Geophysical Monograph Series

Geophysical Monograph Series

184 **Carbon Cycling in Northern Peatlands** *Andrew J. Baird, Lisa R. Belyea, Xavier Comas, A. S. Reeve, and Lee D. Slater (Eds.)*
185 **Indian Ocean Biogeochemical Processes and Ecological Variability** *Jerry D. Wiggert, Raleigh R. Hood, S. Wajih A. Naqvi, Kenneth H. Brink, and Sharon L. Smith (Eds.)*
186 **Amazonia and Global Change Michael Keller, Mercedes** *Bustamante, John Gash, and Pedro Silva Dias (Eds.)*
187 **Surface Ocean–Lower Atmosphere Processes** *Corinne Le Quèrè and Eric S. Saltzman (Eds.)*
188 **Diversity of Hydrothermal Systems on Slow Spreading Ocean Ridges** *Peter A. Rona, Colin W. Devey, Jérôme Dyment, and Bramley J. Murton (Eds.)*
189 **Climate Dynamics: Why Does Climate Vary?** *De-Zheng Sun and Frank Bryan (Eds.)*
190 **The Stratosphere: Dynamics, Transport, and Chemistry** *L. M. Polvani, A. H. Sobel, and D. W. Waugh (Eds.)*
191 **Rainfall: State of the Science** *Firat Y. Testik and Mekonnen Gebremichael (Eds.)*
192 **Antarctic Subglacial Aquatic Environments** *Martin J. Siegert, Mahlon C. Kennicut II, and Robert A. Bindschadler (Eds.)*
193 **Abrupt Climate Change: Mechanisms, Patterns, and Impacts** *Harunur Rashid, Leonid Polyak, and Ellen Mosley-Thompson (Eds.)*
194 **Stream Restoration in Dynamic Fluvial Systems: Scientific Approaches, Analyses, and Tools** *Andrew Simon, Sean J. Bennett, and Janine M. Castro (Eds.)*
195 **Monitoring and Modeling the Deepwater Horizon Oil Spill: A Record-Breaking Enterprise** *Yonggang Liu, Amy MacFadyen, Zhen-Gang Ji, and Robert H. Weisberg (Eds.)*
196 **Extreme Events and Natural Hazards: The Complexity Perspective** *A. Surjalal Sharma, Armin Bunde, Vijay P. Dimri, and Daniel N. Baker (Eds.)*
197 **Auroral Phenomenology and Magnetospheric Processes: Earth and Other Planets** *Andreas Keiling, Eric Donovan, Fran Bagenal, and Tomas Karlsson (Eds.)*
198 **Climates, Landscapes, and Civilizations** *Liviu Giosan, Dorian Q. Fuller, Kathleen Nicoll, Rowan K. Flad, and Peter D. Clift (Eds.)*
199 **Dynamics of the Earth's Radiation Belts and Inner Magnetosphere** *Danny Summers, Ian R. Mann, Daniel N. Baker, and Michael Schulz (Eds.)*
200 **Lagrangian Modeling of the Atmosphere** *John Lin (Ed.)*
201 **Modeling the Ionosphere-Thermosphere** *Jospeh D. Huba, Robert W. Schunk, and George V. Khazanov (Eds.)*
202 **The Mediterranean Sea: Temporal Variability and Spatial Patterns** *Gian Luca Eusebi Borzelli, Miroslav Gacic, Piero Lionello, and Paola Malanotte-Rizzoli (Eds.)*
203 **Future Earth - Advancing Civic Understanding of the Anthropocene** *Diana Dalbotten, Gillian Roehrig, and Patrick Hamilton (Eds.)*
204 **The Galapagos: A Natural Laboratory for the Earth Sciences** *Karen S. Harpp, Eric Mittelstaedt, Noémi d'Ozouville, and David W. Graham (Eds.)*
205 **Modeling Atmospheric and Oceanic Flows: Insightsfrom Laboratory Experiments and Numerical Simulations** *Thomas von Larcher and Paul D. Williams (Eds.)*
206 **Remote Sensing of the Terrestrial Water Cycle** *Venkat Lakshmi (Ed.)*
207 **Magnetotails in the Solar System** *Andreas Keiling, Caitríona Jackman, and Peter Delamere (Eds.)*
208 **Hawaiian Volcanoes: From Source to Surface** *Rebecca Carey, Valerie Cayol, Michael Poland, and Dominique Weis (Eds.)*
209 **Sea Ice: Physics, Mechanics, and Remote Sensing** *Mohammed Shokr and Nirmal Sinha (Eds.)*
210 **Fluid Dynamics in Complex Fractured-Porous Systems** *Boris Faybishenko, Sally M. Benson, and John E. Gale (Eds.)*
211 **Subduction Dynamics: From Mantle Flow to Mega Disasters** *Gabriele Morra, David A. Yuen, Scott King, Sang Mook Lee, and Seth Stein (Eds.)*
212 **The Early Earth: Accretion and Differentiation** *James Badro and Michael Walter (Eds.)*
213 **Global Vegetation Dynamics: Concepts and Applications in the MC1 Model** *Dominique Bachelet and David Turner (Eds.)*
214 **Extreme Events: Observations, Modeling and Economics** *Mario Chavez, Michael Ghil, and Jaime Urrutia-Fucugauchi (Eds.)*
215 **Auroral Dynamics and Space Weather** *Yongliang Zhang and Larry Paxton (Eds.)*
216 **Low-Frequency Waves in Space Plasmas** *Andreas Keiling, Dong-Hun Lee, and Valery Nakariakov (Eds.)*
217 **Deep Earth: Physics and Chemistry of the Lower Mantle and Core** *Hidenori Terasaki and Rebecca A. Fischer (Eds.)*
218 **Integrated Imaging of the Earth: Theory and Applications** *Max Moorkamp, Peter G. Lelievre, Niklas Linde, and Amir Khan (Eds.)*
219 **Plate Boundaries and Natural Hazards** *Joao Duarte and Wouter Schellart (Eds.)*
220 **Ionospheric Space Weather: Longitude and Hemispheric Dependences and Lower Atmosphere Forcing** *Timothy Fuller-Rowell, Endawoke Yizengaw, Patricia H. Doherty, and Sunanda Basu (Eds.)*
221 **Terrestrial Water Cycle and Climate Change Natural and Human-Induced Impacts** *Qiuhong Tang and Taikan Oki (Eds.)*
222 **Magnetosphere-Ionosphere Coupling in the Solar System** *Charles R. Chappell, Robert W. Schunk, Peter M. Banks, James L. Burch, and Richard M. Thorne (Eds.)*
223 **Natural Hazard Uncertainty Assessment: Modeling and Decision Support** *Karin Riley, Peter Webley, and Matthew Thompson (Eds.)*
224 **Hydrodynamics of Time-Periodic Groundwater Flow: Diffusion Waves in Porous Media** *Joe S. Depner and Todd C. Rasmussen (Auth.)*
225 **Active Global Seismology** *Ibrahim Cemen and Yucel Yilmaz (Eds.)*
226 **Climate Extremes** *Simon Wang (Ed.)*
227 **Fault Zone Dynamic Processes** *Marion Thomas (Ed.)*
228 **Flood Damage Survey and Assessment: New Insights from Research and Practice** *Daniela Molinari, Scira Menoni, and Francesco Ballio (Eds.)*
229 **Water-Energy-Food Nexus - Principles and Practices** *P. Abdul Salam, Sangam Shrestha, Vishnu Prasad Pandey, and Anil K. Anal (Eds.)*
230 **Dawn–Dusk Asymmetries in Planetary Plasma Environments** *Stein Haaland, Andrei Rounov, and Colin Forsyth (Eds.)*
231 **Bioenergy and Land Use Change** *Zhangcai Qin, Umakant Mishra, and Astley Hastings (Eds.)*
232 **Microstructural Geochronology: Planetary Records Down to Atom Scale** *Desmond E. Moser, Fernando Corfu, James R. Darling, Steven M. Reddy, and Kimberly Tait (Eds.)*
233 **Global Flood Hazard: Applications in Modeling, Mapping, and Forecasting** *Guy Schuman, Paul D Bates, Giuseppe T. Aronica, and Heiko Apel (Eds.)*
234 **Pre-Earthquake Processes: A Multidisciplinary Approach to Earthquake Prediction Studies** *Dimitar Ouzounov, Sergey Pulinets, Katsumi Hattori, and Patrick Taylor (Eds.)*
235 **Electric Currents in Geospace and Beyond** *Andreas Keiling, Octav Marghitu, and Michael Wheatland (Eds.)*

Geophysical Monograph 235

Electric Currents in Geospace and Beyond

Andreas Keiling
Octav Marghitu
Michael Wheatland
Editors

This Work is a co-publication of the American Geophysical Union and John Wiley and Sons, Inc.

AGU 100
ADVANCING EARTH
AND SPACE SCIENCE

WILEY

This Work is a co-publication of the American Geophysical Union and John Wiley & Sons, Inc.

This edition first published 2018 by John Wiley & Sons, Inc., 111 River Street, Hoboken, NJ 07030, USA and the American Geophysical Union, 2000 Florida Avenue, N.W., Washington, D.C. 20009

Published under the aegis of the AGU Publications Committee
Brooks Hanson, Executive Vice President, Science
Lisa Tauxe, Chair, Publications Committee
For details about the American Geophysical Union visit us at www.agu.org.

Wiley Global Headquarters
111 River Street, Hoboken, NJ 07030, USA

For details of our global editorial offices, customer services, and more information about Wiley products visit us at www.wiley.com.

Library of Congress Cataloging-in-Publication data is available.

ISBN: 978-1-119-32449-2

Cover images: Birkeland currents at Earth derived from AMPERE (Robin J. Barnes, AMPERE Science Data Center, The Johns Hopkins University Applied Physics Laboratory); (Inset, top, left) Current density in the Mars magnetotail; (Inset, top, right) Current density in the equatorial plane at Saturn; (Inset, middle, left) Photospheric continuum emission and respective vertical electric current density of an active region on the Sun; (Inset, middle, right) Current system in the lunar wake; (Inset, bottom) Partial ring current at Earth.

Cover design by Andreas Keiling, Octav Marghitu, and Michael Wheatland

Set in 10/12pt Times by SPi Global, Pondicherry, India

Printed in the United States of America

10 9 8 7 6 5 4 3 2 1

DEDICATION

Olaf Amm was a convener of the Chapman conference that took place in Dubrovnik, Croatia, 2016. Sadly, and far too early, he passed away on 16 December 2014, while in the midst of organizing this conference with his fellow conveners, the editors of this volume. The editors dedicate this volume to Olaf and have asked me, a close colleague of Olaf, to write this dedication note.

Olaf was born in Rendsburg, Germany. He started his studies in geophysics at the University of Münster in the early 1990s and finalized his doctoral thesis at the Technical University of Braunschweig in 1998. Soon after the doctoral defense, Olaf moved to Finland, where he served at the Finnish Meteorological Institute, first as a post-doc researcher, and later as a senior scientist and supervisor for several doctoral students. He recurrently gave lectures on ionospheric physics and potential theory applications in space physics at the Department of Physics of the University of Helsinki and was nominated as a Docent in Space Physics in 2002. During recent years, he had several visiting professorship periods at the Universities of Nagoya (at STELAB) and Kyushu.

Olaf was an accomplished scientist, especially in the fields of ionospheric electrodynamics and magnetosphere-ionosphere coupling processes. He developed several new innovative approaches for creating regional maps of ionospheric currents, conductances and fields with adjustable space resolution on the basis of ground- or space-based magnetic and electric field measurements. The methods that Olaf developed are widely used by other scientists.

Olaf was the principal investigator of the TomoScand network for ionospheric tomography measurements and a member of the PI-Team of the MIRACLE network of magnetometers and auroral cameras. He served also as coinvestigator for the flux-gate magnetometers on the Double Star satellites. The methods that Olaf developed first for the Fennoscandian ground-based instrument networks are nowadays widely used elsewhere when data from multipoint sources are processed.

Olaf was eagerly awaiting the first measurements by the ESA Swarm mission, which would in many ways offer valuable reference material for his theoretical ideas. As preparatory work for Swarm, in 2013 he led an ESA STSE project where a novel analysis method for Swarm electric and magnetic field measurements was developed.

Kirsti Kauristie
Finnish Meteorological Institute, Finland

CONTENTS

CONTRIBUTORS

N. Achilleos
Department of Physics and Astronomy, Centre for Planetary Sciences, University College London, London, UK

Brian J. Anderson
Applied Physics Laboratory, Johns Hopkins University, Laurel, Maryland, USA

V. Angelopoulos
Department of Earth, Planetary and Space Sciences and Institute of Geophysics and Planetary Physics, University of California, Los Angeles, California, USA

Christopher S. Arridge
Department of Physics, Lancaster University, Lancaster, UK

G. Aulanier
LESIA, Observatoire de Paris, PSL Research University, CNRS, Sorbonne Universités, UPMC Univ. Paris 06, Univ. Paris Diderot, Sorbonne Paris Cité, Meudon, France

Graham Barnes
NorthWest Research Associates, Boulder, Colorado, USA

E. S. Belenkaya
Federal State Budget Educational Institution of Higher Education M.V. Lomonosov Moscow State University, Skobeltsyn Institute of Nuclear Physics (SINP MSU), 1(2), Leninskie gory, GSP-1, Moscow, Russian Federation

P. C. Brandt
Applied Physics Laboratory, Johns Hopkins University, Laurel, Maryland, USA

Asgeir Brekke
UiT Norges Arktiske Universitet, Department of Physics and Technology, Tromso, Norway

E. J. Bunce
Department of Physics and Astronomy, University of Leicester, Leicester, UK

J. F. Carbary
Applied Physics Laboratory, Johns Hopkins University, Laurel, Maryland, USA

Xiangning Chu
Department of Atmospheric and Oceanic Sciences, University of California, Los Angeles, California, USA

S. W. H. Cowley
Department of Physics and Astronomy, University of Leicester, Leicester, UK

John C. Coxon
Department of Physics and Astronomy, University of Southampton, Southampton, UK

I. Dandouras
IRAP, Université de Toulouse, CNRS, UPS, CNES, Toulouse, France

R. DeMajistre
Applied Physics Laboratory, Johns Hopkins University, Laurel, Maryland, USA

X-C. Dong
Space Science Institute, School of Astronautics, Beihang University, Beijing, China

M. K. Dougherty
Blackett Laboratory, Imperial College London, London, UK

M. W. Dunlop
Space Science Institute, School of Astronautics, Beihang University, Beijing, China; RAL Space, STFC, Chilton, Oxfordshire, UK

Marius Echim
Institut d'Aeronomie Spatiale de Belgique, Brussels, Belgium; Institute of Space Science, Măgurele, Romania

C. P. Escoubet
ESA/ESTEC, Noordwijk, The Netherlands

S. Fatemi
Space Sciences Laboratory, University of California, Berkeley, California, USA; Solar System Exploration Research Virtual Institute, NASA Ames Research Center, Moffett Field, California, USA

Matthew Fillingim
Space Sciences Laboratory, University of California, Berkeley, California, USA

Gregory D. Fleishman
Physics Department, Center for Solar-Terrestrial Research, New Jersey Institute of Technology, Newark, New Jersey, USA

Ryoichi Fujii
Institute for Space-Earth Environmental Research, Nagoya University

Manolis K. Georgoulis
RCAAM of the Academy of Athens, Athens, Greece

S. Haaland
Birkeland Centre for Space Science, University of Bergen, Norway; Max-Planck Institute for Solar Systems Research, Göttingen, Germany

D. C. Hamilton
Department of Physics, University of Maryland, College Park, Maryland, USA

M. Holmström
Swedish Institute of Space Physics, Kiruna, Sweden

S. Y. Hsieh
Applied Physics Laboratory, Johns Hopkins University, Laurel, Maryland, USA

G. J. Hunt
Department of Physics and Astronomy, University of Leicester, Leicester, UK; Blackett Laboratory, Imperial College London, London, UK

X. Jia
Department of Climate and Space Sciences and Engineering, University of Michigan, USA

Catherine L. Johnson
Department of Earth, Ocean and Atmospheric Sciences, University of British Columbia, Vancouver, BC, Canada; Planetary Science Institute, Tucson, Arizona, USA

Jay R. Johnson
Andrews University, Berrien Spring, Michigan, USA

L. Juusola
Finnish Meteorological Institute, Helsinki, Finland

Haje Korth
Applied Physics Laboratory, Johns Hopkins University, Laurel, Maryland, USA

Krishan K. Khurana
Department of Earth, Planetary and Space Sciences and Institute of Geophysics and Planetary Physics, University of California, Los Angeles, California, USA

S. M. Krimigis
Office of Space Research and Technology, Academy of Athens, Athens, Greece; Applied Physics Laboratory, Johns Hopkins University, Laurel, Maryland, USA

K. D. Leka
NorthWest Research Associates, Boulder, Colorado, USA

Michael W. Liemohn
Department of Climate and Space Sciences and Engineering, University of Michigan, Ann Arbor, Michigan, USA

Jun Lin
Yunnan Observatories, Chinese Academy of Sciences, Kunming, Yunnan P. R. China; Center for Astronomical Mega-Science, Chinese Academy of Sciences, Beijing, P. R. China

Jiang Liu
Department of Earth, Planetary and Space Sciences and Institute of Geophysics and Planetary Physics, University of California, Los Angeles, California, USA

Ramon E. Lopez
Department of Physics, The University of Texas at Arlington, Arlington, Texas, USA

A. T. Y. Lui
JHU/APL, Laurel, Maryland, USA

Carley J. Martin
Department of Physics, Lancaster University, Lancaster, UK

H. R. Middleton
ESA/ESAC, Villanueva de la Cañada, Madrid, Spain

Stephen E. Milan
Department of Physics and Astronomy, University of Leicester, Leicester, UK; Birkeland Centre for Space Science, University of Bergen, Bergen, Norway

D. G. Mitchell
Applied Physics Laboratory, Johns Hopkins University, Laurel, Maryland, USA

Lei Ni
Yunnan Observatories, Chinese Academy of Sciences, Kunming, Yunnan P. R. China; Center for Astronomical Mega-Science, Chinese Academy of Sciences, Beijing, P. R. China

J. D. Nichols
Department of Physics and Astronomy, University of Leicester, Leicester, UK

Clare E. Parnell
School of Mathematics and Statistics, North Haugh, University of St. Andrews, St. Andrews, Fife, Scotland, UK

Alexei A. Pevtsov
National Solar Observatory, Boulder, Colorado, USA; ReSoLVE Centre of Excellence, Space Climate Research Unit, University of Oulu, Finland

Lydia C. Philpott
Department of Earth, Ocean and Atmospheric Sciences, University of British Columbia, Vancouver, BC, Canada

L. C. Ray
Department of Physics, Lancaster University, Lancaster, UK

S. Rochel-Grimald
ONERA-DESP (Département de l'Environnement Spatial), Toulouse, France

A. Runov
Department of Earth, Planetary and Space Sciences and Institute of Geophysics and Planetary Physics, University of California, Los Angeles, California, USA

C. T. Russell
UCLA, Los Angeles, California, USA

B. Schmieder
LESIA, Observatoire de Paris, PSL Research University, CNRS, Sorbonne Universités, UPMC University Paris 06, University Paris Diderot, Sorbonne Paris Cité, Meudon, France

N. Sergis
Office of Space Research and Technology, Academy of Athens, Athens, Greece

J-K. Shi
NSSC, Chinese Academy of Sciences, Beijing, China

R. Slapak
Division of Space Technology, Luleå University of Technology (LTU), Kiruna, Sweden

H. Vanhamäki
Ionospheric Research Unit, University of Oulu, Oulu, Finland; Finnish Meteorological Institute, Helsinki, Finland

C. Vallat
European Space Astronomy Centre, Villanueva de la Cañada, Madrid, Spain

Martin Volwerk
Space Research Institute, Austrian Academy of Sciences, Graz, Austria

Simon Wing
Applied Physics Laboratory, Johns Hopkins University, Laurel, Maryland, USA

Shaosui Xu
Space Sciences Laboratory, University of California, Berkeley, California, USA

M. Yamauchi
Swedish Institute of Space Physics (IRF), Kiruna, Sweden

Y.-Y. Yang
Key Laboratory of Crustal Dynamics, China Earthquake Administration, Beijing, China

Zhonghua Yao
Laboratoire de Physique Atmosphérique et Planétaire, STAR Institute, Université de Liège, Liège, Belgium

Akimasa Yoshikawa
Department of Earth and Planetary Sciences, Kyushu University; International Center for Space Weather Science and Education, Kyushu University

Q.-H. Zhang
Institute of Space Sciences, Shandong University, Weihai, China

Xu-Zhi Zhou
School of Earth and Space Sciences, Peking University, Beijing, China

PREFACE

This book is an international and interdisciplinary collaboration among different space science communities to explore our understanding of electric currents occurring at key locations in our solar system: ionospheres and magnetospheres of Earth, the inner planets, and the outer planets; the Sun's atmosphere; space environments of other heliospheric bodies, such as moons and comets; and the heliopause. Because of this breadth, the book should be of interest to a large readership and is intended to be a reference for years to come.

Electric currents are fundamental to the structure and dynamics of space plasmas, including our own near-Earth space environment (also called "geospace"). This recognition is one of the great achievements in space research, going back to the beginning of the last century, or even earlier; tentative ideas about electric phenomena in geospace were already expressed in the early 18th century (see Chapter 1 for a historic account). One of the great pioneers and a key figure in the study of electric currents is the Norwegian scientist Kristian Birkeland (after whom a type of current system was named). The writing of this book was chosen to coincide with the 150th and 100th anniversaries of his birth and death years (1867 and 1917, respectively). Hence, this book is also intended as a celebration of his achievements and those of the many other pioneers in this research field.

In the 20th century, it was firmly established that a complex system of currents threads the ionosphere and magnetosphere of our near-Earth space environment. While the ionospheres and magnetospheres of other planets are permeated by electric currents as well (see overview in Chapter 2), the plasma properties are quite different from planet to planet, resulting in new phenomena. For one example, while Earth's ring current produces a magnetic field that does not severely change the largely dipolar magnetic field of the inner magnetosphere (Chapters 6 and 7), the Jovian ring current strongly distorts the shape of the magnetic field into a disc-like configuration (Chapter 8). For another example in this book, the high-density plasma of the Sun supports current loops and current sheets in the solar corona, which provide the energy source in active regions for solar flares and coronal mass ejections (Chapters 3, 14, 15, and 23).

These examples, involving the Earth, Jupiter, and the Sun, are only a small selection of the broad range of electric current systems reviewed here. For each and every one of them, the ultimate goal is to understand its generation, its (multiscale) spatial structure, its (short- and long-term) temporal development, as well as its impacts and ramifications on the local space environment. These are common objectives among all researchers studying electric currents in space plasmas.

The book is divided into six parts. Part I provides as introduction a history of the field and several broad overviews of current systems in planetary and solar environments, together with some measuring techniques. The remaining sections contain specialized chapters that focus on the most recent discoveries of the 21st century about ring currents (Part II), current sheets (Part III), field-aligned currents (Part IV), ionospheric currents (Part V), and selected other current systems (Part VI) that do not fall exclusively in any of the other categories. In total, the book comprises 30 chapters, whose quality, we would like to add, also benefited from the enthusiastic efforts of our reviewers, reflected in their thoughtful and constructive criticisms.

Most of the chapters are born out of presentations given at the Chapman conference "Currents in Geospace and Beyond," held in Dubrovnik, Croatia, during the last week of May 2016; an unforgettable event in many respects. The conference's interdisciplinary theme brought together scientists from around the world, and demonstrated that this field is truly a global effort. This book builds on these efforts and shows how, despite politics and gaps in understanding of our own differences, it is possible to bridge gaps and be inclusive across genders, nationalities, and diverse ethnicities. In this spirit, the conference also hosted discussions on equity, diversity, and inclusion, and in particular the goal of achieving a higher representation of women in research and academia. Special thanks to Beverly Smith-Keiling for leading these discussions and also for her role as the chair of the local organizing committee. Every social event and activity was organized with perfection!

Sadly, and far too early, Olaf Amm, our fellow convener, passed away on 16 December 2014. He was enthusiastic about the conference theme and would have significantly contributed to both the conference and this book, which we dedicate to him. His colleague from the Finnish Meteorological Institute, Kirsti Kauristie, wrote the dedication at the beginning of this book.

Andreas Keiling
Octav Marghitu
Michael Wheatland

Part I: Introduction

1
Pioneers of Electric Currents in Geospace

Asgeir Brekke

ABSTRACT

This review shows that the progress of our understanding of the electric currents in geospace has gone through a progressive development from the time of the Enlightenment in the early eighteenth century to the Space Age in the 1970s. When it was found that magnetic field variations were caused by electric currents in the upper atmosphere, important steps were made in the late part of the nineteenth century. The aurora borealis was believed to be an electric phenomenon by several authors as early as the 1750s. The current system linking the creation of the aurora became a main field of interest in the beginning of the twentieth century and has remained so until our time. At present, we have a large variety of instruments and methods such as satellite and ground-based experiments of different kinds and capacities as well as dedicated computer models to study these current systems further. What appears to be lacking, however, is a more detailed knowledge of the variation of the ionospheric conductivities in space and time.

1.1. INTRODUCTION

Geospace is often used as a denotation of the space between the Sun and the Earth. Solar wind plasma is streaming through geospace and due to the omnipresent magnetic field, electrons and positive ions blowing with the wind are forced to move in different directions with electric currents as an outcome. These electric currents forming in the magnetosphere are connected to the ionosphere of the Earth by field-aligned currents that are closed by horizontal currents in the upper atmosphere. The effects on ground caused by these currents have been a challenge for humankind for generations. This chapter will give an overview of the development of our understanding of these electric currents in geospace from the time of the Enlightenment until the 1970s. Important milestones reached in the eighteenth century like the understanding of the relationship between the occurrence of aurora borealis and magnetic field fluctuations, as well as the fact that the aurora is an electrical phenomenon, are elucidated. The important recognition in the last part of the nineteenth century of diurnal magnetic field fluctuations being a result of electric currents in the atmosphere is emphasized. The introduction of cathode rays and electrically charged particles from the Sun being the cause of the aurora, field-aligned currents, and magnetic storms are given a broad coverage including experimental mapping and models of the current systems involved.

It is expected that the multidimensional European Incoherent Scatter Radar EISCAT_3D system in Scandinavia together with modern space-borne technology will improve our ability to better understand the cause and connections of the electric currents in geospace.

1.2. AURORA BOREALIS AND VARIATIONS IN THE EARTH'S MAGNETIC FIELD

We are all standing on the shoulders of somebody. When we are tracing the pioneers of our field, we have to make a choice where to start. *Chapman and Bartles* [1940]

UiT Norges Arktiske Universitet, Department of Physics and Technology, Tromso, Norway

Electric Currents in Geospace and Beyond, Geophysical Monograph 235, First Edition.
Edited by Andreas Keiling, Octav Marghitu, and Michael Wheatland.

claimed that Sir Edmund Halley was the first to publish an observed connection between the aurora borealis and magnetic storms in London on 17 March 1716. From these observations, Halley concluded that the position as well as the shape of the aurora was strongly controlled by the magnetic field. Furthermore, Halley presented the idea that the aurora was due to a magnetic effluvia that streamed out from pores in the Earth's surface and by following the magnetic field lines reached far away from the surface of the Earth, where the ether became luminous [*Halley*, 1716]. Although Halley did not relate his observations to electricity, it might be natural to start this review by mentioning his contributions to the field as they lifted the understanding of the auroral phenomenon from speculative ideas to a frame of reference based on natural science. Furthermore, the aurora borealis is a reasonable introduction to the field as its striking appearance has inspired many scientists throughout history to relate it to electric currents in geospace.

It has been said that the watchmaker George Graham (1673–1751) in London was the first to observe a relationship between the aurora borealis and the variations in the magnetic needle in 1722 [*Chapman and Bartels*, 1940]. In his paper, however, Graham reported peculiar variations in the direction of a magnetic needle. As he described his experiment and dismissed any possible artificial disturbances in his instrument that could explain the remarkable observations, he wrote:

> "I am well assured these Changes in the Direction are owing to some other Cause than Friction of the Needle upon the Pin; but what that Cause is I cannot say, for it seems to depend neither upon Heat nor Cold, a dry or moist Air, clear or cloudy, windy or calm Weather, nor the Height of the Barometer" [*Graham*, 1724–1725, 96–107].

Clearly there was an outside cause of the variations, and in hindsight it is tempting to assume that it must have been at a time of occurrence of northern lights.

Later on, however, the Swedish astronomer and physicist Anders Celsius (1701–1744) probably became the first to have realized a relationship between the aurora borealis and variations in the position of the magnetic needle as he encouraged Olav Peter Hiorter (1696–1750) to look for a possible connection between the occurrence of these two phenomena. Hiorter meticulously made 6638 readings of the position of the magnetic needle from 19 January 1741 until 19 January 1742. When he reported his work in 1747, he had made 10,000 observations and he wrote:

> "But who would have been able to imagine that the northern light had anything in common and a relation to the magnetic needle, and that the northern light when it passes past zenith towards south or accumulating near the western or eastern horizon would cause a considerable perturbation of the magnetic needle amounting to several degrees within a few minutes?" [*Hiorter*, 1747, 27–43].

The first time Hiorter noticed the relation between the northern light and the position of the magnetic needle was 1 March 1741. A few weeks before that, Celsius had written to Graham in London to encourage him to watch the needle, hoping that if a disturbance happened in Uppsala, one could observe whether a similar event occurred at the same time in London, and if so, one could disregard any artificial error source in the local observatories. On 5 April 1741, Graham could report on a variation in London that occurred at the same time in Uppsala and he wrote:

> "The alterations that day were greater, than I had ever met with before. Tho' no alteration of any thing in the Room could occasion it. ...the only thing in which I am certain, is, that there was no change of position of any thing in the Room, that could cause it, being alone the whole day" [*Hiorter*, 1747, 36].

Hiorter certainly reverenced Celsius as he submissively summarized his work in the following way:

> "Thus I am glad to ascribe this discovery completely to the late Professor Celsius alone, ...and who made it possible for me to continue these researches and to publish these discoveries which otherwise would have been buried with him" [*Hiorter*, 1747, 43].

Johan Carl Wilcke (1732–1796) was a German who came to Uppsala as a student in 1749, and devoted himself to studies of electricity and caloric theory. He also followed up on the work by Celsius and Hiorter as he in 1777 wrote a paper about the diurnal and annual variations of the position of the magnetic needle, and stated:

> "The relationship between the magnetic needle and the northern lights is so clear, common and for good that no one that with attention is watching both phenomena, can have any doubt about it....The matter is exactly as Mr. Hiorter has described so clearly. The northern end of the magnetic needle seems to follow the auroras and to be attracted by them" [*Wilcke*, 1777, 274–300].

Then Wilcke discussed the auroral corona or "*Norrskens-Solen*" and stated that:

> "The flames of the northern lights are stretching themselves up along the direction that the Magnetic-force is directing the magnetic needle when it turns freely" [*Wilcke*, 1777, 299].

1.2.1. The Aurora Borealis as an Electric Phenomenon

In 1897, the director of the French Meteorological Institute Alfred Angot (1848–1924) wrote in his book *The Polar Aurora* that the French physicist John Canton (1718–1772) appears to have been the first physicist that "pointed out in 1753 the close analogy which auroras offer with the light of electric discharges produced in very rarefied air" [*Angot*, 1897, 158].

This may well be true, but in 1752, the Norwegian bishop of Bergen Erich Pontoppidan (1698–1764) also discussed the northern lights as being an electrical phenomenon. After reciting contemporary theories like fire from volcanic eruptions in the north or reflections of solar rays from vaporous clouds high above the North Pole, as introduced by famous philosophers at the time like the Russian scientist Leonhard Euler (1707–1783), Pontoppidan expressed himself in a very humble tone as he wrote:

"Are somebody expecting that I should tell my opinion about this problematic matter, then it probably would not be less reasonable than what here is already referred to, if one presented the idea, that the northern light is caused by the electricity of the ether in the air" [*Pontoppidan*, 1752, 7–23].

Pontoppidan based his statement on a work by the French assistant to Isaac Newton (1642–1727), John Theophilus Desaguliers (1683–1744), who with reference to an experiment by Francis Hawksbee (Hauksbee) (1660–1713) wrote: "I suppose Particles of Air to be Electric Bodies always in state of Electricity, and that Vitreous Electricity" (vitreous electricity, that is, positive electricity). Desaguliers then described Hawksbee's experiment:

"Having pumped out all the air from a Glass Globe, he caused it to turn its Axis very swiftly, by means of a Rope with Wheel and Pulley; then rubbing the Glass with his hand during Motion, there appeared a great deal of Light of a purple Colour" [*Desaguliers*, 1742, 140–143].

Pontoppidan compared this rotating globe with the rotating Earth, and wrote:

"One can imagine the terrestrial globe surrounded by air like the glass-globe in an electric machine. When the air is pumped out and the globe is rotating very fast, a purple coloured flame is coming into being, the same colour as exists in the northern light, and this flame must be Ether igneous."

Further Pontoppidan wrote:

"The northern light observed towards the Pole or the axis of the globe can not be caused by ether alone but can also be the very ether itself; which, being aggregated, gives way to the impression of the humid air, and mounts and floats above the clouds, whose motion likewise renders variable" [*Pontoppidan*, 1752, 16–17].

The Russian polyhistor Mikhail Lomonosov (1711–1765) discussed similar experiments as Hawksbee had demonstrated and he claimed:

"During excitation of an electric force in the sphere, from which air is drawn out, sudden rays emitted, and instantly disappeared, and at the same time, new ones in their places popped up, so it looked like continuous glitter. In the northern lights flashes or beams do not suddenly occur to the extent of entire space, but behave similarly. The pillars of the northern lights shining as stripes on the surface of the electric atmosphere in subtle or in pure are very nearly perpendicular to the ether, as in the aforementioned electric sphere from a concave circular surface to the center of converging rays." [*Chernouss*, 2012, 105–107]

In 1779, the American scientist, inventor, and diplomat Dr. Benjamin Franklin (1706–1790) published an article where he assumed that the warm and moist air rising in the tropical zones contains electricity that is transported by winds to the polar regions and that equatorial return winds bring cooler air back from the poles to the tropics. He wrote:

"May not then the great Quantity of Electricity brought into the Polar Regions by the Clouds which are condens'd there & fall in Snow, which Electricity would enter the Earth but cannot penetrate the Ice; May it not, I say, as a bottle overcharg'd, break thro' that low Atmosphere and run along in the Vacuum over the Air towards the Equator, diverging as the degrees of Longitude enlarge, strongly visible where densest, and becoming less visible as it more diverges, till it finds a Passage to the Earth in more temperate Climates; or is mingled with their upper Air?If such an Operation of Nature were really performed, would it not give all the Appearances of an Aurora Borealis?" [*Franklin*, 1779, 291–297]

Definitely, Franklin related the aurora to a global stream of electricity, what we today would call an electric current.

Franklin wrote a letter to Lomonosov where he expressed his opinion about the nature of the aurora borealis. Lomonosov commented on this letter in the following way:

"Franklin's guess about the northern lights, which he refers to by a few words in the same letter of my theories is very different. He attracts electrical matter of the northern lights from the equatorial zone, but I find it in the same place, i.e. in the air and everywhere present. He does not define its place, but I think it is above the atmosphere, ...contains my long standing view that the northern lights arise from the motion of the ether" [*Chernouss*, 2012, 105–107; *Eather*, 1980, 59].

1.2.2. Electric Currents Related to the Northern Lights

In 1820, the Danish physicist Hans Christian Ørsted (1777–1851) discovered electromagnetism, that is, that an electric current could create a magnetic field at a distance. A door was then opened to understand the geomagnetic field variations in terms of global currents.

The French physicist Auguste de la Rive (1801–1873) in 1849 wrote a letter to the French prime minister and physicist M. Arago (1786–1853) with a laudatory phrase in the following way:

"I have only followed the route which you yourself have traced; for more than thirty years ago, you, with indefatigable perseverance, established by your numerous observations the remarkable concordance which exists between the appearance of the Aurora Borealis and the disturbance of the magnetic needle" [*De La Rive*, 1849, 40–46].

Seemingly, Arago had repeated the original work by Celsius, Hiorter, and Graham.

De la Rive had the idea that as an electric current is floating from the hot end of a rod to the cold end, there is a current floating from the warm end at the bottom of a column of air at the equator to the top, that is, a vertical upward current exists from the Earth's surface at the equator, and he wrote:

> "Thus we have a circuit formed; each annular stratum of the atmosphere gives rise to a current, which travels in the upper portion of the stratum towards the pole, re-descends towards the earth through the atmosphere around the pole."

And finally

> "the currents pass also from the equator to the pole in the upper regions of the air, and from the pole to the equator upon the surface of the earth" [*De La Rive*, 1849, 42]. This idea has certainly a lot in common with Franklin's model.

The Scottish physicist Balfour Stewart (1828–1887) in 1886 when discussing observed solar-diurnal variations in the terrestrial magnetism, had to dismiss the Earth-currents as the cause, and maintained:

> "And we are therefore driven to regard electrical currents as being the only conceivable cause, if this cause is to be located in the upper atmospheric regions. In the first place, it may be said that while undoubtedly rarified air is a conductor of electricity, yet it is not a good conductor, and where can we look for sufficient potential to drive current through these upper atmospheric regions? To this I would reply that as a matter of fact we know that there are visible electric currents in the upper atmospheric regions which occur occasionally at ordinary latitudes, and which are very frequent, if not continuous, in certain regions of the Earth. I allude to the aurora which is unquestionably an electric current, and must therefore influence the magnetic field.
>
> While we can with the greatest ease account for it by means of a system of currents in the upper regions of the Earth's atmosphere." [*Stewart*, 1886, 44]

1.3. TIDAL MOTIONS AND THE DYNAMO THEORY

The British physicist Arthur Schuster (1851–1934), who studied the variations of the vertical component of the magnetic field (the *z*-component), stated in 1889:

> "The horizontal movements in the atmosphere which must accompany a tidal action of the Sun or Moon or any periodic variation of the barometer such as is actually observed, would produce electric currents in the atmosphere having magnetic effects similar in character to the observed daily variation" [*Schuster*, 1907, 163–204].

Here is probably one of the first statements in history relating the daily magnetic field variations to the existence of tidal motions, and Schuster continued:

> "If we endeavor to carry the investigation a step further and enquire into the probable origin of these currents, we have at present no alternative to the theory first proposed by Balfour Stewart, that the necessary electromotive force are supplied by the permanent forces of terrestrial magnetism acting on the bodily motion of masses of conducting air which cut through its line of force" [*Schuster*, 1907, 163–204].

Here Schuster introduced for the first time the dynamo theory as an explanation of the diurnal variations of the magnetic field.

Sidney Chapman (1888–1970) later followed up on the studies of tidal motion based on a statistical analysis of magnetic variations from a global network of stations and reached the following conclusion about the *S* (solar) and *L* (lunar) components of the diurnal variations:

> "There is, in any case, little or no reason to doubt that the S and L currents flow in our atmosphere in a layer which is very nearly spherical and concentric with the Earth" [*Chapman*, 1919, 1].

This view resembles De La Rive's ideas but it leaves a system without vertical currents.

1.4. THE BIRKELAND-STØRMER CURRENT SYSTEM

In 1896, a young Norwegian physicist Kristian Birkeland (1867–1917) presented an idea where he assumed that the northern lights where caused by cathode rays streaming out from the Sun, being soaked up by the Earth's magnetic field and forced toward the poles of the globe where they created the northern lights. [*Birkeland*, 1896]. According to his theory there was no strong limitation of how far down in the atmosphere the cathode rays could reach. For Birkeland, the height of the aurora became an important issue.

In 1898, Birkeland received support from the Norwegian government to build an observatory, the Haldde Observatory, and an annex observatory, Talvik, 3.4 km apart at 900 m above sea level in northern Norway in order to study the nature of the northern lights.

Birkeland's initial objective was to measure the height of the northern lights by triangulation from the two observatories connected by a telephone line. He lost his primary goal partly due to the short baseline (3.4 km) and partly because his optical equipment had too low sensitivity. Fortunately, he brought with him magnetometers to the mountain and could combine magnetic records from his observatory with recordings at lower latitudes like Potsdam and Pavlovsk. Birkeland realized that such magnetic-field variations could not be used to derive the full current vector, but only the horizontal component or the equivalent current. From these magnetic data, however, he drew up a map (Fig. 1.1) of a cross-polar horizontal current system that resembles the convection current systems often discussed today.

Birkeland again applied to the Norwegian government for support to discharge a new expedition to the Arctic, and, in 1902–1903, he had 4 stations installed in Iceland, Spitsbergen, Novaia-Zemlja, and Bossekop, respectively. By combining magnetic records from these stations with similar recordings from 23 other stations around the world, he came to the conclusion that field-lined electron precipitation formed vertical currents that connected to a horizontal current along the auroral arcs (Fig. 1.2). He also expanded his model by including two antiparallel current sheets feeding a horizontal part along the auroral arc (Fig. 1.3).

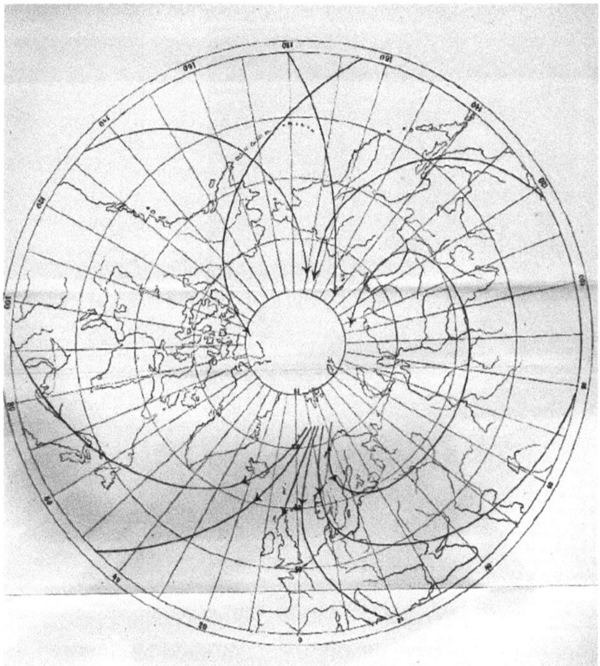

Figure 1.1 Birkeland's convection current system from 1901. The arrows represent the motion of the cathode rays. From *Birkeland* [1902].

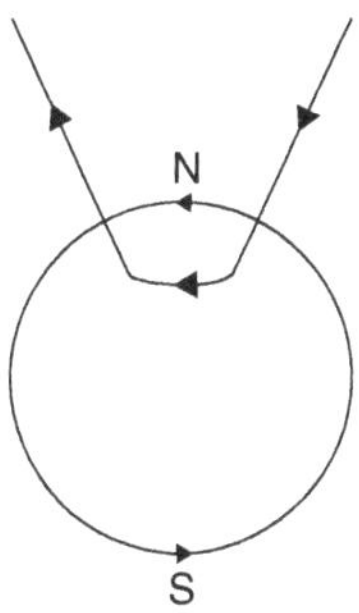

Figure 1.2 Birkeland's line-current model where vertical currents are closed by a current along the auroral arc. After *Boström* [1967].

It is surprising that, when reading Birkeland's large work entitled *The Norwegian Aurora Polaris Expedition 1902–1903, Volume I, On the Cause of Magnetic Storms and the Origin of Terrestrial Magnetism* [*Birkeland*, 1908], actually very little is found in terms of visual observations of the aurora itself. The main part of this gigantic work refers to the interpretations of worldwide magnetic records in terms of global currents. As a matter of fact, Birkeland drew most of his conclusions about the creation of the aurora from experiments in the laboratory, where the Terrella experiment was the central piece. By his small "universe," he could create the auroral rings as well as equatorial ring currents that he

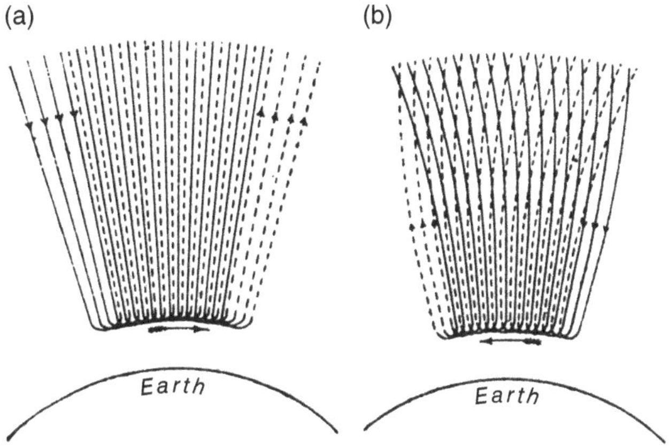

Figure 1.3 Birkeland's expanded current system including two antiparallel current sheets feeding a horizontal current along the auroral arc. The arrows indicate the direction of the motion of cathode rays. From *Birkeland* [1908].

associated with similar phenomena in the real world. Certainly, the experiment was not scaled in true proportions, but, nevertheless, the phenomena discovered gave associations to the reality.

The Norwegian mathematician Carl Størmer (1874–1957) was much inspired by Birkeland's work and set out to calculate the trajectories of electrically charged particles in a magnetic field in an attempt to explain the formation of auroral rays, and so on (Fig. 1.4). He traced the particles from the Sun toward the Earth and found that they often reached the atmosphere at higher latitudes than the 23 degrees from the pole where the auroras most often occurred. In order to improve the agreement, he weakened the magnetic field in his calculations by introducing an eastward (westerly) current around the globe in the equatorial plane at a distance from the Earth. This was probably the first indication of the equatorial ring current ever proposed [*Størmer*, 1955].

1.5. CURRENT SYSTEMS FOR MAGNETIC STORMS

The report from Birkeland's second expedition to the Arctic had, as stated above, a subtitle: "On the cause of magnetic storms and the origin of terrestrial magnetism" [*Birkeland*, 1908].

Birkeland attributed the equatorial storms to flow of electrons at great distances from the Earth: the positive storms were due to electrons deflected away from the Earth, westward, on the sunward side, and the negative storms were due to electrons deflected eastward round the back of the Earth.

Schuster criticized this idea on the grounds that a beam of electrons from the Sun could not hold together against their mutual repulsion, in sufficient strength to provide

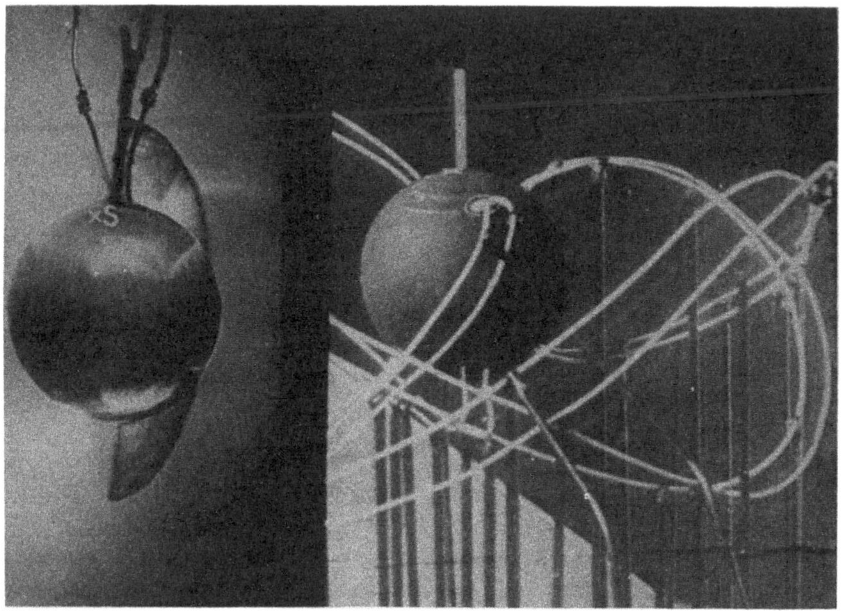

Figure 1.4 A comparison between Birkeland's artificial aurora from the Terrella experiment (left) and Størmer's particle trajectories illustrated by spiral models (right). From *Birkeland* [1908].

the electric currents near the Earth as proposed by Birkeland, and he stated:

> "We must conclude that a swarm of electrons packed with sufficient density to cause a magnetic effect would soon get dissipated laterally into space until its magnetic action becomes negligible."

He continued:

> "The results of the previous investigation conclusively prove that magnetic storms cannot be due to a direct magnetic action of swarms of electrified particles" [*Schuster*, 1911, 44–50].

In 1918, Chapman, however, in contrast to Schuster, supported the idea of precipitating electric particles with the same type of electric charge in order to explain the vertical atmospheric motions that he believed to have observed in connection to a magnetic storm. He wrote:

> "A magnetic storm is generated by the entry into the earth's atmosphere of numbers of electric particles, mainly or entirely of the same sign of charge. They penetrate to a more or less definite level in the upper atmosphere, this level depending on the density and composition of the atmosphere, and upon the physical nature and velocity of the particles."

He continued:

> "On the theory to be described, this cause is a system of electric currents which flow, in more or less horizontal strata, in the upper atmosphere" [*Chapman*, 1918, 78].

He further divided the current system into two subsystems, a latitudinal and a meridional one (Fig. 1.5):

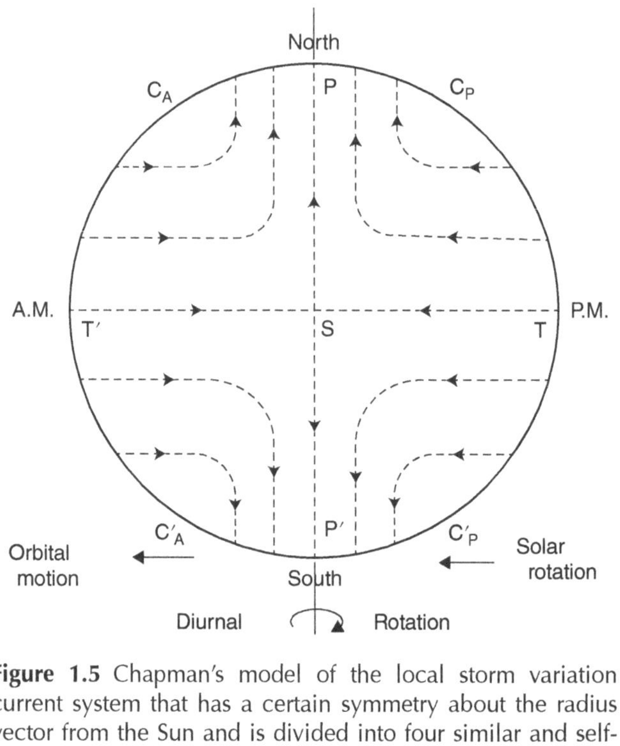

Figure 1.5 Chapman's model of the local storm variation current system that has a certain symmetry about the radius vector from the Sun and is divided into four similar and self-contained quadrantal parts by the "solar meridian." From *Chapman* [1918].

> "The first current system, in which the circulation is round parallels of latitude, is symmetrical about the earth's axis. The electromotive force (E.M.F.) impelling the current arises from inductive action occurring in the plane normal to the E. M. F., i.e., in the meridian plane at each point. The most general action in this plane can be resolved into component parts, in one of which a vertical motion of the atmosphere takes place across the horizontal component of the earth's magnetic field, while in the other a horizontal current of air crosses the vertical magnetic field." [*Chapman*, 1918, 76]

Finally, based on the background of the work by the Norwegian scientists Störmer, Vegard, and Krogness (L. Vegard [1880–1963] and O. A. Krogness [1886–1934] were both students of K. Birkeland), Chapman associated the occurrence of the aurora with these penetrating particles:

> "Auroræ may themselves be the visible manifestation of vertical electric discharges of unusual intensity. Precipitation would seem to be ordinarily confined to high latitudes, extending during a storm over a much wider area, but the injection may not be confined to the regions where auroræ are actually visible" [*Chapman*, 1918, 80].

In 1919, F. A. Lindemann criticized Chapman's theory with the following strategy:

> "The best way to approach the subject is probably by criticising the theory now probably most generally accepted, which has been most elaborately worked out by Dr. Chapman. The recognition of the fact that a radial current on the earth would explain the magnetic phenomena is undoubtedly a most valuable advance, and with this part of his theory it is not proposed to tamper."

And Lindemann continued:

> "Dr. Chapman assumes that beams of α rays are emitted by parts of the solar surface and that the earth is subjected to a magnetic storm when it passes through a beam of this sort" [*Lindemann*, 1919, 669].

The strongest arguments Lindemann put forward against Chapman's theory were the following:

> "There are a number of reasons which show that α rays alone cannot be the true cause of magnetic storms, the main ones being that they cannot be produced on the sun in sufficient quantities, that they cannot proceed as a beam for one or two solar diameters on account of the mutual repulsion on the particles, and finally that they could not approach the earth after the first few seconds on account of the charge the earth would rapidly acquire.... The hypothesis to be examined therefore is that an approximately equal number of positive and negative ions are projected from the sun in something of the form of a cloud and that these are the cause of magnetic storms and auroræ." [*Lindemann*, 1919 673]

Here Lindemann for the first time touched upon what we today call the "Solar Wind."

Lindemann argued against an electromotive force in the solar atmosphere and believed the particles in the cloud were driven out from the Sun by the radiation pressure.

> "It appears certain therefore that such clouds of ionized gas can exist, and that they would be projected radially from the sun at such a speed that they would naturally spread out enough by the time they reached the earth to account for the observed duration of magnetic storms" [*Lindemann*, 1919, 679].

As the ions were heavier than the electrons, the radiation pressure would more effectively drive ions and the electrons would lag behind with the consequences that a radial electric field would be created.

Chapman accepted the criticism by Lindeman and wrote in the 1930s, together with C. V. A. Ferraro, a string of papers related to *A New Theory of Magnetic Storms* [*Chapman and Ferraro*, 1930, 129–130].

> "We have not examined closely the extent to which the stream will cause inflow of ions and electrons into the earth's atmosphere in the polar regions, or how this inflow will give rise to the observed currents along auroral zones; but it seems likely that present theories of the auroræ will need to be modified, because the particles of a neutral stream can approach much closer to the earth, in the equatorial plane, than the single charged particles hitherto considered."

Concerning the currents responsible for the magnetic deviation during a magnetic storm they stated:

> "In the second phase of a magnetic storm the earth's horizontal force is decreased. We attribute this to the formation of a westerly current around the earth, due primarily to the flow of charges across the space "behind" the earth (viewed from the sun)" [*Chapman and Ferraro*, 1930, 129].

Included, in their paper was Figure 1.6, which shows the charged layers along BB' (positive) and CC'(negative) situated in the equatorial plane behind the Earth as seen from the Sun, and being due to the polarization of the stream by the magnetic field. An electric field would be set up to bring the charges between the layers that would enforce the ions from moving from BB' to CC', but the magnetic field would preclude the electrons from moving in the reverse direction. Electrons would instead flow downward and upward along the magnetic lines of force to neutralize the positive ions that traverse from BB' to CC'. From this they maintained that a westerly current could be set up around the Earth, the equatorial ring-current.

Ferraro made a critical survey of their common work until 1933 and stated,

> "We have not been able to make any serious attempt at a mathematical discussion of the processes involved in the main phase of the storm, which we think is due to a westerly current flowing round the Earth at a distance of several Earth radii" [*Ferraro*, 1933, 259].

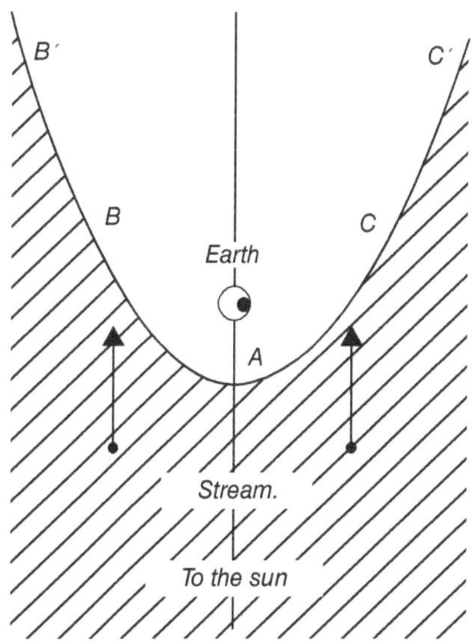

Figure 1.6 A figure presented by *Chapman and Ferraro* [1930] to explain the separation of positive charges along BB′ and negative charges along CC′ in the particle stream behind the Earth seen from the Sun.

Summing up their common work in contrast to Birkeland and Störmer's idea, Ferraro wrote:

> "Our theory differs in many respects from previous corpuscular theories (associated especially with the names of Birkeland and Störmer), especially in that we supposed the solar streams to be neutral but ionized (as had been suggested by F. A. Lindemann) and to approach much nearer to the Earth in the equatorial plane than electric corpuscles considered by Birkeland and Störmer" [*Ferraro*, 1933, 253].

Chapman in 1935 [*Chapman*, 1935] continued on the work on the current system related to magnetic storms and published a model of the complete current system about 16 hours after the outbreak of a magnetic storm. The currents were given as calculated on the assumption that there was no current supply to the zones from the outside.

During the 1960s, global magnetic recordings from the International Geophysical Year (IGY) 1957–1958 were extensively used to infer equivalent currents. Since vertical currents cannot be derived from ground-based geomagnetic recordings only, the equivalent current is a pseudo current in the horizontal plane. *Matsushita* [1967] derived the so-called averaged mean S_q current system for different periods of the day (Fig. 1.7).

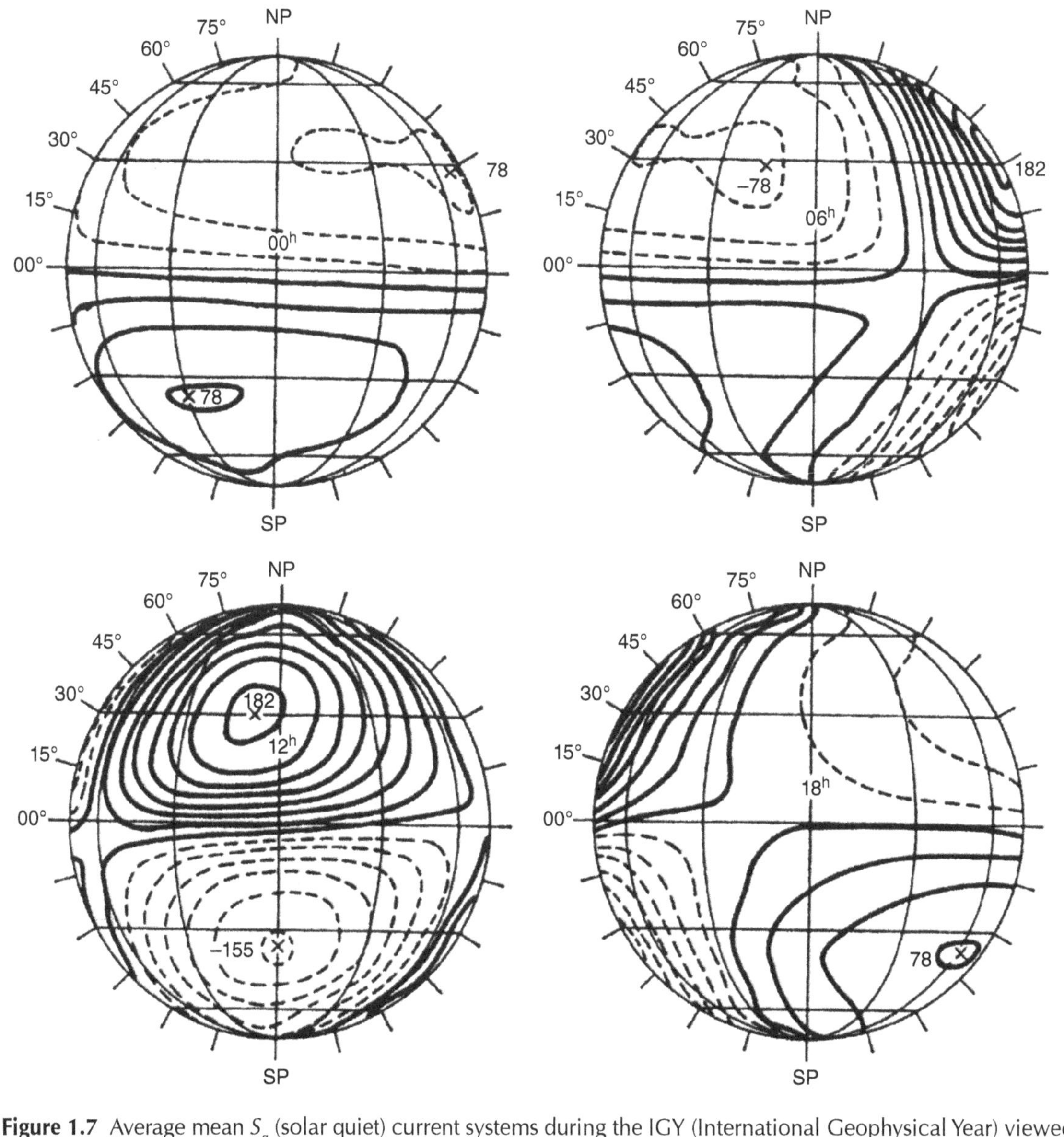

Figure 1.7 Average mean S_q (solar quiet) current systems during the IGY (International Geophysical Year) viewed from the magnetic equator at 00:00, 06:00, 12:00, and 18:00 meridian. The numbers near the crosses indicate vortex current intensity in units of 10^3 A. The distance between the current lines corresponds to 2.5 10^3 A. From *Matsushita* [1967].

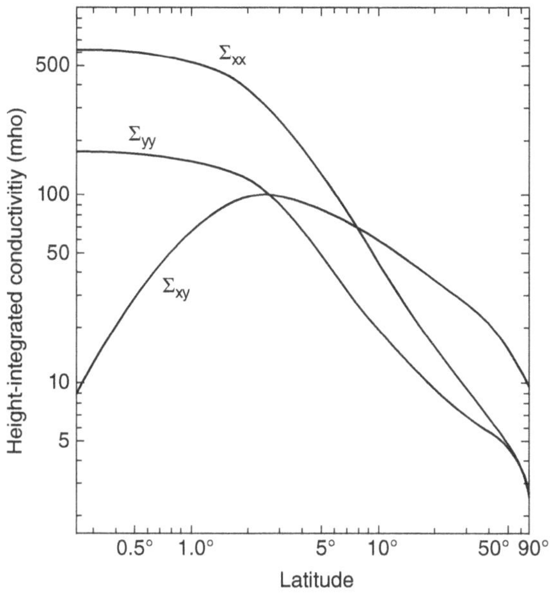

Figure 1.8 Model calculations of the height-integrated conductivities as function of magnetic latitudes derived by *Fejer* [1964].

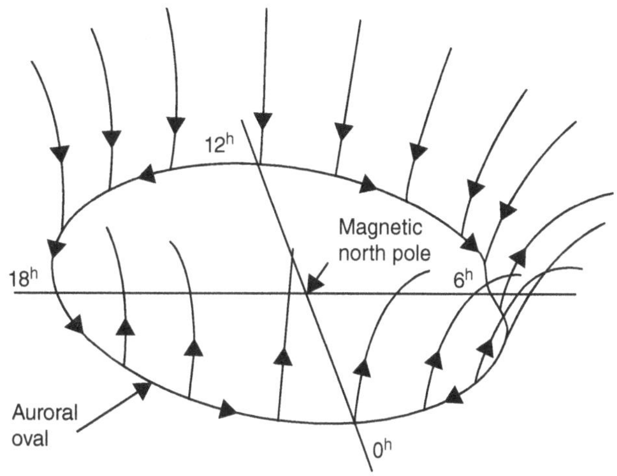

Figure 1.9 The high latitude part of Alfvén's current system showing connection between the auroral zone and field-aligned currents, with a downward current on the dayside and an upward current on the nightside, linked to an eastward current in the evening side and a westward current in the morning side along the auroral oval. From *Alfvén* [1940].

Such data were also widely used to derive electric fields and neutral winds in the upper atmosphere when world-wide models for the ionospheric conductivities were introduced. *Fejer* [1964] calculated such models (Fig. 1.8) that played an important role as input to many of the analyses of the global variations in terms of *F*-region neutral winds and electric fields.

1.6. HANNES ALFVÉN INTRODUCED PLASMA PHYSICS TO MAGNETOSPHERIC RESEARCH

In 1939 the Swedish physicist Hannes Alfvén (1908–1969) published a theory about magnetic storms and the aurora where he criticized the assumption by Chapman and Ferraro that the magnetic field inside the beam is zero [*Alfvén*, 1939]. Later on he offered them an excuse:

> "The assumption that the magnetic field inside the beam is zero was natural at the time when Chapman and Ferraro made it" [*Alfvén*, 1955, 50–64].

Alfvén maintained that the beam possessed a magnetic field that was "frozen in" into the highly conducting matter, and due to the motion of the beam, this magnetic field produced an electric field. He stated, "that, an ionized, but on the average neutral, stream emitted from the sun must be electrically polarized due to its motion in the solar magnetic field. In fact any conductor–and the stream is certainly a good conductor–moving with the velocity v in a magnetic field H becomes polarized so that it posses an electric field E = v/c H. The direction of the field is perpendicular to the magnetic field as well as to the velocity" [*Alfvén*, 1939]. Alfvén introduced a forbidden region in space outside the Earth and argued:

> "Up to the borderline of the forbidden region the ions and electrons neutralize each other so that the resultant space charge is small. On the day side of the borderline the positive ions are no longer neutralized by the electrons. Hence a positive space charge is built up. On the other hand, on the night side near the borderline the electrons are not neutralized by the positive ions. This means that a negative space charge is produced. The positive space charge on the day side and the negative space charge on the night side of the borderline may neutralize each other through a discharge along the magnetic line of force. Charge is transported from the equatorial plane along the magnetic lines of force to the upper atmosphere, which is hit along one curve around each pole, which is the projection along the lines of force of the borderline upon the earth's surface. This curve, over which the discharge takes place, marks the region where the aurora occurs." [*Alfvén*, 1939, ch. 6]

An eastward current in the equatorial plane was also introduced all around the globe, the equatorial ring current, as Störmer proposed more than 20 years before. Field-line transport of electric charges was again connected to the creation of aurora as Birkeland did about 40 years earlier.

Later, Alfvén improved his model and an eastward current was introduced on the evening side along the auroral oval and a westward current on the morning side (Fig. 1.9). Alfvén's work marked a change of paradigm and introduced plasma physics to the understanding of magnetospheric processes.

1.7. INTO THE SPACE AGE

In 1973, *McPherron et al.* [1973] carried out an extensive study of the cause of auroral substorms where they combined ground and geosynchronous magnetic field observations, and introduced a current system including the so-called current wedge (Fig. 1.10) [*McPherron et al.*, 1973]. Like Alfvén, their current system included field-aligned currents between the nightside equatorial plane and the polar latitudes feeding the auroral electrojet. No currents, however, were drawn on the dayside.

Alfvén inspired young scientists in Sweden to engage themselves in studies of magnetospheric physics. One of them was Rolf Boström, who in 1964 wrote a paper about the polar substorm current system where he drew up two models for the currents [*Boström*, 1964]. One was based on the idea of a line current as introduced by Birkeland but with currents flowing along magnetic field lines in contrast to vertically, as Birkeland proposed. In the ionosphere, the current formed the auroral electrojet probably confined to the visible auroral structure. Boström specified the ionospheric current in more details than previously discussed, and wrote:

> "However, since the Hall current must be continues across the boundary of the electrojet a southward polarization electric field will be produced which lowers the northward current component in the electrojet. It will also drive a westward Hall current in the region of enhanced conductivity, the net effect is an intense confined electrojet" [*Boström*, 1967].

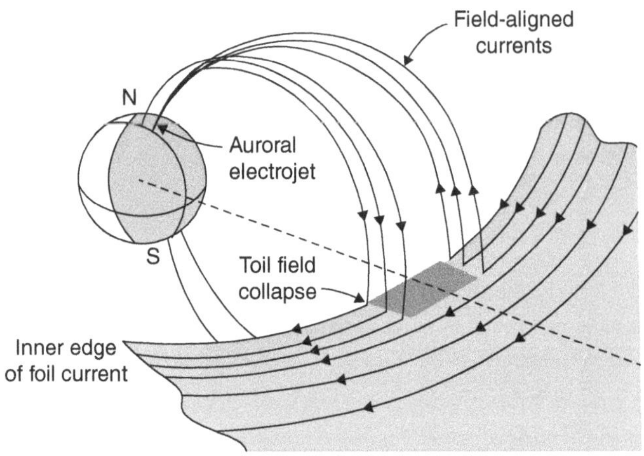

Figure 1.10 The current wedge as introduced by *McPherron et al.* [1973].

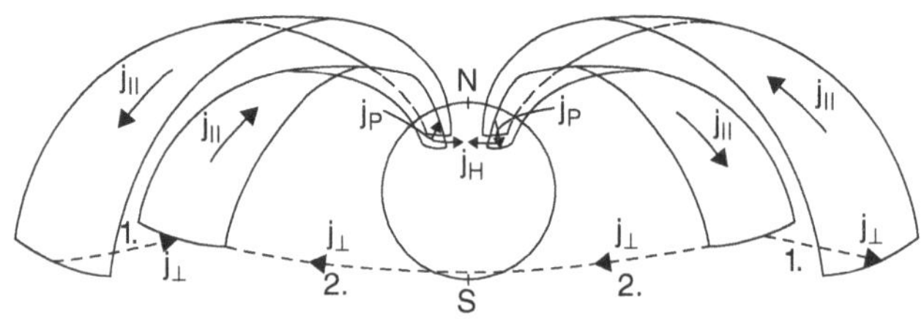

Figure 1.11 The second current system proposed by Boström including the Hall and Pedersen currents in the ionosphere. From *Boström* [1964].

In his second model (Fig. 1.11), Boström assumed that plasma motions in the magnetosphere drove the current system and that Pedersen currents in the ionosphere were linked to sheet currents flowing from the northern and southern edges of the electrojet that is represented by the Hall current. This was actually a further development of Birkeland's original sheet current model. According to Boström, the Pedersen current in the ionosphere represents an energy-loss mechanism since its flow is along the electric field. The corresponding current that flows transverse to the sheet currents in the magnetosphere brakes the plasma motions there, and represents the dynamo, the driving mechanism for the current system [*Boström*, 1964].

In 1970, Armstrong and Zmuda, based on satellite measurements of transverse magnetic disturbances in the auroral oval, presented a field-aligned current system that was "found to fit qualitatively a two-sheet current model proposed by R. Boström. Currents flow into the ionosphere along the higher-latitude sheet and out along the lower-latitude sheet" (7122–7127). An equatorward Pedersen current linked to a downward current from the magnetosphere formed by precipitating positive particles and thermal electrons streaming out of the ionosphere [*Armstrong and Zmuda*, 1970].

Park and Cloutier [1971, 7714–7733] derived from a rocket-borne experiment at Fort Churchill, Canada, a current system adjacent to a quiet auroral arc. Magnetic field variations were measured by a vector magnetometer in vicinity to the arc. "The data were interpreted in terms of a model current system consisting of a northwestward electrojet and two oppositely directed Birkeland's sheet currents, all lying in planes approximately parallel to the auroral arc." A westward electrojet was situated at the position of the arc connected to an upward field-aligned current due to precipitating electrons of 2 to 18 kev. The two field-aligned currents were linked by a Pedersen current in the northward direction driven by a northward field in opposition to the model by Armstrong and Zmuda. According to the authors: "The over–all current configuration differs from theoretical models proposed by Boström and Atkinson." The main reason for this discrepancy was probably the orientation of the arc. Boström's model predicted an eastward current, while the rocket experiment occurred in a westward electrojet. Furthermore, in Boström's model the sheet currents flowed at the edges of the arc where there were horizontal gradients in the conductivity, rather than within the arc, and the electrojet was located between the current sheets. Boström's model also predicted an electric field perpendicular to the arc while the rocket experiment indicated that the field could not be perpendicular to the arc and the electrojet might have been a combination of a Pedersen and a Hall component in contrast to Boström's that indicated that the electrojet was driven by a Hall current only.

1.8. THE RELATION OF THE IONOSPHERIC CONDUCTIVITIES TO THE CURRENTS

So far in the search for a global current system with special emphasis on the auroral region, individual in situ measurements had been performed by rockets to carry out what were basically case studies, and little was known with respect to the time-dependent interplay between electric fields and ionospheric conductivities in forming the details of the currents. When an incoherent scatter radar was introduced to the auroral region in Chatanika, Alaska, in 1971, a new area was opened for ionospheric research at high latitudes. Here *Brekke et al.* [1974], for the first time were able to make long time series measurements of the time development of the electric field and neutral wind components as well as the height-integrated Hall and Pedersen conductivities during auroral substorms. These results (Fig. 1.12) showed that the electric field was strongly polarized, being northward in the evening and southward in the morning. The conductances, however, were most strongly enhanced during southward fields in the morning hours, and the neutral wind played a relative weak role compared with the electric field.

On 13–14 March 1972, the radar observed a very special event (Fig. 1.13). The southward E-field (E_x) was strongly enhanced with respect to the westward field (E_y) with a factor of $E_x / E_y = 3$ at the same time as the Hall conductance Σ_H was enhanced by a factor of 3 with respect to the Pedersen conductance Σ_P. As it turned out that the northward electric current corresponding to fluctuations in the magnetic D-component was close to zero, the event very closely matched the Boström model of the polarization current in an auroral arc. The strength of the westward auroral electrojet was given by $J_y = \Sigma_P (1 + (\Sigma_H/\Sigma_P)^2)E_y \approx 10\Sigma_P E_y$, and the polarization factor was close to 10.

Iijima and Potemra [1976] presented a summary of the distribution and flow directions of large-scale field-aligned currents determined from observations of field-aligned currents at 800 km altitude in the high-latitude region by the satellite Triad (Fig. 1.14).

The currents on the equatorward side are the so-called Region 2 currents, which are into the ionosphere in the evening and out of the ionosphere in the morning. The currents on the poleward side are the so-called Region 1 currents, which are into the ionosphere in the morning and out of the ionosphere in the evening.

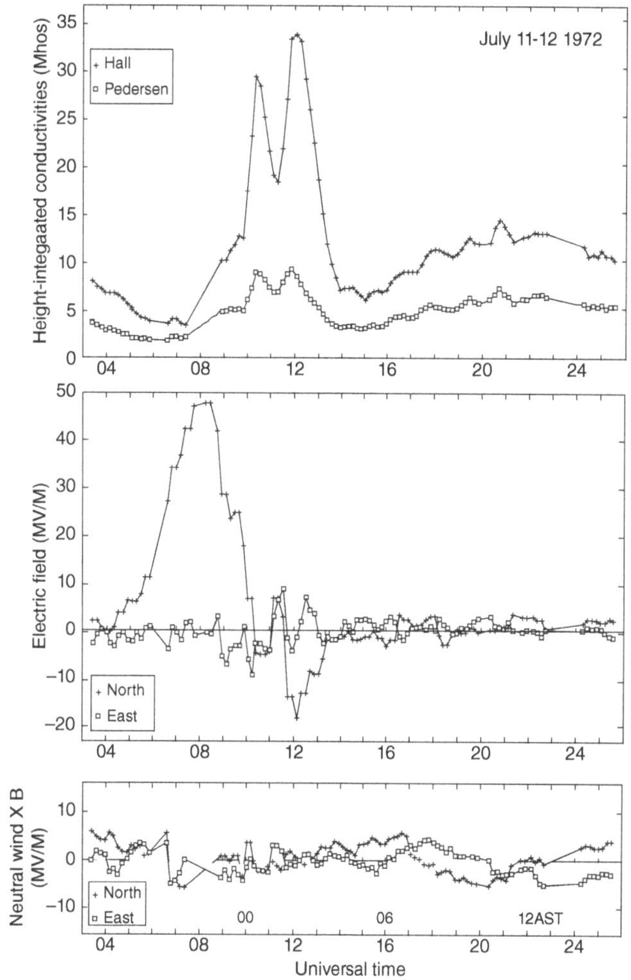

Figure 1.12 (Top) Height-integrated Hall and Pedersen conductivities; (middle) horizontal electric field components; (bottom) horizontal neutral wind components (u x B). All derived in the auroral zone by the incoherent scatter radar Chatanika, Alaska, 11–12 July 1972. From *Brekke et al.*. [1974].

By combining the observations obtained from Triad with the measurements done by Chatanika radar and later confirmed by the EISCAT (The European Incoherent Scatter Association) incoherent scatter radar in Scandinavia (Fig. 1.15), a current system prevails that is much in agreement with the second model by Boström. On evening side, a downward field-aligned current is flowing into the ionosphere on the equatorward side and an upward current on the poleward side of the auroral oval. A Pedersen current aligned with the northward E-field links these currents. On the morning side, the direction of the currents and the electric field are reversed. The ionosphere is therefore a load to the system. In the equatorial plane, however, the currents are directed against the electric field of the magnetotail. The current

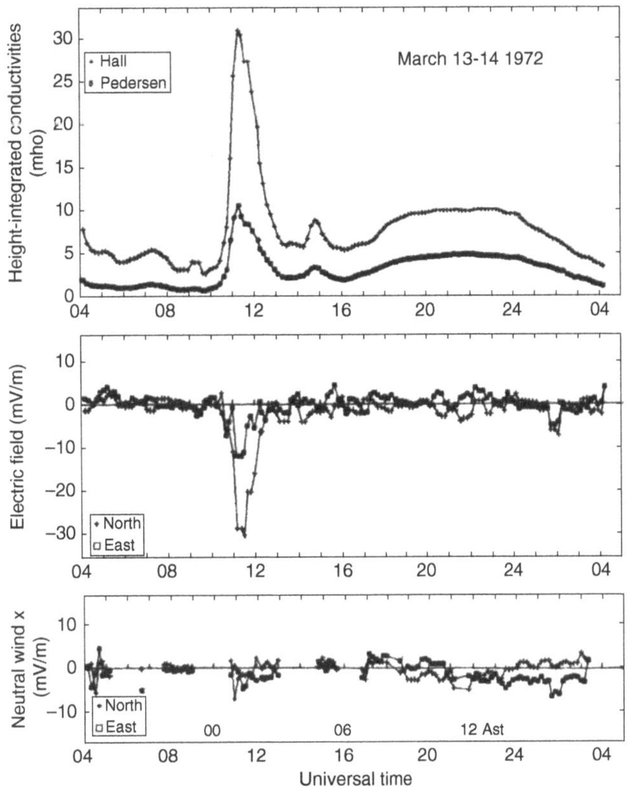

Figure 1.13 (Top) Height-integrated Hall and Pedersen conductivities and (bottom) the horizontal electric field components in the west and southward directions derived by the Chatanika radar on 13–14 March 1972. From *Brekke et al.* [1974].

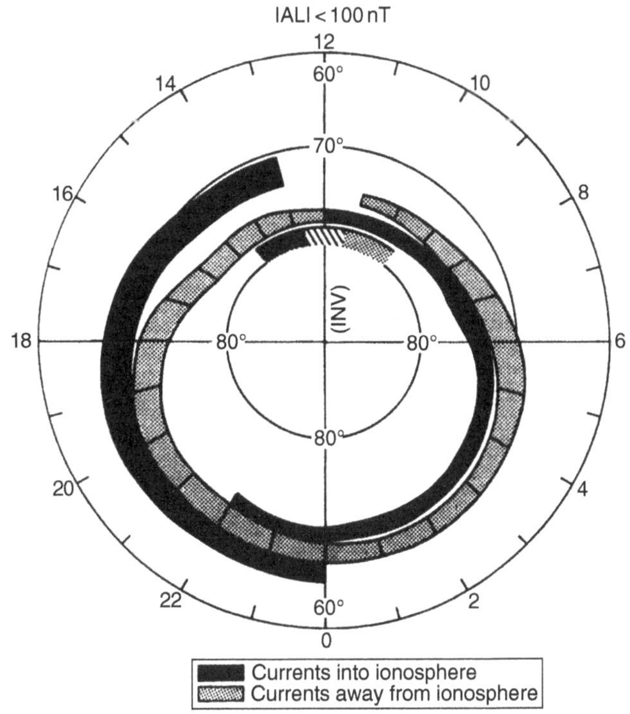

Figure 1.14 Schematic model of observed field-aligned currents as deduced from satellite observations by *Iijima and Potemra* [1976].

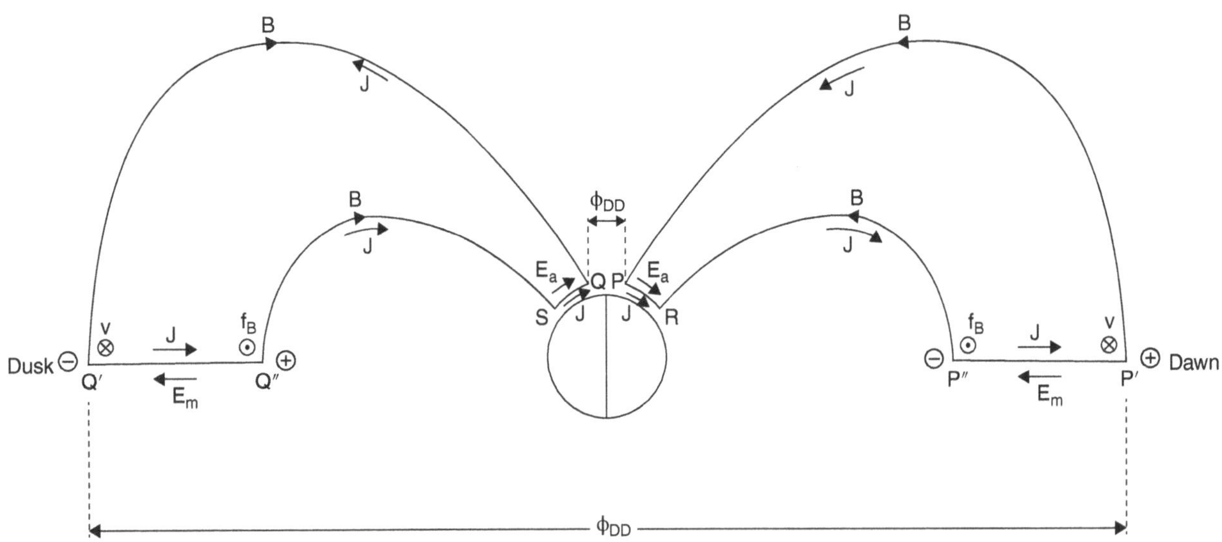

Figure 1.15 A synthesis of the satellite observations of field-aligned currents by the Triad satellite and ground-based observations of electric fields and conductances by the Chatanika and EISCAT incoherent scatter radars showing the current system between the nightside magnetosphere and the auroral zone. From *Brekke* [2013].

force $f_B = j \times B$ is therefore set up as a brake against the magnetospheric convection v.

1.9. CONCLUSION

This review shows that the progress of our understanding of the electric currents in geospace has gone through a progressive development from the time of the Enlightenment in the early eighteenth century to the Space Age in the 1970s. Important steps were made in the late part of the nineteenth century when magnetic field variations were found to be caused by electric currents in the upper atmosphere. The aurora borealis was believed to be an electric phenomenon by several authors as early as the 1750s. The current system linking the creation of the aurora became a main field of interest in the beginning of the twentieth century and has remained so until our time.

At present, we have a large diversity of instruments and methods such as satellite and ground-based experiments of different kinds and capacities as well as dedicated computer models to study these current systems further. What appears to be lacking, however, is a more detailed knowledge of the variations of the interplay between the ionospheric conductivities, electric fields, and neutral winds in space and time, that is, the three most important factors in determining the ionospheric currents. So far, observations of these parameters are available partly integrated over large areas and extended time periods, but that does not satisfy studies of detailed variations of the plasma processes taking place in the upper atmosphere during magnetospheric storms and auroral displays.

The plan to install a phased array incoherent scatter radar system EISCAT_3D in Scandinavia, with unprecedented spatial and time resolutions, is anticipated to lead the science of currents in geospace into a new era.

REFERENCES

Alfvén, H. (1939), A theory of magnetic storms and of the aurorae, Kungl. Svenska Vetenskapsakademien, III, *18* (3), Stockholm.

Alfvén, H. (1940), A theory of magnetic storms and of the aurorae II, The Aurorae III, The magnetic disturbances, Kungl. Svenska Vetenskapsakademien–Akad. Handlingar, II, *18* (9), Stockholm.

Alfvén, H. (1955), On the electric field theory of magnetic storms and aurorae, *Tellus*, *7*(1), 50–64.

Angot, A. (1897), *The Polar Aurora*, Appleton & Company, New York, *158*.

Armstrong, J. C., and A. J. Zmuda (1970), Field-aligned current at 1100 km in the auroral region measured by satellite, *J. Geophys. Res.*, *75*(34), 7122–7127; doi:10.1029/JA075i034p07122.

Birkeland, K. (1896), *Sur les Rayons Cathodiques sous l'Action de Forces Magnétique Intenses*, Archives des Sciences Physique et Naturelle, T. *1*, 497–512.

Birkeland, K. (1902), Expédition norvégienne de 1899–1900 pour l'étude des aurores boréales. Résultats des recherches magnétiques, Skrifter Videnskabsselskabet I Christiania 1901, I, *Mat.- Nat. Klasse. Christiania*, *70*.

Birkeland, K. (1908), The Norwegian Aurora Polaris Expedition, 1902–1903, I, *On the Cause of Magnetic Storms and the Origin of Terrestrial Magnetism. Christiania*, Aschehoug & Co., Kristiania.

Boström, R. (1964), A model of the auroral electrojet, *J. Geophys. Res.*, *69*, 4983–4999.

Boström, R. (1967), Currents in the ionosphere and magnetosphere, *The Birkeland Symposium on Aurora and Magnetic Storms,* Centre National de la Recherche Scientifique., edited by A. Egeland and J. Holtet, Centre national de la recherche scientifique, Paris, 45.

Brekke, A. (2013), *Physics of the Polar Upper Atmosphere*, 2ed., Springer and Praxis Publishing, Chichester, UK, 184.

Brekke, A., J. R. Doupnik, and P.M. Banks (1974), Incoherent scatter measurements of *E* region conductivities and currents, *J. Geophys. Res.*, *79*(25), 3773–3790.

Chapman, S. (1918), An outline of a theory of magnetic storms, *Proc. Roy. Soc. A.*, *95* (666), 78.

Chapman, S. (1919), The Solar and Lunar Diurnal Variations of Terrestrial Magnetism, *Phil. Trans. Roy*, London, A218, 1–118.

Chapman, S. (1935), The electric current-systems of magnetic storms, *J. Geophys. Res.*, *40*(4), 349–370; doi:10.1029/TE040i004p00349.

Chapman, S., and J. Bartles (1940), *Geomagnetism*, Oxford University Press, Oxford, 471.

Chapman, S., and V. C. A. Ferraro (1930), A new theory of magnetic storms, *Nature*, *126*, 129–130.

Chernouss, S. A. (2012), Ideas of Lomonosov in auroral research, *Geophysica*, *48*(1–2), 105–107.

De La Rive, M. A. (1849), On the diurnal variations of the magnetic needle, and the aurora borealis; Extract of a letter to M. Arago, translated for Journal of the Franklin Institute from Annales de Chemie et de Physique, 3me. serie, 25, 310, *J. Frank. Inst.*, *17*, 40–46.

Desaguliers, J. T. (1742), Some conjectures concerning electricity, and the rise of vapours, *Phil. Trans. 1742–1743*, *42*, 140–143.

Eather, R. (1980), *Majestic Lights, The Aurora in Science, History and Arts*, American Geophysical Union, Washington, DC, 59.

Fejer, J. A. (1964), Atmospheric tides and associated magnetic effects, *Rev. Geophys.*, *2*, 275.

Ferraro, C. V. A.(1933), A new theory of magnetic storms, a critical survey, *The Observatory*, August, 253–259.

Franklin, B. (1779), Aurora borealis supposition and conjunctures towards forming an hypothesis for its explanation, read for the French Académie des Sciences, from *Memories of the Life and Writings of Benjamin Franklin,* 2ed, London, printed for Henry Colburn, 1818–1819, Part IV, Philosophical Subjects, 291–297.

Graham, G. (1724–1725), An account of observations made of the variation of horizontal needles at London, in the latter part of year 1722, and beginning of 1723, *Phil. Trans.*, *33*, 96–107.

Halley, E. (1716), An account of the late surprising appearance of the lights seen in the air, *Phil. Trans. Roy. Soc.*, *29*, 406.

Hiorter, O. P. (1747), Om magnet-nålens Åtskillige ändringar, som af framledne Professoren Herr Anders, Celsius blifvit i akt tagne och sedan vidare observerade, Jamt nu framgifne, *Kongl. Swen. Wetenskaps. Acad. Handlingar*, *8*, 27–43,.

Iijima, T., and T. Potemra (1976), Field-aligned currents in the dayside cusp observed by Triad, *J. Geophys. Res.*, *81*(34), 5971–5979.

Lindemann,. F. A. (1919), Note on the theory of magnetic storms, *The London, Edinburgh, and Dublin Philosophical Magazine and Journal of Science*, *38* (228), 669–683.

Matsushita, S. (1967), Solar quiet and lunar daily variations field, Chap. III-1, *Physics of Geomagnetic Phenomena Academic Press*, edited by S. Matsushita and W. H. Campbell, New York, 302–424.

McPherron, R. L., C. T. Russell, M. G. Kivelson, and P. J. Coleman Jr. (1973), Substorms in space: The correlation between ground and satellite observations of the magnetic field, *Radio Sci.*, *8*(11), 1059–1076; doi:10.1029/RS008i011p01059.

Park, R. J., and P. A. Cloutier (1971), Rocket-based measurement of Birkeland currents related to an auroral arc and electrojet, *J. Geophys. Res.*, *76*(31), 7714–7733; doi:10.1029/JA076i031p07714.

Pontoppidan, E. (1752), Det første forsøg paa Norges Naturlige Historie, Kjøbenhavn, 7–23.

Schuster, A. (1907), The Diurnal Variation of Terrestrial Magnetism, *Proc. Roy. Soc. London*, *80*(535), 163–204.

Schuster, A. (1911), The origin of magnetic storms, *Proc. Roy. Soc. Lond. A*, *85*, 44–50.

Stewart, B.(1886), On the cause of the solar-diurnal variation of terrestrial magnetism, *Proc. Roy. Soc. Lond., B*, *1*, 38–49.

Størmer, C. (1955), *The Polar Aurora*, Oxford at the Clarendon Press.

Wilke, J. C. (1777), Rön om Magnet-Nålens årliga och dageliga ändringar i Stockholm, *Kongl. Vetenskps Academiens Handlingar*, *38*, 274–300.

2

Current Systems in Planetary Magnetospheres: A Comparative Overview

Krishan K. Khurana and Jiang Liu

ABSTRACT

Magnetospheric fields and plasmas are confined and guided by various forces such as thermal pressure, magnetic pressure, plasma inertia, and centrifugal forces. Any imbalance between these forces results in a magnetic Lorentz force expressed through the deformation of the ambient field or equivalently the generation of electric currents. A knowledge of large-scale currents therefore provides valuable clues about the physical processes operating in that magnetosphere. In addition, field-aligned currents (FACs) mediate linear and angular momentum over long distances between the ionospheres to the magnetosphere. An understanding of these currents aids in characterizing the magnetosphere-ionosphere coupling processes. In this chapter, we compare and contrast various current systems observed in planetary magnetospheres with those observed at Earth. We show that in the magnetospheres of Jupiter and Saturn, the centrifugal force contributes significantly to the generation of the azimuthal ring current (ARC). Further in these magnetospheres, the corotation enforcement currents (CEC), which flow in the radial direction in the current sheet, rival the strengths of their ARCs. We also review our knowledge of the FACs in these magnetospheres. We pay special attention to local time asymmetries in the distributions and strengths of the major current systems and relate them to internal and external convection drivers.

2.1. INTRODUCTION

The fact that current systems in space create observable field perturbations on the surface of the Earth was recognized long before the advent of the Space Age. The existence of a ring current in space around the Earth was first postulated by *Chapman* in 1918 to explain the reduction of the strength of the magnetic field in the equatorial plane of the Earth during a large magnetic storm. The existence of a magnetopause current system that separates the field lines of the Earth's magnetosphere from those of the solar wind was intuited by *Chapman and Ferraro* [1931] to explain a sudden short-duration enhancement of the field at the commencement of a solar storm. Similarly, guided by observations from polar magnetic observatories, *Birkeland* [1908] postulated the existence of FACs above the atmosphere of the Earth in the polar regions. In situ spacecraft observations beginning in 1957 not only confirmed the existence of these current systems but also elucidated how they are generated, and the role FACs play in transmitting stresses between the solar wind, magnetosphere, and the upper atmosphere (ionosphere/thermosphere). One of the first current systems to be detected in the in situ observations was the Chapman-Ferraro current system (Fig. 2.1), which helps divert the magnetosheath plasma around the Earth's magnetosphere. The currents flowing in a thin current sheet (the magnetopause) keep the plasmas of the

Department of Earth, Planetary and Space Sciences and Institute of Geophysics and Planetary Physics, University of California, Los Angeles, California, USA

Electric Currents in Geospace and Beyond, Geophysical Monograph 235, First Edition.
Edited by Andreas Keiling, Octav Marghitu, and Michael Wheatland.

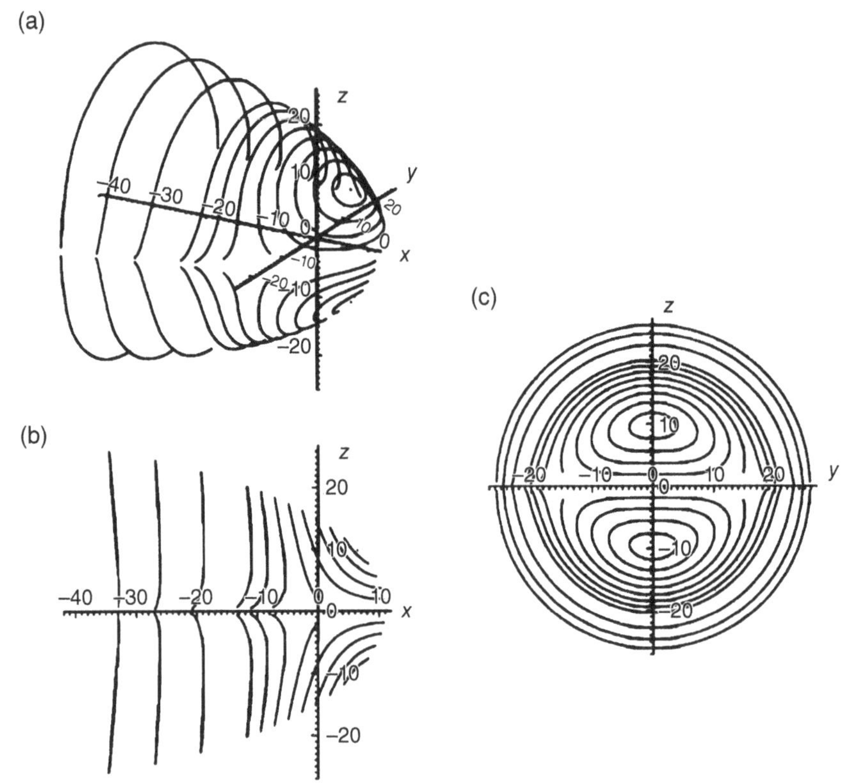

Figure 2.1 Current flow lines on the magnetopause: (a) viewed from the dawn flank of the magnetosphere, (b) projected onto the *x-z* plane, (c) projected onto the *y-z* plane and viewed from the solar direction. Figure from *Sotirelis et al.* [1994].

magnetosheath and the magnetospheres separated and also mediate the stresses between the two systems. For example, on the dayside near the magnetic equator, the current system enhances the field inside the magnetopause and nulls it outside (Fig. 2.1).

The implied $J \times B$ force from the currents within the magnetopause is sunward in the equatorial plane and opposes the dynamic pressure (ρV_{sw}^2) of the solar wind keeping it out of the magnetosphere. In the magnetotail, the Chapman-Ferraro current system merges with the cross-tail current sheet field, keeps the two plasma populations separated, and also stretches the Earth's magnetic field lines into the two magnetotail lobes. When large solar wind dynamic pressure pulses impact the Earth's magnetosphere, rippling of the magnetopause is known to drive twin convection vortices (TCVs) in the ionosphere [*Lanzerotti et al.*, 1987; *Kivelson and Southwood*, 1991]. Other than the TCV-linked phenomena, the effect of Chapman-Ferraro current systems in driving convection or dynamics deep inside the magnetosphere is minimal, and for this reason, we will exclude further discussion of these currents in the outer planet magnetospheres.

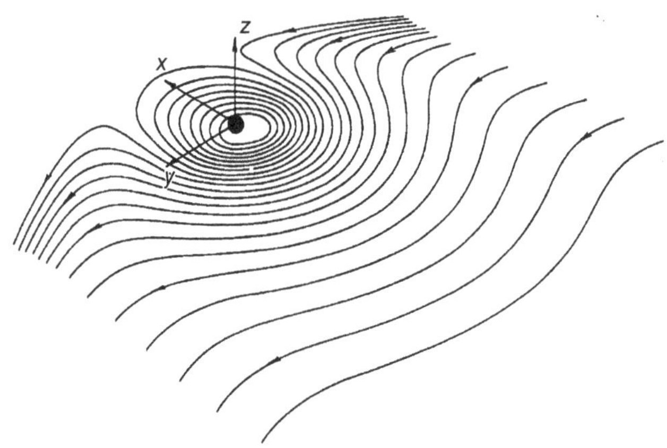

Figure 2.2 A three-dimension view of the current flowing in the equatorial plane of the magnetosphere. Figure from *Tsyganenko* [1990].

The strongest currents in the Earth's magnetosphere flow near the magnetic equator. For example, the ring current, with a strength of 2–4 MA, flows in closed loops in an equatorial current sheet (radial distance, $r < 10\ R_E$) and merges into the magnetotail current on the nightside beyond this distance (Fig. 2.2). The strength of the tail current exceeds 10 MA and is closed at the dawn and dusk flanks of the magnetotail by the Chapman-Ferraro current system of the magnetopause. The origin of both of these currents lies in the diamagnetic nature of plasma because ions and electrons gyrating in the Earth's magnetic field create magnetic fields that oppose the background field. Together, the ring current and the tail current can depress the strength of Earth's magnetic field on its surface by several hundred nT. *Dessler and Parker* [1959] and *Sckopke* [1966] showed that the surface perturbation field is directly proportional to the energy content of the plasma in the ring and tail currents and is given by:

$$\frac{\Delta B}{B} = \frac{2K}{3U_M} \tag{2.1}$$

where ΔB is the magnetic perturbation at the Earth's center from the ring and tail currents, B is the equatorial surface field of the Earth's dipole, K is the total kinetic energy of the particles in the ring and tail currents, and U_M is the energy contained in the Earth's field outside of its surface. The energy content of the ring + tail current particles varies appreciably over time and is strongly influenced by solar wind and IMF conditions. For example, during a major magnetic storm, coronal mass ejections from the Sun are known to depress the Earth's surface field by several hundred nT [*McPherron*, 1991]. The recovery of the Earth's field can take several days because the main loss process for the ring current particles is charge exchange with the Earth's neutral corona, which has a time constant of several days [*McPherron*, 1991].

Another important current system is a partial ring of plasma centered in the midnight sector of the magnetosphere and is accompanied by Region 2 type FACs that close into the ionosphere (Figs. 2.3 and 2.4a). Overall, this current system enhances the strength of the magnetospheric field on the dayside and depresses it on the nightside. The origin of this current system also lies in convection in the magnetosphere driven by the solar wind. Reconnection in the deep magnetotail injects solar wind particles into the magnetotail enhancing particle density and pressure in the near-Earth region on the nightside. The excess plasma is kept confined to the magnetosphere by enhanced ring current in that region depressing the local field. Maintenance of current continuity then requires downward FAC at dusk and upward FAC at dawn, with corresponding closure via westward current flowing in the nightside inner magnetosphere.

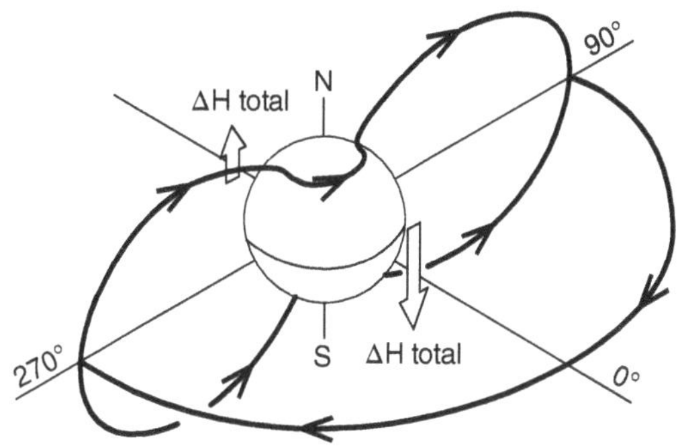

Figure 2.3 Partial ring current centered on the midnight sector of the Earth's magnetosphere and accompanied by field-aligned currents of Region 2 sense. Figure from *Fukushima and Kamide* [1973].

Finally, closely related and having a similar origin are the Region 1 sense FACs that close poleward of Region 2 currents in the ionosphere (Figs. 2.4a and b) and have an opposite sense to the Region 2 currents [*Iijima and Potemra*, 1976]. The origin of Region 1 sense currents lies in the high latitude magnetosphere/magnetosheath "generator," which provides a downward Poynting flux into the ionosphere and imposes dawn to dusk Pedersen currents flowing across the polar cap where momentum captured from the solar wind is transferred into the ionosphere [*Cowley*, 2000]. Region 2 currents ensure continuity of the auroral zone currents in the lower latitude regions and the near-Earth magnetotail (Fig. 2.4a). Region 1 and Region 2 current systems (Fig. 2.4) together drive convection in the Earth's magnetosphere and the ionosphere. In the polar cap, the newly reconnected flux tubes are dragged by the solar wind into the magnetotail, driving the antisunward convection seen in the polar cap. The flux transported to the nightside, returns to the dayside at low latitudes after reconnecting in the magnetotail, driving the sunward convection at low latitudes of the ionosphere. In general, Region 1 sense currents are stronger than Region 2 currents because they are fed by Pedersen currents on both open and closed field lines. For example, during quiet times, the Region 1 current has a strength of 1.6 MA while Region 2 current has a strength of ~1.1 MA [*Iijima and Potemra*, 1978]. During disturbed times, both of the currents are intensified with a strength of 2.7 and 2.5 MA, respectively [*Iijima and Potemra*, 1978]. This discussion makes it clear that most of the current systems in the Earth's magnetosphere owe their existence to the plasma entry and transport processes (abbreviated as plasma convection) induced by the solar wind. Thus, understanding their morphology, strength, and variability provides key insights into how the solar wind populates and drives dynamics in the Earth's magnetosphere.

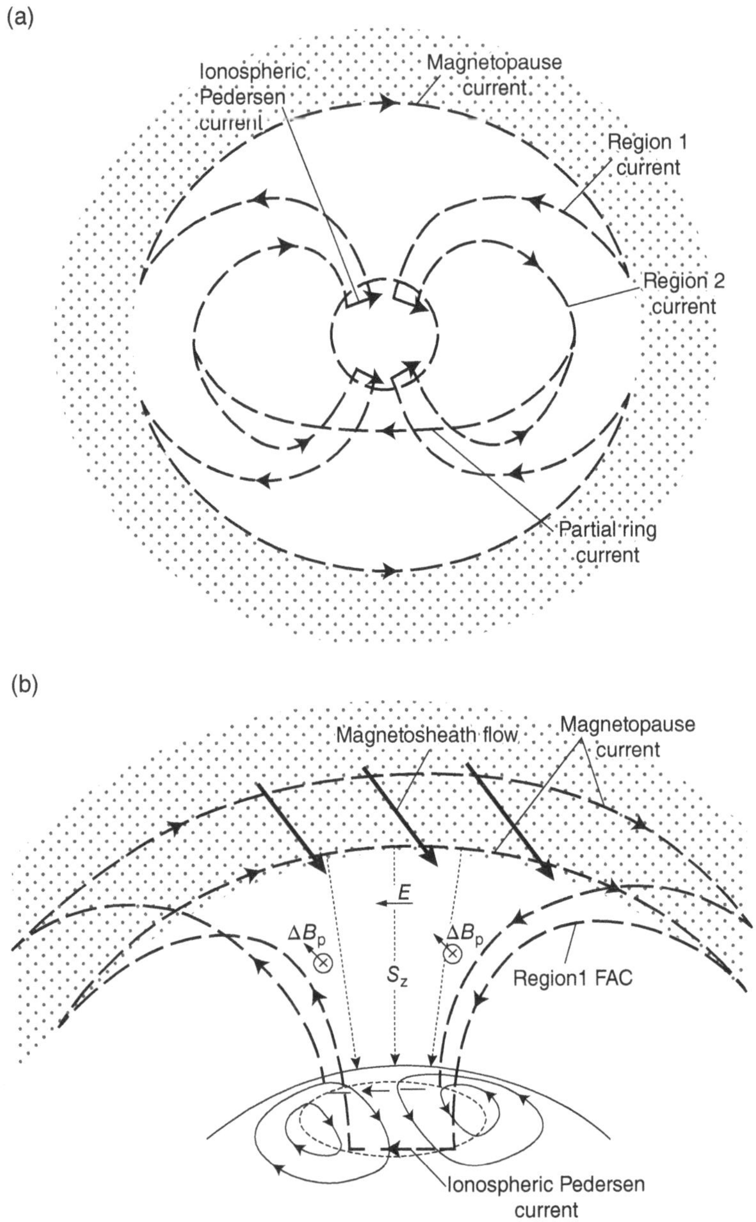

Figure 2.4 (a) The complete current circuit that links both Region 1 and Region 2 current systems and helps transfer momentum and mass from the solar wind into the Earth's magnetosphere from the solar wind. (b) A close-up of the polar cap current circuit closing in the northern ionosphere. The current circuit produces a sunward directed magnetic field in the polar region and a dawn to dusk electric field (from tailward convection of the flux) yielding a net downward Poynting flux of electromagnetic energy into the ionosphere. Both figures from *Cowley* [2000].

2.2. DETERMINING CURRENT STRENGTHS IN OUTER PLANET MAGNETOSPHERES: THE BASICS

The magnetic field at any location in a magnetosphere can be represented as the sum of fields generated by distant currents (such as those in the interior of a planet or in the magnetopause boundary) and those generated by local currents (such as the ring current or the tail current). The former can be expressed by the Laplace equation $\mathbf{J} = \nabla \times \mathbf{B} = -\nabla^2 V = 0$, where V is a scalar potential. In spherical coordinates (r, θ, φ), the Laplace equation admits solutions of the form

$$B = -\nabla\left[V^i + V^e\right]$$
$$= -a\nabla\left[\sum_{n=1}^{N}\sum_{m=0}^{m=n}\left\{\begin{aligned}&\left(\frac{a}{r}\right)^{n+1}\left(g_n^m \cos m\varphi + h_n^m \sin m\varphi\right)\\&+\left(\frac{r}{a}\right)^{n}\left(G_n^m \cos m\varphi + H_n^m \sin m\varphi\right)\end{aligned}\right\}P_n^m\right] \tag{2.2}$$

where a is the planetary radius, P_n^m are associated Legendre functions, g_n^m and h_n^m are the coefficients of the internal scalar potentials, and G_n^m and H_n^m are the coefficients of the external scalar potentials. Both internal and external spherical harmonic series are truncated to low degrees in n. Because, the contributions of higher degree terms of internal harmonics fall off as $(a/r)^{n+2}$ with radial distance, spacecraft that approach the planet the closest, provide the best internal field models. For this reason, at Jupiter, Voyager 1 and Pioneer 11 spacecraft provided the most accurate models of the internal field even though Galileo provided the bulk of local time coverage. The most reliable model of internal field at Jupiter so far is VIP4, which was derived from observations made by Voyager 1 and Pioneer 11 supplemented by information on the auroral location of the footprint of Io flux tube (from ground telescopic observations, *Connerney et al.* [1998]). Saturn's internal magnetic field is extremely axisymmetric (i.e., only terms that have $m = 0$ have finite amplitudes). Currently, the most reliable internal field model derived from Cassini observations is that by *Cao et al.* [2012], which consists of only axisymmetric terms up to $n = 5$. The dipole tilt is constrained to be $< 0.1°$ [*Cao et al.*, 2011]. At Uranus, even though data were available only from a single flyby, well-constrained models (with terms up to $n = 3$) have been presented by *Connerney et al.* [1987] and by *Stanley and Bloxham* [2006] because the large magnetic dipole tilt allowed Voyager 2 to sample a wide range of magnetic latitudes. Neptune's large dipole tilt (47°) has also been exploited to construct well-constrained internal field models (with terms up to $n = 8$) by *Connerney et al.* [1991] and by *Stanley and Bloxham* [2006].

In regions of the magnetosphere where local currents exist, two different approaches have been followed for modeling the magnetic field. *Connerney et al.* [1981] expressed the field in terms of a vector potential $(\nabla \times \mathbf{A} = \mathbf{B})$. Assuming that the strength of the ring current falls as $1/\rho$ with distance, they obtained analytical solutions of the form

$$A^{\pm} = \mu_0 I_0 \int_0^{\infty} J_1(\lambda\rho) J_0(\lambda a) \sinh(\lambda D) e^{\mp\lambda z} \frac{d\lambda}{\lambda^2} \quad \text{for } |z| > D$$

and

$$A^{i} = \mu_0 I_0 \int_0^{\infty} J_1(\lambda\rho) J_1(\lambda a) \left\{1 - e^{-\lambda D} \cosh(\lambda z)\right\} \frac{d\lambda}{\lambda^2} \quad \text{for } |z| < D \tag{2.3}$$

where D is the thickness of the current sheet, I_0 is the strength of the current at the inner edge, and λ is a variable over which the above integrations are evaluated. The magnetic field is obtained using $\mathbf{B} = \nabla \times \mathbf{A}$ and is in general calculated by integrating the equations numerically.

An alternative approach uses Euler potentials (f, g), which are constant along field lines, to define the magnetic field such that $\mathbf{B} = \nabla f \times \nabla g$. In a first application of these functions in Jupiter's magnetosphere, *Barish and Smith* [1975] specified Euler potentials of a form that represented the internal field and the field of Jupiter's current sheet such that inside the magnetosphere, the magnetic field strength fell as $1/r^2$ and the field was confined to a sphere of radius 100 R_J representing the boundaries of the magnetosphere. Later, *Goertz et al.* [1976] and *Khurana* [1997] presented alternative forms of Euler potentials that represented the observed field more accurately and also included the warping of the field lines caused by the sweepback of the field lines.

2.3. SPACECRAFT MISSIONS TO OUTER PLANETS

So far, eight spacecraft have visited the outer planet magnetosphere (Table 2.1). Whereas all eight of these spacecraft have visited Jupiter, three have visited Saturn, and only one spacecraft (Voyager 2) flew by Uranus and Neptune. For Jupiter, most of the information on the magnetosphere was provided by Galileo, which was a low-latitude orbiter and covered all local times and radial distances between 6 and 150 R_J. The spacecraft Juno arrived in the Jovian system in July 2016 and has begun a comprehensive survey of the Jovian system. At Saturn, Cassini recently concluded its 13 yr mission of exploring the Saturn system and provided information on the equatorial and midlatitude regions between the distance of 1.03 and 40 R_S.

2.3.1. Distinguishing Characteristics of the Outer Planet Magnetospheres

Table 2.2 provides the major characteristics of outer planets and their magnetospheres. The giant magnetospheres of Jupiter and Saturn rotate rapidly and are populated by plasma largely supplied by their own internal moons. The strong tilted internal magnetic field of Jupiter creates a large magnetosphere, which is further ballooned by plasma pressure and centrifugal forces exerted by its

Table 2.1 The Orbital Characteristics of the Outer Planet Spacecraft Missions.

Planet	Mission	Date	CA (R_p)	Orbit characteristic (relative to magnetic equator)
Jupiter	Pioneer 10	December 1973	2.9	Equatorial flyby
	Pioneer 11	December 1974	1.6	Midlatitude flyby
	Voyager 1	March 1979	4.89	Equatorial flyby
	Voyager 2	July 1979	10.11	Equatorial flyby
	Ulysses	February 1992	6.3	Midlatitude flyby
	Galileo	December 1995–September 2003	4.0	Equatorial orbiter
	Cassini	December 2000	137	Distant equatorial flyby
	New Horizon	February 2007	32.2	Equatorial magnetotail
Saturn	Pioneer 11	September 1979	1.35	Equatorial flyby
	Voyager 1	November 1980	3.07	Equatorial flyby
	Cassini	June 2004–September 2017	1.03	Equatorial and midlatitude orbiter
Uranus	Voyager 2	January 1986	4.19	Low and high latitude flyby
Neptune	Voyager 2	August 1989	1.18	Low and high latitude flyby

Table 2.2 Major Characteristics of Outer Planets and Their Magnetospheres.

Characteristics	Jupiter	Saturn	Uranus	Neptune
R_{planet} (km)	71,398	60,330	25,559	24,764
Spin period (hr)	9.9	10.7	17.2	16.1
Dipole surf. B (Gauss = 10^5 nT)	4.28	0.22	0.23	0.14
Dipole tilt	10°	0°	59°	47°
n_{sw} (#/cm^3)	0.4	0.1	0.03	0.005
Upstream IMF field (nT)	1.0	0.4	0.2	0.15
Major species	S^+,O^+,H^+	O^+, $H2O^+$, H^+	H^+	N^+, H^+
$R_{Msphere}$ (R_p)	50–100	16–22	18	23–26
Main plasma source	Io	Enceladus	Corona	Triton
Corotation	Near	Partial	Yes	Yes?
Source strength (kg/s)	1,000	300	0.02	0.2

Source: Bagenal [1992], *Kivelson and Bagenal* [2007], *Arridge* [2015].

heavy plasma (mostly ions of S and O). Jupiter's magnetospheric current sheet is seen to rock up and down relative to Jupiter's equator with a period of 9.9 hr. The plasma nearly corotates with Jupiter up to a distance of ~20 R_J [*Belcher*, 1983] but lags from corotation substantially farther outward [*Kane et al.*, 1995]. Saturn's magnetosphere is populated by neutral gas and plasma derived from Enceladus's south-pole plume. One major distinguishing characteristic of Saturn's magnetosphere is that neutral particles are more numerous than charged particles in its inner and middle regions by a factor of ~100 [*Jurac and Richardson*, 2005]. Another puzzling aspect of Saturn's magnetosphere is that despite the very high conductivity of its ionosphere (height integrated conductivity > 10 S; *Moore et al.* [2010]), the plasma lags corotation by more than 20% even at a radial distance of 4 R_S and even more farther out [*Wilson et al.*, 2009]. A comprehensive review of the literature on local-time asymmetries in the magnetodiscs of Jupiter and Saturn was recently presented by *Arridge et al.* [2015]. The magnetosphere of Uranus is quite sparse in its plasma content, which comes mainly from its own atmosphere. Neptune's magnetosphere appears to be populated by ions from its moon Triton with the dominant species being N^+. A large neutral torus was also inferred from the observed density profile of the charged species [*Richardson*, 1993].

2.3.2. Sources of Currents in a Magnetosphere

Currents and accelerations of plasma arise in a magnetosphere when imbalances occur between different forces, such as the pressure gradient force ($-\nabla p$), the centrifugal force ($\rho\boldsymbol{\omega}\times(\boldsymbol{\omega}\times\mathbf{r})$), Coriolis force ($2\rho\boldsymbol{\omega}\times\mathbf{u}$), and so on, according to the momentum equation

$$\frac{\rho du}{dt} = -\nabla\cdot\mathbf{P} + \mathbf{J}\times\mathbf{B} + \rho\boldsymbol{\omega}\times(\boldsymbol{\omega}\times\mathbf{r}) + 2\rho\boldsymbol{\omega}\times\mathbf{u} \quad (2.4)$$

written in a frame rotating with the planet. Here, **P** is the plasma pressure tensor, **ω** is the rotation vector of the planet, **B** is the local magnetic field, **J** is the local current density, and **u** is the plasma velocity. When pressure

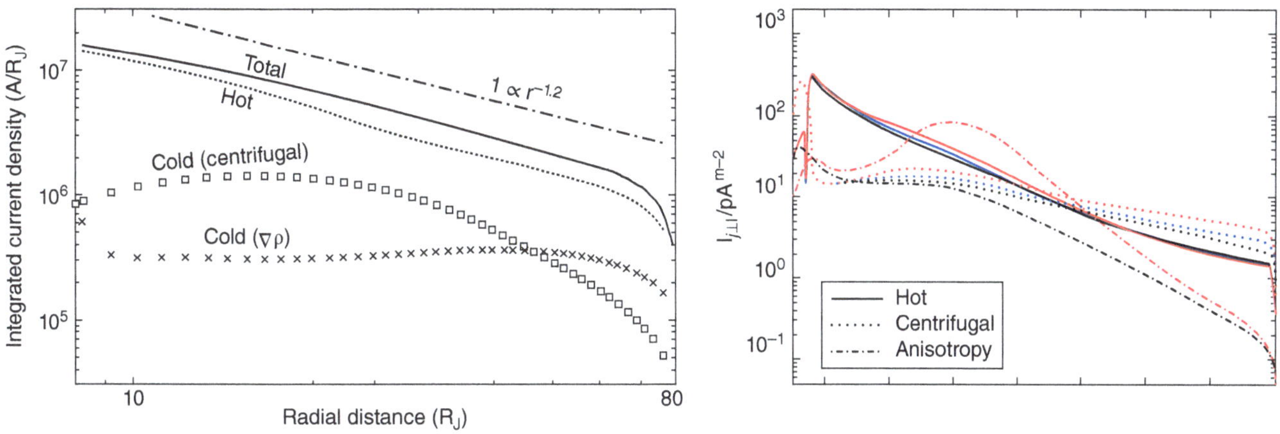

Figure 2.5 (Left) Contributions of pressure (from cold and hot populations) and centrifugal force in generating the ring current in Jupiter's magnetosphere (figure reproduced from *Caudal* [1986]). (Right) Equatorial value of azimuthal current density generated by energetic particles pressure (solid curves), centrifugal force (dashed lines), and plasma anisotropy (broken lines). The black, blue, and red lines correspond to energetic particle anisotropies ($p_{||}/p_{\perp}$) of 0.85, 1.0, and 1.15, respectively. Figure from *Nichols et al.* [2015].

anisotropy is small, the pressure term can be approximated as $-\nabla p$, where p is the scalar plasma pressure. In most regions of planetary magnetospheres, the plasma pressure gradient force and the centrifugal force are directed radially outward. A strong azimuthal current (called the ring current) arises in the magnetosphere such that the field lines apply a net inward Lorentz force ($\mathbf{J}\times\mathbf{B}$) on the plasma to contain it within the magnetosphere. In a steady state, plasma acceleration may not result. *Caudal* [1986] quantitatively assessed the contributions of various forces in generating the ring current. Using magnetic field and particle data from Voyager 1 spacecraft, he showed that in Jupiter's magnetosphere, plasma pressure gradient resulting from the hot particles dominates the generation of the ring current (Fig. 2.5, left). Only beyond a radial distance of 80 R_J, does the centrifugal force become strong enough that its contribution to ring current becomes appreciable. *Nichols* [2011], using more modern data and self-consistent profiles of plasma velocity in the magnetosphere, reached a similar conclusion. *Nichols et al.* [2015] have revisited force balance in Jupiter's magnetodisc by including the effects of hot plasma anisotropy. They show that inward of 20 R_J and outward of 50 R_J, plasma pressure gradient and centrifugal force indeed contribute predominantly to the ring current generation, while plasma anisotropy current can dominate in between (Fig. 2.5, right).

Sergis et al. [2010] have examined the contributions of plasma pressure and centrifugal forces in generating the ring current in Saturn's magnetosphere. They find that because the plasma cools substantially inside of 9 R_S by interactions with the neutral torus of Enceladus and the E-ring dust particles, the centrifugal force dominates over

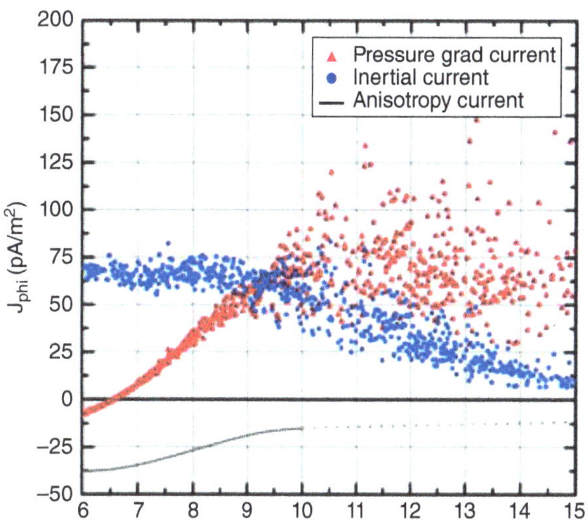

Figure 2.6 Contributions of plasma pressure gradient (red dots), centrifugal force (inertial term, blue dots), and plasma anisotropy (gray line) in generating the ring current in Saturn's magnetosphere. From *Sergis et al.* [2010].

the plasma pressure gradient force in the inner magnetosphere. However, beyond 9 R_S, plasma pressure force plays the dominant role in generating the ring current and inflates the magnetosphere (Fig. 2.6). *Achilleos et al.* [2010] used *Caudal* [1986] formalism with the Cassini spacecraft field and plasma data to obtain stress balance solutions in Saturn's magnetosphere and independently reached the conclusions of *Sergis et al.* [2010]. Using field and plasma observations from 11 equatorial orbits of Cassini spanning September 2005 to July 2006, *Kellett et al.* [2011] also show that plasma pressure force dominates beyond a radial

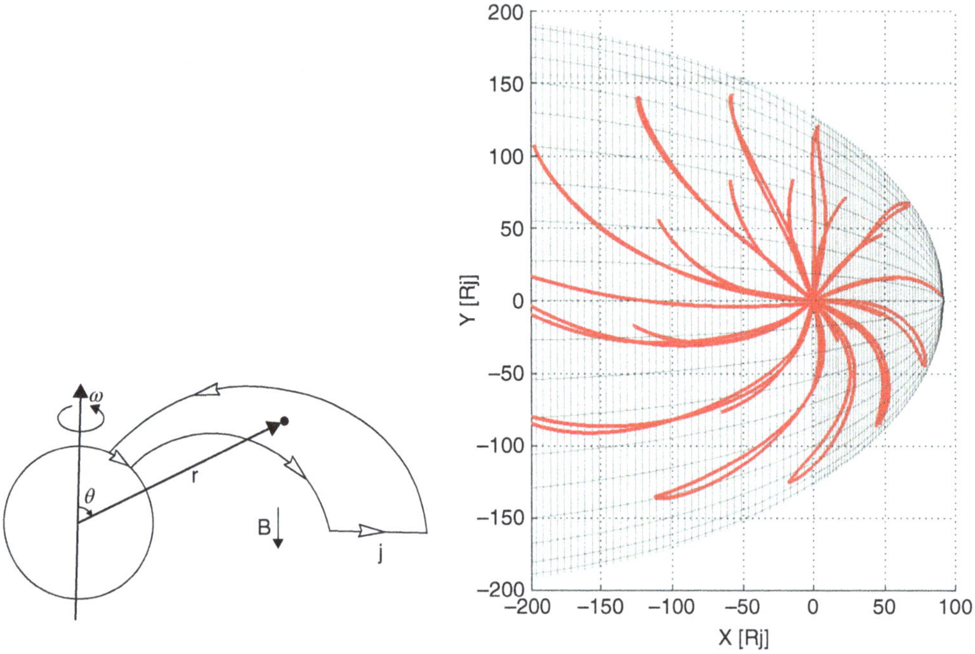

Figure 2.7 (Left) The current system in the magnetosphere and the ionosphere that transmits torque from the ionosphere to the outflowing plasma (figure reproduced from *Hill* [1979]); (right) the resulting configuration of field lines in Jupiter's magnetosphere (from *Khurana* [2002]).

distance of 8–9 R_S whereas in the inner magnetosphere, the centrifugal force dominates over other terms in agreement with previous works. *Kellett et al.* [2011] further find that the dominant pressure gradient current in the outer magnetosphere is strongest in the dusk to midnight sector but declines by a factor of ~2 in the dawn to noon sector. In addition, *Kellett et al.* [2011] find that in the outer magnetosphere, the azimuthal current has a much larger temporal variability, presumably because of the solar wind influence.

In a fast rotating planetary magnetosphere, with an internal plasma source, the plasma is transported outward under the action of a centrifugal driven interchange instability. As the plasma conserves angular momentum while moving outward, it falls behind corotation generating a bend-back of the field lines (azimuthally lagging field). *Hill* [1979] showed that a torque is transmitted to the magnetosphere by FACs that tends to restore corotation in the plasma. In the planetary ionosphere, the viscous torque is exerted by ion-neutral collisions while in the magnetosphere the torque is exerted by an outward radial current J_r and the associated Lorentz force. Figure 2.7 (left) from *Hill* [1979] shows the current system associated with the transmission of torques and the generation of an azimuthal field. *Hill* also showed that the azimuthal velocity of the outflowing plasma is governed by the rate of mass outflow and ionospheric conductivity such that at a large distance L, the plasma velocity is essentially given by

$$\omega = \pi^{\frac{1}{2}} \omega_p \left(\frac{L_0}{L} \right)^2 \quad \text{and} \quad L_0^4 = \pi \Sigma R_p^2 B_p^2 / \dot{M}, \tag{2.5}$$

Here Σ is the height integrated ionospheric conductivity, $\dot{M}$ is the mass outflow rate, and B_p is the planetary magnetic field. The strength of the radial current generated by this process can be as strong as the planetary ring current. Indeed, the swept-back shape of magnetospheric field lines reflects the effects of this large outward transfer of angular momentum (Fig. 2.7, right).

2.3.3. Calculation of Current Density from Magnetic Field Observations of a Single Spacecraft

The electric current density **J**, in any place in a magnetosphere, can be calculated from the derivatives of the magnetic field by the use of Ampère's law ($\nabla \times \mathbf{B} = \mu_0 \mathbf{J}$). Its azimuthal component in a cylindrical coordinate system (r, φ, z) is given by

$$J_\phi = \frac{1}{\mu} \left(\frac{\partial B_r}{\partial z} - \frac{\partial Bz}{\partial r} \right) \tag{2.6}$$

where it is assumed that the z axis is perpendicular to the equatorial current sheet. Each partial derivative term in this equation (2.6) must be computed with respect to one spatial coordinate keeping all the other spatial variables constant. However, along the spacecraft trajectory, all three position variables (r, z, and φ) vary simultaneously and the partial derivatives cannot be obtained from the raw data. One must therefore rely on certain special properties of the current system to compute the partial derivatives correctly [*Khurana*, 2001]. For example, the differenced field ((b_r, b_φ, b_z) = observations–internal field) arises mainly from local and distant sheets of current generated within the magnetosphere. As the magnetic field from distant currents such as the magnetopause does not vary rapidly with position, the local variations in the differenced field arise mainly from local currents. Further, for a thin current sheet in the equatorial plane, $\partial b_r/\partial r$ and $\partial b_z/\partial z$ terms are small, it can be assumed that b_r is a function only of the z coordinate and that b_z is a function only of the r coordinate. Thus, partial derivatives can be performed accurately and the azimuthal current can be obtained from

$$J_\varphi = \frac{1}{\mu}\left(\frac{\partial b_r}{\partial z} - \frac{\partial bz}{\partial r}\right) \tag{2.7}$$

The integration of equation (2.7) over the thickness of the current sheet yields height integrated azimuthal current flowing through the current sheet per unit radial distance.

$$J'_\varphi = \int J_\varphi dz = \frac{1}{\mu}\left(2b_{rl} - 2w\frac{\partial bz}{\partial r}\right) \tag{2.8}$$

where b_{rl} is the radial component of the differenced magnetic field just outside the current sheet, and w is the half-thickness of the current sheet. A similar set of operations can be performed to obtain the height integrated radial current density from

$$J'_r = \int J_r dz = -\frac{2b_{\phi ll}}{\mu} \tag{2.9}$$

where $b_{\varphi l}$ is the azimuthal component of the differenced magnetic field just outside the current sheet. At planets that have moderate to large dipole tilts relative to their spin axes (Jupiter, Uranus, and Neptune), the spacecraft motion relative to the current sheet is caused mainly by the cyclical motion of the current sheet, which is generally located near the magnetic equator (Fig. 2.8). However, in Saturn's magnetosphere, because no appreciable tilt exists between the rotational and dipole axes, the current sheet is generally located in Saturn's geographic equatorial plane for distances below ~15 R_S. One has to rely on high-latitude orbits of the spacecraft to obtain field values in the lobes of Saturn's magnetotail.

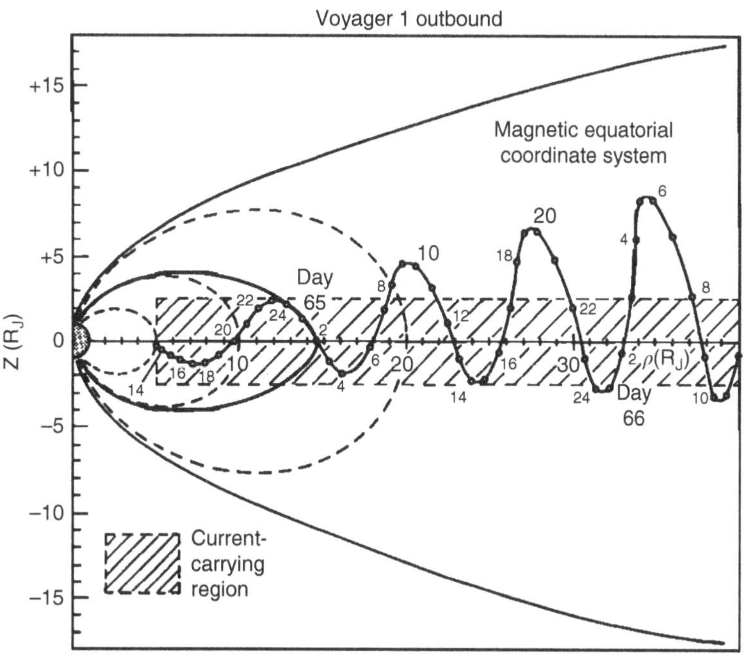

Figure 2.8 The trajectory of Voyager 1 relative to Jupiter's current sheet during the outbound pass. Figure from *Connerney et al.* [1981].

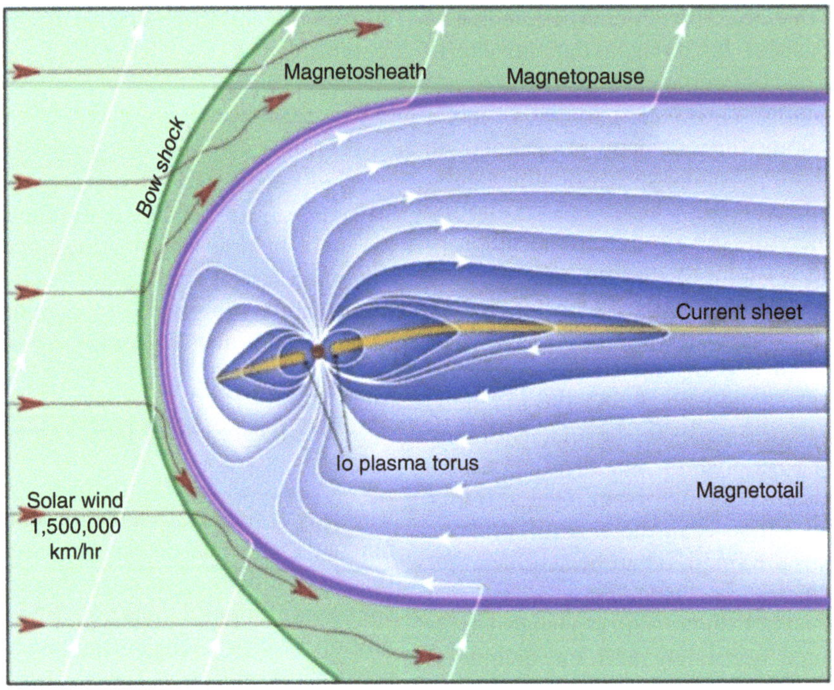

Figure 2.9 A schematic of Jupiter's magnetosphere. The large dayside magnetopause distance (~75 R_J) results from the very strong magnetic field of Jupiter assisted by hot energetic particle and centrifugal force of the fast rotating plasma. Figure courtesy of Francis Bagenal and Steve Bartlett, University of Colorado.

2.4. JUPITER

The five spacecraft that visited Jupiter prior to Galileo (Pioneer 10 and 11, Voyager 1 and 2, and Ulysses) showed that Jupiter's magnetosphere belongs to a class of celestial objects known as rotationally dominant magnetospheres (see Fig. 2.9 for a schematic). They discovered that the large magnetosphere of Jupiter is made possible by the strong internal field of Jupiter (surface strength ~4 Gauss at the equator), the small dynamic pressure of solar wind at 5.2 AU, and the rapid rotation rate of Jupiter (rotation period 9 hr 55 min) that helps inflate the magnetosphere through the action of centrifugal force on charged particles. In Jupiter's magnetosphere, the plasma corotates with Jupiter up to a radial distance of ~20 R_J and even near 100 R_J, the plasma velocity seldom falls below one third of the corotation speed [*Kane et al.*, 1995]. Another distinguishing characteristic found by these spacecraft is that Jupiter's magnetosphere is supplied principally from internal sources because of the strong mass loading near Io (believed to be 1–2 tons/s [*Hill et al.*, 1983]). The mass loading creates and maintains a plasma torus around Io's orbit, which contains more than a million tons of plasma consisting principally of various charged states of low energy O and S ions and electrons [*Belcher*, 1983]. The outward transport of this plasma is responsible for the strong sweepback of Jupiter's magnetic field lines [*Hill*, 1979; *Vasyliunas*, 1983; *Khurana and Kivelson*, 1993]. Radial currents flowing through Jupiter's current sheet accelerate the outflowing plasma toward corotation.

Connerney et al. [1981] provided the first quantitative estimate of ring current flowing in the magnetodisc of Jupiter between the radial distance of 6 and 50 R_J. They showed that good fits to the observations were obtained for an integrated ring current strength of ~60 MA during the Voyager 1 flyby and ~40 MA during the Voyager 2 flyby. These estimates are roughly a factor of 2 lower than those given by *Caudal* [1986].

The five spacecraft before Galileo established the uniqueness of Jupiter's magnetosphere. However, because of their limited local time coverage, they were unable to address a fundamental question in Jovian physics, namely, does solar wind play a role in driving convection in Jupiter's magnetosphere? As already discussed, at Earth, solar wind is the main driver of convection in the magnetosphere, and its influence is manifest in several types of asymmetries in the global magnetic field generated by currents flowing in the magnetosphere/ionosphere system. A hint of a solar wind influence was observed in the Pioneer and Voyager magnetic field data by *Bunce and Cowley* [2001] and *Bunce et al.* [2002] who showed that in the dawn sector, the azimuthal current density indeed varies with local time. However, because of the limited nature of the observations, they could not uniquely determine if the equatorial current was

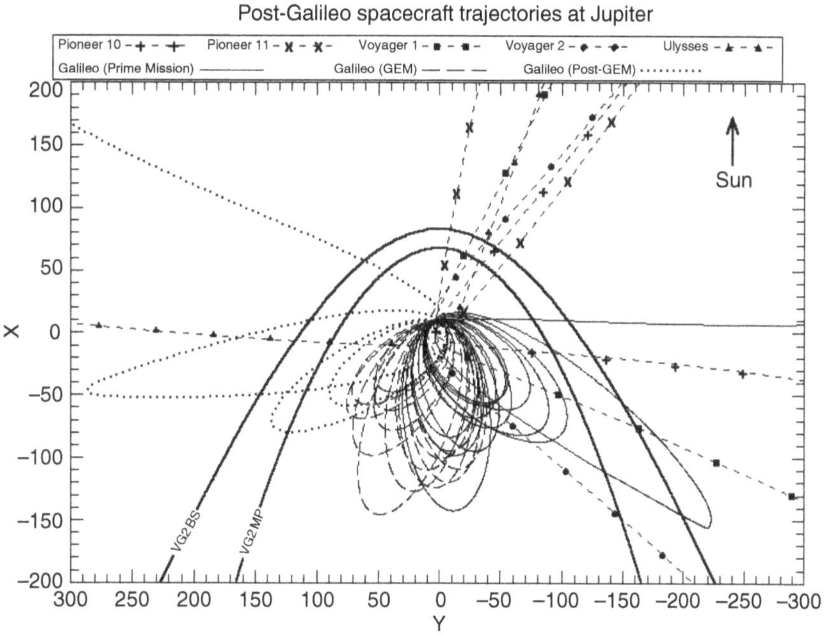

Figure 2.10 The trajectories of six spacecraft used by *Khurana* [2001] in his study. The trajectories for the Galileo Europa Mission (GEM) (dashed lines) provided the midnight/dusk coverage.

divergence-free (i.e., the divergences of azimuthal and radial currents were equal and opposite) or if there was a net divergence of the equatorial current leading to enhanced field-aligned currents in the dawn sector. With the arrival of Galileo, it became possible to map currents throughout the magnetosphere and look for local time asymmetries in various current systems (see Fig. 2.10 for local time coverage by various spacecraft). *Khurana* [2001] used equations (2.8) and (2.9) and the technique described above on Galileo magnetometer data to calculate current densities throughout the Jovian magnetosphere. Figure 2.11 shows height integrated azimuthal current density from that work plotted as a color map on the left and as binned data plots as functions of local time and radial distance on the right. The work clearly showed a day/night asymmetry in the strength of the azimuthal current between 10 and 50 R_J (~144 MA on the nightside versus 88 MA on the dayside).

Figure 2.12 presents a similar set of plots for the height integrated radial current density. The current is seen to be almost uniform with local time up to a radial distance of 25 R_J, but shows marked dawn/dusk asymmetries beyond that distance. It is not yet clear what process or processes are able to maintain this asymmetry, but three different mechanisms have been explored in the literature. *Khurana et al.* [2004] suggested that the radial current asymmetry could be caused by local-time asymmetries in the plasma outflow rate. This would require the outflow to terminate in the dusk sector in the outer magnetosphere but remain continuous in the dawn sector. *Ray et al.* [2014], on the other hand, suggest that the weaker radial current in the noon to dusk sector may result from the stronger magnetic field in that sector such that even a weak current can provide the required $\mathbf{J} \times \mathbf{B}$ force to maintain plasma corotation in that sector. However, the authors note that the stronger field in the dusk sector would also result in faster rotation of the plasma in that sector, a prediction of their model not supported by observations [*Krupp et al.*, 2001]. Finally, *Tao et al.* [2010] have explored the consequences of local-time variations of ionospheric conductance on the radial and field-aligned currents in the magnetosphere. Their numerical model predicts that the strongest field-aligned currents and the power extracted from the ionosphere would be in the noon sector. This result though inconsistent with the observations shown in Figure 2.12, nonetheless, points to yet another variable that must be taken into consideration when assessing force-balance in the Jovian magnetosphere.

Khurana [2001] also calculated the divergence of the equatorial current into FACs and showed that the divergence is consistent with Region 2 type FACs (with its polarity reversed compared to those at the Earth). Thus, an inescapable conclusion of that work is that the solar wind influence reaches deep in the heart of Jupiter's magnetosphere and indicates the presence most likely of a solar wind driven convection system. Figure 2.13 summarizes the strengths and locations of all of the current systems measured or inferred in that work.

The inference of *Khurana* [2001] that a Region 2 FAC system adds to and evacuates the ring current in the equatorial plane is supported by auroral observations, which show that the aurora is 3 times brighter in the dusk

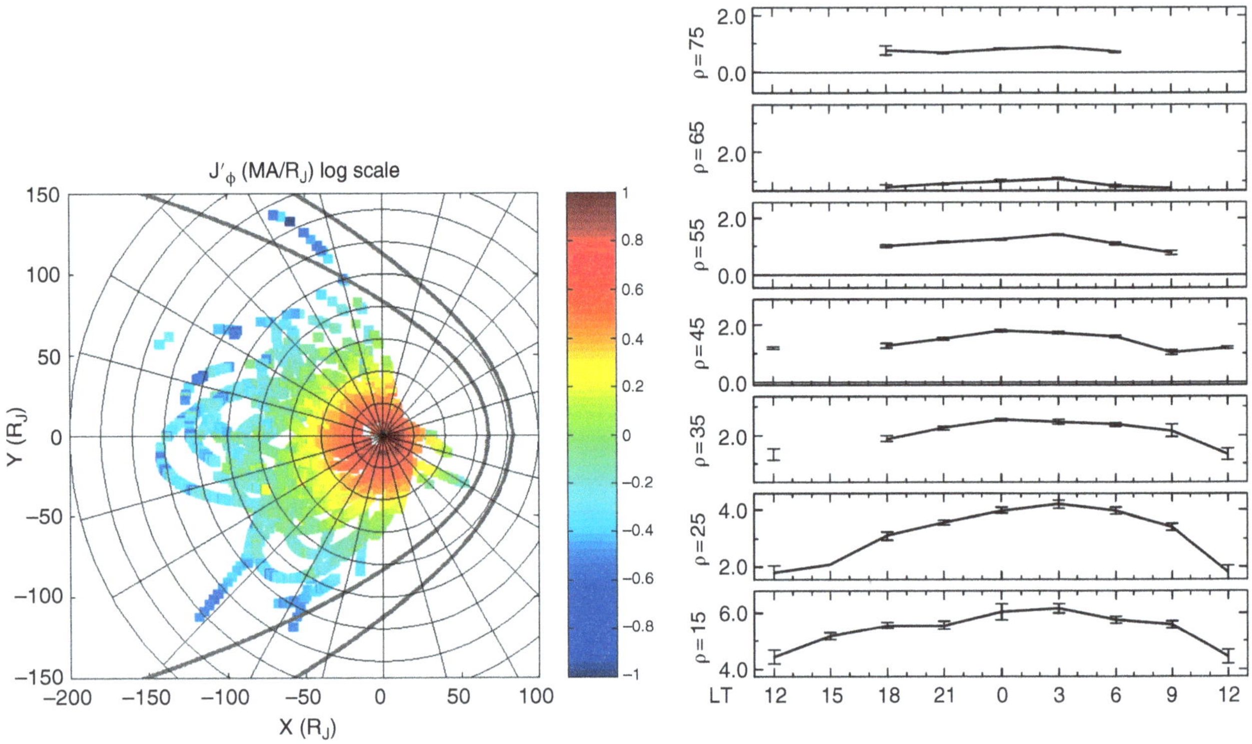

Figure 2.11 (Left) Height-integrated azimuthal current flowing in the equatorial plane of Jupiter; (right) the same binned over local time and radial bins. Figures from *Khurana* [2001].

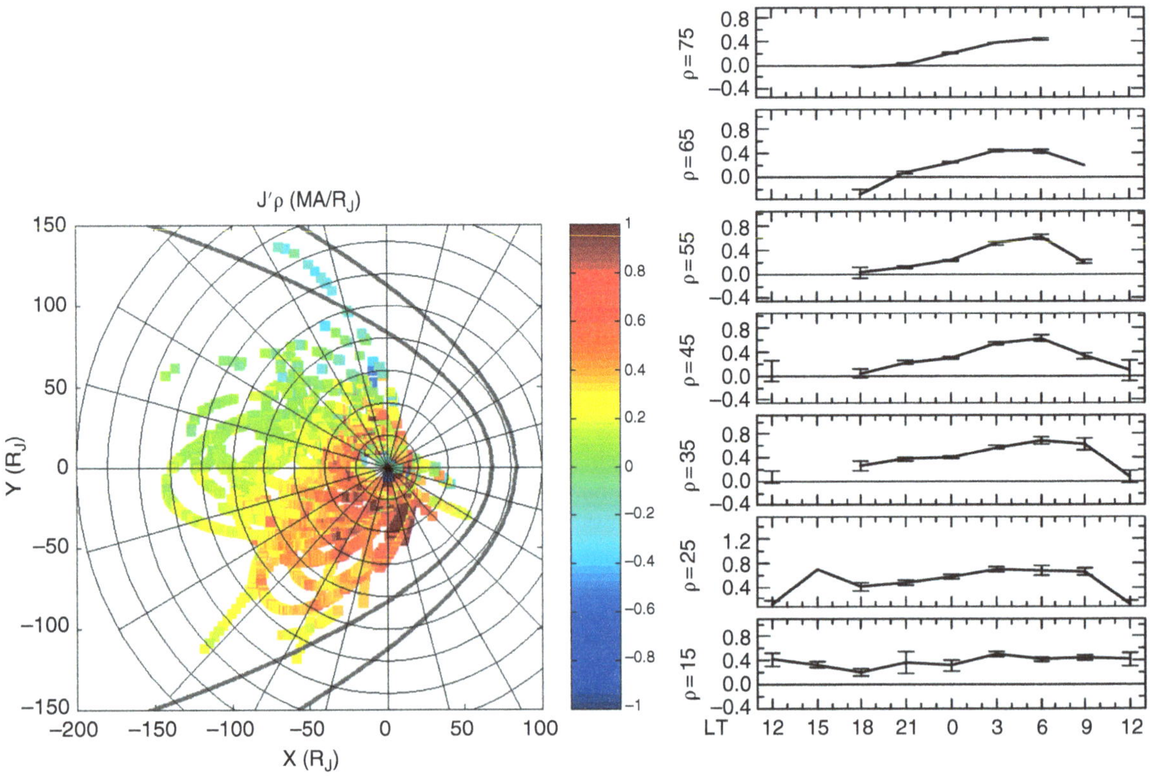

Figure 2.12 (Left) Height-integrated radial current flowing in the equatorial plane of Jupiter; (right) the same binned over local time and radial bins. It can be seen that near Jupiter ($r<25\ R_J$), the radial current is uniform but becomes nonuniform at larger distances. There is a clear dawn-dusk difference in the current strength at large radial distances. Both figures from *Khurana* [2001].

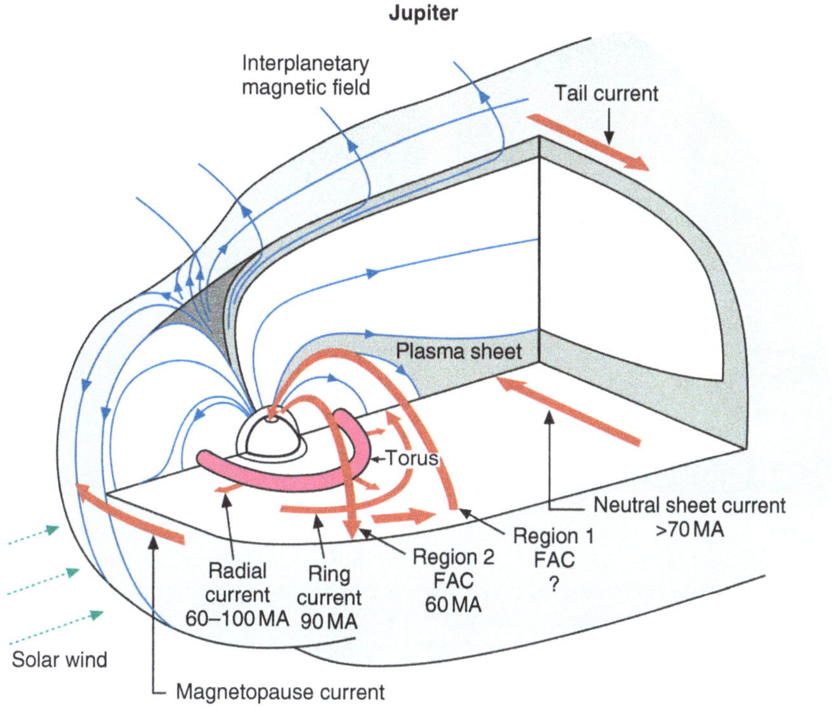

Figure 2.13 A schematic summary of the locations and strengths of various current systems inferred in Jupiter's magnetosphere so far. Figure from *Khurana* [2001].

sector compared with its intensity in the dawn side [*Vogt et al.*, 2011]. *Bonfond et al.* [2015] have suggested that the asymmetry results from the partial ring current located in the nightside magnetosphere (Fig. 2.14).

2.5. SATURN

Saturn is the second largest planet of the solar system (radius = 60,268 km) and possesses a moderately strong internal field (surface equatorial strength = 0.2 Gauss), which is exceedingly axially symmetric and devoid of a measurable tilt [*Acuna and Ness*, 1980; *Burton et al.*, 2009; *Cao et al.*, 2011]. Similar to the other gas giant Jupiter, Saturn is a fast rotator (Saturnian day ~ 10 hr 39 min).

Because of the much reduced solar wind density near Saturn (~0.2 amu/cm^3), the magnetosphere (see Fig. 2.15 for a schematic) is relatively large with the dayside magnetopause known to occur between 16 and 22 R_S [*Arridge et al.*, 2006]. The plasma in Saturn's magnetosphere is derived mostly from Enceladus, an active moon of Saturn. It is estimated that a plume at the south pole of Enceladus injects water group molecules at a rate of 100–300 kg/s [*Hansen et al.*, 2006] into the magnetosphere. Between 1 and 3 kg/s of this material becomes ionized in the immediate vicinity of Enceladus [*Burger et al.*, 2007; *Khurana et al.*, 2007] mainly through charge exchange, while the bulk of neutral molecules escape to populate a neutral torus [*Jurac and Richardson*, 2005]. A small additional contribution to magnetospheric plasma is made by the sputtering of ices from the rings and surfaces of other inner satellites, especially Dione and Tethys [*Burch et al.*, 2007]. The newly picked up plasma consists of charged states of water group molecules [*Richardson*, 1998]. Titan also provides a small source of N^+ near its orbit. The plasma is clearly subcorotational in most of the inner and middle magnetosphere [*Wilson et al.*, 2008; *Thomsen et al.*, 2010; *Arridge et al.*, 2011] and obtains its angular momentum and energy from Saturn's ionosphere (and ultimately from Saturn's atmosphere; see *Gombosi et al.* [2009]). The existence of an equatorial current sheet was revealed only when the data from Voyager 1 and 2 spacecraft became available [*Connerney et al.*, 1983].

There is no consensus yet on the large-scale convection and transport of plasma in the magnetosphere [*Cowley et al.*, 2004; *Hill et al.*, 2005; *Mitchell et al.*, 2009]. *Bunce et al.* [2008] mapped the outer regions of the equatorial current sheet to the ionosphere and showed that this open/closed boundary (the magnetopause on the dayside) maps to the auroral region. They, therefore argue that the Saturnian aurorae are not associated with plasma mass loading and radial transport in the middle magnetosphere. In agreement with this model, *Crary et al.* [2005] showed that the solar wind dynamic pressure and the electric field control the intensity of the Saturnian aurora as would be expected for a magnetosphere driven by solar-wind-induced transport. Recently, using the poleward boundary of the auroral emissions as a proxy for open/closed boundary of the magnetosphere, *Badman et al.* [2014] have shown that between 10 and 50 GWb flux is

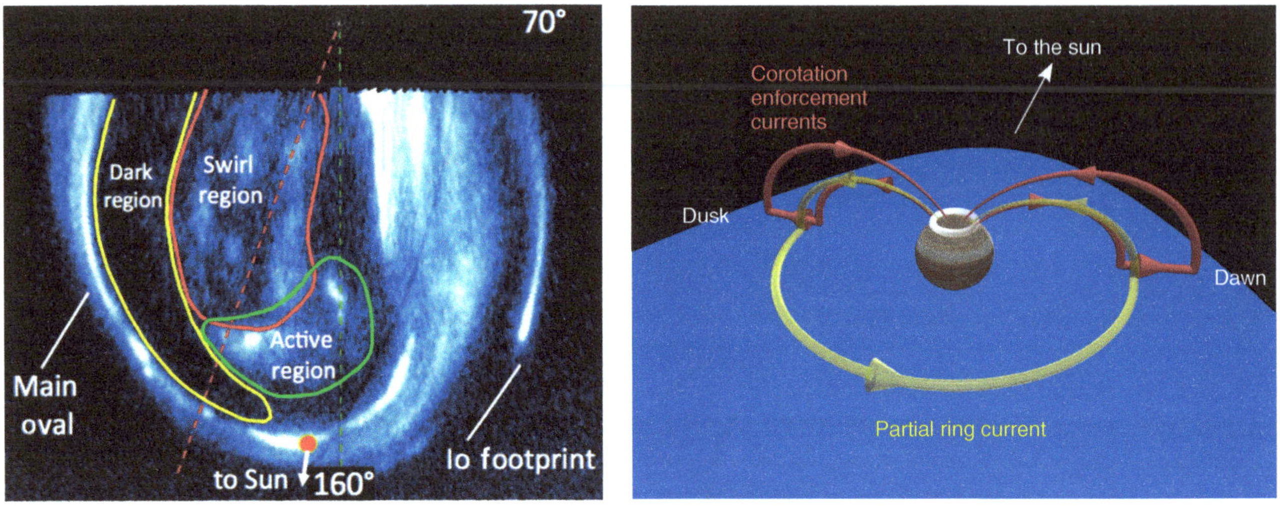

Figure 2.14 (Left) UV auroral emissions in Jupiter's northern hemisphere as imaged by the Hubble Space Telescope. Different features of the emissions are identified in the figure. The main auroral oval is thinner and fainter on the dawn side compared to the dusk side (figure reproduced from *Vogt et al.* [2011]). (Right) A schematic of corotation enforcement current system and the partial ring current system driven by the solar wind induced convection. On the dawn side, at low latitudes, the two current systems (yellow lines for partial ring current system and red lines for corotation enforcement currents) are opposite in sign leading to a dimmer and narrow auroral region. However, on the dusk side, at low latitudes, the two current systems flow in the same direction leading to a brighter aurora (figure from *Bonfond et al.* [2015]).

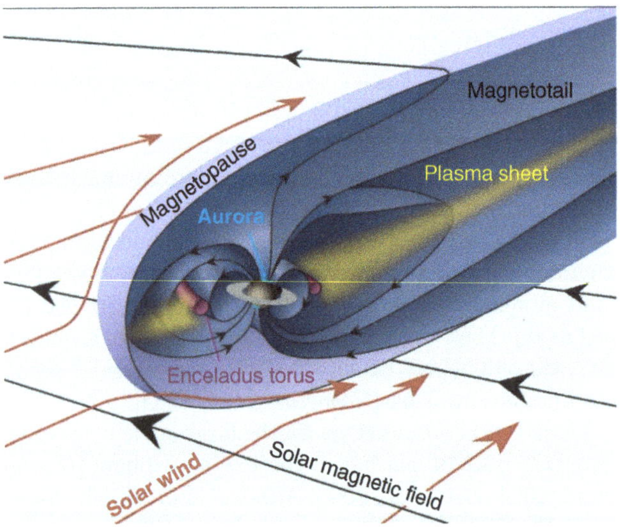

Figure 2.15 The structure of Saturn's magnetosphere. This magnetosphere is filled internally from plasma derived from Enceladus and its neutral torus. The plasma sheet has a shallow bowl shape because the solar wind flow forms a large angle with the magnetic equator of Saturn. Figure courtesy of Fran Bagenal and Steve Bartlett, University of Colorado at Boulder.

open in Saturn's magnetosphere with an average of 35 GWb remaining permanently open. They further calculated the average potential drops by observing variations in polar cap size and found them mostly in the range of a few tens of kV but on a few occasions exceeded ~300 kV, roughly twice the values observed in the Earth's magnetosphere suggesting that at least occasionally solar wind driving is significant. However, *Hill et al.* [2005] and *Liu et al* [2010] have argued that radial plasma transport is accomplished mainly by mesoscale convection cells driven by an internal centrifugal interchange instability. Indeed, centrifugal interchange events have been observed extensively in Saturn's magnetosphere [*Andre et al.*, 2005; *Rymer et al.*, 2009; *Kennelly et al.*, 2013] and the composition and energy dispersion of the interchange injection events indicates that they deliver plasma from the outer magnetosphere and travel inward at least a distance of 6 R_S [*Thomsen et al.*, 2014; *Paranicas et al.*, 2016].

In order to resolve this question, we have now used the technique of *Khurana* [2001] on Cassini magnetic field data obtained between 2004 and 2014 (see Fig. 2.16 for local time coverage provided by Cassini). As discussed earlier, because of the axisymmetric nature of Saturn's magnetic field, the current sheet does not sweep up and down over the spacecraft, and equatorial passes are not able to monitor the strength of the field in the lobes, making the dataset quite sparse in its coverage of the magnetosphere. Figure 2.17 shows the computed height integrated azimuthal and radial current densities in Saturn's magnetosphere in a color map presentation (please notice the different color bar scales for the two plots). It can be clearly seen that even in Saturn's magnetosphere, the ring current is much stronger on the dayside compared with its value in the midnight sector (see also a related chapter [Chapter 9] on Saturn's ring current by *Sergis et al.* in this volume). By integrating the ring

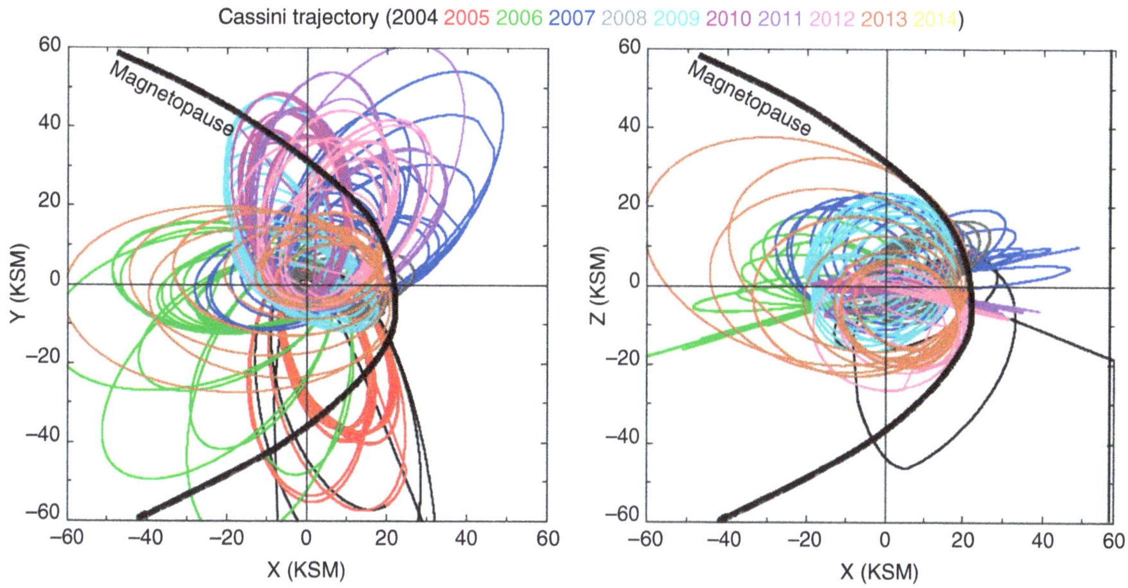

Figure 2.16 The trajectory of Cassini projected onto the *x-y* and *x-z* planes of the KSM coordinate system. In this coordinate system centered at Saturn, the *x* axis points toward the Sun and the *x-z* plane contains Saturn's dipole moment. The trajectories are color coded according to the year of the observations (as shown in the title).

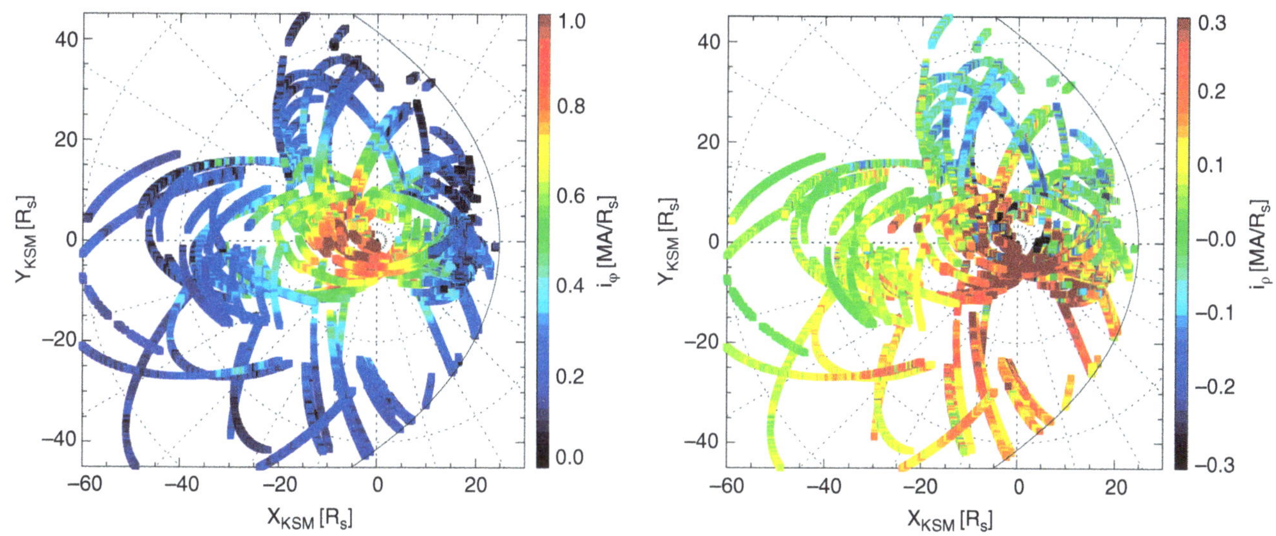

Figure 2.17 Height-integrated azimuthal (left) and radial (right) current densities in Saturn's magnetosphere. The strength of the ring current is much higher in the dawn to midnight sector compared with its value in the noon sector. Similarly, a clear dawn to dusk asymmetry is observed in the radial current profile with the radial currents maximizing in the dawn sector.

current over a radial distance of 4 to 20 RS, we find that its strength is 11 MA on the dayside and 17 MA on the nightside. We attribute this difference to the presence of a partial ring current accompanied by a Region 2 type FAC system. Thus, we infer that Region 2 current system must have a strength of ~6 MA in Saturn's magnetosphere. Next, we find that the radial current system has a strength of 2–4 MA, which is consistent with the values obtained by *Cowley and Bunce* [2003] based on a model of ionosphere/magnetosphere coupling. However, a clear dawn to dusk asymmetry is present in the radial current strength and must have a similar nature to processes operating in the Jovian magnetosphere. Thus, we conclude that Saturn's magnetosphere is also profoundly affected by solar-wind-driven convection.

Next, we dwell on a current system that is truly unique to Saturn's magnetosphere. Saturn's spin and dipole axes are aligned to better than 0.1° [*Cao et al.*, 2011], yet Saturn's SKR wave intensities and many field and plasma parameters (see Fig 2.18, top left) display planetary

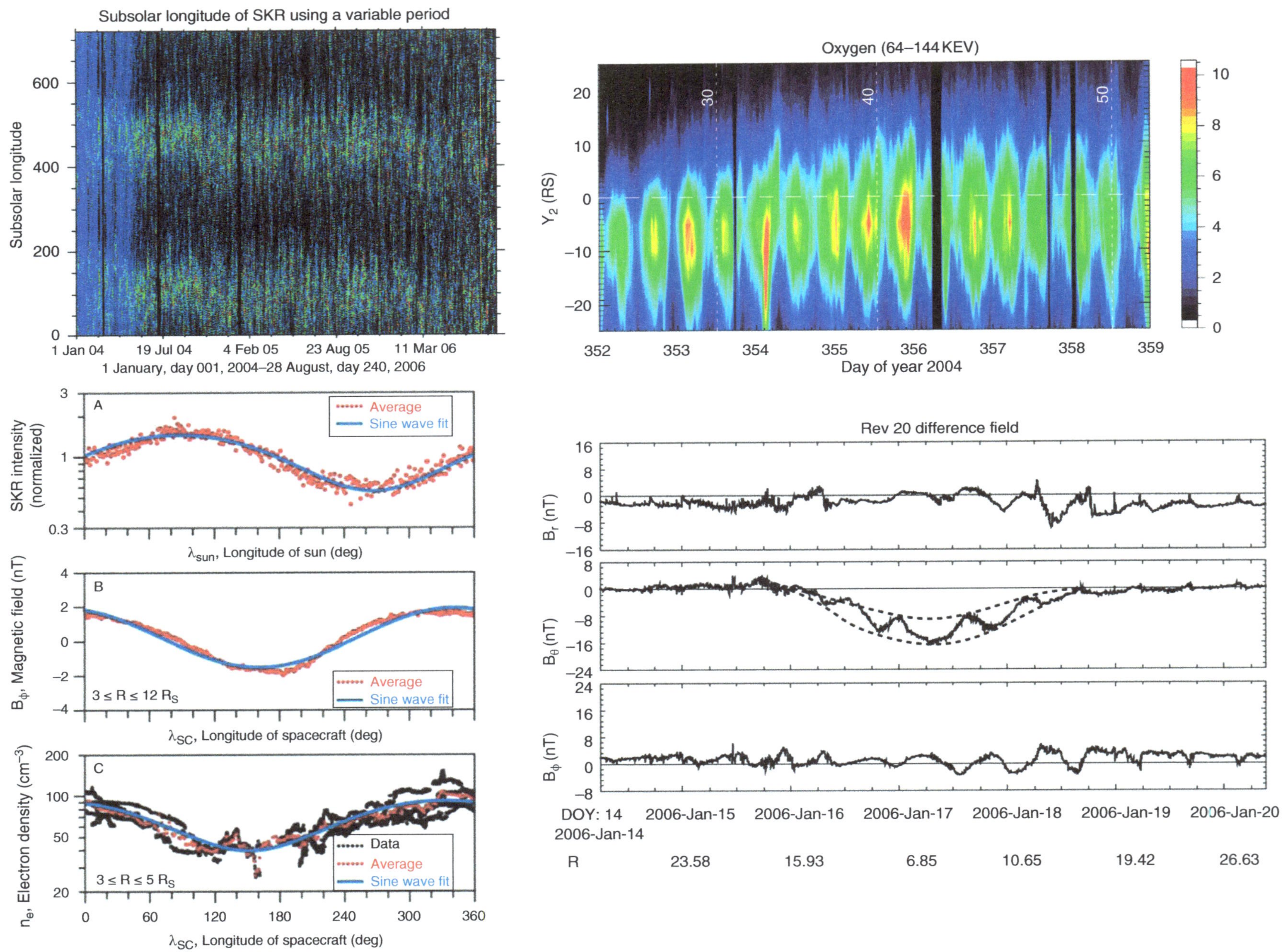

Figure 2.18 Rotational signals in Saturn's magnetosphere (from top left, clockwise) in Saturn Kilometric radiation [*Kurth et al.*, 2007], in the energetic neutral emissions of oxygen [*Carbary et al.*, 2008], in the magnetic field [*Khurana et al.*, 2009], and plasma density [*Gurnett et al.*, 2007].

period oscillations (PPOs). Field and plasma observations from the inner magnetosphere (also called the core region where $r < 12$ R_S) display periodicities [*Southwood and Kivelson*, 2007; *Gurnett et al.* 2007] at the SKR modulation period. The radial and azimuthal components of the magnetic field (B_r and B_ϕ) are seen in quadrature (with B_r leading) when the southern SKR source dominated in strength. *Southwood and Kivelson* [2007] have shown that an interhemispheric current system rotating with the planet and flowing between the northern and southern ionospheres is responsible for the observed periodicities of the magnetic field in the core region. It is not yet known how this current system is generated, but two options have been explored including a global ionospheric conductivity anomaly [*Southwood and Kivelson*, 2007] and a twin vortex system in the ionosphere driven from the atmosphere below [*Jia et al.*, 2012].The electron plasma density is seen to be in phase with B_ϕ and the SKR intensity. *Andrews et al.* [2010a] show that the B_θ component also displays modulations being mostly in phase with the B_r components in the core region and beyond. Magnetic field observations from the middle and outer magnetosphere show that spin periodicities persist both at low latitudes [*Andrews et al.*, 2010a; *Provan et al.*, 2009a], and at high latitudes [*Andrews et al.*, 2010b; *Provan et al.*, 2009b]. A comparison of the SKR modulations with those in the magnetic field reveals that in the predawn to noon sector an intense midmorning SKR source gives the emissions a strobe-like behavior whereas in other local time sectors, the SKR modulations phase depends on local time and gives it a rotating beam character [*Andrews et al.*, 2011]. The magnetic field observations from near the equatorial plane and from high latitude regions in the southern hemisphere display a periodicity closer to that of the southern-SKR emissions whereas magnetic field periodicities from the northern high latitude regions have a period commensurate with the northern-SKR emissions [*Provan et al.*, 2011; *Southwood*, 2011]. The dual periodicities have also been reported in the measured fluxes of energetic ions and electrons [*Carbary et al.*, 2007, 2008, 2009].

It is now known that the SKR period varies at the level of 1% over a timescale of years [*Lecacheux and Aubier*, 1997; *Gurnett et al.*, 2007]. An even more perplexing observation is that the northern and the southern hemispheres modulate SKR radiation at slightly different rotational periods [*Gurnett et al.*, 2009]. These observations cast doubt on the idea that the SKR wave intensities are modulated by a source internal to Saturn (dynamo region) but are supportive of a mechanism resident in the ionosphere or/and the magnetosphere of Saturn.

Two types of models have been advanced to explain the periodic current systems in Saturn's magnetosphere. The first type of models postulate that the source of periodicities exists in the atmosphere of Saturn such as vortical flows arising from the asymmetric heating of the thermosphere from below [*Smith*, 2011; *Jia et al.*, 2012, *Jia and Kivelson* (2012)] while the second type of models place the source such as a rotating plasma mass asymmetry or a ring current asymmetry in the magnetosphere [*Gurnett et al.*, 2007; *Mitchell et al.*, 2009; *Khurana et al.*, 2009; *Brandt et al.*, 2010]. At the time of writing this review, no consensus has emerged on what the source is for these spin period oscillations. Recently, *Hunt et al.* [2014, 2015] have modeled the high latitude FACs associated with these oscillations and they show that part of the current is indeed interhemispheric in nature and has both a northern hemisphere and a southern hemisphere origin (Fig. 2.19) suggesting that the atmosphere/ionosphere is the ultimate driver of these oscillations. The current strengths of these rotating current systems is 1–2 MA/rad for a total current out of each hemisphere ~3–4 MA, similar to the radial current that enforces corotation in Saturn's magnetosphere.

2.6. URANUS

Uranus (radius = 25, 559 km), like its other counterparts in the outer solar system, is a fast rotator (a Uranian day ~17 hr 12 min). However, its rotation axis is aligned very close to its orbital plane giving its magnetosphere some unique characteristics (Fig. 2.20). The surface equatorial field of Uranus is ~0.23 Gauss, just slightly lower than that of the Earth. The magnetic axis of Uranus forms an angle of 59° with its rotation axis, making it the most tilted magnetosphere in the solar system. Only one spacecraft (Voyager 2 in January 1986) has visited this planet and found its dayside magnetopause near 18 R_U [*Ness et al.*, 1986]. No satellite source of plasma was observed, and based on the composition of the plasma in the magnetosphere, it appears that most of the plasma is derived primarily from the Uranian atmosphere and consists mostly of protons [*Bridge et al.*, 1986]. The plasma was found to be near-corotational in most of the magnetosphere and gains its angular momentum from Uranus. The magnetosphere is characterized by a weak current sheet that rotates with the planet (see Fig. 2.21) but is clearly delayed in phase [*Behannon et al.*, 1987]. Because of the large tilt of the current sheet, the spacecraft spent considerable time in the lobes of Uranian magnetotail. However, the residual field (observation–internal field) from the ring current was found to be on average ~1.5 nT, which, when integrated, gives a ring current strength of 1 MA, a value much smaller than that of the Earth's ring current. The residual field is less than 1 nT in the magnetotail. Assuming a magnetotail with a length of ~30 R_U, the integrated tail current is < 2 MA. Clearly, the currents are extremely weak in the magnetosphere of Uranus. For

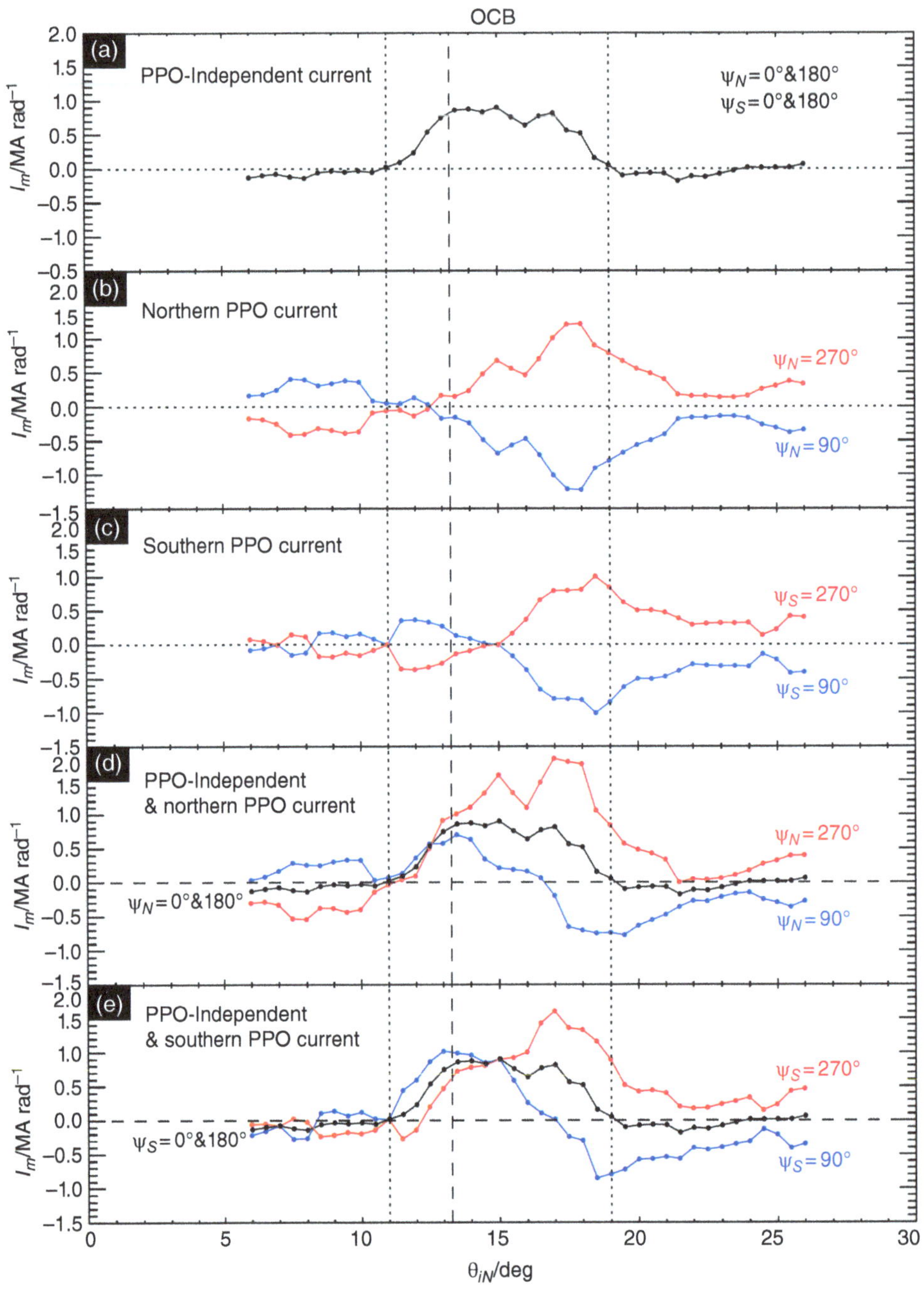

Figure 2.19 Field-aligned currents closing in Saturn's ionosphere: (a) PPO independent radial current, (b) PPO current at the northern period, (c) PPO current at the southern period, (d) PPO independent+northern PPO current, and (e) PPO independent and southern PPO current. Figure from *Hunt et al.* [2015].

this reason, *Connerney et al.* [1987] have termed the Uranian magnetosphere as a "vacuum magnetosphere." Since data from only one flyby are available for Uranus, no conclusions can be drawn about local time differences in the strength of the ring current. Further, no appreciable bend-back of the field lines was observed during the magnetotail flyby, presumably because there is no appreciable mass outflow in the magnetosphere. The magnetotail was however found to be twisted (delayed) by 5.5° because the Alfvén waves that communicate the source location of the dipole from the ionosphere to the magnetotail take finite time to reach the magnetotail.

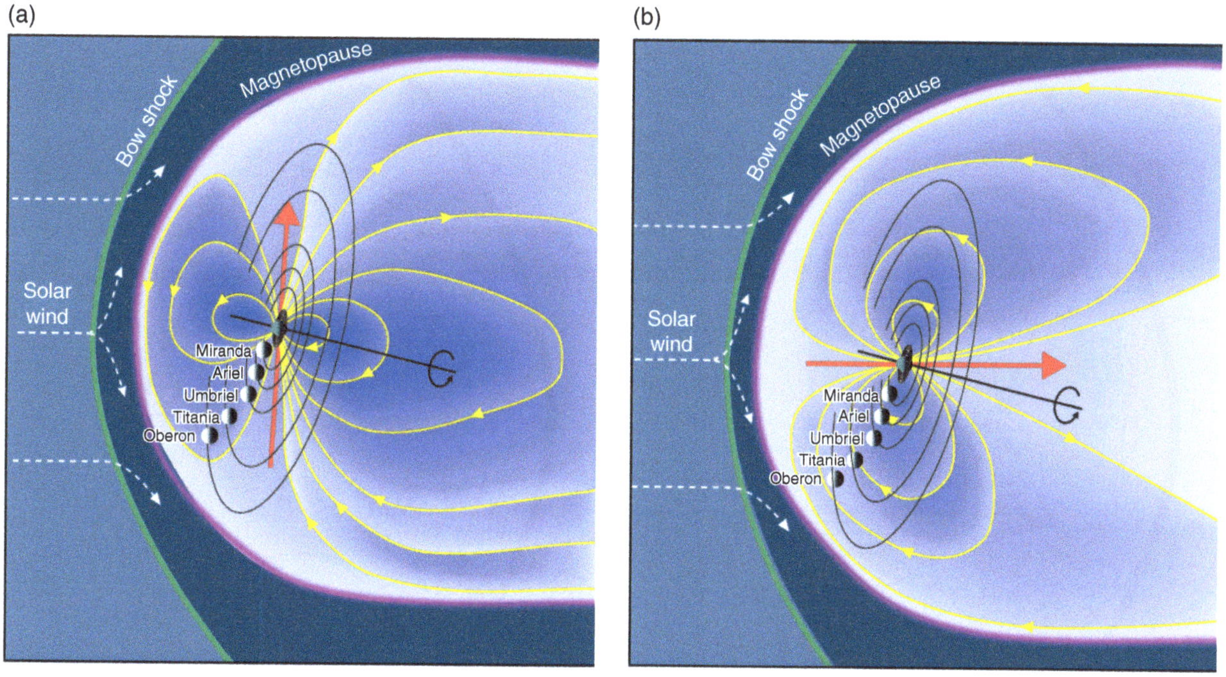

Figure 2.20 The Uranian magnetosphere during the Voyager 2 flyby in January1986 (left panel) and its configuration expected in 2030 (right panel). The magnetosphere had a pole on configuration relative to the Sun during the Voyager flyby. Figure courtesy of *Christopher Arridge, Francis Bagenal, and Steve Bartlett.*

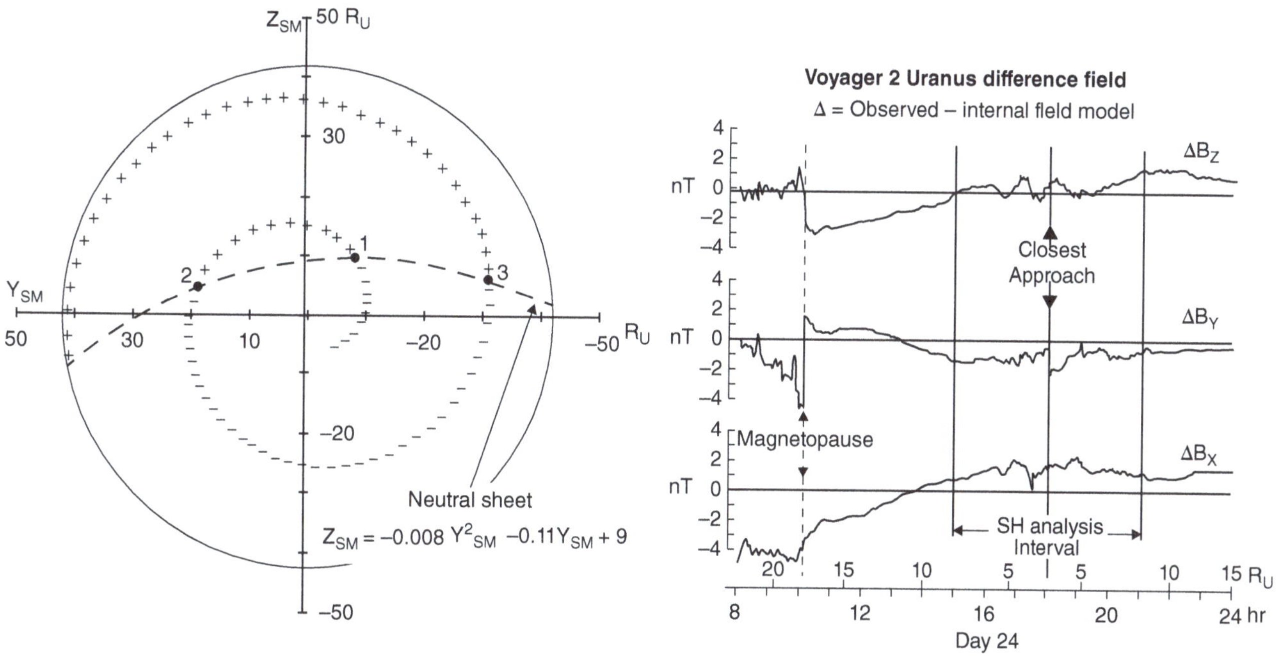

Figure 2.21 (Left) The trajectory of Voyager 2 with respect to the current sheet of Uranus. The current sheet was encountered four times during the Voyager flyby (figure from *Behannon et al.* [1987]); (right) residual field (observed-internal field) in Uranian magnetosphere (figure from *Connerney et al.* [1987]).

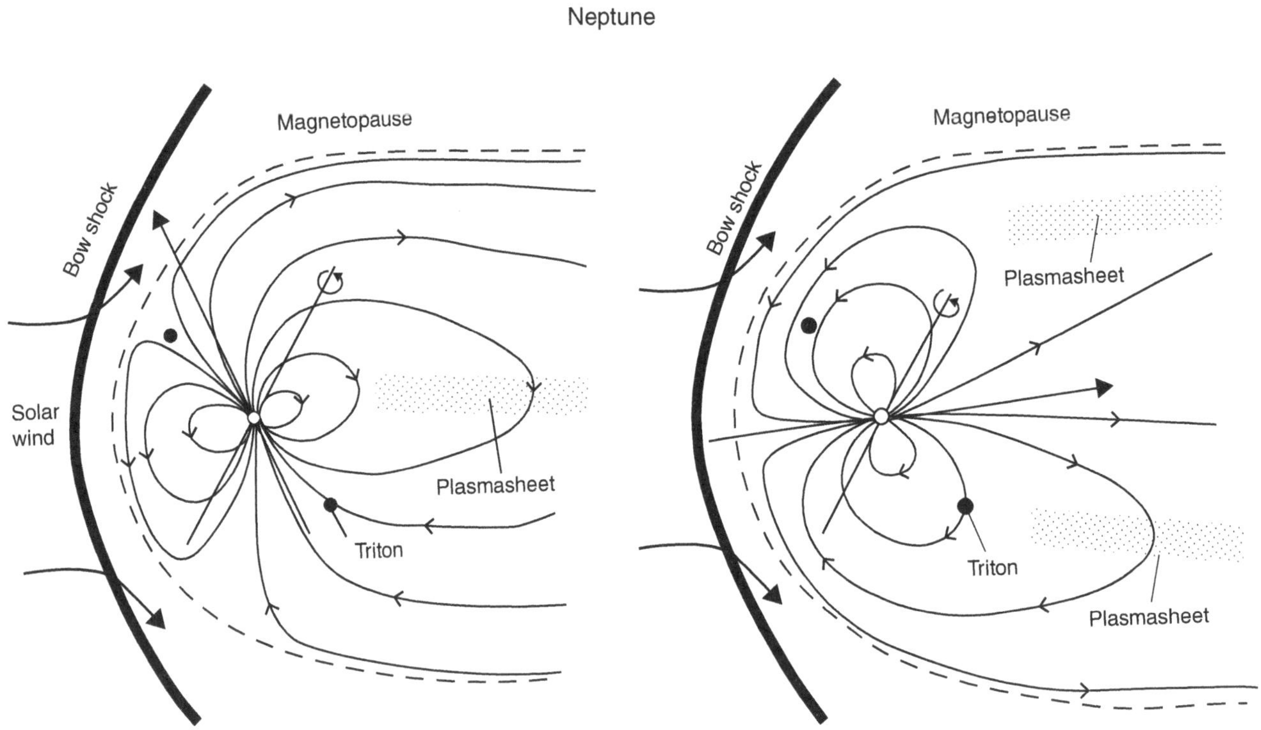

Figure 2.22 Neptune's magnetosphere observed by Voyager 2 during its pass in August 1989. The large dipole tilt of the field leads to major changes in the configuration of Neptune's plasma sheet and even bifurcation for part of the Neptunian day. From *Bagenal* [1992].

2.7. NEPTUNE

Neptune is another gas/ice giant like Uranus and has an equatorial radius of 24,764 km and a rotation period of 16.1 hr. Voyager 2 visited Neptune in August 1989 and found it magnetized (surface equatorial field = 0.14 Gauss, roughly half that of the Earth). Like Uranus, the magnetic dipole axis of this planet is also highly inclined with respect to its rotation axis (tilt = 47°). Even though the magnetic field is quite weak, the extremely low density of the upstream solar wind ensures that the dayside magnetopause is at a distance of 23–26 R_N. Most of the plasma (mainly H^+ and N^+) appears to be supplied by Triton and the atmosphere of Neptune (mainly from a plasma cloud extending outward from the exosphere of Neptune [*Richardson*, 1993]). Because of the extremely large tilt of the magnetic dipole, the equatorial plasma sheet becomes bifurcated for part of a rotation [*Behannon et al.*, 1987] (Fig. 2.22). As a result, the residence time for plasma in the magnetosphere is short and the number densities in the current sheet are very low (of the order of $0.1/cm^3$).

Figure 2.23 shows the trajectory of Voyager 2 during its Neptune flyby in a cylindrical coordinate system in which Z-axis is aligned to the magnetic dipole axis and RHO is the cylindrical radial distance from the Z-axis. The residual data (observations–internal field) are plotted on the right. It can be seen that the residual field is everywhere < 0.2 nT, which places an upper limit on the ring current at < 0.2 MA. There is also no evidence of a radial current system. In addition, in the magnetotail region, no appreciable difference field was observed above the noise level of the instrument.

2.8. SUMMARY AND MAJOR QUESTIONS

Table 2.3 summarizes our knowledge of various current systems operating in outer planet magnetospheres. The review shows that

1. At Jupiter
   a. A strong ring current with a strength of ~90 MA helps keep the plasma confined to the magnetosphere. In the nightside, the ring current morphs into the magnetotail current with a strength exceeding 70 MA.
   b. A strong (60–100 MA strength) magnetosphere/ionosphere coupling current system helps enforce corotation in Jupiter's magnetosphere.
   c. A day/night asymmetry exists in the strength of the ring current, the current being at least 50% stronger in the nightside sector.

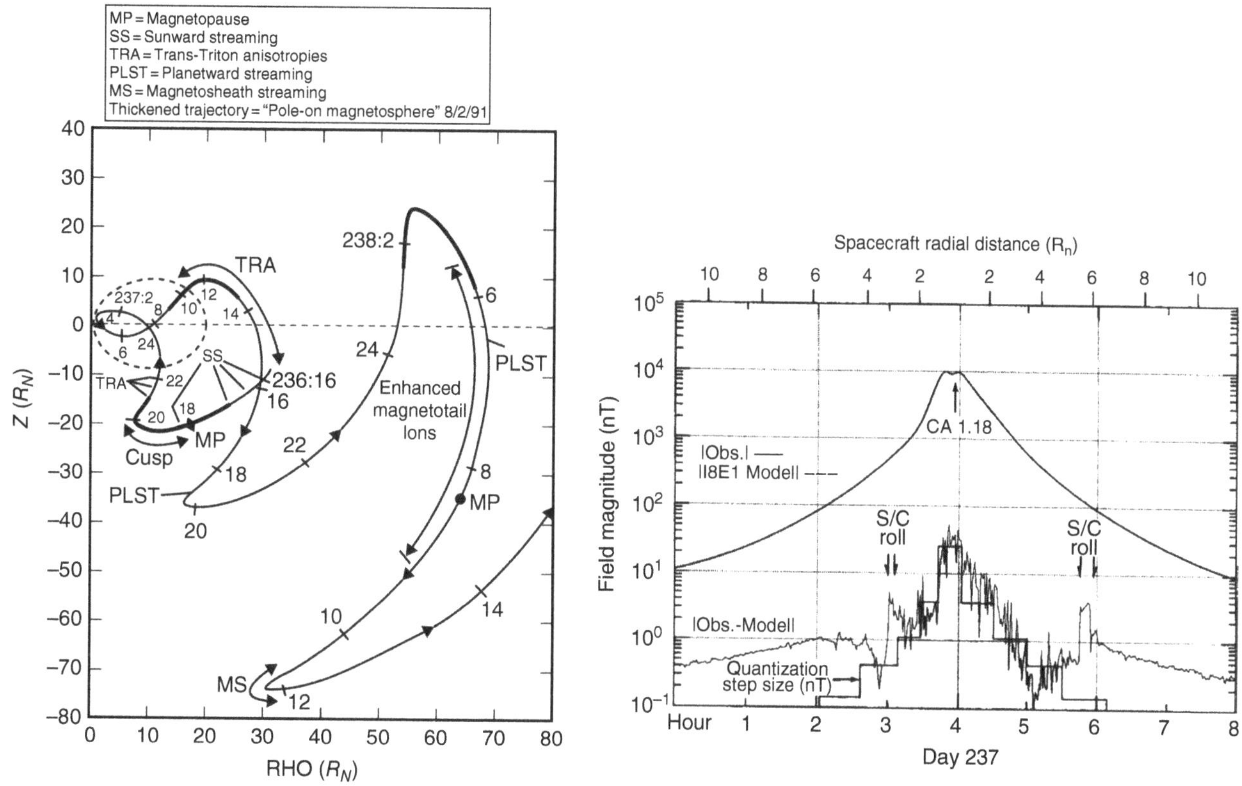

Figure 2.23 (left) The trajectory of Voyager 2 during its August 1989 flyby of Neptune in a magnetic coordinate system. The last magnetopause crossing was encountered near 60 R_N (from *Mauk et al.* [1991]); (right) field strength and residual field in the magnetosphere of Saturn (from *Connerney et al.* [1991]).

Table 2.3 A Comparison of the Strengths of Various Currents Present in the Earth's Magnetosphere With Those of the Outer Planet Magnetospheres

Planet	Ring current (MA)	Radial current (MA)	Tail current (MA)	Region 1 (MA)	Region 2 (MA)
(MA) Earth	2–4	0	>10	3–5	2–4
Jupiter	90	60–100	>70	>24	60
Saturn	10	2–4	>10	?	6
Uranus	<1	0	<2	?	?
Neptune	<0.2	0	?	?	?

d. There is evidence for a partial ring current and associated Region 2 type field-aligned currents in the magnetosphere in magnetic field data and auroral observations.

e. The local-time symmetry of the radial current inside of 20 R_J suggests that the mass outflow is symmetric inside of 20 R_J. However, the absence of an outward radial current beyond 20 R_J on the dayside suggests that mass outflow does not occur in this region.

f. These asymmetries call for a central role for solar wind in driving the dynamics of Jupiter's magnetosphere.

2. At Saturn

a. A modest ring current (strength ~10 MA) helps keep plasma confined in the magnetosphere.

b. There is a clear day/night asymmetry in the strength of the ring current. The ring current strength is enhanced in the nightside sector by at least 50%. A rotating interhemispheric current system facilitates angular momentum transfer between the northern and southern parts of Saturn's ionosphere.

c. Radial current (responsible for field bend-back and reinforcement of plasma corotation) has a modest strength of 2–4 MA in Saturn's magnetosphere.

d. Similar to the Jovian magnetosphere, the radial current is enhanced in the dawn sector.
e. These local-time asymmetries point to the solar wind influence in driving the dynamics of Saturn's magnetosphere as well.

3. At Uranus
   a. The ring current (<1 MA) and tail current (<2 MA) have strengths weaker than those found in the Earth's magnetosphere.
   b. The magnetospheric convection is expected to be dominated by solar-wind-induced electric fields.
4. At Neptune
   a. The ring current strength is extremely small (<0.2 MA).
   b. The magnetospheric convection is expected to be dominated by the solar wind electric field.
   c. The magnetotail is reconfigured in each rotation.

Finally, we are left with an interesting puzzle in both Jupiter's and Saturn's magnetospheres. Observations from the middle and outer magnetospheres of both of these planets show that the sweepback of the field (or equivalently the strength of the radial current) peaks in the dawn sector and minimizes in the dusk sector. As shown by *Hill* [2001] and *Cowley and Bunce* [2001], the strength of the radial current in a magnetosphere is determined mainly by the conductivity of its ionosphere and the mass outflow rate in the equatorial plane. Even though, solar EUV ionization is primarily responsible for the conductivity of the ionosphere, many other factors also affect the conductivity of the ionosphere. *Nichols and Cowley* [2003] have shown that the effective ionospheric conductivity can be much lower than the actual ionospheric conductivity because of the slippage of the ionospheric neutrals. In addition, *Nichols and Cowley* [2004] have shown that auroral precipitation significantly enhances Pedersen conductivity in the ionosphere and cannot be ignored in the upward-flowing current regions of the ionosphere. Thus, one would expect a stronger magnetosphere/ionosphere coupling in the dusk sector (where the auroral intensity peaks) and an enhanced radial current system, but the situation is exactly opposite. Thus, to explain the decrease of the radial currents in the dusk sectors of the Jovian and Saturnian magnetospheres, one is led to conclude that the local plasma outflow rates have been reduced by the convection systems operating in these magnetospheres. The reduction in the outflow rate in the dusk sector would then imply that plasma would be closer to corotation in that sector (see equation [2.5]). However, observations from Jupiter's magnetosphere show precisely an opposite trend, that is, the plasma lags corotation most in the dusk sector [*Krupp et al.*, 2001]. Obviously, our pictures of azimuthally symmetric outflow and local-time symmetric ionospheric conductivity in these magnetospheres needs some major rethinking.

REFERENCES

Achilleos, N., P. Guio, and C. S. Arridge (2010), A model of force balance in Saturn's magnetodisc, *Monthly Not. R. Astron. Soc.*, *401*, 2349–2371; doi:10.1111/j.1365-2966.2009.15865.x.

Acuna, M. H., and N. F. Ness (1980), The magnetic field of Saturn–Pioneer 11 observations. *Science*, *207*, 444–446.

Andre, N. , M. K. Dougherty, C. T. Russell, J. S. Leisner, and K. K. Khurana (2005), Dynamics of the Saturnian inner magnetosphere: First inferences from the Cassini magnetometers about small-scale plasma transport in the magnetosphere, *Geophys. Res. Lett.*, *32*, L14S06; doi:10.1029/2005GL022643.

Andrews, D. J., A. J. Coates, S. W. H. Cowley, M. K. Dougherty, L. Lamy, G. Provan, and P. Zarka (2010b), Magnetospheric period oscillations at Saturn: Comparison of equatorial and high-latitude magnetic field periods with north and south Saturn kilometric radiation periods, *J. Geophys. Res.*, *115*, A12252; doi:10.1029/2010JA015666.

Andrews, D. J., B. Cecconi, S. W. H. Cowley, M. K. Dougherty, L. Lamy, G. Provan, and P. Zarka (2011), Planetary period oscillations in Saturn's magnetosphere: Evidence in magnetic field phase data for rotational modulation of Saturn kilometric radiation emissions, *J. Geophys. Res.*, *116*, A09206; doi:10.1029/2011JA016636.

Andrews, D. J., S. W. H. Cowley, M. K. Dougherty, and G. Provan (2010a), Magnetic field oscillations near the planetary period in Saturn's equatorial magnetosphere: Variation of amplitude and phase with radial distance and local time, *J. Geophys. Res.*, *115*, A04212; doi:10.1029/2009JA014729.

Arridge, C. S. (2015), Magnetotails of Uranus and Neptune, in *Magnetotails in the Solar System*, Geophysical Monograph 207, edited by A. Kelling, C. M. Jackman, and P. A. Delamere, John Wiley & Sons.

Arridge, C. S., et al. (2011), Mapping magnetospheric equatorial regions at Saturn from Cassini prime mission observations, *Space Sci. Rev.*, *164*, 1–83; doi 10.1007/s11214-011-9850-4.

Arridge, C. S., M. Kane, N. Sergis, K. K. Khurana, and C. M. Jackman (2015), Sources of local time asymmetries in magnetodiscs, *Space Sci Rev.*, *187*, 301–333.

Arridge, C. S., N. Achilleos, M. K. Dougherty, K. K. Khurana, and C. T. Russell (2006), Modeling the size and shape of Saturn's magnetopause with variable dynamic pressure, *J. Geophys. Res.*, *111*, A11227; doi:10.1029/2005JA011574.

Badman, S. V., C. M. Jackman, J. D. Nichols, J. T. Clarke, and J.-C. Gerard (2014), Open flux in Saturn's magnetosphere, *Icarus*, *231*, 137–145.

Bagenal, F. (1992), Giant planet magnetospheres, *Ann. Rev. Earth Planet. Sci.*, *20*, 289–328.

Barish, F. D., and R. A. Smith (1975), An analytical model of the Jovian magnetosphere, *Geophys. Res. Lett.*, *2*, 269.

Behannon, K. W., R. P. Lepping, E. C. Sittler, N. F. Ness, B. H. Mauk, S. M. Krimigis, R. L. McNutt Jr. (1987), The magnetotail of Uranus, *J. Geophys. Res. Space Physics*, *92* (A13), 15,354–15,366.

Belcher, J. W. (1983), The low-energy plasma in the Jovian magnetosphere, in *Physics of the Jovian Magnetosphere*, edited by A. J. Dessler, 68–105, Cambridge University Press, New York.

Birkeland, K. (1908), On the cause of magnetic storms and the origin of terrestrial magnetism, The Norwegian aurora POLARIS expedition, 1902–1903, Longmans, Green & Company, London.

Bonfond, B., J. Gustin, J.-C. Gérard, D. Grodent, A. Radioti, B. Palmaerts, S. V. Badman, K. K. Khurana, and C. Tao (2015), The far-ultraviolet main aurora emission at Jupiter, Part 1: Dawn-dusk brightness asymmetries, *Ann. Geophys.*, *33*, 1203–1209.

Brandt, P. C., K. K. Khurana, D. G. Mitchell, N. Sergis, K. Dialynas, J. F. Carbary, E. C. Roelof, C. P. Paranicas, S. M. Krimigis, and B. H. Mauk (2010), Saturn's periodic magnetic field perturbations caused by a rotating partial ring current, *Geophys. Res. Lett.*, *37*, L22103; doi:10.1029/2010GL045285.

Bridge, H. S., et al. (1986), Plasma observations near Uranus: Initial results from Voyager 2, *Science*, *233*, 89–93; doi: 10.1126/science.233.4759.89.

Bunce, E. J., and S. W. H. Cowley (2001), Divergence of the equatorial current in the dawn sector of Jupiter's magnetosphere: Analysis of Pioneer and Voyager magnetic field data, *Planet. Space. Sci.*, *49*, 1089–1113.

Bunce, E. J., C. S. Arridge, S. W. H. Cowley, and M. K. Dougherty (2008), Magnetic field structure of Saturn's dayside magnetosphere and its mapping to the ionosphere: Results from ring current modeling, *J. Geophys. Res.*, *113*, A02207; doi:10.1029/2007JA012538.

Bunce, E. J., P. G. Hanlon, and S. W. H. Cowley (2002), A simple empirical model of the equatorial radial field in Jupiter's middle magnetosphere, based on spacecraft fly-by and Galileo orbiter data, *Planet. Space Sci.*, *50*, 789–806.

Burch, J. L., J. Goldstein, W. S. Lewis, D. T. Young, A. J. Coates, M. K. Dougherty, and N. Andre (2007), Tethys and Dione as sources of outward-flowing plasma in Saturn's magnetosphere, *Nature*, *447*, 14; doi:10.1038.

Burger, M. H., E. C. Sittler, R. E. Johnson, H. T. Smith, O. J. Tucker, and V. I. Shematovich (2007), Understanding the escape of water from Enceladus, *J. Geophys. Res.*, *112*, A06219; doi:10.1029/2006JA012086.

Burton, M. E., M. K. Dougherty, and C. T. Russell (2009), Model of Saturn's internal planetary magnetic field based on Cassini observations, *Planet. Space Sci.*, *57*, 1706–1713.

Cao H., C. T. Russell, J. Wicht, U. R. Christensen, and M. K. Dougherty (2012), Saturn's high degree magnetic moments: Evidence for a unique planetary dynamo, *Icarus*, *221*, 388–395.

Cao H., C. T. Russell, U. R. Christensen, M. K. Dougherty, and M. E. Burton (2011) Saturn's very axisymmetric magnetic field: No detectable secular variation or tilt, *Earth and Planetary Sci. Lett.*, *304*, 22–28, 2011.

Carbary, J. F., D. G. Mitchell, P. Brandt, C. Paranicas, and S. M. Krimigis (2008), ENA periodicities at Saturn, *Geophys. Res. Lett.*, *35*, L07102; doi:10.1029/2008GL033230.

Carbary, J. F., D. G. Mitchell, S. M. Krimigis, and N. Krupp (2009), Dual periodicities in energetic electrons at Saturn, *Geophys. Res. Lett.*, *36*, L20103; doi:10.1029/2009GL040517.

Carbary, J. F., D. G. Mitchell, S. M. Krimigis, D. C. Hamilton, and N. Krupp (2007), Charged particle periodicities in Saturn's outer magnetosphere, *J. Geophys. Res.*, *112*, A06246; doi:10.1029/2007JA012351.

Caudal, G. (1986), A self-consistent model of Jupiter's magnetodisc including the effects of centrifugal force and pressure, *J. Geophys. Res.*, *91*, 4201–4221; doi:10.1029/JA091iA04p04201.

Chapman, S. (1918), An outline of a theory of magnetic storms, *Proc. Phys. Soc. Lond. A*, *95*, 61–83.

Chapman, S., and V. C. A. Ferraro (1931), A new theory of magnetic storms, *Terrestrial magnetism and atmospheric electricity*, *36*, 77–97.

Connerney, J. E. P. M. H. Acuna, and N. F. Ness (1981), Modeling the Jovian current sheet and inner magnetosphere, *J. Geophys. Res.*, *86*, 8370.

Connerney J. E. P., M. H. Acuna, and N. F. Ness (1983), Currents in Saturn's magnetosphere, *J. Geophys. Res.*, *88*, 8779.

Connerney, J. E. P., M. H. Acuna, and N. F. Ness (1987), The magnetic field of Uranus, *J. Geophys. Res.* *92*, 15,329–15,336.

Connerney, J. E. P., M. H. Acuna, and N. F. Ness (1991), The magnetic field of Neptune, *J. Geophys. Res.*, *96*, 19,023–19,042.

Connerney, J. E. P., M. H. Acuna, N. F. Ness, and T. Satoh (1998), New models of Jupiter's magnetic field constrained by the Io flux tube footprint, *J. Geophys. Res.*, *103*, 11929.

Cowley, S. W. H. (2000), Magnetosphere-ionosphere interactions: A tutorial review, in *Magnetospheric Current Systems*, edited by S.-I. Ohtani, R. Fujii, M. Hesse, and R. L. Lysak, American Geophysical Union, Washington, DC.

Cowley S.W.H., and E.J. Bunce (2001), Origin of the main auroral oval in Jupiter's coupled magnetosphere–ionosphere system, *Planetary and Space Science*, *49*, 1067–1088.

Cowley, S. W. H., and E. J. Bunce (2003), Corotation-driven magnetosphere-ionosphere coupling currents in Saturn's magnetosphere and their relation to the auroras, *Ann. Geophys.*, *21*, 1691–1707.

Cowley, S. W. H., E. J. Bunce, and J. M. O'Rourke (2004), A simple quantitative model of plasma flows and currents in Saturn's polar ionosphere, *J. Geophys. Res.*, *109*, A05212; doi:10.1029/2003JA010375.

Crary, F. J., et al. (2005), Solar wind dynamic pressure and electric field as the main factors controlling Saturn's aurorae, *Nature*, *433*, 720–722.

Dessler, A. J., and E. N. Parker (1959), Hydromagnetic theory of geomagnetic storms, *J. Geophys. Res.*, *24*, 2239–2252.

Fukushima, N., and Y. Kamide (1973), Contribution of magnetospheric field-aligned current to geomagnetic bays and Sq fields, A comment on partial ring-current models, *Radio Science*, *8*, 1013–1017.

Goertz, C. K., D. E. Jones, B. A. Randall, E. J. Smith, and M. F. Thomsen (1976), Evidence of open field lines in Jupiter's magnetosphere, *J. Geophys. Res.*, *81*, 3393–3398.

Gombosi T. I., T. P. Armstrong, C. S. Arridge, K. K. Khurana, S. M. Krimigis, N. Krupp, A. M. Persoon, and M. F. Thomsen (2009), Saturn's magnetospheric configuration, in *Saturn*, Cambridge University Press.

Gurnett, D. A., A. M. Persoon, W. S. Kurth, J. B. Groene, T. F. Averkamp, M. K. Dougherty, and D. J. Southwood (2007), The rotation of the inner region of Saturn's plasma disk, *Science*, *309*, 442–445; doi: 10.1126/science.1138562.

Gurnett, D. A., A. Lecacheux, W. S. Kurth, A. M. Persoon, J. B. Groene, L. Lamy, P. Zarka, and J. F. Carbary (2009), Discovery of a north-south asymmetry in Saturn's radio rotation period, *Geophys. Res. Lett.*, *36*, L16102; doi:10.1029/2009GL039621.

Hansen, C. J., L. Esposito, A. I. F. Stewart, J. Colwell, A. Hendrix, W. Pryor, D. Shemansky, and R. West (2006), Enceladus' water vapor plume, *Science*, *311*, 1422–1425.

Hill, T. W. (1979), Inertial limit on corotation, *J. Geophys. Res.*, *84*, 6554.

Hill, T.W. (2001), The Jovian auroral oval, *J. Geophys. Res.*, 106, 8101–8107.

Hill, T. W., A. J. Dessler, and C. K. Goertz (1983), Magnetospheric models, in *Physics of the Jovian Magnetosphere*, 353–394, Cambridge University Press.

Hill, T. W., A. M. Rymer, J. L. Burch, F. J. Crary, D. T. Young, M. F. Thomsen, D. Delapp, N. André, A. J. Coates, and G. R. Lewis (2005), Evidence for rotationally driven plasma transport in Saturn's magnetosphere, *Geophys. Res. Lett.*, *32*, L14S10; doi:10.1029/2005GL022620.

Hunt, G. J., S. W. H. Cowley, G. Provan, E. J. Bunce, I. I. Alexeev, E. S. Belenkaya, V. V. Kalegaev, M. K. Dougherty, and A. J. Coates (2014), Field-aligned currents in Saturn's southern nightside magnetosphere: Subcorotation and planetary period oscillation components, *J. Geophys. Res. Space Physics*, *119*, 9847–9899; doi:10.1002/2014JA020506.

Hunt, G. J., S. W. H. Cowley, G. Provan, E. J. Bunce, I. I. Alexeev, E. S. Belenkaya, V. V. Kalegaev, M. K. Dougherty, and A. J. Coates (2015), Field-aligned currents in Saturn's northern nightside magnetosphere: Evidence for interhemispheric current flow associated with planetary period oscillations, *J. Geophys. Res. Space Physics*, *120*, 7552–7584; doi:10.1002/2015JA021454.

Iijima, T., and T. A. Potemra (1976), The amplitude distribution of field aligned currents at northern high latitudes observed by Triad, *J. Geophys. Res.*, *83*, 1109–1114.

Iijima, T., and T. A. Potemra (1978), Large-scale characteristics of field-aligned currents associated with substorms, *J. Geophys. Res.*, *83*, 599–615.

Jia, X., and M. G. Kivelson (2012), Driving Saturn's magnetospheric periodicities from the upper atmosphere/ionosphere: Magnetotail response to dual sources, *J. Geophys. Res.*, *117*, A11219; doi:10.1029/2012JA018183.

Jia, X., M. G. Kivelson, and T. I. Gombosi (2012), Driving Saturn's magnetospheric periodicities from the upper atmosphere/ionosphere, *J. Geophys. Res.*, *117*, A04215; doi:10.1029/2011JA017367.

Jurac, S., and J. D. Richardson (2005), A self-consistent model of plasma and neutrals at Saturn: Neutral cloud morphology, *J. Geophys. Res.*, *110*, A09220.

Kane, M., B. H. Mauk, E. P. Keath, and S. M. Krimigis (1995), Hot ions in Jupiter's magnetodisc: A model for Voyager 2 low-energy charged particle measurements, *J. Geophys. Res.*, *100*, 19,473.

Kellett, S., C. S. Arridge, E. J. Bunce, A. J. Coates, S. W. H. Cowley, M. K. Dougherty, A. M. Persoon, N. Sergis, and R. J. Wilson (2011), Saturn's ring current: Local time dependence and temporal variability, *J. Geophys. Res.*, *116*, A05220; doi:10.1029/2010JA016216.

Kennelly, T. J., J. S. Leisner, G. B. Hospodarsky, and D. A. Gurnett (2013), Ordering of injection events within Saturnian SLS longitude and local time, *J. Geophys. Res. Space Physics*, *118*, 832–838; doi:10.1002/jgra.50152.

Khurana, K. K. (1997), Euler potential models of Jupiter's magnetospheric field, *J. Geophys. Res.*, *102*, 11,295.

Khurana, K. K. (2001), Influence of solar wind on Jupiter's magnetosphere deduced from currents in the equatorial plane, *J. Geophys. Res.*, *106* (A11): 25999–26016.

Khurana, K. K., and M. G. Kivelson (1993), Inference of the angular velocity of plasma in the Jovian magnetosphere from the sweepback of magnetic field, *J. Geophys. Res.*, *98*, 67.

Khurana, K. K., D. G. Mitchell, C. S. Arridge, M. K. Dougherty, C. T. Russell, C. Paranicas, N. Krupp, and A. J. Coates (2009), Sources of rotational signals in Saturn's magnetosphere, *J. Geophys. Res.*, *114*, A02211; doi:10.1029/2008JA013312.

Khurana, K. K., M. G. Kivelson, V. M. Vasyliunas, N. Krupp, J. Woch, A. Lagg, B. H. Mauk, and W. S. Kurth (2004), The configuration of Jupiter's magnetosphere, in *Jupiter: The Planet, Satellites, and Magnetosphere*, edited by F. Bagenal, T. Dowling, and W. McKinnon, Cambridge University Press, 593–616.

Khurana, K. K., M. K. Dougherty, C. T. Russell, and J. S. Leisner (2007), Mass loading of Saturn's magnetosphere near Enceladus, *J. Geophys. Res. Space Physics*, *112*, A08203; doi:10.1029/2006JA012110.

Kivelson, M. G., and D. J. Southwood (1991), Ionospheric travelling vortex generation by solar wind buffeting of the magnetosphere, *J. Geophys. Res.*, *96*, 1661–1667.

Kivelson, M. G., and F. Bagenal (2007), Planetary magnetospheres, in *Encyclopedia of the Solar System*, edited by L. McFadden, P. R. Weissman, and T. V. Johnson, Academic Press, San Diego.

Krupp, N., A. Lagg, S. Livi, B. Wilken, J. Woch, E. C. Roelof, and D. J. Williams (2001), Global flows of energetic ions in Jupiter's equatorial plane: First-order approximation, *J. Geophys. Res.*, *106*, 26017–26032.

Kurth, W. S., A. Lecacheux, T. F. Averkamp, J. B. Groene, and D. A. Gurnett (2007), A Saturnian longitude system based on a variable kilometric radiation period, *Geophys. Res. Lett.*, *34*, L02201; doi:10.1029/2006GL028336.

Lanzerotti, L. J., R. D. Hunsalter, D. Rice, L.-C. Lee, A. Wolfe, C. G. Maclennan, and L. V. Medford (1987), Ionosphere and ground based response to field-aligned currents near the magnetospheric cusp region, *J. Geophys. Res.*, *92*, 7739.

Lecacheux, A. P., and M. Aubier (1997), Re-visiting Saturnian radiation with Ulysses/URAP, in *Planetary Radio Emissions IV*, edited by H. O. Rucker, S. J. Bauer, and A. Lecacheux, 313–325, Austrian Academy of Sciences Press, Vienna.

Liu, X., T. W. Hill, R. A. Wolf, S. Sazykin, R. W. Spiro, and H. Wu (2010), Numerical simulation of plasma transport in Saturn's inner magnetosphere using the Rice Convection Model, *J. Geophys. Res.*, *115*, A12254; doi:10.1029/2010JA015859.

Mauk, B. H., E. P. Keath, M. Kane, S. M. Krimigis, A. F. Cheng, M. H. Acuna, T. R. Armstrong, and N. F. Ness (1991), The magnetosphere of Neptune: Hot plasmas and energetic particles, *J. Geophys. Res.*, *96*, 19,061–19,084.

McPherron, R. L. (1991), Physica processes producing magnetospheric substorms and magnetic storms, in *Geomagnetism*, 4, edited by J. Jacobs, 593–739, Academic Press Ltd., London.

Mitchell, D. G., et al. (2009), Recurrent energization of plasma in the midnight-to-dawn quadrant of Saturn's magnetosphere, and its relationship to auroral UV and radio emissions, *Planet. Space Sci.*, *57*, 1732–1742; doi:10.1016/j.pss.2009.04.002.

Moore, L., et al. (2010), Latitudinal variations in Saturn's ionosphere: Cassini measurements and model comparisons, *J. Geophys. Res.*, *115*, A11317.

Ness, N. F., M. H. Acuna, K. W. Behannon, L. F. Burlaga, J. E. P. Connerney, R. P. Lepping, and F. M. Neubauer (1986), Magnetic fields at Uranus, *Science*, *233*, 85–89.

Nichols, J. D. (2011), Magnetosphere-ionosphere coupling in Jupiter's middle magnetosphere: Computations including a self-consistent current sheet magnetic field model, *J. Geophys. Res.*, *116*, A10232; doi:10.1029/2011JA016922.

Nichols, J. D., N. Achilleos, and S. W. H. Cowley (2015), A model of force balance in Jupiter's magnetodisc including hot plasma pressure anisotropy, *J. Geophys. Res. Space Physics*, *120*, 10,185–10,206; doi:10.1002/2015JA021807.

Nichols, J.D. and S.W.H. Cowley (2003), Magnetosphere-ionosphere coupling currents in Jupiter's middle magnetosphere: dependence on the effective ionospheric Pedersen conductivity and Iogenic plasma mass outflow rate, *Ann. Geophys.*, *21*, 1419, 2003.

Nichols, J.D. and S.W.H. Cowley (2004) Magnetosphere-ionosphere coupling currents in Jupiter's middle magnetosphere: effect of precipitation-induced enhancement of the ionospheric Pedersen conductivity, *Annales Geophysicae*, *22*, 1799–1827.

Paranicas, C., M. F. Thomsen, N. Achilleos, M. Andriopoulou, S. V. Badman, G. Hospodarsky, C. M. Jackman, et al. (2016), Effects of radial motion on interchange injections at Saturn, *Icarus*, *264*, 342–351.

Provan, G., D. J. Andrews, B. Cecconi, S. W. H. Cowley, M. K. Dougherty, L. Lamy, and P. Zarka (2011), Magnetospheric period magnetic field oscillations at Saturn: Equatorial phase 'jitter' produced by superposition of southern- and northern-period oscillations, *J. Geophys. Res.*, *116*, A04225, doi:10.1029/2010JA016213.

Provan, G., D. J. Andrew, C. S. Arridge, et al. (2009b), Polarization and phase of planetary-period magnetic field oscillations on high-latitude field lines in Saturn's magnetosphere, *J. Geophys. Res.*, *114*, A02225; doi:10.1029/2008JA013782.

Provan, G., S. W. H. Cowley, and J. D. Nichols (2009a), Phase relation of oscillations near the planetary period of Saturn's auroral oval and the equatorial magnetospheric magnetic field, *J. Geophys. Res.*, *114*, A04205; doi:10.1029/2008JA013988.

Ray, L. C., N. A. Achilleos, M. F. Vogt, and J. N. Yates (2014), Local time variations in Jupiter's magnetosphere-ionosphere coupling system, *J. Geophys. Res. Space Physics*, *119*, 4740–4751; doi:10.1002/2014JA019941.

Richardson, J. D. (1993), A quantitative model of plasma in Neptune's magnetosphere, *Geophys. Res. Lett.*, *20*, 1467–1470.

Richardson, J. D. (1998), Thermal plasma and neutral gas in Saturn's magnetosphere, *Rev. Geophys.*, *36*, 501–524.

Rymer, A. M., B. H. Mauk, T. W. Hill, N. Andre, D. G. Mitchell, C. Paranicas, D. T. Young, et al. (2009), Cassini evidence for rapid interchange transport at Saturn, *Planet. Space Sci.*, *57*, 1779–1784.

Sckopke, N. (1966), A general relation between the energy of trapped particles and the disturbance field near the Earth, *J. Geophys. Res.*, *71*, 3125–3130.

Sergis, N., et al. (2010), Particle pressure, inertial force, and ring current density profiles in the magnetosphere of Saturn, based on Cassini measurements, *Geophys. Res. Lett.*, *37*, L02102; doi:10.1029/2009GL041920.

Smith, C. G. A. (2011), A Saturnian cam current system driven by asymmetric thermospheric heating, *Monthly Not. R. Astron. Soc.*, *410*, 2315–2328; doi:10.1111/j.1365-2966.2010.17602.x

Southwood, D. (2011), Direct evidence of differences in magnetic rotation rate between Saturn's northern and southern polar regions, *J. Geophys. Res.*, *116*, A01201; doi:10.1029/2010JA016070.

Southwood, D. J., and M. G. Kivelson (2007), Saturnian magnetospheric dynamics: Elucidation of a camshaft model, *J. Geophys. Res.*, *112*, A12222; doi:10.1029/2007JA012254.

Stanley, S., and J. Bloxham (2006), Numerical dynamo models of Uranus's and Neptune's magnetic fields, *Icarus*, *184*, 556–572.

Tao, C., H. Fujiwara, and Y Kasaba (2010), Jovian magnetosphere-ionosphere current system characterized by diurnal variation of ionospheric conductance, *Planet. Space Sci.*, *58*, 351–364.

Thomsen, M. F., D. B. Reisenfeld, D. M. Delapp, R. L. Tokar, D. T. Young, F. J. Crary, E. C. Sittler, M. A. McGraw, and J. D. Williams (2010), Survey of ion plasma parameters in Saturn's magnetosphere, *J. Geophys. Res.*, *115*, A10220; doi:10.1029/2010JA015267.

Thomsen, M. F., et al. (2014), Ion composition in interchange injection events in Saturn's magnetosphere, *J. Geophys. Res. Space Physics*, *119*, 9761–9772.

Tsyganenko, N. A. (1990), Quantitative models of the magnetospheric magnetic field: Methods and results, *Space Sci. Rev.*, *54*, 75–186.

Vasyliunas, V. M. (1983), Plasma distribution and flow, in *Physics of the Jovian Magnetosphere*, edited by A. J. Dessler, 395–453, Cambridge University Press.

Vogt, M. F., M. G. Kivelson, K. K. Khurana, R. J. Walker, B. Bonfond, D. Grodent, and A. Radioti (2011), Improved mapping of Jupiter's auroral features to magnetospheric sources, *J. Geophys. Res.*, *116*, A03220; doi:10.1029/2010JA016148.

Wilson, R. J., R. L. Tokar, and M. G. Henderson (2009), Thermal ion flow in Saturn's inner magnetosphere measured by the Cassini Plasma Spectrometer: A signature of the Enceladus torus? *Geophys. Res. Lett.*, *36*, L23104; doi:10.1029/2009GL040225.

Wilson, R. J., R. L. Tokar, M. G. Henderson, T. W. Hill, M. F. Thomsen, and D. H. Pontius Jr. (2008), Cassini plasma spectrometer thermal ion measurements in Saturn's inner magnetosphere, *J. Geophys. Res.*, *113*, A12218; doi:10.1029/2008JA013486.

3

Electric Currents in the Solar Atmosphere

Gregory D. Fleishman[1] and Alexei A. Pevtsov[2]

ABSTRACT

We review historical and modern studies pertinent to the measurements, modeling, and role of electric current in the solar atmosphere. We describe how the electric current density is computed from the vector magnetic field measurement at photospheric and chromospheric levels and how it is modeled, using Nonlinear Force-Free Field (NLFFF) extrapolations, in the solar corona. Next, the roles of electric currents in plasma dynamics and heating, energy release, and acceleration and transport of nonthermal particles are discussed. We then consider some interesting properties of electron and ion components forming electric current in a multicomponent plasma, the effect of the electric current on the Alfvén wave properties, current neutralization, and redistribution during solar flares. We discuss present theories about the origin of electric currents from sub-photospheric and photospheric motions. We conclude that evaluation of electric currents in the solar atmosphere has reached a high level of maturity and that accounting for the electric current is highly important for solar physics.

3.1. INTRODUCTION

The Sun's volume can crudely be divided onto its interior and the outer atmosphere. The interior includes the inner core (IC), the radiative zone (RZ), and the convection zone (CZ). While the IC and RZ together compose about 98% of the mass of the Sun, these layers do not have any electric currents with appreciable observable effect on the visible surface. The solar magnetic field observed on the Sun's visible surface and electric currents associated with it are likely created either in a thin transition layer between the RZ and the CZ, or in the bulk of the CZ via the flows of ionized plasma, whose collective action is referred to as a dynamo [*Charbonneau*, 2013]. Once the magnetic field reaches a certain threshold, it becomes buoyant, and rises through the CZ toward the outer layer of solar atmosphere, or corona [*Cheung and Isobe*, 2014; *Fan*, 2009]. The transition between the CZ and corona takes place over two thin layers called the photosphere (about 500 km in average thickness) and the chromosphere (≈2500 km) bounded by a thin transition region (TR). It is in the photosphere and the chromosphere, where the magnetic field vector can be routinely "measured directly," from which the electric currents are inferred.

Practically all major solar features observed in the photosphere are associated with magnetic fields (see Fig. 3.1). Dark sunspots (Fig. 3.1a) correspond to areas of strong magnetic field (see white and black concentrations in Fig. 3.1d). There, the magnetic field inhibits transport of energy via convection, which explains why these areas are darker than their surroundings. In contrast, in the chromosphere (Fig. 3.1b) and corona (Fig. 3.1c), areas around sunspots appear as bright features suggesting either enhanced energy transport (bright plages in the

[1] *Physics Department, Center for Solar-Terrestrial Research, New Jersey Institute of Technology, Newark, New Jersey, USA*
[2] *National Solar Observatory, Boulder, Colorado, USA; ReSoLVE Centre of Excellence, Space Climate Research Unit, University of Oulu, Finland*

Electric Currents in Geospace and Beyond, Geophysical Monograph 235, First Edition.
Edited by Andreas Keiling, Octav Marghitu, and Michael Wheatland.

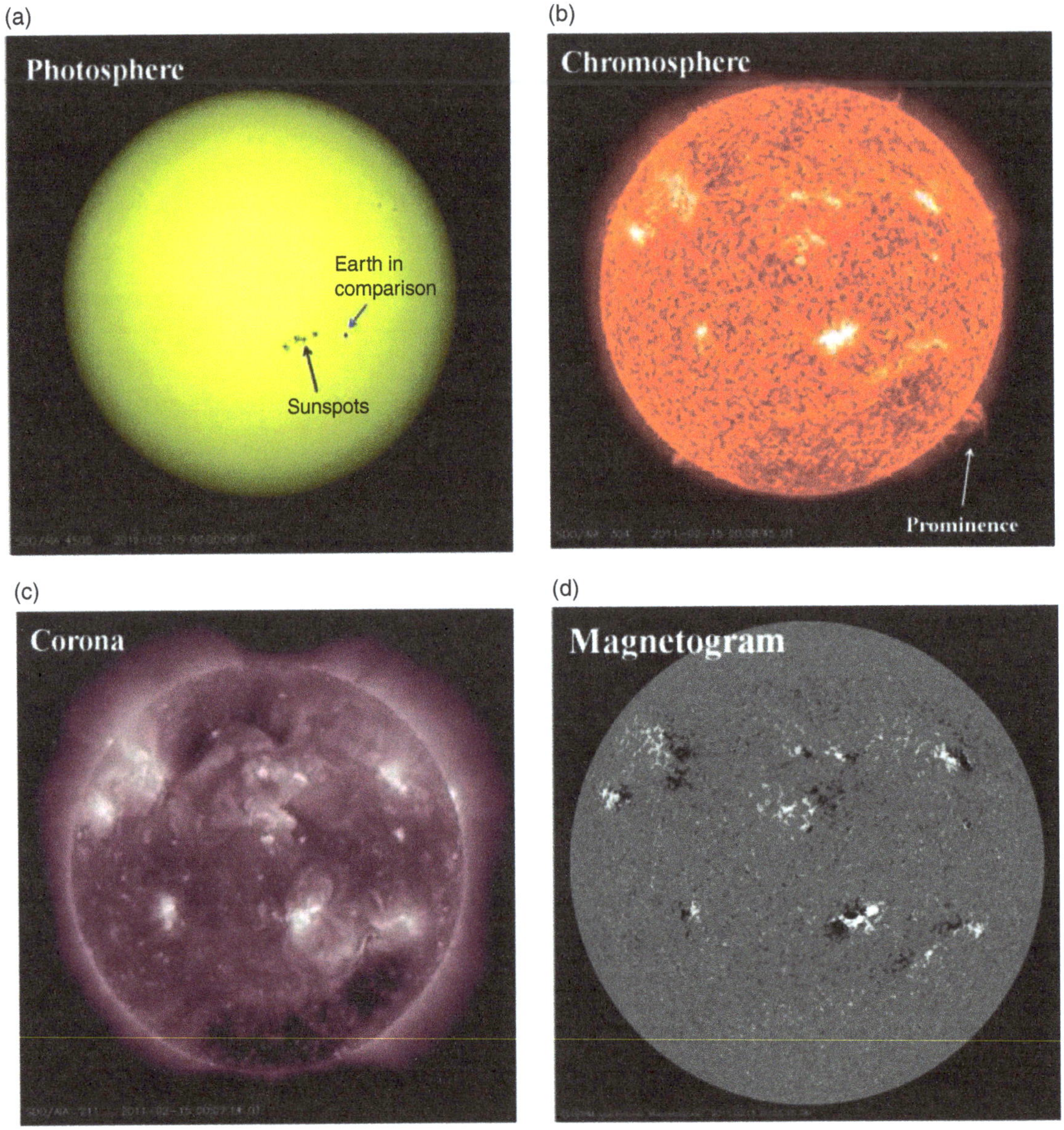

Figure 3.1 Images of the solar photosphere, chromosphere, and corona and the corresponding map of photospheric magnetic fields as observed by the Solar Dynamics Observatory on 15 Feb 2011 around 00:00 UT.

chromosphere) or enhanced heating in the corona (waves, reconnection etc., see, e.g., *Hansteen et al.* [2015]). As active regions decay, their magnetic field spreads over larger areas and weakens. The evolving remnants of active regions form large-scale neutral lines, which may support accumulation of cool chromospheric material forming prominences or chromospheric filaments, when seen against the disk [*Mackay et al.*, 2010; *Parenti*, 2014]. Solar flares and coronal mass ejections, the two major sources of the interplanetary disturbances associated with space weather, are related to groups of sunspots (active regions) and erupting filaments/prominences. Large areas of open, weak magnetic field can be readily recognized in the corona as coronal holes [*Cranmer*, 2009] (see large dark feature in the lower [south] part of the solar disk in Fig. 3.1c). Coronal holes are the areas from which the fast solar wind is launched.

Unlike laboratory measurements, any remote sensing of the electric current relies on measurements of the magnetic field vector and the use of Ampère's law. Even though the determination of magnetic fields on the Sun refers to derivation of the magnetic field components from observations of various polarization states of spectral lines, we call such derivations "measurements," given that the results show a high level of repeatability and consistency when obtained by different observatories. It is true that the complexity in interpretation of the polarized light in terms of magnetic fields produces an uncertainty in the magnetic fields derived that way; however, those uncertainties can be firmly quantified. Similarly, "measurements" of electric currents refer to their computation from the given distribution of magnetic fields. The computation of electric currents involves spatial derivatives and so further increases the uncertainties associated with measurements of magnetic fields.

Currently available techniques of measuring magnetic fields are based on the Zeeman and Hanle effects, and normally can be applied only to the photosphere and the chromosphere. Information about the polarity and the total field strength of the coronal magnetic field can be derived from the observations in radio wavelengths [for reviews, see *Gary et al.*, 2013; *Gelfreikh*, 2004; *Wang et al.*, 2015; *White*, 2005]. In some limited cases, information about the inclination of magnetic field vector relative to the line of sight can also be inferred (e.g., quasi-transverse propagation, QT-surface, *Ryabov et al.* [2005]). The coronal field strength can also be derived via indirect measurements of oscillating coronal structures observed in EUV lines [e.g., *Nakariakov and Ofman*, 2001]. Direct measurements of the fields in low corona are possible via the Hanle effect in EUV resonance lines [*Raouafi*, 2002], or via the Zeeman effect in coronal spectral lines [*Harvey*, 1969; *Lin et al.*, 2000]. Since the corona is optically thin, the observed radiation represents an integral along the line of sight passing through multiple coronal structures. This complicates interpretation of magnetic fields measured this way. In contrast, the corona is optically thick in radio wavelengths, and, thus, coronal fields obtained from radio methods are linked to specific layers in the solar corona, although there may be an uncertainty in the exact height of those layers. Traditionally, the derivation of coronal fields relied on their extrapolation from lower layers of the solar atmosphere. The extrapolation techniques make restrictive assumptions about the magnetic field as being either potential (vacuum or current free) or force free (linear or nonlinear). The applicability of the force-free field approximation to solar atmospheric conditions depends on the ratio of gas (p) and magnetic pressures (commonly called the "plasma beta"): $\beta = \frac{8\pi p}{B^2}$. For $\beta \ll 1$, the pressure gradient can be neglected, and the stationary case of the equation of motion reduces to $\mathbf{j} \times \mathbf{B} = 0$, where j is the electric current density, and $\mathbf{B}$ is the magnetic field. This "force-free field" (FFF) condition implies that the electric currents flow along the magnetic field.

The plasma properties differ significantly between the CZ, the photosphere, chromosphere, and corona. At its base (about 200,000 km below the photosphere), the CZ plasma is very dense ($\rho \simeq 0.1$ gm cm^{-3}) and is completely ionized ($T \simeq 2 \times 10^6$ K). The magnetic field strength is estimated to be in the range of a few dozens kG. In the photosphere, $\rho \simeq 3 \times 10^{-7}$ gm cm^{-3} and $T \simeq 6{,}000$ K. The typical magnetic field strength in sunspots is about a few kilo-Gauss. This implies $\beta \gg 1$ through the convection zone and in the photosphere outside of strong field regions (e.g., sunspot umbrae). In the corona, the gas pressure and the magnetic field decrease significantly $\rho \simeq 10^{-16} - 10^{-14}$ gm cm^{-3}, $T = (1-2) \times 10^6$ K, and $B \simeq 10-100$ G (coronal holes/active regions). This results in $\beta \simeq 0.01 - 0.001 \ll 1$. A more detailed analysis (see Fig. 3.2) indicates that the plasma β is smaller than unity in the range of heights from the mid-chromosphere to mid-upper corona. Evaluation of the plasma β based on observations of magnetic fields at different heights in the solar photosphere and the chromosphere suggests that the magnetic field is not force free in the photosphere, while the FFF conditions are satisfied in the chromosphere [*Metcalf et al.*, 1995]. The reader should understand that these estimates of plasma β refer to average conditions. In various structures, the magnetic fields could be force free or non–force free even if these structures are located at the same height in solar atmosphere (see range of β in Fig. 3.2).

Traditionally, the topological properties and interconnectivity of solar features in the photosphere and corona are presented in terms of their magnetic fields. For example, as a rule the sunspots are organized in groups, whose leading polarity is opposite in sign to the following polarity. The large-scale structure of magnetic field in bipolar active regions resembles the dipole field from a bar magnet. The polarity of leading spots in the Northern Hemisphere is opposite to that in the Southern hemisphere [the so-called Hale or Hale-Nicholson polarity rule; *Hale and Nicholson* [1925]). For the same group, the leading polarity spots are normally situated closer to the equator. The tilt of the main axis of a group relative to the equator varies approximately as sine of the latitude; this functional dependency is referred to as Joy's law. The Hale polarity rule is quite strong: only about 1%–10% of active regions exhibit abnormal polarity orientation. But Joy's law is rather a weaker tendency with a significant scatter (for review, see, *Pevtsov et al.* [2014]).

Solar activity rises and falls with approximately 10–11 year cycles [*Hathaway*, 2010], although cycle-like variations with other periods were reported as well. At the beginning of a sunspot cycle, sunspots emerge at higher latitudes (normally, at about 40°–45°), and as the cycle progresses, the latitude of sunspot emergence moves slowly toward the equator. The leading polarity of sunspots reverses with each new cycle, and, thus, the true magnetic cycle consists of two regular sunspot cycles. The solar cycle variation also includes polar magnetic fields. Polar fields have opposite polarity at the north and

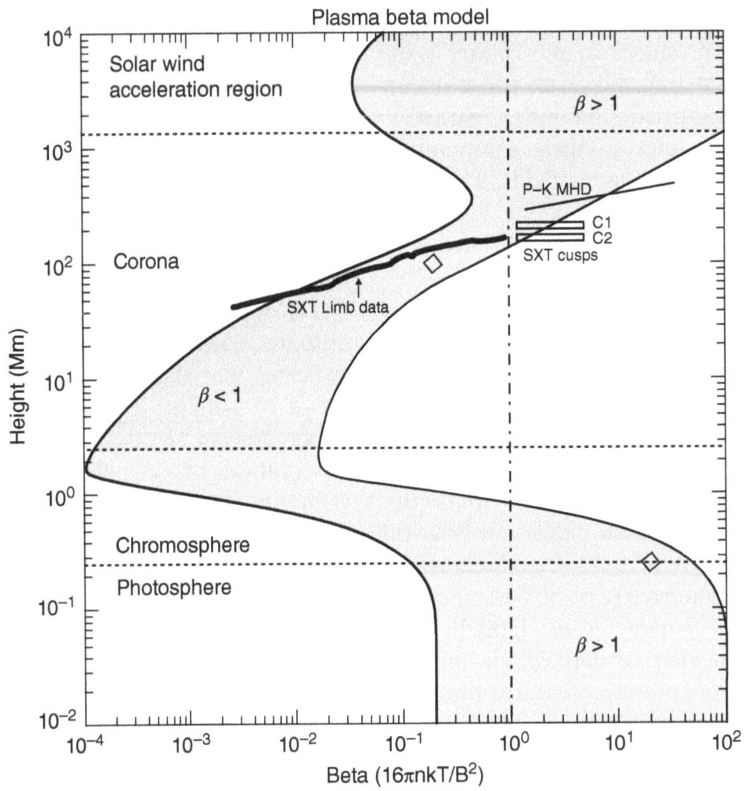

Figure 3.2 Compilation model of plasma β as a function of height in solar atmosphere above active region. Range of magnetic field is taken between (right boundary of shaded area) 150 G (representing plage) and 2500 G (sunspot, thick black line). From *Gary* [2001].

south polar caps. The maximum of the polar field is typically reached around the minimum of the sunspot cycle, and the two polar caps reverse their polarities shortly after the maximum of the sunspot cycle. Thus, during the rising phase of a sunspot cycle, the trailing polarity of sunspots in groups is opposite in sign to the polarity of polar magnetic field in the same hemisphere. During the declining phase of a sunspot cycle, the trailing polarity of sunspots in groups has the same sign as the polar field. This polarity orientation plays an important role in the solar cycle. As active regions (sunspot groups) decay, the magnetic flux of their following polarity is transported by the meridional flows to high latitudes. There it interacts with the existing polar field: first, canceling out the polar field of the previous cycle and, later, rebuilding a new polar field of opposite sign for the next solar cycle. This qualitative representation of the evolution of solar magnetic field forms a foundation for the Babcock-Leighton dynamo model [*Babcock*, 1961].

The traditional "pictures" of solar phenomena are often based on magnetic field representations. Here we specifically refer to general descriptive picture, not MHD models. While the models (even at the dawn of the solar magnetography era) were based on equations that (directly or indirectly) include the electric currents, the descriptive pictures often shaped the interpretation of observations in the framework of these models. Thus, for example, for a long time, the observations did not consider electric currents developing as the result of reconnection, simply because at that time only magnetic field evolution was emphasized by qualitative pictures of this solar phenomenon. Perhaps, responding to that, *Alfvén and Carlqvist* [1967] had noted that a description based on electric currents is "often physically more interesting than a description in terms of magnetic field." The explicit use of electric currents provides another perspective allowing a deeper understanding of the physics of generation (dynamo), topology (connectivity), restructuring (reconnection), and dissipation of solar magnetic fields. Thus, for example, the use of electric currents led to the introduction of the concept of magnetic and current helicity to solar physics in the 1990s [*Abramenko et al.*, 1996; *Pevtsov et al.*, 1994, 1995; *Rust*, 1994; *Seehafer*, 1990; *Zhang and Bao*, 1998]. As a quick reference, the helicity density is defined as a dot product of a vector and its curl; for example, the current helicity density is $h_c = \mathbf{B} \cdot [\nabla \times \mathbf{B}]$. The electric currents were shown to play a role in the reconnection process affecting both the reconnection itself [e.g., *Canfield et al.*, 1996], and the amount of energy released as the result of the reconnection [e.g., *Kusano et al.*, 2004; *Melrose*, 1997]. For more details on current/

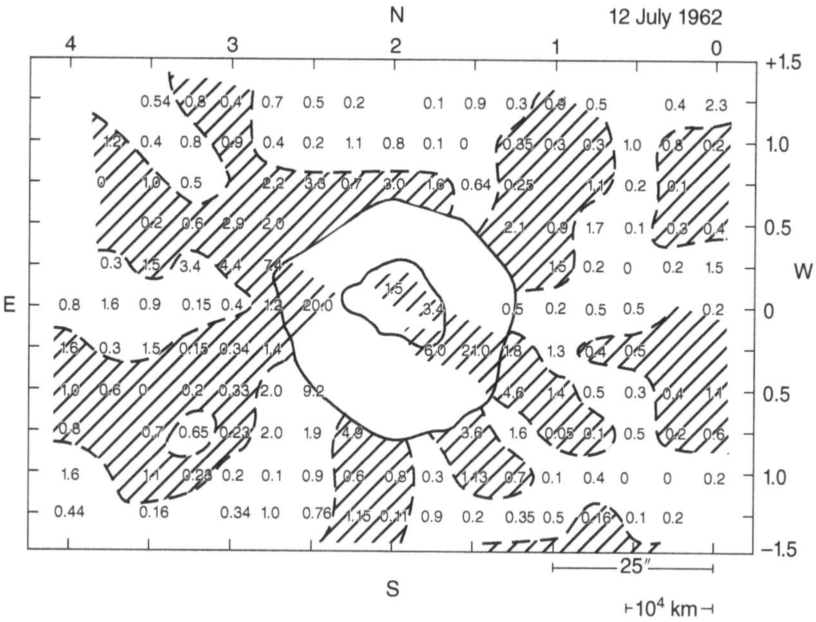

Figure 3.3 Map of vertical electric current density in a sunspot observed on 12 July 1962 at Crimean Astrophysical Observatory. Solid contours outline sunspot umbra and penumbra. *X* and *y* axes indicate heliographic coordinates in degrees. Dashed lines mark boundaries between electric currents of negative (shaded) and positive signs. Note that electric currents of both signs are present in sunspot umbra and penumbra. This is a modified version of Figure 4 from *Severnyi* [1965].

magnetic helicity and the role of electric currents in reconnection, see reviews by *Brown et al.* [1999]; *Buechner and Pevtsov* [2003]; *Pevtsov et al.* [2014].

The earliest observations of electric currents in sunspots (Fig. 3.3) were conducted in 1962 at Crimean Astrophysical Observatory (CrAO, see references to 1964 reports in *Severnyi* [1965]), and were based on a differential form of Ampère's law: $\nabla\times\mathbf{B}=\frac{4\pi}{c}\mathbf{j}+\frac{1}{c}\frac{\partial\mathbf{D}}{\partial t}$ with the displacement term **D** neglected. Since the observations of vector field on the Sun are normally limited to a single height in the atmosphere, only the vertical (z) component of the electric current density could be derived as: $j_z=\frac{c}{4\pi}\left(\frac{\partial B_x}{\partial y}-\frac{\partial B_y}{\partial x}\right)$. Later, the photospheric electric currents in sunspots were also studied by *Harvey* [1969], *Moreton and Severny* [1968], and *Rayrole and Semel* [1970].

In 1968, the first vector magnetograms at the chromospheric heights were observed in CrAO, and the electric currents in the photosphere and the chromosphere were found to be about the same [*Kotov*, 1970, 1972]. Observations taken in two photospheric lines forming at slightly different heights were used to compute horizontal components of electric currents. The horizontal electric currents were found to be larger than the vertical current [*Kotov*, 1971]. *Dravins* [1974] confirmed early findings made by Crimean astronomers. The average current density in sunspots was found to be about 9×10^{-3} A m^{-2} and about 3×10^{-3} A m^{-2} outside the sunspot outer boundary. The vertical currents were smaller than the horizontal ones. Later, *Hofmann and Staude* [1987] estimated the azimuthal component of electric currents assuming a specific geometry of a highly inclined flux tube forming an isolated sunspot to be about $(78\pm36)\times10^{-3}$ A m^{-2}. *Pevtsov and Peregud* [1990] assumed an azimuthal (cylindrical) symmetry to derive the three components of electric current. Both studies confirmed that the horizontal currents in sunspots are typically larger than the vertical currents.

Computation of electric currents based on the differential form of Ampère's equation may be a subject of several uncertainties. First, the 180° ambiguity in the horizontal field direction needs to be resolved. This ambiguity arises from the classical definition of the azimuth of the horizontal magnetic field as the arctangent of the ratio of the Stokes Q and U components. These Stokes components represent properties of the linear polarization of light (see section 3.2). The methods employed for the disambiguation are based on additional assumptions about the properties of the magnetic field in the solar atmosphere [for review, see, *Metcalf et al.*, 2006]. Ambiguity-free determination of total electric current in solar active regions has been formulated by *Semel and Skumanich* [1998], but has not been widely adopted. Second, taking partial derivatives tends to amplify the observing errors. Typically, the errors in the horizontal fields are an order of magnitude larger than the errors in the line-of-sight component. Alternatively, one can rewrite the Ampère's equation in integral form as:

$$\oint_C \mathbf{B}\cdot d\mathbf{l} = \frac{4\pi}{c}\int \mathbf{j}\cdot d\mathbf{S} = \frac{4\pi}{c}\mathbf{J}. \tag{3.1}$$

Integrating horizontal field over the closed contour encompassing a flux element would give the total electric current crossing the surface of this flux element. This approach was briefly discussed in *Rayrole and Semel* [1970], and later used by several researchers (e.g., see, *Gopasyuk* [2015], and references therein]. The total currents found by the two methods seem to agree reasonably well.

3.2. MODERN MEASUREMENTS OF ELECTRIC CURRENT AT THE PHOTOSPHERE

Modern measurements of electric currents in the solar atmosphere follow the early approaches using the differential form of Ampère's law. Measurements of the vector magnetic field, however, have made significant progress. In the early days, the observations were often limited to measuring an integrated polarization over a restricted wavelength range in a wing of spectral line sensitive to the magnetic field in the range of atmospheric heights where this line is formed. To convert these polarization measurements to magnetic field (field strength, inclination, and azimuth) required making additional restrictive assumptions about the solar atmosphere, assuming a functional dependence between the degree of polarization and the magnetic field strength and orientation, as well as simplified radiative transfer modeling. While magnetographs of that type (often referred to as a Babcock-type magnetograph) are still in use, the majority of modern instruments are full-Stokes polarimeters, which sample a full spectral range associated with a spectral line and the neighboring continuum with sufficiently high spectral resolution. Observations are taken to characterize all four Stokes parameters: the total intensity, I, the circularly polarized component, V, and the two linearly polarized components, Q and U. Q and U are defined in coordinate systems rotated by 45° relative to each other. For example, if $+Q$ is maximum at zero and 180° azimuthal orientations (where the reference direction is set by the orientation of the optical element in a magnetograph), and $-Q$ is maximum at 90° and 270°, for $+U(-U)$ the maxima will be at 45° and 225° (135° and 315°) accordingly. The observations of Stokes I, Q, U, and V profiles are inverted to derive a set of thermodynamic and magnetic field parameters best fitting the observed spectro-polarimetry data. This approach has fewer restrictive assumptions, and it mitigates the presence of magneto-optical effects in measurements [*del Toro Iniesta*, 2003]. It also allows determining the gradients of magnetic field with height within the range of formation of the spectral line, and the fractional contribution of polarized and nonpolarized light within each observing pixel. The latter parameter, often referred to as fill-(or filling) factor, provides an estimate of unresolved features with the magnetic field inside a single instrument pixel. A fill-factor of unity corresponds to totally magnetized plasma within the observing pixel, while $f = 0$ means that there is no contribution of magnetic field in any portion of the observed pixel.

Radiative transfer calculations may still use some simplified models of solar atmosphere, although these restrictions have gradually been relaxed. One of the model atmospheres most frequently employed by many inversion techniques is the so-called Milne-Eddington (ME) model. This model makes several assumptions, including: local thermodynamic equilibrium (LTE), no depth dependence of the ratio of the spectral line to the continuum, linearity of the Planck function with depth in the stellar atmosphere, and no gradients in the magnetic field and the line absorption matrix (absorption coefficient, line damping, Doppler width, and velocity) in the spectral line formation region. Under these assumptions, one could derive an analytical solution of the radiative transfer equations, which was done by *Unno* [1956] and *Rachkovsky* [1962, 1967] in the early 1960s; for a recent review, see *del Toro Iniesta and Ruiz Cobo* [2016]. The inversion techniques that were developed later use this Unno-Rachkovsky analytical solution to fit the observed Stokes line profiles with the analytical ones with several (typically, six) free parameters; see *Bommier et al.* (2007), *Bommier* (2016) and references therein.

Full-Stokes inversions based on the ME model are currently in use in such instruments as the Helioseismic and Magnetic Imager (HMI) [*Scherrer et al.*, 2012; *Schou et al.*, 2012]; the Vector Stokes Magnetograph (VSM) at the Synoptic Optical Long-Term Investigations of the Sun facility (SOLIS) [*Balasubramaniam and Pevtsov*, 2011; *Keller et al.*, 2003]; the Imaging Magnetograph eXperiment (IMaX) for the Sunrise Balloon-Borne Solar Observatory [*Martínez Pillet et al.*, 2011]; and the Spectropolarimeter (SP) on the Hinode spacecraft [*Kosugi et al.*, 2007]. For a more complete review of modern measurements of magnetic fields, the reader is referred to *del Toro Iniesta* [2003].

3.2.1. Vertical Currents at the Photosphere

Figure 3.4 shows an example of the maps of the full vector magnetic field and vertical electric current density derived from the HMI/SDO observations. The distribution of electric currents in sunspots is highly structured. The strongest currents flow along the area of the

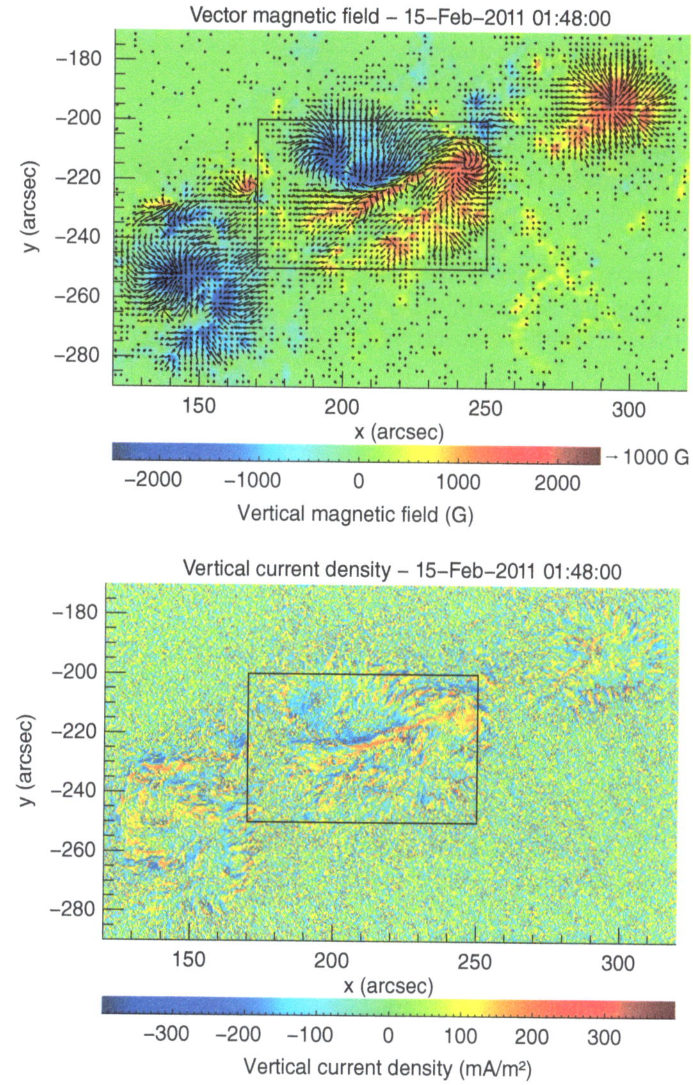

Figure 3.4 Maps of magnetic field (top) and j_z (bottom) derived from SDO/HMI observations of NOAA AR 11158 on 15 February 2011. Note that electric currents of both signs are present in sunspot single polarity areas. The strongest currents are located along the magnetic neutral line separating two sunspots, which is also the area of strongest magnetic shear. From *Musset et al.* [2015], Figure 1.

strongest shear in the magnetic field (see the narrow elongated patterns of positive and negative currents inside the box shown in Fig. 3.4). Currents flowing in umbral areas of sunspots are more disorganized, and both positive and negative currents are present inside each umbra (compare current distribution in darkest (blue or red) areas in Fig. 3.4, top). As each sunspot is composed of a single polarity magnetic field (positive or negative), the presence of oppositely directed currents indicates that the current helicity (h_c) of any single sunspot is sign-alternating. Such a pattern of current helicity is typical for most sunspots [*Gary et al.*, 1987; *Pevtsov and Peregud*, 1990; *Pevtsov et al.*, 1994], although highly twisted δ-spots often exhibit strong electric currents of a single sign inside each umbra [*Pevtsov*, 2005] with opposite currents flowing in opposite magnetic polarities. Individual patches of current helicity inside sunspot umbra have a lifetime of about 27 hours [*Pevtsov et al.*, 1994], which provides an estimate for a lifetime of isolated currents in sunspots in the absence of continuous generation or flare activity.

3.2.2. 3D Picture at the Photosphere

Because of a combination of the narrow height range of the photosphere (≈500 km) and a simplified treatment of the radiative transfer (widespread use of Milne-Eddington stellar atmosphere), the 3D picture of electric currents in the photosphere is not well studied.

High resolution observations of sunspots from instruments on Hinode spacecraft reveal highly structured patterns of electric currents in sunspot penumbra [*Gosain et al.*, 2014]. The pattern is well correlated with the pattern of the penumbral filaments, and it suggests the presence of strong electric current flowing around these filaments. These electric currents are limited to the photosphere, and are unlikely to cross into higher layers of the solar atmosphere.

Puschmann et al. [2010a] applied the Stokes Inversion based on Response functions (SIR) algorithm to invert the Stokes profiles observed by Hinode/SP. The optical depth retrieved from SIR inversions was converted to geometric depth using the genetic algorithm developed by *Puschmann et al.* [2010b]. The algorithm finds a solution for the three components of a vector field by minimizing a merit function composed of the weighted contributions of the magnetic field divergence and the residual force (which includes the pressure gradient, Lorentz force, and gravity). A 3D model of the magnetic field in a sunspot penumbra was constructed on the basis of geometric heights derived from this inversion. The magnetic field and electric currents at the photospheric heights of 200 km resulting from this inversion are shown in Figure 3.5 (top). The topology of the magnetic field in the sunspot penumbra was found to most closely correspond to the so-called uncombed model of sunspot penumbra, where nearly horizontal magnetic flux tubes (so-called intraspines) are embedded in stronger and more vertical magnetic field (so called spines). The pattern can be visualized by putting two hands together at a small angle and inserting fingers on one hand between the fingers on the other hand. In Figure 3.5, the locations of intraspines are outlined by white contours. Strong horizontal currents were found at the boundaries of intraspines. The direction of $\mathbf{j}_\perp$ at intraspine boundaries forms a slight (about 20°) angle relative to the main axis of each horizontal flux tube (intraspine). The overall picture of these currents was interpreted as if they were flowing around a magnetic flux tube. At the low photospheric boundary (0 km), the plasma β in intraspines is larger than unity (corresponding to non-FFF conditions, Fig. 3.5, bottom), while in spines and the photospheric upper boundary, $\beta \approx 0.1$. At the top layer of the photosphere, the magnetic field was found to satisfy a FFF condition; no significant currents were found in spines (the magnetic field was nearly potential there). Amplitudes of the horizontal current densities were about 3–4 times larger than those of the vertical ones, which agrees with previous derivations of electric currents [*Hofmann and Staude*, 1987; *Kotov*, 1971; *Pevtsov and Peregud*, 1990]. The peak amplitude of horizontal currents was $\gtrsim 120$ mA m^{-2}, which also agrees with some early derivations.

The authors also evaluated the amplitude of electric current using a more traditional approach based on

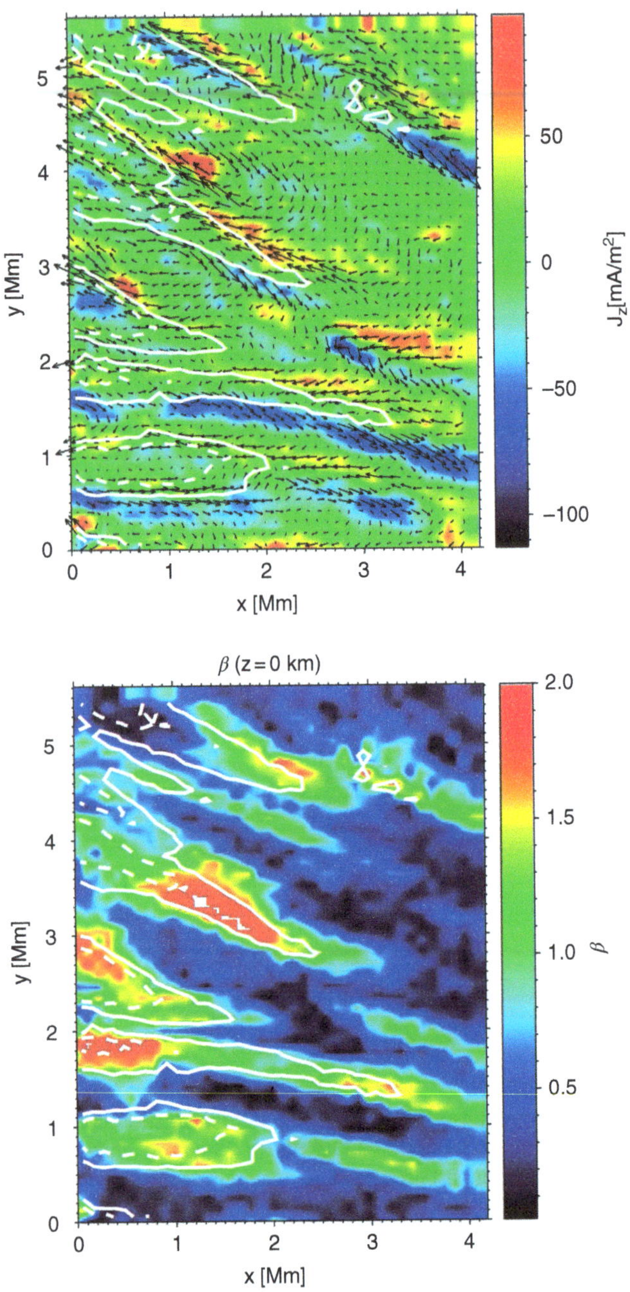

Figure 3.5 Top panel shows color-coded j_z and arrow-size-coded $\mathbf{j}_\perp$ at $z = 200$ km. Bottom panel shows a map of plasma β at $z = 0$ km. In both panels, white contours correspond to $B_z = -650$ G (solid lines) and $B_z = -450$ G (dashed lines) outlining intraspines. This is a modified version of Figures 1 and 4 from *Puschmann et al.* [2010a] © AAS. Reproduced with permission.

inversion of magnetic fields in the framework of the ME atmosphere. This derivation showed that traditional inversions tend to significantly underestimate the total electric current. *Gosain et al.* [2014] also noted that ignoring the fact that the optical depth varies slightly between spines and intraspines (and instead, assigning all observed fields to the same height in atmosphere) may lead to the appearance of false currents on top of the real ones.

3.3. MEASUREMENTS OF ELECTRIC CURRENT AT THE CHROMOSPHERE

Measurement of the magnetic field vector in the chromosphere is a much more challenging task than in the photosphere. The chromosphere cannot be represented well by a uniform spherical layer, which complicates the interpretation of observations greatly; the chromospheric processes are rapid and often small-scale, thus requiring rapid measurements. At the same time, the magnetic field becomes on average weaker. Available measurements employ the few infrared magnetically sensitive lines, whose formation heights are located at the chromospheric levels such as He I 1083.0 nm or Ca II 854.2 nm. In addition to being more challenging observationally, the inversion of these chromospheric measurements is also challenging both theoretically and computationally. Indeed, there is no simple model of the chromosphere, such as the ME photospheric model, and thus there is no simple analytical solution of the radiative transfer equation to be fit to the data. The radiative transfer through the chromosphere is complicated by deviation from LTE and the presence of strong vertical gradients in the involved parameters. Accurate inclusion of all physical processes relevant to the inversion algorithm makes it computationally expensive. But an account of the realistic atmospheric conditions is needed for reliable vector magnetometry [*Lites et al.*, 1994; *Socas-Navarro*, 2002] because the outcome of the radiative transfer depends strongly on the atmospheric conditions.

Here we review some results on the chromospheric vector magnetometry and on the electric currents derived from the magnetic vector data following papers by *Socas-Navarro* [2005a] and *de la Cruz Rodríguez and Socas-Navarro* [2011]. *Socas-Navarro* [2005a] observed an active region (NOAA AR 0634) at the Sun using the Spectro-Polarimeter for Infrared and Optical Regions (SPINOR) instrument simultaneously in two photospheric Iron lines (849.7 and 853.8 nm) and two chromospheric Calcium lines (849.8 and 854.2 nm) to derive the magnetic field vector in a finite volume extending from the photospheric level up to high chromospheric heights.

The outcome of this analysis was a 3D distribution of the magnetic field vector above a sunspot, Figure 3.6. For convenience, the lower portion (0–800 km) of the field lines is shown yellow, while the upper part (800–1600 km) is shown red. Importantly, the upper and lower parts behave consistently with each other: the divergence of the magnetic field is always small (less than 2%) compared with ratio of the magnetic field magnitude B and the box length l, B/l, which validates the 3D vector magnetic field reconstructed from the observations.

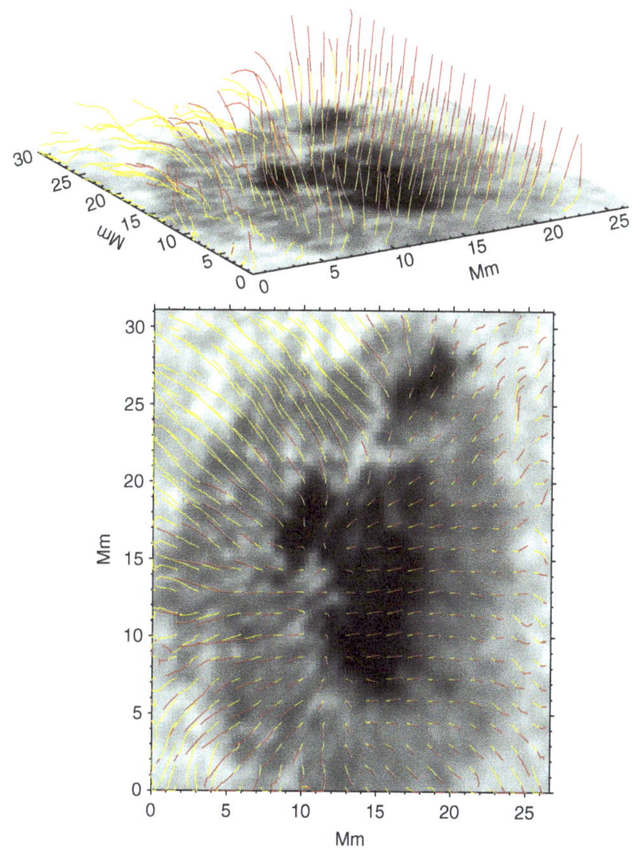

Figure 3.6 Visualization of the 3D magnetic structure above a sunspot in the perspective view (top panel) and top view (bottom panel). The bottom part (0–800 km) of the field lines is shown in yellow, the top part (800–1600 km) is shown in red. From *Socas-Navarro* [2005a] © AAS. Reproduced with permission.

3.3.1. Misalignment Between Vectors of Magnetic Field and Hα Fibrils

Chromospheric fibrils are often considered to be tracers of horizontal magnetic field. Given that the plasma beta is small in the chromosphere, the magnetic configuration is supposed to be not far from a FFF configuration; thus, the electric current vector should be coaligned with the chromospheric fibrils as well. This expectation was, in particular, employed by *Wiegelmann et al.* [2008] to fix the azimuth of the magnetic field at the chromospheric heights while performing nonlinear FFF (NLFFF) extrapolation. *De la Cruz Rodríguez and Socas-Navarro* [2011] challenged this expectation. They tracked a number of Hα fibrils (shown as yellow line segments in Fig. 3.7

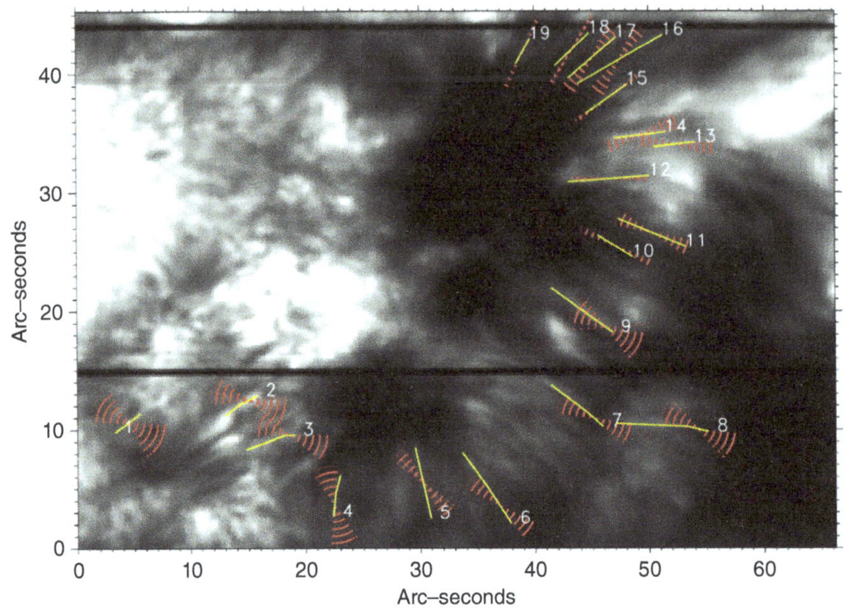

Figure 3.7 Example of misalignment between Hα fibrils (shown as yellow segments) and ranges of the azimuth of the magnetic field vector (red cones that take errors into account). From *de la Cruz Rodríguez and Socas-Navarro* [2011].

and compared them with the magnetic field azimuth shown by red cones, where the cone angle characterizes the uncertainty of the azimuth. It is apparent from the figure that Hα fibrils do not generally align with the magnetic field; thus, the use of the Hα fibrils as tracers of the magnetic field azimuth is not justified in the general case. There are a number of possible interpretations for this misalignment, of which the most plausible is, perhaps, the effect of ambipolar diffusion in the partly ionized plasma of the chromosphere [*Martínez-Sykora et al.*, 2016]. The point is that ions and neutrals will decouple as soon as the ion-neutral collision frequency is small. This means that the magnetic field is also decoupled from the neutral population and allows the field to slip through it.

3.3.2. Misalignment Between Vectors of Magnetic Field and Electric Current

Having the three-dimensional structure of the vector magnetic field, one can compute the vector current density as the curl of the field. *Socas-Navarro* [2005b] used SPINOR data described in the beginning of this section to compute the electric current vector in a finite volume above the sunspot; the results are summarized in Figure 3.8. The top row of this figure shows that the largest currents form filamentary structures, which look like 2D cuts of a current sheet system in 3D.

The low panels in Figure 3.8 show a systematic misalignment between the electric current and the magnetic field vectors. For a truly FFF, the angle between these two vectors should be either 0° or 180°, and thus, the corresponding plots would be a combination of yellow and blue areas, which is clearly not the case. Instead, the bottom panels show considerable misalignment of the electric current vector from the magnetic field vector at all levels. This is indicative of a significant departure of the magnetic field from the force-free configuration even at a relatively high chromospheric level, although the departure is clearly weaker at the chromosphere compared with the photosphere: there are significant areas at the (inner part of the) chromospheric heights, where the field is close to force-free and the vectors are aligned well.

3.3.3. Role of Electric Current in Heating the Chromosphere

In addition to the magnetic field, modern inversions of Stokes profiles allow deriving several thermodynamic parameters of plasma, including its temperature. Figure 3.9 shows the distribution of the plasma temperature at various heights obtained this way. Even visual comparison of the temperature distribution (Fig. 3.9) and electric current density distribution (Fig. 3.8) shows that they are remarkably different with no clear relationship between areas with strong chromospheric currents and the hot patches. This is further quantified by Figure 3.10, which shows spatial correlation between plasma temperature and electric current at various heights. Two observations can be made based on the correlation plots: although (1) the correlation coefficient is small, (2) it is

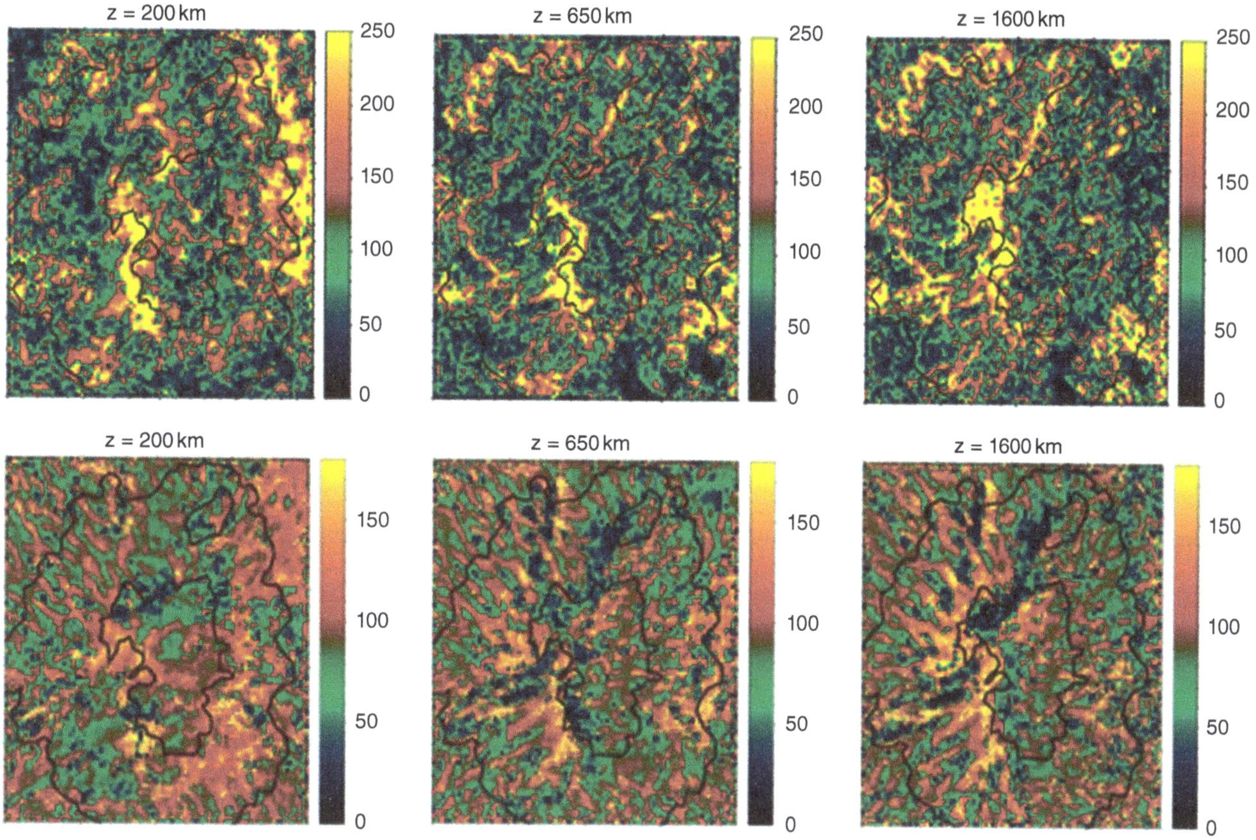

Figure 3.8 Distributions of the current density at different heights $z = 200$, 650, 1600 km (top); the scale is in mA m^{-2} for the sunspot shown in Figure 3.6. The angles between **j** and **B** are given in degrees (bottom). From *Socas-Navarro* [2005b] © AAS. Reproduced with permission.

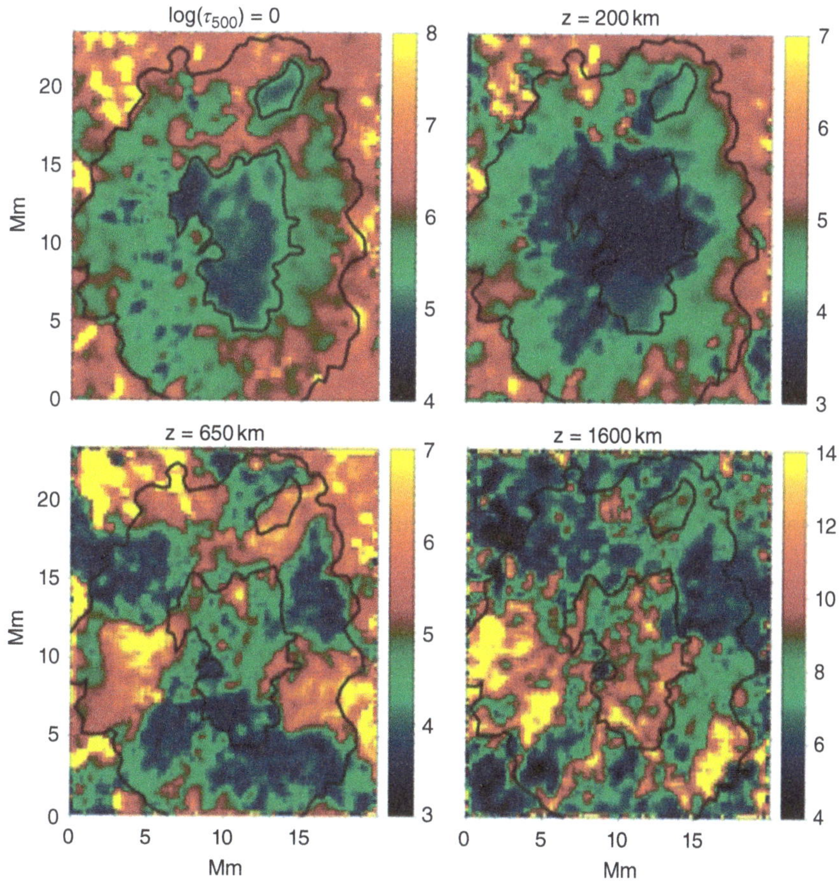

Figure 3.9 Temperature above the sunspot shown in Figure 3.6 derived from the inversion at various heights in the photosphere (0 and 200 km) and chromosphere (650 and 1600 km). Hot patches (≈5 Mm) embedded in a cooler plasma are seen at the chromospheric cuts. From *Socas-Navarro* [2005b] © AAS. Reproduced with permission.

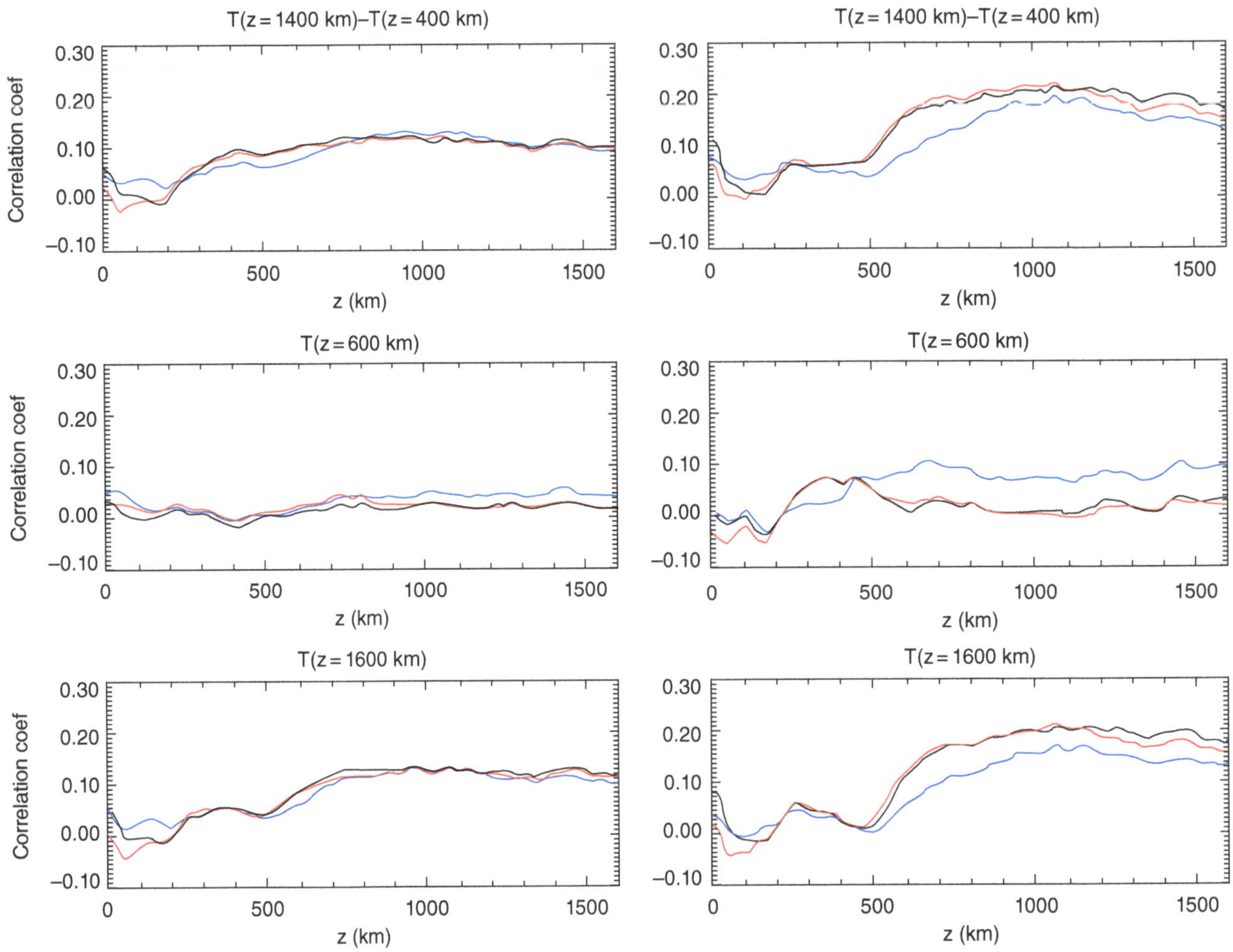

Figure 3.10 Correlation coefficients between temperature at a given height (labeled on top of the panel) and electric current density at the height *z*. Left, for the full area; right, for the penumbra only. Absolute value of the current is shown in black, the perpendicular in red, and parallel in blue. From *Socas-Navarro* [2005b] © AAS. Reproduced with permission.

positive almost everywhere. This is interpreted by *Socas-Navarro* [2005b] as indication that the dissipation of electric currents may contribute to the chromospheric heating, but this cannot be a dominant contribution.

The right panels in Figure 3.10 show the same correlation but restricted to a penumbral area only. Here the correlation coefficient increases by a factor of two indicating that the electric current dissipation may be a more important heating mechanism in the penumbra. Goodman [2004] advocated that the dissipation of Pedersen electric currents (flowing perpendicular to the magnetic field) is highly efficient in heating the solar chromosphere. Although the correlation curves for the entire region (Fig. 3.10, left) do not show a clear distinction between both components, those for the penumbra (Fig. 3.10, right) do show a slightly better correlation with the Pedersen component of the electric current. Contrary to that, *Arber et al.* [2016] argued that the heating provided by the dissipation of high-frequency Alfvén waves through Pedersen resistivity may be insufficient to balance the radiative and conductive losses for a realistic range of field strengths and velocities.

The findings that the electric currents may play only a minor role in chromospheric heating are in qualitative agreement with *Fisher et al.* [1998], who studied a correlation between the total soft X-ray (SXR) brightness of active regions in the corona with several integral parameters and concluded that the magnetic flux is the prime component in this correlation. Once the correlation associated with magnetic flux is subtracted, the residual SXR brightness shows no correlation with the total electric current flowing in an active region.

3.4. MODELING OF ELECTRIC CURRENTS IN THE CORONA

There are a number of coronal magnetic field diagnostics [e.g., *Dalmasse et al.*, 2016; *Gibson et al.*, 2016] including radio methods, which provide either the absolute value of the magnetic field in the case of a gyroresonant process or the line-of-sight (LOS) component in the case of a polarized free-free process during nonflaring periods [e.g., *Lee*, 2007; *Wang et al.*, 2015] or the absolute value and the inclination of the magnetic field vector to the LOS in the case of the gyrosynchrotron process during flares [e.g., *Gary et al.*, 2013]. Only in rare cases are the magnetic field vector diagnostics available for dense, cool structures (e.g., prominences) located at coronal heights [*Levens et al.*, 2016; *Orozco Suárez et al.*, 2014]. The routine observational diagnostics to derive the magnetic field vector in the hot, tenuous corona are still in their infancy. Measurements of the three components of the vector field in the corona off the solar limb based on the Hanle effect are now taken with the Coronal Multichannel Polarimeter (CoMP) [*Tomczyk et al.*, 2008]. The observations represent a LOS column integration, which may be difficult to interpret in terms of the physical 3D structure of the magnetic field. There are recent attempts to disentangle the LOS integration effect using coronal tomography [*Kramar et al.*, 2016].

In the current absence of routine observations, information on the magnetic field *vector* and electric current vector in the corona is derived from static (NLFFF), evolutionary (magnetofrictional), or dynamic (MHD) modeling augmented by one or another coronal diagnostics in some cases [*Wiegelmann and Sakurai*, 2012; *Wiegelmann et al.*, 2014]. Although the electric current, which plays an important role in the corona, is explicitly taken into account in both static NLFFF extrapolations and dynamic MHD models [see, e.g., recent reviews *Chen*, 2011; *Cheung and Isobe*, 2014; *Fan*, 2009; *Raouafi et al.*, 2016; *Schmieder et al.*, 2014; *Shibata and Magara*, 2011], often its role is not carefully considered or is even fully neglected, when interpreting observations.

3.4.1. Nonlinear Force-Free Field Extrapolations

Figure 3.2 shows that the plasma beta in the corona is less or much less than one; thus, there is no force available to balance the Lorentz force $\boldsymbol{j}\times\boldsymbol{B}$ [see, e.g., *Fleishman and Toptygin*, 2013a]. Therefore, a configuration with a significant Lorentz force must evolve toward a new configuration with a much smaller or no Lorentz force, which is called a FFF magnetic configuration. In particular, to have a stationary magnetothermal configuration requires a magnetic field that is reasonably close to the FFF one.

Zero Lorentz force for a nonzero magnetic field and current density implies the electric current density vector directed along the magnetic field vector, $\mathbf{j} \parallel \mathbf{B}$. Recalling that $\mathbf{j} \propto \nabla \times \mathbf{B}$ we conclude $\nabla \times \mathbf{B} = \alpha_{\mathrm{FFF}}\mathbf{B}$, where α_{FFF} is a pseudo-scalar, called the FFF parameter. With the above relationships, we can write down the equations specifying the force-free magnetic field:

$$\nabla\times\mathbf{B}=\alpha_{\mathrm{FFF}}\mathbf{B},\qquad \mathbf{B}\cdot\nabla\alpha_{\mathrm{FFF}}=0. \tag{3.2}$$

A remarkable property apparent from equation (3.2) is that the FFF parameter does not change in the direction of the magnetic field vector; as a result, α_{FFF} does not change along any given field line. This property, at least theoretically, allows computing the electric current density along a given field line if the electric current density is known at the footpoint of this field line and the absolute value of the magnetic field is known along the field line, which requires less information about the magnetic field than for the use of Ampère's law.

Vector magnetic field measurements at the photosphere and/or chromosphere represent a boundary from which the magnetic field in a coronal volume above this boundary can be reconstructed in the force-free regime [*Aly*, 1984, 1989; *McClymont et al.*, 1997; *Molodenskii*, 1969; *Schrijver et al.*, 2006; *Wiegelmann and Inhester*, 2010]. Theoretically, for some boundary conditions an FFF solution exists and it is unique [*Aly*, 1984; *Amari et al.*, 2006], although no solution may exist for an arbitrary boundary condition. Still, the uniqueness of NLFFF solution for given boundary conditions has been established only for small and smooth distributions of the force-free parameter alpha; and the actual meaning of smooth and small has yet to be determined [*Bineau*, 1972; *Boulmezaoud and Amari*, 2000]. Even when this problem has a unique solution, in practice a few reconstruction methods (called NLFFF extrapolations) including optimization [*Rudenko and Myshyakov*, 2009; *Wheatland et al.*, 2000; *Wiegelmann et al.*, 2006], magnetofrictional [e.g., *Valori et al.*, 2007], and Grad-Rubin [e.g., *Wheatland*, 2007] methods yield generally similar, but far from identical, results [*De Rosa et al.*, 2009; *DeRosa et al.*, 2015]. In particular, the required constancy of the force-free parameter α_{FFF} along a field line does not hold equally well in each of the methods; in particular, it does not hold well for various implementations of the optimization method. In contrast, the Grad-Rubin methods specifically attempt to preserve α_{FFF} constancy along all closed field lines.

Figure 3.11 visualizes the 3D magnetic structure obtained with the Grad-Rubin method [*DeRosa et al.*, 2015] in a coronal volume above an AR with a number of reference field lines (red) and flux tubes (green) built around these reference field lines within a 3D modeling

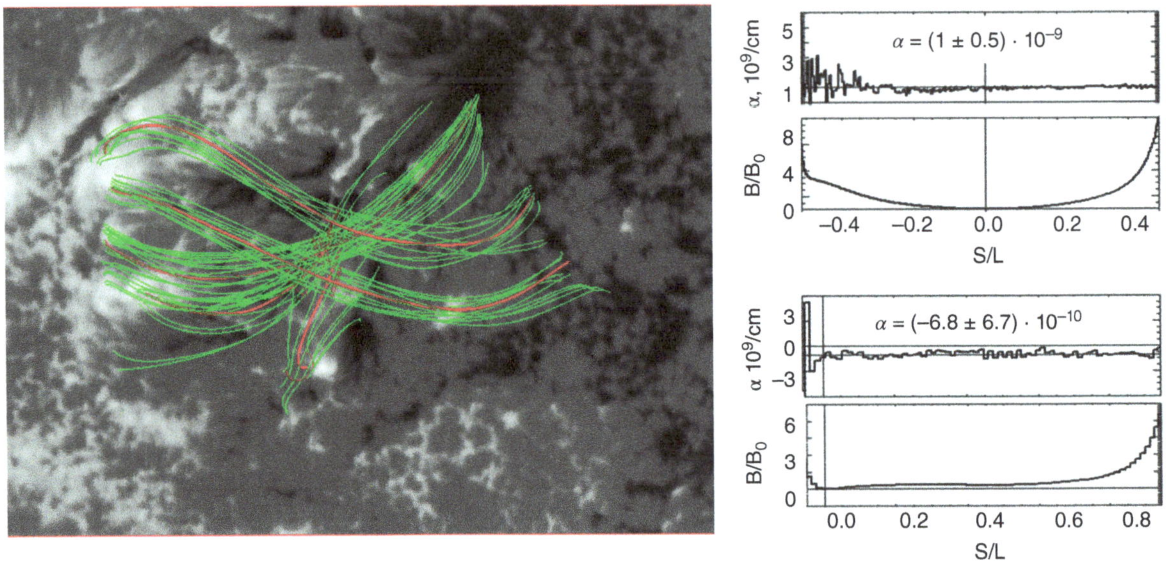

Figure 3.11 Visualization, using GX Simulator tool [*Nita et al.*, 2015], of a NLFFF data cube obtained with a Grad-Rubin method from *DeRosa et al.* [2015]. Grey image represents the bottom photospheric magnetogram; the red lines are the reference field lines; and the green lines are the flux tubes built around those reference field lines to visualize the magnetic field topology. Right panels show distributions of the α_{FFF} parameter and the magnetic field modulus along two of the red field lines shown in the left panel.

tool, GX Simulator [*Nita et al.*, 2015]. The panels to the right show how the force-free parameter α_{FFF} and the absolute value of the magnetic field B vary along two of the red field lines. The plots show that the force-free parameter is reasonably constant along each of these two field lines; thus the electric current density can straightforwardly be computed at any voxel as $\mathbf{j}=\frac{c}{4\pi}\alpha_{FFF}\mathbf{B}$, where c is speed of light, and thus the value of the electric current density drops along the field line proportionally to the magnetic field value. In the plotted field lines, the footpoint-to-loop-top magnetic field ratio is about 5–8, thus, the electric current density at the loop top is, respectively, 5 to 8 times lower than at the corresponding foot point. This tells us that the strongest coronal currents must concentrate at the base of the corona, where the photospheric/chromospheric currents are strong and where the field lines show a reasonably small expansion.

3.4.2. Coronal MHD Modeling

No evolution of electric currents in the corona is captured by static NLFFF extrapolations; thus, modeling of the evolution requires explicit use of the time-dependent MHD approach or possibly the magnetofrictional approach [*Cheung and DeRosa*, 2012; *Fisher et al.*, 2015]. The complete set of MHD equations that would be valid all the way from the (sub-) photospheric up to coronal heights is extremely complicated and expensive to solve numerically. Although advanced codes that take into account realistic physics exist [e.g., *Gudiksen et al.*, 2011], they can be applied only to a very limited spatial domain [*Carlsson et al.*, 2016]. In many cases, when only a coronal domain is studied, the modelers use a simplified set of MHD equations [e.g., *Jiang et al.*, 2016], which explicitly takes into account the condition of low-beta plasma $\beta \ll 1$. In this way, evolution of the magnetic field driven by photospheric motions results in progressively stronger tangling of the magnetic field lines due to freezing of the magnetic flux into the conductive fluid. Thus, the magnetic flux tubes become more strongly twisted, which implies enhancements of the corresponding electric current density along the flux tubes.

Here we briefly describe one example [*Amari et al.*, 2015] of a dynamic model driven by a "flux rope emergence" from the bottom of the modeling data cube. A magnetic structure was injected from the bottom with a significant twist and shear, and then was driven by the bottom (photospheric) velocity field. As a result, strong electric currents flowed in the volume, primarily along the magnetic field in the area of the flux rope emergence, which can be vividly seen in online movies that supplement the paper by *Amari et al.* [2015].

3.5. ELECTRIC CURRENT IN A HOT, MULTICOMPONENT PLASMA

An interesting and nontrivial question is: What is the microscopic composition of electric currents in the hot corona, where atoms are ionized up to various ionization states, while there is no neutral component? It is clear that the electric current is primarily supported by a corresponding bulk motion of electrons, which are light and, thus, easier to accelerate given an electric force. It is also transparent that for a purely hydrogen fully ionized plasma, the protons will move along the electric field vector[1] (i.e., opposite to the electrons) with a bulk velocity consistent with momentum conservation. In the plasma rest frame, we thus have $\mathbf{v}_p = -(m_e/m_p)\mathbf{v}_e \approx 5\times10^{-4}\mathbf{v}_e$. Then, the electron velocity is found from the equation

$$\mathbf{j} = \sum_c e_c n_c \mathbf{v}_c$$

where the summation is performed over all plasma components. In our case, we have two components with the same number densities, opposite charges, and strongly different masses; thus: $\mathbf{j} = -en_e\mathbf{v}_e(1 + m_e/m_p)$, where e is the elementary positive charge, while j can be determined for the given electric field from Ohm's law.

A more complicated picture arises when there are other ions with larger charges (Ze) because momentum conservation alone is insufficient to determine the velocities of the plasma species and explicit consideration of their collisions with each other is needed. It was first performed by *Gurevich* [1961], who demonstrated that "multiply charged" positive ions move, together with negative electrons, in the direction opposite to the protons. A simple explanation of this effect is as follows. A stationary electric current is set up due to balancing the electric force ($\mathbf{F} = eZ\mathbf{E}$) by the dynamic friction forces (proportional to Z^2) between all plasma species. These two forces almost compensate each other in the case of protons ($Z = 1$), which results in a very small bulk velocity of the protons. But this implies that for any $Z > 1$, the dynamic drag force acting on the multiply charged ions by the fast-moving electrons will overcome the net electric force; thus, the multiply charged ions will move in the same direction as the electrons. Quantitative consideration of this effect can be found in *Fleishman and Toptygin* [2013a; Section 11.1.1 and Fig. 11.1 there], from where we estimate the bulk velocity of the protons and ions of abundant ^{4}He and much rarer ^{3}He ions for the typical solar abundances:

$$\mathbf{v}_p = -\frac{2n_4}{n_p + 4n_4}\sqrt{\frac{m_e}{m_p}}\mathbf{v}_e \approx 2.9\times10^{-3}\mathbf{v}_e,$$

$$\mathbf{v}_4 = \frac{n_p}{2(n_p + 4n_4)}\sqrt{\frac{m_e}{m_p}}\mathbf{v}_e \approx 8.8\times10^{-3}\mathbf{v}_e, \qquad (3.3)$$

$$\mathbf{v}_3 = \frac{n_p n_4}{n_e(n_p + 4n_4)}\mathbf{v}_e \approx 5\times10^{-2}\mathbf{v}_e.$$

From here we can draw a few important conclusions about components of the electric current in a multicomponent plasma.

1. Protons move much faster than in a purely hydrogen plasma.
2. Helium ions move at the *same* direction as *electrons*, and move oppositely to *protons*.
3. ^{3}He ions (and other rare ions) move much faster than ^{4}He ions (and other abundant ions).

This phenomenon is likely relevant for understanding enrichment of solar energetic particles by rare elements, most notably ^{3}He, and also for variations of elemental abundances in the corona, reported recently for both flare [*Warren*, 2014] and active region [*Caspi et al.*, 2015] measurements.

3.6. ALFVÉN WAVE ON TOP OF CURRENT-CARRYING MAGNETIC FIELD

Another interesting effect of the electric current present in the plasma is a modification of plasma eigenmodes compared with the current-free case. A fundamental reason behind this modification is the breaking of the plasma's mirror symmetry, which can be quantified by a pseudoscalar quantity: nonzero current helicity density $h_c = \mathbf{B}\cdot[\nabla\times\mathbf{B}] \neq 0$. Here, for simplicity, we consider the corresponding current-induced modification to the Alfvén waves only. To do so, in the full set of MHD equations we neglect all dissipative terms as well as the pressure gradient and any external force, rather than the Lorentz force. The truncated set of equations reads:

$$\nabla\cdot\mathbf{B} = 0, \qquad \frac{\partial\mathbf{B}}{\partial t} = \nabla\times[\mathbf{u}\times\mathbf{B}], \qquad (3.4a)$$

$$\rho\left(\frac{\partial\mathbf{u}}{\partial t} + (\mathbf{u}\cdot\nabla)\mathbf{u}\right) = \frac{1}{4\pi}[\nabla\times\mathbf{B}]\times\mathbf{B}. \qquad (3.4b)$$

To determine the eigenmode, we have to linearize these equations assuming $\boldsymbol{B} = \boldsymbol{B}_0(\boldsymbol{r}) + \boldsymbol{b}$:

$$\frac{\partial\mathbf{b}}{\partial t} = (\mathbf{B}_0\cdot\nabla)\mathbf{u} - (\mathbf{u}\cdot\nabla)\mathbf{B}_0, \qquad (3.5a)$$

$$\frac{\partial\mathbf{u}}{\partial t} = \frac{1}{4\pi\rho}(\nabla\times\mathbf{b})\times\mathbf{B}_0 + \frac{\alpha_{\mathrm{FFF}}}{4\pi\rho}\mathbf{B}_0\times\mathbf{b}, \qquad (3.5b)$$

where $\mathbf{b}$ and $\mathbf{u}$ are the first-order oscillating values. In contrast to the classical case of the uniform mean

[1] For simplicity, we adopt that electric field is elongated along the magnetic field, so no transverse or Hall current is considered, there is no $\boldsymbol{E}\times\boldsymbol{B}$ drift, etc.

magnetic field, in our case we have to take into account that $\nabla \times \mathbf{B}(\mathbf{r}) \neq 0$; thus, the linearized equations have additional terms (the last terms in each row; in the second one, we assume that the field is force free and use $\nabla \times \mathbf{B}_0(\mathrm{r}) = \alpha_{FFF}\mathbf{B}_0$) compared with the classical case. Accordingly, the solution describing properties of the Alfvén waves will also be different from the classical case. A remarkable modification of the Alfvén wave compared with the current-free case (see problem 2.6 in *Fleishman and Toptygin* [2013a]) is that the vectors **u** and **b** are not parallel to each other any longer, which has a number of important implications. In particular, the modified wave possesses a nonzero kinetic helicity (problem 2.6 in *Fleishman and Toptygin* [2013a]) $\langle h_k \rangle = \frac{1}{2}\Re \mathbf{u} \cdot (\nabla \times \mathbf{u}^*)$, which is proportional to the force-free field parameter α_{FFF}.

Then, given that the vectors **b** and **u** are not parallel to each other any longer, the mean electric field of the Alfvén wave, $\mathbf{E}_h \propto \langle \mathbf{u} \times \mathbf{b} \rangle$, turns out to be nonzero. It is easy to estimate from consideration of dimensions (see section 11.2.2 in *Fleishman and Toptygin* [2013a] for quantitative analysis) that this electric field is proportional to both the mean magnetic field and the FFF parameter $\mathbf{E}_h \propto \alpha_{FFF}\mathbf{B}_0$. Remarkably, the direction of this electric field does *not* depend on the **k**-vector, but on the **B**-vector only. An ensemble of such Alfvén waves with random phases will represent a "helical" Alfvénic turbulence with properties distinct from the "standard" Alfvénic turbulence. In particular, given that the mentioned mean electric field has the same direction for all present Alfvén waves independently on their wave-vectors, the magnitude of this electric field will go up for stronger and stronger turbulence (more and more waves). This electric field can become sufficiently large to form runaway particle populations that can be further picked up by stochastic acceleration by the same wave ensemble.

3.7. ELECTRIC CURRENT AND SOLAR FLARES

Various forms of eruptive activity at the Sun are manifestations of excessive magnetic energy stored in the solar corona in a nonpotential magnetic field, which is supported by electric currents. Solar flares, vividly seen as transient brightenings throughout electromagnetic spectrum from radio waves to gamma rays, represent the most rapid release of this excessive magnetic energy; thus, solar flares are closely linked with coronal electric currents. Given that the Reynolds number is overall large in the corona, the rapid energy release in flares requires that the currents are concentrated into small regions such as narrow current sheets. Nonflaring current sheets were inferred observationally [e.g., *Solanki et al.*, 2003] in the system of rising loops low in the corona at the regions with a tangential discontinuity of the field direction.

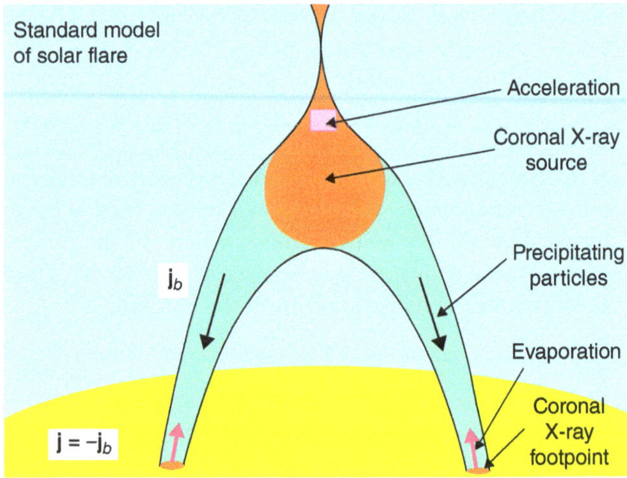

Figure 3.12 A cartoon showing a "standard" model of a solar flare. Precipitating electrons form a direct electric current $\mathbf{j}_b$, which is neutralized by a return current formed by background particles $\mathbf{j} = -\mathbf{j}_b$. A modified version of a figure from *Benz* [2008].

Within a standard model of solar flare, the energy release high in the corona results in acceleration of a fraction of plasma particles to nonthermal energies in an "acceleration region" (Fig. 3.12). Then, these particles precipitate down to the chromospheric footpoints, where they lose their energy in Coulomb collisions and produce X-ray and γ-ray emissions. Besides a variety of multifaceted manifestations, these precipitating particles represent a strong electric current, which would produce an additional magnetic field on top of the coronal one. Estimates with Ampère's law show that this magnetic field can be as large as $\approx 10^5$ G, which would contain more energy than the beam itself. Prompt creation of such an intense magnetic field of $\approx 10^5$ G in the highly conductive corona would require either some type of anomalous resistivity to generate new flux or unrealistically rapid, coherent, and large-scale inflows to compress preexisting flux. The commonly accepted solution for this paradox is creation of the neutralizing return current composed of background particles (Fig. 3.12).

3.7.1. Association Between X-ray Emission and Photospheric Currents

Although theoretically expected, detailed relationships between solar flares and electric currents in the solar atmosphere are not easy to establish observationally. *Janvier et al.* [2014] showed that the photospheric currents are concentrated in the "current ribbons," that is, elongated areas or stripes of an enhanced current density. They presented evidence that the current ribbons, which are also correlated with the flare EUV ribbons, highlight

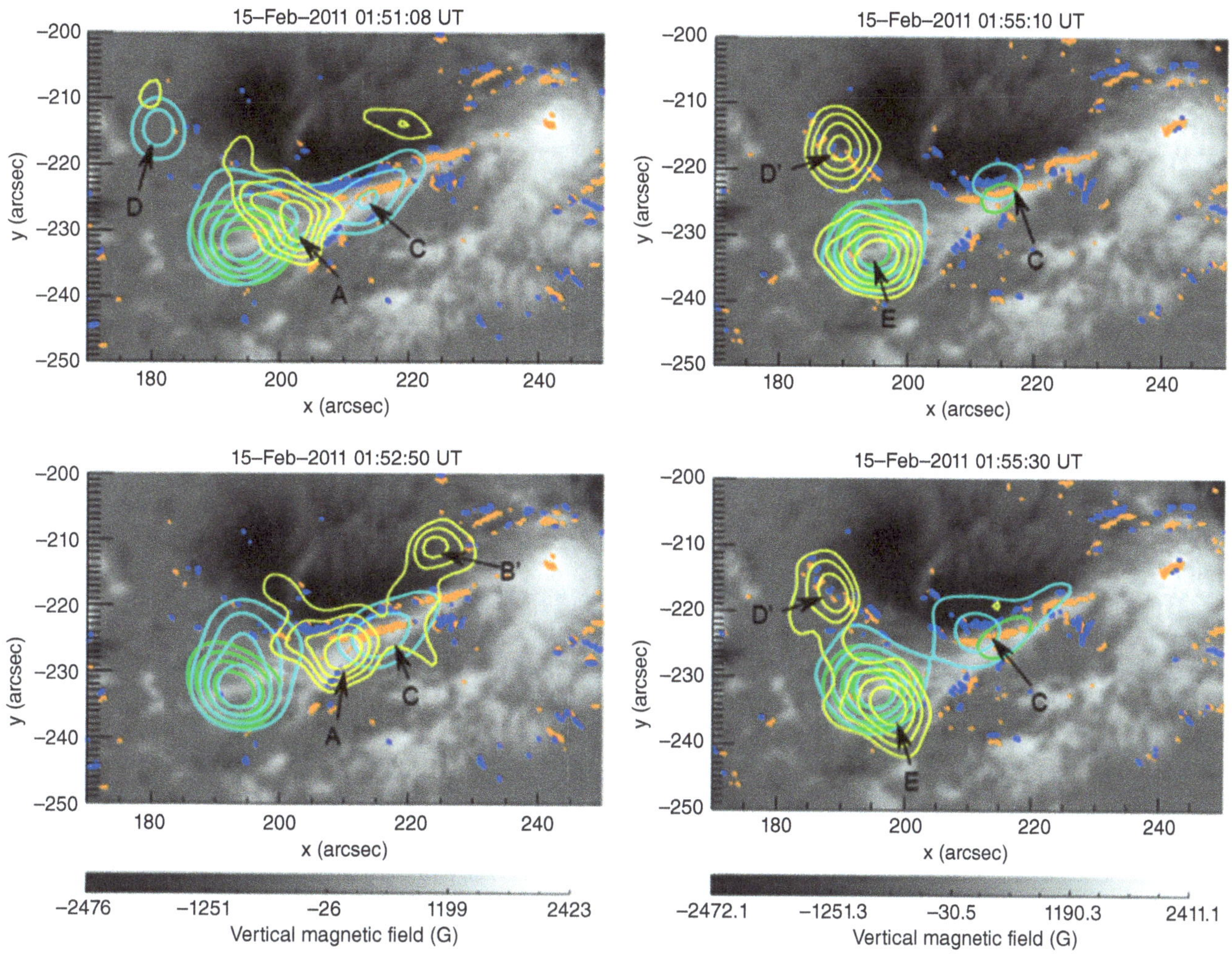

Figure 3.13 Association between photospheric magnetogram (gray image), photospheric positive (yellow dots) and negative (blue dots) currents, and X-ray emission in 12–25 keV (green contours), 25–50 keV (cyan contours), and 50–100 keV (yellow contours) energy ranges at different time frames for AR 11158 at the time of a solar flare (01:48:00 UT). From *Musset et al.* [2015].

the photospheric imprint of coronal electric current sheets. These current sheets may be formed where the gradient of magnetic connectivity is strong. Such areas are known to favor reconnection [see, e.g., *Demoulin et al.*, 1996] that drives the solar flare. Thus, both electric currents and processes of plasma heating and particle acceleration in flares are expected to occur in these same regions. This suggests that the thermal and nonthermal X-ray emissions in solar flares are expected to be closely associated with the photospheric current ribbons.

Musset et al. [2015] reported on a close association between SXR observed with RHESSI and the photospheric electric current distribution derived from the SDO/HMI data, which is summarized in Figure 3.13. Although there is no "pixel-to-pixel" correlation between X-ray brightness and distribution of the electric current density, there are elongated thermal and nonthermal X-ray sources shown in green (12–25 keV) and cyan (25–50 keV) projected onto narrow elongated photospheric current ribbons. The high-energy source (50–100 keV) shown in yellow is supposed to trace the chromospheric footpoints of the flaring loops system. Thus, it is remarkable that the X-ray source A is projected onto the main current ribbon (black arrow in the top panel of Fig. 3.13). The reported spatial relationships confirm the close association between locations of the acceleration region in this flare and electric current sheets.

3.7.2. Evolution of the Photospheric Current at the Course of Flare

It is interesting that in addition to the remarkable spatial relationships between hard X-ray (HXR) emission and photospheric currents, *Musset et al.* [2015] also found a coherent behavior of them in time. Specifically, the appearance of the 50–100 keV X-ray source D′ (upper right panel in Fig. 3.13) is accompanied by simultaneous appearance of cospatial photospheric currents. *Kazachenko et al.* [2015] reported a global change of the magnetic field at the same AR during this flare accompanied by a decrease of magnetic field twist, which implies a corresponding reconfiguration of the electric current system at the course of the flare. It is worthwhile to note that the polarization measurements, on which the magnetic field data are based, can be biased during flares due to a number of known effects [see, e.g., *Hénoux and Karlický*, 2013, and references therein]. For this reason, *Janvier et al.* [2014] and *Musset et al.* [2015] carefully analyzed the changes of the transverse magnetic field during the given flare and concluded that the derived changes of the vertical electric current are real.

This coherence in time evolution of the electric currents and HXR emission is strong evidence that magnetic reconnection is responsible for both these phenomena. Indeed, the observed rapid increase of the photospheric electric current density beneath the source D′ can easily be understood as a result of a reconnection between a flux tube carrying a strong current and another flux tube, perhaps, current free or carrying a weaker current, which was initially rooted in the D′ area. After it reconnects with a current-carrying loop, the current will rapidly become uniform over the coronal portion of the newly formed flux system due to the requirement of the electric current continuity [e.g., *Canfield et al.*, 1996]. Thus, the photospheric electric current must go up at the corresponding footpoint, as observed. Another known outcome of magnetic reconnection is plasma heating and particle acceleration, which is traced by the HXR emission. We note that according to *Petrie* [2012], the relatively rapid changes in the electric current system reported by *Musset et al.* [2015] are not unique to the 15 February 2011 flare but likely typical for other flares as well.

3.8. ORIGIN OF ELECTRIC CURRENTS IN THE SOLAR ATMOSPHERE

Electric current observed in the upper solar atmosphere (photosphere and above) are believed either to be of subphotospheric origin or to be the result of stress applied to the magnetic field after its emergence. Early analysis of different mechanisms for supplying energy to solar flares suggested that while (near-) surface shear flows may contain sufficient energy for small flares, major flares and repeated flare activity require significant electric currents of subphotospheric origin [*McClymont and Fisher*, 1989]. Observations do show that isolated bipolar flux elements emerge already carrying electric currents [e.g., *Leka et al.*, 1996] although the relative motions of footpoints of these flux tubes are also present as part of the flux emergence. The force balance at the interface of sub- and above- photospheric portions of a flux tube dictates a more complex dynamic behavior of the twist (and, therefore, of the electric current). Temporal evolution of electric current in an emerging flux tube at the photospheric level will depend on the speed of emergence: if the emergence is relatively slow, the evolution of magnetic flux and electric current (twist) will follow each other, but for a rapid emergence, propagation of twist (electric current) will be delayed as compared with the magnetic flux [*Longcope and Welsch*, 2000]. Such predicted evolution of magnetic flux and twist (electric currents) was observed in several small bipolar active regions [*Pevtsov et al.*, 2003]. The subphotospheric origin of electric currents is also supported by studies of the multifractality of the magnetic field of active regions [*Abramenko*, 2006] and comparative studies of the current helicity of photospheric and coronal magnetic fields [*Pevtsov et al.*, 1997]. *Gosain et al.* [2014] decomposed the observed vertical density of electric current (derived from high-resolution observations of vector magnetic field) into two components, parallel and orthogonal to the direction of the horizontal field. The second component was used to represent global electric current crossing from the photosphere to the chromosphere and corona. They found these global currents in agreement with the overall twist of the coronal structures above sunspots in support of the idea of continuity of electric current flow between the photosphere and corona. The patterns of electric currents in δ-spots, where the umbra with one polarity field contains upward current and the umbra with the other polarity has downward current, also supports the picture of electric current flowing from the photosphere in one footpoint of a flux tube through the chromosphere, corona, and back to the photosphere in the other footpoint of the same flux tube [*Gopasyuk*, 2015; *Pevtsov*, 2005].

It has been argued that both mechanisms (subphotospheric origin and near-photosphere shear) can create neutralized currents: with main (direct) current flowing in the main body of a magnetic flux tube, and return current flowing in the opposite direction in a boundary layer of the tube [*Melrose*, 1991]. Studies based on low-moderate spatial resolution magnetograms consistently showed that electric currents in active regions are not neutralized in the photosphere. Summarizing findings of several studies, *Pevtsov et al.* [1997] concluded that the pattern and amplitude of j_z derived from vector magnetograms "cannot plausibly be explained on the basis of local currents that close in the photosphere." In a separate study of 21 active regions, *Wheatland* [2000] drew a similar conclusion that the large-scale electric currents flowing in solar active regions are usually unneutralized. References to additional studies addressing the question of observational aspects of neutralization of electric currents can be found in the citation lists of these two papers. Complementing these observations, numerical modeling also indicates that both the flux emergence and the photospheric flows acting on magnetic fields can build up unneutralized currents. Only in some rare circumstances, that is, when the photospheric neutral line does not have any magnetic shear, the action of the photospheric flows may result in completely neutralized currents. For a review of the numerical modeling of unneutralized electric currents, the reader is referred to *Dalmasse et al.* [2015, and references therein].

The subphotospheric origin of electric currents lends support for the hypothesis that the emergence of active region magnetic fields carrying significant electric current can be used to predict the level of flaring activity of a given active region. In line with that hypothesis, *Bobra and Couvidat* [2015] reported that some of the strongest "photospheric" predictors of flare activity are related to electric currents.

3.9. CONCLUSIONS

A brief overview of observational and theoretical studies of electric currents in the solar atmosphere presented here demonstrates that the understanding and qualitative description of many solar processes from magnetic flux emergence [*Cheung and Isobe*, 2014; *Schmieder et al.*, 2014], to reconnection, flaring [*Raouafi et al.*, 2016], and particle acceleration [*Fleishman and Toptygin*, 2013b] may benefit tremendously from explicitly bringing electric currents into the picture. Although the assumption about a force-free field seems justifiable in the chromosphere on the basis of the low plasma β, observed misalignment between magnetic field and electric current vectors seems to contradict this. Observations also suggest that there may be electric current flowing outside and around magnetic flux bundles in sunspot penumbrae, and, at the same time, the currents are continuous along the corresponding current circuit encompassing the photosphere, chromosphere, and corona. In interpretations of observations involving reconnection processes, the interaction of current-carrying magnetic structures and the redistribution of electric currents need to be investigated [see, early work by *Canfield et al.*, 1996].

Finally, one might question if the electric currents derived from the limited spatial resolution magnetographic observations represent a fair approximation of the true electric currents flowing in the solar atmosphere. *Parker* [1996] showed that if a magnetic flux rope is composed of isolated current-carrying magnetic filaments, the currents derived from magnetograph measurements that do not resolve individual threads would show no resemblance to the true currents flowing along the threads. However, we did not see any qualitative change in the overall picture of the electric currents in sunspots as the spatial resolution of magnetographic observations improved from about 1 arc second in 1960 to 0.1–0.15 arc seconds in mid-2010. The newer observations indicate the presence of return currents as predicted by the theory, and the data also support the continuity of electric current from the photosphere, through the chromosphere and corona. Taken together, this justifies that modern magnetographic observations provide a physically correct representation of electric currents in the solar atmosphere. This picture is expected to be improved tremendously with a new generation of instruments including DKIST and FASR. Overall, we conclude that considering the electric current in addition to the magnetic field is essential to address many fundamental physical problems of solar physics.

ACKNOWLEDGMENTS

We are thankful to Dr. Michael Wheatland for critical reading of the manuscripts and his valuable comments. This work was supported in part by NSF grants AGS-1250374 and AGS-1262772, NASA grants NNX14AC87G and NNX16AL67G to New Jersey Institute of Technology, and RFBR grant 15-02-03835. Alexei A. Pevtsov acknowledges the financial support by the Academy of Finland to the ReSoLVE Centre of Excellence (project no. 272157) and NASA grant NNX14AE05G.

REFERENCES

Abramenko, V. I. (2006), Emerging Active Regions: Turbulence in the photosphere versus flaring in the corona, in *Solar MHD Theory and Observations: A High Spatial Resolution Perspective, Astronomical Society of the Pacific Conference Series*, 354, edited by J. Leibacher, R. F. Stein, and H. Uitenbroek, Astronomical Society of the Pacific, 195.

Abramenko, V. I., T. Wang, and V. B. Yurchishin (1996), Analysis of electric current helicity in active regions on the basis of vector magnetograms, *Solar Phys.*, *168*, 75–89; doi:10.1007/BF00145826.

Alfvén, H., and P. Carlqvist (1967), Currents in the solar atmosphere and a theory of solar flares, *Solar Phys.*, *1*, 220–228; doi:10.1007/BF00150857.

Aly, J. J. (1984), On some properties of force-free magnetic fields in infinite regions of space, *Astrophys. J.*, *283*, 349–362; doi:10.1086/162313.

Aly, J. J. (1989), On the reconstruction of the nonlinear force-free coronal magnetic field from boundary data, *Solar Phys.*, *120*, 19–48; doi:10.1007/BF00148533.

Amari, T., J.-F. Luciani, and J.-J. Aly (2015), Small-scale dynamo magnetism as the driver for heating the solar atmosphere, *Nature*, *522*, 188–191; doi:10.1038/nature14478.

Amari, T., T. Z. Boulmezaoud, and J. J. Aly (2006), Well posed reconstruction of the solar coronal magnetic field, *Astron. Astrophys.*, *446*, 691–705; doi:10.1051/0004-6361:20054076.

Arber, T. D., C. S. Brady, and S. Shelyag (2016), Alfvén wave heating of the solar chromosphere: 1.5D models, *Astrophys. J.*, *817*, 94; doi:10.3847/0004-637X/817/2/94.

Babcock, H. W. (1961), The topology of the sun's magnetic field and the 22-YEAR cycle, *Astrophys. J.*, *133*, 572; doi:10.1086/147060.

Balasubramaniam, K. S., and A. Pevtsov (2011), Ground-based synoptic instrumentation for solar observations, in *Solar Physics and Space Weather Instrumentation IV, Proc. SPIE*, *8148*, 814809; doi:10.1117/12.892824.

Benz, A. O. (2008), Flare observations, *Living Rev. Sol. Phys.*, *5*, 1; doi:10.12942/lrsp-2008-1.

Bineau, M. (1972), On the existence of force-free magnetic fields, *Comm. Pure Appl. Math.*, *25*(1), 77–84; doi:10.1002/cpa.3160250107.

Bobra, M. G., and S. Couvidat (2015), Solar flare prediction using SDO/HMI vector magnetic field data with a machine-learning algorithm, *Astrophys. J.*, *798*, 135; doi:10.1088/0004-637X/798/2/135.

Bommier, V., Landi Degl'Innocenti, E., Landolfi, M., and Molodij, G. (2007), UNNOFIT inversion of spectro-polarimetric maps observed with THEMIS, *A&A*, *464*, 323–339.

Bommier, V. (2016), Milne-Eddington inversion for unresolved magnetic structures in the quiet Sun photosphere, *Journal of Geophysical Research (Space Physics)*, *121*, 5025–5040.

Boulmezaoud, T. Z., and T. Amari (2000), On the existence of non-linear force-free fields in three-dimensional domains, *Zeitschrift Angewandte Mathematik und Physik*, *51*, 942–967.

Brown, M. R., R. C. Canfield, and A. A. Pevtsov (1999), *Magnetic Helicity in Space and Laboratory Plasmas*, *111*, American Geophysical Union, Washington, DC; doi:10.1029/GM111.

Buechner, J., and A. A. Pevtsov (2003), Preface, *Adv. Space Res.*, *32*, 1817–1817; doi:10.1016/S0273-1177(03)90615-6.

Canfield, R. C., A. A. Pevtsov, and A. N. McClymont (1996), Magnetic chirality and coronal reconnection., in *Magnetic Reconnection in the Solar Atmosphere, Astronomical Society of the Pacific Conference Series*, 111, edited by R. D. Bentley and J. T. Mariska, 341–346, Astronomical Society of the Pacific.

Carlsson, M., V. H. Hansteen, B. V. Gudiksen, J. Leenaarts, and B. De Pontieu (2016), A publicly available simulation of an enhanced network region of the Sun, *Astron. Astrophys.*, *585*, A4; doi:10.1051/0004-6361/201527226.

Caspi, A., T. N. Woods, and H. P. Warren (2015), New observations of the solar 0.5–5 keV soft x-ray spectrum, *Astrophys. J.*, *802*, L2; doi:10.1088/2041-8205/802/1/L2.

Charbonneau, P. (2013), Where is the solar dynamo?, *J. Phys. Conf. Ser.*, *440*(1), 012014; doi:10.1088/1742-6596/440/1/012014.

Chen, P. F. (2011), Coronal mass ejections: Models and their observational basis, *Living Rev. Sol. Phys.*, *8*, 1; doi:10.12942/lrsp-2011-1.

Cheung, M. C. M., and H. Isobe (2014), Flux emergence (theory), *Living Rev. Sol. Phys.*, *11*, 3; doi:10.12942/lrsp-2014-3.

Cheung, M. C. M., and M. L. DeRosa (2012), A method for data-driven simulations of evolving solar active regions, *Astrophys. J.*, *757*, 147; doi:10.1088/0004-637X/757/2/147.

Cranmer, S. R. (2009), Coronal holes, *Living Rev. Sol. Phys.*, *6*, 3; doi:10.12942/lrsp-2009-3.

Dalmasse, K., D. Nychka, S. Gibson, N. Flyer, and Y. Fan (2016), ROAM: A radial-basis-function optimization approximation method for diagnosing the three-dimensional coronal magnetic field, *Front. Astron. Space Sci.*, *3*, 24; doi:10.3389/fspas.2016.00024.

Dalmasse, K., G. Aulanier, P. Démoulin, B. Kliem, T. Török, and E. Pariat (2015), The origin of net electric currents in solar active regions, *Astrophys. J.*, *810*, 17; doi:10.1088/0004-637X/810/1/17.

de la Cruz Rodríguez, J., and H. Socas-Navarro (2011), Are solar chromospheric fibrils tracing the magnetic field?, *Astron. Astrophys.*, *527*, L8; doi:10.1051/0004-6361/201016018.

De Rosa, M. L., C. J. Schrijver, G. Barnes, K. D. Leka, B. W. Lites, M. J. Aschwanden, T. Amari, et al. (2009), A critical assessment of nonlinear force-free field modeling of the solar corona for active region 10953, *Astrophys. J.*, *696*, 1780–1791; doi:10.1088/0004-637X/696/2/1780.

del Toro Iniesta, J. C. (2003), *Introduction to Spectropolarimetry*, Cambridge University Press, Cambridge.

del Toro Iniesta, J. C., and B. Ruiz Cobo (2016), Inversion of the radiative transfer equation for polarized light, *Living Rev. Sol. Phys.*, *13*, 4; doi:10.1007/s41116-016-0005-2.

Demoulin, P., J. C. Henoux, E. R. Priest, and C. H. Mandrini (1996), Quasi-separatrix layers in solar flares. I. Method., *Astron. Astrophys.*, *308*, 643–655.

DeRosa, M. L., M. S. Wheatland, K. D. Leka, G. Barnes, T. Amari, A. Canou, S. A. Gilchrist, et al. (2015), The influence of spatial resolution on nonlinear force-free modeling, *Astrophys. J.*, *811*, 107; doi:10.1088/0004-637X/811/2/107.

Dravins, D. (1974), Magnetic field and electric current structure in the chromosphere, *Solar Phys.*, *37*, 323–342; doi:10.1007/BF00152492.

Fan, Y. (2009), Magnetic fields in the solar convection zone, *Living Rev. Sol. Phys.*, *6*, 4; doi:10.12942/lrsp-2009-4.

Fisher, G. H., D. W. Longcope, T. R. Metcalf, and A. A. Pevtsov (1998), Coronal heating in active regions as a function of global magnetic variables, *Astrophys. J.*, *508*, 885–898; doi:10.1086/306435.

Fisher, G. H., W. P. Abbett, D. J. Bercik, M. D. Kazachenko, B. J. Lynch, B. T. Welsch, J. T. Hoeksema, et al. (2015), The coronal global evolutionary model: Using HMI vector magnetogram and Doppler data to model the buildup of free magnetic energy in the solar corona, *Space Weather*, *13*, 369–373; doi:10.1002/2015SW001191.

Fleishman, G. D., and I. N. Toptygin (2013a), *Cosmic Electrodynamics: Astrophysics and Magnetic Hydrodynamics of Cosmic Plasmas*, Springer, New York, 388.

Fleishman, G. D., and I. N. Toptygin (2013b), Stochastic particle acceleration by helical turbulence in solar flares, *Mon. Not. R. Astron. Soc.*, *429*, 2515–2526; doi:10.1093/mnras/sts518.

Gary, D. E., G. D. Fleishman, and G. M. Nita (2013), Magnetography of solar flaring loops with microwave imaging spectropolarimetry, *Solar Phys.*, *288*, 549–565; doi:10.1007/s11207-013-0299-3.

Gary, G. A. (2001), Plasma beta above a solar active region: Rethinking the paradigm, *Solar Phys.*, *203*, 71–86; doi:10.1023/A:1012722021820.

Gary, G. A., R. L. Moore, M. J. Hagyard, and B. M. Haisch (1987), Nonpotential features observed in the magnetic field of an active region, *Astrophys. J.*, *314*, 782–794; doi:10.1086/165104.

Gelfreikh, G. B. (2004), Coronal magnetic field measurements through bremsstrahlung emission, in *Solar and Space Weather Radiophysics: Current Status and Future Developments*, edited by D. E. Gary and C. U. Keller, 115–133, Kluwer.

Gibson, S., T. Kucera, S. White, J. Dove, Y. Fan, B. Forland, L. Rachmeler, C. Downs, and K. Reeves (2016), FORWARD: A toolset for multiwavelength coronal magnetometry, *Front. Astron. Space Sci.*, *3*, 8; doi:10.3389/fspas.2016.00008.

Goodman, M. L. (2004), On the creation of the chromospheres of solar type stars, *Astron. Astrophys.*, *424*, 691–712; doi:10.1051/0004-6361:20040310.

Gopasyuk, O. S. (2015), Rotation of sunspots in active region NOAA 10930, *Adv. Space Res.*, *55*, 937–941; doi:10.1016/j.asr.2014.09.005.

Gosain, S., P. Démoulin, and M. López Fuentes (2014), Distribution of electric currents in sunspots from photosphere to corona, *Astrophys. J.*, *793*, 15; doi:10.1088/0004-637X/793/1/15.

Gudiksen, B. V., M. Carlsson, V. H. Hansteen, W. Hayek, J. Leenaarts, and J. Martínez-Sykora (2011), The stellar atmosphere simulation code Bifrost, code description and validation, *Astron. Astrophys.*, *531*, A154; doi:10.1051/0004-6361/201116520.

Gurevich, A. V. (1961), Behavior of multiply charged ions in a plasma, *Sov. Phys. JETP*, *13*, 1282–1286.

Hale, G. E., and S. B. Nicholson (1925), The law of sun-spot polarity, *Astrophys. J.*, *62*, 270; doi:10.1086/142933.

Hansteen, V., N. Guerreiro, B. De Pontieu, and M. Carlsson (2015), Numerical simulations of coronal heating through footpoint braiding, *Astrophys. J.*, *811*, 106; doi:10.1088/0004-637X/811/2/106.

Harvey, J. W. (1969), Magnetic Fields Associated with Solar Active-Region Prominences, Ph.D. Thesis, University of Colorado, Boulder, Colorado.

Hathaway, D. H. (2010), The solar cycle, *Living Rev. Sol. Phys.*, *7*, 1; doi:10.12942/lrsp-2010-1.

Hénoux, J. C., and M. Karlický (2013), Flare line impact polarization. Na D2 589 nm line polarization in the 2001 June 15 flare, *Astron. Astrophys.*, *556*, A95; doi:10.1051/0004-6361/201219478.

Hofmann, A., and J. Staude (1987), Electric current density in the sunspot photosphere derived from vector magnetograms, *Publications of the Astronomical Institute of the Czechoslovak Academy of Sciences*, *66*, 105–107.

Janvier, M., G. Aulanier, V. Bommier, B. Schmieder, P. Démoulin, and E. Pariat (2014), Electric currents in flare ribbons: Observations and three-dimensional standard model, *Astrophys. J.*, *788*, 60; doi:10.1088/0004-637X/788/1/60.

Jiang, C., S. T. Wu, X. Feng, and Q. Hu (2016), Data-driven magnetohydrodynamic modelling of a flux-emerging active region leading to solar eruption, *Nat. Comm.*, *7*, 11522; doi:10.1038/ncomms11522.

Kazachenko, M. D., G. H. Fisher, B. T. Welsch, Y. Liu, and X. Sun (2015), Photospheric electric fields and energy fluxes in the eruptive active region NOAA 11158, *Astrophys. J.*, *811*, 16; doi:10.1088/0004-637X/811/1/16.

Keller, C. U., J. W. Harvey, and M. S. Giampapa (2003), SOLIS: An innovative suite of synoptic instruments, in *Innovative Telescopes and Instrumentation for Solar Astrophysics, Proc. SPIE*, *4853*, edited by S. L. Keil and S. V. Avakyan, 194–204.

Kosugi, T., K. Matsuzaki, T. Sakao, T. Shimizu, Y. Sone, S. Tachikawa, T. Hashimoto, et al. (2007), The Hinode (Solar-B) mission: An overview, *Solar Phys.*, *243*, 3–17; doi:10.1007/s11207-007-9014-6.

Kotov, V. A. (1970), Magnetic field and electric currents of a unipolar sunspot, *Izvestiya Ordena Trudovogo Krasnogo Znameni Krymskoj Astrofizicheskoj Observatorii*, *41*, 67–88.

Kotov, V. A. (1971), On the structure of magnetic field and electric currents of a unipolar sunspot (presented by A. Severny), in *Solar Magnetic Fields, IAU Symposium*, *43*, edited by R. Howard, 212–219, Springer.

Kotov, V. A. (1972), Measurements of the transverse magnetic field in the chromosphere above a sunspot, *Sov. Astron.*, *15*, 687–688.

Kramar, M., H. Lin, and S. Tomczyk (2016), Direct observation of solar coronal magnetic fields by vector tomography of the coronal emission line polarizations, *Astrophys. J.*, *819*, L36; doi:10.3847/2041-8205/819/2/L36.

Kusano, K., T. Maeshiro, T. Yokoyama, and T. Sakurai (2004), The trigger mechanism of solar flares in a coronal arcade with reversed magnetic shear, *Astrophys. J.*, *610*, 537–549; doi:10.1086/421547.

Lee, J. (2007), Radio emissions from solar active regions, *Space Sci. Rev.*, *133*, 73–102; doi:10.1007/s11214-007-9206-2.

Leka, K. D., R. C. Canfield, A. N. McClymont, and L. van Driel-Gesztelyi (1996), Evidence for current-carrying emerging flux, *Astrophys. J.*, *462*, 547; doi:10.1086/177171.

Levens, P. J., B. Schmieder, A. López Ariste, N. Labrosse, K. Dalmasse, and B. Gelly (2016), Magnetic field in atypical prominence structures: Bubble, tornado, and eruption, *Astrophys. J.*, *826*, 164; doi:10.3847/0004-637X/826/2/164.

Lin, H., M. J. Penn, and S. Tomczyk (2000), A new precise measurement of the coronal magnetic field strength, *Astrophys. J.*, *541*, L83–L86; doi:10.1086/312900.

Lites, B. W., V. Martinez Pillet, and A. Skumanich (1994), A quantitative comparison of vector magnetic field measurement and analysis techniques, *Solar Phys.*, *155*, 1–27; doi:10.1007/BF00670727.

Longcope, D. W., and B. T. Welsch (2000), A model for the emergence of a twisted magnetic flux tube, *Astrophys. J.*, *545*, 1089–1100; doi:10.1086/317846.

Mackay, D. H., J. T. Karpen, J. L. Ballester, B. Schmieder, and G. Aulanier (2010), Physics of solar prominences: II. Magnetic structure and dynamics, *Space Sci. Rev.*, *151*, 333–399; doi:10.1007/s11214-010-9628-0.

Martínez Pillet, V., J. C. Del Toro Iniesta, A. Álvarez-Herrero, V. Domingo, J. A. Bonet, L. González Fernández, A. López Jiménez, et al. (2011), The Imaging Magnetograph eXperiment (IMaX) for the Sunrise Balloon-Borne Solar Observatory, *Solar Phys.*, *268*, 57–102; doi:10.1007/s11207-010-9644-y.

Martínez-Sykora, J., B. De Pontieu, M. Carlsson, and V. Hansteen (2016), Misalignment between chromospheric features and magnetic field, *ArXiv e-prints*.

McClymont, A. N., and G. H. Fisher (1989), On the mechanical energy available to drive solar flares., 219–225, *American Geophysical Union, Geophysical Monograph Series*, *54*, Washington, DC.

McClymont, A. N., L. Jiao, and Z. Mikic (1997), Problems and progress in computing three-dimensional coronal active region magnetic fields from boundary data, *Solar Phys.*, *174*, 191–218; doi:10.1023/A:1004976720919.

Melrose, D. B. (1991), Neutralized and unneutralized current patterns in the solar corona, *Astrophys. J.*, *381*, 306–312; doi:10.1086/170652.

Melrose, D. B. (1997), A solar flare model based on magnetic reconnection between current-carrying loops, *Astrophys. J.*, *486*, 521–533.

Metcalf, T. R., K. D. Leka, G. Barnes, B. W. Lites, M. K. Georgoulis, A. A. Pevtsov, K. S. Balasubramaniam, et al. (2006), An overview of existing algorithms for resolving the 180 ambiguity in vector magnetic fields: Quantitative tests with synthetic data, *Solar Phys.*, *237*, 267–296; doi:10.1007/s11207-006-0170-x.

Metcalf, T. R., L. Jiao, A. N. McClymont, R. C. Canfield, and H. Uitenbroek (1995), Is the solar chromospheric magnetic field force-free?, *Astrophys. J.*, *439*, 474–481; doi:10.1086/175188.

Molodenskii, M. M. (1969), Integral properties of force-free fields, *Sov. Astron.*, *12*, 585.

Moreton, G. E., and A. B. Severny (1968), Magnetic fields and flares in the region CMP 20 September 1963, *Solar Phys.*, *3*, 282–297; doi:10.1007/BF00155163.

Musset, S., N. Vilmer, and V. Bommier (2015), Hard X-ray emitting energetic electrons and photospheric electric currents, *Astron. Astrophys.*, *580*, A106; doi:10.1051/0004-6361/201424378.

Nakariakov, V. M., and L. Ofman (2001), Determination of the coronal magnetic field by coronal loop oscillations, *Astron. Astrophys.*, *372*, L53–L56; doi:10.1051/0004-6361:20010607.

Nita, G. M., G. D. Fleishman, A. A. Kuznetsov, E. P. Kontar, and D. E. Gary (2015), Three-dimensional radio and x-ray modeling and data analysis software: Revealing flare complexity, *Astrophys. J.*, *799*, 236; doi:10.1088/0004-637X/799/2/236.

Orozco Suárez, D., A. Asensio Ramos, and J. Trujillo Bueno (2014), The magnetic field configuration of a solar prominence inferred from spectropolarimetric observations in the He i 10 830 Å triplet, *Astron. Astrophys.*, *566*, A46; doi:10.1051/0004-6361/201322903.

Parenti, S. (2014), Solar prominences: observations, *Living Rev. Sol. Phys.*, *11*, 1; doi:10.12942/lrsp-2014-1.

Parker, E. N. (1996), Inferring mean electric currents in unresolved fibril magnetic fields, *Astrophys. J.*, *471*, 485; doi:10.1086/177983.

Petrie, G. J. D. (2012), The abrupt changes in the photospheric magnetic and Lorentz force vectors during six major neutral-line flares, *Astrophys. J.*, *759*, 50; doi:10.1088/0004-637X/759/1/50.

Pevtsov, A. A. (2005), Helicity generation and signature in solar atmosphere, *High. Astron.*, *13*, 89.

Pevtsov, A. A., and N. L. Peregud (1990), Electric currents in a unipolar sunspot, 161–165, *Geophysical Monograph Series, 58, American Geophysical Union*, Washington, DC; doi:10.1029/GM058p0161.

Pevtsov, A. A., M. A. Berger, A. Nindos, A. A. Norton, and L. van Driel-Gesztelyi (2014), Magnetic helicity, tilt, and twist, *Space Sci. Rev.*, *186*, 285–324; doi:10.1007/s11214-014-0082-2.

Pevtsov, A. A., R. C. Canfield, and A. N. McClymont (1997), On the subphotospheric origin of coronal electric currents, *Astrophys. J.*, *481*, 973–977.

Pevtsov, A. A., R. C. Canfield, and T. R. Metcalf (1994), Patterns of helicity in solar active regions, *Astrophys. J.*, *425*, L117–L119; doi:10.1086/187324.

Pevtsov, A. A., R. C. Canfield, and T. R. Metcalf (1995), Latitudinal variation of helicity of photospheric magnetic fields, *Astrophys. J.*, *440*, L109–L112; doi:10.1086/187773.

Pevtsov, A. A., V. M. Maleev, and D. W. Longcope (2003), Helicity evolution in emerging active regions, *Astrophys. J.*, *593*, 1217–1225; doi:10.1086/376733.

Puschmann, K. G., B. Ruiz Cobo, and V. Martínez Pillet (2010a), The electrical current density vector in the inner penumbra of a sunspot, *Astrophys. J.*, *721*, L58–L61; doi:10.1088/2041-8205/721/1/L58.

Puschmann, K. G., B. Ruiz Cobo, and V. Martínez Pillet (2010b), A geometrical height scale for sunspot penumbrae, *Astrophys. J.*, *720*, 1417–1431; doi:10.1088/0004-637X/720/2/1417.

Rachkovsky, D. N. (1962), Magnetic rotation effects in spectral lines, *Izvestiya Ordena Trudovogo Krasnogo Znameni Krymskoj Astrofizicheskoj Observatorii*, *28*, 259–270.

Rachkovsky, D. N. (1967), The reduction for anomalous dispersion in the theory of absorption line formation in a magnetic field, *Izvestiya Ordena Trudovogo Krasnogo Znameni Krymskoj Astrofizicheskoj Observatorii*, *37*, 56–61.

Raouafi, N. E. (2002), Stokes parameters of resonance lines scattered by a moving, magnetic medium: Theory of the two-level atom, *Astron. Astrophys.*, *386*, 721–731; doi:10.1051/0004-6361:20020113.

Raouafi, N. E., S. Patsourakos, E. Pariat, P. R. Young, A. C. Sterling, A. Savcheva, M. Shimojo, et al. (2016), Solar coronal jets: Observations, theory, and modeling, *Space Sci. Rev.*, *201*, 1–53; doi:10.1007/s11214-016-0260-5.

Rayrole, J., and M. Semel (1970), Evaluation of the electric current in a sunspot by the study of the observed transverse component of the magnetic field, *Astron. Astrophys.*, *6*, 288–293.

Rudenko, G. V., and I. I. Myshyakov (2009), Analysis of reconstruction methods for nonlinear force-free fields, *Solar Phys.*, *257*, 287–304; doi:10.1007/s11207-009-9389-7.

Rust, D. M. (1994), Spawning and shedding helical magnetic fields in the solar atmosphere, *Geophys. Res. Lett.*, *21*, 241–244; doi:10.1029/94GL00003.

Ryabov, B. I., V. P. Maksimov, S. V. Lesovoi, K. Shibasaki, A. Nindos, and A. Pevtsov (2005), Coronal magnetography of solar active region 8365 with the SSRT and NoRH radio heliographs, *Solar Phys.*, *226*, 223–237; doi:10.1007/s11207-005-2691-0.

Scherrer, P. H., J. Schou, R. I. Bush, A. G. Kosovichev, R. S. Bogart, J. T. Hoeksema, Y. Liu, et al. (2012), The helioseismic and magnetic imager (HMI) investigation for the Solar Dynamics Observatory (SDO), *Solar Phys.*, *275*, 207–227; doi:10.1007/s11207-011-9834-2.

Schmieder, B., V. Archontis, and E. Pariat (2014), Magnetic flux emergence along the solar cycle, *Space Sci. Rev.*, *186*, 227–250; doi:10.1007/s11214-014-0088-9.

Schou, J., P. H. Scherrer, R. I. Bush, R. Wachter, S. Couvidat, M. C. Rabello-Soares, R. S. Bogart, et al. (2012), Design and ground calibration of the helioseismic and magnetic imager (HMI) instrument on the Solar Dynamics Observatory (SDO), *Solar Phys.*, *275*, 229–259; doi:10.1007/s11207-011-9842-2.

Schrijver, C. J., M. L. De Rosa, T. R. Metcalf, Y. Liu, J. McTiernan, S. Régnier, G. Valori, M. S. Wheatland, and T. Wiegelmann (2006), Nonlinear force-free modeling of coronal magnetic fields, Part I: A quantitative comparison of methods, *Solar Phys.*, *235*, 161–190; doi:10.1007/s11207-006-0068-7.

Seehafer, N. (1990), Electric current helicity in the solar atmosphere, *Solar Phys.*, *125*, 219–232; doi:10.1007/BF00158402.

Semel, M., and A. Skumanich (1998), An ambiguity-free determination of J_Z in solar active regions, *Astron. Astrophys.*, *331*, 383–391.

Severnyi, A. B. (1965), The nature of solar magnetic fields (the fine structure of the field), *Soviet Astron.*, *9*, 171–182.

Shibata, K., and T. Magara (2011), Solar flares: Magnetohydrodynamic processes, *Living Rev. Sol. Phys.*, *8*, 6; doi:10.12942/lrsp-2011-6.

Socas-Navarro, H. (2002), Zeeman diagnostics of solar magnetic fields, in *SOLMAG 2002, Proceedings of the Magnetic Coupling of the Solar Atmosphere Euroconference, ESA Special Publication*, 505, edited by H. Sawaya-Lacoste, 45–51.

Socas-Navarro, H. (2005a), The three-dimensional structure of a sunspot magnetic field, *Astrophys. J.*, *631*, L167–L170; doi:10.1086/497334.

Socas-Navarro, H. (2005b), Are electric currents heating the magnetic chromosphere?, *Astrophys. J.*, *633*, L57–L60; doi:10.1086/498145.

Solanki, S. K., A. Lagg, J. Woch, N. Krupp, and M. Collados (2003), Three-dimensional magnetic field topology in a region of solar coronal heating, *Nature*, *425*, 692–695; doi:10.1038/nature02035.

Tomczyk, S., G. L. Card, T. Darnell, D. F. Elmore, R. Lull, P. G. Nelson, K. V. Streander, J. Burkepile, R. Casini, and P. G. Judge (2008), An instrument to measure coronal emission line polarization, *Solar Phys.*, *247*, 411–428; doi:10.1007/s11207-007-9103-6.

Unno, W. (1956), Line formation of a normal Zeeman triplet, *Phys. Rep.*, *8*, 108.

Valori, G., B. Kliem, and M. Fuhrmann (2007), Magnetofrictional extrapolations of Low and Lou's force-free equilibria, *Solar Phys.*, *245*, 263–285; doi:10.1007/s11207-007-9046-y.

Wang, Z., D. E. Gary, G. D. Fleishman, and S. M. White (2015), Coronal magnetography of a simulated solar active region from microwave imaging spectropolarimetry, *Astrophys. J.*, *805*, 93; doi:10.1088/0004-637X/805/2/93.

Warren, H. P. (2014), Measurements of absolute abundances in solar flares, *Astrophys. J.*, *786*, L2; doi:10.1088/2041-8205/786/1/L2.

Wheatland, M. S. (2000), Are electric currents in solar active regions neutralized?, *Astrophys. J.*, *532*, 616–621; doi:10.1086/308577.

Wheatland, M. S. (2007), Calculating and testing nonlinear force-free fields, *Solar Phys.*, *245*, 251–262; doi:10.1007/s11207-007-9054-y.

Wheatland, M. S., P. A. Sturrock, and G. Roumeliotis (2000), An optimization approach to reconstructing force-free fields, *Astrophys. J.*, *540*, 1150–1155; doi:10.1086/309355.

White, S. M. (2005), Radio measurements of coronal magnetic fields, in *Chromospheric and Coronal Magnetic Fields, ESA Special Publication*, 596, edited by D. E. Innes, A. Lagg, and S. A. Solanki, 10.1.

Wiegelmann, T., and B. Inhester (2010), How to deal with measurement errors and lacking data in nonlinear force-free coronal magnetic field modelling?, *Astron. Astrophys.*, *516*, A107; doi:10.1051/0004-6361/201014391.

Wiegelmann, T., and T. Sakurai (2012), Solar force-free magnetic fields, *Living Rev. Sol. Phys.*, *9*, 5; doi:10.12942/lrsp-2012-5.

Wiegelmann, T., B. Inhester, and T. Sakurai (2006), Preprocessing of vector magnetograph data for a nonlinear force-free magnetic field reconstruction, *Solar Phys.*, *233*, 215–232; doi:10.1007/s11207-006-2092-z.

Wiegelmann, T., J. K. Thalmann, C. J. Schrijver, M. L. De Rosa, and T. R. Metcalf (2008), Can we improve the preprocessing of photospheric vector magnetograms by the inclusion of chromospheric observations?, *Solar Phys.*, *247*, 249–267; doi:10.1007/s11207-008-9130-y.

Wiegelmann, T., J. K. Thalmann, and S. K. Solanki (2014), The magnetic field in the solar atmosphere, *A&A Rev.*, *22*, 78; doi:10.1007/s00159-014-0078-7.

Zhang, H., and S. Bao (1998), Latitudinal distribution of photospheric current helicity and solar activities, *Astron. Astrophys.*, *339*, 880–886.

4
Multipoint Analysis of Electric Currents in Geospace Using the Curlometer Technique

M. W. Dunlop[1,9], S. Haaland[2], X-C. Dong[1], H. R. Middleton[3], C. P. Escoubet[4], Y.-Y. Yang[5], Q.-H. Zhang[6], J-K. Shi[7], and C. T. Russell[8]

ABSTRACT

Four-point magnetic field measurements in space allow estimates of the electric current density through the curlometer technique, which estimates electric current density from Ampère's law, and is relevant to the magnetosphere and surrounding regions, which contain high conductivity plasma. Knowledge of spacecraft separations, magnetic field measurement accuracy, and the form of the current structures sampled (e.g., relative scale size) limits the accuracy of the method. Despite these conditions of application, in many regions of the magnetosphere it has been shown to be robust and reliable. A number of studies have applied the method successfully such as: the ring current; the magnetotail current sheet; the magnetopause currents, and field aligned currents, as well as to other current structures (e.g., flux tubes). Where time stationarity and other special assumptions can be made, or where the spacecraft configuration is highly irregular or less than four spacecraft are available, the method can still be applied to obtain partial components of the current. We discuss the application of the curlometer technique to the four-point observations from the Cluster mission in terms of its adaptability and performance (including the lessons learned) and illustrate its use with recent data from the MMS mission, which covers a much smaller spatial regime.

[1] *Space Science Institute, School of Astronautics, Beihang University, Beijing, China*
[2] *Birkeland Centre for Space Science, University of Bergen, Norway; Max-Planck Institute for Solar Systems Research, Göttingen, Germany*
[3] *ESA/ESAC, Villanueva de la Cañada, Madrid, Spain*
[4] *ESA/ESTEC, Noordwijk, The Netherlands*
[5] *Key Laboratory of Crustal Dynamics, China Earthquake Administration, Beijing, China*
[6] *Institute of Space Sciences, Shandong University, Weihai, China*
[7] *NSSC, Chinese Academy of Sciences, Beijing, China*
[8] *UCLA, Los Angeles, California, USA*
[9] *RAL Space, STFC, Chilton, Oxfordshire, UK*

4.1. INTRODUCTION: CURRENTS IN GEOSPACE

As shown in Chapter 2, a wide range of currents permeates many regions of space and, in particular, currents are ubiquitous in the Earth's magnetosphere (see Fig. 4.1), supporting both large-scale and small-scale plasma structures (in the form of sheets, tubes, and other structures) and connecting different regions (as in the case of field-aligned currents). In the terrestrial magnetosphere, pronounced current systems include the ring current (see Chapter 7), the cross-tail current sheet (see Chapter 10), and the magnetopause current (see Chapter 13). Field-aligned currents (FACs, see section IV of this volume)

Electric Currents in Geospace and Beyond, Geophysical Monograph 235, First Edition.
Edited by Andreas Keiling, Octav Marghitu, and Michael Wheatland.

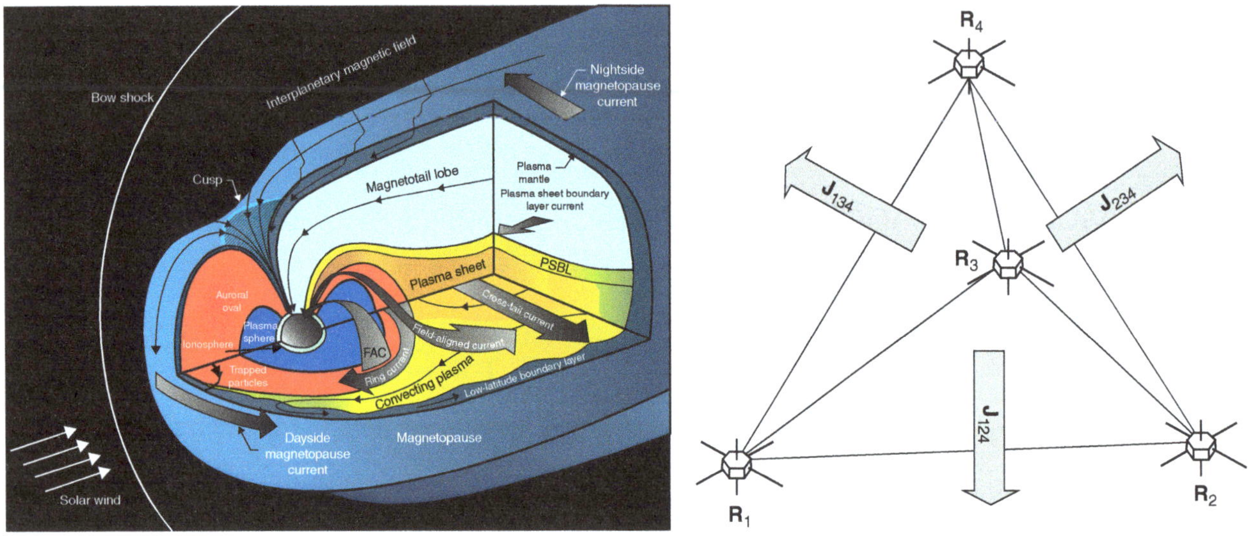

Figure 4.1 (Left) Depiction of the large-scale currents in the Earth's magnetosphere (adapted from *Kivelson and Russell* [1995]), and (right) the curlometer concept (after *Dunlop et al.* [1988]).

connect the collisionless plasma regime in the magnetosphere with the Earth's collision-dominated ionosphere. Other currents may exist in the plasma sheet boundary layer and within the cusp boundaries, and within transient structures such as opened flux tubes caused by magnetic reconnection.

Currents are set up by differential motion of charged particles. In simple terms, the current density can thus be derived from density and velocity moments of ions and electrons:

$$\mathbf{J} = \Sigma q n_s \mathbf{V_s} \sim q n\left(\mathbf{V_i} - \mathbf{V_e}\right) \quad (4.1)$$

where $n = n_i = n_e$ is the density (assuming quasi neutrality), V_i is the ion bulk velocity and V_e is the corresponding electron bulk velocity. In practice, this approach is difficult and will require particle detectors with identical characteristics for all ion species and electrons in terms of angular resolution, energy ranges, sensitivity, and calibration. Since particle moments will have to be based on 3D particle velocity distribution functions, measurements often rely on spacecraft spin (and thus have limited time resolution) or arrays of detectors to obtain full coverage. Examples, using particle moments from the Cluster satellites [*Escoubet et al.*, 2001] can be found in publications by, for example, *Henderson et al.* [2008] and *Petrukovich et al.* [2015]. Recent results from the Magnetospheric Multiscale (MMS) mission [*Burch et al.*, 2016] benefit from far better time resolution, and sampling at much smaller spatial separations, and will provide new opportunities to use particle moments to estimate currents (see section 4.3.2).

In space, either currents are also sources of magnetic fields or they modify existing magnetic fields (e.g., the Earth's dipole magnetic field). The relation between currents and magnetic field perturbations is given by Ampère's law (for highly conducting plasmas where quasi neutrality typically applies, the displacement current, $\mu_0\varepsilon_0\partial E/\partial t$, where E is the electric field, can be neglected in the corresponding Maxwell equation [*Russell, et al.*, 2016]):

$$\mu_0 \mathbf{J} = \mathbf{curl}\left(\mathbf{B}\right)$$

This relation enables us to derive the strength and orientation of currents directly from the magnetic field and its gradients without any particle measurements (or to compare between the two estimates). Magnetic fields can typically be measured with higher accuracy and higher time resolution than particle moments.

Multiple spacecraft flying in formation (a constellation) allow the curlometer technique, using Ampère's law above, to estimate the electric current density from **curl(B)**, suitable for the magnetosphere and surrounding regions, which contain plasma with high conductivity. The method was introduced to utilize four-point measurements in space; anticipating the realization of the four-spacecraft Cluster mission [*Escoubet et al.*, 2001], which began full science operations in February 2001 and provided the spatial measurements from which a linear calculation of the curl of the magnetic field can be made.

Knowledge of spacecraft separations, magnetic field measurement accuracy, and the form of the current structures sampled (e.g., relative scale size) limits the quality of the method. Despite these conditions of application, in many regions of the magnetosphere it has been shown to be robust and reliable. A number of studies have applied the method successfully, such as the ring current; the magnetotail current sheet; the magnetopause currents, and field aligned currents; as well as other current structures (e.g., flux tubes). Where time stationarity and other special assumptions can be made, or where the spacecraft configuration is highly irregular or less than four spacecraft are available, the method can still be applied to obtain partial components of the current.

In this chapter, we discuss the performance, adaptability, and lessons learned from multispacecraft techniques, with emphasis on the curlometer technique, and the adaptability of the method. In particular, the application to the new sampling regime offered by the new four-point observations of the MMS mission (which covers a dramatically different spatial regime to that of Cluster), offers a tool able to address open questions on small-scale current structures.

4.2. MULTISPACECRAFT ANALYSIS

4.2.1. The Curlometer

The curlometer method [*Dunlop et al.*, 1988] provides all components of the electric current density from measurements taken from four spacecraft, based on the assumption that: **curl(B)** = μ_0**J**. Using the individual positions and measured magnetic fields at each spacecraft, the curlometer estimates the average current density in the tetrahedral configuration. This may be done by either estimating linear approximations to the spatial gradients needed for **curl(B)**, or using the linear, integral form of Ampère's law ($\mu_0\int \mathbf{J}.\mathbf{ds} = \int \mathbf{B}.\mathrm{dl}$), which leads to an estimate of the average current density, J_{1ij}, normal to the face 1ij of the tetrahedron (see the right hand-side of Fig. 4.1) from:

$$\mu_0 \langle \mathbf{J} \rangle . (\Delta\mathbf{R}_i \times \Delta\mathbf{R}_j) = \Delta\mathbf{B}_i.\Delta\mathbf{R}_j - \Delta\mathbf{B}_j.\Delta\mathbf{R}_i,$$
$$\text{e.g.} \mu_0 < \mathbf{J} >_{123} (\Delta\mathbf{R}_{12} \times \Delta\mathbf{R}_{13}) = \Delta\mathbf{B}_{12}.\Delta\mathbf{R}_{13} - \Delta\mathbf{B}_{13}.\Delta\mathbf{R}_{12}, \quad (4.2)$$

where here, $\Delta\mathbf{B}_{1i}$, $\Delta\mathbf{R}_{1i}$ have been shortened in notation to $\Delta\mathbf{B}_i$, $\Delta\mathbf{R}_i$, but any reference spacecraft can be used. Three faces of the tetrahedron provide three noncoplanar components of the current, which can then be used to construct a full, average vector current density.

An estimate of the average value of div(**B**) over the volume of the tetrahedron can also be made from $< div(\mathbf{B}) > |\Delta\mathbf{R}_i.\Delta\mathbf{R}_j \times \Delta\mathbf{R}_k| = |\Sigma_{cyclic}\Delta\mathbf{B}_i.\Delta\mathbf{R}_j \times \Delta\mathbf{R}_k|$, which indicates by its value (since div(**B**) = 0) the approximate size of the neglected nonlinear gradients [*Dunlop et al.*, 2002; *Robert et al.*, 1998; and *Haaland et al.*, 2004a], indirectly. The shape and size of the spacecraft configuration compared with the relative orientation and spatial scale of the current structures affect the quality of the linear estimate through the form of the nonlinear gradients in **B**. Usually these may be indirectly monitored by the quality parameter, Q = |div(**B**)|/|**curl(B)**|, although for highly nonregular tetrahedral shapes, the use is less certain. Since div(**B**) should ideally be zero, small values of Q are desirable, and, typically, Q is used only to indicate when the curlometer estimate may be bad. Note, however, that div(**B**) is also calculated from gradients in **B** and, thus, is restricted by the same constraints as the curl calculations. The value of Q therefore can not be used to derive *quantitative* error estimates of the current determination. Finally, measurement errors contribute to the overall uncertainty in the linear approximation of both div(**B**) and **curl(B)**.

In the above equations, $\Delta\mathbf{B}_{ij}$ and $\Delta\mathbf{R}_{ij}$ represent differences in the measured magnetic field and spatial position vectors between spacecraft *i* and *j*, respectively (assumed to be measured simultaneously, so that not only uncertainties in the measured **B** and **R**, but also interspacecraft timing errors affect the estimate). Since the full vector current can be constructed using any three of the four faces of the configuration, there is a redundancy in the calculation (by cycling through the choices of three faces), which can verify the sensitivity of the estimate for each Cartesian component of **J** (thus providing a way to assess uncertainty independently to the estimate of div(**B**), and hence Q). This original form of the curlometer is identical to the process of directly estimating the linear spatial gradients for each current density component (see section 4.2.2), for example, from the dyadic of **B**, but the error handling (and hence relevant quality parameter) is slightly different. In the integral form of the method, the estimate is often self-stabilizing for common current structures such as sheets and tubes.

Although the method can be applied to the measured **B** point by point in time, the nature of the linear approximation suggests a further consideration for the preparation of the input data. If the background field contains strong nonlinear gradients (this is the case for the internal geomagnetic field), then the neglect of these gradients in the linear estimates can imply nonphysical (i.e., not real) currents (the effect is significant in the inner magnetosphere and therefore affects the ring current calculation, and the calculation of FACs at low orbit). Moreover, if the curlometer is applied to a geomagnetic model field (such as the IGRF) in which the current is zero, then implied currents from nonzero values will result from the estimate. This effect was first noted in *Dunlop et al.* [2002] and was pursued more recently by *Grimald et al.* [2012]. The solution is to subtract a zero current model field from the measured magnetic field measurements prior to the application of the curlometer, which is then applied to the residual measured fields [*Dunlop et al.*, 2015b]. It is of course important that any detrending of the data in this way does not remove any real currents, which may be present (and it is usually found to be sufficient to use a defined model field containing no current), unless different current systems (such as large and small scale) need to be separated, and modeled currents are first subtracted.

A further consideration is that nonregular tetrahedral spacecraft configurations preferentially access some components of the current more accurately than others (depending on alignment, or misalignment, to the dominant current direction and its form). This can be a significant pitfall to consider when sampling large-scale currents at the magnetopause or magnetotail, for example; but it can also be a benefit to the measurement of highly directed currents such as field-aligned currents (FACs) where one face of the tetrahedron formed by the spacecraft can be used to estimate the component of the current that is closest to the FAC direction [*Dunlop et al.*, 2015a].

In general terms, accurate computation of any current requires the spatial configuration to be small compared to the characteristic scale size of current structure to minimize the effect of the nonlinear gradients (i.e., nonmeasured gradients in the current density). This requirement, however, is limited by the effect of the measurement errors in **B** and **R** and by timing errors between spacecraft, which become more significant at smaller spatial scales. Thus, smaller tetrahedral scales require higher absolute accuracy in **B** and **R**, and higher temporal behavior requires higher cadence and accuracy of the measurement times. For the smallest Cluster tetrahedron scales (a few 100 km), measurement uncertainty (~0.1 nT in **B**; a few km for **R** and millisecond timing) was sufficiently low (for accessing minimum currents greater than a few nAm^{-2}) that linearization errors typically dominate the lack of knowledge in the estimates. This is why Q was taken as a reasonable quality indicator. At separations of tens of km (as accessed by the MMS mission [*Burch et al.*, 2016]), the curlometer is likely to be more often in the linear regime where errors due to gradients in the current density are small. On these spatial scales, however, the measurement errors could become significant unless the currents measured are large and therefore it is more relevant that all these measurement errors should be minimized. In fact, it is natural that at Cluster separation scales the curlometer misses small-scale structure and therefore that the estimate will typically be lower than actual currents that have forms varying over the spacecraft configuration. We might expect that on MMS scales the estimated currents will be larger than comparative measurements by Cluster (see section 4.3.2). At the magnetopause or in the magnetotail, for example, the boundary layer will often be much larger than the MMS separation scales so that substructures can be resolved.

The preceding outline represents the key issues revealed from the application of the method to Cluster data, following its use in many regions of near-Earth space since the launch of Cluster (see, for instance, *Dunlop and Eastwood*, [2008], for a review, and *Dunlop et al.* [2016], for a summary of the practical experience gained). In section 4.3 we will illustrate the key pitfalls learned through a selection of these applications. Ready to use implementations of the curlometer method can be obtained from the Cluster Science Archive (https://www.cosmos.esa.int/web/csa/multi-spacecraft), and see the technical note by *Middleton and Masson* [2016].

4.2.2. Other Supporting Techniques: Fewer Than Four Spacecraft

There are a number of alternative methods that have been developed relating to the generalized problem of estimating spatial gradients of **B**, and the mapping of these calculations onto the computation of **curl(B)** [*Chanteur*, 1998; *Harvey*, 1998], as well as other gradient-based methods [e.g., *Shi et al.*, 2005, 2006]. For example, standard methods have been developed based on the use of planar reciprocal vectors [*Vogt et al.*, 2008] and the method of least squares [e.g., *DeKeyser et al.*, 2007], and a full calculation of the magnetic rotation rates (MRA, magnetic rotation analysis), leading to estimates of field-line curvature as well as the current density directly has also been developed [see *Shen and Dunlop*, 2008]. The MRA method can also be readily applied to irregular configurations of three to five spacecraft [*Shen et al.*, 2012a].

Recently, the power of these methods has been demonstrated in situations where less than four point measurements are available, or for severely distorted configurations where three spacecraft remain regular or are aligned to the dominant direction of current density. In general, results are returned only for some components of **J** (for example, three spacecraft may still recover one component normal to the plane of the spacecraft), unless other data sets are used in conjunction with the magnetic field, such as the plasma moments, or where assumptions in the behavior of the currents can be made (e.g., stationarity of the field, known FACs, or force-free structures) such as at low Earth orbit recently covered by the multispacecraft Swarm mission [e.g., *Vogt et al.*, 2009, 2013; *Shen et al.*, 2012a, 2012b; *Ritter and Lühr*, 2013].

One application of the method of *Shen et al.* [2012a] has been applied to the group of three magnetospheric THEMIS spacecraft, which came into a close, three-spacecraft configuration at times when they were providing coverage through the Earth's ring current [*Yang et al.*, 2016]. The left-hand side of Figure 4.2 is taken from *Yang et al.* [2016] to illustrate that the current component normal to the configuration can be readily obtained, but

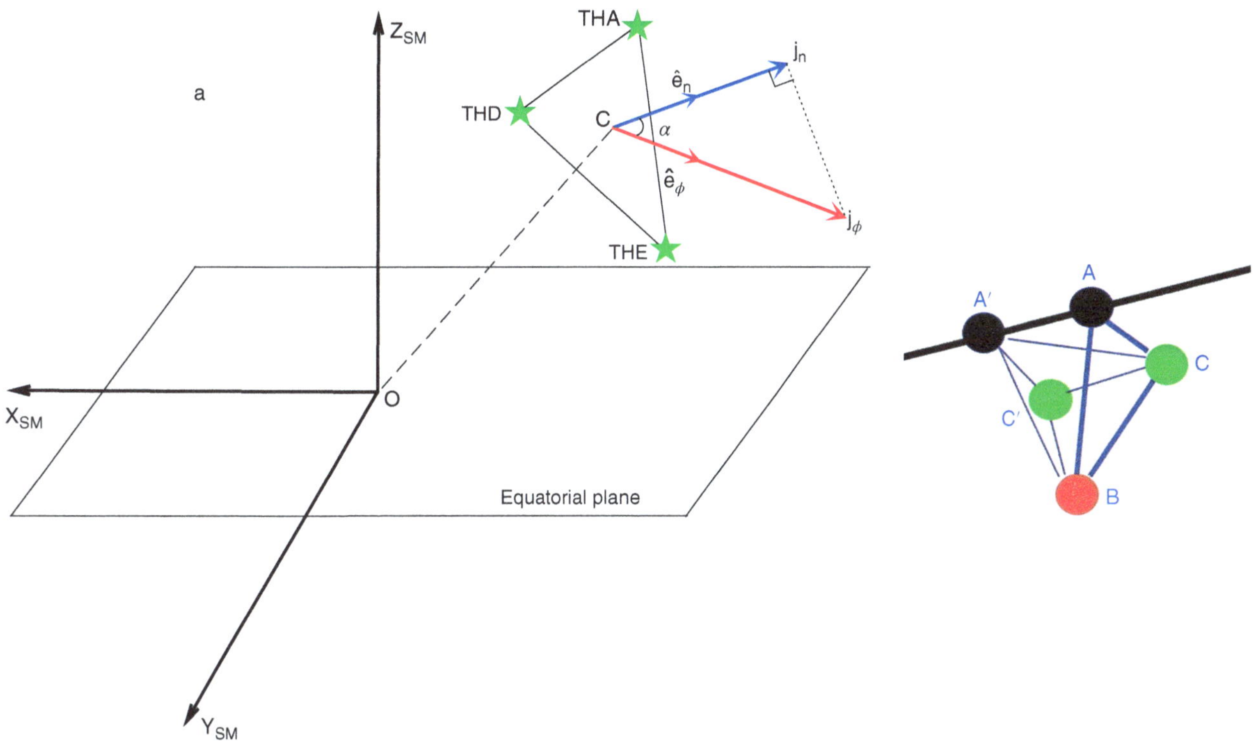

Figure 4.2 (Left) Application of the method of *Shen et al.* [2012a] for the case of three THEMIS spacecraft [from *Yang et al.*, 2016]. (Right) Orbit track of the three Swarm spacecraft (A, B, C) when these are grouped closely together. The positions A′, C′ are of these spacecraft a few seconds earlier. When these are combined, giving up to five positions, then the full curlometer can be applied and all components of the current density, *J*, may be recovered (i.e., by using selected configurations of four spatial positions, e.g., A,A′,B,C and using some mean time reference suitable for each configuration). From *Dunlop et al.* [2015a, 2015b].

generally takes some angle relative to the J_ϕ component (in the X, Y plane of the magnetic dipole equator). For large-scale current systems, such as the ring current, where the direction of the main current can be assumed, or for systems of known FACs, where the dominant current is along the known magnetic field direction, this type of application is useful (see section 4.3.3 and 4.3.5).

At low Earth orbit, assumptions of stationarity of the field often can be made, so that positions of the spacecraft at adjacent times can be combined to produce added measurement points, as depicted on the right-hand side of Figure 4.2, allowing the multipoint curlometer to be applied with fewer than four spacecraft. Different configurations of the combined positions are associated with different effective mean times for the measurement. These considerations have been explored in two recent papers [*Dunlop et al.*, 2015a, 2015b], which have estimated the full current density at Swarm altitudes using two to four spacecraft positions, and have shown coordinated field-aligned current (FAC) signatures at Cluster and Swarm. In this high field region magnetic residuals are computed by first subtracting a high-resolution internal field model before application of the curlometer.

This is by no means an exhaustive list and the continued usefulness of the curlometer and related techniques in fact lies in flexibility of application and formulation (methodology). Indeed the benchmarking of the method can easily be adapted to new regimes, which we briefly describe below.

4.2.3. Technique Implementation

From equation (4.2), and with reference to the right-hand side of Figure 4.1, it can be seen that the size of the current components perpendicular to each face of the tetrahedron can be found from the terms on the right-hand side of equation (4.2), where the normals to each face (and hence the orientation of these current components) are obtained from $(\Delta\mathbf{R}_i\times\Delta\mathbf{R}_j)/|(\Delta\mathbf{R}_i\times\Delta\mathbf{R}_j)|$, on the left-hand side of equation (4.2). Cartesian components can be found from projections of three of these components onto X, Y, Z coordinates in the usual way. In order to make this computation from four spacecraft data the time series need to be interpolated onto a common timeline.

4.3. APPLICATIONS

The curlometer method has been successful in view of its wide applicability and robustness and has been successfully applied in many different regions of the Earth's magnetosphere, such as the magnetopause [e.g., *Dunlop et al.*, 2002; *Dunlop and Balogh*, 2005; *Haaland et al.*, 2004a; *Panov et al.*, 2008]; the magnetotail current sheet [e.g., *Runov et al.*, 2006; *Nakamura et al.*, 2008; *Narita et al.*, 2013]; the ring current and inner magnetosphere [e.g., *Vallat et al.*, 2005; *Shen et al.*, 2014, *Yang et al.*, 2016]; field-aligned currents (FAC) [e.g., *Forsyth et al.*, 2008; *Shi, et al.*, 2010], and other transient signatures [e.g., *Roux, et al.*, 2015; *Xiao et al.*, 2004; *Shen, et al.*, 2008], as well as to structures in the solar wind [e.g., *Eastwood et al.*, 2002]. See also Chapters 8 and 11 in this volume.

Many papers have used this technique with Cluster and other multispacecraft combinations. Table 4.1 gives a summary of the typical current density values found previously in the Earth's magnetosphere and solar wind. As noted above, the values are based on the average current density found for typical large-scale structure and as such are likely to miss peaks of small-scale structure within current layers.

4.3.1. The Magnetopause Current Sheet

The magnetopause (MP) is a current sheet forming the boundary between the planetary magnetic field on one side and the shocked solar wind on the other side. It is thus a key region for transfer of mass, momentum, and energy from the solar wind into the magnetosphere. The location of the magnetopause is largely dictated by the balance between the solar wind (primarily dynamic) pressure on one side and the (primarily magnetic) pressure inside the magnetosphere. Due to pressure variations in the solar wind on all timescales, the magnetopause is constantly moving back and forth [e.g., *Sibeck et al.*, 1991]. Typical thicknesses for the Earth's magnetopause vary from a few hundred kilometers (a few ion gyro radii) to several thousands of kilometers (*Berchem and Russell*, 1982; *Paschmann et al.*, 2005; *Panov et al.*, 2008). Typical current densities are of the order 10–100 nAm^{-2}, but small-scale current layers within the magnetopause can have even higher current densities as shown later in this chapter.

An early direct demonstration of the Chapman-Ferraro current (see Fig. 4.3, left, after *Dunlop and Balogh*, [2005]) used a combination of the curlometer and discontinuity analysis (see *Haaland et al.* [2004b] for a review) to probe the magnetopause thickness and orientation for a number of events showing persistence of the MP current; the presence of boundary ripples tied to the current density orientation, and broad scaling of the current density with thickness. The MP was shown to be as thin as 200 km for some crossings (see Fig. 4.3, right, after *Haaland et al.* [2004a]).

4.3.1.1. Example Curlometer Results From the Flank Magnetopause

Figure 4.4 shows Cluster magnetic field measurements and the derived current density and quality criteria from a dusk flank magnetopause (position [–6 16 –7 Re GSE]) crossing on 4 November 2003, about 19:05 UT. The four Cluster spacecraft were flying in a tetrahedron-like formation with a separation distance of around 200 km. At around 19:05 UT, a sharp rotation in the magnetic field was observed, indicating a crossing of the magnetopause current layer from the magnetosphere to the magnetosheath. The actual crossing only took about 17 seconds. Results from plasma moments and triangulation suggested a magnetopause normal velocity of around 35 km/sec, and thus a magnetopause thickness of the order of 600 km.

Structures inside the magnetopause current layer are difficult to resolve with Cluster's 4-sec spin resolution plasma data, but the current (panel e) derived from the high-resolution magnetic field measurements reveal a

Table 4.1 Typical Current Density Values

Feature/Region	Typical values for **J**
Magnetopause currents	~10 nA m^{-2} [*Dunlop and Eastwood*, 2008], up to several 100 nA m^{-2}. See, e.g., *Panov et al.* [2008], Fig. 3.2.
Currents in flux transfer events	~1 nA m^{-2} [*Dunlop and Eastwood*, 2008], up to 10 nA m^{-2} [*Pu et al.*, 2005].
Current at the cusp boundaries	~20 nA m^{-2} [*Dunlop et al.*, 2002]. See also Chapter 21 of this volume.
Field aligned currents (FAC) near polar cap	~2 μA m^{-2} at 500 km altitude and ~20 nA m^{-2} at 2.5 R_E altitude [*Dunlop et al.*, 2015a]; See also Chapter 17 of this volume.
Magnetotail current sheet	Up to ~30 nA m^{-2} [*Runov et al.*, 2006].
Currents in the plasma sheet boundary layer	~10 nA m^{-2} [*Nakamura et al.*, 2004].
Ring current	9–27 nA m^{-2} at 4–4.5 R_E [*Zhang et al.*, 2011]. See also Chapter 8 of this volume.
Solar wind current sheet	~10 nA m^{-2} [*Eastwood et al.*, 2002].

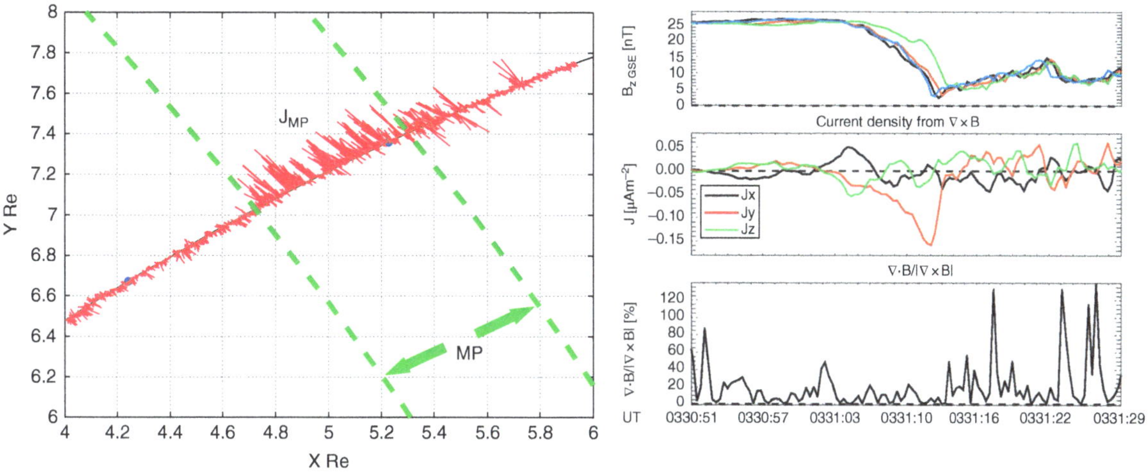

Figure 4.3 (Left) Scan of the curlometer results from the traversal on 26 January 2001, plotted as projections of current orientations as Cluster moves outbound through repeated in-out motions of the MP and into the magnetosheath. (Right) Plot of the current density for a thin magnetopause (2 March 2002) calculated as extending over 200 km, with estimated current densities reaching 150 nAm^{-2}. Here: $\nabla.\mathbf{B} = \text{div}(\mathbf{B})$ and $\nabla \text{x} \mathbf{B}$ = **curl**(**B**).

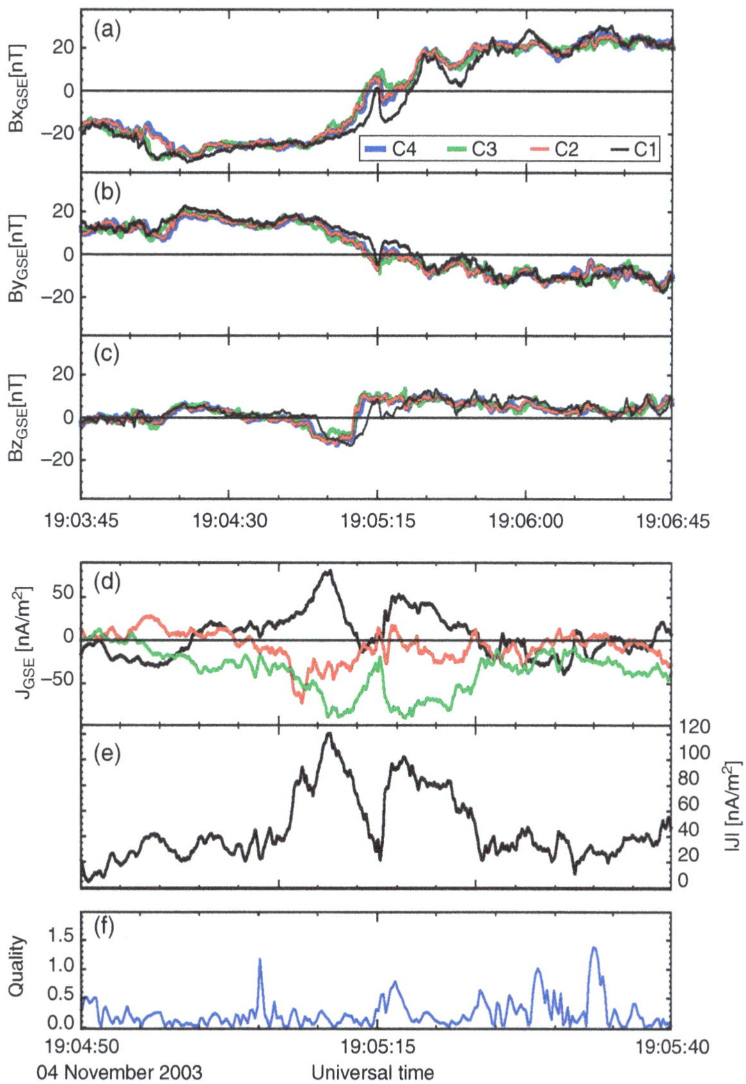

Figure 4.4 Magnetopause current sheet profile during a dusk magnetopause crossing on 4 November 2003. Panels (a), (b), and (c) show high-resolution (at 22.4 vectors/sec) GSE components of the magnetic field; standard Cluster color coding has been used and is indicated in panel (a). Panel (d) shows the individual components of the current (red, X; black, Y, and green, Z). Panel (e) shows the calculated current density, which exhibits a pronounced two-peak structure in this case. Panel (f) shows the quality criteria Q = |div(**B**)|/|**curl**(**B**)| discussed in section 4.2.1. After *Haaland et al.* [2014].

layered structure of the current sheet. The magnetospheric side has significant Xgse and Ygse components (black and red lines in panel d), whereas the current on the magnetosheath side has a much stronger Zgse component (green line in panel e). We interpret this as the signature of two adjacent current sheets, with different directions, where each current sheet is only a few ion inertial lengths thick.

Layered current sheets are in no way unique. Similar features were already seen in the first ever reported curlometer results by *Dunlop et al.* [2002], where one of the investigated events from the cusp region showed two distinct peaks in the current density. Layered magnetopause current sheet structures are also often seen in typical kinetic models of the tangential discontinuity magnetopause [e.g., *Roth et al.*, 1996].

4.3.1.2. Flank Magnetopause Characteristics

One advantage of the curlometer technique is its robustness and ease of use. *Haaland and Gjerloev* [2013] and *Haaland et al.* [2014] calculated characteristic flank magnetopause properties from more than 1700 Cluster magnetopause traversals during the years 2001–2010. For around 500 crossings, the curlometer technique could be applied to determine peak and average current densities (the other crossings were either partial crossings or had spacecraft separation distances too large for reliable curlometer estimates). Results are shown in Figure 4.5. An interesting result of that study was the persistent dawn-dusk asymmetry observed. *Haaland et al.* [2014] noted that the dusk magnetopause was found to be persistently thinner and with a higher current density than its dawn counterpart, and that, conversely, the typical dawn magnetopause was found to be thicker and with a lower current density. The total current carried is similar for the two flanks. The asymmetry remained if thicknesses were normalized to the local gyro radius. *Haaland et al.* [2014] attributed these results to asymmetries in the magnetosheath and bow-shock geometry, and/or asymmetries in the ring current. These asymmetries would cause a larger magnetic shear and thus higher current density at the dusk-side magnetopause.

4.3.2. Small-Scale Current Structures: MMS Measurements

The small-scale size of the MMS configuration has allowed the detailed structure of many small-scale currents to be revealed for the first time. Figure 4.6 shows an MMS magnetopause crossing (top) compared with a crossing seen by Cluster for similar conditions (studied previously by *Panov et al.* [2006]). For thin MP layers, the scale of the Cluster configuration was much greater than the current layer and as such the estimates of the current densities can be underestimated (often by up to a factor of 2 or 3). The thicknesses of the overall current layers for the MP events in Figure 4.6 are similar and are small compared to the Cluster separations, but large compared to those of MMS; so it is expected that the large-scale current measured at Cluster is underestimated and indeed this shows lower values than those for MMS.

In addition, the sharp spikes in current density seen in the MMS data, associated with smaller scale structures within the current layer, are not seen at all by Cluster. The error estimates from Q are similar in this comparison, but typically are smaller for MMS. In contrast, the errors due to instrumentation are insignificant compared to Q at Cluster since the separation scale is large so that ΔB is much larger than for MMS, as discussed in section 4.2.

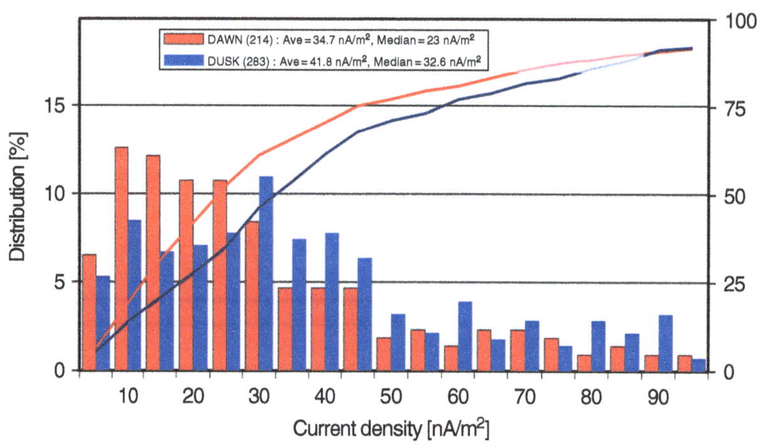

Figure 4.5 (Left) Distribution of magnetopause current densities based on the curlometer results. Each bin in the histograms are 5 nA/m² wide. Lines show cumulated values, and refer to the secondary vertical axis at right. Dawn crossings are characterized by lower current densities than their dusk counterparts. After *Haaland et al.* [2014].

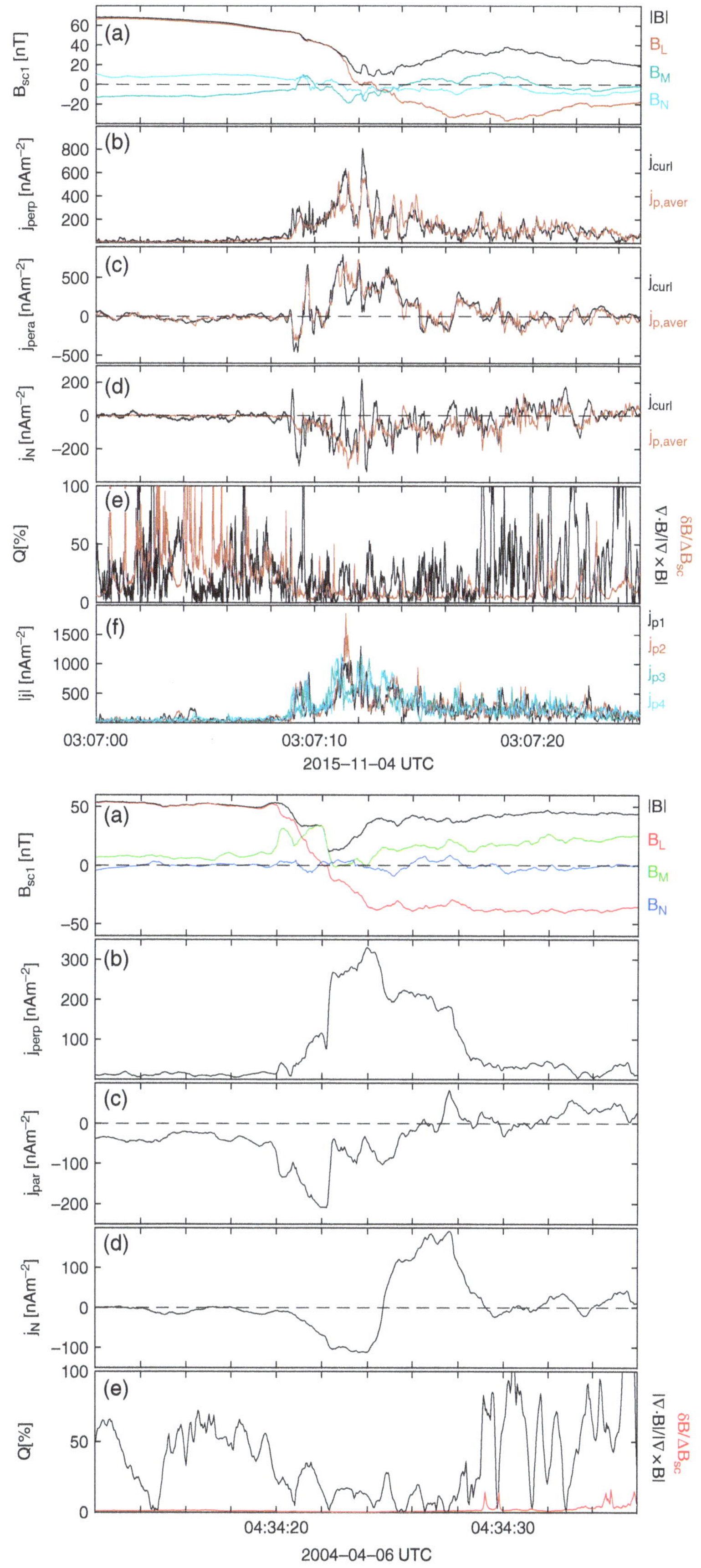

Figure 4.6 Two MP crossings seen by MMS and Cluster for similar thicknesses of ~ 350 km. (top) MMS data showing: (a) magnetic field in boundary normal coordinates; (b) to (d) the total current density from the curlometer and the mean value of the current density from the four spacecraft ion moments (after equation [4.1]) for the perpendicular, parallel, and boundary normal components; (e) comparison of the quality parameter Q and a crude (upper) estimate of the limit on measurement errors from $\delta B/\Delta B$; (f) the individual total current estimated at each spacecraft from the plasma moments. (Bottom) Cluster data showing (a) to (e) similar parameters to the plots for MMS.
Note: In panels 'e' (top and bottom) $Q = |\text{div}(\mathbf{B})|/|\mathbf{curl}(\mathbf{B})|$

It can also be seen that the individual currents at each spacecraft from the plasma moments (panel (f) for the MMS event) show significant differences, while the mean value over all spacecraft is quite consistent with the curlometer result for all current components (here we have shown the field perpendicular (Chapman-Ferraro current) and parallel (field-aligned current) components and the component normal to the MP, separately). This suggests that there is indeed small-scale structure within the MP current layer and that the dominant current carriers are measured by the plasma moments. The most significant differences between the plasma currents and the curlometer arise in the normal component.

4.3.3. The Terrestrial Ring Current

Another region where the curlometer has been applied is the terrestrial ring current. The ring current (RC) is variable in strength, extends significantly in latitude (–30°–30°) and extends radially from about 2–7 R_E (see Chapter 7 for more details of the RC and other studies of this region). Cluster, for example, generally crosses the ring current every perigee pass. During the earlier phase of the mission, the polar orbit passed normally through the ring current, as was first reported by *Vallat et al.* [2005]. Careful selection of high-quality passes allows a full azimuth scan of the ring current density at all local times (*Zhang et al.* [2011], shown on the left-hand side of Fig. 4.7) and for a limited radial extent (~4–4.5 R_E). By checking the stability of the current density for each pass, the orientation of the Cluster configuration typically allows the azimuthal (ring plane) component, J_ϕ, to be estimated accurately. In order to suppress the effect of nonlinear spatial gradients in the Earth's internal field in this region, the IGRF should be subtracted from the measured data so that estimates of the current density are applied to the field residuals. This requirement has been discussed by *Shen et al.* [2014], who also computed the magnetic field line curvature (using the MRA method) for a number of storm-time events, which also shows the strengthening of the RC with storm activity is reflected in the field-line curvature. MMS also covers the ring current and comparisons may be made with Cluster measurements, but on MMS scales.

A further study has been carried out by *Yang et al.* [2016] using coverage achieved by the three magnetospheric THEMIS spacecraft, while in a close configuration (see left-hand side of Fig. 4.2). The coverage is summarized on the right-hand side of Figure 4.7, where radial coverage is achieved from about 4 up to 12 R_E. The azimuthal coverage is unfortunately limited on the dawn side so that LT trends are not well resolved, but the results show a clear radial profile for the recovery phase of each storm period and resolve the Eastward reverse current on the inner edge of the RC. The eastward to westward ring current boundary is found to be at L = 5, on average, but

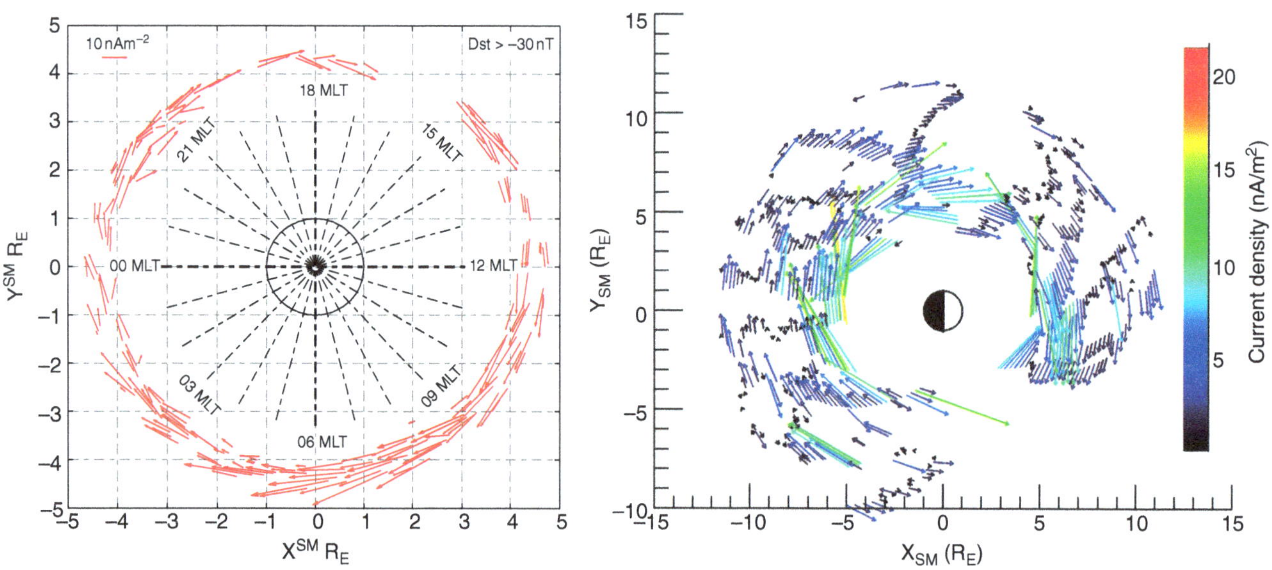

Figure 4.7 (Left) Full azimuth scan (magnetic local time, MLT) of ring current (RC) passes, plotted in solar magnetic (SM) coordinates (dipole aligned), and between –30° to 30° latitude [*Zhang et al.*, 2011]. The length and direction of the vectors represent 10 min averages of the current density obtained from the curlometer. The measurements represent nonstorm (Dst>–30 nT) values of the RC. The RC strength is seen to increase with MLT on the dawn side (03–12 MLT) and is a little suppressed on the dusk side. (Right) THEMIS coverage of the RC for current densities projected into the ring plane and for all storm activities during the recovery phase (i.e., outside the main phase of the storm).

is sensitive to storm activity. There is also a bias to strong storm events near midnight LT.

4.3.4. The Cross-tail Current Sheet

The magnetotail is another region where the curlometer technique has been successfully applied and is reviewed in Chapter 10. Layered current sheet structures similar to those found at the magnetopause have also been observed in the magnetotail. For example, *Runov et al.* [2006] presented a survey of 30 magnetotail current sheet profiles from the first season of Cluster-tail traversals in 2001.

Their study, which was based on the curlometer technique, revealed three distinct tail-current sheet profiles, illustrated in Figure 4.8. Type I current sheet profiles (top panel in Fig 4.8) resemble classic Harris-type current sheet with a symmetric current density in the middle of the neutral sheet (i.e., where the B_Max component of the B-field is zero). Type II current sheets (middle panel) are bifurcated, with two distinct current peaks, one on each side of the neutral sheet and a weaker current in the middle. Finally, type III current sheets are single-peak current sheets, but with the current peak shifted away from the neutral sheet. *Runov et al.* [2006] inferred that layered current sheets were the rule rather than the exception for the magnetotail current sheet. They also noted that type I current sheets were typically narrower (~2000 km) and with a higher peak current density than type II current sheets (~4000 km thickness). Type I current sheets were also found to be more short-lived and transient, possibly because they are more susceptible to various plasma instabilities than the thicker, bifurcated type II current sheets.

4.3.5. Field-Aligned Currents

The curlometer has been applied to a wide number of situations involving FACs, which are covered in detail in Section IV of this volume. As mentioned in section 4.2.2, the technique can be applied to both Cluster sampling at a few to several R_E distances from the Earth, and to low-orbit (400 to several hundred kilometer altitudes) as recently sampled by Swarm (*Dunlop et al.*, 2015b). The technique was applied to a survey of events in the magnetotail [*Shi et al.*, 2010] and to a particular event in the plasma sheet boundary layer [*Cheng et al.*, 2016]. For the latter, the particle and curlometer estimates of current density were shown to agree and be dominated by the

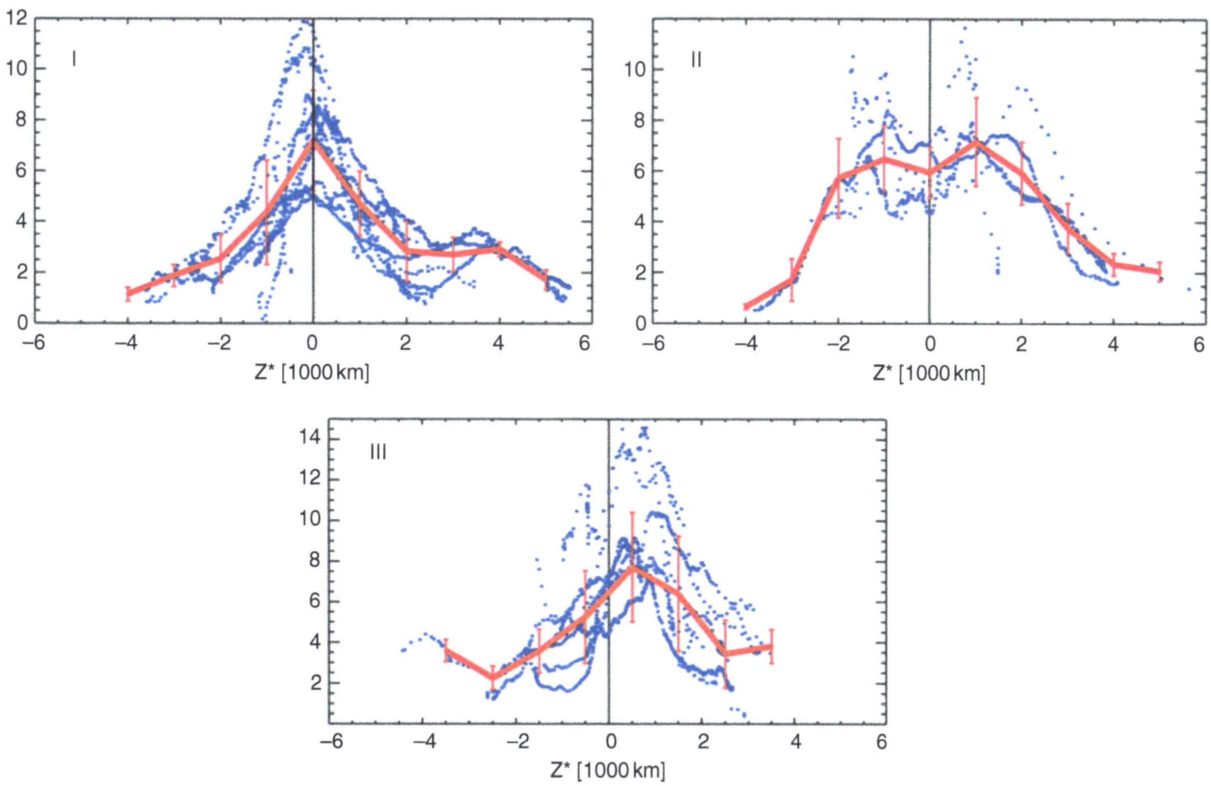

Figure 4.8 Magnetotail current density in μA/m² (vertical axes) versus vertical distance from neutral sheet (horizontal axes) individual crossings (blue dots) and ensemble averages (red lines) for three distinctly different current sheet types. From *Runov et al.* [2006].

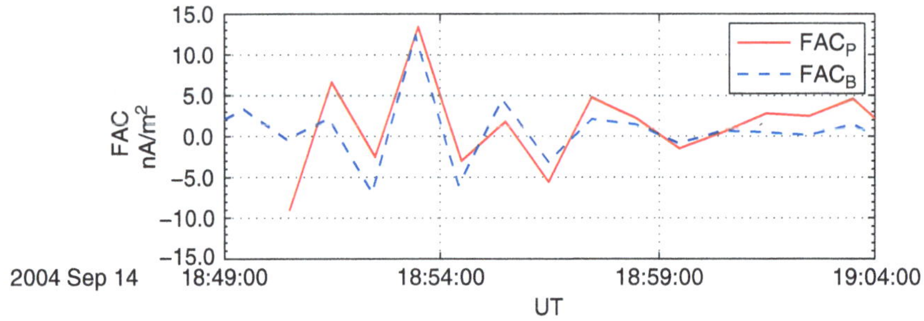

Figure 4.9 The densities of FACs calculated by C3 ions and electron data (FAC_P) and the "curlometer" with the magnetic field data (FAC_B).

electron carriers (see Fig. 4.9). The presence of FACs associated with bursty flows in the magnetotail has also been shown by *Forsyth et al.* [2008].

It is worth mentioning that the curlometer can also measure the currents associated with flux transfer events (FTEs, reconnected flux tubes). These currents are often field aligned in the core of the flux tube, reflecting the force-free nature of the structure. An early demonstration of currents in a large-scale FTE seen by Cluster was recently discussed by *Roux et al.* [2015].

4.4. SUMMARY

Multispacecraft determination of currents, first from Cluster, later from the SWARM and MMS missions, has provided a wealth of new information about current structures in the magnetosphere, and the curlometer has proven to be a reliable and robust tool to determine 3D currents over a wide parameter range in various regions of geospace.

For small-scale current structures, spacecraft separation and configuration, as well as the ability to filter out magnetic field contributions from other sources will necessarily constrain the applicability of the curlometer method [e.g., *Forsyth et al.*, 2011]. In these circumstances, direct current determination from particle moments may provide comparative estimates, but the measured distributions require carefully checking to ensure that the calculated moments contain all particles contributing to currents. As noted, such comparisons have been made using Cluster [e.g., the cited *Henderson et al.*, 2008: *Petrukovich et al.*, 2015, references], but for Cluster this approach has been limited by the availability (ion moments not available from all four spacecraft) and time resolution (Cluster provides particle moments at 4-sec spin resolution) of the plasma measurements. This is a key combination where existing missions like MMS and Swarm (but also proposed missions like THOR and Alfvén) are expected to have plasma instrumentation with the capability for high time-resolution measurements, to provide better estimates and new insight on small-scale current structures. Initial results from MMS [e.g., *Eastwood et al.*, 2016] are very promising.

ACKNOWLEDGMENTS

All Cluster data, including software implementation of the curlometer, can be obtained from the ESA Cluster Science Archive (http://www.cosmos.esa.int/web/csa/access). M.W. Dunlop is partly supported by an STFC in-house research grant and is supported by the NSFC grant 41574155. S. Haaland was supported by the Norwegian Research Council under grant BCSS 223252. J.K. Shi is supported by NSFC with No. 41674145. This work is partly supported through the ISSI Team 'from Cluster to MMS'. We thank the International Space Science Institute in Bern, Switzerland, its staff and directors. We also thank the MMS Science Data Center at LASP, University of Colorado, for provision of MMS data.

REFERENCES

Berchem, J., and C. T. Russell (1982), The thickness of the magnetopause current layer: ISEE 1 and 2 observations, *J. Geophys. Res.*, *87*(A4), 2108–2114; doi:10.1029/JA087iA04p02108.

Burch, J. L., T. E. Moore, R. B. Torbert, and B. L. Giles (2016), Magnetospheric multiscale overview and science objectives, *Space Sci. Rev.*, *5*, 1–17; doi:10.1007/s11214-015-0164-9.

Chanteur, G. (1998), Spatial interpolation for four spacecraft: Theory, in *Analysis Methods for Multispacecraft Data*, ISSI Science Report Series, SR-001, 395–418, Kluwer Academic Pub.

Cheng, Z. W., J. C. Zhang, J. K. Shi, L. M. Kistler, M. Dunlop, I. Dandouras, and A. Fazakerley (2016), The particle carriers of field-aligned currents in the Earth's magnetotail during a substorm, *J. Geophys. Res. Space Physics*, *121*, 3058–3068; doi:10.1002/2015JA022071.

De Keyser, J., F. Darrouzet, M. W. Dunlop, and P. M. E. Decreau (2007), Least-squares gradient calculation from multi-point observations of scalar and vector fields: Methodology and applications with Cluster in the plasmasphere, *Ann. Geo.*, *25*, 971–987; doi: 10.5194/angeo-25-971-2007.

Dunlop, M. W., and A. Balogh (2005), Magnetopause current as seen by Cluster, *Ann. Geophys.*, *23*(3), 901–907; doi:10.5194/angeo-23-901-2005.

Dunlop, M. W., and J. P. Eastwood (2008), The Curlometer and other gradient based methods, in *Multi-Spacecraft Analysis Methods Revisited*, edited by G. Paschmann and P. W. Daly,

ISSI ISSI Science Report Series, SR-008, 17–21, Kluwer Academic Pub.

Dunlop, M. W., D. J. Southwood, K.-H. Glassmeier, and F. M. Neubauer (1988), Analysis of multipoint magnetometer data, *Adv. Space Res.*, *8*, 273–277; doi:10.1016/0273-1177(88)90141-X.

Dunlop, M. W., et al. (2002), Four-point Cluster application of magnetic field analysis tools: The Curlometer, *J. Geophys. Res.*, *107*(A11), 1384; doi:10.1029/2001JA005088.

Dunlop, M.W., J.-Y. Yang, Y.-Y. Yang, C. Xiong, H. Lühr, Y. V. Bogdanova, C. Shen, et al. (2015a), Simultaneous field-aligned currents at Swarm and Cluster satellites, *Geophys. Res. Lett.*, *42*, 3683–3691; doi:10.1002/2015GL063738.

Dunlop, M. W., S. Haaland, C.P. Escoubet, and X.-C. Dong (2016), Commentary on accessing 3-D currents in space: Experiences from Cluster, *J. Geophys. Res.*, *121*, 7881–7886; doi:10.1002/2016JA022668.

Dunlop, M. W., Y.-Y. Yang, J-Y. Yang, H. Luhr, C. Shen, N. Olsen, Q.-H. Zhang, et al. (2015b), Multi-spacecraft current estimates at Swarm, *J. Geophys. Res.*, *120*; doi:10.1002/2015JA021707.

Eastwood, J. P., A. Balogh, M. W. Dunlop, and C. W. Smith (2002), Cluster observations of the heliospheric current sheet and an associated magnetic flux rope and comparisons with ACE, *J. Geophys. Res.*, *107*(A11), 1365; doi:10.1029/2001JA009158.

Eastwood, J. P., et al. (2016), Ion-scale secondary flux ropes generated by magnetopause reconnection as resolved by MMS, *Geophys. Res. Lett.*, *43*; doi:10.1002/2016GL068747.

Escoubet, C. P., M. Fehringer, and M. Goldstein (2001), Introduction: The Cluster mission, *Ann. Geophys.*, *19*, 1197–1200; doi10.5194/angeo-19-1197-2001.

Forsyth, C., et al. (2008), Observed tail current systems associated with bursty bulk flows and auroral streamers during a period of multiple substorms, *Ann. Geophys.*, *26*; doi:10.5194/angeo-26-167-2008.

Forsyth, C., M. Lester, A. Fazakerley, C. Owen, and A. Walsh (2011), On the effect of line current width and relative position on the multi-spacecraft curlometer technique, *Planet. Space Science*, *59*, 598–605; doi:10.1016/j.pss.2009.12.007.

Grimald, S., I. Dandouras, P. Robert, and E. Lucek (2012), Study of the applicability of the curlometer technique with the four Cluster spacecraft in regions close to Earth, *Ann. Geophys.*, *30*, 597–611; doi:10.5194/angeo-30-597-2012.

Haaland, S., and J. Gjerloev (2013), On the relation between asymmetries in the ring current and magnetopause current, *J. Geophys. Res.*, *118*, 7593–7604; doi:10.1002/2013JA019345.

Haaland, S., B. O. Sonnerup, M. Dunlop, E. Georgescu, G. Paschmann, B. Klecker, and A. Vaivads (2004a), Orientation and motion of a discontinuity from Cluster curlometer capability: Minimum variance of current density, *Geophys. Res. Letts.*, *31*, L10804–L10804; doi:10.1029/2004GL020001.

Haaland, S., J. Reistad, P. Tenfjord, J. Gjerloev, L. Maes, J. DeKeyser, R. Maggiolo, C. Anekallu, and N. Dorville (2014), Characteristics of the flank magnetopause: Cluster observations, *J. Geophys. Res.*, *119* (11), 9019–9037; doi:10.1002/2014JA020539.

Haaland, S. E., B. Ö. Sonnerup, M. W. Dunlop, A. Balogh, E. Georgescu, H. Hasegawa, B. Klecker, et al. (2004b), Four-spacecraft determination of magnetopause orientation, motion and thickness: comparison with results from single-spacecraft methods, *Ann. Geophys.*, *22*, 1347–1365.

Harvey, C. C. (1998), Spatial gradients and the volumetric tensor, in *Analysis Methods for Multispacecraft Data*, edited by G. Paschmann and P. W. Daly, ISSI Science Report Series, SR-001, 395–418, Kluwer Academic Pub.

Henderson, P. D., et al. (2008), The relationship between j and Pe in the magnetotail plasma sheet: Cluster observations, *J. Geophys. Res.*, *113*, A07S31; doi:10.1029/2007JA012697; doi:10.1029/2007JA012697.

Kivelson, M., and C. T. Russell (1995), *Introduction to Space Physics*, Cambridge University Press.

Middleton, H. R., and A. Masson (2016), The Curlometer technique: A beginner's guide, *ESA Technical Note*, ESDC-CSA_TN_0001; http://www.cosmos.esa.int/web/csa/multi-spacecraft.

Nakamura, R., W. Baumjohann, M. Fujimoto, Y. Asano, A. Runov, C. Owen, A. Fazakerley, B. Klecker, H. Rème, and E. Lucek (2008), Cluster observations of an ion-scale current sheet in the magnetotail under the presence of a guide field, *J. Geophys. Res.*, 1978–2012, *113*; doi: 10.1029/2007JA012760.

Narita, Y., R. Nakamura, and W. Baumjohann (2013), Cluster as current sheet surveyor in the magnetotail, *Ann. Geophys*, *31*, 1605–1610; doi:10.5194/angeo-31-1605-2013.

Panov, E., J. Büchner, M. Fränz, A. Korth, Y. Khotyaintsev, B. Nikutowski, S. Savin, K.-H. Fornaçon, I. Dandouras, and H. Reme (2006), CLUSTER spacecraft observation of a thin current sheet at the Earth's magnetopause, *Adv. Space Res.*, *37*(7), 1363–1372; doi:10.1029/2006GL026556.

Panov, E. V., J. Buchner, M. Franz, A. Korth, S. P. Savin, H. Reme, and K.-H. Fornacon (2008), High-latitude Earth's magnetopause outside the cusp: Cluster observations, *J. Geophys. Res.*, *113*, A01220; doi:10.1029/2006JA012123.

Paschmann, G., S. Haaland, B. U. Ã. Sonnerup, H. Hasegawa, E. Georgescu, B. Klecker, T. D. Phan, H. Reme, and A. Vaivads (2005), Characteristics of the near-tail dawn magnetopause and boundary layer, *Ann. Geophys.*, *23*, 1481–1497; doi:10.5194/angeo-23-1481-2005.

Petrukovich, A., et al. (2015), Current sheets in the earth magnetotail: Plasma and magnetic field structure with cluster project observations, *Space Sci. Rev.*, *188*, 311–337; doi:10.1007/s11214-014-0126-7.

Ritter, P., and H. Lühr (2013), Determining field-aligned currents with the Swarm constellation mission, *Earth, Planet Space*, *65*, 1285–1294; doi:10.5047/eps.2013.09.006.

Robert, P., M.W. Dunlop, A. Roux, and G. Chanteur (1998), Accuracy of current density determination, in *Analysis Methods for Multispacecraft Data*, edited by G. Paschmann and P. W. Daly, ISSI Science Report Series, SR-001, 395–418, Kluwer Academic Pub.

Roth, M., J. de Keyser, and M. M. Kuznetsova (1996), Vlasov theory of the equilibrium structure of tangential discontinuities in space plasmas, *Space Sci. Rev.*, *76*(3–4), 251–317.

Roux, A., P. Robert, D. Fontaine, O. LeContel, P. Canu, and P. Louarn (2015), What is the nature of magnetosheath FTEs?, *J. Geophys. Res. Space Physices*, *120*(6), 4576–4595; doi:10.1002/2015JA020983.

Runov, A., et al. (2006), Local structure of the magnetotail current sheet: 2001 Cluster observations, *Ann. Geophys.*, *24*, 247–262; doi:10.5194/angeo-24-247-2006.

Russell, C. T., J.G. Luhmann, and R. J. Strangeway (2016). *Space Physics: An Introduction*, Cambridge University Press.

Shen, C., and M. W. Dunlop (2008), Geometric structure analysis of the magnetic field, in *Multi-Spacecraft Analysis Methods Revisited*, edited by G. Paschmann and P.W. Daly, ISSI Science Report Series, SR-008, 27–32, Kluwer Academic Pub.

Shen, C., et al. (2008), Flattened current sheet and its evolution in substorms. *J. Geophys. Res.*, *113*, A07S21; doi:10.1029/2007JA012812.

Shen, C., et al. (2014), Direct calculation of the ring current distribution and magnetic structure seen by Cluster during geomagnetic storms, *J. Geophys. Res.*, *119*, 2458–2465; doi:10.1002/2013JA019460.

Shen, C., J. Rong, M. Dunlop, Y. H. Ma, G. Zeng, and Z. X. Liu (2012a), Spatial gradients from irregular, multiple-point spacecraft configurations, *J. Geophys. Res.*, *117*, A11207; doi:10.1029/2012JA018075.

Shen, C., Z. J. Rong, and M. Dunlop (2012b), Determining the full magnetic field gradient from two spacecraft measurements under special constraints, *J. Geophys. Res.*, *117*, A10217; doi:10.1029/2012JA018063.

Shi, J. K., et al. (2010), South-north asymmetry of field-aligned currents in the magnetotail observed by Cluster, *J. Geophys. Res.*, *115*; doi:10.1029/2009JA014446.

Shi, Q. Q., C. Shen, Z. Y. Pu, M. W. Dunlop, Q.-G. Zong, H. Zhang, C. J. Xiao, Z. X. Liu, and A. Balogh, (2005), Dimensional analysis of observed structures using multipoint magnetic field measurements: Application to Cluster *Geophys. Res. Lett.*, *32*, CiteID L12105; doi:10.1029/2005GL022454.

Shi, Q. Q., C. Shen, M. W. Dunlop, Z. Y. Pu, Q.-G. Zong, Z. X. Liu, E. Lucek, and A. Balogh (2006), Motion of observed structures calculated from multi-point magnetic field measurements: Application to Cluster, *Geophys. Res. Lett.*, *33*, CiteID L08109; doi:10.1029/2005GL025073.

Sibeck, D. G., R. E. Lopez, and E. C. Roelof (1991), Solar wind control of the magnetopause shape, location, and motion, *J. Geophys. Res.*, *96*(A4), 5489–5495; doi:10.1029/90JA02464.

Vallat, C., et al. (2005), First current density measurements in the ring current region using simultaneous multi-spacecraft CLUSTER-FGM data, *Ann. Geophys.*, *23*, 1849–1865; doi:10.5194/angeo-23-1849-2005.

Vogt, J., A. Albert, and O. Marghitu (2009), Analysis of three-spacecraft data using planar reciprocal vectors: methodological framework and spatial gradient estimation, *Ann. Geophys.*, *27*, 3249–3273; doi:10.5194/angeo-27-3249-2009.

Vogt, J., E. Sorbalo, M. He, and A. Blagau (2013), Gradient estimation using configurations of two or three spacecraft, *Ann. Geophys.*, *31*, 1913–1927.

Vogt, J., G. Paschmann, and G. Chanteur (2008), Reciprocal vectors, in *Multi-Spacecraft Analysis Methods Revisited*, edited by G. Paschmann and P. W. Daly, ISSI Science Report Series, SR-008, 27–30, Kluwer Academic Pub.

Xiao, C. J., et al. (2004), Inferring of flux rope orientation with the minimum variance analysis technique, *J. Geophys. Res.*, *109*; doi:10.1029/2004JA010594.

Yang, Y.-Y., C. Shen, M. Dunlop, Z.-J. Rong, X. Li, V. Angelopolous, Z.-Q. Chen, G.-Q. Yan, and Y. Ji (2016), Storm time current distribution in the inner magnetospheric equator: THEMIS observations, *J. Geophys. Res.*, *120*; doi/10.1002/2015JA022145.

Zhang, Q.-H., M. W. Dunlop, M. Lockwood, R. Holme, Y. Kamide, W. Baumjohann, R. -Y. Liu, et al. (2011), The distribution of the ring current: Cluster observations, *Ann. Geo.*, *29*, 1655–1662; doi:10.5194/angeo-29-1655-2011.

5

Inferring Currents from the Zeeman Effect at the Solar Surface

Graham Barnes and K. D. Leka

ABSTRACT

The brightness and morphology of coronal structures and the sharply bound morphology of photospheric magnetic structures indicate that electric currents permeate the solar atmosphere. Quantifying the strength of these currents is, however, difficult. In this introduction to measuring electric currents in the atmosphere of the Sun, we present an overview of the present methods for estimating them near the solar surface, including a brief overview of the steps required to infer a photospheric current density from remotely sensed observations of polarized light. We focus on results from the Zeeman effect at the photosphere, and highlight the limitations of spatial resolution and the challenges of sampling the solar atmosphere at multiple heights, both of particular interest for understanding the true nature of the solar electric currents.

5.1. INTRODUCTION

The solar atmosphere is threaded with magnetic fields and, by extension, electric current systems. Many of the ubiquitous features of the Sun, from the confined character of the magnetic structures to the energy storage and release required for the energetic explosive events, point to the existence of solar electric currents.

However, electric currents in the solar atmosphere are not measured directly: "What we actually measure are electrons on the CCD" [*J. Schou, private communication*]. Slightly less literally, what are measured are photons from the Sun that have been influenced by the magnetized plasma in which they originate. Multiple additional steps are needed just to estimate the magnetic field vector, before any component of the electric current density can be inferred.

In the high-density plasma close to the Sun's surface in the photosphere, the most commonly measured effects are the Zeeman and Hanle effects; the former impacts energy-level splitting and the polarization state in absorption lines and is most relevant for the stronger fields and higher densities of the solar photosphere, while the Hanle depolarization of scattered light is the primary mechanism by which magnetic fields in the solar chromosphere are now being investigated, where the field strength and density are both lower on average than at the photosphere (see details and reviews in *del Toro Iniesta and Ruiz Cobo* [2016]; *Stenflo* [1994]). Readers interested in advances in solar polarimetric observing and analysis techniques are directed to the proceedings of the solar polarization meetings series, published by the Astronomical Society of the Pacific [e.g., *Kuhn et al.*, 2011].

For a typical absorption line sampling the Zeeman effect in the photosphere, the height of formation covers a relatively small range compared to the scale height of the field, so the field is considered to be measured at a single height, and local variations in this height are generally ignored [*Faurobert et al.*, 2009, 2012]. In the low density of the corona, the Sun's outer atmosphere, thermal bremsstrahlung and thermal gyroresonance emission are observed in radio wavelengths, and can be used to determine a line-of-sight integrated value of the magnetic field [*Miyawaki et al.*, 2016; *White and Kundu*, 1997]. The direct analysis of Stokes polarization signals

NorthWest Research Associates, Boulder, Colorado, USA

Electric Currents in Geospace and Beyond, Geophysical Monograph 235, First Edition.
Edited by Andreas Keiling, Octav Marghitu, and Michael Wheatland.

from coronal forbidden lines is also now being used to infer coronal magnetic structures [*Dalmasse et al.*, 2016; *Dove et al.*, 2011; *Judge and Casini*, 2001], although the analysis is challenging due to weak fields, low signal to noise, and again the corona is optically thin [*Dima et al.*, 2016; *Kramar et al.*, 2013, 2016; *Merenda et al.*, 2006; *Schad et al.*, 2016]. The line-of-sight integral for coronal emission in both wavelength regimes means that these observations are generally not useful for determining the current because the lack of information about the spatial variation of the field precludes applying Ampère's law. Still, modeling efforts have demonstrated the possibility of differentiating between current-filled and current-free coronal streamer-type structures when Zeeman and Hanle diagnostics are combined [*Judge et al.*, 2006]. Between the corona and the photosphere, the chromospheric atmosphere can be optically thin or thick depending on the structure being observed and where in the spectral line (wings versus core) the observation is made, providing an expectedly nonplanar surface. This review focuses on the electric current as inferred using the Zeeman effect primarily in the photosphere. Many of the same principles may apply for the solar chromosphere/corona magnetic field as inferred from the Hanle effect, although the geometrical considerations pose additional challenges in that regime.

Figure 5.1 shows photospheric images of a group of sunspots as seen in continuum light and with the line-of-sight component of the magnetic field; three areas of particular solar structure are indicated: quiet Sun (where the field is weak), a sunspot umbra (where the field is strong), and penumbra (where the field is both strong and highly inclined). Figure 5.2 shows an example of the polarization signals in the common Stokes formalism observed by the Hinode Solar Optical Telescope/SpectroPolarimeter (SOT/SP) [*Kosugi et al.*, 2007; *Lites et al.*, 2013; *Tsuneta et al.*, 2008] in these structures. In the quiet Sun, two magnetically sensitive spectral absorption lines are clearly evident in the unpolarized component, while the polarized states are dominated by noise. In the umbra and penumbra, there is a clear signal present in both polarized and unpolarized states.

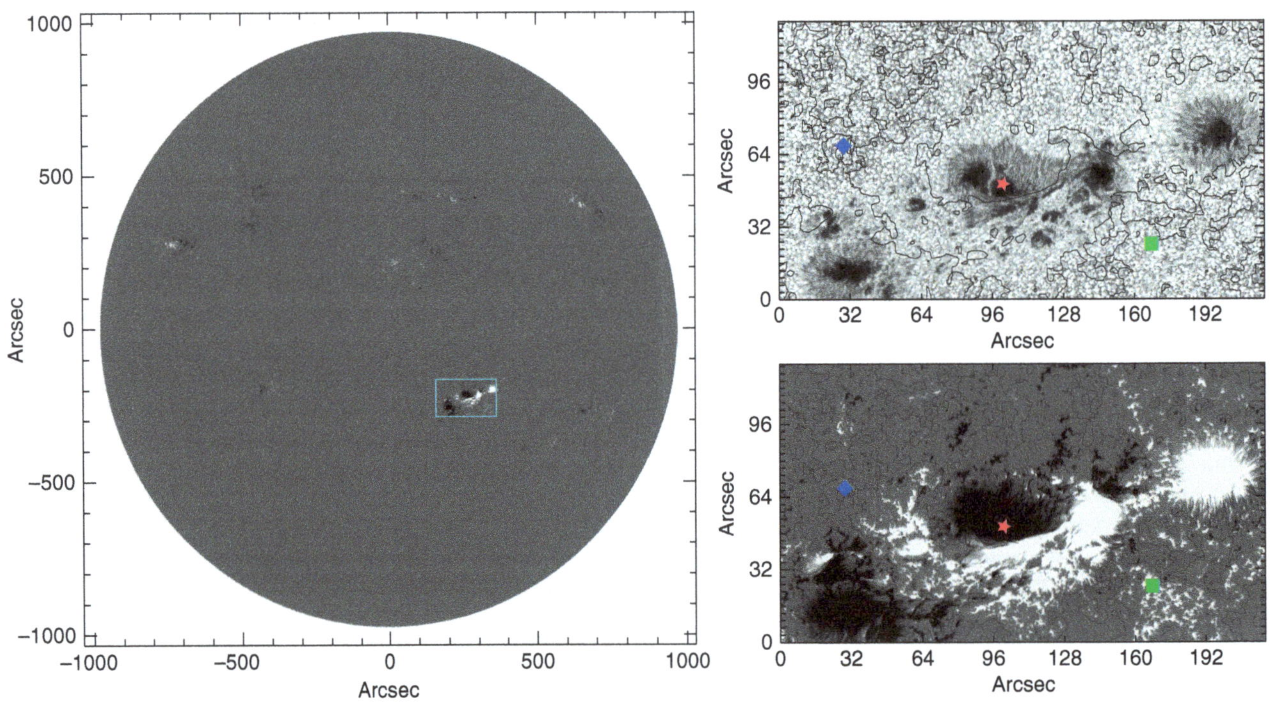

Figure 5.1 The photospheric surface of the Sun as seen on 15 February 2010 at 10:12TAI: (left) The line-of-sight component of the magnetic field, with the field of view of SOT/SP encompassing NOAA Active Region 11158 outlined by the blue box; (right) NOAA AR 11158 as observed by SOT/SP in continuum light (top) and the line-of-sight component of the magnetic field (bottom) inverted using the MERLIN inversion and available online (http://sot.lmsal.com/data/sot/level2d/). A quiet Sun point is indicated with a blue diamond, a plage point is labeled with a green square, and an umbral point is labeled with a red star. Spectra from these locations are shown in Figure 5.2.

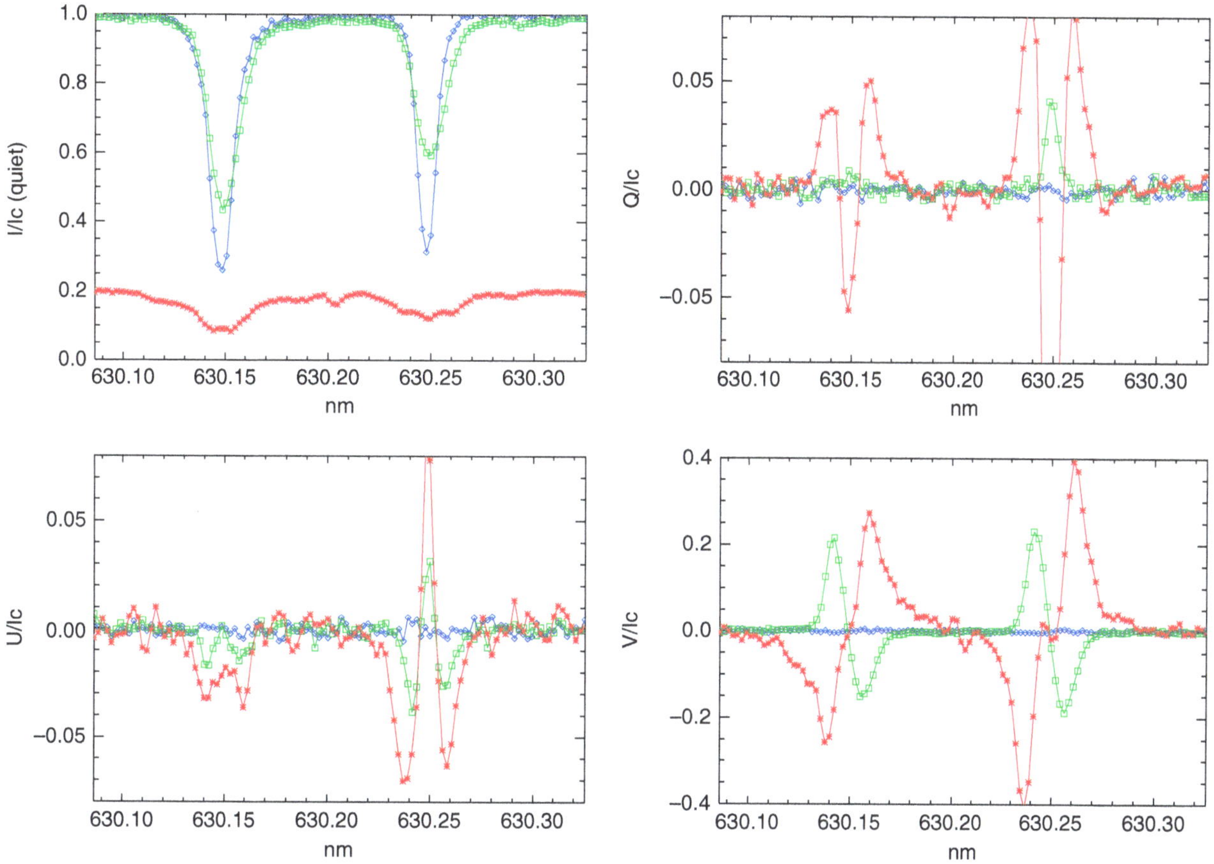

Figure 5.2 Four polarization states as measured by SOT/SP in different solar structures as indicated in Figure 5.1: quiet Sun (blue/diamonds), penumbra (green/squares), and umbra (red/asterisks). (Top, left) The unpolarized state, I; (top right, bottom left) two linear polarization states, *Q* and *U*, which are most sensitive to the component of the field transverse to the line of sight; (bottom, right) the circular polarization, *V*, which is most strongly influenced by the component of the field along the line of sight.

To leading order, a magnetic field can be considered to influence the polarized states through the Zeeman effect according to

$$B_{\|} \propto V, \tag{5.1}$$

$$B_{\perp} \propto \left(Q^2 + U^2\right)^{1/2}, \tag{5.2}$$

$$\phi \approx \tan^{-1}(U/Q), \tag{5.3}$$

where $B_{\|}$ is the component of the field directed along the line of sight, $B_{\perp}$ is the magnitude of the field transverse to the line of sight, and ϕ is the direction of the transverse component of the field, while Q, U, V are the Stokes parameters with Q and U being the linear polarization with orientation differing by 45°, and V being the circular polarization.

With the initial observations of the linear polarization component of magnetically sensitive spectral lines and its interpretation as the transverse component of the solar magnetic field, the existence of electric current systems in the solar atmosphere was inferred [*Rayrole and Semel*, 1970; *Severnyi*, 1965]. Indeed it was quickly recognized that the observed sharp (at 6″) gradients in field strength, especially near the inferred magnetic polarity inversion lines, could be indications of strong current systems. It was also quickly realized that chromospheric fibril structures reflected the nonpotential current-carrying character of the photospheric magnetic field [*Hale*, 1927; *Nakagawa et al.*, 1971]. As spatial resolution, polarimetric sensitivity, and data reduction algorithms improved, including algorithms for implementing coordinate transforms to fully consider the heliographic magnetic components [*Canfield et al.*, 1993; *Gary and Hagyard*, 1990], the understanding of the crucial role that electric current systems play in solar physics increased as well. This understanding spans the role of currents as energy storage and possible trigger mechanisms for solar energetic events, as indications of helicity transfer, and in the organization of solar structures from the smallest penumbral fibrils to the largest coronal loops. Even with the recent advances in observational and interpretive capability, there are many limitations to these measurements; we will highlight here only some of the effects of limited spatial resolution.

5.2. FROM PHOTONS TO CURRENT

There are four main steps needed to infer the (vertical) current density at a given location on the solar surface: (1) Instrumental calibration, (2) inversion, (3) disambiguation, and (4) Ampère's law. Calibration is extremely important, but is instrument-specific and so will not be covered in this review. Common approaches to methods of inversion and disambiguation are outlined below, including the possibility of additional information becoming available from more sophisticated methods that are presently not routinely employed. The various ways to compute the current from Ampère's law are then discussed.

5.2.1. Inversion

Magnetic fields on the Sun are inferred from the field strength and direction needed to reproduce the observed polarization signals. As alluded to above, very rough approximations can relate the two directly. However, in the context of inferring electric currents, these approximations and other approaches that do not rely on good spectral information can be misleading. The magneto-optical rotation of the plane of linear polarization that occurs when the light passes through a magnetized medium can introduce apparent vertical current systems where there are none. Hence, a full inversion of the polarized spectra of magnetically sensitive spectral lines, which accounts for these effects, is required [*del Toro Iniesta*, 2003; *del Toro Iniesta and Ruiz Cobo*, 2016]. The goal of the inversion is to determine the set of parameters describing the field and plasma that result in emergent spectra that best match the observations. The model for the magnetized plasma used in this procedure generally includes simplifying assumptions.

One very common set of assumptions used for producing vector magnetic field maps is the Milne-Eddington atmosphere (ME) [*del Toro Iniesta*, 2003; *Mihalas*, 1978; *Skumanich and Lites*, 1987]): the field strength and direction are constant with depth over the line-forming region, as are the governing thermodynamic parameters, such as temperature, density, velocity, as also manifest in the radiative transfer parameters such as the Voigt-function damping parameter and the line-to-continuum opacity ratios. Only the source function is allowed to vary (linearly) with optical depth. An unmagnetized component of plasma can also be included in this atmospheric model, occupying a fraction of the volume specified by a filling factor determined as part of the inversion. Typically, about 10 parameters are thus needed to characterize the field and the plasma. For forward-model-based inversion methods, the magnetic and thermodynamic behavior of the atmospheric model are incremented by means of some nonlinear least-squares fitting technique to optimize the agreement between the model and the observed Stokes profiles [*del Toro Iniesta and Ruiz Cobo*, 2016]. The resulting maps of the vector magnetic fields on the solar surface have been shown to be a reasonable average of the true magnetic and thermodynamic structures over the ≈ 150 km in height that most photospheric lines are formed [*Westendorp Plaza et al.*, 1998]. Inversions based on the ME approximation can be used for data that are comparatively sparsely sampled spectrally, and are sufficient fast and robust to be routinely applied to large numbers of pixels [e.g., *Borrero et al.*, 2011; *Centeno et al.*, 2014], but contain no information on the variation of the magnetic field along the line of sight.

Without information about the gradient along the line of sight, the full current vector cannot be computed (see section 5.2.3). One way to obtain gradient information is to apply a ME inversion to two spectral lines with different heights of formation. This has been tried by *Bommier* [2014] using the UNNOFIT inversion [*Bommier et al.*, 2007] for the Fe I 630.15 nm and 630.25 nm lines whose height of formation differs by approximately 100 km. The difference in the field from the two inversions in principle can be used to determine the gradient, but when $\nabla \cdot B$ is estimated this way, it gives a systematically nonzero result, so work remains to be done to make this approach reliable for estimating the current.

For instruments with high spectral resolution, more sophisticated inversions that relax some of the assumptions in the ME approximation to allow for gradients along the line of sight are possible. For example, the Stokes Inversion based on Response functions (SIR) [*Ruiz Cobo and del Toro Iniesta*, 1992] includes gradients while assuming local thermodynamic equilibrium (LTE). The Non-LTE Inversion COde using the Lorien Engine (NICOLE) [*Socas Navarro et al.*, 2000; *Socas-Navarro et al.*, 2015] goes a step further and relaxes the assumption of local thermodynamic equilibrium in its multilevel treatment of Zeeman line transfer. This inversion code was developed for application to deep lines that span the photosphere and chromosphere, such as the infrared triplet lines of Ca II, but is well suited for retrieving the gradients for suitable photospheric lines such as the Fe I pair at 630.15 and 630.25 nm used by the SOT/SP. Indeed, an early application of non-LTE codes provided one of the first maps of the current vector in a sunspot [*Socas Navarro*, 2005].

5.2.2. Disambiguation

There is a degeneracy in the Zeeman regime in the polarization states for transverse fields that are oppositely directed (recall that the direction is given by $\phi \approx \tan^{-1}(U/Q)$), so there is an inherent 180° ambiguity in the direction of the transverse field (the Hanle effect and forbidden-line transitions also suffer from different degeneracies). Figure 5.3 schematically illustrates how the choice of the direction of the transverse component of the field affects the horizontal components, and hence the vertical current density. For a single height inversion, some assumption(s) and/or approximation(s) must be made to remove this ambiguity.

There are two main approaches to resolving this inherent ambiguity in the direction of the transverse field. One is to construct a reference field, typically either potential or linear force-free, and pick the selection closest to this reference field. The second is to approximate the vertical derivative needed to compute the divergence of the field, then minimize the value of $|\nabla \cdot \mathbf{B}|$. Again, a potential or linear force-free field is generally used to approximate the vertical derivative. There are many implementations of these methods. An overview including comparison of their performance on synthetic data is given in *Metcalf et al.* [2006], while the effects of noise, spatial resolution, and method assumptions are evaluated in *Leka et al.* [2009].

For inversions that retrieve information about the variation of the field along the line of sight, it is no longer necessary to approximate the vertical derivative from a reference field. Instead, the divergence can be estimated

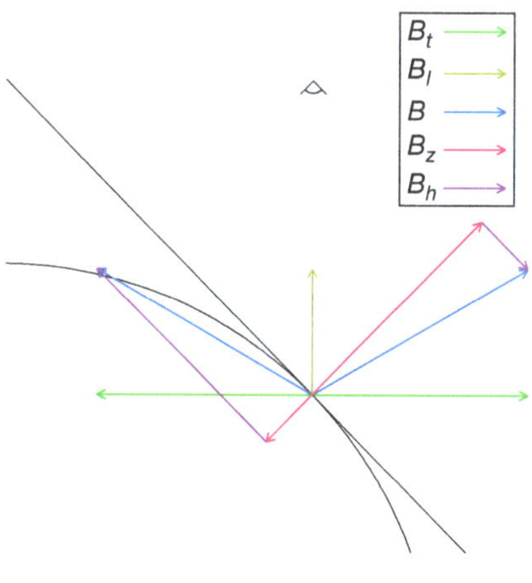

Figure 5.3 Illustration of the ambiguity in vector magnetograms. The direction of the component of the field transverse to the line of sight is ambiguous by 180°, as indicated by the two green arrows. When transforming to heliographic components, the two choices of direction can result in not just a different direction and magnitude for the horizontal field (purple) but even a change in sign of the vertical field (red).

using only inferred values of the field from the inversion [*Crouch and Barnes*, 2008]. Care must be taken in how this is done, since the use of finite differences or similar methods to approximate the horizontal derivatives makes this a nonlocal problem. When applied locally [*Cuperman et al.*, 1992; *Li et al.*, 2007; *Wu and Ai*, 1990] or sequentially [*Boulmezaoud and Amari*, 1999], an incorrect solution can still result. The use of a global optimization method to minimize the magnitude of the divergence is therefore most appropriate [*Crouch et al.*, 2009], although even this can be susceptible to the influence of noise [*Crouch*, 2013]. Nevertheless, the use of gradient information removes the necessity for one assumption in disambiguating the data, and is thus preferable when the necessary information is available.

5.2.3. Ampère's Law

Once the magnetic field vector has been determined at a discrete set of locations, Ampère's law is used to infer the current density. This is most commonly done using the differential form

$$\mathbf{J} = \frac{c}{4\pi} \nabla \times \mathbf{B}. \tag{5.4}$$

Note that for a single height inversion, only the horizontal variation of the field is known, and the vertical derivatives needed to determine the horizontal components of $\mathbf{J}$ cannot be computed. One is left with simply the radial component of the current (shown in spherical coordinates here, but often approximated in planar geometry by ignoring the curvature of the solar surface):

$$J_r = \frac{c}{4\pi r \sin\theta} \left[\frac{\partial}{\partial \theta} \left(B_\phi \sin\theta \right) - \frac{\partial B_\theta}{\partial \phi} \right]. \tag{5.5}$$

It is frequently of interest to know the total current through a solar structure, such as a sunspot. In this case, the integral form of Ampère's law is also sometimes used (e.g., *Georgoulis et al.* [2012]):

$$I = \int_S \mathbf{J} \cdot \mathbf{n} dA = \frac{c}{4\pi} \oint_{\delta S} \mathbf{B} \cdot d\mathbf{r}, \tag{5.6}$$

where the closed curve δS must lie on the surface of the Sun for a typical inversion, so again it is the radial current that is being computed.

There are multiple ways in which the derivatives (integral) needed to compute the vertical current can be approximated. Generally, these either take the form of finite differences or finite elements. However, within each category, different orders of differencing can be used, along with different stencils. When the surface of the Sun

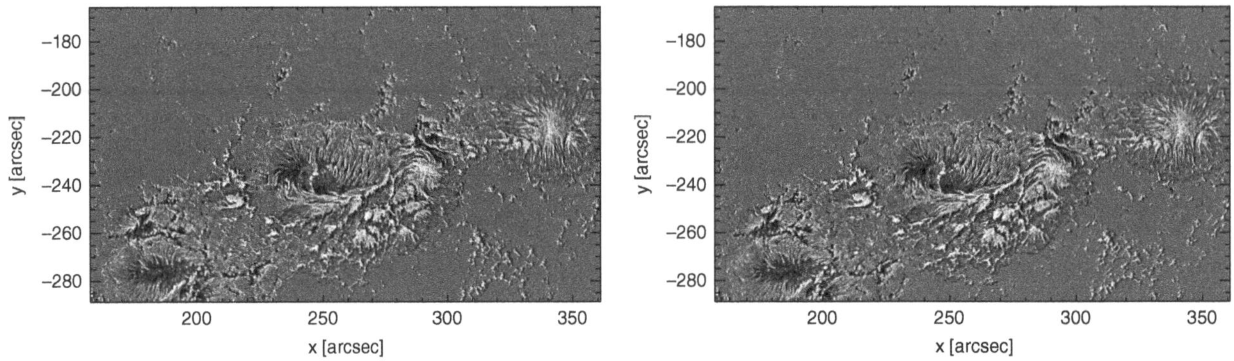

Figure 5.4 The vertical current density from the inversion of SOT/SP data shown in Figure 5.1 for NOAA AR 11158 on 15 February 2011 at 10:11 TAI. The grey scale in both panels is saturated at ± 0.1 A m^{-2}: (left) first-order forward differences; (right) first-order backward differences. Only pixels for which the field is well measured are shown. The results are qualitatively similar.

is approximated as a plane (a substantial approximation, especially in sunspot areas), and the grid is regular in both directions, one common approach is to use a four-point stencil to compute the current on a staggered grid where it is accurate to second order [*Canfield et al.*, 1993].

To illustrate the impact that the implementation has on the result, Figure 5.4 shows the radial current density computed using forward and backward first-order differences from SOT/SP observations of NOAA AR 11158 on 15 February 2011 at 10:11 TAI. Qualitatively, the results are very similar, with, for example, predominantly negative current in the sunspot in the lower left corner of the image. However, when the results of the two differencing schemes are compared point by point (Fig. 5.5), there are substantial differences in the magnitude of the current density at any particular pixel. From this we conclude that it is possible to infer the existence of strong currents at the solar surface, but the magnitude of the current is not generally precisely known.

For a sufficiently spatially resolved field, the forward and backward differences should result in the same value for the current. In the case shown here, the lack of the same value indicates that the grid size is too large for the differences to accurately represent the derivatives. While a higher order differencing scheme will converge faster as the grid size is decreased, in the observational case, the grid size is set by the instrument and cannot be varied, so a higher order scheme is not necessarily more accurate. The presence of unresolved structure becomes apparent in many inversion codes in the value of the dimensionless filling factor, α. In the center of sunspots, where the magnetic field is less variable, inversions generally find $\alpha \approx 1$. For a pixel containing unresolved structure, the best fit to the spectral line consists of a magnetic field that only occupies a fraction of the volume, and α is not unity. When this is the case, there is no physical possibility for

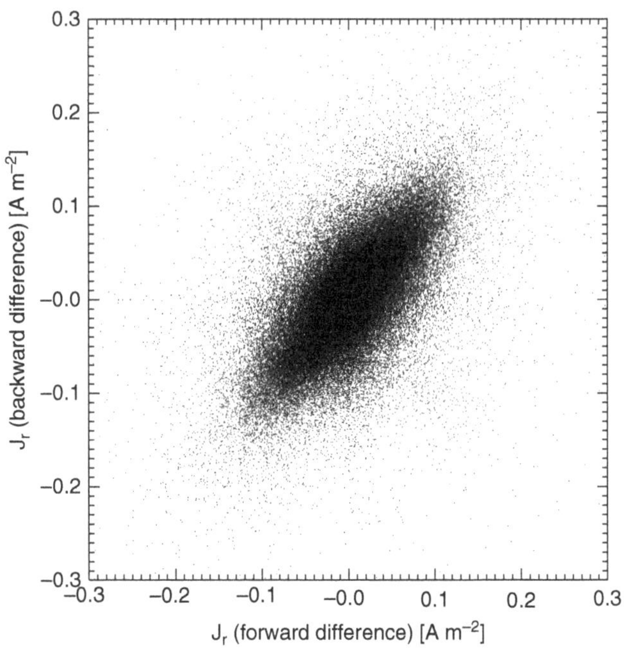

Figure 5.5 The vertical current density from SOT/SP data of NOAA AR 11158 on 15 February 2011 at 10:11 TAI comparing first-order forward differences (horizontal axis) with first-order backward differences (vertical axis). Only pixels for which the field is well measured are used in this scatter plot.

any finite difference (or finite element) scheme to accurately determine the spatial derivative of the field.

An appropriate test is to evaluate the differences in inferred current by comparing the forward and backward differences for all strong field pixels compared with only those with a large filling factor. The correlation coefficient when including all strong field pixels shown in Figure 5.5 is 0.57, but restricting to $\alpha > 0.9$, the correlation coefficient rises to 0.80. The increase in correlation supports the idea that a lack of spatial resolution is impacting the estimated radial current magnitude in these data.

5.3. THE EFFECTS OF RESOLUTION ON ESTIMATES OF THE CURRENT

One of the main impacts on measuring the current is the resolution of the instrument being used, both spectral and spatial. In contrast to the SOT/SP, the Helioseismic and Magnetic Imager (HMI) [*Hoeksema et al.*, 2014; *Scherrer et al.*, 2012; *Schou et al.*, 2012] on board NASA's Solar Dynamics Observatory (SDO) [*Pesnell et al.*, 2012] has lower resolution. It has a spatial resolution of approximately 1.0″ compared with SOT/SP's roughly 0.3″ resolution. For HMI, and in some cases for SP, the observations are critically sampled, resulting in a pixel size that is half the spatial resolution of the telescope. HMI has a spectral resolution of approximately 8 pm and samples at six points across a single Ni I line at 676.8 nm, as compared with a spectral resolution of approximately 2 pm for SP, which samples at 112 points across two Fe I lines at 630.15 nm and 630.25 nm. The reduced resolution of HMI is a trade-off for the improved cadence and a much larger field of view than SOT/SP: HMI typically produces a vector magnetogram of the full solar disk every 12 min as compared to a single active region sized scan observed over roughly an hour for SOT/SP. In many cases, this makes HMI a more suitable instrument for studying solar energetic events like flares, which happen on timescales short compared with a typical SOT/SP scan.

Figure 5.6 shows the Stokes polarization spectra measured by HMI at approximately the same locations as shown in Figure 5.2 for SOT/SP. Qualitatively, the same features are seen in both. In the quiet Sun, there is an obvious absorption line in the unpolarized signal from HMI, but the polarized states are dominated by noise. In the sunspot umbra, in each line in the SOT/SP data and in the HMI data, the V signal shows the same negative-to-positive trend with wavelength, and the Q and U signals show the same patterns between the two data sets. However, the relative magnitudes of the polarization signals are significantly different between the two, as is the level of spectral detail available; SOT/SP data clearly constrain any model atmosphere and field more strongly. For HMI, only a ME inversion is routinely performed, using the Very Fast Inversion of the Stokes Vector (VFISV) [*Borrero et al.*, 2011; *Centeno et al.*, 2014].

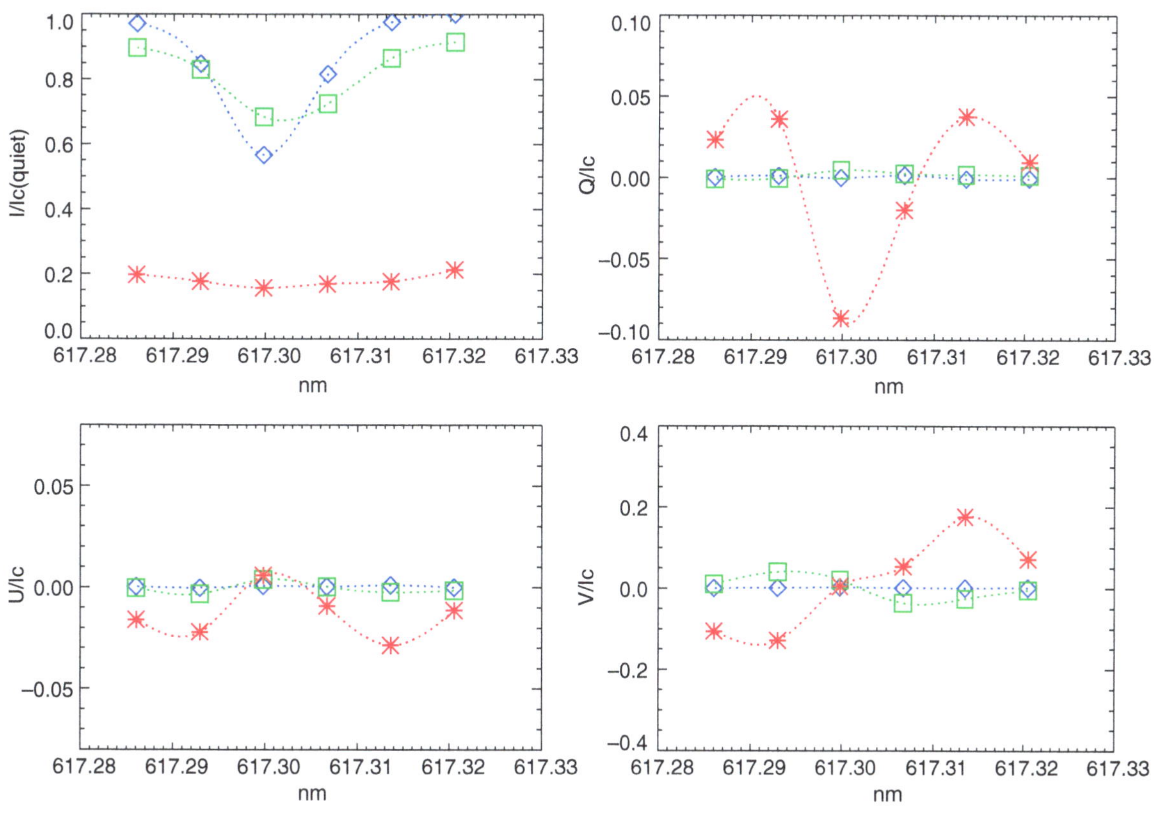

Figure 5.6 Four polarization states as measured by SDO/HMI at approximately the same locations as shown in Figure 5.2. The dotted lines are a cubic spline interpolation of the observed data points intended to guide the eye and not a fit from an inversion.

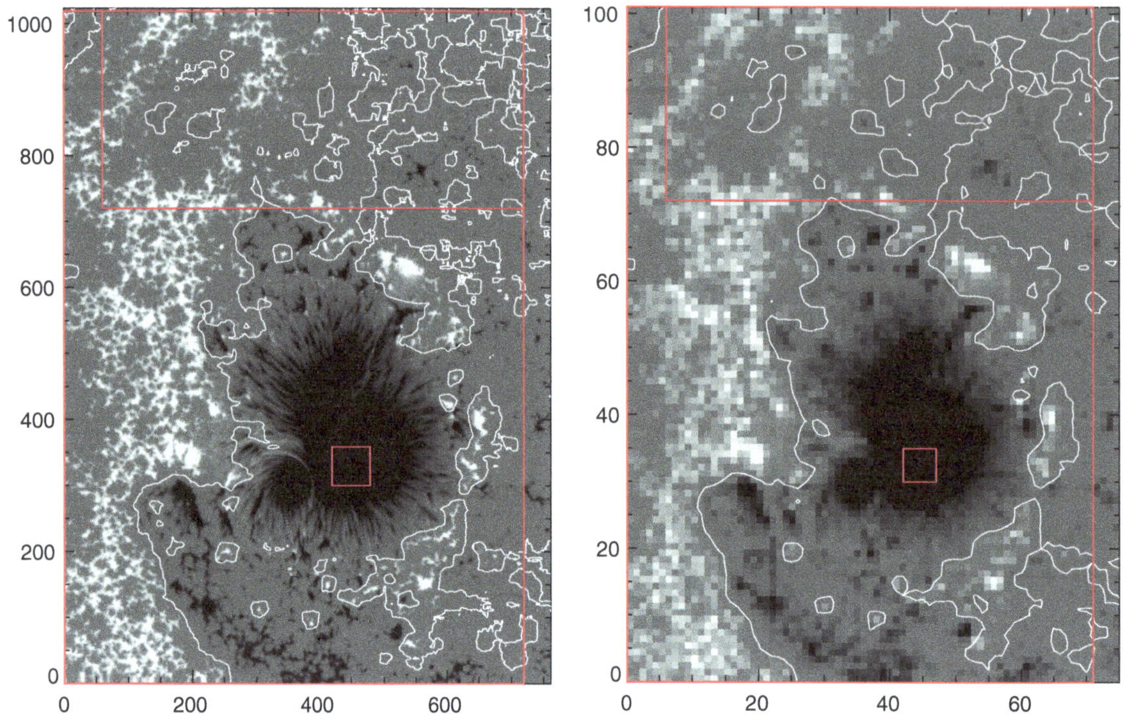

Figure 5.7 The line-of-sight component of the magnetic field, scaled to $\pm 10^3$ G, seen at two spatial resolutions: (left) full resolution SOT/SP data; (right) results from spatially binning the spectra by a factor of 10. Three areas are highlighted: a large full field-of-view box, a smaller area of unresolved plage (near top), and the center of a sunspot umbra [*Leka and Barnes,* 2012].

The data from SOT/SP and HMI are different in spectral and spatial resolution, spectral sampling method, as well as the line that is being observed. To disentangle the effect of spatial resolution from the other effects, consider SOT/SP data that have been spatially degraded without changing the other properties. Figure 5.7 shows the magnetic field seen at two spatial resolutions inferred from SOT/SP for NOAA AR 10953 observed at 18:35 UT on 30 April 2007, inverted using the NCAR/HAO Milne-Eddington inversion [*Lites et al.*, 1993; *Lites and Skumanich*, 1990; *Skumanich and Lites*, 1987]. To construct the lower spatial resolution map, the spectra were spatially binned and then inverted to mimic the effect of a lower spatial resolution instrument [*Leka and Barnes,* 2012]. As expected, the smallest spatial scales are no longer resolved in the binned data.

To see the impact that (only) decreasing spatial resolution has on the current, Figure 5.8 shows the total unsigned current

$$I_{\text{tot}} = \int_S |\mathbf{J} \cdot \mathbf{n}| \, dA, \tag{5.7}$$

in several different solar structures as a function of the pixel size (inverse of resolution). There is a general tendency for the total current to decrease as the spatial resolution decreases, with slightly different trends seen in different solar structures, presumably as a result of the current due to the smallest scale features being lost.

So is the current at the highest resolutions observed by SOT/SP correct? *Parker* [1996] showed that the mean current that would be inferred for unresolved magnetic fibrils bears little resemblance to the actual current. When the net current in a fibril is zero, yet current is present, a nonzero net current will still be inferred if the fibril is not resolved. Conversely, *Leka and Barnes* [2012] considered a potential (current-free) field model for an active region, constructed to include spatial scales smaller than can be resolved by SOT/SP but otherwise include solar-like magnetic field characteristics. Figure 5.9 shows the total current that would be inferred to be present in this potential field as a function of the pixel size. The current at the highest spatial resolution (smallest pixel size) is the lowest, thus the closest to being correct, but then increases, corresponding to the inability of finite differences to accurately recover the horizontal derivatives of the field, especially in the unresolved plage area. For some structures, there is then a decrease in the total current at the lowest resolution, similar to what is seen in the SOT/SP data. Note, however, that the magnitude of the current in the potential field case is substantially smaller than what is seen for the Sun. Thus, it seems likely that real currents are being detected, and indeed the magnitude of the currents

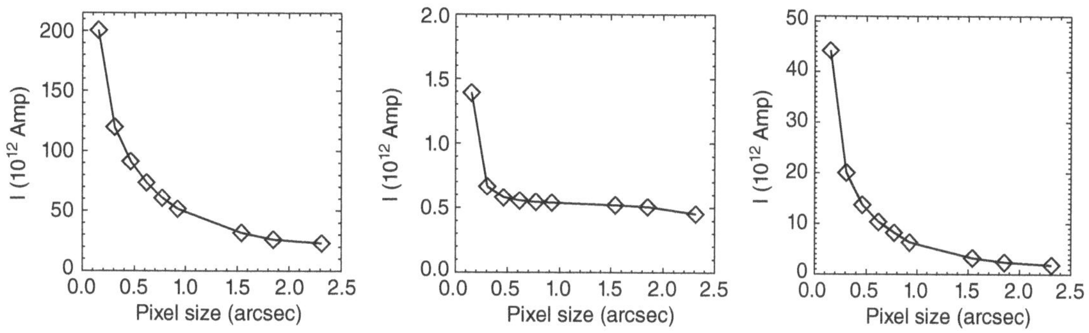

Figure 5.8 The total unsigned current ($\int|\mathbf{J}\cdot\mathbf{n}|$) as a function of pixel size for the three different areas of SOT/SP data, as indicated in Figure 5.7: (left to right) full FOV, umbra, plage.

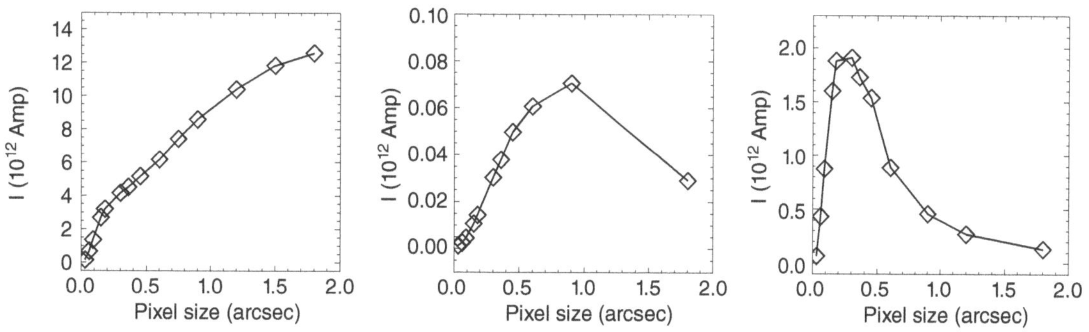

Figure 5.9 The total unsigned current ($\int|\mathbf{J}\cdot\mathbf{n}|$) as a function of pixel size for three different "solar"-structure areas in the constructed potential-field model described in *Leka and Barnes* [2012], areas selected to match the characteristics of observed structures well, such as those in Figure 5.7 and referenced in Figure 5.8: (left to right) full field of view (which contains umbral, penumbral, quiet, and unresolved plage-like structures), just umbra, just plage.

could be substantially larger than what is inferred if there are spatial scales smaller than those presently resolved.

5.4. SUMMARY AND FUTURE DIRECTIONS

Energy stored and then released from the coronal magnetic field, in the form of both major solar energetic events and the continual background brightenings of coronal loops, make it clear that electric currents must be running through much of the solar atmosphere. However, to determine the magnitude of that current at any particular location is very challenging.

Inferring electric currents in the solar atmosphere requires a number of intermediate steps, each of which presents its own challenges. In general, the magnetic field in the solar atmosphere is determined by observing its effect on photons passing through magnetized plasma. From the field estimated at discrete locations, Ampère's law is used to infer the current.

Most often, it is the impact of the Zeeman effect on an absorption line formed in the photosphere that is inverted to determine the magnetic field. For instruments observing this effect, there is a trade-off between measuring the line with high spectral resolution, having high spatial resolution, a large field of view, and high cadence. The instruments presently in use each have perhaps two of these properties.

Knowledge of the full current vector through the solar atmosphere may be the ultimate goal, and steps to achieve it are under way now. As demonstrated, unresolved structure poses problems for estimating even the normal component of the current; hence high-resolution observations (both in the spectral and spatial domains) are need to fully resolve the magnetic field vector and allow finite-difference methods to be employed with better confidence on discretely sampled data. This combination of high spectral and spatial resolution typically limits the cadence and field of view over which the current density can be inferred, but this challenge should be accessible with new facilities such as the Daniel K. Inouye Solar Telescope presently under construction. If the full current density vector is desired, then an inversion method that includes information about gradients in the field along the line of sight is required. Further advances would be made by understanding solar structures to the point of extending all calculations to treat nonplanar surfaces.

Not all topics in solar physics require quantitative absolute knowledge of the solar electric currents.

Well-characterized estimates of the vertical current density, as presently obtained with myriad assumptions applied to data, which do not fully resolve all solar structures, continue to provide insights into general trends and statistical understanding for topics where the strength of the current density is not necessary except in a relative sense.

ACKNOWLEDGMENTS

We thank both referees for clarifying several of the topics presented here. This material is based upon work supported by NASA under award No. NNX16AF83G.

REFERENCES

Bommier, V. (2014), Electromagnetism in a strongly stratified plasma showing an unexpected effect of the Debye shielding, *Comptes Rendus Physique*, *15*, 430–440; doi:10.1016/j.crhy.2014.03.003.

Bommier, V., E. Landi Degl'Innocenti, M. Landolfi, and G. Molodij (2007), UNNOFIT inversion of spectro-polarimetric maps observed with THEMIS, *Astron. Astrophys.*, *464*, 323–339; doi:10.1051/0004-6361:20054576.

Borrero, J. M., S. Tomczyk, M. Kubo, H. Socas-Navarro, J. Schou, S. Couvidat, and R. Bogart (2011), VFISV: Very Fast Inversion of the Stokes Vector for the Helioseismic and Magnetic Imager, *Solar Phys.*, *35*; doi:10.1007/s11207-010-9515-6.

Boulmezaoud, T. Z., and T. Amari (1999), On the removal of the sign ambiguity in the photospheric transverse magnetic field, *Astron. Astrophys.*, *347*, 1005–1008.

Canfield, R. C., J. F. de La Beaujardière, Y. Fan, K. D. Leka, A. N. McClymont, T. Metcalf, D. L. Mickey, J.-P. Wülser, and B. W. Lites (1993), The morphology of flare phenomena, magnetic fields, and electric currents in active regions, I. Introduction and methods, *Astrophys. J.*, *411*, 362–369; doi:10.1086/172836.

Centeno, R., J. Schou, K. Hayashi, A. Norton, J. T. Hoeksema, Y. Liu, K. D. Leka, and G. Barnes (2014), The Helioseismic and Magnetic Imager (HMI) vector magnetic field pipeline: Optimization of the spectral line inversion code, *Solar Phys.*, *289*, 3531–3547; doi:10.1007/s11207-014-0497-7.

Crouch, A. D. (2013), Resolving the azimuthal ambiguity in vector magnetogram data with the divergence-free condition: The effects of noise and limited spatial resolution, *Solar Phys.*, *282*, 107–131; doi:10.1007/ s11207-012-0149-8.

Crouch, A. D., and G. Barnes (2008), Resolving the azimuthal ambiguity in vector magnetogram data with the divergence-free condition: Theoretical examination, *Solar Phys.*, *247*, 25–37; doi:10.1007/s11207-007-9096-1.

Crouch, A. D., G. Barnes, and K. D. Leka (2009), Resolving the azimuthal ambiguity in vector magnetogram data with the divergence-free condition: Application to discrete data, *Solar Phys.*, *260*, 271–287; doi:10.1007/ s11207-009-9454-2.

Cuperman, S., J. Li, and M. Semel (1992), Magnetic shear-based removal of the 180-deg ambiguity in the observed transverse photospheric magnetic field: Feasibility and limitations, *Astron. Astrophys.*, *265*, 296–307.

Dalmasse, K., D. Nychka, S. Gibson, N. Flyer, and Y. Fan (2016), ROAM: A radial-basis-function optimization approximation method for diagnosing the three-dimensional coronal magnetic field, *Front. Astron. Space Sci.*, *3*, 24; doi:10.3389/fspas.2016.00024.

del Toro Iniesta, J. C. (2003), *Introduction to Spectropolarimetry*, Cambridge University Press.

del Toro Iniesta, J. C., and B. Ruiz Cobo (2016), Inversion of the radiative transfer equation for polarized light, *Living Rev. Sol. Phys.*, *13*, 4; doi:10.1007/s41116-016-0005-2.

Dima, G., J. Kuhn, and S. Berdyugina (2016), Infrared dual-line Hanle diagnostic of the coronal vector magnetic field, *Front. Astron. Space Sci.*, *3*, 13; doi:10.3389/fspas.2016.00013.

Dove, J. B., S. E. Gibson, L. A. Rachmeler, S. Tomczyk, and P. Judge (2011), A ring of polarized light: Evidence for twisted coronal magnetism in cavities, *Astrophys. J.*, *731*, L1; doi:10.1088/2041-8205/731/1/L1.

Faurobert, M., C. Aime, C. PÉrini, H. Uitenbroek, C. Grec, J. Arnaud, and G. Ricort (2009), Direct measurement of the formation height difference of the 630nm Fe I solar lines, *Astron. Astrophys.*, *507*, 29–32; doi:10.1051/ 0004-6361/200913330.

Faurobert, M., G. Ricort, and C. Aime (2012), A cross-correlation method for measuring line formation heights in the solar photosphere, *Astron. Astrophys.*, *548*, A80; doi:10.1051/0004-6361/201219640.

Gary, G. A., and M. J. Hagyard (1990), Transformation of vector magnetograms and the problems associated with the effects of perspective and the azimuthal ambiguity, *Solar Phys.*, *126*, 21–36.

Georgoulis, M. K., V. S. Titov, and Z. Mikić (2012), Non-neutralized electric current patterns in solar active regions: Origin of the shear-generating Lorentz force, *Astrophys. J.*, *761*, 61; doi:10.1088/0004-637X/761/1/61.

Hale, G. E. (1927), The fields of force in the atmosphere of the sun, *Nature*, *119*, 708–714; doi:10.1038/119708a0.

Hoeksema, J. T., Y. Liu, K. Hayashi, X. Sun, J. Schou, S. Couvidat, A. Norton, M. Bobra, R. Centeno, K. D. Leka, G. Barnes, and M. Turmon (2014), The Helioseismic and Magnetic Imager (HMI) vector magnetic field pipeline: Overview and performance, *Solar Phys.*, *289*, 3483–3530; doi:10.1007/s11207-014-0516-8.

Judge, P. G., and R. Casini (2001), A synthesis code for forbidden coronal lines, in *Advanced Solar Polarimetry–Theory, Observation, and Instrumentation*, Astronomical Society of the Pacific Conference Series, 236, edited by M. Sigwarth, San Francisco, CA, 503.

Judge, P. G., B. C. Low, and R. Casini (2006), Spectral lines for polarization measurements of the coronal magnetic field, IV. Stokes signals in current-carrying fields, *Astrophys. J.*, *651*, 1229–1237; doi:10.1086/507982.

Kosugi, T., K. Matsuzaki, T. Sakao, T. Shimizu, Y. Sone, S. Tachikawa, T. Hashimoto, et al. (2007), The Hinode (Solar-B) mission: An overview, *Solar Phys.*, *243*, 3–17; doi:10.1007/s11207-007-9014-6.

Kramar, M., B. Inhester, H. Lin, and J. Davila (2013), Vector tomography for the coronal magnetic field. II. Hanle effect measurements, *Astrophys. J.*, *775*, 25; doi:10.1088/0004-637X/775/1/25.

Kramar, M., H. Lin, and S. Tomczyk (2016), Direct observation of solar coronal magnetic fields by vector tomography of the coronal emission line polarizations, *Astrophys. J.*, *819*, L36; doi:10.3847/2041-8205/819/2/L36.

Kuhn, J. R., D. M. Harrington, H. Lin, S. V. Berdyugina, J. Trujillo-Bueno, S. L. Keil, and T. Rimmele, eds. (2011), *Solar Polarization Workshop 6*, Astronomical Society of the Pacific Conference Series, 437, San Francisco, CA.

Leka, K. D., and G. Barnes (2012), Modeling and interpreting the effects of spatial resolution on solar magnetic field maps, *Solar Phys.*, *277*, 89–118; doi:10.1007/s11207-011-9821-7.

Leka, K. D., G. Barnes, A. D. Crouch, T. R. Metcalf, G. A. Gary, J. Jing, and Y. Liu (2009), Resolving the 180° ambiguity in solar vector magnetic fields: Evaluating the effects of noise, spatial resolution, and method assumptions, *Solar Phys.*, *260*, 83–108; doi:10.1007/s11207-009-9440-8.

Li, J., T. Amari, and Y. Fan (2007), Resolution of the 180° ambiguity for inverse horizontal magnetic field configurations, *Astrophys. J.*, *654*, 675–686; doi:10.1086/509062.

Lites, B., D. F. Elmore, P. Seagraves, and A. Skumanich (1993), Stokes profile analysis and vector magnetic fields, VI. Fine scale structure of a sunspot, *Astrophys. J.*, *418*, 928–942.

Lites, B. W., and A. Skumanich (1990), Stokes profile analysis and vector magnetic fields V. The magnetic field structure of large sunspots observed with Stokes II, *Astrophys. J.*, *348*, 747–760.

Lites, B. W., D. L. Akin, G. Card, T. Cruz, D. W. Duncan, C. G. Edwards, D. F. Elmore, et al. (2013), The Hinode spectropolarimeter, *Solar Phys.*, *283*, 579–599; doi:10.1007/s11207-012-0206-3.

Merenda, L., J. Trujillo Bueno, E. Landi Degl'Innocenti, and M. Collados (2006), Determination of the magnetic field vector via the Hanle and Zeeman effects in the He I λ10830 multiplet: Evidence for nearly vertical magnetic fields in a polar crown prominence, *Astrophys. J.*, *642*, 554–561; doi:10.1086/501038.

Metcalf, T. R., K. D. Leka, G. Barnes, B. W. Lites, M. K. Georgoulis, A. A. Pevtsov, G. A. Gary, et al. (2006), An overview of existing algorithms for resolving the 180° ambiguity in vector magnetic fields: Quantitative tests with synthetic data, *Solar Phys.*, *237*, 267–296; doi:10.1007/s11207-006-0170-x.

Mihalas, D. (1978), *Stellar Atmospheres*, W. H. Freeman & Co., New York.

Miyawaki, S., K. iwai, K. Shibasaki, D. Shiota, and S. Nozawa (2016), Coronal magnetic fields derived from simultaneous microwave and EUV observations and comparison with the potential field model, *Astrophys. J.*, *818*, 8; doi:10.3847/0004-637X/818/1/8.

Nakagawa, Y., M. A. Raadu, D. E. Billings, and D. McNamara (1971), On the topology of filaments and chromospheric fibrils near sunspots, *Solar Phys.*, *19*, 72–85; doi:10.1007/BF00148825.

Parker, E. N. (1996), Inferring mean electric currents in unresolved fibril magnetic fields, *Astrophys. J.*, *471*, 485–488; doi:10.1086/177983.

Pesnell, W. D., B. J. Thompson, and P. C. Chamberlin (2012), The Solar Dynamics Observatory (SDO), *Solar Phys.*, *275*, 3–15; doi:10.1007/ s11207-011-9841-3.

Rayrole, J., and M. Semel (1970), Evaluation of the electric current in a sunspot by the study of the observed transverse component of the magnetic field, *Astron. Astrophys.*, *6*, 288.

Ruiz Cobo, B., and J. C. del Toro Iniesta (1992), Inversion of Stokes profiles, *Astrophys. J.*, *398*, 375–385; doi:10.1086/171862.

Schad, T. A., M. J. Penn, H. Lin, and P. G. Judge (2016), Vector magnetic field measurements along a cooled stereo-imaged coronal loop, *Astrophys. J.*, *833*, 5; doi:10.3847/0004-637X/833/1/5.

Scherrer, P. H., J. Schou, R. I. Bush, A. G. Kosovichev, R. S. Bogart, J. T. Hoeksema, Y. Liu, et al. (2012), The Helioseismic and Magnetic Imager (HMI) investigation for the Solar Dynamics Observatory (SDO), *Solar Phys.*, *275*, 207–227; doi:10.1007/s11207-011-9834-2.

Schou, J., J. M. Borrero, A. A. Norton, S. Tomczyk, D. Elmore, and G. L. Card (2012), Polarization calibration of the Helioseismic and Magnetic Imager (HMI) onboard the Solar Dynamics Observatory (SDO), *Solar Phys.*, *275*, 327–355; doi:10.1007/s11207-010-9639-8.

Severnyi, A. B. (1965), The nature of solar magnetic fields (the fine structure of the field), *Sov. Astron.*, *9*, 171.

Skumanich, A., and B. W. Lites (1987), Stokes profile analysis and vector magnetic fields I: Inversion of photospheric lines, *Astrophys. J.*, *322*, 473–482.

Socas Navarro, H. (2005), Are electric currents heating the magnetic chromosphere?, *Astrophys. J.*, *633*, L57–L60; doi:10.1086/498145.

Socas-Navarro, H., J. de la Cruz Rodríguez, A. Asensio Ramos, J. Trujillo Bueno, and B. Ruiz Cobo (2015), An open-source, massively parallel code for non-LTE synthesis and inversion of spectral lines and Zeeman-induced Stokes profiles, *Astron. Astrophys.*, *577*, A7; doi:10.1051/0004-6361/201424860.

Socas Navarro, H., J. Trujillo Bueno, and B. Ruiz Cobo (2000), Non-LTE inversion of Stokes profiles induced by the Zeeman effect, *Astrophys. J.*, *530*, 977–993; doi:10.1086/308414.

Stenflo, J. O. (1994), *Solar Magnetic Fields*, Kluwer Academic Publishers.

Tsuneta, S., K. Ichimoto, Y. Katsukawa, S. Nagata, M. Otsubo, T. Shimizu, Y. Suematsu, et al. (2008), The Solar Optical Telescope for the Hinode mission: An overview, *Solar Phys.*, *249*, 167–196; doi:10.1007/s11207-008-9174-z.

Westendorp Plaza, C., J. C. del Toro Iniesta, B. Ruiz Cobo, V. Martinez Pillet, B. W. Lites, and A. Skumanich (1998), Optical tomography of a sunspot, I. Comparison between two inversion techniques, *Astrophys. J.*, *494*, 453–471.

White, S. M., and M. R. Kundu (1997), Radio observations of gyroresonance emission from coronal magnetic fields, *Solar Phys.*, *174*, 31–52.

Wu, L.-X., and G.-X. Ai (1990), On the determination of the transverse magnetic field and current on photosphere, *Acta Astrophys. Sin.*, *10*, 371–373.

Part II: Ring Currents

6

ENA Imaging of Planetary Ring Currents

P. C. Brandt, S. Y. Hsieh, R. DeMajistre, and D. G. Mitchell

ABSTRACT

Basic techniques of ENA cameras are presented followed by an overview of an inversion technique applied to ENA images obtained by the High Energy Neutral Atom (HENA) camera on board the Imager for Magnetopause-to-Aurora Global Exploration (IMAGE) mission. Observations from HENA are used to demonstrate the storm-time evolution and acceleration of the ring current ions. Plasma pressure retrieved from HENA images are used to illustrate the three-dimensional structure of the partial ring current (PRC) system. ENA observations of the Saturnian magnetosphere by the Ion and Neutral Camera (INCA) on board Cassini are used to contrast its force balance of the Saturnian magnetosphere with that of Earth. The dynamics of Saturn's magnetosphere are reviewed showing Earth-like phenomena such as large-scale injections not unlike those at Earth. A brief review is given of the only ENA observations of the Jovian magnetosphere by INCA during the Cassini flyby showing relatively stable ENA emissions indicating a presence of a Europa torus. A description of the future ENA observations of the Jovian magnetosphere on board the Jupiter Icy Moon Explorer (JUICE) by the European Space Agency (ESA) is given. The paper concludes by contrasting ENA imaging of the three planetary inner magnetospheres and summarizing future directions.

6.1. INTRODUCTION

Energetic Neutral Atom (ENA) imaging has become a powerful technique for remote sensing the interaction between space plasma and neutral gas. After its birth in the 1980s marked by the first dedicated ENA image (Fig. 6.1) of the terrestrial ring current by *Roelof et al.* [1985], ENA imaging has now successfully been used on several missions to image the terrestrial magnetospheric energetic ion distributions of the ring current including Astrid [*Barabash et al.*, 1998; *Brandt et al.*, 2001], IMAGE [*Brandt et al.*, 2002a; *Mitchell et al.*, 2000], Double Star [*Lu et al.*, 2010; *McKenna-Lawlor et al.*, 2005], and TWINS [*McComas et al.*, 2009a]. ENA imaging of the Saturnian magnetosphere has also been carried out with great success from Cassini to image the Saturnian ring current [*Brandt et al.*, 2008a; *Krimigis et al.*, 2007; *Mitchell et al.*, 2009] and its neutral gas distribution around the E-ring [*Dialynas et al.*, 2013a] as well as the global distribution of Titan's exosphere [*Brandt et al.*, 2012]. During its Jupiter flyby, the Cassini mission also obtained distant ENA images of Jupiter's magnetosphere indicating the presence of a neutral Europa torus [*Brandt et al.*, 2017; Mauk *et al.*, 2004; *Mitchell et al.*, 2004] that now have guided the development of the designs of ENA cameras for the European Space Agency's (ESA) Jupiter Icy Moon Explorer (JUICE). ENA imaging has been proven to be an indispensable tool in understanding the global interaction between the heliosphere and the interstellar medium as demonstrated by the Interstellar Boundary Explorer (IBEX) mission [*McComas et al.*, 2009b, 2014], Cassini [*Dialynas et al.*, 2017; *Krimigis et al.*, 2009], and by Mars Express and Venus Express [*Brandt et al.*, 2009; *Galli et al.*, 2013]. Furthermore, ENA detectors on board Mars Express and Venus Express have been used to

Applied Physics Laboratory, Johns Hopkins University, Laurel, Maryland, USA

Electric Currents in Geospace and Beyond, Geophysical Monograph 235, First Edition.
Edited by Andreas Keiling, Octav Marghitu, and Michael Wheatland.

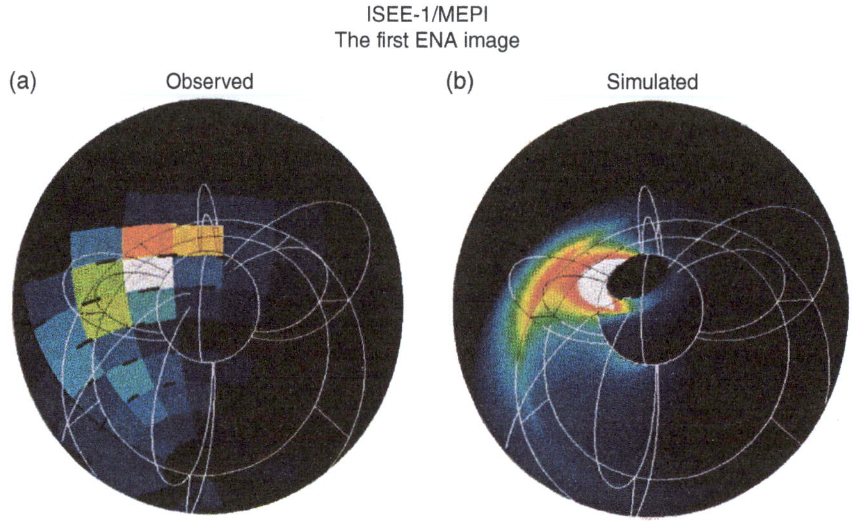

Figure 6.1 (a) The first ENA image reconstructed from measurements by the Medium Energy Particles Instrument on board the ISEE-1 mission. The image shows the hydrogen ENA emissions from the ring current during a geomagnetic storm (Dst=−241 nT) on 29 September 1978. (b) The corresponding simulated image assuming a ring current ion distribution peaking at 2.6 R_E. After *Roelof* [1987].

uncover the solar-wind interaction with Mars and Venus [*Futaana et al.*, 2011]. It is somewhat surprising that ENA imaging has also become an effective tool for studying plasma interactions with surfaces such as between the solar wind and the lunar surface [*McComas et al.*, 2009b; *Wieser et al.*, 2009]. ENA imaging of the heliospheric boundary, Mars and Venus solar-wind interaction, and surface interactions are beyond the scope of this review and will not be discussed further since these targets do not directly involve any ring currents.

As depicted in Figure 6.2, ENAs are produced by charge-exchange between singly charged ions and neutral gas (or plasma) typically surrounding a magnetized planet or a body in space. For sufficiently energetic parent ions, the resulting ENA propagates in a straight trajectory from its origin of creation. The energy loss during the charge exchange reaction is generally in the 10s eV range and, therefore, if the parent ion is sufficiently energetic the ENA is a sample of the source population, carrying spectral, compositional, and pitch-angle information of the source plasma. In general, the charge-exchange cross section decreases with energy, but a few exceptions exist such as energetic He^+ ions colliding with H gas [*Barnett et al.*, 1990; *Lindsay and Stebbings*, 2005].

The charge-exchange medium is usually dominated by neutral gas, such as exospheres and upper atmospheres of planets and moons. As shown in Figure 6.2, the ENA intensity is a line-of-sight (LOS) integral convolving the ion intensities and neutral gas densities. ENA imaging is often associated with imaging of the ion distributions, but if the gradients of the neutral gas distributions are sharper than those of the ion distributions, the ENA image can represent the neutral gas distribution rather than the ion distribution. At Earth, the geocoronal distribution falls exponentially with altitude such that ENA intensities are heavily weighted toward low altitudes, whereas for the Jovian system, the neutral gas left behind by Europa in its orbit results in ENA images that are more representative of the neutral gas distribution. Ionic plasma can act as a charge-exchange medium as well. The ion-ion charge-exchange cross section is generally lower than that of the same neutral species, but, if sufficiently dense, the ionic plasma can produce appreciable intensities of ENAs. This may be the case in the Io plasma torus around Jupiter, where the densities of O^+, S^+, and S^{++} are likely comparable to those of the neutral densities in the Europa torus.

In this chapter, we will review some of the key ENA observations that have demonstrated how large-scale ring current distributions at Earth and Saturn behave and how these have helped us understand the electrical current systems in magnetospheres. The first section provides a brief overview of ENA measurement techniques, followed by a description of how the 3D parent ion distributions are retrieved from ENA images using constrained linear inversion in the second section. The third section, reviews ENA images of the terrestrial ring current and its development through a geomagnetic storm and describes the general characteristics of the 3D Partial Ring Current (PRC) system driven by the azimuthally asymmetric plasma pressure during the main phase of the storm. The fourth section discusses the contrasting force balance in the Saturnian system and the global morphology and

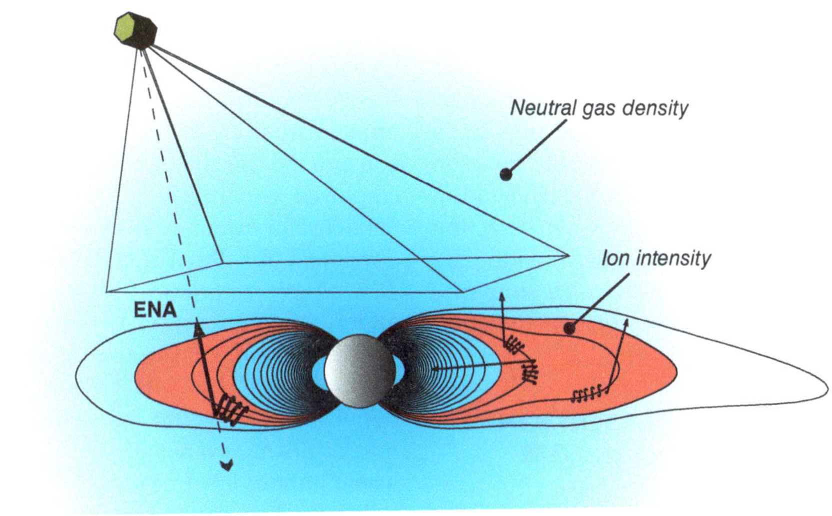

Figure 6.2 ENAs are produced when singly charged energetic ions (red region) charge exchanges with the exosphere (blue shaded region) and the upper atmosphere. The newly formed ENA travels then in straight lines (black arrows) unaffected by electric and magnetic fields and carries with it spectral and compositional information of the parent plasma. An image is formed by detecting the direction, energy, and species of the ENA.

dynamics of energetic ion injections as observed by the Ion and Neutral Camera (INCA) on board the Cassini mission, and how this current system is related to some of the mysterious, but periodic magnetic field perturbations in the magnetosphere. The paper concludes with an overview of ENA imaging of Jupiter and the future science opportunities that the NASA-funded Jovian Energetic Neutrals and Ions (JENI) ENA camera will provide from the JUICE mission.

6.2. ENA MEASUREMENT TECHNIQUES

Two basic techniques exist to detect and image ENAs. A foil-based time-of-flight (TOF) technique is used for ENAs above about 1 keV/nucleon and is the technique used in Cassini/INCA [*Krimigis et al.*, 2004], IMAGE/HENA [*Mitchell et al.*, 2000], and MENA [Pollock *et al.*, 2000], TWINS [*McComas et al.*, 2009a], and now also in JUICE/JENI [*Mitchell et al.*, 2016]. Other applications in development include those using a collimating system to overcome the scattering of the lower energy end of the ENAs [*Westlake et al.*, 2016]. A system of deflection plates in front of a slit aperture prevents charged particles from entering the instrument. The remaining ENAs penetrate a thin foil in the entrance aperture that has to be approximately 1 μg/cm² carbon to transmit particles of this low energy [*Allegrini et al.*, 2015a]. As the ENA penetrates the foil, it produces secondary electrons, both on the front and back of the foil depending on energy [*Allegrini et al.*, 2003] and the primary ENA also scatters in angle, which is one of the sources of "blur" in ENA images. This image blur is mathematically represented by the point-spread function (PSF) and is introduced by convolving it with simulated ENA images or deconvolving it from observed ENA images.

Atomic layer deposition of graphene is a relatively new technique that may eventually replace the carbon foils because of its superior properties in scattering and energy thresholds [*Allegrini et al.*, 2015b]. The electrons from the entrance foil can be used as a start timing and position pulse. After a certain TOF (usually on the order of 100 ns), the stop position and timing pulse are detected by a stop Microchannel Plate (MCP) or Solid-State Detector (SSD) system. The combination of start and stop timing and position pulses then provides the TOF and direction of the primary ENA. The TOF also serves as a coincidence timing window, beyond which events are rejected and therefore can be used to reduce unwanted background such as UV photons and penetrating electrons. Several variations on this basic concept exist, including using backscattered electrons from a stop foil covering the stop MCP as an additional coincidence pulse for improved background rejection. The UV photon intensity impinging on detectors can be reduced by the use of foils, but their increased thickness worsens the angular scattering and increases the energy threshold.

For primary ENA energies much less than 1 keV/nucleon, a conversion surface is introduced instead of an entrance foil to convert the ENA to a charged ion (highest probability for negatively charged). A conventional analysis system for charged particles can then be used after the conversion surface, such as an electrostatic analyzer system. This technique has been used by IMAGE/LENA [*Moore et al.*, 2000], IBEX-Lo [*McComas et al.*, 2009b]. On IBEX-Hi an ultrathin foil was used instead for converting the primary ENAs to positive ions.

In summary, foil-based TOF techniques provide a relatively high-efficiency detection of ENAs, but build on a straight-path concept that requires careful consideration of UV-rejection techniques. Low-energy ENA detection techniques based on the charge-conversion technique can rely on subsequent bent-path analysis, which simplifies rejection of UV, but the conversion efficiency is inherently relatively low (a few percent or less).

6.3. RETRIEVAL OF RING-CURRENT ION DISTRIBUTIONS FROM ENA IMAGES

Global information about the ring-current morphology and dynamics can be retrieved directly from ENA images obtained from suitable vantage points. Because of the relatively azimuthally symmetric geocoronal density distribution, large-scale local time information about the ring-current ion distribution is readily obtained from images obtained from a vantage point above one of the poles as was done by, for example, *Brandt et al.* [2002a]. However, due to the exponentially falling geocoronal density, it is difficult to discern the radial distribution of the ring current directly from images from a polar vantage point. Conversely, from an equatorial vantage, the radial distribution of the ring current can be retrieved with relatively good certainty, whereas the uncertainties of the local-time distribution are higher. Several studies have also made important conclusions about the large-scale temporal evolution of the ring current, for example, how and when substorm injections contribute to the ring current, role of O^+, and the relation to Dst [*Brandt et al.*, 2002b; *Ohtani et al.*, 2001, 2005a,b; *Ohtani et al.*, 2007]. An improved ring-current strength index through the use of ENA observations has been suggested and discussed by several [*Brandt et al.*, 2003; *Jorgensen*, 1997; *Valek et al.*, 2014], but is complicated by the directional dependence of ENA intensities.

In order to retrieve defining quantities, such as the plasma pressure distribution of the ring current, one has to apply retrieval techniques that take into account the neutral gas distribution, magnetic field structure, and charge-exchange cross sections. Several techniques have been developed to accomplish this including forward modeling techniques [*Roelof and Skinner*, 2000] and different versions of inversion (reverse modeling) techniques [*DeMajistre et al.*, 2004; *Perez et al.*, 2012]. Forward modeling techniques rely on parametric models of the parent ion and neutral gas distributions. Therefore, their solutions are restricted to the range of distributions the parametric model can provide, but are often relatively straightforward to construct and run. Inversion techniques, on the other hand, can find solutions of more complex distributions, but require usually a more sophisticated mathematical framework to ensure proper implementation. In the following, we briefly describe a constrained linear inversion technique that has been employed extensively in geophysics and described in detail by, for example, *Twomey* [1977]. This technique was developed for the IMAGE/HENA image analysis by *DeMajistre et al.* [2004]. In all of these techniques, the underlying assumption is that the ENA emissions are "optically thin," meaning that the mean-free path of a newly created ENA is greater than the scale of the population being imaged. This approximation indeed breaks down severely in the dense upper atmosphere, where intense ENA emissions are created from multiple charge-exchange and stripping reactions. To overcome these complication, one can use approximations represented by ENA-emitting low-altitude surfaces.

The basis of any inversion technique starts with the equation of measurement, which for ENA imaging can be stated as the counts in a pixel, C_i represented by

$$C_i = \int_{\Delta t} \int_{\Delta E} \int_0^{2\pi} \int_0^{\pi} \left[A_i(\epsilon, \beta, E, t) \right] j_{ena} \sin\epsilon \, d\epsilon \, d\beta dE dt, \tag{6.1}$$

where A_i is the response of the pixel i to ENA intensity j_{ena} at time t, energy E, and angular position ϵ and β. The instrument response function, A_i, includes the angular response as well as the PSF of the instrument. For foil-based TOF ENA cameras, the PSF is composed of the uncertainty (or "spread") of the primary ENA direction due to the scattering in foils and uncertainty in the positioning system. The HENA camera samples images in $6° \times 6°$ or $3° \times 3°$ bins depending on energy. The task is now to rewrite equation (6.1) into an appropriate and convenient linear equation system that can be solved with linear inversion techniques.

As illustrated in Figure 6.2, the differential intensity in units of $(\text{cm}^2\,\text{s}\,\text{sr}\,\text{keV})^{-1}$ of the ENAs at energy E can be expressed as the LOS integral

$$j_{ena} = \int_0^{s_e} n^H(s) \sigma_{10}^H(E) j_{ion}(s, E) ds + j_{ena}^e(s_e), \tag{6.2}$$

where s is the distance along the line of sight determined by ϵ and β, n^H is the number density of hydrogen, σ_{10}^H is the charge-exchange cross section for protons on hydrogen (1 and 0 denote the initial and final charge states of the projectile), and j_{ion} is the ion intensity. The limits of integration are taken from the detector to infinity or to the exobase for LOSs that intersect it. The ENA emissions from the optically thin regions of the geocorona are separated from those of the optically thick region below the exobase by introducing the ENA intensity term $j^{e}_{ena}(s_e)$ at the exobase, which can then include approximations for the emission intensity from the optically thick region, such as the thick target approximation described by *Bazell et al.* [2010].

The 3D parent ion intensity distribution can be reduced to 2D by assuming that ion pitch angles can be mapped along the magnetic field lines, which in this case is done assuming the magnetic field model by *Tsyganenko and Sitnov* [2005]. We represent therefore the ion intensity as a spatial and pitch-angle distribution only in the magnetic equator as

$$j_{ion}\left(L, \phi, \theta, \mu_{eq}\right) = U\left(\mu_{eq}, \mu_{eqlc}\left(L\right), m\right)\sum_{p=0}^{N} j_{ion}^{(p)} P_p^*\left(\mu_{eq}\right) \tag{6.3}$$

where $P_p^*(\mu_{eq})$ are the Legendre polynomials scaled to the interval (0, 1) of the cosine of the equatorial pitch angle, μ_{eq}. The parameter m is set to 1 if the loss cone is present or 0 if it is not. $U(\mu_{eq}, \mu_{eqlc}(L), m)$ is a binary function governing the behavior of the loss cone, which is considered either completely filled ($m = 0$) or completely empty ($m = 1$).

Given the expressions above, the measurement equation can therefore be expressed as a well-defined linear equation system

$$C_i = \sum_p \sum_k \sum_l b_{ikl}^p j_{ion}^p\left(L_l, \phi_k\right) \tag{6.4}$$

where the quadrature weights b_{ikl}^p contain now all information about the magnetic field, geocorona, geometries, and so on. In more convenient matrix notation, we can instead write

$$\mathbf{C} = \left|\mathbf{K}^{\parallel} \quad \mathbf{K}^{\perp}\right| \begin{vmatrix} \mathbf{J}_{ion}^{\parallel} \\ \mathbf{J}_{ion}^{\perp} \end{vmatrix} \tag{6.5}$$

where

$$\begin{aligned} j_{ion}^{\perp} &\equiv j_{ion}^{(0)} - j_{ion}^{(1)} \\ j_{ion}^{\parallel} &\equiv j_{ion}^{(0)} + j_{ion}^{(1)} \end{aligned} \tag{6.6}$$

which are the equatorial ion intensities perpendicular ($j_{ion}(\mu = 0)$) and parallel ($j_{ion}(\mu = 1)$) to the field lines, respectively.

Because the image in itself does not contain sufficient information to determine the ion distributions uniquely, a least-squares solution with added constraints on the ion distribution has to be sought, which can be written

$$\mathbf{J}_{ion} = \left(\mathbf{K}^T \sigma_C^{-2} \mathbf{K} + \gamma \mathbf{H}\right)^{-1} \mathbf{K}^T \sigma_C^{-2} \mathbf{C} \tag{6.7}$$

where γ is a scalar, H is a constraint matrix, and σ_C is the measurement covariance matrix.The constant γ serves a tuning parameter, determining the relative contribution of the a priori constraints to the solution. As γ becomes larger, the solution is more tightly bound to the constraint. The constraint matrix H can take several forms, for example, the Markhov constraint used here that favors solutions where the mean around any given point is small. For appropriate values of γ and H, there exists a unique solution that has been treated extensively by, for example, *Twomey* [1977], *Rodgers* [2000], and *Menke* [1989]. Therefore, constraint type and its constraint strength are of decisive importance in finding a correct and unique solution. *DeMajistre et al.* [2004] uses a simulation approach to construct synthetic images based on realistic model ring current ion distributions. In this way, the correct solution is already known and the inversion can be applied to the synthetic images with different constraint types and strengths. The optimal solution can then be readily identified as the minimum difference between the model ion distribution and the retrieved ion distribution. However, with real observations, the only available measure of goodness of fit is the difference between the observed image and the simulated image ($C = KJ$).

Figure 6.3 displays an example of an inversion of ENA images obtained by IMAGE/HENA during a typical recovery phase on 23–24 April 2001, and how those compare to simultaneous in situ measurements by the Cluster mission. Because the pitch-angle distributions (PADs) in the recovery phase are peaked around 90°, two IMAGE vantage points were chosen 3 hr apart to achieve an angular separation to retrieve the pitch angles with sufficient accuracy. Figure 6.3a shows one of the hydrogen ENA images in the 27–39 keV range. Dipole field lines are drawn for dipole L shells 4 and 8 for the four MLTs as indicated. From the ENA image alone, we note a decrease of ENA intensity close to Earth, which is a direct indicator of the peak PADs at the magnetic equator. Figure 6.3b shows the ENA image simulated ($C = KJ$) using the best solution of the proton distribution shown in Figures 6.3c (perpendicular equatorial pitch angles) and 6.3d (parallel equatorial pitch angles). Figures 6.3c and 6.3d show these proton distributions in the equator with dashed circles separated by 2 R_E starting at 2 R_E distance from Earth's center. The overlaid black curve with time tick marks represents the projected trajectory of the Cluster spacecraft.

Figure 6.3e shows the comparison between the simultaneous, in situ proton measurements by the Composition and Distribution Function Analyzer (CODIF) [*Reme et al.*, 2001] on board Cluster spacecraft (black lines), and the corresponding proton intensities from the HENA inversions along the Cluster trajectory (red lines). The perpendicular (thick lines) and parallel (thin lines) pitch

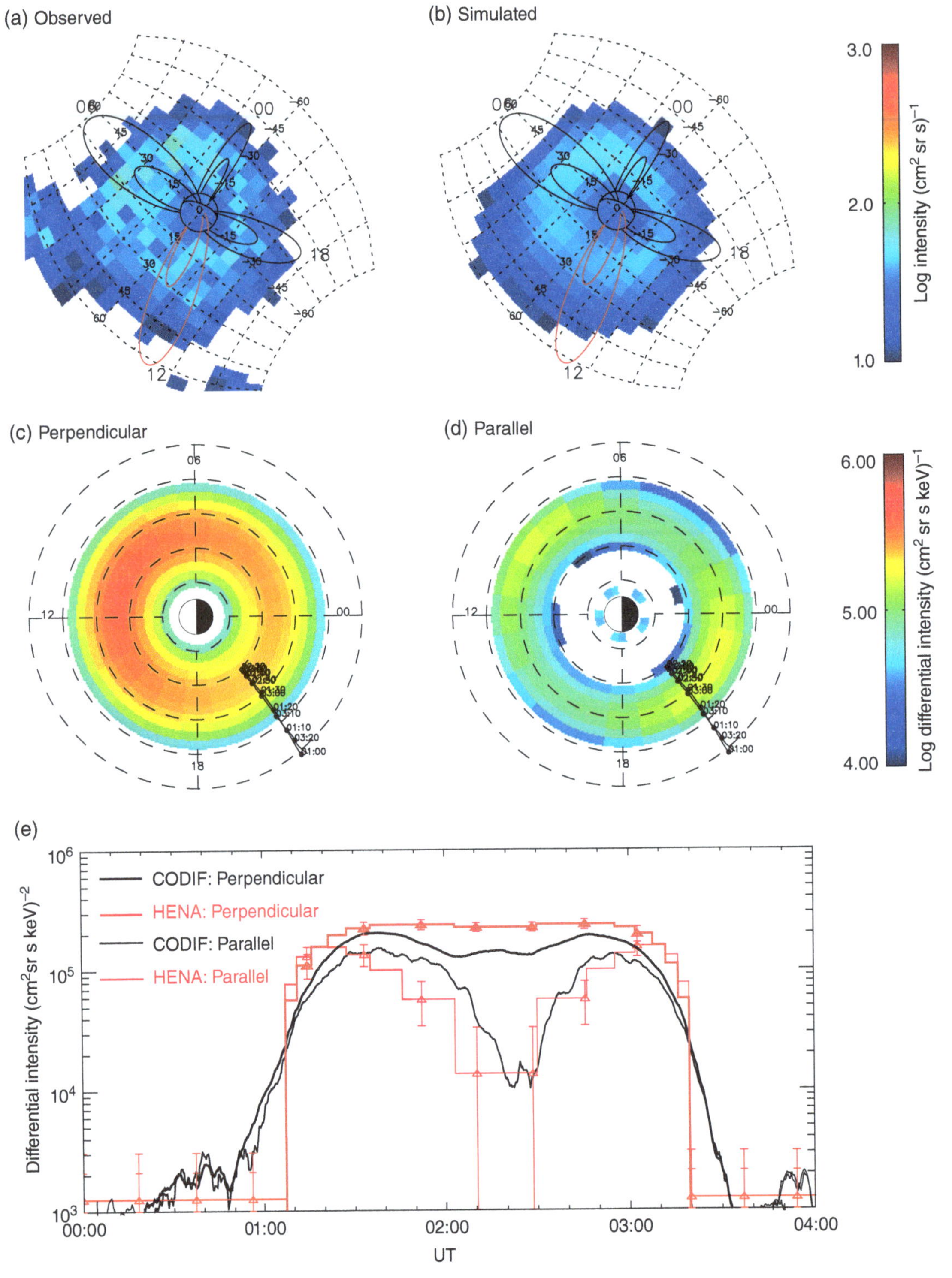

Figure 6.3 Example of a retrieval of a typical recover-phase ring current proton distribution in the 27–39 keV range using a linear inversion technique as described in *DeMajistre et al.* [2004]. The inversion uses two vantage points separated 3 hr apart. First vantage point was at 22:00 UTC 23 April 2001 situated above the north pole (not shown). (a) The observation from the second vantage point at 01:00 UTC 24 April 2001. (b) The simulated image using the best solution of the ion distribution below. (c) The equatorial proton distribution of the best solution for perpendicular pitch angles. The magnetic local time (MLT) is indicated in the figure and dashed circles are separated by 2 R_E, starting at 2 R_E. The black curve with time tick marks is the projected trajectory of the Cluster 4 spacecraft. (d) The same as (c), but for parallel pitch angles. (e) The comparison between in situ proton distribution in the same 27–39 keV range obtained by Cluster/CODIF in black thick (perpendicular pitch angles) and black thin (parallel pitch angles) lines. The red thick (perpendicular pitch angles) and red thin (parallel pitch angles) lines represent the proton intensities retrieved from the HENA image along the same Cluster trajectory.

angles at the Cluster spacecraft position are shown. The decrease of proton intensities centered around 02:20 UTC is a manifestation of the decreasing proton intensities at the inner edge of the ring current, which HENA reproduces fairly well for parallel pitch angles. Note, however, that for perpendicular pitch angles, the inversion shows a presence of protons down to about 2 R_E that may not be real. From these near-polar vantage points, the inner edge of the ring current for near perpendicular pitch angles is obscured by ENA emissions coming from high latitudes of higher L shells and therefore it is difficult from such images to discern the position of the inner edge, regardless of angular resolution. This intrinsic problem can be eliminated by using equatorial vantage points, but at the expense of reduced MLT coverage and resolution.

6.4. ENA IMAGING OF THE TERRESTRIAL RING CURRENT

At Earth, the plasma pressure responsible for the ring current is distributed between about 2.5 to 8 R_E and is carried by protons and heavy ionospheric ions (O^+ and N^+) in the approximate energy range of 10–300 keV/e [*Hamilton et al.*, 1988]. Although protons normally dominate the plasma pressure in quiet and storm conditions [*Gloeckler and Hamilton*, 1987], the plasma pressure carried by the ionospheric ions in the ring current can be comparable to that of protons and sometimes even dominate during strong geomagnetic storms [*Daglis et al.*, 1999; *Nosé et al.*, 2005]. As expected from their much larger charge-exchange cross sections, the O^+ ion content of the ring current decays relatively quickly after a geomagnetic storm main phase [*Krimigis et al.*, 1985]. Conversely, because of the relatively small charge-exchange cross sections of protons at higher energies, it turns out that a high-energy proton ring current survives even quiet times as has been shown from Cluster [*Vallat et al.*, 2005] and Van Allen Probe observations [*Gkioulidou et al.*, 2016], making justice to the historic name the "ring current."

ENAs from the terrestrial ring current are formed by charge exchange between singly charged ions and the neutral hydrogen of the exosphere [*Ostgaard et al.*, 2003] or the neutral O in the upper atmosphere [*Hedin*, 1991]. The fact that the ring current plasma pressure is dominated by ions in the 10–300 keV range as described above, foil-based TOF ENA cameras in the approximate range of 1–300 keV (hydrogen) and 10–300 keV (oxygen) are sufficient to image the majority of the pressure. Therefore, ENA imaging of the Earth's ring current is an efficient technique to diagnose the global morphology and dynamics of the ring-current pressure on short timescales (10–30 min).

It is well known from several studies that the ring-current ion distributions are localized to the nightside during the main phase and then symmetrize in magnetic local time during the recovery phase. Most notably, *Ebihara et al.* [2002] used 3 years of geotail proton measurements in the 1–200 keV range to compute the spatial distribution of energy density in the magnetic equator. In contrast, Figure 6.4 shows a sequence of ENA hydrogen images in the 60–119 keV range, each obtained with a 20 min accumulation time by the HENA camera on board IMAGE from a north polar vantage point looking down on Earth [*Brandt et al.*, 2008b]. The dipole field lines of L = 4 and 8 are drawn for reference at each indicated MLT. The sequence shows ENA intensities centered around the nightside during the main phase and how the ring current proton distribution spreads westward as the recovery phase progresses. Note that even though the morphology of the main phase intensities is relatively stable during the main phase, individual energetic ions still drift westward around the Earth and out the dayside magnetopause. The spatially stable morphology in intensity comes from the fact that the energy of ions maximize at a specific point close to Earth at a certain MLT [*Lyons and Williams*, 1984] and therefore the ion differential intensity also maximizes at that point.

Many studies of O^+ ions involving both in situ and ENA observations have indicated an energization mechanism that appears to be more nonadiabatic than for protons (see *Keika et al.* [2013] and references therein). From the very first mass-separated ENA images from IMAGE/HENA, it was clear that the response of O^+ during storm-time injections was very different than that of protons [*Mitchell et al.*, 2003]. Figures 6.5a–6.5c (top row) show the evolution of a substorm injection in hydrogen ENAs in the 60–119 keV range. Figures 6.5d–6.5f (bottom row) show the evolution of a substorm injection in oxygen ENAs in the 180–264 keV range. Although the SYM-H shows that this storm was in its the recovery phase during this period, the interplanetary magnetic field (IMF) B_z was negative around this period, which enhanced the ring current as can be seen from the ENA intensification in the image at 23:30 UTC. Note that, while the intensification of the off-limb emissions and the emissions from within the limb in the hydrogen images are both strong, the oxygen images show the major intensification from Earth itself. This behavior indicates preferential O^+ acceleration along the magnetic field lines. The observation is consistent with the results of *Mauk* [1986], *Delcourt* [2002], *Fok et al.* [2006], and *Jones et al.* [2006], which conclude that O^+ is differentially and efficiently accelerated during dipolarizations, much more than protons due to the larger gyro period of O^+, in particular in the parallel direction although some arguments still exist that this depends on the duration of the dipolarization.

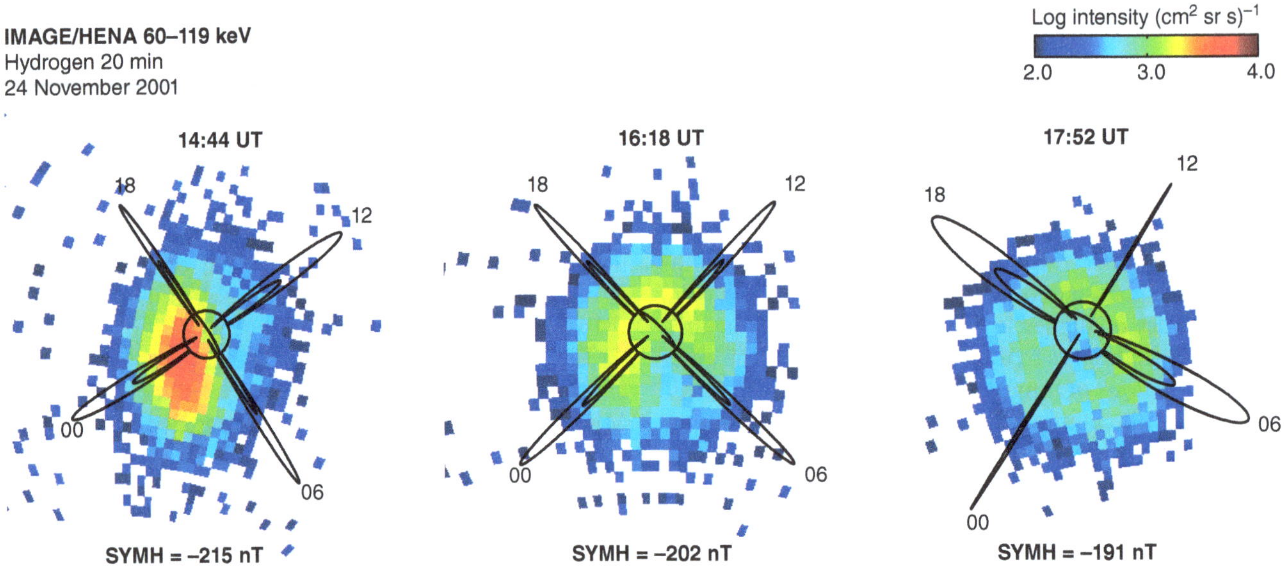

Figure 6.4 Hydrogen ENA image sequence obtained by the HENA imager on board the IMAGE spacecraft. The sequence was obtained during a transition from the storm main phase to the storm recovery phase. The IMF turned from southward to northward at around 15:00 UTC and the ring current responded rapidly by becoming more symmetric. After *Brandt et al.* [2008b].

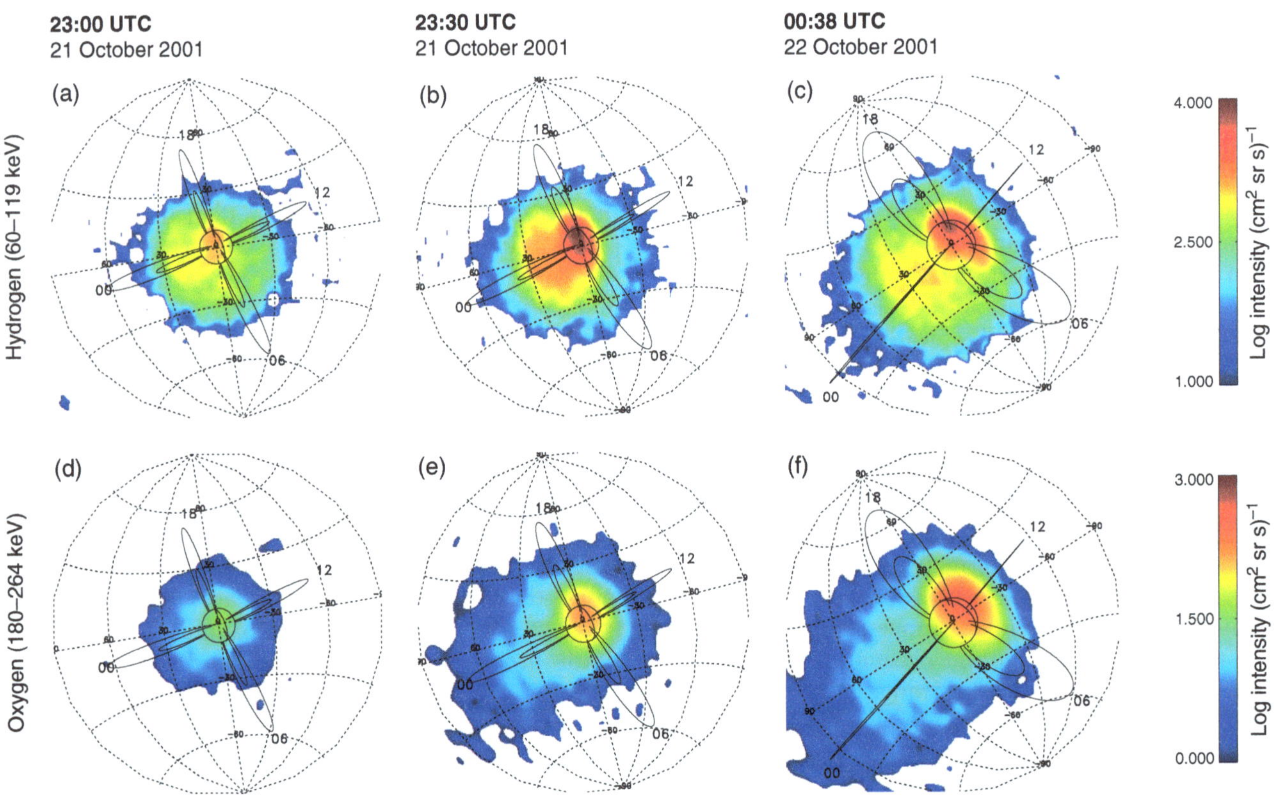

Figure 6.5 Hydrogen and oxygen ENA images display drastically different behaviors that indicate differences in acceleration mechanisms. The displayed sequence shows hydrogen ENAs in the 60–119 keV range (top row) and oxygen ENAs in the 180–264 keV range (bottom row) during a storm-time substorm injection on 21–22 October 2001. Adapted from *Mitchell et al.* [2003].

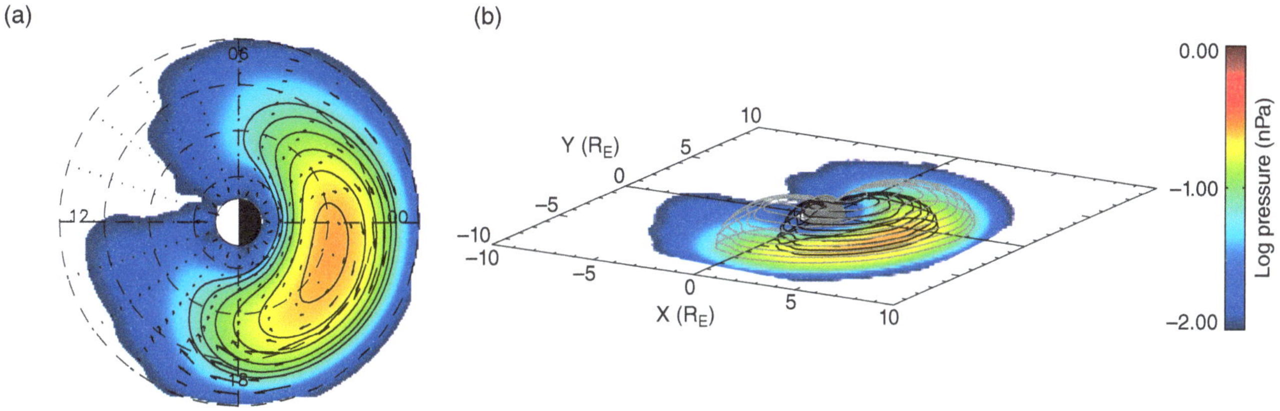

Figure 6.6 (a) The equatorial proton pressure distribution in the 27–60 keV range retrieved from a HENA hydrogen image during the storm main phase of 18 April 2002. For an isotropic pressure, the electrical current follows the isopressure contours. The magnitude of the retrieved current is represented by the length of the arrows. Note that in the absence of a constant interhemispheric FAC, there are no FAC coming out of the magnetic equator and the current closes on itself in the magnetic equator. (b) The resulting three-dimensional distribution of electrical current flow lines. At higher latitudes, the westward current on the outer edge of the pressure distribution deflects into the ionosphere due to the increasing magnitude of the FAC around the location of azimuthal pressure gradients. This entire current system is known as the Partial Ring Current.

Steady-state force balance necessitates the existence of electrical currents to counteract the force of pressure gradients of plasma trapped in a magnetic field [*Daglis et al.*, 1999; *Parker*, 1957]. These "pressure-driven" currents compose the ring current system at Earth. The asymmetric ring-current pressure distribution during main phases is responsible for the Partial Ring Current (PRC), while the more symmetric pressure distribution during recovery phases is responsible for the Symmetric Ring Current (SRC). As we will see below, the low-altitude connection of the PRC system to the ionosphere is a core part of the so-called Region 2 Field-Aligned Current (FAC) system. In order to visualize the electrical current systems associated with pressure distributions, one can use the force-balance equation neglecting the inertial term,

$$\mathbf{J} \times \mathbf{B} = \nabla \cdot P \tag{6.8}$$

where P is in general the pressure tensor and B is the magnetic field. By assuming isotropic pressures and current continuity $\nabla \cdot \mathrm{J} = 0$, one can derive the three-dimensional structure of the current system [*Vasyliunas*, 1984]. *Roelof* [1989] and *Roelof et al.* [2004] employed an effective technique for deriving the current flow lines using an Euler-potential formalism to describe the current density. Figure 6.6a shows the partial equatorial proton pressure distribution in the 27–60 keV range retrieved from an IMAGE/HENA 10 min image accumulation during the geomagnetic storm main phase of 18 April 2002 [*Brandt et al.*, 2008a]. The electrical current density in the equatorial plane is computed in a dipole field for simplicity and is represented by arrows. As expected for an isotropic pressure distribution, the current flows along the pressure contours and therefore the current flows westward on its outer edge and eastward on its inner edge. Because of the stronger magnetic field at the inner edge, this eastward current is much weaker than the stronger westward current, which dominates the PRC. Note that the current flow lines close on themselves in the equator without connecting to the ionosphere. This natural and critical feature of the PRC has been called the "Banana Current" [*Liemohn et al.*, 2011] and was visualized earlier by *Roelof* [1989]. This feature has now also been successfully retrieved by the TS07d empirical magnetic field model [*Sitnov et al.*, 2008; *Stephens et al.*, 2016].

Figure 6.6b illustrates now the current flow lines in three dimensions of the same pressure distribution shown in Figure 6.6a. Current flow lines are shown on two pressure isosurfaces. The outer set of current flow lines in gray is computed on the pressure isosurface of 0.09 nPa, and the inner set of current flow lines in black is computed on the 0.1 nPa isosurface. Just like in Figure 6.6a, the current in the equatorial plane follows the contours of the pressure. Above and below the magnetic equator, the FAC increases such that the westward current on the outer edge of the pressure distribution bends toward higher latitudes where the azimuthal pressure gradients become larger. At first glance, this may seem inconsistent with Kirchoff's law, which is a special case of current continuity equation for current-carrying wires. The resolution to this apparent inconsistency lies in the fact that the PRC system is not a set of wires, but a current system driven by a three-dimensional pressure distribution where the currents can be thought of as distributed on isopressure surfaces. No relevant "wires" can be identified in this case, but the total current (charge per unit time) has to be integrated across the same corresponding areal cross section in the equatorial plane as in the top side ionosphere where all the flow lines come together. The current density going in and out of the ionosphere therefore is an integral along the field line from the equator to the top of the ionosphere. The distribution of the current density going in and out of the ionosphere is known as the Region 2 current system [*Iijima and Potemra*, 1976]. The entire 3D current system comprises the PRC, and newly coined terms such as the Banana Currents are a part of it. During the recovery phase of a geomagnetic storm, the ring current pressure distribution becomes more azimuthally symmetric because magnetospheric convection has weakened to the point where curvature and gradient drift of ions dominates so that the drift paths are closed around the Earth and, therefore, the FACs diminish during this phase due to the lack of strong azimuthal pressure gradients.

Multiple other studies have been conducted using ENA observations to reveal globally important aspects. Most notably, these include a construction of composite ENA images of the storm-time ring current by *Reeves and Henderson* [2001] using the comprehensive energetic particle pitch-angle distribution/imaging proton spectrometer (CEPPAD/IPS) on board the POLAR mission; an observation of a narrow flow channel from a low-altitude vantage point (1000 km) at the equator viewing the ring current from below obtained by the Astrid mission[*Brandt et al.*, 2002c]; analysis of HENA images and model comparison of possible signatures of fine-scale interchange instabilities during sawtooth injections [*Yang et al.*, 2008]; stereo inversions of the ring current using images from the Double Star/NUADU ENA detector [*Lu et al.*, 2010]; and pitch-angle retrievals using stereo inversions of the ring current using images from the TWINS mission (see review by *Goldstein and McComas* [2013]).

6.5. ENA IMAGING OF SATURN'S MAGNETOSPHERE

At Saturn, the force balance is dominated by centrifugal forces exerted by the low-energy (<3 keV) ions inside of about 7 R_S and by radial pressure gradients in the energetic ions (≥3 keV) beyond about 12 R_S [*Sergis et al.*, 2017]. Whereas only in situ statistical studies have shed light on the low-energy population, ENA imaging of the energetic protons and O^+ ions have revealed dynamic large-scale injections into the inner magnetosphere that nearly corotate with Saturn [*Krimigis et al.*, 2007; *Mitchell et al.*, 2005; *Mitchell et al.*, 2009] and are believed to originate from reconnection driven by centrifugal stresses of the dense, low-energy plasma. This implies that plasma is transported onto closed drift trajectories, where the corotational electric field is sufficiently strong to dominate the drift of even 100 keV protons, whereas the opposing magnetic curvature and gradient drift only reduces the drift speed by a few tens of percent at these L shells and energies. Therefore, the large-scale injections give rise to rather localized energetic ion distributions that rotate around Saturn with about 11–13 hour period and disperses in to a spiral shape as it decays, likely due to charge-exchange [*Brandt et al.*, 2008b]. However, many exceptions to this anticipated behavior exist. Furthermore, ENA images recently have been used also to derive the globally warped shape of the plasma sheet [*Carbary and Mitchell*, 2016].

Saturn's magnetosphere is in some respect an ideal ENA object: most of the neutral gas is confined to the equatorial plane and Saturn's spin axis is almost perfectly aligned with its magnetic dipole axis. Therefore, the equatorial neutral gas acts as a screen for displaying magnetospheric ion intensification at all local times. Most of Saturn's neutral gas is hosted by the so-called E-ring and is composed of water products (OH, O, H, H_2O) as demonstrated by Hubble Space Telescope (HST) and Cassini observations combined with neutral-plasma modeling [*Jurac and Richardson*, 2005; *Richardson*, 1998; *Smith and Richardson*, 2017]. The source of gas is most certainly the icy moon Enceladus and its presence of water plumes [*Dougherty et al.*, 2006; *Smith et al.*, 2010]. Of the four water products, the H-density distribution appears to extend and dominate at large distances from Saturn, but is by far the most complicated one to constrain due to its relatively low UV emissions and its difficulty to model. ENA images by INCA have also been used to constrain the neutral density distribution at Saturn [*Dialynas et al.*, 2013a].

Figure 6.7 shows a sequence of ENA images obtained by INCA on board Cassini displaying how a large-scale proton injection propagates into the inner magnetosphere and begins to drift around Saturn. The image was obtained 17 April 2009 (Day of Year 107) in the 24–55 keV range, each accumulated for 60 min. The outer circle represents the orbit of Titan at around 20 R_S, and the next circle inside of that represents Dione's orbit at 6.9 R_S. The *x*-axis lies in the plane spanned by Saturn's rotational axis (*z*-axis) and the Sun vector, and is directed toward the Sun. The *y*-axis completes the right-hand system. As can be seen from the images, the ENA emissions appear to be originating from the equatorial plane, which is a result of the equatorial confinement of the neutral gas as discussed above. During the first hour starting at around 16:23 UTC (Fig. 6.7a), the ENA emissions intensify around Dione's orbit, which could be a new injection around midnight or part of a preexisting injection that

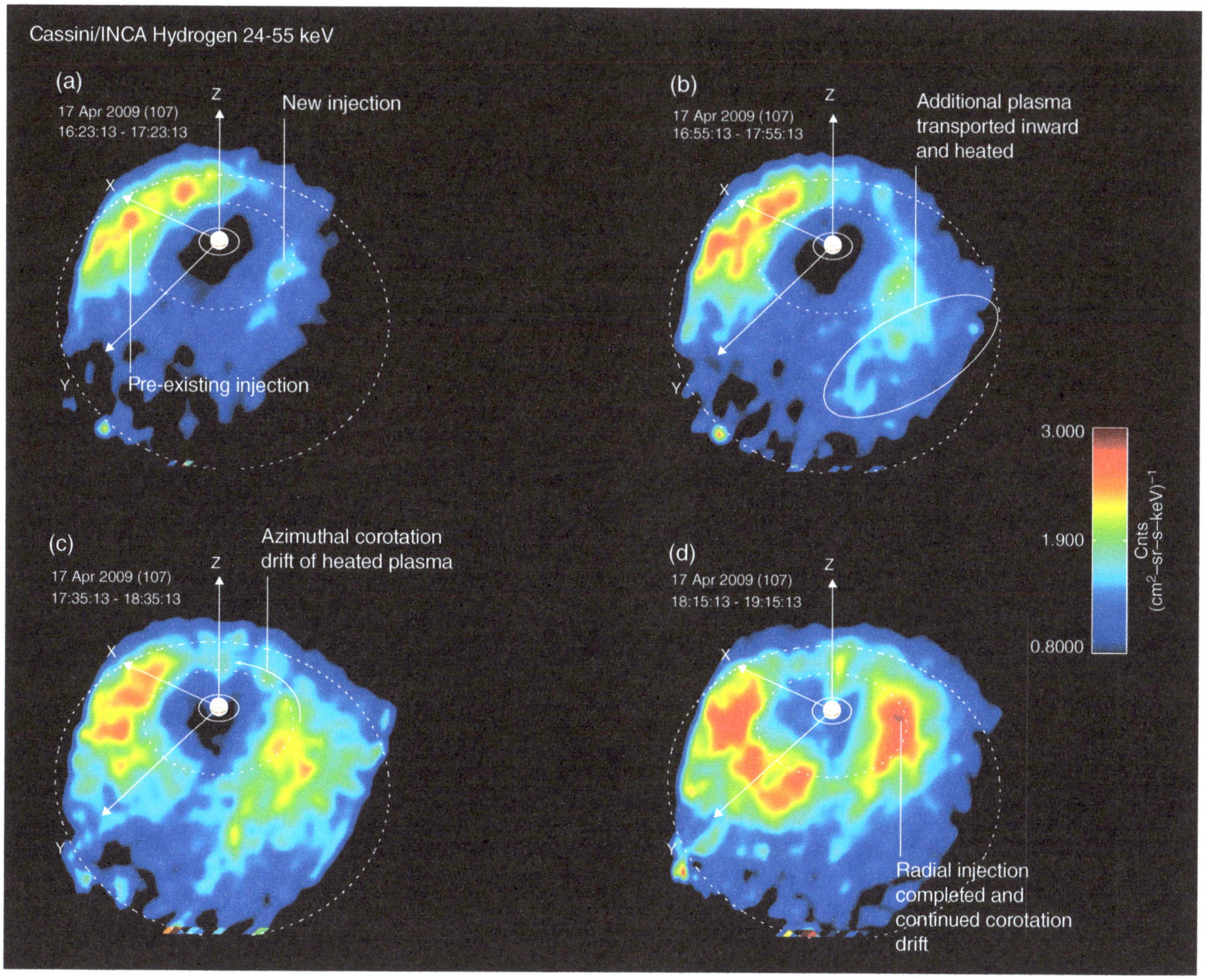

Figure 6.7 Hydrogen ENA-image sequence of a large-scale injection at Saturn in the 24–55 keV range obtained by Cassini/INCA. Outer dashed circle represents Titan's orbital distance (20 R_S), and the inner dashed circle represents Dione's orbital distance (6.3 R_S). The *z*-axis is parallel to Saturn's rotational axis and the *x*-axis points toward the Sun and lies in the plane spanned by the *z*-axis and the Sun vector. (a) A new injection appears around midnight at Dione's orbit. (b) Further ENA intensification beyond Dione's orbit is indicative of additional inward plasma transport and associated heating. (c) The expected azimuthal corotation drift can be observed, while further intensifications take place beyond Dione's orbit. (d) The completion of the injection is marked by the diminishing intensities beyond Dione's orbit.

intensifies as indicated in *Achilleos et al.* [2014]. Nevertheless, in Figure 6.7b, the midnight injection is followed by further ENA intensification between Dione's and Titan's orbit, which can be interpreted as additional plasma propagating radially inward with subsequent plasma heating (adiabatic or nonadiabatic) that increases the ion intensities. As the injection progresses in Figure 6.7c with further intensification between the orbits of Dione and Titan, one can also note how the ENA intensification rotates anticlockwise around Saturn, corresponding to the strong corotation drift, which can generally account for the dawn-dusk asymmetries seen in in situ data [*Dialynas et al.*, 2013b]. Note also how the preexisting injection indicated in Figure 6.7a has azimuthally drifted around Saturn as expected [*Brandt et al.*, 2008b]. In Figure 6.7d, the ENA intensities have reached their maximum and have further drifted azimuthally, but the ENA intensities that were previously present closer to the orbit of Titan have now diminished, which marks the end of the large-scale injection. This is interpreted as a radially inward propagation of heated plasma, with a strong azimuthal drift component. This global behavior is consistent with fast flows of low-density plasma flux tubes and associated dipolarization fronts that heats plasma as they rush into the inner magnetosphere, reported by *Mitchell et al.* [2015], *Rymer et al.* [2009], and

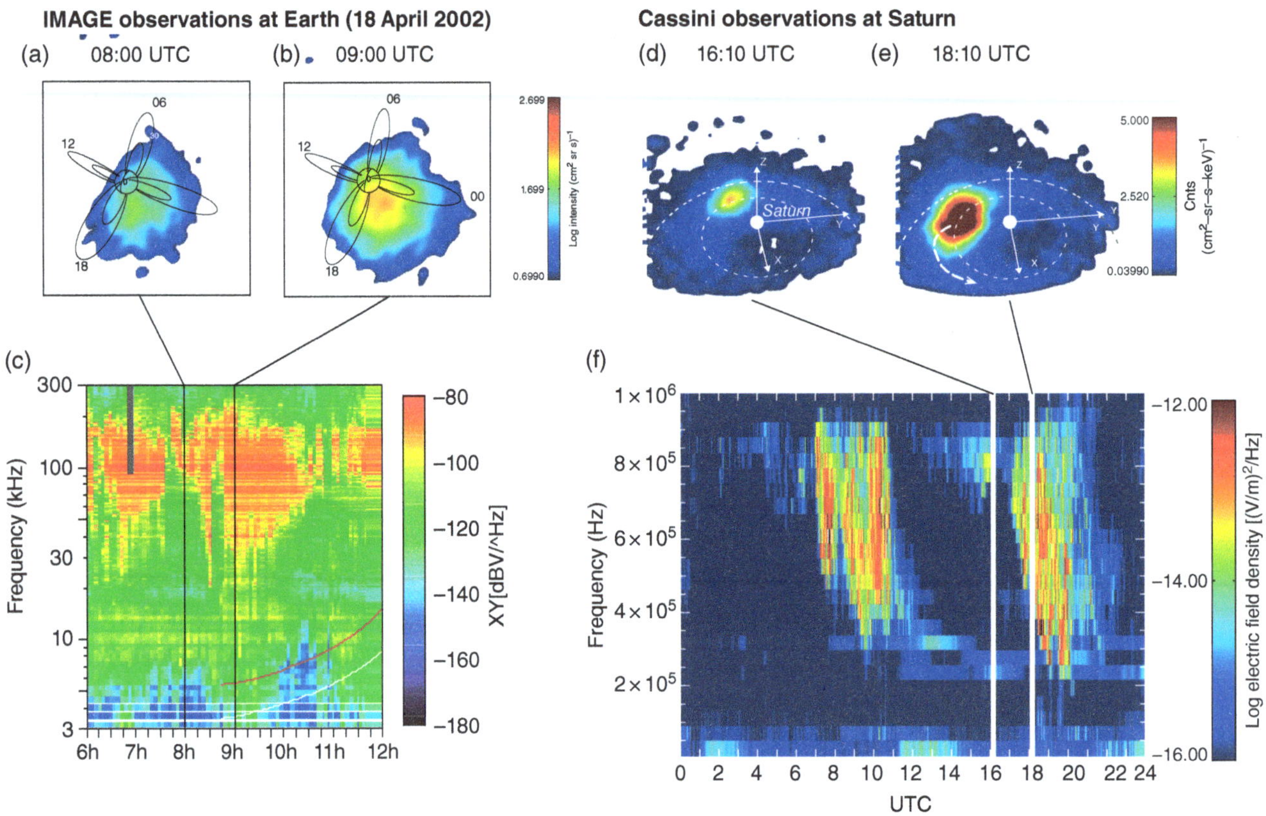

Figure 6.8 Large-scale magnetospheric injections and kilometric radiation are intimately linked phenomena at both Earth and Saturn. (a) Oxygen ENA image of a terrestrial substorm injection obtained by IMAGE/HENA in the 52–180 keV range just before substorm onset on 18 April 2002, and (b) approximately 35 min after onset. (c) Frequency spectrogram obtained by IMAGE/RPI showing the evolution of AKR during substorm injection. (d) Hydrogen ENA image of a large-scale injection at Saturn obtained by Cassini/INCA in the 24–55 keV range at the first onset of the injection on 11 January 2009, and (e) 2 hr into the injection. (f) Frequency spectrogram displays the periodic SKR bursts obtained by Cassini/RPWS during the same time.

its likely auroral counterpart *Jackman et al.* [2013]. While the trigger mechanisms of large-scale injections at Saturn may be very different from that of Earth, the heating and transport mechanisms of ring current plasma appear to be quite similar to what has been observed and modeled in the terrestrial magnetosphere as bursty-bulk flows [*Runov et al.*, 2011].

The commonality between magnetospheric phenomena at Earth and Saturn is further supported by the similar correspondences between ENA, auroral and kilometric radio emissions at both Earth and Saturn (see for example *Ohtani et al.* [2007] and *Mitchell et al.* [2009]), as well as the seemingly differential acceleration of protons and O^+ also seen at Saturn reported by, for example, *Dialynas et al.* [2009]. Saturn displays not only rotationally periodic Saturn Kilometric Radiation (SKR) bursts [*Gurnett et al.*, 2007] that have been used to define Saturn's planetary rotation period, but also large-scale injections observed in ENAs [*Carbary et al.*, 2008; *Mitchell et al.*, 2009; *Paranicas et al.*, 2005], charged-particle and magnetic-field signatures [*Carbary et al.*, 2007; *Provan et al.*, 2009] that all occur at that same period. Figure 6.8 illustrates the striking similarity between the terrestrial and Saturnian magnetospheres on how low-frequency radio bursts appear in direct correlation with large-scale magnetospheric injections. Figures 6.8a and 6.8b show IMAGE/HENA images during a storm-time substorm injection in oxygen 52–180 keV range. Figure 6.8c shows the burst of auroral kilometric radiation (AKR) and its extension to lower frequencies commencing at about 08:30 h obtained by the Radio Plasma Imager (RPI) [*Green et al.*, 2000] on board IMAGE. This onset of the low-frequency AKR has a high correlation with the terrestrial substorm onset [*Liou et al.*, 2000], and, as discussed above, IMAGE/HENA observes clear ENA intensifications during each substorm onset with corresponding AKR bursts. Similarly, Figures 6.8d–6.8e show the hydrogen INCA images in the 24–55 keV range during a typical injection at

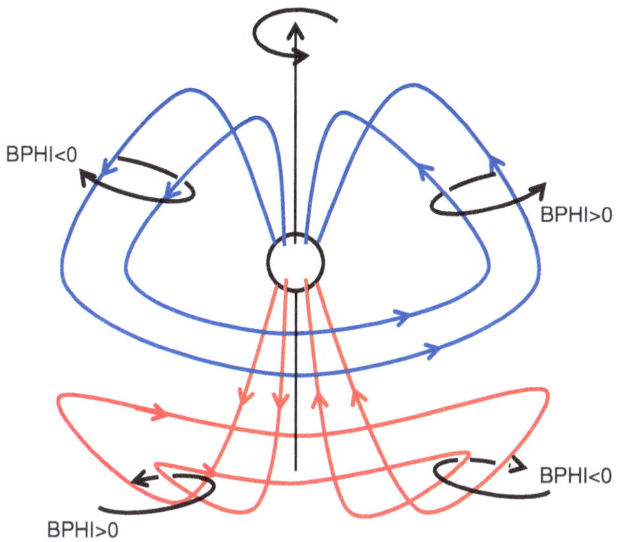

Figure 6.9 At Saturn, data suggest that a PRC system similar to that of Earth exists, but rotates around Saturn as shown by *Provan et al.* [2009] and *Brandt et al.* [2010]. Above and below the equator, the Cassini spacecraft encounter periodic perturbations in azimuthal magnetic field component, ΔB_φ. As the relatively slow-moving Cassini spacecraft moves through the northern half of the current system (blue), it encounters periodic field perturbations in the radial, ΔB_r, and polar, ΔB_θ field components that are in phase with each other. However, in the southern half of the current system (red), these two components appear 180° out of phase.

Saturn obtained on 11 January 2009. Figure 6.8f shows the strongly correlated onset of a periodic SKR burst at about 16:00 UTC and its corresponding low-frequency extension through 20:00 UTC obtained by the Radio Plasma Wave Science (RPWS) instrument on board Cassini [*Gurnett et al.*, 2004]. Note also the previous SKR burst at around 09:00 UTC, which was also closely correlated with a large-scale injection as seen in ENAs.

At least the hot plasma pressure distributions have been shown often to be longitudinally asymmetric [*Krimigis et al.*, 2007]. Therefore, the associated current system is the same as described above for the terrestrial PRC, but in this case it is rotating with the planet at some fraction of its corotation. Figure 6.9a shows a simplistic concept to demonstrate what such current system may look like. As described for Earth above, the pressure distribution would be centered between the two pairs of FAC systems that go in and out of the ionosphere. At higher latitudes, the FACs would therefore produce perturbations in the φ component of the magnetic field, ΔB_φ, as indicated in Figure 6.9a, and closer to the equator, the perturbations would be in the radial, ΔB_r, and polar, ΔB_θ, components of the magnetic field. So, for example, during a north-to-south traversal of the inner magnetosphere by the Cassini spacecraft, the PRC system would rotate at a much faster rate than the spacecraft speed, such that Cassini measures periodic perturbations. At northern latitudes, it measures ΔB_r and ΔB_θ in phase, and at southern latitudes these two components are instead 180° out of phase. The periodic encounters with ΔB_φ are directly indicative of the structure of the FAC system. Note that any interhemispheric FAC can be added to this system without violating the force balance equation discussed above.

During a near-equatorial pass through the Saturnian magnetosphere, periodic ENA intensifications were observed on the nightside for more than 10 days continuously [*Paranicas et al.*, 2005]. Later data revealed that each injection is indeed a new injection from the magnetotail, and that the preexisting injection may be acting as a trigger of the new injection [*Mitchell et al.*, 2009]. During this pass, magnetic field measurements also showed clear periodic signatures in all components. This fortuitous observation provided fields, particles, and remote ENA observations offering an excellent opportunity to probe the cause of magnetic field periodicities at Saturn. As illustrated in Figure 6.10a, the pass started in the dayside magnetosphere with a closest approach near dusk at around 6 R_S, swinging around the nightside and exiting the magnetosphere in the postmidnight sector. To illustrate the scale and geometries of this pass, an ENA image (obtained at a different time) has been overlaid with a model magnetic field and the Cassini trajectory (dashed line), and the rotation direction of the PRC is indicated by the white arrow. Using only magnetic field data and only ENA and particle measurements, *Provan et al.* [2009] and *Brandt et al.* [2010] concluded that the periodic perturbations in the magnetic field were due to a rotating PRC system. Figure 6.10b is adapted from *Brandt et al.* [2010] and shows the comparison between the residual of the total measured magnetic field (black line) and the simulated field (red line) using plasma pressure distributions derived from in situ measurements and ENA observations from Cassini. The pressure model uses in situ plasma measurements below 2 keV compiled by *Sergis et al.* [2010]. Above 2 keV, the "hot" plasma pressure morphology is derived from ENA observations by INCA by matching simulated ENA images using a parametric ion distribution to the observed ENA images. The magnitude of the hot plasma pressure is estimated by spectral extrapolation using the statistical in situ model by *Dialynas et al.* [2009]. Magnetic field perturbations due to the pressure-driven current system are computed by using perturbation theory of a dipole field. Given the simplicity of this approach, the agreement with the measured magnetic field perturbation magnitude and phase is quite remarkable. However, it should be noted that several additional features of periodic phenomena remain unexplained.

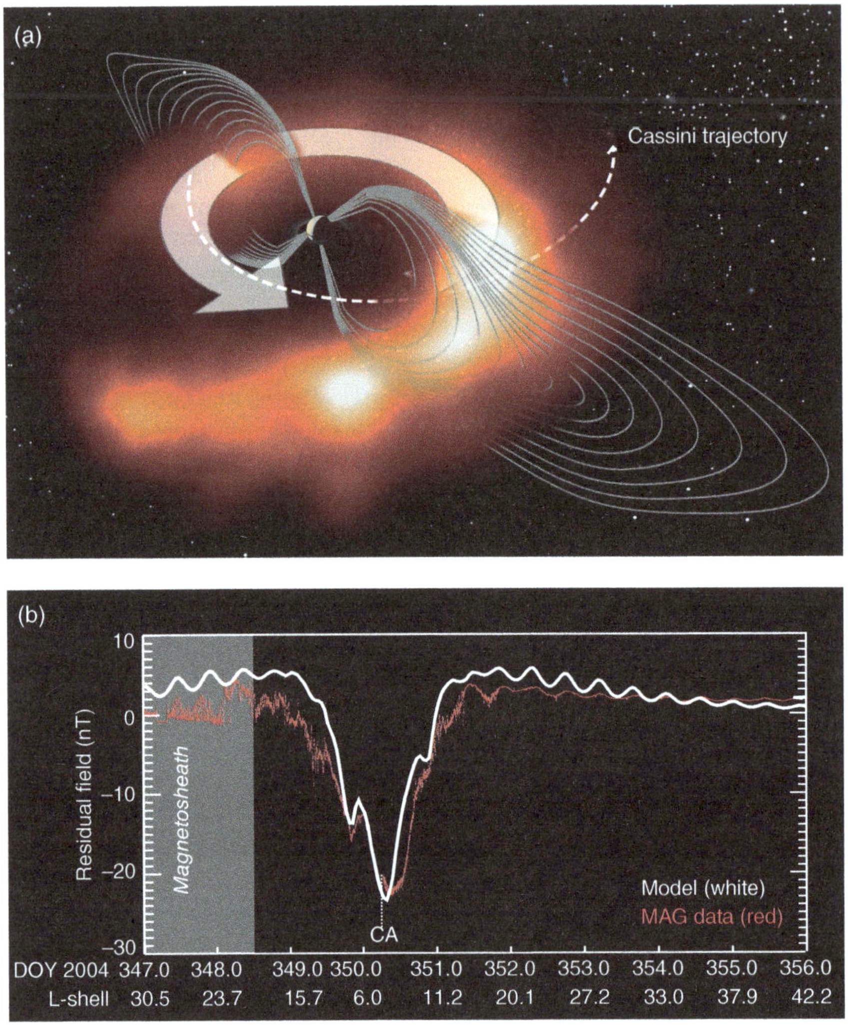

Figure 6.10 Using in situ and remote ENA observations, *Brandt et al.* [2010] showed that asymmetric hot-plasma pressure distributions can account for the periodic magnetic field perturbations. (a) ENA image obtained by INCA overlaid with a model magnetic field, showing the approximate Cassini trajectory during a 10 day pass through the equatorial magnetosphere when periodic ENA brightenings were observed simultaneous with clear periodic magnetic field perturbations. (b) Corresponding magnetic field measurements and resulting simulated field perturbations using plasma pressure distributions derived from data.

6.6. ENA IMAGING OF JUPITER'S RING CURRENT

At Jupiter there exists azimuthal currents that are set up as a consequence by the radially outward plasma flow driven by the centrifugal forces acting on the relatively dense and cold plasma [*Achilleos et al.*, 2014]. This current inflates the magnetic field and contributes to forming the Jovian magnetodisc. However, from Galileo measurements it turns out that the balance between the magnetic stress of the magnetodisc cannot be balanced by the centrifugal force exerted by the plasma alone, but that also currents driven by the radial plasma-pressure gradient and the plasma-pressure anisotropies likely play a significant role in forming the magnetodisc [*Paranicas et al.*, 1991]. A significant part of the plasma pressure of the Jovian inner magnetosphere is believed to come from quasi-periodic large-scale injections [*Kronberg et al.*, 2005; *Krupp et al.*, 1998; *Mauk et al.*, 1999; *Woch et al.*, 1998], perhaps not unlike the ones discovered at Saturn and Earth.

In contrast to Saturn, the Jovian magnetosphere is plasma dominated. Pick up of heavy ions from the ionized upper atmosphere of Io acts as the dominating source of both plasma and neutrals in the Jovian system, resulting in the Io Plasma Torus (IPT) and neutral torus [*Thomas et al.*, 2004]. While it may be intuitive to associate Io's tenuous SO_2 atmosphere with a direct supply of volcanic eruptions, it appears that the atmosphere is sustained by sublimation of SO_2 surface

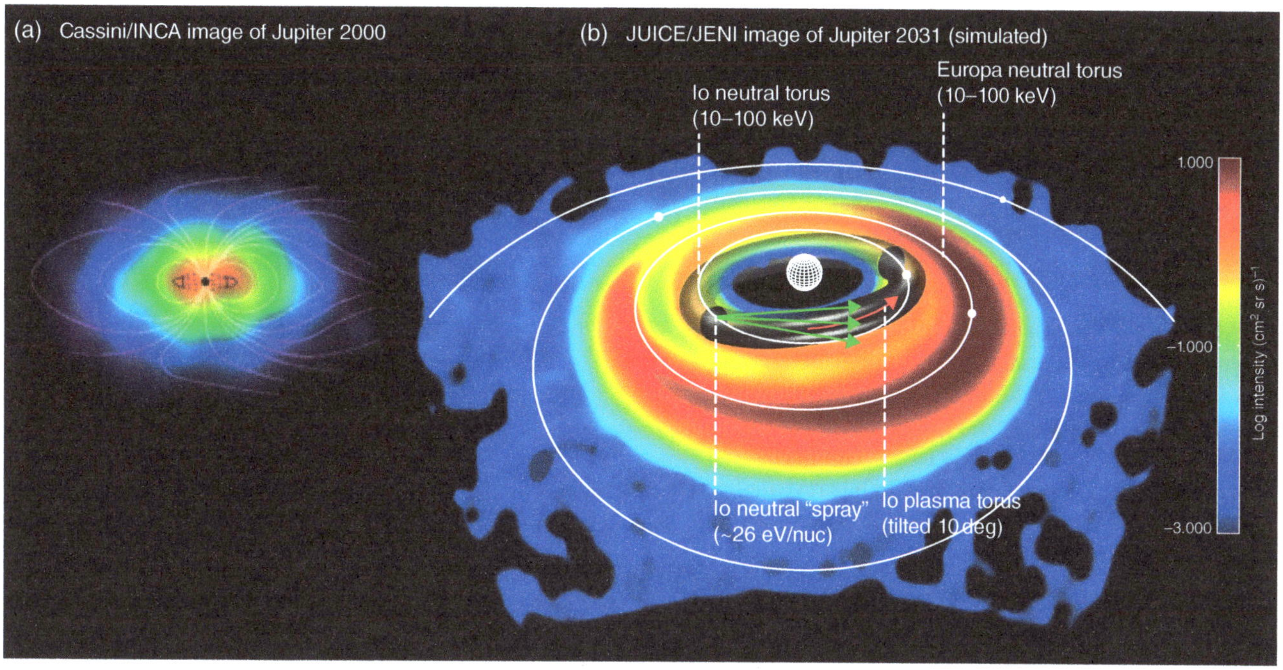

Figure 6.11 (a) ENA image of the Jovian system obtained by Cassini/INCA during the distant Jupiter flyby in December 2000 [*Mitchell et al.*, 2004]. (b) Simulated hydrogen ENA image in the 40 keV range of what the JENI ENA camera may reveal as JUICE is in orbit around Jupiter after 2030.

frost [*Tsang et al.*, 2016]. Europa, on the other hand, releases water products to space creating a cloud of neutrals along its orbit from surface- (sputtering, radiolysis) or subsurface release (plume eruptions) [*Plainaki et al.*, 2012; *Smyth and Marconi*, 2006]. It is reasonable to expect that there is little contribution of Iogenic neutrals around Europa due to the very short neutral lifetime at Io preventing them from reaching Europa's orbital distance through outward diffusion. Furthermore, the ongoing debate regarding the relative roles of these processes at Europa has certainly been fueled by the recent observations of potential plume structures [*Roth et al.*, 2014; *Sparks et al.*, 2016].

At the end of 2000 and beginning of 2001, Cassini flew by Jupiter at a closest approach of about 140 R_J. During this flyby, INCA accumulated several ENA images of the ENA emission from Jupiter (see Figure 6.11a). Analysis by *Mauk et al.* [2003], *Mauk et al.* [2004], and *Mitchell et al.* [2004] concluded that the dominant source of these emissions came from charge-exchange between the energetic ion distribution and the neutral gas of the Europa torus. Because of the relatively large fly-by distance, it was difficult to discern the energetic ion distribution much beyond Europa's distance from these observations. In addition, the neutral gas beyond the Europa torus is not well know and less dense than in the Europa torus, which means that the ENA intensity of large-scale injections is plausibly much weaker than at Europa's orbit.

In June 2022, the European Space Agency is launching the Jupiter Icy Moon Explorer (JUICE), which carries the second-generation ENA camera (after INCA) JENI, designed and built by The Johns Hopkins University Applied Physics Laboratory [*Mitchell et al.*, 2016]. Figure 6.11b shows the expected ENA image of the Jovian magnetosphere using simulations based on Galileo energetic ion measurements reported by *Mauk et al.* [2004] and the model neutral gas distribution by *Smyth and Marconi* [2006]. The circles represent, from the center, the orbits of Io, Europa, Ganymede, and Callisto. The simulation has exaggerated the energetic proton intensities inside of Europa's orbit to illustrate what the possible ENA emissions would look like from the Io neutral torus region, which follows the orbital path of Io. The gray, tilted, doughnut shaped circle represents the IPT demonstrating how low-energy ENAs are produced from charge-exchange between ions in the IPT and neutrals in the neutral Io torus, something that the Jovian Neutral Atom detector on board JUICE will investigate [*Futaana et al.*, 2015]. Because of the 9.4° tilt of Jupiter's magnetic dipole axis with respect to its spin axis, ENA emissions from charge-exchange with energetic ions of the magnetodisc will be modulated by the rotational period as it wobbles up and down through the neutral gas confined to the spin equator.

6.7. SUMMARY

Since its birth in the 1980s, ENA imaging has become a powerful tool for imaging energetic ions in planetary magnetospheres, which is the only technique that can image the ring current and plasma sheet ion distributions, and captures a significant part of the plasma pressure that governs the structure of the inner magnetosphere. At Earth, several missions have successfully used the ENA technique including Astrid-1, Double Star, TWINS, and IMAGE. These have provided global information on how magnetospheric injections build up the ring current, and how it evolves during geomagnetic storms, allowing us to estimate and visualize the 3D current systems associated with the partial and symmetric ring current systems.

At Saturn, INCA on board Cassini has revealed mysteriously periodic, large-scale injections that result in a PRC system rotating around Saturn. While it has been demonstrated that this rotating PRC system is consistent with the periodic magnetic field signatures at Saturn, the cause of the periodicity remains one of the biggest outstanding questions. The only ENA observations of Jupiter have been provided by Cassini during its distant flyby of Jupiter. These observations revealed that the dominant ENA intensities originated from the Europa torus. Any presence of large-scale injections in the INCA observations that previous Galileo in situ measurements have indicated, still remains unclear. In 2030, the first dedicated ENA camera will be inserted into orbit around Jupiter on board the ESA JUICE mission and will provide clarity to many of the unanswered questions.

ENA imaging has moved from being a proof-of-concept technique to a global observational tool that has opened up the frontier to reveal global dynamics of invisible plasma populations. Instrument developments have already made significant improvements in angular resolution nearing one degree. To make ENA imaging more accessible and understandable to the larger science community and to the public, future developments will have to prioritize higher angular resolution moving to less than one degree and efficient background rejection that will enable images approaching photographic qualities that possibly could reveal bursty-bulk flows and their global impacts on plasma heating. Imaging heavy ENAs at low energies (<1 keV) with high angular resolution and sensitivity remains a challenge and is therefore another critical development that would enable making global observations of, for example, the complete circulation of O^+ from the terrestrial ionosphere to the inner magnetosphere. It is of equal importance to ensure that analysis techniques mature and become available so that reliable and well understood data products of global ion distributions can be provided to, and used by, the scientific community.

ACKNOWLEDGMENTS

This work has been supported by NASA Grants NNX10AQ56G, NNX10AF17G, NNX12AG81G, NASA contract NAS5-97271and NSF Grant 1303646. The authors wish to thank Dr. W. S. Kurth for useful comments and for use of data from Cassini/RPWS.

REFERENCES

Achilleos, N., N. Andre, X. Blanco-Cano, P. C. Brandt, P. A. Delamere, and R. Winglee (2014), 1. transport of mass, momentum and energy in planetary magnetodisc regions, *Space Sci. Rev.*; doi:10.1007/s11214-014-0086-y.

Allegrini, F., P. Bedworth, R. W. Ebert, S. A. Fuselier, G. Nicolaou, and S. Sinton (2015b), Energy loss and straggling of 1-50 keV H, He, C, N, and O ions passing through few layer graphene, *Nuc. Inst. Meth. Phys. Res. B*, *358*, 223–228; doi:10.1016/j.nimb.2015.06.028.

Allegrini, F., R. F. Wimmer-Schweingruber, P. Wurz, and P. Bochsler (2003), Determination of low-energy ion-induced electron yields from thin carbon foils, *Nuc. Inst. Meth. Phys. Res. B*, *211*, 487–494; doi:10.1016/S0168-583X(03)01705-1.

Allegrini, F., R. W. Ebert, G. Nicolaou, and G. Grubbs (2015a), Semi-empirical relationships for the energy loss and straggling of 1-50 keV hydrogen ions passing through thin carbon foils, *Nuc. Inst. Meth. Phys. Res. B*, *359*, 115–119; doi:10.1016/j.nimb.2015.07.098.

Barabash, S., O. Norberg, R. Lundin, S. Olsen, K. Lundin, P. C. Brandt, E. C. Roelof, C. Chase, and B. Mauk (1998), Energetic neutral atom imager on the Swedish micro satellite Astrid, in *Measurement Techniques in Space Plasmas*, edited by R. F. Pfaff et al., 257–262, AGU Monograph 103, American Geophysical Union, Washington, DC.

Barnett, C. F., H. T. Hunter, M. I. Fitzpatrick, I. Alvarez, C. Cisneros, and R. A. Phaneuf (1990), Atomic data for fusion, 1: Collisions of H, H2, He, and Li atoms and ions with atoms and molecules, NASA STI/Recon Technical Report N, 91.

Bazell, D., E. C. Roelof, T. Sotirelis, P. C. Brandt, H. Nair, P. Valek, J. Goldstein, and D. McComas (2010), Comparison of TWINS images of low-altitude emission of energetic neutral atoms with DMSP precipitating ion fluxes, *J. Geophys. Res. Space Physics*, *115*, A10204; doi:10.1029/2010JA015644.

Brandt, P. C., B. H. Mauk, J. Westlake, D. G. Mitchell, and H. T. Smith (2017), Cassini/INCA Observations of an Asymmetric Europa Torus, *Planet. Space Sci.*, in preparation.

Brandt, P. C., C. P. Paranicas, J. F. Carbary, D. G. Mitchell, B. H. Mauk, and S. M. Krimigis (2008a), Understanding the global evolution of Saturn's ring current, *Geophys. Res. Lett.*, *35*, 17,101; doi:10.1029/2008GL034969.

Brandt, P. C., D. G. Mitchell, R. DeMajistre, E. C. Roelof, S. Ohtani, J.-M. Jahn, C. Pollock, and G. Reeves (2003), Storm-substorm relationships during the 4 October 2000 storm, IMAGE global ENA imaging results, Storm, in *Disturbances in Geospace: The Substorm Relationship*, 103–118, AGU Monograph 142; doi:10.1029/142GM10.

Brandt, P. C., E. C. Roelof, R. Decker, P. Wurz, S. Barabash, D. Bazell, and T. Sotirelis (2009), A Residual Source of

Energetic Neutral Atoms Across the Sky Obtained by the Neutral Particle Detector Onboard Venus Express, in American Institute of Physics Conference Series, American Institute of Physics Conference Series, 1183, edited by X. Ao and G. Z. R. Burrows, 102–112; doi:10.1063/1.3266764.

Brandt, P. C., K. Dialynas, I. Dandouras, D. G. Mitchell, P. Garnier, and S. M. Krimigis (2012), The distribution of Titan's high-altitude (out to 50,000 km) exosphere from energetic neutral atom (ENA) measurements by Cassini/INCA, *Planet. Space Sci.*, *60*, 107–114; doi:10.1016/j.pss.2011.04.014.

Brandt, P. C., K. K. Khurana, D. G. Mitchell, N. Sergis, K. Dialynas, J. F. Carbary, E. C. Roelof, C. P. Paranicas, S. M. Krimigis, and B. H. Mauk (2010), Saturn's periodic magnetic field perturbations are caused by a rotating partial ring current, *Geophys. Res. Lett.*, *37*, 22,103; doi:10.1029/2010GL045285.

Brandt, P. C., S. Barabash, E. C. Roelof, and C. J. Chase (2001), ENA imaging at low altitudes from the Swedish microsatellite Astrid: Extraction of the equatorial ion distribution, *J. Geophys. Res.*, *106*(A11), 25,731–25,744.

Brandt, P. C., S. Ohtani, D. G. Mitchell, M. C. Fok, E. C. Roelof, and R. DeMajistre (2002a), Global ENA observations of the storm mainphase ring current: Implications for skewed electric fields in the inner magnetosphere, *Geophys. Res. Lett.*, *29*(20), 1954; doi:10.1029/2002GL015160.

Brandt, P. C., S. Ohtani, D. G. Mitchell, R. DeMajistre, and E. C. Roelof (2002b), ENA observations of a global substorm growth phase dropout in the nightside magnetosphere, *Geophys. Res. Lett.*, *29*(20), 1962; doi:10.1029/2002GL015057.

Brandt, P. C., Y. Ebihara, S. Barabash, and E. C. Roelof (2002c), ENA images of a narrow flow channel from the plasma sheet. Astrid-1 observations, *J. Geophys. Res.*, *107*(A10), 1723; doi:10.1029/2001JA000230.

Brandt, P. C., Y. Zheng, T. S. Sotirelis, K. Oksavik, and F. J. Rich (2008b), The linkage between the ring current and the ionosphere system, *AGU Monograph 181*, 135–143; doi:10.1029/181GM13.

Carbary, J. F., and D. G. Mitchell (2016), Seasonal variations in Saturn's plasma sheet warping, *Geophys. Res. Lett.*, *43*, 11; doi:10.1002/2016GL071790.

Carbary, J. F., D. G. Mitchell, P. Brandt, C. Paranicas, and S. M. Krimigis (2008), ENA periodicities at Saturn, *Geophys. Res. Lett.*, *35*, 7102; doi:10.1029/2008GL033230.

Carbary, J. F., D. G. Mitchell, S. M. Krimigis, D. C. Hamilton, and N. Krupp (2007), Charged particle periodicities in Saturn's outer magnetosphere, *J. Geophys. Res.*, *112*(A11), 6246; doi:10.1029/2007JA012351.

Daglis, I. A., R. M. Thorne, W. Baumjohann, and S. Orsini (1999), The terrestrial ring current: Origin, formation, and decay, *Rev. Geophys.*, *37*, 407–438; doi:10.1029/1999RG900009.

Delcourt, D. C. (2002), Particle acceleration by inductive electric fields in the inner magnetosphere, *J. Atmos. Solar Terr. Phys.*, *64*, 551–559.

DeMajistre, R., E. C. Roelof, P. C. Brandt, and D. G. Mitchell (2004), Retrieval of global magnetospheric ion distributions from high energy neutral atom (ENA) measurements by the IMAGE/HENA instrument, *J. Geophys. Res.*, *109*(A04214); doi:10.1029/2003JA010322.

Dialynas, K., P. C. Brandt, S. M. Krimigis, D. G. Mitchell, D. C. Hamilton, N. Krupp, and A. M. Rymer (2013a), The extended Saturnian neutral cloud as revealed by global ENA simulations using Cassini/MIMI measurements, *J. Geophys. Res. Space Phys.*, *118*, 3027–3041; doi:10.1002/jgra.50295.

Dialynas, K., S. M. Krimigis, D. G. Mitchell, D. C. Hamilton, N. Krupp, and P. C. Brandt (2009), Energetic ion spectral characteristics in the Saturnian magnetosphere using Cassini/MIMI measurements, *J. Geophys. Res.*, *114*, 1212; doi:10.1029/2008JA013761.

Dialynas, K., S. M. Krimigis, D. G. Mitchell, E. C. Roelof, and R. B. Decker (2013b), A Three-coordinate system (Ecliptic, Galactic, ISMF) spectral analysis of heliospheric ENA emissions using Cassini/INCA measurements, *Astrophys. Journ.*, *778*, 40; doi:10.1088/0004-637X/778/1/40.

Dialynas, K., S. M. Krimigis, D. G. Mitchell, R. B. Decker, and E. C. Roelof (2017), The bubble-like shape of the heliosphere observed by Voyager and Cassini, *Nat. Astron.*, *1*, 0115; doi:10.1038/s41550-017-0115.

Dougherty, M. K., K. K. Khurana, F. M. Neubauer, C. T. Russell, J. Saur, J. S. Leisner, and M. E. Burton (2006), Identification of a dynamic atmosphere at Enceladus with the Cassini magnetometer, *Science*, *311*, 1406–1409; doi:10.1126/science.1120985.

Ebihara, Y., M. Ejiri, H. Nilsson, I. Sandahl, A. Millillo, M. Grande, J. F. Fennell, and J. L. Roeder (2002), Statistical distribution of the storm-time proton ring current: Polar measurements, *Geophys. Res. Lett.*, *29*(20), 30–1–30–4.

Fok, M. C., T. E. Moore, P. C. Brandt, D. C. Delcourt, S. P. Slinker, and J. A. Fedder (2006), Impulsive enhancements of oxygen ions during substorms, *J. Geophys. Res.*, *111*; doi:10.1029/2006JA011839.

Futaana, Y., J.-Y. Chaufray, H. T. Smith, P. Garnier, H. Lichtenegger, M. Delva, H. Gröller, and A. Mura (2011), Exospheres and Energetic Neutral Atoms of Mars, Venus and Titan, *Space Sci. Rev.*, *162*, 213–266; doi: 10.1007/s11214-011-9834-4.

Futaana, Y., S. Barabash, X.-D. Wang, M. Wieser, G. S. Wieser, P. Wurz, N. Krupp, and P. C. Brandt (2015), Low-energy energetic neutral atom imaging of Io plasma and neutral tori, *Planet. Space Sci.*, *108*, 41–53; doi:10.1016/j.pss.2014.12.022.

Galli, A., P. Wurz, P. Kollmann, P. C. Brandt, M. Bzowski, J. M. Sokól, M. A. Kubiak, A. Grigoriev, and S. Barabash (2013), Heliospheric energetic neutral hydrogen measured with ASPERA-3 and ASPERA-4, *Astrophys. J.*, *775*, 24; doi:10.1088/0004-637X/775/1/24.

Gkioulidou, M., A. Y. Ukhorskiy, D. G. Mitchell, and L. J. Lanzerotti (2016), Storm time dynamics of ring current protons: Implications for the long-term energy budget in the inner magnetosphere, *Geophys. Res. Lett.*, *43*, 47364744; doi:10.1002/2016GL068013.

Gloeckler, G., and D. C. Hamilton (1987), AMPTE ion composition results, *Physica Scripta*, *T18*, 73–84.

Goldstein, J., and D. J. McComas (2013), Five Years of Stereo Magnetospheric Imaging by TWINS, *Space Sci. Rev.*, *180*, 39–70; doi:10.1007/s11214-013-0012-8.

Green, J. L., R. F. Benson, S. F. Fung, W. W. L. Taylor, S. A. Boardsen, B. W. Reinisch, D. M. Haines, et al. (2000), Radio plasma imager simulations and measurements, *Space Sci. Rev.*, *91*, 361–389.

Gurnett, D. A., A. M. Persoon, W. S. Kurth, J. B. Groene, T. F. Averkamp, M. K. Dougherty, and D. J. Southwood (2007),

The variable rotation period of the inner region of Saturn's plasma disk, *Science*, *316*, 442; doi:10.1126/science.1138562.

Gurnett, D. A., W. S. Kurth, D. L. Kirchner, G. B. Hospodarsky, T. F. Averkamp, P. Zarka, A. Lecacheux, et al. (2004), The Cassini radio and plasma wave investigation, *Space Sci. Rev.*, *114*, 395–463; doi:10.1007/s11214-004-1434-0.

Hamilton, D. C., G. Gloeckler, F. M. Ipavich, W. Studemann, B. WIlken, and G. Kremser (1988), Ring current development during the great geomagnetic storm of February 1986, *J. Geophys. Res.*, *93*(A12), 14,343–14,355.

Hedin, A. E. (1991), Extension of the MSIS thermospheric model into the middle and lower atmosphere, *J. Geophys. Res.*, *96*, 1159–1172.

Iijima, T., and T. A. Potemra (1976), The amplitude distribution of field-aligned currents at northern high latitudes observed by TRIAD, *J. Geophys. Res.*, *81*(13), 2165–2174.

Jackman, C. M., N. Achilleos, S. W. Cowley, E. J. Bunce, A. Radioti, D. Grodent, S. V. Badman, M. K. Dougherty, and W. Pryor (2013), Auroral counterpart of magnetic field dipolarizations in Saturn's tail, *Planet. Space Sci.*, *82–83*, 34–42; doi:10.1016/j.pss.2013.03.010.

Jones, S. T., M.-C. Fok, and P. C. Brandt (2006), Modeling global O^+ substorm injection using analytic magnetic field model, *J. Geophys. Res.*, *111*(A11S07); doi:10.1029/2006JA011607.

Jorgensen, A. M. (1997), Global energetic neutral atom (ENA) measurements and their association with the Dst index, *Geophys. Res. Lett.*, *24*, 3173–3176.

Jurac, S., and J. D. Richardson (2005), A self-consistent model of plasma and neutrals at Saturn: Neutral cloud morphology, *J. Geophys. Res. Space Physics*, *110*(9); doi:10.1029/GL2004JA010635.

Keika, K., L. M. Kistler, and P. C. Brandt (2013), Energization of O^+ ions in the Earth's inner magnetosphere and the effects on ring current buildup: A review of previous observations and possible mechanisms, *J. Geophys. Res. Space Physics*, *118*, 4441–4464; doi:10.1002/jgra.50371.

Krimigis, S. M., D. G. Mitchell, D. C. Hamilton, S. Livi, I. Dandouras, S. Jaskulek, T. P. Armstrong, et al. (2004), Magnetosphere Imaging Instrument (MIMI) on the Cassini mission to Saturn, *Space Sci. Rev.*, *114*(1–4), 233–329.

Krimigis, S. M., D. G. Mitchell, E. C. Roelof, K. C. Hsieh, and D. J. McComas (2009), Imaging the interaction of the heliosphere with the interstellar medium from Saturn with Cassini, *Science*, *326*, 971; doi:10.1126/science.1181079.

Krimigis, S. M., G. Gloeckler, R. W. McEntire, T. A. Potemra, F. L. Scarf, and E. G. Shelley (1985), Magnetic storm of September 4, 1984: A synthesis of ring current spectra and energy densities measured with AMPTE/CCE, *Geophys. Res. Lett.*, *12*(5), 329–332.

Krimigis, S. M., N. Sergis, D. G. Mitchell, D. C. Hamilton, and N. Krupp (2007), A dynamic, rotating ring current around Saturn, *Nature*, *450*, 1050–1053; doi:10.1038/nature06425.

Kronberg, E. A., J. Woch, N. Krupp, A. Lagg, K. K. Khurana, and K.-H. Glassmeier (2005), Mass release at jupiter: Substorm-like processes in the Jovian magnetotail, *J. Geophys. Res.*, *110* (A03211); doi:10.1029/2004JA010777.

Krupp, N., J. Woch, A. Lagg, B. Wilken, S. Livi, and D. J. Williams (1998), Energetic particle bursts in the predawn Jovian magnetotail, *Geophys. Res. Lett.*, *25*, 1249–1252; doi:10.1029/98GL00863.

Liemohn, M. W., R. Ilie, N. Y. Ganushkina, A. J. Ridley, J. U. Kozyra, M. F. Thomsen, and J. E. Borovsky (2011), Testing the necessity of transient spikes in the storm time ring current drivers, *J. Geophys. Res. Space Physics*, *116*, A04226; doi:10.1029/2010JA015914.

Lindsay, B. G., and R. F. Stebbings (2005), Charge transfer cross sections for energetic neutral atom data analysis, *J. Geophys. Res.*, *110*, 12,213; doi:10.1029/2005JA011298.

Liou, K., C. Meng, A. T. Y. Lui, P. T. Newell, and R. R. Anderson (2000), Auroral kilometric radiation at substorm onset, *J. Geophys. Res.*, *105*, 25,325–25,332; doi:10.1029/2000JA000038.

Lu, L., S. McKenna-Lawlor, S. Barabash, P. C. Brandt, J. Balaz, Z. X. Liu, Z. H. He, and G. D. Reeves (2010), Comparisons between ion distributions retrieved from ENA images of the ring current and contemporaneous, multipoint ion measurements recorded in situ during the major magnetic storm of 15 May 2005, *J. Geophys. Res.*, *115*(A12218); doi:10.1029/2010JA015770.

Lyons, L. R., and D. J. Williams (1984), *Quantitative Aspects of Magnetospheric Physics*, 2d ed., D. Reidel Publishing Company.

Mauk, B. H. (1986), Quantitative modeling of the "convection surge" mechanism for ion acceleration, *J. Geophys. Res.*, *91*(A12), 13,423–13,431.

Mauk, B. H., D. G. Mitchell, S. M. Krimigis, E. C. Roelof, and C. P. Paranicas (2003), Energetic neutral atoms from a trans-Europa gas torus at Jupiter, *Nature*, *421*, 920–922; doi:10.1038/nature01431.

Mauk, B. H., D. G. Mitchell, R. W. McEntire, C. P. Paranicas, E. C. Roelof, D. J. Williams, S. M. Krimigis, and A. Lagg (2004), Energetic ion characteristics and neutral gas interactions in Jupiter's magnetosphere, *J. Geophys. Res.*, *109*, 9; doi:10.1029/2003JA010270.

Mauk, B. H., D. J. Williams, R. W. McEntire, K. K. Khurana, and J. G. Roederer (1999), Storm-like dynamics of Jupiter's inner and middle magnetosphere, *J. Geophys. Res.*, *104*, 22,759–22,778; doi:10.1029/1999JA900097.

McComas, D. J., F. Allegrini, J. Baldonado, B. Blake, P. C. Brandt, J. Burch, J. Clemmons, et al. (2009a), The Two Wide-angle Imaging Neutral-atom Spectrometers (TWINS) NASA mission-of-opportunity, *Space Sci. Rev.*, *142*, 157–231; doi:10.1007/s11214-008-9467-4.

McComas, D. J., F. Allegrini, M. Bzowski, M. A. Dayeh, R. DeMajistre, H. O. Funsten, S. A. Fuselier, et al. (2014), IBEX: The first five years (2009–2013), *Astrophys. J. Suppl. Ser.*, *213*, 20; doi:10.1088/0067-0049/213/2/20.

McComas, D. J., F. Allegrini, P. Bochsler, M. Bzowski, M. Collier, H. Fahr, H. Fichtner, et al. (2009b), IBEX, interstellar boundary explorer, *Space Sci. Rev.*, *18*; doi:10.1007/s11214-009-9499-4.

McKenna-Lawlor, S., L. Li, S. Barabash, K. Kudela, J. Balaz, I. Strharsky, K. Brinkfeldt, et al. (2005), The NUADU experiment on TC-2 and the first Energetic Neutral Atom (ENA) images recorded by this instrument, *Ann. Geophys.*, *23*(8), 2825–2849.

Menke, W. (1989), *Geophysical Data Analysis: Discrete Inverse Theory*, Academic Press Inc.

Mitchell, D. G., C. P. Paranicas, B. H. Mauk, E. C. Roelof, and S. M. Krimigis (2004), Energetic neutral atoms from Jupiter measured with the Cassini magnetospheric imaging instrument: Time dependence and composition, *J. Geophys. Res.*, *109*, 9; doi:10.1029/2003JA010120.

Mitchell, D. G., P. C. Brandt, E. C. Roelof, D. C. Hamilton, K. C. Retterer, and S. Mende (2003), Global imaging of O^+ from IMAGE/HENA, *Space Sci. Rev.*, *109*(1–4), 63–75.

Mitchell, D. G., P. C. Brandt, E. C. Roelof, J. Dandouras, S. Krimigis, B. H. Mauk, C. P. Paranicas, et al. (2005), Energetic ion acceleration in Saturn's magnetotail: Substorms on Saturn?, *Geophys. Res. Lett.*, *32*(L20S01); doi:10.1029/2005GL022647.

Mitchell, D. G., P. C. Brandt, J. Carbary, W. Kurth, S. Krimigis, C. Paranicas, N. Krupp, et al. (2015), Injection, interchange and reconnection: Energetic particle observations in saturn's magnetotail, *AGU Monograph 207*, 327–342.

Mitchell, D. G., P. C. Brandt, J. H. Westlake, S. E. Jaskulek, G. B. Andrews, and K. S. Nelson (2016), Energetic particle imaging: The evolution of techniques in imaging high-energy neutral atom emissions, *J. Geophys. Res.*, *121*, 88048820; doi:10.1002/2016JA022586.

Mitchell, D. G., S. E. Jaskulek, C. E. Schlemm, E. P. Keath, R. E. Thompson, B. E. Tossman, J. D. Boldt, et al. (2000), High energy neutral atom (HENA) imager for the IMAGE mission, *Space Sci. Rev.*, *91*, 67–112.

Mitchell, D. G., S. M. Krimigis, C. Paranicas, P. C. Brandt, J. F. Carbary, E. C. Roelof, W. S. Kurth, et al. (2009), Recurrent energization of plasma in the midnight-to-dawn quadrant of Saturn's magnetosphere, and its relationship to auroral UV and radio emissions, *Planet. Space Sci.*, *57*, 1732–1742; doi:10.1016/j.pss.2009.04.002.

Moore, T. E., D. J. Chornay, M. R. Collier, F. A. Herrero, J. Johnson, M. A. Johnson, J. W. Keller, et al. (2000), The low-energy neutral atom imager for IMAGE, in *The Image Mission*, Springer, 155–195.

Nosé, M., S. Taguchi, K. Hosokawa, S. P. Christon, R. W. McEntire, T. E. Moore, and M. R. Collier (2005), Overwhelming O^+ contribution to the plasma sheet energy density during the October 2003 superstorm: Geotail/EPIC and IMAGE/LENA observations, *J. Geophys. Res.*, *110*, 9; doi:10.1029/2004JA010930.

Ohtani, S., H. Korth, P. C. Brandt, L. G. Blomberg, H. J. Singer, M. G. Henderson, E. A. Lucek, et al. (2007), Cluster observations in the inner magnetosphere during the 18 April 2002 sawtooth event: Dipolarization and injection at r = 4.6 RE, *J. Geophys. Res. Space Physics*, *112*, A08213; doi:10.1029/2007JA012357.

Ohtani, S., M. Nose, G. Rostoker, H. Singer, A. T. Y. Lui, and M. Nakamura (2001), Storm-substorm relationship: Contribution of the tail current to Dst, *J. Geophys. Res.*, *106*(A10), 21,199–21,209.

Ohtani, S., P. C. Brandt, D. G. Mitchell, H. Singer, M. Nosé, G. D. Reeves, and S. Mende (2005a), Storm-substorm relationship: Variations of the hydrogen and oxygen ENA intensities during storm-time substorms, *J. Geophys. Res.*, *110*(A07219); doi:10.1029/2004JA010954.

Ohtani, S., P. C. Brandt, H. J. Singer, D. G. Mitchell, and E. C. Roelof (2005b), Statistical characteristics of hydrogen and oxygen ENA emission from the storm-time ring current, *J. Geophys. Res.*, *111*(A06209); doi:10.1029/2005JA011201.

Ostgaard, N., S. B. Mende, H. U. Frey, G. R. Gladstone, and H. Lauche (2003), Neutral hydrogen profiles derived from geocoronal imaging, *J. Geophys. Res.*, *108*(A7), SMP 18–1–18–12; doi:10.1029/2002JA009749.

Paranicas, C. P., B. H. Mauk, and S. M. Krimigis (1991), Pressure anisotropy and radial stress balance in the Jovian neutral sheet, *J. Geophys. Res.*, *96*(A12), 21,135–21,140.

Paranicas, C. P., D. G. Mitchell, P. C. Brandt, S. M. Krimigis, and B. H. Mauk (2005), Periodic intensity variations in global ENA images of Saturn, *Geophys. Res. Lett.*, *32*(L21101); doi:10.1029/2005GL023656.

Parker, E. N. (1957), Newtonian development of the dynamical properties of ionized gases of low density, *Phys. Rev.*, *107*(4), 924–933.

Perez, J. D., E. W. Grimes, J. Goldstein, D. J. McComas, P. Valek, and N. Billor (2012), Evolution of CIR storm on 22 July 2009, *J. Geophys. Res.*, *117*(A09221); doi:10.1029/2012JA017572.

Plainaki, C., A. Milillo, A. Mura, S. Orsini, S. Massetti, and T. Cassidy (2012), The role of sputtering and radiolysis in the generation of Europa exosphere, *Icarus*, *218*, 956–966; doi:10.1016/j.icarus.2012.01.023.

Pollock, C. J., K. Asamura, J. Baldonado, M. M. Balkey, P. Barker, J. L. Burch, E. J. Korpela, et al. (2000), Medium energy neutral atom (MENA) imager for the IMAGE mission, *Space Sci. Rev.*, *91*(1), 113–154.

Provan, G., D. J. Andrews, C. S. Arridge, A. J. Coates, S. W. H. Cowley, S. E. Milan, M. K. Dougherty, and D. M. Wright (2009), Polarization and phase of planetary-period magnetic field oscillations on high-latitude field lines in Saturn's magnetosphere, *J. Geophys. Res.*, *114*, 2225; doi:10.1029/2008JA013782.

Reeves, G. D., and M. G. Henderson (2001), The storm-substorm relationship: Ion injections in geosynchronous measurements and composite energetic neutral atom images, *J. Geophys. Res.*, *106*, 5833–5844.

Reme, H., C. Aoustin, J.-M. Bosqued, I. Dandouras, B. Lavraud, J.-A. Sauvaud, A. Barthe et al. (2001), First multispacecraft ion measurements in and near the Earth's magnetosphere with the identical cluster ion spectrometry (CIS) experiment, *Ann. Geophys.*, *19*, 1303.

Richardson, J. D. (1998), Thermal plasma and neutral gas in Saturn's magnetopshere, *Rev. of Geophys.*, *36*(4), 501–524.

Rodgers, C. (2000), *Inverse Methods for Atmospheric Sounding*, World Scientific.

Roelof, E. C. (1987), Energetic neutral atom image of a storm-time ring current, *Geophys. Res. Lett.*, *14*, 652–655.

Roelof, E. C. (1989), Remote sensing of the ring current using energetic neutral atoms, *Adv. Space Res.*, *9*(12), 12,195–12,203.

Roelof, E. C., and A. J. Skinner (2000), Extraction of ion distributions from magnetospheric ENA and EUV images, *Space Sci. Rev.*, *91*, 437–459.

Roelof, E. C., D. G. Mitchell, and D. J. Williams (1985), Energetic neutral atoms ($E \approx 50$ keV) from the ring current: IMP 7/8 and ISEE 1, *J. Geophys. Res.*, *90*(A11), 10,991–11,008.

Roelof, E. C., P. C. Brandt, and D. G. Mitchell (2004), Derivation of currents and diamagnetic effects from global pressure distributions obtained by IMAGE/HENA, *Adv. Space Res.*, *33*/5, 747–751; doi:10.1016/S0273-1177(03) 00633-1.

Roth, L., J. Saur, K. D. Retherford, D. F. Strobel, P. D. Feldman, M. A. McGrath, and F. Nimmo (2014), Transient water vapor at Europa's south pole, *Science*, *343*, 171–174; doi:10.1126/science.1247051.

Runov, A., V. Angelopoulos, X.-Z. Zhou, X.-J. Zhang, S. Li, F. Plaschke, and J. Bonnell (2011), A THEMIS multicase study of dipolarization fronts in the magnetotail plasma sheet, *J. Geophys. Res.Space Physics*, *116*, A05216; doi:10.1029/2010JA016316.

Rymer, A. M., B. H. Mauk, T. W. Hill, N. André, D. G. Mitchell, C. Paranicas, D. T. Young, et al. (2009), Cassini evidence for rapid interchange transport at Saturn, *Planet. Space Sci.*, *57*, 1779–1784; doi:10.1016/j.pss.2009.04.010.

Sergis, N., C. M. Jackman, M. F. Thomsen, S. M. Krimigis, D. G. Mitchell, D. C. Hamilton, M. K. Dougherty, N. Krupp, and R. J. Wilson (2017), Radial and local time structure of the saturnian ring current revealed by cassini, *J. Geophys. Res.*, *122*, 1803–1815; doi:doi:10.1002/2016JA023742.

Sergis, N., S. M. Krimigis, E. C. Roelof, C. S. Arridge, A. M. Rymer, D. G. Mitchell, D. C. Hamilton, et al. (2010), Particle pressure, inertial force, and ring current density profiles in the magnetosphere of Saturn, based on Cassini measurements, *Geophys. Res. Lett.*, *37*(L02102); doi:10.1029/2009GL041920.

Sitnov, M. I., N. A. Tsyganenko, A. Y. Ukhorskiy, and P. C. Brandt (2008), Dynamical data-based modeling of the storm-time geomagnetic field with enhanced spatial resolution, *J. Geophys. Res.*, *113*, 7218; doi:10.1029/2007JA013003.

Smith, H. T., and J. D. Richardson (2017), The spatial and temporal distribution of Enceladus-generated neutral gas in saturn's magnetosphere, *J. Geophys. Res.*, to be submitted.

Smith, H. T., R. E. Johnson, M. E. Perry, D. G. Mitchell, R. L. McNutt, and D. T. Young (2010), Enceladus plume variability and the neutral gas densities in Saturn's magnetosphere, *J. Geophys. Res. Space Physics*, *115*, A10252; doi:10.1029/2009JA015184.

Smyth, W. H., and M. L. Marconi (2006), Europa's atmosphere, gas tori, and magnetospheric implications, *Icarus*, *181*, 510–526; doi:10.1016/j.icarus.2005. 10.019.

Sparks, W. B., K. P. Hand, M. A. McGrath, E. Bergeron, M. Cracraft, and S. E. Deustua (2016), Probing for Evidence of Plumes on Europa with HST/STIS, *Astrophys. J.*, *829*, 121; doi:10.3847/0004-637X/829/2/121.

Stephens, G. K., M. I. Sitnov, A. Y. Ukhorskiy, E. C. Roelof, N. A. Tsyganenko, and G. Le (2016), Empirical modeling of the storm time innermost magnetosphere using Van Allen Probes and THEMIS data: Eastward and banana currents, *J. Geophys. Res. Space Physics*, *121*, 157–170; doi:10.1002/2015JA021700.

Thomas, N., F. Bagenal, T. W. Hill, and J. K. Wilson (2004), The Io neutral clouds and plasma torus, in *Jupiter, the Planet, Satellites, and Magnetosphere*, Cambridge University Press, New York.

Tsang, C. C. C., J. R. Spencer, E. Lellouch, M. A. Lopez-Valverde, and M. J. Richter (2016), The collapse of Io's primary atmosphere in Jupiter eclipse, *J. Geophys. Res. Planets*, *121*, 1400–1410; doi:10.1002/2016JE005025.

Tsyganenko, N. A., and M. I. Sitnov (2005), Modeling the dynamics of the inner magnetosphere during strong geomagnetic storms, *J. Geophys. Res.*, *101*(A03208); doi:10.1029/2004JA010798.

Twomey, S. (1977), *Introduction to the Mathematics in Remote Sensing and Indirect Measurements*, Developments in Geomathematics 3, Elsevier Scientific Publishing Company.

Valek, P. W., J. Goldstein, D. J. McComas, M.-C. Fok, and D. G. Mitchell (2014), Large magnetic storms as viewed by TWINS: A study of the differences in the medium energy ENA composition, *J. Geophys. Res. Space Physics*, *119*, 2819–; doi:10.1002/2014JA019782.

Vallat, C., I. Dandouras, M. Dunlop, A. Balogh, E. Lucek, G. K. Parks, M. Wilber, E. C. Roelof, G. Chanteur, and H. Rème (2005), First current density measurements in the ring current region using simultaneous multi-spacecraft CLUSTER-FGM data, *Ann. Geophys.*, *23*, 1849–1865.

Vasyliunas, V. K. (1984), Fundamentals of current description, in *Magnetospheric Currents*, edited by T. A. Potemra, 63–66, AGU Monograph 28, American Geophysical Union, Washington, DC.

Westlake, J. H., D. G. Mitchell, P. C.-s. Brandt, B. G. Andrews, and G. Clark (2016), The Low-Energy Neutral Imager (LENI), *J. Geophys. Res. Space Physics*, *121*, 8228–8236; doi:10.1002/2016JA022547.

Wieser, M., S. Barabash, Y. Futaana, M. Holmström, A. Bhardwaj, R. Sridharan, M. B. Dhanya, P. Wurz, A. Schaufelberger, and K. Asamura (2009), Extremely high reflection of solar wind protons as neutral hydrogen atoms from regolith in space, *Planet. Space Sci.*, *57*, 2132–2134; doi:10.1016/j.pss.2009.09.012.

Woch, J., N. Krupp, A. Lagg, B. Wilken, S. Livi, and D. J. Williams (1998), Quasi-periodic modulations of the Jovian magnetotail, *Geophys. Res. Lett.*, *25*, 1253–1256; doi:10.1029/98GL00861.

Yang, J., F. R. Toffoletto, R. A. Wolf, S. Sazykin, R. W. Spiro, P. C. Brandt, M. G. Henderson, and H. U. Frey (2008), Rice Convection Model simulation of the 18 April 2002 sawtooth event and evidence for interchange instability, *J. Geophys. Res. Space Physics*, *113*, A11214; doi:10.1029/2008JA013635.

7

Terrestrial Ring Current: A Review of Cluster Results Based on the Curlometer Technique

I. Dandouras[1], S. Rochel-Grimald[2], C. Vallat[3], and M. W. Dunlop[4]

ABSTRACT

The inner magnetosphere electric currents configuration and mapping is one of the key elements for understanding current loop closure inside the entire magnetosphere. The ring current is toroidal and flows in the near-Earth region, where the magnetic field is dipole-like. This current system is driven by the pressure gradients, and it is formed by the drift of the energetic charged particles that are injected from the magnetotail toward the Earth during the magnetospheric storms and substorms. A method for directly computing currents in space is the multispacecraft curlometer technique, which is based on the Maxwell-Ampère law application. This requires the use of four-point simultaneous magnetic field measurements. The FGM (Fluxgate Magnetometer) experiment on board the four Cluster spacecraft allowed for the first time the instantaneous calculation of the magnetic field gradients and thus the measurement of the local current density. This technique requires, however, a careful analysis concerning all the factors that can affect the accuracy of the current density calculation. The CIS (Cluster Ion Spectrometry) experiment on board these spacecraft provides the ion distribution functions of the current carriers within its energy range. Earlier and more recent results, based on Cluster data acquired during passes in the ring current region at different perigee altitudes, are reviewed in the light of recent progress on the accuracy of the method.

7.1. INTRODUCTION

The terrestrial ring current is a toroidal current system flowing in the near-Earth region, where the magnetic field is mainly dipole-like (but is also disturbed by the field generated by the currents flowing in space [cf. *Ganushkina et al.*, 2010; *Lühr et al.*, 2017]). The ring current is one of the oldest concepts in magnetospheric physics. In 1907, Störmer first introduced a ring-shaped current flowing around the Earth [*Störmer*, 1907]. *Chapman and Ferraro* [1931, 1941] used a ring current concept for their geomagnetic storm model. The existence of a ring current forming around the Earth was then more developed by *Singer* [1957], who showed that a westward electric current was produced by the energetic particles (~1 keV to a few hundred keV) trapped in the geomagnetic field, which are drifting due to the gradient of the field. The equivalent current can be schematized as toroidal and flowing around the Earth at geocentric distances from about 2 R_E to 9 R_E. The development of an intense ring current is an important feature of the geomagnetic storm [*Daglis et al.*, 1999].

[1] *IRAP, Université de Toulouse, CNRS, UPS, CNES, Toulouse, France*

[2] *ONERA-DESP (Département de l'Environnement Spatial), Toulouse, France*

[3] *European Space Astronomy Centre, Villanueva de la Cañada, Madrid, Spain*

[4] *Space Science Institute, School of Astronautics, Beihang University, Beijing, China; RAL Space, STFC, Chilton, Oxfordshire, UK*

Electric Currents in Geospace and Beyond, Geophysical Monograph 235, First Edition.
Edited by Andreas Keiling, Octav Marghitu, and Michael Wheatland.

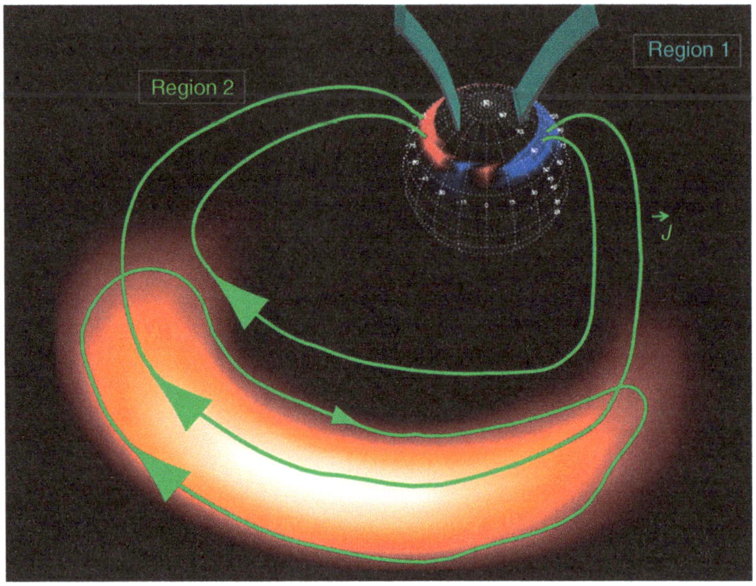

Figure 7.1 A schematic representation of the partial ring current system, as seen from the nightside. The orange-colored structure represents the plasma pressure distribution and the green lines represent the current flow. Courtesy of P. C. Brandt.

The global current system in the inner magnetosphere is more complex (cf. Chapter 2, this volume), but it can be described as a current system driven mainly by pressure gradients. Plasma convected earthward from the magnetotail, or injected earthward during geomagnetic storms or substorms, builds up a pressure distribution with a peak around 3 R_E. Considering the MHD momentum equation under static conditions, simplifying for isotropic plasma pressure and ignoring gravity and collisions [*Parker*, 1957]:

$$\mathbf{J} = \frac{\mathbf{B} \times \nabla P}{B^2} \tag{7.1}$$

where **J** represents the current density, **B** represents the magnetic field, and *P* is the plasma pressure. This plasma pressure profile corresponds to a current circulation with a westward current outside (tailward in the nightside) of the pressure peak and an eastward current inside of the pressure peak (cf. Figure 7.1). These represent the westward and eastward ring currents, respectively.

The ring current system is mostly a partial ring current. The current closure around the pressure peak is represented by a banana current [*Liemohn et al.*, 2013], whereas the westward-eastward current asymmetry results in a part of the westward perpendicular current $J_\perp$ closing through field-aligned currents $J_\parallel$ into and out of the ionosphere [*Roelof et al.*, 2004; *Liemohn et al.*, 2015; *Ganushkina et al.*, 2015]. This is shown schematically in Figure 7.1, where the orange-colored structure represents the plasma pressure distribution, as seen from the nightside, and the green lines, the current flow. The partial westward ring current (left-pointing arrows), eastward ring current (small right-pointing arrow), banana current (contour around the plasma pressure maximum), and the Region 2 field-aligned currents are shown.

7.2. CLUSTER MISSION AND INSTRUMENTATION

The Cluster mission consists of four identical spacecraft launched in 2000 on similar elliptical polar orbits, with an initial perigee at about 4 R_E and an apogee at 19.6 R_E [*Escoubet et al.*, 2001]. This allows the Cluster spacecraft to cross the ring current region from south to north, during every perigee pass, and to obtain the latitudinal profile following almost the same magnetic flux tube [*Vallat et al.*, 2005; *Dandouras et al.*, 2009]. In addition, due to the precession of its orbit, over a year Cluster is crossing the equator successively at all MLT ranges.

During the mission, the interspacecraft separation strategy has been planned in order to allow the study of the various plasma structures encountered by Cluster along the orbit. The maneuvers to adapt the interspacecraft separation take place once or twice a year, depending on the spatial scales of the plasma structures to be studied by the mission. The tetrahedron formed by the four spacecraft can thus have characteristic sizes ranging between typically 100 km and a few R_E, whereas ~10 km interspacecraft distances have been also achieved. Moreover, the tetrahedron geometry evolves along the orbit. As a consequence, the mission allows the analysis of the ring current only for limited time periods, during which the spacecraft constellation is suitable.

Onboard each spacecraft, 11 experiments allow a wide set of measurements of the plasma parameters (particles and fields). Among them, the Fluxgate Magnetometer (FGM), the Cluster Ion Spectrometer (CIS), as well as the high-energy instrument Research with Adaptive Particle Imaging Detectors (RAPID) are present.

The FGM experiment on board Cluster consists of two triaxial fluxgate magnetometers and an onboard data-processing unit on each spacecraft [*Balogh et al.*, 2001]. It allows an accurate measurement of the ambient magnetic field (up to 8 pT resolution), while the background interference from the spacecraft has been minimized by placing the magnetometers on 5 m booms, avoiding interference with the spacecraft.

The Cluster Ion Spectrometer experiment consists of two complementary ion sensors, the Composition and Distribution Function analyzer (CODIF) and the Hot Ion Analyzer (HIA). CODIF is a mass spectrometer that gives the ion composition with a 22.5° angular resolution whereas HIA offers a better angular resolution (~5°) but with no mass discrimination. The CIS experiment is capable of measuring the full three-dimensional ion distribution functions of the major ion species, from thermal energies (~1 eV/e) up to about 40 keV/e, with a one spacecraft spin (4 sec) time resolution [*Rème et al.*, 2001].

The RAPID experiment is an imaging spectrometer for ions and electrons on board the four Cluster spacecraft, providing the complete particle parameters, that is, velocity and atomic mass number, over an energy range from ~30 keV up to 1.5 MeV [*Wilken et al.*, 2001].

7.3. THE CURLOMETER TECHNIQUE

Currents in space can be measured by three different methods [e.g., *Ganushkina et al.*, 2015]: (1) direct measurements of the current carriers; (2) deriving the current density from plasma pressure measurements, through equation (7.1); and (3) magnetic field measurements. A method for directly computing current density from magnetic field measurements is the multispacecraft curlometer technique, which is based on the application of the Maxwell-Ampère law. This technique requires the use of four-point simultaneous magnetic field measurements. The curlometer technique has been described in detail by *Dunlop et al.* [1988; 2002] and *Robert et al.* [1998]. Here we only briefly outline the method.

According to Maxwell-Ampère's law, and assuming stationarity,

$$\mu_0 \mathbf{J} = \nabla \times \mathbf{B} \tag{7.2}$$

Thus, in a discrete Cartesian coordinate system, we have:

$$\begin{aligned}
(\mathrm{curl}\,\mathbf{B})_x &\approx \frac{\Delta Bz}{\Delta y} - \frac{\Delta By}{\Delta z} \\
(\mathrm{curl}\,\mathbf{B})_y &\approx \frac{\Delta Bx}{\Delta z} - \frac{\Delta Bz}{\Delta x} \\
(\mathrm{curl}\,\mathbf{B})_z &\approx \frac{\Delta By}{\Delta x} - \frac{\Delta Bx}{\Delta y}
\end{aligned} \tag{7.3}$$

These equations can be applied to the Cluster constellation data (four simultaneous points for the magnetic field measurements, from the four spacecraft) to evaluate the magnetic field gradients within the constellation and thus the current density through the tetrahedron formed by the four spacecraft.

Considering that the ring current is carried by energetic charged particles drifting around the Earth, and that it is centered close to the plane of the magnetic equator and it has a finite latitudinal extent above and below this plane, a cylindrical coordinate system is preferable to a Cartesian. The local cylindrical coordinate system, derived from the Solar-Magnetic (SM) system, is thus usually adopted.

The main assumption required for the application of the curlometer technique is that the variation of the magnetic field inside the tetrahedron is linear, which implies that the current density inside the tetrahedron is constant. This is equivalent to stating that the four spacecraft are passing simultaneously through the same current sheet. This condition is well satisfied as long as the inter-spacecraft separation is small enough compared to the plasma structures. For the ring current region, this implies interspacecraft distances typically no larger than ~500 km [*Vallat et al.*, 2005].

In addition to the size scale of the tetrahedron, two other parameters that have been used to characterize the tetrahedron geometry are the elongation and the planarity. The relationship between these two parameters and the current density error has been extensively discussed by *Robert et al.* [1998] and by *Chanteur* [2000]. Due to orbital dynamics, severe distortions of the tetrahedron can occur in some parts of the orbit, particularly close to perigee, adversely affecting the accuracy of the current density calculation in these cases.

Other parameters that can affect the accuracy of the results are the contribution of the Earth's dipolar field (if not subtracted from the magnetic field measurements) and associated truncation errors induced by the curlometer method, and any eventual inaccuracies in the magnetic field measurements or in the spacecraft positions knowledge. A review of the possible error sources, and limitations in the accuracy of the curlometer method when applied to the ring current region, is given in *Vallat et al.* [2005] and in *Grimald et al.* [2012].

Based on the fact that the magnetic field is divergence free, a simple criterion that has been often used to check the validity of the curlometer calculations is the ratio div **B**/|curl **B**|, required to be typically less than 0.5 [*Dunlop et al.*, 2002; *Zhang et al.*, 2011].

7.4. OBSERVATIONS AND ANALYSIS

7.4.1. The 2002 Ring Current Observations

During the period February to June 2002, the four Cluster spacecraft achieved the closest overall spacing, with a 100 km typical separation in a regular tetrahedron formation while crossing the cusp region. This spacecraft constellation resulted in an elongated tetrahedron when traversing the ring current region, in the along-the-orbit direction, with up to ~250 km separations in the z-axis direction of the Solar-Magnetic (SM) coordinate system and around 100 km separations in the magnetic equator plane. Nineteen perigee passes during this period, for which FGM data were available, have been studied in detail by *Vallat et al.* [2005], corresponding to the MLT sector from 17 to 01 hours (dusk to postmidnight sector) and covering a variety of magnetospheric activity conditions, from quiet to storm main phase and to recovery. Here we will briefly review two of them, one corresponding to quiet conditions and another that occurred during a geomagnetic storm.

Figure 7.2 shows the Cluster ion data and the calculated current density using the curlometer method for a quiet-time perigee pass: 18 March 2002, with $K_p = 1+$ and $D_{st} \sim 10$ nT. The ring current region is between the two vertical dashed lines. When Cluster was entering the ring current region, the CIS-CODIF data revealed a clear decrease of the energetic ion fluxes at energy ranges up to 40 keV, that is, up to the instrument upper limit. The ring current region is flanked at higher latitudes by the inner plasma sheet horns (nose structures), where very narrow "filamentations" are observed in the current density, due to the fine-scale structure of the currents there.

The "filamentations," which are of a scale smaller than the Cluster interspacecraft separations, disappeared when the constellation entered the ring current region. The current density profile there became extremely smooth, almost linear all along the traversal until the entry into the northern inner plasma sheet (at 11:28 UT). The J_φ component in particular (bottom panel), which is the azimuthal component of the current, was remarkably stable and about –20 nA m^{-2} ± 20%. The mean value of the total current J_{total} was about 20.9 nA m^{-2} ± 20% (cf. middle panel). The negative value of J_φ is in good agreement with the expected ring current configuration, flowing in this region from dawn to dusk (cf. Fig. 7.1), that is, a westward current. The ring current is observed to have a large latitudinal extent, between I_{lat} –65° and 65°, and at the magnetic equator it was almost fully azimuthal.

Figure 7.3 shows the application of the curlometer technique on a storm-time event, where the D_{st} index reached a value of –101 nT. Even if this event is more disturbed than the one previously analyzed, the transition from the plasma sheet to the ring current is still characterized by a change from a strongly oscillating profile of the current components, until 17:43 UT, related to narrowly structured field-aligned currents, to a more regular one. Once the filamentations disappeared (after 17: 43 UT), the current profile became smoother, even if it still showed a few small amplitude perturbations related to the high activity level, possibly due to multiple particle injections. The dominant current component in the ring current region is also here, as expected, the J_φ component, with a mean value of J_φ = –24.6 nA m^{-2} ± 20%. The mean value of J_{total} is about 25.6 nA m^{-2} ± 20%. In spite of a much stronger geomagnetic activity level, compared to the previous event, the increase of the current density along the Cluster trajectory did not exceed 25% with respect to the quiet-time conditions.

7.4.2. Extension to Cluster Ring Current Observations Covering All MLT Sectors

Using Cluster observations up to April 2004, *Zhang et al.* [2011] extended the analysis of the ring current density to cover all MLT sectors (cf. also Chapter 4, this volume). The events during which the four spacecraft had small separations and close to regular tetrahedral configurations, and the D_{st} index was greater than –30 nT (nonstorm events), were examined. The div **B**/|curl **B**| criterion has been used to control the accuracy of the results. Figure 7.4 shows the distribution of the current density vectors projected on the magnetic equator plane.

The results obtained by *Zhang et al.* [2011] confirm the existence of a substantial quiet-time westward ring current, shown also by *Vallat et al.* [2005]. As indicated earlier, the medium-energy (<40 keV) ion fluxes can be very low during such events (cf. Figure 7.2), which raises the question of which are the current carriers. Higher-energy (above ~95 keV) ion data from the RAPID instrument on board Cluster, showing increases of the proton fluxes in the ring current region during such quiet-time ring current observations, point toward these very energetic ion populations as the main contributors to the permanent ring current. These data, for the 18 March 2002 quiet-time event, are shown in Figure 7.5a, which combines ion measurements from the CIS-CODIF instrument, for low to medium energies (up to 40 keV), separately for H^+, He^+, and O^+ ions, and from the RAPID

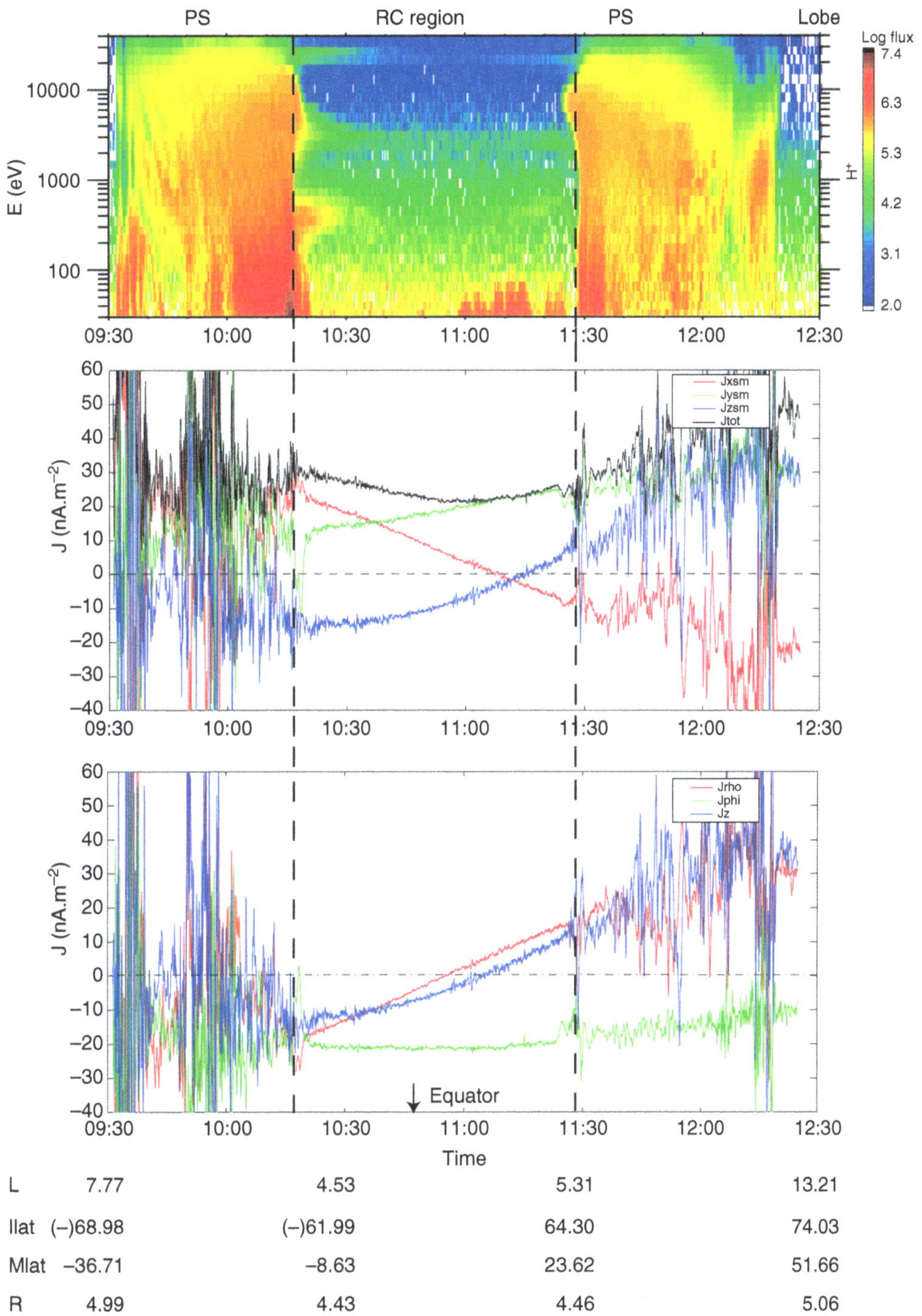

Figure 7.2 Cluster data for the 18 March 2002 ring current event: H^+ energy-time spectrogram (CIS-CODIF instrument) for spacecraft 4 (SC4) in particle flux units (ions/cm^2 s sr keV), current density components in the SM coordinate system and in nA/m^2 (second panel), and transformed into cylindrical coordinates (bottom panel). The black dashed lines demarcate the ring current region. *L*-shell, invariant latitude, magnetic latitude, and geocentric distance are indicated at the bottom. From *Vallat et al.* [2005].

instrument for high-energy (28–4000 keV) H^+ ions. The presence here of ~100–1000 keV protons as the main ring current carriers contrasts with the 20 April 2002 geomagnetic storm event data, shown in Figure 7.5.b, which reveal an intense ion population starting from as low as 4 keV.

These observations are consistent with the results of *Gkioulidou et al.* [2016], showing that the high-energy

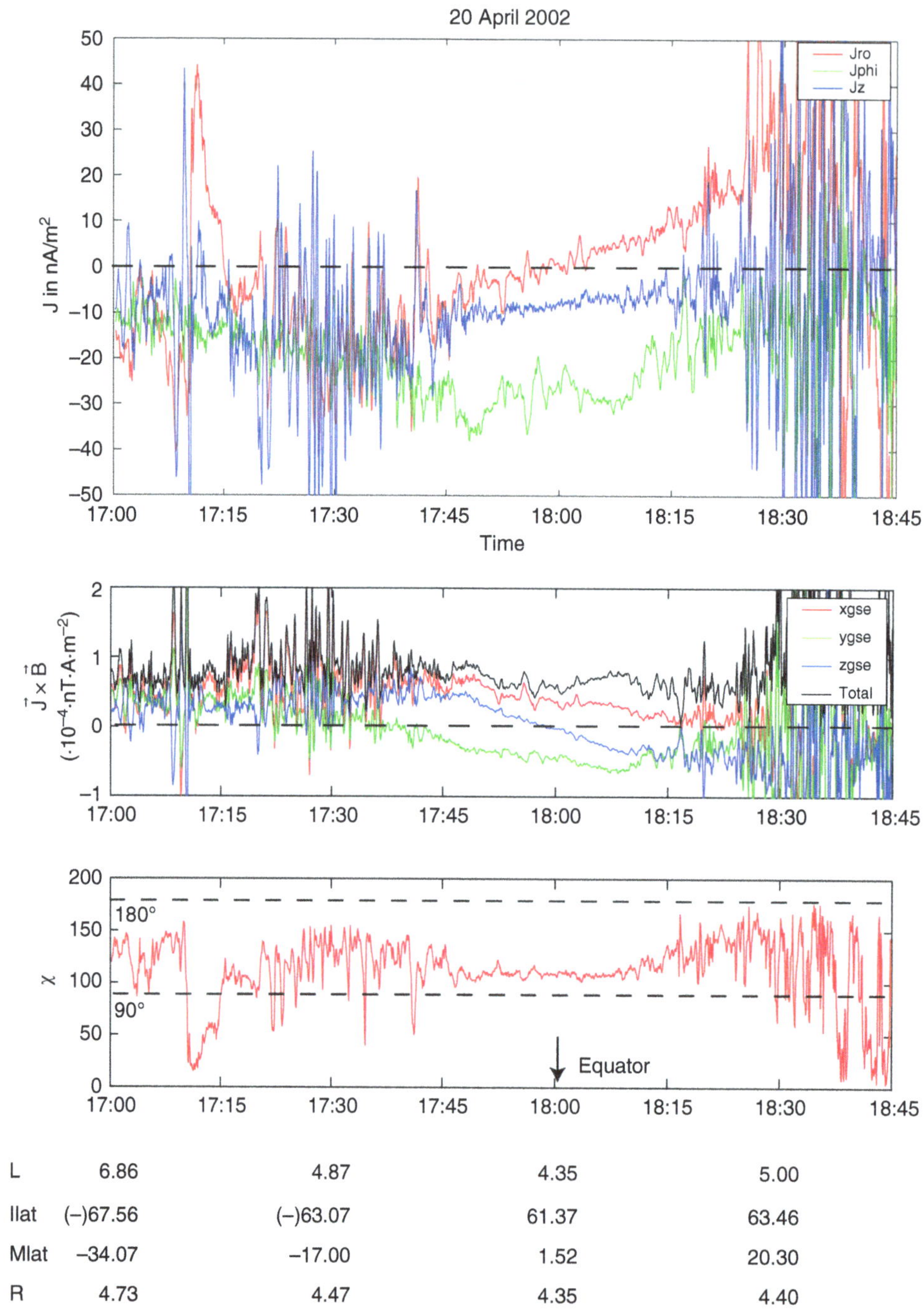

Figure 7.3 Curlometer results for the 20 April 2002 event. From top to bottom: current density in the local cylindrical coordinate system and in nA m^{-2}, pressure gradient (computed from **J** × **B**), and χ angle between local current flow and magnetic field. *L*-shell, invariant latitude, magnetic latitude and geocentric distance are indicated at the bottom of the plot. From *Vallat et al.* [2005].

proton component (>100 keV) varies on much longer timescales and shows either no correlation or even an anti-correlation with the absolute value of the *SYM-H* index, while the low-energy component of the protons (<80 keV) is highly governed by convective timescales and correlates very well with the absolute value of the *SYM-H* index.

The results obtained from Cluster, showing a substantial quiet-time westward ring current, were obtained during the period 2002–2004, that is, close to the maximum/early declining phase of Solar Cycle 23. They are consistent with the analysis of *Lühr and Mauss* [2010], showing: (1) a pronounced dependence of the quiet-time ring current on the

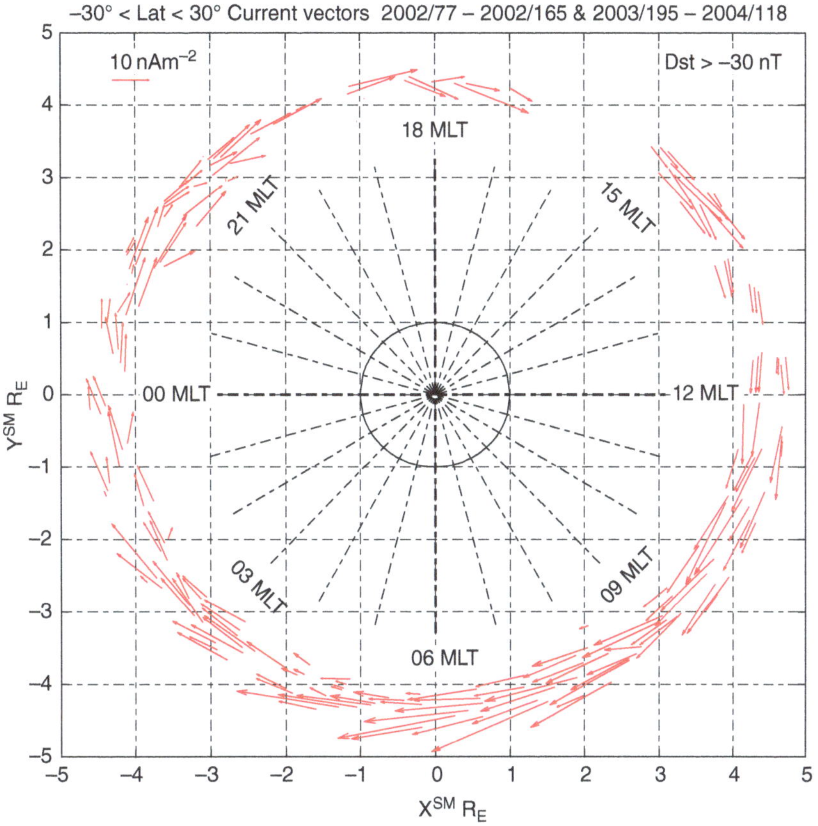

Figure 7.4 The distributions of ring current density vectors around the equator plane in SM coordinates, around Cluster perigee crossings between 18 March 2002 and 14 June 2002 and between 14 July 2003 and 27 April 2004. The vectors represent the directions and the magnitude of the current density in the X-Y plane. From *Zhang et al.* [2011].

solar cycle phase, and particularly on the $F_{10.7a}$ index; (2) a time-dependent bias of the D_{st} index, that underestimates by a varying amount the magnetic field produced by the ring current, which can be up to 15 nT. The baseline field, which is subtracted from the magnetic observatories measurements in order to calculate the D_{st} index, is thus contaminated by the contribution of the quiet-time westward ring current, which is slowly varying as a function of the solar cycle. This suggests that the D_{st} index cannot be used in order to estimate the quiet-time ring current. It also shows the limits of applicability of the Dessler-Parker-Sckopke relation [*Dessler and Parker*, 1959; *Sckopke*, 1966] between the energy content of the inner-magnetosphere trapped energetic particles and the average ground magnetic perturbations.

7.4.3. Cluster Ring Current Observations Close to the Earth

During the period 2007–2013, the Cluster perigee geocentric distances had decreased to ~2 R_E, compared with ~4 R_E during the early years. These allowed exploring the innermost part of the ring current region, where the eastward ring current is expected [*Le et al.*, 2004]. However, the increased size of the spacecraft tetrahedron during this period (typical spacecraft separations were ~1000 km) and the strong gradients of the magnetic field in this region pose a particular challenge in the applicability of the curlometer technique, and at the same time allow testing the limits of its applicability.

Figure 7.6 shows the curlometer analysis for the 15 May 2009 moderate activity conditions (D_{st} ~−18 nT) event. The blue shaded areas indicate the time intervals during which the ring current population is observed in the SC1 (Cluster spacecraft 1) ion data. Since the current density is calculated using the four satellites, and given the large (~1000 km) spacecraft separations, these values give only an indication and an average position. It is first noticed that the average current density J_{av} and its components went up almost to infinity before the spacecraft entered the ring current region. This sharp increase is linked with a strong increase of div B, which also reached infinity during the same time intervals. This observation suggests a very poor accuracy of the J_{av} calculation in this

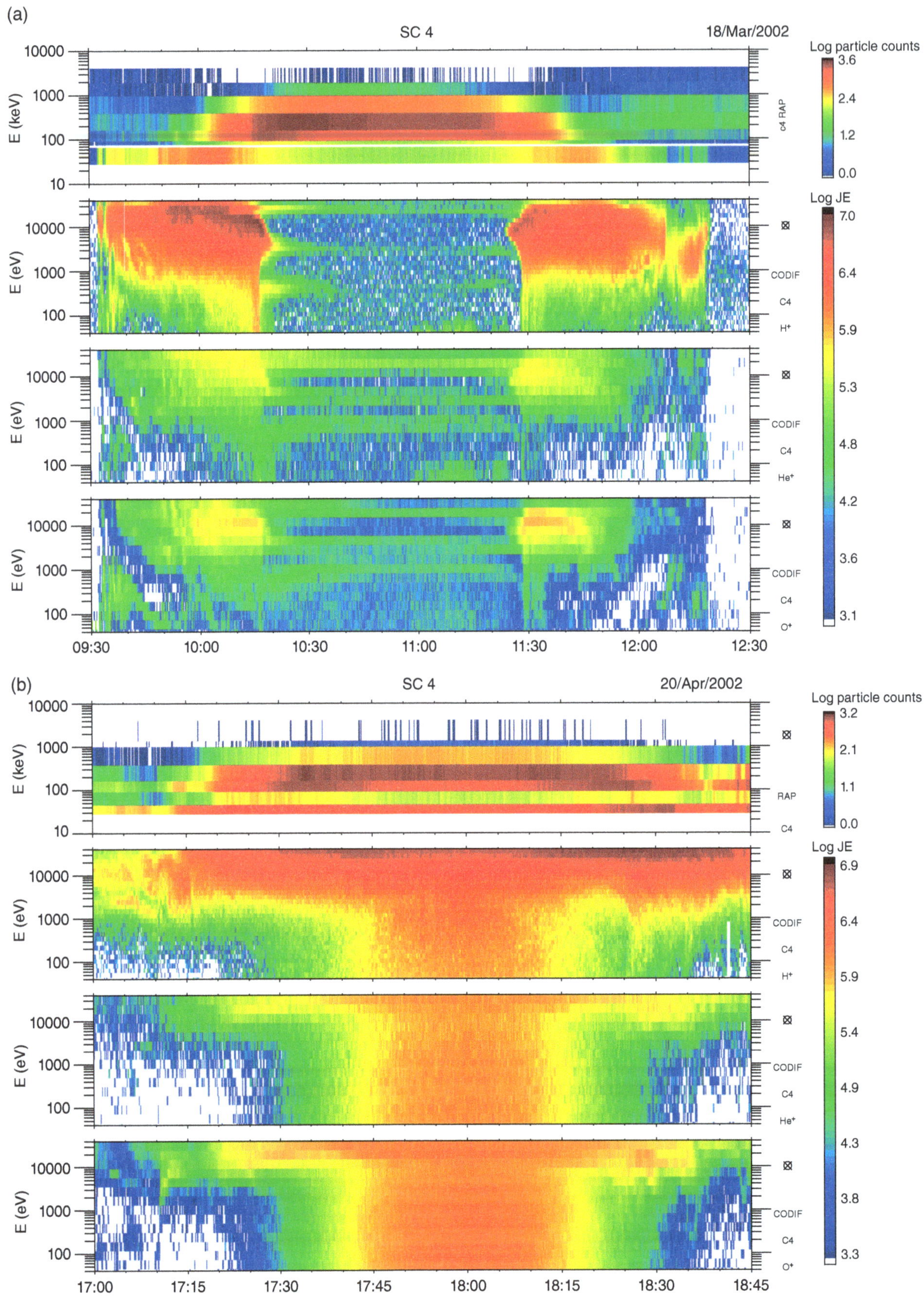

Figure 7.5 (a) Ion energy-time spectrograms from the RAPID instrument for high-energy (28–4000 keV) H^+ ions, in particle count rates (proportional to particle energy flux), top panel, and from the CIS-CODIF instrument, for the low-to-medium energies (up to 40 keV), separately for H^+, He^+, and O^+ ions, in particle energy flux units (keV/cm^2 s sr keV). (b) Same as (a) but for the 20 April 2002 geomagnetic storm event. The diffuse red vertical feature, appearing in the CIS-CODIF data between ~17:45 and ~18:15 UT, corresponds to background counts due to penetrating particles from the outer radiation belt (cf. *Ganushkina et al.* [2011]).

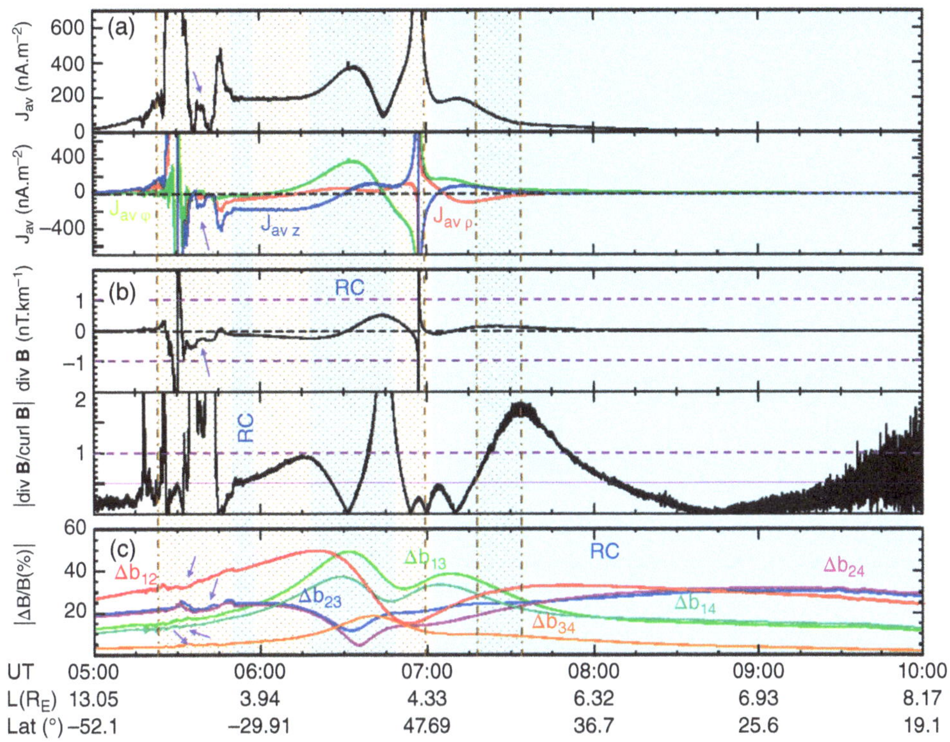

Figure 7.6 Estimation of the current density along the Cluster orbit during the 15 May 2009 ring current region crossing. Panel (a): average current density (total and components in cylindrical coordinates); panel (b): div **B** and div **B**/|curl **B**|; panel (c): $B_i - B_j/B$, where *i* and *j* are the satellite numbers. In each panel, the blue shaded areas indicate the time intervals during which the ring current was observed in the spacecraft 1 data. The brown dashed lines are located where |div **B**| = 1 nT km^{-1}. From *Grimald et al.* [2012].

part of the orbit. Starting from 05:48 UT, the four satellites were located inside the ring current region. The average current density in the azimuthal direction, obtained using the curlometer method, was negative until 06:03 UT, consistent with a westward ring current. At 06:03 UT the azimuthal current direction changed to become eastward at $L = 3.6$ R_E. This result is in agreement with past studies, which located the current reversal boundary between 3 R_E and 4 R_E [*Le et al.*, 2004]. However, we note that div **B**/|curl **B**| was between 0.5 and 1.0 during this westward-to-eastward reversal, suggesting only a marginal accuracy of this calculation result.

The analysis of the accuracy of the curlometer method in this innermost part of the magnetosphere shows that most of the error comes from (1) the interspacecraft distances being larger than the characteristic scales of the ring current structure and (2) the linearity hypothesis on the magnetic field. The subtraction of the Earth's dipolar magnetic field, from the magnetic field measurements, is thus essential. The requirement for div **B**/|curl **B**| < 0.5 is a necessary but not an adequate condition for obtaining accurate results.

The subtraction of the geomagnetic field from the Cluster spacecraft measurements, in order to calculate the current density, has been also used by *Shen et al.* [2014] to study the ring current distribution and magnetic structure during geomagnetic storms. Their analysis confirmed the location of the eastward ring current at smaller geocentric distances than the westward ring current, and showed also that during storm times the magnetic field topology is significantly changing. The curvature radius of the magnetic field lines decreases as a function of storm intensity and, for big storms, it can become as small as 0.4 R_E at geocentric distances of ~3.9 R_E.

7.4.4. Calculation of the Ring Current Density by Other Missions

Direct current density calculation through magnetic field gradient measurements requires, as explained in section 7.3, four-point simultaneous magnetic field measurements at appropriate locations. In order to overcome these restrictions and to extend the use of the curlometer technique to some cases where only two or three spacecraft were available, or the tetrahedron was strongly distorted, methods have been developed for determining the magnetic field gradient from only two or three spacecraft. These methods use statistical average magnetic field data

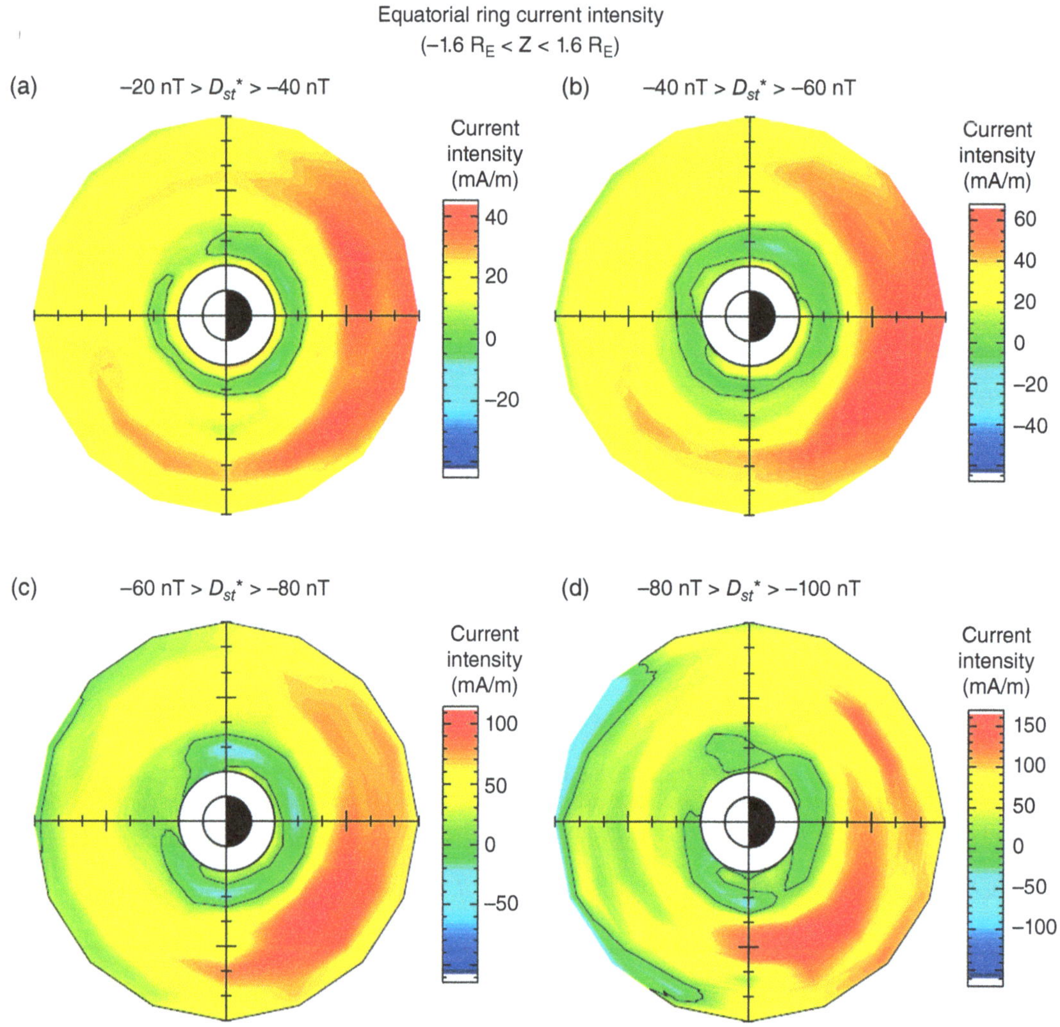

Figure 7.7 Equatorial ring current intensity as a function of magnetic local time and distance from the dipole axis, for four D_{st}* levels. The local current intensity is calculated by integrating the azimuthal component of the current density over $-1.6<Z<1.6$ R_E. From *Le et al.* [2004].

[*Le et al.*, 2004], or assume that the magnetic field is approximately stationary and force free [*Shen et al.*, 2012a], or assume that the ring current around the magnetospheric equator is along the azimuthal direction [*Shen et al.*, 2012b; *Yang et al.*, 2016].

Le et al. [2004] used 20 yr of intercalibrated magnetic field data from the ISEE, AMPTE/CCE, and Polar missions to produce statistical magnetic field maps for different D_{st} levels. They then derived the local current density by calculating the curl of the statistical magnetic field. The results are summarized in Figure 7.7, where the azimuthal component of the local current density has been integrated over a ±1.6 R_E column to produce the local ring current intensity. These results clearly show the local time asymmetry of the ring current, and how it evolves as a function of the activity level, and also the transition between the eastward and the westward ring current.

Yang et al. [2016] used THEMIS three-spacecraft data to analyze the azimuthal current density distribution in the equatorial magnetosphere during storm events, at geocentric distances ranging from 4 to 12 R_E. They showed that the transition between eastward and westward ring currents occurs generally at geocentric distances less than 5 R_E, and that during the main storm phase this transition distance moves closer to the Earth. During the main phase, the current also presents more fluctuations, apparently linked to the dynamics of the storm development in the magnetosphere. During the following recovery phase, however, the profile of the westward ring current is smoother and has almost no MLT dependence.

7.5. CONCLUSIONS

The curlometer technique has been extensively used to analyze the ring current, and it reproduces very well its structure. However, several precautions have to be taken in order to insure the accuracy of the method, in particular for observations close to the Earth (geocentric distances less than ~3.5 R_E), due to the strong nonlinearity of the magnetic field there, or when the spacecraft tetrahedron is too large or highly deformed.

The results show a large latitudinal extent of the ring current, with a mostly azimuthal current close to the magnetic equator and a discontinuous evolution of its orientation close to the boundary between the ring current and the "horns" of the plasma sheet, where there is an abrupt change from a perpendicular current to a field-aligned current system. In the equatorial plane, the transition between the eastward and the westward ring current is established.

The results also show the existence of a substantial quiet-time ring current, carried mostly by high-energy (>100 keV) ions. During magnetic storm conditions, and for the events analyzed, the current density measured can increase by 25%–50% and is accompanied by the presence of carriers covering a broader energy range, starting from energies as low as a few keV.

ACKNOWLEDGMENTS

I. Dandouras thanks CNES for its support to the Cluster project at IRAP. M. W. Dunlop is partly supported by an STFC in-house research grant and is supported by the NSFC grant 41574155. The authors are grateful to the CIS, FGM, and RAPID instrument teams, and to the CSA (Cluster Science Archive), for the data availability.

REFERENCES

Balogh, A., C. M. Carr, M. H. Acuña, M. W. Dunlop, T. J. Beek, P. Brown, K.-H. Fornacon, et al. (2001), The Cluster Magnetic Field investigation: Overview of in-flight performance and initial results, *Ann. Geophys.*, *19*, 1207–1217.

Chanteur, G. (2000), Accuracy of field gradient estimations by Cluster: Explanation of its dependency upon elongation and planarity of the tetrahedron, in *Proceedings of the Cluster-II Workshop Multiscale/Multipoint Plasma Measurements*, Imperial College, London, 1999, ESA SP-449, 265–268, European Space Agency.

Chapman, S., and V. C. A. Ferraro (1931), A new theory of magnetic storms, *Terr. Mag.*, *36*, 77–97.

Chapman S., and V. C. A. Ferraro (1941), The geomagnetic ring-current: Its radial stability, *Terr. Mag.*, *46*, 1–6.

Daglis, I. A., R. M. Thorne, W. Baumjohann, and S. Orsini (1999), The terrestrial ring current: Origin, formation and decay, *Rev. Geophys.*, *37*, 407–438.

Dandouras, I., J. Cao, and C. Vallat (2009), Energetic ion dynamics of the inner magnetosphere revealed in coordinated Cluster-Double Star observations, *J. Geophys. Res.*, *114*, A01S90; doi:10.1029/2007JA012757.

Dessler, A. J., and E. N. Parker (1959), Hydromagnetic theory of geomagnetic storms, *J. Geophys. Res.*, *64*, 2239–2252.

Dunlop, M. W., D. J. Southwood, K.-H. Glassmeier, and F. M. Neubauer (1988), Analysis of multipoint magnetometer data, *Adv. Space Res.*, *8* (9–10), 273–277.

Dunlop, M. W., A. Balogh, K.-H. Glassmeier, and P. Robert (2002), Four-point Cluster application of magnetic field analysis tools: The curlometer, *J. Geophys. Res.*, *107*, 1384–1387.

Escoubet, C. P., M. Fehringer, and M. Goldstein (2001), The Cluster mission, *Ann. Geophys.*, *19*, 1197–1200.

Ganushkina, N. Y., M. W. Liemohn, M. V. Kubyshkina, R. Ilie, and H. J. Singer (2010), Distortions of the magnetic field by storm-time current systems in Earth's magnetosphere, *Ann. Geophys.*, *28*, 123–140.

Ganushkina, N. Yu., I. Dandouras, Y. Y. Shprits, and J. Cao (2011), Locations of boundaries of outer and inner radiation belts as observed by Cluster and Double Star, *J. Geophys. Res.*, *116*, A09234; doi:10.1029/2010JA016376.

Ganushkina, N. Y., M. W. Liemohn, O. Amariutei, I. A. Daglis, I. Dandouras, S. Dubyagin, D. L. De Zeeuw, et al. (2015), Defining and resolving current systems in geospace, *Ann. Geophys.*, *33;* doi:10.5194/angeo-33-1369-2015.

Gkioulidou, M., A. Y. Ukhorskiy, D. G. Mitchell1, and L. J. Lanzerotti (2016), Storm time dynamics of ring current protons: Implications for the long-term energy budget in the inner magnetosphere, *Geophys. Res. Lett.*, *43;* doi:10.1002/2016GL068013.

Grimald, S., I. Dandouras, P. Robert, and E. Lucek (2012), Study of the applicability of the curlometer technique with the four Cluster spacecraft in regions close to Earth, *Ann. Geophys.*, *30*, 597–611; doi:10.5194/angeo-30-597-2012.

Le, G., C. T. Russell, and K. Takahashi (2004), Morphology of the ring current derived from magnetic field observations, *Ann. Geophys.*, *22*, 1267–1295; doi:10.5194/angeo-22-1267-2004.

Liemohn, M. W., N. Y. Ganushkina, R. M. Katus, D. L. De Zeeuw, and D. T. Welling (2013), The magnetospheric banana current, *J. Geophys. Res.*, *118*, 1009–1021; doi:10.1002/jgra.50153.

Liemohn, M. W., R. M. Katus, and R. Ilie (2015), Statistical analysis of storm-time near-Earth current systems, *Ann. Geophys.*, *33*, 965–982.

Lühr, H., and S. Maus (2010), Solar cycle dependence of quiet-time magnetospheric currents and a model of their near-Earth magnetic fields, *Earth Planet Space*, *62*, 843–848.

Lühr, H., C. Xiong, N. Olsen, and G. Le (2017), Near-Earth magnetic field effects of large-scale magnetospheric currents, *Space Sci. Rev.*, *206*, 521–545.

Parker, E.N. (1957), Newtonian development of the dynamical properties of ionized gases of low density, *Phys. Rev.*, *107*, 924–933.

Rème, H., C. Aoustin, J. M. Bosqued, I. Dandouras, B. Lavraud, J. A. Sauvaud, A. Barthe, et al. (2001), First multispacecraft ion measurements in and near the Earth's magnetosphere with the identical Cluster ion spectrometry (CIS) experiment, *Ann. Geophys.*, *19*, 1303–1354.

Robert, P., A. Roux, C. C. Harvey, M. Dunlop, P. W. Daly, and K.-H Glassmeier (1998), Tetrahedron Geometric Factors, in *Analysis Methods for Multi-Spacecraft data*, ISSI Sci. Rep. SR-001, 323–348.

Roelof, E. C., P. C:son Brandt, and D. G. Mitchell (2004), Derivation of currents and diamagnetic effects from global plasma pressure distributions obtained by IMAGE/HENA, *Adv. Space, Res.*, *33*, 5, 747–751.

Sckopke, N. (1966), A general relation between the energy of trapped particles and the disturbance field near the Earth, *J. Geophys. Res.*, *71*, 3125–3130.

Shen, C., Z. J. Rong, and M. Dunlop (2012a), Determining the full magnetic field gradient from two spacecraft measurements under special constraints, *J. Geophys. Res.*, *117*, A10217; doi:10.1029/2012JA018063.

Shen, C., Z. J. Rong, M. W. Dunlop, Y. H. Ma, X. Li, G. Zeng, G. Q. Yan, W. X. Wan, Z. X. Liu, C. M. Carr, and H. Rème (2012b), Spatial gradients from irregular, multiple-point spacecraft configurations, *J. Geophys. Res.*, *117*, A11207; doi:10.1029/2012JA018075.

Shen, C., Y. Y. Yang, Z. J. Rong, X. Li, M. Dunlop, C. M. Carr, Z. X. Liu, et al. (2014), Direct calculation of the ring current distribution and magnetic structure seen by Cluster during geomagnetic storms, *J. Geophys. Res.*, *119*, 2458–2465; doi:10.1002/2013JA019460.

Singer, S. F. (1957), A new model of magnetic storms and aurorae, *Eos Trans. AGU*, *38*, 175–190.

Störmer, C. (1907), Sur les trajectoires des corpuscules électrisés dans l'espace sous l'action du magnétisme terrestre avec application aux aurores boréales, *Arch. Sci. Phys. Nat.*, *24*, 317–364.

Vallat, C., I. Dandouras, M. Dunlop, A. Balogh, E. Lucek, G. K. Parks, M. Wilber, E. C. Roelof, G. Chanteur, and H. Rème (2005), First current density measurements in the ring current region using simultaneous multi-spacecraft CLUSTER-FGM data, *Ann. Geophys.*, *23*, 1849–1865, 1432-0576/ag/2005-23-1849.

Wilken, B, P. W. Daly, U. Mall, K. Aarsnes, D. N. Baker, R. D. Belian, J. B. Blake, et al. (2001), First results from the RAPID imaging energetic particle spectrometer on board Cluster, *Ann. Geophys.*, *19*, 1355–1366.

Yang, Y. Y., C. Shen, M. Dunlop, Z. J. Rong, X. Li, V. Angelopoulos, Z. Q. Chen, G. Q. Yan, and Y. Ji (2016), Storm time current distribution in the inner equatorial magnetosphere: THEMIS observations, *J. Geophys. Res.*, *121*, 5250–5259; doi:10.1002/2015JA022145.

Zhang, Q.-H., M. W. Dunlop, M. Lockwood, R. Holme, Y. Kamide, W. Baumjohann, R.-Y. Liu, et al. (2011), The distribution of the ring current: Cluster observations, *Ann. Geophys.*, *29*, 1655–1662.

8

The Nature of Jupiter's Magnetodisk Current System

N. Achilleos

ABSTRACT

This chapter gives an overview of the properties of the magnetodisk current in the Jovian system. We describe the global morphology of the current sheet embedded in the plasmadisk / magnetodisk and the observational signatures of currents in this structure. We then consider the role of disk currents in force balance and plasmasheet structure in an axisymmetric, rotating system. We also describe the dependence of current density on spatial location, global size of the magnetosphere, and asymmetries plausibly associated with the influence of solar wind. We conclude with a simplified description of the microscopic nature of the particle motions in the magnetospheric plasma, whose collective action produces the currents themselves.

8.1. INTRODUCTION

A prominent feature of the giant, rapidly rotating magnetospheres of planets such as Jupiter and Saturn is the presence of a disklike body of plasma lying close to the magnetic equator of the planet itself. In the middle magnetosphere region (radial distances ~10–40 R_J, 1 $R_J \approx 71500$ km = Jupiter radius), we may appeal to the approximation of the disk as a structure that has symmetry about the magnetic (dipole) equator of the planet. Since Jupiter's dipole moment is tilted at an angle of ~10° with respect to its rotation axis, it follows that the magnetic field in this region takes the form of a tilted, rotating structure. The plasma situated at different radial distances within this disk rotates at different angular velocities, the general trend being a decline in angular velocity from the initial, planetary value as one moves outward in the radial direction. The forces on this plasma disk are related to the currents that flow both radially and azimuthally throughout it.

This chapter gives an overview of the properties of the magnetodisk current in the Jovian system. We begin with a description of the global morphology of the current sheet embedded in the plasmadisk/magnetodisk and the observational signatures of both azimuthal and radial flowing currents in this structure. We then consider the conceptual link between disk current, force balance, and plasmasheet structure in an axisymmetric, rotating plasmadisk. As part of these descriptions, we will also build a picture of the general behavior of the current density and its dependence on factors such as spatial location with respect to the planet, global size of the magnetospheric system, and asymmetries plausibly associated with the influence of the solar wind. We conclude with a description of the microscopic nature of the particle motions in the magnetospheric plasma, whose collective action produces the currents themselves. The specific descriptions in this chapter are also relevant to the comparison of planetary current systems by *Khurana* (Chapter 2, this

Department of Physics and Astronomy, Centre for Planetary Sciences, University College London, London, UK

Electric Currents in Geospace and Beyond, Geophysical Monograph 235, First Edition.
Edited by Andreas Keiling, Octav Marghitu, and Michael Wheatland.

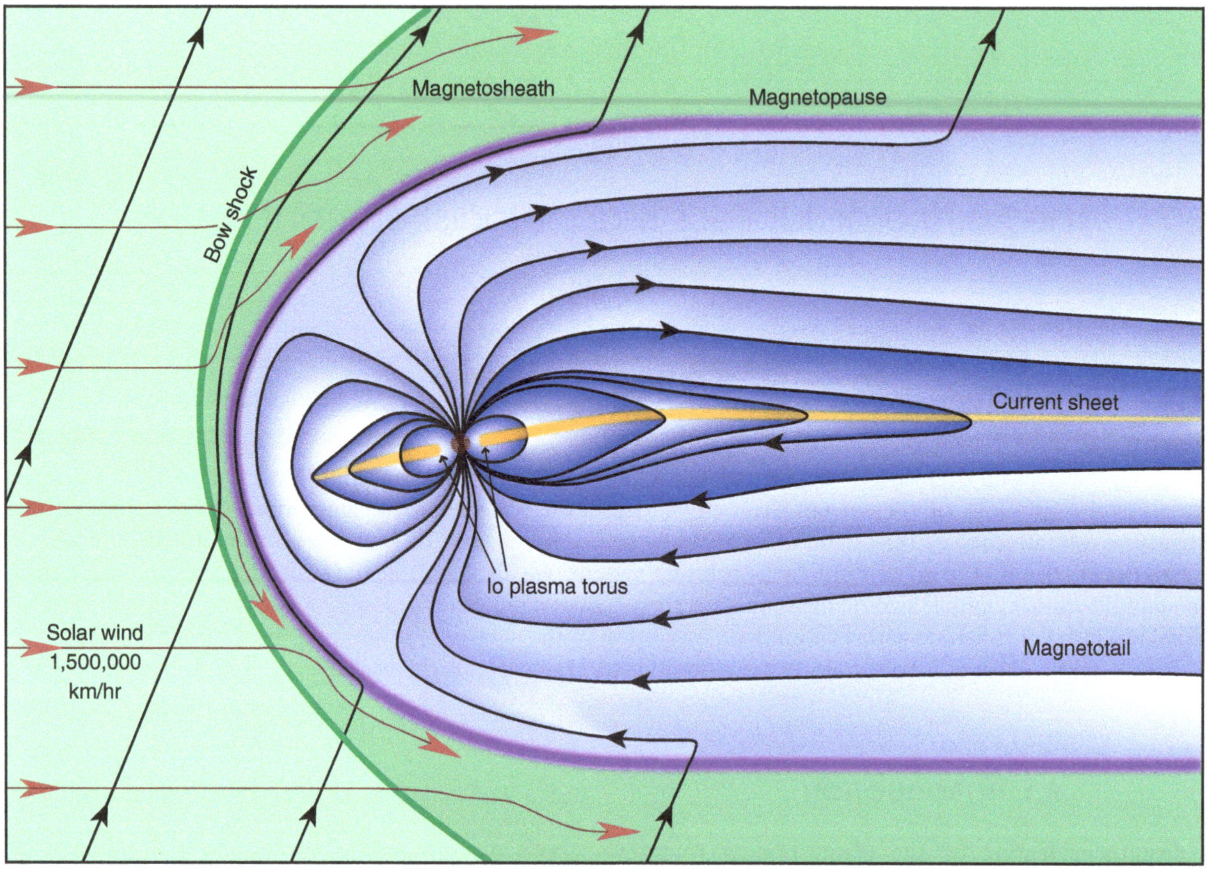

Figure 8.1 Noon-midnight cross section of Jupiter's magnetosphere. Image credit: Steve Bartlett and Fran Bagenal.

volume) and the discussion of giant planet current sheets by *Arridge and Martin* (Chapter 12, this volume).

8.2. GLOBAL MORPHOLOGY OF THE JOVIAN CURRENT DISK

The global structure of the main current system with which we are concerned is illustrated by the diagram of the Jovian magnetosphere in Figure 8.1. The highlighted current sheet region emphasizes some important aspects of the structure and orientation of the plasmasheet/magnetodisk, which carries this current. Figure 8.1 also shows the enormity of Jupiter's magnetosphere, which extends on the dayside to common distances near ~62 and ~93 R_J [*Joy et al.*, 2002]. We may investigate the equatorial magnetosphere, which is dominated over much of its radial extent by the magnetodisk current sheet, in terms of three broad regions, as follows [e.g., *Khurana et al.*, 2004].

8.2.1. Io Torus

The Jovian magnetosphere's principal source of plasma is the volcanic satellite Io, which orbits the planet at a distance of ~6 R_J, and adds on the order of ~1000 kg s^{-1} of sulphur and oxygen plasma to the system [*Bagenal and Delamere*, 2011; *Hill et al.*, 1983]. The ensuing cold/warm plasma torus near Io's orbit, containing ions of temperatures ~2–60 eV [*Bagenal and Sullivan*, 1981; *Bagenal et al.*, 1985; *Scudder et al.*, 1981], is aligned along the centrifugal equator (those points along the local dipolar field lines farthest from the planet's rotation axis). The torus contains on the order of ~10^6 tonnes of plasma and extends between ~5–10 R_J with scale height on the order of ~1 R_J. Plasma is radially transported outward from the torus through processes such as flux tube interchange and magnetic field ballooning/reconfiguration (e.g., *Thomas et al.* [2004]; *Kivelson and Southwood* [2005]; review by *Achilleos et al.* [2015]; and references therein).

8.2.2. Middle Magnetosphere

The middle magnetosphere is the main focus for our description, and extends between ~10–40 R_J in the equatorial plane. It is the region within which the magnetodisk plasma generally lies close to the dipole equator and also shows significant departure from corotation with the planet (i.e., the internal field). The latter behavior arises because of the finite conductance of 'that part of' the ionosphere which is magnetically conjugate to this region, and the decline of equatorial magnetospheric field strength with radial distance. Finite conductance is a principal factor that limits the radial disk current [e.g., *Cowley and Bunce*, 2001; *Hill*, 1979]. Since the equatorial field strength decreases with increasing radial distance, these combined factors limit the azimuthal Lorentz (**J** × **B**) force, which acts to accelerate outflowing plasma toward planetary corotation. Beyond the distances mentioned, the plasma thus lags the planet with respect to angular velocity. This subcorotation is a key concept in the theory of magnetosphere-ionosphere coupling of *Hill* [1979]. The radial current system is closed by ionospheric currents as well as field-aligned currents, some of which correspond to auroral particle precipitation (e.g., *Cowley and Bunce* [2001]; *Nichols and Cowley* [2004]; *Ray et al.* [2012]; *Ray and Ergun* [2012]; Chapter 21 in this volume by *Ray* and references therein).

The plasma in the middle magnetosphere generally consists of a low-energy "warm" component ($\lesssim 100$ eV) originating from outward diffusion of Io torus plasma, and an energetic "hot" component ($\gtrsim 20$ keV), which is thought to be accelerated/heated torus plasma interspersed with plasma from the outer magnetosphere moving inward (e.g., *Siscoe et al.* [1981]; *Kivelson et al.* [1997]; *Thorne et al.* [1997]; *Bagenal and Delamere* [2011]).

Let us now consider the azimuthal currents in the plasmadisk. In the middle magnetosphere, these currents generate perturbation magnetic fields, which become comparable to the internal field of the planet beyond ~20 R_J. In *Connerney et al.* [1981]'s magnetic field model for an axisymmetric current disk, the current flows azimuthally within an equatorial disk of uniform thickness 2D, extending between radii ρ_1 and ρ_2, and in which the current density J_φ is assumed positive everywhere (i.e., in the direction of planetary rotation) with radial dependence $J_\varphi \propto \frac{1}{\rho}$. By tilting the equatorial plane of the model to match the position of the Jovian dipole equator, *Connerney et al.* [1981] were able to reasonably fit some of the Voyager and Pioneer magnetic observations of the perturbation field (measured field minus a model internal field), for radial distances $\lesssim 30\,R_J$ (Fig. 8.2 shows an example).

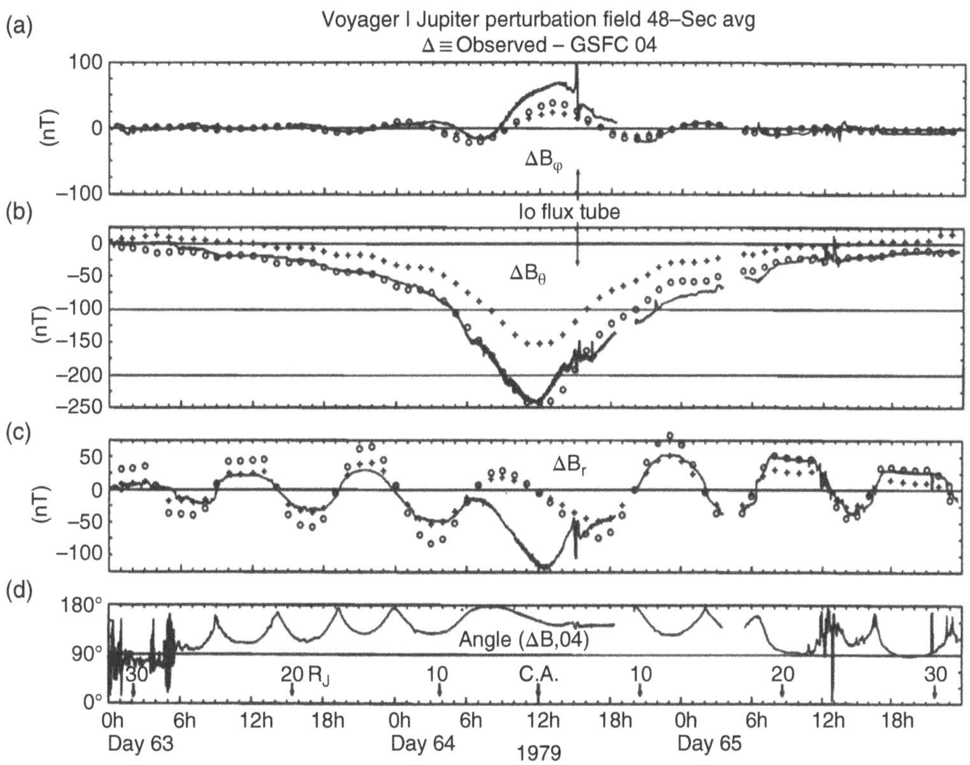

Figure 8.2 Voyager 1 observations of the Jovian perturbation field (estimated using the O4 internal field model), with corresponding predictions from two disk current models of the same vertical thickness (2 D = 5 R_J) and inner radius (ρ_1 = 5 R_J), but differing outer radius and current density (circles: $\rho_2 = 50R_J, J_1 = 5.1MAR_J^{-2}$ at ρ_1; crosses: $\rho_2 = 30R_J, J_1 = 3.42MAR_J^{-2}$). The field components (top three panels: (a) azimuthal, (b) meridional, (c) radial) are in a spherical polar system. The bottom panel (d) shows time-dependent spacecraft radial distance and angle between the perturbation field vector ΔB and O4 model. From *Connerney et al.* [1981].

The observed oscillating radial (ΔB_r) and azimuthal (ΔB_φ) field components in Figure 8.2 are signatures of the rotating, tilted current sheet moving over the spacecraft twice per planetary rotation. Both of these components pass through zero as the spacecraft moves relative to the sheet, from the northern ($\Delta B_r > 0$, $\Delta B_\varphi < 0$) to the southern ($\Delta B_r < 0$, $\Delta B_\varphi > 0$) magnetic hemisphere. The ΔB_r reflects the radially stretched field (Fig. 8.1), in which the inward magnetic curvature force balances the outward forces in this region from the radial plasma pressure gradient and the centrifugal force (see section 8.3). The ΔB_φ observations generally display the azimuthal bendback of field lines from the poloidal geometry of the *Connerney et al.* [1981] model; this is the effect of the radial corotation enforcement currents that flow in the disk. Appropriate detailed analysis of ΔB_φ observations may be used to estimate the Io mass-loading rate [*Khurana and Kivelson*, 1993].

8.2.2.1. Physical Origin of Asymmetry in the Current Disk

The planar, axisymmetric approximation is less suitable for describing the morphology of the current disk at distances beyond ~20–30 R_J. At these larger distances, centrifugal and pressure-related forces become important influences on the stress balance in the system (see section 8.3). As a result, the equilibrium current sheet is located between the dipole and centrifugal equators. *Khurana* [1992] also demonstrated that the sheet in the distant magnetotail ($\gtrsim 60 R_J$) develops a hinging and becomes parallel to the direction of solar wind flow outside the magnetosphere. Stress associated with the solar wind dynamic pressure is communicated to the plasma inside the magnetosphere via the fringing field produced by forces acting at the magnetopause boundary.

A very simple example of this type of communication is illustrated by the equatorial magnetic field ΔB generated by magnetopause currents in a simplified, axisymmetric model of the magnetodisk region (see section 8.3). The model considered later with radius 60 R_J has an equatorial ratio $\Delta B/B_{eq}$ of magnetopause field to total field, which varies with radial distance. The $\Delta B/B_{eq}$ increases with distance from the planet, exceeds ~20% beyond ~20 R_J, and attains a value ~40% just inside the magnetopause. In this example, the magnetopause fringing field thus contributes significantly to the total field strength even in the middle magnetosphere, and becomes stronger for a more compressed system. In the more general case, the fringing field may also support strong rotation or hinging of the magnetospheric field in the neighborhood of the magnetopause itself.

In addition to this hinging effect, systematic delays in the expected arrival times of a planar current sheet are observed in this region. This phenomenon arises partly because of the finite propagation time of the Alfvén waves launched from the ionosphere, which communicate information about the rocking of the rotating, tilted planetary dipole to the distant magnetosphere. Current sheet delays are also controlled by the strong bending of the field lines out of meridian planes, against the direction of planetary rotation, which arises from the azimuthal stresses on the plasma. Field lines at Jovian dawn, for example, can be bent out of their near-planet meridian planes by angles $\lesssim 40°$ for distances $\lesssim 100$ R_J [*Khurana et al.*, 2004]. It is important to note that the frequency of current sheet encounters at any given distance is controlled by the period of the planet. The rotating dipole ultimately imposes a modulation pattern that influences the vertical location of the magnetospheric plasma with respect to the rotational equator. Thus, while outer magnetospheric disk plasma subcorotates with respect to the planet, the vertical location of maximum density along a subcorotating flux tube is variable, and partly determined by the rotational phase of the planetary dipole.

To represent these types of hinging and delay effects, *Khurana* [1992] developed a "hinged-magnetodisk" model. In an earlier study, *Behannon et al.* [1981] introduced hinging and delay effects into their axisymmetric model of the current sheet surface. *Khurana* [1992] presented a model in which the hinging was linked to solar wind forcing, thus breaking axial symmetry, and was able to improve the quality of the model fit to Pioneer and Voyager observations. His formula for the altitude Z_{CS} of the current sheet with respect to the Jovigraphic equator was:

$$Z_{CS} = \rho \tan\left(\theta_{dip}\right)\left(\frac{x_o}{x}\tanh\left(\frac{x}{x_o}\right)\cos\left(\lambda - \delta\right)\right) \quad (8.1)$$

In this equation, ρ is cylindrical radial distance (with respect to the planet's rotational axis) and θ_{dip} is the dipole tilt. The hinging is implemented through the coordinate x along the Jupiter-Sun axis, and the hinge distance x_o indicates where hinging produces significant departure from a planar structure. The λ represents longitude, and the harmonic term introduces a delay in the azimuthal current sheet position by making use of $\delta = \delta_o - \left(\Omega_J \rho_o / v_o\right)\ln\cosh\left(\rho/\rho_o\right)$, where δ_o is the longitude of the plane containing the dipole, Ω_J is the planet's angular velocity of rotation, ρ_o is the distance at which delay effects cause significant departure with respect to a planar current sheet, and v_o can be interpreted as a wave velocity that carries information about dipole orientation into the outer magnetosphere. Figure 8.3 illustrates some of these effects by presenting maps of Z_{CS} from the point of view looking from the north onto the Jovigraphic equatorial plane. We show two examples of the model

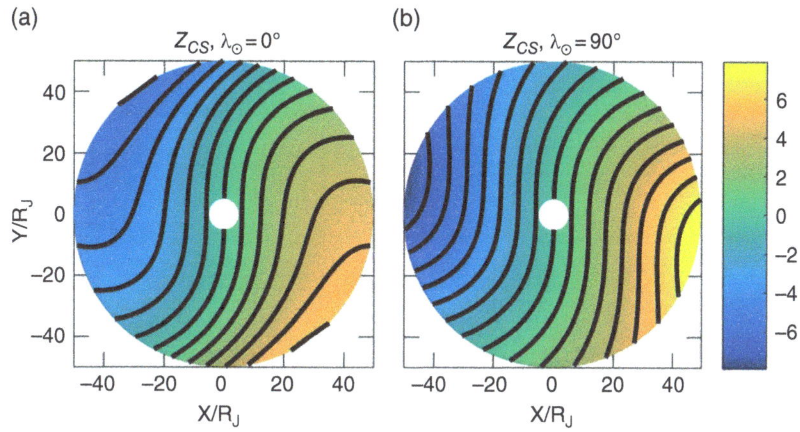

Figure 8.3 Color-coded plots of current sheet altitude Z_{CS}, with respect to the Jovigraphic equator (XY plane), calculated using the hinged-magnetodisk formulae of *Khurana* [1992] for projected radial distances $\rho = 5$–50 R$_J$. The projection of the northern planetary dipole lies along the −X axis in both panels, while the planet-Sun direction is at a longitude of $\lambda_\odot = 0°$ (positive X direction) and 90° (positive Y direction) in the respective left (a) and right (b) panels. The planet's rotational axis is assumed, for simplicity, to be orthogonal to the XY plane (the actual obliquity of this axis is small, ~3°). The black curves are contours of constant Z_{CS} spaced by 1 R$_J$ (with $Z_{CS} = 0$ contours passing through plot centers), and illustrate the strong hinging associated with the $\lambda_\odot = 0°$ configuration. The azimuthal bending of these contours in the outer region of the sheet produces the observed current sheet delay effects described in the text. Values for the model parameters (see text) used for these plots were taken from *Khurana* [1992].

current sheet, for configurations where the Jupiter-Sun direction is parallel and orthogonal to the dipole longitude meridian. The former case shows the most pronounced hinging. We used the following geometrical parameters for the plots in Figure 8.3, identical to those from the fit of *Khurana* [1992]: $x_o = 33.5$ R$_J$, $\rho_o = 33.2$ R$_J$, $v_o = 37.4$ R$_J$/hr.

Magnetic field models that incorporate azimuthal bendback into the field lines can be constructed using the approach where the field is expressed in terms of two Euler potential functions, f and g:

$$\mathbf{B} = \nabla f \times \nabla g. \tag{8.2}$$

Without loss of generality, if the Euler potential g is, for example, a linear function of azimuthal angle φ (magnetic longitude), we obtain a poloidal field structure, where all field lines lie in a single plane. This poloidal class of Euler-potential field models has been combined with the force-balance condition to obtain equilibrium models for both field and plasma distribution (see section 8.3 for further details).

Goertz et al. [1976] used an Euler potential with dependence on both ρ and φ to create nonpoloidal field lines with bendback. *Khurana* [1997] further developed this type of field model for the current sheet by using Euler potentials suitable for coupling the current sheet field to a model of Jupiter's internal magnetic field. He included current sheet delay and hinging by making the Euler potentials dependent on Z_{CS}, the altitude of the current sheet (see equation [8.1]).

We conclude this section on Jovian current sheet asymmetry by summarizing the local-time asymmetry in the intensity of the magnetodisk currents, which was found by *Khurana* [2001], from an analysis of magnetometer data from six spacecraft to have visited the Jovian system, including Galileo. The height-integrated radial and azimuthal current densities flowing in the current sheet (J'_ρ and J'_φ) were calculated in this study according to the following formulae, which essentially arise from Ampère's law applied across the vertical thickness of the sheet:

$$\mu_o J'_\rho = -2\Delta B_{\varphi,l} / \mu_o \tag{8.3}$$

$$\mu_o J'_\varphi = \left(2\Delta B_{\rho,l} - 2W \frac{\partial}{\partial \rho}(\Delta B_z) \right) \tag{8.4}$$

where $\Delta B_{\varphi,l}$ and $\Delta B_{\rho,l}$ are the respective perturbation fields in the cylindrical radial and azimuthal directions, measured just outside the current sheet (lobe region); ΔB_z is the vertical perturbation field; and W is the half-thickness of the sheet. W is not constant; it may vary according to solar wind dynamic pressure (see section 8.3) and local time. For example, *Khurana et al.* [2004] present Galileo magnetic data in the outer magnetosphere (35–55 R$_J$) that indicate that the dawn-side half-thickness is ~1.5 R$_J$ - thin compared to the inferred duskside value of ~7.6 R$_J$.

The resulting profiles of J'_ρ and J'_φ as a function of local time are shown in Figure 8.4, from *Khurana et al.*

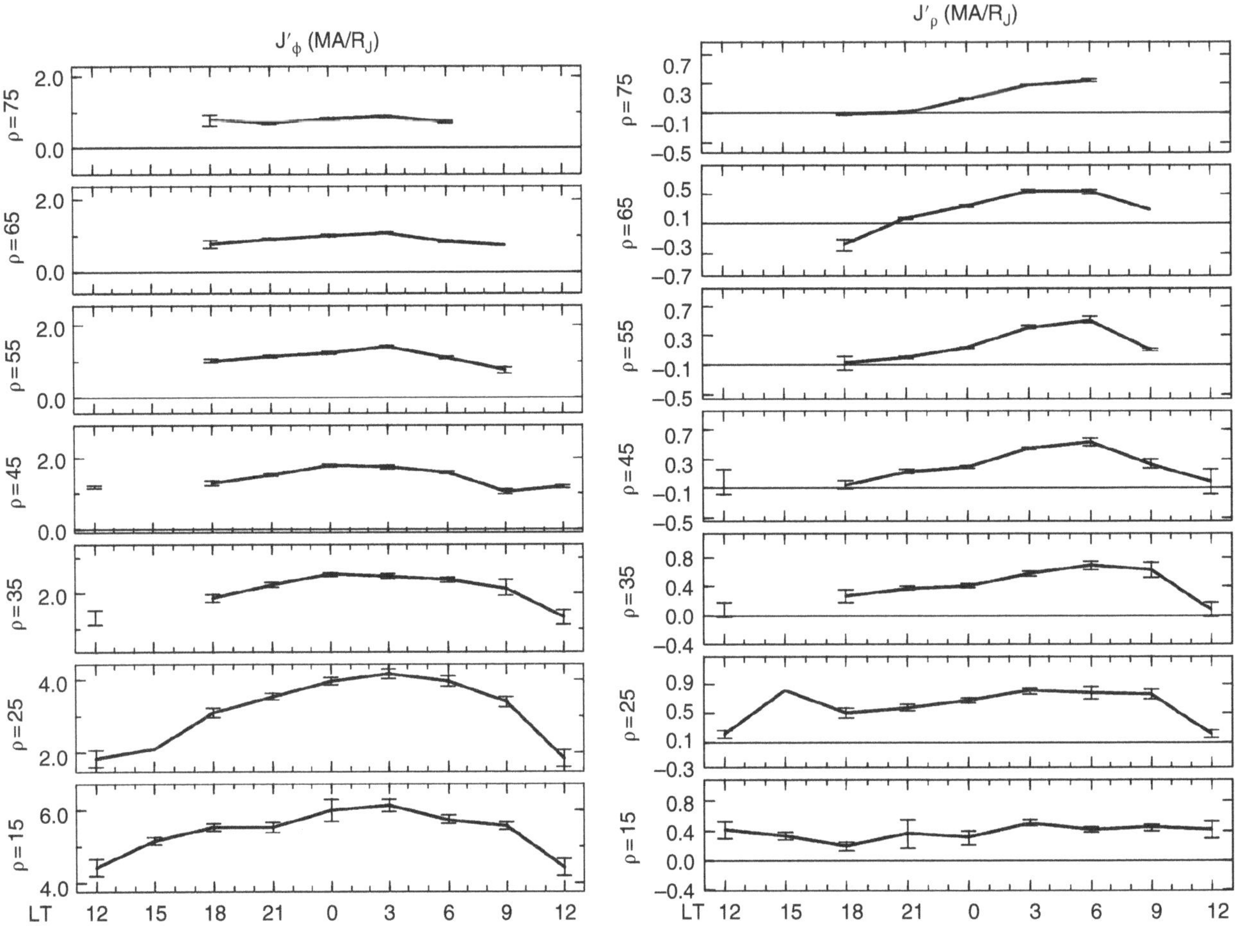

Figure 8.4 Binned profiles, derived from magnetic data, of the height-integrated radial and azimuthal current density in the current sheet, as a function of local time. Profiles for different radial distance intervals are shown, according to the vertical axis labels. From *Khurana et al.* [2004], adapted from *Khurana* [2001].

[2004] (adapted from *Khurana* [2001]). There is a day-night asymmetry in the azimuthal current density: the total azimuthal current flowing on the nightside between distances 10–50 R_J is ~144 MA, compared with the corresponding dayside current, integrated over the same distances, of ~88 MA. Current continuity thus requires the existence of field-aligned current systems at dusk and dawn that respectively feed and drain the nightside partial ring current. This type of current system is analogous to the terrestrial Region 2 currents, associated with solar wind-driven convection. It thus seems plausible that the solar wind may be modulating the local-time structure of the middle magnetosphere at Jupiter.

Another important property of the current profiles in Figure 8.4 is the occurrence of negative (planetward) values of J'_ρ, especially at distances near ~65 R_J in the dusk sector. This finding indicates that, beyond ~65 R_J, the relatively simple picture of a subcorotating magnetodisk breaks down. Excellent reviews of global plasma distribution and flow may be found in, for example, *Vasyliūnas* [1983], *Kivelson and Southwood* [2005], and *Bagenal and Delamere* [2011].

8.2.3. Outer Dayside Magnetosphere

Jupiter's dayside current sheet must, of course, be finite in radial extent, but to what distances does it reach? The magnetopause certainly represents a generous outer boundary for the magnetodisk. In fact, spacecraft observations have revealed that the magnetodisk current sheet does not extend all the way to the magnetopause. In the morning-to-postnoon sector, there is a region known as the "cushion," with radial extent of order ~20 R_J (near noon), lying between the outer edge of the plasmadisk and the magnetopause, where the field shows strong fluctuations and is mainly southward directed [*Kivelson and Southwood*, 2005; *Smith et al.*, 1974]. Magnetic nulls are sometimes present in the cushion: these are a likely signature of bubbles of plasma detaching from the edge of the plasmadisk [*Kivelson and Southwood*, 2005; *Southwood et al.*, 1993].

8.3. FORCE-BALANCE MODELS OF THE MAGNETODISK

Caudal [1986] developed a model of the Jovian magnetodisk, based on the assumption of global force balance between the Lorentz force, plasma (isotropic) pressure gradient, and centrifugal force in an axisymmetric system with parallel rotational and magnetic (planetary dipole direction) axes. This condition may be written as:

$$\mathbf{J} \times \mathbf{B} - \nabla P + n_i m_i \rho \omega^2 \mathbf{e}_\rho \approx 0 \qquad (8.5)$$

where the symbols have the following meanings: current density **J**, magnetic field **B**, plasma pressure P, ion number density n_i, ion mass m_i, cylindrical radial distance ρ, plasma angular velocity ω, and cylindrical radial unit vector $\mathbf{e}_\rho$. *Caudal* [1986] showed that this physical condition was equivalent to a differential equation relating the spatial gradients of the magnetic field (radial and meridional) to a "source function" $g(r, \theta)$, which represents the influence of the plasma properties on the field structure. The $g(r, \theta)$ can then be determined throughout the magnetospheric volume from its equatorial values (constructed from observations) by using the force-balance formalism and model assumptions (e.g., uniform temperature and effective ion mass along a field line). Recent adaptations of *Caudal* [1986]'s model have been presented for the Kronian magnetosphere [*Achilleos et al.*, 2010a; *Achilleos et al.*, 2010b] and for the Jovian magnetosphere [*Nichols*, 2011; *Nichols et al.*, 2015].

This type of model is useful for illustrating how the plasmadisk structure responds to changes in magnetospheric size. In Figure 8.5, we show the computed pressure of the cold plasma population from an implementation of *Caudal* [1986]'s Jovian magnetodisk model, computed for two different equatorial radii 60 and 90 R_J. Magnetic field lines anchored to the same planetary surface footpoints are also shown in both panels, as well as the $\beta_c = 0.2$ contour of cold plasma beta, which represents a plasmasheet boundary. The plasmasheet in the middle magnetosphere is thinner for the expanded model. This behavior is linked to the corresponding change in the shape of the field lines, which become significantly more disklike (radially distended) in the expanded case. In these models, cold plasma pressure along a given field line decreases, relative to the equatorial value P_o, according to:

$$P = P_o \exp\left(\frac{\rho^2 - \rho_o^2}{2l^2}\right) \qquad (8.6)$$

where ρ_o is the cylindrical radial distance of the equatorial point of the field line and l is a pressure scale length given by $l^2 = 2\,k_B T/(m_i \omega^2)$, where k_B is Boltzmann's constant and T is temperature. The points on the field line with cylindrical distance $\rho_o - l$ are thus located at a smaller vertical distance from the equator for a more disklike field, leading to a thinner plasmasheet. A centrifugally confined plasma, situated on flux tubes with equatorial radii ρ_o, would satisfy the condition $l^2 \ll \rho_o^2$, which, using the definition of l^2, is equivalent to $k_B T \ll \frac{1}{2} m_i \omega^2 \rho_o^2$, that is, the typical thermal energy of the cold ions in the disk is small compared to their kinetic energy of bulk rotation. Some representative, equatorial values of l according to the 60 R_J magnetodisk model (which incorporates observed values of relevant plasma parameters) are $l = \sim 3$, ~ 8, and ~ 40 R_J at respective radial distances of 10, 30, and 60 R_J.

Figure 8.6 shows the azimuthal current density for the same magnetodisk models used for Figure 8.5. We show the total current density from both hot and cold plasmas. The total azimuthal current that flows across the entire meridian is ~270 MA and ~310 MA for the respective compressed and expanded models. The corresponding magnetic moments of these extensive current systems are ~1.7 μ_J and ~3.3 μ_J, which exceed the magnetic moment μ_J of the planet's dipole field (defined $\mu_J = 4\pi B_J R_J^3/\mu_o$, where $B_J \approx 428000$ nT is the equatorial magnetic field strength at the planet's surface). These total currents are similar in magnitude to those flowing through the

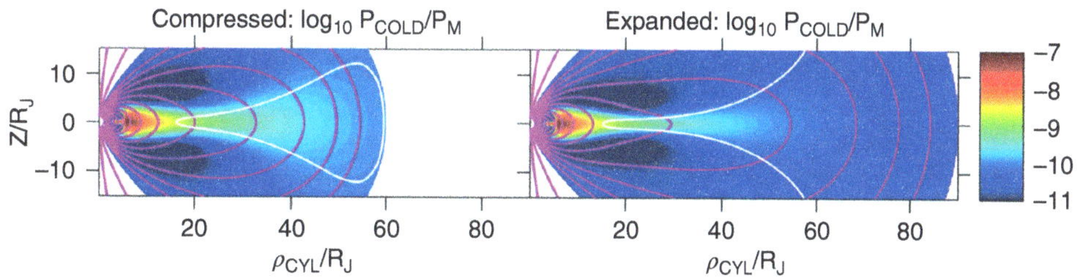

Figure 8.5 Magnetic meridians from compressed and expanded force-balance models of the Jovian magnetodisk (see text). Logarithm of normalized cold plasma pressure is indicated by the color scale (the pressure normalization factor is $P_M = B_J^2/\mu_o$, where $B_J = 428000$ nT is the approximate equatorial surface field strength at Jupiter). The solid magenta curves are magnetic field lines and the white curves are the loci defined by cold plasma beta $\beta_c = 0.2$.

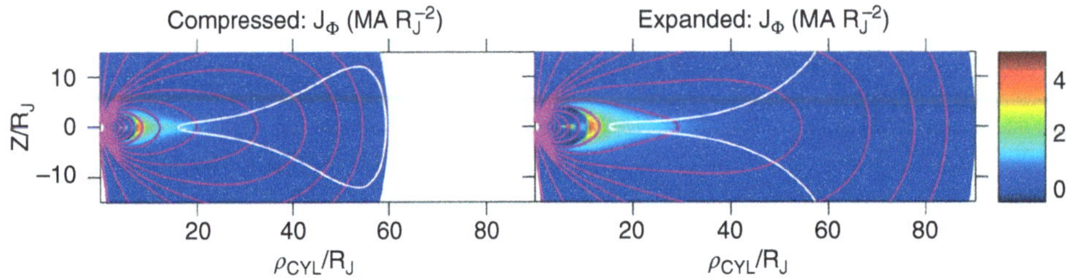

Figure 8.6 As for Figure 8.5, but with the color scale now showing total azimuthal current density (from both cold and hot plasmas), in units of MAR_J^{-2}.

Connerney et al. [1981] current disk models described in Figure 8.2 (~150–290 MA). Not visible at the scale of these plots is the region of weak, negative (i.e., flowing against the direction of planetary rotation) current found in the inner magnetosphere (at distances $\lesssim 7R_J$). Negative current arises in regions where the component of force acting on the plasma in the direction orthogonal to the magnetic field, due to pressure gradients plus centrifugal force, is inward directed.

Nichols et al. [2015] has recently developed the first Caudalian model of the Jovian magnetodisk to include the effects of anisotropic hot plasma pressure. Observations by *Paranicas et al.* [1991] of the anisotropic particle pressure (for particle energies $\gtrsim 28$ keV) were used to constrain the ratio $P_\parallel/P_\perp$ of pressures parallel and perpendicular to the magnetic field. This ratio reaches values up to ~1.2 at the model equator. Force balance equations were then numerically integrated to produce models of the magnetodisk as a function of a hot plasma parameter K_h, which is the product of hot plasma pressure and flux tube volume in the outer magnetosphere, and constrained using spacecraft observations. Figure 8.7 shows some of the results of this study: specifically, maps of the contributions to current density, which balance the different forces on the plasma, and the geometry of the magnetic field lines. The hot plasma pressure gradient is balanced by the strongest contribution to the current overall but note also that the hot plasma anisotropy current also rapidly increases with K_h. The increased radial stretching of the field with K_h generally confines the currents closer to the equatorial plane, and also reduces the magnitude of the local hot plasma pressure gradient.

The calculations by *Nichols et al.* [2015] also have important implications for auroral morphology and magnetodisk stability. With the inclusion of hot plasma pressure anisotropy, the observed few-degree latitudinal shifts of the modeled main oval and Ganymede footprint, arising from realistic changes in K_h, are comparable to the analogous shifts observed in auroral images (e.g., *Grodent et al.* [2008]; *Bonfond et al.* [2012]). In addition, the model's middle magnetosphere attains equatorial values of $\beta_{h,\parallel}-\beta_{h,\perp}$ (difference between parallel and perpendicular hot plasma beta) of ~1–2, for $K_h = 2.0–2.5 \times 10^7$ Pa m T^{-1}. These values suggest that this region may be susceptible to the firehose instability (see also *Kivelson and Southwood* [2005], who discuss the "centrifugal parallel heating" mechanism by which this instability may arise at the equatorial portions of rotating flux tubes, which simultaneously expand as they rotate into the dusk sector). Indeed, for K_h values larger than this range, $\beta_{h,\parallel}-\beta_{h,\perp}$ can exceed 2 near ~25 R_J and the model does not converge. *Nichols et al.* [2015] point out that this result suggests that small-scale release, or drizzle, of Iogenic plasma may commonly arise from the Jovian middle magnetosphere under conditions of adequately high hot plasma pressure (see also *Bagenal* [2007]).

In summary, the inclusion of pressure anisotropy in a Jovian magnetodisk model, as described by *Nichols et al.* [2015], results in the following main modifications when compared with the predictions of isotropic-pressure models: (1) Radial force associated with anisotropy plausibly promotes firehose instability in the middle magnetosphere, possibly associated with an additional channel of mass loss. (2) Anisotropy enhances the distortion of the magnetic field lines, and greatly facilitates few-degree shifts in the main auroral latitude, compatible with those observed.

8.4. MICROSCOPIC NATURE OF THE DISK CURRENT

In general terms, the relation between the magnetic field flux density **B** (commonly referred to in space physics as the "magnetic field"), the free current density $\mathbf{J_f}$, and the plasma magnetization **M** is:

$$\nabla \times \mathbf{B} = \mu_o \left(\mathbf{J_f} + \nabla \times \mathbf{M}\right) \quad (8.7)$$

where we recognize that the total source of the vector field **B** comprises both free and magnetization current density, the latter being equal to ∇ × **M**. Free current is associated with drift motions, which displace the guiding centers of particles, while magnetization current can arise from the collective action of many particle

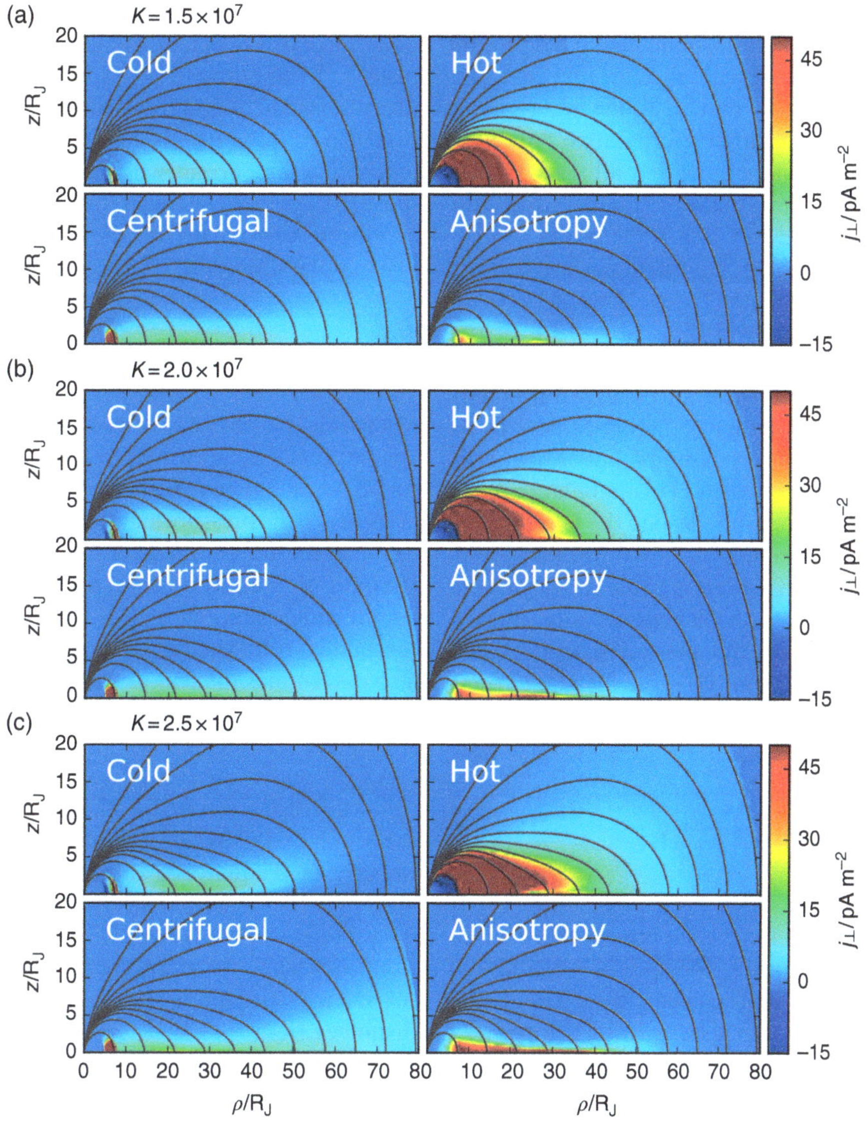

Figure 8.7 Magnetodisk models showing current density on a color scale and magnetic field lines as solid curves. Model outputs associated with the different forces on the plasma (pressure gradients for cold and hot populations, centrifugal force, anisotropy force) are shown. Three sets of models are shown, labeled with the corresponding hot plasma parameter K_h (see text) used for the calculations. From *Nichols et al.* [2015].

gyrations, even in a plasma where the guiding centers are stationary (provided the curl of the magnetization is nonzero).

We may further investigate the relative intensities of the two types of current in a steady-state system by adapting the example described by *Gurnett and Bhattacharjee* [2005]. For simplicity, we consider a cylindrically symmetric system with magnetic field everywhere directed parallel to the axis of symmetry. The magnetic moment of the plasma per unit volume is:

$$\mathbf{M} = \left(-NW_g/B\right)\hat{\mathbf{B}} \tag{8.8}$$

where N is the total number density of particles in the isotropic plasma with mean kinetic energy of gyration W_g per particle, and $\hat{\mathbf{B}}$ denotes a constant unit vector along the direction of the magnetic field. The magnetization current density, $\mathbf{J}_m = \nabla \times \mathbf{M}$, can then be evaluated in this picture:

$$\mathbf{J}_m = \left(\frac{-1}{B^2}\right)\nabla P \times \mathbf{B} + \left(\frac{P}{B^3}\right)\nabla B \times \mathbf{B} \tag{8.9}$$

where we have used the definition for isotropic plasma pressure $P = NW_g$.

Consider a simple plasma with particles of equal gyroradius r_g and threaded by a uniform magnetic field in the z direction. Equation (8.9) reveals that, in this case, the magnetization current density reduces to the first term on the right-hand side, which is often referred to as "diamagnetic current density." Assume now that all the particles in our fictitious plasma have the same temperature, and all gyrate in the xy plane, but that the guiding centers themselves are stationary. Along a reference line defined by $x = x_o$, the current depends on the motions of particles with guiding centers located within a distance $\sim r_g$ of the reference line. If the plasma has a density gradient in the x direction, this implies that the number of gyrating particles crossing the line, which have guiding centers to its right, will not equal the corresponding number for particles with guiding centers to its left. This imbalance leads to a net current in the y direction, despite the fact that the guiding centers are stationary.

The final term in equation (8.9) is equal to $-J_g$, where J_g is the contribution to the current density due to magnetic gradient drift of the plasma particles. The other drift current in this example is the inertial current J_Ω associated with the bulk rotation of the plasma, which is given by:

$$\mathbf{J}_\Omega \approx \left(\frac{NW_\Omega}{\rho B^2} \right) \mathbf{e}_\rho \times \mathbf{B} \tag{8.10}$$

where W_Ω is the rotational kinetic energy per plasma ion, ρ is cylindrical radial distance, and $\mathbf{e}_\rho$ is the cylindrical radial unit vector. This current may be pictured as arising from the drift associated with the centrifugal force in a corotating frame of reference. For a single ion of mass m_i and charge q_i, the centrifugal force per unit charge is simply $(2\ W_\Omega/(q_i\rho))\mathbf{e}_\rho$. The corresponding guiding center drift velocity will thus be $v_{d\Omega} = (2\ W_\Omega/(q_i\rho B^2))\mathbf{e}_\rho \times \mathbf{B}$. Hence, the inertial current density carried by the rotating ions will be, for our quasi-neutral plasma, $J_\Omega = (Nq_i/2)\ v_{d\Omega}$, which is equal to the expression in equation (8.10). This will be a good approximation for the total inertial current density, since the mass of the electrons is negligible when compared with that of the ions.

Using the above expressions for current densities, we note that the total current density in this example is:

$$\mathbf{J_t} = \mathbf{J_m} + \mathbf{J_g} + \mathbf{J}_\Omega = \left(\left(\frac{-1}{B^2} \right) \nabla P + \frac{NW_\Omega}{\rho B^2} \mathbf{e}_\rho \right) \times \mathbf{B} \tag{8.11}$$

The expression for $\mathbf{J_t}$ indicates an important general principle: the drift and magnetization currents for a plasma in magnetostatic equilibrium combine to produce a total current density $\mathbf{J_t}$ for which the associated magnetic body force $\mathbf{J_t} \times \mathbf{B}$ exactly balances the other forces on the plasma. In the present example, the other forces are the pressure gradient and the centrifugal volume force.

It also follows, in this case, that the magnitude of the magnetization current will be large compared to that of the drift current wherever both of the following local conditions are satisfied by the radial length scale l_P of the plasma pressure:

$$l_P << l_B, \quad l_P << \frac{W_g}{W_\Omega} \rho \tag{8.12}$$

where l_B is the length scale of the field strength. In a more general context, we would expect magnetization current to be dominant in a rotating plasma at natural boundaries where the pressure length scale is small in comparison to the minimum of: the length scale of the magnetic field strength, the radius of curvature of the field lines, and a length scale (see equation [8.12]) involving the ratio of the thermal to rotational energy of a plasma ion. This ratio also determines the length scale for pressure in a centrifugally confined plasma according to the theory of *Caudal* [1986] (see section 8.3).

A fully realistic quantitative comparison of magnetization and drift current for the Jovian system is beyond the scope of this review. Nevertheless, we may appeal to the predicted equatorial values of the ratio l_P/l_B from our 60 R_J magnetodisk model in order to obtain a preliminary indication of where we would expect magnetization current to be dominant. When considering total plasma pressure, we find $l_P/l_B < 0.1$, for example, at radial distances from ~43 R_J to just inside the magnetopause. This behavior arises due to the very gradual (compared to pure dipole) decrease in equatorial field strength of the disklike field in the outer magnetosphere. When considering the cold plasma alone, we find that $0.5 < l_P/l_B < 1$ within the inner and outer edges of the Io plasma torus, themselves natural plasma boundaries with relatively strong spatial gradients in plasma pressure.

8.5. SUMMARY

A combination of data analysis and theoretical modeling related to Jupiter's magnetodisk has produced an ever more detailed picture of the system. Elements of this picture include the magnetodisk/current sheet morphology and its response to the various stresses on the disk plasma, and how these stresses vary in response to changes in the drivers of magnetospheric structure, both external (solar wind) and internal (Io mass-loading rate, global change in energetic particle population).

Further quantifying the relative importance of planetary rotation versus solar wind influence on the field and plasma configuration of the magnetodisk is an ongoing endeavor in the magnetospheric community. To support this aim, we require in situ observations of the system under a comprehensive range of solar wind conditions, as well as reliable tools for propagating solar wind parameters, observed in near-Earth space, out to the orbit of Jupiter (e.g., the mSWiM model, *Zieger and Hansen* [2008]). Further modeling and extensive field/particle observations are also strongly required to fully characterize and understand the influence of plasma pressure anisotropy on the magnetodisk structure and stability. On the observational side, missions such as Juno (currently in orbit around Jupiter) and JUICE (JUpiter ICy moon Explorer, scheduled for launch in 2022) will be of great value in these pursuits.

ACKNOWLEDGMENTS

N. Achilleos was supported by the UK STFC Consolidated Grant (UCL/MSSL Solar and Planetary Physics, ST/N000722/1), and also acknowledges the support of the International Space Science Institute (ISSI), following useful discussions at ISSI Team Meetings "Coordinated Numerical Modeling of the Global Jovian and Saturnian Systems" and "Modes of Radial Plasma Transport in Planetary Systems." N. Achilleos is very grateful to both referees for their comments and suggestions.

REFERENCES

Achilleos, N., P. Guio, and C. S. Arridge (2010a), A MNRAS, model of force balance in Saturn's magnetodisc, *Monthly Not. R. Astron. Soc.*, *401*, 2349–2371.

Achilleos, N., P. Guio, C. S. Arridge, N. Sergis, R. J. Wilson, M. F. Thomsen, and A. J. Coates (2010b), Influence of hot plasma pressure on the global structure of Saturn's magnetodisk, *Geophys. Res. Lett.*, *37*, L20,201.

Achilleos, N., N. André, X. Blanco-Cano, P. C. Brandt, P. A. Delamere, and R. Winglee (2015), 1. Transport of mass, momentum and energy in planetary magnetodisc regions, *Space Sci. Rev.*, *187*, 229–299.

Bagenal, F. (2007), The magnetosphere of Jupiter: Coupling the equator to the poles, *J. Atmos. Solar Terr. Phys.*, *69*, 387–402.

Bagenal, F., and J. D. Sullivan (1981), Direct plasma measurements in the Io torus and inner magnetosphere of Jupiter, *J. Geophys. Res.*, *86*, 8447–8466.

Bagenal, F., and P. A. Delamere (2011), Flow of mass and energy in the magnetospheres of Jupiter and Saturn, *J. Geophys. Res.*, *116*, 5209.

Bagenal, F., R. L. McNutt, J. W. Belcher, H. S. Bridge, and J. D. Sullivan (1985), Revised ion temperatures for Voyager plasma measurements in the Io plasma torus, *J. Geophys. Res.*, *90*, 1755–1757.

Behannon, K. W., L. F. Burlaga, and N. F. Ness (1981), The Jovian magnetotail and its current sheet, *J. Geophys. Res.*, *86*, 8385–8401.

Bonfond, B., D. Grodent, J.-C. Gérard, T. Stallard, J. T. Clarke, M. Yoneda, A. Radioti, and J. Gustin (2012), Auroral evidence of Io's control over the magnetosphere of Jupiter, *Geophys. Res. Lett.*, *39*, L01, 105.

Caudal, G. (1986), A self-consistent model of Jupiter's magnetodisc including the effects of centrifugal force and pressure, *J. Geophys. Res.*, *91*, 4201–4221.

Connerney, J. E. P., M. H. Acuña, and N. F. Ness(1981), Modeling the Jovian current sheet and inner magnetosphere, *J. Geophys. Res.*, *86*, 8370–8384.

Cowley, S. W. H., and E. J. Bunce (2001), Origin of the main auroral oval in Jupiter's coupled magnetosphere-ionosphere system, *Planet. Space Sci.*, *49*, 1067–1088.

Goertz, C. K., B. A. Randall, M. F. Thomsen, D. E. Jones, and E. J. Smith (1976), Evidence for open field lines in Jupiter's magnetosphere, *J. Geophys. Res.*, *81*, 3393–3397.

Grodent, D., J.-C. Gérard, A. Radioti, B. Bonfond, and A. Saglam (2008), Jupiter's changing auroral location, *J. Geophys. Res. Space Physics*, *113*, A01, 206.

Gurnett, D. A., and A. Bhattacharjee (2005), *Introduction to Plasma Physics*, Cambridge University Press, chap. 3.

Hill, T. W. (1979), Inertial limit on corotation, *J. Geophys. Res.*, *25*, 6554–6558.

Hill, T. W., A. J. Dessler, and C. K. Goertz (1983), Magnetospheric models, in *Physics of the Jovian Magnetosphere*, Cambridge University Press, 353–394.

Joy, S. P., M. G. Kivelson, R. J. Walker, K. K. Khurana, C. T. Russell, and T. Ogino (2002), Probabilistic models of the Jovian magnetopause and bow shock locations, *J. Geophys. Res. Space Physics*, *107*, 1309.

Khurana, K. K. (1992), A generalized hinged-magnetodisc model of Jupiter's nightside current sheet, *J. Geophys. Res.*, *97*, 6269–6276.

Khurana, K. K. (1997), Euler potential models of Jupiter's magnetospheric field, *J. Geophys. Res.*, *102*, 11,295–11,306.

Khurana, K. K. (2001), Influence of solar wind on Jupiter's magnetosphere deduced from currents in the equatorial plane, *J. Geophys. Res.*, *106*, 25,999–26,016.

Khurana, K. K., and M. G. Kivelson (1993), Inference of the angular velocity of plasma in the Jovian magnetosphere from the sweepback of magnetic field, *J. Geophys. Res.*, *98*, 67–79.

Khurana, K. K., M. G. Kivelson, V. M. Vasyliunas, N. Krupp, J. Woch, A. Lagg, B. H. Mauk, and W. S. Kurth (2004), The configuration of Jupiter's magnetosphere, in *Jupiter: The Planet, Satellites and Magnetosphere*, edited by F. Bagenal, T. E. Dowling, and W. B. McKinnon, Cambridge University Press, 593–616.

Kivelson, M. G., and D. J. Southwood (2005), Dynamical consequences of two modes of centrifugal instability in Jupiter's outer magnetosphere, *J. Geophys. Res.*, *110*, A12,209.

Kivelson, M. G., et al (1997), Galileo at Jupiter: Changing states of the magnetosphere and first looks at Io and Ganymede, *Adv. Space Res.*, *20*, 193–204.

Nichols, J., and S. Cowley (2004), Magnetosphere-ionosphere coupling currents in Jupiter's middle magnetosphere: effect of precipitation-induced enhancement of the ionospheric Pedersen conductivity, *Ann. Geophys.*, *22*, 1799–1827.

Nichols, J. D. (2011), Magnetosphere-ionosphere coupling in Jupiter's middle magnetosphere: Computations including a self-consistent current sheet magnetic field model, *J. Geophys. Res. Space Physics*, *116*, A10, 232.

Nichols, J. D., N. Achilleos, and S. W. H. Cowley (2015), A model of force balance in Jupiter's magnetodisc including hot plasma pressure anisotropy, *J. Geophys. Res. Space Physics*, *120*, 10.

Paranicas, C. P., B. H. Mauk, and S. M. Krimigis (1991), Pressure anisotropy and radial stress balance in the Jovian neutral sheet, *J. Geophys. Res.*, *96*, 21.

Ray, L. C., and R. E. Ergun (2012), *Auroral Signatures of Ionosphere-Magnetosphere Coupling at Jupiter and Saturn*, American Geophysical Union, Geophysical Monograph Series 197, Washington, DC.

Ray, L. C., R. E. Ergun, P. A. Delamere, and F. Bagenal (2012), Magnetosphere-ionosphere coupling at Jupiter: A parameter space study, *J. Geophys. Res. Space Physics*, *117*, A01, 205.

Scudder, J. D., E. C. Sittler, and H. S. Bridge (1981), A survey of the plasma electron environment of Jupiter: A view from Voyager, *J. Geophys. Res.*, *86*, 8157–8179.

Siscoe, G. L., A. Eviatar, R. M. Thorne, J. D. Richardson, F. Bagenal, and J. D. Sullivan (1981), Ring current impoundment of the Io plasma torus, *J. Geophys. Res.*, *86*, 8480–8484.

Smith, E. J., L. Davis Jr., D. E. Jones, P. J. Coleman Jr., D. S. Colburn, P. Dyal, C. P. Sonett, and A. M. A. Frandsen (1974), The planetary magnetic field and magnetosphere of Jupiter: Pioneer 10, *J. Geophys. Res.*, *79*, 3501–3513.

Southwood, D. J., M. K. Dougherty, P. Canu, A. Balogh, and P. J. Kellogg (1993), Correlations between magnetic field and electron density observations during the inbound Ulysses Jupiter flyby, *Planet. Space Sci.*, *41*, 919–930.

Thomas, N., F. Bagenal, T. W. Hill, and J. K. Wilson (2004), The Io neutral clouds and plasma torus, in *Jupiter, The Planet, Satellites and Magnetosphere*, Cambridge University Press, 561–591.

Thorne, R. M., T. P. Armstrong, S. Stone, D. J. Williams, R. W. McEntire, S. J. Bolton, D. A. Gurnett, and M. G. Kivelson (1997), Galileo evidence for rapid interchange transport in the Io torus, *Geophys. Res. Lett.*, *24*, 2131.

Vasyliūnas, V. M. (1983), Plasma distribution and flow, in *Physics of the Jovian Magnetosphere*, edited by A. J. Dessler, Cambridge University Press, New York, 395–453.

Zieger, B., and K. C. Hansen (2008), Statistical validation of a solar wind propagation model from 1 to 10 AU, *J. Geophys. Res. Space Physics*, *113*, A08, 107.

9
The Ring Current of Saturn

N. Sergis[1], E. J. Bunce[2], J. F. Carbary[3], S. W. H. Cowley[2], X. Jia[4], D. C. Hamilton[5], S. M. Krimigis[1,3], D. G. Mitchell[3], and M. K. Dougherty[6]

ABSTRACT

Saturn is a rapidly rotating planet with a relatively strong intrinsic magnetic field and significant internal plasma sources, that produce the required internal pressure to deflect the solar wind and form a large magnetosphere, which holds a complex and dynamic electric current system. A very important part of this global system is a current carried by magnetospheric charged particles flowing near the equatorial plane, in the azimuthal direction around the planet at radial distances between ~ 8 R_S and ~ 15 R_S, and is often referred to as the "ring current." As in most magnetized planets, the ring current plays a key role in the magnetosphere-ionosphere coupling at Saturn. In particular, its azimuthal asymmetry is associated with the generation and topology of the field-aligned currents, which are believed to affect, or to some extent drive, the periodic variability that has been well observed in several magnetospheric properties but is not fully understood. In this chapter, we provide an up-to-date overview of the Kronian ring current, including the most recent advances. Following a brief historical review, the chapter focuses on results from the Cassini spacecraft, which spent over 13 yr in orbit around the planet (2004–2017) and returned a large volume of observational data, which have not been fully exploited yet. The text summarizes results that have been published over the last decades, and is organized in three subsections that cover complementary approaches on the subject: in situ observations, remote imaging, and modeling.

9.1. INTRODUCTION

The concept of a planetary ring current was well established and fairly understood for Earth, before any spacecraft reached Saturn. When Pioneer 11 encountered Saturn (closest approach 21,000 km) in September 1979, it became evident that Saturn satisfied the conditions required to maintain an azimuthal (ring) current. The planet has a dipole magnetic field that is strong enough to form a sizeable magnetosphere, and also has several internal plasma sources (for example, Enceladus, rings, ionosphere, other moons) that continuously supply the system with electron and ion plasma, forming an inhomogeneous and at times quite variable plasma sheet.

Unlike the case of Earth, the Saturnian magnetic field protects quite effectively the magnetosphere from the solar wind plasma, which enters the outer magnetosphere during magnetic reconnection that occurs in the tail and more rarely in the dayside. In addition, Saturn is a fast rotator (period of ~10.8 hr). As the nonuniform magnetospheric field rotates rapidly, ions and electrons drift at different speeds and directions around the planet (due to

[1] *Office of Space Research and Technology, Academy of Athens, Athens, Greece*

[2] *Department of Physics and Astronomy, University of Leicester, Leicester, UK*

[3] *Applied Physics Laboratory, Johns Hopkins University, Laurel, Maryland, USA*

[4] *Department of Climate and Space Sciences and Engineering University of Michigan, USA*

[5] *Department of Physics, University of Maryland, College Park, Maryland, USA*

[6] *Blackett Laboratory, Imperial College London, London, UK*

Electric Currents in Geospace and Beyond, Geophysical Monograph 235, First Edition.
Edited by Andreas Keiling, Octav Marghitu, and Michael Wheatland.

the "grad **B**" drift, the "curvature" drift, and the inertia drift associated with the plasma corotation with the planet). In combination with the established pressure gradient, predominantly along the radial direction, they produce a fairly strong azimuthal electric current that can be viewed as the sum of the drift current mentioned before, plus the magnetization current due to the gyration of the particles about their guiding centers given as curl **M**, where magnetization **M** is the gyration magnetic dipole moment per unit volume (see detailed discussion in *Bunce et al.* [2007]).

The ring current constitutes an essential part of the magnetospheric current system and the global magnetospheric-ionospheric coupling. Its azimuthal asymmetries are closely related to the structure and strength of the field-aligned currents and thus affects the observed, and so far imperfectly understood, periodicities in several of Saturn's magnetospheric phenomena, all of which show a variability with a period very close to that of the planetary rotation [*Carbary and Mitchell*, 2013]. As *Galopeau and Lecacheux* [2000] initially suggested based on Ulysses observations, and *Gurnett et al.* [2009] confirmed using data from the Cassini Radio and Plasma Wave Science Instrument (RPWS), the period of radio emission changes slowly, suggesting that the source of the observed periodicities should be related to a type of asymmetric atmospheric-ionospheric coupling, rather than the interior of the planet. Hence, the study and further understanding of the interacting current components that constitute the three-dimensional magnetospheric current system that, apart from the ring current includes the field-aligned currents, becomes increasingly important.

9.2. EARLY OBSERVATIONS

The presence of a ring current in Saturn's magnetosphere was first inferred from magnetic field data obtained during the flybys of Pioneer 10 in September 1979 [*Smith et al.*, 1980] and Voyager 1 and 2 in November 1980 and August 1981 [*Ness et al.*, 1981, 1982], from a depression in the equatorial field strength compared with that expected for the planetary field alone at radial distances ~5–10 R_S (R_S is Saturn's 1 bar equatorial radius, equal to 60,268 km). Apart from its manifestation in the magnetic field, the ring current was also partially measured by the Low Energy Charged Particle (LECP) sensor onboard Voyager 1 and 2 [*Krimigis et al.*, 1981, 1983; *Mauk et al.*, 1985]. However, as only a limited part of the particle energy spectrum was measured (>30 keV), the nature of the azimuthal current could not be fully revealed.

The field perturbations in the Voyager data were fitted with a four-parameter current disk model previously developed for Jupiter [*Connerney et al.*, 1981a], in which an eastward-directed current is taken to flow in a cylindrically symmetric disk with inner and outer radii R_1 and R_2, respectively, half-thickness D, with the current density J_φ varying inversely with the cylindrical radial distance ρ from the axis

$$J_\varphi = \frac{I_o}{\rho} \tag{9.1}$$

The associated fields are in general obtained from an integral over a product of Bessel functions, but can, in some circumstances, be approximated by analytic functions [*Edwards et al.*, 2001; *Giampieri and Dougherty*, 2004]. Following initial fits to the Voyager 1 data reported by *Ness et al.* [1981] and *Connerney et al.* [1981b], *Connerney et al.* [1983] showed that both Voyager 1 and 2 magnetic field measurements could be well fitted by a disk with $R_1 = 8$ R_S, $R_2 = 15.5$ R_S, $D = 3$ R_S, and a current parameter $\mu_o I_o = 64$ nT. The total eastward disk current of ~11.5 MA produces a northward-directed quasi-uniform field inside the inner edge of the disk, opposite to the southward planetary field, of ~10 nT. The magnetic moment of the disk corresponds to ~40% of the planetary dipole.

Equivalent fits to the Pioneer 11 data were undertaken by *Bunce and Cowley* [2003], who found in this case $R_1 = 6.5$ R_S, $R_2 = 12.5$ R_S, $D = 2$ R_S, and $\mu_o I_o = 76.5$ nT, thus with a total current of ~9.6 MA and a magnetic moment of ~20% of the planetary dipole. The inward radial displacement of the disk compared with the Voyager results was attributed to the compressed nature of Saturn's magnetosphere during the Pioneer 11 flyby due to high solar wind dynamic pressure, with a subsolar magnetopause at ~17 R_S compared with ~23.5 R_S for Voyager 1.

Combining magnetic field and energetic particle data returned by Voyager 1, *Mauk et al.* [1985] argued that pressure gradient currents are important in more central regions. They showed that the ring current has a radial structure somehow different than the one assumed by the Connerney model (see comparison in Fig. 7 of *Mauk et al.* [1985]), with a distinct maximum of ~100 pA/m^2 at 11–12 R_S. However, they pointed out that any direct comparison is of limited value and that the deviations can be, at least in part, justified by the fact that the model is more sensitive to the current content at a given radial distance, than to local current density.

It is, nevertheless, noteworthy that in spite of the substantially different techniques employed and the limited available data, all of the above studies produced maximum current densities and total current estimations that compared favorably and were further proven to be in good agreement with what Cassini extensively measured nearly 20 yr later.

9.3. THE CASSINI ERA

Cassini began orbiting Saturn in mid 2004 (Saturn Orbit Insertion [SOI] on 4 July 2004) and by the end of 2016 had completed 255 full orbits around the giant planet, covering a large part of its magnetosphere in the equatorial plane as well as in high latitudes. Undeniably, the detailed sampling provided by Cassini is beyond comparison with the limited amount of measurements returned by Pioneer 11 and Voyager 1 and 2 encounters, in terms of both spatial and temporal coverage. With the Cassini Plasma Spectrometer (CAPS) [*Young et al.*, 2004] and the Magnetospheric Imaging Instrument (MIMI) [*Krimigis et al.*, 2004], Cassini is capable of measuring essentially the full energy spectrum of charged particles, from few eV to few MeV, with the additional ability of separating ion species and charge states. Combined with high-resolution magnetic field measurements provided by the Cassini magnetometer (MAG) [*Dougherty et al.*, 2004], the particle data revealed the global average plasma flow regime in the Saturnian magnetosphere, together with a strong temporal variability, demonstrated primarily in the form of hot plasma "injections" (i.e., energetic particles that are periodically injected inward and subsequently drift around the planet), or rotating, hot plasma "blobs" [*Krimigis et al.*, 2007; *Mitchell et al.*, 2015].

The Saturnian ring current, in particular, was extensively examined using in situ data (CAPS, MIMI, and MAG) and remote monitoring through Energetic Neutral Atom (ENA) emissions captured by the MIMI Ion Neutral Camera (INCA). Questions that remained unanswered since the early 1980s regarding the composition, energy extent, variability, and, most important, the physical mechanisms that control the azimuthal current at Saturn, have now been addressed through much more sophisticated approaches.

In what follows, we present an overview of the Saturnian ring current, including some of the most recent advances, organized in three subsections (i.e., in situ observations, remote imaging, and modeling).

9.3.1. In Situ Observations

With a much larger data volume available from Cassini, initially from its equatorial plane orbits, *Bunce et al.* [2007] selected 17 (out of the first 27) such orbits, for which an inbound magnetopause occurred within 4 days of closest approach, so that modulation of the current with system size could be examined further. Additional indication of such modulation had previously been provided by *Alexeev et al.* [2006], based on modeling of the SOI field data, during which the magnetosphere was further expanded to an estimated subsolar distance of ~24.8 R_S. Due to the near-equatorial nature of these orbits, D could not be determined, but was assumed by *Bunce et al.* [2007] to be 2.5 R_S based on the flyby results, while R_1, R_2, and $\mu_o I_o$ were then determined from the radial profile of the northward perturbation field (the fit in essence determining $\mu_o I_o D$ with D assumed). It was found that R_1 varies only weakly between ~6.5 and ~8 R_S as the subsolar magnetopause distance increases from ~17 to ~26 R_S, while R_2 varies between ~15.5 and ~21 R_S over the same range, thus tracking within a few R_S of the subsolar magnetopause. The more scattered $\mu_o I_o$ values typically vary from ~45 to ~55 nT over this range, such that the northward perturbation field in the inner region varies between ~9.5 and ~13 nT, the total current between ~9 and ~15 MA, and the dipole moment between ~25% and ~65% of the planetary dipole. *Bunce et al.* [2008] employed these results to examine the consequent deformation of the field lines in Saturn's magnetosphere, showing that they retain a quasi-dipolar form when it is strongly compressed, but becoming more magnetodisk-like when expanded, in agreement with the findings of *Arridge et al.* [2008a], *Brandt et al.* [2008], and the subsequent modeling work of *Achilleos et al.* [2010].

When thermal plasma and energetic particle measurements were included in the analysis, it became clear that, in analogy with the Terrestrial ring current, we can distinguish a "quiescent" ring current and a "disturbed" ring current, with the latter controlled by the hot plasma [*Sergis et al.*, 2007]. As Cassini returned data from multiple passes through the ring current, a picture emerged that was much more than a single flyby snapshot. The high-energy ion population imposes a strong variability on the system, demonstrating the importance of hot plasma and in particular the role of the heavier water group ions (W^+) as pressure and current carriers. The energy profile of the Saturnian ring current and the relative contribution of thermal plasma and hot particles were elucidated soon after.

Brandt et al. [2010] combined thermal and suprathermal pressures measured by Cassini and computed the pressure-driven currents and the expected magnetic field depression. Separating a "symmetric" and an "asymmetric" ring current component, in analogy to the "quiescent" and "disturbed" current, they suggested that the energetic particles regulate the asymmetric component and are associated with the magnetic field periodicities observed in Saturn's magnetosphere, while thermal plasma constitutes the underlying symmetric ring current (see further discussion in section 9.3.3 on modeling approaches).

The current density perpendicular to the magnetic field can be expressed in terms of the plasma parameters, by applying Newton's second law considering local stress balance [*Vasyliunas*, 1984; *Kellett et al.*, 2010, 2011]:

$$J_{\perp} = \frac{\mathbf{B}}{B} \times \left[\rho \frac{d\mathbf{V}}{dt} + \nabla P_{\perp} + \left(P_{\parallel} - P_{\perp}\right)\left(\mathbf{b} \cdot \nabla\right)\mathbf{b} \right] \quad (9.2)$$

where ρ is the plasma mass density, **V** is the bulk velocity, **b** is the unit vector along the magnetic field, and $P_{\parallel}$ and $P_{\perp}$ are the "field-parallel" and "field-perpendicular" plasma pressures, respectively. The three separate terms on the right-hand side of equation (9.2) are the inertial current, the perpendicular pressure gradient current, and the anisotropy current.

If we consider the equatorial part of the Saturnian plasma sheet (for example, within ±1 R_S from the rotational plane), we can assume that there is local symmetry about the planet's spin and magnetic axis, that the plasma bulk velocity is purely azimuthal, and that the radial and azimuthal components of the magnetic field are small compared to the polar component. Equation (9.2) can then be reduced to:

$$J_{\varphi} \approx \frac{1}{B}\left[\frac{\rho V_{\varphi}^{2}}{r} - \frac{\partial P_{\perp}}{\partial r} - \frac{(P_{\perp} - P_{\parallel})}{R}\right] \qquad (9.3)$$

where r is the radial distance, and R is the local radius of curvature of the field lines. In the above expression, all physical quantities can be directly determined by Cassini measurements, with the exception of R. Observations, however, show that at Saturn significant pressure anisotropy is present only inside ~10 R_S, where $P_{\perp} > P_{\parallel}$. In this region, the equatorial field does not usually depart strongly from the dipole, therefore R can be well approximated by $r/3$.

Another underlying assumption is that of the gyrotropic distribution of the energetic particles, required for equations (9.2) and (9.3) to apply. Extensive MIMI/CHEMS and INCA comparisons of energetic particle spectra obtained along different pointing directions in the plasma sheet of Saturn, show that almost all of the times we can safely assume gyrotropic particle distributions (see Figs. 4 and 5 from *Sergis et al.* [2009]). We should note, however, that nongyrotropic distributions may appear close to boundaries such as the magnetopause or the plasma sheet-lobe boundary, but for large radial distances (>15 R_S) where particles obtain large gyroradii due to the much weaker magnetic field and can bypass the boundary leaking from one region to the adjacent [*Sergis et al.*, 2013].

It becomes obvious that the average radial profiles of plasma properties such as the mass density, the flow velocity, and the total particle pressure can lead to a data-based computation of the ring current density, under the assumptions mentioned earlier. Furthermore, the separate computation of each of the three terms of equation (9.3) becomes possible, and can help us explore deeper the nature of the ring current and learn more about its driving mechanism.

The average distribution of the basic plasma properties was computed soon after Cassini completed its first set of equatorial orbits. *Wilson et al.* [2008], *McAndrews et al.* [2009], and *Thomsen et al.* [2010] presented radial profiles of the plasma number and mass density, temperature (both parallel and perpendicular to the magnetic field), and flow velocity from the Cassini/CAPS instrument. *Sergis et al.* [2007, 2009] provided the corresponding radial distribution for the energetic ion population. In panels (a) and (b) of Figure 9.1, we present the radial distribution of the median ion plasma temperature and azimuthal velocity and the mean pressure in the equatorial magnetosphere of Saturn for the thermal plasma (energies below 45 keV, CAPS data), as computed from data collected during the first 4.5 yr of the mission (October 2004 to April 2009). Similarly, panels (c) and (d) show the corresponding radial profiles of the hot (>3 keV) ion pressure and the total plasma pressure, from 11 equatorial orbits of Cassini in 2005 and 2006, which are in addition compared with the measured magnetic pressure. From Figure 9.1 it becomes evident that:

1. The thermal plasma pressure steadily drops with increasing radial distance over the range shown in the figure, and dominates the total pressure inside 8 R_S, while the hot plasma pressure develops a broad maximum around 10 R_S and becomes the major contributor to the total particle pressure beyond 12 R_S. For both thermal and hot plasma, the heavier W^+ ions are the most important pressure carriers.
2. The hot plasma pressure is characterized by increased scatter, indicative of the dynamic temporal behavior of the energetic ions, commonly observed in the form of hot plasma injection events in the middle magnetosphere of Saturn, as a result of local plasma interchange instability.
3. The plasma temperature increases with increasing radial distance, reaching 600 eV near 15 R_S, while the plasma bulk velocity remains within 50%–70% of corotation.
4. Beyond 9 R_S, the particle pressure becomes clearly higher than the local magnetic pressure, resulting a high plasma beta (ratio of particle to magnetic pressure) that remains in the range of 2 to 10 as far as the outer parts of the dayside magnetosphere.

With plasma and magnetic field properties either measured or estimated based on in situ data, it was possible to compute the ring current density from equation (9.3). Since 2008, several independent studies attempted to estimate the magnitude of the ring current and describe its nature and global distribution.

Sergis et al. [2010] combined magnetic field and particle data to examine the radial distribution of the azimuthal current density. The resulting radial profiles for each ring current component, for example, inertial, pressure gradient, and anisotropy, are shown in blue, red, and black respectively in Figure 9.2a, while the total

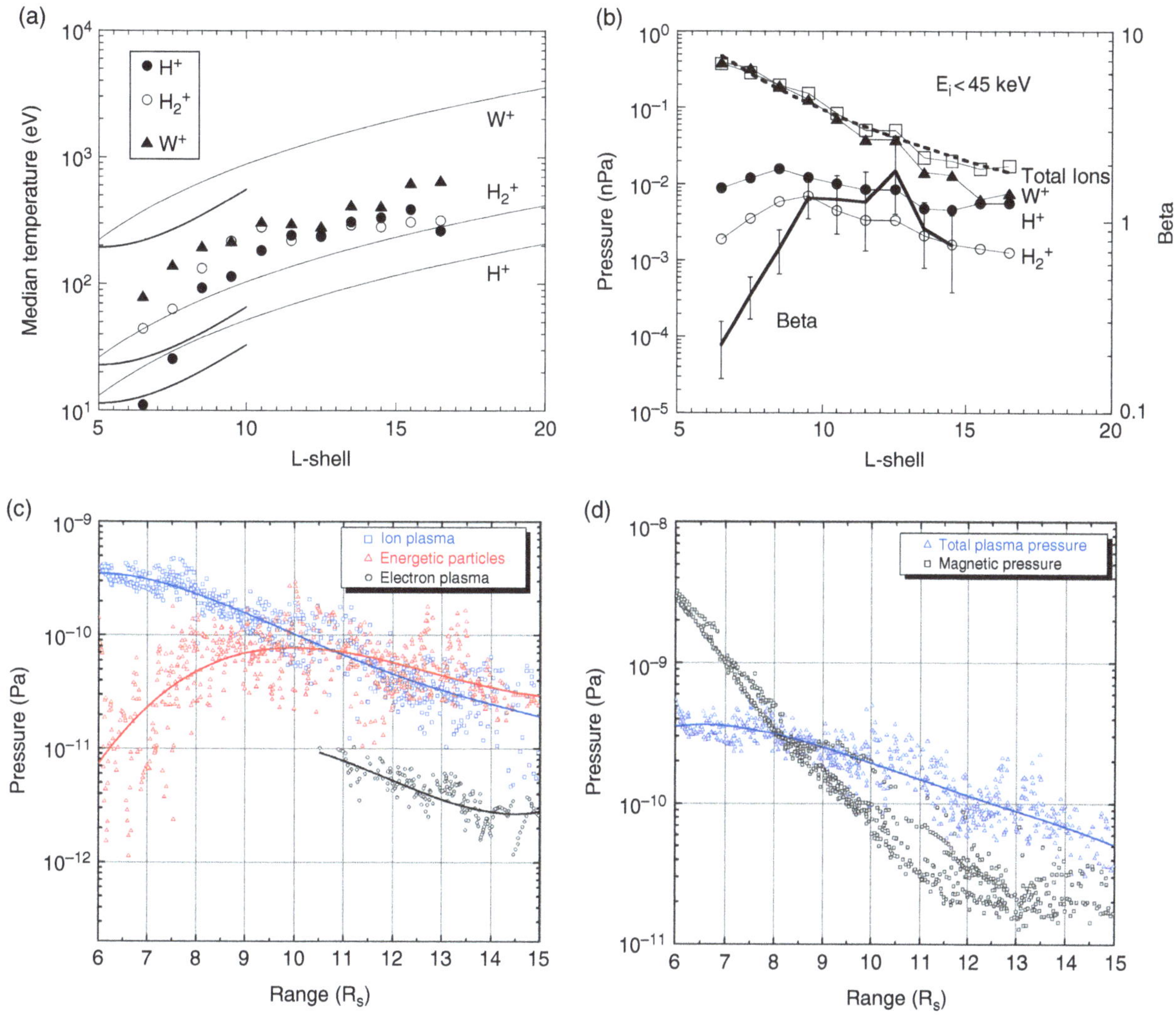

Figure 9.1 (a) Median temperatures of three ion species, compared with the pick-up gyrational energy for full corotation (thin solid lines) and the bulk flow velocity derived by *Wilson et al.* [2008] (heavy solid lines) (adapted from *Thomsen et al.* [2010]; based on Cassini data from July 2004 [SOI] to the end of 2009). (b) Mean values of the pressures (for E < 45 keV) of the three species and their sum, plotted as a function of L for only low-latitude points with corotation within the CAPS field of view. The dashed line is a power law fit to the total ion pressure in this energy range, given as $P_{ion}(nPa) = 550\ L^{-3.77}$ (from *Thomsen et al.* [2010]; Cassini data, July 2004 to end of 2009). Also shown is the contribution to the plasma beta from this energy range, using observed equatorial magnetic field values as presented by *Sergis et al.* [2010]. (c) Radial pressure profile for thermal ion plasma (blue), energetic particles (red), and thermal electron plasma (black), together with polynomial fits of the same color (Cassini data from 2005 and 2006). The apparent scatter is indicative of the intense dynamics present in the Saturnian magnetosphere. Electron moments are not available inside 10 Rs due to the spacecraft potential noise [after *Sergis et al.*, 2010]. (d) Radial profiles for the magnetic pressure (black) and the total particle pressure (blue), with a polynomial fit of the same colors, based on Cassini data from 2005 and 2006. Adapted from *Sergis et al.* [2010].

azimuthal current is presented in Figure 9.2b, together with the corresponding profiles from previous studies.

The total ring current J_φ, shown in panel (b), forms a maximum between 8 and 12 R_S reaching ~100 to 150 pA/m^2, in the same range of distances where maximum ENA intensities have been observed [*Carbary et al.*, 2008a], although MIMI/INCA samples ENAs produced only by the hot (keV) ions. This nonmonotonic dependence on radial distance was also suggested by *Mauk et al.* [1985] and by *Beard and Gast* [1987] long before the arrival of Cassini. Beyond its maximum region, J_φ drops with increasing radial distance as $\sim r^{-2.2}$, that is, faster than the $1/r$ rate earlier suggested by disk current models (blue and orange lines). The anisotropy current (black line in panel [a]

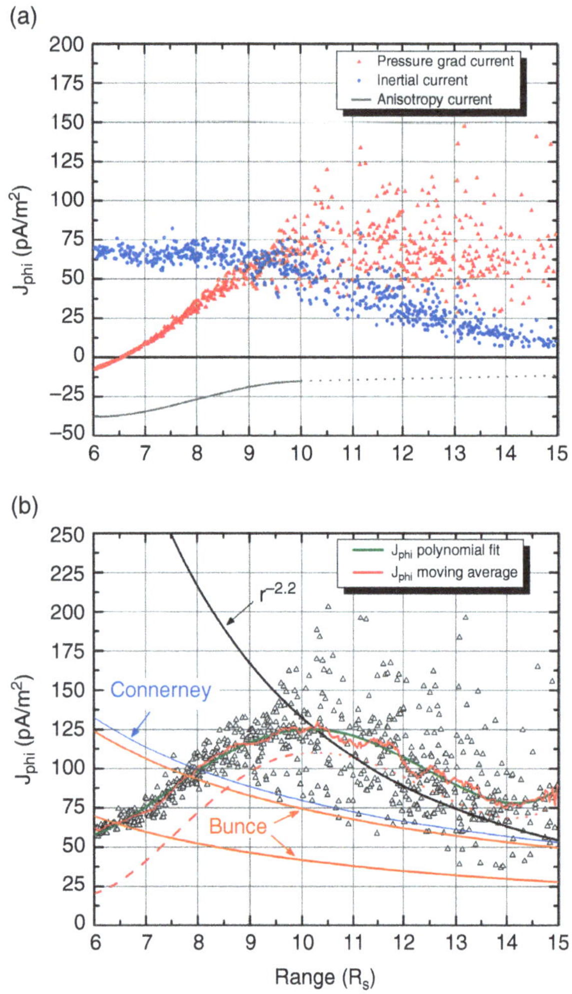

Figure 9.2 (a) Radial profiles for the inertial, the pressure gradient, and the pressure anisotropy contribution to the total current density in blue, red, and gray, respectively. (b) Radial profile of the total ring current density J_φ. The red line is a moving average, the green line is a polynomial fit to the data, and the black line represents an $r^{-2.2}$ power law. The blue line represents the output of the *Connerney et al.* [1983] model (Voyager data) and the orange lines correspond to the min and max J_φ profiles produced by *Bunce et al.* [2007] (Cassini data). The red dashed line (dotted for $r > 10$ R_S) is the total current density if the pressure anisotropy current is included. Adapted from *Sergis et al.* [2010].

of Fig. 9.2) does not seem to affect much the presence or the height of the total current maximum.

While earlier studies suggested that the azimuthal current is primarily inertial (for example, carried mostly by thermal ions that nearly corotate with the planet at ranges between 6 and 15 R_S), the inclusion of the hot plasma pressure component revealed that the Saturnian ring current exhibits a changing nature with radial distance. As Figure 9.2a indicates, it is inertial inside of ~8 R_S, and pressure gradient driven beyond 12 R_S. In its maximum region (8–12 R_S), it appears mixed in nature. Its variability becomes stronger with increasing radial distance, as the mass density drops and the hot plasma and associated pressure gradient current begins to dominate.

In a complementary approach, *Kellett et al.* [2010, 2011] derived plasma parameters on 22 near-equatorial Cassini passes, and used these to determine pass-by-pass profiles of the inertia, pressure gradient, and anisotropy current densities. A typical example is displayed in Figure 9.3, which shows results from the inbound pass of Cassini Rev 15 in 2005, which spanned the dayside sector between ~0900 hr LT and dusk. The last inbound magnetopause crossing occurred at ~22.6 R_S just off the plot, corresponding to an estimated subsolar magnetopause at ~20.1 R_S. Figure 9.3a shows profiles of particle number density, where the yellow data show the electron number density determined from upper hybrid resonance waves and the blue data similarly show the electron number density obtained by integration of the electron spectrum. The red data show the hot ion (water plus proton) number density obtained by integrating the corresponding spectra between 3 and >200 keV. Assuming predominantly singly charged ions, the difference between the hot ion number density and the electron number density then corresponds (through overall charge neutrality) to the warm ion number density below energies ~3 keV (mainly water ions with energies ~100 eV originating and transported outward from Enceladus). The magenta data similarly show the electron number density with energies above 20 eV.

Figure 9.3b shows the hot ion (red) and electron (blue) pressure profiles, determined from the measured spectra. The warm ion pressures (green plus yellow) are determined from their number density inferred from Figure 9.3a combined with models of the warm ion composition and temperature determined from specially selected Voyager and Cassini data when instrument viewing was favorable [*Richardson*, 1995; *Wilson et al.*, 2008; *McAndrews et al.*, 2009]. Inside ~12 R_S, the warm ions are significantly anisotropic with $P_\perp > P_\parallel$ arising from pick-up by the near-corotating plasma flow of ions newly created from Kepler-orbiting neutrals. Green lines show the water ion pressures and yellow the (insignificant) proton pressures, with the solid lines showing the perpendicular pressure and the dot-dashed lines the parallel pressure. The electron pressure (blue) in the inner region is similarly determined as the sum of >20 eV pressure and the cool electron pressure, from the density seen in Figure 9.3a and an electron temperature model [*Schippers et al.*, 2008; *Persoon et al.*, 2009]. Overall, the pressure is dominated in the inner region by the warm water ion pressure, while beyond ~12 R_S the hot ion pressure becomes comparable, typically exceeding the warm ion pressure by a factor of

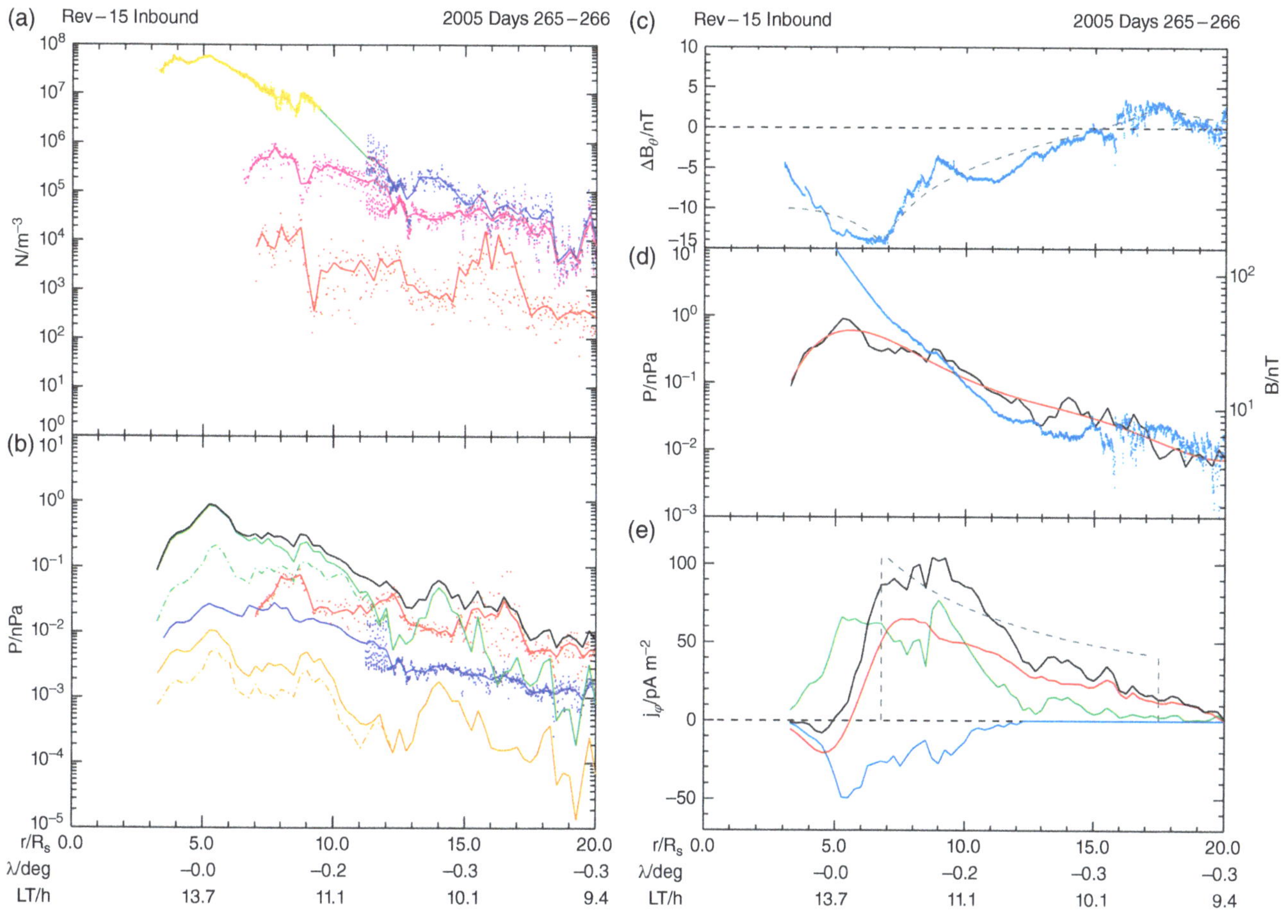

Figure 9.3 Radial profiles for the near-equatorial inbound Cassini pass of Rev 15 showing (a) particle number density (m^{-3}), (b) plasma pressure (nPa), (c) perturbation B_θ field component (nT), (d) total perpendicular pressure and field pressure (nPa), and (e) azimuthal current density (pA/m^2). See text for details. Adapted from *Kellett et al.* [2010].

~2 in the outer region. The electron pressure is found to be ~10% of the ion pressure, or less, throughout.

The total perpendicular pressure, shown by the black line in Figure 9.3b, is reproduced in Figure 9.3d together with the magnetic pressure (blue data). The equatorial plasma beta is small in the inner region, but typically exceeds unity by modest factors beyond ~9–10 R_S. The red line shows a fitted fifth order polynomial from which the pressure gradient in equation (9.3) is estimated. Combined with magnetic parameters, this yields the pressure gradient current shown by the red line in Figure 9.3e. The blue and green lines in this figure similarly show the pressure anisotropy and inertia currents, the latter requiring a model of the plasma velocity [*Kane et al.*, 2008; *Wilson et al.*, 2009], while the black line shows the total current density. All three components are seen to contribute significantly, with the anisotropy current canceling a significant fraction of the inertia current (both terms dominated throughout by warm water ions) inside ~12 R_S. *Kellett et al.* [2010, 2011] showed that this relation follows from the ion pick-up process in the inner region, in which the newly picked-up ions gyrate perpendicular to the field with a speed equal to the difference between the plasma corotation speed and the slower Kepler speed of the neutrals. In agreement with the *Sergis et al.* [2010] results, the total current profile for this particular pass (black) is similar to that of the pressure gradient current (red), dominated by warm water group ions inside ~10–12 R_S, and by hot ions at larger distances, augmented by typical factors of ~1.5–2 by the difference between the warm water ion inertia and anisotropy currents.

Figure 9.3c shows (in blue) the colatitudinal perturbation field, given as $\Delta B_\theta = B_\theta(\text{obs}) - B_\theta(\text{planetary})$, while the black dashed curve shows a fitted current disk model, as in sections 9.2 and 9.3.1, with $R_1 = 6.75$ R_S, $R_2 = 17.5$ R_S, $D = 1.5$ R_S, and $\mu_o I_o = 55$ nT. The corresponding model current density profile is shown by the black dashed line in Figure 9.3e, which is seen to provide a good rough approximation to the current density profile determined from the plasma and field data (black solid line), although the decrease of J_φ appears steeper in the data than in the model. These results are typical of those obtained over the ensemble of 22 passes examined by *Kellett et al.* [2011], which also indicated that pass-by-pass variations are less than a factor of ~2 in the region inside ~10 R_S, increasing to factors of ~2–3 in the outer region where hot ion effects are more important.

The ring current J_φ can also be deduced, at least statistically, by taking the curl of the magnetic field vector **B** through Maxwell's equation $\nabla \times \mathbf{B} = \mu_0 \mathbf{J}$. In this case, **B** can be constructed by bin-averaging the magnetic field observations within a regular cylindrical grid. Using Cassini magnetometer observations from 2004 through 2010, such a field map was constructed and the curl of **B** readily obtained in cylindrical coordinates [*Carbary et al.*, 2012]. This statistical compilation showed that the ring current flows eastward (in the corotation sense), extends from ~3 R_S to ~20 R_S, and develops a maximum reaching ~75 pA/m^2 at ~9.5 R_S. Its half-width along the vertical direction z is estimated ~1.5 R_S. The total azimuthal ring current is 9.2±1.0 MA, which is in very good agreement with other estimates made from charged particles [e.g., *Sergis et al.*, 2010] and from the field modeling outlined in sections 9.2 and 9.3.1.

In a more recent study, *Sergis et al.* [2017] combined magnetic field and particle data obtained from a much larger part of the mission (2004 to 2013) and showed that a significant local time asymmetry is present in the strength and radial behavior of the ring current, in agreement to the remote imaging results (see section 9.3.2). More specifically, the average azimuthal current density maximizes in the dusk-to-dawn local time sector (1700 to 0400 hr) where it reaches values of >100 pA/m^2, contrary to the dawn-to-afternoon sector (0400 to 1700 hr) where J_φ varies between 60 and 80 pA/m^2. This asymmetry of nearly a factor of 2 is almost entirely imposed by the hot plasma distribution, as the radial distributions of the thermal plasma and the magnetic field do not seem to vary strongly with local time [*Kellett et al.*, 2011; *M. F. Thomsen*, private communication]. This conclusion is also supported by the very good agreement between the global ring current distribution and the wide-view ENA images in the keV energy range (see also section 9.3.2). We should note, however, that the J_φ estimations stated above carry an uncertainty of ~50%, mostly due to the energetic particle dynamics and especially in the dawn local time sector, which is less completely sampled.

Although in the present review we do not focus on the vertical structure of the plasma sheet, since the majority of the ions that carry the ring current (and especially the heavier W^+ ions) are confined to the rotational equator, we should briefly mention that several studies, based either on analyses of in situ and remote data or on model fits to measurements agree that the thickness (or scale height) of the plasma sheet is ~2–3 R_S [e.g., *Carbary et al.*, 2008d; *Kellett et al.*, 2009; *Sergis et al.*, 2011]. *Krimigis et al.* [2007] further showed that the distribution of the hot plasma is quite thicker in the dayside, thinning out toward the flanks and fading away forming a nothward tilted hot plasma tail (data concerned the Northern Hemisphere winter). This picture was soon after verified by long-term ENA imaging [*Carbary et al.*, 2008e].

When the high inclination orbits of Cassini revealed the vertical structure and dynamics of the Saturnian plasma sheet in more detail [*Arridge et al.*, 2008b; *Sergis et al.*, 2011], it became clear that its central part (in terms of vertical distance) does not always coincide with the rotational equatorial plane, although the planetary magnetic moment vector is aligned with the planet's rotational axis within 0.3°. The plasma sheet is bowl shaped and oscillates vertically (i.e., along the planetary spin axis), exhibiting distinct modes of periodic variation, linked to the planetary rotation (spinning mode), to global magnetospheric compression and expansion (breathing mode), and to short timescale north-south motion (the so-called flapping) [*Ramer et al.*, 2016]. In addition, the plasma sheet displays a seasonal change of its vertical displacement, being increasingly pushed away from the rotational equator as the solar wind attack angle changes, reaching ±26.7° at the solstices. As a result, measurements taken near the rotational plane do not always characterize the current sheet, but can at times include also the outer (uppermost/lowermost) parts of the plasma sheet, depending on the radial distance, the rotational phase, and the Saturnian season. This is an issue that should be addressed in future analyses, so that the computed pressure gradients are not overestimated. A suggestion could be to filter the particle measurements, either using the in situ magnetic field vector as a diagnostic tool to determine how close to the center of the current sheet Cassini actually is, or by adapting the vertical displacement of the plasma sheet as estimated by models [e.g., *Arridge et al.*, 2008b].

9.3.2. Remote Imaging

Single spacecraft measurements cannot distinguish between spatial and temporal variability (i.e., the spacecraft can only provide information for a specific position and time). However, energetic neutral imaging (ENA), available with the wide field-of-view camera of Cassini/INCA, allows the remote observation of large parts of the magnetosphere through the detection of fast-moving neutral particles, which are the product charge exchange between energetic ions and slow neutral atoms.

Charge exchange occurs when an energetic ion collides with a low-energy neutral particle, generating an ENA and a low-energy ion. The general reaction is $X_*^+ + X \Rightarrow X_* + X^+$ (X_* indicating particle X in a high-energy state), and nearly all of the energy is transferred to the outgoing ENA. This synergy between the ENAs and their parent ions allows mapping of the ion population by observation of the ENAs from long range, much as a conventional imager images a photonic source. An ENA imager can image the energetic ions in a magnetospheric ring current, provided sufficient neutral atoms are present [*Gruntman*, 1997].

In Saturn's magnetosphere, outgassing from the moon Enceladus produces copious amounts of neutral atoms from the water species hydrogen (H) or oxygen (O) [e.g., *Porco et al.*, 2006], leading to large fluxes of energetic H and O neutrals. These neutral species have readily been observed by the Ion Neutral Camera (INCA), which is part of the Magnetospheric Imaging Instrument (MIMI) on Cassini [*Krimigis et al.*, 2004]. INCA images Saturn's H_* and O_* between the energies of ~5 keV and ~3 MeV within a field of view of 120° x 90° at a time resolution of ~6 min.

Through energetic H and O atoms, INCA has observed the hot ion component of Saturn's ring current from orbit insertion in mid-2004 through the present era (late 2016). The initial long-range views of the ring current (or plasma sheet in this context) showed it to be concentrated approximately on the equator and centered on Saturn [*Krimigis et al.*, 2005]. The ring current (hot plasma component) is readily captured in ENA fluxes when viewed from high latitude, as shown in Figure 9.4. The figure displays equatorial maps of energetic hydrogen fluxes that have been corrected for viewing geometry and the relative rotational motion of the source ions, and filtered for observer range (>20 R_S) and latitude (>25°). The maps represent accumulations during four different seasons: (A) southern summer, (B) vernal equinox, (C) northern spring, and (D) northern summer. The four epochs also correspond to different phases of the solar cycle, with (A) and (B) near solar minimum, and (C) and (D) near solar maximum. In the first three cases, the ring current appears as the doughnut-shaped enhancement centered on Saturn and lying roughly between the orbits of the moons Enceladus and Titan. In the final epoch, the ENA fluxes appear only as a maximum in the early morning, suggesting the presence of a "quiescent" ring current (see also discussion in section 9.3.1), although what is sampled by the INCA camera is the hot and more dynamic part of the current.

When a complete torus is evident, it generally peaks in radial distance near the orbit of the moon Rhea (~8.7 R_S). This distance corresponds roughly to the outer edge of the tenuous E-ring, the outermost ring of Saturn that is produced by Enceladus outgassing. The energetic ions producing the ENA have limited penetration into the E-ring neutral cloud, whose outer edge represents a sort of boundary.

The ENA ring current clearly manifests local time asymmetries [*Carbary et al.*, 2008a, 2016]. Early observations of the ENA suggested a maximum in the midnight sector [*Carbary et al.*, 2008a], although this asymmetry was later found to change throughout the course of the seasons and/or solar cycle. This is in very good agreement with the asymmetry that has been observed in the local time distribution of hot plasma injections [*Muller et al.*, 2010] and in suprathermal pressure and total plasma beta [*Sergis et al.*, 2017]. The ring current asymmetry may be related to a dawn-dusk electric field detected in charged particle observations [*Andriopoulou et al.*, 2012; *Thomsen et al.*, 2012].

One of the major advantages of ENA imaging is that the temporal variability of the system can be monitored through continuous global viewing. The ring current displays interesting dynamics that are captured by MIMI/INCA in time-lapse ENA movies. Viewed from high latitude at sufficiently high time resolution, the maximum ENA emission concentrates into "blobs" that generally move in a counterclockwise corotational sense, but at speeds much less than full corotation [*Carbary et al.*, 2008b; *Carbary and Mitchell*, 2014a].

The ENA intensities also exhibit strong periodicities near the ~10.7 hr planetary period [*Carbary et al.*, 2008c]. Oxygen ENA emissions display dual periodicities much like those discovered in the radio emissions and also have a local time dependence [*Carbary et al.*, 2014b]. These ring current periodicities have definite phase relations to the radio emission periodicities [*Carbary et al.*, 2010a, 2011], and the ENA fluxes display a localized structure when mapped into a longitude system based on the radio emissions [*Kurth et al.*, 2008; *Gurnett et al.*, 2011; *Carbary et al.*, 2010b]. This structure may be related to spirals hypothesized by various theories and may contribute to other periodicities in the magnetic field and charged particles [*Brandt et al.*, 2008, 2010]. The blobs in the ring current are also intimately associated with injection events arising from the centrifugal interchange that characterize rapidly rotating magnetospheres such as Saturn's [*Mitchell et al.*, 2015].

The resulting images and movies revealed a partial and highly variable, rotating ring current. We should note, however, that as the INCA camera captures ENAs in the range of 5 keV and above, only the high-energy and far more variable part of the ring current can be made globally visible.

9.3.3. Modeling

The structure and variability of the ring current have also been modeled in global magnetohydrodynamics (MHD) simulations of Saturn's magnetosphere. It should be noted, however, that since an MHD model does not

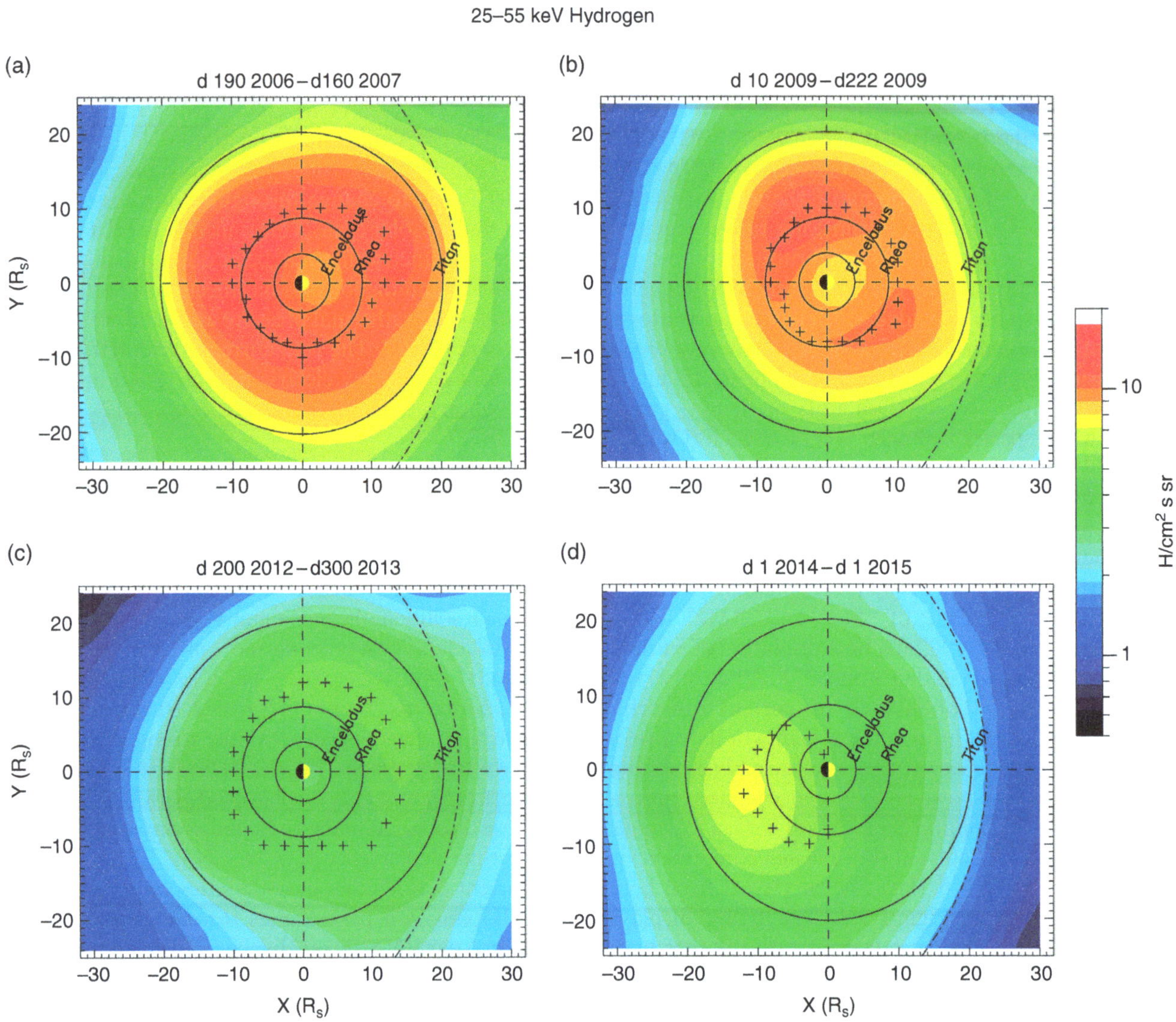

Figure 9.4 Equatorial maps of energetic *H* atom fluxes in four epochs when the spacecraft was observing from high latitudes. Epoch (a) was during Saturn's southern summer, and (b) was just before the vernal equinox; epoch (c) was during northern spring, and epoch (d) was just before northern summer. The view is from the north pole looking down on the equatorial plane, with the Sun to the right. The nominal magnetopause appears as a dot-dash line [*Kanani et al.*, 2010], and the orbits of three major moons appear as circles. Crosses show the maxima in radial profiles at 24 equally spaced local times.

simulate effects of energetic particles and assumes isotropic plasma pressure, current contributions from energetic particle pressure and anisotropy are not included. Nevertheless, an MHD model can provide useful insights into the large-scale current systems present in a magnetosphere. *Jia et al.* [2012b] have developed an atmospheric vortex model based on their global MHD model [e.g., *Hansen et al.*, 2005; *Jia et al.*, 2012a] to understand the magnetospheric periodicities observed ubiquitously in Saturn's magnetosphere. Figure 9.5 shows a series of snapshots of the simulated azimuthal current density in the equatorial plane through roughly one Saturn rotation. Throughout the equatorial plane, the azimuthal current density is positive, consistent with the outward ballooning of the dipole field. The global average ring current corresponds to the disklike geometry of the equatorial plasma sheet and the highly stretched field structure of the outer magnetosphere. Also present in the model are localized enhancements of azimuthal current, which form the asymmetric ring current.

The azimuthal current appears to consist of two components, one fixed in local time and one rotating. The structure fixed in local time has the most intense azimuthal current located near midnight, in agreement with the

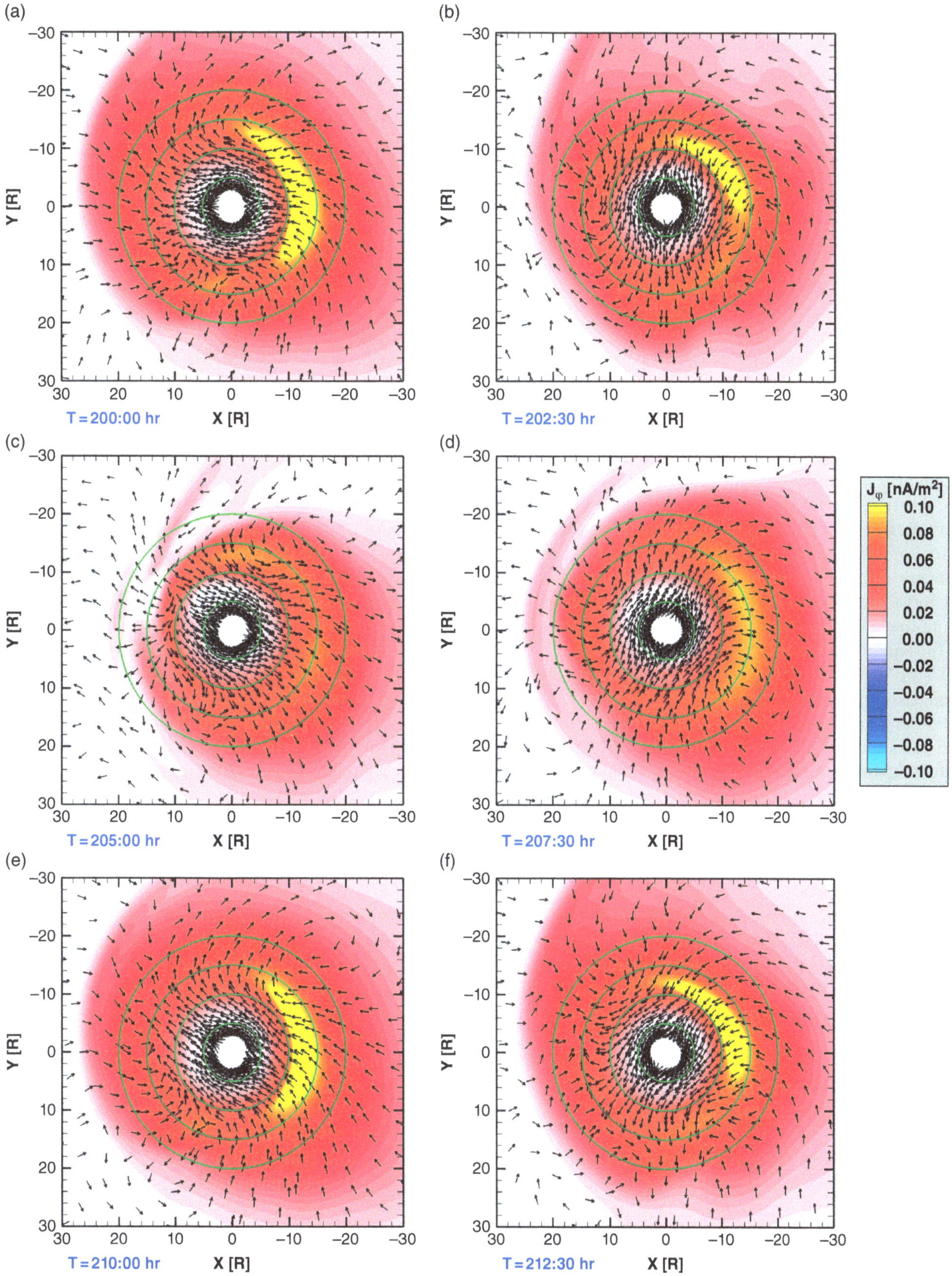

Figure 9.5 Snapshots of the azimuthal current through roughly one Saturn rotation from the atmospheric vortex simulation. The azimuthal current density (J_φ) is shown in color in the equatorial plane. Superimposed on the contours are black arrows representing unit vectors along the perturbation magnetic field that are produced by the rotating current system arising from the imposed atmospheric vortex. Adapted from *Jia et al.* [2012b].

Kellett et al. [2010] result discussed earlier. The rotating structure has a weaker intensity and is only seen between ~12 and ~15 R_S during certain rotational phases (panels [a], [b], [e], and [f] in Fig. 9.5). As the rotating component encounters the local-time-fixed component, the latter weakens and then intensifies again a few hours after the rotating part passes through it. However, it remains unclear as to what causes this complex response because of many other changes of the system that take place concurrently with the interaction between the fixed and rotating parts of the ring current.

Brandt et al. [2010] have modeled the magnetic perturbations resulting from the rotating partial ring current. Based on the pressure of energetic particles (>3 keV) derived by forward modeling to simulate ENA images, and assuming a dipole geometry, they determined the current system corresponding to the partial ring current and then calculated the resultant magnetic perturbations. Comparison with Cassini magnetometer data show that the rotating partial ring current is intimately connected with periodic modulations in the magnetic field with features consistent with those observed in the Cassini data. Our present understanding, however, is that these effects are most likely driven by thermospheric-ionospheric coupling that produces the observed associated phenomena in the equatorial plane [*Jia et al.*, 2012b].

Using a model of force balance in Saturn's disklike magnetosphere, and equatorial observations as a boundary condition, *Achilleos et al.* [2010] described how the variations in hot plasma pressure affect the global magnetic field configuration. They concluded that the suprathermal pressure variations change the location of the magnetopause, even at fixed solar wind dynamic pressure, and affect the magnetic mapping between ionosphere and disk. They further produced radial profiles of the separate components (inertial and pressure gradient) of the azimuthal current density. Although sensitive to the plasma angular velocity input (from other models or Cassini observations), the total current exhibits a maximum of ~80 pA/m^2 between 10 R_S and 12 R_S in very good agreement with previous data-based results [e.g., *Sergis et al.*, 2010; *Kellett et al.*, 2011].

9.4. CONCLUDING REMARKS AND OPEN QUESTIONS

Nearly 40 yr ago, the first spacecraft encounters with the Saturnian magnetosphere (Pioneer 11 and Voyager 1 and 2) were too brief to provide insight into its dynamics or to identify the dominant, internal sources of plasma, yet, they managed to confirm the presence of an intense azimuthal current through both particle and magnetic field measurements, and even allowed initial modeling of its global structure. Cassini arrived at Saturn on 4 July 2004. Spending over a decade orbiting the planet and flying by its moons, it covered a large part of the magnetosphere and, with a state-of-the-art payload, was able to capture the ring current distribution and dynamics, as presented in this chapter.

While the Cassini in situ measurements helped the broad outer planet community to uncover the characteristics of the Saturnian ring current, it was the remote ENA imaging that offered a more appealing and informative overview of the global azimuthal current, capturing the high-energy and most dynamically active component. Figure 9.6 provides an illustration of the average ring current of Saturn, as seen by the ENA camera on board Cassini (MIMI/INCA). Although the measured quantity presented is the intensity of the energetic particle emission (energies of few keV), which depends on the distribution of the neutral atoms, this picture can be considered as a reliable global illustration of the suprathermal part of the ring current. The main features seen here, such as the presence of a maximum at 8–11 R_S and the local time asymmetry, are in very good agreement with the conclusions from studies based on analysis of in situ measurements.

Before the Cassini arrival, it was not absolutely clear in what degree the Saturnian azimuthal current would turn out to be closer in nature to the Terrestrial or the Jovian one, although most expressed predictions favored a Jovian-like ring current. Ground measurements, combined with data from the small fleet of spacecraft that monitor Earth, have shown that the Terrestrial magnetosphere (and ring current) is primarily modulated by the solar wind, in contrast to the magnetospheres of Jupiter and Saturn where rapid rotation and mass loading essentially drive, or at least strongly affect, almost all of the observed magnetospheric phenomena.

As mentioned by *Kivelson* [2005], although the enforced plasma rotation forms an azimuthal charged particle flow and therefore a ring current around Saturn (as also at Jupiter), this current cannot be viewed as an analogue of the Terrestrial ring current, which becomes important mostly during and following magnetic storms. In a rotation-dominated magnetosphere that is well protected from the solar wind, and with a major internal ion source (Enceladus), the conditions are, as Cassini revealed, quite different, and a direct comparison between the two ring currents is not easy to attempt.

We now know that, with a major source of plasma localized close to the rapidly rotating planet, the Saturnian ring current and the planetary semipermanent plasmadisk resembles the disk current of Jupiter's middle magnetosphere much more than the Terrestrial. However, when the two giant magnetospheres are viewed using appropriate dimensionless variables, certain differences emerge, and undoubtedly reflect to the azimuthal current. In such

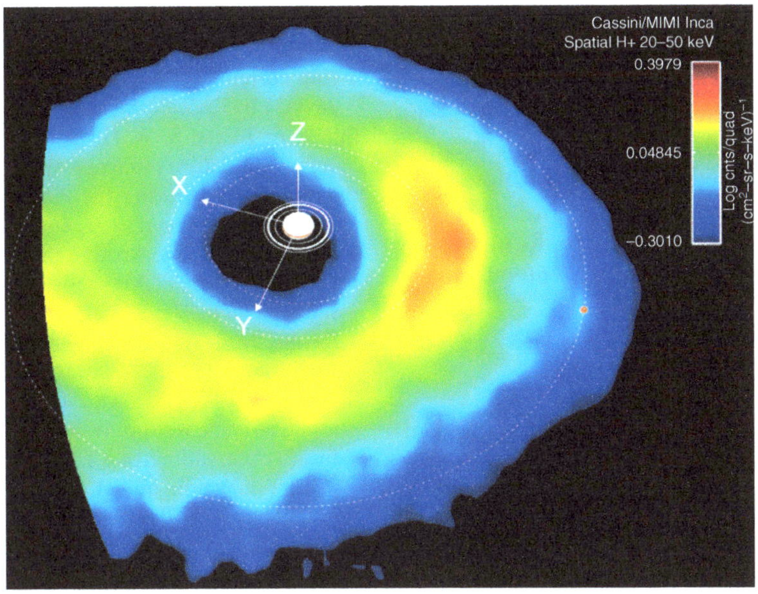

Figure 9.6 ENA image of the hot ion component of the Saturnian ring current, as viewed from above the northern hemisphere. This image, in the range 20 to 50 keV, was obtained on 28 February 2007 with MIMI/INCA while Cassini was moving at latitude of 58°, at local time of 1902 hr and radial distance of 27.9 Rs. Saturn is shown at the center, and the dotted circles represent the orbits of Dione (6.29 Rs), Rhea (8.74 Rs), and Titan (20.2 Rs). The *z* axis is parallel to Saturn's spin axis, the *x* axis points roughly sunward in the Sun-spin axis plane, and the *y* axis completes the system pointing roughly toward the dusk. The INCA field of view is seen as the curved image cutoff on the left. The image is a cospatial average of several frames over the period between 13:42 and 00:22 UTC (following day). Adapted from *Sergis et al.* [2009].

an approach, *Vasyliunas* [2008] showed that the mass input at Saturn could be larger than that at Jupiter, because the scale quantities (for example, the solar wind mass flux incident on the projected magnetospheric area or the critical mass flux that would completely prevent corotation) are also much smaller at Saturn than at Jupiter. In that perspective, and given the relative effect on the planetary magnetic field, the Saturnian ring current could be considered stronger than the Jovian.

Though Cassini provided a comprehensive description of the extent, the composition, the origin, the driving mechanism, the strength, and the effects of azimuthal current, certain issues remain yet to be explored:

1. Despite the fact that Cassini orbited Saturn for over 12 years, certain parts this huge magnetosphere have not been sampled in the detail that would allow us to reach definitive conclusions. In particular, the dawn local time sector is significantly less covered at ranges and latitudes where the azimuthal current flows. That limits our ability to determine in what degree the ring current is, at least at times, affected by a plasma flow pattern that follows the so-called Vasyliunas cycle [*Vasyliunas*, 1983] or by the analogous and often concurrently present Dungey cycle [*Dungey*, 1961]. In addition, as pointed out by *Kivelson and Bagenal* [2007], it is not yet fully explained how particle transport operates and how inertial effects due to the fast rotation compare to convection processes, primarily imposed by interaction with the solar wind. In the same context, the ring current response to solar wind extreme conditions and phenomena that are in some cases triggered subsequently (for example, magnetic reconnection, plasma sheet collapse) has not yet been entirely explored.

2. Magnetotail reconnection and plasmoid formation produce global impacts on the magnetosphere as hot, tenuous plasma is carried inward in rapidly moving flux tubes returning from the tail reconnection region to the dayside. Observations [*Bunce et al.*, 2005; *Hill et al.*, 2008; *Jackman et al.*, 2011; *Mitchell et al.*, 2015] and models [*Jia et al.*, 2012a] have shown that such flux tubes generate significant disturbances in the magnetosphere and the ionosphere, particularly on the (unfortunately less sampled) dawn side, intensifying field-aligned currents that would in turn cause auroral brightening. More recently, *Thomsen et al.* [2015] identified in both data and MHD simulations the formation of a sharp boundary (a plasmapause), which separates flux tubes that have successfully traversed the nightside without significant mass loss, from those that were involved in magnetic reconnection in the tail and return at super-corotational speed, containing low density and high temperature O^+ rich plasma, suggestive of Vasyliunas-type reconnection. They further argued that tail reconnection can strip off inner magnetospheric

plasma into at least a dipole L of 8.6 in the dawn local time sector, that is, where the bulk of the ring current flows. Given that the structure and intensity of the field-aligned current system is in large modified by the asymmetric azimuthal current [*Cummings*, 1966], it becomes very important to map the equatorial current flow in the best available detail, especially in a region as sensitive as the dawn local time sector, in order to form a more complete picture of the field-aligned current system and the overall current topology and closure.

ACKNOWLEDGMENTS

The authors thank MIMI team colleagues for comments that improved this study, and M. Kusterer and J. Vandegriff (JHU/APL) for assistance with the MIMI data. Work at JHU/APL was supported by NASA and by subcontracts at the University of Maryland and the Academy of Athens. E. J. Bunce and S. W. H. Cowley were supported by STFC Consolidated Grant ST/N000749/1.

REFERENCES

Achilleos, N., P. Guio, C. S. Arridge, N. Sergis, R. J. Wilson, M. F. Thomsen, and A. J. Coates (2010), Influence of hot plasma pressure on the global structure of Saturn's magnetodisk, *Geophys. Res. Lett.*, *37*, L20201; doi: 10.1029/2010GL045159.

Alexeev, I. I., V. V. Kalegaev, E. S. Belenkaya, S. Y. Bobrovnikov, E. J. Bunce, S. W. H. Cowley, and J. D. Nichols (2006), A global magnetic model of Saturn's magnetosphere, and a comparison with Cassini SOI data, *Geophys. Res. Lett.*, *33*, L08101; doi: 10.1029/2006GL025896.

Andriopoulou, N., M. Roussos, N. Krupp, C. Paranicas, M. Thomsen, S. Krimigis, M. K. Dougherty, K. H. Glassmeier (2012), A noon-to-midnight electric field and nightside dynamics in Saturn's inner magnetosphere using microsignature observations, *Icarus*, *220*, 503–513; doi: 10.1016/j.icarus.2012.05.010.

Arridge, C. S., C. T. Russell, K. K. Khurana, N. Achilleos, S. W. H. Cowley, M. K. Dougherty, D. J. Southwood, and E. J. Bunce (2008a), Saturn's magnetodisc current sheet, *J. Geophys. Res.*, *113*, A04214; doi: 10.1029/2007JA012540.

Arridge, C. S., K. K. Khurana, C. T. Russell, D. J. Southwood, N. Achilleos, M. K. Dougherty, A. J. Coates, and H. K. Leinweber (2008b), Warping of Saturn's magnetospheric and magnetotail current sheets, *J. Geophys. Res.*, *113*, A08217; doi: 10.1029/2007JA012963.

Beard, D. B., and M. A. Gast (1987), The magnetosphere of Saturn, *J. Geophys. Res.*, *92*, 5763–5767; doi: 10.1029/JA092iA06p05763.

Brandt, P., et al. (2008), Understanding the global evolution of Saturn's ring current, *Geophys. Res. Lett.*, *35* (17), CiteIDL17101; doi: 10.1029/2008GL034969.

Brandt, P. C., K. K. Khurana, D. G. Mitchell, N. Sergis, K. Dialynas, J. F. Carbary, E. C. Roelof, C. P. Paranicas, S. M. Krimigis, and B. H. Mauk (2010), Saturn's periodic magnetic field perturbations caused by a rotating partial ring current, *Geophys. Res. Lett.*, *37*, L22103; doi: 10.1029/2010GL045285.

Bunce, E. J., and S. W. H. Cowley (2003), A note on the ring current in Saturn's magnetosphere: Comparison of magnetic data obtained during the Pioneer-11 and Voyager-1 and -2 fly-bys, *Ann. Geophys.*, *21*, 661–669.

Bunce, E. J., C. S. Arridge, S. W. H. Cowley, and M. K. Dougherty (2008), Magnetic field structure of Saturn's dayside magnetosphere and its mapping to the ionosphere: Results from ring current modeling, *J. Geophys. Res.*, *113*, A02207; doi: 10.1029/2007JA012538.

Bunce, E. J., S. W. H. Cowley, D. M. Wright, A. J. Coates, M. K. Dougherty, N. Krupp, W. S. Kurth, and A. M. Rymer (2005), In situ observations of a solar wind compression-induced hot plasma injection in Saturn's tail, *Geophys. Res. Lett.*, *32*, L20S04; doi:10.1029/2005GL022888.

Bunce, E. J., S. W. H. Cowley, I. I. Alexeev, C. S. Arridge, M. K. Dougherty, J. D. Nichols, and C. T. Russell (2007), Cassini observations of the variation of Saturn's ring current parameters with system size, *J. Geophys. Res.*, *112*, A10202; doi: 10.1029/2007JA012275.

Carbary, J. F., and D. G. Mitchell (2013), Periodicities in Saturn's magnetosphere, *Rev. Geophys.*, *51*,1–30; doi: 10.1002/rog.20006.

Carbary, J. F., and D. G. Mitchell (2014a), Keogram analysis of ENA images at Saturn, *J. Geophys. Res. Space Physics*, *119;* doi: 10.1002/2014JA019784.

Carbary, J. F., D. C. Hamilton, S. P. Christon, D. G. Mitchell, and S. M. Krimigis (2010b), Longitude dependences of energetic H+ and O+ at Saturn, *J. Geophys. Res.*, *115*, A07226; doi: 10.1029/2009JA015133.

Carbary, J. F., D. G. Mitchell, A. M. Rymer, N. Krupp, D. Hamilton, S. M. Krimigis, and S. V. Badman (2016), Local time asymmetries in Saturn's magnetosphere, in *Dawn–Dusk Asymmetries in Planetary Magnetospheres*, edited by S. Haaland, A. Runov, and C. Forsyth, AGU monograph, Wiley, in press.

Carbary, J. F., D. G. Mitchell, and P. C. Brandt (2014b), Local time dependences of oxygen ENA periodicities at Saturn, *J. Geophys. Res. Space Physics*, *119*, 6577–6585; doi: 10.1002/2014JA020214.

Carbary, J. F., D. G. Mitchell, C. Paranicas, E. C. Roelof, and S. M. Krimigis (2008e), Direct observation of warping in the plasma sheet of Saturn, *Geophys. Res. Lett.*, *35*, L24201; doi: 10.1029/2008GL035970.

Carbary, J. F., D. G. Mitchell, P. Brandt, C. Paranicas, and S. M. Krimigis (2008c), ENA periodicities at Saturn, *Geophys. Res. Lett.*, *35*, L07102; doi: 10.1029/2008GL033230.

Carbary, J. F., D. G. Mitchell, P. Brandt, E. C. Roelof, and S. M. Krimigis (2008a), Statistical morphology of ENA emissions at Saturn, *J. Geophys. Res.*, *113*, A05120; doi: 10.1029/2007JA012873.

Carbary, J. F., D. G. Mitchell, P. Brandt, E. C. Roelof, and S. M. Krimigis (2008b), Track analysis of energetic neutral atom blobs at Saturn, *J. Geophys. Res.*, *113*, A01209; doi: 10.1029/2007JA012708.

Carbary, J. F., D. G. Mitchell, P. Brandt, E. C. Roelof, and S. M. Krimigis (2008d), Periodic tilting of Saturn's plasma sheet, *Geophys. Res. Lett.*, *35*, L24101; doi: 10.1029/2008GL036339.

Carbary, J. F., D. G. Mitchell, P. C. Brandt, S. M. Krimigis, and D. A. Gurnett (2011), ENA periodicities and their phase relations to SKR emissions at Saturn, *Geophys. Res. Lett.*, *36*, L16106; doi:10.1029/2011GL048418.

Carbary, J. F., D. G. Mitchell, S. M. Krimigis, D. A. Gurnett, and W. S. Kurth (2010a), Phase relations between energetic neutral atom intensities and kilometric radio emissions at Saturn, *J. Geophys. Res.*, *115*, A01203; doi: 10.1029/2009JA014519.

Carbary, J. F., N. Achilleos, and C. S. Arridge (2012), The statistical ring current of Saturn, *J. Geophys. Res.*, *117*, A06223; doi: 10.1029/2011JA017472.

Connerney, J. E. P., M. H. Acuña, and N. F. Ness (1981a), Modeling the Jovian current sheet and inner magnetosphere, *J. Geophys. Res.*, *86*, 8370–8384.

Connerney, J. E. P., M. H. Acuña, and N. F. Ness (1981b), Saturn's ring current and inner magnetosphere, *Nature*, *292*, 724–726.

Connerney, J. E. P., M. H. Acuña, and N. F. Ness (1983), Currents in Saturn's magnetosphere, *J. Geophys. Res.*, *88*, 8779–8789.

Cummings, W. D. (1966), Particles and fields in the Earth's magnetosphere: Magnetic storms, particles and fields in the ionosphere: Low-latitude ionospheric currents, Earth's main magnetic field (geomagnetic field): Time variations, diurnal to secular, *J. Geophys. Res.*, *71*, 19, 4495; doi: 10.1029/JZ071i019p04495.

Dougherty, M. K., et al. (2004), The Cassini magnetic field investigation, *Space Sci. Rev.*, *114*, 331–383.

Dungey, J. W. (1961), Interplanetary magnetic field and the auroral zones, *Phys. Rev. Lett.*, *6*, 47–48; doi:10.1103/PhysRevLett.6.47.

Edwards, T. M., E. J. Bunce, and S. W. H. Cowley (2001), A note on the vector potential of Connerney et al.'s model of the equatorial current sheet in Jupiter's magnetosphere, *Planet. Space Sci.*, *49*, 1115–1123.

Galopeau, P. H. M., and A. Lecacheux (2000), Variations of Saturn's radio rotation period measured at kilometer wavelengths, *J. Geophys. Res.*, *105, 13*, 089–13, 101; doi: 10.1029/1999JA005089.

Giampieri, G., and M. K. Dougherty (2004), Modeling of the ring current in Saturn's magnetosphere, *Ann. Geophys.*, *22*, 653–659.

Gruntman, M. (1997), Energetic neutral atom imaging of space plasmas, *Rev. Sci. Instr.*, *68* (10), 3617–3656.

Gurnett, D. A., A. Lecacheux, W. S. Kurth, A. M. Persoon, J. B. Groene, L. Lamy, P. Zarka, and J. F. Carbary (2009), Discovery of a north-south asymmetry in Saturn's radio rotation period, *Geophys. Res. Lett.*, *36*, L16102; doi: 10.1029/2009GL039621.

Gurnett, D. A., J. B. Groene, T. F. Averkamp, W. S. Kurth, S.-Y. Ye, and G. Fischer (2011), An SLS4 longitude system based on a tracking filter analysis of Saturn kilometric radiation, in *Planetary Radio Emissions VII*, edited by H. O. Rucker, W.S. Kurth, P. Louarn, and G. Fischer, pp. 51–63, Austrian Academy of Sciences, Vienna, Austria.

Hansen, K. C., A. J. Ridley, G. B. Hospodarsky, N. Achilleos, M. K. Dougherty, T. I. Gombosi, and G. Tóth (2005), Global MHD simulations of Saturn's magnetosphere at the time of Cassini approach, *Geophys. Res. Lett.*, *32*, L20S06; doi: 10.1029/2005GL022835.

Hill, T. W., et al. (2008), Plasmoids in Saturn's magnetotail, *J. Geophys. Res.*, *113*, A01214; doi:10.1029/2007JA012626.

Jackman, C. M., J. A. Slavin, and S. W. H. Cowley (2011), Cassini observations of plasmoid structure and dynamics: Implications for the role of magnetic reconnection in magnetospheric circulation at Saturn, *J. Geophys. Res.*, *116*, A10212; doi:10.1029/2011JA016682.

Jia, X., K. C. Hansen, T. I. Gombosi, M. G. Kivelson, G. Tóth, D. L. DeZeeuw, and A. J. Ridley (2012a), Magnetospheric configuration and dynamics of Saturn's magnetosphere: A global MHD simulation, *J. Geophys. Res.*, *117*, A05225; doi: 10.1029/2012JA017575.

Jia, X., M. G. Kivelson, and T. I. Gombosi (2012b), Driving Saturn's magnetospheric periodicities from the upper atmosphere/ionosphere, *J. Geophys. Res.*, *117*, A04215; doi: 10.1029/2011JA017367.

Kanani, S. J., et al. (2010), A new form of Saturn's magnetopause using a dynamic pressure balance model, based on in situ, multi-instrument Cassini measurements, *J. Geophys. Res.*, *115*, A06207; doi: 10.1029/2009JA014262.

Kane, M., D. G. Mitchell, J. F. Carbary, S. M. Krimigis, and F. J. Crary (2008), Plasma convection in Saturn's outer magnetosphere determined from ions detected by the Cassini INCA experiment, *Geophys. Res. Lett.*, *35*, L04102; doi:10.1029/2007GL032342.

Kellett, S., C. S. Arridge, E. J. Bunce, A. J. Coates, S. W. H. Cowley, M. K. Dougherty, A. M. Persoon, N. Sergis, and R. J. Wilson (2010), Nature of the ring current in Saturn's dayside magnetosphere, *J. Geophys. Res.*, *115*, A08201; doi:10.1029/2009JA015146.

Kellett, S., C. S. Arridge, E. J. Bunce, A. J. Coates, S. W. H. Cowley, M. K. Dougherty, A. M. Persoon, N. Sergis, and R. J. Wilson (2011), Saturn's ring current: Local time dependence and temporal variability, *J. Geophys. Res.*, *116*, A05220; doi: 10.1029/2010JA016216.

Kellett, S., E. J. Bunce, A. J. Coates, and S. W. H. Cowley (2009), Thickness of Saturn's ring current determined from north-south Cassini passes through the current layer, *J. Geophys. Res.*, *114*, A04209; doi: 10.1029/2008JA013942.

Kivelson, M. G. (2005), The Current Systems of the Jovian Magnetosphere and Ionosphere and Predictions for Saturn, *Space Science Reviews*, *116*, 299–318; doi: 10.1007/s11214-005-1959-x.

Kivelson, M. G., and F. Bagenal (2007), Encyclopedia of the Solar System, 2d ed., *Elsevier Academic Press*, *519*.

Krimigis, S. M., et al. (1981), Low-energy charged particles in Saturn's magnetosphere: Results from Voyager 1, *Science*, *212*, 225–231; doi: 10.1126/science.212.4491.225.

Krimigis, S. M. et al. (2004), Magnetosphere Imaging Instrument (MIMI) on the Cassini mission to Saturn/Titan, *Space Sci. Rev.*, *114*, 233–329.

Krimigis, S. M., et al. (2005), Dynamics of Saturn's magnetosphere from MIMI during Cassini's orbital insertion, *Science*, *307*, 1270–1273; doi: 10.1126/science110578.

Krimigis, S. M., J. F. Carbary, E. P. Keath, T. P. Armstrong, L. J. Lanzerotti, and G. Gloeckler (1983), General characteristics of hot plasma and energetic particles in the Saturnian magnetosphere: Results from the Voyager spacecraft, *J. Geophys. Res.*, *88*, 8871–8892; doi: 10.1029/JA088iA11p08871.

Krimigis, S. M., N. Sergis, D. G. Mitchell, D. C. Hamilton, and N. Krupp (2007), A dynamic, rotating ring current around Saturn, *Nature*, *450* (7172), 1050–1053; doi: 10.1038/nature06425.

Kurth, W. S., T. F. Averkamp, D. A. Gurnett, J. B. Groene, and A. Lecacheux (2008), An update to a Saturnian longitude system based on *kilometric* radio emissions, *J. Geophys. Res.* *113*, A05222; doi:10.1029/2007JA012861.

Mauk, B. H., S. M. Krimigis, and R. P. Lepping (1985), Particle and field stress balance within a planetary magnetosphere, *J. Geophys. Res.*, *90*, 8253–8264; doi:10.1029/JA090iA09p08253.

McAndrews, H. J., et al. (2009), Plasma in Saturn's nightside magnetosphere and the implications for global circulation, *Planet Space Sci.*; doi: 10.1016/j.pss.2009.03.003.

Mitchell, D. G., P. C. Brandt, J. F. Carbary, W. S. Kurth, S. M. Krimigis, C. Paranicas, N. Krupp, et al. (2015), Injection, interchange, and reconnection, in *Magnetotails in the Solar System*, edited by A. Keiling, C. M. Jackman, and P. A. Delamere, John Wiley & Sons, Inc, Hoboken, NJ; doi: 10.1002/9781118842324.ch19.

Müller, A. L., J. Saur, N. Krupp, E. Roussos, B. H. Mauk, A. M. Rymer, D. G. Mitchell, and S. M. Krimigis (2010), Azimuthal plasma flow in the Kronian magnetosphere, *J. Geophys. Res.*, *115*, A08203; doi:10.1029/2009JA015122.

Ness, N. F., M. H. Acuña, K. W. Behannon, L. F. Burlaga, J. E. P. Connerney, R. P. Lepping, and F. Neubauer (1982), Magnetic field studies by Voyager 2: Preliminary results at Saturn, *Science*, *215*, 558.

Ness, N. F., M. H. Acuña, R. P. Lepping, J. E. P. Connerney, K. W. Behannon, L.F. Burlaga, and F. Neubauer (1981), Magnetic field studies by Voyager 1: Preliminary results at Saturn, *Science*, *212*, 211.

Persoon, A. M., D. A. Gurnett, O. Santolik, W. S. Kurth, J. B. Faden, J. B. Groene, G. R. Lewis, et al. (2009), A diffusive equilibrium model for the plasma density in Saturn's magnetosphere, *J. Geophys. Res.*, *114*, A04211; doi: 10.1029/2008JA013912.

Porco, C. C., et al. (2006), Cassini observes the active south pole of Enceladus, *Science*, *311*, 1383–1401; doi:10.1126/science.1123013.

Ramer, K. M., M. G. Kivelson, N. Sergis, K. K. Khurana, and X. Jia (2016), Spinning, breathing, and flapping: Periodicities in Saturn's middle magnetosphere, *J. Geophys. Res.*; doi: 10.1002/2016JA023126

Richardson, J. D. (1995), An extended plasma model for Saturn, *Geophys. Res. Lett.*, *22*, 1177–1180.

Schippers, P., M. Blanc, N. André, I. Dandouras, G. R. Lewis, L. K. Gilbert, A. M. Persoon, et al. (2008), Multi-instrument analysis of electron populations in Saturn's magnetosphere, *J. Geophys. Res.*, *113*, A07208; doi: 10.1029/2008JA013098.

Sergis, N., C. M. Jackman, A. Masters, S. M. Krimigis, M. F. Thomsen, D. C. Hamilton, D. G. Mitchell, M. K. Dougherty, and A. J. Coates (2013), Particle and magnetic field properties of the Saturnian magnetosheath: Presence and upstream escape of hot magnetospheric plasma, *J. Geophys. Res. Space Physics*, *118*, 1620–1634; doi:10.1002/jgra.50164.

Sergis, N., C. S. Arridge, S. M. Krimigis, D. G. Mitchell, A. M. Rymer, D. C. Hamilton, N. Krupp, M. K. Dougherty, and A. J. Coates (2011), Dynamics and seasonal variations in Saturn's magnetospheric plasma sheet, as measured by Cassini, *J. Geophys. Res.*, *116*, A04203; doi:10.1029/2010JA016180.

Sergis, N., et al. (2007), Ring current at Saturn: Energetic particle pressure in Saturn's equatorial magnetosphere measured with Cassini/MIMI, *Geophys. Res. Lett.*, *34* (9), CiteID L09102; doi:10.1029/2006GL029223.

Sergis, N., et al. (2009), Energetic particle pressure in Saturn's magnetosphere, measured with Magnetospheric Imaging Instrument on Cassini, *J. Geophys. Res.*, *114, A2*, CiteID A02214; doi:10.1029/2008JA013774.

Sergis, N., et al. (2010), Particle pressure, inertial force, and ring current density profiles in the magnetosphere of Saturn, based on Cassini measurements, *Geophys. Res. Lett.*, *37*, L02102; doi:10.1029/2009GL041920.

Sergis, N., C. M. Jackman, M. F. Thomsen, S. M. Krimigis, D. G. Mitchell, D. C. Hamilton, M. K. Dougherty, N. Krupp, and R. J. Wilson (2017), Radial and local time structure of the Saturnian ring current, revealed by Cassini, *J. Geophys. Res. Space Physics*, *122*, doi:10.1002/ 2016JA023742.

Smith, E. J., L. Davis Jr., D. Jones, P. J. Coleman Jr., D. S. Colburn, P. Dyal, C. P. Sonett (1980), Saturn's magnetic field and magnetosphere, *Science*, *207*, 407.

Thomsen, M. F., D. B. Reisenfeld, D. M. Delapp, R. L. Tokar, D. T. Young, F. J. Crary, E. C. Sittler, M. A. McGraw, and J. D. Williams (2010), Survey of ion plasma parameters in Saturn's magnetosphere, *J. Geophys. Res.*, *115*, A10220; doi: 10.1029/2010JA015267.

Thomsen, M. F., D. G. Mitchell, X. Jia, C. M. Jackman, G. Hospodarsky, and A. J. Coates (2015), Plasmapause formation at Saturn, *J. Geophys. Res. Space Physics*, *120*, 2571–2583; doi: 10.1002/2015JA021008.

Thomsen, M. F., E. Roussos, M. Andriopoulou, P. Kollman, C. S. Arridge, C. P. Paranicas, D. A. Gurnett, R. L. Powell, R. L. Tokar, and D. T. Young (2012), Saturn's inner magnetospheric convection pattern: Further evidence, *J. Geophys. Res.*, *117*, A09208; doi:10.1029/2011JA017482.

Vasyliunas, V. M. (1983), Plasma distribution and flow, in *Physics of the Jovian Magnetosphere*, edited by A. J. Dessler, 395–453, Cambridge University Press, New York.

Vasyliunas, V. M. (1984), Fundamentals of current description, in *Magnetospheric Currents*, edited by T. A. Potemra, 63–66, Geophysical Monograph Series, 28, AGU, Washington, DC.

Vasyliunas, V. M. (2008), Comparing Jupiter and Saturn: Dimensionless input rates from plasma sources within the magnetosphere, *Ann. Geophys.*, *26*, 1341–1343.

Wilson, R. J., R. L. Tokar, and M. G. Henderson (2009), Thermal ion flow in Saturn's inner magnetosphere measured by the Cassini plasma spectrometer: A signature of the Enceladus torus? *Geophys. Res. Lett.*, *36*, L23104; doi:10.1029/2009GL040225.

Wilson, R. J., R. L. Tokar, M. G. Henderson, T. W. Hill, M. F. Thomsen, D. H. Pontius Jr. (2008), Cassini plasma spectrometer thermal ion measurements in Saturn's inner magnetosphere, *J. Geophys. Res.*, *113*, CiteID A112218; doi: 10.1029/2008JA013486.

Young, D. T., et al. (2004), Cassini plasma spectrometer investigation, *Space Sci. Rev.*, *114* (1–4), 1–112; doi: 10.1007/s11214-004-1406-4.

Part III: Current Sheets

10

Review on the Characteristics of the Current Sheet in the Earth's Magnetotail

A. T. Y. Lui

ABSTRACT

The current sheet in the Earth's magnetotail is a plasma region where many dynamic phenomena occur and has been attributed to be the origin of many magnetospheric disturbances. In addition, it is often viewed as a prototype of current sheets in other planets in our solar system as well as in astrophysical systems. The Earth's magnetotail has been surveyed by more satellite missions than at other planets, thus providing valuable knowledge that can be utilized in studies of other plasma systems. In this review, the basic characteristics of the current sheet are briefly discussed, followed by descriptions of some prominent structures and its motions. Its internal properties are also found to be influenced by the external solar wind magnetic field. A fundamentally important link between the Earth's magnetotail and the ionosphere is via magnetic-field-aligned currents, especially during substorm periods. Evaluation of this link brings to light the importance of kinetic processes in magnetospheric dynamics and some major discrepancies of the usually adopted fluid description of these phenomena. In particular, observations on the changes in current density and magnetic field during substorm activities in the magnetotail reveal very short timescales, indicating kinetic processes at play. The frozen-in condition that is implicitly assumed in fluid treatments of plasma dynamics is found to be invalid in these dynamic episodes. In addition, there is compelling evidence showing that observed features of the substorm current system are in agreement with the predictions by a kinetic approach and contradict the predictions by the fluid approach.

10.1. INTRODUCTION

Matter in space is almost all in the plasma state, also known as the fourth state of matter. In simple terms, plasma is a gas so hot that its constituents are electrically charged particles of electrons and ions. This characteristic makes it significantly different from the other three states by its ability to embody electric currents within it. As our common daily experience shows, electric currents transmit power from one region to another. What is so different from our daily experience is that electric currents in space plasmas are carried by the entire medium rather than through conducting wires. In other words, the entire volume of plasma can act as conducting wires. *Parks* [2004a] has provided a comprehensive fundamental understanding of electric currents in space plasma.

The solar wind interacts with the Earth's magnetic field and creates a magnetic cavity around the Earth known as the magnetosphere. This interaction creates several plasma domains around and within the magnetosphere. Some of these domains contain intense electric currents, such as the bow shock, magnetopause, and magnetotail. In particular, the magnetotail has a reservoir of energetic charged particles called the plasma sheet estimated to

JHU/APL, Laurel, Maryland, USA

Electric Currents in Geospace and Beyond, Geophysical Monograph 235, First Edition.
Edited by Andreas Keiling, Octav Marghitu, and Michael Wheatland.

be 4–6 R_E thick normally [*Bame et al.*, 1967]. Embedded within the plasma sheet is a thinner central region called the current sheet that has a significant electric current directed mostly in the dawn-to-dusk direction. Its midplane is often referred to as the neutral sheet where the magnetic field is directed mostly sunward above it and mostly antisunward below it. This change in the magnetic field direction, accompanied in general by a decrease in the magnetic field magnitude, is a common means to identify the neutral sheet from observations of satellites that carry a magnetometer. The neutral sheet captures a lot of attention from space researchers because it harbors very dynamic processes with occasional explosive energy releases.

The following topics are covered for this brief review: the current sheet basics, current sheet structures, current sheet motions, the influence of interplanetary magnetic field (IMF) B_y on the tail B_y, field-aligned currents from the current sheet, and the substorm current system.

10.2. CURRENT SHEET BASICS

Early measurements of magnetic field from the IMP-1 satellite provide a broad overview of the current sheet characteristics in the magnetotail [*Speiser and Ness*, 1967]. From this early survey, the current sheet was found to begin forming near the magnetic equatorial plane at a geocentric distance of 10 ± 3 R_E. Its thickness was observed to be larger with proximity to the Earth and in the dawn side of the magnetotail. The current sheet thickness was estimated to be in the range of ~500–5000 km. Observations from ISEE and Cluster confirmed that a thin current sheet can have a thickness comparable to or less than the ion inertial length near a current disruption or magnetic reconnection site (~400–800 km) [*Sergeev et al.*, 1993; *Sanny et al.*, 1994; *Nakamura et al.*, 2002, 2006; *Asano et al.*, 2004].

There is a nonzero magnetic field component threading through the neutral sheet, that is, the midplane of the current sheet is not ideally neutral. From IMP-1 measurements at the geocentric distances of 20–33 R_E in the magnetotail, the magnetic field normal to the current sheet B_z was found to be 1–4 nT in the dawn side of the tail and < 1 nT near the noon-midnight meridian plane. Later measurements from IMP-6 showed that the average B_z at the neutral sheet to be ~3.5 nT at the dawn and dusk flanks and ~1.8 nT near midnight [*Fairfield*, 1979]. Three years of Cluster observations at downtail distances of ~20 R_E indicated that the B_z at the dawn-dusk flank was even larger, at ~8 nT (see Fig. 5c in *Petrukovich et al.* [2005]). The peak current density for thin neutral sheets typically is in the range of ~5–60 nA/m^2 [*Nakamura et al.*, 2006]. The neutral sheet current normally is a diamagnetic current in which the ions, having the higher pressure and pressure gradient than that of the electrons, are the dominant current carriers [e.g., *Parks*, 2004a]. While protons are the main species during usual conditions, O^+ ions can dominate the plasma pressure during magnetic storms [e.g., *Krimigis et al.*, 1985] and substorms [e.g., *Daglis et al.*, 1991]. At such times, O^+ ions can contribute ~5%–10% or higher of the neutral sheet current [e.g., *Kistler et al.*, 2005; *Artemyev et al.*, 2009]. In addition, electrons may become the major current carriers when the current sheet becomes very thin [*Asano et al.*, 2004; *Artemyev et al.*, 2009].

It is noted that the neutral sheet projected on the *YZ*-plane is often tilted from the *Y*-axis [*Speiser and Ness*, 1967; *Mihalov et al.*, 1968]. In addition, the tail field projected on the equatorial plane deviates from the *X*-axis by ~$3.1 \pm 1.5°$ due to solar wind aberration [*Behannon*, 1970].

10.3. CURRENT SHEET STRUCTURES

In the classification of the Earth's current sheet signatures using 3 yr of IMP-6 data, an occasional feature is a decrease in the elevation angle at the neutral sheet crossing [*Lui*, 1984]. This feature has been noted previously and interpreted as the signature of a magnetic island [*Mihalov et al.*, 1968; *Schindler and Ness*, 1972; *Speiser*, 1973; *Lui and Meng*, 1979]. It is often accompanied by a southward magnetic field component at the neutral sheet, a signature regarded as an indicator of magnetic reconnection. However, a statistical survey of this structure shows its presence during both quiet (verified with global images from DMSP satellites) and substorm times [*Lui and Meng*, 1979]. Therefore, magnetic islands that are associated with filamentation of the current sheet [*Speiser*, 1973] can be viewed as an intrinsic feature of the neutral sheet unrelated to substorm activities.

Some magnetic islands are formed during substorms to play a role in particle energization. For example, a large-scale magnetic island, called alternatively as plasmoid, is proposed as a repeatable signature of magnetic reconnection in the midtail region (beyond ~20 R_E downtail) during substorms [*Hones*, 1979; *Hones et al.*, 1984]. This single plasmoid picture has been extended to include multiple plasmoid and flux rope structures [*Zong et al.*, 1997; *Slavin et al.*, 2003]. In the tail lobe, a plasmoid can be remotely sensed as a traveling compression region (TCR) by ISEE 3 magnetic observations [*Slavin et al.*, 1984]. These various magnetic structures are illustrated in the schematic diagram in Figure 10.1 [*Zong et al.*, 2004].

Another structure that is also intimately related to particle energization is the presence of highly disordered and/or rapid magnetic fluctuations of the neutral sheet. This feature was noticed in the early investigation of magnetic field in the magnetotail even before

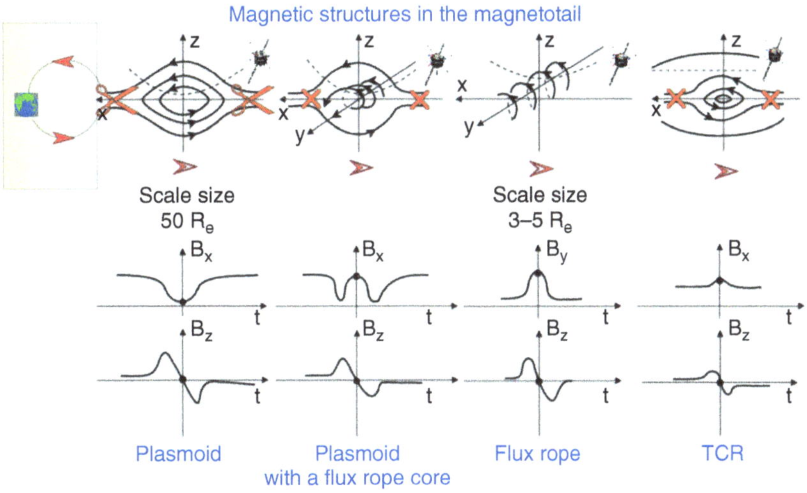

Figure 10.1 Various magnetic structures found in the magnetotail: (a) a single plasmoid, (b) a plasmoid with a flux rope core, (c) a flux rope, and (d) a TCR [*Zong et al.*, 2004].

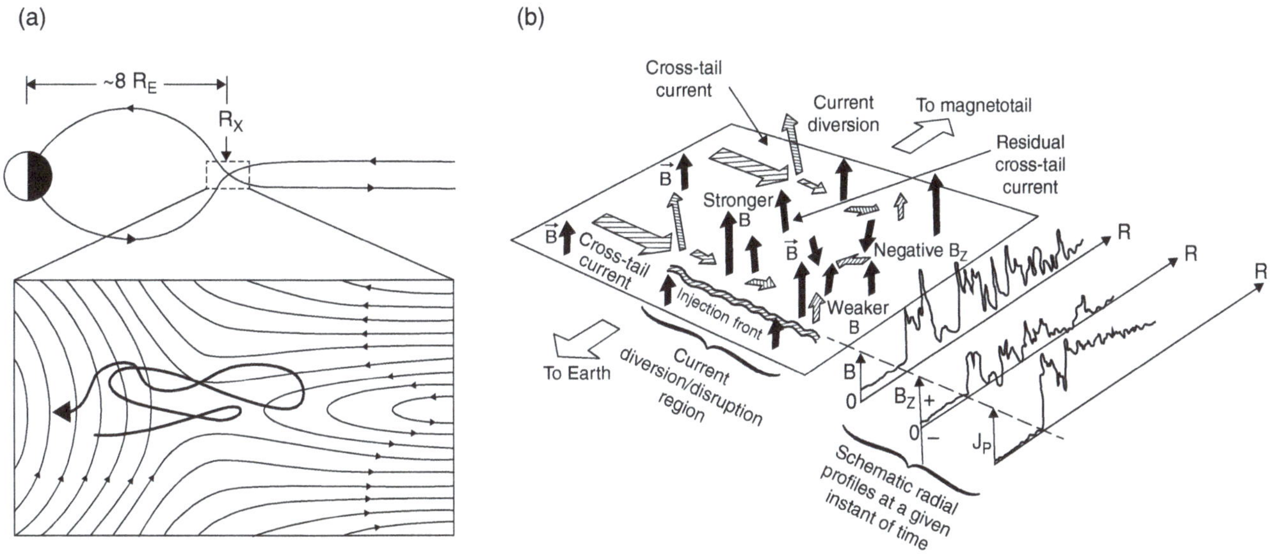

Figure 10.2 Two different interpretations of highly disordered and/or rapid magnetic fluctuations at the neutral sheet: (a) multiple encounters of a reconnection site [*Takahashi et al.*, 1987]; (b) magnetic turbulence due to current disruption [*Lui et al.*, 1988].

accompanying particle measurements were available [*Hruska and Hruskova*, 1970; *Bowling*, 1975]. These magnetic fluctuations were described as transient turbulence by *Bowling* [1975] and were found during enhanced geomagnetic activity indicated by the Kp index.

The occurrence of such turbulence during substorms was reported by *Takahashi et al.* [1987] and by *Lui et al.* [1988] with each reporting a different event and a different interpretation. The former report interpreted the observations as multiple encounters of a magnetic reconnection site by the satellite while the latter report interpreted the observations as detecting magnetic turbulence arising from current disruption. These two interpretations are illustrated in Figure 10.2.

The current disruption (CD) phenomenon has been identified by the occurrence of large magnetic fluctuations predominantly around the neutral sheet of the magnetotail where $B_z >> B_x$, B_y. During CD, magnetic fluctuations reaches the level in which $\delta B_z/B_{zo}$ is of the order one or larger, where B_{zo} is the B_z value before CD onset. It typically lasts for several minutes. During CD, the B_z component could become negative at times in spite of a strong background positive B_z component from the dipole magnetic field [e.g., *Takahashi et al.*, 1987; *Lui et al.*, 1988]. It is accompanied by particle energization (e.g., *Lui et al.*, 1988) and intense fluctuating electric fields [e.g., *Lui*, 2011]. The cross-tail current breaks up into filaments and may reverse its direction [e.g., *Lui*, 2011, 2004]. The associated plasma flow pattern is not organized by the B_z polarity, unlike magnetic reconnection. In other words, tailward (earthward) plasma flow can occur in the presence of positive (negative) B_z. This is one of the distinctions between CD and magnetic reconnection [*Lui*, 2015a]. The frozen-in condition assumed by the magnetohydrodynamic (MHD) approach is broken [e.g., *Lui et al.*, 2007; *Lui*, 2011, 2013]. The B_z component is substantially increased after CD, indicating the relaxation of a stretched tail-like magnetic field configuration before CD to a more dipolarized configuration. A schematic diagram of the CD phenomenon is illustrated in Figure 10.3. The cross-field current instability is a kinetic instability that generates oblique whistler waves to disrupt the cross-tail current [*Lui et al.*, 1991; *Lui*, 2004]. The multiscale nature of plasma waves excited at CD onset is demonstrated well by the wavelet analysis [e.g., *Lui and Najmi*, 1997; *Lui et al.*, 2008; *Yoon et al.*, 2009; *Le Contel et al.* 2009; *Zhou et al.*, 2009]. More detailed analysis of CD events shows CD to be non-MHD turbulence [*Consolini et al.*, 2005]. The validity of the term *turbulence* to describe the large magnetic and electric fluctuations during CD is also demonstrated by the multiscale and multifractal nature of these disturbances [*Lui*, 2002].

10.4. CURRENT SHEET MOTIONS

Satellite measurements often reveal multiple crossings of the neutral sheet and are interpreted as due to its flapping motion in which the distortion of the current sheet surface is on the *XZ*-plane [*Speiser and Ness*, 1967; *Speiser*, 1973].

Flapping of the neutral sheet on the *YZ*-plane has also been inferred by noting that there are multiple neutral sheet crossings that show a continuous rotation of the azimuthal angle of the magnetic field [*Lui et al.*, 1978]. Figure 10.4 shows some cases of this continuous rotation from IMP-5 20s averaged measurements. For reference, multiple neutral sheet crossings that indicate a normal flapping motion on *XZ*-plane is shown in Figure 10.4a. The satellite location in the GSM coordinates is given at the bottom of the panel. The neutral sheet in this event can be identified by the change of the azimuthal component ϕ_{SM}, which is the azimuth angle in the GSM coordinate, from ~180° to ~360° and vice versa. The changes occurred within the range of 180° and 360°, indicating that the B_y component had the same sign for each crossing. Distinct from this normal flapping signature, events in Figure 10.4b–10.4d showed a continuous

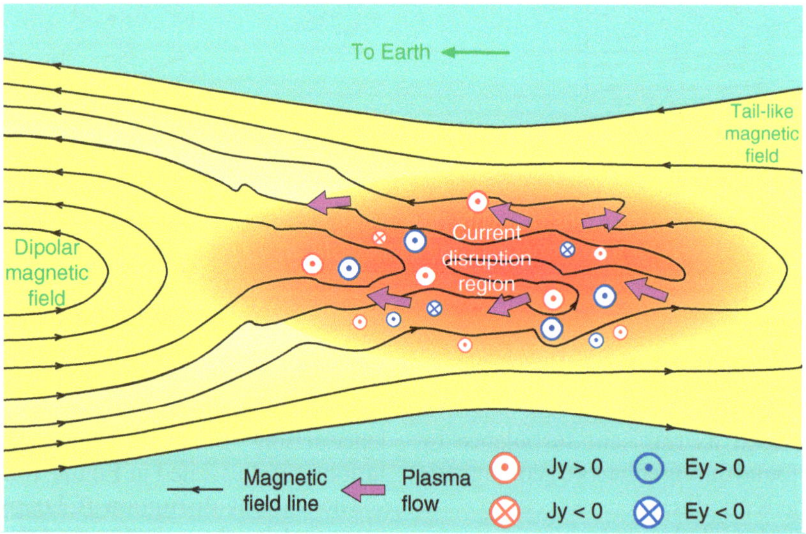

Figure 10.3 A schematic diagram to illustrate the complexities of a current disruption region where large magnetic and electric field fluctuations occur, particles are energized, a plasma flow pattern occurs without being governed by the B_z pattern, the frozen-in condition is broken, and the cross-tail current breaks up into several current filaments [*Lui*, 2011].

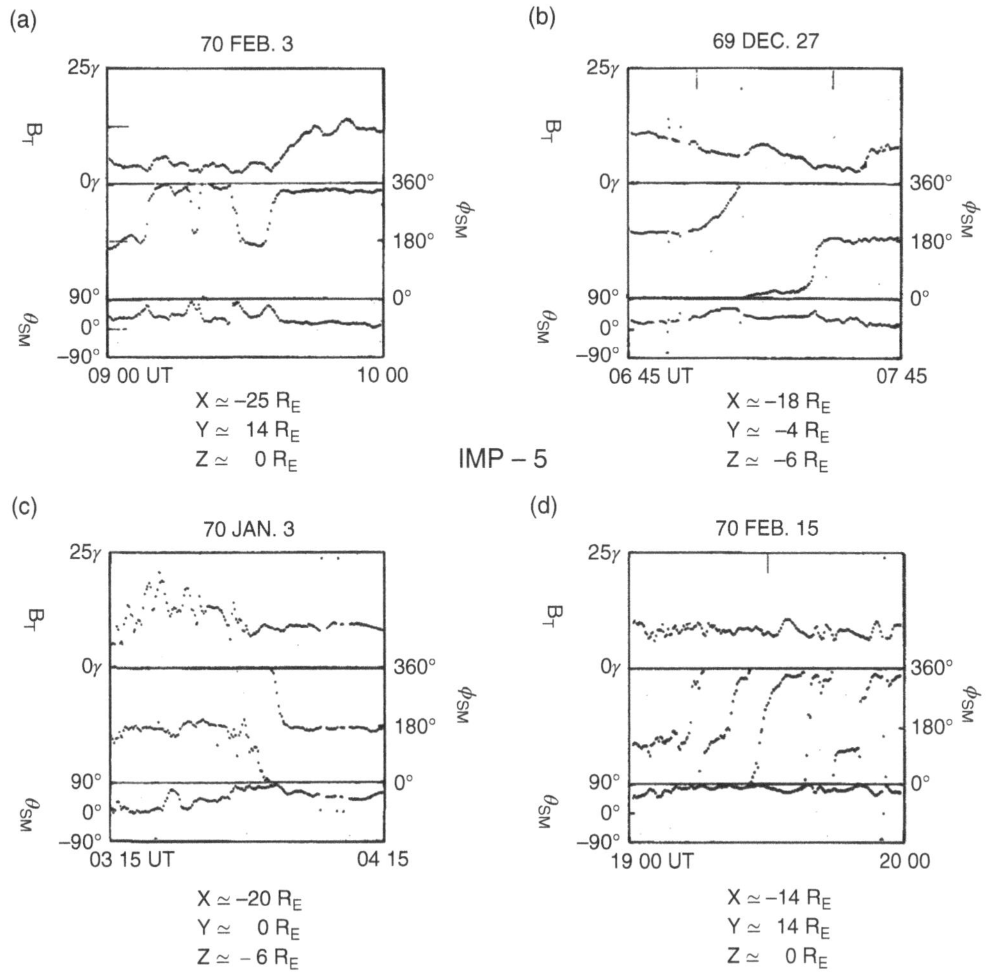

Figure 10.4 (a) Multiple neutral sheet crossings due to flapping on *XZ*-plane; (b–d) multiple neutral sheet crossings showing continuous rotation in ϕ_{SM}, indicating flapping on the *YZ*-plane [*Lui et al.*, 1978]. The unit γ used in the old literature is equivalent to nT.

rotation of ϕ_{SM}. For example, in Figure 10.4b, the first neutral sheet crossing occurred with ϕ_{SM} changing from ~180° to ~360° and the second neutral sheet crossing occurred with ϕ_{SM} changing from ~0° to ~180°, indicating that the B_y component had the opposite sign for the two adjacent crossings. Similar continuous rotation of ϕ_{SM} can be seen in the multiple neutral sheet crossings for the other events in Figure 10.4c–10.4d, indicating the B_y component alternating its sign for successive neutral sheet crossings.

It is shown that a simple model of a current sheet with a hump on the *YZ*-plane can reproduce this continuous rotation as illustrated in Figure 10.5. The model represents the current at the neutral sheet by a set of current wires extending to infinity in the dawn and dusk directions. The wires, spaced 0.01 R_E apart, have an equivalent strength of 0.03 A/m at its earthward edge with an exponential decrease over a distance of 60 R_E. The radius and height of the hump in the equatorial plane is taken to be 1 R_E and 0.6 R_E, respectively.

The early investigation was made without simultaneous measurement of plasma flow to confirm the movement of the current sheet plasma as a whole with the occurrence of multiple neutral sheet crossings. Later satellite missions carried multiple science payloads that can provide the opportunity for such a check. AMPTE/IRM is such a mission and this check was made by correlating the rate of change in the measured B_x component (dB_x/dt) with simultaneous plasma flow component in the *Z*-direction [*Sergeev et al.*, 1998]. A high-correlation coefficient, often better than 0.8, was found for several events of multiple neutral sheet crossings.

The flapping surface of the neutral sheet can have a rather unusual shape. In a case study based on observations from four Cluster satellites, *Runov et al.* [2005] showed that the flapping neutral sheet surface had a

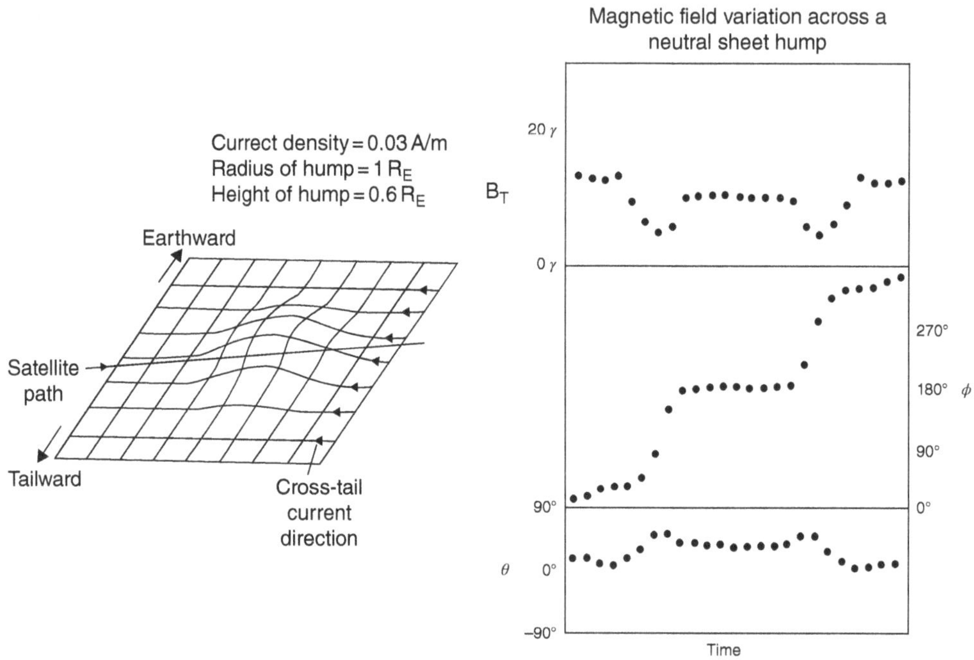

Figure 10.5 (Left) A neutral sheet surface with flapping on the *YZ*-plane represented by a hump of the surface with a satellite passing through it in the dawn-dusk direction; (right) the expected magnetic field components to be observed by a satellite passing through this flapping surface [*Lui et al.*, 1978].

quasi-rectangular shape on the *YZ*-plane as schematically drawn in Figure 10.6a. Both the *Y*- and *Z**-dimensions of this shape are estimated to be ~6000 km, where Z* is the estimated vertical scale, defined as $Z^*(t) = \int(\partial B_l/\partial t)[\nabla_n B_l]^{-1}dt$. The parameter B_l is the maximum variance component of the magnetic field and ∇_n is the gradient along the current sheet normal (see equation (2) in *Runov et al.* [2005]). The barycenter location with respect to the neutral sheet $Z^* = 0$ for each crossing is given in Figure 10.6b. Current density profile around the neutral sheet is reconstructed with the estimated *Z**-coordinate for these crossings in Figure 10.6c. As indicated, the peak current density at these neutral sheet crossings was ~9 nA/m². A study of 78 neutral sheet crossings showed the current density varying between 5 and 25 nA/m² and half-thickness varying between 1 and 20 ion thermal gyroradii. In addition to the center-peaked current sheet distribution, there are two other types [*Runov et al.*, 2006]. The peak current density often deviates from the center of the current sheet, a signature of an asymmetric sheet. Closer to the Earth at the downtail distances between 8 and 12 R_E where the dipolar field configuration becomes more prominent, the current sheet is found to split into two, reaching to higher latitudes in both hemispheres [*Antonova*, 2003; *Antonova et al.*, 2009; *Saito*, 2015]. This feature is called current sheet bifurcation, the third type of current sheet distribution. Current sheet bifurcation can be found occasionally at other radial distances as well [*Hoshino et al.*, 1996; *Runov et al.*, 2006].

Current sheet flapping motions can extend over a considerable distance along the tail axis. This is revealed by simultaneous observations of neutral sheet crossings at two different radial distances by Cluster and Double Star [*Zhang et al.*, 2005]. For the reported conjunction study on 5 August 2004, Cluster was in the radial range of 16–19 R_E while Double Star was in the radial range of 10–13 R_E. During the eight neutral sheet crossings identified almost simultaneously at both satellites, the separation between these two satellite missions was ~5 R_E. The flapping amplitude was found to be 4000–8000 km. Later Cluster observations showed that the amplitude could be as large as a few R_E [*Yushkov et al.*, 2016].

The flapping motion may actually be ultralow frequency (ULF) waves propagating along the current sheet. By combining electric and magnetic field measurements from THEMIS (Time History of Events and Macroscale Interactions during Substorms), *Shen et al.* [2015] showed that strong toroidal mode ULF waves that could also be interpreted as current sheet flapping were induced by solar wind pressure pulses.

A different type of current sheet flapping, called slippage, was proposed in which the magnetic field component joining the two sides of the neutral sheet maintains its orientation in the *Z*-direction instead of being tilted to

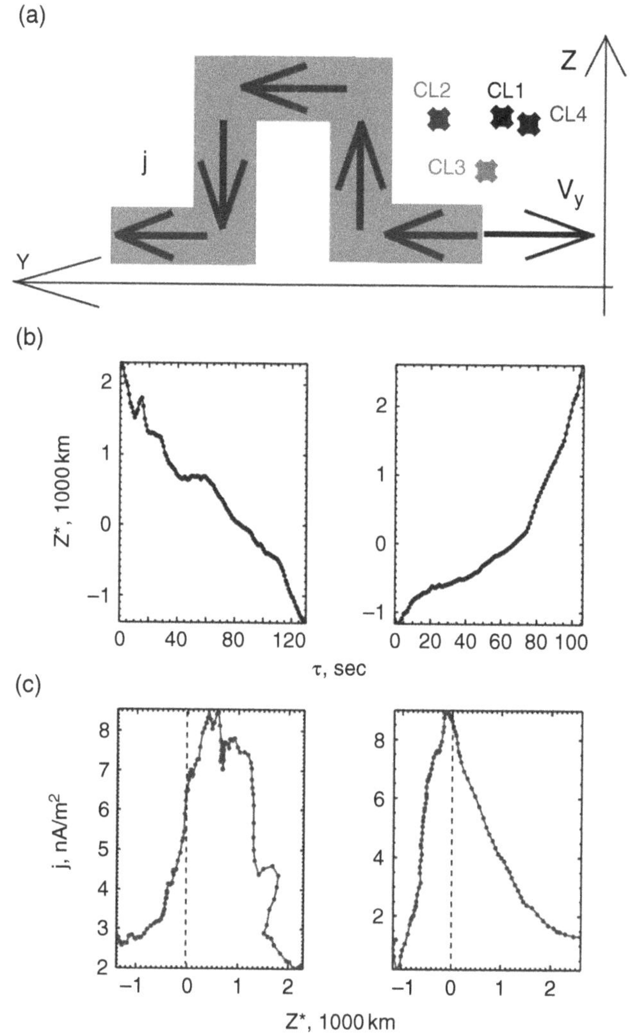

Figure 10.6 (a) An interpretation of the neutral sheet shape for the 12 August 2001 event; (b) the Cluster barycenter location during the two crossings; (c) reconstructed current density profiles for the two crossings [*Runov et al.*, 2005].

be normal to the neutral sheet surface [*Petrukovich et al.*, 2006, 2008; *Rong et al.*, 2010]. This slippage explanation does not give the alternating sign of the B_y component for successive neutral sheet crossings in current sheet flapping as originally reported by *Lui et al.* [1978] and may represent a different type of flapping.

It has been noted [e.g., *Lui*, 2004; *Asano et al.*, 2005) that the observed current sheet does not conform to the well-known Harris current sheet [*Harris*, 1962]. It has important non-fluid ion kinetic effects such as ions with transient Speiser-type orbits being the main current carriers [*Artemyev et al.*, 2010; *Artemyev and Zelenyi*, 2013], hot asymmetric wings in the ion population, and thin embedded current sheet where electrons carry significant current [*Artemyev et al.*, 2009; *Petrukovich et al.*, 2011]. Further details on the plasma and magnetic structures of the current sheet in the Earth's magnetotail can be found in recent reviews [*Petrukovich et al.*, 2015; *Zelenyi et al.*, 2015].

10.5. INFLUENCE OF THE IMF B_Y ON THE TAIL B_Y

As there is strong evidence of magnetic reconnection on the dayside magnetopause, it is anticipated that there may be association between the B_y component in the tail and the IMF B_y component. The first such effort was made by *Fairfield* [1979] using hourly averaged B_y component measured by IMP-6 at geocentric distances of 20–33 R_E. After removing the tail flaring and solar wind aberration effect, this component was compared with the hourly averaged IMF B_y provided in *King* [1977]. The result from over 3000 hours of data showed a relation of B_y (tail) = 0.13 B_y (IMF)–0.30 in nT. This relation was interpreted as penetration of IMF B_y in the magnetotail.

A different approach to this issue was taken by *Lui* [1984] who used measurements from IMP-5 and IMP-6 at neutral sheet crossings only so that possible contamination by tail flaring and solar wind aberration could be eliminated. The result showed a slightly different relation of B_y (tail) = 0.5 ± 0.1 B_y (IMF)–0.5 ± 0.2 in nT at the neutral sheet, suggesting that the association is stronger at the neutral sheet than at other parts of the magnetotail as reported by *Fairfield* [1979].

The two previous statistical studies were made in the magnetotail where the neutral sheet is prominent. A different statistic at downstream distances closer to the Earth, that is, the geosynchronous altitude, using 5 yr of GOES-6 and 6 yr of GOES-7 was performed by *Wing et al.* [1995]. The result showed also a different slope of 0.79 ± 0.04 in the midnight sector for the relationship between the two B_y components with a correlation coefficient of 0.50. The correlation coefficient was slightly higher at 0.61 for the local noon sector but the slope was less, namely, 0.29 ± 0.01. The correlation coefficient for the dawn sector was only 0.23 and the slope had the intermediate value of 0.52 ± 0.05.

The variability in the amount of influence on the magnetospheric B_y component by the IMF B_y is further exemplified in the work of *Petrukovich et al.* [2005]. From 3 yr of Cluster observations at ~20 R_E downtail, the typical slope was 0.4–0.6, but was below 0.20 at the dawn and dusk flanks.

The influence of the IMF B_y on the magnetospheric B_y is often cast as the "penetration" of IMF B_y in the magnetosphere. However, the nonuniform influence shown in these observations casts severe doubt on the validity of this simple notion from the expectation of dayside magnetic reconnection or from a superposition of IMF and geomagnetic field. Based on 11 yr of Geotail measurements, it is suggested that there are several factors contributing to the value of the B_y component in the magnetotail [*Petrukovich*, 2011].

10.6. FIELD-ALIGNED CURRENT FROM THE CURRENT SHEET

As discussed in the introduction section, currents have the capacity to carry power from one plasma regime to another. In particular, many dynamic processes in the ionosphere are in fact powered by magnetospheric processes via field-aligned current connection between the two plasma regimes.

A case in point is the multipoint analysis of magnetospheric and ionospheric flow vortices during a substorm on 19 February 2008 [*Keiling et al.*, 2009]. For this event, four of the THEMIS satellites were close to the neutral sheet between 9 and 12 R_E downtail. Three of them concentrated at ~23 MLT while another one was at ~21 MLT. A counterclockwise plasma flow vortex was detected by the three close-by satellites while the isolated satellite recorded a clockwise plasma flow vortex. In the ionosphere, a pair of conjugate clockwise and counterclockwise flow vortices, inferred from equivalent ionospheric currents (EICs), appeared simultaneously. Snapshots of the plasma flows from the three THEMIS satellites projected on the GSM *XY*-plane and the conjugate ionospheric flows inferred from EICs are shown in Figure 10.7. Red and yellow circles in the ionosphere indicate inferred downward and upward field-aligned currents (FACs), respectively. The generation mechanism for the FACs was attributed to an increase in the plasma flow vorticity in the magnetosphere.

For the flow vorticity term to generate FACs, the flow vortex is required to change in time continuously. If the flow vorticity decreases with time, then the sense of the FACs reverses. A case of FAC generation from a decrease in plasma flow vorticity in the magnetotail was reported by *Lui et al.* [2010]. A transient narrow auroral feature extruding from the poleward boundary of the diffuse aurora was seen on 19 March 2009. It moved westward-poleward initially to form part of a vortex pattern, followed by its equatorward-dawnward retreat later, as shown in Figure 10.8. This auroral activity did not occur during a substorm, as indicated by the AU/AL indices.

During this auroral activity, THEMIS satellites, which were projected near the same MLT of the auroral feature, detected appreciable plasma flows that form part of a flow vortex pattern. The plasma flows were initially duskward-earthward and changed to duskward-tailward later, as shown in Figure 10.9.

This plasma flow pattern was detected during the equatorward-dawnward retreat of the auroral feature. These

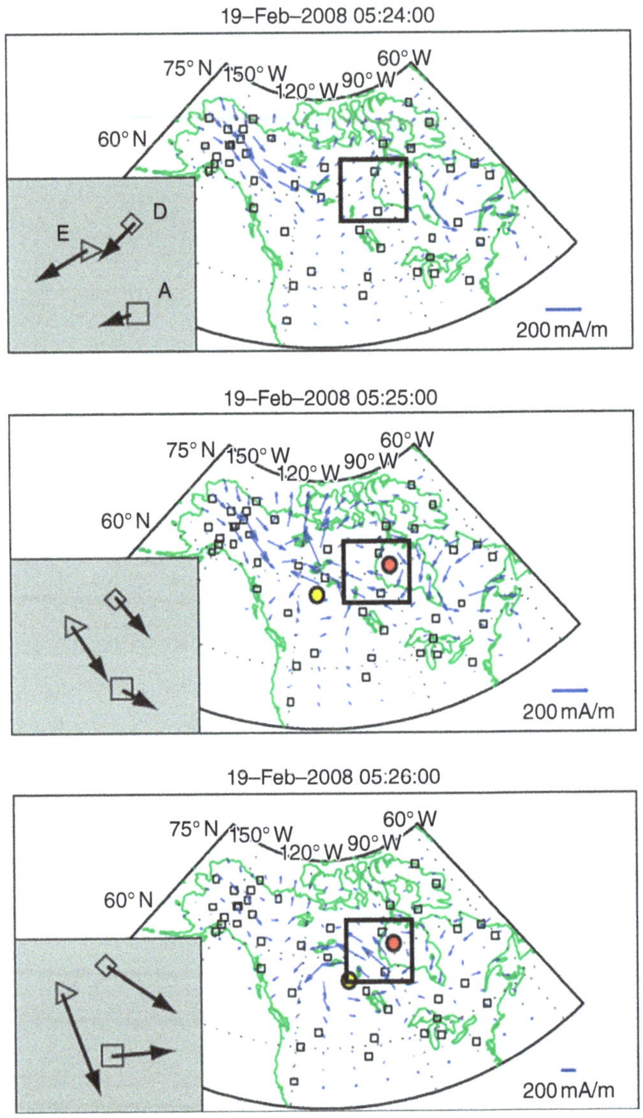

Figure 10.7 Comparison of conjugate flows in the ionosphere with plasma flows in the magnetotail shown in the bottom left insert for each panel. The ionospheric flows were inferred from the equivalent ionospheric currents [*Keiling et al.*, 2009].

observations suggest that THEMIS satellites remotely sensed a counterclockwise flow vortex (viewed from above the equatorial plane) in the magnetotail with decreasing flow vortex strength, which was considered to be the generation mechanism for the auroral feature seen simultaneously.

In examining the overshoot and rebound of a bursty bulk flow event from THEMIS observations, *Panov et al.* [2010] noted that the oscillatory flow led to flow vortices with opposite sense of rotation. However, they did not relate these vortices to FAC. In a later study of the same event by correlating the flow observations with the ground observations of auroral forms and magnetic

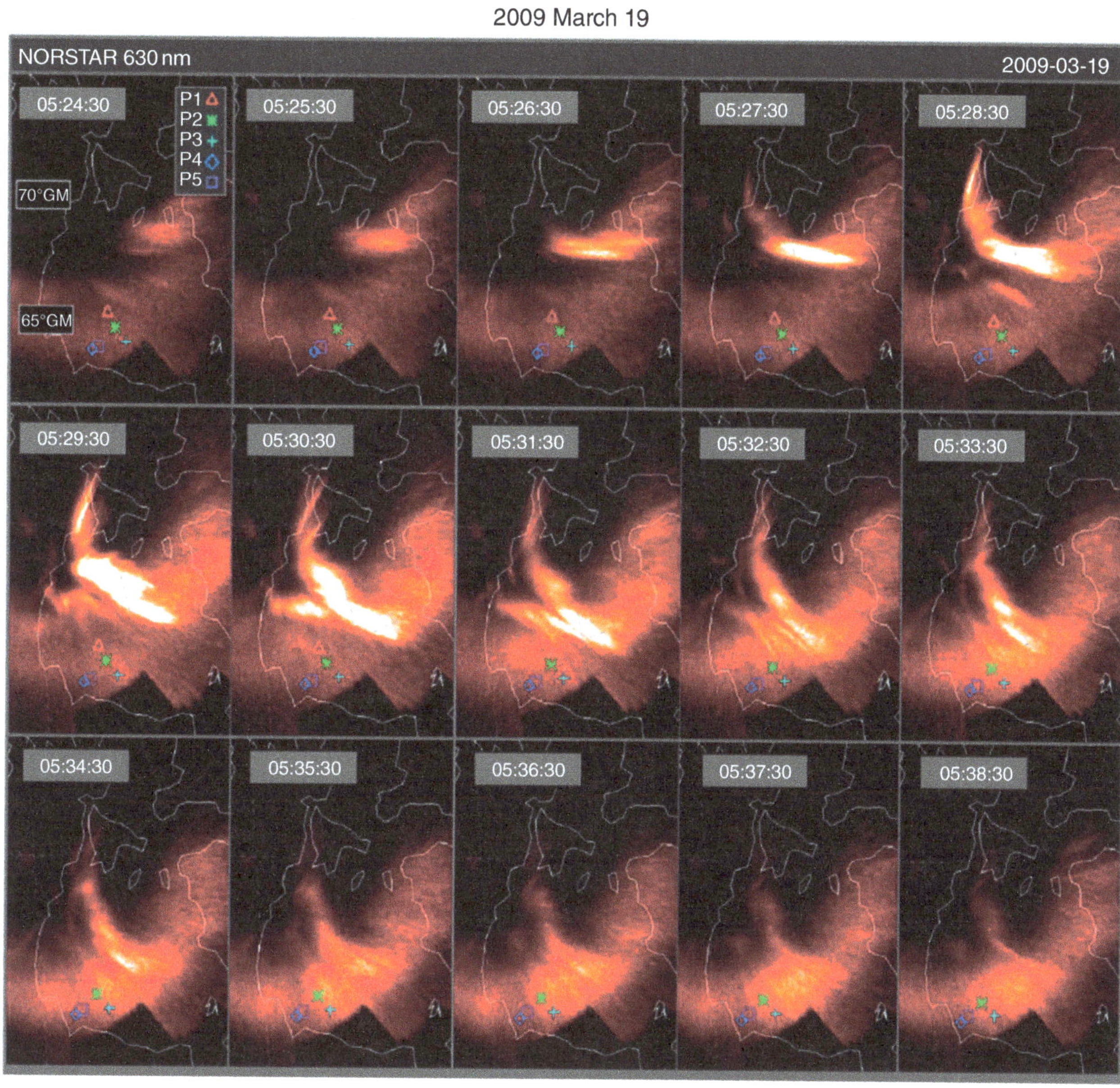

Figure 10.8 A time sequence of auroral images at 630.0 nm emission from the All-Sky-Imager in Rankin Inlet on 19 March 2009 showing an auroral vortex pattern when THEMIS satellites were in the same local time sector and detected a plasma flow vortex pattern [*Lui et al.*, 2010].

perturbations, *Panov et al.* [2013b] concluded that the oscillating currents associated with flow oscillations consisted of polarization current and the oscillatory part of the pressure gradient current. Another plausible interpretation of the flow braking in this event is CD in which the cross-tail current was disrupted by the cross-field current instability [*Lui et al.*, 1991] with the generation of whistler waves that were observed for this event [*Panov et al.*, 2013a].

Recently, multipoint observations indicate that magnetospheric vortices could be produced by solar wind dynamic pressure changes. *Shi et al.* [2014] showed a plasma vortex with a scale of 5–10 R_E and propagating downtail while expanding in size. It was induced by a solar wind dynamic pressure increase. The observed vortex had features consistent with the prediction of a model proposed by *Sibeck* [1990]. Simultaneously, a poleward propagating vortex in the ionosphere was

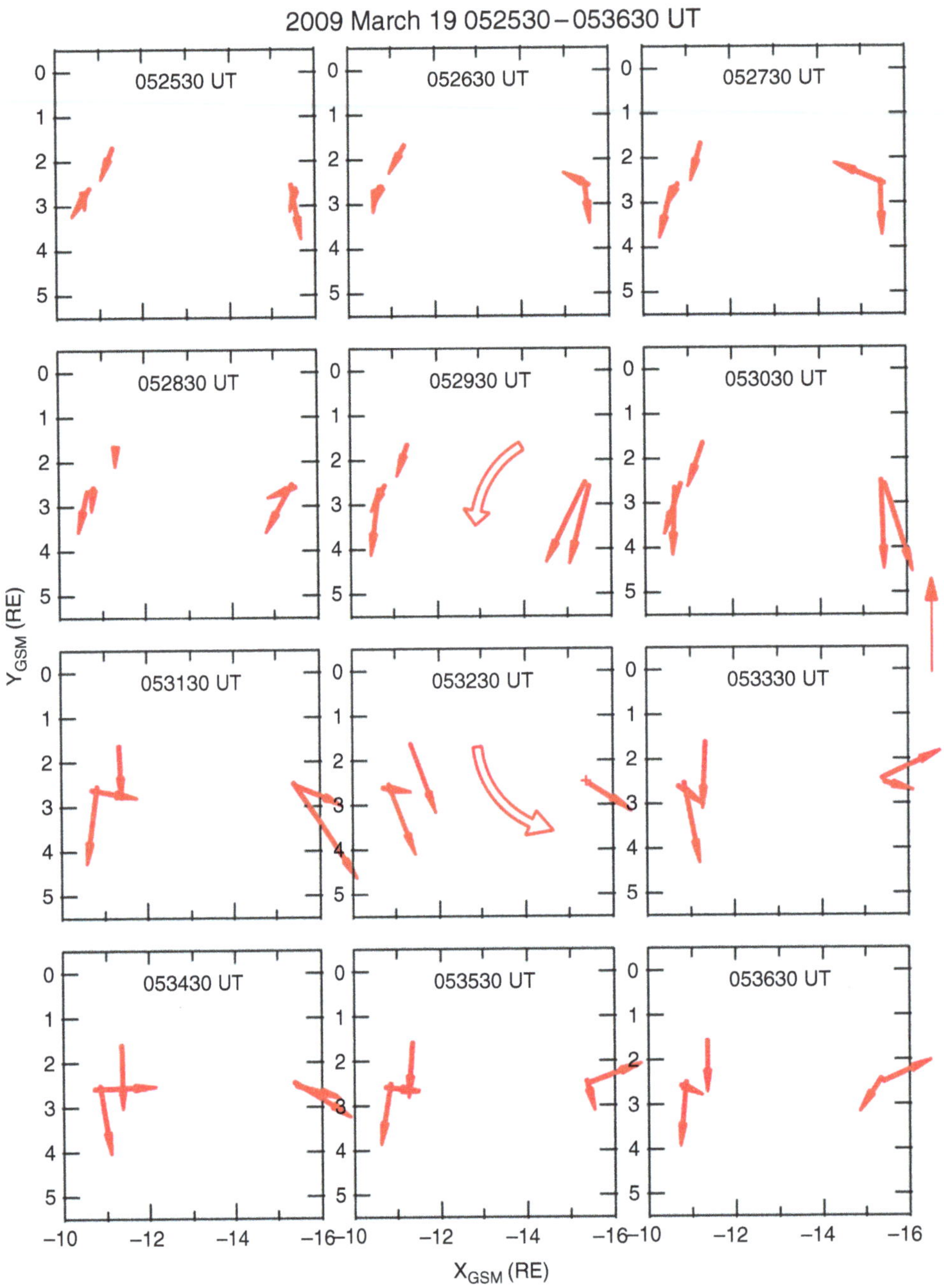

Figure 10.9 Snapshots of plasma flow pattern in the magnetotail showing the flow rotation indicative of a plasma flow vortex located mostly tailward of the satellites in the magnetotail [*Lui et al.*, 2010].

inferred from ground magnetometer observations, consistent with the expectation of the vortex in the magnetotail.

A decrease in solar wind dynamic pressure also produces a flow vortex in the magnetosphere, as reported by *Zhao et al.* [2016]. The observed vortex had a rotation opposite to that associated with an increase in solar wind dynamic pressure. The vortex was propagating tailward. Near the ionospheric projection of the observing satellite location, a vortex inferred from the ground magnetometer data was also found with the sense consistent with that in the magnetotail. The estimated FAC density from the three satellites was about 0.15 nA/m^2 with the total FAC of the vortex as 0.15–0.3 MA, similar to the total FAC in pseudo-breakup but an order of magnitude less than the total FAC in moderate substorms, which is ~1 MA.

10.7. SUBSTORM CURRENT SYSTEM

Magnetospheric substorms are major disturbances that capture the attention of many space researchers [*Akasofu*, 1964, 1968]. An intrinsic substorm phenomenon is the appearance of a new current system in the magnetosphere. Two types of current circuit were proposed [*Boström*, 1964]. The first type consists of azimuthal current paths in the ionosphere and the magnetosphere, linked by FACs. The second type consists of meridian current paths in both regions and linked by FACs also. More specifically, FAC in type 2 circuit closes approximately in the radial direction on the magnetospheric equatorial plane and approximately in the north-south direction in the ionosphere. Based on ground and space magnetic perturbations during substorms, the type 1 circuit was emphasized as the primary component and coined the substorm current wedge (SCW) [*McPherron et al.*, 1973]. However, in the ionosphere, the SCW is largely a Hall current and does not dissipate much power during substorms while the meridian current circuit (type 2) is largely a Pedersen current and dissipates much power during substorms, thus requiring a dynamo to drive it. Therefore, in terms of energy consideration, this latter circuit is a more important one [*Kamide and Kokubun*, 1996; *Lui and Kamide*, 2003; *Akasofu*, 2003, 2013].

Substorm expansion typically lasts for tens of minutes. It is quite unlikely that the SCW is mainly driven by large-scale plasma flow vortices in the magnetosphere with continuous vorticity increase throughout the disturbance period. Other mechanisms to generate the substorm current system must be considered [e.g., *Hasegawa and Sato*, 1979; *Lui*, 1996; *Birn et al.*, 1999]. In the latest version [*Birn and Hesse*, 2013, 2014; *Sergeev et al.*, 2014; *Kepko et al.*, 2015], the SCW arises mainly from the accumulation of multiple wedgelets related to dipolarization fronts (DFs), also called dipolarizing flux bundles (DFBs) [*Liu et al.*, 2013]. An additional azimuthal circuit with currents opposite to that in the SCW occurs at the earthward edge of the SCW in this latest version. This additional circuit arises from the plasma flow shear related to the flow vorticity for the SCW. Incoming plasma flow is deflected toward dusk (dawn) in the evening (morning) portion of the flow braking region. This scenario for SCW is mainly based on MHD simulations and the current generation is schematically illustrated in Figure 10.10.

This scenario has several problems. First of all, the ionospheric counterpart of DFs and DFBs are streamers [*Sergeev et al.*, 2000; *Nakamura et al.*, 2001]. Streamers do not always precede substorm onset [*Yahnin et al.*, 2000; *Mende et al.*, 2011] and streamers are frequently observed without substorm onset [*Xing et al.*, 2010]. Therefore, the role of DFs and DFBs in triggering substorm onset is questionable.

There is also a fundamental problem in describing the SCW by the MHD approach. There are many examples showing that substorm dynamics (e.g., substorm dipolarization) exhibit important non-MHD behaviors [e.g., *Lui et al.*, 1988, 1999; *Ohtani et al.*, 1992, 1998; *Cheng and*

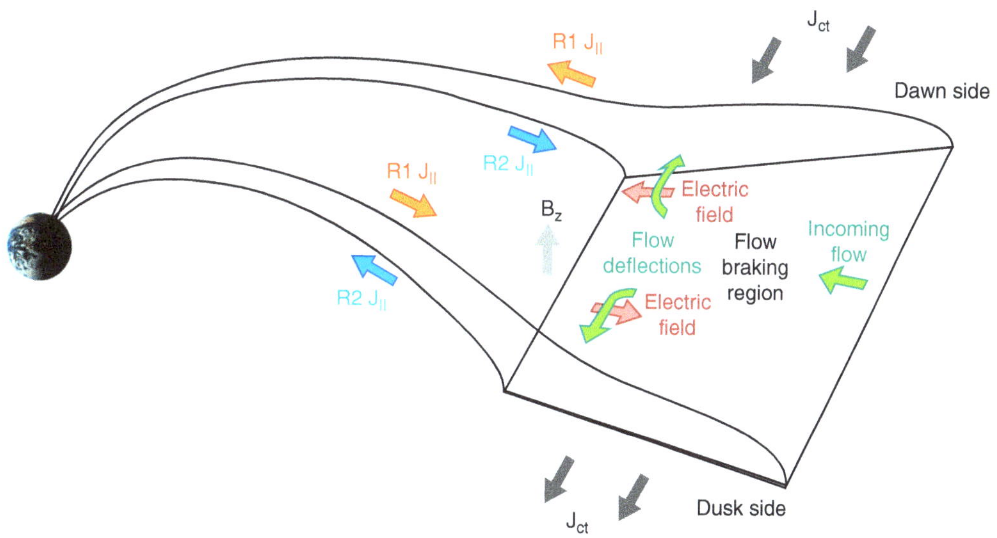

Figure 10.10 A schematic illustration to show the generation of a pair of field-aligned currents (denoted as R1 and R2) by plasma flow braking. The R2 field-aligned current is related to the shear flow produced by flow deflection at the flow braking region. The electric field associated with this flow deflection is also shown. Note that the electric field is tailward-pointing on the dusk side of this flow braking region. The cross-tail current (J_{ct}) is also indicated.

Lui, 1998; *Chen et al.*, 2000, 2003; *Parks et al.*, 2001; *Parks*, 2004b; *Liang et al.*, 2008; *Yoon et al.*, 2009; *Mühlbachler et al.*, 2009; *Mok et al.*, 2010; *Cao et al.*, 2013; *Kozelov and Kozelova*, 2013; *Lui*, 2013, 2015b; *Hwang et al.*, 2014; *Li et al.*, 2014]. In particular, it is demonstrated that the frozen-in condition required to associate DFs and DFBs as magnetic flux carriers from the midtail region to the near-Earth region is invalid [*Lui*, 2015b]. In fact, DFs and DFBs are distinct from the substorm dipolarization in the near-Earth region [*Lui*, 2014]. DFs and DFBs are local features of limited spatial extent in the magnetosphere whereas near-Earth dipolarization is a global phenomenon covering a substantial dawn-dusk spatial extent like that for the SCW. Furthermore, several analyses of substorm injections into the inner magnetosphere showed that most injections at geosynchronous altitude and deeper in do not originate in the midtail plasma sheet because most DFs have flux tube entropy too high [*Dubyagin et al.*, 2011; *Lui*, 2011; *Kim et al.*, 2012; *Sergeev et al.*, 2012; *Turner et al.*, 2015]. Instead, these deep injections are suggested to be energization of local populations deep inside the magnetosphere by magnetosonic waves launched possibly from the flow braking sites [*Turner et al.*, 2015].

There are some substorm features in this MHD scenario that are inconsistent with observations. For instance, substorm expansion onsets typically occur in the premidnight sector and observations in the ionosphere indicate that the electric field equatorward of the onset auroral arc points southward [*Akasofu*, 2003]. For the scenario in Figure 10.10, the shear flow in the dusk side of the flow braking region is directed duskward so that the associated electric field from MHD consideration ($E=-V\times B$) is directed tailward. The corresponding electric field projected in the ionosphere is then northward-pointing and the closure meridian current in the magnetotail is directed earthward. These predictions can be compared with observations in the magnetotail for the substorm current system.

The direction of the meridian current in the magnetotail during substorms can be extracted with a fortuitous satellite configuration in which two THEMIS satellites have almost identical projections on the equatorial plane and are separated only in the north-south direction. One such case is shown in Figure 10.11 in which the evolutions of the magnetic field and the embedded electric current as well as the corresponding load/dynamo parameter ($J\bullet E$) were obtained [*Lui*, 2013]. In more specific details, based on the integral form of the Ampère's law, $J_i = \int(\nabla\times B)dl/\mu_o \approx (B_{k1}-B_{k2})/\mu_o$, where subscript $i=y$, x when $k=x$, y, respectively, and subscripts 1, 2 refer to the two satellites having almost identical equatorial projections. The electric field E is the averaged value of the corresponding component (i.e., E_y for J_y and E_x for J_x) at the two satellites. For the calculation of $J\bullet E$, the GSE coordinate is used because E components are more precisely known in that coordinate system. The satellites were in the premidnight sector at the dusk side of the CD region. It is found that the perturbed meridian current J_x during the CD interval (02:24–02:36 UT) was mainly negative relative to the predisruption level (marked by the blue dashed line in Figure 10.11c), indicating a tailward current instead. Furthermore, the dynamo/load parameter for the meridian current path was mainly a dynamo ($J\bullet E<0$), implying that the electric field associated with the meridian current circuit was positive. The electric field projected in the ionosphere was then southward, consistent with observations [*Akasofu*, 2003]. Interestingly, in the MHD scenario, both the current and the electric field are opposite to observations in this sector of the magnetotail, thus also implying a dynamo effect for the meridian closure. It appears that non-MHD behavior with ions and electrons decoupled is necessary to create earthward electric field in this region [*Zelenyi et al.*, 2010; *Artemyev et al.*, 2011].

Another noteworthy point from this study is that the changes in magnetic field and current densities had the fast kinetic timescale, that is, in seconds, implying substorm dipolarization having the timescale of a kinetic process and not an MHD process.

The CD model for substorms [*Lui*, 1991] is quite compatible with these characteristics. For this model, CD and the associated substorm dipolarization are caused by a kinetic plasma instability with unmagnetized ions and magnetized electrons [*Lui and Kamide*, 2003]. As the tail-like field changes to dipolar-like field in dipolarization, the magnetized electrons convect earthward with the duskward induced electric field from dipolarization. The unmagnetized ions, on the other hand, do not follow the convection as closely as the magnetized electrons and so lag behind them. This differential motion between the two species generates an earthward-directed electric field simultaneously with a tailward-directed current. This constitutes the dynamo for the meridian current circuit. The cross-tail current ($J_\perp$) is interrupted by the turbulence associated with the kinetic plasma instability and constitutes the azimuthal current circuit. In terms of the physical process, it is the magnetic field gradient (∇B) in the dipolarizing region that diverts the cross-tail current to form the FACs for this circuit. FAC from the magnetosphere to the ionosphere is due to $J_\perp\bullet\nabla B>0$ in the dawn side of the dipolarizing region and the opposite sense for the dusk side where $J_\perp\bullet\nabla B<0$. The two current circuits proposed by the CD model are illustrated in Figure 10.12. The collapse of the magnetic field in dipolarization by CD can produce an earthward

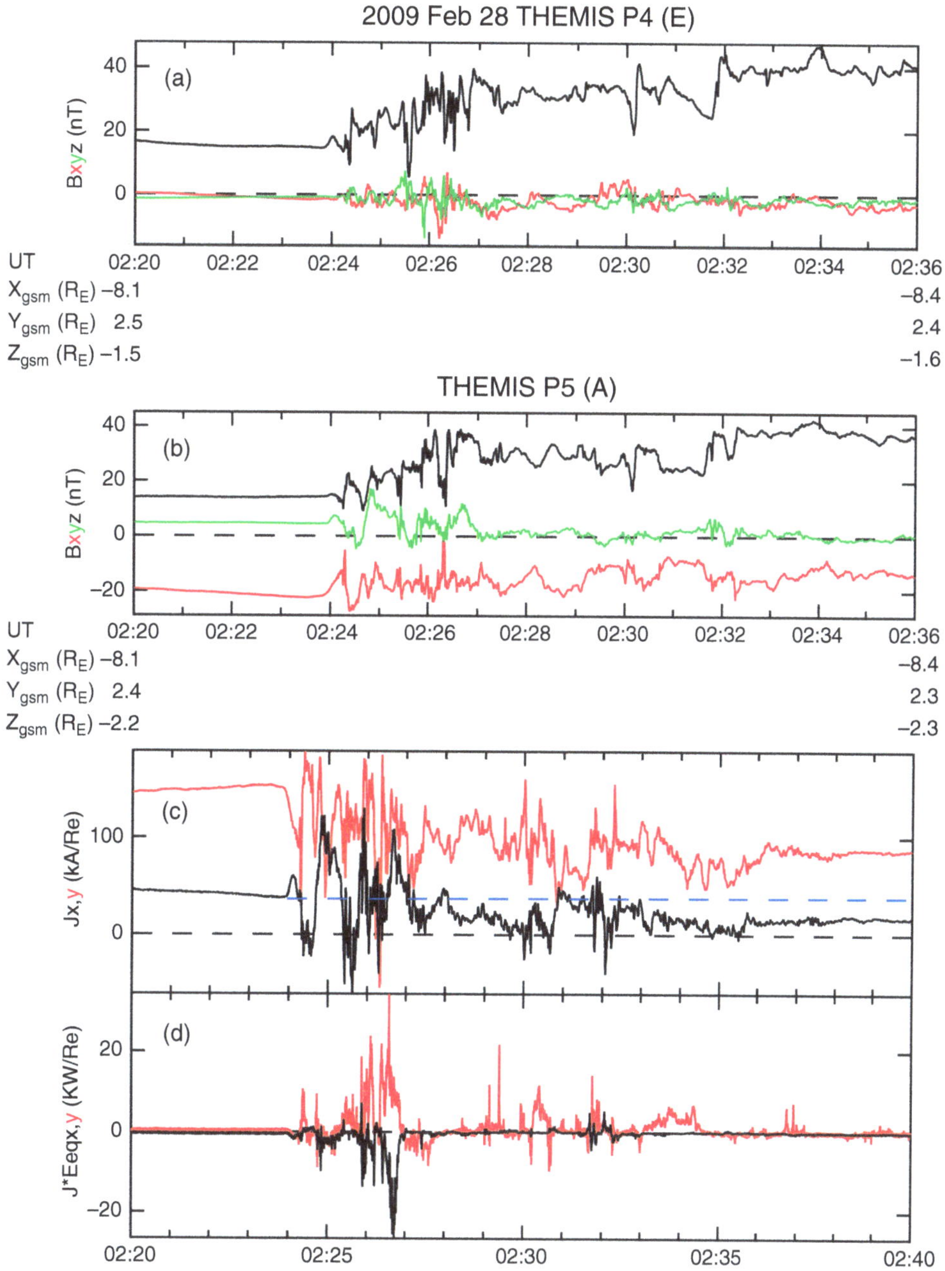

Figure 10.11 (a,b) The magnetic field measurements from two THEMIS satellites in which their equatorial projections were almost identical (their GSM locations are given at the bottom of each panel); (c,d) the inferred current and power parameters based on the Ampère's law [*Lui*, 2013].

propagating magnetosonic wave spatially localized in the dawn-dusk direction. The magnetosonic wave is generated by the sudden plasma pressure increase earthward of the dipolarizing plasma. This is a feature that *Turner et al.* [2015] suggested to be responsible for dipolarization deep inside the magnetosphere. Therefore, the cause of the earthward propagating magnetosonic wave that produces deep particle injections in the inner magnetosphere is not necessarily from the flow braking region but could well be from the CD region instead.

In the CD model, there are multiple localized CD sites that constitute the global evolution of the substorm current system. This scenario is likely to produce multiple localized meridian current circuits during substorm expansion as indicated by previous detailed

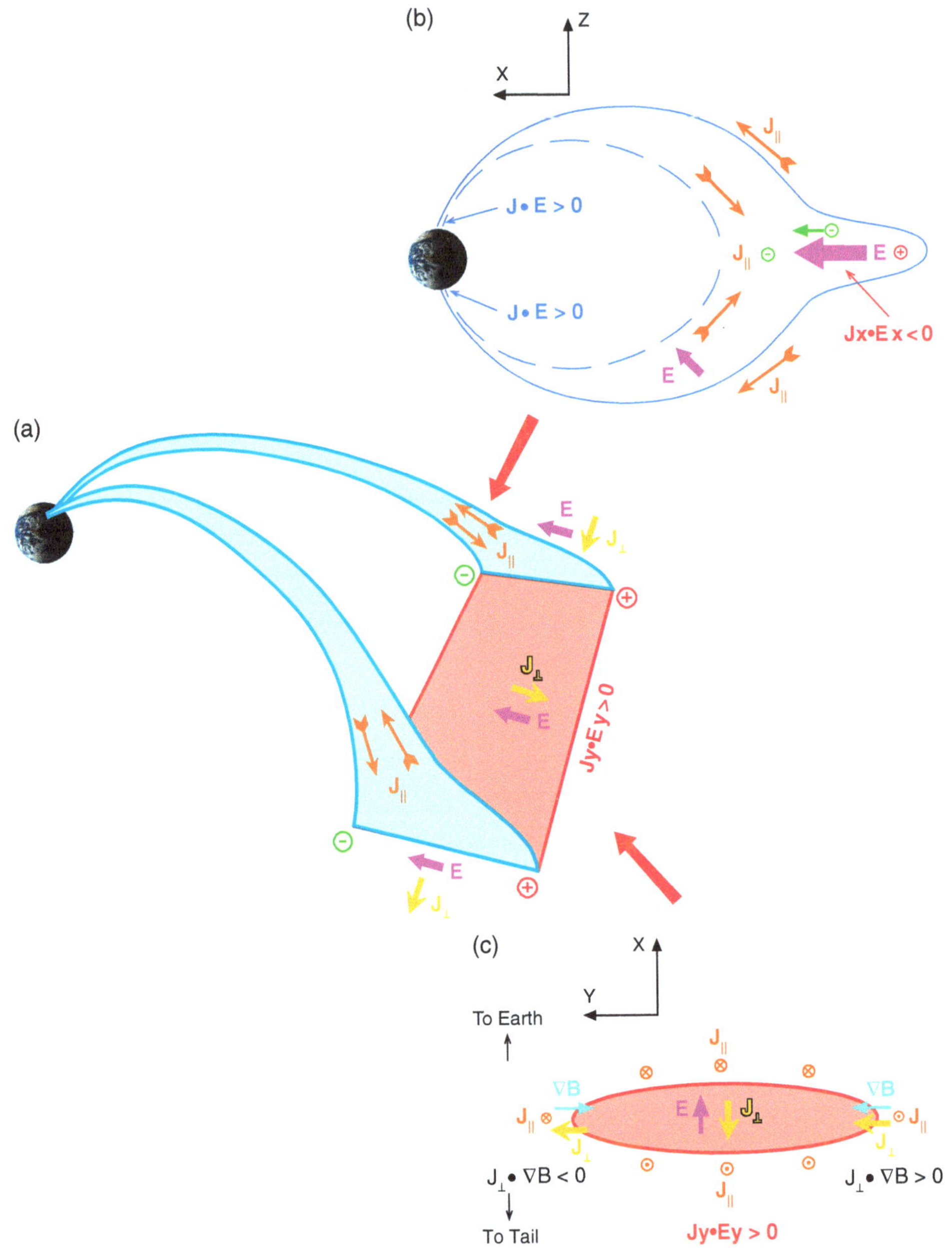

Figure 10.12 Schematic diagrams to illustrate the dynamo action associated with a dipolarizing region for the substorm current system from a single current disruption site (colored red). (a) A three-dimensional view of the substorm current system; (b) the meridian current circuit with the dynamo at the dipolarizing region in the magnetosphere and the load in the ionosphere; (c) the meridian and azimuthal current circuits in the equatorial plan where the dynamo for the meridian current circuit and the load for the azimuthal current circuit are both in the dipolarizing region.

investigations of the substorm current system [*Fujii et al.*, 1994; *Gjerloev and Hoffman*, 2002, 2014]. Furthermore, magnetic reconnection is invoked in the late expansion phase that could set up another current system that may be more global in nature than those by localized CD sites. This additional current system may be distinct from the CD current system and could well be the substorm current generated by the boundary layer model as a consequence of magnetic reconnection that develops farther down the magnetotail [*Rostoker*, 1996, 2007]. This scenario is likely to produce two relatively independent SCW as indicated by the analysis of *Gjerlov and Hoffman* [2014].

The observations shown in Figure 10.11 as well as the non-MHD nature of DFs and DFBs constitute compelling evidence for the CD model. The link between CD and substorm injection was described in detail previously by the change of the Lorenz force at the CD site [*Lui*, 1996]. The substorm injection can also be inferred from the consideration of low entropy created by CD as investigated in several reports [*Wolf et al.*, 2009; *Yang et al.*, 2010, 2011, 2012]. Note that the physical process to produce entropy depletion can be magnetic reconnection or CD [*Wolf et al.*, 2009; *Yang et al.*, 2010]. Modeling entropy depletion with the Rice Convection model reproduces the major features observed in substorms, including substorm injection [*Yang et al.*, 2012]. Therefore, the CD model for SCW has strong observational support from the THEMIS mission and provides convincing explanation for substorm injection as well, contradicting the perception that it does not provide explanation nor is supported by observations for substorm developments related to the SCW formation, as stated in *Kepko et al.* [2015].

10.8. SUMMARY AND DISCUSSION

There are several interesting and puzzling features of the current sheet in the Earth's magnetotail that are revealed more clearly in recent multipoint observations.

1. Occasional magnetic turbulence is present at the neutral sheet during substorms.
2. The neutral sheet surface often exhibits wavy profiles on the *XZ*- and *YZ*-planes.
3. The IMF B_y component has a nonuniform influence on the tail B_y component.
4. Increasing and decreasing strengths of plasma flow vortex can generate field-aligned currents with opposite sense of direction.
5. Substorm current system consists of two current circuits, with a dynamo for the meridian current circuit and a load for the azimuthal current circuit in the magnetosphere.

The phenomenon (1) and the responsible physical process are probably substorm-related and are playing a significant role in particle energization. The physical processes responsible for the features in (2) and (3) are still unknown. Feature (3) indicates that the simple notion of IMF penetrating to the magnetosphere is probably incorrect. Feature (4) probably has important implications on magnetosphere-ionosphere coupling. Feature (5) will remain controversial if the non-MHD aspects of the substorm current system are ignored. The resolution of these features remains to be challenging problems for future investigation.

ACKNOWLEDGMENTS

This work was supported by the NSF grant AGS-1250634 to The Johns Hopkins University Applied Physics Laboratory.

REFERENCES

Akasofu, S.-I. (1964), The development of the auroral substorm, *Planet. Space Sci.*, *12*, 273–282.

Akasofu, S.-I. (1968), *Polar and Magnetospheric Substorms*, D. Reidel, Norwell, Massachusetts.

Akasofu, S.-I. (2003), A source of auroral electrons and the magnetospheric substorm current systems, *J. Geophys. Res.*, *108 (A4)*, 8006; doi:10.1029/2002JA009547.

Akasofu, S.-I. (2013), The relationship between the magnetosphere and magnetospheric/auroral substorms, *Ann. Geophys.*, *31*, 387–394.

Antonova, E. E. (2003), Investigation of the hot plasma pressure gradients and the configuration of magnetospheric currents from INTERBALL, *Adv. Space Res.* *31*(5), 1157–1166.

Antonova, E. E., I. P. Kirpichev, I. L. Ovchinnikov, K. G. Orlova, and M. V. Stepanova (2009), High latitude magnetospheric topology and magnetospheric substorm, *Ann. Geophys.*, *27*, 4069–4073.

Artemyev, A. V., and L. M. Zelenyi (2013), Kinetic nature of current sheets in the Earth magnetotail, *Space Sci. Rev.*, *178*, 419–440.

Artemyev, A. V., A. A. Petrukovich, R. Nakamura, and L. M. Zelenyi (2010), Proton velocity distribution in thin current sheets: Cluster observations and theory of transient trajectories, *J. Geophys. Res.*, *115*, A12255; doi:10.1029/2010JA015702.

Artemyev, A. V., A. A. Petrukovich, R. Nakamura, and L. M. Zelenyi (2011), Cluster statistics of thin current sheets in the Earth magnetotail: Specifics of the dawn flank, proton temperature profiles and electrostatic effects, *J. Geophys. Res.*, *116*, A09233; doi:10.1029/2011JA016801.

Artemyev, A. V., L. M. Zelenyi, H. V. Malova, G. Zimbardo, and D. Delcourt (2009), Acceleration and transport of ions in turbulent current sheets: formation of non-Maxwellian energy distribution, *Nonlin. Processes Geophys.*, *16*, 631–639.

Asano, Y., T. Mukai, M. Hoshino, Y. Saito, H. Hayakawa, and T. Nagai (2004), Statistical study of thin current sheet evolution around substorm onset, *J. Geophys. Res.*, *109*, A05213; doi:10.1029/2004JA010413.

Asano, Y., R. Nakamura, W. Baumjohann, A. Runov, Z. Vörös, M. Volwerk, T. L. Zhang, A. Balogh, B. Klecker, and H. Rème (2005), How typical are atypical current sheets, *Geophys. Res. Lett.*, *32*, L03108; doi:10.1029/2004GL021834.

Bame, S. J., J. R. Asbridge, H. E. Felthauser, E. W. Hones Jr., and I. B. Strong (1967), Characteristics of the plasma sheet in the Earth's magnetotail, *J. Geophys. Res.*, *72*, 113–129.

Behannon, K. W. (1970), Geometry of the geomagnetic tail, *J. Geophys. Res.*, *75*, 743–753.

Birn, J., and M. Hesse (2013), The substorm current wedge in MHD simulations, *J. Geophys. Res.*, *118*, 3364–3376; doi:10.1002/jgra.50187.

Birn, J., and M. Hesse (2014), The substorm current wedge: Further insights from MHD simulations, *J. Geophys. Res.*, *119*, 3503–3513; doi:10.1002/2014JA019863.

Birn, J., M. Hesse, G. Haerendel, W. Baumjohann, and K. Shiokawa (1999), Flow braking and the substorm current wedge, *J. Geophys. Res.*, *104*, 19895–19903.

Boström, R. (1964), A model of the auroral electrojets, *J. Geophys. Res.*, *69*, 4983–4999.

Bowling, S. B. (1975), Transient occurrence of magnetic loops in the magnetotail, *J. Geophys. Res.*, *80*, 4741–4745.

Cao, J., Y. Ma, G. Parks, H. Reme, I. Dandouras, and T. Zhang (2013), Kinetic analysis of the energy transport of bursty bulk flows in the plasma sheet, *J. Geophys. Res.*, *118*; doi:10.1029/2012JA018351.

Chen, L.-J., A. Bhattacharjee, K. Sigsbee, G. Parks, M. Fillingim, and R. Lin (2003), Wind observations pertaining to current disruption and ballooning instability during substorms, *Geophys. Res. Lett.*, *30*, 68–1; doi 10.1029/2002GL016317.

Chen, L.-J., G. K. Parks, M. McCarthy, D. Larson, R. P. Lin (2000), Kinetic properties of bursty bulk flow events, *Geophys. Res. Lett.*, *27*, 1847–1850.

Cheng, C. Z., and A. T. Y. Lui (1998), Kinetic ballooning instability for substorm onset and current disruption observed by AMPTE/CCE, *Geophys. Res. Lett.*, *25*, 4091–4094.

Consolini, G., M. Kretzschmar, A. T. Y. Lui, G. Zimbardo, and W. M. Macek (2005), On the magnetic field fluctuations during magnetospheric tail current disruption: A statistical approach, *J. Geophys. Res.*, *110*, A07202; doi:10.1029/2004JA010947.

Daglis, I. A., E. T. Sarris, and G. Kremser (1991), Ionospheric contribution to the cross-tail current enhancement during the substorm growth phase, *J. Atmos. Terr. Phys.*, *53*, 1091–1098.

Dubyagin, S., V. Sergeev, S. Apatenkov, V. Angelopoulos, A. Runov, R. Nakamura, W. Baumjohann, J. McFadden, and D. Larson (2011), Can flow burst penetrate into the inner magnetosphere? *Geophys. Res. Lett.*, *38*, L08102; doi:10.1029/2011GL047016.

Fairfield, D. H. (1979), On the average configuration of the geomagnetic tail, *J. Geophys. Res.*, *84*, 1950–1558.

Fujii, R., R. A. Hoffman, P. C. Anderson, J. D. Craven, M. Sugiura, L. A. Frank, and N. C. Maynard (1994), Electrodynamic parameters in the nighttime sector during auroral substorms, *J. Geophys. Res.*, *99*, 6093–6112.

Gjerloev, J. W., and R. A. Hoffman (2002), Currents in auroral substorms, *J. Geophys. Res.*, *107* (A8), SMP 5; doi: 10.1029/2001JA000194.

Gjerloev, J. W., and R. A. Hoffman (2014), The large-scale current system during auroral substorms, *J. Geophys. Res. Space Physics*, *119*, 4591–4606; doi: 10.1002/2013JA019176.

Harris, E. G. (1962), The equilibrium of oppositely directed magnetic fields, *Nuovo Cimento*, *23*, 115–121.

Hasegawa, A., and T. Sato (1979), Generation of field aligned current during substorm, *Dynamics of the Magnetosphere*, edited by S.-I. Akasofu, D. Reidel Publ. Co., Hingham, Massachusetts, 529–542.

Hones, E. W., Jr. (1979), Transient phenomena in the magnetotail and their relation to substorms, *Space Sci. Rev.*, *23*, 393–410.

Hones, E. W., Jr., J. Birn, D. N. Baker, S. J. Bame, W. C. Feldman, D. J. McComas, R. D. Zwickl, J. A. Slavin, E. J. Smith, and B. T. Tsurutani (1984), Detailed examination of a plasmoid in the distant magnetotail with ISEE 3, *Geophys. Res. Lett.*, *11*, 1046–1049.

Hoshino, M., A. Nishida, T. Mukai, Y. Saito, T. Yamamoto, and S. Kokubun (1996), Structure of plasma sheet in magnetotail: Double-peaked electric current sheet, *J. Geophys. Res.*, *101*, 24775–24786.

Hruska, A., and J. Hruskova (1970), Transverse structure of the Earth's magnetotail and fluctuations of the tail magnetic field, *J. Geophys. Res.*, *75*, 2449–2457.

Hwang, K.-J., M. L. Goldstein, T. E. Moore, B. M. Walsh, D. G. Baishev, A. V. Moiseyev, B. M. Shevtsov, and K. Yumoto (2014), A tailward moving current sheet normal magnetic field front followed by an earthward moving dipolarization front, *J. Geophys. Res. Space Physics*, *119*, 5316–5327; doi:10.1002/2013JA019657.

Kamide, Y., and S. Kokubun (1996), Two component auroral electrojet: Importance for substorm studies, *J. Geophys. Res.*, *101*, 13027–13046.

Keiling, A., V. Angelopoulos, A. Runov, J. Weygand, S. V. Apatenkov, S. Mende, J. McFadden, et al. (2009), Substorm current wedge driven by plasma flow vortices: THEMIS observations, *J. Geophys. Res.*, *114*, A00C22; doi:10.1029/2009JA014114.

Kepko, L., R. L. McPherron, O. Amm, S. Apatenkov, W. Baumjohann, J. Birn, M. Lester, R. Nakamura, T. I. Pulkkinen, V. Sergeev (2015), Substorm current wedge revisited, *Space Sci. Rev.*, *190*:1–46; doi: 10.1007/s11214-014-0124-9.

Kim, H.-S., D.-Y. Lee, S. Ohtani, M.-Y. Park, and B.-H. Ahn (2012), On near-tail bubble penetration into geosynchronous altitude, *J. Geophys. Res.*, *117*, A07205; doi:10.1029/2012JA017749.

King, J. H. (1977), Interplanetary medium data book, Publ. 77-04, NASA Space Sci. Data Center, Greenbelt, Maryland.

Kistler, L. M., C. Mouikis, E. Möbius, B. Klecker, J. A. Sauvaud, H. Réme, A. Korth, M. F. Marcucci, R. Lundin, G. K. Parks, and A. Balogh (2005), Contribution of nonadiabatic ions to the cross-tail current in an O^+ dominated thin current sheet, *J. Geophys. Res.*, *110*, A06214; doi:10.1029/2004JA010653.

Kozelov, T. V., and B. V. Kozelova (2013), Substorm-associated explosive magnetic field stretching near the earthward edge of the plasma sheet, *J. Geophys. Res. Space Phys.*, *118*, 3323–3335; doi:10.1002/jgra.50344.

Krimigis, S. M., G. Gloeckler, R. W. McEntire, T. A. Potemra, F. L. Scarf, and E. G. Shelley (1985), Magnetic storm of September 4, 1984: A synthesis of ring current spectra and energy densities measured with AMPTE/CCE, *Geophys. Res. Lett.*, *12*, 329–332.

Le Contel, O., A. Roux, C. Jacquey, P. Robert, M. Berthomier, T. Chust, B. Grison, et al. (2009), Quasi-parallel whistler mode waves observed by THEMIS during near-Earth dipolarizations, *Ann. Geophys.*, *27*, 2259–2275.

Li, H., C. Wang, and S. Y. Fu (2014), Classification of fast flows in central plasma sheet: Superposed epoch analysis based on THEMIS observations, *J. Geophys. Res. Space Physics*, *119*, 7199–7213; doi:10.1002/2014JA020105.

Liang, J., E. F. Donovan, W. W. Liu, B. Jackel, M. Syrjasuo, S. B. Mende, H. U. Frey, V. Angelopoulos, and M. Connors (2008), Intensification of pre-existing auroral arc at substorm expansion phase onset: Wave-like disruption during the first tens of seconds, *Geophys. Res. Lett.*, *35*, L17S19; doi:10.1029/2008GL033666.

Liu, J., V. Angelopoulos, A. Runov, and X.-Z. Zhou (2013), On the current sheets surrounding dipolarizing flux bundles in the magnetotail: The case for wedgelets, *J. Geophy. Res. Space Physics*, *118*, 2000–2020; doi:10.1002/2013JA019395.

Lui, A. T. Y. (1984), Characteristics of the cross-tail current in the Earth's magnetotail, in *Magnetospheric Currents*, AGU Geophysical Monograph 28, 158–170.

Lui, A. T. Y. (1991), A synthesis of magnetospheric substorm models, *J. Geophys. Res.*, *96*, 1849–1856.

Lui, A. T. Y. (1996), Current disruption in the Earth's magnetosphere: Observations and models, *J. Geophys. Res.*, *101*, 13067–13088.

Lui, A. T. Y. (2002), Multiscale phenomena in the near-Earth magnetosphere, *J. Atmos. Solar Terr. Phys.*, *64*, 125–143.

Lui, A. T. Y. (2004), Potential plasma instabilities for substorm expansion onset, *Space Sci. Rev.*, *113*, 127–206.

Lui, A. T. Y. (2011), Reduction of the cross-tail current during near-Earth dipolarization with multi-satellite observations, *J. Geophys. Res.*, *116*, A12, A12239; doi:10.1029/2011JA17107.

Lui, A. T. Y. (2013), Cross-tail current evolution during substorm dipolarization, *Ann. Geophys.*, *31*, 1131–1142; doi:10.5194/angeo-31-1131-2013.

Lui, A. T. Y. (2014), Evidence for two types of dipolarization in the Earth's magnetotail, *J. Atmos. Solar Terr. Phys.*, *115–116*, 17–24.

Lui, A. T. Y. (2015a), Comparison of current disruption and magnetic reconnection, *Geosci. Lett.*, *2*:14; doi 10.1186/s40562-015-0031-2.

Lui, A. T. Y. (2015b), Dipolarization fronts and magnetic flux transport, *Geosci. Lett.*, *2*:15; doi 10.1186/s40562-015-0032-1.

Lui, A. T. Y. and A.-H. Najmi (1997), Time-frequency decomposition of signals in a current disruption event, *Geophys. Res. Lett.*, *24*, 3157–3160.

Lui, A. T. Y. and C.-I. Meng (1979), Relevance of southward magnetic fields in the neutral sheet to anisotropic distribution of energetic electrons and substorm activity, *J. Geophys. Res.*, *84*, 5817–5827.

Lui, A. T. Y., and Y. Kamide (2003), A fresh perspective of the substorm current system and its dynamo, *Geophys. Res. Lett.*, *30*, (18), 1958; doi:10.1029/2003GL017835.

Lui, A. T. Y., C.-I. Meng and S.-I. Akasofu (1978), Wavy nature of the magnetotail neutral sheet, *Geophys. Res. Lett.*, *5*, 279–282.

Lui, A. T. Y., C.-L. Chang, A. Mankofsky, H.-K. Wong, and D. Winske (1991), A cross-field current instability for substorm expansions, *J. Geophys. Res.*, *96*, 11389–11401.

Lui, A. T. Y., E. Spanswick, E. F. Donovan, J. Liang, W. W. Liu, O. LeContel, and Q.-G. Zong (2010), An active auroral streamer and the concurrent magnetotail activity, *J. Geophys. Res.*, *115*, A10210; doi:10.1029/2010JA015449.

Lui, A. T. Y., K. Liou, M. Nosé, S. Ohtani, D. J. Williams, T. Mukai, K. Tsuruda, S. Kokubun (1999), Near-Earth dipolarization: Evidence for a non-MHD process, *Geophys. Res. Lett.*, *26*, 2905–2908.

Lui, A. T. Y., P. H. Yoon, C. Mok, and C.-M. Ryu (2008), Inverse cascade feature in current disruption, *J. Geophys. Res.*, *113*, A00C06; doi:10.1029/2008JA013521.

Lui, A. T. Y., R. E. Lopez, S. M. Krimigis, R. W. McEntire, L. J. Zanetti, T. A. Potemra (1988), A case study of magnetotail current sheet disruption and diversion, *Geophys. Res. Lett.*, *15*, 721–724.

Lui, A. T. Y., Y. Zheng, H. Rème, M. W. Dunlop, G. Gustafsson, and C. J. Owen (2007), Breakdown of the frozen-in condition in the Earth's magnetotail, *J. Geophys. Res.*, *112*, A04215; doi:10.1029/2006JA012000.

McPherron, R. L., C. T. Russell, and M. Aubry (1973), Satellite studies of magnetospheric substorms on August 15, 1968, 9, Phenomenological model for substorms, *J. Geophys. Res.*, *78*, 3131–3149.

Mende, S. B., H. U. Frey, V. Angelopoulos, and Y. Nishimura (2011), Substorm triggering by poleward boundary intensification and related equatorward propagation, *J. Geophys. Res.*, *116*, A00I31; doi:10.1029/2010JA015733.

Mihalov, J. D., D. S. Colburn, R. G. Currie, and C. P. Sonnett (1968), Configuration and reconnection of the geomagnetic tail, *J. Geophys. Res.*, *73*, 943–959.

Mok, C., C.-M. Ryu, P. H. Yoon, and A. T. Y. Lui (2010), Obliquely propagating electromagnetic drift ion-cyclotron instability, *J. Geophys. Res.*, *115*, A04218; doi:10.1029/2009JA014871.

Mühlbachler, S., D. Langmayr, A. T. Y. Lui, N. V. Erkaev, I. V. Alexeev, P. W. Daly, and H. K. Biermat (2009), Cluster observations showing the indication of the formation of a modified-two-stream instability in the geomagnetic tail, *Adv. Space Res.*, *43* (10), 1588–1593; doi:10.1016/j.asr.2009.01.012.

Nakamura, R., W. Baumjohann, R. Schödel, M. Brittnacher, V. A. Sergeev, M. Kubyshkina, T. Mukai, and K. Liou (2001), Earthward flow bursts, auroral streamers, and small expansions, *J. Geophys. Res.*, *106*, 10,791–10,802.

Nakamura, R., W. Baumjohann, A. Runov, M. Volwerk, T. L. Zhang, B. Klecker, Y. Bogdanova, et al. (2002), Fast flow during current sheet thinning, *Geophys. Res. Lett.*, *29* (23), 10.1029/2002GL016200.

Nakamura, R., W. Baumjohann, Y. Asano, A. Runov, A. Balogh, C. J. Owen, A. N. Fazakerley, M. Fujimoto, B. Klecker, and H. Rème (2006), Dynamics of thin current sheets associated with magnetotail reconnection, *J. Geophys. Res.*, *111*, A11206; doi:10.1029/2006JA011706.

Ohtani, S., K. Takahashi, C. T. Russell (1992), Radial expansion of the tail current disruption during substorms: A new approach to substorm onset region, *J. Geophys. Res.*, *97*, 3129–3136.

Ohtani, S., K. Takahashi, T. Higuchi, A. T. Y. Lui, and H. E. Spence, J. F. Fennell (1998), AMPTE/CCE - SCATHA simultaneous observations of substorm-associated magnetic fluctuations, *J. Geophys. Res.*, *103*, 4671–4682.

Panov, E. V. , A. V. Artemyev, W. Baumjohann, R. Nakamura, and V. Angelopoulos (2013a), Transient electron precipitation during oscillatory BBF braking: THEMIS observations and theoretical estimates, *J. Geophys. Res.*, *118*, 3065–3076; doi:10.1002/jgra.50203.

Panov, E. V., R. Nakamura, W. Baumjohann, V. Angelopoulos, A. A. Petrukovich, A. Retinò, M. Volwerk, et al. (2010), Multiple overshoot and rebound of a bursty bulk flow, *Geophys. Res. Lett.*, *37*, L08103; doi:10.1029/2009GL041971.

Panov, E. V., W. Baumjohann, R. Nakamura, O. Amm, M. V. Kubyshkina, K.-H. Glassmeier, J. M. Weygand,

V. Angelopoulos, A. A. Petrukovich, and V. A. Sergeev (2013b), Ionospheric response to oscillatory flow braking in the magnetotail, *J. Geophys. Res.*, *118*, 1529–1544; doi:10.1002/jgra.50190.

Parks, G. K. (2004a), Physics of Space Plasmas: An Introduction, 2d ed., Addison-Wesley Publishing Co., Redwood City California.

Parks, G. K. (2004b), Why space physics need to go beyond the MHD box, *Space Sci. Rev.*, *113*, 97–125.

Parks, G. K., L. J. Chen, M. Fillingim, and M. McCarthy (2001), Kinetic characterization of plasma sheet dynamics, *Space Sci. Rev.*, *95*, 237–255.

Petrukovich, A., A. Artemyev, I. Vasko, R. Nakamura, and L. Zelenyi (2015), Current sheets in the Earth magnetotail: Plasma and magnetic field structure with Cluster project observations, *Space Sci. Rev.*, *188*, 311–337; doi 10.1007/s11214-014-0126-7.

Petrukovich, A. A. (2011), Origins of plasma sheet *By*, *J. Geophys. Res.*, *116*, A07217; doi: 10.1029/2010JA016386.

Petrukovich, A. A., A. V. Artemyev, H. V. Malova, V. Y. Popov, R. Nakamura, and L. M. Zelenyi (2011), Embedded current sheets in the Earth's magnetotail, *J. Geophys. Res.*, *116*, A00125; doi:10.1029/2010JA015749.

Petrukovich, A. A., T. L. Zhang, W. Baumjohann, R. Nakamura, A. Runov, A. Balogh, and C. Carr (2006), Oscillatory magnetic flux tube slippage in the plasma sheet, *Ann. Geophys.*, *24*, 1695–1704.

Petrukovich, A. A., W. Baumjohann, R. Nakamura, A. Runov, and A. Balogh (2005), Cluster vision of the magnetotail current sheet on a macroscale, *J. Geophys. Res.*, *110*, A06204; doi:10.1029/2004JA010825.

Petrukovich, A. A., W. Baumjohann, R. Nakamura, and A. Runov (2008), Formation of current density profile in tilted current sheets, *Ann. Geophys.*, *26*, 3669–3676.

Rong, Z. J., C. Shen, A. A. Petrukovich, W. X. Wan, and Z. X. Liu (2010), The analytic properties of the flapping current sheets in the earth magnetotail, *Planet. Space Sci.*, *58*, 1215–1229.

Rostoker, G. (1996), Phenomenology and physics of magnetospheric substorms, *J. Geophys. Res.*, *101*, 12955–12973.

Rostoker, G. (2007), Substorms, in *Handbook of the Solar-Terrestrial Environment*, ed. by Y. Kamide and A. Chian, 376–395, Springer-Verlag, Berlin, Heidelberg.

Runov, A., V. A. Sergeev, R. Nakamura, W. Baumjohann, S. Apatenkov, Y. Asano, T. Takada, et al. (2006), Local structure of the magnetotail current sheet: 2001 Cluster observations, *Ann. Geophys.*, *24*, 247–262.

Runov, A., V. A. Sergeev, W. Baumjohann, R. Nakamura, S. Apatenkov, Y. Asano, M. Volwerk, et al. (2005), Electric current and magnetic field geometry in flapping magnetotail current sheets, *Ann. Geophys.*, *23*, 1391–1403.

Saito, M. (2015), THEMIS two-point measurements of the cross-tail current density: A thick bifurcated current sheet in the near-Earth plasma sheet, *J. Geophys. Res. Space Physics*, *120*, 6258–6275; doi:10.1002/2015JA021142.

Sanny, J., R. L. McPherron, C. T. Russell, D. N. Baker, T. I. Pulkkinen, and A. Nishida (1994), Growth phase thinning of the near-Earth current sheet during the CDAW-6 substorm, *J. Geophys. Res.*, *99*, 5805–5816.

Schindler, K., and N. F. Ness (1972), Internal structure of the geomagnetic neutral sheet, *J. Geophys. Res.*, *77*, 91–100.

Sergeev, V., A. V. Nikolaev, N. A. Tsyganenko, V. Angelopoulos, A. V. Runov, H. J. Singer, and J. Yang (2014), Testing a two-loop pattern of the substorm current wedge (SCW2L), *J. Geophys. Res.*, *119*, 947–963; doi:10.1002/2013JA019629.

Sergeev, V., Y. Nishimura, M. Kubyshkina, V. Angelopoulos, R. Nakamura, and H. Singer (2012), Magnetospheric location of the equatorward prebreakup arc, *J. Geophys. Res.*, *117*, A01212; doi:10.1029/2011JA017154.

Sergeev, V. A., D. G. Mitchell, C. T. Russell, and D. J. Williams (1993), Structure of the tail plasma/current sheet at ~11 R_E and its changes in the course of a substorm, *J. Geophys. Res.*, *98*, 17345–17365.

Sergeev, V. A., J.-A. Sauvaud, D. Popescu, R. A. Kovrazhkin, K. Liou, P. T. Newell, M. Brittnacher, G. Parks, R. Nakamura, T. Mukai, and G. D. Reeves (2000), Multi-spacecraft observation of a narrow transient plasma jet in the Earth's plasma sheet, *Geophys. Res. Lett.*, *27*, 851–854.

Sergeev, V. A., V. Angelopoulos, C. Carlson, and P. Sutcliffe (1998), Current sheet measurements within a flapping plasma sheet, *J. Geophys. Res.*, *103*, 9177–9187.

Shen, X. C., Q.-G. Zong, Q. Q. Shi, A. M. Tian, W. J. Sun, Y. F. Wang, X. Z. Zhou, S. Y. Fu, M. D. Hartinger, and V. Angelopoulos (2015), Magnetospheric ULF waves with increasing amplitude related to solar wind dynamic pressure changes: The time history of events and macroscale interactions during substorms (THEMIS) observations, *J. Geophys. Res. Space Physics*, *120*, 7179–7190; doi:10.1002/2014JA020913.

Shi, Q. Q., M. D. Hartinger, V. Angelopoulos, A. M. Tian, S. Y. Fu, Q.-G. Zong, J. M. Weygand, et al. (2014), Solar wind pressure pulse-driven magnetospheric vortices and their global consequences, *J. Geophys. Res. Space Physics*, *119*, 4274–4280; doi:10.1002/2013JA019551.

Sibeck, D. G. (1990), A model for the transient magnetospheric response to sudden solar wind dynamic pressure variations, *J. Geophys. Res.*, *95*, 3755–3771.

Slavin, J. A., E. J. Smith, B. T. Tsurutani, D. G. Sibeck, H. J. Singer, D. N. Baker, J. T. Gosling, E. W. Hones, and F. L. Scarf (1984), Substorm associated traveling compression regions in the distant tail: ISEE-3 geotail observations, *Geophys. Res. Lett.*, *11*, 657–660.

Slavin, J. A., R. P. Lepping, J. Gjerloev, M. L. Goldstein, D. H. Fairfield, M. H, Acuna, A. Balogh, et al. (2003), Cluster electric current density measurements within a magnetic flux rope in the plasma sheet, *Geophys. Res. Lett.*, *30*(7), 1362; doi:10.1029/2002GL016411.

Speiser, T. W. (1973), Magnetospheric current sheets, *Radio Sci.*, *8*, 973–977.

Speiser, T. W., and N. F. Ness (1967), The neutral sheet in the geomagnetic tail: Its motion, equivalent currents, and field line connection through it, *J. Geophys. Res.*, *72*, 131–141.

Takahashi, K., L. J. Zanetti, R. E. Lopez, R. W. McEntire, T. A. Potemra, and K. Yumoto (1987), Disruption of the magnetotail current sheet observed by AMPTE/CCE, *Geophys. Res. Lett.*, *14*, 1019–1022.

Turner, D. L., S. G. Claudepierre, J. F. Fennell, T. P. O'Brien, J. B. Blake, C. Lemon, M. Gkioulidou, et al. (2015), Energetic electron injections deep into the inner magnetosphere associated with substorm activity, *Geophys. Res. Lett.*, *42*, 2079–2087; doi:10.1002/2015GL063225.

Wing, S., P. T. Newell, D. G. Sibeck, and K. B. Baker (1995), Large statistical study of the entry of interplanetary magnetic field Y-component into the magnetosphere, *Geophys. Res. Lett.*, *22*, 2083–2086.

Wolf, R. A., Y. Wan, X. Xing, J.-C. Zhang, and S. Sazykin (2009), Entropy and plasma sheet transport, *J. Geophys. Res.*, *114*, A00D05; doi:10.1029/2009JA014044.

Xing, X., L. Lyons, Y. Nishimura, V. Angelopoulos, D. Larson, C. Carlson, J. Bonnell, and U. Auster (2010), Substorm onset by new plasma intrusion: THEMIS spacecraft observations, *J. Geophys. Res.*, *115*, A10246; doi:10.1029/2010JA015528.

Yahnin, A. G., I. A. Kornilov, T. A. Kornilova, V. A. Sergeev, A. T. Y. Lui, K. Liou, C.-I. Meng, and A. Pajunpaa (2000), Do the observations confirm the high-speed flow braking model for substorms?, *Proceedings of the Fifth International Conference on Substorms*, *ESA SP-443*, 345–348.

Yang, J., F. R. Toffoletto, and Y. Song (2010), Role of depleted flux tubes in steady magnetospheric convection: Results of RCM-E simulations, *J. Geophys. Res.*, *115*, A00I11; doi:10.10 29/2010JA015731.

Yang, J., F. R. Toffoletto, R. A. Wolf, S. Sazykin, P. A. Ontiveros, and J. M. Weygand (2012), Large-scale current systems and ground magnetic disturbance during deep substorm injections, *J. Geophys. Res.*, *117*, A04223; doi:10.1029/2011 JA017415.

Yang, J., R. A. Wolf, and F. R. Toffoletto (2011), Accelerated thinning of the near-Earth plasma sheet caused by a bubble-blob pair, *Geophys. Res. Lett.*, *38*, L01107; doi:10.1029/2010 GL045993.

Yoon, P. H., A. T. Y. Lui, and J. W. Bonnell (2009), Identification of plasma instability from wavelet spectra in a current disruption event, *J. Geophys. Res.*, *114*, A04207; doi:10.1029/ 2008JA013816.

Yushkov, E. V., A. V. Artemyev, A. A. Petrukovich, and R. Nakamura (2016), Current sheet flapping in the near-Earth magnetotail: peculiarities of propagation and parallel currents, *Ann. Geophys.*, *34*, 739–750.

Zelenyi, L. M., A. Artemyev, and A. Petrukovich (2010), Earthward electric field in the magnetotail: Cluster observations and theoretical estimates, *Geophys. Res. Lett.*, *37*, L06105; doi:10.1029/2009GL042099.

Zelenyi, L. M., A. Artemyev, and A. Petrukovich (2015), Properties of magnetic field fluctuations in the Earth's magnetotail and implications for the general problem of structure formation in hot plasmas, *Space Sci. Rev.*, *188*, 287–310; doi: 10.1007/s11214-014-0037-7.

Zhang, T. L., R. Nakamura, M. Volwerk, A. Runov, W. Baumjohann, H. U. Exchelberger, C. Carr, et al. (2005), Double Star/Cluster observation of neutral sheet oscillations on 5 August 2004, *Ann. Geophys.*, *23*, 2909–2914.

Zhao, H. Y., X. C. Shen, B. B. Tang, A. M. Tian, Q. Q. Shi, J. M. Weygand, Z. H. Yao, et al. (2016), Magnetospheric vortices and their global effect after a solar wind dynamic pressure decrease, *J. Geophys. Res. Space Physics*, *121*, 1071–1077; doi:10.1002/2015JA021646.

Zhou, M., M. Ashour-Abdalla, X. Deng, D. Schiver, M. El-Alaoui, and Y. Pang (2009), THEMIS observation of multiple dipolarization fronts and associated wave characteristics in the near-Earth magnetotail, *Geophys. Res. Lett.*, *36*, L20107; doi:10.1029/2009GL040663.

Zong, Q.-G., B. Wilken, G. D. Reeves, I. A. Daglis, T. Doke, T. Iyemori, S. Livi, et al. (1997), Geotail observations of energetic ion species and magnetic field in plasmoid-like structures in the course of an isolated substorm event, *J. Geophys. Res.*, *102*, 11409–11428.

Zong, Q.-G., T. A. Fritz, Z. Y. Pu, S. Y. Fu, D. N. Baker, H. Zhang, A. T. Lui, et al. (2004), Cluster observations of earthward flowing plasmoid in the tail, *Geophys. Res. Lett.*, *31*, L18803; doi:10.1029/2004GL020692.

11

Recent Advances Regarding the Mars Magnetotail Current Sheet

Michael W. Liemohn[1] and Shaosui Xu[2]

ABSTRACT

The Mars induced magnetotail, including the lobes, plasma sheet, and current sheet, is reviewed and some recent studies are highlighted. In particular, three issues of the Mars magnetotail are examined for which there is apparent conflict in the published results but for which resolution exists. The Mars tail region shows some similarities to that of Venus but not only some striking differences. Reviewed here is the structure of the Mars tail, the ion composition in the tail, and the influence of the crustal fields on the tail. The tail plasma sheet at Mars not only twists in response to the rotation of the interplanetary magnetic field clock angle but also shifts depending on the solar flux and crustal field locations. At Mars, the plasma sheet and lobes are dominated by the loss of planetary ions to deep space, out to the induced magnetosphere boundary. In addition, the crustal fields have a significant impact on the structure of the magnetic field in the tail and the flow of planetary ions in the plasma sheet, with the pattern being simple when the strong crustal fields are on the dayside and very complicated when the strong crustal fields are on the nightside.

11.1. INTRODUCTION

Mars's interaction with the solar wind has been described as being very similar to that of Venus [*Luhmann et al.*, 1987, 1992; *Nagy et al.*, 1990, 2004]. That is, without the presence of a strong internally generated dipole magnetic field, the solar wind can impinge directly on the topside ionosphere of Mars, creating a relatively small obstacle in the solar wind flow [*Luhmann and Brace*, 1991]. Figure 11.1 shows the "classic view" of the Mars space environment, in the absence of strong crustal fields [*Nagy et al.*, 2004]. The subsolar bow shock location is only about a planetary radius from the surface and an induced magnetosphere of draped interplanetary magnetic field (IMF) forms a long tail behind the planet, with a thin plasma sheet between the induced magnetotail lobes. Note that it shows the ionopause extending many Mars radii downtail, implying that the planetary ions dominate the density along the draped IMF lines.

The Mars magnetotail plasma sheet carries a current. Since the first drawing of a schematic like Figure 11.1, the magnetic shear presumed to exist behind the planet specified that a current sheet should exist in this region. The exact nature of this current sheet, however, was not well known until recently. Several studies examining observations from Mars Global Surveyor (MGS), Mars Express (MEX), and more recently the Mars Atmosphere and Volatile Evolution (MAVEN) missions, in conjunction with numerical modeling from global magnetohydrodynamic (MHD), hybrid, and test particle codes have revealed many of the important properties and dynamics of the Mars tail current sheet.

Two recent reviews have covered the magnetotail of Mars. *Bertucci et al.* [2011] examined the upstream plasma interaction with Venus, Mars, and Saturn's moon Titan. While this review mostly focused on the dayside

[1] *Department of Climate and Space Sciences and Engineering, University of Michigan, Ann Arbor, Michigan, USA*
[2] *Space Sciences Laboratory, University of California, Berkeley, California, USA*

Electric Currents in Geospace and Beyond, Geophysical Monograph 235, First Edition.
Edited by Andreas Keiling, Octav Marghitu, and Michael Wheatland.

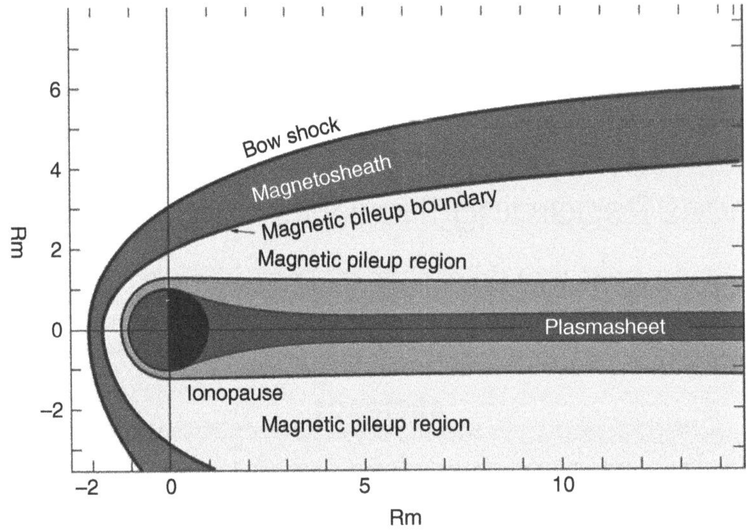

Figure 11.1 Schematic of the structure of the Mars space environment, illustrating the major plasma and magnetic boundaries and regions in the equatorial plane [*Nagy et al.*, 2004].

interaction regions with these unmagnetized celestial bodies, brief mention was included about the processes in the tail regions. *Dubinin and Frănz* [2015] is a review specifically about the magnetotail regions of Venus and Mars. While it covered the compositional aspects of the Mars magnetotail and the loss of planetary ions to deep space, the review was rather brief regarding the structure and dynamics of the magnetotail, especially that of Mars. With the arrival of MAVEN at Mars, numerous studies have been published in the last year on this topic.

This report summarizes the key findings about the Mars tail current sheet in recent years, in particular, the new understanding since the comprehensive review of *Nagy et al.* [2004] and what isn't fully covered in the other recent reviews. In particular, three specific issues of the Mars tail region are covered for which there are apparent discrepancies in the published literature. The first topic is the structure of the tail, in particular the location of the tail current sheet. The tail current sheet rotates within the y-z plane as the IMF clock angle changes orientation, with the current sheet typically aligned with the solar wind motional electric field (E_{SW}). In the E_{SW} frame, however, the tail current sheet is often not directly downtail of the planet but instead offset to the left or right. The second topic is ion composition in the tail. Planetary ions are expected to dominate in the current sheet region of the central tail, but there is some uncertainty about which plasma source, the solar wind or the Mars upper atmosphere, contributes the most to other regions of the Mars tail. The third topic is the influence of crustal fields on the Mars current sheet. Studies find different, sometimes opposing, relative contribution of the crustal fields to planetary ion escape. These studies are reviewed and explanations for the differences are postulated.

11.2. COMPARISONS WITH VENUS

Mars's interaction has been described as being similar to solar minimum conditions at Venus, when the solar wind plasma pressure exceeds the maximum thermal pressure of the planetary ionosphere most of the time [e.g., *Luhmann et al.*, 1987; *Cloutier et al.*, 1999; *Fox*, 2009]. This means that the draped IMF must magnetize the top layers of the ionosphere to increase the pressure and stand off the dynamic pressure of the oncoming solar wind. This ionospheric magnetization slows the IMF progression past the planetary obstacle, resulting in a draping of the IMF around the planet. The IMF is typically not perpendicular to the Sun-Mars line but rather at an angle, with the nominal Parker spiral orientation of the upstream IMF being 50°–60° away from radial. The draping effect and this incoming angle lead to a standard magnetic field configuration around the two unmagnetized inner planets. Figure 11.2 illustrates this phenomenon, showing the typical configuration of the IMF around Venus [*McComas et al.*, 1986]. The same regions of upstream solar wind, magnetosheath, and induced magnetosphere are seen in this schematic, as in Figure 11.1. The view is from over the North Pole, and it is seen that the upstream IMF lines are close to parallel with the bow shock normal vector on the dawn side and nearly perpendicular to the bow shock normal vector on the dusk side. Because of the relatively slow passage of the IMF around the planet, the field line, therefore, bends

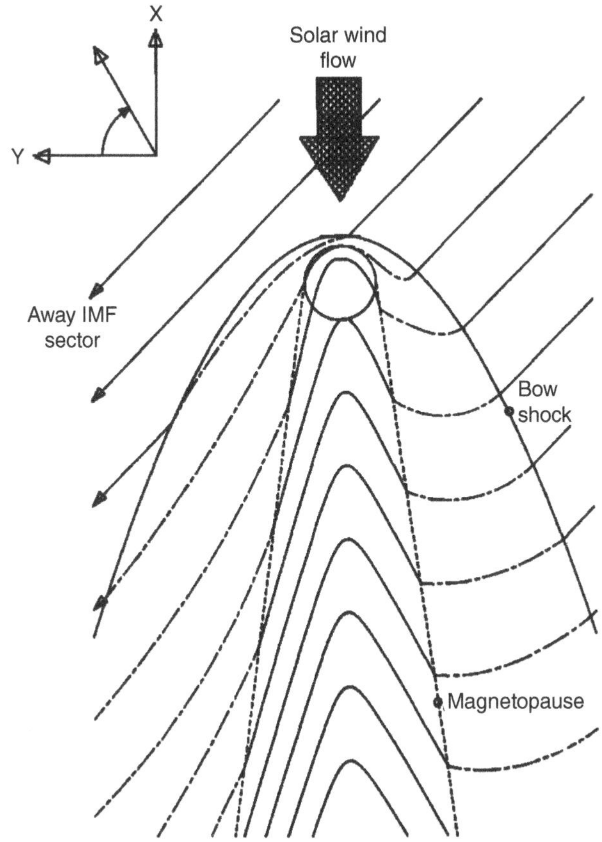

Figure 11.2 Schematic of a Parker-spiral IMF draping around an unmagnetized planet, like Mars [*McComas et al.*, 1986].

back toward the planet. Kinks in the IMF lines occur at the bow shock, at the induced magnetosphere boundary (IMB), and close to the x-z meridian beginning in the subsolar region through the central tail.

Figure 11.2 deserves some extra attention because it depicts a consequence of this standard draping scenario with regard to the current sheet. The field lines are drawn such that the magnetic field flux is represented by the spacing between the field lines. The spacing is closer inside the IMB, thus giving it the name of "induced magnetosphere." The magnetic flux is not uniform in this region but rather has two distinct values on the dawn and dusk tail lobes, with the field strength on the dusk side being larger. This results in extra magnetic pressure in the dusk side lobe relative to the dawn lobe, and a systematic shift of the magnetotail current sheet toward dawn. This was noted by *McComas et al.* [1986], who hypothesized this explanation and confirmed by MHD simulations [e.g., *Ma et al.*, 2013]. *Rong et al.* [2016] showed that the exact structure of the near-Venus tail asymmetry is actually insensitive to the value of the IMF cone angle.

Another issue is the plasma population in the tail current sheet. *McComas et al.* [1986] noted that both H^+ and O^+ were regularly observed near the Pioneer Venus Orbiter apoapse distance at 8–12 R_V, and they conducted calculations to place an upper bound on the planetary ion mass loss rate to deep space of ~6×10^{24} O^+/s, most of which was cold and flowing within the magnetotail plasma sheet. Using Venus Express data, *Barabash et al.* [2007b] found a number density composition ratio in the tail plasma sheet for planetary H^+ to O^+ of roughly two to one. This was confirmed by other studies using the Venus Express data set [e.g., *Fedorov et al.*, 2011; *Edberg et al.*, 2011]. While planetary H^+ dominates over other planetary species, *Fedorov et al.* [2011] showed that the dominant proton population in the Venus tail was from the solar wind, implying a relatively small contribution from the planet to the total ion density and/or pressure in the tail.

Is this the case at Mars, regarding both the current sheet location and the composition of the plasma sheet? This question could not be answered at the time of the *Nagy et al.* [2004] review, but it now can be answered with a review of recent studies.

11.3. MARS TAIL AND CURRENT SHEET CONFIGURATION

While data analysis is useful for a statistical picture of the Mars tail region, numerical modeling provides a complete solution for a particular solar wind input and planetary condition. One particular study to mention here is that of *Brain et al.* [2010], who presented a robust comparison of many of the numerical models that solve the Mars interaction with the solar wind. Framed as a modeling challenge, the codes were run with close to identical inputs for direct comparison of the results with each other. The paper focused on the induced magnetospheric structure in the x-z plane and on the total ion loss rate. While this shows the north-south extent of the plasma sheet, it does not show the location of the sheet, in particular this offset found by *McComas et al.* [1986] for Venus.

Xu et al. [2016] conducted an analysis of boundaries from multifluid MHD simulation results, both pressure boundaries as well as compositional boundaries. Figure 11.3 shows two of those boundaries, for the quantities of plasma β (thermal plasma pressure divided by magnetic pressure) and a related quantity which they called β^* (with the addition of plasma dynamic pressure to the numerator of the ratio). The green surface is the unity isocontour of each of these quantities, in Figures 11.3a and 11.3b, respectively, with values less than unity inside the surfaces. This defines the induced magnetosphere, and the green surfaces are one measure of the IMB. Barely visible within the outer surface is an inner surface; this is the lower boundary of the induced magnetosphere located at or near the inner boundary of the MHD simulation

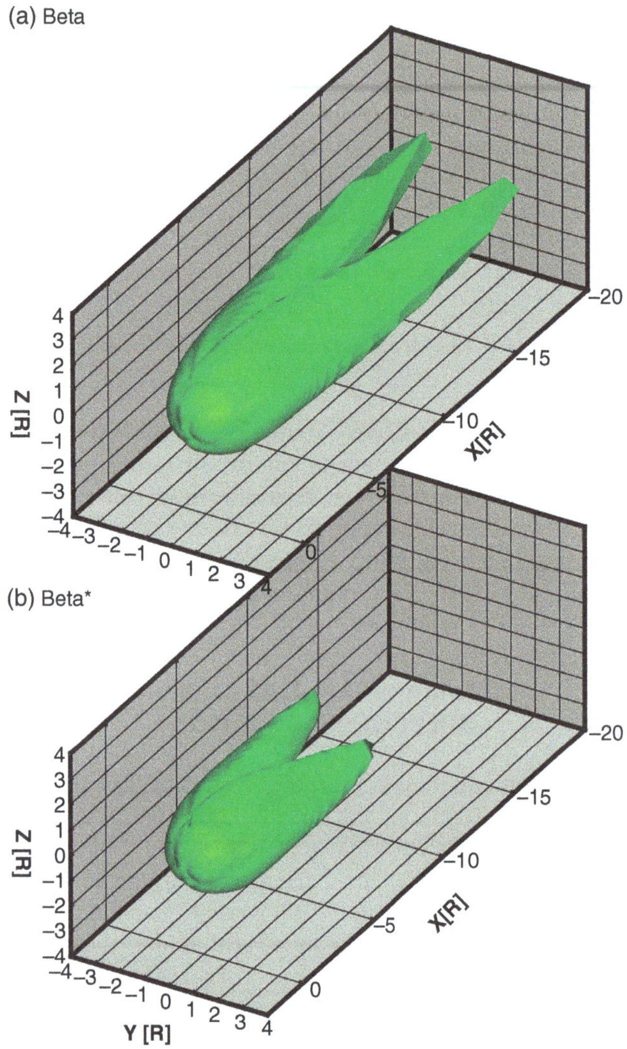

Figure 11.3 Contours of (a) plasma $\beta = 1$ and (b) plasma $\beta^* = 1$ around Mars for nominal away-sector Parker spiral IMF conditions [*Xu et al.*, 2016]. The view is from above the equatorial plane in the afternoon sector, with distances given in Mars radii.

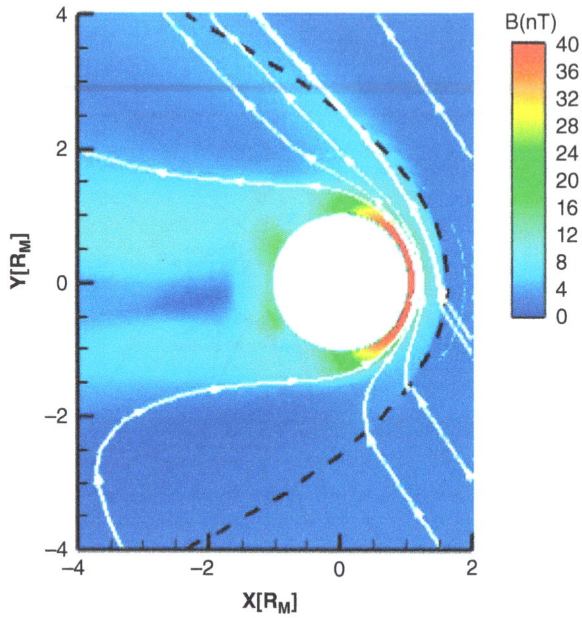

Figure 11.4 A view of the near-Mars space equatorial plane, showing magnetic field intensity (color) and direction (white traces), from a multifluid MHD simulation [*Najib et al.*, 2011]. Note that the Bz component of the magnetic field is neglected in the field traces on this plot.

domain of 100 km altitude. The plasma sheet is between the elongated isosurfaces extending into the tail.

For this simulation (solar maximum, perihelion, nominal solar wind with an away-sector Parker spiral IMF), the lobes extend at least 15–20 R_M downtail. The lobe region is bifurcated, as expected with the draped magnetic field configuration, and the plasma sheet begins as a very narrow channel near the planet and slowly widens to ~3 R_M across near the end of the lobes. These images give an overall understanding of the shape and size of the Mars magnetotail.

Another perspective of the structure of the Mars magnetotail is presented in Figure 11.4, from *Najib et al.* [2011]. This image, from a multifluid MHD simulation, shows the equatorial plane view of the magnetotail. Shown is the magnetic field intensity as well as a few selected magnetic field traces of the x-y field components in this plane. The black dashed line is an empirical location for the Mars bow shock, for reference. This simulation, for solar maximum (like those in Figure 11.3), clearly shows the relatively larger dusk-side magnetotail lobe and the dawnward shift of the magnetotail current sheet, in the same direction as that found by *McComas et al.* [1986] for Venus. Another key feature of the tail shown in this image is the S-shape of the draped IMF, as in the schematic of Figure 11.2. Remember, however, that the z-component of the magnetic field was omitted from the field line trace, so these white lines in Figure 11.4 may not be actual field lines because they do account for the out-of-plane portion of the tracing. Similar to the *Najib et al.* [2011] results, the study by *Luhmann et al.* [2015], in examining multifluid MHD results from the very near Mars tail region, showed a shift of the Mars tail current sheet to the dawn side.

A robust examination of current densities from MHD results was conducted by *Li et al.* [2013]. Figure 11.5 presents a key result of that study, showing current density and direction in the tail region. The y-z plane slice 1.1 R_M shows not only a region of ionospheric current directly behind the planet, but also a well-formed current around the two magnetotail lobes. Similarly, at 2 R_M downtail, the larger coverage of the plot includes not only the cross-tail and magnetopause currents around the lobes but also

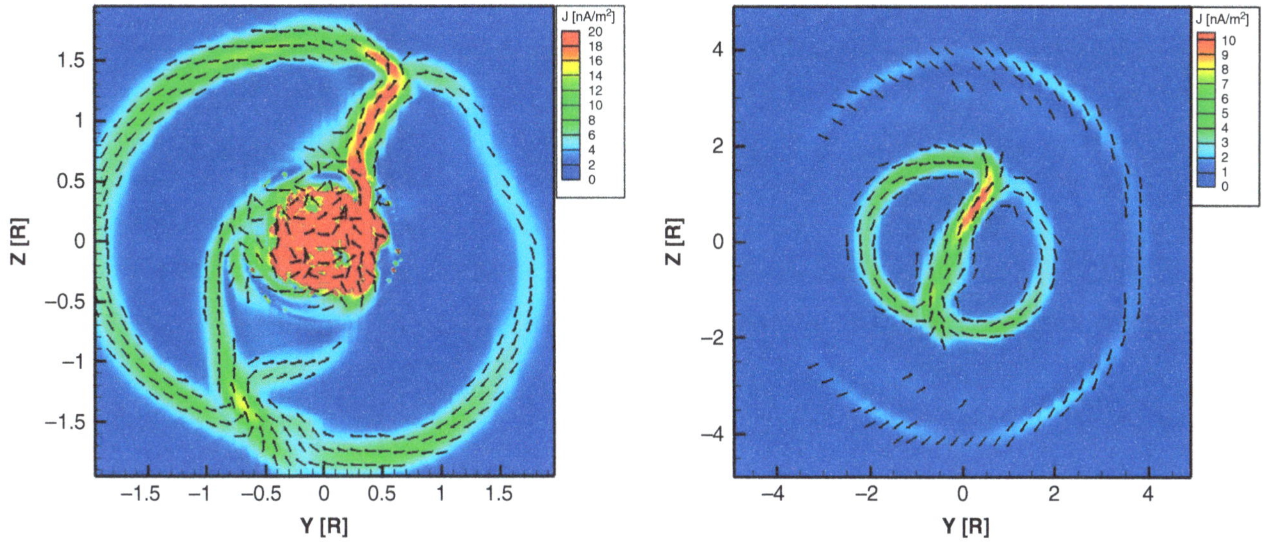

Figure 11.5 Current density amplitude in the Mars magnetotail at two different locations, (a) X = –1.1 and (b) X = –2 R_M. The color shows the magnitude and the arrows show the 2-d component direction in the y-z plane [*Li et al.*, 2013].

the bow shock currents associated with the kink in the IMF field lines at this boundary. Note that the bow shock currents flow in a similar double-loop pattern as the magnetotail currents within the IMB, closing through a + Z-oriented current in the central tail region. This closure current between the IMB and bow shock is too weak to be seen with the color scale used in Figure 11.5b. Rather than an offset, they found a clockwise (as viewed from the Sun) rotation of the tail current sheet and lobes. Note that this is not expected given the IMF having a Parker spiral configuration with no Z component.

In Figure 11.5, the currents are stronger in the northern hemisphere because this is the direction of + E_{SW} and the lack of crustal fields allows the draped IMF field lines to better scavenge this hemisphere's ionosphere more than in the southern hemisphere. Particles from the dayside are preferentially pulled into this hemisphere before making their way downstream. Therefore, the intensity of the northern hemisphere tail current sheet is stronger than the southern hemisphere. The bifurcated current sheet in the southern hemisphere in Figure 11.5a is an effect of planetary ions from the dayside that missed bombarding the planet's upper atmosphere and instead skirted around the planet.

Other studies, however, have found a different configuration to the tail. Specifically, an examination of results from other studies shows that the location of the tail current sheet is not always on the dawn side of the midnight meridian, as was found by *McComas et al.* [1986] and the other studies mentioned above. Solving a multispecies MHD equation set (different continuity equations but a single momentum and energy equation that combines all ion species), *Ma et al.* [2002] found an opposite shift, with the tail current sheet moved over to the dusk side of the tail. A similar shift was shown in the figures of *Ma et al.* [2004] and the multifluid MHD results of *Xu et al.* [2016]. An example of this duskward shift of the plasma sheet is shown in Figure 11.6, from *Xu et al.* [2016]. The black line in these panels shows the β unity isocontour location from Figure 11.3a, above, and the color background shows the derivative of the logarithm of β in these 2-d slices. As with the early studies, +E_{SW} in the upstream solar wind is directed in the + Z direction. Figure 11.6 reveals that the dawn lobe is larger than the dusk lobe in these simulation results, and the current sheet is correspondingly shifted toward the dusk. That is, several studies have found an opposite dawn-dusk asymmetry, calling into question the explanation of the shift given by *McComas et al.* [1986].

These differences in the results from these various numerical modeling studies are not because of a systematic difference in the code. All of these Mars global magnetosphere studies are examining results from the same MHD model, except with different upstream and planetary conditions. So, what controls the dawn-dusk location of the tail current sheet?

Two answers have been posited for this question. The first is the location of the crustal fields. Examining multispecies MHD results, *Luhmann et al.* [2015] showed that nearly all of the very near Mars tail was magnetically connected to the planet (or, at least, to the inner boundary of the simulation, on which the magnetic field is specified by the *Cain et al.* [2003] crustal field empirical model). Therefore, the location of the crustal fields, and in

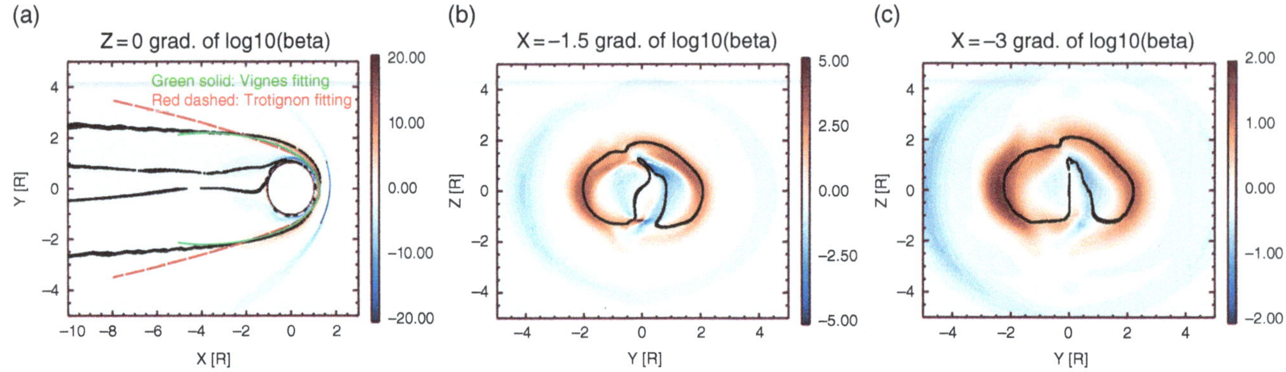

Figure 11.6 Gradient of the logarithm of β, shown for (a) Z = 0 plane, (b) X = −1.5 R_M, and (c) X = −3 R_M [*Xu et al.*, 2016]. The green and red lines in (a) show the MPB fitting from *Vignes et al.* [2000] and *Trotignon et al.* [2006], respectively.

particular the net polarity of the nightside crustal fields with respect to the IMF direction, could be controlling the shape and size of the magnetotail lobes and therefore the location of the magnetotail current sheet. For example, *Fang et al.* [2010] found that crustal fields inhibit the loss of planetary ions to deep space, and if the crustal fields are on the dayside, the loss rate drops by nearly a factor of two. This loss is largely controlled by the accessibility of the planetary ions to the open or draped magnetic fields, on which they will slowly drift downtail or be picked up by the motional electric field of the solar wind. Similarly, *Andersson and Ergun* [2012] showed that the combined effect of crustal fields and neutral winds could change the dayside equatorial ionospheric outflow of particular species by a factor of two or so throughout a Mars day. Therefore, this hypothesis that the crustal fields control the size and shape of the magnetotail lobes is very reasonable.

The second answer is given by *Liemohn et al.* [2017]. They examined multifluid MHD simulation results in the absence of crustal field sources, thereby removing this influence from the simulation results. They found that the location of the tail current sheet still shifted from the dawn side during solar maximum conditions to the dusk side during solar minimum conditions. Their explanation for this shift was either the draped IMF slipping relatively easily past the planet when the ionosphere was strong or the draped IMF getting hung up on the dayside, hindered in its slippage by the planet when the ionosphere was relatively weak.

Beyond these numerical modeling studies, observations have revealed the structure of the tail, the relative size of the magnetotail lobes, and the dawn-dusk location of the tail current sheet. Because the IMF can rotate to any clock angle, observations are often presented and discussed in terms of Mars-Solar-Electric (MSE) coordinates, in which the + Z axis points in the direction of the solar wind motional electric field. For instance, *Halekas et al.* [2006], in their survey of MGS magnetometer data, found a large number of current sheet crossings. This is perhaps unusual because the MGS orbit was gravitationally locked into a 2 am–2 pm local time configuration. Therefore, all of the nightside passes of MGS were in the postmidnight sector. Furthermore, they state that most of the current sheet observations were over regions of very weak crustal magnetic field and that the configuration is consistent with a draped IMF configuration. The implication is that, at least for a significant fraction of the data set, the dawn lobe is small and that the current sheet is shifted in that direction. Some data show the opposite result though. *DiBraccio et al.* [2015], in their analysis of a case study of MAVEN data, found a premidnight location for the current sheet. Yet another finding is inconclusive, specifically that of *Harada et al.* [2015], who conducted a survey of MAVEN magnetometer data in the Mars tail region. Figure 11.7 presents a key result from that study, showing the magnetic field components and total intensity, in MSE coordinates. Figure 11.7a contains no systematic shift of the Bx sign change. There is some hint in Figure 11.7d, the total magnetic field intensity, about a possible shift in favor of a larger dawn-side lobe than the dusk-side lobe. It should be noted, however, that this survey only included values within 1.5 R_M downtail, which is quite close to the planet. Moreover, for a nominal Parker spiral IMF, the E_{SW} direction is opposite for toward and away sector. It means that, even if there is an asymmetry, it will be averaged in MSE coordinates as dawn and dusk flip with the IMF direction and their contribution to MSE "dawn" and "dusk" are about equal.

These numerical and observational studies show that the Mars current sheet location and dynamic variability is not exactly like Venus. Other factors, especially the strong crustal fields and the weaker ionosphere, play a critical role in the structure of the Mars magnetotail.

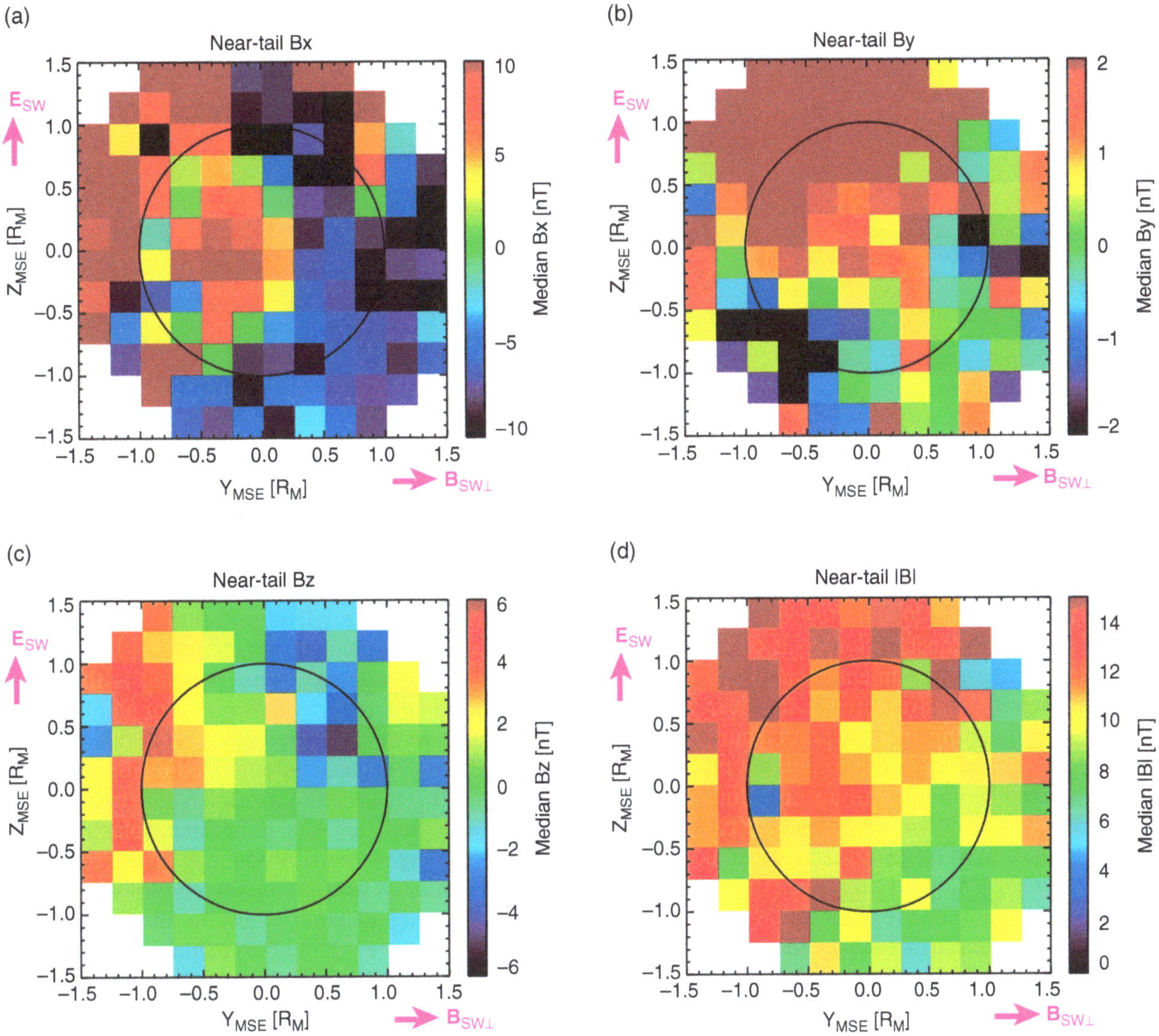

Figure 11.7 Statistics of magnetic field components and magnitude in the near-Mars tail ($-1.5\ R_M < X < -1.0\ R_M$), with each measurement rotated into MSE coordinates [*Harada et al.*, 2015].

Specifically, the dawnward shift of the tail current sheet that is seen at Venus is sometimes evident at Mars as well, but at Mars it can sometimes shift to the dusk side of the midnight meridian, especially during solar minimum conditions and when the strong crustal fields are in a preferential location relative to the IMF orientation to cause such a shift.

11.4. MARS TAIL PLASMA SHEET COMPOSITION

The ion composition of the tail of Venus is a mix of protons and heavy ions, presumably indicating a mix of solar wind and planetary contributions to the plasma sheet between the magnetotail lobes. Let's examine the evidence for Mars.

There have been a number of observational studies that reveal planetary ion dominance in the Mars magnetotail. *Barabash et al.* [2007a] presented statistics of ion fluxes in the Mars tail from MEX. Figure 11.8 shows the key set of figures from that study, planetary ions seen at 2 R_M downtail. The observations were rotated into MSE coordinates and this allows the plasma sheet to be clearly seen in these results. Not only is there O^+ moving downtail but also there are molecular ions, notably O_2^+ and CO_2^+. *Dubinin et al.* [2011] compiled a review of ion observations around Mars and Venus, and *Dubinin and Fränz* [2015] provide an updated perspective on ion composition in the tail. MAVEN has further contributed to understanding of composition in the Mars tail region [e.g., *Jakosky et al.*, 2015; *Dong et al.*, 2015b; *Harada et al.*, 2015; *Halekas*

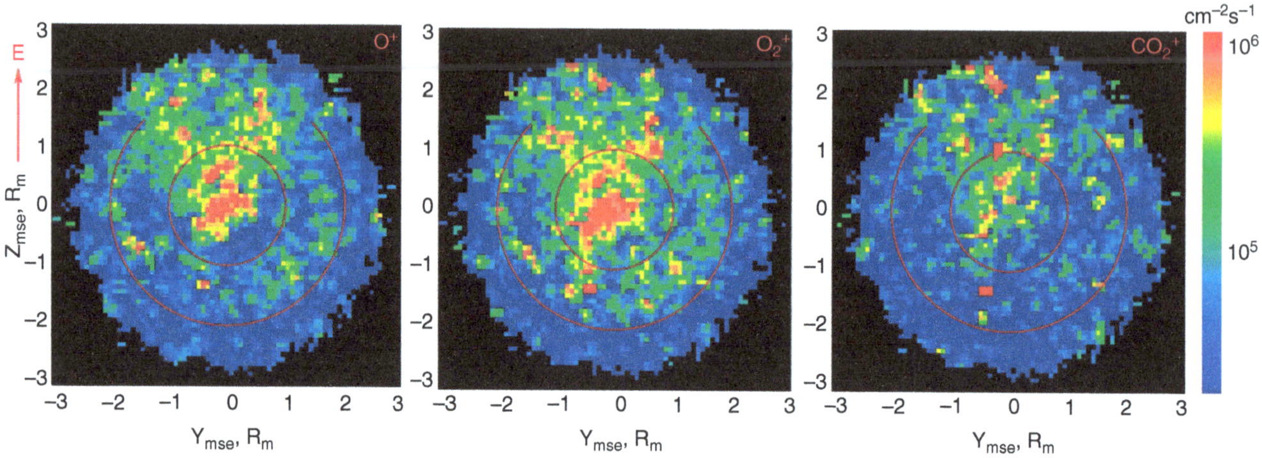

Figure 11.8 Statistics of observed ion fluxes (O^+, O_2^+, and CO_2^+, respectively) in the near-Mars tail (X = −2 R_M), organized by the estimated direction of the solar wind electric field along the +Z axis [*Barabash et al.*, 2007a].

et al., 2015]. *Dong et al.* [2015b] showed that the energetic plume of escaping planetary ions is a constant feature of the Mars space environment, *Halekas et al.* [2015] revealed dispersed ion structures in the near-Mars tail, presumably originating from a point source near the planet, and *Jakosky et al.* [2015] found that planetary ion escape increased by more than an order of magnitude during the passage of an interplanetary coronal mass ejection. Interestingly, a figure was not produced for H^+ in the *Barabash et al.* [2007a] study, even though the MEX ion mass analyzer is sensitive to H^+ and regularly sees the solar wind. The reason is because there isn't much H^+ in the tail.

Others examined this issue of H^+ in the tail. Specifically, *Nilsson et al.* [2010] conducted a robust survey of H^+ and O^+ fluxes seen by MEX. Figure 11.9 shows the results for the proton statistics from that study, split by the local time of the strong crustal field region. In both panels, though, it is clearly seen that the H^+ fluxes dramatically decrease inside of the IMB. Solar wind protons are greatly reduced within the IMB and the tail region is dominated by planetary ions. *Harada et al.* [2015] showed statistics of ion fluxes in the very-near Mars tail, revealing low fluxes for both O^+ and H^+ in the region directly behind the planet.

In support of these observations, numerical modeling routinely reveals the dominance of the planetary ions inside the IMB. Unfortunately, most do not calculate an ion density ratio. *Najib et al.* [2011] shows densities of H^+ and planetary ions in the x-z plane, showing H^+ densities dropping by a factor of several inside the tail, a place where the planetary ion densities rise. It is difficult to extract an exact ratio from these values, though. One numerical study that did this ratio calculation is *Xu et al.* [2016], who calculated the location of ion composition boundaries. They found the locations where either the number density or mass density flips from H^+ dominance to heavy ion dominance. Their finding is that the ion composition boundary, by either mass or number density, is very close to the pressure boundary as defined by β or β^*, but that the planetary-dominated region includes the plasma sheet. That is, the entire tail region inside the IMB is filled with planetary ions rather than solar wind ions.

The Mars nightside induced magnetosphere has a distinct energetic ion feature: the presence of a strong sheet of fast-beaming planetary ions in the $+E_{SW}$ direction within and "above" the tail. Planetary ions are accelerated to keV energies when they enter, or are "born" within, the magnetosheath or upstream solar wind [e.g., *Luhmann*, 1990]. Several observational studies have noted the presence of this escape channel for Mars [e.g., *McKenna-Lawlor et al.*, 1993; *Carlsson et al.*, 2008; *Dubinin et al.*, 2011; *Liemohn et al.*, 2014; *Jakosky et al.*, 2015; *Dong et al.*, 2015b], and quite a few numerical modeling studies predict that it could be up to 50% of the total ion loss from the planet [e.g., *Cravens et al.*, 2002; *Fang et al.*, 2008, 2015; *Kallio et al.*, 2010; *Modolo et al.*, 2012; *Curry et al.*, 2013; *Liemohn et al.*, 2013]. *Curry et al.* [2015], who conducted identical simulations for Venus and Mars, noted that this sheet of fast-moving planetary ions is considerably stronger at Mars than at Venus.

An interesting comparison can be made between the observational spatial distribution of planetary ions in the near-Mars tail region, seen in Figure 11.8, and an analogous numerically derived plot in Figure 11.6b. The plasma sheet seems to be wide in the $+E_{SW}$ hemisphere in the observations of Figure 11.8 and narrow in that hemisphere in the numerical model results in Figure 11.6. There are two explanations for this apparent discrepancy. The first is that the *Barabash et al.* [2007a] study uses

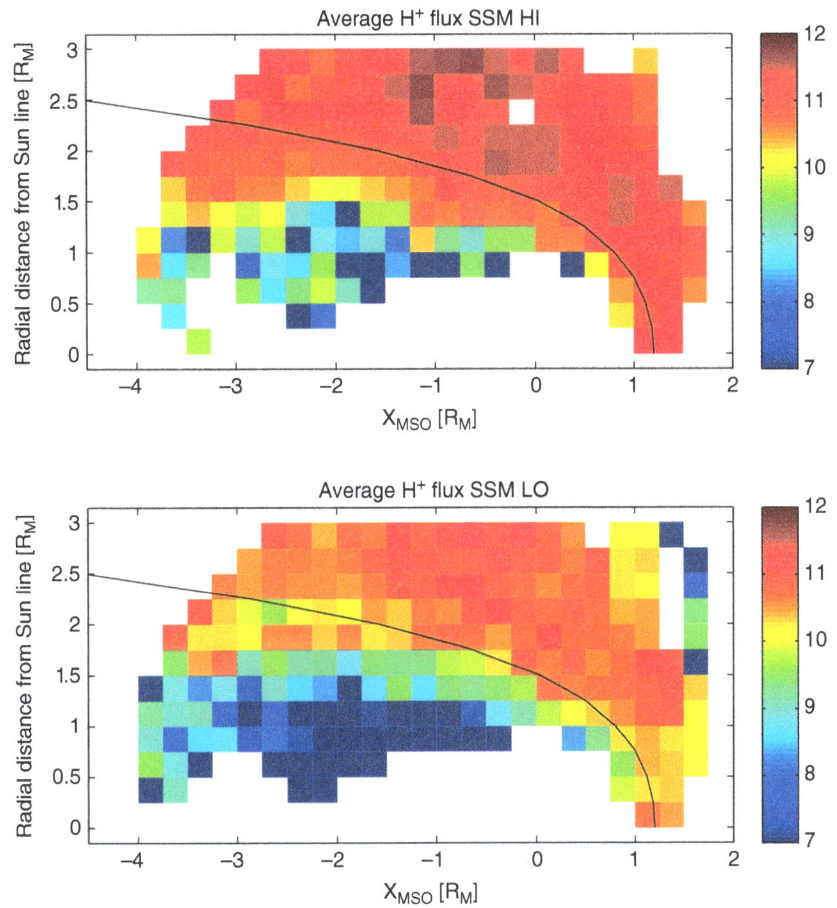

Figure 11.9 H^+ fluxes (cm^{-2} s^{-1}) in near-Mars space as seen by MEX [*Nilsson et al.*, 2010]. The top panel shows fluxes when the strong crustal field region is on the dayside and the lower panel shows fluxes when that region is on the nightside. The black line shows an empirical location for the IMB.

MGS magnetic field observations as a proxy for the IMF direction and $+E_{SW}$ determination. This proxy is only good within ~90°, and this ambiguity yields a wide plasma sheet in the MEX ion statistics. Another explanation is that the model results in Figure 11.6 are not showing ion composition but rather plasma β (and its gradient). The *Xu et al.* [2016] study showed that the planetary ions dominate throughout the tail region, out to the IMB.

Beyond the structure of the tail discussed in the previous section, this composition dominance by planetary ions is another key difference between Mars and Venus. Venus has a mix of both planetary and solar wind particles throughout the tail while Mars does not show this mix, instead having a plasma sheet and lobe region dominated by planetary ions.

11.5. INFLUENCE OF THE MARS CRUSTAL FIELDS

MGS greatly enhanced our understanding of the solar wind-Mars interaction by quantifying the intense crustal magnetic fields, both from direct magnetic measurements [*Acuña et al.*, 1998, 1999; *Connerney et al.*, 1999, 2001, 2005] and from electron reflectometry [*Mitchell et al.*, 2001, 2007; *Lillis et al.*, 2004, 2008; *Brain, et al.*, 2006, 2007]. Alternating east-west oriented magnetic features of strong field source have been identified in the southern hemisphere, as seen in Figure 11.4 [*Connerney et al.*, 2001]. These fields create a complex interaction between Mars and the solar wind. MHD results reveal a lumpy magnetic field topology over the strong crustal fields [e.g., *Ma et al.*, 2004; *Najib et al.*, 2011; *Dong et al.*, 2015a]. The crustal fields do not have a big effect on the bow shock location, but they have a large influence on the magnetic pile-up region and the sheath-ionosphere boundary across the dayside. The crustal fields at Mars are at least 100 times stronger than those on the Moon [cf., *Lin et al.*, 1998] and are 10 times stronger than those at Earth [cf., *Langel et al.*, 1982]. Several good models have been developed to quantify the crustal magnetic field of Mars [e.g., *Purucker et al.*, 2000; *Arkani-Hamed*, 2001, 2002; *Cain et al.*, 2003; *Morschhauser et al.*, 2014].

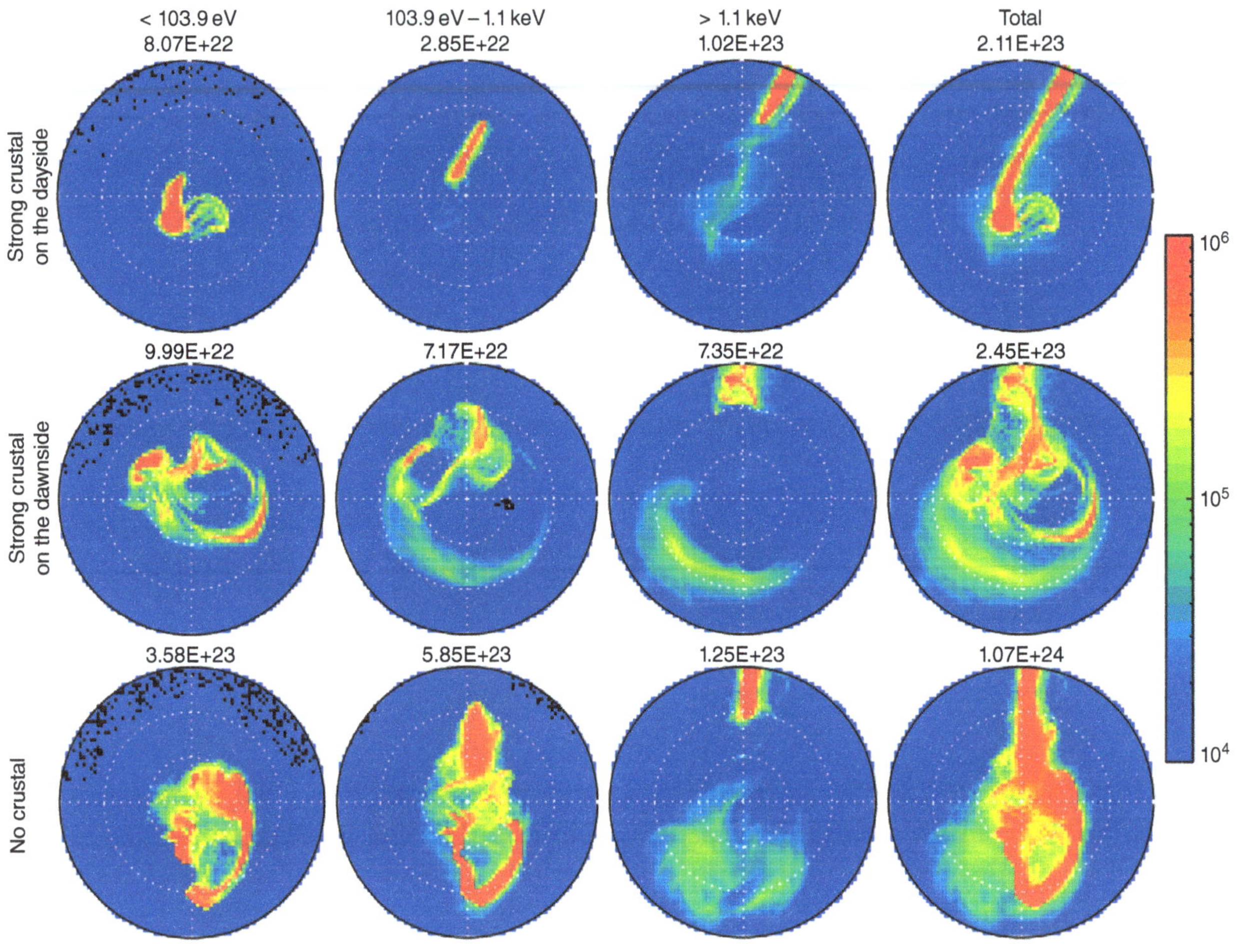

Figure 11.10 O^+ fluxes in the y-z plane (out to a radius of 3 R_M) at 4 R_M in the Mars tail from test particle tracing through MHD fields, partitioned by particle energy (columns) for three simulations: the strong crustal field region on the dayside (top row), on the nightside (middle row), and without crustal fields in the MHD model results (lower row) [*Fang et al.*, 2010].

Many studies have examined the influence of the strong crustal fields on the tail, including the reviews and studies mentioned above. Several recent numerical studies examining the influence of crustal fields on total ion loss from the planet are the hybrid code analyses of *Brecht and Ledvina* [2012, 2014] and *Brecht et al.* [2016] and the MHD assessment by *Ma et al.* [2014]. All find that locating the strong crustal fields on the dayside lowers the total ion loss rate from Mars. This is in contrast to some observational studies, however, like *Nilsson et al.* [2010], who found that the O^+ fluxes in the tail were increased when the strong crustal fields were on the dayside.

The answer to this conundrum of whether the crustal fields increase or decrease planetary ion loss is found in another study, *Fang et al.* [2010], which presented ion fluxes in the tail for three simulations of O^+ test particle traces through MHD fields. Figure 11.10 shows a key set of figures from these simulations, presenting the ion fluxes at 4 R_M downtail partitioned by their energy and the total ion fluxes in the rightmost column. The top row shows results with the strong crustal fields on the dayside. The ions are confined to a well-structured and narrow plasma sheet, with just a little structure of the lowest-energy particles in the central tail. The middle row shows results for the case when the strong crustal field region is on the nightside. The tail plasma sheet loses coherency, developing dramatic structure in the location of the high ion fluxes. The lower row shows the test particle results with no crustal fields included in the MHD simulation. The structure seen in the -z hemisphere is a natural feature of how the planetary ions are accelerated in the magnetosheath.

The difference between the three simulation results in Figure 11.10 is stark. The crustal field local time has a dramatic influence on the magnetic and electric field structure of the Mars space environment, and therefore a substantial difference in the flow of planetary ions through the magnetotail. When the strong crustal field region is on the dayside, it inhibits planetary ion loss, as seen in the difference between the -z hemispheres in the top and bottom rows of Figure 11.9. When on the dayside, it does not perturb the structure of the lobes and leaves the plasma in a thin sheet. When the strong crustal field region is on the nightside, the tail region is highly perturbed.

The difference seen in top and middle rows can explain the difference between the MHD and hybrid results and the observations. Specifically, when the strong crustal fields are on the dayside, the planetary ions form a very high flux region down the middle of the tail. When the strong crustal fields are on the nightside, the total loss might be higher but the fluxes are spread across a much larger area, lowering the fluxes at any one point. So, one explanation for the mode-data discrepancy is that the *Nilsson et al.* [2010] study, in selecting and averaging MEX observations, preferentially removed the low O^+ fluxes in the lobes, because they state that they omitted from the O^+ statistics those times when the H^+ flux was above a threshold value of $10^9 cm^{-2} s^{-1}$ (the cyan-green color change in Fig. 11.9). This would have the effect of increasing the average observed O^+ flux values inside the IMB when the strong crustal fields were on the dayside.

11.6. SUMMARY

This chapter reviewed some of the key features of the Mars magnetotail, defined as the lobes, plasma sheet, and current sheet, including its structure, ion composition, and the influence of crustal fields. Not only does the IMF twist the plasma sheet to align with the solar wind motional electric field, but also the current sheet at Mars moves depending on the solar flux and crustal field locations. The plasma sheet and lobes are dominated by the loss of planetary ions to deep space, with densities larger than the solar wind at least out to the IMB. In addition, the crustal fields have a significant impact on the structure of the magnetic field in the tail and the flow of planetary ions in the plasma sheet, with the pattern being simple when the strong crustal fields are on the dayside and very complicated when the strong crustal fields are on the nightside.

The Mars-induced magnetotail shows some similarities to that of Venus and also some striking differences. One difference is the dawn-dusk motion of the plasma sheet. At Venus, this motion is small, with the tail current sheet typically shifted toward the dawn side. At Mars, the current sheet shifts to the dusk side during solar minimum and with the relative location of the strong crustal fields. Another difference is the ion composition of the tail. At Venus, the ions exhibit a mixture of solar and planetary sources, while at Mars the planetary ions dominate the composition of the tail.

The Mars tail is an interesting current system in the solar system and offers a unique perspective on plasma physics. It is a large-scale natural laboratory, and the Sun regularly conducts experiments on the Mars space environment. Slowly but surely, we are learning about it. This review summarized some of the new advances in our understanding of the Mars tail region, in particular the tail current sheet, from both observational and numerical modeling studies since the last round of such papers [e.g., *Nagy et al.*, 2004; *Bertucci et al.*, 2011; *Dubinin and Fränz*, 2015]. Both of these approaches, however, are limited. There are very few spacecraft orbiting Mars right now, each with their own orbital characteristics and instrument complements. Specifically, MEX does not have a magnetometer and MAVEN's apoapsis is ~6000 km altitude, which is within 2 R_M of the surface and therefore does not sample particularly far downtail. Full knowledge of the Mars space environment is difficult to extract from one or two measurement locations, and therefore ambiguities exist. Similarly, numerical models each have assumptions built into the adopted equation sets, applied numerical techniques, and imposed initial and boundary conditions. For example, MHD models assume a drifting Maxwellian velocity space distribution, while hybrid models have an inherent trade-off between grid resolution and counting statistics accuracy. Even with these limitations, significant advancements have been achieved in recent years, as highlighted above.

As for future studies of the Mars tail region, the combination of MAVEN and MEX at Mars offers the chance for tail observations by one satellite with the other spacecraft in the upstream solar wind. Both spacecraft precess around Mars and so this should occur several times for year, with each interval lasting weeks. Such an orbit configuration should definitively answer questions about solar wind control of the magnetotail structure and dynamics.

ACKNOWLEDGMENTS

The authors thank NASA for supporting this work, particularly under grants NNX16AQ04G, NNX13AF26G, and NNX11AD80G and a contract from the Mars Scout program. No new data or model results were used in this paper; all figures, tables, and values are from referenced sources. The authors thank the two anonymous reviewers for helpful suggestions on improving the clarity and comprehensiveness of the presentation.

REFERENCES

Acuña, M. H., et al. (1998), Magnetic field and plasma observations at Mars: Initial results of the Mars Global Surveyor Mission, *Science*, *279*, 1676.

Acuña, M. H., et al. (1999), Global distribution of crustal magnetization discovered by the Mars Global Surveyor MAG/ER experiment, *Science*, *284*, 790.

Andersson, L., and R. E. Ergun (2012), Neutral wind effects on ion outflow at Mars, *Earth Planet Space*, *64*, 105–112; doi: 10.5047/eps.2011.06.047.

Arkani-Hamed, J. (2001), A 50-degree spherical harmonic model of the magnetic field of Mars, *J. Geophys. Res.*, *106*, 23,197.

Arkani-Hamed, J. (2002), An improved 50-degree spherical harmonic model of the magnetic field of Mars derived from both high-latitude and low-latitude data, *J. Geophys. Res.*, *107*(E10), 5083; doi: 10.1029/2001JE001835.

Barabash, S., A. Fedorov, R. Lundin, and J.-A. Sauvaud (2007a), Martian atmospheric erosion rates, *Science*, *315*, 501–503; doi:10.1126/science.1134358.

Barabash, S., et al. (2007b), The loss of ions from Venus through the plasma wake, *Nature*, *450*, 650–653; doi: 10.1038/nature06434.

Bertucci, C., F. Duru, N. Edberg, M. Fraenz, C. Martinecz, K. Szego, and O. Vaisberg (2011), The induced magnetospheres of Mars, Venus, and Titan, *Space Sci. Rev.*, *162*, 113–171; doi: 10.1007/s11214-011-9845-1.

Brain, D., et al. (2010), First results from the SWIM model challenge, *Icarus*, *206*, 139; doi:10.1016/j.icarus.2009.06.030.

Brain, D. A., D. L. Mitchell, and J. S. Halekas (2006), The magnetic field draping direction at Mars from April 1999 through August 2004, *Icarus*, *182*(2), 464–473.

Brain, D. A., R. J. Lillis, D. L. Mitchell, J. S. Halekas, and R. P. Lin (2007), Electron pitch angle distributions as indicators of magnetic field topology near Mars, *J. Geophys. Res. 112*, A09201; doi:10.1029/2007JA012435.

Brecht, S. H., and S. A. Ledvina (2012), Control of ion loss from Mars during solar minimum, *Earth Planet Space*, *64*(12); doi:10.5047/eps.2011.05.037.

Brecht, S. H., and S. A. Ledvina (2014), The role of the Martian crustal magnetic fields in controlling ionospheric loss, *Geophys. Res. Lett.*, *41*, 5340–5346; doi:10.1002/2014GL060841.

Brecht, S. H., S. A. Ledvina, and S. W. Bougher (2016), Ionospheric loss from Mars as predicted by hybrid particle simulations, *J. Geophys. Res. Space Physics*, *121*; doi:10.1002/2016JA022548.

Cain, J. C., B. B. Ferguson, and D. Mozzoni (2003), An n = 90 internal potential function of the Martian crustal field, *J. Geophys. Res.*, *108*(E2), 5008; doi: 10.1029/2002JE001487.

Carlsson, E., D. Brain, J. Luhmann, S. Barabash, A. Grigoriev, H. Nilsson, and R. Lundin (2008), Influence of IMF draping direction and crustal magnetic field location on Martian ion beams, *Planet. Space Sci.*, *56*, 861–867; doi: 10.1016/j.pss.2007.12.016.

Cloutier, P. A., et al. (1999), Venus-like interaction of the solar wind with Mars, *Geophys. Res. Lett.*, *26*, 2685.

Connerney, J. E. P., M. H. Acuña, P. J. Wasilewski, G. Kleteschka, N. F. Ness, H. Reme, R. P. Lin, and D. L. Mitchell (2001), The global magnetic field of Mars and implications for crustal evolution, *Geophys. Res. Lett.*, *28*, 4015.

Connerney, J. E. P., M. H. Acuña, N. F. Ness, G. Kletetschka, D. L. Mitchell, R. P. Lin, and H. Reme (2005), Tectonic implications of Mars crustal magnetism, *Proc. Natl. Acad. Sci.*, *102*(42), 14,970–14,975.

Connerney, J. E. P., M. H. Acuña, P. J. Wasilewski, N. F. Ness, H. Reme, C. Mazelle, D. Vignes, R. P. Lin, D. L. Mitchell, and P. A. Cloutier (1999), Magnetic lineations in the ancient crust of Mars, *Science*, *284*, 794.

Cravens, T. E., A. Hoppe, S. A. Ledvina, and S. McKenna-Lawlor (2002), Pickup ions near Mars associated with escaping oxygen atoms, *J. Geophys. Res.*, *107*(A8), 1170; doi:10.1029/2001JA000125.

Curry, S. M., J. Luhmann, Y. Ma, M. Liemohn, C. Dong, and T. Hara (2015), Comparative pick-up ion distributions at Mars and Venus: Consequences for atmospheric deposition and escape, *Planet. Space Sci.*, *115*, 35–47; doi: 10.1016/j.pss.2015.03.026.

Curry, S. M., M. W. Liemohn, X. Fang, Y. Ma, A. F. Nagy, and J. Espley (2013), The influence of production mechanisms on pickup ion loss at Mars, *J. Geophys. Res. Space Physics*, *118*; doi: 10.1029/2012JA017665.

DiBraccio, G. A., et al. (2015), Magnetotail dynamics at Mars: Initial MAVEN observations, *Geophys. Res. Lett.*, *42*, 8828–8837; doi:10.1002/2015GL065248.

Dong, C., S. W. Bougher, Y. Ma, G. Toth, Y. Lee, A. F. Nagy, V. Tenishev, D. J. Pawlowski, M. R. Combi, and D. Najib (2015a), Solar wind interaction with the Martian upper atmosphere: Crustal field orientation, solar cycle, and seasonal variations, *J. Geophys. Res. Space Physics*, *120*, 7857–7872; doi:10.1002/2015JA020990.

Dong, Y., X. Fang, D. A. Brain, J. P. McFadden, J. S. Halekas, J. E. Connerney, S. M. Curry, Y. Harada, J. G. Luhmann, and B. M. Jakosky (2015b), Strong plume fluxes at Mars observed by MAVEN: An important planetary ion escape channel, *Geophys. Res. Lett.*, *42*, 8942–8950; doi:10.1002/2015GL065346.

Dubinin, E., and M. Fränz (2015), Magnetotails of Mars and Venus, in *Magnetotails in the Solar System, Geophysical Monograph Series*, *207*, 43–59, American Geophysical Union, John Wiley and Sons.

Dubinin, E., M. Fraenz, A. Fedorov, R. Lundin, N. Edberg, F. Duru, and O. Vaisberg (2011), Ion energization and escape on Mars and Venus, *Space Sci. Rev.*, *162*, 173–211; doi: 10.1007/s11214-011-9831-7.

Edberg, N. J. T., et al. (2011), Atmospheric erosion of Venus during stormy space weather, *J. Geophys. Res.*, *116*, A09308; *doi:*10.1029/2011JA016749.

Fang, X., M. W. Liemohn, A. F. Nagy, J. G. Luhmann, and Y. Ma (2010), On the effect of the Martian crustal magnetic field on atmospheric erosion, *Icarus*, *206*, 130; doi: 10.1016/j.icarus.2009.01.012.

Fang, X., M. W. Liemohn, A. F. Nagy, Y. Ma, D. L. De Zeeuw, J. U. Kozyra, and T. Zurbuchen (2008), Pickup oxygen ion distribution around Mars, *J. Geophys. Res.*, *113*, A02210; doi: 10.1029/2007JA012736.

Fang, X., Y. Ma, D. Brain, Y. Dong, and R. Lillis (2015), Control of Mars global atmospheric loss by the continuous rotation of the crustal magnetic field: A time-dependent MHD study, *J. Geophys. Res. Space Physics*, *120*, 10,926–10,944; doi:10.1002/2015JA021605.

Fedorov, A., S. Barabash, J.-A. Sauvaud, Y. Futaana, T. L. Zhang, R. Lundin, and C. Ferrier (2011), Measurements of the ion escape rates from Venus for solar minimum, *J. Geophys. Res.*, *116*, A07220; *doi:*10.1029/2011JA016427.

Fox, J. L. (2009), Morphology of the dayside ionosphere of Mars: Implications for ion outflows, *J. Geophys. Res.*, *114*, E12005; doi:10.1029/2009JE003432.

Halekas, J. S., D. A. Brain, R. J. Lillis, M. O. Fillingim, D. L. Mitchell, and R. P. Lin (2006), Current sheets at low altitudes in the Martian magnetotail, *Geophys. Res. Lett.*, *33*, L13101; doi:10.1029/2006GL026229.

Halekas, J. S., J. P. McFadden, J. E. P. Connerney, J. R. Espley, D. A. Brain, D. L. Mitchell, D. E. Larson, et al. (2015), Time-dispersed ion signatures observed in the Martian magnetosphere by MAVEN, *Geophys. Res. Lett.*, *42*, 8910–8916; doi:10.1002/2015GL064781.

Harada, Y., et al. (2015), Marsward and tailward ions in the near-Mars magnetotail: MAVEN observations, *Geophys. Res. Lett.*, *42*, 8925–8932; doi:10.1002/2015GL065005.

Jakosky, B. M., et al. (2015), MAVEN observations of the response of Mars to an interplanetary coronal mass ejection, *Science*, *350* (6261), aad0210.

Kallio, E., K. Liu, R. Jarvinen, V. Pohjola, and P. Janhunen (2010), Oxygen ion escape at Mars in a hybrid model: High energy and low energy ions, *Icarus*, *206*(1), 152–163.

Langel, R. A., C. C. Schnetzler, J. D. Phillips, and R. J. Horner (1982), Initial vector magnetic anomaly map from MAGSAT, *Geophys. Res. Lett.*, *9*, 273.

Li, L., L. Xie, Y. Zhang, T. Liu (2013), Model investigation of current system and influence of the crustal fields on the large scale structure of current sheets at Mars, *Planet. Space Sci.*, *86*, 80–85; doi: 10.1016/j.pss.2013.07.004.

Liemohn, M. W., B. C. Johnson, M. Fraenz, and S. Barabash (2014), Mars Express observations of high altitude planetary ion beams and their relation to the "energetic plume" loss channel, *J. Geophys. Res. Space Physics*, *119*, 9702–9713; doi: 10.1002/2014JA019994.

Liemohn, M. W., S. M. Curry, X. Fang, and Y. Ma (2013), Comparison of high-altitude production and ionospheric outflow contributions to O^+ loss at Mars, *J. Geophys. Res. Space Physics*, *118*, 4093–4107; doi: 10.1002/jgra.50388.

Liemohn, M. W., S. Xu, C. Dong, S. W. Bougher, B. C. Johnson, R. Ilie and D. L. De Zeeuw (2017), Ionospheric control of the dawn-dusk asymmetry of the Mars magnetotail current sheet, *J. Geophys. Res. Space Physics*, *122*, 6397–6414; doi: 10.1002/2016JA023707.

Lillis, R. J., D. L. Mitchell, R. P. Lin, and M. H. Acuña (2008), Electron reflectometry in the Martian atmosphere, *Icarus*, *194*, 544–561; doi:10.1016/j.icarus.2007. 09.030.

Lillis, R. J., D. L. Mitchell, R. P. Lin, J. E. P. Connerney, and M. H. Acuña (2004), Mapping crustal magnetic fields at Mars using electron reflectometry, *Geophys. Res. Lett.*, *31*, L15702; doi: 10.1029/2004GL020189.

Lin, R. P., D. L. Mitchell, D. W. Curtis, K. A. Anderson, C. W. Carlson, J. McFadden, M. H. Acuña, L. L. Hood, and A. Binder (1998), Lunar surface magnetic fields and their interaction with the solar wind: Results from Lunar Prospector, *Science*, *281*, 1480.

Luhmann, J. G. (1990), A model of the ion wake of Mars, *Geophys. Res. Lett.*, *17*, 869–872.

Luhmann, J. G., and L. H. Brace (1991), Near-Mars space, *Rev. Geophys.*, *29*, 121.

Luhmann, J. G., C. T. Russell, F. L. Scarf, L. H. Brace, and W. C. Knudsen (1987), Characteristics of the Mars-like limit of the Venus-solar wind interaction, *J. Geophys. Res.*, *92*, 8545.

Luhmann, J. G., C. T. Russell, L. H. Brace, and O. L. Vaisberg (1992), The intrinsic magnetic field and solar wind interaction of Mars, in *Mars*, edited by H. H. Kieffer, B. M. Jakosky, C. W. Snyder, and M. S. Matthews, University of Arizona Press, Tuscon, 1054.

Luhmann, J. G., et al. (2015), Implications of MAVEN Mars near-wake measurements and models, *Geophys. Res. Lett.*, *42*, 9087–9094; doi:10.1002/2015GL066122.

Ma, Y., A. F. Nagy, I. V. Sokolov, and K. C. Hansen (2004), Three-dimensional, multispecies, high spatial resolution MHD studies of the solar wind interaction with Mars, *J. Geophys. Res.*, *109*, A07211; doi: 10.1029/2003JA010367.

Ma, Y., A. F. Nagy, K. C. Hansen, D. L. DeZeeuw, and T. I. Gombosi (2002), Three-dimensional multispecies MHD studies of the solar wind interaction with Mars in the presence of crustal fields, *J. Geophys. Res.*, *107*(A10), 1282; doi: 10.1029/2002JA009293.

Ma, Y., X. Fang, C. T. Russell, A. F. Nagy, G. Toth, J. G. Luhmann, D. A. Brain, and C. Dong (2014), Effects of crustal field rotation on the solar wind plasma interaction with Mars, *Geophys. Res. Lett.*, *41*, 6563–6569; doi:10.1002/2014GL060785.

Ma, Y. J., A. F. Nagy, C. T. Russell, R. J. Strangeway, H. Y. Wei, and G. Toth (2013), A global multispecies single-fluid MHD study of the plasma interaction around Venus, *J. Geophys. Res. Space Physics*, *118*, 321–330; doi: 10.1029/2012JA018265.

McComas, D. J., H. E. Spence, C. T. Russell, and M. A. Saunders (1986), Magnetic field draping and consistent plasma properties of the Venus magnetotail, *J. Geophys. Res.*, *91*, 7939–7953.

McKenna-Lawlor, S. M. P., V. Afonin, Y. Yeroshenko, E. Keppler, E. Kirsch, and K. Schwingenschuh (1993), First identification in energetic particles of characteristic plasma boundaries at Mars and an account of various energetic particle populations close to the planet, *Planet. Space Sci.*, *41*, 373–380.

Mitchell, D. L., R. J. Lillis, R. P. Lin, J. E. P. Connerney, and M. H. Acuña (2007), A global map of Mars' crustal magnetic field based on electron reflectometry, *J. Geophys. Res.*, *112*, E01002, doi:10.1029/2005JE002564.

Mitchell, D. L., R. P. Lin, C. Mazelle, H. Re`me, P. A. Cloutier, J. E. P. Connerney, M. H. Acuña, and N. F. Ness (2001), Probing Mars' crustal magnetic field and ionosphere with the MGS Electron Reflectometer, *J. Geophys. Res.*, *106*, 23,419–23,428.

Modolo, R., G. M. Chanteur, and E. Dubinin (2012), Dynamic Martian magnetosphere: Transient twist induced by a rotation of the IMF, *Geophys. Res. Lett.*, *39*, L01106; doi:10.1029/2011GL049895.

Morschhauser, A., V. Lesur, and M. Grott (2014), A spherical harmonic model of the lithospheric magnetic field of Mars, *J. Geophys. Res. Planets*, *119*, 1162–1188; doi:10.1002/2013JE004555.

Nagy, A., et al. (2004), The plasma environment of Mars, in *Mars' Magnetism and Its Interaction with the Solar Wind*, edited by D. Winterhalter, M. Acuna, and A. Zakharov, 33–114, Springer, Netherlands.

Nagy, A. F., T. I. Gombosi, K. Szego, R. Z. Sagdeev, V. D. Shapiro, and V. I. Shevchenko (1990), Venus mantle-Mars planetosphere: What are the similarities and differences?, *Geophys. Res. Lett.*, *17*, 865.

Najib, D., A. F. Nagy, G. Tóth, and Y. Ma (2011), Three-dimensional, multifluid, high spatial resolution MHD model studies of the solar wind interaction with Mars, *J. Geophys. Res.*, *116*, A05204; doi:10.1029/2010JA016272.

Nilsson, H., E. Carlsson, D. A. Brain, M. Yamauchi, M. Holmstrom, S. Barabash, R. Lundin, and Y. Futaana (2010), Ion escape from Mars as a function of solar wind conditions: A statistical study, *Icarus*, *206*, 40–49; doi: 10.1016/j.icarus.2009.03.006.

Purucker, M., D. Ravat, H. Frey, C. Voorhies, T. Sabaka, and M. Acuña (2000), An altitude-normalized magnetic map of Mars and its interpretation, *Geophys. Res. Lett.*, *27*, 2449.

Rong, Z. J., G. Stenberg, Y. Wei, L. H. Chai, Y. Futaana, S. Barabash, W. X. Wan, and C. Shen (2016), Is the flow-aligned component of IMF really able to impact the magnetic field structure of Venusian magnetotail?, *J. Geophys. Res. Space Physics*, *121*, 10,978–10,993; doi:10.1002/2016JA022413.

Trotignon, J., C. Mazelle, C. Bertucci, and M. Acuña (2006), Martian shock and magnetic pile-up boundary positions and shapes determined from the Phobos 2 and Mars Global Surveyor data sets, *Planet. Space Sci.*, *54*(4), 357–369; doi:10.1016/j.pss.2006.01.003.

Vignes, D., C. Mazelle, H. Rme, M. H. Acuña, J. E. P. Connerney, R. P. Lin, D. L. Mitchell, P. Cloutier, D. H. Crider, and N. F. Ness (2000), The solar wind interaction with Mars: Locations and shapes of the bow shock and the magnetic pile-up boundary from the observations of the MAG/ER experiment onboard Mars Global Surveyor, *Geophys. Res. Lett.*, *27*, 49–52; doi:10.1029/1999GL010703.

Xu, S., M. W. Liemohn, C. Dong, D. L. Mitchell, and S. M. Bougher (2016), Pressure and ion composition boundaries at Mars, *J. Geophys. Res. Space Physics*, *121*, 6417–6429; doi: 10.1002/2016JA022644.

12

Current Sheets at the Giant Planets

Christopher S. Arridge and Carley J. Martin

ABSTRACT

In this review we examine the equatorial current sheets of the giant planets, studying their equilibrium structure, stress balance, geometry, and time-dependent motion in response to the solar wind and coupling to the planet. These equatorial current sheets exhibit a wide range of dynamic behaviors, which include magnetic reconnection, and temporal variations in the strength of the current sheet and associated changes in stress balance. Current sheets also appear to be in constant motion and these small-scale motions have been interpreted as surface modes on the current sheet.

12.1. INTRODUCTION

From an observational perspective, many current systems in space plasma environments take the form of a sheet, where the volume currents are confined to a thin surface. Current sheets are also theoretically and empirically convenient. It is also useful to differentiate between a neutral current sheet (or neutral sheet), where there is no magnetic flux crossing the plane of the current sheet, and a current sheet, where there may be flux crossing the sheet.

The structure of current systems in Earth's magnetosphere has been studied since the start of the space age. The magnetotail or cross-tail current sheet was surveyed in detail in the mid-1960s and shown to have asymmetries and a nonplanar structure. The geometry of this thin current sheet contrasts clearly with the more toroidal ring current in Earth's inner magnetosphere as a consequence of the magnitude of the pressure gradient stresses associated with the ring current compared to the $\mathbf{j} \times \mathbf{B}$ force in the ring current. For a recent review of the Earth's magnetotail, see *McPherron* [2015] and *Khurana* (Chapter 2, this volume).

Department of Physics, Lancaster University, Lancaster, UK

Russell and Brody [1967] were the first to remark that the position of the current sheet in Earth's magnetotail was dependent on the orientation of Earth's dipole. If no other effects were present then the current sheet would be a planar structure. However, numerous studies have examined how the tail current sheet is deformed by the dynamic pressure of the solar wind forcing the tail current sheet to asymptotically become parallel to the solar wind flow, and the orientation of the interplanetary magnetic field which rotates the cross-tail current sheet around the Earth-Sun line [e.g., *Tsyganenko et al.*, 1998].

The flyby of Pioneers 10 and 11 with Jupiter in 1973 and 1974 revealed that such a thin near-equatorial current sheet also extended around to the noon and dawn sectors of the Jovian magnetosphere [*Smith et al.*, 1974], similar to that described by *Piddington* [1969] as a wrapped-up magnetotail, and now referred to as the magnetodisc. Figure 12.1 shows a wire-frame model of the magnetodisc and Jupiter's magnetosphere as revealed to the public at a Pioneer 10/11 press conference in 1973. In the late 1970s, both Voyager 1 and 2 made observations of Jupiter's magnetic field and confirmed this very different magnetic field configuration to Earth [*Connerney et al.*, 1982]. Since then, Galileo took magnetic field measurements at the equator of Jupiter from 1995 to 2003 thoroughly

Electric Currents in Geospace and Beyond, Geophysical Monograph 235, First Edition.
Edited by Andreas Keiling, Octav Marghitu, and Michael Wheatland.

Figure 12.1 John Simpson and James Van Allen illustrating the configuration of the Jovian magnetosphere and its equatorial current sheet using a wire model during the Pioneer 10/11 news conference, 27 January 1973. Credit: NASA.

mapping the current sheet as a washer-shaped disk [e.g., *Khurana and Schwarzl*, 2005].

This is one of the main geometrical differences: the current sheet distorts the field into a disk-like configuration over a wide range of local times. Other differences include the presence of both azimuthal and radial currents in the sheet and the source of the mechanical stresses that give rise to the currents. A similar thin equatorial current sheet was first observed at Saturn on the dawn flank of the magnetosphere by Pioneer 11 [*Smith et al.*, 1980]. However, it wasn't possible to determine the extent of this thin current sheet, and therefore demonstrate the presence of a magnetodisc at Saturn, until more extensive Cassini data were available [e.g., *Arridge et al.*, 2007, 2008a].

Figure 12.2 illustrates the main current systems in the magnetospheres of Jupiter and Saturn. Azimuthal and radial currents, exerting radial and azimuthal $\mathbf{j} \times \mathbf{B}$ forces, flow in the equatorial current sheet and are revealed by radial stretching of the field and sweep-back of magnetic field lines out of their meridional planes. As emphasized by *Kivelson* [2000], there is no special emphasis of the current in $\mathbf{j}$ in $\mathbf{j} \times \mathbf{B}$ and in the magnetohydrodynamic (MHD) momentum equation. Of equal importance are flows and plasma stresses in the magnetosphere. As discussed in section 12.2, these currents are asymmetrical in local time and vary with radial distance and so current flows in and out of the equatorial current sheet to balance $\nabla \cdot \mathbf{j}_\perp$. Field-aligned currents also flow in the inner/middle magnetosphere at Saturn and are associated with planetary period oscillations (see chapter by *Hunt, Cowley and Nichols*, this volume).

At Earth, the magnetotail currents close via the magnetopause, forming a theta-shaped current loop in the magnetotail. Such a configuration is expected in the magnetotails of Jupiter and Saturn, however, it is not clear at what distance the azimuthal currents deviate away from flowing azimuthally around the planet and close on the magnetopause. This may be at the point where wave travel time from the ionosphere to the equatorial magnetosphere is so long that the ionosphere and magnetosphere are effectively decoupled [*Vasyliunas*, 1994]. In this limit, the ionosphere cannot exert any significant forces on the magnetospheric plasma and so the magnetospheric flows are almost entirely determined by processes in the magnetosphere. Flows in the ionosphere are then determined by mapping these magnetospheric flows. However, the distance at which this occurs, and the effects on the $\mathbf{j} \times \mathbf{B}$ force and then the inferred currents is not well constrained.

At Uranus and Neptune, Voyager 2 revealed the presence of Chapman-Ferraro and cross-tail currents, however, the orientation of the magnetic dipole to the solar wind creates a highly asymmetrical magnetosphere. At solstice, Uranus's magnetosphere is reminiscent of the terrestrial magnetosphere in that the dipole is approximately perpendicular to the incoming solar wind flow. However, the key difference is that the planetary rotation

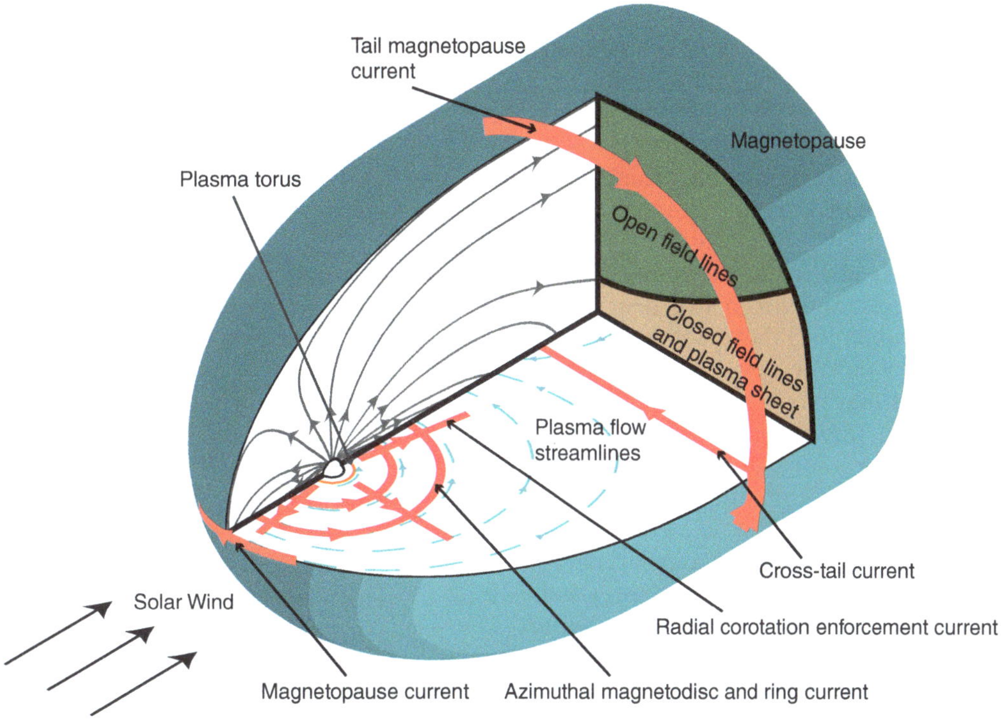

Figure 12.2 Schematic illustration of current systems in the magnetospheres of Jupiter and Saturn. Field-aligned currents, closing the corotation enforcement currents and partial ring current via the ionosphere, are omitted for clarity. After *Khurana* [2001].

axis is parallel to the incoming solar wind flow so that the magnetotail is wound up like a corkscrew. At Neptune, and during equinox conditions at Uranus, the magnetosphere oscillates between a pole-on and perpendicular (Earth-like) configuration. Simple models suggest that in a pole-on configuration, the tail current sheet is expected to take the form of a cylinder (see Fig. 7.8 in *Arridge* [2015]) with currents closing entirely within the tail. However, recent numerical MHD modeling suggests that this might not be the case [*Mejnertsen et al.*, 2016]. For further reading we direct the reader to reviews by *Lepping* [1994] and *Arridge* [2015].

In this review, we specifically examine the equatorial current sheet in the Jovian and Saturnian magnetospheres and their dynamics. We briefly examine the physical origin, geometry, and asymmetries in these current sheets. However, rather than focusing on the details of the physics of the equatorial current sheet, we use this description to lead to a discussion of dynamics. Dynamic processes include small-scale ripples and flapping, time-dependence in the stresses and currents, tearing, and magnetic reconnection.

Other recent reviews of these current systems include *Jackman et al.* [2014a], *Achilleos et al.* [2015], *Arridge et al.* [2015], *Arridge* [2015], *Kivelson* [2015], and *Louarn et al.* [2015]. Comprehensive discussions of this topic can also be found in *Vasyliunas* [1983] and *Kivelson* [2000]. Other chapters in this volume also address these current systems, including Khurana and Liu (Chapter 2) for an overview and synthesis of these current systems at different planets; chapters by Sergis et al. (Chapter 9) and Achilleos (Chapter 8) cover the ring current systems at Saturn and Jupiter, respectively; Belenkaya (Chapter 13) discusses Chapman-Ferraro current systems; Hunt, Cowley, and Nichols (Chapter 27) present observational work on the magnetosphere-ionosphere coupling current systems at Saturn; and Ray (Chapter 21) presents a theoretical discussion of such coupling currents at Jupiter.

Section 12.2 focuses on the physics of the equatorial current sheet in the magnetospheres of Jupiter and Saturn and its driving mechanisms. Section 12.3 then examines observed dynamic behavior and its theoretical interpretation.

12.2. EQUATORIAL CURRENT SHEET

The equatorial current sheet at Jupiter and Saturn is typically studied in a spherical polar coordinate system (B_r, B_θ, B_φ), where B_r is the field component along a line from the center of the planet to the observer, B_θ points in a colatitudinal direction from the observer, and B_φ completes the right-handed set and points in approximately the direction of corotation at Jupiter and Saturn.

In this coordinate system, the flyby of Pioneers 10 and 11 with Jupiter in 1973 and 1974 showed that the magnetic field in the middle and outer magnetosphere had large B_r and B_φ components compared to the B_θ component, and that the B_r and B_φ components had an antiphase relationship. Hence, the field is radially stretched away from the planet where field lines were pulled out of their meridional planes against corotation, thus adopting a swept-back configuration. This radially extended and swept-back configuration is found over most of the magnetosphere beyond around 20 R_J [*Khurana*, 2001; *Khurana and Schwarzl*, 2005]. A similar configuration of Saturn's field was first observed using data from the Pioneer 11 flyby [*Smith et al.*, 1980] although the disk-like geometry was not clear as the noon sector was more dipolar. *Arridge et al.* [2008a] surveyed the current sheet using Cassini magnetometer data and found that the current sheet could also be present on the dayside, except that the magnetopause had to be at larger subsolar distances in order for the sheet to be detected on the dayside. Hence, *Arridge et al.* [2008a] concluded that a magnetodisc was present at all local time distances sampled by Cassini at that time.

Both the current sheets at Jupiter and Saturn have a thickness on the order of a few planetary radii, which increases with radial distance and has local-time asymmetries [*Arridge et al.*, 2015; *Khurana and Schwarzl*, 2005]. Figure 12.3 presents Voyager 1 data showing current sheet crossings at Jupiter. When the spacecraft is in the lobes, the field is steady in direction and strength and rapidly rotates during current sheet crossings, associated with diamagnetic depressions in the current sheet and an increase in magnetic fluctuations.

Jupiter's field has been found to be swept-back over nearly all local-time sectors of the magnetosphere, except for some modest sweep-forward in the dusk sector [*Khurana and Schwarzl*, 2005]. As discussed below, and by *Khurana* [2001], this is indicative of an internal process associated with the outward transport of plasma. The geometry of the field at Saturn is swept back in the dawn, noon, and dusk sectors [*Arridge et al.*, 2015], strongly so in the dawn sector, but the case for the geometry of the field lines in the dusk sector has received less attention in the literature. *Bunce et al.* [2003] have argued that one might expect more sweep-forward in the dusk sector at Saturn.

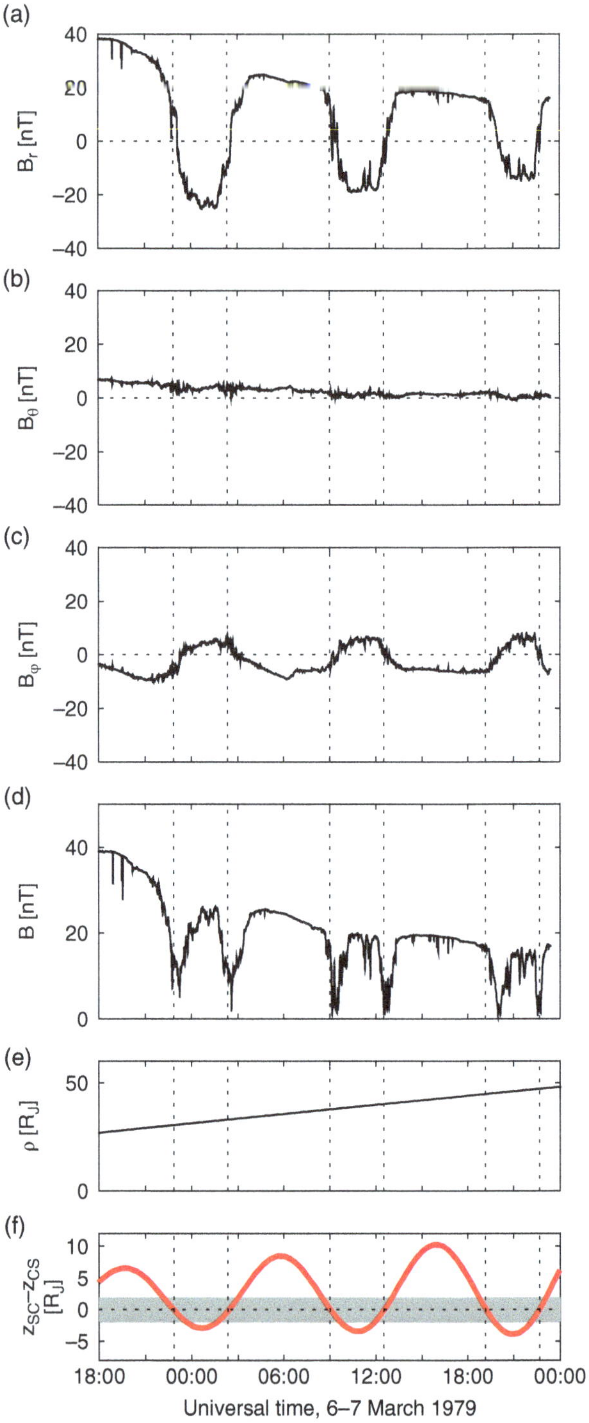

Figure 12.3 Voyager 1 magnetometer data showing current sheet crossings at Jupiter. Panels (a–c) show the magnetic field in spherical polar coordinates; (d) shows the field strength; (e) the distance from Jupiter, and (f) the trajectory of the spacecraft relative to the current sheet (indicated by the grey region).

12.2.1. Physics of the Current Sheet: Azimuthal and Radial Currents

The magnetospheres of Jupiter and Saturn have significant internal plasma sources originating in the interiors, surfaces, and atmospheres of natural satellites. In the Jovian magnetosphere, this source is dominated by the satellite Io, buried deep within the magnetosphere. In the Saturnian magnetosphere, Enceladus plays an important role, but additional sources are more distributed in the Saturnian magnetosphere. Due to rapid rotation of the magnetosphere, the added mass is unstable to the centrifugal interchange instability, a Rayleigh-Taylor-like instability where the centrifugal force plays the role of gravity. Under the action of this instability, the added mass is transported outward in the magnetosphere, filling the plasma sheet, and is ultimately lost to the solar wind [e.g., *Thomsen*, 2013].

The equatorial current and plasma sheet arise from mechanical stresses in the rotating plasma, which must be balanced by magnetic stresses (the $\mathbf{j}\times\mathbf{B}$ force). The identification of these local stresses and their global importance is a nontrivial problem that has recently been discussed in detail by *Vasyliunas* [2015]. Our starting point is the MHD momentum balance equation [e.g., *Vasyliunas*, 1983]:

$$\rho\frac{D\mathbf{u}}{Dt} = -\nabla\cdot\mathrm{P} + \mathbf{j}\times\mathbf{B} \tag{12.1}$$

where $\mathrm{P} = P_{\perp}\mathrm{I} + \left(P_{\parallel} - P_{\perp}\right)\frac{\mathbf{BB}}{B^2}$ is the (in general anisotropic) pressure tensor, with parallel $P_{\parallel}$ and perpendicular $P_{\perp}$ pressures, ρ is the mass density, $\mathbf{u}$ is the plasma velocity, $\mathbf{j}$ is the volume current density, and $\mathbf{B}$ is the magnetic field vector. Often in the magnetospheres of Jupiter and Saturn, the advective derivative is replaced with terms including the centrifugal force.

Understanding radial stress balance, particularly the relative roles of the centrifugal force, pressure gradients, and anisotropy from different particle populations, has been the focus of observational and theoretical work since Voyager. Early studies at both planets argued that centrifugal forces associated with cold internally produced plasma were important mechanical stresses, if not the dominant stresses [e.g., *Arridge et al.*, 2007; *McNutt*, 1984; *Russell et al.*, 1999a and 1999b]. However, additional work has shown that plasma anisotropy, both in low-energy plasma and energetic particles, are important contributors [e.g., *Kellett et al.*, 2010; *Nichols et al.*, 2015; *Paranicas et al.*, 1991]. Figure 12.4 shows a semi-empirical model of the currents associated with plasma pressure, the centrifugal force, and plasma anisotropy from Nichols *et al.* (2015). These results show that anisotropy can provide an important source of mechanical stress in the middle magnetosphere. As the quality and extent of observational data improves, a detailed picture of radial stress balance is emerging. The radial currents, associated with azimuthal fields, are associated with azimuthal stress balance exerting a torque on the equatorial plasma to enforce (partial) corotation. The reader is referred to the chapters by Ray and Hunt et al. (Chapters 21 and 27 this volume) for more detailed discussion of magnetosphere-ionosphere coupling.

In applying this equation to the magnetodisc, we often adopt a geometrical thin-sheet, which takes into account of the fact that the spatial length scales normal to the sheet are much smaller than those in the tangential direction along the surface of the sheet. Following *Vasyliunas* [1983], we write introduce a new coordinate system decomposed into a direction normal to the sheet $\mathbf{e}_z$ and vectors with a subscript t are tangential to the sheet. Thus the magnetic field is written as $\mathbf{B} = \mathbf{B}_z\mathbf{e}_z + \mathbf{B}_t$ and gradients become $\nabla = e_z\frac{\partial}{\partial z} + \nabla_t$. Then the thin-sheet approximation can be formally stated by the statement $\nabla_t \ll \partial/\partial z$.

With this definition, Maxwell's equations become:

$$0 = \frac{\partial B_z}{\partial z} + \nabla_t\cdot\mathbf{B}_t \tag{12.2}$$

$$\mu_0\mathbf{j} = \mathbf{e}_z\times\left(\frac{\partial\mathbf{B}_t}{\partial z} - \nabla_t B_z\right) + \nabla_t\times\mathbf{B}_t \tag{12.3}$$

One particular implication for the magnetic field in the thin sheet approximation is obtained from equation (12.2), which shows that the gradient of the field normal to the current sheet depends on the tangential gradients of the field tangential to the current sheet; azimuthally symmetrical and radially constant fields imply that the field normal to the sheet is constant with distance from the sheet. However, a more detailed understanding of the thin sheet approximation is found when studying stress balance normal to the sheet. A textbook analysis typically results in pressure balance between the lobes and center of the current sheet. However, a more detailed analysis in a rotating system reveals that there are two thin sheet approximations *Vasyliunas* [1983]. The first implies pressure balance between the lobes and center of the sheet. This always holds for highly stretched field configurations, but additionally holds for almost dipolar fields when the plasma is cold (i.e., the thermal plasma speed is smaller than the azimuthal bulk flow speed). The second is encapsulated in the constancy of the normal field relative to the current sheet, which only holds when the plasma is cold.

Although the current and plasma sheet are conceptually viewed as a colocated uniform sheet, theoretical and observational studies show that these sheets have a complex structure in the normal direction [e.g., *Artemyev et al.*, 2014; *Kellett et al.*, 2009; *Sergis et al.*, 2011; *Yin et al.*, 2000]. At a simplified level this consists of a denser, cold heavy ion disk surrounded by a hotter sheet. In particular, *Artemyev et al.* [2014] have shown for crossings of Jupiter's current sheet that the central current sheet is only a few sulphur ion gyroradii (~0.25 R_J) in thickness. Figure 12.5 illustrates the central results showing the embedding of a thin current sheet inside a broader current sheet. *Artemyev et al.* [2014] have also found that a significant part of the vertical stress balance is mediated

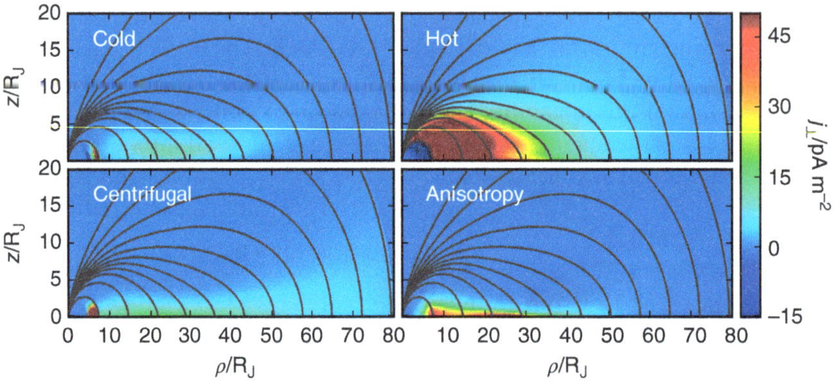

Figure 12.4 Semi-empirical Euler potential model showing the contribution of different plasma stresses to the total stress balance in Jupiter's magnetodisc. Magnetic field lines are indicated by the grey lines. The top panels show the pressure gradient force associated with cold and hot plasma. The bottom left panel shows the stress associated with the centrifugal force, and the bottom right panel shows the stress associated with hot plasma pressure anisotropy. From *Nichols et al.* [2015].

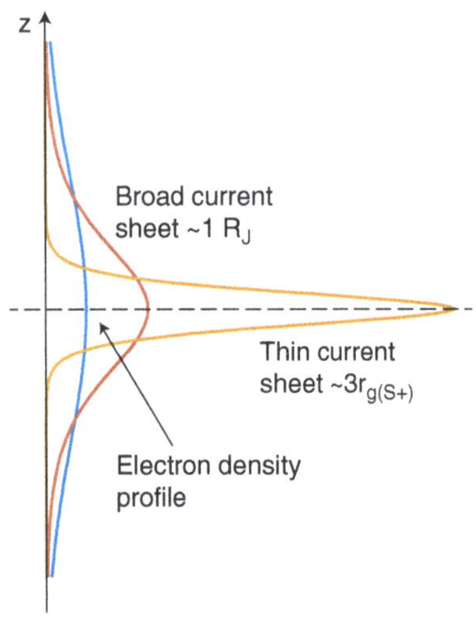

Figure 12.5 Kinetic results from *Artemyev et al.* [2014].

by the magnetic field in the azimuthal direction (which peaks in the center of the sheet).

Kinetic studies have shown that different layers in the current sheet can support different current systems, for example *Yin et al.* [2000] have shown radially outward currents in the center of the sheet and radially inward currents at the edges. These studies highlight the limitations of the magnetohydrodynamic approach and that for detailed studies a kinetic approach is required. This is particularly relevant for understanding bifurcated current sheets where the current density maximizes away from the center of the current sheet. These have been observed in Earth's magnetotail current sheet [*Sergeev et al.*, 1993] and also Jupiter's magnetodisc [*Israelevich et al.*, 2007]. The physical origin of these bifurcated current sheets remains unclear but their equilibria can be modeled using a kinetic approach [e.g., *Génot et al.*, 2005].

We can define a coordinate system where $\mathbf{e}_a$ is in the direction of the largest change in magnetic field (usually in the radial direction), and $\mathbf{e}_b$ is in the direction of the current, and $\mathbf{e}_c$ is normal to the current sheet. The variation in B_a as the spacecraft crosses the current sheet is distinctive and if, for example, the current sheet were behaving as a Harris current sheet [*Harris*, 1962], then the derivative of the magnetic field with time would show a much larger change at the center of the current sheet. More generally, from Ampère's $j_b = \partial B_a/\partial z$ and we can write the total derivative for the change in B_a as $dB_a = (\partial B_a/\partial z)dz$ and hence $dB_a/dt = (\partial B_a/\partial z)(dz/dt)$. Thus the detailed variation of j_b can be recovered from measurements of the time derivative of B_a. *Israelevich et al.* [2007] use this method to show that bifurcation of the Jovian current sheet is possible. Unlike the peak of the differential at $B_a = 0$, *Israelevich et al.* [2007] found evidence for a local minima at the center of the current sheet, with maxima either side, showing the majority of the current flowing off the center of the current sheet. Although *Israelevich et al.* [2007] found few examples of this structure at Jupiter but related the bifurcation to the presence of pressure anisotropies in the current sheet. *Volwerk et al.* [2013] have reinvestigated the presence of bifurcations in the current sheets at Jupiter and Saturn and have found weak evidence for bifurcations at either planet, in accordance with the limited detections reported by *Israelevich et al.* [2007].

12.2.2. Geometry and Motion of the Current Sheet

Geometrical deformations in the shape and position of the Earth's magnetotail, as a result of the solar wind dynamic pressure, interplanetary magnetic field, and diurnal motion of Earth's magnetic dipole [e.g., *Tsyganenko et al.*, 1998] are also present in the current sheets of Jupiter [e.g., *Khurana and Schwarzl*, 2005] and Saturn [e.g., *Arridge et al.*, 2008b]. The offset of the magnetic axis to the rotational axis produces a diurnal up-down flapping motion of the current sheet, at roughly the same period as Jupiter's rotation period. Although present in Earth's magnetotail, the effect is very pronounced in magnetometer data from Jupiter and is visible as a "square wave" feature in the radial and azimuthal components of the field [e.g., *Smith et al.*, 1974], and can be seen in Figure 12.3. This is known as the "wavy magnetodisc" model.

To move or deform the current sheet, the current carriers must also move. In the case of the diurnal effect of planetary rotation with a tilted dipole, the motion of the dipole is transmitted outward into the magnetosphere via Alfvén waves. Since these waves propagate at a finite speed, the current sheet will not remain as a rigid structure [*Northrop et al.*, 1974]. Observationally, this leads to delays, or a lag, in the observed position of the current sheet compared with a rigid wavy magnetodisc model. Figure 12.6 illustrates the wavy magnetodisc relative to the orbital plane.

In the presence of azimuthal plasma flow **u**, the actual group velocity for the waves is given by $\mathbf{u}+v_A\mathbf{B}/B$, where v_A is the Alfvén speed. By considering the trajectory of an Alfvén wave packet into the magnetosphere, *Northrop et al.* [1974] derived a general expression for the incremental delay of the deformation of the current sheet with increasing radial distance. Using solutions to the MHD equations that relate the plasma flow and magnetic field [*Chandrasekhar*, 1956; *Mestel*, 1961, 1968], the incremental delay [*Goertz*, 1981; *Khurana and Schwarzl*, 2005; *Northrop et al.*, 1974] can be shown to be

$$\frac{d\delta}{d\rho}=\frac{B_\phi}{\rho B_\rho}-\frac{\Omega_J-\Omega_m}{u_\rho+v_A B_\rho/B} \quad (12.4)$$

where δ is the delay, Ω_J is the rotation period of Jupiter, Ω_m is the rotation period of the magnetosphere, and the position and field are specified in cylindrical coordinates ρ, φ, z and B_ρ, B_φ, B_z, respectively. There are two physical effects that produce the incremental delay. The first is the sweep-back of the field lines, the first term on the right-hand side in equation (12.4). The second is the wave travel time in a subcorotating magnetosphere (second term), where there may also be a radial outflow of plasma ($u_\rho>0$). Figure 12.7 shows the delays due to these terms as determined from current sheet crossings at Jupiter [*Khurana and Schwarzl*, 2005]. Figure 12.7a shows the measured bend-back and then the resulting delay in figure Figure 12.7b. The remaining delays were assumed to be due to wave delay (Fig. 12.7c), but the contribution from outflow and subcorotation was not determined.

Although at the time of writing, Saturn does not have a significant dipole tilt, flapping motions have been detected in the equatorial current sheet and these have also been found to be systematically delayed [*Arridge et al.*, 2011]. However, the contribution of the two terms in equation (12.4) to this delay has not been determined.

Uranus's magnetosphere is asymmetrical with a tilt of around 60° between the dipole and spin axes. Similar to Jupiter and Saturn, the orientation of the dipole is communicated into the distant magnetosphere by Alfvén waves. At equinox, Uranus's spin axis is parallel (or antiparallel) to the planet-Sun line. As the planet rotates the magnetotail lobes coil up into a double helix. The pitch of the resulting helix and the orientation of the current

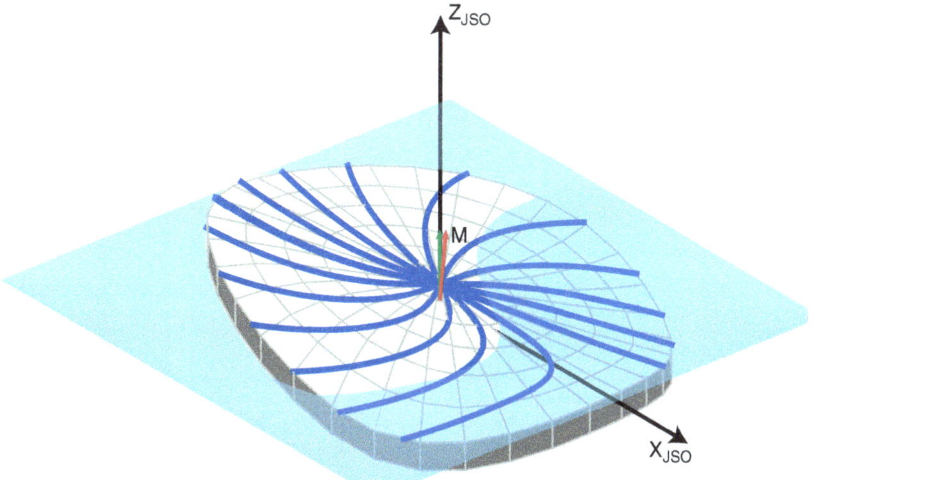

Figure 12.6 Diagram showing wavy non-rigid magnetodisc. Blue lines show Alfvén wave packet trajectories carrying information about the rotating dipole to the distant plasma sheet. The blue plane is the rotational equator of the planet. The green and red arrows show the rotation axis and magnetic axis of Jupiter, respectively.

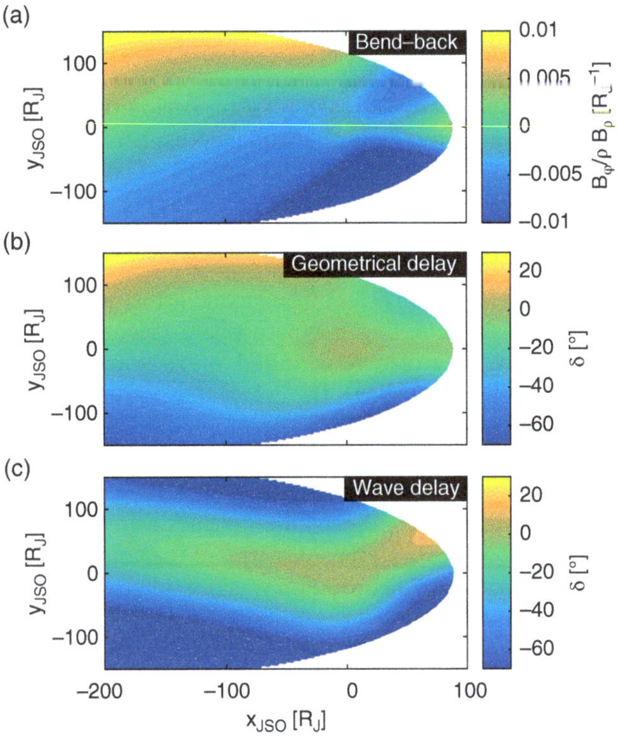

Figure 12.7 Diagram showing the contributions from field line bend-back (a,b) and wave delay to observed lags in Jupiter's wavy magnetosphere. Panel (a) shows the bend-back or bend-forward of field lines from their meridional planes. Panel (b) shows the delay associated with this geometry. Panel (c) shows the wave delay associated with wave packets traveling in the magnetosphere. After *Khurana and Schwarzl* [2005].

sheet are determined by the wave propagation speed along the tail and the rotation rate of the planet. The propagation speed was found to be around 0.4°/R_u, slower than the 0.28°/R_u expected for wave propagation at the Alfvén speed. *Tóth et al.* [2004] interpreted as the result of friction between the lobes and the solar wind (Fig. 12.8).

In addition to these nonrigid flapping motions, at a given distance from the planet, the current sheet has been observed to deviate from the magnetic axis due to a combination of centrifugal forces pushing the dense cold ions toward the rotational equator [*Hill et al.*, 1974; *Khurana*, 1992; *Khurana and Kivelson*, 1989] and solar wind pressure forcing the current sheet to become parallel with the solar wind flow. In the inner magnetosphere, the current and plasma sheet lies roughly in two-thirds of the way from the rotational equator to the magnetic equator, at an angle of roughly 6.4° [*Hill et al.*, 1974], although *Vasyliunas* [1983] has shown that the angle depends on the ratio of thermal speed to bulk flow speed. At Saturn, the magnetic equator is offset from the rotational equator due to the quadrupole component of the internal magnetic field. This results in a small shift of the centrifugal equator from the magnetic equator of less than 1°, becoming larger with increasing radial distance so that the tilt is less than 0.2° at 10 R_S.

Solar-wind-driven distortions of the magnetosphere are usually characterized by a hinging distance, a characteristic distance over which the current sheet asymptotically becomes parallel to the solar wind flow. The hinging distance at Jupiter and Saturn has been

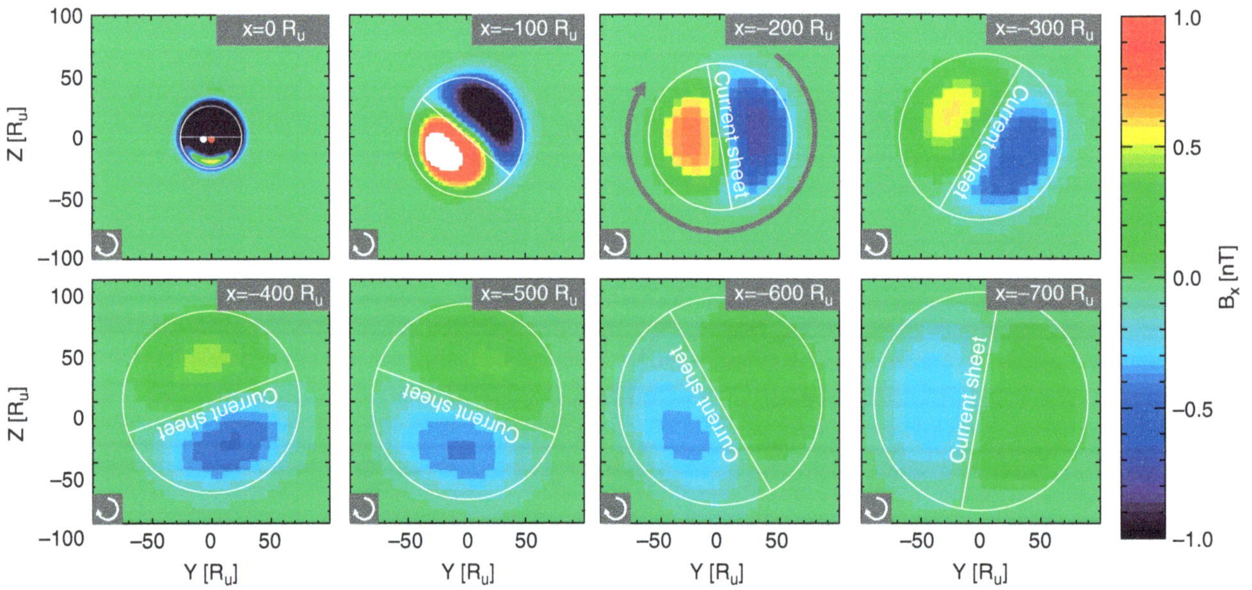

Figure 12.8 MHD simulation results from *Tóth et al.* [2004] showing the slices through the magnetotail at different distances downtail, showing the double helix formation of the magnetotail lobes and the orientation of the tail current sheet.

empirically determined to be 47 R_J [*Khurana and Schwarzl*, 2005] and 16–30 R_S [*Arridge et al.*, 2008b], respectively.

Models of solar wind-driven distortion at Jupiter and Earth have assumed that the current sheet becomes parallel to the solar wind flow on the dayside as well as the nightside [*Khurana and Schwarzl*, 2005; *Tsyganenko*, 2002]. The solar wind-driven distortion in the Saturnian system was found to pull the current sheet into a bowl shape [*Arridge et al.*, 2008b]. During southern hemisphere summer and upward distortion of the current sheet was observed on the dayside and nightside. Subsequently this dayside distortion has also been detected in the terrestrial magnetosphere and theorized to be a universal effect when there is a north-south asymmetry due to dipole tilt and day-night magnetospheric asymmetries imposed by the solar wind [*Tsyganenko and Andreeva*, 2014; *Tsyganenko et al.*, 2015]. Figure 12.9 shows that when a planetary internal field is confined inside the magnetopause then the region of minimum field strength is then deformed away from the equator [*Tsyganenko and Andreeva*, 2014]. Such an effect has yet to be reported for Jupiter, although the small obliquity of Jupiter may make it difficult to detect.

As mentioned above, even though Saturn's dipole tilt is very small, periodic global flapping motions have also been detected at Saturn [e.g., *Arridge et al.*, 2008; *Arridge et al.*, 2011; *Carbary et al.*, 2008]. *Burch et al.* [2008] interpreted these periodicities not as flapping motions, but as periodic reconnection and release of plasmoids down the tail along a spiral path. Perturbations in the north-south component of the magnetic field were used as evidence to support this interpretation. However, few examples of plasmoids have been seen at Saturn and these are usually accompanied by a fast antiplanetward flow. *Jackman et al.* [2009] showed that the north-south perturbations discussed by *Burch et al.* [2008] could be produced by a flapping motion of the current sheet. This is discussed in more detail in the next section. The driver of these periodicities is still the subject of vigorous debate, but appears to be driven by current systems in the inner and middle magnetosphere (see Chapter 27 by Hunt et al. in this volume). From the results of an MHD simulation of Saturn's magnetosphere, *Kivelson and Jia* [2014] argue that the flapping is controlled by compressional waves that weaken in the evening sector and strengthen in the morning sector. This simulation shows that the magnetopause position is also dependent on the compressional waves.

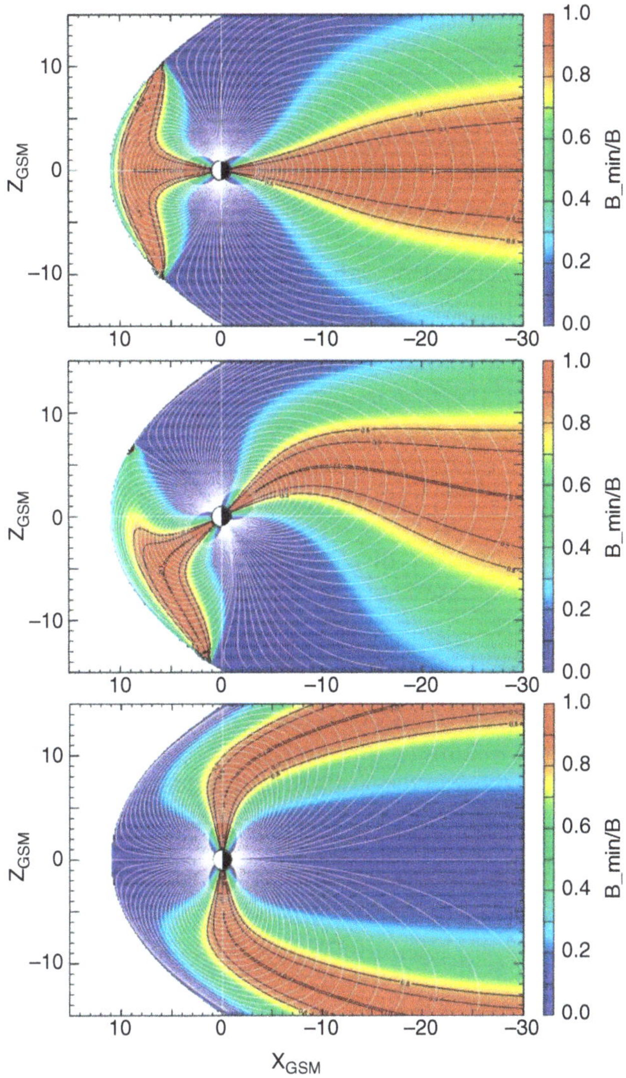

Figure 12.9 Dipole magnetic fields confined inside the magnetosphere. The color scale shows the inverse mirror ratio where the minimum field strength along the field is deformed away from the equator.

12.3. DYNAMICS

Here we use the observational and theoretical background from the previous sections to understand dynamics in global current sheets at Jupiter and Saturn. *Sharma et al.* [2008] have reviewed dynamic processes in the terrestrial magnetotail and examples of many of these transients have been reported in current sheets at the giant planets. We can group some of these transients into a set that are associated with magnetic reconnection in the magnetotail/magnetodisc, for example, plasmoids, traveling compression regions, magnetic reconnection ion diffusion regions, and dipolarizations. Another set of transients can be identified as due to flapping motions of the current/plasma sheet. In this section, we focus on these two broad groupings, along with a third grouping, which are known as magnetic nulls.

The current sheet itself is also the subject of time variability. The physical stresses discussed in section 12.2.1 and encapsulated in equation (12.1) are subject to variability, which produces variations in the current density. Such variations have been detected at both Jupiter [*Russell et al.*, 2001] and Saturn [*Kellett et al.*, 2011; *Leisner et al.*, 2007], and also discussed at Earth [e.g., *Daglis*, 2006]. *Kellett et al.* [2011] examined the plasma stresses in Saturn's magnetosphere and found that the current density varied within a factor of two, principally in the plasma pressure gradient in the middle and outer magnetosphere, with associated variations in the B_θ component of the magnetic field (related to Dst at Earth) up to a factor of two. *Russell et al.* [2001] and *Leisner et al.* [2007] adopted a different approach in that they subtracted a model magnetic field and examined the residual magnetic field to search for perturbations. At Saturn, numerous small amplitude perturbations were found and were interpreted as the effect of the solar wind, with larger more infrequent perturbations. The magnitude of these perturbations was generally consistent with *Kellett et al.* [2011].

12.3.1. Magnetic Reconnection

A variety of transient structures have been detected in the magnetotails and magnetodiscs of the giant planets, similar to those in the terrestrial magnetosphere [*Sharma et al.*, 2008], including plasmoids [*Russell et al.*, 1998; *Smith et al.*, 2016] for example, traveling compression regions [e.g., *Jackman et al.*, 2014b], magnetic reconnection ion diffusion regions [*Arridge et al.*, 2016], and dipolarizations [e.g., *Arridge et al.*, 2016; *Kasahara et al.*, 2011]. A current system analogous to the substorm current wedge was proposed by *Jackman et al.* [2013] to explore the auroral consequences of dipolarizations. Evidence for the hall current system was found in a diffusion region at Saturn [*Arridge et al.*, 2016].)

Evidence for weak tearing in Jupiter's current sheet beyond 40 R_J has been presented by *Russell et al.* [1999a and 1999b]. Although plasmoids have been seen at a distance of 50 R_J, most plasmoids are seen between 80 and 100 R_J [*Vogt et al.*, 2014] with a statistical X-line between 90 and 120 R_J [*Vogt et al.*, 2010], although this is dependent on local time. In comparison, a theoretical analysis of tearing presented by *Zimbardo* [1991] shows that the instability is expected around 60 R_J. In data from Saturn, *Smith et al.* [2016] showed that more planetward moving structures were observed inside 30 R_S, suggesting an average X-line around this distance. This is supported by evidence for direct observation of a diffusion region at 29 R_S was presented by *Arridge et al.* [2016].

The presence of multiple ion species and rapid rotation introduces original effects into the process of magnetic reconnection at the giant planets. The presence of significant quantities of heavy ions introduces multiple scales to the ion diffusion region as ions with different masses and charge states demagnetize on different scales [*Shay and Swisdak*, 2004]. The presence of heavy ions also depresses the reconnection rate, perhaps providing some explanation for the weak tearing found by *Russell et al.* [1999].

Heavy ions and rapid rotation also add a significant centrifugal force that stretches magnetic field lines eventually making the current sheet so thin it can reconnect. In the absence of a solar wind interaction, this provides a method of ejecting plasma from the magnetosphere and cycling magnetic flux, known as the Vasyliunas cycle. This mechanism is thought to be superposed onto the Dungey cycle. Various authors have claimed to observe the effects of reconnection due to these two cycles [e.g., *Delamere et al.*, 2015; *Thomsen et al.*, 2015a; *Yao et al.*, 2017, submitted manuscript], which can affect the composition of the plasma sheet [*Thomsen et al.*, 2015b]. There is also evidence that reconnection can trigger the outflow of plasma from Saturn's ionosphere [*Felici et al.*, 2016].

12.3.2. Flapping Motions

Increases in solar wind dynamic pressure are thought to decrease the hinging distance to force the current sheet to become parallel to the solar wind at a smaller radial distance. Variations in the mean position of the current/plasma sheet have been detected in Galileo fields and particles data at Jupiter [*Vasyliunas et al.*, 1997; *Waldrop et al.*, 2005], however it is difficult to directly attribute these to variations in the solar wind dynamic pressure. *Vasyliunas et al.* [1997] also suggests that variability of the internal plasma source from the Io torus and drifting longitudinal asymmetries may also be responsible for some of the observed variability. Similar variability is also observed at Saturn [*Arridge et al.*, 2011] and at least one event has been associated with an abrupt increase in the solar wind dynamic pressure [*André et al.*, unpublished manuscript]. However, these are large-scale distortions.

Flapping motions of the current sheet represent small-scale variations superimposed on the larger global-scale current sheet motions discussed in section 12.2.2. Such warps and corrugations are well known in the heliospheric current sheet [*Shodan et al.*, 1994] and have been detected at Earth, Jupiter, and Saturn [e.g., *Russell et al.*, 1999; *Sharma et al.*, 2008; *Volwerk et al.*, 2013; *Martin and Arridge*, 2017]. Figure 12.10 shows Galileo data from a radial distance of 54.6 R_J where the current sheet is observed flapping during a north-to-south traversal of the current sheet. These observations led *Russell et al.* [1999] to describe Jupiter's magnetodisc current sheet to be constantly in motion, not only at the rotational period of the planet but also with a period of around 10 min about its equilibrium position. They found that the amplitude of the perturbation varied from one current sheet encounter to the next but was small inside 40 R_J.

Similar structures were also identified in Cassini magnetometer data [*Arridge et al.*, 2007] and have been studied in more detail by [*Martin and Arridge*, 2017]. In this more recent study, the waves were generally found to propagate outward in the magnetosphere and their amplitude was found to increase with radial distance. At Jupiter, oscillatory motions of the current sheet are found beyond around 30 R_J and the current sheet was reported to be in almost chaotic motion beyond 50 R_J *Russell et al.* [1999]. At Saturn such flapping motions are found beyond around 15 R_S [*Martin and Arridge*, 2017].

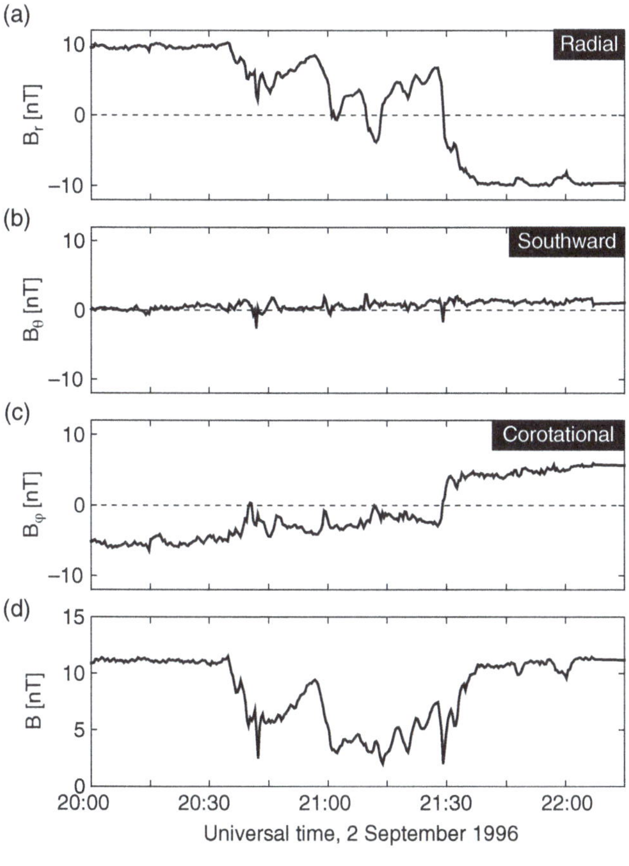

Figure 12.10 Example of flapping behavior observed in Jupiter's magnetodisc at 54.6 R_J. After *Russell et al.* [1999].

Small-scale flapping motions are usually analyzed in the MHD limit and are interpreted as surface waves where the current sheet acts as a waveguide, bounded by the high Alfvén speed in the lobes. Two fundamental modes are found: the sausage mode, which changes in current density and pressure, and the kink mode, which transverse oscillations. Figure 12.11 illustrates the kink mode and the sausage mode in a current sheet. The MHD equations are linearized and combined with a simple current sheet model. This is often a Harris or modified Harris sheet (the additional of a north-south or B_z component). From this basic formulation, a wide range of authors have investigated different physical aspects. *Cramer* [1994] examined the case where the field exhibits different levels of shear across the current sheet. *Tirry et al.* [1997] found that when the waves propagated obliquely to the equilibrium magnetic field, the wave eigenmodes resonantly coupled to localized Alfvén waves, leading to damping. *Yamauchi and Lui* [1997] examined the relationship between the waves and the tearing mode. *Fruit et al.* [2002] and *Louarn et al.* [2004] have performed detailed analyses and comparisons with Cluster data in the Earth's magnetotail. *Erkaev et al.* [2009] computed wave modes for a 2D current sheet with gradients associated with the current and a gradient of B_z along the current sheet. *Fruit and Louarn* [2011] also examined propagation in a 2D current sheet but also numerically solved the full wave equation, including propagation into the lobes rather than assuming the waves are evanescent in the lobes.

Lachin [1998] is the only author to offer a theoretical interpretation of surface modes in the magnetospheres of the giant planets and predicted a period of around 10 min, compatible with the observations of *Russell et al.* [1999]. A number of studies have shown that the damping these waves can also lead to heating of the plasmasheet. However, their analysis was highly idealized. Further work is required to examine the propagation of MHD surface modes, and their coupling with Alfvén, slow and fast modes in the magnetosphere.

The origin of these flapping motions are not well understood. In Earth's magnetotail, they are found propagating from the center of the tail to the flanks of the magnetosphere and thought to be associated with magnetic reconnection. *Volwerk et al.* [2013] examined

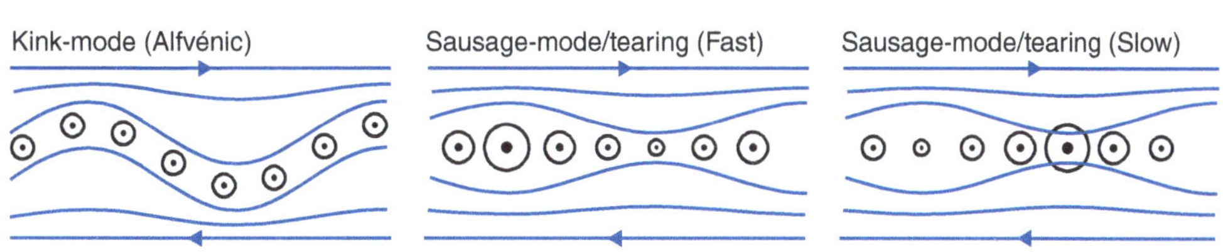

Figure 12.11 Current sheet surface modes. After *Yamauchi and Lui* [1997].

propagation directions for a small set of events at Jupiter and Saturn and found evidence of propagation in the direction of the magnetic field (composed of both azimuthal and radial components). They also found that the periods were not in agreement with the double-gradient model [*Erkaev et al.*, 2009], but this might not be expected since this does not include all the plasma stresses present at Saturn and Jupiter. Further work is required to examine these motions. In the absence of regular plasma data, it is also difficult to fully separate kink- and sausage-mode motions, especially if both are present together [*Kubyshkina et al.*, 2014] and work is required to fully understand these wave modes at the giant planets.

12.3.3. Nulls

Complicating the analysis of flapping perturbations at Jupiter are structures known as magnetic nulls. These were first described using magnetometer data from the Ulysses spacecraft [*Haynes et al.*, 1994] but were also detected by magnetometers on the Pioneer and Voyager spacecraft [*Leamon et al.*, 1995]. The main distinguishing characteristic of a null is that the B_θ component falls to very low values in the center of the null and the transverse B_r and B_φ components also fall in antiphase, hence the field strength drops to a very low magnitude inside the event leading to the term null. The typical duration of a null is of the order of 5–10 min. The leading physical interpretation for nulls is that they are pieces of the plasma sheet that have detected from its outer edge and lie in a region of quasi-dipolar field known as the cushion. These nulls should be filled with iogenic plasma, or at least a mixture of iogenic and plasma of solar wind origin, sufficient to provide the observed diamagnetic effect. Unfortunately, the plasma instrumentation on previous missions to Jupiter has not been sufficient (due to a combination of sampling rate, pointing, and mission local time coverage) to study the interior of these nulls.)

Saturn fields and particles data have been searched for evidence of similar structures in the Saturnian magnetosphere. Although some candidate events have been discussed, no directly analogous structures have been indentified. Part of this reason may lie in the suggestion by *Went et al.* [2011] that the cushion region at Jupiter is much thicker than the equivalent structure at Saturn, where the magnetodisc essentially extends right to the magnetopause [*Arridge et al.*, 2008a].

12.4. SUMMARY

In this review we have examined the equatorial currents that dominate the magnetospheres of Jupiter and Saturn and we have examined their equilibria, geometrical properties, and dynamics.

Attempts to understand stress balance in the sheet have been carried out using observations and semiempirical modeling methods. These studies have highlighted the general importance of pressure gradients, anisotropy, and centrifugal force in the current sheet. Multiple species and populations of the same species can make contributions to these terms. This has led to a thorough picture of the time-independent equilibrium of the sheet. However, numerous studies have revealed bifurcations in the current sheet and that kinetic structure is important and more work is required in this area.

The geometry and motion of Jupiter's magnetodisc is fairly well constrained. The discovery of bowl-like geometries on the dayside at Saturn and Earth has not yet motivated a search for a similar geometry at Jupiter, but this may be hard to recover due to the small obliquity of Jupiter. A more detailed understanding of wave propagation in the magnetospheres of Jupiter and Saturn would help to understand nonrigid motion of the magnetodisc in more detail.

Small-scale flapping motions of current sheets appear to be an ubiquitous feature of planetary current sheets. There is a rich array of theoretical work that models these as MHD surface modes, and this is an area for growth at the giant planets.

ACKNOWLEDGMENTS

CSA was funded by a Royal Society Research Fellowship.

REFERENCES

Achilleos, N., N. André, X. Blanco-Cano, P. C. Brandt, P. A. Delamere, and R. Winglee (2015), 1. Transport of mass, momentum and energy in planetary magnetodisc regions, *Space Sci. Rev.*, *187*(1), 229–299; doi:10.1007/s11214-014-0086-y.

Arridge, C., C. Russell, K. Khurana, N. Achilleos, N. André, A. Rymer, M. Dougherty, and A. Coates (2007), Mass of Saturn's magnetodisc: Cassini observations, *Geophys. Res. Lett.*, *34*(9); doi:10.1029/2006GL028921.

Arridge, C., C. Russell, K. Khurana, N. Achilleos, S. Cowley, M. Dougherty, D. Southwood, and E. Bunce (2008a), Saturn's magnetodisc current sheet, *J. Geophys. Res. Space Physics*, *113*(A4).

Arridge, C., K. Khurana, C. Russell, D. Southwood, N. Achilleos, M. Dougherty, A. Coates, and H. Leinweber (2008b), Warping of Saturn's magnetospheric and magnetotail current sheets, *J. Geophys. Res. Space Physics*, *113*(A8).

Arridge, C. S. (2015), *Magnetotails of Uranus and Neptune*, John Wiley & Sons, Inc., 119–133.

Arridge, C. S., J. P. Eastwood, C. M. Jackman, G.-K. Poh, J. A. Slavin, M. F. Thomsen, N. André, et al. (2016), Cassini in situ observations of long-duration magnetic reconnection in

Saturn's magnetotail, *Nature Phys.*, *12*, 268–271; doi:10.1038/nphys3565.

Arridge, C. S., M. Kane, N. Sergis, K. K. Khurana, and C. M. Jackman (2015), Sources of local time asymmetries in magnetodiscs, *Space Sci. Rev.*, *187*(1), 301–333; doi:10.1007/s11214-015-0145-z.

Arridge, C. S., N. André, K. Khurana, C. Russell, S. Cowley, G. Provan, D. Andrews, et al. (2011), Periodic motion of Saturn's nightside plasma sheet, *J. Geophys. Res. Space Physics*, *116* (A11).

Arridge, C. S., N. André, N. Achilleos, K. K. Khurana, C. L. Bertucci, L. K. Gilbert, G. R. Lewis, A. J. Coates, and M. K. Dougherty (2008), Thermal electron periodicities at 20RS in Saturn's magnetosphere, *Geophys. Res. Lett.*, *35*, L15107; doi:10.1029/2008GL034132.

Artemyev, A. V., I. Y. Vasko, and S. Kasahara (2014), Thin current sheets in the Jovian magnetotail, *Planet. Space Sci.*, *96*, 133–145; doi:10. 1016/j.pss.2014.03.012.

Bunce, E., S. Cowley, and J. Wild (2003), Azimuthal magnetic fields in Saturn's magnetosphere: Effects associated with plasma subcorotation and the magnetopause-tail current system, *Ann. Geophys.*, *21*(8), 1709–1722.

Burch, J., J. Goldstein, P. Mokashi, W. Lewis, C. Paty, D. Young, A. Coates, M. Dougherty, and N. André (2008), On the cause of Saturn's plasma periodicity, *Geophys. Res. Lett.*, *35*(14).

Carbary, J. F., D. G. Mitchell, P. Brandt, E. C. Roelof, and S. M. Krimigis (2008), Periodic tilting of Saturn's plasma sheet, *Geophys. Res. Lett.*, *35*, L24101; doi:10.1029/2008GL036339.

Chandrasekhar, S. (1956), Axisymmetric magnetic fields and fluid motions, *Astrophys. J.*, *124*, 232; doi:10.1086/146217.

Connerney, J., M. Acuna, and N. Ness (1982), Voyager 1 assessment of Jupiter's planetary magnetic field, *J. Geophys. Res. Space Physics*, *87*(A5), 3623–3627.

Cramer, N. F. (1994), Magneto-acoustic surface waves on current sheets, *J. Plasma Phys.*, *51*, 221–232.

Daglis, I. A. (2006), Ring current dynamics, *Space Sci. Rev.*, *124*, 183–202; doi:10.1007/s11214-006-9104-z.

Delamere, P. A., A. Otto, X. Ma, F. Bagenal, and R. J. Wilson (2015), Magnetic flux circulation in the rotationally driven giant magnetospheres, *J. Geophys. Res. Space Physics*, *120*, 4229–4245; doi:10.1002/2015JA021036.

Erkaev, N. V., V. S. Semenov, I. V. Kubyshkin, M. V. Kubyshkina, and H. K. Biernat (2009), Mhd model of the flapping motions in the magnetotail current sheet, *J. Geophys. Res.*, *114*, A03206.

Felici, M., C. S. Arridge, A. J. Coates, S. V. Badman, M. K. Dougherty, C. M. Jackman, W. S. Kurth, et al. (2016), Cassini observations of ionospheric plasma in Saturn's magnetotail lobes, *J. Geophys. Res. Space Physics*, *121*, 338–357; doi:10.1002/2015JA021648.

Fruit, G., and P. Louarn (2011), On the propagation of MHD eigenmodes in a 2-d-magnetotail, *Ann. Geophys.*, *29*, 19–30.

Fruit, G., P. Louarn, A. Tur, and D. Le Quéau (2002), On the propagation of magnetohydrodynamic perturbations in a harris-type current sheet 1. propagation on discrete modes and signal reconstruction, *J. Geophys. Res.*, *107*, 1411.

Génot, V., F. Mottez, G. Fruit, P. Louarn, J.-A. Sauvaud, and A. Balogh (2005), Bifurcated current sheet: model and Cluster observations, *Planet. Space Sci.*, *53*, 229–235; doi:10.1016/j.pss.2004.09.048.

Goertz, C. (1981), The orientation and motion of the predawn current sheet and jupiter's magnetotail, *J. Geophys. Res. Space Physics*, *86*(A10), 8429–8434.

Harris, E. G. (1962), On a plasma sheath separating regions of oppositely directed magnetic field, *Il Nuovo Cimento* (1955–1965), *23*(1), 115–121.

Haynes, P. L., A. Balogh, M. K. Dougherty, D. J. Southwood, and A. Fazakerley (1994), Null fields in the outer jovian magnetosphere: Ulysses observations, *Geophys. Res. Lett.*, *21*, 405–408.

Hill, T. W., A. J. Dessler, and F. C. Michel (1974), Configuration of the Jovian magnetosphere, *Geophys. Res. Lett.*, *1*, 3–6; doi:10.1029/GL001i001p00003.

Israelevich, P., A. Ershkovich, and R. Oran (2007), Bifurcation of the tail current sheet in Jovian magnetosphere, *Planet. Space Sci.*, *55*(15), 2261–2266.

Jackman, C., C. Arridge, H. McAndrews, M. Henderson, and R. Wilson (2009), Northward field excursions in Saturn's magnetotail and their relationship to magnetospheric periodicities, *Geophys. Res. Lett.*, *36*(16).

Jackman, C. M., C. S. Arridge, N. André, F. Bagenal, J. Birn, M. P. Freeman, X. Jia, et al. (2014a), Large-scale structure and dynamics of the magnetotails of Mercury, Earth, Jupiter and Saturn, *Space Sci. Rev.*, *182*(1), 85–154; doi:10.1007/s11214-014-0060-8.

Jackman, C. M., J. A. Slavin, M. G. Kivelson, D. J. Southwood, N. Achilleos, M. F. Thomsen, G. A. DiBraccio, et al. (2014b), Saturn's dynamic magnetotail: A comprehensive magnetic field and plasma survey of plasmoids and traveling compression regions and their role in global magnetospheric dynamics, *J. Geophys. Res. Space Physics*, *119*, 5465–5494; doi:10. 1002/2013JA019388.

Jackman, C. M., N. Achilleos, S. W. H. Cowley, E. J. Bunce, A. Radioti, D. Grodent, S. V. Badman, M. K. Dougherty, and W. Pryor (2013), Auroral counterpart of magnetic field dipolarizations in Saturn's tail, *Planet. Space Sci.*, *82*, 34–42; doi:10.1016/j.pss.2013.03.010.

Kasahara, S., E. A. Kronberg, N. Krupp, T. Kimura, C. Tao, S. V. Badman, A. Retinò, and M. Fujimoto (2011), Magnetic reconnection in the Jovian tail: X-line evolution and consequent plasma sheet structures, *J. Geophys. Res. Space Physics*, *116*(A15), A11219; doi:10.1029/2011JA016892.

Kellett, S., C. S. Arridge, E. J. Bunce, A. J. Coates, S. W. H. Cowley, M. K. Dougherty, A. M. Persoon, N. Sergis, and R. J. Wilson (2010), Nature of the ring current in Saturn's dayside magnetosphere, *J. Geophys. Res. Space Physics*, *115*, A08201; doi:10.1029/2009JA015146.

Kellett, S., C. S. Arridge, E. J. Bunce, A. J. Coates, S. W. H. Cowley, M. K. Dougherty, A. M. Persoon, N. Sergis, and R. J. Wilson (2011), Saturn's ring current: Local time dependence and temporal variability, *J. Geophys. Res. Space Physics*, *116*, A05220; doi:10.1029/2010JA016216.

Kellett, S., E. J. Bunce, A. J. Coates, and S. W. H. Cowley (2009), Thickness of Saturn's ring current determined from north-south Cassini passes through the current layer, *J. Geophys. Res. Space Physics*, *114*, A04209; doi:10.1029/2008JA013942.

Khurana, K. (2001), Influence of solar wind on jupiter's magnetosphere deduced from currents in the equatorial plane, *J. Geophys. Res. Space Physics*, *106*, 25.

Khurana, K. K. (1992), A generalized hinged-magnetodisc model of Jupiter's nightside current sheet, *J. Geophys. Res. Space Physics*, *97*(A5), 6269–6276.

Khurana, K. K., and H. K. Schwarzl (2005), Global structure of Jupiter's magnetospheric current sheet, *J. Geophys. Res. Space Physics*, *110*(A7).

Khurana, K. K., and M. G. Kivelson (1989), On Jovian plasma sheet structure, *J. Geophys. Res. Space Physics*, *94*(A9), 11,791–11,803; doi:10.1029/JA094iA09p11791.

Kivelson, M. G. (2000), *Currents and Flows in Distant Magnetospheres*, Geophysical Monograph Series 118, American Geophysical Union, Washington, DC, 339–352.

Kivelson, M. G. (2015), Planetary magnetodiscs: Some unanswered questions, *Space Sci. Rev.*, *187*(1), 5–21; doi:10.1007/s11214-014-0046-6.

Kivelson, M. G., and X. Jia (2014), Control of periodic variations in saturn's magnetosphere by compressional waves, *J. Geophys. Res. Space Physics*, *119*(10), 8030–8045.

Kubyshkina, D. I., D. A. Sormakov, V. A. Sergeev, V. S. Semenov, N. V. Erkaev, I. V. Kubyshkin, N. Y. Ganushkina, and S. V. Dubyagin (2014), How to distinguish between kink and sausage modes in flapping oscillations?, *J. Geophys. Res. Space Physics*, *119*(4), 3002–3015; doi:10.1002/2013JA019477.

Lachin, A. (1998), Low frequency waves in the solar system, Ph.D. thesis, Imperial College London.

Leamon, R. J., M. K. Dougherty, D. J. Southwood, and P. L. Haynes (1995), Magnetic nulls in the outer magnetosphere of Jupiter: Detections by Pioneer and Voyager spacecraft, *J. Geophys. Res.*, *100*, 1829–1835; doi:10.1029/94JA01963.

Leisner, J. S., C. T. Russell, K. K. Khurana, and M. K. Dougherty (2007), Measuring the stress state of the Saturnian magnetosphere, *Geophys. Res. Lett.*, *34*, L12103; doi:10.1029/2007GL029315.

Lepping, R. P. (1994), Comparison of the field configurations of the magnetotails of Uranus and Neptune, *Planet. Space Sci.*, *42*, 847–857; doi:10.1016/0032-0633(94)90065-5.

Louarn, P., G. Fruit, E. Budnik, J. A. Sauvaud, C. Jacquey, D. Le Quéau, H. Rème, E. Lucek, and A. Balogh (2004), On the propagation of low-frequency fluctuations in the plasma sheet: 1. cluster observations and magnetohydrodynamic analysis, *J. Geophys. Res.*, *109*, A03216.

Louarn, P., N. Andre, C. M. Jackman, S. Kasahara, E. A. Kronberg, and M. F. Vogt (2015), Magnetic reconnection and associated transient phenomena within the magnetospheres of Jupiter and Saturn, *Space Sci. Rev.*, *187*(1), 181–227; doi:10.1007/s11214-014-0047-5.

Martin, C. J., and C. S. Arridge (2017), Cassini observations of aperiodic waves on Saturn's magnetodisc, *J. Geophys. Res. Space Physics*, 122, 8063–8077, doi:10.1002/2017JA024293.

McNutt, R.L., Jr. (1984), Force balance in outer planet magnetospheres, in *Physics of Space Plasmas, Proceedings of the 1982-4 MIT Symposia*, edited by J. Belcher, H. Bridge, T. Change, B. Coppi, and J.R. Jasperse, 179, SPI Conference Proceedings and Reprint Series, #5. Scientific Publishers, Inc.

Mejnertsen, L., J. P. Eastwood, J. P. Chittenden, and A. Masters (2016), Global mhd simulations of Neptune's magnetosphere, *J. Geophys. Res. Space Physics*, *121*(8), 7497–7513; doi:10.1002/2015JA022272, 2015JA022272.

Mestel, L. (1961), A note on equatorial acceleration in a magnetic star, *Monthly Not. R. Astron. Soc.*, *122*, 473; doi:10.1093/mnras/122. 6.473.

Mestel, L. (1968), Magnetic braking by a stellar wind-I, *Monthly Not. R. Astron. Soc.*, *138*, 359; doi:10.1093/mnras/138.3.359.

Nichols, J. D., N. Achilleos, and S. W. H. Cowley (2015), A model of force balance in Jupiter's magnetodisc including hot plasma pressure anisotropy, *J. Geophys. Res. Space Physics*, *120*(A9), 10; doi:10.1002/2015JA021807.

Northrop, T. G., C. K. Goertz, and M. F. Thomsen (1974), The magnetosphere of Jupiter as observed with Pioneer 10: 2. Nonrigid rotation of the magnetodisc, *J. Geophys. Res.*, *79*, 3579; doi:10.1029/JA079i025p03579.

Paranicas, C. P., B. H. Mauk, and S. M. Krimigis (1991), Pressure anisotropy and radial stress balance in the Jovian neutral sheet, *J. Geophys. Res.*, *96*, 21; doi:10.1029/91JA01647.

Piddington, J. H. (1969), Cosmic electrodynamics. Interscience Monographs and Texts in Physics and Astronomy. New York: Wiley-Interscience, and Huntington: Krieger.

Russell, C., D. Huddleston, K. Khurana, and M. Kivelson (1999a), Structure of the Jovian magnetodisk current sheet: Initial Galileo observations, *Planet. Space Sci.*, *47*(8), 1101–1109.

Russell, C. T., and K. I. Brody (1967), Some remarks on the position and shape of the neutral sheet, *J. Geophys. Res.*, *72*, 6104–6106; doi:10.1029/JZ072i023p06104.

Russell, C. T., D. E. Huddleston, K. K. Khurana, and M. G. Kivelson (1999b), Observations at the inner edge of the Jovian current sheet: Evidence for a dynamic magnetosphere, *Planet. Space Sci.*, *47*, 521–527; doi:10.1016/S0032-0633(98)00119-6.

Russell, C. T., K. K. Khurana, D. E. Huddleston, and M. G. Kivelson (1998), Localized reconnection in the near Jovian Mmagnetotail, *Science*, *280*, 1061; doi:10.1126/science.280.5366.1061.

Russell, C. T., Z. J. Yu, K. K. Khurana, and M. G. Kivelson (2001), Magnetic field changes in the inner magnetosphere of Jupiter, *Adv. Space Res.*, *28*, 897–902; doi:10.1016/S0273-1177(01)00513-0.

Sergeev, V. A., D. G. Mitchell, C. T. Russell, and D. J. Williams (1993), Structure of the tail plasma/current sheet at 11 re and its changes in the course of a substorm, *J. Geophys. Res. Space Physics*, *98*(A10), 17,345–17,365; doi:10.1029/93JA01151.

Sergis, N., C. S. Arridge, S. M. Krimigis, D. G. Mitchell, A. M. Rymer, D. C. Hamilton, N. Krupp, M. K. Dougherty, and A. J. Coates (2011), Dynamics and seasonal variations in Saturn's magnetospheric plasma sheet, as measured by Cassini, *J. Geophys. Res. Space Physics*, *116*, A04203; doi:10.1029/2010JA016180.

Sharma, A. S., R. Nakamura, A. Runov, E. E. Grigorenko, H. Hasegawa, M. Hoshino, P. Louarn, et al. (2008), Transient and localized processes in the magnetotail: A review, *Ann. Geophys.*, *26*, 955–1006; doi:10.5194/angeo-26-955-2008.

Shay, M. A., and M. Swisdak (2004), Three-species collisionless reconnection: Effect of O+ on magnetotail reconnection, *Phys. Rev. Lett.*, *93*, 175001; doi:10.1103/PhysRevLett.93.175001.

Shodan, S., N. U. Crooker, W. J. Hughes, and G. L. Siscoe (1994), Heliospheric current sheet inclinations predicted from source surface maps, *J. Geophys. Res.*, *99*, 2531–2536.

Smith, A. W., C. M. Jackman, and M. F. Thomsen (2016), Magnetic reconnection in Saturn's magnetotail: A comprehensive magnetic field survey, *J. Geophys. Res. Space Physics*, *121*(4), 2984–3005; doi:10.1002/2015JA022005, 2015JA022005.

Smith, E., L. Davis, D. Jones, P. Coleman, D. Colburn, P. Dyal, C. Sonett, and A. Frandsen (1974), The planetary magnetic field and magnetosphere of Jupiter: Pioneer 10, *J. Geophys. Res.*, *79*(25), 3501–3513.

Smith, E., L. Davis, D. Jones, P. Coleman, D. Colburn, P. Dyal, and C. Sonett (1980), Saturn's magnetic field and magnetosphere, *Science*, *207*(4429), 407–410.

Thomsen, M. F. (2013), Saturn's magnetospheric dynamics, *Geophys. Res. Lett.*, *40*, 5337–5344; doi:10.1002/2013GL057967.

Thomsen, M. F., C. M. Jackman, D. G. Mitchell, G. Hospodarsky, W. S. Kurth, and K. C. Hansen (2015b), Sustained lobe reconnection in Saturn's magnetotail, *J. Geophys. Res. Space Physics*, *120*, 10,257–10,274; doi:10.1002/2015JA021768.

Thomsen, M. F., D. G. Mitchell, X. Jia, C. M. Jackman, G. Hospodarsky, and A. J. Coates (2015a), Plasmapause formation at Saturn, *J. Geophys. Res. Space Physics*, *120*, 2571–2583; doi:10.1002/2015JA021008.

Tirry, W. J., V. M. Cadez, and M. Goossens (1997), MHD surface type quasi-modes of a current sheet model, *Astron. Astrophysics*, *324*, 1170–1178.

Tóth, G., D. Kovács, K. C. Hansen, and T. I. Gombosi (2004), Three-dimensional MHD simulations of the magnetosphere of Uranus, *J. Geophys. Res. Space Physics*, *109*(A18), A11210; doi:10.1029/2004JA010406.

Tsyganenko, N. A. (2002), A model of the near magnetosphere with a dawn-dusk asymmetry 1. mathematical structure, *J. Geophys. Res. Space Physics*, *107*(A8), SMP 12-1–SMP 12-15; doi:10.1029/2001JA000219.

Tsyganenko, N. A., and V. A. Andreeva (2014), On the "bowl-shaped" deformation of planetary equatorial current sheets, *Geophys. Res. Lett.*, *41*, 1079–1084; doi:10.1002/2014GL059295.

Tsyganenko, N. A., S. B. P. Karlsson, S. Kokubun, T. Yamamoto, A. J. Lazarus, K. W. Ogilvie, C. T. Russell, and J. A. Slavin (1998), Global configuration of the magnetotail current sheet as derived from Geotail, Wind, IMP 8 and ISEE 1/2 data, *J. Geophys. Res.*, *103*, 6827–6842; doi:10.1029/97JA03621.

Tsyganenko, N. A., V. A. Andreeva, and E. I. Gordeev (2015), Internally and externally induced deformations of the magnetospheric equatorial current as inferred from spacecraft data, *Ann. Geophys.*, *33*, 1–11; doi:10.5194/angeo-33-1-2015.

Vasyliunas, V. M. (1983), Plasma distribution and flow, in *Physics of the Jovian Magnetosphere*, edited by A. J. Dessler, Cambridge University Press, 395–453.

Vasyliunas, V. M. (1994), Role of the plasma acceleration time in the dynamics of the Jovian magnetosphere, *Geophys. Res. Lett.*, *21*(6), 401–404; doi:10.1029/94GL00192.

Vasyliunas, V. M. (2015), *Magnetotail: Unsolved Fundamental Problem of Magnetospheric Physics*, John Wiley & Sons, Inc., 1–19.

Vasyliunas, V. M., L. A. Frank, K. L. Ackerson, and W. R. Paterson (1997), Geometry of the plasma sheet in the midnight-to-dawn sector of the Jovian magnetosphere: Plasma observations with the Galileo spacecraft, *Geophys. Res. Lett.*, *24*, 869–872; doi:10.1029/97GL00757.

Vogt, M. F., C. M. Jackman, J. A. Slavin, E. J. Bunce, S. W. H. Cowley, M. G. Kivelson, and K. K. Khurana (2014), Structure and statistical properties of plasmoids in Jupiter's magnetotail, *J. Geophys. Res. Space Physics*, *119*(2), 821–843; doi:10.1002/2013JA019393, 2013JA019393.

Vogt, M. F., M. G. Kivelson, K. K. Khurana, S. P. Joy, and R. J. Walker (2010), Reconnection and flows in the Jovian magnetotail as inferred from magnetometer observations, *J. Geophys. Res. Space Physics*, *115*(A6); doi:10.1029/2009JA015098, a06219.

Volwerk, M., N. André, C. S. Arridge, C. M. Jackman, X. Jia, S. E. Milan, A. Radioti, et al. (2013), Comparative magnetotail flapping: An overview of selected events at Earth, Jupiter and Saturn, *Ann. Geophys.*, *31*, 817–833; doi:10.5194/angeo-31-817-2013.

Waldrop, L. S., T. A. Fritz, M. G. Kivelson, K. Khurana, N. Krupp, and A. Lagg (2005), Jovian plasma sheet morphology: particle and field observations by the Galileo spacecraft, *Planet. Space Sci.*, *53*, 681–692; doi:10.1016/j.pss.2004.11.003.

Went, D. R., M. G. Kivelson, N. Achilleos, C. S. Arridge, and M. K. Dougherty (2011), Outer magnetospheric structure: Jupiter and Saturn compared, *J. Geophys. Res. Space Physics*, *116*, A04224; doi:10.1029/2010JA016045.

Yamauchi, M., and A. T. Y. Lui (1997), Modified magnetohydrodynamic waves in a current sheet in space, *Phys. Plasmas*, *4*, 4382–4387; doi:10.1063/1. 872600.

Yin, L., F. V. Coroniti, P. L. Pritchett, L. A. Frank, and W. R. Paterson (2000), Kinetic aspects of the Jovian current sheet, *J. Geophys. Res. Space Physics*, *105*(A11), 25,345–25,361; doi:10.1029/2000JA000016.

Zimbardo, G. (1991), Collisionless reconnection in Jupiter's magnetotail, *Geophys. Res. Lett.*, *18*, 741–744.

13
Planetary Magnetopause and Heliopause Current Sheets

E. S. Belenkaya

ABSTRACT

The interaction of two kinds of plasma at the boundary between them leads to formation of two current sheets. Each current is carried by the ions of the outer plasma. We consider a few aspects connected with the magnetopause/heliopause influence on the surrounding space. We investigate a fundamental property of the magnetopause, as a multilevel structure, its reaction to the sudden solar wind dynamic pressure pulse, and the possibility to amplify the magnetic field in a passing by plasma flow at the magnetopause of a rapidly rotating magnetized planet. We show that FTEs could arise in one of the Mercury cusps dependent on the sign of B_x IMF component. The dynamo mechanism, suggested by us for Jupiter's magnetopause, was applied to explain the unexpected observations of the magnetic field strength drop at the heliopause without change of field's direction. Here we consider the Voyager 2 observations in the outer heliosphere to clarify our estimations. Although Voyager 2 did not intersect the heliopause yet, it directly measures the plasma speed, which allows us to use its measurements for extrapolation of the velocity field.

13.1. INTRODUCTION

Magnetopause is a boundary of the magnetized planet magnetosphere, the cavity, in which magnetic fields of the planet and the magnetospheric current systems are contained. Magnetopause plays a significant role for the magnetosheath and the magnetosphere, between which it is located, mainly due to the reconnection processes acting at it or in its vicinity [e.g., *Ergun et al.*, 2016]. During these processes, the transport of mass, momentum, and energy takes place through the boundary. In the last years, attention of the scientific community was focused on the comparative investigation of the magnetospheres of the solar system planets [e.g., *Keiling et al.*, 2014].

Federal State Budget Educational Institution of Higher Education M. V. Lomonosov Moscow State University, Skobeltsyn Institute of Nuclear Physics (SINP MSU), 1(2), Leninskie gory, GSP-1, Moscow, Russian Federation

In a rough approximation, magnetopause is not permeable for magnetized plasmas located on both sides of it. However, there are some exceptions. First, the zones of the magnetopause, where the magnetospheric magnetic field is very weak: the cusp regions and flanges of the tail neutral current sheet. When magnetospheric magnetic field is close to zero, it does not act as an obstacle for the outer plasma flow, which can partially penetrate through these regions. Second, the reconnection process, which is very significant for magnetospheric dynamics, allows partial penetration of plasma and fields. As a result, the inner and outer boundary layers arise on both sides of the magnetopause, in which plasmas from the magnetosheath and magnetosphere are present [*Cowley*, 2015].

Magnetopause can be considered as a current sheet (or sheets) separating two different sorts of plasma and their magnetic fields [*Song et al.*, 1993]. Due to this, magnetopause manifests itself by the magnetic field drop and/or the change of its orientation.

Electric Currents in Geospace and Beyond, *Geophysical Monograph 235*, First Edition.
Edited by Andreas Keiling, Octav Marghitu, and Michael Wheatland.

Heliopause is a boundary of the solar system in which the interplanetary magnetic field and solar wind plasma dominate. It also divides two sorts of plasma: the interstellar and heliospheric, and their magnetic fields. In this sense, it is similar to the magnetopause. However, at this boundary, very complicated processes, specific for the interaction with the interstellar medium, take place, including those that are associated with ENAs (energetic neutral atoms).

Both, magnetopause and heliopause are formed when a flow of magnetized plasma passes by the obstacle, whose strong magnetic field stops it. For this reason, the magnetospheric/heliospheric magnetic field does not allow the particles of the outer flow to penetrate inside the magnetosphere/heliosphere. As a result, the flow's particles reflecting from the boundary between two mediums generate a magnetopause/heliopause current. In turn, the magnetospheric/heliospheric charged particles should also carry their magnetopause/heliopause current. However, it is necessary to take into account a lot of complex processes, specific for the heliopause, connected with the interstellar neutral atoms, charge-exchange interactions, creation of the new heliospheric energetic neutrals, and so on [e.g., *Avinash et al.*, 2014]. In section 13.2, we will consider the magnetopause current sheets in more detail.

In collisionless plasma, the diffusion of charged particles across the magnetic field at the magnetopause acts on a length smaller than a few Larmor radii or the inertia lengths, however, this narrow region plays a crucial role in forming and dynamics of the magnetosphere [e.g., *Hasegawa*, 2012]. In section 13.3, we consider how disturbance of the magnetopause current under the action of the sudden jump of the solar wind dynamic pressure influences the magnetic field of the terrestrial and Kronian magnetospheres. Magnetopause of each planet has its own specific features determined by its position in the solar system and the planet's characteristics. On the example of Mercury (section 13.4) and Jupiter (section 13.5), we demonstrate how the magnetopauses of these smallest and greatest planets of the solar system affect the environment of their magnetospheres. In section 13.6, a principal role of the heliopause current sheet in the interaction of the interstellar flow with the heliosphere is shown, and in section 13.7, the summary is given.

13.2. CURRENT SHEETS AT THE MAGNETOPAUSE

Starting in the 1930s, many models of the magnetopause structure were created [e.g., *Chapman and Ferraro*, 1931a, b, 1932, 1933, 1940]. In these models, interplanetary magnetic field (IMF) was ignored, as well as the existence of the magnetospheric plasma. *Chapman and Ferraro* [1931a, b] showed that for the infinitely conducting magnetopause its electric current field fully screens the dipole field. *Ferraro* [1952] considered several plane sheets at the magnetopause. Later, after the spacecraft investigations, it was found that considering the magnetopause currents it is necessary to take into account two magnetized plasmas (solar wind and magnetospheric). In several models [*Sestero*, 1964, 1966; *Lee and Kan*, 1979a; *Roth*, 1978, 1979], solutions of Vlasov equation were searched as integrals of motion. The integrals of motion are the total energy and the generalized momentum along the direction of velocity shift. *Lee and Kan* [1979b] included the ratio of the ion to electron temperatures, considering the magnetopause properties. It occurred that the hotter ions determine the thickness of the magnetopause.

The terrestrial magnetopause vicinity was analyzed earlier, for example, by such spacecraft as ISEE, AMPTE, and Geotail. In recent years, due to the Cluster, THEMIS, and MMS (Magnetospheric Multiscale mission) data, the magnetopause structure has been investigated in more detail [e.g., *Hasegawa*, 2012; *Eastwood et al.*, 2015; *Burch et al.*, 2016]. *Hasegawa* [2012] paid attention to a new possibility to study 2D and 3D magnetopause structures having several spacecraft data, and to the reconnection processes acting at the magnetopause and investigated by the in situ and remote observations. He emphasized also the significant role of the magnetopause Kelvin-Helmholtz instability and considered its evolution. *Eastwood et al.* [2015] describe the basic structure of the magnetopause from the modern point of view. *Burch et al.* [2016], using the MMS measurements, found the experimental support of the idea that the minimum space scale of the reconnection diffusion region is the character electron scale.

Considering the plasma behavior on a scale of the Larmor radii, we should apply the kinetic approach, contrary to the MHD one appropriate for larger sizes. *Belenkaya* [2001] presented a simple model of the magnetopause current sheets describing explicitly the role of each of them. The magnetopause is considered as a plane stationary boundary. This rough assumption is commonly used in zero approximation.

Near the magnetopause, the density of the solar wind/magnetosheath plasma is 5 to 10 times higher than the density of the magnetospheric plasma, while the temperature of plasma in the magnetosheath is 6 to10 times lower than in the magnetosphere. Figure 13.1 shows the ion orbits (Larmor circles) on both sides of the magnetopause, which is presented as a thin line along the Y-axis in the equatorial plane (X, Y). X is the direction to the Sun. Larmor radius ρ is given by equation (13.1):

$$\rho = mV_{\perp} / eB, \qquad (13.1)$$

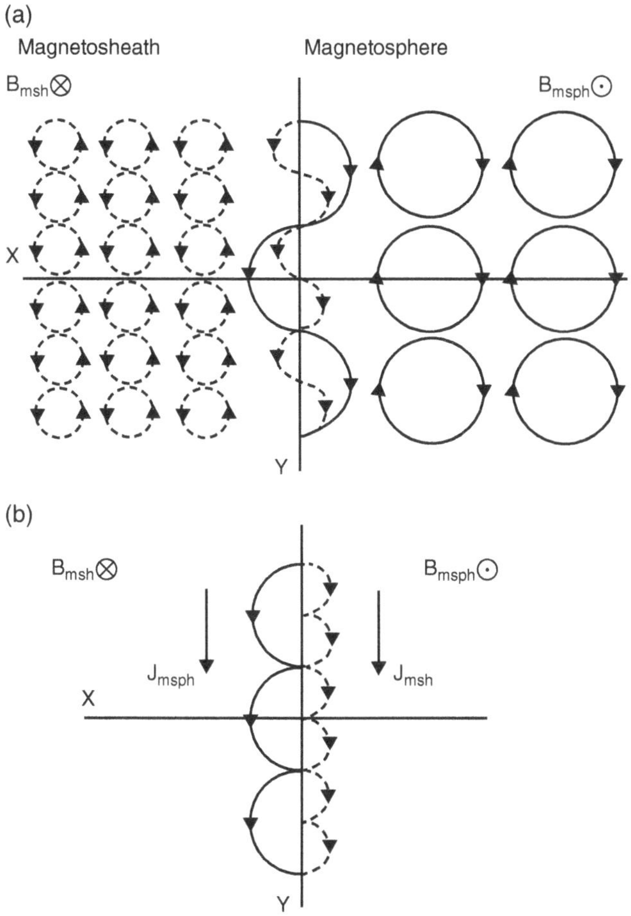

Figure 13.1 Southward IMF, equatorial plane (X, Y); magnetopause is located along Y axis; X axis is directed to the Sun. (a) The magnetospheric ion Larmor orbits are marked by solid curves and magnetosheath ion Larmor circles are marked by dashed curves. (b) Each sort of ions does not penetrate through the boundary on the length exceeding two Larmor radii. Resulting current at the boundary caused by the magnetospheric ions is shown by solid curve, and resulting current created by the magnetosheath ions is shown by dashed curve. Directions of the currents are shown by arrows.

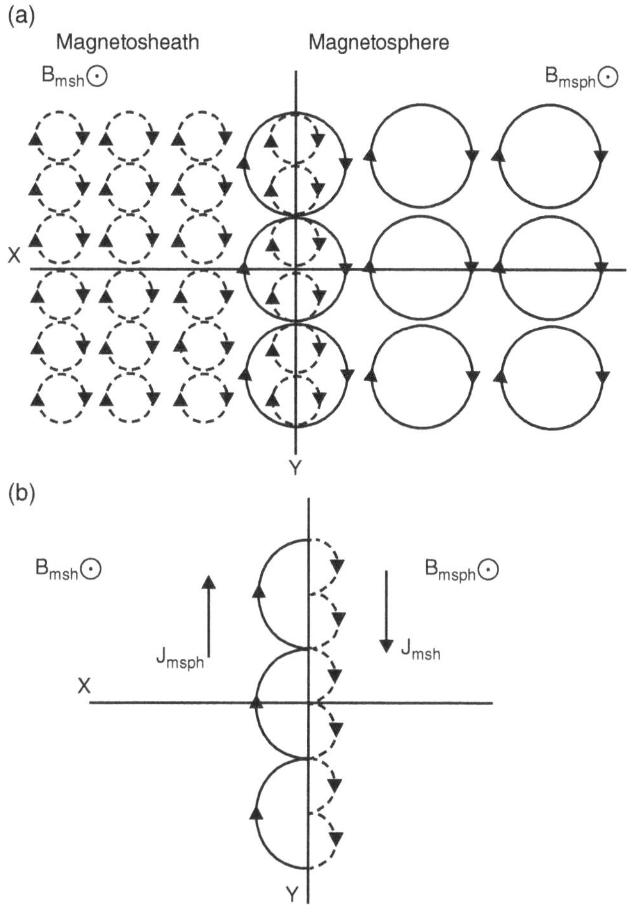

Figure 13.2 In the same format as Figure 13.1, but for northward IMF.

where m is the particle's mass, e is its charge, and $V_\perp$ is its velocity perpendicular to the magnetic field, **B**. In Figure 13.1, a case of the southward IMF is shown. Cyclotron circles of the magnetospheric (magnetosheath) ions are drawn by solid (dashed) curves. As the magnetic field changes its direction on 180° at the boundary, the sense of the ion's rotation also changes. The current on the arcs of the magnetospheric cyclotron circles intersecting the magnetopause, which are located in the magnetosphere, are annihilated by the neighboring ions' current. As the Larmor radii on both sides of the boundary are different due to the different plasma parameters, the current on the arcs of the magnetospheric cyclotron circles located in the magnetosheath, is not annihilated by the surrounding ions. This residual current is marked as J_{msph} in Figure 13.1. By analogy, the current on the arcs of the magnetosheath ion orbits placed in the magnetosphere, remains and is marked as J_{msh} in Figure 13.1. Thus, two current sheets are formed on both sides of the boundary: the current carried by the magnetospheric ions is placed on the magnetosheath side, and the current of the magnetosheath ions is located on the magnetospheric side. These currents are parallel to each other in the case of southward IMF and directed to the dusk. They form the magnetopause current zone.

For northward IMF (see Fig. 13.2), the magnetic field does not change its direction on the boundary, and, consequently, the sense of the ion's rotation conserves on the both sides of it. The remaining current carried by the magnetospheric (magnetosheath) ions is located in the magnetosheath (magnetosphere) and is directed to the dawn (dusk). Therefore, for northward IMF, these currents are antiparallel.

Two current layers considered by us for southward and northward IMF, are the magnetopause currents with the total thickness ~$2(\rho_{msh} + \rho_{msph})$, where ρ_{msh} and ρ_{msph} are the magnetosheath and magnetospheric ion cyclotron radii, respectively. According to observations, the Earth's low-latitude magnetopause has a thickness of a few to about 10 times the proton Larmor radius or inertia length (~800 km) [e.g., *Hasegawa*, 2012]. The currents considered here are created by plasma space inhomogeneity at the boundary and are the diamagnetic currents. This means that magnetospheric ions create the current whose magnetic field tries to prevent the IMF penetration into the magnetosphere, while the magnetosheath ion current does not allow the magnetospheric field to penetrate into the magnetosheath. As the current is proportional to the density of its carriers, the magnetosheath ions create more effective shielding than the magnetospheric ions. In the magnetospheres of other planets, the similar magnetopause current sheets could be generated by the same reason.

13.3. REACTION OF THE TERRESTRIAL AND KRONIAN MAGNETOPAUSE CURRENT SHEETS ON THE SUDDEN DROP OF THE SOLAR WIND DYNAMIC PRESSURE

Knowing the nature of the magnetopause currents allows us to find their disturbance under the action of the solar wind dynamic pressure jump. We showed above that the magnetosheath charged particles create the current located on the magnetospheric side of the magnetopause. This current is directed from dawn to dusk independent of the IMF orientation and screens the magnetospheric magnetic field. The magnetopause current carried by magnetosheath ions is controlled by the solar wind dynamic pressure. For this reason, variations in the solar wind pressure cause the disturbance in the magnetospheric magnetic field due to their influence on the magnetopause current. *Belenkaya et al.* [2006] and *Belenkaya* [2014] considered the action of the dynamic pressure pulse on the terrestrial and Kronian magnetospheres. Disturbance of the magnetopause current is the beginning of the magnetospheric reaction on the pressure pulse. After it, the MHD waves are generated. The propagation tailward of the disturbance caused by the magnetopause current forms the transition current system. *Belenkaya et al.* [2006, 2014b] showed that this system contains the increased magnetopause diamagnetic current and the polarization current closing it at the tailward edge of the perturbed magnetospheric zone. These two currents form some kind of a ring in the magnetospheric equatorial plane, which exists during the pulse action; that is why we called this current system the "transient ring current." The situation arising when the solar wind pressure jump is accompanied by a northward IMF turning is described by *Belenkaya et al.* [2004, 2007].

Considering the Kronian magnetosphere, it is necessary to take into account that plasma motion in it is controlled not only by the solar wind, as in the case of the Earth, but also by the planetary fast rotation [e.g., *Cowley et al.*, 2004]. *Cowley et al.* [2005] investigated how the solar wind pressure pulse is connected with disturbance of the Kronian aurora. *Masters et al.* [2012] noted that the reconnection at the Kronian magnetopause could be significantly different from that one for the Earth (probably, less effective).

Following *Tamao* [1975] and *Wilken et al.* [1982], studying the terrestrial magnetosphere under the action of a strong pressure jump, *Belenkaya et al.* [2006] discussed the development of the transient ring current in the Kronian magnetosphere. Due to the antiparallel directions of the Earth's and Saturn's dipole moments, the directions of currents in the transient ring current systems of these two planets are also opposite. For this reason, magnetic fields generated by these two transient systems are antiparallel: southward for Saturn and northward for the Earth. *Belenkaya* [2014] calculated enhancement of the Saturn's magnetopause current during the solar wind pressure pulses on 16 and 26 January 2004. It was shown that the magnetopause current was increased three times on 16 January and by the order of magnitude on 26 January 2004.

For the Earth, expressions for the enhanced magnetopause current and magnetic field of the transient ring current were obtained in *Belenkaya et al.* [2014a]. On the basis of these expressions, the disturbed magnetic field was calculated for several cases (17 April 1993, 4 June 2000, and 19 May 2005). Here we investigate the case of 6 November 2000, considered, for example, by *Shi et al.* [2005]. The solar wind parameters and IMF were measured by ACE. At ~18:00 UT, the pressure pulse was registered. This pulse was caused mainly by the sharp increase of the density, while IMF was almost constant, as well as velocity (~600 km s^{-1}). Density increased from 6.7 cm^{-3} to 20 cm^{-3}. In *Belenkaya et al.* [2014a], the maximum magnetopause current density increase, δj_{sw}, due to the solar wind density drop, δn_{sw}, and velocity change, δV_{sw}, (when it exists) is given as follows:

$$\delta j_{sw} = (e/\pi) n_{sw} V_{sw} \left[(1 + \delta n_{sw}/n_{sw})(1 + \delta V_{sw}/V_{sw}) - 1 \right] \tag{13.2}$$

The corresponding northward magnetic field of the arising transient ring current is

$$\delta B_z = (2\mu_0 n_{sw} m_{sw})^{1/2} V_{sw}[(1 + \delta n_{sw}/n_{sw})(1 + \delta V_{sw}/V_{sw}) - 1] / [\pi (1 + \delta n_{sw}/n_{sw})^{1/2}] \tag{13.3}$$

(μ_0 is the magnetic permeability of the vacuum). As in the studied case δV_{sw}~0, the last expression transfers to

$$\delta B_z = \left(2\mu_0 m_{sw} / n_{sw}\right)^{1/2} V_{sw} \delta n_{sw} / [\pi \left(1 + \delta n_{sw} / n_{sw}\right)^{1/2}] \tag{13.4}$$

For $n_{sw} = 6.7$ cm^{-3}, $\delta n_{sw} = 13.3$ cm^{-3}, and $V_{sw} = 600$ km s^{-1}, we obtain the northward δB_z~37 nT in the low-latitude magnetosphere. The SYM-H index given, for example, in *Shi et al.* [2005], shows at 18 UT the positive sharp increase by ~33 nT, which is in good accordance with our calculations. This peak is considered by us as caused by the solar wind pressure pulse, which induced the magnetopause current enhancement followed by the development of the transient ring current with magnetic field δB_z.

13.4. CHARACTER FEATURE OF THE MERCURY'S MAGNETOPAUSE STRUCTURE

Mercury has a rather small magnetic field. It is located at the heliocentric distance ~0.3 AU. At Mercury's orbit, IMF is mainly radially directed, which is supported by both the IMF description by the Parker spiral and the MESSENGER observations. These conditions determine the specific features of the IMF interaction with the Hermean magnetosphere. *Belenkaya et al.* [2013] show that, as in the case of the Earth, Mercury's magnetospheric structure for strong radial IMF is rather complex and different from those for the other IMF orientations. This was demonstrated on the base of the Hermean paraboloid magnetospheric magnetic field model introduced in *Alexeev et al.* [2008]. In the paraboloid model, magnetopause is described by a paraboloid of revolution. Using this model, *Alexeev et al.* [2010] found from the MESSENGER data the value of the magnetic dipole moment ~196 nT R_M^3 and its northward offset ~0.17 R_M.

When IMF is oriented close to the direction of the Sun-planet line, as is common for Mercury, the reconnection with the magnetospheric magnetic field takes place in two cusps. In one cusp, the reconnection has a 3D character and occurs in a magnetic field neutral point, while in the other cusp the 2D reconnection takes place at a quasi-neutral line located at the magnetopause. In which cusp what kind of reconnection occurs depends on a sign of the dominant radial IMF. For sunward (antisunward) IMF, the neutral point is in the northern (southern) cusp, and the quasi-neutral line lies on the southern (northern) cusp's magnetopause.

At the planetary magnetopauses, the flux transfer events (FTEs), generated as a result of unsteady reconnection, are often observed [e.g., *Keiling et al.*, 2014]. Knowledge of the magnetospheric and magnetopause structures for the radial IMF orientation, typical for Mercury's environment, allows us to suggest that at the quasi-neutral line in the cusp, the FTEs could arise in a similar way, as they are generated at the dayside terrestrial magnetopause neutral line for IMF with southward component [*Belenkaya et al.*, 2013]. This suggestion was verified using the MESSENGER's measurements of magnetic field [see, e.g., *Slavin et al.*, 2012].

Belenkaya et al. [2013] selected the cases with similar IMF orientation, with large sunward radial component. For this orientation, we should expect arising of FTEs at the quasi-neutral line on the magnetopause of the southern cusp and the absence of FTEs in the northern cusp, where magnetic null exists. This was supported by observations. For example, on 11 April 2011, when the measured IMF was {35; 0; 20 nT}, the FTEs were registered by MESSENGER downstream of the southern cusp, and on 21 February 2012, when dominated B_x was ~25 nT, no FTEs were seen in the northern cusp region in accordance with our analysis. Thus, at the Mercury magnetosphere, FTEs can arise not only at the low-latitude dayside magnetopause for southward IMF, but also in one of the cusps for strong radial IMF dependent on its sign.

13.5. POSSIBLE MECHANISM OF IMF AMPLIFICATION AT THE JOVIAN MAGNETOPAUSE

Belenkaya [1996] considered a special kind of amplification of the solar wind magnetic field at Jupiter's magnetopause. This amplification is in principle different from the magnetic barrier arising when the solar wind flow breaks dawn by the planet's magnetosphere. The new considered mechanism is connected with braking of rotation transferred from the planet just outside the magnetopause.

Jupiter is the largest planet of the solar system, possessing the strongest magnetic field (magnetic dipole moment M ~1.6 · 10^{20} T m^3) and powerful magnetosphere. It is a fast-spinning planet with high-conducting ionosphere. The main source of magnetospheric energy is the energy of Jupiter's spinning (period of rotation is ~10 hr). Plasma corotation is transferred along open field lines to the magnetopause and stops just out of it, in a thin layer with thickness δ_V. Relation δ_V/δ_B is equal to $(\mu_0\sigma\nu)^{1/2}$, where δ_B is the thickness of a magnetopause layer, where magnetic field changes, and σ and ν are the magnetopause conductivity and effective viscosity, respectively [*Belenkaya*, 1996, and references therein]. According to our scenario

$$\delta_V / \delta_B = \left(\mu_0 \sigma \nu\right)^{1/2} > 1 \tag{13.5}$$

which means that $\sigma > 1/(\mu_0 \nu)$. This determines the lower limit of magnetopause conductivity, which is necessary for providing the action of the suggested mechanism, a hydromagnetic dynamo. Expression (13.5) describes one of the necessary conditions [e.g., *Landau and Lifshitz*, 1959], stating that the thickness of the magnetopause layer, where velocity brakes, should be larger than the thickness of a layer, where magnetic field changes (for example, for Earth, this ratio ≥ 3).

A dynamo can generate a magnetic field in a conducting media by flows, which kinetic energy transfers to the magnetic energy. During this process, the toroidal field is stretched from the small background poloidal field by differential rotation, and α-effect creates poloidal field from the toroidal one due to the Coriolis force (αΩ dynamo) [e.g., *Moffat*, 1978]. For the dynamo action, the enhanced magnetic field diffusion is necessary; otherwise, the increased toroidal field stops the dynamo development.

Just out of the Jovian magnetopause, in the thin layer where rotation stops, the magnetic diffusion $\lambda = (\mu_0 \sigma)^{-1}$ exists due to the magnetopause conductivity σ, the lower limit of which was determined above by equation (13.5), with the upper limit determined by field-aligned conductivity. *Belenkaya* [1996] showed that for Jupiter such enhancement of the solar wind magnetic field could occur out of the magnetopause at the distance l from the rotation axis: $l_0 \leq l \leq R_1$, where $l_0 \sim 30\ R_J$ and $R_1 \sim 90\ R_J$ ($R_J \sim 71.5 \cdot 10^3$ km is the Jupiter's radius). Here R_1 is the distance to the subsolar magnetopause from the planetary center and l_0 is the minimum distance from the rotation axis, for which $\Omega_0 l / V_0 \geq 1$ ($l_0 = V_0/\Omega_0$), where $V_0 \sim 400$ km s^{-1} is the solar wind velocity and $\Omega_0 = 2 \cdot 10^{-4}$ s^{-1} is Jupiter's angular velocity. This means that for such values of l, kinetic rotation energy exceeds the kinetic solar wind flow energy. Only from more powerful sources of kinetic energy, the additional magnetic energy of the solar wind may be obtained.

Belenkaya [1996] estimated maximum amplification of IMF at Jupiter's magnetopause by a factor of ~10. However, as the real value of Ω at the magnetopause is less than that for rigid corotation (Ω_0), l_0 is greater than 30 R_J. Consequently, the length of the dynamo layer (from l_0 to R_1) decreases, and decreases the field amplification factor $b/b_0 \approx \Omega\, l / V_A$, where $V_A = b_0/(\mu_0 n m_p)^{1/2}$ is the Alfvén velocity, n and m_p are the solar wind/magnetosheath proton density and proton mass, respectively. According to *Belenkaya* [2004], $\Omega/\Omega_0 \sim 0.2$, and consequently, $b/b_0 \approx 2$.

This approach can be applied not only to Jupiter, but also to any rapidly spinning planet with strong magnetic field and large conductivities of the ionospheric and magnetopause layers. These conductivities are less than the field-aligned conductivity due to collisions in the ionosphere and existence of magnetic diffusion at the magnetopause. The interrelation between the planetary magnetic field and the angular velocity should provide the condition: $R_1 > l_0 = V_0/\Omega_0$. In the solar system, this is valid only for Jupiter [*Belenkaya*, 1996]. However, among exoplanets, the appropriate objects could be found. Moreover, in the following section we show that this mechanism can work also at the heliopause.

13.6. DYNAMO AT THE HELIOPAUSE

In principle, the causes of the magnetopause and heliopause formation are similar. Flow of the magnetized plasma passes the obstacle, whose strong magnetic field stops it, creating the draping. For this reason, change of magnetic field strength and direction is expected at the heliopause, similar to the case of a planetary magnetopause. However, when Voyager 1 (V1) intersected the heliopause on 25 August 2012 at 121.7 AU, the magnetic field strength increased twice (up to 0.4 nT), while the field direction did not change and retained the spiral direction observed inside the heliosphere [*Burlaga et al.*, 2013]. The conclusion that V1 really intersected the heliopause was supported by simultaneous decrease of the heliospheric particle number, increase of the galactic cosmic ray intensity, growth of plasma density by the order of magnitude and diminishment of the temperature. On the base of the numerical observations, the local interstellar magnetic field (LIMF) orientation was expected to be different from the spiral form. LIMF was suggested to be directed to the center of the ribbon of strongest flux ENAs emission detected by Interstellar Boundary Explorer (IBEX). Ulysses and SOHO/SWAN data gave their estimations of the LIMF direction, which is measured by V1 more and more accurately, as the spacecraft goes farther from the heliopause. All these data were combined by *Schwadron et al.* [2015] using a triangulation method to obtain the modern understanding of the LIMF orientation. It was confirmed that LIMF is a poloidal vector. For this reason, not only the field strength drop, but the change in the magnetic field direction at the heliopause should exist, however, was not observed. One of the possible explanations of this unexpected phenomenon was proposed by *Belenkaya* [2015].

Belenkaya [2015] suggested that a dynamo could operate in a thin layer in the closest outer heliosheath (area between a possible interstellar bow shock and heliopause) and showed that all necessary conditions for dynamo action are fulfilled there. For example, comparison of δ_V/δ_B with 1 gives the following result. In a layer with thickness δ_V just out of the heliopause, plasma rotation driven by the Sun due to the strong open magnetic flux, should stop. *Schwadron et al.* [2015] predict that at ~164.6 AU V1 will see the poloidal LIMF, so rotation will disappear maximum on the length $\delta_v \approx 164.6–121.7 \approx 43$ AU, if we assume that the azimuthal velocity along the magnetic field is not zero (see explanation below). *Opher and Drake* [2013] on the base of the MHD modeling give the estimation of the thickness of the sheet with strong rotation ~10 AU. The thickness of a layer where magnetic field strength changes according to the V1 observations is $\delta_B \approx 0.01–0.02$ AU [*Webber and McDonald*, 2013]. Thus, this necessary condition for dynamo $\delta_v/\delta_B \gg 1$ is fulfilled.

Here we use the Voyager 2 (V2) data to clarify the situation in the outer heliosphere and to check once more the hypothesis of dynamo at the heliopause. In spite of V2 did not intersect the heliopause yet, it provides directly measured plasma density and velocity, contrary to V1, for which the density and velocity values in the outer heliosphere were derived indirectly from the suggested hypothesis, because its plasma science (PLS) instrument didn't work from 1980.

The solar wind velocity was estimated using the V1 Low-Energy Charged Particle (LECP) detector working with ion energies from 30 keV to 10 MeV. *Kane et al.* [1998] noted that the work with this instrument allows us to estimate only radial velocity. It was suggested that the measured anisotropies are mainly due to convection as a result of Compton-Getting effect. The first-order anisotropies are

$$\mathbf{A}_1 = \rho \mathbf{b} \times (\nabla f / f) + (1/f)(\partial f / \partial v)\mathbf{V} \quad (13.6)$$

[e.g., *Northrop et al.*, 1979]. Here v is the particle velocity, $\mathbf{V}$ is the convection velocity relative to the spacecraft, f is the distribution function, $\mathbf{b}$ is the unit vector in the direction of magnetic field, ρ is the gyroradius (see equation (13.1)). The Compton-Getting convection anisotropy is described by the second term, while the first term is the density gradient anisotropy perpendicular to the magnetic field.

Consequently, using this indirect method of the velocity determination by V1, nothing can be told about the azimuthal velocity and field-aligned velocity. *Fisk and Gloekler* [2013] noted that the evidence made from V1 observations beyond 113 AU that the anisotropy of the solar wind ions is very small and therefore the velocity is small too, is not correct. Instead, *Fisk and Gloekler* [2013] suggested that very small low-energy anisotropy means that velocity is not small, but is oriented along a magnetic field, which is observed to be in the azimuthal direction, and small anisotropy is a measure of velocity deviation from the magnetic field. *Richardson and Decker* [2015] also stated that "convective flow does not cause all the observed anisotropy and the Compton-Getting analysis is unreliable" for V1. According to this analysis, the radial velocity calculated for V1 decreased from 70 km s^{-1} to 0 in the inner heliosheath (zone between the termination shock (TS) and heliopause) [*Burlaga and Ness*, 2012].

Richardson and Burlaga [2013] and *Richardson and Decker* [2015] showed the radial, azimuthal and normal components of the solar wind velocity in the inner heliosheath directly measured by V2. No disappearance of the radial velocity (as it was indirectly obtained for V1 in the inner heliosheath) was seen. The radial velocity measured by V2 was ~130 km s^{-1} at 84–89 AU and decreased to 100 km s^{-1} at ~90 AU and to ~70 km s^{-1} at ~91 AU. The azimuthal velocity, which was ~0 at the TS, beyond it increased significantly up to ~100 km s^{-1} (intersection with TS was at ~84 AU for V2 and at ~94 for V1). We suppose that in the same regions of the outer heliosphere for V1 and V2 locations in the sense that both spacecraft are placed at the same distances from TS, the solar wind rotation velocities are similar. From the V2 direct measurements, it follows that in the inner heliosheath the plasma flows in the azimuthal direction are significant.

In Figure 14 from *Richardson and Burlaga* [2013], the calculations fulfilled using the *Pogorelov et al.* [2009] model are also shown. For the V2 azimuthal velocity, these model calculations give ~30 km s^{-1}, which can be considered as a turning of the solar wind. So for plasma rotation at 90 AU, we obtain $V = 100 - 30 = 70$ km s^{-1} for V2. If we suggest that for V1, rotation with the same velocity (70 km s^{-1}) was in the corresponding place of the inner heliosheath (determined by the same distance from TS: $\Delta l = 90 - 84 = 6$ AU), then we could calculate the total azimuthal velocity at $l_{i\text{-}h} = 94 + 6 = 100$ AU. At $l_{i\text{-}h} = 100$ AU, the *Pogorelov et al.* [2009] model gives for V1 the turning velocity –20 km s^{-1}, so the azimuthal velocity there was $70 - 20 = 50$ km s^{-1}. Thus, for our further estimations for V1, we will use the azimuthal field-aligned velocity $V_{f\text{-}a}$~50 km s^{-1} at $l_{i\text{-}h} = 100$ AU, as a character value for the inner heliosheath.

As in the outer heliosheath, the spiral ***B*** was observed by V1, we suppose that, at least partially, plasma continue to move along magnetic field lines there. We consider that in the inner heliosheath, the solar wind plasma moves along IMF, while in the outer heliosheath, the interstellar plasma flows along the LIMF. *Opher and Drake* [2013] using 3D multifluid modeling showed that for IMF $\neq 0$ the LIMF rotates in the Sun rotation sense just beyond the heliopause near the stagnation point. The authors stated that in the intersections of V1 and V2 with the heliopause (in the past and in future, respectively), magnetic field direction should remain.

Figure 13.3 schematically demonstrates the location of dynamo region just out of the heliopause according to our scenario. A possible interstellar bow shock is shown by a thin curve, the heliopause is drawn by a thick curve, the outer heliosheath is located between

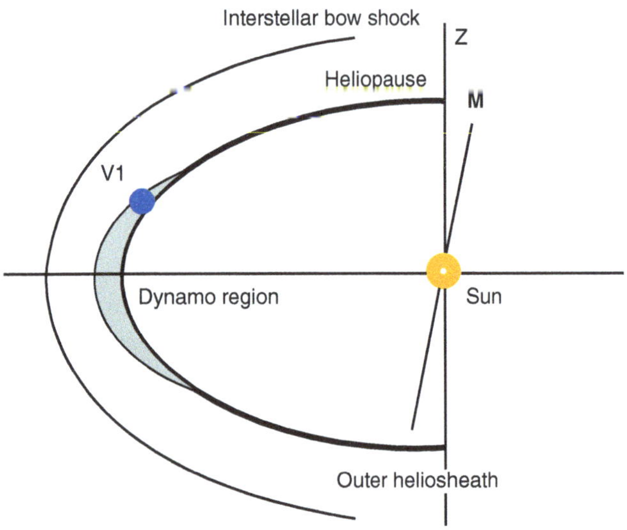

Figure 13.3 Interstellar bow shock (thin black curve) and heliopause (thick black curve); dynamo layer is the grey shaded area. Sun is the yellow circle, V1 location is the blue circle. **M** and *Z* are the magnetic and rotation Sun's axes.

them, and the dynamo layer is shaded in grey. The Sun is marked by a yellow circle and the place of intersection of the heliopause by V1 is marked by a blue circle. The magnetic dipole (**M**) is tilted with respect to the rotational axis of the Sun (*Z*).

The nonlocal character of the processes in the space plasma is caused by the high field-aligned conductivity. The Sun rotation is transferred along open field lines to the heliopause and beyond it and brakes in the dynamo layer giving its energy to the magnetic field amplification there. For the dynamo action, it is necessary that the conductivity at the heliopause current sheet, σ_{eff}, was sufficiently high, but less than the field-aligned conductivity out of this sheet. We found that $\sigma_{eff} = 0.5 \cdot 10^9$ s$^{-1} \ll \sigma_{sw}$, σ_{VLISM}, where $\sigma_{VLISM} \sim 10^{14} - 10^{16}$ s^{-1} is the solar wind conductivity, and $\sigma_{VLISM} = 3.5 \cdot 10^{12}$ s^{-1} is the conductivity of the very local interstellar medium (VLISM) [see *Belenkaya*, 2015, and references therein].

By analogy with the method described in section 13.5, we can conclude that the character distance from the Sun's rotation axis to the dynamo layer out of the heliopause, l, changes from l_0 to R_1, where now R_1 is the heliocentric distance to the front point of the heliopause, and at l_0 the azimuthal velocity, $\Omega_{eff} l_0$, is equal to the interstellar plasma flux velocity V_{VLISM} ~26 km s^{-1}. The enhancement of the magnetic field should be provided by kinetic energy of azimuthal motion, which should exceed the energy of the interstellar plasma flow. We assume that the angular velocities in the inner and outer heliosheaths are equal to the effective angular velocity Ω_{eff}. Here we take into account that Ω_{eff} is not equal to the Sun's angular velocity $\Omega_{Sun} = 2.9 \cdot 10^{-6}$ rad s^{-1}. We showed above that the azimuthal field-aligned velocity character for the heliosheath is $V_{f-a} \sim 50$ km s^{-1}, as we determined for $l_{i-h} = 100$ AU. So, $\Omega_{eff} = V_{f-a}/l_{i-h} = V_{VLISM}/l_0$, or $l_0 = l_{i-h}\ V_{VLISM}/V_{f-a} \sim 52$ AU. Thus, the dynamo layer beyond the heliopause is located at distances from the rotation axis from ~52 to ~122 AU.

To clarify this estimation, we consider the *Landau and Lifshitz* [1959] and *Vainshtein and Zeldovitz* [1972] statement that if a dynamo acts, the amplified magnetic field **b** energy density $b^2/2\mu_0$ is of the order of the energy density of the main flow, generating the dynamo. In our case, it is the kinetic energy density of the azimuthal flow $\wp(\Omega_{eff} l)^2/2$, where $\wp$ is the density of plasma. Thus,

$$\wp\left(\Omega_{eff} l_{eff}\right)^2/2 = b^2/2\mu_0 \text{ or } \Omega_{eff} l_{eff} = b/\left(\mu_0 \wp\right)^{1/2} = U_A, \quad (13.7)$$

where $U_A \sim 30$ km s^{-1} is the Alfvén velocity in the VLISM. The effective distance from the rotation axis, l_{eff}, where the dynamo layer can be developed and where the magnetic and kinetic energy densities are equal, is $l_{eff} = U_A/\Omega_{eff} = l_{i-h} U_A/V_{f-a} = 60$ AU. This clarification allows us to estimate more accurately the distances from the rotation axis for the dynamo zone: from 60 to 121.7 AU. The point of intersection of V1 with the heliopause has the distance from the rotation axis $l_h = 121.7$ AU $\times \cos(34.5°) = 100.2$ AU, if the latitude of this point is the same as for the crossing of V1 with TS. Thus, the point where V1 exited into the VLISM belongs to the dynamo region, in accordance with Figure 13.3. *Linde et al.* [1998] on the base of the MHD calculations stated that out of the heliopause, the magnetic field growths up to ~0.4 nT in a thin layer adjacent to the heliopause at the distances from ~70 to 130 AU from the Sun, which is in good accordance with our estimation.

13.7. SUMMARY

The interaction of two kinds of plasma at the boundary between them leads to formation of two current sheets. Each current is carried by the ions of the outer plasma. These currents are diamagnetic. Observations show that in the case of Earth, the number of the solar wind ions is enough for creation of the current, which magnetic field totally shields the magnetospheric field, while the number of the magnetospheric ions is not sufficient for total screening, and a small part of the solar wind magnetic field (0.1–0.2) penetrates into the magnetosphere. In the magnetospheres of other planets, similar magnetopause currents could arise due to the same mechanism.

Investigation of the space plasma shows that many aspects of magnetospheric physics can be transferred to other astrophysical objects, and vice versa. In particular, some processes acting at the magnetopause occur at the heliopause. However, there are many specific phenomena that should be taken into account studying the heliopause. They are associated only with this object and do not exist somewhere else. Here we pay attention to some features of the current sheets of planetary magnetopauses and heliopause.

A relatively thin boundary (a few Larmor ion radii) of the magnetosphere/heliosphere plays a significant role in the dynamics of its environment. We consider a few aspects connected with the magnetopause/heliopause influence on the space inside and outside the magnetosphere/heliosphere. We investigate a fundamental property of the magnetopause, as a multilevel structure, its reaction to the sudden solar wind dynamic pressure pulse, and a possibility to amplify the magnetic field in passing by plasma flow at the magnetopause of a rapidly spinning magnetized planet. Studying the sudden pressure pulse impact, we introduced a short-living transient current system, existing during the pulse action [*Belenkaya et al.*, 2014c], and estimate here its magnetic field for the case of 6 November 2000 (δB_z ~37 nT).

Considering the Mercury's magnetosphere, we show that FTEs could arise in the subsolar magnetopause under the southward IMF conditions and at the magnetopause quasi-neutral line in one of the cusps depending on the sign of the dominant radial IMF component. For sunward (antisunward) IMF, the neutral point is in the northern (southern) cusp, and the quasi-neutral line, associated with the FTEs arising, lies on the southern (northern) cusp's magnetopause.

The dynamo mechanism, suggested by us earlier for Jupiter's magnetopause [*Belenkaya*, 1996], was applied to explain the unexpected observations of the magnetic field strength drop at the heliopause without changing the field's direction [*Belenkaya*, 2015]. Here we consider the Voyager 2 observations in the outer heliosphere to clarify our estimations. In spite of Voyager 2 did not intersect the heliopause yet, it directly measures the plasma speed (contrary to Voyager 1), which allows us to use its measurements for extrapolation of the velocity field vectors.

ACKNOWLEDGMENTS

Work at the Federal State Budget Educational Institution of Higher Education M.V. Lomonosov Moscow State University, Skobeltsyn Institute of Nuclear Physics (SINP MSU) was partially supported by RFBR grant 16-05-00760 and the European FP7 project IMPEx (no. 262863).

REFERENCES

Alexeev, I., E. Belenkaya, J. Slavin, H. Korth, B. Anderson, D. Baker, S. Boardsen, et al. (2010), Mercury's magnetospheric magnetic field after the first MESSENGER flybys, *Icarus*, *209*, 23–39.

Alexeev, I. I., E. S. Belenkaya, S. Y. Bobrovnikov, J. A. Slavin, and M. Sarantos (2008), Paraboloid model of Mercury's magnetosphere, *J. Geophys. Res.*, *113*, A12210; doi: 10.1029/2008JA013368.

Avinash, K., G. P. Zank, B. Dasgupta, and S. Bhadoria (2014), Instability of the heliopause driven by charge exchange interactions, *Astrophys. J.*, *791*(102); doi:10.1088/0004-637X/791/2/102.

Belenkaya, E. S. (1996), Generation of the magnetic field at the magnetopauses of the rapidly rotating planets, *J. Geophys. Res.*, *101*(A1), 41–47.

Belenkaya, E. S. (2001), Currents at the subsolar low shear magnetopause, *J. Geophys. Res.*, *106*(A11), 25, 437–25, 450.

Belenkaya, E. S. (2004), The Jovian magnetospheric magnetic and electric fields: Effects of the interplanetary magnetic field, *Planet. Space Sci.*, *52*, 499–511.

Belenkaya, E. S. (2014), Response of the currents in Earth's and Saturn's dayside magnetopause to sudden change in the solar wind, *Geomag. Aeron.*, *54*(3), 287–291.

Belenkaya, E. S. (2015), Dynamo in the outer heliosheath: Necessary conditions, *Solar Phys.*, *290* (7), 2077–2092.

Belenkaya, E. S., D. A. Parunakyan, I. I. Alexeev, V. V. Kalegaev, M. S. Blokhina, and M. S. Grigoryan (2014a), Low-latitude variations in the geomagnetic field caused by solar wind disturbances, *Geomag. Aeron.*, *54*(2), 203–206.

Belenkaya E. S., I. I. Alexeev, and C. R. Clauer (2004), Field-aligned current distribution in the transition current system, *J. Geophys. Res.*, *109*, A11207; doi:10.1029/2004JA010484.

Belenkaya, E. S., I. I. Alexeev, and C. R. Clauer Jr. (2007), Magnetic field of the transition current system: dawn-dusk asymmetry, *Ann. Geophys.*, *25*, 1899–1911.

Belenkaya, E.S., I. I. Alexeev, J. A. Slavin, and M. S. Blokhina (2013), Influence of the solar wind magnetic field on the Earth and Mercury magnetospheres in the paraboloidal model, *Planet. Space Sci.*, *75*, 46–55, doi:10.1016/j.pss.2012.10.013 DOI:10.1016/j.pss.2012.10.013#doilink.

Belenkaya, E. S., S. W. H. Cowley, and I. I. Alexeev (2006), Saturn's aurora in the January 2004 events, *Ann. Geophys.*, *24*, 1649–1663.

Belenkaya, E. S., S. W. H. Cowley, and V. V. Kalegaev (2014b), The response of the high-latitude ionosphere to the solar wind pressure jump with a southward IMF on January 10, 1997, *Geomag. Aeron.*, *54*(6), 702–711.

Belenkaya, E. S., V. V. Kalegaev, and M. S. Blokhina (2014c), Polar cap response to the solar wind density jump under constant southward IMF, *Geomag. Aeron.*, *54*(2), 203–206.

Burch, J. L, R. B. Torbert, T. D. Phan, L. J. Chen, T. E. Moore, R. E. Ergun, J. P. Eastwood, et al. (2016), *Electron-scale measurements of magnetic reconnection in space, Science, 352(6290):aaf2939*; doi: 10.1126/science.aaf2939.

Burlaga, L. F., and N. F. Ness (2012), Heliosheath magnetic fields between 104 and 113 Au in a region of declining speeds and a stagnation region, *Astrophys. J.*, *749*(13); doi:10.1088/0004-637X/749/1/13.

Burlaga, L. F., N. F. Ness, and E. C. Stone (2013), Magnetic field observations as Voyager 1 entered the heliosheath depletion region, *Science*, *341*(6142), 147–150; doi: 10.1126/science.1235451.

Chapman, S., and V. C. A. Ferraro (1931a), A new theory of magnetic storms, *Terr. Magn. Atmos. Electr.*, *36*, 77–91.

Chapman, S., and V. C. A. Ferraro (1931b), A new theory of magnetic storms, *Terr. Magn. Atmos. Electr.*, *36*, 171–186.

Chapman, S., and V. C. A. Ferraro (1932), A new theory of magnetic storms, *Terr. Magn. Atmos. Electr. 37*, 147–156.

Chapman, S., and V. C. A. Ferraro (1933), A new theory of magnetic storms, *Terr. Magn. Atmos. Electr.*, *38*, 79–96.

Chapman, S., and V. C. A. Ferraro (1940), A new theory of magnetic storms, *Terr. Magn. Atmos. Electr.*, *45*, 245–268.

Cowley, S. W. H. (2015), Dungey's reconnection model of the Earth's magnetosphere: The first forty years, in *Magnetospheric Plasma Physics: The Impact of Jim Dungey's Research*, edited by D. J. Southwood, S. W. H. Cowley, and S. Mitton, Springer Publishers, 1–32; doi:10.1007/978-3-319.18359-6.

Cowley, S.W. H., E. J. Bunce, and R Prangé (2004), Saturn's polar ionospheric flows and their relation to the main auroral oval, *Ann. Geophys.*, *22*, 1379–1394.

Cowley, S.W. H., S. V. Badman, E. J. Bunce, J. T. Clarke, J.-C. Gérard, C. M. Jackman, S. E. Milan, and T. K. Yeoman (2005), Reconnection in a rotation-dominated magnetosphere and its relation to Saturn's auroral dynamics, *J. Geophys. Res.*, *110*, A02201; doi:10.1029/2004JA010796.

Eastwood, J. P., H. Hietala, G. Toth, T. D. Phan, and M. Fujimoto (2015), What controls the structure and dynamics of Earth's magnetosphere? *Space Sci Rev.*, *188*, 251–286; doi 10.1007/s11214-014-0050-x.

Ergun, R. E., K. A. Goodrich, F. D. Wilder, J. C. Holmes, J. E. Stawarz, S. Eriksson, A. P. Sturner, et al. (2016), Magnetospheric multiscale satellites observations of parallel electric fields associated with magnetic reconnection, *Phys. Rev. Lett.*, *116*(23), 235102; doi:10.1103/PhysRevLett.116.235102.

Ferraro, V. C. A. (1952), On the theory of the first phase of a geomagnetic storm: A new illustrative calculation based on a idealized (plane, not cylindrical) model field distribution, *J. Geophys. Res.*, *57*(1), 15–49.

Fisk, L. A., and G. Gloeckler (2013), The global configuration of the heliosheath inferred from recent Voyager 1 observations, *Astrophys. J.*, *776*(79); doi:10.1088/0004-637X/776/2/79.

Hasegawa, H. (2012), Structure and dynamics of the magnetopause and its boundary layers, *Monogr. Environ. Earth Planets*, *1*, 71–119; doi:10.5047/meep.2012.00102.0071.

Kane, M., R. B. Decker, B. H. Mauk, and S. M. Krimigis (1998), The solar wind velocity determined from Voyager 1 and 2: Low-Energy Charged Particle measurements in the outer heliosphere, *J. Geophys. Res.*, *103*(A1), 267–276.

Keiling A., C. M. Jackman, and P. T. Delamere (2014), *Magnetotails in the Solar System*, American Geophysical Union, Washington, DC.

Landau, L. D., and E. M. Lifshitz (1959), *Electrodynamics of Continuous Media*, Pergamon, New York.

Lee, L. C., and J. R. Kan (1979a), A unified kinetic model of the tangential magnetopause structure, *J. Geophys. Res.*, *84*(11), 6417–6426.

Lee, L. C., and J. R. Kan (1979b), Transition layer between two magnetized plasmas, *J. Plasma. Phys.*, *22*, 515–524.

Linde T.J., Gombosi T.I., Roe P.L., Powell K.G., and DeZeeuw D.L. (1998), Heliosphere in the magnetized local interstellar medium: Results of a three-dimensional MHD simulation, *J. geophys. Res.*, *103*, 1889–1904.

Masters, A., J. P. Eastwood, M. Swisdak, M. F. Masters, A., J. P. Eastwood, M. Swisdak, M. F. Thomsen, et al. (2012), The importance of plasma conditions for magnetic reconnection at Saturn's magnetopause, *Geophys. Res. Lett.*, *39*, L08103; doi:10.1029/2012GL051372.

Moffatt, H. K. (1978), *Magnetic Field Generation in Electrically Conducting Fluids*, Cambridge University Press, New York.

Northrop, T. G., T. J. Birmington, and A. W. Schardt (1979), Anisotropies in the fluxes of Pioneer 10 protons, *J. Geophys. Res.*, *84*, 47–55; doi: 10.1029/JA084iA01p00047.

Opher, M., and J. F. Drake (2013), On the rotation of the magnetic field across the heliopause, *Astrophys. J. Lett.*, *778*(L26); doi:10.1088/2041-8205/778/2/L26.

Pogorelov, N. V., J. Heerikhuisen, J. J. Mitchell, I. H. Cairns, and G. P. Zank (2009), Heliospheric asymmetries and 2–3 kHz radio emission under strong interstellar magnetic field conditions, *Astrophys. J.*, *695*, L31–L34; doi: 10.1088/0004-637X/695/1/L31.

Richardson, J. D., and L. F. Burlaga (2013), The solar wind in the outer heliosphere and heliosheath, *Space Sci. Rev.*, *176* (1–4), 217–235; doi: 10.1007/s11214-011-9825-5.

Richardson, J. D., and R. B. Decker (2015), Plasma and flows in the heliosheath, 13th Annual International Astrophysics Conference: Voyager, IBEX, and the Interstellar Medium IOP Publishing, *J. Phys. Conf. Ser. 577*, 012021; doi:10.1088/1742–6596/577/1/012021.

Roth, M. (1978), Structure of tangential discontinuities at the magnetopause: The nose of the magnetopause, *J. Atmos. Ter. Phys.*, *40*, 323–329.

Roth, M. (1979), A microscopic description of interpenetrated plasma regions, in *Proceedings of Conference of Magnetospheric Boundary Layers, ESA SP-148*, edited by B. Battric and J. Mort, European Space Research and Technology, Noordwijk, The Netherlands.

Schwadron, N. A., J. D. Richardson, L. F. Burlaga, D. J. McComas, and E. Moebius (2015), Triangulation of the interstellar magnetic field, *Astrophys. J. Lett.*, *813*(L20); doi:10.1088/2041-8205/813/1/L20.

Sestero, A. (1964), Structure of plasma sheets, *Phys. Fluids*, *7* (1), 44–51.

Sestero, A. (1966), Vlasov equation study of plasma motion across magnetic fields, *Phys. Fluids*, *9*(10), 2006–2013.

Shi, Y., E. Zesta, L. R. Lyons, A. Boundouridis, K. Yumoto, and K. Kitamura (2005), Effect of solar wind pressure enhancements on storm time ring current asymmetry, *J. Geophys. Res.*, *110*, A10205; doi:10.1029/2005JA011019.

Slavin, J. A., S.M Imber, S. A. Boardsen, G. A. DiBraccio, T. Sundberg, M. Sarantos, T. Nieves-Cnchilla, et al. (2012), MESSENGER observations of a flux transfer event shower at Mercury, *J. Geophys. Res.*, *117*, A00M06, 10.1029/2012JA017926.

Song, P., C. T. Russell, R. J. Fitzenreiter, J. T. Gosling, M. F. Tomsen, D. G. Mitchell, S. A. Fuselier, G. K. Parks, R. R. Anderson, and D. Hubert (1993), Structure and properties of the subsolar magnetopause for northward interplanetary magnetic field: Multiple-instrument particle observations, *J. Geophys. Res.*, *98*, 11, 319–11,337.

Tamao, T. (1975), Unsteady interactions of solar wind disturbances with the magnetosphere, *J. Geophys. Res.*, *80*, 4230–4236.

Vainshtein, S. I., and Y. Zel'dovich (1972), Origin of magnetic fields in astrophysics, *Sov. Phys. Usp.*, *15*, 159–172.

Webber, W. R., and F. B. McDonald (2013), Recent Voyager 1 data indicate that on 25 August 2012 at a distance of 121.7 AU from the Sun, sudden and unprecedented intensity changes were observed in anomalous and galactic cosmic rays, *Geophys. Res. Lett.*, *40*(9), 1665–1668; doi:10.1002/grl.50383.

Wilken, B., C. K. Goertz, D. N. Baker, P. R. Higbie, and T. A. Fritz (1982), The SSC on July 29, 1977 and its propagation within the magnetosphere, *J. Geophys. Res.*, *87*(A8), 5901–5910.

14

MHS Models of Current Layers in the Solar Atmosphere

Clare E. Parnell

ABSTRACT

Electric currents are pervasive throughout the solar atmosphere. However, when intense localized concentrations of currents (current sheets or current layers) arise, the dynamic behavior of the magnetic field and plasma can completely change giving rise to significant releases of magnetic energy. As such, understanding how and where current layers form is essential to both understanding and predicting magnetic-energy release events. In this review, the creation of current layers in two dimensions, both force-free and magnetohydrostatic, will be discussed. Then the formation of three-dimensional current layers will be discussed, including where and how they may form. Also the differences between current layers in low and high beta magnetohydrostatic equilibria will be discussed. Finally, I will briefly touch on the evolution of a current layer once magnetic reconnection is initiated.

14.1. INTRODUCTION

The solar atmosphere is a plasma with a large magnetic Reynolds number, $R_m = L_0 v_0/\eta$, where L_0 and v_0 are typical length scales and speeds in the system, $\eta = 1/\mu\sigma$ is the magnetic diffusivity, μ is the permeability of free space, and σ is the conductivity. In the solar corona, typical features, such as magnetic loops and sunspots, have $L_0 \approx 10^6$–10^8 m and $v_0 \approx 10^4$–3×10^4 ms^{-1}. The magnetic diffusivity, which depends on temperature, is approximately 1 m^{2}s^{-1}, hence $R_m \approx 10^{10}$–10^{12} in the solar corona. The dimensionless magnetic Reynolds number measures the ratio of the advection and diffusion terms in the magnetic induction equation, which governs the behavior of the magnetic field [e.g., *Priest*, 2014].

If $R_m > 1$, then advection dominates and the magnetic field is said to be frozen-into the plasma. In the solar corona, which is known to be a low plasma beta environment (that is, its magnetic pressure is much greater than its plasma pressure), the magnetic field drags the plasma with it. In the photosphere, the plasma beta is greater than one and the plasma pushes the magnetic field around. Either way, plasma elements are fixed on field lines and cannot move from one field line to another. In such a situation, magnetic structures cannot change topology. Instead, the photospheric foot points of the solar corona's magnetic field lines are moved about by overshoots of convections cells from below the solar surface injecting magnetic energy into the atmosphere via the Poynting flux. These foot point motions can cause field lines in the corona to become tangled, as they cannot reconfigure/reconnect when $R_m > 1$, generating electric currents as magnetic energy is built up in the solar atmosphere.

For reconnection to occur, localized regions of strong electric currents are required. This is because in such regions the gradients of the magnetic field change significantly over a short distance (i.e., become locally large) and the relevant length-scale for the system is no longer the global size of a feature, but the short distance over which the magnetic field gradients change. When the current becomes very strong in a localized region with a typical length scale of the order of a few ion inertial lengths, then a loss of thermal equilibrium in the plasma may occur leading to instabilities [*Cairns*, 1985]. These can generate turbulence, disrupting the flow of electrons,

School of Mathematics and Statistics, North Haugh, University of St. Andrews, St. Andrews, Fife, Scotland, UK

Electric Currents in Geospace and Beyond, Geophysical Monograph 235, First Edition.
Edited by Andreas Keiling, Octav Marghitu, and Michael Wheatland.

and causing the conductivity of the plasma to fall by several orders of magnitude and the magnetic diffusivity to rise by the same factor.

For instance, in the solar corona, if the typical length is of the order of an ion inertial length in a hydrogen plasma ($L_0 \approx 10$ m) and the local plasma velocity reduces to $v_0 \approx 10$ ms^{-1} with an enhanced magnetic diffusivity of $\eta = 10^2$ m^2 s^{-1}, then the magnetic Reynolds number becomes 1. This means diffusion becomes important, and plasma elements may, within the region $R_m \leq 1$, move from field line to field line. Thus, magnetic reconnection can occur (magnetic structures can reconfigure) and a global release of magnetic energy into thermal, kinetic and particle energy can occur.

Although for reconnection to occur current layers must have very short length-scales in at least one direction, the accumulations of current could extend over large distances. For instance, if current layers reside in a helmet streamer, they could have lengths of several solar radii or arise above a region of emerging flux before it reconnects with the overlying field, they could extend over many millions of meters [e.g., *Archontis et al.*, 2005].

The solar corona, which makes up the largest fraction of the solar atmosphere, is not only a low plasma beta environment, but, for many features, the typical length scale in the system is less than the pressure scale height and the typical velocity is less than the sound, Alfvén and gravitational free-fall speeds. This means that the equilibrium momentum equation for the solar corona is given by

$$\mathbf{j} \times \mathbf{B} = \mathbf{0},$$

where **j** is the electric current density and **B** is the magnetic field. A magnetic field that satisfies this equation is said to be force-free and, if $\mathbf{j} \neq \mathbf{0}$, then the electric currents will be parallel to **B** [e.g., *Priest,* 2014].

In the chromosphere or photosphere, the plasma pressure, p, is much larger and thus the pressure force cannot be ignored and a magnetohydrostatic equilibrium, where

$$\nabla p = \mathbf{j} \times \mathbf{B},$$

will be formed instead, if any (static) equilibrium exists at all; the photosphere and chromosphere are highly dynamic, so it is not certain whether equilibria of any form arise.

A necessary and sufficient condition for reconnection in three dimensions [*Hesse and Schindler,* 1988; *Schindler et al.*, 1988] is that there exists a localized region in which

$$\left| \int_l \frac{\mathbf{E} \cdot \mathbf{B}}{B} dl \right| = \left| \int_l \frac{\mathbf{j} \cdot \mathbf{B}}{\sigma B} dl \right| > \varepsilon, \quad \text{where } \varepsilon \ll 1, \qquad (14.1)$$

where, in MHD, the electric field $\mathbf{E} = \mathbf{j}/\sigma - \mathbf{v} \times \mathbf{B}$, with **v** the plasma velocity and l the length along a field line. Note that in a plasma, the conductivity σ is never actually infinite, but, in general, it is large. As mentioned earlier, σ becomes small (i.e., the resistivity $1/\sigma$ becomes large) in localized regions associated with large electric currents as a result of microinstabilities. Thus, large, localized electric currents are essential for magnetic reconnection.

One might imagine that reconnection should readily occur everywhere throughout a force-free field since $\mathbf{j} \cdot \mathbf{B} \neq 0$. This, however, is not the case since force-free fields are typically associated with large length scales and thus R_m is large. As mentioned above, an intense localized accumulation of current density is required to locally reduce $R_m \leq 1$ and permit magnetic reconnection to occur. Here, we define a localized intense current accumulation of finite width as a current layer while a current sheet is one that is infinitesimally thin (singular).

The focus of this chapter is the nature of the magnetic field in the vicinity of these important local features. More specifically, we concentrate on MHS equilibria containing current sheets or layers. In a real plasma, the thinnest dimension of the current layer will approach scales of 10 or so ion (and electron) inertial lengths. To study the interior of the current layers that have reached these scales, the dynamics of particles must be followed, which is outside the scope of this chapter. MHD, however, is a valid regime in which to study the formation of current layers in the solar corona since the effect of particles does not become important until the current layer has collapsed to scales much smaller than those typically found in the system.

For instance, if we assume a domain of length 1 Mm in two dimensions (2D), the thinnest current layer found, for instance, by *Fuentes-Fernández and Parnell* [2012] using a 2D nonresistive MHD experiment reaches 500 m. This length is equivalent to approximately 70 ion inertial lengths in a purely hydrogen plasma and some 7150 times the Debye shielding length in the corona and thus still within the MHD regime. In their experiment, the interior of the current layer was still resolved at this scale and thus discussing the behavior of the magnetic field and forces inside the current layers in these experiments is also valid.

In 2D magnetic fields, MHS current sheets/layers form at null points. Section 14.2.1 considers why this is the case and section 14.2.2 reviews the range of possible local force-free magnetic field solutions containing current sheets. Since the advent of computers powerful enough to study the collapse of magnetic null points, various investigations have been undertaken to ascertain the effects of pressure on the formation of current layers. In Section 14.2.3, magnetohydrostatic (MHS) equilibria involving current layers are discussed.

In three dimensions (3D), current layers can arise not only at null points, but also at a range of other locations such as at spines and separatrix surfaces (which are special lines and surfaces made up of field lines that extend out-of or into null points and bald patches in the case of separatrix surfaces) and separators (special field lines that connect pairs of null points; the only nongeneric separators are created by the intersection of separatrix surfaces). Current layers have also been found to form at geometrical features called quasi-separatrix layers (QSLs). Sections 14.3.1 and 14.3.2 review some of the types of current layers that form at both topological and geometrical three-dimensional features, respectively. Finally, in section 14.4, we consider the implications for reconnection of the different types of current layers and consider how prevalent these types of current layers are likely to be in the solar atmosphere.

First, note that here null points in n-dimensional fields are called nD null points. If the magnetic field of a null point is differentiable, then it is possible to expand the field about the null as a Taylor series. In n-dimensions, the local magnetic field about such a null is given by the linear terms of the Taylor series, which can be written as

$$\mathbf{B}_{\text{null}} = \mathbf{M} \cdot \mathbf{r}^{\text{T}} + \ldots,$$

where $\mathbf{M}$ is an $n \times n$ matrix and $\mathbf{r}$ is the position vector relative to the null. The nature of the local field about the null is defined by the eigenvalues of the matrix $\mathbf{M}$ [see *Parnell et al.*, 1996, for a review]. In particular, if $\mathbf{M}$ is singular (i.e., it has m eigenvalues that are zero, where $m < n$), then the local field about the null looks like a null of $n - m$ dimensions, with the higher-order terms in the Taylor series ensuring that the field is only zero at the null point. For instance, we will see that a 2D null may collapse to form a 1D null (i.e., the local field may change from a hyperbolic [X-like] nature to antiparallel field lines) and that a 3D null may collapse to form a 2D null with an X-point like structure or even antiparallel field (1D null). Note that if $m = n$, then the null is of second order or higher: such nulls are not generic.

Not all null points are differentiable and thus the magnetic field about them cannot be expanded as a Taylor series. Nondifferentiable nulls may arise, for instance, where there are tangential discontinuities. Since current sheets (singular current layers) are tangential discontinuities, we will come across nondifferentiable nulls when discussing them.

14.2. TWO-DIMENSIONAL CURRENT SHEETS

14.2.1. Collapse of 2D Current Sheets

As already mentioned, in two dimensions, current sheets form at null points. *Uberoi* [1963] and *Imshennik and Syrovatskiĭ* [1967] found sets of exact nonlinear 2D solutions to the incompressible and compressible MHD equations, respectively, describing the local collapse of a 2D X-point to form a current sheet.

It is easy to show that a potential 2D X-point (i.e., a linear magnetic field of the form $\mathbf{B} = y\hat{\mathbf{x}} + x\hat{\mathbf{y}}$, which is in force balance, Fig. 14.1a) will collapse following any perturbation that squashes it, creating a local magnetic field with a uniform current density of the form

$$\mathbf{B} = y\hat{\mathbf{x}} + (1-\alpha)x\hat{\mathbf{y}} \quad \text{and} \quad \mathbf{j} = -\frac{\alpha}{\mu}\hat{\mathbf{z}},$$

where $0 < \alpha < 1$. This perturbed magnetic field has a Lorentz force,

$$\mathbf{j} \times \mathbf{B} = \frac{\alpha}{\mu}(1-\alpha)x\hat{\mathbf{x}} - \frac{\alpha}{\mu}y\hat{\mathbf{y}},$$

that acts to further the collapse of the X-type null, as shown by the large arrows in Figure 14.1b.

A 2D O-point has a nonzero Lorentz force and, hence, is not a force-free field. It can be represented by a flux function of the form $A(x, y) = 0.5((1-\alpha)x^2 + y^2)$, which produces an O-point with a uniform current density,

$$\mathbf{B} = y\hat{\mathbf{x}} - (1-\alpha)x\hat{\mathbf{y}} \quad \text{and} \quad \mathbf{j} = -\frac{2-\alpha}{\mu}\hat{\mathbf{z}},$$

where $0 < \alpha < 1$. Thus, the Lorentz force acting on this field is

$$\mathbf{j} \times \mathbf{B} = -\frac{2-\alpha}{\mu}\left[(1-\alpha)x\hat{\mathbf{x}} + y\hat{\mathbf{y}}\right].$$

When $\alpha = 0$, the field lines are perfectly circular and the Lorentz force acts radially inward to try to collapse the magnetic field down to the origin (Fig. 14.1c), which, in the absence of pressure, would produce a current singularity.

If the O-point is perturbed such that it becomes elliptical, then the magnitude of the Lorentz force along a field line is not constant ($|\mathbf{j} \times \mathbf{B}| = -(2-\alpha)\sqrt{(1-\alpha)^2 x^2 + y^2}/\mu$). The force always acts inward toward a line elongated parallel to the major axis of the elliptical field lines (Fig. 14.1d). This means in one direction it acts to increase the perturbation, but in the other direction it acts to counter the collapse. Hence, the Lorentz force now acts to collapse the magnetic field into a line rather than a point. Thus, technically a current sheet could be produced provided the Lorentz force is not balanced by a pressure force (which in reality it is).

These local models, in principal, permit the possibility of unbounded current growth if energy is continually supplied to the system and the field lines are free to move

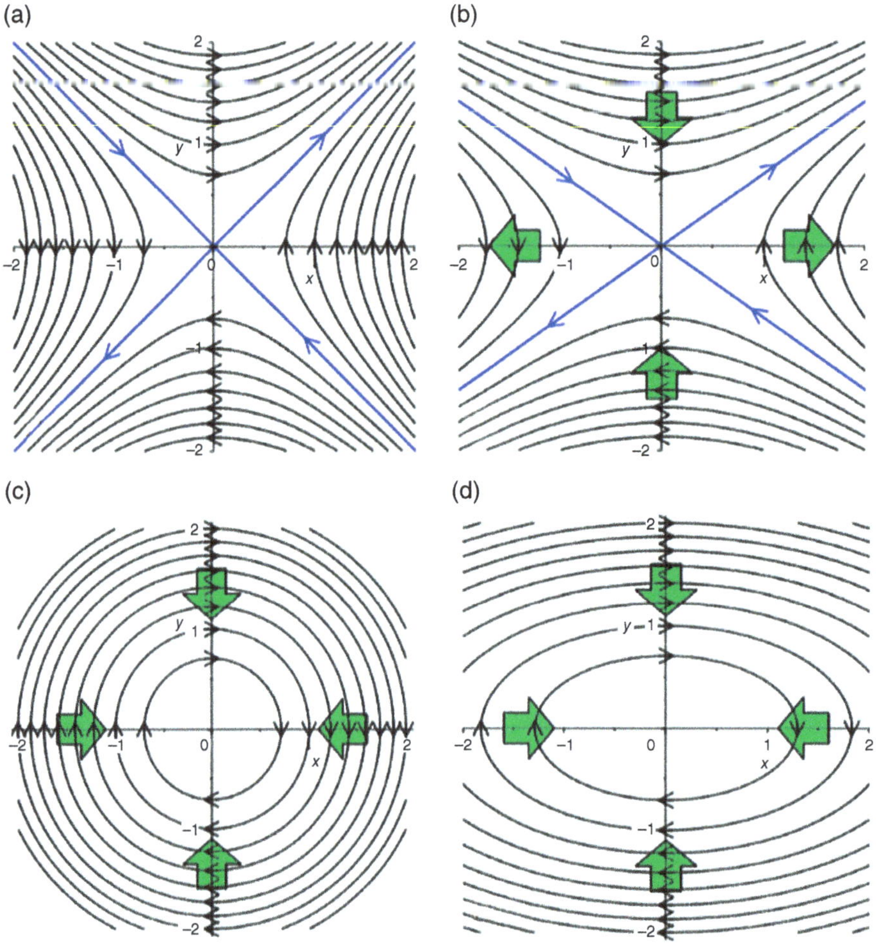

Figure 14.1 2D null point magnetic fields: (a) potential X-point; (b) X-point with $\alpha = 0.5$; (c) O-point with $\alpha = 0$; (d) O-point with $\alpha = 0.7$. The separatrices of the X-points are plotted as blue lines. The direction of the Lorentz force, if nonzero, is indicated by the large green arrows.

such that the X-point can continue collapsing until the field lines are essentially antiparallel. In reality, the field lines are not completely free to move and the energy in the system cannot increase indefinitely. Instead, under nonresistive MHD, the system seeks a state where most of the field is in force balance. This process continues until resistive effects become important. To do this, the magnetic field lines bend such that the current is focused into one location about the null point, which can be modeled analytically as a magnetic field with a singular current sheet (section 14.2.2) or numerically as a magnetic field with a current layer of finite thickness (section 14.2.3).

14.2.2. 2D Force-Free Current Sheets

When an X-type null point collapses, in the absence of a meaningful plasma pressure, the magnetic field tends toward a state that is force free everywhere outside the current sheet/layer that forms at the null. The magnetic field about the current sheet may have zero current or just current parallel to the magnetic field.

Analytical models of local force-free field configurations in 2D typically represent a current sheet as a tangential discontinuity. One of the most well known models involves a current sheet that has nondifferentiable Y-type null points at its ends. A model for a global field enclosing such a current sheet was described by *Green* [1965]. Here, we do not write down Green's full magnetic field expression, but instead consider just the local magnetic field about his type of current sheet, which (in complex notation with $Z = x + iy$) is given by the expression

$$B_y + iB_x = \sqrt{Z^2 - a^2}, \tag{14.2}$$

where the current sheet is represented by a cut in the complex plane lying between $-a < Z < a$. Everywhere outside the current sheet the magnetic field is potential (Fig. 14.2a).

This is not the only potential 2D magnetic field containing a current sheet. *Syrovatskiĭ* [1971] derived a more general form

$$B_y + iB_x = \frac{Z^2 - l^2}{\sqrt{Z^2 - a^2}}. \tag{14.3}$$

Here, if $l = a$, then this magnetic field takes the same form as equation (14.2). If $l < a$ then the ends of the current sheet have infinite magnetic field strength and there are nondifferentiable null points at $(x, y) = (\pm l, 0)$ on the current sheet, which has length $2a$ (Fig. 14.2b). These magnetic fields were derived by considering the nonresistive MHD relaxation of an X-point. However, if $l > a$, then a pair of X-type null points are found lying beyond either end of the current sheet. The creation of these new null points requires reconnection and, thus, such configurations cannot be reached by the nonresistive MHD evolution of a single X-point.

Bungey and Priest [1995] came up with an even more general form for the local 2D magnetic field about a current sheet encompassing the previously mentioned magnetic fields, given by

$$B_y + iB_x = B_0 \left[\frac{Z^2 - 2acZ - ba^2 - \frac{1}{2}a^2}{\sqrt{Z^2 - a^2}} \right], \tag{14.4}$$

where a defines the half length of the current sheet, as above. This formula was not derived by considering the collapse of an X-point, and as a result *Bungey and Priest* [1995] focus on many configurations that are not accessible via nonresistive X-point collapse.

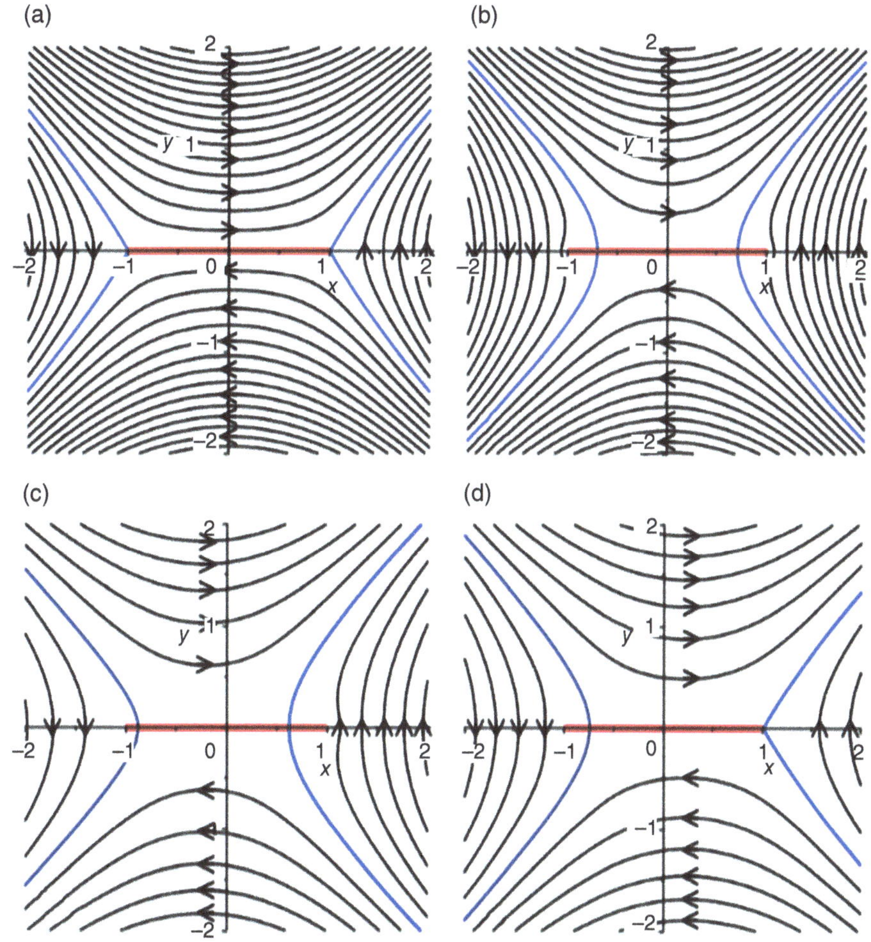

Figure 14.2 Examples of 2D potential magnetic fields with a singular current sheet extending from $-1 < x < 1$ (shown in red): (a) Green's field with $a = 1$; (b) a Syrovatskiĭ-type field with $a = 1$ and $l = 1/\sqrt{2}$ (or in equation [14.4], $B_0 = a = 1$, $b = c = 0$); (c) asymmetric Syrovatskiĭ-type field with nulls located at $Z = -0.75$ and $Z = 1$ ($B_0 = a = 1$, $b = 3/64$, $c = -1/8$); (d) a hybrid field ($B_0 = a = 1$, $b = 1/4$, $c = 1/8$) with a null located on the current sheet at $Z = -0.75$ and a Y-point at $Z = 1$. The separatrices (field lines that start or end at a null) are shown in blue.

There are a few new magnetic field configurations that may be reached by the collapse of an X-point in a nonsymmetric manner. For example, an asymmetric generalization of Syrovatskii's field may be found in which the nondifferentiable null points are located on the current sheet at $-a < Z_n = ac \pm a\sqrt{b+c^2+0.5} < a$ (Fig. 14.2c).

Also a hybrid field may be constructed (Fig. 14.2d), which looks like Green's current sheet on the right and Syrovatskii's on the left. That is, the current sheet has a nondifferentiable Y-type null at one end and a singularity (the magnetic field is infinite) at the other with one nondifferentiable null along its length.

Additionally, if in the Syrovatskii field the two nondifferentiable nulls coincide, then a higher-order nondifferentiable null arises on the current sheet. This nondifferentiable null has four separatrices and the extensions of the branch cuts either side. Thus, about this nondifferentiable null, the magnetic field is divided into six topologically distinct regions (not shown).

Other authors have looked beyond the very local field about the current sheet. For instance, an early model for a solar flare due to emerging magnetic fields was presented by *Tur and Priest* [1976] and involved the formation of a current sheet between the emerging and overlying field. *Heyvaerts et al.* [1977] built on this model and since then there have been a host of papers exploring the consequences of 2D magnetic reconnection as a result of emerging magnetic field [e.g., *Yokoyama and Shibata*, 1995, and many others]. Reconnection in a current sheet below a flux rope has been proposed as a mechanism for a coronal mass ejection (CME) [e.g., *Forbes and Priest*, 1995]. Also *Priest et al.* [1994] and *Parnell et al.* [1994] proposed that X-ray bright points may result from reconnection at a current layer formed due to cancellation of converging flux features.

14.2.3. Two-Dimensional Magnetohydrostatic Current Layers

About all null points, even in regions such as the solar corona where the typical plasma beta is small, the plasma pressure will become important since $|\mathbf{B}|$ approaches zero as the null is approached. *Klapper* [1998] showed analytically that the formation of a current singularity at a 2D X-point would take an infinite time to achieve in an ideal incompressible plasma, but what happens during the formation of a current layer in a compressible plasma? Do the gradients in the plasma pressure end up balancing the Lorentz force and thus prevent the formation of a singularity?

The effect of compressibility on the formation of a current layer at an X-point in 2D was considered for the first time by *Rastätter et al.* [1994]. They used an Eulerian magnetofrictional code (with a nonlinear grid), which damped out kinetic energy by adding a linear damping term dependent on viscosity to the momentum equation, but without considering an energy equation. The system was perturbed by an enhancement in pressure, which led to a current layer forming that had cusp-shaped separatrices. A nonzero current density extended along the separatrices that formed the cusp shape (e.g., Fig. 14.3a) and a finite jump in plasma pressure was seen between the inside and outside of the cusp regions. The current layer they discovered had a finite width, which they claimed was due to numerical, rather than physical, reasons. The integral of the current density over the layer width remained constant along each extended separatrix.

Over a decade later, *Craig and Litvinenko* [2005] and *Pontin and Craig* [2005] revisited this problem, using a Lagrangian magnetofrictional code in order to achieve finer scales than those found by *Rastätter et al.* [1994]. By investigating the effect of changing the grid resolution, they confirmed that the current layer continued to strengthen and narrow as the grid resolution increased. In particular, the peak current density j_{max}, which lies at the null, grows at a logarithmic rate proportional to the size of the numerical grid, N, such that $j_{max} \propto N^{\mu}$, where $0<\mu<2$ depending on the plasma beta of the system [*Pontin and Craig*, 2005]. This implies that the current is attempting to become singular. By looking at the effects of increasing the initial plasma pressure in the system, they found that an increase in plasma pressure reduced the rate of formation of the current singularity, but did not prevent its formation. The current extended beyond the current layer along the separatrices in agreement with *Rastätter et al.* [1994].

The same problem was then studied using an Eulerian nonresistive MHD code, allowing the energy transfer from kinetic energy to internal energy due to viscous damping to be properly accounted for [*Fuentes-Fernández et al.*, 2011, 2012a]. The aim of these studies was to investigate the MHD evolution during the formation of the MHS current-layer in high-beta [*Fuentes-Fernández et al.*, 2011] and low-beta [*Fuentes-Fernández et al.*, 2012a] systems, as well as studying the resulting force-balance in the field. Many aspects of the previous studies were replicated, such as enhanced currents found not only in the current layer, but also along the cusplike separatrices extending from the ends of the current layer, the existence of the singularity at the null point (Fig. 14.3c) and enhanced pressure within the cusps (Fig. 14.3b).

In order to study the growth rate of the peak current at the null, *Fuentes-Fernández et al.* [2011] ran their experiments for almost 200 fast-mode crossing times. After the initial rapid formation of the current layer, the peak current density grew very slowly at a rate $j_{max} \propto t^{\alpha}$, where α lay

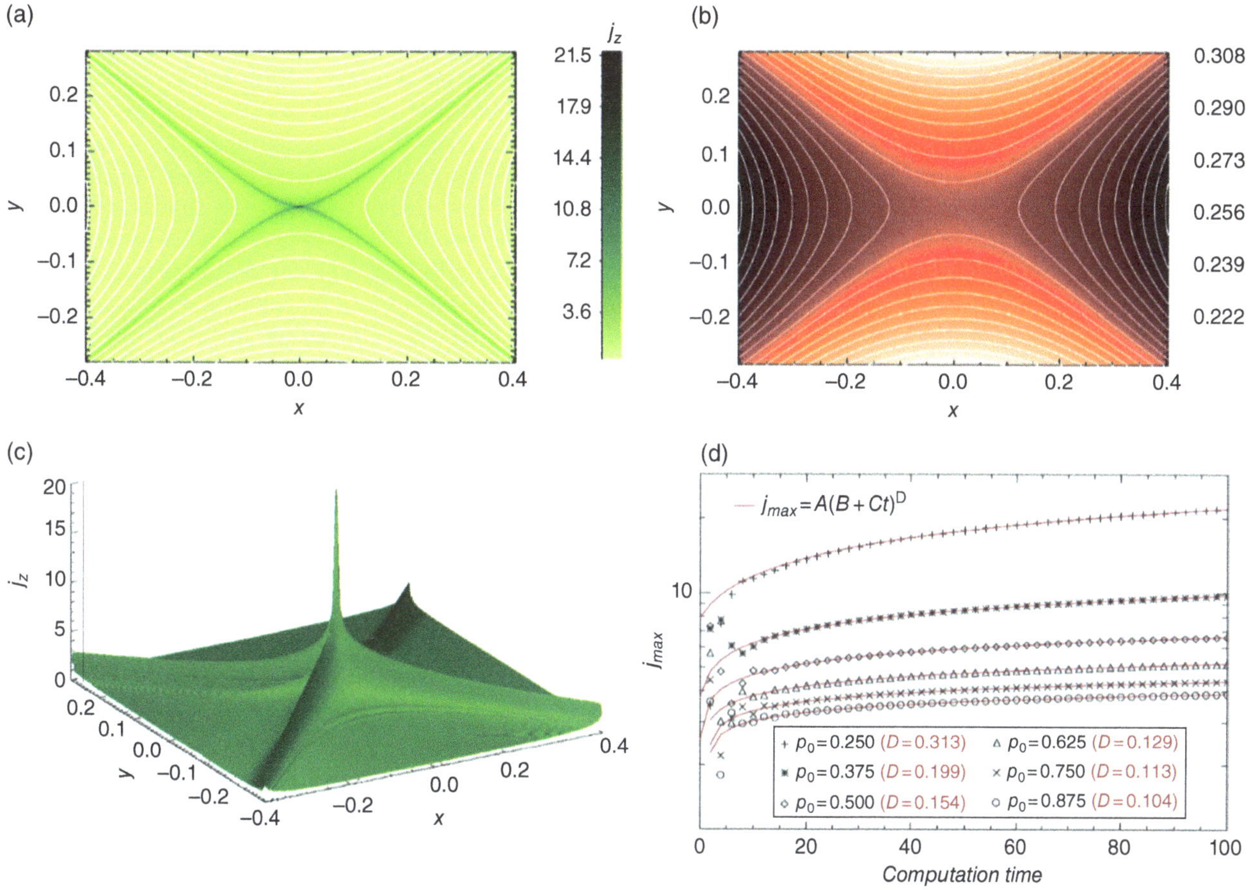

Figure 14.3 Contour plot of (a) current density and (b) pressure in a 2D MHS equilibrium; field lines are overplotted. (c) Plot of current density highlighting the strength of the current at the null relative to that on the separatrices and elsewhere. (d) Growth rate of peak current for different initial pressures. From *Fuentes-Fernández et al.* [2011], reproduced with permission © ESO.

between 0.1 and 0.33 depending on the value of the plasma beta. Such a logarithmic growth rate is an indication of an infinite-time singularity (Fig. 14.3d). The consequences of this type of behavior are that the final magnetic field is in force balance everywhere except within the very localized enhanced current regions both within the layer and along the separatrices. In fact, no matter what resolution, and no matter how long these experiments are run, these regions would never be in force balance, since the forces there are acting to try to collapse the current layer to form a singularity at the origin.

Fuentes-Fernández et al. [2011] demonstrated that, due to the line-tied boundary conditions and the symmetry of the system, the total current in the numerical domain is conserved. This means that as the peak current density at the null continues to grow, the current layer both narrows and shortens. If the field lines were not line tied on the boundary, as would be the case in a driven MHD experiment, then extra energy could enter the system. The current layer could then grow both in strength and length as it collapses provided the additional energy enters the system fast enough.

14.3 THREE-DIMENSIONAL CURRENT ACCUMULATIONS

14.3.1. Current Layers at Topological Features

As already mentioned, in 3D there are many locations at which current layers may form. In this section, we concentrate on topological features, namely 3D null points, their associated separatrix surfaces and spine lines, as well as separators. A separatrix surface of a 3D null is formed by an infinite number of field lines that all extend into/out of the null, whereas the spine lines are a pair of field lines that extend out of/into the null (see, for example, Fig. 14.4a,b).

It is possible that a field line that forms the spine of one null or lies in its separatrix surface may connect to another null: such a field line is called a separator. The only generic type of separator connecting two null points is formed by a field line lying in the separatrix surfaces from the two different nulls and thus lies at the boundary between four topologically distinct domains, just like a null point does in 2D. Separators can also be formed by the intersection of a separatrix surface arising from a bald patch [*Titov et al.*, 1993], with another separatrix surface from either a null or different bald patch.

14.3.1.1. Current Accumulations About 3D Null Points

It has been shown that a 3D magnetic null point is prone to local collapse, in a similar way to a 2D magnetic null point [*Bulanov and Olshanetskii*, 1984; *Parnell et al.*, 1997] and currents may accumulate in the vicinity of the null. Furthermore, *Parnell et al.* [1997] showed that the only 3D differentiable null points that may exist within an MHS equilibrium are potential. First, note that the curl of the MHS force balance equation, in the absence of gravity or a viscous force, implies

$$\nabla \times (\mathbf{j} \times \mathbf{B}) = \nabla \times \nabla p \equiv \mathbf{0}.$$

The local magnetic field about a 3D differentiable null point may be expanded as a Taylor series. Thus, the magnetic field close to the null point may be written as $\mathbf{B} = \mathbf{M} \cdot \mathbf{r}^{\mathrm{T}}$, where $\mathbf{M}$ is a 3×3 nonsingular matrix containing constant coefficients. Thus, the current $\mathbf{j}$ associated with the local field about the null is constant and

$$\nabla \times (\mathbf{j} \times \mathbf{B}) = -(\mathbf{j} \cdot \nabla)(\mathbf{M} \cdot \mathbf{r}) = -\mathbf{M} \cdot \mathbf{j}.$$

Hence, it is only possible to find a pressure force to balance the Lorentz force of a 3D differentiable null point if the null point has zero current density, that is, if it is potential [*Parnell et al.*, 1997]. Moreover, this shows that all 3D nonpotential differentiable null points will collapse (with or without pressure). Their collapse can only be constrained by a velocity flow [*Titov and Hornig*, 2000]. In contrast, all 2D differentiable null points can reside in an MHS equilibrium, whether potential or nonpotential, since in 2D $\mathbf{M} \cdot \mathbf{j}$ is always zero. This implies that to form a 3D MHS equilibrium the system must transform any 3D null to a potential null or collapse it to either a 2D null or a pair of 3D nondifferentiable nulls associated with the 3D equivalent of a tangential discontinuity.

Rickard and Titov [1996] looked at the linear response to perturbations of the magnetic field in the vicinity of a radial (axisymmetric) potential null modeled using cylindrical coordinates with the spine of the null aligned along the z-axis. An $m=0$ axisymmetric perturbation, corresponding to a twisting of the field about the spine of the null, led to the creation of current parallel/antiparallel to the spine lines, while an $m=1$ perturbation led to a folding of the spine toward the separatrix surface and the creation of current perpendicular to the spine.

Many authors have investigated the nature of reconnection at 3D nulls, with most starting from a situation involving a potential nongeneric radial magnetic null point in which two eigenvalues are identical. A few authors start from a generic form for a potential null, such as $\mathbf{B} = x\hat{\mathbf{x}} + py\hat{\mathbf{y}} - (p+1)z\hat{\mathbf{z}}$ where $p \neq 1$ (for details of all the different types of 3D null points, see Parnell *et al.* [1996]). The field is then driven to create an enhanced current region within which reconnection occurs. As such, the nature of the driver plays a major role in dictating the nature of current accumulated, which is dissipated, via reconnection, before it forms a genuine current layer. Here, we focus on current layers, formed through ideal (nonresistive) relaxation, that reside in 3D MHS equilibria.

To study the formation of current layers at 3D null points, *Pontin and Craig* [2005] used the same Lagrangian magneto-frictional relaxation code that they had used to study current layers at 2D X-points. In particular, they studied the nature of current accumulations at 3D nulls created by the folding of the separatrix surface and spine toward each other, and thus associated with a current perpendicular to the spine (e.g., Fig. 14.4).

They were able to demonstrate that the peak current, which lies at the null, scales with the resolution of the numerical grid, just like in 2D, implying that the 3D null point is a current singularity. The similarity between a current sheet formed at a 2D X-point and at a 3D null due to the folding of the spine and separatrix surface and viewed in the plane perpendicular to the current direction was highlighted by *Pontin and Craig* [2005]. The main difference was that, in 3D, enhanced current was only found to extend along the separatrix surface, rather than the spine lines (e.g., Fig. 14.4e,f), whereas in 2D the extensions of enhanced current occurred evenly along all four separatrices (e.g., Fig. 14.3a).

Fuentes-Fernández and Parnell [2012, 2013] also considered the formation of MHS equilibria containing current layers formed at 3D nulls and, like Fuentes-Fernández *et al.* [2011], they used a nonresistive MHD code as opposed to a magnetofrictional relaxation code. As found in 2D, the MHS equilibria created were in force balance everywhere except in the localized current accumulations themselves.

Fuentes-Fernández and Parnell [2013] studied the same sort of current accumulations as *Pontin and Craig* [2005]. In particular, they followed the MHD relaxation of an initially non-force-free field (Fig. 14.4a,b) with a uniform current density throughout the domain directed perpendicular to the spine, of the form

$$\mathbf{B} = x\hat{\mathbf{x}} + by\hat{\mathbf{y}} + \left(j_{\perp}y - (b+1)z\right)\hat{\mathbf{z}} \quad \text{and} \quad \mathbf{j} = j_{\perp}/\mu_0\hat{\mathbf{x}}.$$

The current preferentially accumulates in the separatrix surface with the peak current found at the null and a band of enhanced current aligned with the axis of folding of the spine and separatrix surface (which is also the

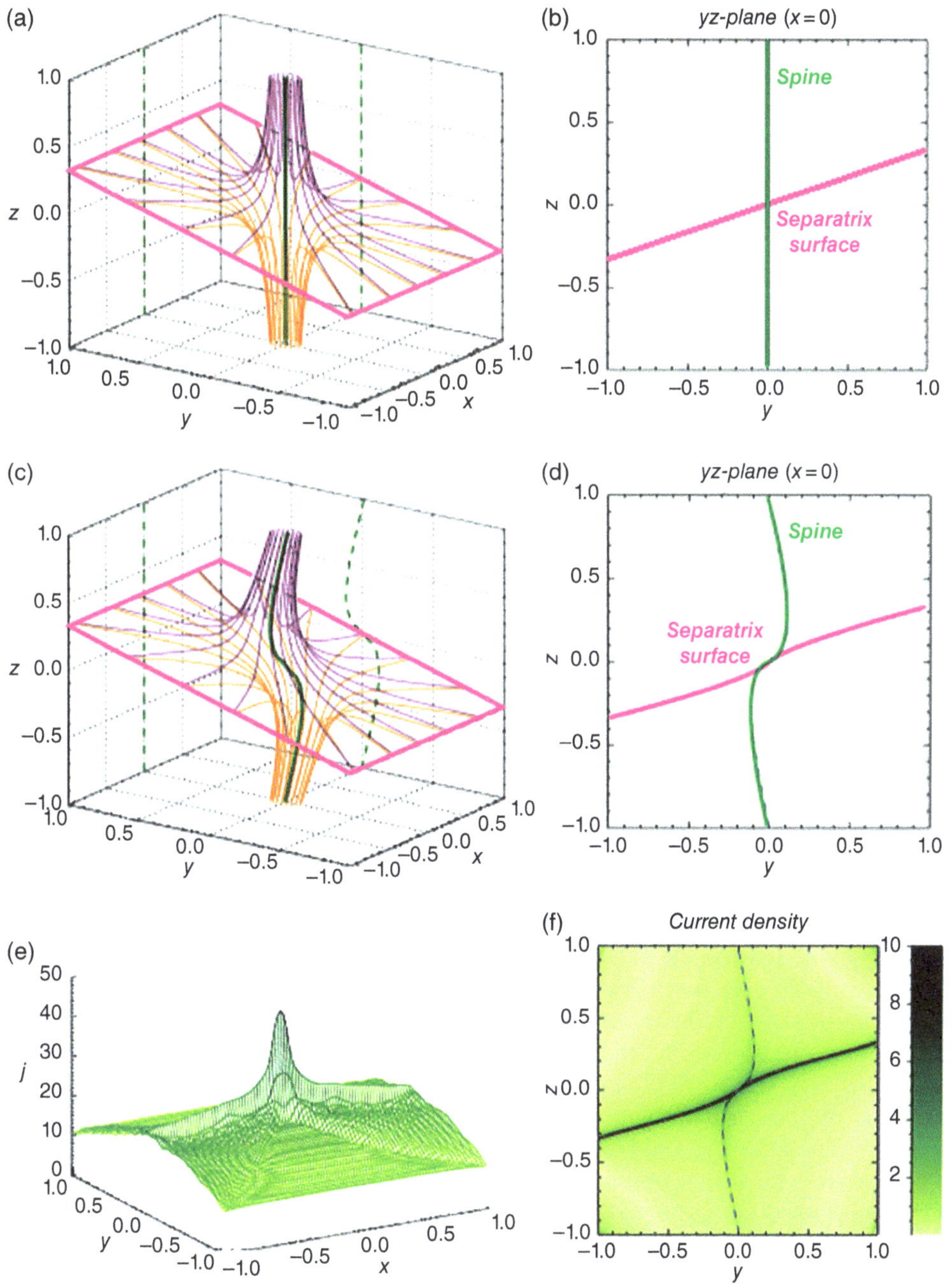

Figure 14.4 (a) Initial 3D null-point magnetic field with a uniform current in the x-direction. The orange and purple lines are sample field lines. The pink rectangle bounds the separatrix surface and the green line follows the spine lines. (b) Cut showing the location of the spine and separatrix surface in the initial magnetic field in the $x = 0$ plane. (c) Final MHS magnetic equilibrium with current layer at the origin (lines as above). (d) Cut showing the location of the spine and separatrix surface in the final MHS equilibrium in the $x = 0$ plane. (e) Plot of current density on the MHS separatrix surface. (f) Contour plot of current density in the final MHS equilibrium. All images taken from *Fuentes-Fernández and Parnell* [2013], reproduced with permission © ESO.

direction of the current). The separatrix surface and spine form a cusp shape (Fig. 14.4d,f), just like the separatrices of the 2D-point. The growth rate of the peak current at the null was once again found to be very slow with a logarithmic dependency on time.

Fuentes-Fernández and Parnell [2013] noted that since only potential 3D differentiable nulls may appear in MHS equilibria, then, due to the folding of the spine and separatrix surface toward each other, all the field lines in the separatrix surface must also fold in an attempt to lie in the same plane (perpendicular to the direction of the current) such that the magnetic field very close to the null forms a 2D X-point-like field (not pictured). In other words, in a Taylor series expansion of the magnetic field about this null the 3×3 matrix associated with the linear terms would be singular with only two nonzero eigenvalues.

Fuentes-Fernández and Parnell [2012] investigated the characteristics of the current accumulations formed following the nonresistive MHD relaxation of an initial nonlinear force-free field with a uniform current density throughout the domain directed parallel to the spine (Fig. 14.5a-b), of the form

$$\mathbf{B} = \left(x - \frac{j_{\|}}{2} y\right)\hat{\mathbf{x}} + \left(\frac{j_{\|}}{2} x + by\right)\hat{\mathbf{y}} - (b+1)z\hat{\mathbf{z}}$$

$$\text{and} \quad \mathbf{j} = j_{\|}/\mu_0 \ \hat{\mathbf{z}}.$$

They found that the current localizes itself about the spine lines of the null, with the width of the current layer tapering as it heads toward the null point such that at the null itself $\mathbf{j}=\mathbf{0}$, as predicted by *Parnell et al.* [1997] (Fig. 14.5e,f). All that changes from the initial field is the concentration of field-line twist, which, instead of being uniform, is now located within the segments of the field lines that lie close to the spine lines. Near to (and in) the separatrix surface, the field lines are as close to radial as the line-tied boundary conditions permit (Fig. 14.5d).

There are known to be three different reconnection regimes at 3D nulls [e.g., *Priest and Pontin*, 2009], namely, spine-fan, torsional spine and torsional fan. The first type of 3D null MHS equilibria discussed above, where the spine and fan fold in toward each other, would result in the most generic 3D reconnection regime, that is spine-fan reconnection, while the latter MHS equilibrium involving a localization of the current about the spine would permit torsional-spine reconnection. However, no MHS equilibria have been created that permit the torsional-fan reconnection regime.

14.3.1.2. Current Accumulations at Separators

Generic magnetic separators, formed from the intersection of two separatrix surfaces, lie at the boundary between four topologically distinct flux domains. Thus, they are arguably the 3D equivalent to 2D nulls, which are also bounded by four topologically distinct domains. So separators are possible locations for the formation of current layers.

Longcope and Cowley [1996] considered a situation involving two null points of opposite sign (A and B) connected by a pair of separators (Fig. 14.6a). They noted that if a current parallel to the separator were applied, it would generate a Lorentz force that would cause the separatrix surfaces forming the separator to fold toward each other, just as the separatrices do in 2D. They predicted that the separators and associated 3D nulls collapse to form an annular disc of current and a set of four nondifferentiable 3D nulls (Fig. 14.6b). In this scenario, both the null points and separators are affected by the formation of the current ribbon and any resulting reconnection will involve all these topological features.

In global potential field models of the solar atmosphere, pairs of nulls are often connected by one, as opposed to two, separators [*Platten et al.*, 2014]. In order to investigate the nature of single separator current layers, *Stevenson et al.* [2015] used a nonresistive MHD code to study their formation following the same approach as *Fuentes-Fernández et al.* [2011]. That is they started with a non-force-free magnetic configuration and allowed it to relax with line-tied boundary conditions. The single separator magnetic field studied (Fig. 14.7a) had the form

$$\begin{aligned}
B_x &= x + 2cxz + byz - \frac{j_{\text{sep}}}{2} y, \\
B_y &= (2a - 2c)yz - (1 + aL)y + bxz + \frac{j_{\text{sep}}}{2} x, \\
B_z &= -a(z^2 - Lz) + cx^2 + (a - c)y^2 + bxy
\end{aligned}$$

where L is the length of the current sheet and a, b, and c are constants that must satisfy the constraints

$$a > 0, \quad b^2 > \frac{(aL - cL - 1)(cL + 1)}{L^2}$$

$$\text{and} \quad b^2 > \frac{a^2(cL+2)^2}{4(1+aL)},$$

to ensure that there are only two null points in the field situated at (0, 0, 0) and (0, 0, L) connected by a single separator that runs from the lower null to the upper null. The electric current associated with this field

$$\mathbf{j} = \frac{j_{\text{sep}}}{\mu}\hat{\mathbf{z}},$$

lies parallel to the separator and is initially uniform throughout the region.

Following the nonresistive MHD relaxation of this field with line-tied boundary conditions, current accumulates about the separator, away from the null points, forming a twisted ribbon-like current layer (Fig. 14.7b). Outside the current layer the magnetic field is potential except above and below the separator where there is also enhanced current about the separatrix surface of the nearest null. This current arises due to the requirement that globally the current must form a closed loop; in this case the actual closure of the current loop is achieved outside the domain since here we are focusing on the local

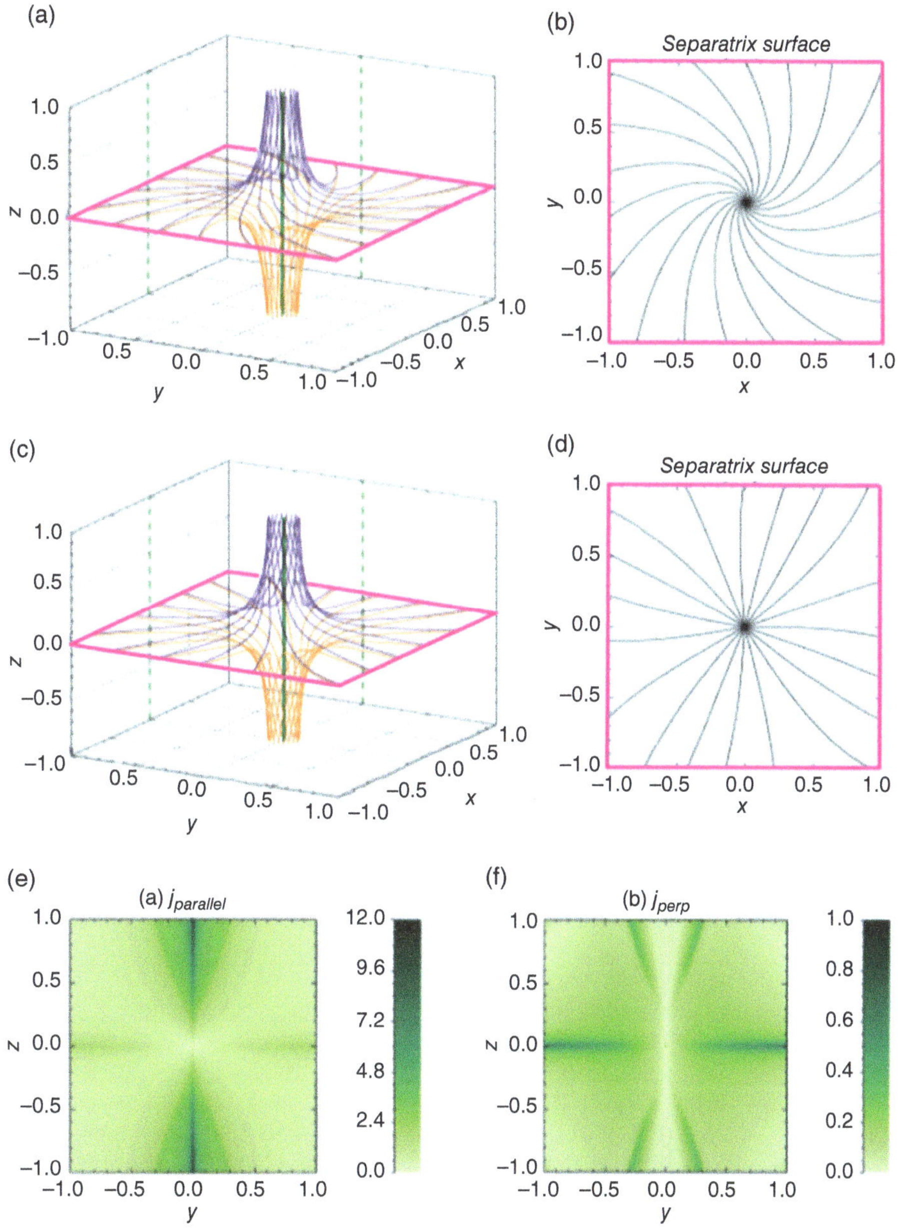

Figure 14.5 (a) Initial 3D null point magnetic field with a uniform current in the z-direction (lines as in Fig. 14.4a). (b) Sample initial magnetic field lines in the separatrix surface of the initial magnetic field. (c) Final MHS magnetic equilibrium with current focused on the two spine lines away from the null (lines as above). (d) The same sample field lines as (b), in the separatrix surface of the final MHS equilibrium. Contour plot of current density (e) parallel to **B** and (f) perpendicular to **B** in the $x = 0$ plane of the MHS equilibrium. All images taken from *Fuentes-Fernández and Parnell* [2012], reproduced with permission © ESO.

field about the separator. The current is antiparallel to the magnetic field outside the separator (Fig. 14.7f) where it spreads out over the separatrix surface and so does not form a current layer.

The separator current layer contains strong electric current parallel to the magnetic field (separator). It forms a concentrated ribbon-like structure, which is twisted about the separator and has enhanced wings of current on the separatrix surfaces (Fig. 14.7d,e). It arises due to the closing up of the two separatrix surfaces, as predicted by *Longcope and Cowley* [1996], with a cusp-type configuration forming in cuts through the separatrix surfaces perpendicular to the separator.

In cuts perpendicular to the separator, there are a number of similarities with the 2D X-point MHS equilibria found by *Fuentes-Fernández et al.* [2011].

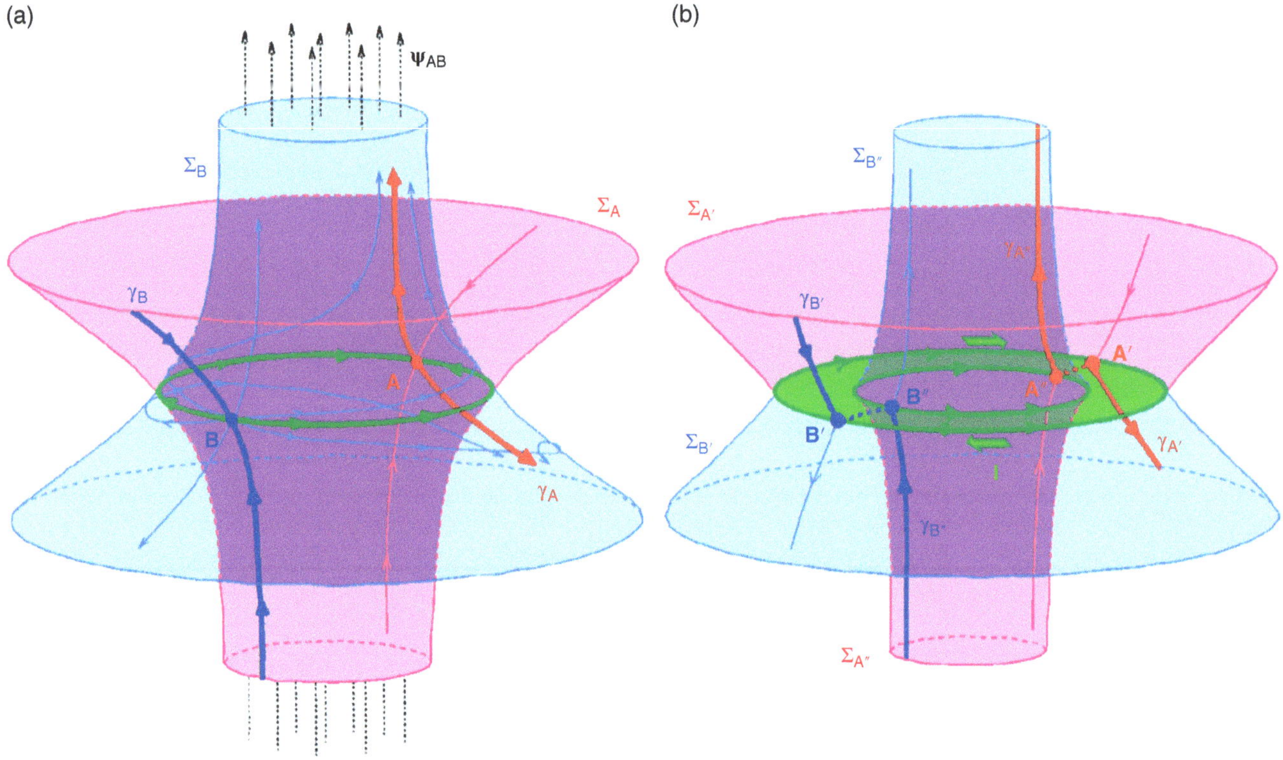

Figure 14.6 Sketches of (a) a 3D double separator magnetic field with two oppositely signed nulls and (b) after the collapse of the field about these separators with a current sheet and the two nulls replaced by four nondifferentiable nulls. The red and dark blue spheres denote negative (A-type) and positive (B-type) nulls, respectively. The spines (γ-lines) associated with these nulls are the thick lines of the same color. The separatrix surfaces (Σ-surfaces) associated with the negative nulls are shaded light pink while field lines from these nulls are pink lines. The separatrix surfaces (Σ-surfaces) and field lines associated with the positive nulls are shaded/colored light blue, respectively. The separators are the thick dark green lines and the current sheet, which is bounded by two separators, is colored light green. The light green arrows indicate the clockwise flow of the electric current. Images based upon sketches in *Longcope and Cowley* [1996].

For instance, Figure 14.7c shows that within these cusps the pressure is enhanced and outside them it is decreased, relative to the initial uniform plasma pressure. *Stevenson et al.* [2015] found that the growth rate of the maximum current, which lies on the separator, follows a slow rate that depends logarithmically on time (Fig. 14.7g). They also determined that the MHS configuration is in equilibrium everywhere except at the separator current layer, which continues to collapse trying to focus all the current on the separator itself. This suggests the collapse of a separator in 3D is also an infinite time singularity.

A study of the reconnection in this single separator scenario [*Stevenson and Parnell*, 2015 a,b] showed that the diffusion region is a twisted ribbon-like structure centered on the separator away from the nulls and that the null points did not play a role in the reconnection. In a plane perpendicular to the separator, the reconnection looked like 2D reconnection, although, unlike in 2D, the major magnetic field component in the diffusion region was out of the 2D plane, parallel to the separator.

14.3.2. Current Layers at Geometrical Features

In 3D, current layers form in the absence of null points and other topological features such as separators. Since the 1980s there have been a series of numerical experiments in which simple (null-free), typically uniform, initial magnetic fields undergo complex boundary driving motions in order to try to form current sheets or current layers.

The first of these sets of experiments was run by *van Ballegooijen* [1988 a,b]. He found that following five shearing motions of varying phase, after each of which the field relaxed to form a force-free state, current accumulations formed. The length-scales of these accumulations exponentially decreased after each shear. Similar experiments using MHD, reduced MHD or magnetofrictional codes have been run over the years with either

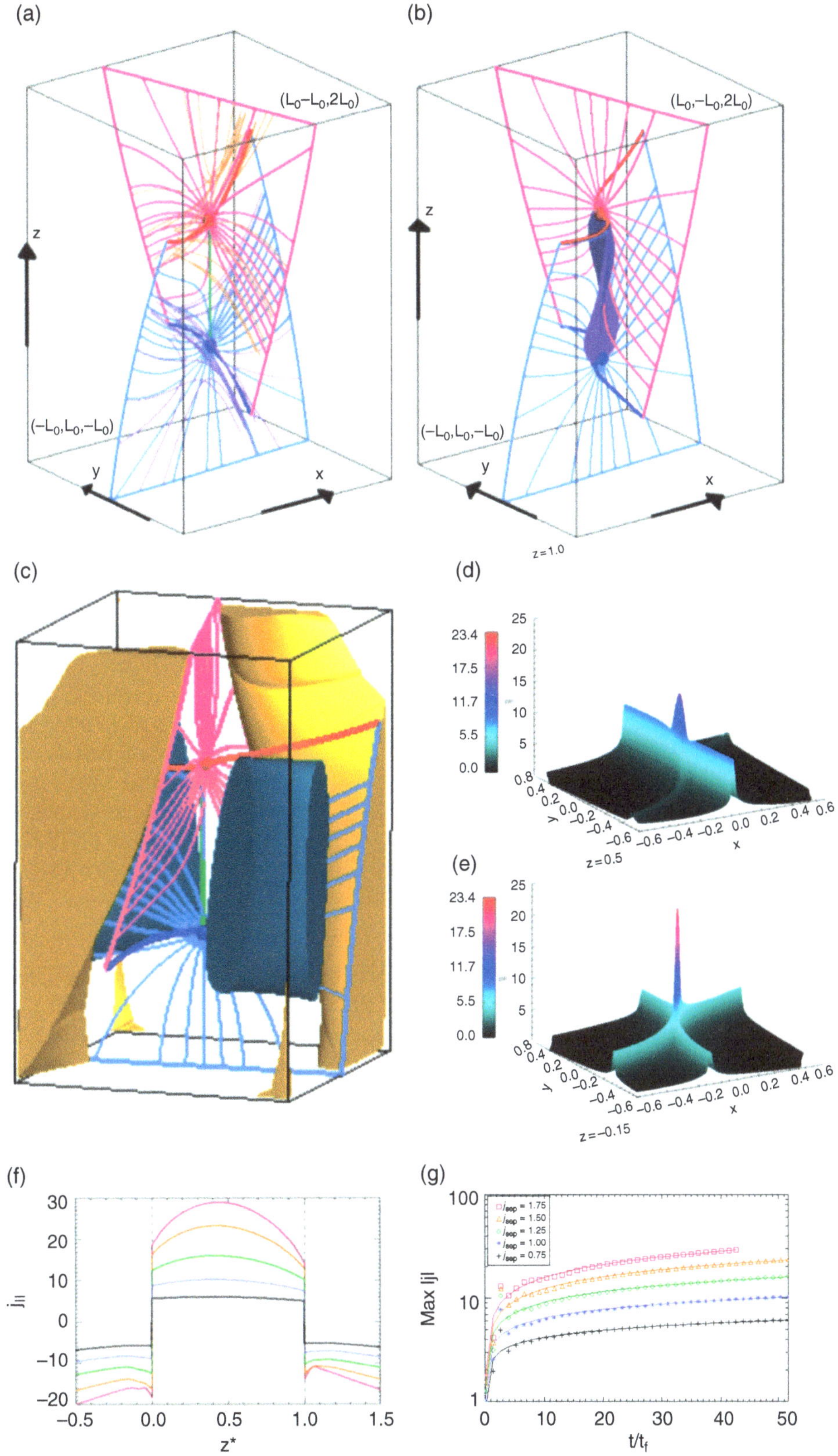

Figure 14.7 Single separator (a) initial force-free and (b) final MHS equilibrium magnetic field configurations analyzed by *Stevenson et al.* [2015]. The color coding is the same as that used in Figure 14.6 except here in (b) an isosurface of $j_{||} > 0$ is plotted in purple. (c) Regions of plasma pressure that are greater (yellow) and less than (blue) the initial uniform pressure in the final MHS equilibrium. Plots of current density in (d) the plane of the upper null and (e) half way along the separator. (f) Plot of the current along the *z*-axis for various MHS equilibria formed from the same initial setup except for their initial uniform current. (g) The growth rate of the maximum current, which lies on the separator for the same set of experiments shown in (f). All images taken from *Stevenson et al.* [2015], reproduced with permission © ESO.

a greater number of shears or with fewer, but larger, shears. In each case, the results basically confirm the above results. That is, the complex driving of a simple (null-free) magnetic field can lead to the formation of current layers that will eventually be thin enough for reconnection to occur. A comprehensive review of all of these driven experiments has been written by *Wilmot-Smith* [2015].

Here, we are interested in the formation of MHS current layers due to nonresistive relaxation following a perturbation, as opposed to boundary driving. In particular, can strong current layers and/or current sheets be created in 3D MHS null-free configurations?

Janse and Low [2009] conducted a series of calculations that demonstrated that a line-tied 3D magnetic field that is nonpotential and non-force-free will relax to a state that must contain current accumulations. However, they did not show where these currents would accumulate and so were not able to say whether these current accumulations were large-scale or formed current layers similar to those about null points and separators.

Wilmot-Smith et al. [2009, 2010] considered the relaxation of a null-free non-force-free magnetic field configuration involving a uniformly directed magnetic field that had been braided in a so-called pigtail braid. Their initial magnetic field involved a uniform magnetic field with the addition of six flux rings whose placement was staggered in a zigzag manner within the uniform field,

$$\mathbf{B} = b_0\hat{\mathbf{z}} + \sum_{i=1}^{6}\sqrt{2}\left(-1\right)^i b_0 E_i\left(y\hat{\mathbf{x}} - \left(x - x_i\right)\hat{\mathbf{y}}\right),$$

where

$$E_i = \exp\left(-\frac{\left(x-x_i\right)^2}{2} - \frac{y^2}{2} - \frac{\left(z-z_i\right)^2}{4}\right),$$

with the parameters $x_i = (-1)^{i+1}$ and z_i the ith element of $\{-20,-12,-4,4,12,20\}$. The flux rings, located at $(x_i, 0, z_i)$, have radius $\sqrt{2}$ and length 2. They all have the same strength, but alternate rings are directed in opposite directions introducing alternating right-handed and left-handed crossings of the field. Sample field lines from the initial (pale colored lines) and final equilibrium (dark coloured lines) magnetic configurations are plotted in Figure 14.8a. In the final MHS equilibrium, large-scale current structures exist all the way along the field lines associated with the braiding. There is no indication of the formation of a current singularity (Fig. 14.8b).

The MHS equilibrium formed by *Wilmot-Smith et al.* [2010] had no electric field, but in the situation of a uniform conductivity, σ, the parallel electric field is $\mathbf{E}_{\parallel} = \mathbf{j}_{\parallel}/\sigma$. As already mentioned, a necessary and sufficient condition in 3D for reconnection is given by equation (14.1), thus, as a proxy, *Wilmot-Smith et al.* [2010] calculated $\int_l j_{\parallel} dl$ along many field lines. This provides one value per field line and can be plotted on a 2D slice (in an xy-plane) through their braided flux tube (Fig. 14.8c). This plot, which clearly shows small-scale structures, will look different in every slice taken through the flux tubes due to the braiding of the field lines. *Wilmot-Smith et al.* [2010] argued that, despite the lack of a true current layer, due to the presence of these small-scale structures, this magnetic configuration is actually an ideal place for reconnection to occur.

Pontin and Hornig [2015] extended the above work by considering not just one but a series of pigtail braids, one after the other, such that the magnetic field is even more twisted. They used an improved version of the Lagrangian relaxation code, in combination with a nonresistive MHD code, to form MHS equilibria. In their final MHS equilibria, they found that increasing the number of braids caused the thickness of the current layers to decrease and the current intensity to increase exponentially. In particular, they found three end-to-end pigtail braids were enough to form current layers, thus confirming the results of *van Ballegooijen* [1988b] and other authors.

The formation of a current accumulation in a much simpler null-free magnetic field was considered by *Craig and Pontin* [2014]. They used their original Lagrangian relaxation code to study the MHS equilibria formed after perturbing an initial magnetic field of the form

$$\mathbf{B} = B_0 x\hat{\mathbf{x}} - B_0 y\hat{\mathbf{y}} + B_z\hat{\mathbf{z}},$$

where B_0 and B_z are constants. This magnetic field was studied by *Priest and Démoulin* [1995] and it may be thought of as the magnetic field about a separator, but with one null point below the base and the other above the top of the box. The locations of what would be the separatrix surfaces of these external nulls are identified as regions of high Q. The squashing factor, Q, is a geometrical measure of the magnetic field [*Titov*, 2007; *Titov and Hornig*, 2002]. Field lines that have foot points that are close by at one end and highly spread out at the other end will have a high Q value and form a quasi-separatrix layer (QSL).

The location of the current accumulations found by *Craig and Pontin* [2014] were dependent on the length (in the z direction) of the box. A shorter box produced current layers near the top of the box off to the sides, while a longer box produced a more central intense current near the base of the box. These regions of intense current coincided with the regions of high Q. They again considered the behavior of the peak current on varying the grid resolution and found a similar logarithmic dependence as that found by *Pontin and Craig* [2005] for the collapse of 2D and 3D null points. Thus, they

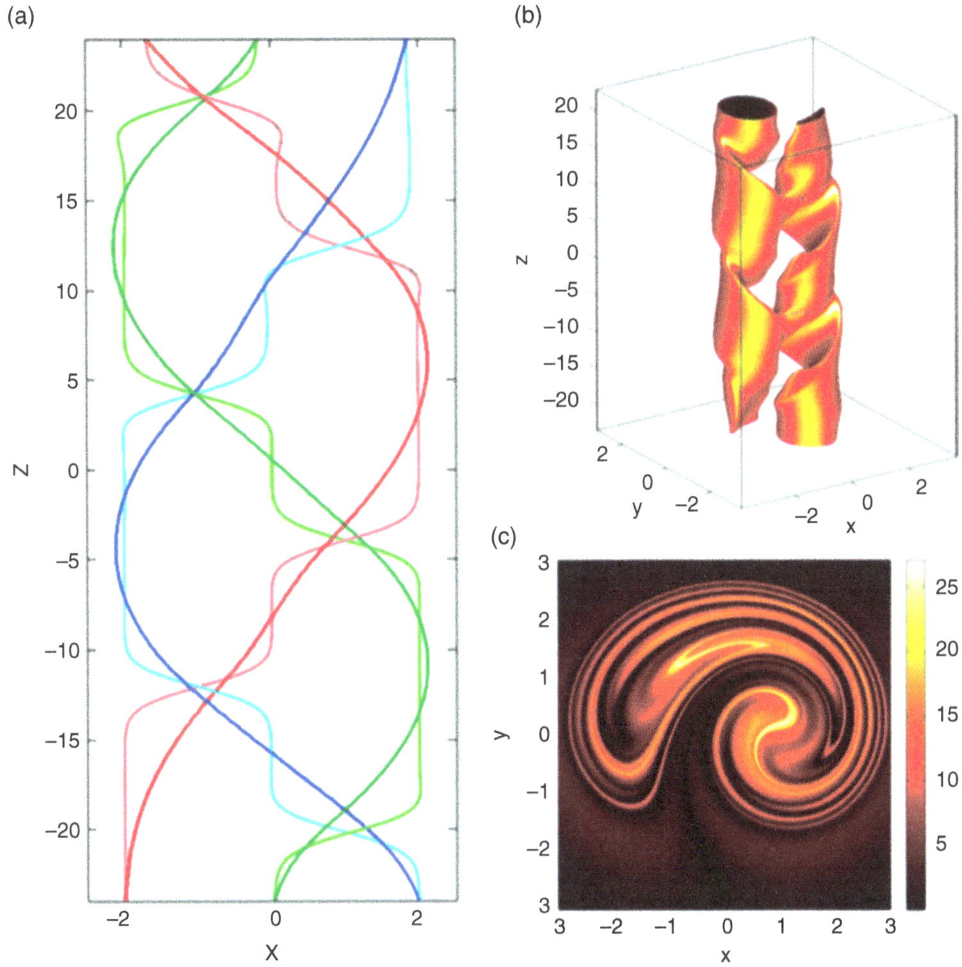

Figure 14.8 (a) Example field lines in the initial 3D braided magnetic field (pale colored lines) with the corresponding field lines after relaxation of the 3D MHS magnetic field (dark colored lines). (b) Isosurfaces of electric current density in the relaxed MHS configuration. (c) A plot of the $\int_l E_{\parallel} dl$ revealing considerable fine scale structure. All images are taken from or in the case of (a) based on figures from *Wilmot-Smith et al.* [2010]. © AAS., reproduced with permission.

concluded that current singularities could be formed following the collapse of certain null-free 3D magnetic fields.

Both *Wilmot-Smith et al.* [2009] and *Pontin and Hornig* [2015] also calculated Q in their braiding experiments. They found that the QSLs were not good predictors of the locations of either the current features or the $\int_l j_{\parallel} dl$, although regions of high Q did form similar fine-scale structures.

These studies confirmed that finite width current layers may form with simple null-free magnetic fields, but it is less certain whether singular current sheets can form. Indeed, it may be that current sheets can form only in some null-free configurations, but the characteristics of the plasma flow or magnetic field that permit or prevent a current sheet from forming are not yet known. Answers to these questions will require further research, both analytically and numerically.

14.4. DISCUSSION

In this review, we consider the creation and nature of MHS equilibria involving singular current sheets and finite-width current layers in both 2D and 3D scenarios. MHD is a valid regime in which to consider the formation of current layers in the solar corona since the effect of particles does not become important until scales of the order of 10 or so ion inertial lengths are reached and these will not arise until the current layer has collapsed to scales much smaller than those typically found in the system.

The creation of finite-width current layers in both 2D and 3D MHS equilibria about topological features such as null points, spines, and separators are all associated with infinite time singularities. In each case, it is found that the current density shows a significant spike at the null or on the spine or separator, which continues to grow logarithmically with respect to time as the magnetic field tries to attain a state where the current is located solely on the topological feature. In each of these situations, nonresistive MHD experiments that are run for hundreds of Alfvén/fast-mode crossing times tend to states where the field outside the current layer is in an MHS equilibrium, but inside the current layer the Lorentz and pressure forces remain unbalanced and the peak current continues to grow.

The 2D and 3D MHS nonresistive MHD, or magnetofrictional, relaxation experiments in which a finite current layer forms at a single null point with all fields retaining just the single null all behave similarly. In the case of a 2D X-point, the null collapses to form a configuration closely resembling a 1D null, namely, antiparallel field about a null line. In the 3D MHS equilibria where there is a component of current perpendicular to the spine of the null, the field about the null collapses such that it resembles either that of a 1D null line or 2D X-point, depending on the initial eigenvalues of the null. In both of these cases, current accumulates about (at) the null and, local to the null, on both the separatrices in 2D, but only on the separatrix surface in 3D. If the current about the initial 3D null is parallel to the spines, then, in the final equilibrium, the 3D null does not collapse, but is potential and all the current accumulates on the spines away from the null.

Analytical expressions for force-free magnetic fields about collapsed 2D X-type null points represent the current accumulation as a singular current sheet, that is, a tangential discontinuity across which the field is not continuous [*Bungey and Priest*, 1995; *Green*, 1965; *Syrovatskiĭ*, 1971]. In these models, the original X-point null is replaced by a pair of nulls that lie either at the ends of the current sheet or somewhere along its length and as a result are nondifferentiable. If the end of the current sheet is not a null point, then it will be a singular point instead where the field is infinite. Such configurations have not been found by nonresistive MHD experiments since, as already mentioned, such states would take an infinite time to reach.

In 3D, current layers have been shown to form about separators. In one case, the collapse of a pair of 3D nulls linked by two separators was considered and an analytical model analogous to that of *Green* [1965] was proposed. In this analytical model, put forward by *Longcope and Cowley* [1996], both separators and both nulls collapse and in the resulting potential magnetic configuration, they are replaced by a singular current sheet in the form of an annular disc with four nondifferentiable nulls.

In another case, the formation of a current layer at a single separator was considered by *Stevenson et al.* [2015]. They formed a MHS equilibrium via a nonresistive MHD experiment in which the separator and both 3D nulls are retained. The current accumulates in a finite current layer about the separator (away from the nulls). On the separatrix surfaces that lie outside the nulls, rather than in between where the separator is, the current accumulations are more dispersed and do not form current layers.

Two models showing current layers in topologically simple (null-free) 3D MHS equilibria were considered [*Craig and Pontin*, 2014; *Pontin and Hornig*, 2015]. In one, the accumulations were of finite width whereas in the other they were found to show properties that suggested they could be singular. In both situations regions of high Q, QSLs were found. This suggests that in some situations, but not all, topologically simple (null-free) 3D magnetic fields may form singular current layers. However, further studies are required to determine what the properties of the initial field and the perturbation must be to create a singular current region.

Given the relatively long formation time of MHS current layers it is not clear how prevalent such current layers will be in dynamic scenarios where plasma flows are nonnegligible, such as those found in the solar atmosphere. It is possible that dynamically driven current enhancements are more important than those developed through MHD relaxation. This, of course, needs to be investigated properly before any firm conclusions can be made. However, since the plasma parameter regimes in the latest numerical experiments are vastly different from those found in the solar atmosphere, this is hard to do.

The MHS equilibria involving current layers appear to be stable to ideal instabilities (although some that are twisted are unstable to the kink instability [*Wilmot-Smith et al.*, 2010]). However, the equilibria are unstable to resistive instabilities, although exactly what type is not clear.

One interesting finding from resistive experiments that start from the MHS equilibria involving a current layer formed at a topological feature is that the reconnection follows a similar pattern. That is, as soon at the resistive instability is initiated, for instance by prescribing an anomalous resistivity that occurs above a given critical current, there is very rapid reconnection that dissipates the current above the critical value. Reconnection then stops and the magnetic field is left to evolve in a nonresistive MHD manner, building up current in the current layer, as it attempts to reach an MHS equilibrium again. As soon as the current in the layer reaches the critical value, this again leads to reconnection. Thus, the initial phase of rapid reconnection is followed by a phase of slower impulsive bursty reconnection [e.g., *Fuentes-Fernández et al.*, 2012b; Stevenson and Parnell, 2015a,b], which will continue until the MHD relaxation process can no longer build up enough current to reach the critical value required to trigger reconnection. This pattern of energy release is reminiscent of a solar flare, but the amount of energy released in current models is very small, less than that of a nanoflare, due to numerical constraints and the domain size considered.

Once reconnection actually starts, the topology of the magnetic field may change, and additional null points and separators can, and do tend to, be created inside the current layers. In fact, in the case of current layers at both 3D null points and separators, the onset of reconnection leads to an initial cascade in scales as the current layer breaks up into smaller and smaller localized current layers. This is directly coupled to an increase in the number of topological features [e.g., *Haynes et al.*, 2007; *Parnell et al.*, 2008, 2010; *Wyper and Pontin*, 2014]. Similarly, in the case of reconnection in the pigtail braid [*Pontin et al.*, 2011], there is a cascade in scales with multiple small current layers being formed. As the current dissipates and the magnetic energy in the field decreases, the number of null points and separators returns to the number in the field before being perturbed (or less), while in the case of the pigtail braid, a state is achieved that is neither potential nor constant-alpha force-free due to the constraints of line-tied boundary conditions and particular properties of the initial magnetic field (for more details see *Pontin et al.* [2011] and *Yeates et al.* [2010]).

In the solar atmosphere, it is interesting to consider how prevalent the different topological features are in order to postulate which of the above types of current layers (and associated reconnection) are likely to be most common in the solar atmosphere. The abundance of magnetic null points has been estimated using magnetic field extrapolations from a range of different observed photospheric magnetograms.

A recent paper by *Edwards and Parnell* [2015] considered global potential field source-surface models extrapolated from a range of different synoptic magnetograms from both solar maximum and solar minimum. Such models provide a very conservative estimate on the number of topological features with many more expected in dynamical models [e.g., *Parnell et al.*, 2010]. Null points are found to follow a butterfly diagram with fewer nulls located above active regions than above quiet-Sun regions. The numbers of nulls are, as a consequence, anticorrelated with the solar cycle [*Edwards and Parnell*, 2015]. Extrapolations with 301 harmonics from MDI synoptic magnetograms produced between 2000 and 3000 nulls in the solar atmosphere (a lower limit due to model limitations). This equates to a smaller null density than that estimated by *Longcope and Parnell* [2009], who found one null point lying above 1.5 Mm for every 322 Mm^2 patch of quiet Sun using high-resolution partial frame MDI data. *Longcope et al.* [2009] found more than twice as many nulls above 1.5 Mm using data from Hinode.

Of these null points, approximately half are isolated, that is, they are not connected by any separators. All the other nulls are connected to at least one separator whose length can extend to more than a solar radius [*Platten et al.*, 2014]. So there are thousands of separators through the solar atmosphere, especially above quiet-Sun regions.

However, much of the solar atmosphere is devoid of topological features, especially one megameter or more above the solar surface. These magnetic field regions are typically associated with large-scale features on the solar surface, such as active regions or unipolar regions at the poles. Whether the flow patterns in these regions are sufficiently complex to generate current layers is an open question.

It is therefore difficult to assess the relative contributions of reconnection at the different types of current layers to either heating the solar atmosphere or to powering solar events such as solar flares. The very small size of current layers means that they cannot be observed using the current suite of solar missions. Thus much work remains to be done to bridge the gap between kinetic and observational scales before the importance of current layers can be properly appreciated.

ACKNOWLEDGMENTS

The author is grateful to Dr. A. Wilmot-Smith for useful discussions on aspects of this review and thanks the referees for constructive feedback that has helped improve the paper. The author acknowledges the support of STFC through the Solar & Magnetospheric Theory Group's 2016 Consolidated Grant.

REFERENCES

Archontis, V., F. Moreno-Insertis, K. Galsgaard, and A. W. Hood (2005), The three-dimensional interaction between emerging magnetic flux and a large-scale coronal field: Reconnection, current sheets, and jets, *Astrophys. J.*, *635*, 1299–1318; doi:10.1086/497533.

Bulanov, S. V., and M. A. Olshanetskii (1984), Magnetic collapse near zero points of the magnetic field, *Phys. Lett. A*, *100*, 35–38; doi:10.1016/0375-9601(84)90349-9.

Bungey, T. N., and E. R. Priest (1995), Current sheet configurations in potential and force-free fields., *Astron. Astrophys.*, *293*, 215–224.

Cairns, R. A. (1985), *Plasma Physics*, Blackie & Sons Ltd, Glasgow.

Craig, I. J. D., and D. I. Pontin (2014), Current singularities in line-tied three-dimensional magnetic fields, *Astrophys. J.*, *788*, 177; doi:10.1088/0004-637X/788/2/177.

Craig, I. J. D., and Y. E. Litvinenko (2005), Current singularities in planar magnetic X points of finite compressibility, *Phys. Plasmas*, *12*(3), 032301; doi:10.1063/1.1854154.

Edwards, S. J., and C. E. Parnell (2015), Null point distribution in global coronal potential field extrapolations, *Solar Phys.*, *290*, 2055–2076; doi:10.1007/s11207-015-0727-7.

Forbes, T. G., and E. R. Priest (1995), Photospheric magnetic field evolution and eruptive flares, *Astrophys. J.*, *446*, 377; doi:10.1086/175797.

Fuentes-Fernández, J., and C. E. Parnell (2012), Magnetohydrodynamics dynamical relaxation of coronal magnetic fields. III. 3D spiral nulls, *Astron. Astrophys.*, *544*, A77; doi:10.1051/0004-6361/201219190.

Fuentes-Fernández, J., and C. E. Parnell (2013), Magnetohydrodynamics dynamical relaxation of coronal magnetic fields. IV. 3D tilted nulls, *Astron. Astrophys.*, *554*, A145; doi:10.1051/0004-6361/201220346.

Fuentes-Fernández, J., C. E. Parnell, and A. W. Hood (2011), Magnetohydrodynamics dynamical relaxation of coronal magnetic fields. II. 2D magnetic X-points, *Astron. Astrophys.*, *536*, A32; doi:10.1051/0004-6361/201117156.

Fuentes-Fernández, J., C. E. Parnell, and E. R. Priest (2012a), The onset of impulsive bursty reconnection at a two-dimensional current layer, *Phys. Plasmas*, *19*(7), 072901; doi:10.1063/1.4729334.

Fuentes-Fernández, J., C. E. Parnell, A. W. Hood, E. R. Priest, and D. W. Longcope (2012b), Consequences of spontaneous reconnection at a two-dimensional non-force-free current layer, *Phys. Plasmas*, *19*(2), 022,901–022,901; doi:10.1063/1.3683002.

Green, R. M. (1965), Modes of annihilation and reconnection of magnetic fields, in *Stellar and Solar Magnetic Fields*, IAU Symposium, 22, edited by R. Lust, *398*.

Haynes, A. L., C. E. Parnell, K. Galsgaard, and E. R. Priest (2007), Magnetohydrodynamic evolution of magnetic skeletons, *Proc. Roy. Soc. Lond. A*, *463*; doi:10.1098/rspa.2007.1815.

Hesse, M., and K. Schindler (1988), A theoretical foundation of general magnetic reconnection, *J. Geophys. Rev.*, *93*, 5559–5567; doi:10.1029/JA093iA06p05559.

Heyvaerts, J., E. R. Priest, and D. M. Rust (1977), An emerging flux model for the solar flare phenomenon, *Astrophys. J.*, *216*, 123–137; doi:10.1086/155453.

Imshennik, V. S., and S. I. Syrovatskiĭ (1967), Two-dimensional flow of an ideally conducting gas in the vicinity of the zero line of a magnetic field, *Sov. J. Exper. Theo. Phys.*, *25*, 656.

Janse, Å. M., and B. C. Low (2009), The topological changes of solar coronal magnetic fields; I. Spontaneous current sheets in three-dimensional fields, *Astrophys. J.*, *690*, 1089–1104; doi:10.1088/0004-637X/690/2/1089.

Klapper, I. (1998), Constraints on finite-time current sheet formation at null points in two-dimensional ideal incompressible magnetohydrodynamics, *Phys. Plasmas*, *5*, 910–914; doi:10.1063/1.872659.

Longcope, D., C. Parnell, and C. DeForest (2009), The Density of Coronal Null Points from Hinode and MDI, in *The Second Hinode Science Meeting: Beyond Discovery-Toward Understanding*, Astronomical Society of the Pacific Conference Series, 415, edited by B. Lites, M. Cheung, T. Magara, J. Mariska, and K. Reeves, *178*.

Longcope, D. W., and C. E. Parnell (2009), The number of magnetic null points in the quiet sun corona, *Solar Phys.*, *254*, 51–75; doi:10.1007/s11207-008-9281-x.

Longcope, D. W., and S. C. Cowley (1996), Current sheet formation along three-dimensional magnetic separators, *Phys. Plasmas*, *3*, 2885–2897; doi:10.1063/1.871627.

Parnell, C. E., A. L. Haynes, and K. Galsgaard (2008), Recursive reconnection and magnetic skeletons, *Astrophys. J.*, *675*, 1656–1665; doi:10.1086/527532.

Parnell, C. E., E. R. Priest, and V. S. Titov (1994), A model for X-ray bright points due to unequal cancelling flux sources, *Solar Phys.*, *153*, 217–235; doi:10.1007/BF00712503.

Parnell, C. E., J. M. Smith, T. Neukirch, and E. R. Priest (1996), The structure of three-dimensional magnetic neutral points, *Phys. Plasmas*, *3*, 759–770; doi:10.1063/1.871810.

Parnell, C. E., R. C. Maclean, and A. L. Haynes (2010), The Detection of Numerous Magnetic Separators in a Three-Dimensional Magnetohydrodynamic Model of Solar Emerging Flux, *Astrophys. Lett.*, *725*, L214–L218; doi:10.1088/2041-8205/725/2/L214.

Parnell, C. E., T. Neukirch, J. M. Smith, and E. R. Priest (1997), Structure and collapse of three-dimensional magnetic neutral points, *Geo. Astro. Fluid Dyn.*, *84*, 245–271; doi:10.1080/03091929708208979.

Platten, S. J., C. E. Parnell, A. L. Haynes, E. R. Priest, and D. H. Mackay (2014), The solar cycle variation of topological structures in the global solar corona, *Astron. Astrophys.*, *565*, A44; doi:10.1051/0004-6361/201323048.

Pontin, D. I., A. L. Wilmot-Smith, G. Hornig, and K. Galsgaard (2011), Dynamics of braided coronal loops. II. Cascade to multiple small-scale reconnection events, *Astron. Astrophys.*, *525*, A57; doi:10.1051/0004-6361/201014544.

Pontin, D. I., and G. Hornig (2015), The Structure of Current Layers and Degree of Field-line Braiding in Coronal Loops, *Astrophys. J.*, *805*, 47; doi:10.1088/0004-637X/805/1/47.

Pontin, D. I., and I. J. D. Craig (2005), Current singularities at finitely compressible three-dimensional magnetic null points, *Phys. Plasmas*, *12*(7), 072112; doi:10.1063/1.1987379.

Priest, E. (2014), *Magnetohydrodynamics of the Sun*, Cambridge University Press, New York.

Priest, E. R., and D. I. Pontin (2009), Three-dimensional null point reconnection regimes, *Phys. Plasmas*, *16*(12), 122,101–122,101; doi:10. 1063/1.3257901.

Priest, E. R., and P. Démoulin (1995), Three-dimensional magnetic reconnection without null points, 1. Basic theory of magnetic flipping, *J. Geophys. Rev.*, *100*, 23,443–23,464; doi:10.1029/95JA02740.

Priest, E. R., C. E. Parnell, and S. F. Martin (1994), A converging flux model of an X-ray bright point and an associated canceling magnetic feature, *Astrophys. J.*, *427*, 459–474; doi:10.1086/174157.

Rastätter, L., A. Voge, and K. Schindler (1994), On current sheets in two-dimensional ideal magnetohydrodynamics caused by pressure perturbations, *Phys. Plasmas*, *1*, 3414–3424; doi:10.1063/1.870490.

Rickard, G. J., and V. S. Titov (1996), Current Accumulation at a Three-dimensional Magnetic Null, *Astrophys. J.*, *472*, 840; doi:10.1086/178114.

Schindler, K., M. Hesse, and J. Birn (1988), General magnetic reconnection, parallel electric fields, and helicity, *J. Geophys. Rev.*, *93*, 5547–5557; doi:10. 1029/JA093iA06p05547.

Stevenson, J. E. H., and C. E. Parnell (2015a), Spontaneous reconnection at a separator current layer: 1. Nature of the reconnection, *J. Geophys. Res. Space Physics*, *120*, 10; doi:10.1002/2015JA021730.

Stevenson, J. E. H., and C. E. Parnell (2015b), Spontaneous reconnection at a separator current layer: 2. Nature of the waves and flows, *J. Geophys. Res. Space Physics*, *120*, 10; doi:10.1002/2015JA021736.

Stevenson, J. E. H., C. E. Parnell, E. R. Priest, and A. L. Haynes (2015), The nature of separator current layers in MHS equilibria. I. Current parallel to the separator, *Astron. Astrophys.*, *573*, A44; doi:10.1051/0004-6361/201424348.

Syrovatskiĭ, S. I. (1971), Formation of current sheets in a plasma with a frozen-in strong magnetic field, *Sov. J. Exper. Theo. Phys.*, *33*, 933.

Titov, V. S. (2007), Generalized squashing factors for covariant description of magnetic connectivity in the solar corona, *Astrophys. J.*, *660*, 863–873; doi:10.1086/512671.

Titov, V. S., and G. Hornig (2000), Magnetohydrodynamic flows sustaining stationary magnetic nulls, *Phys. Plasmas*, *7*, 3542–3550; doi:10.1063/1. 1286862.

Titov, V. S., and G. Hornig (2002), Magnetic connectivity of coronal fields: geometrical versus topological description, *Adv. Space Res.*, *29*, 1087–1092; doi:10.1016/S0273-1177(02)00021-2.

Titov, V. S., E. R. Priest, and P. Demoulin (1993), Conditions for the appearance of "bald patches" at the solar surface, *Astron. Astrophys.*, *276*, 564.

Tur, T. J., and E. R. Priest (1976), The formation of current sheets during the emergence of new magnetic flux from below the photosphere, *Solar Phys.*, *48*, 89–100; doi:10.1007/BF00153334.

Uberoi, M. S. (1963), Some Exact Solutions of Magnetohydrodynamics, *Phys. Fluids*, *6*, 1379–1381; doi:10.1063/1.1710956.

van Ballegooijen, A. A. (1988a), Force free fields and coronal heating part I. The formation of current sheets, *Geo. Astro. Fluid Dyn.*, *41*, 181–211; doi:10.1080/03091928808208850.

van Ballegooijen, A. A. (1988b), Magnetic fine structure of solar coronal loops, in *Solar and Stellar Coronal Structure and Dynamics*, edited by R. C. Altrock, *115–124*.

Wilmot-Smith, A. L. (2015), An overview of flux braiding experiments, *Phil. Trans. R. Soc. Lond. A*, *373*, 20140,265–20140,265; doi:10.1098/rsta.2014.0265.

Wilmot-Smith, A. L., D. I. Pontin, and G. Hornig (2010), Dynamics of braided coronal loops, I. Onset of magnetic reconnection, *Astron. Astrophys.*, *516*, A5; doi:10.1051/0004-6361/201014041.

Wilmot-Smith, A. L., G. Hornig, and D. I. Pontin (2009), Magnetic braiding and quasi-separatrix layers, *Astrophys. J.*, *704*, 1288–1295; doi:10.1088/0004-637X/704/2/1288.

Wyper, P. F., and D. I. Pontin (2014), Non-linear tearing of 3D null point current sheets, *Phys. Plasmas*, *21*(8), 082114; doi:10.1063/1.4893149.

Yeates, A. R., G. Hornig, and A. L. Wilmot-Smith (2010), Topological constraints on magnetic relaxation, *Phys. Rev. Lett.*, *105*(8), 085002; doi:10.1103/PhysRevLett.105.085002.

Yokoyama, T., and K. Shibata (1995), Magnetic reconnection as the origin of X-ray jets and Hα surges on the Sun, *Nature*, *375*, 42–44; doi:10.1038/375042a0.

15

Large-Scale Current Sheets in Flares and CMEs

Jun Lin and Lei Ni

ABSTRACT

The catastrophe model of the solar eruption predicted that a long current sheet (CS) develops in the major eruption. The CS length and the dynamics of the ejecta are dominated by both the rate of magnetic reconnection and the background magnetic field. Formation and development of the long CS were confirmed by observations soon after it was predicted. Furthermore, the CS was found much thicker than expected by the traditional theory. Works in both theories and observations regarding the physics behind the unexpectedly thick CS have been performed for many years, but the question is still open. Recent numerical experiments suggest that the turbulence due to various instability inside CS accounts for the thick CS and fast reconnection that drives the major eruption. In this review, we first introduce the model and its prediction of the long CS, and the corresponding confirmations by observations. Then, we discuss studies of various CS features performed so far, including investigations of its geometric scale and internal fine structures, together with the possible mechanisms for observations. Finally, we list the differences between the classical understanding of the CS and the facts revealed by both the observations and numerical simulations, and discuss possible reasons why the differences occur.

15.1. INTRODUCTION

Magnetic field possesses the property that it tends to maintain its initial topological connectivity unchanged unless diffusion takes place, namely, any change in the magnetic topology must be accompanied by magnetic diffusion. The universe is filled with magnetic field and plasma, and interactions between them result in twisting, tangling, writhing, stretching, compressing, breaking, and reconnecting of the magnetic field, leading to the conversion of magnetic energy into plasma heating and kinetic energy, and vice versa. In the universe, however, both the length scale and the electrical conductivity are usually very large, with the magnetic Reynolds number, R_m (or equivalently, the Lundquist number, S), varying from 10^6 to 10^{12} in the solar corona [*Priest*, 2014], and from 10^{15} to 10^{21} in the interstellar medium [*Heitsch and Zweibel*, 2003], respectively. The magnetic field diffusion in such an environment is thus confined to very local regions including X-type neutral points (or X-points for short), current sheets (CS), quasi-separatrix layers (QSL), null points, and magnetic separators.

Usually, an X-point, null point, QSL, and magnetic separator could exist in a complex magnetic structure in the corona, no matter whether this structure is static, quasi-static, or dynamic (see discussions of *Titov and Démoulin* [1999], *Lin et al.* [2002], *Qiu et al.* [2007], *Roussev et al.* [2012], *Zhang et al.* [2017], and references therein). But a CS cannot exist statically or quasi-statically in the corona because of the diffusion caused by the finite electrical resistivity of the coronal plasma. Instead, a CS could form only as a result of the collapse of an X-point due to either stretching or squeezing of the

Yunnan Observatories, Chinese Academy of Sciences, Kunming, Yunnan P. R. China; Center for Astronomical Mega-Science, Chinese Academy of Sciences, Beijing, P. R. China

Electric Currents in Geospace and Beyond, Geophysical Monograph 235, First Edition.
Edited by Andreas Keiling, Octav Marghitu, and Michael Wheatland.

magnetic field nearby (e.g., see Fig. 2.6 of *Priest* [2014] and the related discussions).

A long CS in the corona is produced by the eruption due to severe stretching of a closed magnetic field in the corona (e.g., see *Forbes and Isenberg* [1991]; *Isenberg et al.* [1993]; *Forbes et al.* [1994]; *Forbes and Priest* [1995]; and *Lin and Forbes* [2000]; and *Lin et al.* [2003]). It appears below the coronal mass ejection (CME) and above the associated flare, separating two plasmas associated with magnetic fields of opposite polarity (see Fig. 15.1). Magnetic reconnection in a CS converts magnetic energy into heating (accounting for the flare) and kinetic energy of the bulk plasma (accounting for the CME), associated with the production of energetic particles. (We note here that, in this work, we are focusing on such large-scale CSs forming in the major eruption.)

Before *Lin and Forbes* [2000], the well-known CSHKP model provided the standard description of a two-ribbon flare in a major eruption [*Švestka and Cliver*, 1992]. In this model, a long CS extends from the top of the flare to infinity, and magnetic reconnection occurring in the CS produces growing flare loops in the corona and separating flare ribbons on the solar disk. *Martens and Kuin* [1989] pointed out that the magnetic structure in the CSHKP model is actually the lower part of the disrupting magnetic configuration in the major event, the upper part of this configuration is an eruptive prominence (or CME), and a CS connects them to one another. But the CS in the *Martens and Kuin* [1989] scenario was short, so their model is not suitable for describing the long duration evolution in the configuration.

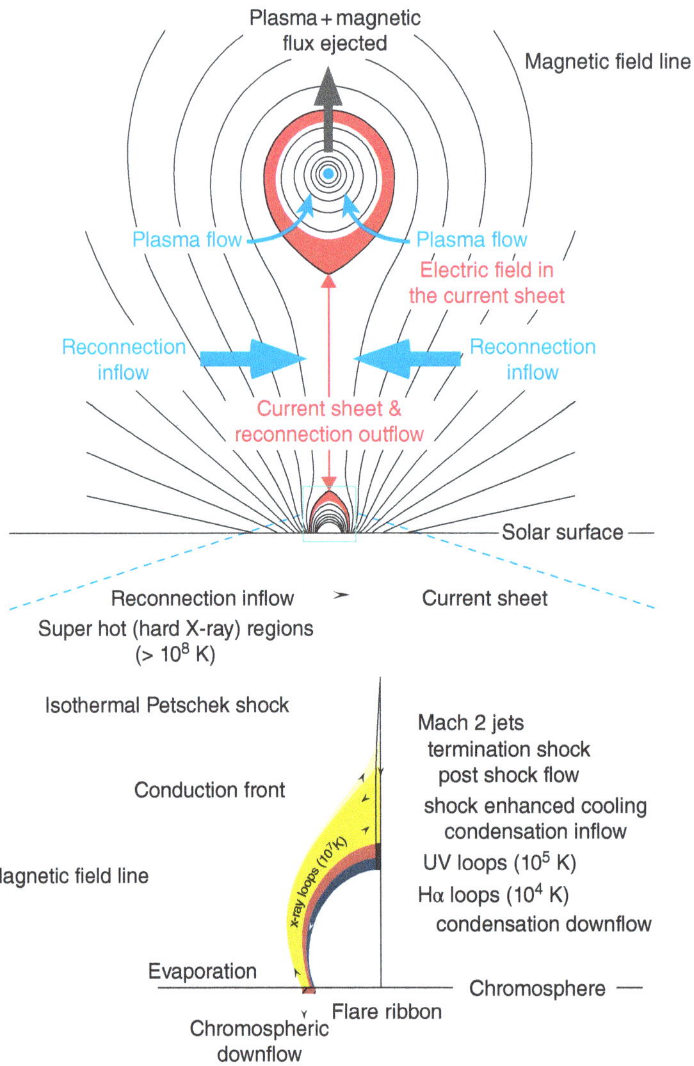

Figure 15.1 Schematic diagram of the magnetic configuration appearing in the eruptive process described in the model of *Lin and Forbes* [2000]. From *Lin et al.* [2004].

In the Yohkoh era, on the other hand, the Martens-Kuin scenario of the eruption was gradually pieced together. With the direct and the indirect observational evidence that could be collected [*Klimchuk et al.*, 1994; *McAllister et al.*, 1992, 1994; *Hanaoka et al.*, 1994; and *Khan et al.*, 1994], a common feature of the magnetic structure that forms in the eruptive event emerged: the flare loop/arcade observed in X-ray is below the eruptive prominence/CME. The sample provided by *Hanaoka et al.* [1994] was believed more than the others because they observed the arcade not only in X-rays but also in microwave. *Gopalswamy et al.* [1997] looked into the evolutionary features of the X-ray emission source in a flare arcade with a bright knot at the apex, and interpreted this bright knot as a reconnection region that separates the flare arcade/loop from the eruptive prominence/CME.

Hanaoka et al. [1994] figured out that the eruption should create an elongated/stretched magnetic configuration similar to that shown in Figure 15.1 (see also their Fig. 13 for comparison). After having compared the kinematic features of plasma blobs and fine structures of magnetic field inside the eruptive prominence with those of the flare loops/arcades, *Hanaoka et al.* [1994] concluded that the eruption created a severely stretched magnetic configuration in which the flare is confined to a very local region near the solar surface and the eruptive prominence is high above it, and the elongated magnetic field is a necessity for reconnection in the eruption.

But there was not a suitable theoretical description for this scenario at that time. With the development of the catastrophe model for the solar eruption by *Lin and Forbes* [2000], the scenario of *Hanaoka et al.* [1994] evolved from a qualitative conjecture into a quantitative description, and the most important progress made in this evolution was the success in implanting self-consistently a long CS in the disrupting magnetic structure.

Formation of the long CS in the eruption was confirmed by observations [*Ciaravella et al.*, 2002; *Webb et al.*, 2003] soon after it was predicted by the theory [*Lin and Forbes*, 2000]. Later, *Lin et al.* [2007, 2009] realized that a CME/flare CS could as well possess a considerably large thickness according to observations, and their results have continuously been confirmed by the following observations (e.g, see *Lin et al.* [2015] for more discussion). This constitutes a serious challenge to the traditional picture and theory about magnetic reconnection and the CS, and is also an open question to both the theory and the observation.

In the next section, we will go through the work in both theory and observation about the long CS developed by the major eruption. Geometrical scales and the related studies of the CS are introduced in section 15.3. (We need to note here that major eruptions in the SDO era have been much less frequent than those in the SOHO era, so we do not expect as many papers on the CME/flare CS in the SDO era as those in the SOHO era.) The physics behind these observations will be discussed in section 15.4, and we finally summarize this work in section 15.5.

15.2. CME-FLARE CURRENT SHEETS

When developing the catastrophe model for the solar eruption, *Lin and Forbes* [2000] and *Lin* [2002] realized that a long CS may form in the eruption. They noticed that if the rate of reconnection is not high enough, the increasing tension force associated with the CS formation may become strong enough to prevent the flux rope from escaping from the Sun, and that the rate of magnetic reconnection is governed by the local Alfvén speed near the CS such that the rate of magnetic reconnection, $M_A = v_i/v_A$, must be less than unity. Here, v_i and v_A are the reconnection inflow speed and the Alfvén speed near the CS, respectively. As pointed out by *Priest and Forbes* [2000], the dissipation of the magnetic field in the reconnection process is controlled by the diffusive rate of the magnetic field in the reconnection region (or CS), and the timescale at which the diffusion erases the magnetic field in the CS, τ_d, is much longer than the time scale at which Aflvén waves traverse the CS, τ_A.

So, if the reconnection inflow speed, v_i, is too fast, say exceeding v_A, the magnetic field brought by the reconnection inflow to the CS cannot be dissipated fast enough, and the magnetic field will pile up outside the CS, preventing reconnection from proceeding further with v_i being suppressed consequently. Therefore, for a smoothly proceeding magnetic reconnection process, $M_A < 1$ usually holds. Thus, if v_A decreases with height very fast, the reconnection process can hardly weaken effectively the magnetic tension that tends to stop the motion of the CME, and the loss of equilibrium may eventually cease to develop to an eruption.

Lin and Forbes [2000] and *Lin* [2002] showed that, in an isothermal corona, the plasma decreases in density exponentially with height *h* (from the surface of the Sun), but the magnetic field usually decreases quadratically with *h* at higher altitudes, so v_A may decrease with *h* at lower altitudes, but increase at higher altitudes, which allows reconnection to erase the CS quickly enough, such that no long CS is expected (see Fig. 6 of *Lin and Forbes* [2000]). In a realistic corona, on the other hand, the plasma decreases in density exponentially for $h < 0.7$ solar radii ($R_\odot$), and decreases quadratically as $h > 0.7\ R_\odot$ (e.g., see *Lin* [2002] and the references therein). So the resultant Alfvén speed decreases at higher altitudes as $1/h$ if the magnetic field still decreases with *h* quadratically, and the reconnection rate may be suppressed in a certain way, but not very seriously, and hence the flux rope still could escape in the eruption if $M_A > 0.03$, and a long CS could be expected (see detailed discussions by *Lin* [2002]). This explains why most of the eruptive prominences observed developed to CMEs smoothly, and only a few were the so-called failed eruptions (e.g., see *Ji et al.* [2003]; *Myers et al.* [2015]; and references therein).

A reconnection rate with a value of a few percent has been continuously obtained in observations. *Ko et al.* [2003] obtained M_A between 0.01 and 0.03, *Lin et al.* [2005, 2007] found M_A varied from 0.008 to 0.25, and *Takasao et al.* [2012] gave $0.055 < M_A < 0.2$, for major eruptions during which the CS formed as a result of stretching of the coronal magnetic field. *Su et al.* [2013] revealed that M_A was between 0.05 and 0.5 measured near the CS formed in a compact flare developed due to the collision of two coronal magnetic loops, and *Zhu et al.* [2016] noticed for a similar compact flare that M_A equaled 0.05 or ranged from 0.01 to 0.02, depending on the method of estimating the Alfvén speed near the CS.

Forbes and Lin [2000] and *Lin and Forbes* [2000] pointed out that the behavior of the CS itself depends on both the reconnection rate and the flux rope motion such that fast reconnection erases the CS fast and helps the flux rope escape fast, but fast motion of the flux rope stretches the magnetic field severely, allowing the CS to develop fast as well. At the early stage of the eruption, the disrupting magnetic configuration evolves catastrophically and is governed by the Alfvén timescale (τ_A, about 10–30 min in the corona), which is shorter than the reconnection timescale, so the CS development dominates the CS erosion, and a long CS is expected in the major eruption (see also numerical simulations by *Linker et al.* [2003] and *Reeves et al.* [2010]). After time τ_A, magnetic reconnection starts governing the evolution with the erosion of the CS dominating the development, and the flare is apparently produced with growing flare loops (interested readers should refer to *Forbes and Acton* [1996] and *Lin* [2002] for detailed discussion), and the flux rope quickly escapes from the Sun developing to a CME with a fast expanding CME-bubble [e.g., see *Lin et al.*, 2004; *Lin and Soon*, 2004, for details].

The scenario of a long CS developed between CME and flare in the solar eruption was quickly confirmed by *Ciaravella et al.* [2002]. They noticed a very narrow feature in the spectral line of [Fe XVIII] 974 Å that forms at a temperature around $10^{6.8}$ K at the location between the flare and the core of the CME at the altitude of $r = 1.55$ $R_\odot$ above the surface, suggesting an edge-on view of the CS as shown in Figure 15.1. A composite of EIT and LASCO C2 images with the UVCS slit image superposed is given in Figure 15.2. Several other high temperature lines of [Ca XIV], Fe XVII, and Ne IX could be recognized at the same location, especially when the UVCS entrance slit was placed 0.15 $R_\odot$ lower. Cooler lines did not show the same peak, but they were detectable at

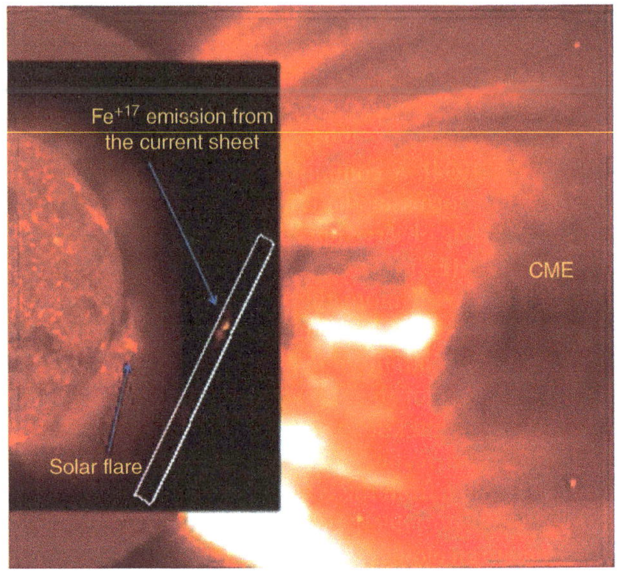

Figure 15.2 Composite images taken by EIT and LASCO on board SOHO that displayed the flare loops and CME core. The white rectangle including a bright point between the bigger images contains the UVCS slit image where the position of the emission from the spectral line [Fe XVIII] 974 Å can be identified. Several important features of the disrupting magnetic configuration are specified. Original picture template taken from *Kohl et al.* [2006].

nearby locations along the slit. A temperature of $10^{6.8}$ K to $10^{6.9}$ K may be deduced from the [Fe XVIII] to Fe XVII intensity ratio. A good review of this issue was given by *Kohl et al.* [2006].

Here the brackets indicate that the relevant spectral line is a so-called forbidden emission line. This sort of spectral line is produced by atoms, ions, and molecules, which undergo a transition between two energy levels that does not take place in a particular condition like that in the terrestrial environment where the gas density is fairly high and the emission is usually dominated by the electric-dipole transition. In a very tenuous medium like the corona, on the other hand, the other kinds of transitions, such as the magneto-dipole and/or electric-quadruple transitions, can occur but at a much lower rate. The spectral lines, like [Fe XVIII] and [Ca XIV], are produced this way.

When investigating 26 CME events that were observed in white-light by the Solar Maximum Mission satellite, *Webb et al.* [2003] noticed that magnetic disconnection features were displayed by these events. Such events are usually named "candidate disconnection events" (CDEs). Some structures that look like rays formed and developed a few hours following these CDEs, and they were also often observed to connect to the streamers above the limb. *Webb et al.* [2003] identified these structures as CSs behind the CMEs.

Then, the long CS connecting the CME to the associated flare in the major eruption, and the observational consequences of the reconnection process inside this CS, was successively identified and studied by *Ko et al.* [2003], *Raymond et al.* [2003], *Lin et al.* [2005], *Bemporad* [2008], *Ciaravella and Raymond* [2008], *Vršnak et al.* [2009], *Ko et al.* [2010], *Schettino et al.* [2010], *Takasao et al.* [2012], *Ciaravella et al.* [2013], and *Ling et al.* [2014] in EUV and visible wave bands; and by *Innes et al.* [2003a, 2003b], *Asai et al.* [2004], *Kolománski and Karlický* [2007], *Milligan et al.* [2010], *Liu* [2013], and *Liu et al.* [2013] in EUV, X-ray, and visible wave bands, respectively.

In the works by *Poletto et al.* [2004] and *Bemporad et al.* [2006], in addition to the remote sensing data, the in situ data from Ulysses were also used. The detection of Fe ions with high ionization states implied that magnetic reconnection had occurred or was occurring. This could be considered the first evidence obtained via in situ measurements for magnetic reconnection occurring in the solar eruption although we have long known that reconnection must be happening in the CS during the solar eruption (e.g., see *Klimchuk* [1996], *Gallagher et al.* [2002], and references therein for more details).

Bemporad et al. [2006] noticed the elevation of the Fe ionization state with apparent deviations from +10 to +16 inside an ICME bubble. They concluded that the high Fe ionization state detected by Ulysses confirmed the structure of the ICME as described by Figure 15.1, and that Ulysses passed through the outer layer of the ICME.

They realized that the related magnetic structure was smooth as well and therefore, that the fluctuation of the Fe ionization state should result from a filamentary and bursty reconnection process in the CS. Soft X-ray emission was observed by RHESSI in this event, and low temperature (from a few times 10^4 K to a few times 10^5 K) flare loops were found located below the coronal X-ray source (e.g., see *Saint-Hilaire et al.* [2009]). These results are consistent with those of *Sui and Holman* [2003], *Sui et al.* [2004], and *Susino et al.* [2013], and were further confirmed by *Song et al.* [2016] recently.

In addition to the above results deduced from the combination of remote sensing and in situ measurement, which directly indicated the persistent connection of the flare to the associated CME (ICME) by a long CS, the results deduced from in situ measurement alone indirectly suggested the similar structure in the disrupting magnetic configuration as shown by Figure 15.1. The common feature deduced from these indirect results is the bidirectional electrons streaming throughout the ICME at the location around 1 AU, implying that the closed magnetic field lines still thread the entire event at the distance of 1 AU, and magnetic reconnection taking place in the CS is believed to account for the bidirectionally streaming electrons (e.g., see also *Bame et al.* [1981], *Gosling et al.* [1987], *de Koning et al.* 2005, *Wimmer-Schweingruber et al.* [2006], and references therein for details).

The CS was also observed by *Aurass et al.* [2009] in radio wavelengths in the event of 17 May 2002. The EUV spectral data obtained by UVCS helped deduce the plasma temperature (5×10^6 K) and density (between 7.3×10^7 cm^{-3} and 1.09×10^8 cm^{-3}) of the CS, confirming the existence of a CS where magnetic reconnection was taking place, and the CS was found to extend from, at least, the heliocentrical distance of 1.5 $R_\odot$ to 2.1 $R_\odot$. Similar results for the plasma temperature and density were also obtained in the Yohkoh era with a temperature of about 9×10^6 K and a density of about 10^8 cm^{-3} (e.g., see also *McKenzie and Hudson* [1999] for details). The drifting pulsating structures (DPSs) at high frequency (>2 GHz) were found to be colocated with a source of hard X-ray emission on the top of a rising flare loop system (e.g., see *Karlický et al.* [2004] as well) if the frequency of the radio source was related to the height of the source via a given model of the plasma density distribution in the corona following *Sittler and Guhathakurta* [1999]. The low frequency (40–170 MHz) DPS and a type III burst in the same frequency range were also observed in this event, which allows us to deduce that the location where the energetic particles were mainly accelerated was in the region of the middle and high corona at a heliocentrical distance ranging from 1.43 $R_\odot$ to 2.19 $R_\odot$ in which the CS was usually located.

Recently, *Gao et al.* (2016) reported a type II radio burst with a broken lane in the metric band. Combining the information collected in the other wavelengths, they realized that the broken lane was caused by the collision of a CME-driven shock with a CS produced by the previous eruption. Two eruptions took place successively from the same source region or two different regions but very close to each other. A CS was created by the first eruption, and the fast mode shock driven by the second CME collided with the CS, which turned the shock from a perpendicular one into a quasi-parallel one near the CS. Energetic particles bound around the shock then escaped from the shock in that region, and the type II burst disappeared as a result of the lack of energetic particles around the shock. After passing through the CS, the shock returned to a perpendicular configuration, and the type II radio burst resumed. Hence, a type II radio burst with a broken lane appeared on the dynamic spectrum. A similar type II radio burst was also reported earlier due to the collision of a CME-driven shock with a helmet streamer [*Feng et al.*, 2012].

With the CME/flare CS being identified by observations continuously, numerical simulations have also been finding and revealing various important behaviors and physical properties of the CS. *Linker et al.* [2003], *Riley et al.* [2007], and *Reeves et al.* [2010] successively confirmed the formation and development of a long CS in a major eruption. *Linker et al.* [2003] demonstrated that the CS behaved almost in the same way described by the analytic model of *Forbes and Lin* [2000], *Lin and Forbes* [2000], and *Lin* [2002]. *Riley et al.* [2007] noticed the appearance of magnetic islands in the CS, and suggested that they resulted from the tearing mode instability that usually takes place in a long CS.

Reeves et al. [2010] performed a 2.5-D simulation for the eruption in a spherical domain extending from 1 to 20 $R_\odot$, and a long CS was found to form outward quickly behind the CME, and the Poynting flux was the dominant carrier of energy into the CS. Their results manifested a CME bubble with a V-like hot shell and a cool interior as indicated by the observations of *Webb et al.* [2003], *Raymond et al.* [2003], *Bemparod et al.* [2006], *Song et al.* [2016], and *Webb and Vourlidas* [2016], and as suggested by the theories of *Lin et al.* [2004] and *Lin and Soon* [2004]. The work of *Mei et al.* [2012] revealed that the CS is created as the flux rope quickly moves upward, and plasma instabilities produce many medium and small size structures inside the CS.

15.3. GEOMETRICAL SCALE OF THE CURRENT SHEET

Before the long CS was definitely identified as the common feature of the magnetic structure in the eruption, several important characteristics of the reconnection region above the flare loops had been noticed and studied. These characteristics include supra-arcade down-flow, a fan of spikelike "rays," and the retraction of flare loops. The related works were performed based on the observations of Yohkoh, TRACE, RHESSI, as well as SOHO [*Švestka*, 1996; *Švestka et al.*, 1998; *McKenzie and Hudson*, 1999; *McKenzie*, 2000; *Gallagher et al.*, 2002; *Khan et al.*, 2007]. Consulting the features of the relevant structures shown in Figure 15.1, we conclude that observations of those characteristics suggested that the CS was seen face-on (namely the line of sight [LOS] is along the two thick horizontal arrows shown in Fig. 15.1) although it was not easy to observe a CME/flare CS face-on at that time.

In the SDO era, on the other hand, the SDO/AIA can perform coronal observations with a higher resolution, and identifying a CS face-on becomes easier. For example, *Warren et al.* [2011], *Savage and McKenzie* [2011], *Savage et al.* [2012a, 2012b], *Takasao et al.* [2012], *Mckenzie* [2013], *Doschek et al.* [2014], *Hanneman and Reeves* [2014], and *Innes et al.* [2014] studied the supra-arcade plasma flows, the internal structures, and plasma temperatures inside CSs identified in various events that were observed by HINODE and SDO. In order to deduce the CS thickness from observations, on the other hand, we do need to observe the CS edge-on before the in situ measurement becomes available.

Following *Ciaravella et al.* [2002] and *Webb et al.* [2003], *Ko et al.* [2003] identified, in a single event, the fast CME in the LASCO data, the growing flare loops in the EIT, CDS, and MK4 data, the CS connecting the flare to the CME in the UVCS, LASCO, and MK4 data, as well as the antisunward reconnection outflow in the LASCO data. Before *Ko et al.* [2003], no one had reported similar studies of the eruption observed by multiple instruments both in orbit and on the ground covering a large wavelength range. Then, *Lin et al.* [2005] recognized in a different event the merging motion of the two legs of the disrupting magnetic configuration right above the flare loop system as specified by the two thick blue arrows in Figure 15.1, directly measured the speed of the reconnection inflow near the CS in the direction normal to the sheet, and further deduced the reconnection rate, M_A, that ranged from 0.01 to 0.23. The first tentative measurement of M_A was performed by *Yokoyama et al.* [2001] who obtained M_A varying from 0.001 to 0.03; *Chen et al.* [2004] pointed out the weakness of the method of *Yokoyama et al.* [2001] in determining the speed of the reconnection inflow near the CS; and *Narukage and Shibata* [2006] repeated this measurement for a larger number of events, and eventually found that M_A was between 0.001 and 0.07.

Lin et al. [2007] studied the CS in the 18 November 2003 event. They noticed that the reconnection inflow disappeared or stopped where the CS was presumably located, and a gap showed between two bright spots in a Lyα image of the UVCS slit correspondingly. This gap was similar to that reported by *Yokoyama et al.* [2001]. Figure 15.3 shows a composite of LASCO/C2, UVCS, and EIT 195 Å images, which clearly manifests several important objects developed in this eruption. These objects include a CME, a flare, and a CS between the CME and flare, with their relative positions arranged according to Figure 15.1, if we ignore the error in aligning the images that were not taken simultaneously.

The two UVCS composite images in Figure 15.3 were made by stacking up several individual UVCS slit exposures (images) taken at different times from right to left in a chronological order. The left image of the UVCS slit consists of five exposures, and the right one consists of four. The dark gap in the left image suggests the disappearance of the emission in the Lyα spectral line as a result of either the high temperature ($>2\times10^4$ K), or the lower density of H I particles, or the high speed ($>200\,\mathrm{km\,s^{-1}}$) flow of H I particles perpendicular to LOS in that region due to the Doppler dimming (e.g., see *Noci et al.* [1987] and references

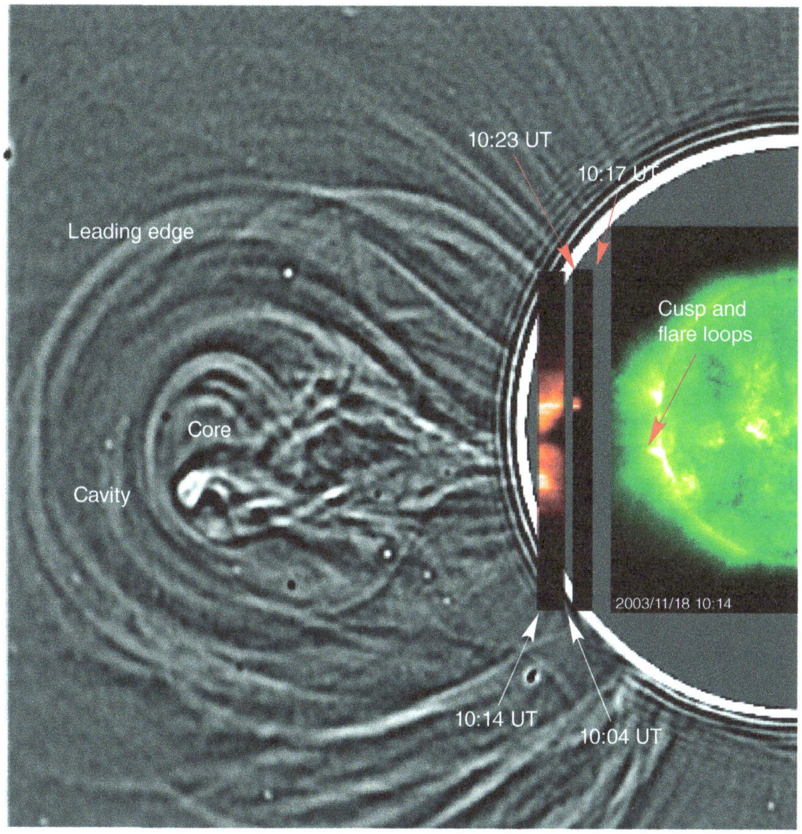

Figure 15.3 Composite C2-UVCS-EIT image of the 18 November 2003 event. The C2 image was taken at 10:26 UT and has been enhanced by the wavelet technique in order to emphasize the morphological features at various positions of the CME. The EIT 195 Å image was taken at 10:14 UT, and the left UVCS image was made by stacking chronologically five UVCS exposures (images) in Lyα that were taken at a heliocentric distance of 1.70 $R_\odot$ covering the time interval from 10:04 UT (right edge) to 10:14 UT (left edge), and the right UVCS image was made by stacking chronologically four UVCS exposures in Lyα that were taken at a heliocentric distance of 1.52 $R_\odot$ covering the time interval from 10:17 UT (right edge) to 10:23 UT (left edge). From *Lin et al.* [2005].

therein). In this particular event, the low emission measure in the Lyα probably results from the latter two causes [*Lin et al.*, 2007]. The change in the size of the gap thus indicates the reconnection inflow and the coincidence of the gap with the CS in space.

The time profile of the gap width shows an asymptotic behavior such that the inflow slows down as it approaches the sheet (e.g., see Fig. 1 of *Lin et al.* [2007]), implying that the reconnection inflow stops somewhere near the edges of the CS, which yields a reconnection inflow speed near the CS ranging from $8\,\mathrm{km\,s^{-1}}$ to $85\,\mathrm{km\,s^{-1}}$ and an apparent value of the CS thickness, d, of about 6.8×10^4 km. The scales of the CSs in the 23 March 1998 event [*Ciaravella et al.*, 2002] and in the 8–10 January 2002 event [*Ko et al.*, 2003] were also investigated by *Lin et al.* [2007], and the apparent values of d for these two events were found to be up to 1.1×10^5 km and 1.3×10^5 km, respectively.

In the image of the UVCS slit at right in Figure 15.3, we can hardly see the dark gap. Only a bright feature appears at the location on one leg of the stretched magnetic structure seen in the five exposures of the UVCS slit at left. *Lin et al.* [2005] found this feature about 10 times brighter in Lyα than the nearby region, but only about twice as bright in O VI. The lack of low temperature material in the region around this bright feature when the four images were taken is probably because the slit was located right below the CS tip, where the temperature is high. This is suggestive of significant accumulation of the material of low temperature spanning from 7×10^4 K to 3×10^5 K at the leg of CME right outside the CS, and the slit at that time was likely to have been located in the flare loop region where the temperature was high (refer to the lower panel in Fig. 15.1). Therefore, the temperature analysis performed by *Lin et al.* [2005] provided a more in-depth probe of the condition inside the CS

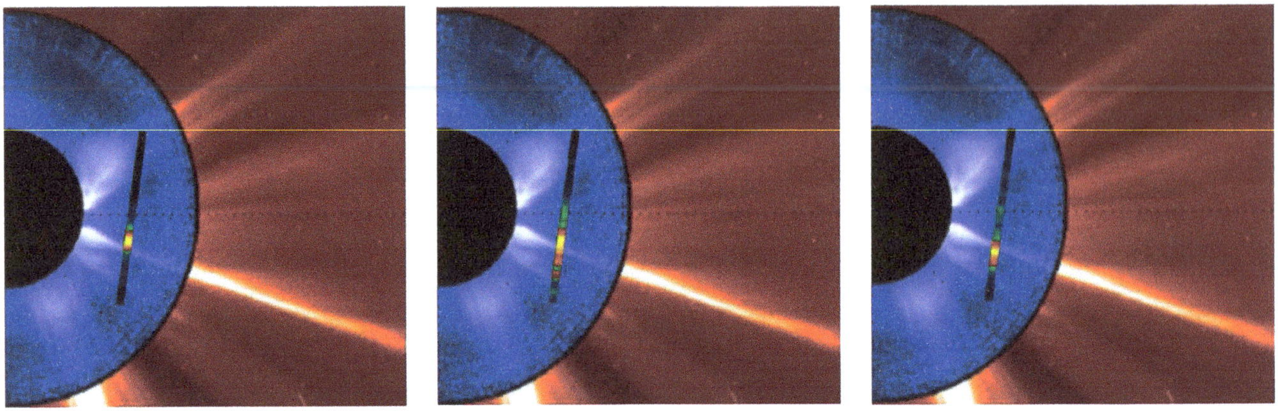

Figure 15.4 Composite image of the current sheet and the structures nearby observed on 4 November 2003 by LASCO C2 at 23:54 UT, MK IV at 23:57, and UVCS from 20:12 to 20:51 UT. The three panels show UVCS images of [Fe XVIII] (left panel), Si XII (middle panel), and O VI lines (right panel). From *Ciaravella et al.* [2013].

than previous similar works by *Yokoyama et al.* [2001], *Chen et al.* [2004], and *Narukage and Shibata* [2006].

Ciaravella and Raymond [2008] studied a CS on 4 November 2003, produced by the most energetic event of Solar Cycle 23. This event created a fast CME at speed of 2657 km s^{-1}, and a flare of GOES class > X28 (*Kiplinger and Garcia* [2004] estimated that the X-ray flare from GOES was most probably class X-30.6). In the event, shortly after the CME crossed the UVCS slit, a narrow and bright feature appeared in the [Fe XVIII] line and lasted for about 4 hr. Figure 15.4 displays a composite image of the ray-like structures in the wake of the CME that were observed in [Fe XVIII], Si XII, and O VI spectral lines. Similar to those seen in Figure 15.2, the low temperature features are spread over a large area, and the high temperature feature appears in a local and confined area only. Thus, several important features, like the flare, CME, and CS connecting the flare to CME, that are frequently observed in a major eruption as described by Figure 15.1 are confirmed again in this event. The apparent value of d was about 1.392×10^5 km.

Savage et al. [2010] studied the cart-wheel CME, which occurred on 9 April 2008, and found the apparent thickness of the CS behind the CME d was between 4×10^3 km and 5×10^3 km. Later, *Landi et al.* [2012] studied this event again, and measured d along the spectrograph slit that crossed the CS diagonally. They obtained a value of about 1.3×10^5 km, which was apparently too large by a geometric factor of about 50%. Dividing by this factor gives an apparent value of d of 8.7×10^4 km. Considering the projection effects further, we believe that this value is the upper limit to the true value of d, namely $d \leq 6.5 \times 10^4$ km. Therefore, the results of *Savage et al.* [2010] and *Landi et al.* [2012] can actually be considered to agree with one another pretty well.

Ling et al. [2014] investigated a major eruption occurring on 7 September 2005. A fast CME at a speed of 2500 km s^{-1} and the last extraordinarily energetic flare of the previous solar cycle were produced by this event. *Ling et al.* [2014] studied the data obtained by a ground-based instrument (namely the MLSO MK4 coronameter in Hawaii [*Elmore et al.*, 2003]) only because SOHO did not work then. The field of view (FOV) of MK4 extends from 1.08 $R_\odot$ to 2.86 $R_\odot$ in heliocentric distance, which allows us to observe the top part of flare loops through the coronagraph since the height of flare loops in a major eruption can reach up to 1.10 $R_\odot$ in heliocentric distance (e.g., see discussions by *Lin* [2002] as well). This event developed a long CS of which the lower tip could be seen clearly in the FOV of MK4. *Ling et al.* [2014] noticed that the CS possessed a thickness of $\approx 7 \times 10^3$ km near the lower tip of the sheet, and d was typically around 2.8×10^4 km for the CS in the MK4 FOV. This is consistent with the result deduced from both the observations (e.g., see *Savage et al.* [2010] and *Landi et al.* [2012]) and numerical experiments (e.g., see *Shen et al.* [2011] and *Mei et al.* [2012]) that d becomes larger with height and time.

Ciaravella et al. [2013] analyzed 28 events that took place in the period from 1997 to 2004 observed by various instruments in various wavelengths, and found that the apparent values of d varied from 4.0×10^4 km to 3.5×10^5 km, and the temperature in the CS ranged from 2.63×10^6 K to 7.94×10^6 K. Most recently, *Seaton et al.* [2017] investigated a major eruption observed by SDO/AIA on 25 February 2004. This event produced an X4.5 flare and a fast CME at a speed of 2100 km s^{-1}, and a CS was seen with a temperature ranging from 8×10^6 K to 10^7 K, which connected the CME to the flare. The CS had a thickness of a few time 10^3 km for most of its lifetime, and the other phenomena were also displayed by the event, including supra-arcade plasma downflows and shrinking flare loops that are believed to result from magnetic reconnection in the CS (e.g., see also *Lin et al.* [1995], *Forbes and Acton* [1996], *Lin* [2004], *Savage et al.* [2012b], and references therein). Furthermore, the rate of magnetic reconnection, M_A, was determined to be between 0.004 and 0.007.

Studying the information that we can collect about the CME/flare CS, especially about the value of *d*, we notice the fact that the apparent value of *d* deduced from observations generally ranges from a few times 10^3 km to a few times 10^5 km, no matter by what instrument, in which wavelength, by which approach, and in which event these observations were made. Although the observed values of *d* span three orders of magnitudes, the range of the difference in these values can still be considered small compared to that between the traditionally expected value of *d* (see discussions below shortly) and the observed ones, which span seven to eight orders of magnitudes. The instruments obtaining the data include those in orbit (e.g., SOHO, HINODE, STEREO, and SDO) and on the ground (MK4 of MLSO); the wavelengths in which these observations were performed include UV, EUV, and white light; the approaches include analyzing white light data, spectral data, and filtergrams in various wavelengths; and the events include those occurring in the SOHO, HINODE, and SDO era. *Ciaravella and Raymond* [2008] provided a good example regarding this issue such that they obtained the same value for *d* from the data in different wavelengths observed in the same event.

All these results have revealed the fact that the CME/flare CS can possess a very large apparent thickness. But it is very hard for the existing theory to explain how the fast reconnection process could take place in such a large-scale region to allow the magnetic energy to be released rapidly enough to drive the fast CMEs and major flares that are usually observed in a typical solar eruptive process. Traditionally, the CS is thought to be so thin that it cannot be observed because its thickness is not allowed to exceed the proton Larmor radius as pointed out by *Litvinenko* [1996], *Wood and Neukirch*, [2005], and so on. The Larmor radius in the coronal environment is a few tens of meters. Thus, the difference in *d* between the theoretical expectation and the observational result is huge and could result from some issues that can significantly affect our observations. These issues include projection effects, complex internal structure, as well as the thermal halo around the CS because of the leak of the thermal energy.

Lin et al. [2007 and 2009] discussed the projection effect on measuring *d* directly according to the observational data, and concluded that the projection effect is not very apparent. The projection effect always exists, and the observed value of *d* is no doubt the upper limit to the true value of *d*, but the difference between tens of meters and 10,000 km is too huge to be explained by the projection effect. To specify this point, we take the observations in Lyα shown in Figure 15.3 as an example. Because the thickness of a CS in reality is finite, broadening the dark gap seen in Lyα emission by projection occurs inevitably at the cost of reducing the contrast of the image due to the fact that the intensity of Lyα emission is proportional to both H I density and roughly the depth of the observed object in LOS. Therefore, a dark sheet appearing 10 times wider than its natural *d* will have only about 0.1 contrast [*Lin et al.*, 2009].

Figure 11a of *Lin et al.* [2005] indicated that the Lyα gap was two to four times fainter than the adjacent emission. This gives a contrast ranging from 0.5 to 0.75, which suggests that the projection effect is not significant for this event. For the data in which the CS looks bright, furthermore, *Lin et al.* [2009] also found that there is a trade-off to increasing the observed value of *d* by the projection effects that may cause negative effects on the brightness of the CS in other ways. Considering the fact that there are usually other bright features like the CME and helmet streamer around, and that *d* is always small compared to the length of the sheet, we realize that usually the CS developed in an eruption cannot be easily recognized (see also discussions by *Ko et al.* [2003]) unless it is roughly observed edge-on. *Lin et al.* [2009] thus estimated that the increase in the value of *d* due to the impact of the projection effect is less than a factor of 10.

Complex structures inside the CS may also lead to an extra increase in the observed value of *d*. *Vršnak et al.* [2009] studied the events of 26 June 2005, 8 January 2002, and 18 November 2003. They performed the relevant plasma diagnoses, and directly measured values of d that ranged from 7×10^4 km to 2.1×10^5 km. Their results indicated that the global structure of the CS was highly stretched though its interior could be complex, including multiple ray-like features, and may help us understand why a CS is mostly observed as a straight feature, and a curved CS has never been observed. This is similar to the extended CS case in the solar wind at 1 AU. Because of the stretching by the solar wind, an extended CS is typically highly planar and no significant warping occurs in the reconnection process [*Phan et al.*, 2009].

So, both theories and observations indicated that the eruption stretches the coronal magnetic field severely in a certain direction and the CS develops mainly in this direction (e.g., see also *Titov and Démolin* [1999], *Lin et al.* [2002], *Ciaravella and Raymond* [2008], *Ling et al.* [2014], and references therein). Because its emission measure is low compared with that of the nearby structures like the CME and flare loops, the CS is not easy to detect if the LOS is in the paper-plane of Figure 15.1 due to either the block by the CME and/or by the occulting disk (roughly corresponding to the case of the halo CME) or its low emission measure if the observation is made face-on.

Although observing and identifying the CS face-on is no longer impossible in the SDO era as we mentioned earlier, it is still more difficult than observing the CS edge-on. In the case of edge-on, no matter how complex the CS structure is, what we can observe is a thick CS with multiple ray-like features as pointed out by *Vršnak et al.* [2009]. For the time being, on the other hand, we do not know whether the whole ray-like bright structure is the CS. This is an open question to which the Solar Probe [*Fox et al.*, 2016] may give the answer in the future.

The thermal halo could be another issue that causes an extra broadening of the CS observed in high temperature spectral lines such as [Fe XVIII] and [Ca XIV]. *Yokoyama and Shibata* [1998, 2001] found that the thermal energy inside the CS may leak to the inflow region due to thermal conduction because of the difference in temperature between the sheet and the inflow region, so as to heat the plasma in the inflow region and produce a thermal halo around the CS. *Seaton and Forbes* [2009] and *Reeves et al.* [2010] confirmed this result in numerical experiments of higher resolution, which suggests that the CS is actually embedded in the thermal halo, and what we would observe directly is actually the halo and the apparent value of d is in fact the scale of the halo instead of the CS itself. In addition to the observations made by UVCS, the thermal halo may also affect the results deduced from the HINODE data (e.g., see also *Landi et al.* [2012], *Savage et al.* [2010], as well as discussions of *Lin et al.* [2015]).

However, the following two issues indicate the limited impact of the thermal halo. First of all, d has been deduced from the white-light data as well, in which brightness of the observed object is dependent on the density of the free electron only, and the thermal behavior of the object does not affect its manifestations in white light. Furthermore, *Ciaravella and Raymond* [2008] measured d for the same event from both white-light and UVCS data obtained simultaneously at the same location, and they got the same results for d. In this case, if the thermal halo did occur, the region where the thermal halo happened should be less dense than that inside the CS so that the emission measure due to the thermal halo in this region is negligible. The recent work of *Seaton et al.* [2017] further confirmed this point.

Second, numerical experiments by *Ni et al.* [2016] also indicated the limited impact of the thermal halo on measuring d (see Fig. 15.5). Their simulations started with the Harris CS configuration in which the magnetic field component normal to the CS exists, heat conductions both parallel and perpendicular to the magnetic field were included, and the thermal halo surrounding the CS could be recognized. The colored shading contours in Figure 15.5a are for the electric current density, J_z, inside and around the CS, those in Figure 15.5b show the distribution of the temperature, T, and the continuous curves in both panels are magnetic field lines. Such a distribution of magnetic field, T, and J_z near the CS is directly governed by the reconnection inflow near the CS, and somewhat depends on the rate of reconnection, M_A.

Figures 15.5a and 15.5b indicate that the concentration of high-temperature plasma roughly occurs at the location with an apparent concentration of J_z, and comparison of the colored shading contours suggests that the distributions of J_z and T in the CS are approximately colocated in space except in the region where the large plasmoid at left is located. Furthermore, in Figure 15.5c, we plot the distribution of both J_z and T across the sheet at the location of $x = 0.51$ in Figures 15.5a and 15.5b, respectively, from which we notice that the two profiles have almost the same shape, and occupy almost the same extension in space. This confirms the results of *Mei et al.* [2012] (see also their Fig. 9f). We also compare the profiles of T and J_z at the right plasmoid, and obtain similar conclusions on their colocation in space, but obviously, we get a larger value for d in this case.

According to the standard definition of the CS thickness, the true value of d is the full width at the half-maximum of the profile of the electrical current density across the sheet. So, results of *Ni et al.* [2016] indicated that the accumulation of the hot plasma occurs in the region of high values of J_z. This could be ascribed to the fact that the Joule heating is proportional to J_z^2 although the slow (or even the fast) mode shock may also heat the plasma in the CS (e.g., see *Mei et al.* [2012] and *Ni et al.* [2016] for more details). Thus, the identification of profiles of J_z and T across the CS implies that, in a real CME/flare event, the true value of d by the standard definition may not be very different from that determined by the distribution of a plasma parameter sensitive to the plasma temperature and/or density.

15.4. REASONABLENESS OF THE THICK CURRENT SHEET

Discussions in previous sections indicate that the large value of d should result from the intrinsic properties of magnetic reconnection and the other related processes occurring in the CME/flare CS. Both theories [*Shibata and Tanuma*, 2001; *Tanuma et al.*, 2001] and observations [*Bemporad*, [2008]; *Susino et al.*, 2013] suggested the role of the tearing mode instability in broadening the CS and in enhancing the electrical resistivity in a thick CS. Before we look into these properties and processes, it is worthwhile to discuss the lower limit to the observed values of d.

When studying the event of 4 November 2003, *Ciaravella and Raymond* [2008] analyzed the extension

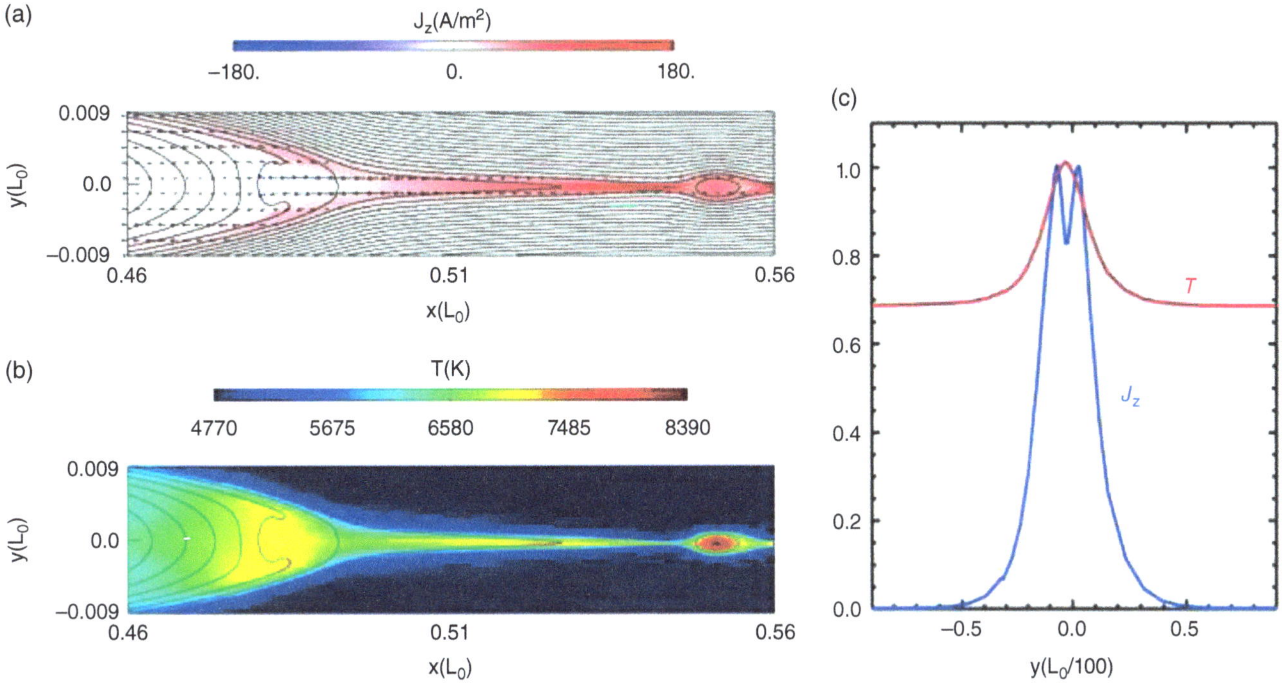

Figure 15.5 Distributions of the electric current density (J_z) inside a reconnection CS (a) as well as the temperature (T) in the same region, (b) as shown by the colored shading contours obtained in the numerical experiments by *Ni et al.* [2016]. The continuous curves manifest contours of the magnetic field. Panel (c) displays the profiles of J_z (blue curve) and T (red curve) across the CS at the location of $x = 0.51$ in panels (a) and (b) for comparison. Values of the two parameters for the plot have been normalized to their own maxima in the region, respectively.

of the post-flare loops, the LOS depth of the CS, as well as the relative positions and orientations of these features. They obtained that the true value of d was between 2.784×10^4 km and 5.568×10^4 km, namely the actual thickness could be two and a half to five times narrower than the apparent thickness. This is the only case so far where a lower limit to d was deduced directly from observations.

For the other events, on the other hand, the relative positions and orientations of various features near the CS could not be fully determined, so we cannot deduce the true value of d for these events using only the observational data collected. However, we might be able to determine the lower limit to d for the specific event by applying reasonable theoretical arguments.

Numerical simulations indicated that the thickness of the CS may decrease continuously at the beginning of the reconnection process due to the pressure of the reconnection inflow, but the decrease in the thickness will cease after the turbulence commences inside the CS [*Shen et al.*, 2011; *Mei et al.*, 2012]. It is now widely accepted that the turbulence results from the tearing mode instability occurring in the long CS [*Furth et al.*, 1963; *Priest and Forbes*, 2000], and the formation of turbulent structures produces an extra pressure to neutralize the impact of the reconnection inflow.

Another role of these structures in reconnection is that their formation tears the original large-scale CS into many small-scale islands or blobs (see also *Guidoni and Longcope* [2011]), which significantly suppresses the local magnetic Reynolds number in the CS, and dramatically enhances the magnetic diffusion inside the CS such that the ratio of the resultant resistivity (also known as the hyper-resistivity, see *Strauss* [1988] and *Bhattacharjee and Yuan* [1995]) to the conventional anomalous resistivity produced by the collision of electrons with ion-acoustic waves could reach up to 10^9 in the quiet coronal environment, and up to 10^6 in the CME/flare CS (e.g., see discussions by Strauss [1988] and Lin *et al.* [2009] for more details). Combining these two consequences of the tearing mode, we conclude that magnetic reconnection could take place at a reasonably fast rate in a thick CME/flare CS.

In a CS where magnetic reconnection takes place, the timescale of reconnection, τ_R, is usually long compared to the Alfvén timescale, $\tau_A = l/v_A$, and is short compared to the diffusion time scale, $\tau_d = l^2/\eta$, of the Spitzer resistivity (see also discussions of *Priest and Forbes* [2000]). Here, $l = d/2$, and η is the magnetic diffusivity of the plasma in the sheet. So

$$\tau_A \ll \tau_R \ll \tau_d. \tag{15.1}$$

For the tearing mode in different situations with different boundary conditions, the inequalities in equation (15.1) become

$$S^{-\alpha} \ll kl \ll S^{\gamma} \tag{15.2}$$

as S and kl are not extremely large, say exceeding 10^3 [*Priest*, 1985]. Here $S=\tau_d/\tau_A$ is the local Lundquist number of the CS, not that of the whole system [*Priest and Forbes*, 2000, p. 180], k is the wave number of the turbulence due to the tearing mode, α and γ are constants between 0 and 1 for different cases (e.g., see more discussions in detail by *Lin et al.* [2015] and references therein).

The relation in equation (15.2) indicates that kl could possess a finite lower limit. *Lin et al.* [2007] showed that $S = M_A^{-1}$, and *Shen et al.* [2011] further summarized the relation of $l_{\min}$, the lower limit to l, to the other observables such as

$$l_{\min} = M_A^{\alpha}/k = M_A^{\alpha}\lambda/2\pi, \tag{15.3}$$

where $\lambda = 2\pi/k$ is the wavelength of the turbulence, which is approximated by the distance between two adjacent blobs observed to flow along the CS (refer to Fig. 6.3 of *Priest and Forbes* [2000], and Figs. 3c and 3d of *Lin et al.* [2005]). *Strauss* [1988] obtained a scaling similar to equation (15.3) via a different approach based on the argument that the reconnection outflow can suppress the tearing modes by convecting them away unless they grow faster than they are convected away.

For the events that have been studied so far, M_A ranges from 10^{-3} to 10^{-1} [*Yokoyama et al.*, 2001; *Ko et al.*, 2003; *Lin et al.*, 2005; *Vršnak et al.*, 2009; *Liu et al.*, 2010; *Savage et al.*, 2010; *Su et al.*, 2013]. So S for the CS varies from 10 to 10^3, which is small compared with the value of S for the quiet corona ($\approx 10^8$–10^{12}; e.g., see *Priest* [2014]). The values of S of interest were also within the same range for the sheets occurring in the laboratory [*Furth et al.*, 1963; *Priest and Forbes*, 2000]. Summarizing the results of previous works, *Lin et al.* [2015] found that $l_{\min}$ ($d_{\min}$) ranges from 7.4×10^2 km to 4.55×10^3 km (from 1.48×10^3 km to 9.10×10^3 km), which is ten to one hundred times smaller than the apparent values of l (d). As we mentioned earlier, although this difference spans two orders of magnitude, and is surely larger than the factor of two and a half to five found by *Ciaravella and Raymond* [2008], it is apparently much less than the difference between 10^4 km and a few meters! This suggests that the true value of d should not differ very much from the apparent value as long as d can be measured directly in observations.

We note here that observations and equation (15.3) give the upper and the lower limits to the true value of d, respectively, not the true value of d itself, and that it is not very unreasonable for these two limits to differ from one another by one or two orders of magnitude. For the time being, there is not an accepted approach to determining d observationally, and the best we can do is just to figure out these two limits. Completely solving this problem should be possible with the Solar Probe Plus (e.g., see *Fox et al.* [2016]).

The above results indicate that the reasonably fast reconnection process does not necessarily have to take place in a very thin CS in the solar eruption. Both increasing the resistivity and decreasing the size of the diffusion region can speed up magnetic reconnection. In a given environment, the resistivity does not change too much, but the size can change dramatically either through decreasing the scale of the diffusion region or via producing small-scale structures inside the region (see also *Guidoni and Longcope* [2011]). Obviously, the occurrence of the tearing mode belongs to the second case.

As the tearing mode develops, a large-scale CS is "torn" into many small pieces like magnetic islands or plasmoids, and the original large-scale CS becomes an assembly of many small structures (e.g., see also *Loureiro et al.* [2007]), which allows the dissipation of a large amount of magnetic energy to occur in many small pieces simultaneously. Therefore, the diffusion in the CS is significantly enhanced although the global scale, especially the thickness, of the sheet is still large. This is analogous to what occurs in the process of parallel computing that is widely used in modern numerical experiments, in which large problems are often divided into smaller ones that can be solved simultaneously in a shorter time.

The enhanced diffusion is equivalent to adding an extra resistivity in the reconnection region, which is also known as the "hyper-resistivity" (e.g., see also *Strauss* [1988] and *Bhattacharjee and Yuan* [1995] for details). It is worth pointing out that this resistivity was studied and deduced by both *Strauss* [1988] and *Bhattacharjee and Yuan* [1995] in the framework of MHD when the impact of the turbulence on diffusing the magnetic field was investigated. Therefore, the hyper-resistivity is an MHD property of the turbulent plasma, which occurs on the MHD scale on which no kinetic feature of the plasma shows yet.

In fact, the other diffusive structures and processes may also occur inside the CME/flare CS in reality. *Mei et al.* [2012] showed that a CME/flare CS actually possesses complex structures, in which three types of magnetic reconnection: Sweet-Parker, Petschek, and turbulence, could even take place simultaneously. The work by *Ni et al.* [2015a] further suggested that the dissipation at the Petschek-shock and the fast mode shock could be much more efficient than that at the other parts of the CS.

In the turbulent region, on the other hand, coalescence between two adjacent plasma blobs produced by the tearing mode has been observed frequently in numerical experiments (e.g., see *Bárta et al.* [2011]; *Shen et al.* [2011]; *Mei et al.* [2012]; *Shen et al.* [2013]; *Ni et al.* [2015a]). Further reconnection takes place when the two blobs merge into one, yielding more dissipation of the magnetic field. *Shen et al.* [2013] and *Ni et al.* [2012, 2013, 2015b] even found that the spectrum of the kinetic energy of the plasma flow in the sheet displays a power-law shape that indicates a cascading behavior even shown by the plasma flow. Therefore, the occurrence of small structures and multiple styles of diffusion, which appears to be supported by the observations of *McKenzie* [2013], *Doschek et al.* [2014], and *Reeves et al.* [2017], makes magnetic reconnection in the large-scale CME/flare CS a very efficient process of energy release, and the extra pressure produced by these small structures prevents the global reconnection inflow from squeezing the sheet to a very small thickness.

15.5. SUMMARY

Because of the low resistivity of the coronal plasma, it is traditionally believed that magnetic reconnection needs to take place in a very confined region so that the rate of reconnection is high enough to drive solar eruptions as energetic as those observed. In the case that the CS could be stretched to be very long, "confined" therefore means "very thin." In the coronal environment, d is believed to be roughly the proton Larmor radius, which is of order tens of meters [*Litvinenko*, 1996; *Wood and Neukirch*, 2005; and references therein]. Since the first direct measurements of d were performed by *Lin et al.* [2007], on the other hand, more and more observations have revealed that the apparent value of d may reach up to more than 10^4 km, and the impact of projection effects, complex structures, and the thermal halo on the measurement is limited and cannot yield a difference between several meters and several 10^4 km.

The above inconsistency of the theoretical expectation and the observational results might be resolved via the following considerations.

1. The traditional knowledge about magnetic reconnection and the CS is based on an implicit assumption that the diffusion occurs in a region of which both internal and external structures are simple, but both the observational results (e.g., see also *McKenzie* [2013], *Doschek et al.* [2014], and *Reeves et al.* [2017] for more details) and theoretical results (e.g., see also *Shibata and Tanuma* [2001], *Bárta et al.* [2011], *Shen et al.* [2011, 2013], *Mei et al.* [2012], and *Ni et al.* [2015b] for more details) we have obtained so far have revealed the simple fact that the reconnecting CS could actually possess a very complex internal feature and include many multiscale structures.

2. The traditional knowledge also implicitly assumes that magnetic reconnection commences in the magnetic configuration including a preexisting CS, and the CS itself is static, but now we know that the CME/flare CS does not exist prior to the eruption, but instead it is produced in a highly dynamic process.

3. Behaviors of individual particles cannot govern dynamic properties of the coronal magnetic structure because of the force-free condition in the corona. Furthermore, the CME/flare CS is always in a highly dynamic state, especially at the beginning of the eruption, at which time its length increases at a speed of several times 10^2 km s^{-1} [*Forbes and Lin*, 2000; *Lin and Forbes*, 2000; *Lin*, 2002; *Linker et al.*, 2003; *Webb et al.*, 2003; *Webb and Vourlidas*, 2016] and at an acceleration of a few m s^{-2} [*Webb and Vourlidas*, 2016]. It is impossible for individual particles to influence the scale of a rapidly evolving magnetic structure in a force-free environment. Instead the value of d is determined by the turbulence invoked by various plasma instabilities inside the sheet [*Strauss*, 1986, 1988; *Lazarian* and *Vishniac*, 1999; *Lazarian et al.*, 2012; *Vishniac et al.*, 2012], and it has nothing to do with the Larmor radius of the proton in the corona.

4. Traditionally, the CS is implicitly defined as the region where the Joule heating takes place as a result of either the Spitzer resistivity or the conventional anomalous one or both, but numerical experiments have clearly demonstrated that the reconnecting CS eventually develops to a very complicated structure although it could possess a simple structure at the very beginning. The slow mode shock, plasmoids, and merging of plasmoids are very popular features of the sheet (e.g., see *Skender and Lapenta* [2010]; *Bárta et al.* [2011]; *Shen et al.* [2011, 2013]; *Mei et al.* [2012]; *Ni et al.* [2012, 2013, 2015a, 2015b]). Therefore, it is actually impossible to distinguish a specific diffusion mechanism from the others.

5. Overall, we have used *diffusion*, *dissipation*, and *current sheet* more generally here than were used traditionally. These terms here refer to any process leading to magnetic diffusion and any region where such diffusion occurs (e.g., see also discussions of *Lin and Forbes* [2000]). But the questions of the geometric scale and the fine structures inside the CME/flare CS, together with interactions of these structures, still remains open. The true value of d and the correct way of measuring d constitutes a big challenge to future work in both observations and theories.

ACKNOWLEDGMENTS

We are grateful for the referees' valuable and constructive advice, comments, and suggestions for improving this work, which was supported by Program 973 grant 2013CBA01503, NSFC grants U1631130, 11573064, and 11333007, and CAS grants QYZDJ-SSW-SLH012 and

XDA15010900, the Western Light 2014, and the Youth Innovation Promotion Association CAS 2017. Part of calculations in this work were performed at the Supercomputing Platform Tianhe-2 with the support from the NSFC-Guangdong Joint Funds (the 2nd Phase) of NSFC2015-460 and NSFC2015-463. This work was also supported by the grant associated with the Project of the Group for Innovation of Yunnan Province.

REFERENCES

Asai, A., T. Yokoyama, M. Shimojo, and K. Shibata (2004), Downflow motions associated with impulsive nonthermal emissions observed in the 2002 July 23 solar flare, *Astrophys. J.*, *605*, L77.

Aurass, H., F. Landini, S. Braune, et al. (2009), Coronal current sheet signatures during the 17 May 2002 CME-flare, *Astron. Astrophys.*, *506*, 901.

Bame, S. J., J. R. Asbridge, W. C. Feldman, et al. (1981), Bi-directional streaming of solar wind electrons greater than 80 eV-ISEE evidence for a closed-field structure within the driver gas of an interplanetary shock, *Geophys. Res. Lett.*, *8*, 173.

Bárta, M., J. Büchner, M. Karlický, et al. (2011), Spontaneous current-layer fragmentation and cascading reconnection in solar flares, I. Model and analysis, *Astrophys. J.*, *737*, 24.

Bemporad, A. (2008), Spectroscopic detection of turbulence in post-CME current sheets, *Astrophys. J.*, *689*, 572.

Bemporad, A., G. Poletto, S. T. Suess, et al. (2006), Current sheet evolution in the aftermath of a CME event, *Astrophys. J.*, *638*, 1110.

Bhattacharjee, A., and Y. Yuan (1995), Self-consistency constraints on the dynamo mechanism, *Astrophys. J.*, *449*, 739.

Chen, P., K. Shibata, D. H. Brooks, and H. Isobe (2004), A re-examination of the evidence for reconnection inflow, *Astrophys. J.*, *602*, L61.

Ciaravella, A., and J. C. Raymond (2008), The current sheet associated with the 2003 November 4 coronal mass ejection: Density, temperature, thickness, and line width, *Astrophys. J.*, *686*, 1372.

Ciaravella, A., J. C. Raymond, J. Li, et al. (2002), Elemental abundances and post- coronal mass ejection current sheet in a very hot active region, *Astrophys. J.*, *575*, 1116.

Ciaravella, A., D. F. Webb, S. Giordano, et al. (2013), Bright ray-like features in the aftermath of coronal mass ejections: White light versus ultraviolet spectra, *Astrophys. J.*, *766*, 65.

de Koning, C. A., et al. (2005), An unusually fast interplanetary coronal mass ejection observed by Ulysses at 5 AU on 15 November 2003, *J. Geophys. Res.*, *110*, A01102.

Doschek, G. A., D. E. McKenzie, and H. P. Warren (2014), Plasma dynamics above solar flare soft x-ray loop tops, *Astrophys. J.*, *788*, 26.

Elmore, D. F., J. T. Burkepile, J. A. Darnell, et al. (2003), Calibration of a ground-based solar coronal polarimeter in polarimetry in *Astronomy*, ed. by Silvano Fineschi, Proceedings of the SPIE, *4843*, 66.

Feng, S., Y. Chen, X. Kong, G. Li, H. Song, X. Feng, and Y. Liu (2012), Radio Signatures of Coronal-mass-ejection-Streamer Interaction and Source Diagnostics of Type II Radio Burst, *Astrophys. J.*, *753*, 21.

Forbes, T. G., and E. R. Priest (1995), Photospheric magnetic field evolution and eruptive flares, *Astrophys. J.*, *446*, 377.

Forbes, T. G., and J. Lin (2000), What can we learn about reconnection from coronal mass ejections? *J. Atmos. Solar Terr. Phys.*, *62*, 1499.

Forbes, T. G., and L. W. Acton (1996), Reconnection and field line shrinkage in solar flares, *Astrophys. J.*, *459*, 330.

Forbes, T. G., and P. A. Isenberg (1991), A catastrophe mechanism for coronal mass ejections, *Astrophys. J.*, *373*, 294.

Forbes, T. G., E. R. Priest, and P. A. Isenberg (1994), On the maximum energy release in flux-rope models of eruptive flares, *Solar Phys.*, *150*, 245.

Fox, N. J., M. C. Velli, S. D. Bale, et al. (2016), The Solar Probe Plus Mission: Humanity's First Visit to Our Star, *Space Sci. Rev.*, *204*, 7.

Furth, H. P., J. Killeen, and M. N. Rosenbluth (1963), Finite-resistivity instabilities of a sheet pinch, *Phys. Fluids*, *6*, 459.

Gallagher, P. T., B. R. Dennis, S. Krucker, et al. (2002), Rhessi and trace observations of the 21 April 2002 x1.5 flare, *Solar Phys.*, *210*, 341.

Gao, G., M. Wang, N. Wu, J. Lin, E. Ebenezer, and B. Tan (2016), The Broken Lane of a Type II Radio Burst Caused by Collision of a Coronal Shock with a Flare Current Sheet, *Solar Phys.*, *291*, 3369.

Gopalswamy, N., Y. Hanaoka, and M. R. Kundu, et al. (1997), Radio and x-ray studies of a coronal mass ejection associated with a very slow prominence eruption, *Astrophys. J.*, *475*, 348.

Gosling, J. T., D. N. Baker, S. J. Bame, W. C. Feldman, R. D. Zwickl, and E. J. Smith (1987), Bidirectional solar wind electron heat flux events, *J. Geophys. Res.*, *92*, 8519.

Guidoni, S. E., and D. W. Longcope (2011), Density enhancements and voids following patchy reconnection, *Astrophys. J.*, *730*, 90.

Hanaoka, Y., H. Kurokawa, S. Enome, et al. (1994), Simultaneous observations of a prominence eruption followed by a coronal arcade formation in radio, soft x-rays, and H(alpha), *Pub. Astron. Soc. Japan*, *46*, 205.

Hanneman, W. J., and K. K. Reeves (2014), Thermal Structure of Current Sheets and Supra-arcade Downflows in the Solar Corona, *Astrophys. J.*, *786*, 95.

Heitsch, F., and E. G. Zweibel (2003), Suppression of fast reconnection by magnetic shear, *Astrophys. J.*, *590*, 291.

Innes, D. E., D. E. McKenzie, and T. Wang (2003a), SUMER spectral observations of post-flare supra-arcade inflows, *Solar Phys.*, *217*, 247.

Innes, D. E., D. E. McKenzie, and T. Wang (2003b), Observations of 1000 $km\,s^{-1}$ Doppler shifts in 10^7 K solar flare supra-arcade, *Solar Phys.*, *217*, 267.

Innes, D. E., L. Guo, A. Bhattacharjee, et al. (2014), Observations of supra-arcade fans: Instabilities at the head of reconnection jets, *Astrophys. J.*, *796*, 27.

Isenberg, P. A., T. G. Forbes, and P. Démoulin (1993), Catastrophic evolution of a force-free flux rope: A model for eruptive flares, *Astrophys. J.*, *417*, 368.

Karlický, M., F. Fárník, and S. Krucker (2004), High-frequency slowly drifting structures and X-ray sources observed by RHESSI, *Astron. Astrophys.*, *419*, 365.

Khan, J. I., H. M. Bain, and L. Fletcher (2007), The relative timing of supra-arcade downflows in solar flares, *Astron. Astrophys.*, *475*, 333.

Khan, J. I., Y. Uchida, A. H. McAllister, and T. Watanabe (1994), YOHKOH soft x-ray observations related to a prominence eruption and arcade flare on 7 May 1992, in *X-Ray Solar Physics from Yohkoh*, edited by Y. Uchida et al., Univ. Acad. Press, Tokyo, 201.

Kiplinger, A. L., and H. A. Garcia (2004), Soft X-ray parameters of the great flares of active region 486, *Bull. AAS*, *36*, 739.

Klimchuk, J. A. (1996), Post-eruption arcades and 3-D magnetic reconnection, *Astron. Soc. Pac. Seri. 111*, 319.

Klimchuk, J. A., L. W. Acton, K. L. Harvey, et al. (1994), Coronal eruptions observed by YOHKOH, in *X-Ray Solar Physics from Yohkoh*, edited by Y. Uchida et al., Tokyo Univ. Acad. Press, 181.

Ko, Y.-K., J. C. Raymond, J. Lin, et al. (2003), Modeling UV and X-ray emission in a post-coronal mass ejection current sheet, *Astrophys. J.*, *594*, 1068.

Ko, Y.-K., J. C. Raymond, B. Vršnak, et al. (2010), Modeling UV and X-ray emission in a post-coronal mass ejection current sheet, *Astrophys. J.*, *722*, 625.

Kohl, J. L., G. Noci, S. R. Cranmer, et al. (2006), Ultraviolet spectroscopy of the extended solar corona, *Astron. Astrophys. Rev.*, *13*, 31.

Kolománski, S., and M. Karlický (2007), The interaction of a plasmoid with a loop-top kernel, *Astron. Astrophys.*, *475*, 685.

Landi, E., J. C. Raymond, M. P. Miralles, et al. (2012), Post-coronal mass ejection plasma observed by Hinode, *Astrophys. J.*, *751*, 21.

Lazarian, A., and E. T. Vishniac (1999), Reconnection in a weakly stochastic field, *Astrophys. J.*, *517*, 700.

Lazarian, A., et al. (2012), Turbulence, magnetic reconnection in turbulent fluids and energetic particle acceleration, *Space Sci. Rev.*, *173*, 557.

Lin, J. (2002), Energetics and propagation of coronal mass ejections in different plasma environments, *ChJAA*, *2*, 539.

Lin, J. (2004), Motions of flare ribbons and loops in various magnetic configurations, *Solar Phys.*, *222*, 115.

Lin, J., and T. G. Forbes (2000), Effects of reconnection on the coronal mass ejection process, *J. Geophys. Res.*, *105*, 2375.

Lin, J., and W. Soon (2004), Evolution of morphological features of CMEs deduced from catastrophe model of solar eruptions, *New Astron.*, *9*, 611.

Lin, J., A. A. van Ballegooijen, and T. G. Forbes (2002), Evolution of a semicircular flux rope with two ends anchored in the photosphere, *J. Geophys. Res.*, *107*, 1438.

Lin, J., J. C. Raymond, and A. A. van Ballegooijen (2004), The role of magnetic reconnection in the observable features of solar eruptions, *Astrophys. J.*, *602*, 422.

Lin, J., J. Li, T. G. Forbes, et al. (2007), Features and properties of coronal mass ejection/flare current sheets, *Astrophys. J.*, *658*, L123.

Lin, J., J. Li, Y.-K. Ko, et al. (2009), Investigation of thickness and electrical resistivity of the current sheets in solar eruptions, *Astrophys. J.*, *693*, 1666.

Lin, J., N. A. Murphy, C. Shen, et al., (2015), Review on current sheets in CME development: Theories and observations, *Space Sci. Rev.*, *194*, 237.

Lin, J., T. G. Forbes, E. P. Priest, et al. (1995), Models for the motions of flare loops and ribbons, *Solar Phys.*, *159*, 275.

Lin, J., W. Soon, and S. L. Baliunas (2003), Theories of solar eruptions: A review, *New Astron. Rev.*, *47*(2), 53.

Lin, J., Y.-K. Ko, L. Sui, et al. (2005), Direct observations of the magnetic reconnection site of an eruption on 2003 November 18, *Astrophys. J.*, *622*, 1251.

Ling, A. G., D. F. Webb, J. T. Burkepile, et al. (2014), Development of a current sheet in the wake of a fast coronal mass ejection, *Astrophys. J.*, *784*, 91.

Linker, J. A., Z. Mikić, R. Lionello, et al. (2003), Flux cancellation and coronal mass ejections, *Phys. Plasma*, *10*(5), 1971.

Litvinenko, Y. (1996), Particle acceleration in reconnecting current sheets with a nonzero magnetic field, *Astrophys. J.*, *462*, 997.

Liu, R. (2013), Dynamical processes at the vertical current sheet behind an erupting flux rope, *Mon. Not. R. Astron. Soc.*, *434*, 1309.

Liu, R., J. Lee, T. Wang, et al. (2010), A reconnecting current sheet imaged in a solar flare, *Astrophys. J.*, *723*, L28.

Liu, W., Q. Chen, and V. Petrosian (2013), Plasmoid ejections and loop contractions in an eruptive M7.7 solar flare: Evidence of particle acceleration and heating in magnetic reconnection outflows, *Astrophys. J.*, *773*, 166.

Loureiro, N. F., A. A. Schekochihin, and S. C. Cowley (2007), Instability of current sheets and formation of plasmoid chains, *Phys. Plasma*, *14*, 100703.

Martens, P. C. H., and N. P. M. Kuin (1989), A circuit model for filament eruptions and two-ribbon flares, *Solar Phys.*, *122*, 263.

McAllister, A., Y. Uchida, J. Khan, et al. (1994), Coronal magnetic fields parallel to magnetic polarity inversion lines, in *X-Ray Solar Physics from Yohkoh*, edited by Y. Uchida, et al., Univ. Acad. Press, Tokyo, 189.

McAllister, A., Y. Uchida, S. Tsuneta, et al. (1992), The structure of the coronal soft x-ray source associated with the dark filament disappearance of 1991 September 28 using the YOHKOH soft x-ray telescope, *Publ. Astron. Soc. Japan*, *44*, L205.

McKenzie, D. E. (2000), Supra-arcade downflows in long-duration solar flare events, *Solar Phys.*, *195*, 381.

McKenzie, D. E. (2013), Turbulent dynamics in solar flare sheet structures measured with local correlation tracking, *Astrophys. J.*, *766*, 39.

McKenzie, D. E., and H. S. Hudson (1999), X-ray observations of motions and structure above a solar flare arcade, *Astrophys. J.*, *519*, L93.

Mei, Z., C. Shen, N. Wu, et al. (2012), Numerical experiments on magnetic reconnection in solar flare and coronal mass ejection current sheets, *Mon. Not. R. Astron. Soc.*, *425*, 2824.

Milligan, R. O., R. T. J. McAteer, B. R. Dennis, et al. (2010), Evidence of a plasmoid-looptop interaction and magnetic inflows during a solar flare/coronal mass ejection eruptive event, *Astrophys. J.*, *713*, 1292.

Myers, C. E., M. Yamada, H. Ji, et al. (2015), A dynamic magnetic tension force as the cause of failed solar eruptions, *Nature*, *528*, 526.

Narukage, N., and K. Shibata (2006), Statistical analysis of reconnection inflows in solar flares observed with SOHO EIT, *Astrophys. J.*, *637*, 1122.

Ni, L., B. Kliem, J. Lin, et al. (2015a), Fast magnetic reconnection in the solar chromosphere mediated by the plasmoid instability, *Astrophys. J.*, *799*, 79.

Ni, L., J. Lin, I. I. Roussev, et al. (2016), Heating mechanisms in the low solar atmosphere through magnetic reconnection in current sheets, *Astrophys. J.*, *832*, 195.

Ni, L., J. Lin, N. A. Murphy, et al. (2013), Effects of the non-uniform initial environment and the guide field on the plasmoid instability, *Phys. Plasma*, *20*, 061206.

Ni, L., J. Lin, Z. Mei, et al. (2015b), Numerical experiments on the detailed energy conversion and spectrum studies in a corona current sheet, *Astrophys. J.*, *812*, 92.

Ni, L., U. Ziegler, Y. Huang, et al. (2012), Effects of plasma beta on the plasmoid instability, *Phys. Plasma*, *19*, 072902.

Noci, G., J. L. Kohl, and G. L. Withbroe (1987), Solar wind diagnostics from Doppler enhanced scattering, *Astrophys. J.*, *315*, 706.

Phan, T. D., J. T. Gosling, and M. S. Davis (2009), Prevalence of extended reconnection X-lines in the solar wind at 1 AU, *Geophys. Res. Lett.*, *36*, L09108.

Poletto, G., S. T. Suess, A. Bemporad, et al. (2004), Evidence for the same hot plasma after coronal mass ejection events, in both remote and in situ observations, *Astrophys. J.*, *613*, L173.

Priest. E. R. (1985), The magnetohydrodynamics of current sheets, *Rep. Prog. Phys.*, *48*, 955.

Priest, E. R. (2014), *Magnetoydrodynamics of the Sun*, Cambridge University Press, Cambridge, UK.

Priest, E. R., and T. G. Forbes (2000), *Magnetic Reconnection: MHD Theory and Applications*, Cambridge University Press, Cambridge, UK.

Qiu, J., Q. Hu, T. A. Howard, et al. (2007), On the magnetic flux budget in low-corona magnetic reconnection and interplanetary coronal mass ejections, *Astrophys. J.*, *659*, 758.

Raymond, J. C., A. Ciaravella, D. Dobrzycka, et al. (2003), Far-ultraviolet spectra of fast coronal mass ejections associated with x-class flares, *Astrophys. J.*, *597*, 1106.

Reeves, K. K., J. A. Linker, Z. Mikić, et al. (2010), Current sheet energetics, flare emissions, and energy partition in a simulated solar eruption, *Astrophys. J.*, *721*, 1547.

Reeves, K. K., M. S. Freed, D. E. McKenzie, et al. (2017), An exploration of heating mechanisms in a supra-arcade plasma sheet formed after a coronal mass ejection, *Astrophys. J.*, *836*, 55.

Riley, P., R. Lionello, Z. Mikić, et al. (2007), "Bursty" reconnection following solar eruptions: MHD simulations and comparison with observations, *Astrophys. J.*, *655*, 591.

Roussev, I. I., K. Galsgaard, C. Down, et al. (2012), Explaining fast ejections of plasma and exotic X-ray emission from the solar corona, *Nature Phys.*, *8*, 845.

Saint-Hilaire, P., S. Krucker, S. Christe, et al. (2009), The X-ray detectability of electron beams escaping from the Sun, *Astrophys. J.*, *696*, 941.

Savage, S. L., and D. E. McKenzie (2011), Quantitative examination of a large sample of supra-arcade downflows in eruptive solar flares, *Astrophys. J.*, *730*, 98.

Savage, S. L., D. E. McKenzie, and K. K. Reeves (2012b), Re-interpretation of supra-arcade downflows in solar flares, *Astrophys. J.*, *747*, L40.

Savage, S. L., D. E. McKenzie, K. K. Reeves, et al. (2010), Reconnection outflows and current sheet observed with Hinode/XRT in the 2008 April 9 "Cartwheel CME" flare, *Astrophys. J.*, *722*, 329.

Savage, S. L., G. Holman, K. K. Reeves, et al. (2012a), Low-altitude reconnection inflow-outflow observations during a 2010 November 3 solar eruption, *Astrophys. J.*, *754*, 13.

Schettino, G., G. Poletto, and M. Romoli (2010), CMEs from AR 10365: Morphology and physical parameters of the ejections and of the associated current sheet, *Astrophys. J.*, *708*, 1135.

Seaton, D. B., A. E. Bartz, and J. M. Darnel (2017), Observations of the formation, development, and structure of a current sheet in an eruptive solar flare, *Astrophys. J.*, *835*, 139.

Seaton, D. B., and T. G. Forbes (2009), An analytical model for reconnection outflow jets including thermal conduction, *Astrophys. J.*, *701*, 348.

Shen, C., J. Lin, and N. A. Murphy (2011), Numerical experiments on fine structure within reconnecting current sheets in solar flares, *Astrophys. J.*, *737*, 14.

Shen, C., J. Lin, N. A. Murphy, et al. (2013), Statistical and spectral properties of magnetic islands in reconnecting current sheets during two-ribbon flares, *Phys. Plasma*, *20*, 072114.

Shibata, K., and S. Tanuma (2001), Plasmoid-induced-reconnection and fractal reconnection, *Earth Planet Space*, *53*, 473.

Sittler, E. C., Jr. and M. Guhathakurta (1999), Semiempirical Two-dimensional MagnetoHydrodynamic Model of the Solar Corona and Interplanetary Medium, *Astrophys. J.*, *523*, 812.

Skender, M., and G. Lapenta (2010), On the instability of a quasiequilibrium current sheet and the onset of impulsive bursty reconnection, *Phys. Plasma 17*, 2905.

Song, H., Z. Zhang, Y. Chen, et al. (2016), A statistical study of the average iron charge state distributions inside magnetic clouds for solar cycle 23, *Astrophys. J.*, Sup. Seri., *224*, 27.

Strauss, H. R. (1986), Hyper-resistivity produced by tearing mode turbulence, *Phys. Fluid*, *29*, 3668.

Strauss, H. R. (1988), Turbulent reconnection, *Astrophys. J.*, *326*, 412.

Su, Y., A. M. Veronig, G. D. Holman, et al. (2013), Imaging coronal magnetic-field reconnection in a solar flare, *Nature Phys.*, *9*, 489.

Sui, L., and G. D. Holman (2003), Evidence for the formation of a large-scale current sheet in a solar flare, *Astrophys. J.*, *596*, L251.

Sui, L., G. D. Holman, and B. R. Dennis (2004), Evidence for magnetic reconnection in three homologous solar flares observed by RHESSI, *Astrophys. J.*, *612*, 546.

Susino, R., A. Bemporad, and S. Krucker (2013), Plasma heating in a post eruption current sheet: A case study based on ultraviolet, soft, and hard x-ray data, *Astrophys. J.*, *777*, 93.

Švestka, Z. (1996), *Solar Phys.*, Speeds of rising post-flare structures, *169*, 403.

Švestka, Z., and E. W. Cliver (1992), History and Basic Characteristics of Eruptive Flares, in *Eruptive Solar Flares*, Proc. of Col. #133 IAU, held at Iguazu, Argentina, August

2-6, 1991. Ed. by Z. Švestka, B.V. Jackson, M.E. Machado; Publisher, Springer-Verlag, New York, 1.

Švestka, Z., F. Fárník, H. S. Hudson, et al. (1998), Large-scale active coronal phenomena in Yohkoh SXT Images IV, solar wind streams from flaring active regions, *Solar Phys.*, *182*, 179.

Takasao, S., A. Asai, H. Isobe, et al. (2012), Simultaneous observation of reconnection inflow and outflow associated with the 2010 August 18 solar flare, *Astrophys. J.*, *745*, L6.

Tanuma, S., T. Yokoyama, T. Kudoh, et al. (2001), Two-dimensional magnetohydrodynamic numerical simulations of magnetic reconnection triggered by a supernova shock in the interstellar medium: Generation of X-ray gas in the galaxy, *Astrophys. J.*, *551*, 312.

Titov, V. S., and P. Démoulin (1999), Basic topology of twisted magnetic configurations in solar flares, *Astron. Astrophys.*, *351*, 707.

Vishniac, E. T., et al. (2012), Reconnection current sheet structure in a turbulent medium, *Nonlinear Proc. Geophys.*, *19*, 605.

Vršnak, B., G. Poletto, and E. Vuji (2009), Morphology and density structure of post-CME current sheets, *Astron. Astrophys.*, *499*, 905.

Warren, H. P., C. M. O'Brien, and N. R. Sheeley Jr. (2011), Observations of reconnecting flare loops with the atmospheric imaging assembly, *Astrophys. J.*, *742*, 92.

Webb, D. F., et al. (2003), Observational evidence of new current sheets trailing coronal mass ejections, *J. Geophys. Res.*, *108*(A12), 1440.

Webb, D. F., and A. Vourlidas (2016), LASCO white light observations of eruptive current sheets trailing CMEs, *Solar Phys.*, *291*, 3725.

Wimmer-Schweingruber, R. F., et al. (2006), Understanding interplanetary coronal mass ejection signatures, Report of Working Group B, *Space Sci. Rev.*, *123*, 177.

Wood, P., and T. Neukirch (2005), Electron acceleration in reconnecting current sheets, *Solar Phys.*, *226*, 73.

Yokoyama, T., and K. Shibata (1998), A two-dimensional magnetohydrodynamic simulation of chromospheric evaporation in a solar flare based on a magnetic reconnection model, *Astrophys. J.*, *494*, L113.

Yokoyama, T., and K. Shibata (2001), Magnetohydrodynamic simulation of a solar flare with chromospheric evaporation effect based on the magnetic reconnection model, *Astrophys. J.*, *549*, 1160.

Yokoyama, T., K. Akita, T. Morimoto, et al. (2001), Clear evidence of reconnection inflow of a solar flare, *Astrophys. J.*, *546*, L69.

Zhang, Q., Y. Su, and H. Ji (2017), Pre-flare coronal dimmings, *Astron. Astrophys.*, *598*, 3.

Zhu, C., R. Liu, D. Alexander, et al. (2016), Observation of the evolution of a current sheet in a solar flare, *Astrophys. J.*, *821*, L29.

Part IV: Field-Aligned Currents

16

A Review of Birkeland Current Research Using AMPERE

John C. Coxon[1], Stephen E. Milan[2], and Brian J. Anderson[3]

ABSTRACT

We review research into the Birkeland currents (also known as field-aligned currents) that has been conducted using the Active Magnetosphere and Planetary Electrodynamics Response Experiment (AMPERE). We open with a short review of the history of research into the Birkeland current systems, before describing the conception of AMPERE and its roots in Iridium telecommunications satellite engineering telemetry data. We describe the difference between Iridium engineering telemetry data and AMPERE data, and review the work that has been done using both datasets. We review research into Regions 1 and 2 Birkeland current during geomagnetic storms and consequently the ways in which the currents are driven by the solar wind, before moving onto the substorm current wedge and discussing the present controversy over this phenomenon in AMPERE data. Ways in which AMPERE data can be used to examine ionospheric conductivity are detailed alongside ways in which AMPERE has contributed to empirical models of the coupled solar wind-magnetosphere-ionosphere system. Finally, we look to the future and speculate on the manner in which AMPERE may yet unlock the secrets of the magnetosphere.

16.1. INTRODUCTION

At the turn of the twentieth century, the existence of field-aligned currents (FACs) was proposed by the Norwegian scientist Kristian Birkeland [*Birkeland*, 1908, 1913]. They were proposed as a mechanism to link the ionosphere to the solar wind, in order to explain observations of magnetic perturbations and aurorae that he made at an observatory in the north of Norway. The existence of field-aligned currents was not immediately accepted by the scientific community, and proved controversial until their eventual detection some 50 years after Birkeland's proposal [*Zmuda et al.*, 1966]; as a result, field-aligned currents are often referred to as Birkeland currents, in his honor. An overview of the debate surrounding Birkeland currents can be found within the detailed historical account that opens this monograph [*Brekke*, Chapter 1, this volume].

The first detailed study of Birkeland current structure was conducted by *Iijima and Potemra* [1976, 1978], who split the Birkeland current system into two regions, both of

[1] *Department of Physics and Astronomy, University of Southampton, Southampton, UK*

[2] *Department of Physics and Astronomy, University of Leicester, Leicester, UK; Birkeland Centre for Space Science, University of Bergen, Bergen, Norway*

[3] *Applied Physics Laboratory, Johns Hopkins University, Laurel, Maryland, USA*

Electric Currents in Geospace and Beyond, Geophysical Monograph 235, First Edition.
Edited by Andreas Keiling, Octav Marghitu, and Michael Wheatland.

which flow in an oval above the ionosphere. The Region 1 (R1) current connects the ionosphere to the magnetopause (on the dayside and flanks), and to the current sheet (on the nightside); the R1 current oval encircles the polar cap. The Region 2 (R2) current connects to the partial ring current in the inner magnetosphere, and the oval lies equatorward of R1. These two regions of Birkeland current are often found during periods of magnetic reconnection on the dayside, and they are consistent with reconnection-driven ionospheric convection [e.g., *Cowley*, 2000; *Coxon et al.*, 2014a]. During substorms, when reconnection occurs in the magnetotail, a magnetic bay is observed in the AL index [*Akasofu and Meng*, 1969], which is a result of a current flowing west in the ionosphere; this current is known as the substorm electrojet. Field-aligned currents (FACs) flow into and out of the ionosphere at either end of the electrojet, and these currents are collectively referred to as the substorm current wedge, or SCW [McPherron et al., 1973; *Clauer and McPherron*].

Results from the Active Magnetosphere and Planetary Electrodynamics Response Experiment (AMPERE) on the R1/R2 current system and the SCW will be reviewed in sections 16.3 and 16.4, respectively. In addition to research on these large-scale current systems, AMPERE has also contributed to the study of other current systems, which are briefly reviewed in section 16.5. We discuss the contributions of AMPERE to studies of ionospheric conductivity in section 16.6 and to modeling studies (both as a contribution to empirical models as a comparison to theoretical models) in section 16.7. Finally, in section 16.8, we look to the future.

16.2. ABOUT AMPERE

AMPERE was conceived to investigate the dynamics of the Birkeland currents using magnetometer data from the attitude control systems of the Iridium telecommunications satellite network [*Anderson et al.*, 2014]. AMPERE allows for the currents during dayside driving and during a substorm to be examined globally, and an example of the progression of the current systems is shown in Figure 16.1, which shows R1 and R2 currents flowing in Figure 16.1a,b and which shows the presence of additional current in the 21-03 MLT sector after substorm onset (Fig. 16.1c), which is an example of the SCW.

Data from the Iridium spacecraft constellation were first used to investigate the Birkeland current density by *Anderson et al.* [2000], and following initiation of AMPERE by the National Science Foundation, modifications to the Iridium flight software enabled acquisition of the higher rate AMPERE data, which were first made available in December 2010 [*Burke et al.*, 2011]. The Iridium constellation comprises 66 satellites in the communications network together with on-orbit spares, which orbit the Earth in six circular, polar, orbital planes at an altitude of 780 km and a longitudinal spacing of ~2 hr of MLT, although as the orbits do not pass directly above the poles, this can be 1–3 hr of MLT [*Murphy et al.*, 2013]. There are 11 satellites per orbital plane in the communications network, and each orbit takes 104 min, such that it takes a satellite 9.5 min to move from its current position to the current position of the next satellite.

The use of magnetometer data from the Iridium avionics magnetometers [*Anderson et al.*, 2000; Waters *et al.*, 2001] used the original engineering telemetry (TLM) data from the satellites, which were originally for monitoring satellite system performance. The TLM packets were nominally acquired every ~200 s, so magnetometer samples were acquired with a separation of ~1400 km in the along-track direction. The latitude spacing between samples in the TLM magnetometer data was therefore almost 15°, at least three times the nominal latitude extent of the R1/R2 current system [cf. *Iijima and Potemra*, 1976]. As a result, studies with these TLM data required at least one hour of magnetometer samples to obtain even a coarse estimate of the magnetic signals of the Birkeland current system [cf. *Korth et al.*, 2010]. Even at this low capability, the data were used successfully for a variety of studies and it was established that the current systems derived from Iridium TLM data are consistent with R1 and R2 currents flowing during southward IMF and NBZ currents flowing during northward IMF, with rotations due to B_Y; examples of the large-scale current systems are shown later in Figure 16.3 [*Anderson et al.*, 2008].

The time intervals between the TLM data packets could not be decreased within the bandwidth or onboard storage constraints of the space vehicles. Yet, only by increasing the data sampling by at least an order of magnitude could the Iridium magnetometer data acquire enough data to resolve the large-scale currents on each pass to take advantage of the satellite constellation revisit time of only 9.5 min. To solve this problem, the Active Magnetosphere and Planetary Electrodynamics Response Experiment (AMPERE) was conceived to support the purchase of a new data stream of avionics magnetometer and vehicle attitude and position data. Figure 16.2 shows examples of data acquired in 10 min intervals for the Iridium TLM data and AMPERE standard and high rate sampling. Unlike the Iridium TLM data, full vector data are used in the AMPERE data processing and inversions: the examples illustrate the vast increase in data density achieved by AMPERE. The spherical harmonic fit uses a longitude order of 5 and a latitude order of 20 in the range of colatitudes $0° \leq \theta \leq 60°$ [*Anderson et al.*, 2014], and Figure 16.2 shows that under standard rate sampling, the data are sufficient to resolve signals with scales of ~1.5° in latitude [*Anderson et al.*, 2017].

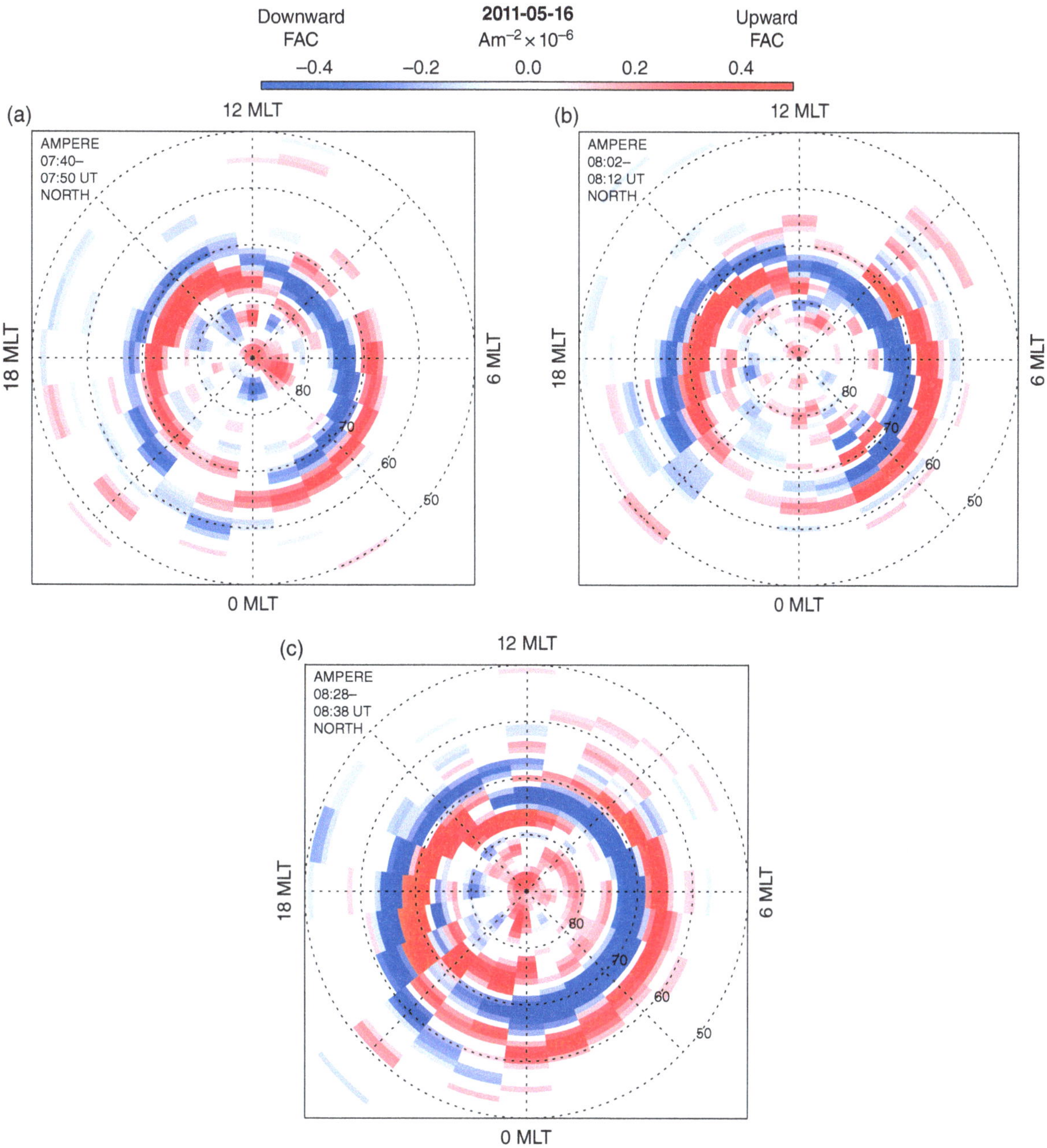

Figure 16.1 AMPERE current densities from (a) 07:45 UT, (b) 08:07 UT, and (c) 08:33 UT, 16 May 2011 in the Northern Hemisphere. The densities are plotted in polar coordinates with noon at the top of the page; dotted lines represent a grid of 3 hr MLT and 10° latitude, respectively. Upward current is defined as positive, with the color scale at the top indicating the densities in each panel. The panels show the progression of a substorm. From *Murphy et al.* [2013].

It is important to note that the TLM data were incapable of properly resolving magnetospheric reconfigurations or dynamics because at least 1 hr of data had to be accumulated to derive even a crude estimate of the currents. Only with the availability of AMPERE data was it possible to derive new global estimates of the Birkeland currents on 10 min timescales, shorter than a nominal magnetosphere-ionosphere reconfiguration time. With this advance, it finally became possible to observe the global dynamics of the Birkeland currents and truly resolve spatial and temporal dynamics. This review covers studies using both the Iridium-TLM and AMPERE data and the type of data used by each study are distinguished by the notations TLM or AMPERE data.

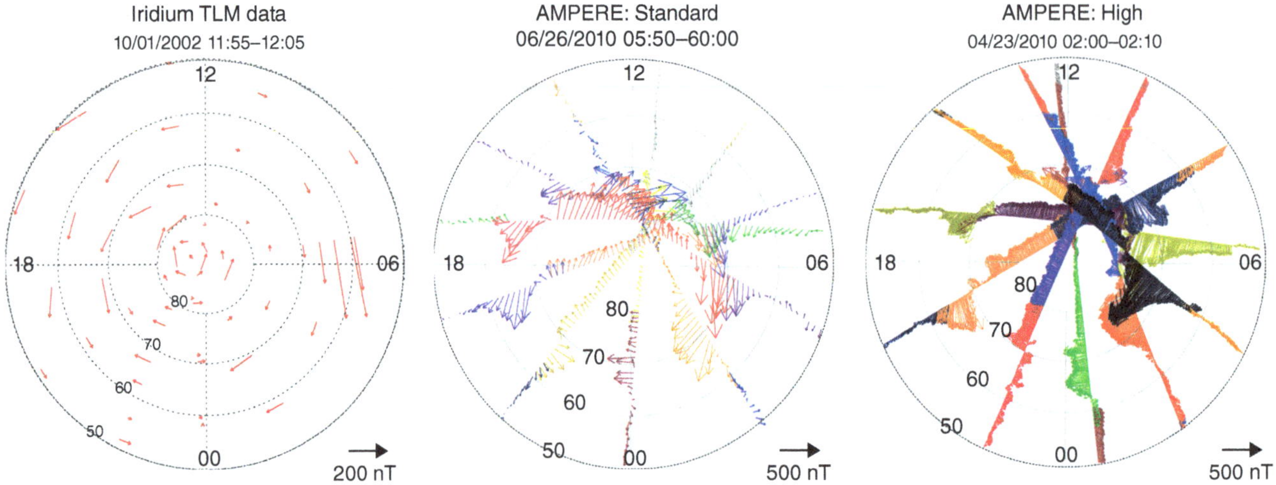

Figure 16.2 Comparison of magnetic field data returned from the Iridium Communications constellation of satellites in 10 min intervals with the engineering telemetry data (Iridium TLM) on the left and after the AMPERE development for standard or low rate (19.44 s between samples) and high rate (2.16 s between samples). All panels show reduced data from which the background field has been subtracted and further baseline removal has been applied. The magnetic perturbations are plotted as arrows originating at the point of measurement in the local horizontal plane and projected into AACGM latitude and local time. Arrow length scaling is shown by reference arrows at the lower right of each panel. For AMPERE data, different colors denote data from different satellites. For an example of the current densities derived from the magnetic data, see Figure 16.3.

While the Iridium TLM data were useful only for characterizing the large-scale average configuration of the Birkeland currents [*Anderson et al.*, 2000], the increased data rate achieved by AMPERE allowed studies of relatively small-scale features, which were previously only detectable with spacecraft constellations as Cluster [*Dunlop et al.*, Chapter 4, this volume] and Swarm [*Ritter and Lühr*, 2006]. With the global distribution of the AMPERE data, it became possible to identify the dynamics in the magnitudes of both R1 and R2 current, as well as the shape and distribution of the R1 and R2 current ovals and a method to do this algorithmically is described in *Clausen et al.* [2012] and *Coxon et al.* [2014a]. This review will focus on research conducted into Birkeland currents using the Iridium TLM and AMPERE datasets. Reviews of the current systems affecting Earth, and of magnetosphere-ionosphere coupling have previously been conducted [*Cowley*, 2000; *Milan et al.*, 2017], as have reviews of the different current systems [*Ganushkina et al.*, 2015, as well as the reviews found throughout this volume such as *Khurana and Liu*, Chapter 2, this volume].

The magnetic field data returned by AMPERE are fit to a continuous spherical harmonic fit and the curl of this fit is used to derive the radial current density making use of Ampère's law. Since it is the radial current density that is derived, equating this with the Birkeland current density is equivalent to assuming that the Birkeland current (and therefore the magnetic field) is radial. Due to the spacing of the satellites in each orbital plane, data are composed of a series of 10 min windows, evaluated every 2 min, and the data are on a grid of resolution 1° latitude and 1 hr of MLT. The data are presented in altitude adjusted geomagnetic (AAGCM) coordinates at an altitude of 780 km. We note that for ease of description, we adopt a convention in which the temporal midpoint of the 10 min window is used to describe an interval of data (such that an interval of data between 19:42 and 19:52 UT would be described as occurring at 19:47 UT).

16.3. REGIONS 1 AND 2 BIRKELAND CURRENT DURING GEOMAGNETIC STORMS

The first scientific study to use data from the Iridium spacecraft examined two geomagnetic storms, one on 22–23 September 1999 and the other on 21–22 October 1999 [*Anderson et al.*, 2002], Subsequent studies of Birkeland currents being driven by the solar wind have included case studies of geomagnetic storms from 2000–2015 using Iridium TLM data [*Anderson and Korth*, 2007; *Anderson et al.*, 2005; *Brandt et al.*, 2004; *Eriksson et al.*, 2008; *Liou et al.*, 2005; *Sitnov et al.*, 2010; *Stephens et al.*, 2013] and AMPERE data [*Knipp et al.*, 2014; *Le et al.*, 2016; *Lu et al.*, 2014; *Lyons et al.*, 2016; *Matsui et al.*, 2016; *Wilder et al.*, 2012] as well as case studies specifically of the dynamical evolution of the currents at activity onsets [*Anderson et al.*, 2014] or in conjunction with MMS observations in the magnetopause [*Anderson et al.*, 2016; *Le et al.*, 2016; *Matsui et al.*, 2016].

Maps of the currents during different IMF conditions have enhanced our knowledge of how the IMF changes Birkeland current patterns, as demonstrated by Figure 16.3. The ability to automatically detect the magnitude and location of the R1 and R2 currents [*Clausen et al.*, 2012; *Coxon et al.*, 2014a] has enabled statistical studies that confirm the magnitude of the Birkeland currents is consistent with driving by magnetic reconnection [*Coxon et al.*, 2014a,b] and the location of the Birkeland current ovals are consistent with expansions and contractions of the polar cap [*Anderson et al.*, 2014; *Clausen et al.*, 2012; *Coxon et al.*, 2014a,b]. Work has also shown that more R1 and R2 current flows when the current ovals are expanded during periods of dayside reconnection [*Anderson et al.*, 2014; *Coxon et al.*, 2014b], whereas Steady Magnetospheric Convection (SMC) events do not cause an expansion of the current ovals, but do cause more current to flow in the R1/R2 system [*Stephens et al.*, 2013]. Expansions of the current ovals observed with AMPERE are beginning to be used to identify periods of dayside reconnection [e.g., *Anderson et al.*, 2016], providing a new tool for identifying periods of driving. A more detailed review of the work that has been done on these topics follows in the rest of this section.

There is much to be done using AMPERE data to characterize and understand the Birkeland currents and their reaction to solar wind driving. Iridium TLM data have shown that the current ovals are in fact somewhat displaced toward dusk [*Anderson et al.*, 2005], although this has not yet been studied with AMPERE data. Comparisons of the aurora with Birkeland currents have been performed with Iridium TLM data [*Korth et al.*, 2004a] and AMPERE [*Carter et al.*, 2016; *Korth et al.*, 2014], although there is disagreement on whether upward current is colocated with auroral displays on the dusk flank of the polar cap, and the exact nature of the link between aurora and Birkeland current is still very much an open question. It has been suggested that the Birkeland current system can hold only a limited amount of current, and Iridium TLM data have been presented to support the idea that R1 current saturates during geomagnetic storms [*Anderson and Korth*, 2007], but more work needs to be done to substantiate this discovery. Work has demonstrated that the upward and downward currents in the Northern Hemisphere appear to be systematically stronger than those in the Southern Hemisphere [*Coxon et al.*, 2016a], and the nature of the physics underpinning this observation is also an open question.

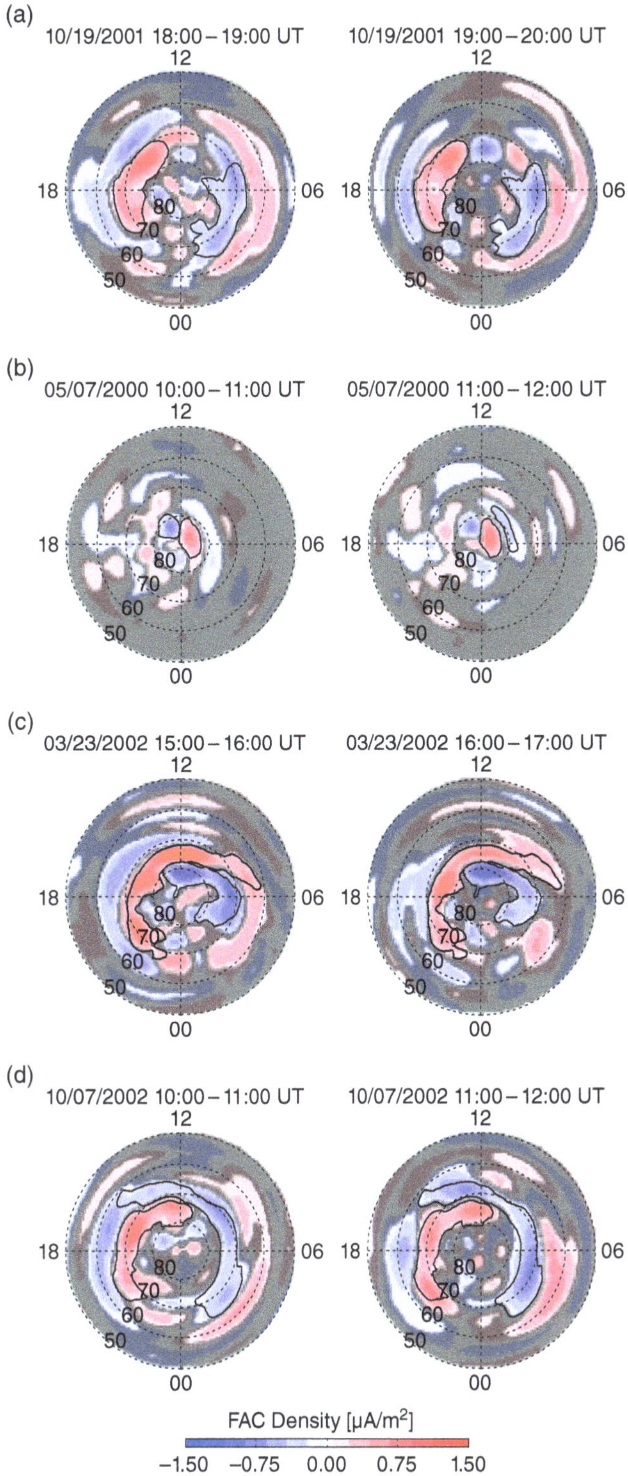

Figure 16.3 A plot showing examples of (a) southward, (b) northward, (c) dawnward, and (d) duskward IMF in current maps using Iridium TLM data. Upward currents are shown in red; downward currents are in blue; and grey shading indicates currents below the 2σ confidence level. Contours in the plots show the areas that stay constant between the time intervals plotted (above the 2σ confidence level). From *Anderson et al.* [2008].

16.3.1. Location

Looking at the global distribution of the currents is one of the key abilities of AMPERE that cannot be achieved via other methods. During southward IMF,

Birkeland currents are expected to be distributed in the Regions 1 and 2 (R1 and R2) current systems, and indeed this has been observed statistically with Iridium data [*Anderson et al.*, 2008] and subsequently AMPERE [*Carter et al.*, 2016]. Empirical Orthogonal Function (EOF) analysis and Principal Component Analysis (PCA) have been used to decompose AMPERE data into eigenfunctions that describe the modes of variability of the dataset; the first 12 eigenfunctions found by PCA are shown in Figure 16.4 [*Milan et al.*, 2015]. These eigenfunctions can be multiplied by scaling factors and

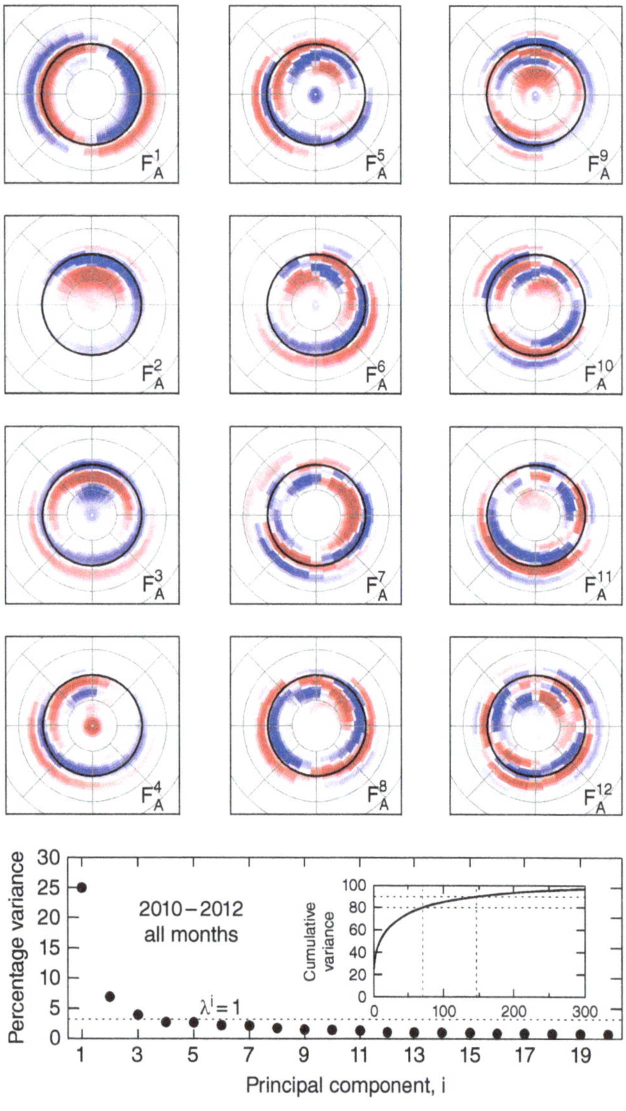

Figure 16.4 The 12 eigenfunctions responsible for the most variability in AMPERE data. F_A^1 shows the R1/R2 current system and is responsible for 25% of the variability in a current map, F_A^2 and F_A^3 are responsible for < 10% of the variability each, and the other eigenfunctions are responsible for < 1% each. There is no obvious candidate for the SCW in any of the eigenfunctions. From *Milan et al.* [2015].

combined to reconstruct maps of the current density (in this sense they are similar to spherical harmonics), and the single eigenfunction that describes the most variability in the dataset (called the primary component) is the R1 and R2 current system, with other components accounting for <10% [*Milan et al.*, 2015]. The first observations of the R1 and R2 current systems showed that they had, on average, a spiral structure with overlapping currents near noon [*Iijima and Potemra*, 1978]; this structure has been shown during intervals in which there is a nonzero B_Y component in the IMF and is seen in Figure 16.3c,d [*Anderson et al.*, 2008].

The latitude of the equatorward edges of the R1 and R2 current ovals have been shown to be anticorrelated with *Kp* ($R = -0.68$), such that during periods of higher *Kp*, or during periods of southward IMF, the current ovals are observed to move to lower latitudes [*Anderson et al.*, 2002]. The location of the Birkeland currents in AMPERE data was also observed to be consistent with the evolution of SYM-H during six storms examined by *Zou et al.* [2014], with the current ovals moving between 5° and 10° equatorward during the geomagnetic storm. A storm on 17 April 2002 evaluated using Iridium data showed that the Birkeland currents moved 5° equatorward during the initial phase of the storm and then moved a further 5°–6° in the main phase [*Liou et al.*, 2005]. During the St. Patrick's Day storm in 2013, the R1 and R2 system was observed as southward IMF began and then expanded equatorward by ~1°–2° during a second interval of southward IMF; another expansion of ~2°–3° was observed during a third period of southward IMF, alongside developing smaller-scale currents [*Lyons et al.*, 2016]. A case study of the 17 March 2015 geomagnetic storm revealed that a prolonged interval of southward IMF led to observations of Birkeland currents moving equatorward by ~8°, with the most equatorward currents occurring at the most southward IMF [*Le et al.*, 2016]. Modeling of a CIR-driven magnetic storm showed that the field-aligned currents reduced in density and contracted when tail currents dominated in the model run, whereas they increased in density and expanded when the partial ring current was dominant [*Sitnov et al.*, 2010].

For comparison, the current ovals are also observed to expand during the growth phase of a substorm, prior to the onset of the expansion phase; this can be seen in Figure 16.1a,b, which shows the R1 and R2 currents moving 2°–3° to lower latitudes prior to the expansion phase onset [*Murphy et al.*, 2013]; this was corroborated by *Coxon et al.* [2014b], who employed a superposed epoch analysis to show that the average latitude of the current ovals moved toward the equator by 1°–2° over the 2 hr before substorm expansion phase onset. Although these increases in the size of the current ovals lie close to the threshold of AMPERE's ability to resolve latitudinal features, they are consistent with increases observed in the latitude of the auroral ovals prior to expansion phase onset [e.g., *Milan et al.*, 2009]. In contrast, steady magnetospheric convection (SMC) events do not result in significant equatorward expansion of the R1 current system, but the R2 current system does thicken, extending farther equatorward [*Stephens et al.*, 2013], which suggests that the polar cap is not expanding in size but that the amount of current flowing increases.

The currents have been shown statistically to move to lower latitudes during periods of southward IMF [*Anderson et al.*, 2008; Matsuo *et al.*, 2015] and also during periods of enhanced solar wind electric field [*Korth et al.*, 2010], which is consistent with magnetic flux being opened at the magnetopause by dayside reconnection. In Figure 16.3, the currents in the example of southward IMF are farther equatorward than in any other IMF condition. EOF analysis of AMPERE data showed that the function that explained the second most variability in the AMPERE dataset was associated with movement of R1 and R2 currents, but was only weakly correlated with solar wind driving [*Cousins et al.*, 2015a]; however, contractions of the polar cap are not expected to be driven by the solar wind, which may explain why the correlation was not high. A substorm growth phase magnetic field model [*Yue et al.*, 2015] was developed using empirical plasma sheet pressures from Geotail and THEMIS. Comparison with AMPERE showed a relatively poor correspondence, with the simulation only yielding downward currents in a very small part of the ionosphere relative to observations; however, the general pattern of the simulation results, if enlarged, did resemble the AMPERE observations. The model showed the equatorward motion of the auroras in the growth phase. Superposed epoch analyses show that the R1 current oval expands through the growth phase of a substorm and reaches maximum size ~ 15 min after expansion phase onset, before beginning to contract again [*Clausen et al.*, 2012, 2013a,b; *Coxon et al.*, 2014b]. This observation was interpreted as evidence for the current ovals expanding with dominant dayside reconnection and contracting with dominant nightside reconnection, as part of the expanding/contracting polar cap (ECPC) paradigm [*Cowley and Lockwood*, 1992; *Milan et al.*, 2015; *Siscoe and Huang*, 1985]. The other results reviewed here also appear to be consistent with the ECPC model.

Expansions and contractions of the current ovals during geomagnetic storms are sometimes accompanied by a dawn-dusk asymmetry in the currents, which is also observed during storms. The geomagnetic storm of 1 October 2002 was found to show good agreement in R2 current between modeled currents and Iridium measurements [*Brandt et al.*, 2004], and *Anderson et al.* [2005] presented an analysis of this storm and the subsequent storm, with minimum Dst = −163 nT. As the storm progressed, the Birkeland currents developed a dawn-dusk asymmetry during the main phase, with the currents lying at more equatorial latitudes on the dusk flank of the polar cap than on the dawn flank. During the recovery phase, the currents on both sides returned to approximately equal latitudes; a similar phenomenon was seen in a storm on 11–12 August 2000 [*Anderson and Korth*, 2007].

A statistical survey of geomagnetic storms found that the dusk currents were shifted equatorward, relative to the dawn currents, by an average of 2.4°, but that the extent to which the currents were asymmetrical was controlled in part by B_Y; dawn currents could be equatorward of dusk currents at large enough values of B_Y [*Anderson et al.*, 2005]. B_Y is known to have a significant effect on the morphology of the Birkeland currents observed by AMPERE [*Carter et al.*, 2016], with observations that there is a ~1.25 hr delay between a pulse of B_Y and a peak in dawnward current [*Wilder et al.*, 2013] seemingly consistent with currents flowing as a result of forces acting to restore symmetry in cases of asymmetrical field lines [*Reistad et al.*, 2016; *Tenfjord et al.*, 2015], although results indicate that this effect is stronger in the summer hemisphere [*Green et al.*, 2009]. Removing the effect of B_Y to look at the dawn-dusk asymmetry during geomagnetic storms confirmed that larger asymmetries occurred during the storm main phase; for each 100 nT in ASY-H, the dusk currents strengthened and moved approximately 1° equatorward relative to those on the dawn flank, attributed to ion pressure associated with the partial ring current [*Anderson et al.*, 2005].

The knowledge that the current ovals observed by AMPERE expand and contract consistent with the ECPC has led to the data being used to verify periods of dayside (and nightside) magnetic reconnection. *Anderson et al.* [2016] used observations of the expanding R1 oval to show that dayside reconnection was occurring and thus demonstrated that MMS observations of a deviation in magnetopause location was consistent with magnetopause erosion due to dayside reconnection. Expansions of the current ovals were also used to identify dayside reconnection intervals in an investigation of GPS scintillations by *Clausen et al.* [2016].

Field-aligned currents facilitate the travel of particles from elsewhere in the magnetosphere into the ionosphere, and are therefore intimately related to auroral displays. Currents tend to be stronger when the current ovals are expanded [*Coxon et al.*, 2014b], consistent with the dynamic behavior of emissions previously established by global auroral imaging using such missions as IMAGE and Polar. Upward Birkeland currents are thought to be colocated with auroral precipitation [*Waters et al.*, 2001]; however, recent results differ on whether upward current on the dusk side of the polar cap are well colocated with aurora in this region [*Carter et al.*, 2016; *Korth et al.*, 2014]. This is an open question that needs to be resolved in future studies using AMPERE (and perhaps other Birkeland current measurements) in conjunction with auroral observations.

16.3.2. Integrated Total Current

As well as enabling hitherto unprecedented views of the location of the currents flowing in the magnetosphere, AMPERE provides for analysis of the total amount of current flowing, both by using current density and by calculating the magnitude of the current flowing [e.g., *Anderson et al.*, 2014]. Some studies have created indices that describe the maximum magnetic perturbations detected by Iridium spacecraft, and found that these measures of current activity were correlated with the AE index calculated using ground magnetometers, indicating that the magnetic perturbations and consequently the Birkeland currents increased with geomagnetic activity [*Anderson et al.*, 2002] and after the commencement of geomagnetic storms [*Liou et al.*, 2005]. This has been substantiated by AMPERE, which shows that more current flows in the R1 and R2 current systems at times of more southward IMF [*Anderson et al.*, 2014; *Cousins et al.*, 2015a; *Matsuo et al.*, 2015], with current density increasing as the current ovals move to equatorward latitudes in the geomagnetic storm of 17 March 2015 [*Le et al.*, 2016]. Superposed epoch analyses also appear to show that the current magnitudes peak approximately coincident with the current oval reaching its equatormost point [*Coxon et al.*, 2014b].

Steady magnetospheric convection (SMC) events have been found to increase the density and thickness of the R1 and R2 current systems [*Stephens et al.*, 2013], and the solar wind electric field is the most important factor driving Birkeland current intensities during southward IMF [*Korth et al.*, 2010], which suggests that the currents are driven by magnetic reconnection, especially at the magnetopause. Indeed, it has recently been explicitly demonstrated that R1 and R2 current magnitudes are consistent with driving by both dayside and nightside reconnection [*Coxon et al.*, 2014a,b; *Milan et al.*, 2015]. Furthermore, examination of upward and downward Birkeland current flowing as a function of both reconnection and conductivity shows that the amount of current (averaged over 27 days to remove solar effects) is well represented by a combination of conductivity enhancements due to solar illumination and driving by dayside reconnection [*Coxon et al.*, 2016a]. In the same paper, the magnitudes of the upward and downward current in the Northern Hemisphere were systematically larger than the Southern Hemisphere even when seasonal and diurnal effects were taken into consideration, implying that an asymmetry might be present in the Birkeland currents in each hemisphere. Although the confluence of the orbital tracks of the Iridium constellation lies farther from the pole in the Southern Hemisphere than in the Northern Hemisphere, this asymmetry seems likely to be a physical phenomenon, since observations show that there are differences in ionospheric convection [*Förster and Haaland*, 2015] associated with asymmetries in Earth's magnetic field [*Cnossen and Förster*, 2016; *Laundal et al.*, 2016], so it is reasonable that such an asymmetry may also be present in the Birkeland currents.

Estimates of the energy input to the ionosphere have also been made in two studies, which combined data from the Iridium constellation with data from SuperDARN to estimate the total electromagnetic energy contribution by calculating the Poynting flux [*Korth et al.*, 2008a; *Waters et al.*, 2004]. In the first study, the total energy input during the event was estimated to be 50 GW and the Iridium and SuperDARN estimate showed good agreement with Defense Meteorological Satellite Program (DMSP) calculations of Poynting flux [*Waters et al.*, 2004]. In the second study, the Iridium and SuperDARN calculation was corrected for a perceived underestimate compared to the DMSP calculation, and the total energy input was estimated at 188 GW, which was found to be an order of magnitude greater than the 20 GW from particle precipitation calculated using auroral imaging [*Korth et al.*, 2008a].

Two geomagnetic storms during 10–13 August 2000 were found to exhibit evidence for R1 current saturation, a phenomenon in which magnetic perturbations associated with R1 current act to reduce the subsolar magnetic field and therefore also reconnection rate [*Anderson and Korth*, 2007, and references therein]. The second storm in this period showed a maximum interplanetary electric field E_Y, which was 3.6 times greater than that of the first storm, but the Birkeland currents in the second storm were only twice as big. A sample of 23 events was used to fit a power law of current $I = 4.2 + 2.6E_Y^{0.48}$, indicating that the currents do indeed show the theorized saturation. However, no margins of error were calculated and the conclusion mostly comes from a small number of data at $E_Y > 10\,\mathrm{mV\,m^{-1}}$. Similar analyses have not yet been performed with AMPERE data to establish whether the currents do saturate.

16.4. THE SUBSTORM CURRENT WEDGE

Substorms are a key mechanism driving nightside Birkeland currents, and AMPERE enables, for the first time, snapshots in which the current density is visible across the entire substorm current wedge (SCW) region [*Clauer and McPherron*, 1974; *McPherron et al.*, 1973]. As a result, much work has been done using AMPERE to examine the SCW, including several case studies of substorms [*Connors et al.*, 2014, 2015; *Murphy et al.*, 2012, 2013; *Sergeev et al.*, 2014a]. Other studies have used a statistical approach [*Clausen et al.*, 2012, 2013a,b; *Coxon et al.*, 2014b, 2017]. It has been found that the onset of nightside currents, presumably reflecting onset of return convection in the magnetotail, is critical to the development of the global R1/R2 current system [*Anderson et al.*, 2014]. While most authors agree that Birkeland currents on the nightside are intimately related to substorms, the exact nature of the SCW is one of the biggest open questions in the field, and studies using AMPERE are critical to developing our understanding of substorms. The traditional picture of the SCW involves a single current flowing into the ionosphere on the postmidnight side and a single current flowing out of the ionosphere on the premidnight side, in the same sense as R1 [*Clauer and McPherron*, 1974]. There is debate as to whether this picture is accurate, or whether there is also a current in the sense of R2, which flows as a result of a substorm as part of a "two loop" substorm current wedge [*Sergeev et al.*, 2011, 2014b]. Another topic of debate is whether the currents that compose the SCW should be considered as R1 and R2 currents, or whether they should be considered as distinct currents that happen to flow in a similar sense. To differentiate between the models of *Clauer and McPherron* [1974] and *Sergeev et al.* [2014a], we refer to them as "single loop" and "two loop" current wedges in the rest of this section while reviewing the work to date in more detail.

Although the large-scale structure of the SCW is an open question that is perfect for AMPERE, the SCW is also thought to have small-scale structure, which may be beyond the limitations of the AMPERE dataset. Authors have used Cluster data to show that the SCW may comprise tens of filamentary current sheets, rather than broad regions of current [*Forsyth et al.*, 2014] and MMS observations found current sheets with a thickness of 25 km in the plasma sheet during substorms [*Nakamura et al.*, 2016]. As such, work with AMPERE focuses on large-scale structure.

Case studies have contributed evidence to this debate. *Murphy et al.* [2013] used AMPERE data from events in which the footprint of the Iridium satellite contributing to the AMPERE data was close to the onset of the substorms in question, identifying three events in 2010 and 2011 for which this was the case. Observations for two of the events showed R1 and R2 current flow, but such flow was flanked by additional current systems in both of those events (an example is shown in Fig. 16.1); the third event showed no sign of the R1 and R2 current system in the expected sense. Despite the structure observed in the cases chosen, they concluded that averaging latitudinally could replicate the single loop SCW of *Clauer and McPherron* [1974]. *Connors et al.* [2014] showed a case study of a substorm in AMPERE data, concluding that the two loop substorm current wedge was not supported by their observations. However, the AMPERE data they present appear to show an enhancement in R2-sense currents flowing equatorward of the R1-sense currents after expansion phase onset, consistent with the two loop model of *Sergeev et al.* [2011]; in subsequent analysis of the event, they noted that the substorm had involved regions both near Earth and farther down the magnetotail [*Connors et al.*, 2015]. Finally, a case study employing AMPERE data in addition to four GOES spacecraft and three THEMIS spacecraft [*Sergeev et al.*, 2014a] found that the current wedge more closely resembled the two loop substorm current wedge [*Sergeev et al.*, 2014b]. As such, all the case studies done using AMPERE data appear to support the two loop SCW, rather than the single loop model.

A number of statistical analyses of substorm current development have also been performed. *Clausen et al.* [2013a,b] examined substorm expansion phase onsets in the first four months of 2010, looking first at the morphology of the currents during substorms [*Clausen et al.*, 2013a]. The authors found that the R1 current density on the nightside was enhanced after substorm expansion phase, and that R1 currents were dominant on the dayside prior to onset and on the nightside after onset. They found no corresponding enhancement in R2 current, concluding that the currents were in agreement with the single loop SCW [*Clauer and McPherron*, 1974]. Although these statistical studies have not shown any evidence of the two loop SCW, it should be noted that they did not rotate or scale the current maps from AMPERE with the latitude or MLT of substorm onset, which might have allowed some spatial smoothing to occur.

Clausen et al. [2013b] then turned to examining the influence of the amount of open magnetic flux on the substorms observed, concluding that when more the magnetosphere contained more open magnetic flux at the time of substorm onset, the substorm was more intense, presenting evidence that the R1 current density (among other parameters) was more enhanced during substorms occurring at times of higher magnetic flux. Another superposed epoch analysis was performed by *Coxon et al.* [2014b], who examined AMPERE data for 3 yr between 2010–2012, looking at the size of the R1 and R2 current ovals and the R1 and R2 current magnitudes. As well as finding that the R1 and R2 current ovals increased in size until ~10 min after substorm onset, they also found that R1 and R2 current were both enhanced after onset and that the ratio of R1 current to R2 current increased after substorm onset. They concluded that the auroral electrojet at the ionospheric footprint of the SCW was closing R1 current in the dawn flank through the R1 current system on the dusk flank of the polar cap, but that R2 current was also enhanced, providing insight into the larger structure of the SCW [*Coxon et al.*, 2014b]. A follow-up study of the morphology of the current systems (2010–2015) showed that if the substorms were rotated to the MLT of substorm onset and binned by substorm onset latitude, the R1 and R2 current systems both showed enhancement at substorm onset [*Coxon et al.*, 2017], consistent with the two loop SCW [*Sergeev et al.*, 2014a]. The 12 eigenfunctions that describe the most variability in the AMPERE dataset (Fig. 16.4) were found by principal component analysis [*Milan et al.*, 2015], but none of these eigenfunctions appear to be associated with the SCW and *Milan et al.* [2015] concluded that there was no eigenfunction representing the SCW. The fact that the SCW can be well described in terms of R1/R2 sense currents gives an insight as to why no such eigenfunction was found.

Murphy et al. [2012] investigated a somewhat different aspect of Birkeland current variation during substorms, investigating the phenomenon of auroral dimming first observed by *Pellinen and Heikkila* [1978]. They found evidence for auroral dimming in a substorm that was located under the track of an Iridium satellite, noting that the upward Birkeland current density in the MLT of substorm onset reduced close to the latitude of substorm onset consistent with observations of a decrease in auroral intensity prior to onset. This was expanded upon by *Coxon et al.* [2017], who found that a superposed epoch analysis of current density close to the location of substorm onset showed a similar decrease in density prior to substorm onset.

In summary, the recent history of the substorm current wedge is one of an ever more complex system. Authors have observed smaller-scale structures within the traditional, large-scale substorm current wedge [*Murphy et al.*, 2013; *Nakamura et al.*, 2016] and more complicated large-scale structures have been proposed [*Sergeev et al.*, 2011, 2014b]. Although AMPERE has shed significant light on the behavior of the Birkeland currents around substorm expansion phase onset, results have been interpreted differently by different authors. Case studies appear to demonstrate the existence of a two loop current system, and despite statistical studies which do not concur, the most recent statistical study is consistent with this view [*Coxon et al.*, 2017]. The exact morphology of the currents close to substorm onset, and the behaviour of currents prior to onset, remain open questions.

16.5. OTHER CURRENT SYSTEMS

While we have reviewed the largest Birkeland current systems (sections 16.3 and 16.4), it is also important to review the work that has been done with AMPERE on current systems outside of these. One of these current systems is the banana current system, which is driven by pressure gradients in the magnetosphere and flows eastward on the earthward side of a peak in pressure located within the inner magnetosphere [*Liemohn et al.*, 2013]. Consequently, a westward current is required to close it, either through the ionosphere (path 1) or through a current on the far side of the same pressure peak in the inner magnetosphere (path 2). *Liemohn et al.* [2013] concluded that the current would close in the magnetosphere rather than in the ionosphere, and the resulting shape gave rise to its moniker. Observations with AMPERE showed no sign of a pair of upward and downward Birkeland currents of the required sense for path 1 at low latitudes [*Stephens et al.*, 2016], which is consistent with the original conclusion that the current closed through path 2 [*Liemohn et al.*, 2013].

Observations of an auroral arc made before a substorm showed that the arc was located ~4° equatorward of the R1/R2 interface in the early growth phase but this decreased to ~1° shortly before onset [*Motoba et al.*, 2015]. AMPERE data were used to infer the location of the R1/R2 boundary, and RBSP-B data when the spacecraft was very close to the arc showed that the arc was connected to the magnetotail by an upward Birkeland current on the tailward side of an enhancement in plasma pressure, which was interpreted as evidence for localized pressure gradients driving Birkeland current and therefore the auroral arc [*Motoba et al.*, 2015].

Another current system beyond the structures of Regions 1 and 2 current and the SCW has been observed by AMPERE. During intervals of northward IMF, the R1 and R2 current systems are weakened, and a different dominant current system is observed; this system is called NBZ (because it occurs coincident with northward B_Z). Periods of northward IMF both before and during the recovery phase of geomagnetic storms have resulted in current distributions in AMPERE consistent with the expected NBZ current system [*Eriksson et al.*, 2008; *Le et al.*, 2016] and statistical analysis of Iridium current density also shows these distributions when binned by IMF conditions [*Anderson et al.*, 2008]. Birkeland current densities higher than 0.7 mA m^{-2} and solar wind proton densities lower than 4 cm^3 are associated with the occurrence of high-latitude dayside aurora [*Korth et al.*, 2004a], and the motions of upward currents within an NBZ current system on 15 May 2005 were linked to the motion of a transpolar arc in the same geomagnetic storm [*Eriksson et al.*, 2008]. Averaged Iridium TLM data suggest that the NBZ current system does not occur in the winter hemisphere and that Birkeland current is reconfigured to the nightside in the winter hemisphere, in comparison to the summer hemisphere [*Green et al.*, 2009]. Periods of sustained northward IMF allow for the identification of currents driven by a viscous

Kelvin-Helmholtz interaction, and such an interaction is thought to account for <0.2 MA [*Korth et al.*, 2005].

16.6. CONDUCTIVITY AND ITS EFFECT ON BIRKELAND CURRENTS

Conductivity is a relatively underrepresented topic for AMPERE studies so far. The large spatial scale afforded by AMPERE data seemingly makes AMPERE a natural choice for any large-scale study of conductivity, or such phenomena as Joule heating, but not many authors have used AMPERE data to tackle these open questions. Two of the main studies to examine the issue used Iridium TLM data, with one study comparing estimates of Birkeland currents using SuperDARN and conductivity estimates to data [*Green et al.*, 2006] and another directly estimating the conductivity from Iridium TLM data [*Green et al.*, 2007]. *Coxon et al.* [2016a] used a model with dayside reconnection and conductivity inputs [*Milan*, 2013] to compare the monthly averaged modeled Birkeland currents to monthly averaged AMPERE observations, finding good agreement between modeled results and AMPERE, but a distinct asymmetry between the current flowing in the Northern and Southern Hemispheres.

Ionospheric vorticity has been used as a proxy for Birkeland current flow, which has been shown to be a reasonable proxy in an average sense [e.g., *Chisham et al.*, 2009], although variations in conductivity lead to disagreements between the vorticity and the actual currents. Some authors have combined radar data with models of ionospheric conductivity to estimate the Birkeland currents directly, and these estimates have been compared with Iridium TLM data; this technique does well in sunlit conditions where SuperDARN covers the area of interest, but poorly on Earth's nightside and where SuperDARN data were not available [*Green et al.*, 2006]. This is consistent with underestimates of Birkeland currents from the Lyon-Fedder-Mobarry magnetohydrodynamic model on the nightside [*Korth et al.*, 2004b], and *Green et al.* [2006] suggested that estimates of conductivity due to particle precipitation were to blame. A technique to estimate the conductivity from Iridium TLM data showed enhanced conductivity between the R1 and R2 current systems and estimated the ratio between Hall and Pedersen conductivity at close to unity [*Green et al.*, 2007].

The Birkeland currents determine the curl-free component of the height-integrated horizontal current in the ionosphere $J_{\perp,\,cf}$. Fukushima's theorem states that, from the Earth's surface, the magnetic perturbations due to Birkeland current flow are canceled completely by the magnetic perturbations due to these curl-free currents, in the case where the magnetic field is radial and the conductivity is uniform [*Fukushima*, 1969, 1976], such that ground magnetic perturbations are directly associated only with the divergence-free component of the horizontal ionospheric current $J_{\perp,\,df}$. *Laundal et al.* [2015] took advantage of the fact that the total horizontal current $J_\perp$ could be decomposed into $J_{\perp,\,cf}$ and $J_{\perp,\,df}$, and as a result argued that where $J_\perp = 0$, $J_{\perp,\,df}$ will be entirely determined by $J_{\perp,\,cf}$, such that the global Birkeland current distribution alone determines the local magnetic perturbation. This decomposition has proven useful in evaluating the relative contributions within the polar cap at different times of day, showing that during sunlit hours, the Hall current system dominates measurements made on the ground, whereas in darkness, the Birkeland current contribution can be measured using ground magnetometer data. This was attributed to conductivity in the polar cap approaching zero when the polar cap is in darkness [*Laundal et al.*, 2015, 2016].

Birkeland currents themselves vary with conductivity, relying primarily on solar illumination on the dayside and particle precipitation at premidnight magnetic local times [*Laundal et al.*, 2016]. It is possible to model the amount of Birkeland current flowing in the magnetosphere with a model that takes magnetic reconnection rates as input, combined with estimates of the ionospheric conductivity [*Milan*, 2013], which can in turn also be modeled using functions of the solar zenith angle [*Moen and Brekke*, 1993; *Robinson and Vondrak*, 1984]. Such a model has been constructed, and has been tested by using the dayside reconnection rate, averaged over a 27 day period. Then, the magnitudes of upward and downward current observed by AMPERE for each hemisphere can be calculated and summed, and the 27 day average of that quantity can be compared to the modeled current magnitudes [*Coxon et al.*, 2016a], showing excellent agreement between such a model and observations. This is consistent with the view that Birkeland currents are driven by the convection of the magnetosphere. However, as mentioned in section 16.3.2, the amount of current flowing was found to be asymmetrical between the Northern and Southern Hemispheres [*Coxon et al.*, 2016a]. The reasons for this remain open to question.

16.7. MODELING, ASSIMILATION, AND SIMULATION

Data from AMPERE have also been used to develop or test statistical models, as inputs to ionospheric electrodynamics assimilation, and in comparisons with physical simulations of the magnetosphere-ionosphere system. In this section, we will give an overview of the models that have been validated using AMPERE data, in order to demonstrate how useful this dataset is to those trying to model the electrodynamics of the ionosphere. The global distribution and continuous availability make these data and products particularly amenable to assimilation analyses. In addition, because the Birkeland current density is conserved from low to high altitudes, and because the current is a fundamental quantity evaluated within MHD, as distinct from auroral emission for example, the AMPERE results have proven to be particularly useful for testing results from MHD magnetospheric simulations.

16.7.1. Model Development and Validation

Various model results have been validated using AMPERE data over the course of its availability, and AMPERE data have also been used as the underpinning for empirical models [*Yue et al.*, 2015]. The first such comparison of AMPERE data to model results was conducted by *Papitashvili et al.* [2002], who constructed a model of the Birkeland current morphology with IMF direction and season using Ørsted and Magsat data. They found agreement with the results reviewed in section 16.3 and compared their model with Iridium TLM data, concluding that the two corresponded well. Subsequent work based on Iridium TLM data that decimated the intervals used for the statistics using only those for which the Birkeland current distributions were stable found that the southward and northward IMF distributions were much more distinct than were derived from statistical models using single satellite data that could not be as selective in the intervals included in the analysis [cf. *Anderson et al.*, 2008; *Korth et al.*, 2010].

The amount of open flux in the magnetosphere, also known as the magnetotail magnetic flux (MTF), can be modeled using spacecraft measurements of the magnetic field in the magnetotail in conjunction with solar wind measurements. This procedure has been performed using Cluster data in the magnetotail and solar wind data from OMNI to estimate the MTF in 2001 to 2009 using automated identifications of the different regions of the magnetotail [*Boakes et al.*, 2014] and also in October 2010 using visual identifications [*Shukhtina et al.*, 2016]. It would in principle be possible to extend this using more recent identifications of Cluster in the magnetotail lobes [*Coxon et al.*, 2016b]. MTF estimates were compared to IMAGE data in August/September 2005 and AMPERE data in October 2010, achieving correlations of 0.7.

An Automated Forward Modeling (AFM) technique has been developed that allows for the total current and the location of the auroral electrojet to be output, using a nonlinear optimization algorithm (based on the Levenberg-Marquardt method) with a model of auroral current systems and taking ground magnetometer data as input [*Connors and Rostoker*, 2015; *Connors et al.*, 2014, 2015]. This technique can be used globally (AGM), regionally (ARM), and meridionally (AMM). Results from the ARM technique indicated that three-dimensional current system parameters could be inferred from ground magnetometer data [*Connors et al.*, 2014], and AMPERE was used to identify intervals at which an auroral electrojet should be visible such that AMM could be compared to the meridional ionospheric currents estimated using the Spherical Elementary Current Systems (SECS) technique [*Pulkkinen et al.*, 2003], finding agreement within 10% [*Connors and Rostoker*, 2015]. The technique of gridded integration used by *Connors et al.* [2014] was used to identify the difference between the upward and downward field-aligned currents in each sector, which was inferred to be the meridional current; comparison of this AMPERE-derived quantity to the AMM technique showed reasonable (if unquantified) agreement between the two estimates [*Connors and Rostoker*, 2015].

16.7.2. Assimilation Studies

Assimilation analysis for ionospheric electrodynamics has an extensive history [*Knipp et al.*, 1989; *Richmond and Kamide*, 1988] and AMPERE data have proven to be a particularly useful input to the Assimilative Mapping of Ionospheric Electrodynamics (AMIE) algorithm [*Lu*, 2016; *Matsuo et al.*, 2015; *Wilder et al.*, 2012]. The AMIE procedure can ingest various datasets in addition to AMPERE and can yield spatial patterns that are more distinct and that lack small-scale, rapidly oscillating features that can occur in the spherical harmonic fitting due to disparities in the latitude and longitude resolution of the input data [*Matsuo et al.*, 2015]. The AMIE outputs relying solely on measurements taken from the ground grossly underestimate the cross-cap potential in comparison with satellite data [*Lu*, 2016; *Marsal et al.*, 2012]. Figure 16.5 shows results from *Wilder et al.* [2012] using AMIE for the 5 April 2010 storm at 00:30 UT. The panel on the left shows the electromagnetic energy flux derived from ground magnetometer data together with SuperDARN flows and DMSP satellite data. The panel on the left is for the same interval but after adding the AMPERE data. The magnitude and the distribution of the energy flux are dramatically different between the two. AMIE has also been used with AMPERE to show that during the recovery phase of a geomagnetic storm, when the IMF was northward, intense Joule heating associated with NBZ currents took place in the dayside polar regions [*Wilder et al.*, 2012]. Another method for assimilating satellite magnetic perturbations in estimating ionospheric properties is the Utah State University data assimilation model, which when using Iridium magnetic perturbations together with other measurements exhibits good agreement with SuperDARN returns [*Zhu et al.*, 2012].

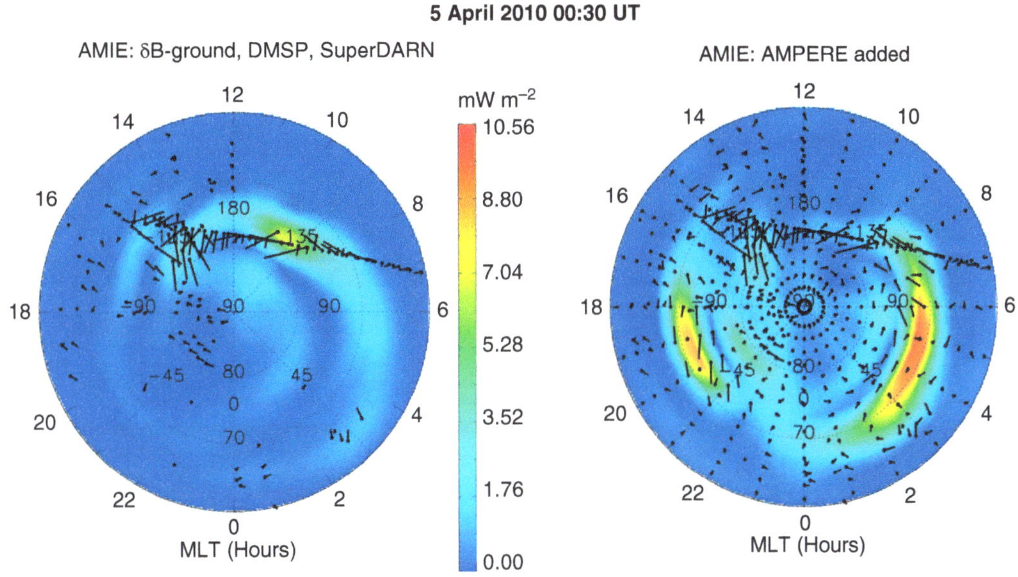

Figure 16.5 Results using AMIE of assimilative determination of ionospheric Joule heating derived for 00:30 UT on 5 April 2010 using ground magnetometer, SuperDARN, and DMSP data only (left) and when also using AMPERE results (right). The dramatic difference in distribution and intensity of the heating reflects the importance of including these additional data in such analysis. Adapted from *Wilder et al.* [2012].

Perhaps the most stringent test whether the electrodynamics using AMIE are valid is given by comparisons with storm-time thermospheric density dynamics. The AMIE procedure using AMPERE has been used as an input to the Thermosphere-Ionosphere (-Mesosphere)-Electrodynamics General Circulation Model, known as TI(M)EGCM. Initial results showed that TIEGCM was underestimating the induction currents in the Earth's surface, but improved the model's minimum timescale of variation from 6 hr to 10 min [*Marsal et al.*, 2012]; simulations of ground observations only improved modestly when the conductivities in the model were improved [*Marsal*, 2015]. Traveling atmospheric disturbances were modeled with TI(M)EGCM [*Lu et al.*, 2014], and comparison to data showed that neutral winds over the polar region were well reproduced. In addition, the densities at the GOCE and CHAMP satellites and those derived from the thermospheric simulation showed remarkably similar dynamics as shown in Figure 16.6 indicating that the assimilated electrodynamics closely reproduced the thermospheric response even though the model underestimated the neutral mass density and overestimated the storm recovery time somewhat [*Lu et al.*, 2014].

An even more ambitious assimilation challenge is to use data to adjust MHD simulations. An initial study of the feasibility was conducted by *Merkin et al.* [2016]. They found that rather than attempting to assimilate the currents, it was actually computationally more stable and hence more feasible to assimilate the low altitude magnetic perturbations as a boundary condition. By deriving the covariance matrix between the low altitude signals in the closed field region, attributed to R2 currents and the simulation, their initial analysis demonstrated that the technique could be used to reduce the discrepancy between the simulation and the observations. Future work to actually ingest the low altitude boundary into the simulation would require resolving potentially competing effects of the solar wind/IMF drivers for the simulation and the signals associated with the R1 currents. Given the close linkage between the R1 and R2 currents [cf. *Carter et al.*, 2016] allowing potentially conflicting drivers in the simulation assimilation scheme is a problem that has yet to be solved.

16.7.3. Simulation Comparisons

Magnetohydrodynamical models have been a subject of research in the community since the 1970s [e.g., *Merkin et al.*, 2016, and references therein], but for most of the history of the field, it has been difficult to compare their global predictions of Birkeland current to data. AMPERE allows for MHD modeling to be directly compared to global measurements of the Birkeland current systems. The Lyon-Fedder-Mobarry (LFM) model is an established MHD model, and has been well compared with Iridium and AMPERE data. High resolution simulations with this model generated R1 currents that exceeded results from the Iridium TLM derived field-aligned currents and that were attributed to underestimates in Iridium estimates of Birkeland current [*Korth*

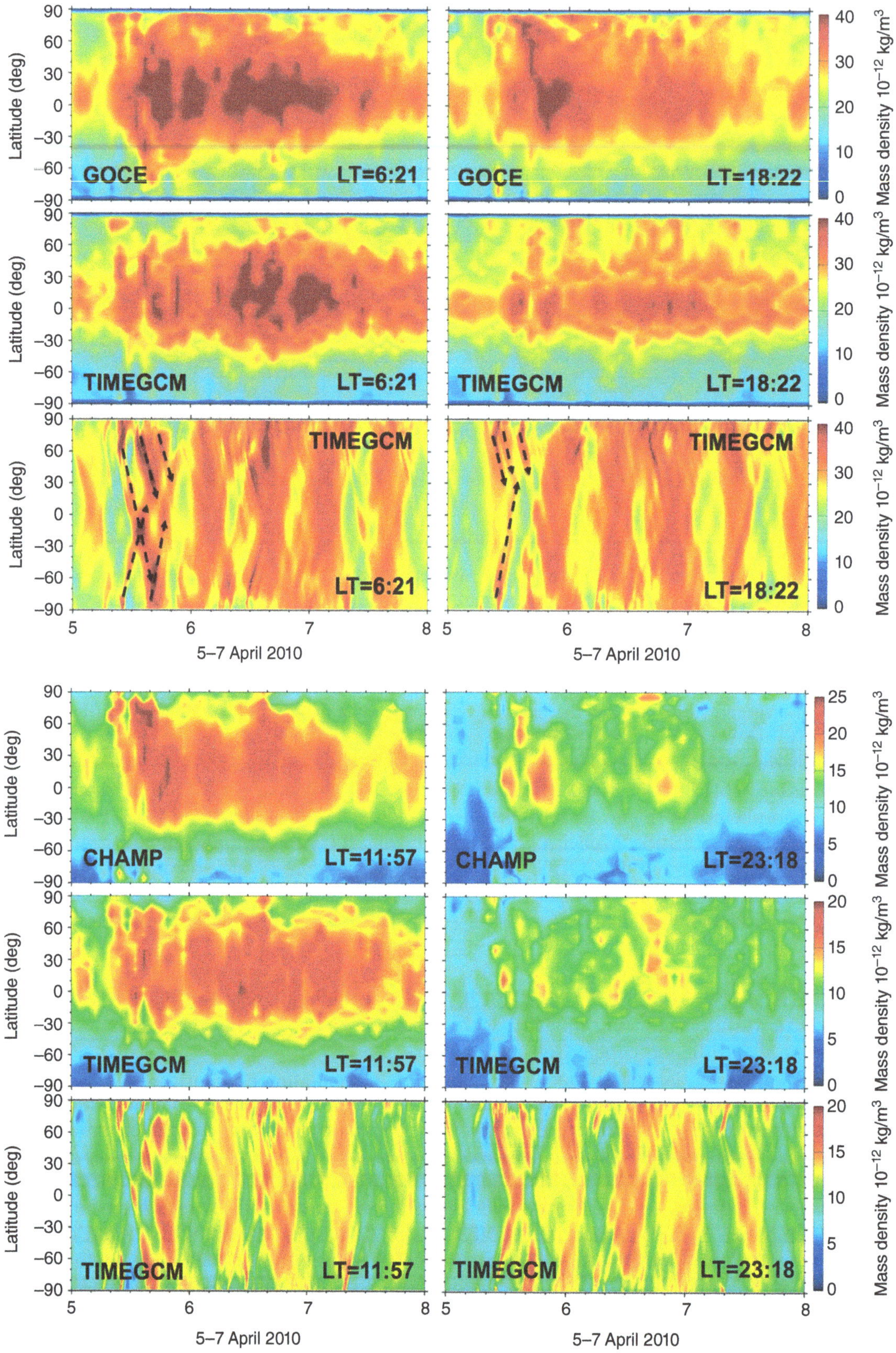

Figure 16.6 Thermospheric circulation results for the 5–7 April 2010 geomagnetic storm compared against neutral densities derived from accelerometer data from the GOCE and CHAMP satellites [*Lu et al.*, 2014]. Neutral density determinations from the GOCE and CHAMP satellites are shown in the top panels and reflect the elliptical orbits of the satellites. The neutral densities evaluated along the GOCE and CHAMP orbits from the thermosphere, ionosphere (mesosphere), exosphere, general circulation model, TI(M)EGCM, are shown in the middle panels. The TI(M)EGCM results were obtained using the AMIE procedure inputs for ionospheric electrodynamics using AMPERE data. Bottom panels show the neutral density at a fixed altitude clearly showing the traveling density disturbances between the polar regions. Adapted from *Lu et al.* [2014].

et al., 2004b, 2008b] and/or overestimates in the simulated electric field [*Korth et al.*, 2008b]. In the same comparisons, the simulated R2 currents and nightside currents underestimated the observations, indicating that the simulation ionospheric conductivity on the nightside was too low [*Korth et al.*, 2004b]. A similar problem was exhibited in the simulations run by *Korth et al.* [2008a], and both simulations underestimated the R2 current.

In further work, a case study of an event that demonstrated low R2 current and weak solar wind driving was conducted and the LFM model reproduced the currents derived from Iridium TLM data [*Merkin et al.*, 2007]. Another case study performing comparisons with AMPERE results and a higher resolution run of the LFM model for the 3–5 August 2010 geomagnetic storm found that the simulated magnetic perturbations well reproduced AMPERE observations except for some peaks in current density that exceeded the AMPERE measurements. The simulation yields filamentary currents with longitudinal extent smaller than the spacing between Iridium orbit tracks the discrepancy was attributed to an underestimate in the AMPERE inversions [*Merkin et al.*, 2013]. *Kleiber et al.* [2016] developed a framework to compare simulations to satellite products by automatically detecting R1 and R2 current regions and found that high-resolution simulations corresponded well to AMPERE observations in an example event.

A study of the Whole Heliosphere Interval (Carrington Rotation 2068) was conducted, in which the results of the LFM model were compared with AMPERE observations, both parameterized by eight IMF clock angles: this showed that the patterns agreed well, and that the ratio R1/R2 decreased as the resolution of the model increased [*Wiltberger et al.*, 2016], suggesting that the low R2 currents shown in previous simulations [*Korth et al.*, 2004b, 2008a] were due to underestimation of the R2 currents in the lower resolution simulations. A previous statistical study was conducted by *Korth et al.* [2011], comparing Iridium TLM results with the Space Weather Modeling Framework (SWMF) comprising a magnetosphere model coupled to the Rice Convection Model (RCM). This simulation showed that modeled currents and currents observed with AMPERE had a correlation of ~0.8, and demonstrated that increased solar wind electric field and pressure resulted in larger currents, with a larger electric field also being associated with an equatorward expansion of the current ovals, consistent with observations using AMPERE data (see section 16.3). In this work, the Iridium TLM data were heavily averaged and therefore were considered to considerably underestimate the true currents whereas the SWMF simulation was a specific run for IMF and solar wind conditions of the statistical binning used for the Iridium TLM data. The LFM model has also been combined with the RCM, in order to test how well magnetic perturbations from datasets like AMPERE can be assimilated into the LFM-RCM model, with the results indicating that such an approach is feasible for the future [*Merkin et al.*, 2016].

The SWMF was used in a case study of a geomagnetic storm, giving good agreement with currents derived from Iridium TLM observations and reproducing a rotation in R2 current toward noon with a rotation of the ring current toward the dusk during the storm main phase [*Zaharia et al.*, 2010]. The R2 current system had previously been modeled in the storm main phase, using ion and pressure distributions computed from the High Energy Neutral Atom (HENA) data from the IMAGE spacecraft, with magnetic perturbations found to be well modeled on the nightside, as opposed to results from MHD models [*Brandt et al.*, 2004]. Further research on the R2 current system was conducted using the Comprehensive Ring Current Model (CRCM), which reproduced the rotation in R2 current during geomagnetically active times and which reported that the distribution of R2 current in latitude and local time changed depending on whether polar cap potential, ionospheric conductivity, or plasma sheet density was enhanced [*Zheng et al.*, 2006].

A recent study comparing results of the three MHD simulations run for the Space Weather Prediction Center (SWPC) storm challenge events has also been done to critically assess the relative capabilities of different models to reflect the storm-time dynamics as represented by AMPERE results [*Anderson et al.*, 2017]. Two of the five SWPC challenge storms occurred after January 2010 and the AMPERE results for these storms, 4–5 April 2010 and 5–6 August 2011, showed that the total Birkeland current in the simulations could range from a factor of three below to a factor of three greater than the AMPERE results, with LFM and SWMF generally underestimating the current whereas the Open Geospace General Circulation Model (OpenGGCM) generally overestimated the current. Detailed comparison with the distributions of current revealed that, in general, the models and AMPERE differed substantially as indicated by linear regression between the radial current densities evaluated on the same latitude/longitude grids. The regression coefficients were generally between 0 and 0.5 and averaged around 0.25. The SWPC challenge simulations used low resolution codes and the results of *Merkin et al.* [2013] with the higher resolution codes demonstrate that the low resolution codes are not generally capable of reproducing the currents as determined by AMPERE.

16.8. THE FUTURE WITH AMPERE

AMPERE data have been used with SuperDARN data to look at global effects of solar wind and magnetosphere phenomena many times since Iridium data became available [e.g., *Waters et al.*, 2004], and further work combining these two datasets will give the community unparalleled views of the behavior of magnetosphere-ionosphere coupling, but these are not the only two global datasets that can be used in combination. The recent development of the SuperMAG dataset [*Gjerloev*, 2009, 2012] has already unlocked new science, both alone and in combination with AMPERE [e.g., *Laundal et al.*, 2015], and the new development of a global fit to SuperMAG data similar to that of the original work on AMPERE will enable science to be done with a complete view of the magnetic perturbations above and below the aurora [*Waters et al.*, 2015]. Furthermore, combinations of data both from the ground and above the aurora will lead to more reliable maps of the ionospheric electrodynamics, meaning that the last 6 yr of global Birkeland current maps may be the precursor to a new age of reliable, global views of the electrodynamic ionosphere [*Cousins et al.*, 2015b; *Lu*, 2016; *Matsuo et al.*, 2015].

Recent work with AMPERE has suggested that there may be an asymmetry between Birkeland current flow in the two hemispheres [*Coxon et al.*, 2016a], and work will be needed in the future both to better quantify this asymmetry and to explore phenomena related to Birkeland current in order to examine the origin of such an asymmetry [*Laundal et al.*, 2016]. Additionally, it is clear that the resolution of the current systems at work on Earth's nightside is a pressing concern for the community, with no real consensus on the morphology of the current systems in this area (as evidenced by section 16.4), and authors have noted that this should be a priority [*Liemohn et al.*, 2016]. Open questions abound concerning the nature of the exact morphology of the Birkeland currents [e.g., *Anderson et al.*, 2005; *Milan et al.*, 2015], the link between Birkeland current and aurora [*Carter et al.*, 2016], and the nature of Birkeland current saturation [*Anderson and Korth*, 2007].

Combining AMPERE data with other datasets, whether via AMIE or similar procedures, or using new techniques of inverting the electrodynamics of the high-latitude ionosphere, promises to be of great utility for a wide range of investigations of Earth's space environment. The availability of nearly continuous 24/7 data coverage offers unprecedented opportunities for coordination with ongoing missions, including Cluster, MMS, and Van Allen Probes making observations in the magnetosphere as well as TIMED and the DMSP satellites making complementary low-altitude observations. The future NASA missions ICON and GOLD missions target middle- and low-latitude thermospheric dynamics and require knowledge of the high-latitude drivers of the global ionosphere and thermosphere that AMPERE can play a key role in providing, and missions making measurements of currents on smaller scales will provide vital information at the other end of the scale. The future of AMPERE beyond the 2017 time frame rests with the Iridium NEXT constellation, which is the planned second generation of the Iridium Constellation. We expect it to yield higher quality input data and hence higher fidelity data products for an even wider array of scientific studies than those to which it has already been applied.

ACKNOWLEDGMENTS

John C. Coxon was funded on NERC grant NE/L007177/1 and STFC Ernest Rutherford grant ST/L002809/1, while Stephen E. Milan was funded on STFC consolidated grant ST/N000749/1. Support for AMPERE has been provided under NSF sponsorship under grants ATM-0739864 and AGS-1420184. We are greatly indebted to the contribution of Iridium Communications for providing the data for AMPERE. The AMPERE data used in this review and the papers cited herein are available at http://ampere.jhuapl.edu/. The authors would like to thank the AGU and the conveners of the AGU Chapman Conference on Currents in Geospace in Dubrovnik, Croatia, and John C. Coxon acknowledges the NSF for a travel grant that enabled him to attend.

REFERENCES

Akasofu, S. I., and C. I. Meng (1969), A study of polar magnetic substorms, *J. Geophys. Res.*, *74* (1), 293–313; doi:10.1029/JA074i001p00293.

Anderson, B., and H. Korth (2007), Saturation of global field aligned currents observed during storms by the Iridium satellite constellation, *J. Atmos. Solar Terr. Phys.*, *69* (1–2), 166–169; doi: 10.1016/j.jastp.2006.06.013.

Anderson, B. J., C. T. Russell, R. J. Strangeway, F. Plaschke, W. Magnes, D. Fischer, H. Korth, et al. (2016), Electrodynamic context of magnetopause dynamics observed by magnetospheric multiscale, *Geophys. Res. Lett.*, *43* (12), 5988–5996; doi: 10.1002/2016GL069577.

Anderson, B. J., H. Korth, C. L. Waters, D. L. Green, and P. Stauning (2008), Statistical Birkeland current distributions from magnetic field observations by the Iridium constellation, *Ann. Geophys.*, *26* (3), 671–687; doi:10.5194/angeo-26-671-2008.

Anderson, B. J., H. Korth, C. L. Waters, D. L. Green, V. G. Merkin, R. J. Barnes, and L. P. Dyrud (2014), Development of large-scale Birkeland currents determined from the Active Magnetosphere and Planetary Electrodynamics Response Experiment, *Geophys. Res. Lett.*, *41* (9), 3017–3025; doi:10.1002/2014GL059941.

Anderson, B. J., H. Korth, D. T. Welling, V. G. Merkin, M. J. Wiltberger, J. Raeder, R. J. Barnes, et al. (2017), Comparison of predictive estimates of high-latitude electrodynamics with observations of global-scale Birkeland currents, *Space Weather*; doi:10.1002/2016SW001529.

Anderson, B. J., K. Takahashi, and B. A. Toth (2000), Sensing global Birkeland currents with IridiumR engineering magnetometer data, *Geophys. Res. Lett.*, *27* (24), 4045–4048; doi:10.1029/2000GL000094.

Anderson, B. J., K. Takahashi, T. Kamei, C. L. Waters, and B. A. Toth (2002), Birkeland current system key parameters derived from Iridium observations: Method and initial validation results, *J. Geophys. Res. Space Physics*, *107* (A6), SMP 11-1–SMP 11-13; doi:10.1029/2001JA000080.

Anderson, B. J., S.-I. Ohtani, H. Korth, and A. Ukhorskiy (2005), Storm time dawn dusk asymmetry of the large-scale Birkeland currents, *J. Geophys. Res. Space Physics*, *110* (A12); doi:10.1029/2005JA011246.

Birkeland, K. (1908), *The Norwegian Aurora Polaris Expedition 1902–1903*, vol. *1*, H. Aschelhoug & Co., Christiania, Norway.

Birkeland, K. (1913), *The Norwegian Aurora Polaris Expedition 1902–1903*, vol. *2*, H. Aschelhoug & Co., Christiania, Norway.

Boakes, P. D., R. Nakamura, M. Volwerk, and S. E. Milan (2014), ECLAT Cluster spacecraft magnetotail plasma region identifications (2001–2009), *Dataset Papers in Science*, *2014* (11), 1–13.

Brandt, P., E. C. Roelof, S.-I. Ohtani, D. G. Mitchell, and B. J. Anderson (2004), IMAGE/HENA: Pressure and current distributions during the 1 October 2002 storm, *Adv. Space Res.*, *33* (5), 719–722; doi:https://doi.org/10.1016/S0273-1177(03)00633-1.

Burke, W. J., G. R. Wilson, C. S. Lin, F. J. Rich, J. O. Wise, and M. P. Hagan (2011), Estimating Dst indices and exospheric temperatures from equatorial magnetic fields mea sured by DMSP satellites, *J. Geophys. Res. Space Physics*, *116* (A1); doi:10.1029/2010JA015310, a01205.

Carter, J. A., S. E. Milan, J. C. Coxon, M.-T. Walach, and B. J. Anderson (2016), Average field-aligned current configuration parameterized by solar wind conditions, *J. Geophys. Res. Space Physics*, *121* (2), 1294–1307; doi:10.1002/2015JA021567.

Chisham, G., M. P. Freeman, G. A. Abel, W. A. Bristow, A. Marchaudon, J. M. Ruohoniemi, and G. J. Sofko (2009), Spatial distribution of average vorticity in the high-latitude ionosphere and its variation with interplanetary magnetic field direction and season, *J. Geophys. Res. Space Physics*, *114* (A9); doi: 10.1029/2009JA014263.

Clauer, C. R., and R. L. McPherron (1974), Mapping the local time-universal time development of magnetospheric substorms using mid-latitude magnetic observations, *J. Geophys. Res.*, *79* (19), 2811–2820; doi:10.1029/JA079i019p02811.

Clausen, L. B. N., J. B. H. Baker, J. M. Ruohoniemi, S. E. Milan, and B. J. Anderson (2012), Dynamics of the region 1 Birkeland current oval derived from the Active Magnetosphere and Planetary Electrodynamics Response Experiment (AMPERE), *J. Geophys. Res.*, *117*, A06,233.

Clausen, L. B. N., J. B. H. Baker, J. M. Ruohoniemi, S. E. Milan, J. C. Coxon, S. Wing, S. Ohtani, and B. J. Anderson (2013a), Temporal and spatial dynamics of the regions 1 and 2 Birkeland currents during substorms, *J. Geophys. Res. Space Physics*, *118* (6), 3007–3016; doi:10.1002/jgra.50288.

Clausen, L. B. N., J. I. Moen, K. Hosokawa, and J. M. Holmes (2016), GPS scintillations in the high latitudes during periods of dayside and nightside reconnection, *J. Geophys. Res. Space Physics*, *121* (4), 3293–3309; doi:10.1002/2015JA022199.

Clausen, L. B. N., S. E. Milan, J. B. H. Baker, J. M. Ruohoniemi, K.-H. Glassmeier, J. C. Coxon, and B. J. Anderson (2013b), On the influence of open magnetic flux on substorm intensity: Ground- and space-based observations, *J. Geophys. Res. Space Physics*, *118* (6), 2958–2969; doi:10.1002/jgra.50308.

Cnossen, I., and M. Förster (2016), North-south asymmetries in the polar thermosphere ionosphere system: Solar cycle and seasonal influences, *J. Geophys. Res. Space Physics*, *121* (1), 612–627; doi:10.1002/2015JA021750.

Connors, M., and G. Rostoker (2015), Inverting magnetic meridian data using nonlinear optimization, *Earth Planet Space*, *67* (1), 155; doi:10.1186/s40623-015-0315-y.

Connors, M., C. T. Russell, X. Chu, and R. L. McPherron (2015), The February 24, 2010 substorm: A refined view involving a pseudo breakup/expansive phase/poleward bound ary intensification sequence, *Earth Planet Space*, *67* (1), 195; doi:10.1186/s40623-015-0363-3.

Connors, M., R. L. McPherron, B. J. Anderson, H. Korth, C. T. Russell, and X. Chu (2014), Electric currents of a substorm current wedge on 24 February 2010, *Geophys. Res. Lett.*, *41* (13), 4449–4455; doi:10.1002/2014GL060604.

Cousins, E. D. P., T. Matsuo, A. D. Richmond, and B. J. Anderson (2015a), Dominant modes of variability in large-scale Birkeland currents, *J. Geophys. Res. Space Physics*; doi:10.1002/2014JA020462.

Cousins, E. D. P., T. Matsuo, and A. D. Richmond (2015b), Mapping high-latitude ionospheric electrodynamics with SuperDARN and AMPERE, *J. Geophys. Res. Space Physics*; doi:10.1002/2014JA020463.

Cowley, S. W. H. (2000), Magnetosphere-ionosphere interactions: A tutorial review, in *Magnetospheric Current Systems*, Geophysical Monograph Series, 118, edited by S.-I. Ohtani, R. Fujii, M. Hesse, and R. L. Lysak, 91–106, American Geophysical Union, Washington, DC, USA.

Cowley, S. W. H., and M. Lockwood (1992), Excitation and decay of solar wind-driven flows in the magnetosphere-ionosphere system, *Ann. Geophys.*, *10*, 103–115.

Coxon, J. C., C. M. Jackman, M. P. Freeman, C. Forsyth, and I. J. Rae (2016b), Identifying the magnetotail lobes with Cluster magnetometer data, *J. Geophys. Res. Space Physics*, *121* (2), 1436–1446; doi:10.1002/2015JA022020.

Coxon, J. C., I. J. Rae, C. Forsyth, C. M. Jackman, R. C. Fear, and B. J. Anderson (2017), Birkeland currents during substorms: Statistical evidence for a region 2 intensification after onset and a local reduction in density before onset, *J. Geophys. Res. Space Physics*; doi:10.1002/2017JA023967.

Coxon, J. C., S. E. Milan, J. A. Carter, L. B. N. Clausen, B. J. Anderson, and H. Korth (2016a), Seasonal and diurnal variations in AMPERE observations of the Birkeland currents compared to modeled results, *J. Geophys. Res. Space Physics*, *121* (5), 4027–4040; doi:10.1002/2015JA022050.

Coxon, J. C., S. E. Milan, L. B. N. Clausen, B. J. Anderson, and H. Korth (2014a), The magnitudes of the regions 1 and 2 Birkeland currents observed by AMPERE and their role in solar wind-magnetosphere-ionosphere coupling, *J. Geophys. Res. Space Physics*, *119* (12), 9804–9815; doi:10.1002/2014JA020138.

Coxon, J. C., S. E. Milan, L. B. N. Clausen, B. J. Anderson, and H. Korth (2014b), A superposed epoch analysis of the regions

1 and 2 Birkeland currents observed by AMPERE during substorms, *J. Geophys. Res. Space Physics*, *119* (12), 9834–9846; doi:10.1002/2014JA020500.

Eriksson, S., M. R. Hairston, F. J. Rich, H. Korth, Y. Zhang, and B. J. Anderson (2008), High-latitude ionosphere convection and Birkeland current response for the 15 May 2005 magnetic storm recovery phase, *J. Geophys. Res. Space Physics*, *113* (A3); doi:10.1029/2008JA013139,a00A08.

Förster, M., and S. Haaland (2015), Interhemispheric differences in ionospheric convection: Cluster EDI observations revisited, *J. Geophys. Res. Space Physics*, *120* (7), 5805–5823; doi:10.1002/2014JA020774.

Forsyth, C., A. N. Fazakerley, I. J. Rae, C. E. J. Watt, K. Murphy, J. A. Wild, T. Karlsson, et al. (2014), In situ spatiotemporal measurements of the detailed azimuthal substructure of the substorm current wedge, *J. Geophys. Res. Space Physics*, *119* (2), 927–946; doi:10.1002/2013JA019302.

Fukushima, N. (1969), Equivalence in ground geomagnetic effect of Chapman-Vestine's and Birkeland-Alfven's electric current-systems for polar magnetic storms, *Report of Ionosphere and Space Research in Japan*, *23*, 219–227.

Fukushima, N. (1976), Generalized theorem for no ground magnetic effect of vertical currents connected with Pedersen currents in the uniform-conductivity ionosphere, *Report of Ionosphere and Space Research in Japan*, *30* (1), 35–40.

Ganushkina, N. Y., M. W. Liemohn, S. Dubyagin, I. A. Daglis, I. Dandouras, D. L. De Zeeuw, Y. Ebihara, et al. (2015), Defining and resolving current systems in geospace, *Ann. Geophys.*, *33* (11), 1369–1402; doi:10.5194/angeo-33-1369-2015.

Gjerloev, J. W. (2009), A global ground-based magnetometer initiative, *Eos, Transactions American Geophysical Union*, *90* (27), 230–231; doi:10.1029/2009EO270002.

Gjerloev, J. W. (2012), The SuperMAG data processing technique, *J. Geophys. Res. Space Physics*, *117* (A9); doi:10.1029/2012JA017683.

Green, D. L., C. L. Waters, B. J. Anderson, and H. Korth (2009), Seasonal and inter915 planetary magnetic field dependence of the field-aligned currents for both northern and southern hemispheres, *Ann. Geophys.*, *27* (4), 1701–1715; doi:10.5194/angeo-27-1701-2009.

Green, D. L., C. L. Waters, B. J. Anderson, H. Korth, and R. J. Barnes (2006), Comparison of large-scale Birkeland currents determined from Iridium and SuperDARN data, *Ann. Geophys.*, *24* (3), 941–959; doi:10.5194/angeo-24-941-2006.

Green, D. L., C. L. Waters, H. Korth, B. J. Anderson, A. J. Ridley, and R. J. Barnes (2007), Technique: Large-scale ionospheric conductance estimated from combined satellite and ground-based electromagnetic data, *J. Geophys. Res. Space Physics*, *112* (A5); doi:10.1029/2006JA012069,a05303.

Iijima, T., and T. A. Potemra (1976), The amplitude distribution of field-aligned currents at northern high latitudes observed by Triad, *J. Geophys. Res.*, *81*, 2165–2174.

Iijima, T., and T. A. Potemra (1978), Large-scale characteristics of field-aligned currents associated with substorms, *J. Geophys. Res.*, *83*, 599–615.

Kleiber, W., B. Hendershott, S. R. Sain, and M. Wiltberger (2016), Feature-based validation of the Lyon-Fedder-Mobarry magnetohydrodynamical model, *J. Geophys. Res. Space Physics*, *121* (2), 1192–1200; doi:10.1002/2015JA021825.

Knipp, D. J., A. D. Richmond, G. Crowley, O. de la Beaujardire, E. Friis-Christensen, D. S. Evans, J. C. Foster, I. W. McCrea, F. J. Rich, and J. A. Waldock (1989), Electrodynamic patterns for September 19, 1984, *J. Geophys. Res. Space Physics*, *94* (A12), 16,913–16,923; doi:10.1029/JA094iA12p16913.

Knipp, D. J., T. Matsuo, L. Kilcommons, A. Richmond, B. Anderson, H. Korth, R. Redmon, B. Mero, and N. Parrish (2014), Comparison of magnetic perturbation data from LEO satellite constellations: Statistics of DMSP and AMPERE, *Space Weather*, *12* (1), 2–23; doi:10.1002/2013SW000987.

Korth, H., B. J. Anderson, and C. L. Waters (2010), Statistical analysis of the dependence of large-scale Birkeland currents on solar wind parameters, *Ann. Geophys.*, *28* (2), 515–530; doi:10.5194/angeo-28-515-2010.

Korth, H., B. J. Anderson, H. U. Frey, and C. L. Waters (2005), High-latitude electromagnetic and particle energy flux during an event with sustained strongly northward IMF, *Ann. Geophys.*, *23* (4), 1295–1310; doi:10.5194/angeo-23-1295-2005.

Korth, H., B. J. Anderson, H. U. Frey, T. J. Immel, and S. B. Mende (2004a), Conditions governing localized high-latitude dayside aurora, *Geophys. Res. Lett.*, *31* (4); doi:10.1029/2003GL018911,l04806.

Korth, H., B. J. Anderson, J. G. Lyon, and M. Wiltberger (2008b), Comparison of Birkeland current observations during two magnetic cloud events with mhd simulations, *Ann. Geophys.*, *26* (3), 499–516; doi:10.5194/angeo-26-499-2008.

Korth, H., B. J. Anderson, J. M. Ruohoniemi, H. U. Frey, C. L. Waters, T. J. Immel, and D. L. Green (2008a), Global observations of electromagnetic and particle energy flux for an event during northern winter with southward interplanetary magnetic field, *Ann. Geophys.*, *26* (6), 1415–1430; doi:10.5194/angeo-26-1415-2008.

Korth, H., B. J. Anderson, M. J. Wiltberger, J. G. Lyon, and P. C. Anderson (2004b), Intercomparison of ionospheric electrodynamics from the iridium constellation with global MHD simulations, *J. Geophys. Res. Space Physics*, *109* (A7); doi:10.1029/2004JA010428, a07307.

Korth, H., L. Rastätter, B. J. Anderson, and A. J. Ridley (2011), Comparison of the observed dependence of large-scale Birkeland currents on solar wind parameters with that obtained from global simulations, *Ann. Geophys.*, *29* (10), 1809–1826; doi: 10.5194/angeo-29-1809-2011.

Korth, H., Y. Zhang, B. J. Anderson, T. Sotirelis, and C. L. Waters (2014), Statistical relationship between large-scale upward field-aligned currents and electron precipitation, *J. Geophys. Res. Space Physics*, *119* (8), 6715–6731; doi: 10.1002/2014JA019961.

Laundal, K. M., C. C. Finlay, and N. Olsen (2016), Sunlight effects on the 3D polar current system determined from low Earth orbit measurements, *Earth Planet Space*, *68* (1), 142; doi:10.1186/s40623-016-0518-x.

Laundal, K. M., S. E. Haaland, N. Lehtinen, J. W. Gjerloev, N. Østgaard, P. Tenfjord, J. P. Reistad, et al. (2015), Birkeland current effects on high-latitude ground magnetic field perturbations, *Geophys. Res. Lett.*, *42* (18), 7248–7254; doi:10.1002/2015GL065776.

Le, G., H. Lühr, B. J. Anderson, R. J. Strangeway, C. T. Russell, H. Singer, J. A. Slavin, et al. (2016), Magnetopause erosion during the 17 March 2015 magnetic storm: Combined field-aligned currents, auroral oval, and magnetopause observations, *Geophys. Res. Lett.*, *43* (6), 2396–2404; doi:10.1002/2016GL068257.

Liemohn, M. W., N. Y. Ganushkina, R. M. Katus, D. L. De Zeeuw, and D. T. Welling (2013), The magnetospheric banana current, *J. Geophys. Res. Space Physics*, *118* (3), 1009–1021; doi:10.1002/jgra.50153.

Liemohn, M. W., N. Y. Ganushkina, R. Ilie, and D. T. Welling (2016), Challenges associated with near-Earth nightside current, *J. Geophys. Res. Space Physics*, *121* (7), 6763–6768; doi:10.1002/2016JA022948.

Liou, K., P. T. Newell, B. J. Anderson, L. Zanetti, and C.-I. Meng (2005), Neutral composition effects on ionospheric storms at middle and low latitudes, *J. Geophys. Res. Space Physics*, *110* (A5); doi:10.1029/2004JA010840, a05309.

Lu, G. (2016), Large scale high-latitude ionospheric electrodynamic fields and currents, *Space Sci. Rev.*, 1–20; doi:10.1007/s11214-016-0269-9.

Lu, G., M. E. Hagan, K. Husler, E. Doornbos, S. Bruinsma, B. J. Anderson, and H. Korth (2014), Global ionospheric and thermospheric response to the 5 April 2010 geomagnetic storm: An integrated data-model investigation, *J. Geophys. Res. Space Physics*, *119* (12), 10,358–10,375; doi:10.1002/2014JA020555.

Lyons, L. R., B. Gallardo-Lacourt, S. Zou, J. M. Weygand, Y. Nishimura, W. Li, M. Gkioulidou, et al. (2016), The 17 March 2013 storm: Synergy of observations related to electric field modes and their ionospheric and magnetospheric effects, *J. Geophys. Res. Space Physics*, *121* (11), 10,880–10,897; doi: 10.1002/2016JA023237.

Marsal, S. (2015), Conductivities consistent with Birkeland currents in the ampere-driven tie-gcm, *J. Geophys. Res. Space Physics*, *120* (9), 8045–8065; doi: 10.1002/2015JA021385.

Marsal, S., A. D. Richmond, A. Maute, and B. J. Anderson (2012), Forcing the tie-gcm model with Birkeland currents from the active magnetosphere and planetary electrodynamics response experiment, *J. Geophys. Res. Space Physics*, *117* (A6); doi:10.1029/2011JA017416.

Matsui, H., P. J. Erickson, J. C. Foster, R. B. Torbert, M. R. Argall, B. J. Anderson, J. B. Blake, et al. (2016), Dipolarization in the inner magnetosphere during a geomagnetic storm on 7 October 2015, *Geophys. Res. Lett.*, *43* (18), 9397–9405; doi:10.1002/2016GL070677.

Matsuo, T., D. J. Knipp, A. D. Richmond, L. Kilcommons, and B. J. Anderson (2015), Inverse procedure for high-latitude ionospheric electrodynamics: Analysis of satellite borne magnetometer data, *J. Geophys. Res. Space Physics*, *120* (6), 5241–5251; doi:10.1002/2014JA020565.

McPherron, R. L., C. T. Russell, and M. P. Aubry (1973), Satellite studies of magnetospheric substorms on August 15, 1968: 9. phenomenological model for substorms, *J. Geophys. Res.*, *78* (16), 3131–3149; doi:10.1029/JA078i016p03131.

Merkin, V. G., B. J. Anderson, J. G. Lyon, H. Korth, M. Wiltberger, and T. Motoba (2013), Global evolution of Birkeland currents on 10 min timescales: Mhd simulations and observations, *J. Geophys. Res. Space Physics*, *118* (8), 4977–4997; doi:10.1002/jgra.50466.

Merkin, V. G., D. Kondrashov, M. Ghil, and B. J. Anderson (2016), Data assimilation of low-altitude magnetic perturbations into a global magnetosphere model, *Space Weather*, *14* (2), 165–184; doi:10.1002/2015SW001330.

Merkin, V. G., J. G. Lyon, B. J. Anderson, H. Korth, C. C. Goodrich, and K. Papadopoulos (2007), A global MHD simulation of an event with a quasi-steady northward IMF component, *Ann. Geophys.*, *25* (6), 1345–1358; doi:10.5194/angeo-25-1345-2007.

Milan, S. E. (2013), Modeling Birkeland currents in the expanding/contracting polar cap paradigm, *J. Geophys. Res. Space Physics*, *118* (9), 5532–5542; doi: 10.1002/jgra.50393.

Milan, S. E., A. Grocott, C. Forsyth, S. M. Imber, P. D. Boakes, and B. Hubert (2009), A superposed epoch analysis of auroral evolution during substorm growth, onset and recovery: Open magnetic flux control of substorm intensity, *Ann. Geophys.*, *27* (2), 659–668; doi:10.5194/angeo-27-659-2009.

Milan, S. E., J. A. Carter, H. Korth, and B. J. Anderson (2015), Principal component analysis of Birkeland currents determined by the Active Magnetosphere and Planetary Electrodynamics Response Experiment, *J. Geophys. Res. Space Physics*, *120* (12), 10,415–10,424; doi:10.1002/2015JA021680.

Milan, S. E., L. B. N. Clausen, J. C. Coxon, J. A. Carter, M.-T. Walach, K. Laundal, N. Østgaard, et al. (2017), Overview of solar wind-magnetosphere-ionosphere-atmosphere coupling and the generation of magnetospheric currents, *Space Sci. Rev.*, *206* (1), 547–573; doi:10.1007/s11214-017-0333-0.

Moen, J., and A. Brekke (1993), The solar flux influence on quiet time conductances in the auroral ionosphere, *Geophys. Res. Lett.*, *20* (10), 971–974; doi: 10.1029/92GL02109.

Motoba, T., S. Ohtani, B. J. Anderson, H. Korth, D. Mitchell, L. J. Lanzerotti, K. Shiokawa, M. Connors, C. A. Kletzing, and G. D. Reeves (2015), On the formation and ori1051 gin of substorm growth phase/onset auroral arcs inferred from conjugate space-ground observations, *J. Geophys. Res. Space Physics*, *120* (10), 8707–8722; doi: 10.1002/2015JA021676.

Murphy, K. R., I. R. Mann, I. J. Rae, C. L. Waters, B. J. Anderson, D. K. Milling, H. J. Singer, and H. Korth (2012), Reduction in field-aligned currents preceding and local to auroral substorm onset, *Geophys. Res. Lett.*, *39* (15); doi: 10.1029/2012GL052798.

Murphy, K. R., I. R. Man, I. Jonathan Rae, C. L. Waters, H. U. Frey, A. Kale, H. J. Singer, B. J. Anderson, and H. Korth (2013), The detailed spatial structure of field aligned currents comprising the substorm current wedge, *J. Geophys. Res. Space Physics*; doi:10.1002/2013JA018979.

Nakamura, R., V. A. Sergeev, W. Baumjohann, F. Plaschke, W. Magnes, D. Fischer, A. Varsani, et al. (2016), Transient, small-scale field-aligned currents in the plasma sheet boundary layer during storm time substorms, *Geophys. Res. Lett.*, *43* (10), 4841–4849; doi: 10.1002/2016GL068768.

Papitashvili, V. O., F. Christiansen, and T. Neubert (2002), A new model of field-aligned currents derived from high-precision satellite magnetic field data, *Geophys. Res. Lett.*, *29* (14), 28-1–28-4; doi:10.1029/2001GL014207.

Pellinen, R. J., and W. J. Heikkila (1978), Observations of auroral fading before breakup, *J. Geophys. Res. Space Physics*, *83* (A9), 4207–4217; doi: 10.1029/JA083iA09p04207.

Pulkkinen, A., O. Amm, and A. Viljanen (2003), Ionospheric equivalent current distributions determined with the method of spherical elementary current systems, *J. Geophys. Res. Space Physics*, *108* (A2); doi:10.1029/2001JA005085, 1053.

Reistad, J. P., N. Østgaard, P. Tenfjord, K. M. Laundal, K. Snekvik, S. Haaland, S. E. Milan, K. Oksavik, H. U. Frey, and A. Grocott (2016), Dynamic effects of restoring footpoint symmetry on closed magnetic field lines, *J. Geophys. Res. Space Physics*, *121* (5), 3963–3977; doi:10.1002/2015JA022058.

Richmond, A. D., and Y. Kamide (1988), Mapping electrodynamic features of the high latitude ionosphere from localized observations: Technique, *J. Geophys. Res. Space Physics*, *93* (A6), 5741–5759; doi:10.1029/JA093iA06p05741.

Ritter, P., and H. Lühr (2006), Curl-B technique applied to Swarm constellation for determining field-aligned currents, *Earth Planet Space*, *58* (4), 463–476; doi: 10.1186/BF03351942.

Robinson, R. M., and R. R. Vondrak (1984), Measurements of E region ionization and conductivity produced by solar illumination at high latitudes, *J. Geophys. Res. Space Physics*, *89* (A6), 3951–3956; doi:10.1029/JA089iA06p03951.

Sergeev, V. A., A. V. Nikolaev, M. V. Kubyshkina, N. A. Tsyganenko, H. J. Singer, J. V. Rodriguez, V. Angelopoulos, et al. (2014a), Event study combining magnetospheric and ionospheric perspectives of the substorm current wedge modeling, *J. Geophys. Res. Space Physics*, *119* (12), 9714–9728; doi:10.1002/2014JA020522.

Sergeev, V. A., A. V. Nikolaev, N. A. Tsyganenko, V. Angelopoulos, A. V. Runov, H. J. Singer, and J. Yang (2014b), Testing a two-loop pattern of the substorm current wedge (SCW2L), *J. Geophys. Res. Space Physics*, *119* (2), 947–963; doi:10.1002/2013JA019629.

Sergeev, V. A., N. A. Tsyganenko, M. V. Smirnov, A. V. Nikolaev, H. J. Singer, and W. Baumjohann (2011), Magnetic effects of the substorm current wedge in a spread-out wire model and their comparison with ground, geosynchronous, and tail lobe data, *J. Geophys. Res. Space Physics*, *116* (A7); doi:10.1029/2011JA016471.

Shukhtina, M. A., E. I. Gordeev, V. A. Sergeev, N. A. Tsyganenko, L. B. N. Clausen, and S. E. Milan (2016), Magnetotail magnetic flux monitoring based on simultaneous solar wind and magnetotail observations, *J. Geophys. Res. Space Physics*, *121* (9), 8821–8839; doi:10.1002/2016JA022911.

Siscoe, G. L., and T. S. Huang (1985), Polar cap ination and deation, *J. Geophys. Res. Space Physics*, *90* (A1), 543–547; doi:10.1029/JA090iA01p00543.

Sitnov, M. I., N. A. Tsyganenko, A. Y. Ukhorskiy, B. J. Anderson, H. Korth, A. T. Y. Lui, and P. C. Brandt (2010), Empirical modeling of a CIR-driven magnetic storm, *J. Geophys. Res. Space Physics*, *115* (A7); doi:10.1029/2009JA015169, a07231.

Stephens, G. K., M. I. Sitnov, A. Y. Ukhorskiy, E. C. Roelof, N. A. Tsyganenko, and G. Le (2016), Empirical modeling of the storm time innermost magnetosphere using Van Allen probes and THEMIS data: Eastward and banana currents, *J. Geophys. Res. Space Physics*, *121* (1), 157–170; doi:10.1002/2015JA021700.

Stephens, G. K., M. I. Sitnov, J. Kissinger, N. A. Tsyganenko, R. L. McPherron, H. Korth, and B. J. Anderson (2013), Empirical reconstruction of storm time steady magnetospheric convection events, *J. Geophys. Res. Space Physics*, *118* (10), 6434–6456; doi:10.1002/jgra.50592.

Tenfjord, P., N. Østgaard, K. Snekvik, K. M. Laundal, J. P. Reistad, S. Haaland, and S. E. Milan (2015), How the IMF BY induces a BY component in the closed magnetosphere and how it leads to asymmetric currents and convection patterns in the two hemispheres, *J. Geophys. Res. Space Physics*, *120* (11), 9368–9384; doi:10.1002/2015JA021579.

Waters, C. L., B. J. Anderson, and K. Liou (2001), Estimation of global field-aligned currents using the Iridium R System magnetometer data, *Geophys. Res. Lett.*, *28* (11), 2165–2168; doi:10.1029/2000GL012725.

Waters, C. L., B. J. Anderson, R. A. Greenwald, R. J. Barnes, and J. M. Ruohoniemi (2004), High-latitude Poynting flux from combined Iridium and SuperDARN data, *Ann. Geophys.*, *22* (8), 2861–2875; doi:10.5194/angeo-22-2861-2004.

Waters, C. L., J. W. Gjerloev, M. Dupont, and R. J. Barnes (2015), Global maps of ground magnetometer data, *J. Geophys. Res. Space Physics*, *120* (11), 9651–9660; doi:10.1002/2015JA021596.

Wilder, F., D. G. Crowley, B. J. Anderson, and A. D. Richmond (2012), Intense dayside Joule heating during the 5 April 2010 geomagnetic storm recovery phase observed by AMIE and AMPERE, *J. Geophys. Res. Space Physics*, *117* (A5); doi: 10.1029/2011JA017262.

Wilder, F. D., S. Eriksson, H. Korth, J. B. H. Baker, M. R. Hairston, C. Heinselman, and B. J. Anderson (2013), Field-aligned current reconfiguration and magnetospheric response to an impulse in the interplanetary magnetic field BY component, *Geophys. Res. Lett.*, *40* (11), 2489–2494; doi:10.1002/grl.50505.

Wiltberger, M., E. J. Rigler, V. Merkin, and J. G. Lyon (2016), Structure of high latitude currents in magnetosphere-ionosphere models, *Space Sci. Rev.*, 1–24; doi: 10.1007/s11214-016-0271-2.

Yue, C., C.-P. Wang, Y. Nishimura, K. R. Murphy, X. Xing, L. Lyons, M. Henderson, V. Angelopoulos, A. T. Y. Lui, and T. Nagai (2015), Empirical modeling of 3d force balanced plasma and magnetic field structures during substorm growth phase, *J. Geophys. Res. Space Physics*, *120* (8), 6496–6513; doi:10.1002/2015JA021226.

Zaharia, S., V. K. Jordanova, D. Welling, and G. Tóth (2010), Self-consistent inner magnetosphere simulation driven by a global mhd model, *J. Geophys. Res. Space Physics*, *115* (A12); doi:10.1029/2010JA015915, a12228.

Zheng, Y., A. T. Y. Lui, M.-C. Fok, B. J. Anderson, P. C. Brandt, T. J. Immel, and D. G. Mitchell (2006), Relationship between Region 2 field-aligned current and the ring current: Model results, *J. Geophys. Res. Space Physics*, *111* (A11); doi:10.1029/2006JA011603.

Zhu, L., R. Schunk, L. Scherliess, and V. Eccles (2012), Importance of data assimilation technique in defining the model drivers for the space weather specification of the high latitude ionosphere, *Radio Science*, *47* (4); doi:10.1029/2011RS004936,rS0L24.

Zmuda, A. J., J. H. Martin, and F. T. Heuring (1966), Transverse magnetic disturbances at 1100 kilometers in the auroral region, *J. Geophys. Res.*, *71*, 5033–5045.

Zou, S., M. B. Moldwin, A. J. Ridley, M. J. Nicolls, A. J. Coster, E. G. Thomas, and J. M. Ruohoniemi (2014), On the generation/decay of the storm-enhanced density plumes: Role of the convection flow and field-aligned ion flow, *J. Geophys. Res. Space Physics*, *119* (10), 8543–8559; doi:10.1002/2014JA020408.

17

Birkeland Currents at Mercury: Review and Comparison With Earth

Brian J. Anderson[1], Catherine L. Johnson[2,3], Haje Korth[1], and Lydia C. Philpott[2]

ABSTRACT

Birkeland currents are present at both Earth and Mercury, and differences in these currents reflect fundamental differences in the convection dynamics of these systems. At Earth, the conductive ionosphere and the high altitude magnetosphere are linked with these currents through the Pedersen conductance, which closes the currents in the ionosphere, and through magnetospheric plasma, which carries the Region 2 currents between the evening and morning sectors while shielding the inner magnetosphere from the convection potential. The currents reflect the convection of magnetic flux from the dayside, over the polar caps into the magnetotail, and subsequently sunward in the tail to the dayside. At Mercury, the analysis indicates that the Birkeland currents close through the planet itself with at least half of the current closing through the core. There is no evidence of Region 2 currents at Mercury, implying that the magnetosphere and planet itself are not shielded from the convection cycle and suggesting that the planet also participates in the return convection although it is not known how this is actually accomplished. Future exploration of Mercury making measurements of the plasma flow and electric fields may resolve some of the basic questions about how the convection cycle closes at Mercury.

17.1. INTRODUCTION

Magnetic field signatures at high latitudes attributed to quasi-steady field-aligned currents were recently reported in observations obtained in orbit at Mercury [*Anderson et al.*, 2014a], making Mercury the only planet in our solar system with a global magnetosphere and Birkeland currents but without an ionosphere. The Birkeland current system at Earth is well studied [e.g., *Coxon et al.*, Chapter 16, this volume], and its configuration and dynamics yield a wealth of insight into the physics of Earth's magnetosphere and coupling electrodynamics with the ionosphere [cf. *Cowley*, 2000; *Richmond and Thayer*, 2000]. Their observation at Mercury was surprising because the absence of a permanent atmosphere implied the lack of a highly electrically conducting ionosphere that would facilitate their closure [*Russell et al.*, 1988; *Slavin et al.*, 2007]. It is therefore worthwhile to examine the ways in which the Birkeland currents at Earth and Mercury differ and to consider the implications for corresponding basic differences in the dynamics and configurations of the magnetospheres and coupling to the planet or ionosphere.

To illustrate how the Birkeland currents relate to differences in these two magnetospheres, Figure 17.1 shows a schematic illustrating the relationship of field-aligned currents to the magnetosphere and the planet. The figure shows a view of the dawn-dusk meridian plane and three representative open magnetic field lines (dark blue), that cross the magnetopause boundary and two closed equatorward field lines that thread magnetospheric plasma (if present). Because particle motions are not significantly impeded parallel to the magnetic field, the motional potential imposed on the open magnetic field lines as a

[1] *Applied Physics Laboratory, Johns Hopkins University, Laurel, Maryland, USA*
[2] *Department of Earth, Ocean and Atmospheric Sciences, University of British Columbia, Vancouver, BC, Canada*
[3] *Planetary Science Institute, Tucson, Arizona, USA*

Electric Currents in Geospace and Beyond, Geophysical Monograph 235, First Edition.
Edited by Andreas Keiling, Octav Marghitu, and Michael Wheatland.

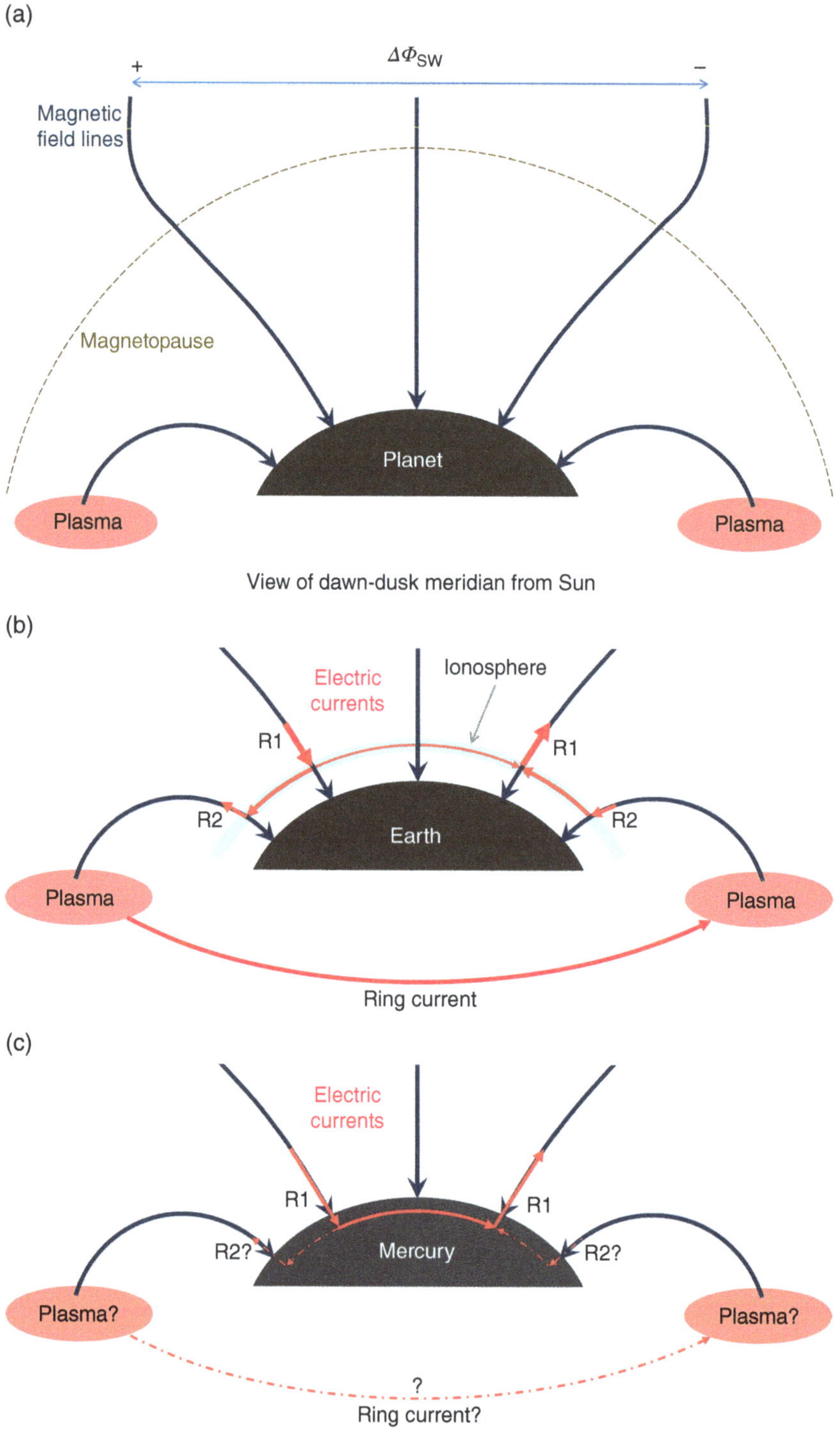

Figure 17.1 Schematic of a magnetized planet's electrodynamic connection to the solar wind motional dynamo and net potential. The potential maps along magnetic lines of force to the planet. (a) The magnetopause and planetary magnetic field in the northern polar regions in the dawn-dusk meridian plane. Views showing electric currents (red) (b) at Earth and (c) at Mercury within their respective magnetospheres. Birkeland currents are shown as currents parallel to the magnetic lines of force, and closure currents are shown flowing in the ionosphere at Earth and within the planet at Mercury. The Region 1 and Region 2 currents are denoted as R1 and R2, respectively.

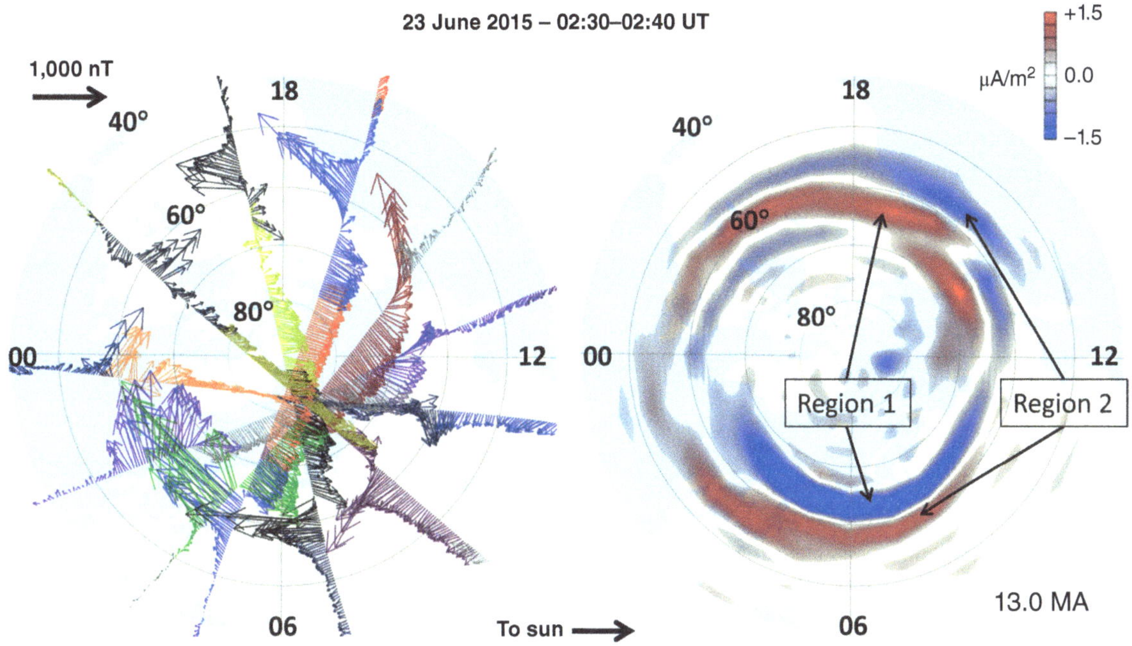

Figure 17.2 Example determination of Birkeland currents at Earth from the Active Magnetosphere and Planetary Electrodynamics Response Experiment (AMPERE). View is from above the Earth's north magnetic pole and registered in magnetic latitude and local time. Left panel shows magnetic perturbations perpendicular to the radial direction, $\delta\mathbf{B}_{\perp}$, for a 10 min interval from 02:30 to 02:40 UT on 23 June 2015 derived from magnetometer data acquired from the Iridium Communications constellation. Vectors show $\delta\mathbf{B}_{\perp}$ starting at the location point of the measurement plotted in the direction of $\delta\mathbf{B}_{\perp}$ with the length scaled as shown by the 1000 nT reference vector. Data from different satellites are plotted with different colors. For this interval, data were acquired at 8 s intervals from all satellites. Right panel shows radial current densities derived from the $\delta\mathbf{B}_{\perp}$ data via spherical harmonic fit to the $\delta\mathbf{B}_{\perp}$ data and Ampére's law to derive J_r. Radial current is shown by the color bar with upward current in red and downward current in blue. For consistency with displays of MESSENGER results, the direction to the Sun is to the right with dusk at the top and midnight at the left. The Region 1 (poleward) and Region 2 (equatorward) currents are indicated in the right panel.

result of their connection to the solar wind by magnetic reconnection on the dayside magnetopause, is mapped along the lines of force to low altitudes (Fig. 17.1a). For both Earth and Mercury, the planetary dipole moment is directed southward so that in the north polar region, the magnetic field is pointed radially inward, southward, as shown. In the south polar region, the field is directed radially outward, that is, still southward. Views of the currents near the planet for Earth (Fig. 17.1b) and Mercury (Fig. 17.1c) illustrate the electric currents (red arrows) that flow at low altitudes resulting from the imposed electric potential. For both Earth and Mercury, currents flow down to the planet from the most positive potential and upward on field lines linked to the most negative potential. These correspond to the higher latitude currents termed Region 1 [cf. *Iijima and Potemra*, 1976; *Korth et al.*, 2010]. At Earth, the ionosphere (gray layer), embedded within Earth's thermosphere, provides substantial electrical conductivity across the magnetic field, and closes the downward Birkeland current at dawn with the upward current at dusk. Some of this current closes across the polar cap (thin red arrow) but much of this current closes equatorward (thicker red lines in the ionosphere) to the opposite polarity Birkeland currents that flow to plasma within the magnetosphere, termed the Region 2 currents [cf. *Iijima and Potemra*, 1976; *Brandt et al.*, 2004]. At Earth, this plasma supports a ring current that flows azimuthally in the near equatorial magnetosphere without electrical dissipation so that the net resistance to closure of the Region 1 currents through Region 2 is lower than it is over the polar cap. Hence, more current flows through the Region 2 system than flows over the polar cap [cf. *Richmond and Thayer*, 2000]. The net effect is that the closure of Region 1 to Region 2 in the ionosphere is a lower resistance path than closure of Region 1 over the polar cap.

To illustrate the distribution of the Region 1 and Region 2 currents at Earth, Figure 17.2 shows an example of observed horizontal magnetic perturbations from which the background main field has been subtracted,

$\delta\mathbf{B}_{\perp}$, on the left and corresponding Birkeland currents on the right at Earth from the Active Magnetosphere and Planetary Electrodynamics Response Experiment (AMPERE) for a 10 min interval from 02:30 to 02:40 UT on 23 June 2015 during a geomagnetic storm. AMPERE uses avionics magnetometer data from the constellation of 66 Iridium Communications satellites to obtain global, continuous sampling of Earth's magnetic field at 780 km in six near-polar orbit planes that are equally spaced in longitude [cf. *Anderson et al.*, 2000, 2014b; *Claussen et al.*, 2012]. With 11 satellites in each orbit plane, the temporal spacing between satellites is 9 min, and data are processed using 10 min data windows to yield estimates for the radial current through the orbit altitude surface (see details of processing approach in *Waters et al.*, [2001] and *Anderson et al.* [2014b]). The right-hand panel gives the radial current density, J_r, derived from a spherical harmonic fit to the data on the left yielding $\delta\mathbf{B}_{\perp\mathrm{Fit}}$, and the curl of $\delta\mathbf{B}_{\perp\mathrm{Fit}}$ to obtain the radial current from Ampère's law, $\mu_0 J_r = (\nabla \times \delta\mathbf{B}_{\perp\mathrm{Fit}}) \cdot \hat{\mathbf{r}}$, where μ_0 is the permeability of free space and $\hat{\mathbf{r}}$ is the radial unit vector [*Waters et al.*, 2001]. This interval prominently displays the Region 1 and Region 2 currents originally identified statistically [*Iijima and Potemra*, 1976; *Papitashvilli et al.*, 2002; *Weimer*, 2005; *Anderson et al.*, 2008; *Korth et al.*, 2010]. The largest magnetic perturbations are westward centered near 60° magnetic latitude (MLAT) in the dawn sector and eastward centered near 55° MLAT in the dusk sector and span about 5° MLAT. The latitudes of the currents during the storm are roughly 10° farther equatorward than during moderate geomagnetic activity, but the distributions are essentially the same [e.g., *Anderson et al.*, 2017]. The equatorward gradients of these features correspond to a downward current in the dusk sector and an upward current in the dawn sector and are the Region 2 currents. The magnetic signals over the pole are generally sunward and relatively uniform with a magnitude about one-third of the maximum signals at dawn and dusk. The gradients between the dawn and dusk perturbations and those over the pole correspond to upward currents at dusk and downward currents at dawn and are the Region 1 currents. The Region 2 currents close without dissipation in the high altitude hot magnetospheric plasmas. As a result, the ionospheric resistance for closure of the Region 1 currents is lower via the short path to the Region 2 currents than via the longer path over polar cap. Hence, most of the total ionospheric closure current flows from Region 1 to Region 2 via Pedersen currents.

At Mercury, there are two major departures from the situation at Earth. First, Mercury's transient exosphere is so tenuous that it is not collisional and thus does not facilitate any substantial cross-field motion of charged particles so the conductivity perpendicular to the field is extremely low. Hence, the closure of the Region 1 current must occur either at the surface [*Cheng et al.*, 1987] or within the planet itself. For this reason, it had been thought that perhaps only transient field-aligned currents, carried by Alfvén waves, would occur at Mercury [*Slavin et al.*, 1997; *Glassmeier*, 1997; *Lyatsky et al.*, 2010]. A second major difference at Mercury is that the magnetosphere is so much smaller relative to the planet itself [cf. *Russell et al.*, 1988; *Slavin et al.*, 2007] that plasma convection/drift paths, which at Earth close on themselves around the planet within the magnetosphere, are not supported within Mercury's equatorial magnetosphere [*Glassmeier*, 2000]. Plasma that could support a steady-state ring current is not present at Mercury [*Slavin et al.*, 2007; *Korth et al.*, 2014] and, hence, there is no mechanism to close current in the magnetospheric plasma. The existence of a Region 2 current system is problematic at Mercury. Indeed, the presence or absence of a Region 2 polarity Birkeland current serves as indirect evidence for or against a ring current at Mercury. Because of this uncertainty, the Region 2 system, closure to Region 1, and closure within the magnetosphere are all shown in Figure 17.1c as dashed red arrows with question marks. The presence or absence of Region 2 currents is significant with respect to both the equatorial magnetospheric plasma and the convection path for magnetic flux within the magnetosphere from the nightside magnetosphere, through the meridian plane shown in Figure 17.1, to the dayside to supply the sunward magnetospheric flows required by dayside magnetic reconnection. The relationships between Birkeland currents and magnetospheric convection are discussed in more detail below.

In section 17.2, we present a brief overview of the theoretical concepts of Birkeland current sources and their relationships to magnetospheric convection. The present state of understanding of steady-state Birkeland currents at Mercury is then reviewed in section 17.3. Implications of the differences between the currents at Mercury and Earth are discussed in section 17.4, and prospects for future exploration of currents and convection are covered in section 17.5.

17.2. PHYSICS OF BIRKELAND CURRENTS: SOURCES AND CONVECTION

Field-aligned currents in plasmas are intimately linked to the distribution, pressures, and motions of plasma where the currents are generated, and the spatial distributions of the Birkeland currents are therefore closely related to the convection dynamics of the magnetosphere. To understand the implications of the Birkeland current distributions and polarities for magnetospheric dynamics, we present a brief overview of the theory of sources of field-aligned currents in space plasmas and of the relationship to convection where the Birkeland currents close at low altitudes.

17.2.1. Field-Aligned Current Generation in Space Plasmas

Formally, the origins of Birkeland currents in space plasmas can be described in terms of the field-aligned integral of the divergence of perpendicular plasma current [*Vasyliunas*, 1970; *Strangeway*, 2012], which in turn is determined by plasma flows and pressures. It is convenient to write $\mathbf{J} = \mathbf{J}_\perp + J_\parallel \mathbf{b}$, where $\mathbf{J}$ is the current density, and $\mathbf{b} = \mathbf{B}/B$, so that $J_\parallel$ and $\mathbf{J}_\perp$ are the current densities parallel and perpendicular to $\mathbf{B}$, respectively. If $\mathbf{J}_\perp = 0$, the total current is conserved along a flux tube, that is, $J_\parallel/B$ is constant along a field line. The gradient of $J_\parallel/B$ along $\mathbf{b}$ at a given point on the field line can be written as

$$\frac{d}{ds}\left(\frac{J_\parallel}{B}\right) = \mathbf{b}\cdot\nabla\left(\frac{\mathbf{J}\cdot\mathbf{b}}{B}\right) \tag{17.1}$$

where s is distance along the field. Integrating equation (17.1) along the magnetic field line from the equator to low polar altitudes yields the total field-aligned current. The right-hand side of equation (17.1) can conveniently be expressed in terms of plasma pressure gradients and flow accelerations.

The current density is related to plasma dynamics through the magnetohydrodynamics (MHD) momentum equation

$$\rho\left(\frac{\partial \mathbf{u}}{\partial t} + (\mathbf{u}\cdot\nabla)\mathbf{u}\right) = \mathbf{J}\times\mathbf{B} - \nabla P \tag{17.2}$$

where ρ is the plasma mass density, $\mathbf{u}$ is the plasma flow velocity, and P is the scalar thermal pressure, which, for simplicity, is assumed isotropic in this formulation. Following the treatment of *Strangeway* [2012], equation (17.2) yields an expression for $d(J_\parallel/B)/ds$. Taking the curl of equation (17.2), dotting with the magnetic field, and rearranging terms yields

$$\begin{aligned}(\mathbf{b}\cdot\nabla)\left(\frac{\mathbf{J}\cdot\mathbf{b}}{B}\right) = {} & \frac{2}{B^2}\mathbf{b}\cdot(\nabla P\times\nabla B) \\ & +\frac{1}{B}\mathbf{b}\cdot\left(\rho\left(\frac{\partial\mathbf{u}}{\partial t}+(\mathbf{u}\cdot\nabla)\mathbf{u}\right)\times\frac{\nabla V_A^2}{V_A^2}\right) \\ & +\frac{\rho}{B}\mathbf{b}\cdot\left(\nabla\times\left(\frac{\partial\mathbf{u}}{\partial t}+(\mathbf{u}\cdot\nabla)\mathbf{u}\right)\right)\end{aligned} \tag{17.3}$$

where V_A is the Alfvén speed, $V_A = B^2/(\rho\mu_0)$, and $\nabla\cdot\mathbf{J} = 0$ is assumed, which is valid for low frequencies such that the displacement current can be ignored. The right-hand side of equation (17.3) gives the sources of field-aligned currents, which have been conveniently grouped into three distinct mechanisms corresponding to each term. The first term gives the currents driven by pressure gradients orthogonal to gradients in the magnetic field intensity, corresponding to a divergent plasma magnetization current. The second term represents the currents driven by stresses due to plasma accelerations perpendicular to gradients in the Alfvén velocity and corresponds to flow accelerations. The third term gives currents driven by intensification of flow vorticity corresponding to changes in divergence of the electric field.

It is important to bear in mind that the derivative operator in the second and third term is both the partial time derivative, $\partial/\partial t$, and the advection operator, $\mathbf{u}\cdot\nabla$, so that currents are driven by changes in the frame of a plasma parcel. The import of this is that currents can be generated by time stationary flows that accelerate the plasma as it flows through the standing flow pattern. Currents driven by magnetopause processes arise primarily from the second (flow accelerations) and third (flow gradients) terms and correspond to the drivers for the current from the solar wind dynamo along field lines to the polar regions within the magnetosphere (cf. Fig. 17.1a). Currents driven within the magnetosphere correspond primarily, though not exclusively, to the first (gradients in the plasma pressure) and second (flow breaking) terms (cf. Fig. 17.1b).

17.2.2. Convection

The relationship between Birkeland currents and magnetospheric convection as mapped to the ionosphere or planetary surface can be viewed to a first approximation in terms of the electrostatic potential at the altitude where the currents close. For the purposes of this discussion, the electrical conductance is very high along the magnetic field above the ionosphere at Earth, or the planetary surface at Mercury, and field lines can be considered equipotentials. Particle motion from high to low altitude, from low to high magnetic field intensity, must overcome the mirror force. Thus, to carry an upward current requires a filed-aligned potential drop, the magnitude of which was first estimated in terms of the high altitude electron density and temperature by *Knight* [1973]. The field-aligned potentials associated with aurora are typically ~10% of the total potential so, for the purposes of this discussion, they are ignored. In regions without dissipation, that is where the conductivity is infinite, one has the simple relation

$$\mathbf{E} + \mathbf{u}\times\mathbf{B} = 0 \tag{17.4}$$

Thus, the electric field patterns at low altitude correspond to flow patterns, which in turn can be mapped along the magnetic field to estimate the broad features of high altitude convection.

The flow of steady-state Birkeland currents through the low altitude resistive medium, be it the planet or the ionosphere, can be related to the corresponding electric fields by considering an electrostatic treatment. We assume a constitutive relation between the electric field and current perpendicular to **B** in the resistive layer, $\mathbf{J}_\perp$ and $\mathbf{E}_\perp$, as $\mathbf{J}_\perp = \underline{\sigma}_\perp \cdot \mathbf{E}_\perp$, where $\underline{\sigma}_\perp$ is the conductivity tensor, which at Earth includes both the Pedersen and Hall conductivities. To derive an estimate for the electric field, we consider the simple case of a scalar conductivity, σ, which is roughly equivalent to ignoring the contribution of the Hall conductivity at Earth to closing the Birkeland currents. The Birkeland currents are given by $J_\parallel = \nabla \cdot \mathbf{J}_\perp = \sigma \nabla \cdot \mathbf{E}_\perp + \mathbf{E}_\perp \cdot \nabla \sigma$. So they can be treated formally as a source term for the electric field. For the even simpler case of uniform conductivity, this reduces to $\nabla \cdot \mathbf{E}_\perp = J_\parallel / \sigma$.

Figure 17.3 shows the solution for the electric field and the electrostatic potential in the case of a uniform height-integrated conductivity given the Birkeland currents shown in Figure 17.2. From equation (17.4), recognizing that the background magnetic field is nearly radially inward in the high latitudes at Earth, the contours of the electric potential correspond also to flow lines in the ionosphere. This potential pattern corresponds to antisunward flows over the polar cap, where the electric field is directed nominally from dawn to dusk, and sunward flows around dawn and dusk at lower latitudes, approximately from 50° to 65° MLAT for this storm-time interval, where the electric field is directed roughly dusk to dawn. The flow therefore consists of two cells of circulation.

The sequence of processes at high latitudes responsible for this convection was first discussed more than 50 years ago by *Dungey* [1961], reproduced in Figure 17.4 (with the Sun to the right), and is one of the foundational concepts in Earth-magnetospheric physics. Magnetic lines of force are labeled "1" for those that are anchored at both ends at Earth, and "2" for those that are entirely in the solar wind. Lines of force that are open to the solar wind on one end but anchored at Earth in the north or south are labeled "3" and "4," respectively. The magnetized solar wind flows from right to left. The convection cycle

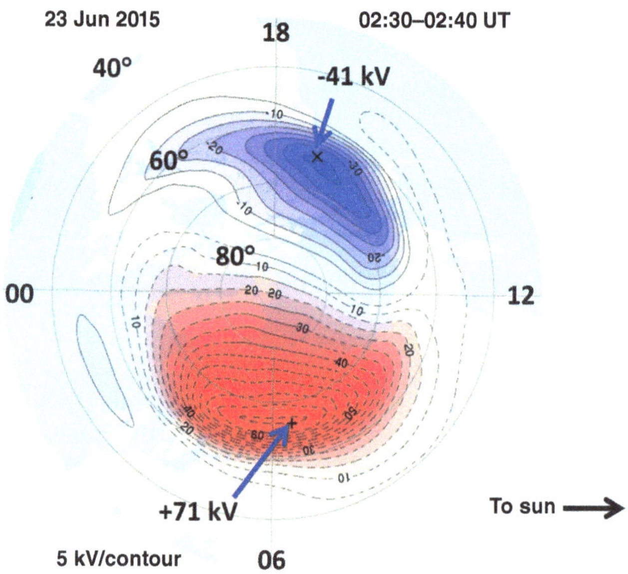

Figure 17.3 Representative electrostatic potential distribution at Earth derived for the Birkeland current distribution of Figure 17.2 under the assumption of a uniform 8S electrical conductance in the ionosphere to close the field-aligned currents. Positive and negative potentials are color coded red and blue, respectively. Major (labeled) contours are every 10 kV.

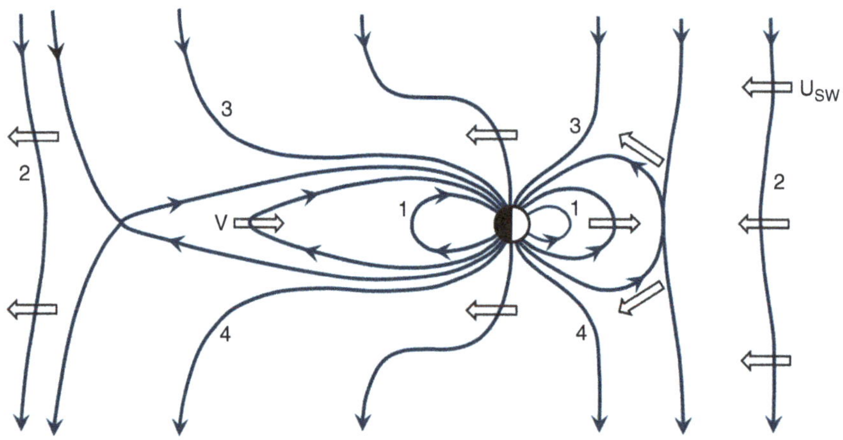

Figure 17.4 Schematic of the Dungey magnetic convection cycle for Earth in the noon-midnight meridian [adapted from *Dungey*, 1961]. Lines show notional magnetic lines of force with the magnetic field direction indicated by the dark blue arrows. Double arrows indicate the direction of plasma flow. Diagram shows the convection for southward IMF with the solar wind flowing from right to left.

begins with magnetic reconnection in the subsolar dayside magnetopause region. This launches flows away from the subsolar region and initiates the transport of magnetic field lines over the north and south polar caps into the tail forming the magnetotail. This leads to accumulation of magnetic flux in the magnetotail lobes with oppositely directed magnetic fields separated by the cross-tail current sheet. Magnetic reconnection at the neutral sheet both expels flux tailward and initiates the planetward convection of freshly closed field lines that form the dawn and dusk return flows, ultimately returning flux to the dayside. Thus, the potential pattern of Figure 17.3 reflects the large-scale magnetospheric convection by representing the complete magnetic convection cycle.

For purposes of comparison with Mercury, a key feature of the convection and Birkeland current pattern at Earth is that the return flows from the nightside to the dayside occur poleward of the equatorward extent of the Region 2 current and extend into the Region 1 currents spanning that portion of the Region 1 current that closes through Region 2. With reference to Figure 17.1b, these return flows occur in the region occupied by magnetospheric plasma in the dawn and dusk at high altitudes. The Region 2 currents and the plasma that supports them effectively shield lower altitudes and lower latitudes from the convection potential shown in Figure 17.3 such that the potential and convection is essentially confined to the polar region. A key question for Mercury therefore is whether Region 2 currents are present and accomplish a similar shielding and if they are not present, where and how the return convection occurs.

17.3. BIRKELAND CURRENTS AT MERCURY

The absence of an ionosphere at Mercury implies that electrodynamic coupling between the magnetosphere and the planet differs from that at Earth [*Glassmeier*, 2000]. Birkeland currents are present in magnetospheric simulations for models with a conducting planetary body [*Janhunen and Kallio*, 2004; *Ip and Kopp*, 2004; *Jia et al.*, 2015]. There are several mechanisms for current closure at Mercury. *Glassmeier* [1997] estimated the conductance of the planet at ~10 s timescales by integrating the material conductivity over the skin depth. Transient currents, treated as Alfvén waves, are reflected from the surface and do not require closure [*Lyatsky et al.*, 2010]. At longer timescales, the low conductance of the regolith inhibits closure near the surface [*Glassmeier*, 2000]. Neutral exospheric densities give negligible Pedersen conductance via ion-neutral collisions [cf. *Slavin et al.*, 2007]. A conductance as high as 0.1 S might result from charge transport via a chain of collisions with the regolith; such collisions create secondary emissions, which in turn collide with the regolith producing additional secondary emissions [*Cheng et al.*, 1987]. *Janhunen and Kallio* [2004] proposed that currents could close through the planet, radially through the low conductance outer layers of the crust and upper mantle and laterally at depths where temperature and conductance are higher.

17.3.1. Dynamic Currents

The first report of field-aligned currents at Mercury was by *Slavin et al.* [1997] using data from the second Mariner 10 flyby in 1974. Figure 17.5 is the key figure of their paper, which shows a region from about 20:50:55 UT to 20:51:30 of rapid, few second, changes in the dawn-dusk (Y) component of the magnetic field before and after a somewhat slower gradient in B_Y. These signals were interpreted as field-aligned currents, potentially arising from substorm dynamics, and the inferred total current was ~100 kA. An alternative to a steady state current, is that the signal indicates the passage of an Alfvén wave over the spacecraft propagating along the magnetic field toward the planet, which in that case corresponds to a time-dependent current [*Slavin et al.*, 1997]. This observation, as well as the confirmation by Mariner 10 that Mercury's exosphere was very tenuous, prompted consideration of time variable currents that might be supported at Mercury but that did not require closure in the steady-state circuit element analog as illustrated in Figure 17.1. *Glassmeier* [1997] pointed out that the skin depth, determined by the wave frequency and material conductivity, of waves incident on the surface would determine the energy loss and reflection of waves and thereby control the current that would flow. Waves propagating transverse to the magnetic field could lead to a transient field-aligned current because the incident and reflected waves would be spatially displaced and hence the field-aligned current of the wave would appear as a transient Birkeland current even though the currents did not close through the material of the planet [*Lyatsky et al.*, 2010]. One challenge to confirming these proposals is that the timescale for the waves, on the order of seconds, is comparable to a number of other physical processes at Mercury, including ion cyclotron wave periods, fast-mode MHD transit times across the dayside magnetosphere, as well Alfvén wave transit times between the northern and southern hemispheres [*Glassmeier*, 1997].

Indeed, the magnetosphere of Mercury turned out to be permeated by magnetic fluctuations over a broad frequency range [e.g., *Boardsen et al.*, 2012; *Anderson et al.*, 2013], which made unambiguous identification of transient Birkeland currents associated with these mechanisms challenging. Nonetheless, the occurrence of sudden changes in magnetotail convection, identified as Mercury substorm signatures, were clearly confirmed [*Slavin et al.*, 2009, 2010; *Sundberg et al.*, 2012; *Sun et al.*, 2015].

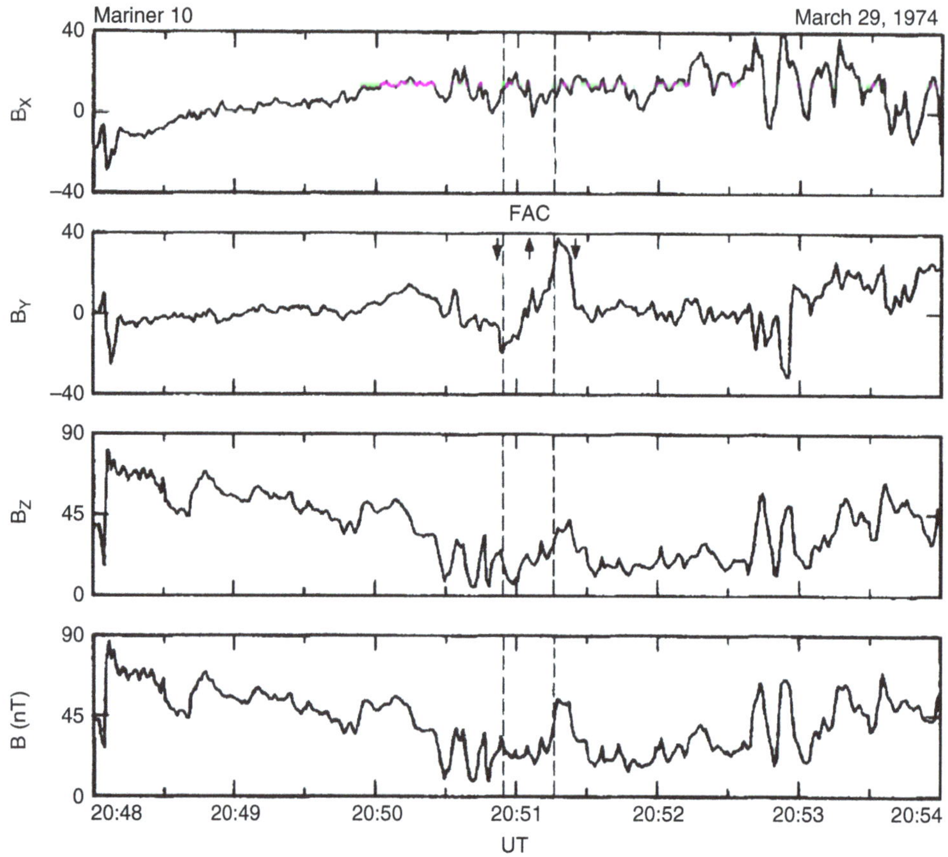

Figure 17.5 Magnetic field from the second Mariner 10 Mercury flyby in Mercury ecliptic (ME) coordinates where X is sunward, Z is north and perpendicular to the ecliptic plane, and Y completes the right-hand system in the direction approximately opposite the planetary orbital motion. Closest approach occurred near 20:47 UT and the signal interpreted as a field-aligned current is indicated by the dashed lines and arrows from ~20:50:55 to ~20:51:15 UT. From *Slavin et al.* [1997].

Thus, transient current systems almost certainly occur, and the observational challenge at low altitudes is to distinguish their signals from the other sources of similar magnetic signals. Future, simultaneous observations of electric fields, magnetic fields, and plasma flows will be of enormous diagnostic benefit in this regard.

17.3.2. Steady-State Currents at Mercury: Polar Magnetic Signatures

The finding of steady-state Birkeland currents at Mercury was made only after extensive observations from orbit were made by the MESSENGER spacecraft [*Solomon et al.*, 2001; *Anderson et al.*, 2014a]. The polar pass of Mariner 10 on the third and final flyby in 1975 did not reveal clear signatures of field-aligned currents [*Ness et al.*, 1975, 1976; *Philpott et al.*, 2014]. In retrospect, the magnetic field on the third Mariner 10 flyby was quiet relative to typical fluctuation levels that are more generally present even over the pole [cf. *Anderson et al.*, 2013], so it may be that Mercury's magnetosphere was in a relatively quiescent state during the third Mariner 10 flyby.

Indeed the signature of Mercury's Birkeland currents is highly variable from one MESSENGER orbit to the next. Figure 17.6 shows examples from six different orbits of MESSENGER at different times during the four-year orbital mission. In all cases, the magnetic field residuals are plotted as line segments in the direction of the residual signal starting at the point along the orbit where they were observed. As in *Anderson et al.* [2014a], the residuals were derived by subtracting both the internal field, represented as an offset dipole [*Anderson et al.*, 2012; *Johnson et al.*, 2012], denoted as $\mathbf{B}_{Int}(\mathbf{r})$, and an external field model including signals from the magnetopause and tail currents [*Johnson et al.*, 2012; *Korth et al.*, 2015], denoted $\mathbf{B}_{Ext}(\mathbf{r},t)$, which uses a variable subsolar standoff distance to capture the effects of Mercury's eccentric orbit about the Sun. The residual signals are then $\delta\mathbf{B} = \mathbf{B}_{obs} - \mathbf{B}_M(\mathbf{r},t)$ where $\mathbf{B}_M(\mathbf{r},t) = \mathbf{B}_{Int}(\mathbf{r}) + \mathbf{B}_{Ext}(\mathbf{r},t)$. Perturbations along the net model magnetic field, in the $\mathbf{b}_M = \mathbf{B}_M/|\mathbf{B}_M|$ direction,

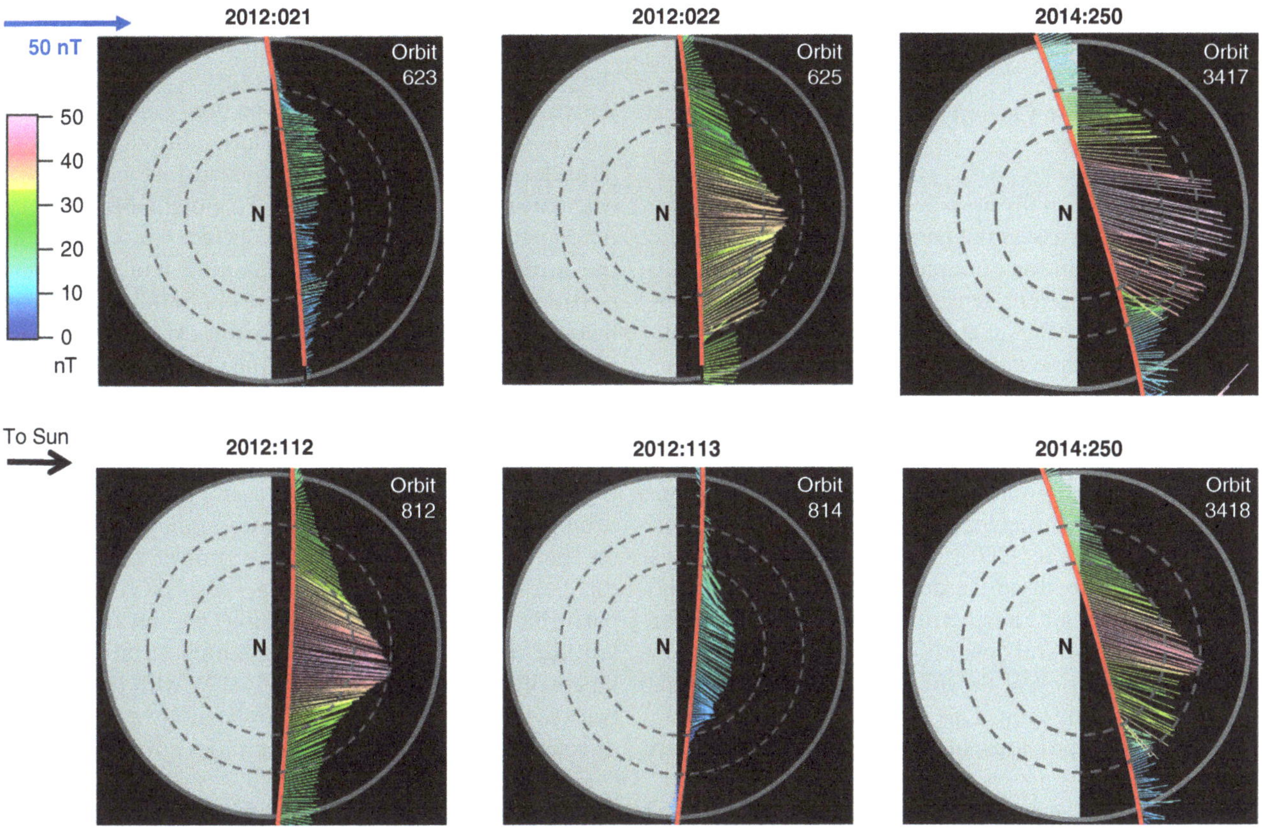

Figure 17.6 MESSENGER Magnetometer magnetic residuals perpendicular to the radial direction, $\mathbf{\delta B}_{\perp}$, after subtracting models for the internal and external field as viewed from above the north pole. The Sun is to the right and dusk is at the top of each frame and the $\mathbf{\delta B}_{\perp}$ are shown as lines starting at the observation point in the $\mathbf{\delta B}_{\perp}$ direction color coded by magnitude and with the length proportional to magnitude as well, scaled according to the 50 nT reference arrow. The sunlit portion of the planet is shaded black to highlight the $\mathbf{\delta B}_{\perp}$ lines. The orbit trajectory is shown in red below 1000 km altitude and black above 1000 km altitude (only evident at the bottom portion of the tracks in the upper left and center panels). The top left two panels show data from two orbits near Mercury perihelion with MESSENGER in a dusk-to-dawn orbit. Bottom left two panels show data from two orbits 90 days (180 orbits) later after one Mercury orbit about the sun and one and one-half planetary rotations. Periapsis altitude of MESSENGER was ~400 km for the four panels on the left. Right two panels show data from two successive orbits after MESSENGER's periapsis altitude had been adjusted to ~100 km altitude.

attributed to local plasma pressure were also removed as described in *Anderson et al.* [2014a] to obtain the net perpendicular perturbation, $\mathbf{\delta B}_{\perp} = \mathbf{\delta B} - (\mathbf{\delta B} \cdot \mathbf{b}_{M})\mathbf{b}_{M}$. For this work, we used Mercury solar orbital (MSO) coordinates aberrated for Mercury's orbital motion [cf. *Johnson et al.*, 2012]. The MSO origin is the center of Mercury, and the Z axis is parallel to the planetary rotation axis, positive northward. Positive X is opposite to the solar wind flow in Mercury's frame, the Y axis completes a right-handed system and is positive duskward.

The upper left and center panels show data from two orbits from days 21 and 22 of 2012 in an approximately dawn to dusk orbit trajectory. The lower left and center panels show data from days 112 and 113 of 2012, 90 days later with nearly identical orbit trajectories but for which the planet has rotated by 180°. The key points of comparison with these four panels are: the signals are generally less than 50 nT and often only 20 nT, roughly 10 times smaller than those observed at Earth; they are highly variable from pass to pass; and they do not rotate with the planet. The latter two points are important in establishing that these signals are driven by dynamics of external origin. The right two panels of the figure show data from two passes after the orbit periapsis altitude had been lowered to ~100 km so that these observations were made substantially closer to the planet than those in the other panels. These passes display stronger signals, ~70 nT on orbit 3417 and over 40 nT on orbit 3418, consistent with signals arising from Birkeland currents since the magnetic signal increases as the flux tube cross-sectional area decreases with decreasing altitude. In all of these cases, the signals consist of a sunward perturbation but do not exhibit the antisunward signals at dawn and dusk that one expects from analogy with Earth (cf. Fig. 17.2).

17.3.3. Steady-State Currents at Mercury: Statistical Analysis

Constructing spatially averaged magnetic maps of the $\delta\mathbf{B}_\perp$ allows one to assess the persistence of the magnetic signatures shown in the examples and is also required to evaluate average current density distributions and intensities. The average $\delta\mathbf{B}_\perp$ was constructed by adding all data together for the same portion of the orbit (ascending or descending) for a given year, corresponding to an altitude range of less than about 1000 km for a given point along the orbit. It is important to bear in mind that the MESSENGER orbit was highly eccentric and that its line of apsides rotated during the course of the orbital mission [cf. *McAdams et al.*, 2007, 2012]. Initially, periapsis occurred in the north on the descending segment of the orbit after reaching its northernmost latitude whereas for the final year of orbital operations, periapsis occurred on the ascending portion of the orbit, prior to the northernmost latitude of the orbit. This implies that the ascending and descending portions of the orbit correspond to substantially different altitudes at the same latitude and also that different years of operations lead to different ranges of altitude on the same portion of the orbit. To derive a field-aligned current, one needs an average $\delta\mathbf{B}_\perp$ over a surface with a single-valued altitude at a given latitude. The current through this surface can then be calculated from Ampere's law. To obtain $\delta\mathbf{B}_\perp$ averages on something close to a single surface, we constructed the averages separately for ascending and descending portions of each year of orbital operations. Only data below 1000 km altitude and northward of 30°N were used. Data in each bin, 1° in latitude by 1 hr in local time, were averaged together.

The result of the average of all of the data for the second year of orbit operations, April 2012 through March 2013, is shown in Figure 17.7, with the average for the ascending orbit portion on the left and the descending orbit portion on the right. The plots show the projection of $\delta\mathbf{B}_\perp$ on the *X-Y* MSO plane. The averages also show a sunward perturbation over the polar region, northward of ~60° latitude that reaches 50 nT at the highest latitudes and tapers off to less than 20 nT by 55°. As with the example tracks, there is no region of oppositely, antisunward, directed $\delta\mathbf{B}_\perp$ over the range of latitudes covered.

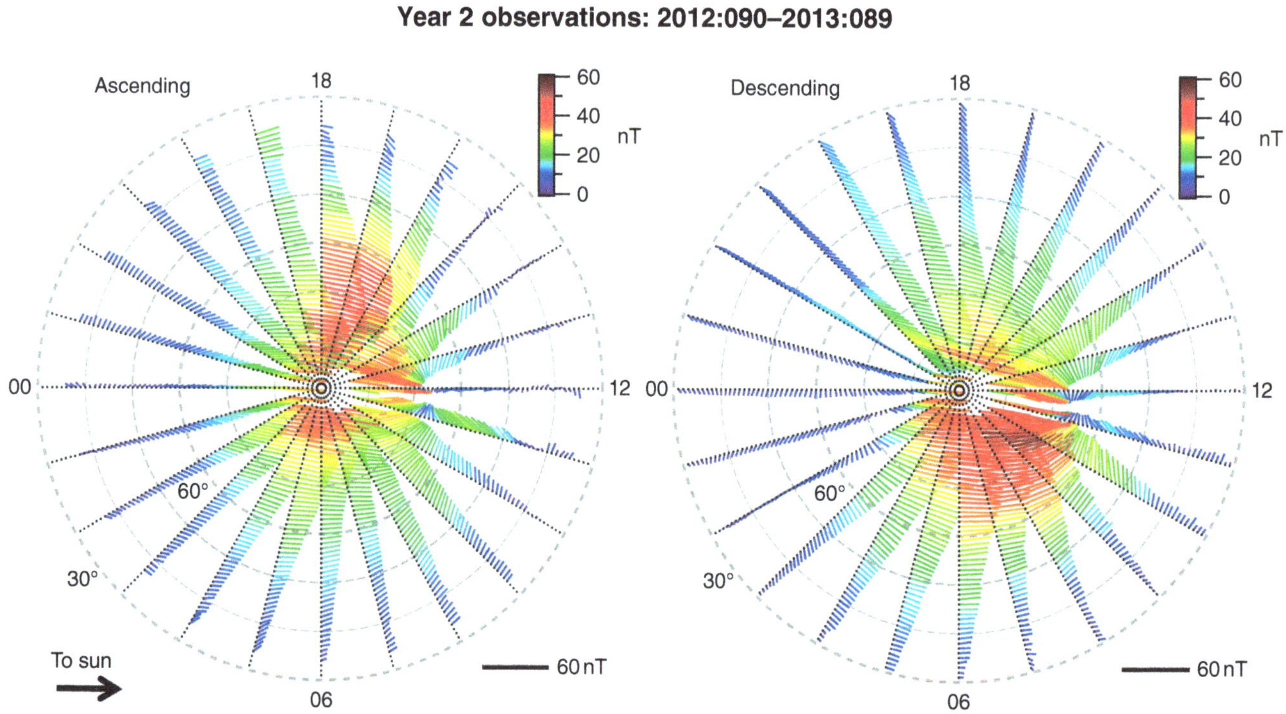

Figure 17.7 Averaged horizontal $\delta\mathbf{B}_\perp$ for the second year of MESSENGER orbital operations at Mercury. Averages were accumulated in Mercury Solar Orbital (MSO) coordinates. Panels show perturbations in 1° MSO latitude and 1 hr MSO local time bins with +*X* (noon) to the right and +*Y* (dusk) at the top. Average $\delta\mathbf{B}_\perp$ are plotted as line segments originating at the averaging latitude-local time bin center and in the direction of the average $\delta\mathbf{B}_\perp$. The magnitude of $\delta\mathbf{B}_\perp$ is indicated both by the line segment length, scaled according to the 60 nT reference, and by the line segment color according to the color bar.

There are opposite dawn-dusk biases in the ascending and descending patterns and they are attributed to the correlation of orbit local time and Mercury heliocentric distance. The ascending average $\delta\mathbf{B}_\perp$ are larger in the dusk sector than in the dawn sector, and the opposite holds for the descending averages. MESSENGER orbit local time and Mercury's distance to the Sun vary in lockstep: at Mercury aphelion, the ascending node was near dusk but at perihelion the ascending node was near dawn [cf. *McAdams et al.*, 2007]. Both the solar wind ram pressure and the solar wind/IMF dynamo are more intense when Mercury is closer to the Sun [cf. *Anderson et al.*, 2013; *Korth et al.*, 2012; *Johnson et al.*, 2016]. Hence, we expect stronger external driving of Mercury's magnetosphere and correspondingly greater $\delta\mathbf{B}_\perp$ at perihelion when the ascending orbit segment was at dusk (descending orbit segment at dawn) than at aphelion when the ascending orbit segment was at dawn (descending orbit segment at dusk). Thus, the dawn-dusk differences reflect this observational effect rather than a true dawn-dusk asymmetry at a given Mercury-Sun distance.

17.3.4. Average Current Densities

Because the altitudes of surfaces on which $\delta\mathbf{B}_\perp$ is averaged vary with latitude and are different between ascending and descending portions of the orbit at the same latitude, it is necessary to evaluate the currents through the surfaces traced out by the satellite orbit, calculate the current densities on these surfaces, and then map these current densities along the model magnetic field to the planetary surface. As described in *Anderson et al.* [2014a], the orbit surfaces were divided into latitude-longitude bins and the total current through each bin was evaluated from the integral of $\delta\mathbf{B}_\perp$ around each bin. The field-aligned current density was then calculated by dividing this by the bin area and by the cosine of the angle between each bin area normal and the model magnetic field, $\mathbf{B}_M$, at the bin's center. This current density was then mapped to the planetary surface along $\mathbf{B}_M$ and converted to a surface current density using the ratio of B_M at the orbit and planetary surfaces. Multiplying the surface field-aligned current by the cosine of the angle between $\mathbf{B}_M$ at the surface point and the radial direction gives the radial current density at the surface, J_{rS}. Finally, each surface point latitude was converted to invariant latitude, Λ, corresponding to the latitude of the field line where it crosses the 1 R_M surface centered on the offset dipole [cf. *Korth et al.*, 2014; *Anderson et al.*, 2014a]. One limitation of this approach is that the model magnetic field, $\mathbf{B}_M$, does not include the effects of the field-aligned currents. Accounting for the effects of these currents on the mapping would require an iterative solution, but since no model exists for the field-aligned currents there is no way to avoid this, and the results must be considered an initial estimate for the currents.

The distributions of J_{rS} for the $\delta\mathbf{B}_\perp$ averages of Figure 17.7 are shown in Figure 17.8. The thick black outlines indicate the minimum Λ sampled below 1000 km altitude and north of 30° MSO latitude. For both ascending and descending portions of the orbit, the J_{rS} distributions consist of an upward current in the dusk/afternoon sector and a downward current in the dawn/morning sector, consistent with the polarity of Region 1 currents at Earth. Integrating the total current from 60° Λ to the pole, the average upward and downward currents are 35 kA and −41 kA, respectively, and the average absolute value is 38 kA, which is a factor of ~200 smaller than the corresponding current at Earth for the storm-time interval shown in Figure 17.2. Even considering that typical total currents at Earth for moderate activity are in the range of 3 to 6 MA [e.g., *Anderson et al.*, 2017], the total current at Mercury is more than 100 times less than at Earth. There is no indication of an equatorward current of opposite polarity despite the fact that the observations extend to Λ more than 20° equatorward of the low-latitude edge of the Region 1 current. The statistical analysis therefore also leads to the conclusion that Mercury's magnetosphere–solar wind interaction does not result in significant Region 2 currents. The consistency of this result for different levels of magnetospheric disturbance at Mercury and its implications are discussed below.

17.3.5. Dependence on Magnetospheric Activity

At Earth, enhanced geomagnetic activity correlates with enhanced Birkeland currents [e.g. *Iijima and Potemra*, 1976], and with the advent of in situ solar wind observations, it was possible to characterize how the Birkeland currents vary in both intensity and distribution with conditions in the solar wind and the interplanetary magnetic field (IMF) [*Weimer*, 2005; *Papitashvilli et al.*, 2002; *Anderson et al.*, 2008; *Korth et al.*, 2010; and *Coxon et al.*, Chapter 16, this volume]. Given that Mercury appears on average to display only Region 1 currents, it is important to explore whether this average might be a convolution of different distributions present for different external conditions. While it is possible to characterize the level of magnetospheric disturbance from the MESSENGER observations [*Anderson et al.*, 2013], the significant variability of the IMF at Mercury renders extrapolations from IMF measurements in the solar wind to conditions during MESSENGER's passage through the magnetosphere unreliable [e.g., *Winslow et al.*, 2012]. Until solar wind/IMF observations are obtained at the same time as measurements at low altitudes over Mercury's polar regions, we will need to learn what we can by distinguishing

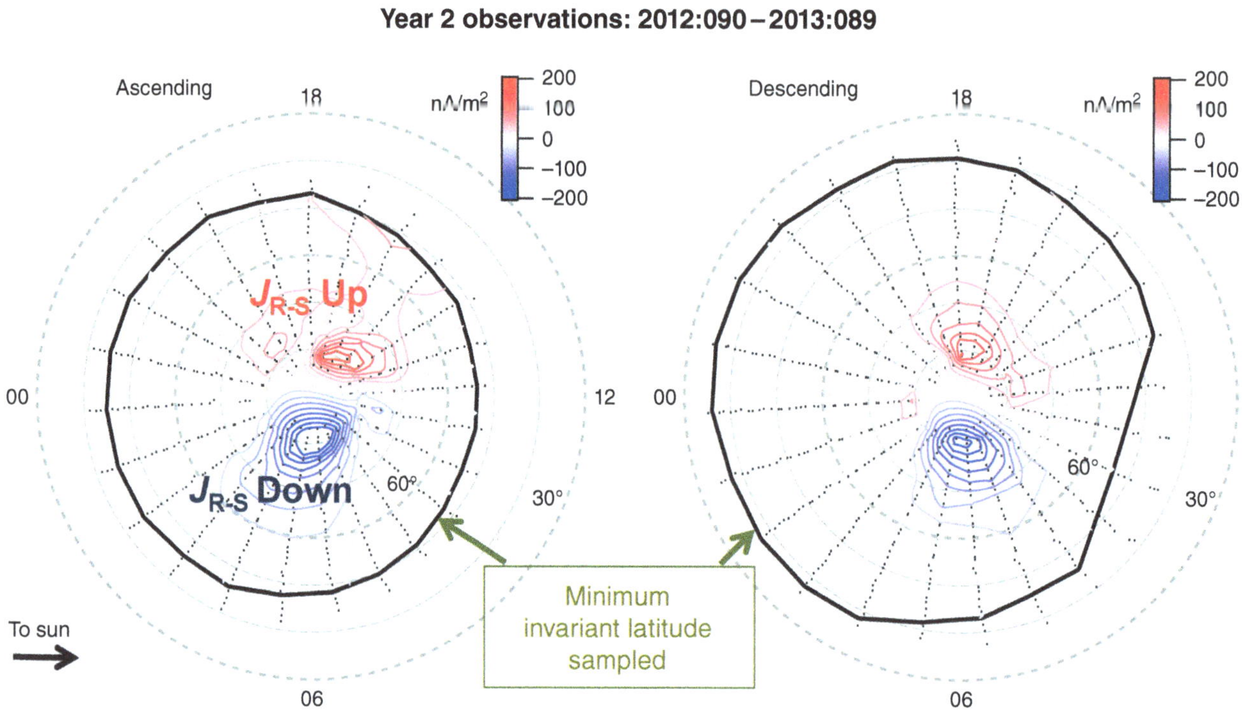

Figure 17.8 Field-aligned currents mapped to Mercury's surface and expressed as surface radial current density derived from the averaged $\mathbf{\delta B}_\perp$ maps of Figure 17.7. Left-hand panel shows the radial current density, J_{rS}, at the surface derived from ascending orbit leg data and right-hand panel shows the result derived from descending orbit legs. The Sun is to the right and dusk at the top with current density contours at 20 nA/m² intervals and shaded with the scale shown in the color bar with upward currents shown as positive (red) and negative (blue). Black contours show the minimum invariant latitude sampled at MESSENGER orbit altitude, which extends to lower invariant latitudes for the descending orbit segment owing to the lower altitudes of the MESSENGER orbit on the descending than on the ascending orbit leg at this time during the mission.

observations based on the magnetic disturbance level, analogous to measures of geomagnetic activity.

To assess the Birkeland currents' dependence on magnetospheric disturbance, the $\mathbf{\delta B}_\perp$ data were analyzed as described above, but they were also binned by magnetic disturbance level [*Anderson et al.*, 2013] in five percentile bins with 0%–20% being the quietest and 80%–100% being the most disturbed. Figure 17.9 shows the average $\mathbf{\delta B}_\perp$ for the four activity ranges: 0%–20%, 20%–40%, 60%–80%, and 80%–100%, again using the second year of orbital observations, for consistency with the data shown above. (Each of the 4 yr of observations gives the same results.) Results for the 40%–60% activity level range were essentially the same as those for all events shown in Figure 17.7 so they are omitted here. The pattern of sunward perturbations maximizing poleward of 60° Λ is present for all levels of activity and the perturbations are stronger for higher levels of disturbance. For the lowest activity range, $\delta\mathbf{B}_\perp$ barely reaches 40 nT, whereas for the highest activity range, $\delta B_\perp$ often exceeds 80 nT (saturated on the color scale shown). The spatial distribution of the currents seems to be preserved and the current intensity grows stronger with increasing activity. It is difficult to tell from these displays whether the pattern only grows in intensity at the same latitudes or expands in latitude as well, but the currents do not dramatically expanding equatorward as they do at Earth.

The corresponding maps of surface current density, J_{rS}, for the same four ranges of magnetic activity are shown in Figure 17.10. For quiet conditions, the currents reach only ~120 nA/m², whereas for the most active conditions the average J_{rS} can exceed 300 nA/m². The integrated total current, average of the absolute values of the upward and downward current, are 20.3 kA, 30.0 kA, 36.3 kA, 42.0 kA, and 60.3 kA for all five activity percentile bins from lowest to highest, consistent with the factor of three increase in maximum J_{rS}, from the lowest to highest activity range. In all cases, the upward currents are in the evening and downward currents in the morning with no indications of a change to this basic pattern with activity. There is no indication of an equatorward set of opposite polarity currents nor is there a switch in the polarity of

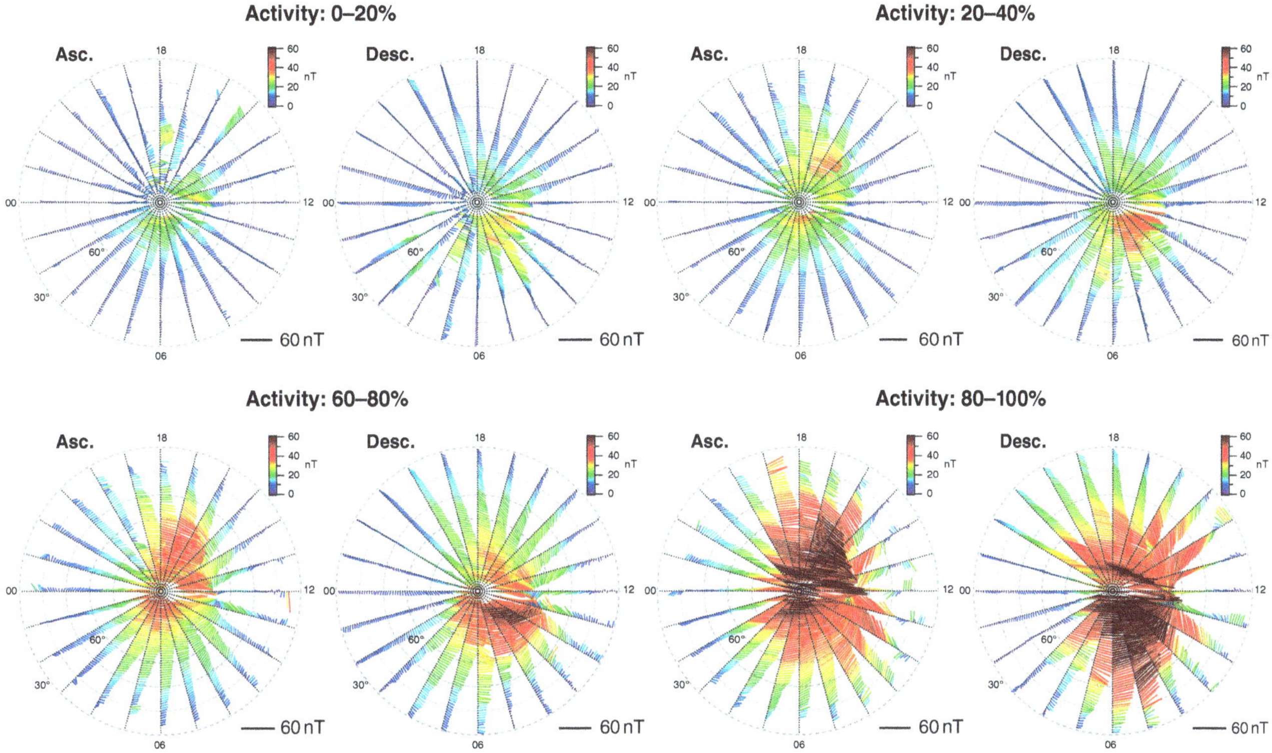

Figure 17.9 Average $\delta\mathbf{B}_\perp$ from ascending and descending orbit legs of the MESSENGER orbit derived from the second year of observations in Mercury orbit from the MESSENGER Magnetometer. Data are divided according to magnetospheric disturbance level by percentile levels for the lowest (upper left, 0%–20%), second (upper right, 20%–40%), fourth (lower left, 60%–80%), and highest (lower right, 80%–100%) quintiles. The Sun is to the right and $\delta\mathbf{B}_\perp$ averages in 1° invariant latitude and 1 hr local time bins are shown as line segments starting at the location of the bin center and in the direction of the average $\delta\mathbf{B}_\perp$. The length of the segments are proportional to $\delta B_\perp$ as shown by the 60 nT reference segments and the segment color also indicates $\delta B_\perp$ as given by the color bar.

the currents for very quiet conditions. Table 17.1 lists the integrated currents and maximum magnetic signals from the statistical patterns for each activity level averaged over the entire mission.

At Earth for northward IMF, the geomagnetic conditions are very quiet yet the Birkeland currents show a prominent reverse polarity set of currents at the highest latitudes. The absence of the northward IMF (often abbreviated as NBZ) current system [*Iijima et al.*, 1984; *Zanetti et al.*, 1984] at Mercury could result from plasma depletion in Mercury's magnetosheath [*Gershman et al.*, 2013], which leads to prominent subsolar reconnection for nearly all orientations of the IMF [*DiBraccio et al.*, 2013].

The patterns in Figure 17.10 also seem to show that the latitude of the maximum current intensity is almost invariant with activity level. At Earth, the auroral oval and the Birkeland currents expand equatorward dramatically during geomagnetic storms, which is attributed to erosion of magnetic flux from the dayside at a rate that exceeds the return transport of flux from the nightside. A lack of equatorward expansion of the Birkeland currents at Mercury may indicate that return convection is always sufficiently rapid to sustain the magnetopause reconnection. It may also reflect the fact that the nominal magnetopause subsolar standoff distance at Mercury is less than 0.5 R_M above the planetary surface [e.g. *Winslow et al.*, 2013], so that erosion cannot progress far before the magnetopause encounters the planet (which will occur in the southern hemisphere for a subsolar standoff distance about 0.2 R_M above the surface) [cf. *Slavin et al.*, 2014]. How magnetopause erosion would progress beyond this point is not clear since the electrodynamics of the dielectric planetary upper material would then couple to the magnetospheric electrodynamics, and this coupling has not been studied quantitatively although the possibility that all of Mercury's magnetic flux could at times be convected into the tail has been suggested [cf. *Slavin et al.*, 2009, 2010]. The role of magnetic fields induced in the core and mantle during extreme conditions may play a role in mitigating such extreme conditions since the planetary moment is not actually fixed [cf. *Heyner et al.*, 2016].

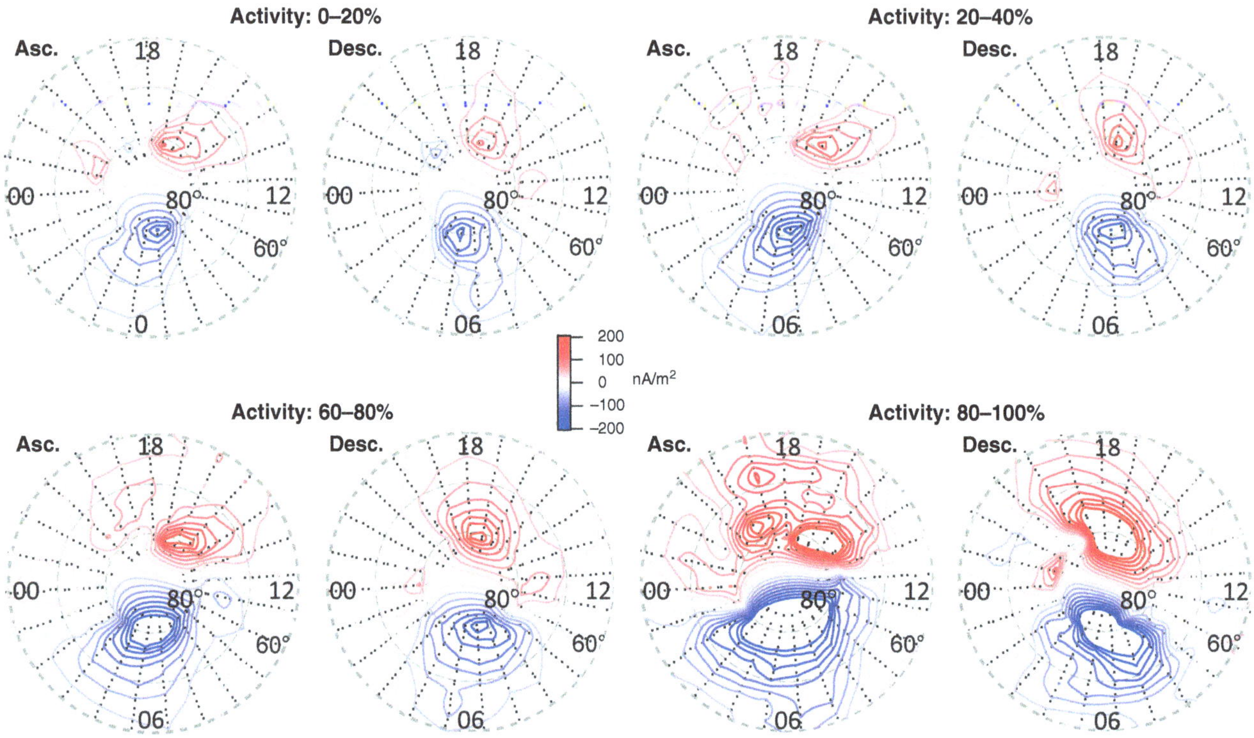

Figure 17.10 Surface radial current density at Mercury derived from MESSENGER magnetometer observations from 4 yr of orbital observations segregated by magnetic disturbance level in quintiles 0%–20%, 20%–40%, 60%–80%, and 80%–100%, as indicated. The median quintile, 40%–60%, is essentially the same as the distributions of all data shown in Figure 17.8. Each pair of panels shows the average current density for ascending (Asc.) and descending (Desc.) orbit segments. The interval between contours is 20 nA/m² and the current density is also indicated by the color bar with upward current shown as red and downward current shown as blue. Plots show the latitude range from the pole to 60° invariant latitude (Λ) with noon at the right and dusk at the top. Coverage extends down to 30° Λ with no indication of opposite polarity currents, as shown in Figure 17.8, so the views here are expanded to better show the currents that are observed.

Table 17.1 Average Birkeland Current Signatures and Total Currents at Mercury for Different Magnetic Disturbance Levels.

Activity level	$\delta B_\perp$–A 90% (nT)*	$\delta B_\perp$–D 90% (nT)	I_{Tot} (kA)**
All	36.1 ± 2.0†	33.3 ± 0.3	37.7 ± 0.9
0–20	22.4 ± 2.2	19.8 ± 0.9	19.0 ± 5.8
20–40	26.8 ± 2.3	25.8 ± 2.4	28.7 ± 2.0
40–60	35.1 ± 2.7	33.1 ± 4.0	34.7 ± 1.5
60–80	39.5 ± 1.3	37.7 ± 2.1	38.8 ± 3.0
80–100	61.8 ± 9.5	57.7 ± 5.4	62.3 ± 13.1

* Average 90 percentile value of the perpendicular magnetic perturbation magnitude computed as the average over the first 3 yr of orbital observations.

** Total surface radial current for ascending and descending orbit segments and averaged over the first 3 yr of orbital observations.

† Uncertainties are standard deviations of the averages of the first 3 yr of orbital observations.

17.3.6. Current Closure

The result that field-aligned currents are present at Mercury necessarily raises the issue of their closure. Observations of magnetopause reconnection yield estimates of the potential, $\Delta\Phi_{SW}$ applied to the magnetosphere of between 15 and 30 kV under nominal conditions [*Slavin et al.*, 2009; *DiBraccio et al.*, 2013]. Although *Cheng et al.* [1987] suggested that charge transport could occur at the surface, the total current is about 10 times too high to be explained by this mechanism at these potentials [cf. *Anderson et al.*, 2014b]. *Janhunen and Kallio* [2004] proposed that Birkeland currents at Mercury could close through the planet itself, possibly passing through the low conductivity but thin upper layers of the semiconductor materials of the crust and mantle to the highly conducting metallic (iron) core below. This idea was tested in a simple shell model shown schematically in Figure 17.11 [*Anderson et al.*, 2014a]. In this calculation,

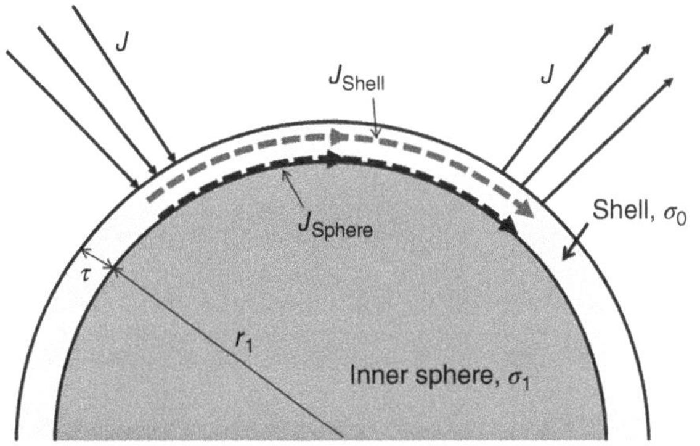

Figure 17.11 Schematic closure path of Birkeland currents at Mercury through the planet partially closing in the outer shell (crust and mantle) and through the core (inner sphere). Owing to the absence of an atmosphere at Mercury, there is no analog of an ionosphere and currents close through the planet. The electrostatic solution for the potential, proportional to the current, is straightforward to obtain using the observed distributions of radial surface current. From *Anderson et al.* [2014b].

the low conducting outer layers were represented as a single layer over a perfectly conducting sphere. Taking the J_{rS} distribution as the upper boundary condition, they derived the electric potential on the surface and throughout the volume of the outer layer recognizing that the potential satisfies $\nabla^2\varphi = 0$ within a layer of constant conductivity. Using the potential estimates from magnetopause reconnection of ~30 kV, and under the assumptions that the potential is dominated by the resistive drop across the layer to the core and that the conductivity increases exponentially with depth, $\sigma(d) = \sigma_s \exp(d/\lambda_\sigma)$, where d is depth below the surface, λ_σ is the conductivity scale length, and σ_s is the surface conductivity, they estimated $\sigma_s \sim 2.5 \pm 1.2 \times 10^{-8}$ S/m using a nominal value for λ_σ of 5 km valid down to ~100 km depth. This value is consistent with the range of values estimated from nominal composition and thermal models for Mercury's crust and mantle [*Verhoeven et al.*, 2009]. The result suggests that the currents most likely close through the planet.

This work was generalized in *Anderson et al.* [2016] where the simple single layer model was replaced with an arbitrary profile of conductivity with depth. Their results are summarized in Figure 17.12, which shows the form for the conductivity profile, results for the fraction of current closing through the core, and the normalized total potential difference versus conductivity scale length. The core was represented by perfect conductivity and placed at depths, denoted by d_{core}, of 370 km, 420 km, or 470 km consistent with results from gravity and topography [*Hauck et al.*, 2013]. The composition and thermal models considered by *Verhoeven et al.* [2009] give a plausible range of conductivity depth profiles for the crust and mantle that can be represented by an exponential increase with depth to a maximum value, σ_{max}, in the range of 0.01 to 1.0 S/m. In addition, the regolith may have a substantially lower conductivity with a depth of ~100 m. These features were represented as shown in Figure 17.12a, and solutions for the three dimensional potential and current versus λ_σ were obtained for different sets of core depth, maximum crust/mantle conductivity, and with or without a regolith layer as indicated in the legends of Figures 17.12b and 17.12c labeled as cases K through P.

Most of the current flows through the upper layers to the core and the fraction closing through the core reaches a maximum of nearly 90% for $\lambda_\sigma = 20$ to 30 km (Fig. 17.12b). For small λ_σ, < 5 km, the conductivity rises so rapidly with depth that there is a large thickness at the maximum crust/mantle conductivity to carry current without reaching the core. However, as λ_σ increases, the thickness of the high conductivity region in the crust/mantle decreases and more current flows through the core. The peak in current closure through the core occurs when the crust/mantle conductivity reaches σ_{max} just at the core. Increases in λ_σ beyond this value lead to progressively greater resistance in current flow to the core and the fraction that closes in the crust/mantle gradually increases.

Figure 17.12c shows that the maximum potential difference is highly linear with increasing λ_σ. This is due to the fact that almost all of the potential drop occurs in the outer few scale lengths in the crust/mantle and is insensitive to how the current closes at depth. This implies that the relation between the conductivity scale length, surface conductivity, and maximum potential difference, $\Delta\psi$, can be written as

$$\Delta\psi = C\frac{\lambda_\sigma}{\sigma_s} \tag{17.5}$$

where C is a constant for a given intensity of current and geometry of surface current distribution. If one takes an average potential of 15 to 30 kV [cf. *DiBraccio et al.*, 2013] corresponding to the currents for the average over all events, one obtains $\lambda_\sigma/\sigma_s = 5.0 \pm 2.4 \times 10^{-11}$ S/m² for all plausible ranges of σ_s and λ_σ. Also, the presence or absence of a 100 m deep regolith has almost no effect since the layer is thin relative to the underlying resistive crust/mantle where most of the potential drop occurs. The model parameters span the range of values for expected materials and thermal structure, λ_σ ~10 to 20 km and $\sigma_s \sim 10^{-10}$ S/m near the surface rising to a maximum ~0.1 S/m in the mantle [*Verhoeven et al.*, 2009], so it seems safe to conclude that the Birkeland current closes through the planet at Mercury.

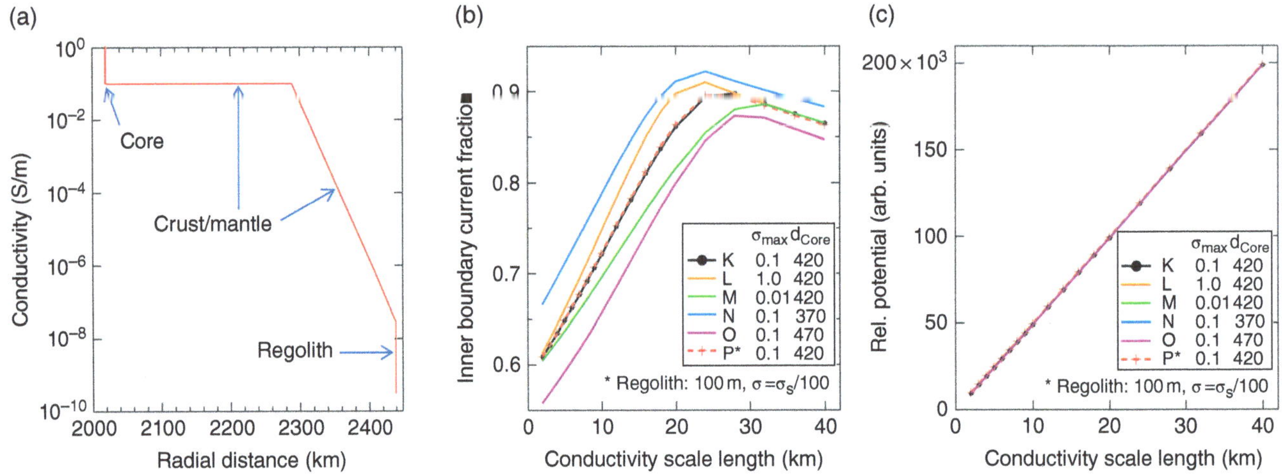

Figure 17.12 Results from an electrostatic model for Birkeland current closure within Mercury for a radially variable conductivity. From left to right, panels show (a) form of the model conductivity profile with radial distance; (b) fraction of current closed through the core as a function of conductivity scale length (λ_σ); and (c) normalized maximum potential difference at the surface versus conductivity scale length. For panels (b) and (c), results are shown for six different variations of the conductivity profile for different maximum conductivities (σ_{max}) and core depths (d_{core}) as indicated by the line colors and patterns. In panel (c), the variations in relative potential are smaller than the thickness of the lines. From *Anderson et al.* [2016].

17.4. IMPLICATIONS IN COMPARISON WITH EARTH

There are two key results for Birkeland currents at Mercury vis-à-vis those at Earth. First, the currents are much weaker at Mercury, only about 1% of those at Earth, owing to the greater electrical resistance of the current closure through the crust/mantle than through the ionosphere at Earth. That the currents are comparatively weak at Mercury is not surprising given the absence of an ionosphere, but they are consistently present. Second, there seems to be no evidence of Region-2-polarity currents at Mercury even though the observations extend equatorward of the Region 1 currents. This latter result has implications for differences in magnetospheric convection and energy dissipation at Mercury in comparison to Earth. Before discussing these, however, it is important to recognize the limitations in our present understanding of the results for Mercury owing to the statistical nature of the analyses required to resolve even these smoothed signatures of the currents.

17.4.1. Statistical and "Instantaneous" Views

When comparing the Birkeland current distributions from Earth (e.g., Fig. 17.2) obtained in as little as 10 min with the statistical results from Mercury (e.g., Fig. 17.8), it is important to bear in mind the limitations of the statistical analysis approach used to characterize the currents at Mercury. The results for Mercury are analogous to those obtained from single satellite studies for Earth [cf. *Iijima and Potemra*, 1976; *Papitashvilli et al.*, 2002; *Weimer*, 2005]. For these studies, to obtain a statistically significant average signal, it was essential to use all of the data irrespective of stability of geomagnetic activity or solar wind and IMF conditions. In addition, to identify how the currents depend on average on different geomagnetic conditions, the data were binned in some way, either by geomagnetic activity indices, for example, Kp, prior to the advent of continuous upstream solar wind and IMF observations, or with solar wind and IMF observations at the L1 point about 230 R_E upstream of Earth making the best possible estimation for the appropriate time delay for imposition of the observed conditions at Earth and then for the manifestation of the corresponding Birkeland currents. The present results for Mercury, binned by magnetic disturbance index, are most analogous to the results for Earth when binned by Kp or similar indices.

The use of solar wind and IMF data to sort observations of the currents from low Earth orbit (LEO) yielded substantial insight into the marked differences in the currents for different external conditions. Most significantly, this analysis led to the discovery of the current system poleward of Region 1 that is present only for northward IMF conditions often termed the northward IMF B_z (NBZ) system [*Iijima et al.*, 1984; *Zanetti et al.*, 1984]. The NBZ current system could not be identified in data organized by geomagnetic activity index. Thus, it is possible that there is an NBZ-type distribution that occurs at

Mercury but is not evident in the present results (cf. Fig. 17.10). Alternatively, we note that magnetic reconnection near the subsolar region appears to occur at Mercury over a much broader range of magnetic shear angles than at Earth [*DiBraccio et al.*, 2013] so that corresponding forward magnetic convection may be predominant at Mercury for all IMF orientations. Reverse convection and the corresponding NBZ currents occur at Earth when reconnection is driven preferentially poleward of the magnetic cusps. Thus, the prevalence of subsolar reconnection at Mercury may imply that reverse convection occurs only infrequently if ever at Mercury, and hence that NBZ currents are rarely present at Mercury.

The analyses that are organized by the solar wind and IMF conditions are subject to uncertainties as well, however. There are two sources of error in these analyses: uncertainties in the delay time from L1 to Earth; and discrepancies between L1 and Earth arising from coherence lengths in the IMF shorter than the L1-Earth distance. The uncertainties in the L1-to-Earth delay estimates are on the order of 20 min out of a nominal average delay of ~60 min so that observations attributed to one set of measurements at L1 have about a one-third probability of actually corresponding to a different time.

How much these intrinsic uncertainties in extrapolations from L1 to Earth affected the statistical results was not known initially but recent results using constellation magnetometer data from LEO provide some guidance to how the single satellite results are affected. To illustrate, Figure 17.13 shows statistical results for Birkeland currents binned by the IMF orientation for strongly northward and strongly southward conditions using two different approaches. The results on the left are those of *Papitashvilli et al.* [2002] obtained from Ørsted data and those on the right are from AMPERE observations. The key difference is that the AMPERE results use only periods for which the IMF was stable in direction to within ± 30° for at least 3 hr to ensure both that advection timing errors from L1 to Earth could be ignored and that the conditions at L1 closely correspond to those at Earth [cf. *Anderson et al.*, 2008; *Korth et al.*, 2010]. Only 3% of the observations met this restrictive criterion for IMF stability, but the constellation provided sufficient data to determine the Birkeland current distribution for every 10 min interval meeting these conditions.

Comparing the northward (top panels) and southward (bottom panels) IMF results, the AMPERE results show almost no features in common between northward and southward IMF, whereas the Ørsted results have essentially the same regions of upward and downward current for both categories even though the intensities of the most poleward currents are greater for northward IMF, and the intensities of the equatorward currents are more intense for southward IMF. The key point for the present discussion is that the statistical results are not able to resolve the very different distributions of currents. In comparison with the AMPERE results, one can see that the single satellite results consist of a mixture of a range of conditions revealing the impact of the uncertainties noted above for these analyses.

With these comparisons in mind, we consider what features of the Birkeland current results for Mercury are the most reliable. The mixing of signals from different conditions would not lead to omission of a Region-2-sense set of currents, and these currents are clearly evident in the single satellite analysis when binned by geomagnetic activity so the nondetection of these currents in the analysis of the MESSENGER data should be reliable. That the NBZ currents were not resolved at Earth when binning data by activity level and were even not well distinguished in statistical analyses organized by IMF suggests that the MESSENGER results should not be taken as ruling out an analog of NBZ currents for northward IMF at Mercury. If present, their signature might be quite weak because there is no hint of a reverse-polarity set of currents either in individual orbits or in the quiet-case statistical results (cf. Fig. 17.10).

17.4.2. Intensity

It is clear from the magnetic signals that the total current at Mercury is much smaller than at Earth. The $\delta\mathbf{B}_\perp$ signatures in Figures 17.2 and 17.6 are not subject to averaging and the signals at Mercury are about one-tenth as large as those at Earth. For currents extended in local time, the $\delta\mathbf{B}_\perp$ observed across the current sheets is proportional to the net current per unit length along the current sheet, for example, in A/m. The difference in total current will therefore scale roughly with the planetary radius, so that the total Birkeland current at Mercury is slightly smaller than the ratio of the magnetic signatures. At Earth, $\delta\mathbf{B}_\perp$ are typically 100 to 1000 nT, while at Mercury the $\delta B_\perp$ range from 10 to 100 nT. Given the factor of ~3 difference in planetary radii, this rough comparison indicates that the total current at Mercury is about a factor of 30 lower than at Earth.

This is consistent with average $\delta B_\perp$ and total currents at Mercury calculated from the statistical results that are given in Table 17.1. For the magnetic signals, the table reports the 90% level of $\delta B_\perp$ evaluated for each of the first 3 yr of orbital observations by MESSENGER and separated by ascending (−A) and descending (−D) orbit segments. The 90% level was chosen to represent the near-maximum $\delta B_\perp$ while discriminating against outliers. The total current, I_{Tot}, is defined as half of the integral of the absolute value of the current so that it represents the current flowing through the system. Results were evaluated for all events averaged together and also segregated

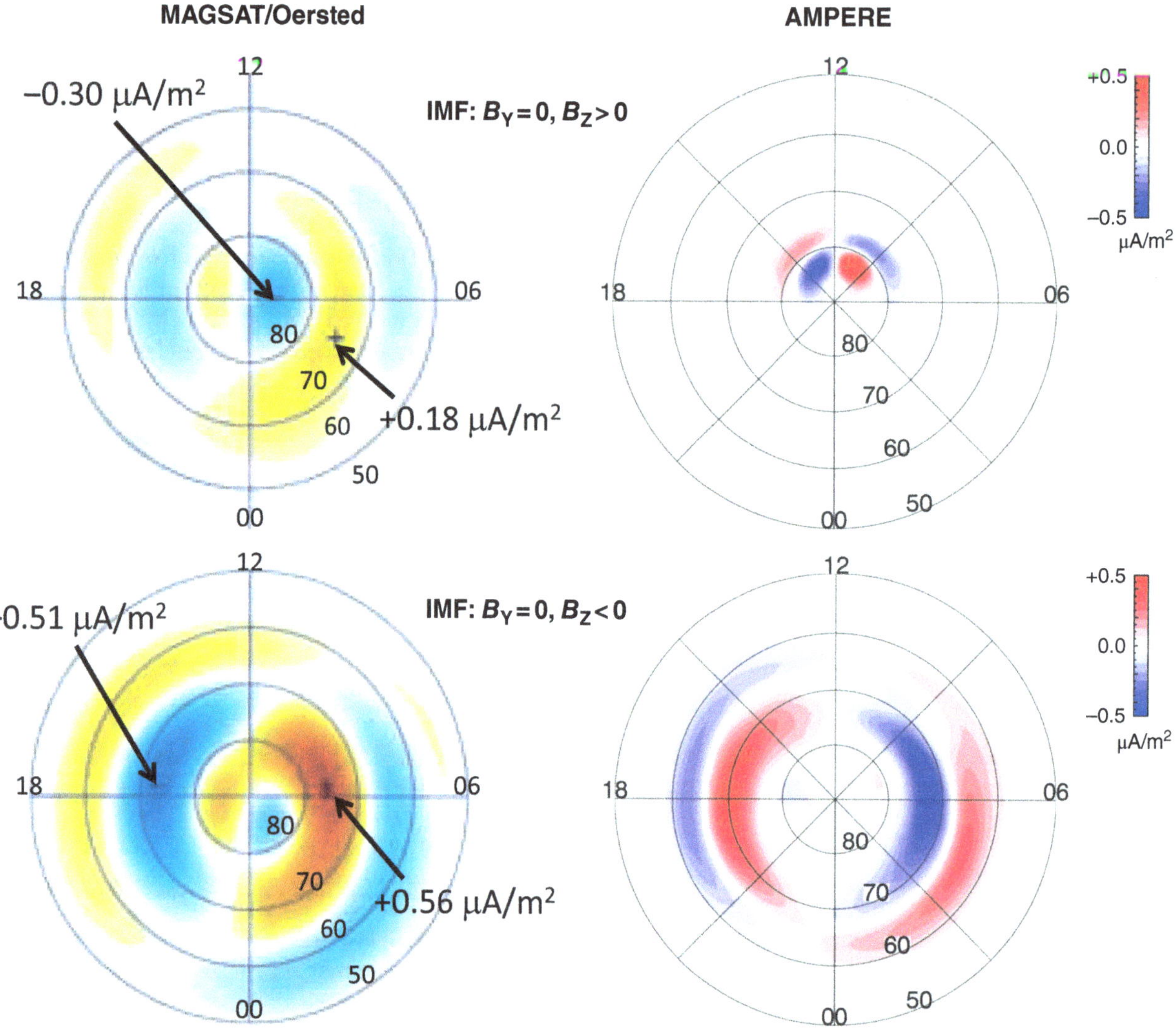

Figure 17.13 Comparison of two different approaches to deriving Birkeland current distributions at Earth for northward (top) and southward (bottom) directed interplanetary magnetic field (IMF). The latitude scales are the same even though the color scales differ. Panels on the left are from *Papitashvilli et al.* [2002] and show statistical averages from the Ørsted high-precision magnetic field mapping mission in which all data are used and binned by solar wind and IMF conditions estimated to correspond to Earth magnetospheric conditions from measurements at L1, ~230 R_E upstream of Earth in the solar wind. The results are shown as originally published with noon at the top and dusk at the left. Panels on the right are obtained from AMPERE and are averages of 10 min data windows using data from ~30 satellites in each 10 min window to derive a complete global estimate of the 2D Birkeland current distribution. The global nature of the AMPERE observations allowed selection of intervals only with very stable IMF and solar wind to ensure high confidence that the L1 observations correspond to estimated conditions at Earth. The IMF stability requirements were satisfied by only ~3% of the observations.

by activity level in five percentile bins. The ninetieth percentile $\delta B_\perp$ increase with magnetic disturbance level from 20 nT for the quietest conditions to ~60 nT for the most disturbed quintile and are about 35 nT on average. The total currents at Mercury range from 19 kA to just over 60 kA from the lowest to highest disturbance quintile and are 38 kA on average. This compares to a typical total Birkeland current at Earth of a 5 MA for moderately disturbed conditions and over 10 MA for storm conditions [cf. *Anderson et al.*, 2017] so that the total current is a factor of about 20 to 50 lower than at Earth.

The difference in the applied electrical potential is not the dominant factor in reducing the currents at Mercury relative to Earth. The stronger IMF at Mercury orbit partially makes up for the fact that Mercury's magnetosphere is much smaller than Earth's, so that the net potential resulting from the solar wind electric field potential at Mercury is on the order of 15 to 30 kV on average, compared with 50 to 100 kV at Earth. Thus, the primary reason for the dramatically lower current at Mercury is that the resistance for current closure through the planet is significantly greater than it is for closure via the ionosphere where they close at Earth.

17.4.3. Region 2 Currents

The apparent absence of Region 2 current is perhaps the most striking feature of the Birkeland currents at Mercury. Not only are Region-2-polarity currents missing in the statistical analyses (cf. Figs. 17.8 and 17.10), but the antisunward perturbations equatorward of the Region 1 currents that are so prominent at Earth (cf. Fig. 17.2) are absent in the observations at Mercury (cf. Figs. 17.6, 17.7, and 17.9). At Earth, the Region 2 currents are immediately adjacent to the Region 1 currents, a configuration that minimizes the energy dissipation of the corresponding closure and electrodynamics [*Barbosa*, 1988], so it is unlikely that there are Region 2 currents at Mercury equatorward of the observations because this would require a latitudinal gap of 20° or more and would be energetically unfavorable.

The Region 2 currents at Earth correspond primarily to pressure gradients (the first term of equation [17.3]) associated with the sunward convection of the plasma sheet and its diversion by the magnetic field nearer the Earth, to form the ring current and partial ring current. The source of these currents at Earth extends to within 5 or 6 R_E geocentric distance where the magnetic field is roughly 10 times its intensity in the more distant magnetotail. At Mercury, the surface magnetic field at midnight is not substantially greater than it is in the magnetotail [cf. *Korth et al.*, 2015; *Winslow et al.*, 2014] so that the inner edge of the plasma sheet at Mercury is evidently the planet itself. This marked difference between Mercury and Earth is illustrated in Figure 17.14, which shows Mercury and Earth to scale on the top and then the planets within their respective magnetospheres on the bottom where the magnetospheres are scaled to the same size. Comparing the schematics, the entire inner magnetosphere at Earth, including the ring current, radiation belts, and plasmasphere, occur in the region of space which at Mercury would be within the planet. More accurately, the closed drift paths that trap the ring current and radiation belts at Earth do not exist in a permanent sense at Mercury so that the plasma pressure populations responsible for the Region 2 currents (cf. Fig. 17.1) are not present except possibly very transiently [cf. *Baumjohann et al.*, 2010b; *Baker et al.*, 2016] consistent with the absence of the Region 2 currents at Mercury.

17.4.4. Convection and Plasma Flow

The absence of the Region 2 current has significant implications for Mercury's magnetosphere. At Earth, the Region 2 currents act to shield the middle and lower latitudes from the high latitude convection (cf. Fig. 17.3) by confining the potential to latitudes poleward of the Birkeland currents. At Mercury, this evidently does not happen, and the potential pattern is not confined. Figure 17.15 shows the electrostatic potential from case K of the current closure solution of Figure 17.12 for Mercury scaled for a 30 kV total potential difference. In comparison with the result of Figure 17.3, it is evident that the potential gradient equatorward of the Birkeland currents is much more gradual at Mercury that at Earth reflecting this lack of the shielding Region 2 currents.

At Earth, the strong potential gradients between the Region 1 and 2 currents correspond to the sunward return flows within the magetosphere to complete the sunward transport of magnetic flux to the dayside required by the global convection (cf. Fig. 17.4). Because the Region 2 currents are linked to plasmas at high altitudes (~4 to 8 R_E), this return convection flow occurs at high altitudes in the equatorial magnetosphere and, to first order, only affects the high latitude ionosphere.

At Mercury, however, the return flows are not confined to high latitudes. Where they actually occur is not known, but it is possible that the sunward flows in the plasma sheet on the nightside impact the planetary surface. This would imply that the planet itself, above the core, must be considered in the convection. The relationship between the external convection electric field and the electric field below the surface is given by the electric field boundary conditions. For a sunward convection velocity $\mathbf{U}_{conv}$ impacting the planet where the magnetic field is $\mathbf{B}$, the convection electric field imposed just above the surface would be $\mathbf{E}_{conv} = -\mathbf{U}_{conv} \times \mathbf{B}/B^2$. The boundary conditions at the surface constrain the internal electric field just below the surface, $\mathbf{E}_{int\text{-}S}$. The tangential electric field at the surface must be continuous so $\mathbf{E}_{int\text{-}S\perp} = \mathbf{E}_{conv\perp}$, where $\mathbf{E}_{\perp} = \mathbf{E} - \mathbf{r}(\mathbf{E}{\cdot}\mathbf{r})/r^2$ and $\mathbf{r}$ is the radial vector. The difference in the radial components of the electric fields is related to the surface charge density, ρ_s, by $(\mathbf{E}_{conv} - \mathbf{E}_{int\text{-}S}){\cdot}(\mathbf{r}/r) = \rho_s/\varepsilon_0$, where ε_0 is the permeability of free space. Solution of the electric field within the planet but above the core, where the very high conductivity imposes $\mathbf{E} = 0$ and $\rho = 0$, determines the distribution of the radial electric field and hence ρ_s.

The electric field imposed at the surface on the nightside by $\mathbf{E}_{conv}$ would be conveyed very rapidly within the planet and could appear on the dayside where it would need to satisfy corresponding boundary conditions with sunward flows, that is $\mathbf{U}_{conv}$ on the dayside just above the surface is also sunward. But this immediately raises the

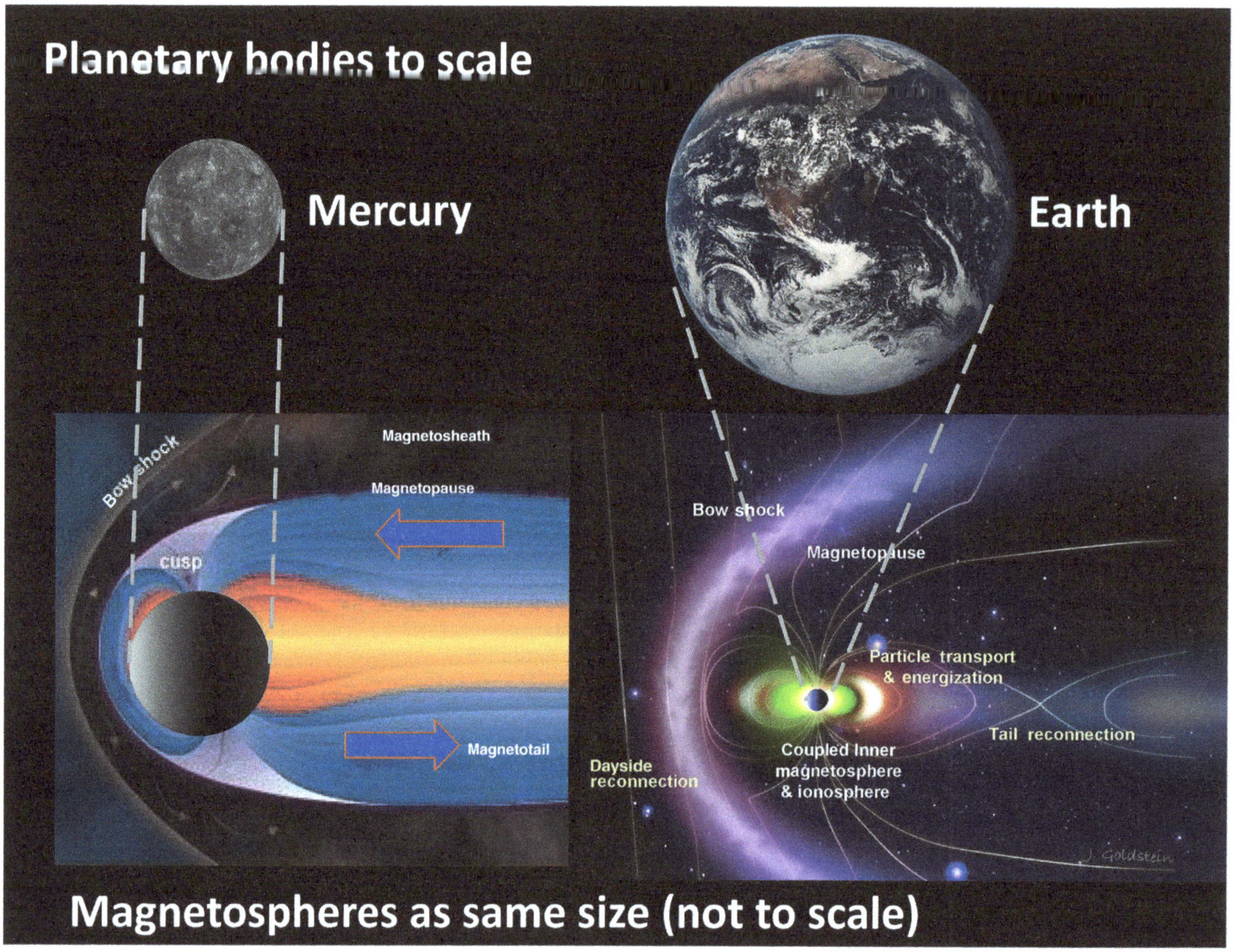

Figure 17.14 Mercury and Earth to scale (top) and in the context of their magnetospheres (bottom) where the magnetospheres are scaled to have the same size. Scaling at bottom illustrates the large fraction of the magnetospheric volume that Mercury occupies relative to Earth in its magnetosphere. Adapted from *Slavin et al.* [2007] and *Burch et al.* [2014].

following question: Where does the plasma come from to support flows lifting sunward off the surface? Without a source of plasma, sunward flows lifting off of the planet on the dayside cannot occur. Thus, how a duskward electric field at the subsolar surface of Mercury could be satisfied is an open question. The lack of Region 2 currents at Mercury therefore raises a question about how magnetic flux is returned to the dayside magnetopause at Mercury and this is a key open issue in the dynamics of Mercury's magnetosphere.

17.5. FUTURE DIRECTIONS

The magnetosphere of Mercury offers surprising contrasts with our home planetary system, and the contrasts provide opportunities to test our theories of planetary magnetospheres and improve our understanding of the basic processes governing these systems. Observationally, there are several avenues to be pursued including more extensive analysis of existing observations and the exciting prospects for new, dual-point, observations at Mercury in the not too distant future. Significant advances will also be possible with advanced physics-based simulations of the solar wind–IMF interaction with Mercury. There are several major challenges in these analyses for Mercury that need to be addressed if we are to fully understand the convection and electrodynamics of this comparatively tiny magnetosphere.

17.5.1. Analyses and Observations

Further analyses of the MESSENGER data could progress along various lines. It would be interesting to understand how well the average distributions of currents

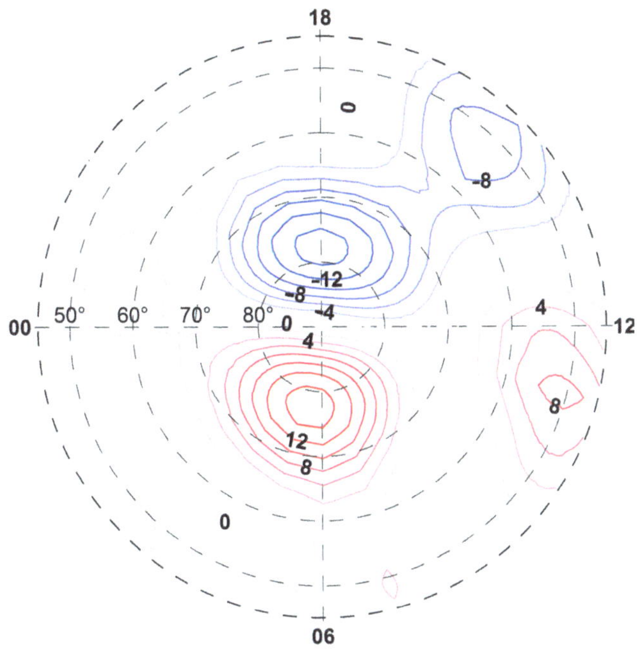

Figure 17.15 Surface potential in kV at Mercury for the average current distributions derived from MESSENGER observations for an assumed 30 kV total potential difference. Pattern is for the current closure solution case K of Figure 17.12 although the surface potential solution is essentially invariant to the conductivity structure assumed for the planet. Positive potential is shown in red contours and negative in blue contours. Contours are shown every 2 kV and labeled at 4 kV intervals. Noon is to the right and dawn at bottom. Invariant latitudes are shown from 45° to the pole.

derived here reflect the state of the system more generally, in particular, if very disturbed or very quiet conditions exhibit extensions of the averages or whether qualitatively new features in the currents and convection emerge under these extremes. Some work has already been done on specific occurrences of extreme states of the system [*Slavin et al.*, 2014], but how these states are reflected in the Birkeland currents is not yet known. The plasma flows and dynamical signatures of plasma convection toward the planet on the nightside and sunward on the dayside should be investigated to see whether constraints can be placed on the return convection to better understand how the convection closes.

The most exciting advances in observations, however, will come from new measurements to be made by the joint European and Japanese BepiColombo mission to Mercury [*Novara*, 2002; *Balogh et al.*, 2007]. This mission offers a number of fundamental advances relative to the measurements made by MESSENGER. First, BepiColombo is a two-spacecraft mission with a low-altitude orbiter, the Mercury Planetary Orbiter (MPO), that will orbit below ~1500 km altitude, remaining within Mercury's magnetosphere for most of its orbit and will obtain the first observations of magnetic fields from relatively low altitudes in the southern hemisphere [*Glassmeier et al.*, 2010]. In addition, the Mercury Magnetosphere Orbiter (MMO) will provide simultaneous observations at higher altitudes [e.g., *Baumjohann et al.*, 2010a], at times in the magnetotail and at other times in the solar wind, allowing a much better understanding of the dynamical processes linking high and low altitudes. Moreover, the MMO will carry instruments to measure the plasma flows and electric fields, providing direct determinations of the convection transport throughout the system [*Delcourt et al.*, 2009; *Miyake et al.*, 2009; *Kasaba et al.*, 2010; *Saito et al.*, 2010; *Sauvaud et al.*, 2010]. The advances both in plasma and field measurements as well as the dual-point nature of the observations promise to resolve many of the enigmas raised by the MESSENGER observations.

17.5.2. Modeling and Simulation

Physical simulations of Mercury's magnetosphere and its interaction with the solar wind and IMF have already led to major advances. As anticipated [*Russell et al.*, 1988; *Glassmeier*, 1997; *Slavin et al.*, 2007; *Müller et al.*, 2012, *Jia et al.*, 2015], ion kinetic processes are fundamentally important and govern many features of the system. The prevalence of quasi-parallel bow shock conditions in the subsolar region for instance imply that the intrinsic dynamics of the shock lead to particle energization and turbulence in the magnetosheath, which affects the magnetopause [e.g., *Trávníček et al.*, 2010]. Given the implications of the Birkeland current results, it will be particularly important to include boundary conditions with the planet itself to understand how the planet participates in the interaction, by closing the Birkeland currents, by conveying electric field dynamics from the nightside to the dayside and thereby perhaps participating in the convection cycle, and by reacting to imposed changes in the magnetic field via induction in the core and mantle [*Jia et al.*, 2015; *Johnson et al.*, 2016]. It is to be expected that interpretation of the dual-point observations from BepiColombo will be greatly facilitated by the availability of such simulations to quantitatively test their consistency with the new observations.

17.5.3. Power of Comparative Magnetospheres

The prospects for continued analyses, new observations, and sophisticated new physical models and simulations suggest that the future for studies of Mercury's solar wind interaction is very bright. Moreover, as our understanding of the dynamics of Mercury's magnetosphere improve, the contrast with Earth's system and those of other planets will most likely become more clear.

Even the initial comparisons of the field-aligned currents at Earth and Mercury presented here reveal fundamental differences between the two systems. The ways that Mercury's planetary body itself reacts to and perhaps governs the magnetosphere is profoundly interesting and reflects a novel aspect in magnetosphere-planet interactions. The prevalence of exoplanets increases the interest in expanding our predictive understanding of the electromagnetic effects on planets due to interactions with their local interplanetary environment and quantitative comparison of the magnetospheres in our own solar system remains one of the most fruitful avenues for these inquiries.

ACKNOWLEDGMENTS

The MESSENGER project was supported by the NASA Discovery Program under contracts NAS5-97271 to the Johns Hopkins University Applied Physics Laboratory and NASW-00002 to the Carnegie Institution of Washington. C. L. Johnson is supported by MESSENGER Participating Scientist grants NNX11AB84G and NNH08CC05C. L. C. Philpott and C. L. Johnson acknowledge support from the Natural Sciences and Engineering Research Council of Canada. Support for AMPERE has been provided under NSF sponsorship under grants ATM-0739864 and AGS-1420184. We are greatly indebted to the contribution of Iridium Communications for providing the data for AMPERE. Data products from AMPERE are available via http://ampere.jhuapl.edu. The MESSENGER data are available via the Planetary Data System and additional data products derived for this work may be obtained by request from the lead author.

REFERENCES

Anderson, B. J., C. L. Johnson, and H. Korth (2013), A magnetic disturbance index for Mercury's magnetic field derived from MESSENGER Magnetometer data, *Geochem. Geophys. Geosyst.*, *14*, 3875–3886; doi:10.1002/ggge.20242.

Anderson, B. J., C. L. Johnson, H. Korth, J. A. Slavin, R. M. Winslow, R. J. Phillips, R. L. McNutt Jr., and S. C. Solomon (2014a), Steady-state field-aligned currents at Mercury, *Geophys. Res. Lett.*, *41*, 7444–7452; doi:10.1002/2014GL061677.

Anderson, B. J., C. L. Johnson, H. Korth, R. M. Winslow, J. E. Borovsky, M. E. Purucker, J. A. Slavin, et al.. (2012), Low-degree structure in Mercury's planetary magnetic field, *J. Geophys. Res.*, *117*, E00l12; doi:10.1029/2012JE004159.

Anderson, B. J., H. Korth, C. L. Johnson, R. J. Phillips, L. C. Philpott, and S. C. Solomon (2016), Closure of Birkeland currents at Mercury: Constraints on the electrical conductivity of the crust and mantle, *Lunar and Planet. Sci.*, *47*, abstract 1243.

Anderson, B. J., H. Korth, C. L. Waters, D. L. Green, and P. Stauning (2008), Statistical Birkeland current distributions from magnetic field observations by the Iridium constellation, *Annales Geophys.*, *26*, 671–687; doi:10.5194/angeo-26-671-2008.

Anderson, B. J., H. Korth, C. L. Waters, D. L. Green, V. G. Merkin, R. J. Barnes, and L. P. Dryud (2014b), Development of large-scale Birkeland currents determined from the Active Magnetosphere and Planetary Electrodynamics Response Experiment, *Geophys. Res. Letters*, *41*, 3017–3025; doi:10.1002/2014GL059941.

Anderson, B. J., H. Korth, D. T. Welling, V. G. Merkin, M. J. Wiltberger, J. Raeder, R. J. Barnes, C. L. Waters, A. A. Pulkkinen, and L. Rastaetter (2017), Comparison of predictive estimates of high latitude electrodynamics with observations of global scale Birkeland currents, *Space Weather*, *15*; doi:10.1002/2016SW001529.

Anderson, B. J., K. Takahashi, and B. A. Toth (2000), Sensing global Birkeland currents with Iridium® engineering magnetometer data, *Geophys. Res. Lett.*, *27*, 4045–4048.

Baker, D. N., R. M. Dewey, D. J. Lawrence, J. O. Goldsten, P. N. Peplowski, H. Korth, J. A. Slavin, et al. (2016), Intense energetic electron flux enhancements in Mercury's magnetosphere: An integrated view with high-resolution observations from MESSENGER, *J. Geophys. Res.: Space Phys.*, *121*, 2171–2184; doi:10.1002/2015JA021778.

Balogh, A., R. Grard, S. C. Solomon, R. Schulz, Y. Langevin, Y. Kasaba, and M. Fujimoto (2007), Missions to Mercury, *Space Sci. Rev.*, *132*, 611–645; doi:10.1007/s11214-007-9212-4.

Barbosa, D. D. (1988), Minimal joule dissipation models of magnetospheric convection, *Pure Appl. Geophys.*, *127*, 473–489; doi:10.1007/BF00879822.

Baumjohann, W., A. Matsuoka, W. Magnes, K.-H. Glassmeier, R. Nakamura, H. Biernat, M. Delva, et al. (2010a), Magnetic field investigation of Mercury's magnetosphere and the inner heliosphere by MMO/MGF, *Planet. Space Sci.*, *58*, 279–286; doi:10.1016/j.pss.2008.05.019.

Baumjohann, W., M. Blanc, A. Fedorov, and K.-H. Glassmeier (2010b), Current systems in planetary magnetospheres and ionospheres, *Space. Sci. Rev.*, *152*, 99–134; doi:10.1007/s11214-010-9629-z.

Boardsen, S. A., J. A. Slavin, B. J. Anderson, H. Korth, D. Schriver, and S. C. Solomon (2012), Survey of coherent ~1 Hz waves in Mercury's inner magnetosphere, *J. Geophys. Res.*, *117*, A00M05; doi:10.1029/2012JA017822.

Brandt, P. C.:Son, E. C. Roelof, S. Ohtani, D. G. Mitchell, and B. Anderson (2004), IMAGE/HENA: Pressure and current distributions during the 1 October 2002 storm, *Adv. Space Res.*, *33*, 719–722; doi:10.1016/S0273-1177(03)00633-1.

Cheng, A. F., R. E. Johnson, S. M. Krimigis, and L. J. Lanzerotti (1987), Magnetosphere, exosphere, and surface of Mercury, *Icarus*, *71*, 430–440; doi:10.1016/0019-1035(87)90038-8.

Clausen, L. B. N., J. B. H. Baker, J. M. Ruohoniemi, S. E. Milan, and B. J. Anderson (2012), Dynamics of the region 1 Birkeland current oval derived from the Active Magnetosphere and Planetary Electrodynamics Response Experiment (AMPERE), *J. Geophys. Res.*, *117*, A06233; doi:10.1029/2012JA017666.

Cowley, S. W. H. (2000), Magnetosphere-ionosphere interactions: A tutorial review, in *Magnetospheric Current Systems, Geophysical Monograph Series*, *118*, edited by S. Ohtani, R. Fujii, M. Hesse, and R. L Lysak, 91–106, American Geophysical Union, Washington, DC.; doi:10.1029/GM118p0091.

Delcourt, D., Y. Saito, J.-M. Illiano, N. Krupp, J.-J. Berthelier, D. Fontaine, M. Fraenz, et al. (2009), The mass spectrum analyzer (MSA) onboard BEPI COLOMBO MMO: Scientific objectives and prototype results, *Adv. Space Res.*, *43*, 869–874; doi:10.1016/j.asr.2008.12.002.

DiBraccio, G. A., J. A. Slavin, S. A. Boardsen,, B. J. Anderson, H. Korth, T. H. Zuburchen, J. M. Raines, D. N. Baker, R. L. McNutt Jr., and S. C. Solomon (2013), MESSENGER observations of magnetopause structure and dynamics at Mercury, *J. Geophys. Res. Space Physics*, *118*, 997–1008; doi: 10.1002/jgra.50123.

Dungey, J. W. (1961), Interplanetary magnetic field and the auroral zones, *Phys. Rev. Lett.*, *6*, 47–48; doi:10.1103/PhysRevLett.6.47.

Gershman, D. J., J. A. Slavin, J. M. Raines, T. H. Zurbuchen, B. J. Anderson, H. Korth, D. N. Baker, and S. C. Solomon (2013), Magnetic flux pileup and plasma depletion in Mercury's subsolar magnetosheath, *J. Geophys. Res.*, *118*, 7181–7199; doi:10.1002/2013JA019244.

Glassmeier, K.-H. (1997), The Hermean magnetosphere and its ionosphere–magnetosphere coupling, *Planet. Space Sci.*, *45*, 119–125; doi:10.1016/S0032-0633(96)00095-5.

Glassmeier, K.-H. (2000), Currents in Mercury's magnetosphere, in *Magnetospheric Current Systems, Geophysical Monograph Series*, *118*, edited by S.-I. Ohtani, R. Fujii, M. Hesse, and R. L. Lysak, 371–380, American Geophysical Union, Washington, DC; doi:10.1029/GM118p0371.

Glassmeier, K.-H., H.-U. Auster, D. Heyner, K. Okrafka, C. Carr, G. Berghofer, B. J. Anderson, et al. (2010), The fluxgate magnetometer of the BepiColombo Mercury Planetary Orbiter, *Planet. Space Sci.*, *58*, 287–299; doi:10.1016/j.pss.2008.06.018.

Hauck, S. A., II, J.-L. Margot, S. C. Solomon, R. J. Phillips, C. L. Johnson, F. G. Lemoine, E. Mazarico, et al. (2013), The curious case of Mercury's internal structure, *J. Geophys. Res. Planets*, *118*, 1204–1220; doi:10.1002/jgre.20091.

Heyner, D., C. Nabert, E. Liebert, and K.-H. Glassmeier (2016), Concerning reconnection-induction balance at the magnetopause of Mercury, *J. Geophys. Res.: Space Phys.*, *121*, 2935–2061; doi:10.1002/2015JA021484.

Iijima, T., and T. A. Potemra (1976), The amplitude distribution of field-aligned currents at northern high latitudes observed by Triad, *J. Geophys. Res.*, *81*, 2165–2174; doi:10.1029/JA081i013p02165.

Iijima, T., T. A. Potemra, L. J. Zanetti, and P. F. Bythrow (1984), Large-scale Birkeland currents in the dayside polar region during strongly northward IMF, A new Birkeland current system, *J. Geophys. Res.*, *89*, 7441–7452; doi:10.1029/JA089iA09p07441.

Ip, W.-H., and A. Kopp (2004), Mercury's Birkeland current system, *Adv. Space Res.*, *33*, 2171–2175; doi:10.1016/S0273-1177(03)00444-7.

Janhunen, P., and E. Kallio (2004), Surface conductivity of Mercury provides current closure and may affect magnetospheric symmetry, *Ann. Geophys.*, *22*, 1829–1837; doi:10.5194/angeo-22-1829-2004.

Jia, X., J. A. Slavin, T. I. Gombosi, L. K. S. Daldorff, G. Toth, and B. Holst (2015), Global MHD simulations of Mercury's magnetosphere with coupled planetary interior: Induction effect of the planetary conducting core on the global interaction, *J. Geophys. Res.: Space Phys.*, *120*, 4763–4775; doi:10.1002/2015JA021143.

Johnson, C. L., L. C. Philpott, B. J. Anderson, H. Korth, S. A. Hauck II, D. Heyner, R. J. Phillips, R. M. Winslow, and, S. C. Solomon (2016), MESSENGER observations of induced magnetic fields in Mercury's core, *Geophys. Res. Lett.*, *43*; doi:10.1002/2015GL067370.

Johnson, C. L., M. E. Purucker, H. Korth, B. J. Anderson, R. M. Winslow, M. M. H. Al Asad, J. A. Slavin, et al. (2012), MESSENGER observations of Mercury's magnetic field structure, *J. Geophys. Res.*, *117*, E00L14. doi:10.1029/2012JE004217.

Kasaba, Y., J.-L. Bougeret, L. G. Blomberg, H. Kojima, S. Yagitani, M. Moncuquet, J.-G. Trotignon, et al. (2010), The Plasma Wave Investigation (PWI) on board the BepiColombo/MMO: First measurement of electric fields, electromagnetic waves, and radio waves around Mercury, *Planet. Space Sci.*, *58*, 238–278; doi:10.1016/j.pss.2008.07.017.

Knight, S. (1973), Parallel electric fields, *Planet. Space Sci.*, *21*, 741–750; doi:10.1016/0032-0633(73)90093-7.

Korth, H., B. J. Anderson, and C. L. Waters (2010), Statistical analysis of the dependence of large-scale Birkeland currents on solar wind parameters, *Ann. Geophys.*, *28*, 515–530.

Korth, H., B. J. Anderson, C. L. Johnson, R. M. Winslow, J. A. Slavin, M. E. Purucker, S. C. Solomon, and R. L. McNutt Jr. (2012), Characteristics of the plasma distribution in Mercury's equatorial magnetosphere derived from MESSENGER Magnetometer observations, *J. Geophys. Res.*, *117*, A00M07; doi: 10.1029/2012JA018052.

Korth, H., B. J. Anderson, D. J. Gershmann, J. M. Raines, J. A. Slavin, T. H. Zurbuchen, S. C. Solomon, and R. L. McNutt Jr. (2014), Plasma distribution in Mercury's magnetosphere derived from MESSENGER Magnetometer and Fast Imaging Plasma Spectrometer observations, *J. Geeophys. Res. Space Phys.*, *119*, 2917–2932; doi:10.1002/2013JA019567.

Korth, H., N. A. Tsyganenko, C. L. Johnson, L. C. Philpott, B. J. Anderson, M. M. Al Asad, S. C. Solomon, and, R. L. McNutt Jr. (2015), Modular model for Mercury's magnetospheric magnetic field confined within the average observed magnetopause, *J. Geophys. Res. Space Phys.*, *120*, 4503–4518. doi:10.1002/2015JA021022.

Lyatsky, W., G. V. Khazanov, and J. A. Slavin (2010), Alfvén wave reflection model of field-aligned currents at Mercury, *Icarus*, *209*, 40–45; doi:10.1016/j.icarus.2009.11.039.

McAdams, J. V., R. W. Farquhar,, A. H. Taylor, and, B. G. Williams (2007), MESSENGER mission design and navigation, *Space Sci. Rev.*, *131*, 219–246.

McAdams, J. V., S. C. Solomon, P. D. Bedini, E. J. Finnegan, R. L. McNutt Jr., A. B. Calloway, D. P. Moessner, et al. (2012), MESSENGER at Mercury: From orbit insertion to first extended mission, *63rd International Astronautical Congress*, Paper IAC-12-C1.5.6, Naples, Italy, October 1–5.

Miyake, W., Y. Saito, M. Harada, M. Saito, H. Hasegawa, A. Ieda, S. Machida, et al. (2009), Mercury Ion Analyzer (MIA) on board Mercury Magnetospheric Orbiter: MMO, *Adv. Space Res.*, *43*, 1986–1992; doi:10.1016/j.asr.2009.03.011.

Müller, J., S. Simon, Y.-C. Wang, U. Motschmann, D. Heyner, J. Schüle, W.-H. Ip, G. Kleindienst, and G. J. Pringle (2012), Origin of Mercury's double magnetopause: 3D hybrid simulation study with A.I.K.E.F., *Icarus*, *218*, 666–687; doi:10.1016/j.icarus.2011.12.028.

Ness, N. F., K. W. Behannon, R. P. Lepping, and, Y. C. Whang (1975), Magnetic field of Mercury confirmed, *Nature*, *255*, 204–205; doi:10.1038/255204a0.

Ness, N. F., K. W. Behannon, R. P. Lepping, and Y. C. Whang (1976), Observations of Mercury's magnetic field, *Icarus*, *28*, 479–488, doi:10.1016/0019-1035(76)90121-4.

Novara, M. (2002), The BepiColombo ESA cornerstone mission to Mercury, *Acta Astronautica*, *51*, 387–395; doi:10.1016/S0094-5765(02)00065-6.

Papitashvili, V. O., F. Christiansen, and T. Neubert (2002), A new model of field-aligned currents derived from high-precision satellite magnetic field data, *Geophys. Res. Lett.*, *29*, 1683; doi:10.1029/2001GL014207.

Philpott, L. C., C. L. Johnson, R. M. Winslow, B. J. Anderson, H. Korth, M. E. Purucker, and S. C. Solomon (2014), Constraints on the secular variation of Mercury's magnetic field from the combined analysis of MESSENGER and Mariner 10 data, *Geophys. Res. Lett.*, *41*, 6627–6634; doi:10.1002/2014GL061401.

Richmond, A. D., and J. P. Thayer (2000), Ionospheric dynamics: A tutorial, in *Magnetospheric Current Systems, Geophysical Monograph Series*, *118*, edited by S. Ohtani, R. Fujii, M. Hesse, and R. L. Lysak, 131–146, American Geophysical Union, Washington, DC; doi:10.1029/GM118p0131.

Russell, C. T., D. N. Baker, and J. A. Slavin (1988), The magnetosphere of Mercury, in *Mercury*, edited by F. Vilas, C. R. Chapman, and M. S. Matthews, University of Arizona Press, Tuscon, 514–561.

Saito, Y., J. A. Sauvaud, M. Hirahara, S. Barabash, D. Delcourt, T. Takashima, and K. Asamura (2010), Scientific objectives and instrumentation of Mercury Plasma Particle Experiment (MPPE) on board MMO, *Planet. Space Sci.*, *58*, 182–200; doi:10.1016/j.pss.2008.06.003.

Sauvaud, J. A., A. Fedorov, C. Aoustin, H.-C. Seran,, E. Le Comte, M. Petiot, J. Rouzaud, et al. (2010), The Mercury Electron Analyzers for the Bepi Colombo mission, *Adv. Space Res.*, *46*, 1139–1148; doi:10.1016/j.asr.2010.05.022.

Slavin, J. A., B. J. Anderson, D. N. Baker, M. Benna, S. A. Boardsen, G. Gloecker, R. E. Gold, et al. (2010), MESSENGER observations of extreme loading and unloading of Mercury's magnetic tail, *Science*, *329*, 665–668; doi:10.1126/science.1188067.

Slavin, J. A., G. A. DiBraccio, D. J. Gershman, S. M. Imber, G. K. Poh, J. M. Raines, T. H. Zurbuchen, et al. (2014), MESSENGER observations of Mercury's dayside magnetosphere under extreme solar wind conditions, *J. Geophys. Res. Space Phys.*, *119*(10), 8087–8116. doi:10.1002/2014ja020319.

Slavin, J. A., J. C. J. Owen, J. E. P. Connerney, and S. P. Christon (1997), Mariner 10 observations of field-aligned currents at Mercury, *Planet. Space Sci.*, *45*, 133–141.

Slavin, J. A., M. H. Acuña, B. J. Anderson, D. N. Baker, M. Benna, S. A. Boardsen, G. Gloecker, et al. (2009), MESSENGER observations of magnetic reconnection in Mercury's magnetosphere, *Science*, *324*, 606–610; doi:10.1126/science.1172011.

Slavin, J. A., S. M. Krimigis, M. H. Acuña, B. J. Anderson, D. N. Baker, P. L. Koehn, H. Korth, S. Livi, B. H. Mauk, S. C. Solomon, and T. H. Zurbuchen (2007). MESSENGER: Exploring Mercury's magnetosphere. *Space Sci. Rev.*, *131*, 133–160, doi:10.1007/s11214-007-9154-x.

Solomon, S. C., R. L. McNutt, R. E. Gold, M. H. Acuna, D. N. Baker, W. V. Boynton, C. R. Chapman, et al. (2001), The MESSENGER mission to Mercury: Scientific objectives and implementation, *Planet. Space Sci.*, *49*, 1445–1465.

Strangeway, R. J. (2012), The relationship between magnetospheric processes and auroral field-aligned current morphology, in *Auroral Phenomenology and Magnetospheric Processes: Earth and Other Planets, Geophysical. Monograph Series*, *197*, edited by A. Keiling, 355–364; doi:10.1029/2012GM001211.

Sun, W.-J., B. J. Anderson, D. N. Baker, S. Fu, D. J. Gershman, S. M. Imber, et al. (2015), MESSENGER observations of magnetospheric substorm activity in Mercury's near magnetotail, *Geophys. Res. Lett.*, *42*, 3692–3699; doi:10.1002/2015GL064052.

Sundberg, T., B. J. Anderson, D. N. Baker, S. A. Boardsen, G. C. Ho, S. M. Krimigis, H. Korth, et al. (2012), MESSENGER observations of dipolarization events in Mercury's magnetotail, *J. Geophys. Res.*, *117*; doi:10.1029/2012JA017756.

Trávníček, P. M., D. Schriver, P. Helliner, D. Herčík, B. J. Anderson, M. Sarantos, and J. A. Slavin (2010), Mercury's magnetosphere–solar wind interaction for northward and southward interplanetary magnetic field: Hybrid simulation results, *Icarus*, *209*, 11–22; doi:10.1016/j.icarus.2010.01.008.

Vasyliūnas, V. M. (1970), Mathematical models of magnetospheric convection and its coupling to the ionosphere, in Particles and Fields in the Magnetosphere, edited by B. M. McCormack, pp. 60–71, D. Reidel, Dordrecht, The Netherlands.

Verhoeven, O., P. Tarits, P. Vacher, A. Rivoldini, and T. Van Hoolst (2009), Composition and formation of Mercury: Constraints from future electrical conductivity measurements, *Planet. Space Sci.*, *57*, 296–305; doi:10.1016/j.pss.2008.11.015.

Waters, C. L., B. J. Anderson, and K. Liou (2001), Estimation of global field aligned currents using Iridium magnetometer data, *Geophys. Res. Lett.*, *28*, 2165–2168.

Weimer, D. R. (2005), Improved ionospheric electrodynamic models and application to calculating Joule heating rates, *J. Geophys. Res.*, *110*, A05 306; doi:10.1029/2004JA010884.

Winslow, R. M., B. J. Anderson, C. L. Johnson, J. A. Slavin, H. Korth, M. E. Purucker, D. N. Baker, and S. C. Solomon (2013), Mercury's magnetopause and bow shock from MESSENGER Magnetometer observations, *J. Geophys. Res. Space Physics*, *118*, 2213–2227; doi:10.1002/jgra.50237.

Winslow, R. M., C. L. Johnson, B. J. Anderson, D. J. Gershman, J. M. Raines, R. J. Lillis, H. Korth, et al. (2014), Mercury's surface magnetic field determined from proton-reflection magnetometry, *Geophys. Res. Lett.*, *41*, 4463–4470; doi:10.1002/2014GL060258.

Winslow, R. M., C. L. Johnson, B. J. Anderson, H. Korth, J. A. Slavin, M. E. Purucker, and S. C. Solomon (2012), Observations of Mercury's northern cusp region with MESSENGER's Magnetometer, *Geophys. Res. Lett.*, *39*, L08112; doi:10.1029/2012GL051472.

Zanetti, L. J., T. A. Potemra, P. F. Bythrow, T. Iijima, and W. Baumjohann (1984), Ionospheric and Birkeland current distributions for northward interplanetary magnetic field: Inferred polar convection, *J. Geophys. Res.*, *89*, 7453–7458; doi:10.1029/JA089iA09p07453.

18

Recent Advances in the Study of Upward Field-aligned Currents Generated Near the Earth's Magnetopause Boundary

Simon Wing[1], Jay R. Johnson[2], and Marius Echim[3]

ABSTRACT

Sheared plasma flows in the boundary layer have been associated with field-aligned currents (FACs) and particle precipitation. While the currents are generated by the potential difference across the magnetopause, the ionosphere plays a key role in controlling the currents. A magnetosphere-ionosphere coupling model based on a Vlasov equilibrium solution for the generator is discussed. A simple analytical theory based on velocity shear generator and a quasi-stationary magnetosphere-ionosphere coupling model are reviewed. The comparisons of the theory with DMSP upward afternoon R1 FAC density ($J_{||}$) and FAC width (Λ) that are located at the low-latitude boundary layer (LLBL) show remarkably good scaling of R1 with solar wind and ionospheric parameters. It is found that $\Lambda \sim n_{sw}^{-0.5}$ and $\Lambda \sim L$ when $\Lambda/L < 5$ where L is the auroral electrostatic scale length. The average LLBL thickness (Δ_m) is inferred to be around 3000 km, which appears to have weak dependence on Vsw. $J_{||}$ has dependencies on Δ_m, Σ_p, n_{sw}, and V_{sw}. The analytical theory provides a simple way to organize data and to infer boundary layer structures from ionospheric data. The calculated and observed $J_{||}$ are in good agreement. Excluding data from noon, $\log(J_{||,max\_cal}) = (0.96 \pm 0.04) \log(J_{||\_obs}) + (0.03 \pm 0.01)$ where $J_{||,max\_cal}$ = calculated $J_{||,max}$ and $J_{||\_obs}$ = observed $J_{||}$.

18.1. INTRODUCTION

A region where solar wind constantly transfers mass, momentum, and energy to the Earth's magnetosphere is the low-latitude boundary layer (LLBL) near the magnetopause [e.g., *Hones et al.*, 1972; *Eastman et al.*, 1976; *Rosenbauer et al.*, 1975; *Crooker*, 1977; *Haerendel et al.*, 1978; *Paschmann et al.*, 1976; *Wing et al.*, 2005]. At the outer layer of the LLBL, the magnetosheath plasma flows in the antisunward direction whereas at the inner layer, the magnetospheric plasma flow is mildly sunward or relatively stagnant, leading to a velocity shear. The velocity shear at the boundary layer gives rise to a potential difference across the boundary, which can drive upward and downward field-aligned currents (FACs) [e.g., *Sonnerup*, 1980; *Stern*, 1983; *Lotko et al.*, 1987; *Siscoe et al.*, 1991; *Iijima and Potemra*, 1976]. Moreover, plasma structures and processes in the LLBL have been linked to discrete auroral arcs at high- latitude polar region in the early afternoon. Plasma interfaces with velocity shears can often be described as tangential structures/discontinuities for which theoretical models based on kinetic theory provide the relationship between the electric field normal to the interface and the velocity shear [e.g., *Roth et al.*, 1996; *Echim and Lemaire*, 2005].

At the low-latitude magnetosheath and the outer layer of the LLBL, the plasma flow is antisunward and as a result, the convection electric field **E** is dawnward. On the other hand, at the low-latitude magnetosphere and the inner layer of LLBL, the plasma flow is mildly sunward, resulting in a duskward electric field **E**. This has an

[1] *Applied Physics Laboratory, Johns Hopkins University, Laurel, Maryland, USA*
[2] *Andrews University, Berrien Spring, Michigan, USA*
[3] *Institut d'Aeronomie Spatiale de Belgique, Brussels, Belgium; Institute of Space Science, Măgurele, Romania*

Electric Currents in Geospace and Beyond, Geophysical Monograph 235, First Edition.
Edited by Andreas Keiling, Octav Marghitu, and Michael Wheatland.

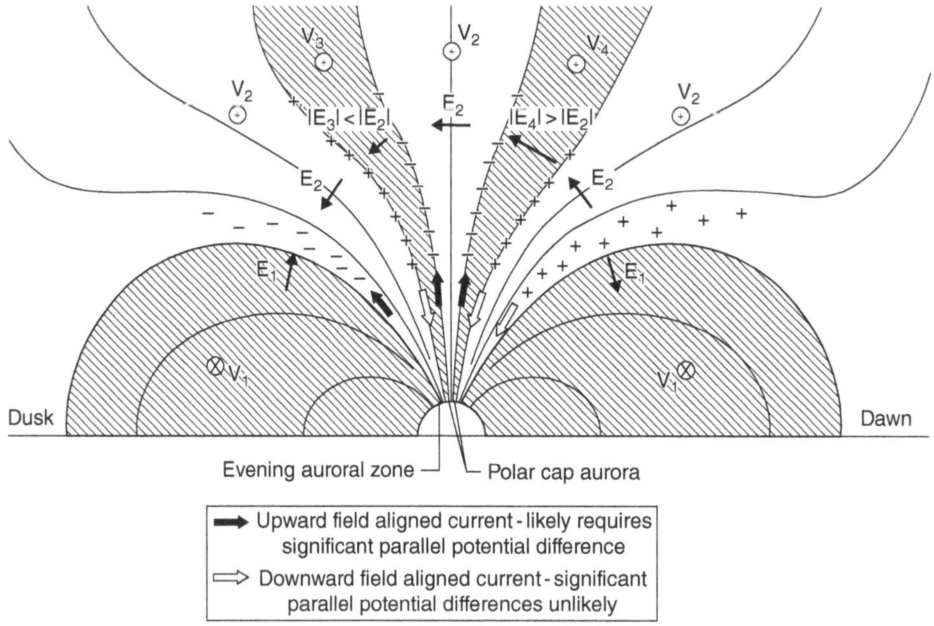

Figure 18.1 Schematic illustration of magnetic field lines emanating from the dusk-dawn meridian plane as viewed sunward from the magnetotail, under the assumption that the polar cap magnetic field lines open into the magnetosheath. A converging electric field exists along the evening boundary between sunward and antisunward convection and along any flow-aligned discontinuity in the magnetosheath flow, with flow speed increasing from left to right in the figure. Such converging electric fields are associated with negatively charged regions that drive upward-field-aligned currents that are likely to require significant field-aligned electric potential differences. As diverging electric field exists along the morning boundary between sunward and antisunward convection and along magnetosheath flow discontinuities, flow speed decreases from left to right in the figure. Such diverging electric fields are associated with positively charged regions, which drive downward field-aligned currents that generally will not require significant field-aligned potential differences. From *Lyons* [1980].

interesting consequence. In the afternoon and near dusk, the electric field at the boundary layer converges, $\nabla \cdot \mathbf{E} < 0$, suggesting a negatively charged discontinuity that leads to upward field-aligned current (FAC), as shown in Figure 18.1 [*Lyons*, 1980]. On the other hand, in the morning and near dawn, the electric field at the boundary layer diverges, $\nabla \cdot \mathbf{E} < 0$, suggesting a positively charged discontinuity that leads to downward FAC. Observationally, Region 1 (R1) FAC in the morning flows downward while R1 in the afternoon flows upward [e.g., *Iijima and Potemra*, 1976], consistent with the expected FAC generated at the LLBL, which led researchers to associate R1 generation with the velocity shear at LLBL. However, studies have shown that not all R1 is located entirely at LLBL or on open-field lines. A portion of R1, perhaps the lower latitude portion, can be located on the closed-field lines well inside the magnetosphere where the convection is sunward and the current generation mechanism may not be the solar wind–magnetospheric plasma velocity shear [e.g., *Bythrow et al.*, 1981; *Vennerstrom et al.*, 2005; *Haraguchi et al.*, 2004; *Xu and Kivelson*, 1994; *Wing et al.*, 2010]. The source regions of R1 are shown in Figure 18.2.

Sonnerup [1980] developed an elaborate one-dimensional steady-state fluid mechanical model that illustrates the FAC generation on the dayside LLBL due to the velocity shear. The model is quite general as it applies to downward and upward FAC in the morning and afternoon Region 1 (R1), respectively, which are assumed to be located at the LLBL, and their adjacent Region 2 (R2). The model does not take into account any kind of turbulence.

A series of papers followed the work of *Sonnerup* [1980] and *Lyons* [1980] exploring how the properties of the boundary layer are controlled by the solar wind and ionospheric parameters. *Lotko et al.* [1987] extended the steady-state model of *Sonnerup* [1980] to include time dependence in a two-dimensional model of the low-latitude boundary layer, which described how nonsteady, incompressible, viscous flows couple to the ionosphere through field-aligned currents. In the model, perturbations in the boundary layer decay away in time through (1) ionospheric drag at large scales, (2) electron acceleration at mesoscales, and (3) viscous damping at small scales. At large scales, the damping is anisotropic in the latitudinal and longitudinal directions with a faster

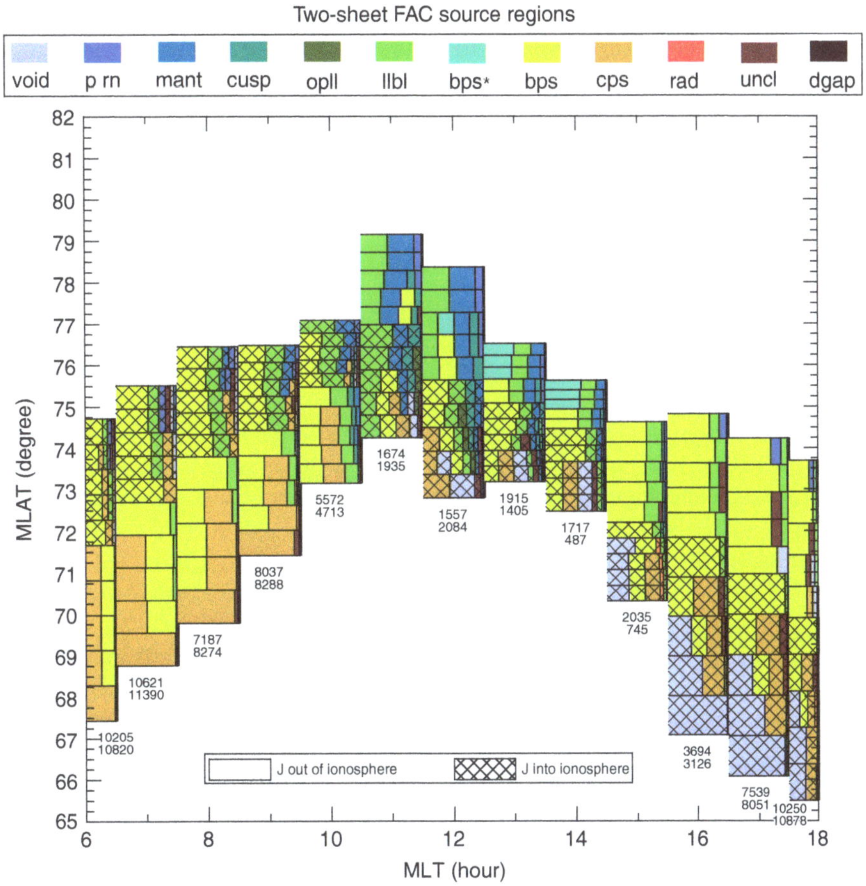

Figure 18.2 The source regions of the dayside R1 and R2. In each MLT, each current sheet is divided into five equally spaced latitudinal regions and within each region, the area of the source (color/grey scale) region is proportional to the frequency of occurrence. The source regions are sorted from left to right in the order of descending frequency of occurrence. The source regions are color/grey scale coded using the scheme depicted in the boxes at the top. The nomenclatures used are: p rn = polar rain, mant = mantle, opll = open-LLBL, bps = boundary plasma sheet, cps = central plasma sheet, rad = radiation belt, uncl = unclassified, and dgap = data gap. bps* is an open-field line or boundary layer region, although the particle signature looks like bps [*Wing et al.*, 2010] (one scenario is presented in section 18.4.4). The hatched and nonhatched regions indicate currents flowing into and out of the ionosphere, respectively. The numbers below R2 indicate the number of events in each current sheet: the top and bottom numbers indicate the number of events in R1 and R2, respectively. Adapted from *Wing et al.* [2010].

damping of latitudinal structure leading to East-West striations. On smaller scales, the damping is isotropic and weaker so that structures may persist over a longer time-scale. These structures have length scales comparable to "inverted V shocks" that are commonly associated with particle precipitation [*Mozer et al.*, 1977; *Temerin et al.*, 1981]. The time-asymptotic solutions provide insight concerning the steady-state development of the boundary layer. These profiles have a strong dependence on Hartmann number with sharper boundary layer flows, but weaker currents for larger viscosity. A key property of this model is that the profiles are constrained by five integration constants (three in the low-latitude boundary layer and two in the ionosphere) indicating the importance of both solar wind and ionosphere constraints on the boundary layer. An integration constant is generally determined by boundary conditions, which may lead to different classes of solutions. Those solutions may also depend on physical parameters such as Pederson conductivity or viscosity.

The work of *Siscoe et al.* [1991] recognized that this boundary layer model could be used to link the high-latitude boundary layer (which generates voltage) and the low-latitude boundary layer (which generates currents) using a different set of boundary conditions. This work imposed a boundary condition based on the polar cap potential drop and allowed for currents at the polar cap boundary (different from *Lotko et al.*, [1987]).

The steady-state solutions produced typical boundary layer features such as reversal in convection from high to low latitude and generally matched observations better, but underestimated the thickness of the low-latitude boundary layer. A general feature of these boundary layer solutions is that the fields have typical variation on the order of 100 km at the ionosphere, which is comparable with the electrostatic auroral scale length (described in section 18.2). When variations in the boundary layer are on this scale or larger, they will generally be associated with field-aligned currents, potential drops, and particle precipitation in the ionosphere. In the shear layer generator discussed in this chapter, boundary layer structures that are smaller than this scale would not map to the ionosphere.

The intense interest in boundary layer-ionosphere coupling led to further development of time-dependent models, which also addressed coupling of Kelvin-Helmholtz (KH) structures generated by velocity shear at the boundary layer. The development of these structures generally broadens the layer [*Matsumoto and Seki*, 2010; *Nakamura et al.*, 2011]. The ionosphere should be taken into account because it can act as a drag to the KH wave development. *Wei et al.* [1993] developed a model for the plasma dynamics at LLBL, which interacts with ionosphere via FAC. In their model, Kelvin-Helmholtz instability (KHI) develops along the flank due to high shear flow, but Pedersen conductivity (Σ_p) can lead to the decay of KH vortices. Due to the competition between shear flow (formation) and Σ_p (decay) of vortices, strong vortices can only form in a limited region. In their simulations, several regions of enhanced field-aligned power density associated with KH vortices and upward FAC filaments were found in the afternoon. The observed auroral bright spots in the afternoon sector may be linked to this enhanced field-aligned power density [e.g., *Evans*, 1985]. Interestingly, in their simulations, the FACs associated with the KH structures are weaker in the postnoon sector than in the prenoon sector due to large parallel electric potential drop ($\Delta\phi_{\|}$) in the postnoon sector, consistent with observations of Region 1 FACs [*Iijima and Potemra*, 1978; *Wing et al.*, 2010; 2011].

Overall, this body of work suggests that the solar wind and ionosphere play key roles in determining the structures in the boundary layer. In this chapter, we focus only on a subset of the field-aligned currents considered in *Sonnerup* [1980], namely the upward field-aligned current in the afternoon R1 that is located entirely at the LLBL or on open-field lines.

The theory of upward field-aligned current generation was significantly advanced by the work of *Lyons* [1980]. *Lyons* [1980] derived an expression for upward field-aligned current density ($J_{\|}$) at the boundary layer and open-field lines, but it does not include velocity shear layer thickness, Δ_m. *Roth et al.* [1993] argued that magnetospheric plasma interfaces can play the role of magnetospheric generators feeding the auroral electric circuit. The arguments of *Roth et al.* [1993] and *Lyons* [1980, 1981] were taken into account in the self-consistent quasi-stationary magnetosphere–ionosphere model developed in *Echim et al.* [2007, 2008].

Echim et al. [2007, 2008] model is based on a Vlasov equilibrium solution for the magnetospheric/LLBL generator and is adapted for velocity shears and/or pressure gradients; the thickness of the sheared layer is derived self-consistently and depends on the asymptotic conditions adopted for the Vlasov equilibrium. Vlasov solutions can be found for a range of velocity shears and pressure gradients as discussed by *De Keyser and Roth* [1998] (see also *De Keyser and Echim* [2010]). This MI coupling model has been tested against observational evidence at the plasma sheet boundary layer generator and in the vicinity of a stable auroral arc [*Echim et al.*, 2009] for field-aligned currents observed at the dayside and connected to the LLBL [*Wing et al.*, 2011] and for observations of polar cap arcs [*Maggiolo et al.*, 2012].

Recently, *Johnson and Wing* [2015] considered a similar approach as *Lyons* [1980], but they included Δ_m, which turns out to be a crucial parameter for calculating $J_{\|}$. They found that $J_{\|}$ varies inversely with Δ_m, for a given magnetic field and velocity shear at the boundary (and hence the potential drop across the boundary). Additionally, FAC latitudinal thickness (Λ) scales with Δ_m. Because $J_{\|}$ depends on Δ_m, we may be able to use observed ionospheric and solar wind parameters to remote sense the boundary layer structures. As in *Sonnerup* [1980], *Johnson and Wing* [2015] did not consider any turbulence at the boundary layer.

One of the main attractions of *Johnson and Wing* [2015] theory is that $J_{\|}$ and Δ_m can be predicted from regularly observed solar and ionospheric parameters. Indeed, *Wing and Johnson* [2015] showed that using these solar and ionospheric parameters, *Johnson and Wing* [2015] prediction of $J_{\|}$ agrees with observations quite well. More recently, *Wing et al.* [2015] found that the $\Delta\phi_{\|}$ predicted by the *Johnson and Wing* [2015] theory was in good agreement with observations. Many features of the *Echim et al.* [2007, 2008] model are captured by [*Johnson and Wing*, 2015] analytic theory.

The rest of this chapter reviews the recent advances in analytical theory, kinetic modeling, and observations of the afternoon R1 upward field-aligned currents generated at the low-latitude boundary layer [*Johnson and Wing*, 2015; *Wing and Johnson*, 2015; *Wing et al.*, 2015; *Echim et al.*, 2007; 2008].

18.2. THEORY FOR UPWARD FIELD-ALIGNED CURRENTS AT LLBL

We begin with a review of *Johnson and Wing* [2015] analytical theory, which serves as a foundation of this chapter. Rather than presenting the full set of equations in *Johnson and Wing* [2015], we present only a subset of the equations that is observationally testable or

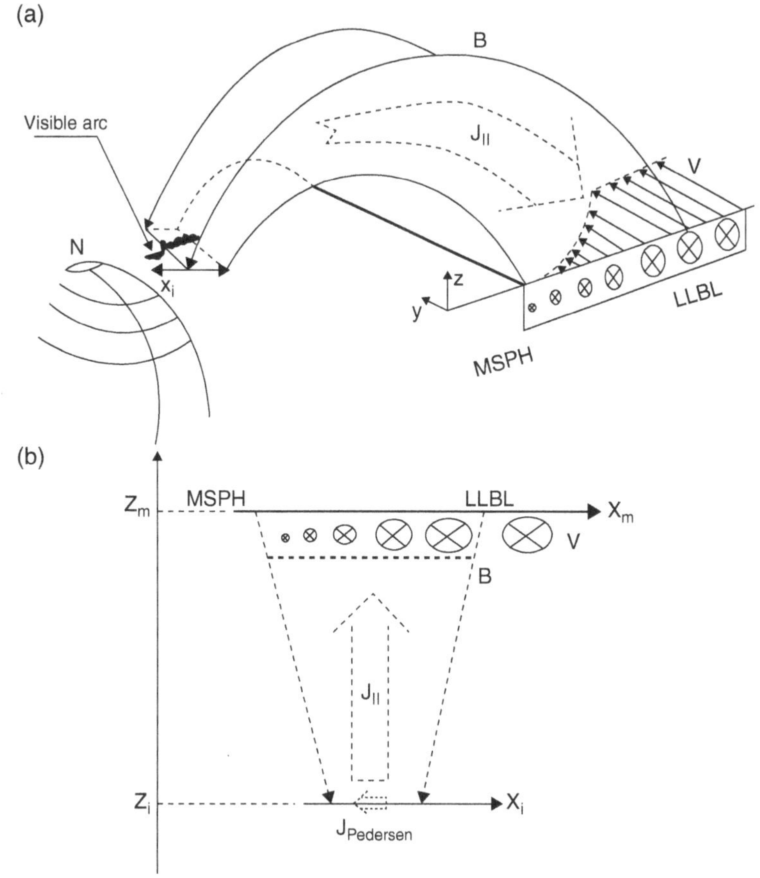

Figure 18.3 The geometry of the upward FAC region between a sheared flow LLBL and the ionosphere. (a) A schematic 3D view of the magnetosphere flank; (b) a simpler, conical geometry adopted to describe a flux tube extended from ionospheric altitudes (z_i) to the magnetosphere (z_m). The velocity profile is illustrated by circles with radius proportional to the local value of the shear velocity. Adapted from *Echim et al.* [2007].

observationally useful. We refer the readers to *Johnson and Wing* [2015] for the full derivations.

Figure 18.3 illustrates the schematic for the geometry that exists at the LLBL in the afternoon where the FAC flows upward. We considered the current continuity equation of the ionosphere

$$\frac{d}{dx}\Sigma_p \frac{d\phi_i}{dx} = J_{\parallel}(\phi_m, \phi_i) \tag{18.1}$$

where Σ_p = height integrated Pedersen conductivity, ϕ_i = electric potential in the ionosphere, ϕ_m = electric potential in the magnetosphere [*Lyons*, 1980; 1981; *Echim et al.* 2007, 2008; and *Johnson and Wing*, 2015]. Throughout the chapter, the subscript m and i denote magnetosphere/LLBL and ionosphere, respectively. The profile of ϕ_m is determined primarily by the solar wind–magnetosphere interaction at the magnetopause [*Echim et al.*, 2008]. The potential drop between the magnetosphere and ionosphere ($\Delta\phi_{\parallel}$) drives FAC out of the ionosphere and their relationship is described by linear Knight relation [*Knight*, 1973],

$$J_{\parallel} = \kappa(\phi_i - \phi_m) \tag{18.2}$$

where the Knight conductivity $\kappa = n_e e^2 / (2\pi m_e k_b T_e)^{0.5}$, n_e = electron number density at LLBL, e = electron charge, m_e = electron mass, T_e = electron temperature at LLBL, and k_b = Boltzman's constant. The linear Knight relation is satisfied in the nonlinear current-voltage relation when $1 \ll e(\phi_i - \phi_m)/T_e \ll B_i/B_m$, where B_m and B_i are the magnetic field strength at the magnetosphere/boundary layer and in the ionosphere at the bottom of the potential drop, respectively.

Assuming κ is constant in the sheared layer and combining equations (18.1) and (18.2),

$$L^2 \frac{d^2\phi_i}{dx^2} = (\phi_i - \phi_m) \tag{18.3}$$

where $L = \sqrt{\Sigma_p / k}$ is the electrostatic auroral scale length [*Lyons*, 1980]. Equation (18.3) can be solved for the ionospheric potential, ϕ_i, with a Fourier transform of ϕ

$$\hat{\phi}(q)=\frac{1}{2\pi}\int_{-\infty}^{\infty}\phi(x)e^{-iqx}\,dx \quad (18.4)$$

Its inverse transform is

$$\phi(x)=\int_{-\infty}^{\infty}\hat{\phi}(q)e^{iqx}\,dq \quad (18.5)$$

The Fourier transform of equation (18.3) is

$$\widehat{\phi_i}(q)=\left(\frac{1}{1+q^2L^2}\right)\widehat{\phi_m}(q) \quad (18.6)$$

$J_{\|}$ and $\Delta\phi_{\|}$ can be obtained similarly by the inversion of their Fourier transforms,

$$\Delta\hat{\phi}(q)=\widehat{\phi_i}(q)-\widehat{\phi_m}(q)=-\left(\frac{q^2L^2}{1+q^2L^2}\right)\widehat{\phi_m}(q) \quad (18.7)$$

$$\widehat{J_{\|}}(q)=-\kappa\left(\frac{q^2L^2}{1+q^2L^2}\right)\widehat{\phi_m}(q) \quad (18.8)$$

Equation (18.8) is analyzed to determine how FAC depends on the boundary layer conditions (controlled by solar wind) and ionospheric conditions (controlled by solar radiation and particle precipitation).

The velocity profile in the LLBL typically varies from an asymptotic magnetosheath flow, V_0, at the outer edge to little or no flow on the inner edge over the thickness of the boundary layer, Δ_m. So, a reasonable velocity profile may be

$$V_y(x_m)=\frac{V_0}{2}\left(1+\tanh\left(x_m/\Delta_m\right)\right) \quad (18.9)$$

where x_m is the magnetospheric coordinate across the magnetopause boundary layer. Using this profile, *Johnson and Wing* [2015] obtained the following magnetospheric potential function

$$\phi_m(x_i)=\frac{V_0\,B_m\sqrt{b}}{2}\left[x_i+\Delta_i\log\left(2\cosh\left(x_i/\Delta_i\right)\right)\right] \quad (18.10)$$

where $\Delta_i=\Delta_m/\sqrt{b}$ is the ionospheric scale length obtained by mapping the boundary layer thickness to ionospheric altitude (for simplicity), $x_m=x_i\sqrt{B_i/B_m}=x_i\sqrt{b}$, $b=B_i/B_m$, and at 200 km, $\sqrt{b}=32$ [e.g., *Echim et al.*, 2007, 2008]. Note that for a dipole geometry, the geometrical scaling factor is different in the latitudinal and longitudinal directions. In the longitudinal direction, $x_m=x_i\sqrt{b}$ as stated above, but in the latitudinal direction, there is an additional factor of $\sqrt{2}$, so that $x_m=x_i\sqrt{2b}$. As noted in section 18.4, if the data were plotted in log-log scales, this geometrical scaling factor would become a constant offset in the y-intercept.

The Fourier transform of ϕ_m is

$$\widehat{\phi_m}(q)=\frac{1}{2\pi}\int\phi_m(x_i)e^{-iqx_i}\,dx_i$$
$$=\frac{V_0\,B_m\sqrt{b}}{2}\left[i\delta'(q)-\frac{\Delta_i}{2q\sinh\left(\pi\Delta_i q/2\right)}\right] \quad (18.11)$$

The current can be derived from the inverse transform of equation (18.8),

$$J_{\|}(x)=\kappa\frac{V_0\,B_m\,\Delta_m}{2}\left(\frac{\pi e^{-|x|/L}}{2\sin(\pi\alpha)}+\sum_{n=1}^{\infty}(-1)^n\frac{n e^{-2n|x|/\Delta_i}}{n^2-\alpha^2}\right) \quad (18.12)$$

where $\alpha=\Delta_i/2L$.

In *Johnson and Wing* [2015] theory, currents are driven by the potential difference across the boundary layer. This potential difference can map to the ionosphere, which, in turn, can drive a Pedersen current in the negative x direction (see Fig. 18.3). In order to maintain current continuity, the ionospheric current must be diverted upward because the electric field at the boundary layer vanishes asymptotically as $x\to-\infty$. The next subsections examine the current maximum and width.

18.2.1. Maximum Current

The current maximizes at the center of the shear layer at $x_i=0$. The maximum current is obtained by evaluating $J_{\|}(0)$,

$$J_{\|,max}=\kappa\frac{V_0\,B_m\,\Delta_m}{2}\left(\frac{\pi}{2\sin(\pi\alpha)}+\sum_{n=1}^{\infty}(-1)^n\frac{n}{n^2-\alpha^2}\right) \quad (18.13)$$

Due to space constraint, detailed derivations are not provided, but they can be found in *Johnson and Wing* [2015]. Moreover, they considered the cases for small and large α and reduced equation (18.13) to a simple expression

$$J_{\|,max}\approx\kappa\frac{V_0\,B_m\,\Delta_m}{4\alpha(1+2\alpha)}=\frac{\Sigma_P V_0\,B_m\,b}{2\left(\Delta_m+\sqrt{b}\,L\right)} \quad (18.14)$$

18.2.2. Maximum Field-Aligned Potential Drop

The potential drop maximizes where the current maximizes at $x_i=0$. *Johnson and Wing* [2015] showed that,

$$\Delta\phi_{\|,max}=\frac{J_{\|}}{\kappa}\approx\frac{V_0\,B_m\,\Delta_m}{4\alpha(1+2\alpha)}=\frac{V_0\,B_m\sqrt{b}\,L}{2\left(1+\Delta_i/L\right)} \quad (18.15)$$

For $L \gg \Delta_i$, $\Delta\phi_{\|,max} \sim L \sim n_{sw}^{-0.5}$, while for $L \ll \Delta_i$, $\Delta\phi_{\|,max} \sim L^2 \sim n_{sw}^{-1}$, which is consistent with the numerical solutions presented in *Echim et al.* [2008]. The dependence of the current and voltage on solar wind velocity is linear (see Fig. 6 of *Echim et al.* [2008]).

18.2.3. Width of the Field-Aligned Current Layer

The width of the field-aligned current layer (Λ) is defined as full width at half maximum. The profile of the width gives a measure of the global extent of the FAC. One may derive the full width at half maximum numerically from equation (18.12), but *Johnson and Wing* [2015] derived an analytical solution

$$\Lambda = \frac{(2\ln 2)L}{1+0.29\alpha} + 2\ln\left(1+\sqrt{2}\right)\Delta_i \qquad (18.16)$$

The width of the boundary layer, Δ_i, can also be obtained from measurement of L and Λ by solving

$$1.0224\alpha^2 + \left(3.5255 - 0.29\frac{\Lambda}{L}\right)\alpha - \left(\frac{\Lambda}{L} - 1.3863\right) = 0 \qquad (18.17)$$

for $\Lambda > 2\ln 2\, L$.

The current width depends on the solar wind density because $L \sim n_{sw}^{-0.5}$ so that for $\Delta_i \ll L$ the width $\Lambda \sim n_{sw}^{-0.5}$.

18.3. A MAGNETOSPHERE-IONOSPHERE QUASI-STATIONARY COUPLING MODEL BASED ON FIELD-ALIGNED CURRENTS WITH A NONLINEAR CURRENT-VOLTAGE RELATIONSHIP

Starting from the kinetic description of magnetospheric generators by *Roth et al.* [1993] and the pioneering work of *Lyons* [1980,1981] and *Echim et al.* [2007, 2008] (see also *De Keyser and Echim* [2010, 2013], *De Keyser et al.* [2010]) developed a magnetosphere-ionosphere (MI) coupling model that includes: (1) a self-consistent description of the magnetospheric generator based on a Vlasov equilibrium solution for tangential plasma interfaces, (2) a coupling of the generator with the auroral ionospheric load via field-aligned currents described by a kinetic current-voltage relationship, and (3) a circuit closure in the ionosphere based on the current continuity (equation [18.1]) that considers the feedback given by the variation of the ionospheric conductivity with the flux of precipitating electrons. Below we review the three components of the MI coupling model and its main predictions as described by *Echim et al.* [2007, 2008].

18.3.1. The LLBL Generator

As noted by *Lyons* [1980, 1981], magnetospheric discontinuities leading to convergent electric fields can generate regions of magnetic field-aligned electric field, upward field-aligned currents, and electron precipitation. The auroral electrodynamic effects were investigated by *Lyons* [1980, 1981] for ad hoc profiles of electric potential. *Roth et al.* [1993] showed that convergent/divergent electric fields occur naturally in magnetospheric tangential plasma interfaces and they illustrated several instances of such Vlasov equilibrium solutions.

Echim et al. [2007, 2008] adapted the tangential interface model of *Roth et al.* [1996] to the inner edge of the low-latitude boundary layer. Figure 18.4 shows a family of such solutions and illustrates the effects on the layer properties of (1) different bulk speeds in the LLBL (from 200 km/s to 500 km/s) and (2) different LLBL plasma densities (from 7 cm^{-3} to 12.5 cm^{-3}). The plasma and electromagnetic field profiles are derived as function of the spatial coordinate (x_m) normal to the interface, from a solution of the Vlasov equation coupled to the Maxwell equations and the quasi-neutrality equation. The solution depends on the kinetic and macroscopic asymptotic values specified in the magnetosphere and magnetosheath. Note that a Vlasov equilibrium can be obtained when the following assumptions are satisfied: (1) the normal component of the magnetic field is equal to zero, and (2) the fields are stationary.

The thickness of tangential plasma interfaces, like the inner edge of the LLBL, derived from Vlasov equilibrium solutions scales as the electron or ion Larmor radius, depending on the relative orientation of the magnetic field and plasma bulk velocity [*Sestero*, 1966; *Lemaire and Burlaga*, 1976; *Roth et al.*, 1996; *De Keyser and Roth*, 1997, 1998; *Echim et al.*, 2005]. The two types of solutions are called electron- and ion-dominated layers [*Sestero*, 1966]. The thickness of the transition layer is of the order of the ion Larmor radius if the convection electric field, $\mathbf{E} = \mathbf{V} \times \mathbf{B}$, is polarized such that the electrons are prevented to enter the transition regions and the main current carriers are the ions; the situation is reversed for electron-dominated layers [*Sestero*, 1966; *Lemaire and Burlaga*, 1976; *De Keyser and Roth*, 1998]. In the example given in Figure 18.4, *Echim et al.* [2008] considered an ion layer at the inner edge of the LLBL and investigated the effects due to increased velocity shear and kinetic pressure gradient (increased LLBL density or electron temperature). The electric potential across the interface is controlled mainly by the velocity shear. Nevertheless, the spatial scale of variations depends also on other parameters: the scale of the density gradient is fixed by the LLBL ion temperature while the scale of the velocity shear is sharper for larger velocities in the LLBL.

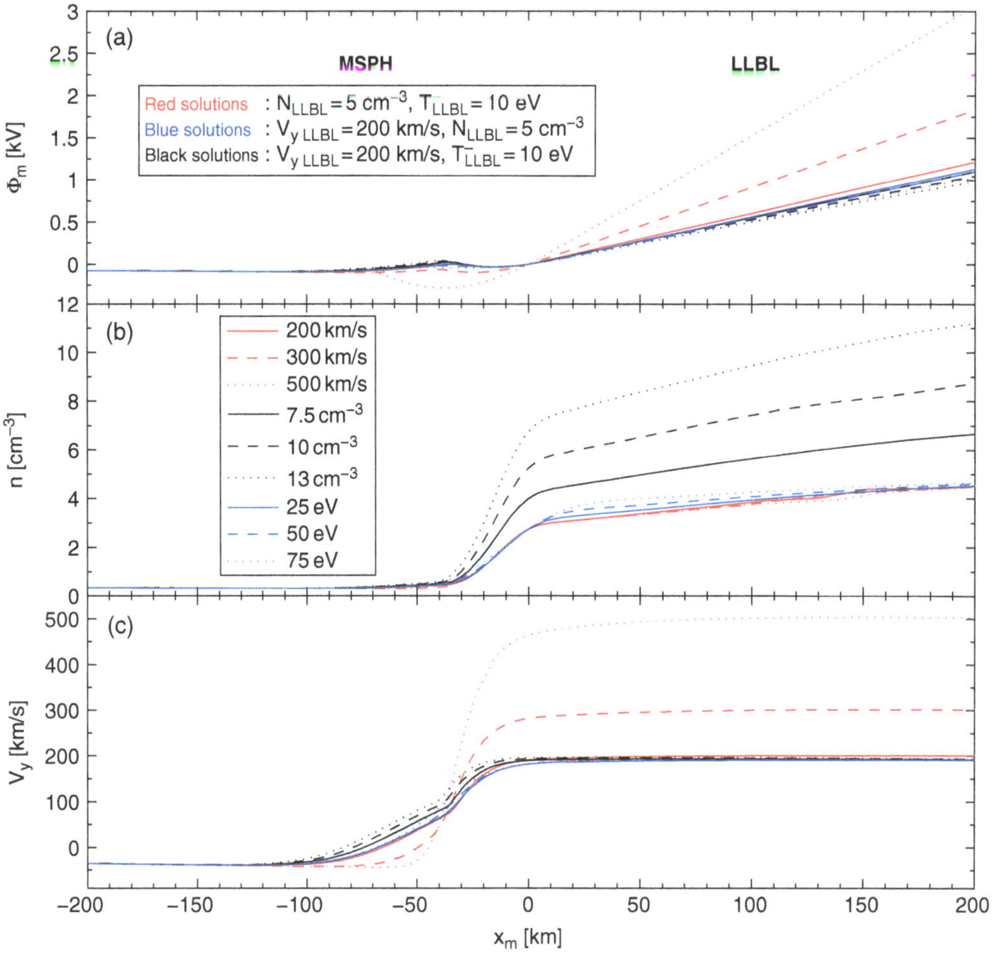

Figure 18.4 Vlasov equilibrium solutions for the inner edge of the LLBL at the interface with the magnetosphere (MSPH). The panels show (a) the electric potential, (b) the plasma density, (c) the antisunward component of the plasma bulk velocity. The figure shows nine different Vlasov solutions corresponding to three different classes of asymptotic solutions that quantify three different types of kinetic effects. All solutions are computed for the same asymptotic plasma parameters at the magnetospheric side: plasma density equal to $N_{MSPH} = 0.5\,cm^{-3}$, plasma velocity equal to zero, electron and ion temperature equal to 200 eV and 1 keV, respectively. The red profiles illustrate the effect of increased velocity shear, blue solutions correspond to increasing electron LLBL temperatures, black profiles illustrate the effect of increased LLBL plasma density. From *Echim et al.* [2008].

18.3.2. Coupling of LLBL to the Ionosphere via Field-Aligned Currents

Assuming that the generator is connected to the ionospheric load by magnetic field lines, the adiabatic motion of particles between the generator and the load contributes to field-aligned currents carried by electrons and ions. The current carried by each species is computed from the first order moment of the velocity distribution function (VDF) as in the kinetic exospheric models of polar and solar wind [e.g., *Lemaire and Scherer*, 1971]. The partial current density due to each population is computed analytically taking into account electron and ion precipitation, outflows and trapped population.

Under the assumption that the electric potential ϕ varies monotonically with the altitude z, one determines the field-aligned current density of each species, s, as a nonlinear relationship between $J_{s\|}$ and $\Delta\phi_{\|}(z) = \phi_i(z) - \phi_m$ at each altitude z, as described by the analytic expressions (4)–(8) in *Echim et al* [2007]. The sum of all partial densities gives the net field-aligned current density at an altitude z.

$$J_{\|} = \sum_s J_{s\|} \tag{18.18}$$

The expression for the FAC density carried by precipitating electrons, $j_{e\|}$, as a function of $\Delta\phi_{\|}$ is given by the current-voltage relation [*Knight, 1973*], whose linearized expression is given by equation (18.2).

18.3.3. Current Closure, Ionospheric Feedback and Auroral Electrodynamics Coupled to the LLBL Generator

The coupling between the generator electric potential, ϕ_m, the field-aligned current density $J_{\|}$ and the ionospheric currents and Pedersen conductivity is described by the current continuity equation at the ionospheric altitude z_i, as given by equation (18.1). In the MI coupling model by *Echim et al.* [2007, 2008] one considers all the partial contributions to the field-aligned density as well as an energy-dependent expression of the height integrated Pedersen conductivity Σ_p [*Harel et al.*, 1977]. *Echim et al.* [2008] computed solutions of the nonlinear differential equation (18.1) for the different profiles of the LLBL electric potential, bulk velocity, and density illustrated in Figure 18.4. The field-aligned current density as well as the thickness of the field-aligned current sheet increases with increased velocity shear at the LLBL, as illustrated in Figure 18.5. One notes that when the nonlinear couplings are considered in equation (18.1), the thickness of the the electron precipitation region is in general slightly narrower than the thickness of the field-aligned current sheet. The precipitating energy flux is computed from the moments of the distribution function of incoming electrons while the field-aligned current density includes contribution from all species and populations [*Echim et al.*, 2007, 2008].

The results of a parametric study of *Echim et al.* [2008] on the electrodynamic properties of the auroral ionosphere are illustrated by Figure 18.6. It shows that the maximum of the field-aligned potential drop $\Delta\phi_{\|,max}$ has a linear dependence on the velocity shear regardless the model adapted for the ionospheric feedback. However, the maximum of the field-aligned current density, $J_{\|,max}$, has a nonlinear variation with the LLBL velocity shear when the Pedersen conductance depends on the energy flux of precipitating electrons, ε_m. The peak value of ε_m depends nonlinearly on the velocity shear. The thickness of the field-aligned current sheet is systematically larger than the thickness of the associated auroral arc (and Pedersen conductance perturbation). This effect is enhanced when Σ_p depends nonlinearly on ε_m. Figure 18.6 compares well with the DMSP observations in *Wing et al.* [2011].

Echim et al [2008] investigated also the effects of LLBL density and found that $\Delta\phi_{\|}$ generally decreases and $J_{\|}$ increases with increasing density, which is qualitatively consistent with DMSP observations [*Wing et al.*, 2011]. However, $\Delta\phi_{\|}$ increases and $J_{\|}$ decreases with increasing T_e at LLBL.

18.4. OBSERVATIONS OF UPWARD FIELD-ALIGNED CURRENTS (THEORY VALIDATION)

Johnson and Wing [2015] and *Wing and Johnson* [2015] used over 20 yr of simultaneous DMSP satellite magnetic field and particle precipitation (SSJ4/5) observations, 1983–2006, to calculate FAC and its source region. They applied the automated algorithm developed by *Higuchi and Ohtani* [2000] to calculate large-scale FACs based on infinite current sheet approximation. To determine the source region of the field-aligned current, they used particle precipitation signatures that were observed simultaneously with FAC. *Newell and Meng* [1988] and *Newell et al.* [1991a, 1991b, and 1991c] developed an automated algorithm for identifying nine magnetospheric source regions of DMSP particle precipitation, but here we only need to know if the particles are located on closed-field lines, open-field lines, or LLBL (see Fig. 18.2). It should be noted that previous studies have used particle precipitation to determine the source region of FAC. For example, *Bythrow et al.* [1981] used particle precipitation to determine the source region of R1 is LLBL. However, a portion of R1, perhaps the lower latitude portion, can be located in the closed-field lines well inside the magnetosphere where the convection is sunward and the current generation mechanism may not be the solar wind–magnetospheric plasma velocity shear [e.g., *Bythrow et al.*, 1981; *Vennerstrom et al.*, 2005; *Haraguchi et al.*, 2004; *Wing et al.*, 2010]. This is shown in Figure 18.2. Hence, *Johnson and Wing* [2015] and *Wing and Johnson* [2015] carefully selected only the afternoon upward R1 that is located entirely in the LLBL, cusp, mantle, and polar rain.

18.4.1. Field-Aligned Current Density ($J_{\|}$)

Wing and Johnson [2015] calculated $J_{\|,\max}$ using equation (18.14) and parameters that were obtained from observations and empirical formulas. They used the approximations $V_0 = 0.15\ V_{sw}$ and $B_m = 30$ nT, which are similar to observations at LLBL between noon and dusk [e.g., *Fujimoto et al.*, 1998; *Vaisberg et al.*, 2001; *Dimmock and Nykyri*, 2013]. The magnetic field ratio was estimated at $b = B_i/B_m \sim 1000$ at DMSP altitude. $L = (\Sigma p/\kappa)^{0.5}$ was obtained using empirical formulas for Σp [*Robinson and Vondrak*, 1984; *Robinson et al.*, 1987] and solar wind observations to infer κ. κ was computed using $n_e = n_{sw}$ [e.g., *Scudder et al.*, 1973; *Phan and Paschman*, 1996] and $T_e = 1 \times 10^6$ K [e.g., *Phan and Paschmann*, 1996].

Figure 18.7 plots the value of $J_{\|,\max}$ predicted by equation (18.14) using Δ_i obtained from the observed Λ and equation (18.17) versus observed $J_{\|}$. If the observed $J_{\|}$ is approximately $J_{\|,\max}$ or $J_{\|,\max} \sim cJ_{\|}$, then on the log-log plot, (1) the slope would be one and (2) any nonunity

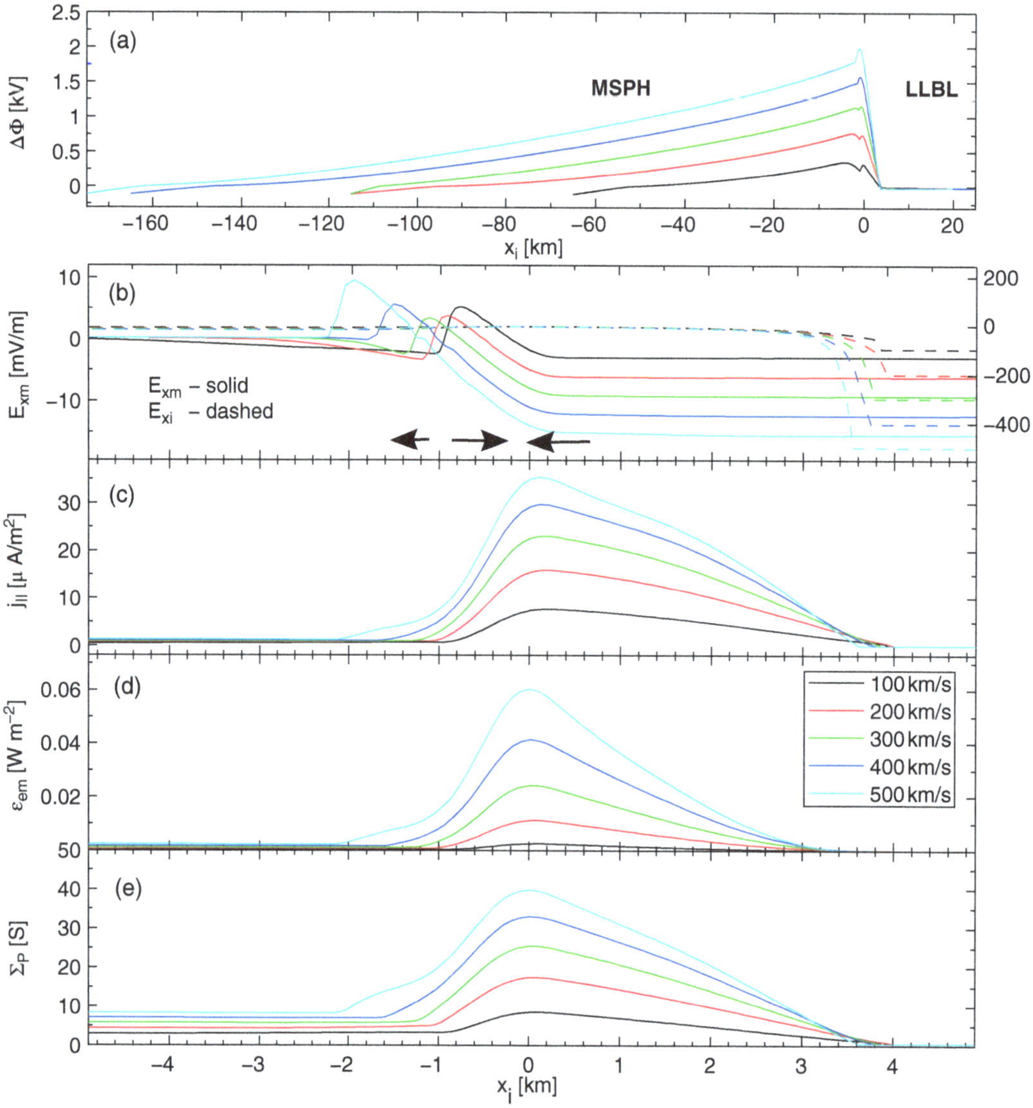

Figure 18.5 Ionospheric electrodynamic parameters as a function of x_i, the coordinate normal to the auroral arc. The panels illustrate solutions of equation (18.1) derived by *Echim et al.* [2008] for a nonlinear dependence on the parallel potential drop, $\Delta\phi_\parallel$, of the field-aligned current density and height integrated Pedersen conductivity. The different ionospheric solutions correspond to different LLBL generator configurations derived for five different velocity shears at the inner edge of the LLBL. The panels show (a) the potential difference between the LLBL generator and the ionosphere, (b) the perpendicular electric field intensity (in the magnetosphere and ionosphere), (c) the field-aligned current density, (d) the energy flux of precipitating electrons, and (e) the height integrated Pedersen conductivity. From *Echim et al.* [2008].

proportionality constant would shift the points in y direction by log(c). In Figure 18.7, the points tend to align along the dashed line of slope = 1, suggesting the validity of this scenario. The observed $J_{\parallel,max}$ is highly correlated with the calculated $J_\parallel$ with correlation coefficient (r) of 0.81. The probability of two uncorrelated random variables to yield $|r| \geq 0.81$ is < 0.01 ($P < 0.01$), suggesting that the correlation is highly significant. *Wing and Johnson* [2015] performed a least square fit and obtained $\log(J_{\parallel,max\_cal}) = (0.96 \pm 0.04)\log(J_{\parallel\_obs}) + (0.03 \pm 0.01)$ where $J_{\parallel,max\_cal}$ = calculated maximum $J_{\parallel,max}$ from equation (18.14) and $J_{\parallel\_obs} = J_\parallel$ observed by DMSP magnetometer. The proportionality constant is nearly unity (y-intercept = 0.03) and $J_{\parallel,max\_cal} \sim 1.07 J_{\parallel\_obs}$.

The points near noon were removed in Figure 18.7 because $J_\parallel$ responds slightly differently at this location. Near noon, the magnetic field lines map to the vicinity of the subsolar region where (1) the thermal currents (see section 18.4.5) may be high due to high magnetosheath densities and temperatures and (2) the velocity shear is

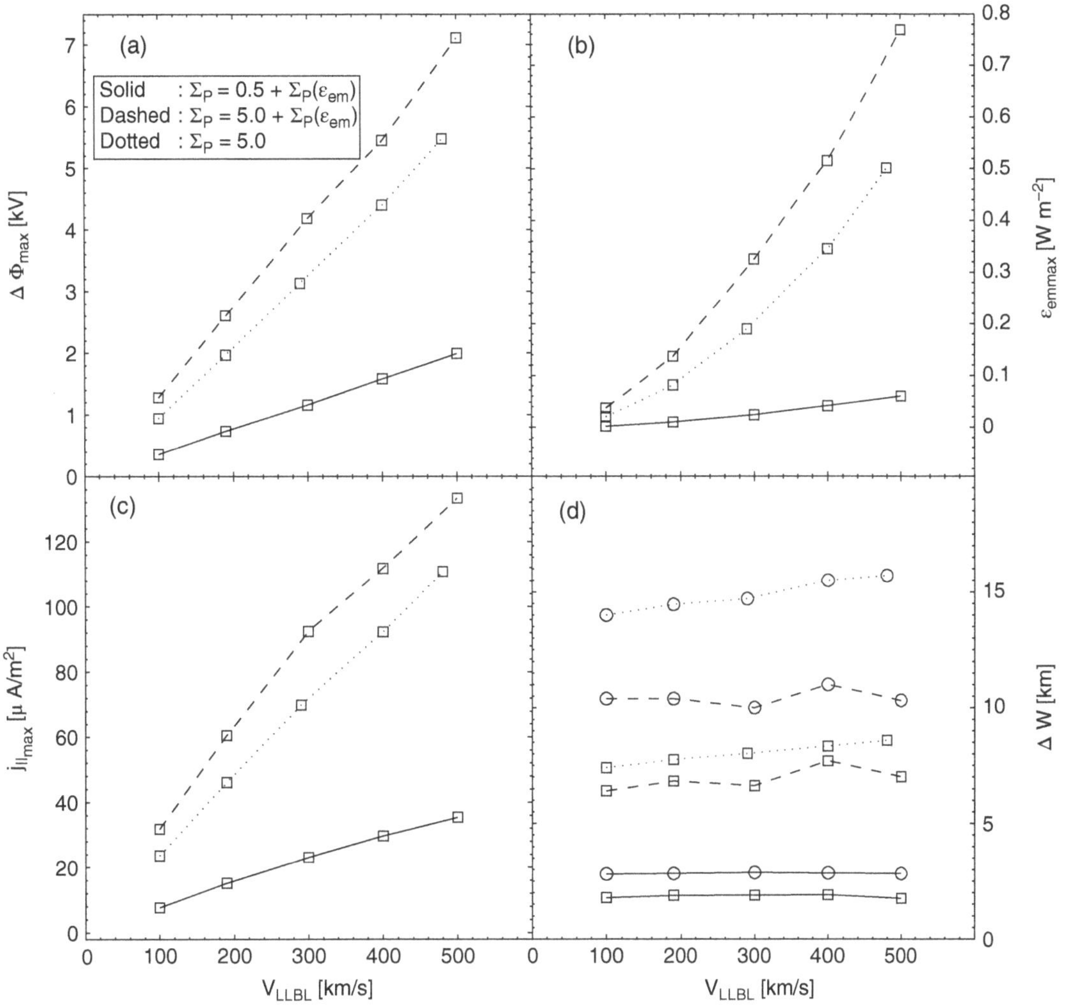

Figure 18.6 Parametric studies on the dependence of ionospheric electrodynamic parameters on the velocity shear at the coupled LLBL generator for three types of ionospheric coupling [*Echim et al.*, 2008]. The three classes of solutions illustrated by solid, dotted and dashed profiles correspond to three models of the height integrated Pedersen conductivity as indicated in panel (a). Panel (a) shows the maximum value of the field-aligned potential difference $\Delta\phi_{\|}$; panel (b) shows the maximum value of the energy flux of precipitating electrons; panel (c) shows the maximum density of field-aligned currents; and panel (d) illustrates with circles and squares the width of the field-aligned current sheet and of the auroral arc at the ionospheric altitude of 400 km, respectively.

minimal and hence the plasma pressure gradient may contribute more significantly to $J_{\|}$ [e.g., *De Keyser and Echim*, 2013].

As discussed in *Wing and Johnson* [2015], a log-log plot is useful to identify scaling relationships and the power law dependence. An unknown constant of proportionality is eliminated from the scaling relationship and become an offset in a log-log plot. The precise values of many of the model input parameters are not known, but they are only known to be proportional to observed solar wind parameters. The usage of log-log plot is illuminating in such cases.

Next, we show how $J_{\|}$ varies with the thickness of the boundary layer, Δ_m. This relationship is shown in Figure 18.8, which plots log $J_{\|}$ versus log Δ_m. $J_{\|}$ was obtained directly from magnetometer observation, while Δ_m was obtained from equation (18.16) and observed values of Λ and L. Figure 18.8 shows that the points tend to align with lines with slope of –1, which is the theoretical prediction for large α limit in equation (18.14). For $\Lambda/L > 5$, the least square returned slope of -0.8 ± 0.2, which is within the theoretical prediction of –1. In contrast, for $\Lambda/L < 5$ (the small α limit, green dots), the slope is closer to 0 because $J_{\|}$ is independent of Δ_m (see equation [19] in *Johnson and Wing* [2015]).

Figure 18.9 plots log $J_{\|}$ versus log n_{sw} for small α ($\Lambda/L < 3$). $J_{\|}$ was obtained from DMSP magnetometer observations while n_{sw} was obtained from solar wind

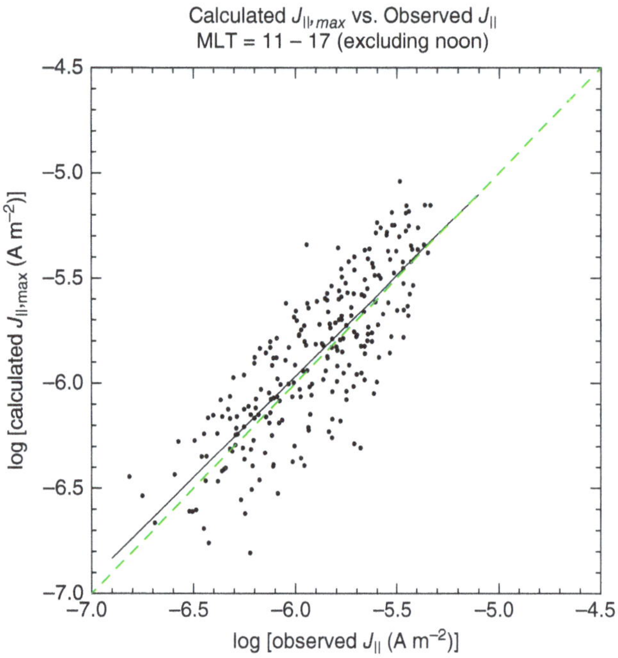

Figure 18.7 Calculated $J_{\|,max}$ versus observed $J_{\|}$ for MLT = 1100–1700, excluding data near noon (MLT = 1200). The calculated $J_{\|,max}$ is obtained from equation (18.14). The least square fitted line (solid line) has a slope of 0.96 ± 0.04, which agrees very well with that of the dashed line, which has a slope of one. From *Wing and Johnson* [2015].

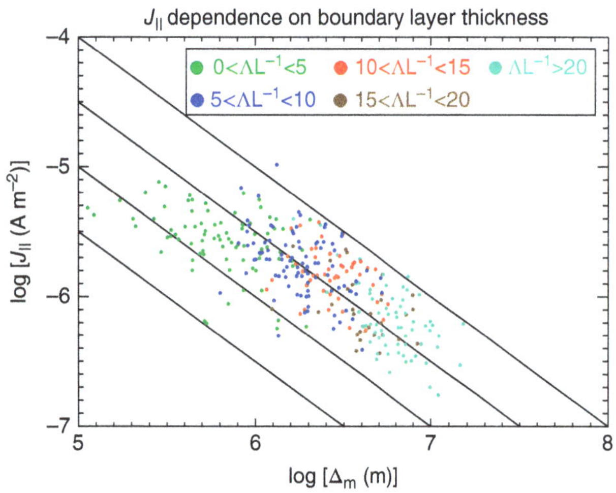

Figure 18.8 Field-aligned current density $J_{\|}$ decreases with increasing Δ_m for data points from 1100 to 1700 MLT. The solid lines have slope = –1, which is expected from equation (18.14). The lines do not fit green dots ($\Lambda/L < 5$) as well because for small α, $J_{\|}$ should be independent of Δ_m (see equation [19] in *Johnson and Wing* [2015]). The large scatter can be attributed partly to the large number of free parameters relating the two parameters as expressed in equation (18.14). From *Johnson and Wing* [2015].

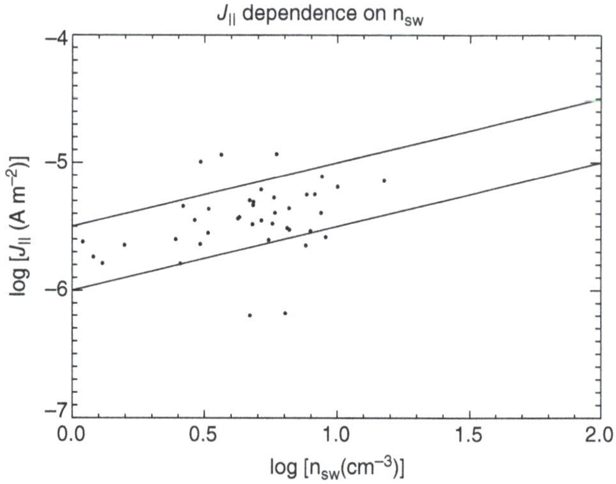

Figure 18.9 Field-aligned current density ($J_{\|}$) increases with n_{sw} for small α ($\Lambda/L < 3$) at 1100–1700 MLT. The solid black lines have slope = 0.5, which is expected from equation (18.14). From *Johnson and Wing* [2015].

observations. Equation (18.14) suggests that the slope should be 0.5, which is similar to Figure 8 of *Echim et al.* [2008]. Figure 18.9 shows that the points are consistent with the predicted value, albeit with a large scatter. The least square fit returned $J_{\|} = n_{sw}^{(0.3\pm0.2)}$, $r = 0.31$, and $P < 0.03$ [*Johnson and Wing*, 2015]. From equation (18.14) and $L = (\Sigma p/\kappa)^{0.5} \sim n_{sw}^{-0.5}$, we can see that for large α (large n_{sw}), $L \ll \Delta_i$, $J_{\|}$ depends mostly on Σ_p, V_{sw}, and Δ_m (or Δ_i).

Figure 18.10 plots log $J_{\|}$ versus log V_{sw} for 1300–1700 MLT. Near noon, FAC would map to near the subsolar magnetopause where the V would be small and, hence, the points near noon were excluded. The theory predicts a slope of 1 (equation [18.14]). Figure 18.10 shows that the points tend to align with the lines of slope = 1, but the scatter is quite large. The least square fit results in $J_{\|} = V_{sw}^{(0.8\pm0.5)}$, $r = 0.18$, and $P < 0.13$ [*Johnson and Wing*, 2015].

The large scatter in Figures 18.9 and 18.10 may result from the anticorrelation between V_{sw} and n_{sw} [e.g., *Richardson et al.*, 1996; *Hundhausen et al.*, 1970; *Wing et al.*, 2016]. This and other sources of errors are discussed in section 18.4.5 and in *Johnson and Wing* [2015].

18.4.2. FAC Latitudinal Width (Λ)

Although theoretical models have been developed for the boundary layer and field-aligned current profiles [e.g., *Lyons*, 1980; *Lotko et al.*, 1987; *Phan et al.*, 1989; *Drakou et al.*, 1994; *Echim et al.*, 2007, 2008], the dependence of the thickness of the boundary layer (Δ_m) and the FAC latitudinal width (Λ) on solar wind parameters have not been investigated observationally at great depth. *Johnson and Wing* [2015] investigated these two parameters using the simple framework presented in section 18.2.3. The results are discussed next.

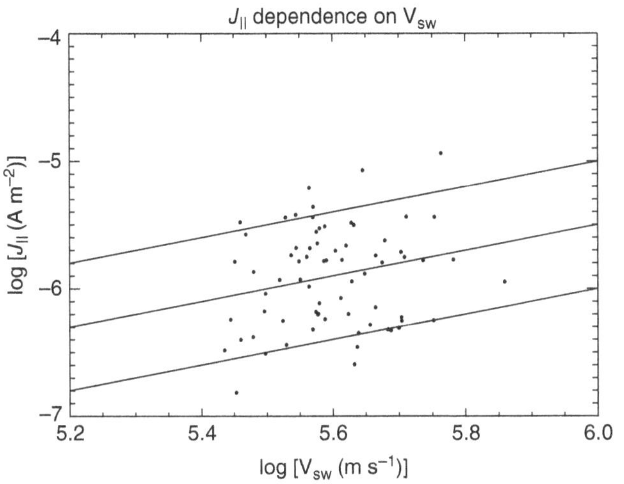

Figure 18.10 Field-aligned current density ($J_{||}$) increases with V_{sw} at 1300–1700 MLT. The solid black lines have slope = 1, which is expected from equation (18.14). From *Johnson and Wing* [2015].

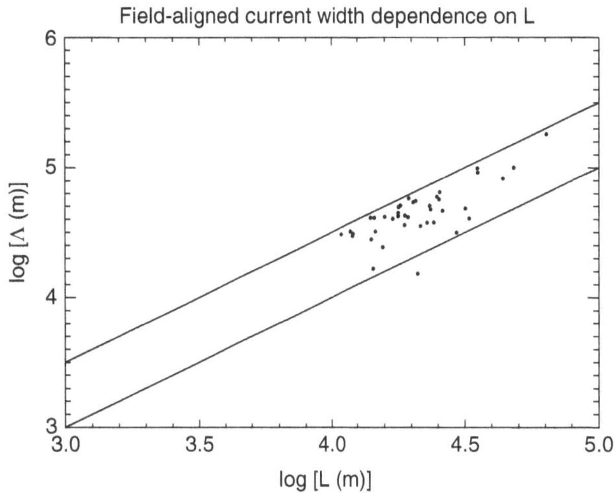

Figure 18.12 Field-aligned current width (Λ) as a function of L for small α ($\Lambda/L < 5$) at 1100–1700 MLT. The solid black lines have a slope = 1, which is the expected slope from equation (18.16). From *Johnson and Wing* [2015].

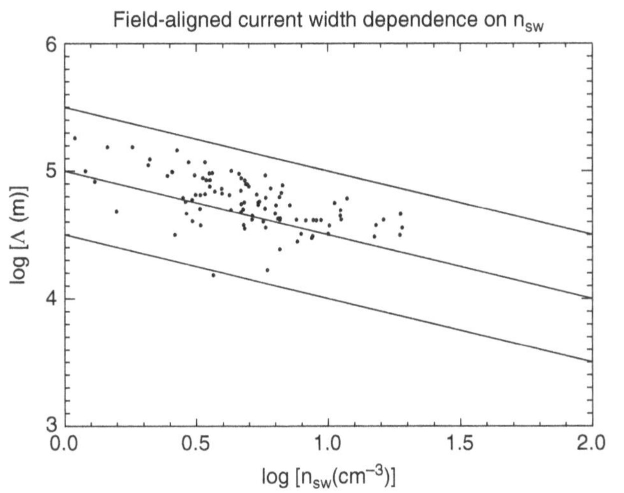

Figure 18.11 Field-aligned current width (Λ) decreases with increasing n_{sw} for small α ($\Lambda/L < 5$) at 1100–1700 MLT. The solid black lines have slope = −0.5, which is the expected slope from equation (18.16) for small α. From *Johnson and Wing* [2015].

First, we examine how Λ varies with n_{sw}. As discussed in section 18.2.3, for $\alpha \ll 1$, equation (18.16) suggests that $\Lambda \sim L \sim n_{sw}^{-0.5}$. Figure 18.11 plots log Λ versus log n_{sw}, for $\Lambda/L < 5$ (small α) at 1100–1700 MLT. The figure shows that all the points tend to align with the lines of slope = −0.5 ($\Lambda \sim n_{sw}^{-0.5}$), consistent with equation (18.16). *Johnson and Wing* [2015] obtained the least square fit of $\log(\Lambda) = -(0.47 \pm 0.06)\ \log(n_{sw}) + (5.1 \pm 0.05)$ or $\Lambda \sim n_{SW}^{-(0.47 \pm 0.06)}$. This result is consistent with the *Echim et al.* [2008] model, which shows that Λ anticorrelates with n_{sw}.

Next, we examine how Λ varies with L. $L = \sqrt{\Sigma_p/\kappa}$ was calculated using empirical formulas for Σ_p and solar wind parameters to infer κ. From equation (18.16), $\Lambda \sim L$ for $\alpha \ll 1$ (or equivalently $\Delta_i \ll L$). Figure 18.12 plots log Λ versus log L for $\Lambda/L < 5$ for 1100–1700 MLT, as in Figure 18.11. Figure 18.12 also draws theoretical lines with a slope of 1, which fit the points well. *Johnson and Wing* [2015] obtained $\log(\Lambda) = (0.9 \pm 0.1)\ \log(L) + (0.9 \pm 0.5)$ or $\Lambda \sim L^{(0.9 \pm 0.1)}$. The uncertainties in the estimates of Σ_p and κ contribute to the rather large scatter. The estimation of Σ_p relied on the accuracies of the *Robinson et al.* [1987] and *Robinson and Vondrak* [1984] empirical formulas and the accuracies of electron mean energy ($<E_e>$), electron energy flux (ε), and solar EUV flux. *Johnson and Wing* [2015] discussed the uncertainties in more details.

18.4.3. Thickness of the Boundary Layer (Δ_m)

Section 18.2 discusses how one can obtain Δ_m from the observed L and Λ using equations (18.16) and (18.17) and the relation $\Delta_m/\Delta_i \sim (B_i/B_m)^{0.5}$ (assume $B_i/B_m \sim 1000$). However, Δ_m can also be obtained from equation (18.14), which relates Δ_m to $J_{||}$, Σ_p, L, V_0, and B_m, and which can be obtained observationally and with empirical formulas.

Figures 18.13a and 18.13b plot log Δ_m versus log V_{sw} where Δ_m is obtained independently from equations (18.16) and (18.14), respectively. The Δ_m obtained from either method has roughly the same value – mean $\Delta_m \sim 3 \times 10^6$ m (~0.5 R_E). The scatter in Figure 18.13 is quite large, but in situ observations at the boundary layer also show similar mean and scatter [e.g., *Eastman and Hones*, 1979; *Phan and Paschmann*, 1996; *Šafránková et al.*, 2007]. Some of the sources of these uncertainties are discussed in section 18.4.5 and in *Johnson and Wing* [2015].

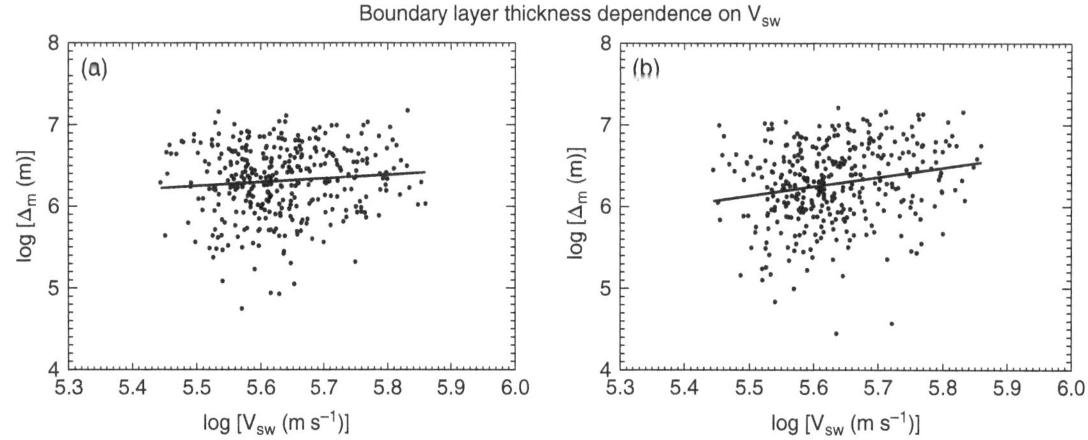

Figure 18.13 The magnetospheric boundary layer (Δ_m) has some dependence on V_{sw}. Δ_m is calculated using two methods: (a) from equation (18.16) and $\Delta_m/\Delta_i \sim (B_i/B_m)^{0.5}$ and (b) from equation (18.14). Both methods return a mean $\Delta_m \sim 3 \times 10^6$ m (~0.5 R_E). The least square fit is plotted as a black line. All data points come from 1100–1700 MLT. The x axis ranges from $V_{sw} \sim 200$ km s^{-1} to 1000 km s^{-1}. From *Johnson and Wing* [2015].

The data in Figure 18.13a have $r = 0.09$ and $P < 0.09$, whereas those in Figure 18.13b have $r = 0.2$ and $P < 0.01$. The low r results from the large scatter or variability in the data, but the low P suggests the trend is statistically significant, particularly in Figure 18.13b. In Figure 18.13a (from equation [18.16]), the least square fit yields $\log(\Delta_m) = (0.5 \pm 0.3)\log(V_{sw}) + (0.4 \pm 0.1)$ or $\Delta_m \sim V_{sw}^{(0.5 \pm 0.3)}$. In Figure 18.13b (from equation [18.14]), the least square fit yields $\log(\Delta_m) = (1.1 \pm 0.3)\ \log(V_{sw}) + (-0.2 \pm 0.2)$ or $\Delta_m \sim V_{sw}^{(1.1 \pm 0.3)}$. The trend that Δ_m increases with V_{sw} seen in Figure 18.13b is meaningful (the slope has small uncertainty). This trend may be attributed to KH mode that becomes more unstable as V_{sw} increases. Moreover, *Foullon et al.* [2008] found that there is observational evidence in Cluster data that suggest that the electron LLBL widens with increasing V_{sw} due to KH mechanism. The large scatter in the data suggests that V_{sw} is only one of many parameters that can affect Δ_m. That is, there are other parameters besides velocity that control the boundary layer thickness.

18.4.4. Field-Aligned Potential Drop ($\Delta\phi_{\parallel}$)

In the upward FAC region slightly poleward of the open-closed boundary, the low-latitude polar rain sometimes shows a ramping up of the electron energy with increasing latitude before reaching a plateau. *Wing et al.* [2015] presented two examples, which are presented in Figure 18.14 and discussed herein. (*Wing et al.* [2010] shows more examples.)

DMSP F7 observed a double cusp [e.g., *Wing et al.*, 2001] at AACGM [*Baker and Wing*, 1989] magnetic latitude ~(–74.7°) – (–76.4°), a mantle at AACGM latitude ~(–76.4°) – (–77.5°), and polar rain (prior to 01:10:09 UT) on 1985 March 27, as shown in Figure 18.14a. Immediately poleward of the mantle, the polar rain electron energies were typical, but at AACGM latitude ~(–77.7°) – (–78.1°) (between the two red vertical dashed lines in the middle panel), the electron differential energy flux peak energy increased with increasing latitude before plateauing at just below 1 keV. The polar rain electron energy does not typically reach this high. From the magnetometer data, it was determined that the polar rain in this interval was located near the poleward edge of the upward FAC region. Poleward of latitude ~–78.1°, the current suddenly weakened and the corresponding polar rain electron energies and fluxes also suddenly dropped to their more typical values.

Figure 18.14b shows another pass exhibiting the same polar rain feature. The DMSP observed cusp at AACGM latitude ~(–77.4°–) – (–80.3°), mantle at AACGM latitude ~(–80.3°) – (–83.2°), and polar rain (prior to 03:27:44 UT) on 20 November 1984. At AACGM latitude ~(–83.2°) – (–85.1°) (between the two red vertical dashed lines in the middle panel), the polar rain electron differential energy flux peak energy increased with increasing latitude before plateauing around 1 keV. As in Figure 18.14a, the polar rain in this interval was located in the upward FAC region. The current changed abruptly from upward to slightly downward at AACGM latitude ~–85.1° and the corresponding polar rain electron energies and fluxes abruptly dropped. Poleward of latitude ~–85.1°, the polar rain electrons looked like the typical polar rain electrons and the current was small or slightly downward.

Wing et al. [2015] suggested that this phenomenon can be explained by electron acceleration in the upward FAC and evolution of the magnetosheath V. Following the reconnection at the dayside magnetopause, the field lines

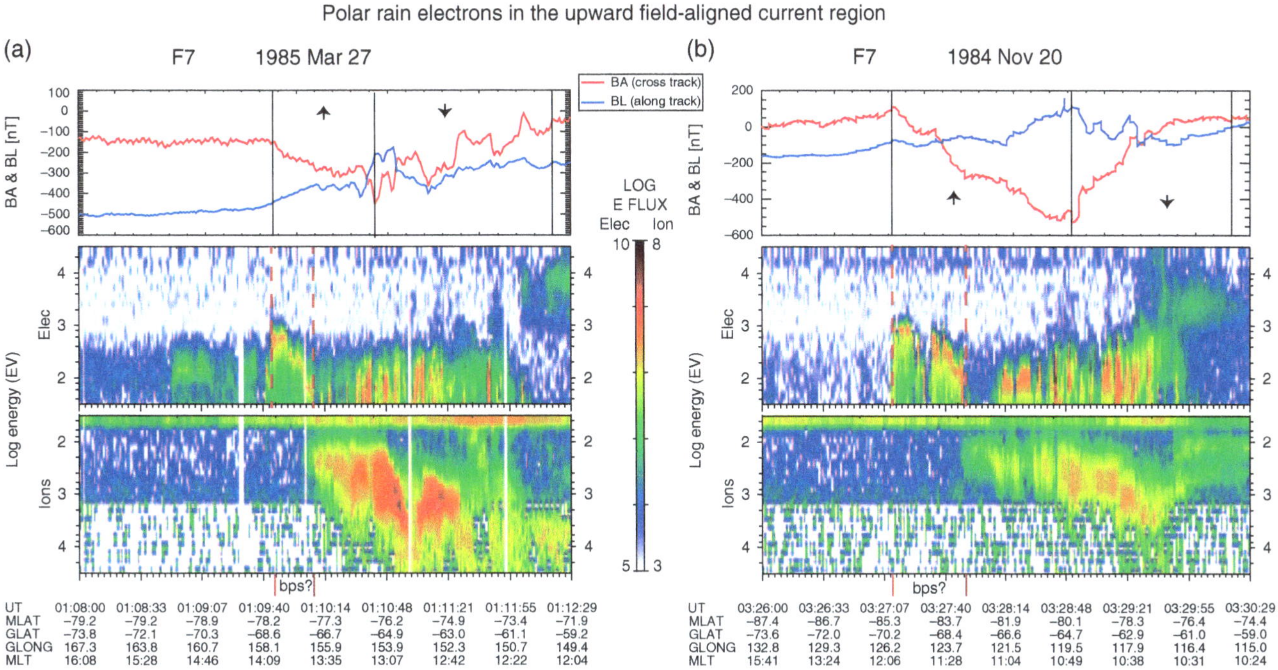

Figure 18.14 Two examples showing at the poleward edge of the upward FAC region, between the two red vertical dashed lines, the polar rain electron energy ramps up with increasing latitude before reaching a plateau. The top panel shows the DMSP magnetic field cross-track (BA, red) and along the track (BL, blue) components. The upward and downward field-aligned currents are indicated by the up and down arrows, respectively. The middle and bottom panels show the electron and ion spectrograms showing log differential energy flux, in units of eV/cm^2 s sr eV, from 32 eV to 30 keV. The ion energy scale is inverted. The white vertical bands in (a) indicate data gaps. The magnetic coordinates are obtained from Altitude Adjusted Corrected Geo-Magnetic (AACGM) coordinate [*Baker and Wing*, 1989]. From *Wing et al.* [2015].

would convect in the antisunward direction to the nightside while the magnetosheath plasma continuously enters the magnetosphere (see for example *Wing et al.* [1996; 2001]). Due to the magnetosheath plasma evolution [e.g., *Spreiter and Stahara*, 1985], polar rain electrons at progressively higher latitude originate from the magnetosheath region that progressively has lower density and higher speed. Figure 18.6a shows that higher V_{sw} can increase $\Delta\phi_{\|}$ (see also observations in *Wing et al.* [2011]). Lower magnetosheath or LLBL density can also increase $\Delta\phi_{\|}$ (see equation [18.15]) and *Wing et al.* [2011]). As the magnetosheath speed increases and density decreases, the $\Delta\phi_{\|}$ and polar rain electron differential flux peak energy progressively increase. Eventually, the magnetosheath density and velocity reach their solar wind values asymptotically, at which point the polar rain electron differential flux peak energy would plateau. As an aside, the particles in the region between two red vertical dashed lines are often classified as bps, which is a closed-field line region, by *Newell et al.* [1991a, 1991b,1991c] algorithm, but this region is most likely open and the electrons are polar rain electrons (see discussion in *Wing et al.* [2015]).

Equation (18.15) shows the expression for $\Delta\phi_{\|}$ at the boundary layer. Assuming solar wind suprathermal $n_e = 0.4\,cm^{-3}$, $T_e = 1 \times 10^6\,K$, $V_0 = 400\,km\,s^{-1}$, $B_0 = 3$ nT, $B_i = 4 \times 10^{-5}$ T, $\Delta_m = 6000\,km$, $\Sigma p = 2S$, equation (18.15) predicts $\Delta\phi_{\|} \sim 1.3\,keV$. Figure 18.14 shows that in both cases, the electron differential energy flux peak energy plateau to about 1 keV, which is close to the theoretical prediction. DMSP satellites do not measure $\Delta\phi_{\|}$ directly, but here, the peak electron energy was used as a proxy for $\Delta\phi_{\|}$ [e.g., *Lyons et al.*, 1979; *Lyons*, 1980]. In both events shown in Figure 18.14, the upward FAC regions are R1, but the phenomenon and mechanism involved would work with any upward FAC region, for example, smaller scale upward FAC that is driven by the magnetosheath-magnetosphere shear flow.

18.4.5. Sources of Uncertainties

Figures 18.7–18.14 show that the data scales well with the expected power law dependence from the analytical theory. However, the scatter in the data is quite large in some cases. Therefore, it would be useful to briefly discuss the origins of the scatter. There are many factors that can contribute to the scatter. For example, there are uncertainties associated with magnetosheath V_0, n_e, T_e, and B_m, ionospheric Σ_p, and the effects of kinetic pressure gradient on $J_{\|}$. The scatter also comes from averaging over all magnetic activities and solar wind conditions.

The theory assumes a linear Knight relation, which ignores thermal current (J_t) and nonlinear saturation. It also assumes Maxwellian distribution in the LLBL electron. Near the subsolar region, higher magnetosheath temperatures and densities may increase the thermal current, $J_t = n_e e V_t / \sqrt{2\pi}$, where J_t = thermal current, n_e = electron density at the boundary layer, V_t = electron thermal velocity at the boundary layer. The $\Delta\phi_{\|}$ can be overestimated when J_t is larger or comparable to $J_{\|}$. Most of the currents presented in Figures 18.7–18.14 exceed typical J_t, but the weaker currents may be comparable (J_t~0.1–1 μA m^{-2} for n_e~0.5–10 cm^{-3} and Te~100 eV) (even when the currents are comparable, the scaling relations may still apply). Finally, the magnetopause processes such as KHI, flux transfer events (FTE), Alfvénic currents (FACs associated with Alfvén waves), and reconnection may introduce small-scale FACs [*Miura and Pritchett*, 1982; *Lotko et al.*, 1987; *Kozlovsky et al.*, 2009; *Hull et al.*, 2010; *Damiano et al.*, 2016], which would introduce noise or scatter to the large-scale FACs computed in Figures 18.7–18.14.

Johnson and Wing [2015] and *Wing and Johnson* [2015] discussed these and other uncertainties in more details.

18.5. SUMMARY AND CONCLUSION

This chapter presents the recent advances in the analytical theory, kinetic modeling, and observations of the afternoon upward R1 current that is generated at the low-latitude boundary layer. The kinetic MI coupling model provides numerical solutions for the electrodynamic properties of the auroral ionosphere as a function of the latitude. It considers a model generator described by Vlasov equilibrium solutions. The theory examines the dependence of the upward field-aligned currents and their structure on ionospheric and solar wind parameters. A main advantage of this framework is that it provides a simple formula for calculating $J_{\|}$ and boundary layer structures using routinely observed parameters in the ionosphere and solar wind and empirical formulas for conductivities.

Using simultaneous measurements of solar wind and DMSP particle and magnetometer data, we show how the MI coupling parameters $J_{\|}$, Δ, Λ, Σ_p, and κ vary with each other and with solar wind parameters. These observations are well organized by the simple analytical expressions. The good agreement between the theory and observations suggests that the theory captures the essential physics governing the upward field-aligned current and the solar wind–magnetosphere–ionosphere coupling at the low-latitude boundary layer.

One the most remarkable success of the analytical theory is that it can predict the scaling of the observed $J_{\|}$ quite well. The least square fit of calculated $J_{\|,max}$ versus observed $J_{\|}$ returned $\log(J_{\|,max\_cal}) = (0.96 \pm 0.04) \log(J_{\|\_obs}) + (0.03 \pm 0.01)$ for data points from 1100–1700 MLT, excluding data from noon. Near noon (MLT = 1200), it was found that $\log(J_{\|,max\_cal}) = (0.83 \pm 0.07) \log(J_{\|\_obs}) + (0.05 \pm 0.03)$.

The analytical theory also provides an insight into how boundary layer structure, Δ_m, maps to ionospheric FAC latitudinal width, Λ, and how the mapping depends on the auroral electrostatic scale length, L. In particular, it was shown that $\Lambda \sim n_{sw}^{-0.5}$ and $\Lambda \sim L$ when $\Lambda/L < 5$. Using ionospheric and solar wind observations, the theory predicted boundary layer thickness (Δ_m)~3000 km, which is consistent with in situ observations. Δ_m appears to have some dependence on V_{sw} such that larger V_{sw} would increase Δ_m. Thus, the simple analytical theory can serve as the basis for inferring boundary layer structures from ionospheric and solar observations. While there can be significant advantages to using this simplified analytical theory for data analysis and modeling, it is also important to recognize it has limitations, which are summarized in section 18.4.5. In some or many cases, a more comprehensive numerical modeling such as the one presented in section 18.3 may be necessary.

For the future work, the analytical theory and model can be expanded to include the small- scale currents generated by the velocity shear in the KH vortices. During periods of northward IMF, there are smaller scale FACs known as northward B_z (NBZ) currents. *Vennerstrom et al.* [2005] suggested while the lower latitude R1 that is located at the boundary layer is generated by the shear flow at LLBL, the higher latitude NBZ currents are generated by the shear flow between newly reconnected field lines and the lobe field lines. It would be interesting to apply the theory and model described in this chapter to upward NBZ currents.

This chapter reviews the recent analytical theory, model, and observations of the upward FACs at the LLBL that are typically found on the dusk flank. At the LLBL on the dawn flank, FACs are usually downward and *Johnson and Wing* [2015] theory and *Echim et al.* [2007, 2008] model may not apply because Knight relation [*Knight*, 1973] may not apply. For the future work, it would be interesting to develop a theory or model that would work for downward FACs, which would be complementary to the works reviewed in this chapter. We would like to note, there have been recent theoretical works and modeling work that focus on current-voltage relations in the downward FAC region [e.g., *Cran-McGreehin and Wright*, 2005a, 2005b; *Temerin and Carlson*, 1998; *De Keyser and Echim*, 2010], although these works did not focus exclusively for the LLBL conditions.

ACKNOWLEDGMENTS

Simon Wing acknowledges support of NASA Grants NNX13AE12G, NNX15AJ01G, NNX16AR10G, and NNX16AQ87G. Jay R. Johnson acknowledges support from NASA Grants (NNH11AR07I, NNX14AM27G, NNH14AY20I, NNX16AC39G), NSF Grants

(ATM0902730, AGS-1203299), and DOE contract DE-AC02-09CH11466. Marius Echim acknowledges support from the Belgian Solar Terrestrial Center of Excellence and the BRAIN-BE project BR/175/A2/MOMA.

REFERENCES

Baker, K. B., and S. Wing (1989), A new magnetic coordinate system for conjugate studies at high latitudes, *J. Geophys. Res.*, *94*(A7), 9139–9143; doi:10.1029/JA094iA07p09139.

Bythrow, P. F., R. A. Heelis, W.B. Hanson, R. A. Power, and R. A. Hoffman (1981), Observational evidence for a boundary layer source of dayside region 1 field-aligned currents, *J. Geophys. Res.*, *86*(A7), 5577–5589; doi:10.1029/JA086iA07p05577.

Cran-McGreehin, A. P., and A. N. Wright (2005), Electron acceleration in downward auroral field-aligned currents, *J. Geophys. Res.*, *110*, A10S15; doi:10.1029/2004JA010898.

Cran-McGreehin, A. P., and A. N. Wright (2005b), Current-voltage relationship in downward field-aligned current region, *J. Geophys. Res.*, *110*, A10S10; doi:10.1029/2004JA010870.

Crooker, N. U. (1977), Explorer 33 entry layer observations, *J. Geophys. Res.*, *82*(4), 515–522; doi:10.1029/JA082i004p00515.

Damiano, P. A., J. R. Johnson, and C. C. Chaston (2016), Ion gyroradius effects on particle trapping in kinetic Alfvén waves along auroral field lines, *J. Geophys. Res. Space Physics*, *121*, 10,831–10,844; doi:10.1002/2016JA022566.

De Keyser, J., and M. Echim (2010), Auroral and sub-auroral phenomena: an electrostatic picture, *Ann. Geophys.*, *28*(2), 633–650.

De Keyser, J., and M. Roth (1997), Equilibrium conditions for the tangential discontinuity magnetopause, *J. Geophys. Res. Space Physics*, *102*, A5, 9513–9530; doi:10.1029/97JA00335.

De Keyser, J., and M. Roth (1998), Equilibrium conditions and magnetic field rotation at the tangential discontinuity magnetopause, *J. Geophys. Res. Space Physics*, *103*, A4, 6653–6662; doi: 10.1029/97JA03710.

De Keyser, J., and M. Echim (2013), Electric potential differences across auroral generator interfaces, *Ann. Geophys.*, *31*, 251–261; doi:10.5194/angeo-31-251-2013.

De Keyser, J., R. Maggiolo, and M. Echim (2010), Monopolar and bipolar auroral electric fields and their effects, *Ann. Geophys.*, *28* (11), 2027–2046.

Dimmock, A. P., and K. Nykyri (2013), The statistical mapping of magnetosheath plasma properties based on THEMIS measurements in the magnetosheath interplanetary medium reference frame, *J. Geophys. Res. Space Physics*, *118*, 4963–4976; doi:10.1002/jgra.50465.

Drakou, E., B. U. O. Sonnerup, and W. Lotko (1994), Self-consistent steady state model of the low-latitude boundary layer, *J. Geophys. Res.*, *99*, 2351–2364; doi:10.1029/93JA02094.

Eastman, T. E., and E. W. Hones Jr. (1979), Characteristics of the magnetospheric boundary layer and magnetopause layer as observed by Imp 6, *J. Geophys. Res.*, *84*, 2019–2028; doi:10.1029/JA084iA05p02019.

Eastman, T. E., E. W. Hones Jr., S. J. Bame, and J. R. Asbridge (1976), The magnetospheric boundary layer: Site of plasma, momentum and energy transfer from the magnetosheath into the magnetosphere, *Geophys. Res. Lett.*, *3*, 685–688; doi:10.1029/GL003i011p00685.

Echim, M. M., J. F. Lemaire, and M. Roth (2005), Self-consistent solution for a collisionless plasma slab in motion across a magnetic field, *Phys. Plasmas*, *7*, 072904; doi:10.1063/1.1943848.

Echim, M. M., M. Roth, and J. De Keyser (2007), Sheared magnetospheric plasma flows and discrete auroral arcs: A quasi-static coupling model, *Ann. Geophys.*, *25*, 317–330; doi:10.5194/angeo-25-317-2007.

Echim, M. M., M. Roth, and J. De Keyser (2008), Ionospheric feedback effects on the quasi-stationary coupling between LLBL and postnoon/evening discrete auroral arcs, *Ann. Geophys.*, *26*, 913–928; doi:10.5194/angeo-26-913-2008.

Echim, M. M., R. Maggiolo, M. Roth, and J. De Keyser (2009), A magnetospheric generator driving ion and electron acceleration and electric currents in a discrete auroral arc observed by Cluster and DMSP, *Geophys. Res. Lett.*, *36*(12), CiteID L12111.

Evans, D. S. (1985), The characteristics of a persistent auroral arc at high latitudes in the 1400 MLT sector, in *The Polar Cusp*, edited by J. Holtet and A. Egeland, 99–109, D. Reidel, Hingham, Massachusetts.

Foullon, C., C. J. Farrugia, A. N. Fazakerley, C. J. Owen, F. T. Gratton, and R. B. Torbert (2008), Evolution of Kelvin-Helmholtz activity on the dusk flank magnetopause, *J. Geophys. Res.*, *113*, A11203; doi:10.1029/2008JA013175.

Fujimoto, M., T. Mukai, H. Kawano, M. Nakamura, A. Nishida, Y. Saito, T. Yamamoto, and S. Kokubun (1998), Structure of the low-latitude boundary layer: A case study with Geotail data, *J. Geophys. Res.*, *103*, 2297–2308; doi:10.1029/97JA02946.

Haerendel, G., G. Paschmann, N. Sckopke, H. Rosenbauer, and P. C. Hedgecock (1978), The frontside boundary layer of the magnetosphere and the problem of reconnection, *J. Geophys. Res.*, *83*(A7), 3195–3216; doi:10.1029/JA083iA07p03195.

Haraguchi, K., H. Kawano, K. Yumoto, S. Ohtani, T. Higuchi, and, G. Ueno (2004), Ionospheric conductivity dependence of dayside region-0, 1, and 2 field-aligned current systems: statistical study with DMSP-F7, *Ann. Geophys.*, *22*, 2775–2783; doi:10.5194/angeo-22-2775-2004.

Harel, M., Wolf, R. A., Reiff, P. H., and Hillis, H. K. (1977), Study of plasma flow near the Earth's plasmapause, Rep. AFGL-TR-77-286, U.S. Air Force Geophys. Lab, Hanscom, AFB, Mass.

Higuchi, T., and S.-I. Ohtani (2000), Automatic identification of large-scale field-aligned current structures, *J. Geophys. Res.*, *105* (25), 305–25; doi:10.1029/2000JA900073.

Hones, E. W., J. R. Asbridge, S. J. Bame, M. D. Montgomery, S. Singer, and S.-I. Akasofu (1972), Measurements of magnetotail plasma flow made with vela 4b, *J. Geophys. Res.*, *77*(28), 5503–5522; doi:10.1029/JA077i028p05503.

Hull, A. J., M. Wilber, C. C. Chaston, J. W. Bonnell, J. P. McFadden, F. S. Mozer, M. Fillingim, and M. L. Goldstein (2010), Time development of field-aligned currents, potential drops, and plasma associated with an auroral poleward boundary intensification, *J. Geophys. Res.*, *115*, A06211; doi:10.1029/2009JA014651.

Hundhausen, A. J., S. J. Bame, J. R. Asbridge, and S. J. Sydoriak (1970), Solar wind proton properties: Vela 3 observations from July 1965 to June 1967, *J. Geophys. Res.*, *75*(25), 4643–4657; doi:10.1029/JA075i025p04643

Iijima, T., and T. A. Potemra (1976), The amplitude distribution of field-aligned currents at northern high latitudes observed by Triad, *J. Geophys. Res.*, *81*, 2165–2174; doi:10.1029/JA081i013p02165.

Iijima, T., and T. A. Potemra (1978), Large-scale characteristics of field-aligned currents associated with substorms, *J. Geophys. Res.*, *83*(A2), 599–615; doi:10.1029/JA083iA02p00599

Johnson, J. R., and S. Wing (2015), The dependence of the strength and thickness of field-aligned currents on solar wind and ionospheric parameters, *J. Geophys. Res. Space Physics*, *120*, 3987–4008; doi: 10.1002/2014JA020312

Knight, S. (1973), Parallel electric fields, *Planet. Space Sci.*, *21*, 741–750.

Kozlovsky, A., T. Turunen, and S. Massetti (2009), Field-aligned currents of postnoon auroral arcs, *J. Geophys. Res.*, *114*, A03301; doi:10.1029/2008JA013666.

Lemaire, J., and L. F. Burlaga (1976), Diamagnetic boundary layers: A kinetic theory, *Astrophys. Space Sci.*, *45*, 303–325.

Lemaire, J., and M. Scherer (1971), Simple model for an ion-exosphere in an open magnetic field, *Phys. Fluids*, *14*, 1683–1694.

Lotko, W., B. U. O. Sonnerup, and R. L. Lysak (1987), Nonsteady boundary layer flow including ionospheric drag and parallel electric fields, *J. Geophys. Res.*, *92*, 8635–8648; doi:10.1029/JA092iA08p08635.

Lyons, L. R. (1980), Generation of large-scale regions of auroral currents, electric potentials, and precipitation by the divergence of the convection electric field, *J. Geophys. Res.*, *85*, 17–24; doi:10.1029/JA085iA01p00017.

Lyons, L. R. (1981), Discrete aurora as the direct result of an inferred high-altitude generating potential distribution, *Journal of Geophysical Research*, *86*, 1-8.

Lyons, L. R., D. S. Evans, and R. Lundin (1979), An observed relation between magnetic field aligned electric fields and downward electron energy fluxes in the vicinity of auroral forms, *J. Geophys. Res.*, *84*(A2), 457–461; doi:10.1029/JA084iA02p00457.

Maggiolo, R., M. Echim, S. C. Wedlund, Y. Zhang, D. Fontaine, G. Lointier, and J.-G. Trotignon (2012), Polar cap arcs from the magnetosphere to the ionosphere: kinetic modelling and observations by Cluster and TIMED, *Ann. Geophys.*, *30*, 283–302; doi:10.5194/angeo-30-283-2012.

Matsumoto, Y., and K. Seki (2010), Formation of a broad plasma turbulent layer by forward and inverse energy cascades of the Kelvin–Helmholtz instability, —DOI:10.1029/2009JA014637.

Miura, A., and P. L. Pritchett (1982), Nonlocal stability analysis of the MHD Kelvin-Helmholtz instability in a compressible plasma, *J. Geophys. Res.*, *87*(A9), 7431–7444; doi:10.1029/JA087iA09p07431.

Mozer, F. S., C. W. Carlson, M. K. Hudson, R. B. Torbert, B. Parady, J. Yatteau, and M. C. Kelley (1977), Observations of paired electrostatic shocks in the polar magnetosphere, *Phys. Rev. Lett.*, *38*, 292.

Nakamura, T. K. M., H. Hasegawa, I. Shinohara, and M. Fujimoto (2011), Evolution of an MHD-scale Kelvin-Helmholtz vortex accompanied by magnetic reconnection: Two-dimensional particle simulations, *J. Geophys. Res.*, *116*, A03227; doi:10.1029/2010JA016046

Newell, P. T., and C.-I. Meng (1988), The cusp and the cleft/boundary layer: Low-altitude identification and statistical local time variation, *J. Geophys. Res.*, *931*, 14,549–14,556; doi:10.1029/JA093iA12p14549.

Newell, P. T., C.-I. Meng, E. R. Sanchez, W. J. Burke, and M. E. Greenspan (1991a), Identification and observations of the plasma mantle at low altitude, *J. Geophys. Res.*, *96*, 35–45; doi:10.1029/90JA01760.

Newell, P. T., E. R. Sanchez, C.-I. Meng, W. J. Burke, M. E. Greenspan, and C. R. Clauer (1991c), The low-latitude boundary layer and the boundary plasma sheet at low altitude: Prenoon precipitation regions and convection reversal boundaries, *J. Geophys. Res.*, *962*, 21,013–21,023; doi:10.1029/91JA01818.

Newell, P. T., S. Wing, C.-I. Meng, and V. Sigillito (1991b), The auroral oval position, structure, and intensity of precipitation from 1984 onward: An automated on-line data base, *J. Geophys. Res.*, *96*, 5877–5882; doi:10.1029/90JA02450.

Paschmann, G., G. Haerendel, N. Sckopke, H. Rosenbauer, and P. C. Hedgecock (1976), Plasma and magnetic field characteristics of the distant polar cusp near local noon: The entry layer, *J. Geophys. Res.*, *81*(16), 2883–2899; doi:10.1029/JA081i016p02883.

Phan, T. D., and G. Paschmann (1996), Low-latitude dayside magnetopause and boundary layer for high magnetic shear, 1. Structure and motion, *J. Geophys. Res.*, *101*, 7801–7816; doi:10.1029/95JA03752.

Phan, T. D., B. U. O. Sonnerup, and W. Lotko (1989), Self-consistent model of the low-latitude boundary layer, *J. Geophys. Res.*, *94*, 1281–1293; doi:10.1029/JA094iA02p01281.

Richardson, J. D., J. W. Belcher, A. J. Lazarus, K. I. Paularena, and P. R. Gazis (1996), Statistical properties of the solar wind, in *Solar Wind Eight*, AIP Conf. Proc., 382, edited by D. Winterhalter et al., 483–486, American Institute of Physics, Melville, New York; doi:10.1063/1.51433.

Robinson, R. M., and R. R. Vondrak (1984), Measurements of E region ionization and conductivity produced by solar illumination at high latitudes, *J. Geophys. Res.*, *89*, 3951–3956; doi:10.1029/JA089iA06p03951.

Robinson, R. M., R. R. Vondrak, K. Miller, T. Dabbs, and D. Hardy (1987), On calculating ionospheric conductances from the flux and energy of precipitating electrons, *J. Geophys. Res.*, *92*, 2565–2569; doi:10.1029/JA092iA03p02565.

Rosenbauer, H., H. Grünwaldt, M. D. Montgomery, G. Paschmann, and N. ySckopke (1975), Heos 2 plasma observations in the distant polar magnetosphere: The plasma mantle, *J. Geophys. Res.*, *80* (19), 2723–2737; doi:10.1029/JA080i019p02723.

Roth, M., D. S. Evans, and J. Lemaire (1993), Theoretical structure of a magnetospheric plasma boundary: Application to the formation of discrete auroral arcs, *J.Geophys. Res.*, *98*(A7), 11, 411–11,423.

Roth, M., J. De Keyser, and M. M. Kuznetsova (1996), Vlasov theory of the equilibrium structure of tangential discontinuities in space plasmas, *Space Sci. Rev.*, *76*(3–4), 251–317; doi: 10.1007/BF00197842.

Safránková, J., Z. Nemecek, L. Prech, J. Simunek, D. Sibeck, and J.-A. Sauvaud (2007), Variations of the flank LLBL thickness as response to the solar wind dynamic pressure and IMF orientation, *J. Geophys. Res.*, *112*, A07201; doi:10.1029/2006JA011889.

Scudder, J. D., D. L. Lind, and K. W. Ogilvie (1973), Electron observations in the solar wind and magnetosheath, *J. Geophys. Res.*, *78*, 6535–6548; doi:10.1029/JA078i028p06535.

Sestero, A. (1966), Structure of plasma sheaths, *Phys. Fluids*, *9*, 2006–2013.

Siscoe, G. L., W. Lotko, and B. U. Sonnerup (1991), A high-latitude, low-latitude boundary layer model of the convection current system, *J. Geophys. Res.*, *96*(A3), 3487–3495; doi:10.1029/90JA02362.

Sonnerup, B. U. O. (1980), Theory of the low-latitude boundary layer, *J. Geophys. Res.*, *85*, 2017–2026; doi:10.1029/JA085iA05p02017.

Spreiter, J. R., and S. S. Stahara (1985), Magnetohydrodynamic and gasdynamic theories for planetary bow waves, in *Collisionless Shocks in the Heliosphere: Reviews of Current Research*, Geophysical Monograph Series, 35, edited by B. T. Tsurutani and R. G. Stone, 85–107, American Geophysical Union, Washington, DC.

Stern, D. P. (1983), The origins of Birkeland currents, *Rev. Geophys. Space Phys.*, *21*, 125–138.

Temerin, M., and C. W. Carlson (1998), Current-voltage relationship in the downward current region, *Geophys. Res. Lett.*, *25*, 2365.

Temerin, M., C. Cattell, R. Lysak, M. Hudson, R. B. Torbert, F. S. Mozer, R. D. Sharp, and P. M. Kintner (1981), The small-scale structure of electrostatic shocks, *J. Geophys. Res.*, *86* (A13), 11278–11298; doi:10.1029/JA086iA13p11278.

Vaisberg, O. L., V. N. Smirnov, L. A. Avanov, J. H. Waite, J. L. Burch, D. L. Gallagher, and N. L. Borodkova (2001), Different types of low-latitude boundary layer as observed by interball tail probe, *J. Geophys. Res.*, *106*(A7), 13,067–13,090; doi:10.1029/2000JA000154.

Vennerstrom, S., T. Moretto, L. Rastätter, and J. Raeder (2005), Field-aligned currents during northward interplanetary magnetic field: Morphology and causes, *J. Geophys. Res.*, *110*, A06205; doi:10.1029/2004JA010802.

Wei, C. Q., and L. C. Lee (1993), Coupling of magnetopause-boundary layer to the polar ionosphere, *J. Geophys. Res.*, *98*(A4), 5707–5725; doi:10.1029/92JA02232.

Wing, S., and J. R. Johnson (2015), Theory and observations of upward field-aligned currents at the magnetopause boundary layer, *Geophys. Res. Lett.*, *42*, 9149–9155; doi:10.1002/2015GL065464.

Wing, S., D. H. Fairfield, J. R. Johnson, and S-I. Ohtani (2015), On the field-aligned electric field in the polar cap, *Geophys. Res. Lett.*, *42*, 5090–5099; doi:10.1002/2015GL064229.

Wing, S., J. R. Johnson, E. Camporeale, and G. D. Reeves (2016), Information theoretical approach to discovering solar wind drivers of the outer radiation belt, *J. Geophys. Res. Space Physics*, *121*, 9378–9399; doi:10.1002/2016JA022711.

Wing, S., J. R. Johnson, P. T. Newell, and C.-I. Meng (2005), Dawn-dusk asymmetry in the northward IMF plasma sheet, *J. Geophys. Res.*, *110*, A08205; doi:10.1029/2005JA011086.

Wing, S., P. T. Newell, and J. M. Ruohoniemi (2001), Double cusp: Model prediction and observational verification, *J. Geophys. Res.*, *106*(A11), 25571–25593; doi:10.1029/2000JA000402.

Wing, S., P. T. Newell, and T. G. Onsager (1996), Modeling the entry of magnetosheath electrons into the dayside ionosphere, *J. Geophys. Res.*, *101*(A6), 13155–13167; doi:10.1029/96JA00395.

Wing, S., S.-I. Ohtani, J. R. Johnson, M. Echim, P. T. Newell, T. Higuchi, G. Ueno, and G. R. Wilson (2011), Solar wind driving of dayside field-aligned currents, *J. Geophys. Res.*, *116*, A08208; doi:10.1029/2011JA016579.

Wing, S., S.-I. Ohtani, P. T. Newell, T. Higuchi, G. Ueno, and J. M. Weygand (2010), Dayside field-aligned current source regions, *J. Geophys. Res.*, *115*, A12215; doi:10.1029/2010JA015837.

Xu, D., and M. G. Kivelson (1994), Polar cap field-aligned currents for southward interplanetary magnetic fields, *J. Geophys. Res.*, *99*(A4), 6067–6078; doi:10.1029/93JA02697.

19

The Current System of Dipolarizing Flux Bundles and Their Role as Wedgelets in the Substorm Current Wedge

Jiang Liu[1], V. Angelopoulos[1], Zhonghua Yao[2], Xiangning Chu[3], Xu-Zhi Zhou[4], and A. Runov[1]

ABSTRACT

In this chapter, we review recent observations of the current system of dipolarizing flux bundles (DFBs), small magnetotail flux tubes (typically < ~3 R_E in X_{GSM} and Y_{GSM}) with a significantly more dipolar magnetic field than their background. When propagating, a DFB or a group of several consecutive DFBs is known as a bursty bulk flow. A DFB's current system consists of a duskward current at the dipolarization front (DF, the leading edge of a DFB), a dawnward current immediately earthward of the DFB, field-aligned currents (FACs) at the DF and earthward of the DFB, and cross-tail current reduction inside the DFB. The FACs and cross-tail current reduction resemble a substorm current wedge (SCW), but are of much smaller scale. Therefore, each DFB possibly serves as a wedgelet, the building block of an SCW. We review a suggested way in which many wedgelets may collectively form a large-scale SCW. We also present some new results that help us to understand the DFB current system.

19.1. INTRODUCTION

An important player in near-Earth magnetotail dynamics is the "bursty bulk flow" (BBF), a 1–10 min timescale, > 100 km/s plasma flow [*Angelopoulos et al.*, 1992]. Likely generated by near-Earth reconnection around or beyond 20 R_E downtail [*Nagai et al.*, 1998; *Sitnov et al.*, 2009; *Angelopoulos et al.*, 2013; *Fu et al.*, 2013], BBFs propagate earthward after generation [*Chen and Wolf*, 1993; *Baumjohann et al.*, 1999; *Forsyth et al.*, 2008; *Runov et al.*, 2009; *Runov et al.*, 2011b]. After propagating for 2–3 min, BBFs stop near the inner edge of the plasma sheet (6–10 R_E downtail) [*Ohtani et al.*, 2006; *Dubyagin et al.*, 2011]. Each BBF is a narrow channel of 2–3 R_E dawn-dusk width [*Nakamura et al.*, 2004] and consists of one or multiple ~1 min timescale flow bursts [*Angelopoulos et al.*, 1992]. Usually coexisting with each flow burst is a localized (<1 min on temporal data series) enhancement in the northward magnetic field B_z [e.g., *Angelopoulos et al.*, 1992; *Ohtani et al.*, 2004]. These enhanced B_z regions correspond to flux tubes with a stronger and more dipolar magnetic field than the background. As they propagate earthward, these flux tubes deform greatly, but they retain their identity as bundles containing a strong magnetic field that becomes increasingly more dipolar [*Runov et al.*, 2009; *Liu et al.*, 2014]. These flux tubes were hence termed "dipolarizing flux bundles" (DFBs) by *Liu et al.* [2013a]. The strong dipolar magnetic field of DFBs result in a curvature force

[1] *Department of Earth, Planetary and Space Sciences and Institute of Geophysics and Planetary Physics, University of California, Los Angeles, California, USA*

[2] *Laboratoire de Physique Atmosphérique et Planétaire, STAR Institute, Université de Liège, Liège, Belgium*

[3] *Department of Atmospheric and Oceanic Sciences, University of California, Los Angeles, California, USA*

[4] *School of Earth and Space Sciences, Peking University, Beijing, China*

Electric Currents in Geospace and Beyond, Geophysical Monograph 235, First Edition.
Edited by Andreas Keiling, Octav Marghitu, and Michael Wheatland.

accelerating plasma earthward, so DFBs are drivers of BBFs [*Li et al.*, 2011; *Liu et al.*, 2014].

In addition to BBFs, dipolarizing flux bundles are associated with other well-known magnetotail structures: The leading edge of a DFB is characterized by a sharp enhancement of B_z (in GSM coordinates, which are used throughout this chapter) known as a "dipolarization front" (DF) [*Nakamura et al.*, 2002] or a reconnection front [*Angelopoulos et al.*, 2013]; inside a DFB, the plasma pressure usually drops, consistent with a "plasma bubble," a theoretical low-entropy flux tube, which is expected to have lower plasma pressure than the ambient plasma at distances not too far from the neutral sheet [*Pontius and Wolf*, 1990; *Chen and Wolf*, 1993; *Birn et al.*, 2004].

Figure 19.1 illustrates the topology of a DFB and a typical DFB observation (adapted from *Liu et al.* [2013a]). The meridional cross section of most dipolarizing flux bundles has the shape of a dipolar flux tube (Figs. 19.1a,b; see the next section for discussion of those with different shapes) [*Liu et al.*, 2013a]. When viewed from Earth, a DFB's transverse cross section (parallel to the *XY* plane) is convex (Fig. 19.1c) [*Sergeev et al.*, 1996; *Guzdar et al.*, 2010; *Liu et al.*, 2013b]. A DFB's interior (red regions in Figs. 19.1b,c) corresponds to the enhanced B_z interval in the magnetotail in situ observation (Fig. 19.1d). The leading edge of a DFB, the dipolarization front, a layer ~1000 km thick [*Runov et al.*, 2009; *Schmid et al.*, 2011; *Liu et al.*, 2013a], corresponds to a sharp B_z enhancement at the beginning of the DFB interval in the in situ observation. At the beginning of a DF, there is usually a minimum B_z (B_z dip; Fig. 19.1d) [*Ohtani et al.*, 2004; *Yao et al.*, 2015] that marks the outer boundary of the DF and

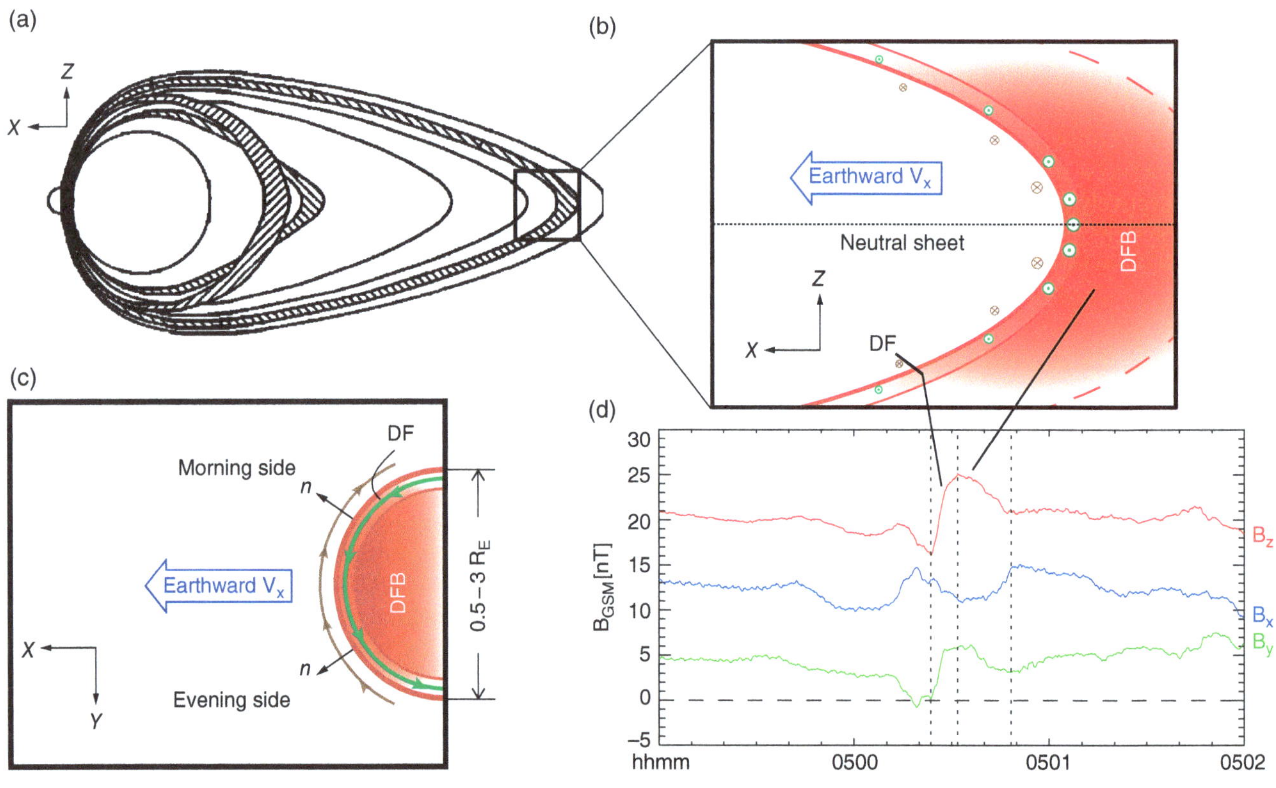

Figure 19.1 (a) Global view of dipolarizing flux bundles in the meridional plane (adapted from *Chen and Wolf* [1993]). The lines represent magnetic field lines, and the shaded flux tubes are DFBs at different stages. (b) Enlargement of a DFB's meridional cross section. (c) DFB's transverse cross section (parallel to the *XY* plane). The **n** arrows are the normal directions of the dipolarization front, which indicate whether an observation is on the morning or evening side of a DFB. Here morning or evening is relative to the localized DFB (local coordinates); the entire DFB flux tube can be in either the dawn or dusk sector of the magnetotail (global coordinates). In (b) and (c), the green arrows inside the DF represent a duskward current responsible for the sharp B_z enhancement at the DF. The tan arrows earthward of the DF are dip currents, which exist only for DFBs with a significant B_z dip (see section 19.2.1). These are only two of the many components of the DFB current system; more are illustrated in Figure 19.7b. Panels (b) and (c) are adopted from *Liu et al.* [2013a]. (d) In situ observation of a DFB in the magnetotail. The first vertical dashed line marks the outer boundary (start) of the DF and DFB, the second marks the inner boundary (end) of the DF, and the third marks the end of the DFB.

the DFB and forms an asymmetric bipolar signature with the B_z maximum at the end of the DF (see more discussion in the next section). Since BBFs are narrow, DFBs are also localized with width in Y of typically 0.5–3 R_E (Fig. 19.1b) [*Sergeev et al.*, 1996; *Liu et al.*, 2013b].

Although localized, DFBs play an important role in global magnetotail dynamics, including magnetic/energy flux transport (they are responsible for > 65% of BBFs' flux transport) [*Angelopoulos et al.*, 1994; *Liu et al.*, 2014] and particle acceleration [e.g., *Fu et al.*, 2011; *Gabrielse et al.*, 2012; *Birn et al.*, 2013; *Runov et al.*, 2015; *Liu et al.*, 2016]. Because these processes are critical aspects of substorms [e.g., *Arnoldy and Chan*, 1969; *Milan et al.*, 2009] and DFBs appear mostly during substorms [*Liu et al.*, 2014], understanding DFBs is crucial to solving the substorm mystery.

Dipolarizing flux bundles may also contribute to another critical aspect of the substorm: its major current system, the substorm current wedge (SCW) [*McPherron et al.*, 1973]. An SCW is created by disruption of the cross-tail currents near the inner boundary of the plasma sheet (6–10 R_E downtail), which may result from pressure gradient change and flow vortices [e.g., *Birn et al.*, 1999; *Birn et al.*, 2004; *Keiling et al.*, 2009; *Yao et al.*, 2012; *Birn and Hesse*, 2013]. This disruption reduces the cross-tail currents and gives rise to field-aligned currents (FACs). A pair of FACs in the Region 1 sense, toward (away from) the ionosphere at the SCW's dawn (dusk) part, makes up most of an SCW. Earthward of the Region-1-sense pair is a weaker Region-2-sense (opposite to Region 1 sense) FAC pair with 20%–60% of the current content of the Region-1-sense pair (Fig. 19.2 is a cartoon illustrating this situation) [*Ritter and Lühr*, 2008; *Sergeev et al.*, 2011; *Sergeev et al.*, 2014]. The net effect of these two current pairs is a Region-1-sense loop that typically contains ~0.3 MA current (this net current varies between 0.1 and 1 MA from event to event) [*Newell and Gjerloev*, 2011; *Sergeev et al.*, 2011; *Chu et al.*, 2014; *Chu et al.*, 2015]. As inferred from ground magnetic activities and auroral brightenings associated with SCWs, an SCW is typically centered at 23:00 magnetic local time (MLT) and spans 2–6 hr of MLT (corresponding to 5–20 R_E at $X = -10$ R_E) [e.g., *Clauer and McPherron*, 1974; *Chu et al.*, 2014].

Recent observations have suggested that an SCW may be formed collectively by the small-scale current system of DFBs. Each DFB serves as a "wedgelet," a building block of a large-scale substorm current wedge. In this chapter, we review these observations (focusing on magnetotail in situ observations) to summarize the findings on the DFB current system and how it is related to the SCW. We also present some new results about DFB currents (specifically, Figs. 19.4 and 19.5, shown later).

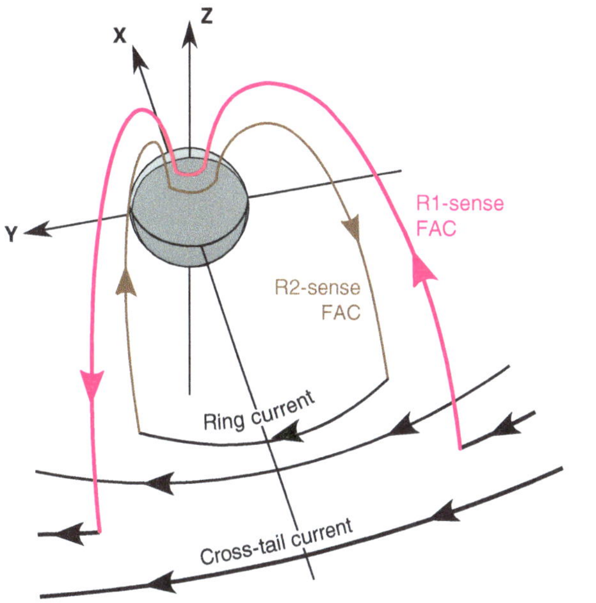

Figure 19.2 A substorm current wedge, modified from Figure 1 of *Sergeev et al.* [2014]. The tan and magenta curves represent the Region-2-sense and Region-1-sense FAC, respectively, of the SCW. The black curves represent the cross-tail current (when farther away from Earth) or the ring current (when closer to Earth). This cartoon illustrates only the Northern Hemisphere; the situation in the Southern Hemisphere is a mirror image of this cartoon about the equatorial plane.

19.2. THE DFB CURRENT SYSTEM

A DFB's current system can be inferred from its magnetic signatures. The most notable magnetic signature of a DFB is a sharp magnetic field change across the dipolarization front (Fig. 19.1d). The most intense DFB current is therefore a concentrated current sheet inside the thin DF layer called a dipolarization front current sheet (DFCS) by *Liu et al.* [2013a] (see also, e.g., *Zhou et al.* [2009]; *Runov et al.* [2011b]). The existence of DFCS has been confirmed by the curlometer technique [e.g., *Hwang et al.*, 2011; *Schmid et al.*, 2011]. Because most of the magnetic field change across a DF is in B_z, a duskward current makes up most of a dipolarization front current sheet (Figs. 19.1b,c). Near the DFB's central meridian (where $n_y = 0$, **n** being a unit vector representing the DF normal direction pointing toward the ambient plasma; see Fig. 19.1c), the duskward current is mainly in $+Y$ direction. Near the DFB's morning/evening flank (where $n_y \sim -1/+1$), the duskward current has a dominant $+X/-X$ component (Fig. 19.1c). For simplicity, however, we still refer to the current as "duskward" because it always has a duskward component. The duskward current has been suggested to be contributed by two factors: the ion finite gyroradius effect and the ion pressure gradient across the

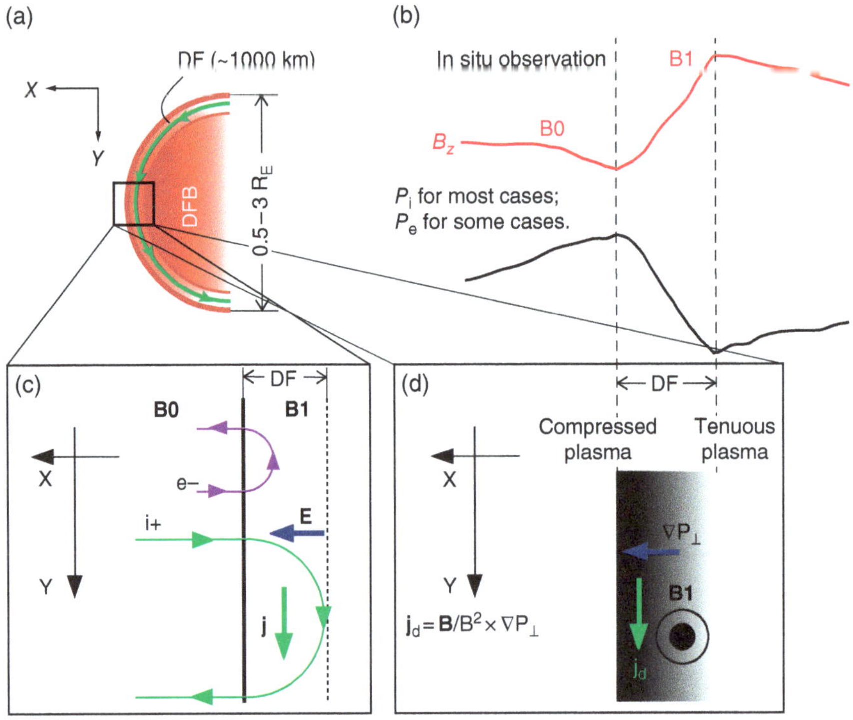

Figure 19.3 The contributors to the duskward current inside the dipolarization front current sheet. (c) and (d) Enlargements of the small square in panel (a). B0 and B1 are the magnetic fields in the ambient plasma ahead of the DFB and inside the DFB, respectively. (c) Finite gyroradius effect of ions leading to the duskward DF current. (d) Duskward DF current as a diamagnetic current (j_d) caused by a thermal pressure gradient. P_i and P_e in panel (b) stand for ion pressure and electron pressure, respectively. Panels (c) and (d) are adapted from *Runov et al.* [2011b].

DF, both of which are illustrated in Figure 19.3 (adapted from [*Runov et al.*, 2011b]). The ion finite gyroradius effect arises from the thinness of the DF layer. Because the DF thickness is on the scale of the ion thermal gyroradius (but much larger than the electron gyroradius) and B_z changes dramatically across the DF, ambient ions and electrons encountering the DFB will have separate gyration paths (Fig. 19.3c). With much larger gyroradii than electrons, ions can penetrate more deeply inside the DF layer, where the ions' gyromotion is duskward. This duskward motion gives rise to a duskward current inside the DF layer [*Li et al.*, 2016]. The other contributor to the duskward current is the ion pressure gradient across the DF. The ion thermal pressure inside the DFB is much lower than that outside, so a duskward diamagnetic current will arise inside the DF layer (Fig. 19.3d). Although for most DFs the ion pressure drop across the DF is much more significant than the electron pressure drop, for some DFs the electron pressure gradient is more significant and dominates diamagnetic current generation [*Zhang et al.*, 2011].

Another notable magnetic signature of most DFBs is an asymmetric bipolar B_z profile with a shallow dip at the beginning of the DF and a high peak at the end of the DF (Fig. 19.1d). This signature can be explained with the aforementioned duskward DF current sheet when considering the DFCS's unique topology, as illustrated by the cartoons in Figure 19.4. If a current sheet is an infinite plane, duskward current inside it can only cause a steplike B_z enhancement to the in situ observation of a spacecraft transecting the sheet from its earthward side (Fig. 19.4a). If the plane sheet is finite (imitating a localized DF), its resultant B_z profile has a bipolar signature similar to that of a DFB, but unlike a DFB's asymmetric B_z profile (Fig. 19.1d), the plane sheet's B_z profile is symmetric (Fig. 19.4b). We arrived at this conclusion using a simple model. First, we use small duskward current segments to fill the thin layer illustrated in the upper part of Figure 19.4b, and then compute the resultant B_z of all segments at any given point using the Biot-Savart law. Adding all segments' resultant B_z gives the B_z profile in the lower part of Figure 19.4b. This method ignores currents outside the finite sheet (or DF), assuming that they are much weaker than those inside. Using the same method, we next consider how the B_z profile is affected by DF topology. As Figure 19.1a,b illustrates, a DF's meridional cross section resembles a dipolar field line, and its transverse cross section can be approximated by a semicircle, so the DF in 3D resembles a saddle (Fig. 19.4c, upper part). Duskward currents flowing along this saddle make the B_z dip less

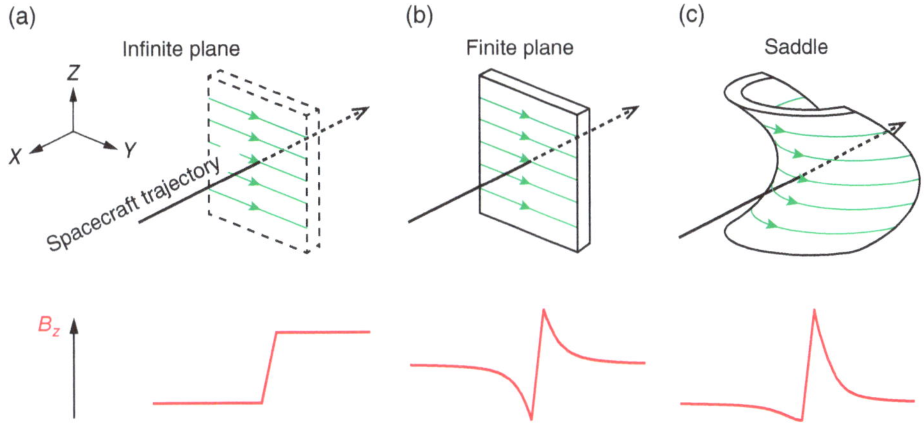

Figure 19.4 Current sheets of different topologies and their resultant B_z profile on a virtual spacecraft transecting them. For an earthward-moving current sheet that imitates a DF, the spacecraft trajectory relative to the current sheet is in the –*X* direction. Green arrows and red lines represent the current direction and B_z profiles, respectively. The B_z profiles in (b) and (c) are computed using a method described in section 19.2.1.

significant than the B_z peak (Fig. 19.4c, lower part), consistent with observations.

Although our model shows that a single current sheet (DFCS) can produce a DFB's bipolar B_z signature (including the B_z dip), current density observations suggest that the cause for the B_z dip is not so simple. *Yao et al.* [2013] investigated the current density ahead of DFBs with and without a significant dip (to be a significant dip, B_L should drop by more than 4 nT within 10 s before DF arrival. B_L, which is close to B_z, is the maximum variation component of B). They found a distinct difference between these two groups. For DFBs with no significant dip, the only clearly visible current is the DFCS. For DFBs with a significant dip, a dawnward current, termed a "dip current," exists earthward of the DFCS (tan arrows in Fig. 19.1b,c). This dip current can be understood as a diamagnetic current arising from the plasma pressure enhancement immediately earthward of a DF (see *Yao et al.* [2015] and the drawing above Fig. 19.3b). The dip current can also be explained as dawnward-moving kinetic ions immediately earthward of the DF, which have been reflected by the DF's sharp B_z enhancement [*Zhou et al.*, 2010; *Zhou et al.*, 2014a; *Eastwood et al.*, 2015]. According to simulations, the ion reflection and thus the dip current should be stronger when closer to the DFB's central meridian, when the DFB moves faster, and when the ambient plasma is denser [*Greco et al.*, 2014; *Zhou et al.*, 2014a]. Observations show that under these conditions, the B_z dip is indeed deeper, indicating stronger dip currents [*Pan et al.*, 2015]. In addition to directly carrying a current, the reflected ions may produce an electric field pointing away from the neutral sheet, which in turn induces a dawnward electron drift current, which may make up part of the dip current [*Lu et al.*, 2016].

Because DFB-related currents supplement background magnetotail currents, a DFB's magnetic variations (e.g., the B_z dip) are superimposed on the background field. Therefore, the sign of the B_z minimum at the dip, which determines the topology of the DFB, should be affected by the background B_z. To determine whether this is true, we examine the B_z minima of DFBs in the *Liu et al.* [2013a] event list (compiled from THEMIS observations). Figure 19.5 shows how the B_z minima vary with downtail distance: the B_z minima are more likely to go negative farther downtail. This is consistent with background control of the B_z minimum, because the background B_z is smaller farther away from Earth, the superimposed B_z reduction is more likely to make B_z negative (assuming that the DFB currents do not decrease with downtail distance as much as the background B_z does). Despite this factor, at any downtail location, the dip B_z minimum is positive for most DFBs. These DFBs as well as their surrounding flux tubes have a dipolar shape (Fig. 19.1b), although the DFB is more dipolar than the flux tubes surrounding it. For the few DFBs that have negative B_z dips, the field line ahead of the DFB is southward. Figure 19.6 is a suggested topology for such DFBs [*Yao et al.*, 2013] (see also, e.g., *Sergeev et al.* [1992]; *Semenov et al.* [2005]; *Sormakov and Sergeev* [2008]).

Next we move deeper inside a DFB to explore the currents behind a DF (the region between the second and third vertical dotted lines in Fig. 19.1d). In this region, the magnetic field change is much more gradual than that at the DF, so the current density can be inferred more accurately from the curlometer technique than from the 1D approximation (i.e., $j_y = -\mu_0 \frac{\partial B_x}{v_x \partial t}$; see, e.g., *Zhou et al.* [2009]; *Runov et al.* [2011a]). Curlometer results show that the current density behind the DF is comparable to that in the background plasma [e.g., *Hwang et al.*, 2011; *Yao et al.*, 2013]. Both currents are much weaker than the DFCS and the dip current. Most often the

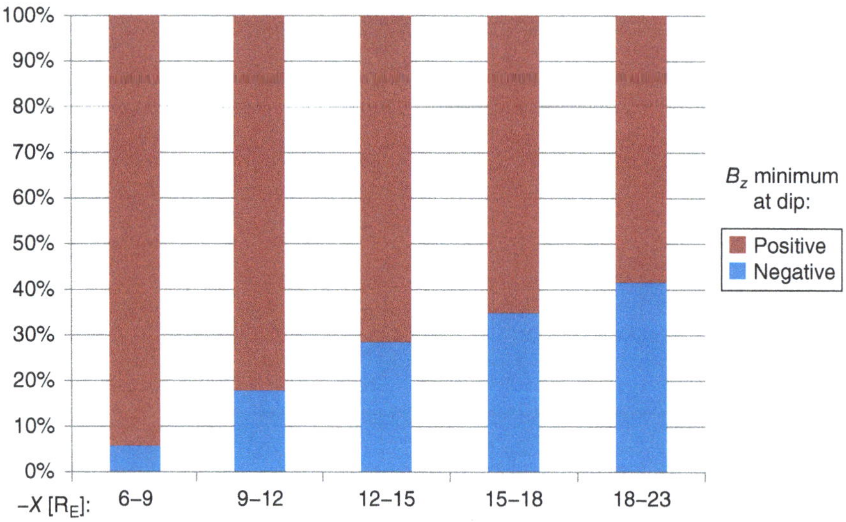

Figure 19.5 Statistical results of the sign of DFBs' dip B_z minimum as a function of downtail distance. The statistical database is from *Liu et al.* [2013a]. The dip B_z minimum is selected as the minimum within the interval $t_0 - 15$ s to t_{in}, where t_0 is the time of the first abrupt jump in B_z and t_{in} is the time of the peak B_z (see Appendix A of *Liu et al.* [2013a] for more details).

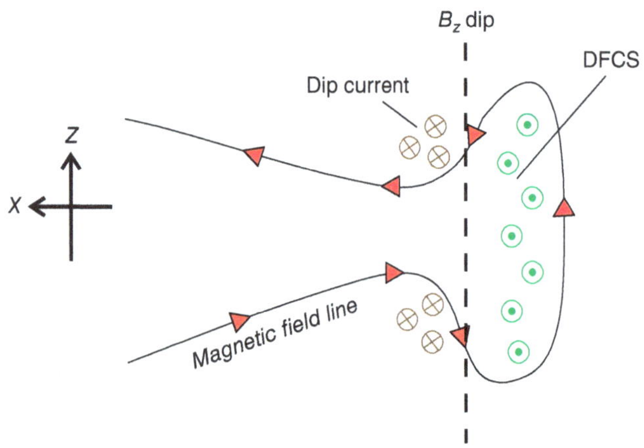

Figure 19.6 Suggested topology for a DFB whose B_z dip goes negative (southward). Adapted from *Yao et al.* [2013].

current behind the DF and the current in the background plasma are both duskward, consistent with the background cross-tail current [e.g., *Hwang et al.*, 2011; *Lu et al.*, 2016]. Because the densities of these two currents are very low, it is difficult to compare their relative magnitudes directly, especially when the curlometer results is affected by instrumental uncertainties. Nevertheless, we can compare them indirectly. The cross-tail current density behind the DF should be smaller than that in the background plasma because $|B_x|$ decreases across a DF [*Zhang et al.*, 2011; *Liu et al.*, 2013a]. To find whether the amount of the cross-tail current (integrated over Z; in Amperes per unit length along X) is reduced behind the DF, *Liu et al.* [2013b] investigated the distribution of the total vertical pressure $P_{\mathrm{ttl,v}} = \dfrac{B_x^2 + B_y^2}{2\mu_0} + P$ around DFBs (P is the thermal pressure). The total vertical pressure approximately equals the lobe magnetic pressure, assuming force balance along Z [*Xing et al.*, 2009], so $P_{\mathrm{ttl,v}}$ can indicate the lobe field strength and thus the amount of cross-tail current. Figure 19.7a displays the statistical distribution of $P_{\mathrm{ttl,v}}$ (adapted from *Liu et al.* [2013b]). The statistical $P_{\mathrm{ttl,v}}$ drops across the DF, suggesting a reduction of cross-tail current behind the DF. The reduced cross-tail current is likely fed into the DFCS, as illustrated by the cartoon Figure 19.7b.

19.2.1. The Field-Aligned Currents of a DFB

To discover the link between a DFB and a substorm current wedge (a field-aligned current system), we focus on the field-aligned component of the DFB current system. *Chen and Wolf* [1993] suggested that a Region-1-sense FAC pair exists on both flanks of a DFB. (See their Fig. 3. Here Region 1 sense means toward/away from the ionosphere in the DFB's morning/evening side.) Simultaneous ground observations of auroral streamers and tail observations of DFBs indicate that a DFB indeed has FACs connected to the ionosphere and that streamers are ground footprints of DFBs [*Lyons et al.*, 1999; *Zesta et al.*, 2000; *Nakamura et al.*, 2001; *Sergeev et al.*, 2004]. The existence of FACs carried by a DFB was confirmed by applying the curlometer technique to Cluster's tail observations, and these FACs coexisted with ionospheric

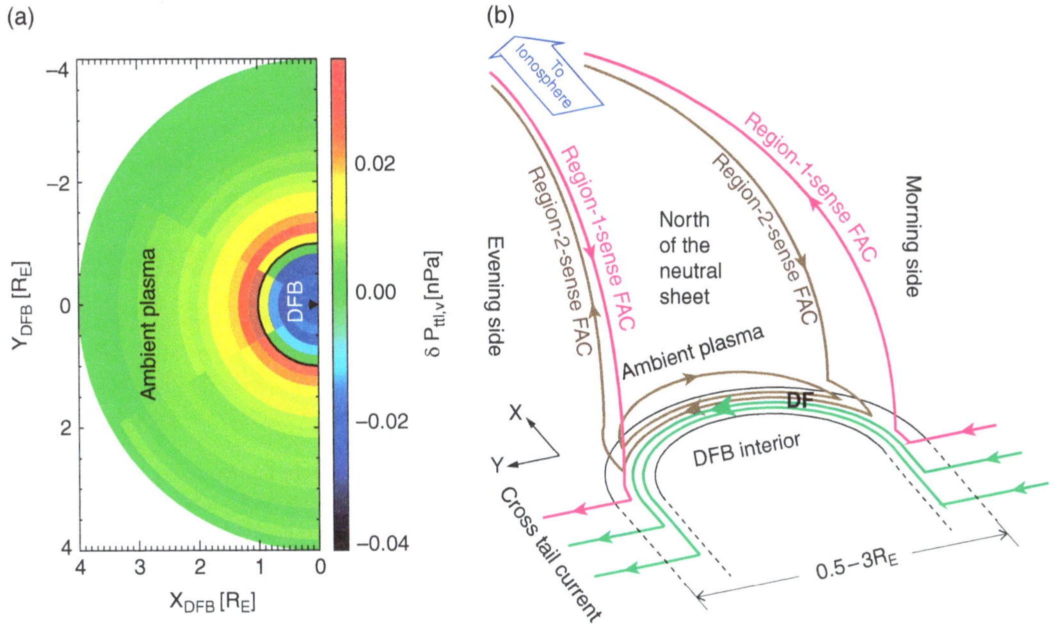

Figure 19.7 (a) Statistical distribution of the total vertical pressure $P_{ttl,v}$ near a DFB (adapted from *Liu et al.* [2013a]; see more details there). The presented $\delta P_{ttl,v}$ has been detrended with the quiet time value at ~3 min before DFB arrival. The color in each bin represents the median value of all data points falling inside that bin. The ambient plasma and the DFB interior are separated by the black semicircle. This figure is in the DFB's local coordinates: X_{DFB} and Y_{DFB} have the same directions as X and Y, respectively, but the entire area can be located in either the dawn or dusk sector of the magnetotail. (b) A possible closure scheme for the DFB current system. The two black semicircles represent the outer and inner boundaries of the DF's transverse cross section, respectively. All currents except field-aligned currents are drawn in the horizontal (XY) plane containing the DF's cross section. This figure illustrates only the Northern Hemisphere ($Z>0$); the situation in the Southern Hemisphere is a mirror image, about the equatorial plane.

effects that suggest they were linked to the ionosphere [*Forsyth et al.*, 2008].

More comprehensive studies reveal additional details about a DFB's field-aligned current system. Around a DFB, the change in B_y depends on the observation point's location relative to both the neutral sheet and the DFB's central meridian. In the ambient plasma immediately earthward of the DF, B_y turns more duskward (dawnward) in two locations: (1) northward of the neutral sheet and at the morning (evening) side of the DFB; (2) southward of the neutral sheet and at the evening (morning) side of the DFB. Behind the B_z dip, B_y turns in the opposite direction [*Sergeev et al.*, 1996; *Liu et al.*, 2013a]. The B_y variations earthward of a DFB are consistent with a pair of FAC sheets in the Region 2 sense (opposite to the Region 1 sense); those behind the dip are consistent with a pair of Region-1-sense FAC sheets. Figure 19.8a is a cartoon illustrating such FAC distribution. The Region-2-sense sheets are typically ~0.8 R_E wide, and the Region-1-sense sheets, which typically coincide with the DF layer, are the field-aligned component of the DFCS [*Liu et al.*, 2013a]. At large distances from the neutral sheet (i.e., when B_x is more than half the lobe field), the reversed B_y behind the dip overshoots the B_y turning earthward of the DFB. This suggests that the Region-1-sense sheet carries more FAC than the Region-2-sense sheet, so the net FACs are in the Region 1 sense. Although the results of *Liu et al.* [2013a] pertain mainly to DFBs at 8–12 R_E downtail, the Region-1-sense and Region-2-sense FAC sheet are also found to exist for DFBs at 17–19 R_E downtail, where the DFBs are in an earlier stage of development [*Sun et al.*, 2013].

The next question is, what gives rise to a DFB's field-aligned currents? According to theories, an FAC can arise from a thermal pressure gradient and/or a changing flow vortex [*Vasyliunas*, 1970; *Sato and Iijima*, 1979]. *Liu et al.* [2013b] created a statistical map of thermal pressure distribution around DFBs (Fig. 19.8b), which shows that the pressure maximizes immediately ahead of a DFB's leading point (where $n_y=0$) and decreases toward the ambient plasma, the DFB's interior, and the DFB's flanks. When combined with estimated flux tube volume distribution around the DFB, such pressure gradients lead to the FAC distribution in Figure 19.8a through the

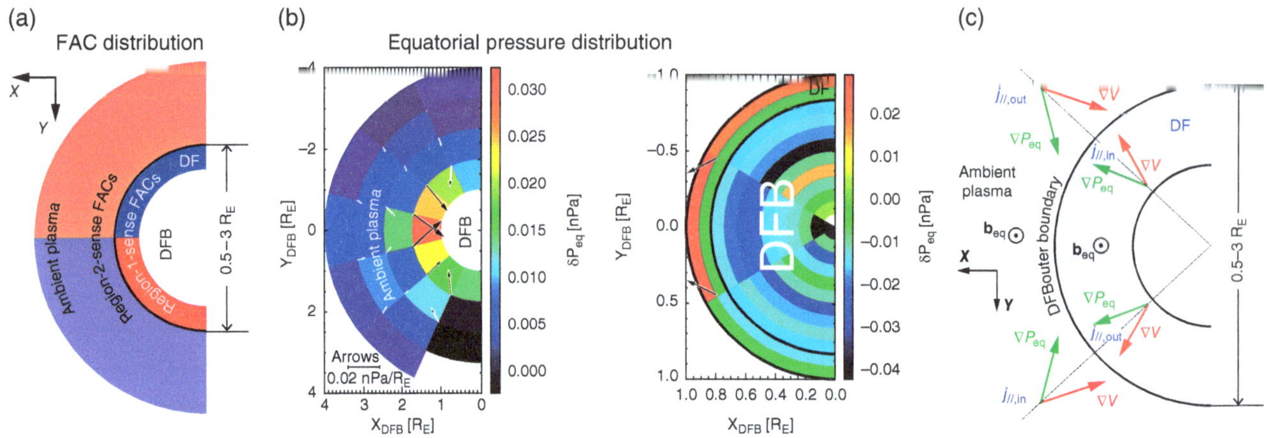

Figure 19.8 (a) Distribution of field-aligned currents around a DFB. Red (blue) indicates FACs away from (toward) the ionosphere. (b) The statistical distribution of the equatorial thermal pressure P_{eq} around a DFB. Near the neutral sheet, $P_{ttl,v}$ can approximate P_{eq}. The color in each bin represents the median value of all data points falling inside that bin. The arrows represent the pressure gradients. The left (right) -hand side shows the distribution outside (inside) the DFB. (c) An illustration of how the pressure gradients lead to FACs. The green (red) arrows represent the directions of P_{eq} (V, the volume of a unit magnetic flux tube) gradients. According to *Vasyliunas* [1970], a positive (negative) $\mathbf{b}_{eq} \cdot (\nabla V \times \nabla P_{eq})$ will lead to an FAC toward (away from) the ionosphere, where b_{eq} is the magnetic field direction at the equator. The FACs toward and away from the ionosphere are represented by $j_{//,in}$ and $j_{//,out}$, respectively. For more information regarding Figures 19.8b and 19.8c, see *Liu et al.* [2013b].

Vasyliunas [1970] equation (see the schematic Figure 19.8c, which is adapted from *Liu et al.* [2013b]. See also *Xing et al.* [2011]; *Xing et al.* [2012]). The cause of such a pressure distribution can be understood in two ways. Under the fluid approximation, the pressure distribution results from the force exerted by the DFB ramming the ambient plasma (and the plasma's counterforce); the pressure is expected to maximize at the DFB's leading point [*Yang et al.*, 2011; *Birn and Hesse*, 2013]. In the kinetic regime, DF-reflected ions contribute to the pressure buildup ahead of the DF [*Zhou et al.*, 2011; *Lu et al.*, 2016]. In either way, the pressure gradients and thus the FACs are supported by the DFB.

The pressure gradients work well in explaining the FACs in the ambient plasma ahead of a DFB. Whether they can account for the FACs inside the DF layer is questionable, however, although the gradients are consistent with the FAC directions there. This is because the DF is too thin for the ions to be treated as fluid, but the pressure gradient part of the *Vasyliunas* [1970] equation was derived from fluid description. The electrons, however, still behave like a fluid around most DFs because the electron gyroradius and inertial length are much smaller than the typical DF thickness [e.g., *Liu et al.*, 2013a]. In addition, for most DFs, the electron fluid is frozen in with the magnetic field because the electron pressure gradient across the DF is usually insignificant [*Runov et al.*, 2011b; *Sun et al.*, 2014a]. As long as a fluid is frozen in with the magnetic field, FACs can arise from the field line twist caused by a changing flow vortex or shear, which is the other part of the *Vasyliunas* [1970] equation. Although flow vortices and shears have been observed around DFBs [*Keiling et al.*, 2009; *Walsh et al.*, 2009; *Panov et al.*, 2010], there is still no convincing evidence of changing electron flow shear across the DF, which can lead to the Region-1-sense FACs inside the DF layer.

Field-aligned currents around DFBs coexist with the aforementioned duskward and dawnward currents. Figure 19.7b illustrates a possible connection between these currents. Inside the DF layer, the duskward current and Region-1-sense FACs together form the DFCS. Near the neutral sheet, the duskward current dominates, but farther away from the neutral sheet, FACs become more important [*Liu et al.*, 2013a]. The current directions suggest that both the duskward current and FACs come from the reduced cross-tail current behind the DF (Fig. 19.7b). In the ambient plasma earthward of the DFB, both the dawnward current and the Region-2-sense FACs are likely connected to the DF current, gaining current continuity.

Our discussion of the DFB current system has focused on the area around the DFB's leading part. The DFB's trailing part has received much less research attention, because observations there are usually too noisy to reveal clean current structures. Occasionally a DFB has a sharp trailing edge; in such cases the trailing edge is also a concentrated current sheet like the DF. This current sheet contains duskward currents and Region-1-sense FACs [*Zhou et al.*, 2013; *Sun et al.*, 2014b]. Because sharp trailing edges rarely accompany DFBs, we still focus on the leading part of the DFB current system in the following.

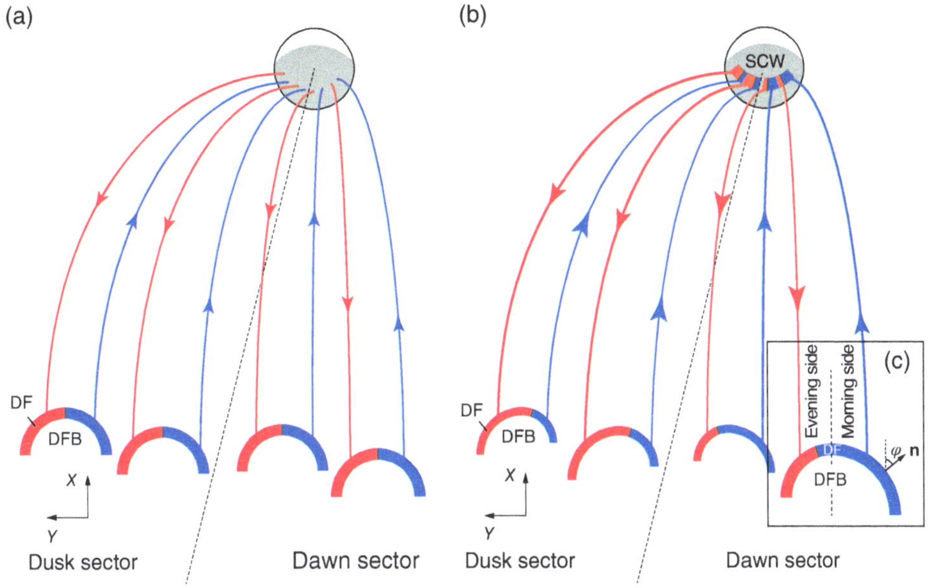

Figure 19.9 Global views from the Northern Hemisphere showing many DFBs (wedgelets) coexisting in the tail. For simplicity, each panel contains four DFBs, but during a realistic substorm, there should be more (see Fig. 19.10). The DFs are illustrated as transverse cross sections. The colored areas and lines with arrows represent FACs; the blue (red) color represents the FACs toward (away from) Earth. (a) The case for symmetric wedgelets. (b) The case for asymmetric wedgelets (adopted from *Liu et al.* [2015]). Figure 19.9b also illustrates the ground footprints of the wedgelet FACs, which are inferred from the *Forsyth et al.* [2014] observations. For clarity, we drew ground FACs much wider than reality. (c) An enlargement of an asymmetric wedgelet.

19.3. THE DFB CURRENT SYSTEM SERVING AS A WEDGELET: HOW DO WEDGELETS FORM A SCW?

As discussed in the last section, a DFB carries a pair of Region-1-sense FACs and a weaker pair of Region-2-sense FACs earthward of the Region-1-sense pair. In addition, a DFB reduces the cross-tail current (Fig. 19.7b). Such a current configuration is similar to that of a substorm current wedge (Fig. 19.2) except that the DFB's field-aligned currents are weaker and more localized (of DFB scale size; see Fig. 19.7b). Because most DFBs appear during substorms, this similarity suggests that DFBs and SCWs could be related. As a DFB's width corresponds to less than 1 hr of MLT, its FACs likely make up a wedgelet, a building block of a several hours of MLT-wide substorm current wedge [*Rostoker*, 1991; *Nakamura et al.*, 1994; *Marklund et al.*, 1998; *Liu et al.*, 2013a; *Liu et al.*, 2013b].

How, then, do weak, localized wedgelets form a strong, global SCW? Auroral observations provide a hint in answering this question. During substorms, several coexisting auroral streamers are often observed to be longitudinally distributed over a width of several hours of MLTs [*Zesta et al.*, 2006; *Lyons et al.*, 2012]. Because each streamer is the ionospheric footprint of a DFB or a train of DFBs, the longitudinal streamer distribution indicates that coexisting DFBs, and thus wedgelets, are distributed in the dawn-dusk direction. Such DFB distribution is commonly seen in global simulations of the substorm-time magnetotail [*Ge et al.*, 2011; *Lu et al.*, 2015; *Wiltberger et al.*, 2015]. Because coexisting wedgelets can be distributed over a wide MLT range, they can collectively form a wide SCW, as illustrated in the schematic Figure 19.9a. In Figure 19.9a and the following, we focus only on Region-1-sense FACs because they are the dominant component of a wedgelet and an SCW. Although the wedgelets in Figure 19.9a can match SCW width, they have a problem matching the typical SCW magnitude. Because each wedgelet has oppositely directed FACs at its morning and evening sides, side-by-side wedgelets could largely cancel each other's FACs. In the case of Figure 19.9a, the net field-aligned currents of many wedgelets will equal those of only one wedgelet. The magnitude of these net currents is much smaller than that of an SCW.

Recent Cluster observations imply a possible solution to this problem. From Cluster data collected ~1 R_E above the ground, *Forsyth et al.* [2014] found ~15 pairs of oppositely directed FAC sheets that filled an SCW region of ~3 hr of MLT width. In the west (east) part of the SCW region, FAC sheets away from (toward) the Earth were wider than those toward (away from) the Earth, so the net FAC was away from (toward) Earth. Therefore, an SCW is effectively a pair of large-scale, Region-1-sense field-aligned currents made up of small, asymmetric pairs of FACs. Each pair of the small FAC sheets is less than half an hour of MLT wide, comparable to the size of a typical DFB: a DFB 0.5–3 R_E wide in the 9–12 R_E downtail region corresponds to ~0.3 hr of MLT. In addition, *Forsyth et al.* [2014] observed that the FAC sheets drift dawnward, which is also a characteristic motion of DFBs in the tail [*Zhou et al.*, 2014b]. The similarity in size and motion of the FAC sheets and DFBs strongly suggests that the FAC sheets are the footprints of wedgelets. If so, each wedgelet in the magnetotail should contain an asymmetric pair of FACs, as illustrated in Figure 19.9b (adapted from *Liu et al.* [2015]).

To determine whether magnetotail wedgelets are indeed asymmetric, *Liu et al.* [2015] examined Region-1-sense field-aligned currents inside the DF layer using the *Liu et al.* [2013a] event list. The FAC density is indeed asymmetric: for a DF in the dawn (dusk) sector of the tail, the FACs toward (away from) Earth are more intense and cover a larger azimuthal range of the DF than the FACs away from (toward) Earth. (Here the azimuth is represented by an angle $\varphi = \text{asin} n_y$; see Fig. 19.9c.) The FAC density must be multiplied by the azimuthal extent of the DF's transverse cross section to become the amount of FAC. The azimuthal extent of the DF is also found to be asymmetric: in the dawn (dusk) sector of the tail, a DF's morning (evening) side is typically twice as wide as its evening (morning) side. Here the evening and morning sides are separated by $n_y = 0$, and this asymmetry pertains only to the leading edge of the DFBs ($n_x > 0$) [*Liu et al.*, 2015]. The asymmetric DF shapes are illustrated in Figure 19.9b. Because of the asymmetries in both DF shape and FAC density, a DF in the dawn (dusk) sector of the tail typically contains two times more FACs toward (away from) Earth than away from (toward) Earth, so the net FAC is toward (away from) Earth. The net FAC of many such DFs will thus be consistent with the SCW's large-scale Region-1-sense FAC. On average, an ~2-R_E-wide DF in the dawn (dusk) sector of the tail contains a net FAC of ~0.1 MA toward (away from) Earth [*Liu et al.*, 2015]. Therefore, it takes ~5 DFBs, or wedgelets, in each sector (dawn or dusk) of the tail to provide the ~0.5 MA Region-1-sense FAC pair of a typical SCW (we estimate the Region-1-sense pair to be ~0.5 MA because the Region-2-sense FAC pair of an SCW typically contains 20%–60% of the current in the Region-1-sense pair and the net SCW is typically ~0.3 MA [e.g., *Sergeev et al.*, 2014]). About five wedgelets in each sector of the tail means a total of ~10 wedgelets across the entire tail; this number is comparable to the ~15 FAC pairs observed by *Forsyth et al.* [2014] (each FAC pair should correspond to a wedgelet).

To verify the ability of asymmetric wedgelets to produce a substorm current wedge, *Liu et al.* [2015] used a simple model containing global, realistic magnetic field lines that was built using the *Chu et al.* [2014] method. This simple model is illustrated in Figure 19.10 (taken from *Liu et al.* [2015]). Figure 19.10a shows the equatorial

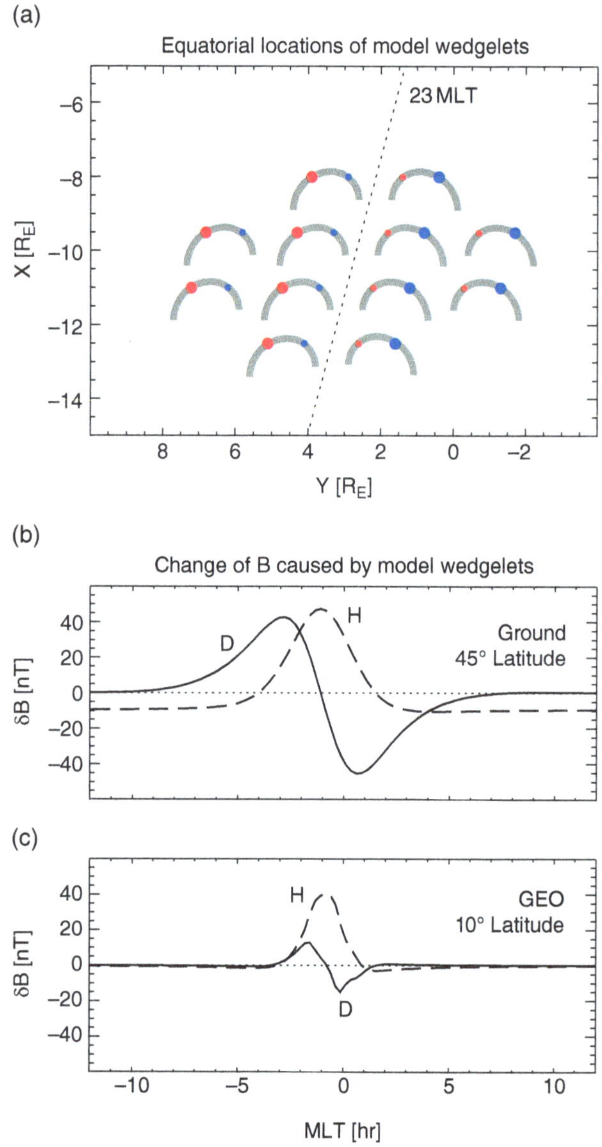

Figure 19.10 Model of asymmetric wedgelets. (a) Equatorial cross section of DFs (grey) and their Region-1-sense FACs. The blue and red circles represent FACs toward and away from Earth, respectively. The FACs are line currents with infinitesimal cross section; the areas of the circles indicate only the FACs' magnitude. The big (small) circles represent 0.15 (0.05) MA. (b) The modification by the modeled FACs to the magnetic field's H and D components at ground (45° latitude in the Northern Hemisphere). (c) The modification at the geosynchronous orbit (10° latitude in the Northern Hemisphere). Adopted from *Liu et al.* [2015].

cross section of the model setting. Each modeled DFB has a typical width of ~2 R_E; the DFBs cover a total width of ~3.5 hr of MLT, a typical SCW width. The DFBs are distributed around 23:00 MLT, the typical central meridian of a SCW. To be consistent with observations, the DFBs are set to carry asymmetric FACs: in the dawn (dusk) sector of the tail each DFB's FAC toward Earth is 0.15 (0.05) MA; the FAC away from Earth is 0.05 (0.15) MA. The modeled wedgelets produce magnetic field variations in both midlatitude ground (Fig. 19.10b) and geosynchronous orbit (Fig. 19.10c). The variations are a bipolar signature in the D component and a peak in the H component, both centered at 23:00 MLT. Such variations are classic SCW signatures [e.g., *Clauer and McPherron*, 1974]. Therefore, the asymmetric wedgelet scenario explains the substorm current wedge very well. The quantity of wedgelets may affect whether a full-scale SCW (when there are many wedgelets) or a pseudo breakup (when there are few wedgelets) will occur [*Yao et al.*, 2014].

19.4. QUESTIONS TO BE ANSWERED

Although the wedgelet scenario is a plausible SCW formation scenario, many questions about it have yet to be answered.

What percentage of a DFB's FACs can eventually reach the ionosphere? *Yao et al.* [2015] suggested that part of the DF current can close locally through the dawnward current ahead of the DF, forming a local banana current (see Fig. 19.7b). If part of the FAC closes locally, its fraction reaching the ionosphere will be smaller than that observed in the tail. Furthermore, the ionospheric FACs observed by *Forsyth et al.* [2014] (as indicated from observations ~1 R_E above ground) were more complex than simple current sheet pairs, and the FACs in the dusk sector were less-well defined as upward/downward pairs than those in the dawn sector. Such complexities suggest that the connection between ionospheric FACs and the DFBs' FACs is not straightforward. A better understanding of this connection requires further investigation.

What mechanism produces asymmetric wedgelets? As mentioned in section 19.2.1, FACs inside a thin DF layer can arise from flow shears across the DF. Therefore, the asymmetric FACs in Figure 19.9 may come from asymmetrically distributed flow shears. To determine whether such asymmetric flow shears actually exist and what ultimately drives them is a worthwhile topic for future observational and theoretical studies.

How do asymmetric wedgelets currents close? Because one arm of each wedgelet's Region-1-sense FAC pair is stronger than the other, a wedgelet cannot achieve current closure by itself. Current closure can be achieved globally, however, as illustrated by the cartoon Figure 19.11

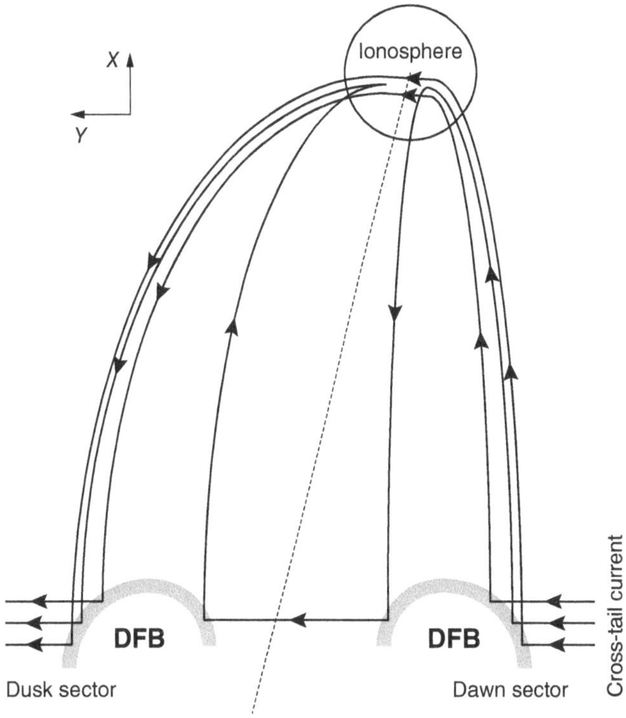

Figure 19.11 A possible way of current closure for our wedgelet scenario, in the same style as Figure 19.9b. For simplicity, we draw only two DFBs (wedgelets); in reality there should be more. Adapted from *Liu et al.* [2015].

(adapted from *Liu et al.* [2015]). In this picture, the redundant FACs of a wedgelet in one tail sector will be closed by wedgelets in the other sector. This current closure scenario will result in a net westward current in the ionosphere, consistent with an SCW's westward electrojet. Such current closure also leads to a net reduction of the cross-tail current in the space between DFBs. This reduction, as well as the cross-tail current reduction inside DFBs (Fig. 19.7), is consistent with a substorm current wedge. The closure scenario in Figure 19.11 requires future studies to verify.

Our discussion of asymmetric wedgelets has focused on Region-1-sense FACs inside the DF layer. Immediately earthward of a DF, there is also a pair of Region-2-sense FACs, which we discussed in section 19.2.1. How, if at all, this pair of FACs is potentially related to an SCW's Region-2-sense FACs is unknown. To fully understand how wedgelets may form an SCW, this relationship must be resolved.

Finally, it is unclear whether wedgelets are responsible for the substorm current wedge during the entire substorm expansion and recovery phases, which typically last for ~30 min. If DFBs emerge continually during this entire interval, a long-lasting SCW can be sustained by wedgelets. If DFBs are present during only the first few minutes of the substorm expansion, however, the wedgelet scenario may account for only the beginning of a long-lasting SCW; a large-scale pressure buildup caused by the accumulation of many stopped DFBs is responsible for the remainder [e.g., *Birn et al.*, 1999; *Yang et al.*, 2011; *Birn and Hesse*, 2014]. Previous observations gave contradictory results on DFB occurrence during different substorm phases [e.g., *Juusola et al.*, 2011; *Chu*, 2015], so it is difficult to tell during which phases the wedgelet scenario can work. We need more studies on DFB occurrence to better understand the role of wedgelets in the substorm process, which is crucial in solving the substorm mystery.

ACKNOWLEDGMENTS

Jiang Liu is supported by NASA's THEMIS contract NAS5-02099, grants NNX14AC17G and NNX16AI51G, and NSF grant 1401822. Zhonghua Yao is a Marie-Curie COFUND postdoctoral fellow at the University of Liege, cofunded by the European Union. We thank Judith Hohl for editing. The only new data in this chapter is in Figure 19.5. These data came from the THEMIS data depository: http://themis.ssl.berkeley.edu/data/themis/. We thank the THEMIS instrument team for the use of data.

REFERENCES

Angelopoulos, V., A. Runov, X.-Z. Zhou, D. L. Turner, S. A. Kiehas, S.-S. Li, and I. Shinohara (2013), Electromagnetic energy conversion at reconnection fronts, *Science*, *341*, 1478–1482; doi:10.1126/science.1236992.

Angelopoulos, V., C. F. Kennel, F. V. Coroniti, R. Pellat, M. G. Kivelson, R. J. Walker, C. T. Russell, W. Baumjohann, W. C. Feldman, and J. T. Gosling (1994), Statistical characteristics of bursty bulk flow events, *J. Geophys. Res.*, *99*, 21257–21280.

Angelopoulos, V., W. Baumjohann, C. F. Kennel, F. V. Coroniti, M. G. Kivelson, R. Pellat, R. J. Walker, H. Luehr, and G. Paschmann (1992), Bursty bulk flows in the inner central plasma sheet, *J. Geophys. Res.*, *97*, 4027–4039; doi:10.1029/91JA02701.

Arnoldy, R. L., and K. W. Chan (1969), Particle substorms observed at the geostationary orbit, *J. Geophys. Res.*, *74*(21), 5019–5028; doi:10.1029/JA074i021p05019.

Baumjohann, W., M. Hesse, S. Kokubun, T. Mukai, T. Nagai, and A. A. Petrukovich (1999), Substorm dipolarization and recovery, *J. Geophys. Res.*, *104*, 24995–25000; doi:10.1029/1999JA900282.

Birn, J., and M. Hesse (2013), The substorm current wedge in MHD simulations, *J. Geophys. Res. Space Physics*, *118*; doi:10.1002/jgra.50187.

Birn, J., and M. Hesse (2014), The substorm current wedge: Further insights from MHD simulations, *J. Geophys. Res. Space Physics*, *119*, 3503–3513; doi:10.1002/2014JA019863.

Birn, J., J. Raeder, Y. Wang, R. Wolf, and M. Hesse (2004), On the propagation of bubbles in the geomagnetic tail, *Ann. Geophys.*, *22*, 1773–1786; doi:10.5194/angeo-22-1773-2004.

Birn, J., M. Hesse, G. Haerendel, W. Baumjohann, and K. Shiokawa (1999), Flow braking and the substorm current wedge, *J. Geophys. Res.*, *104*, 19895–19904; doi:10.1029/1999JA900173.

Birn, J., M. Hesse, R. Nakamura, and S. Zaharia (2013), Particle acceleration in dipolarization events, *J. Geophys. Res. Space Physics*, *118*, 1960–1971; doi:10.1002/jgra.50132.

Chen, C. X., and R. A. Wolf (1993), Interpretation of high-speed flows in the plasma sheet, *J. Geophys. Res.*, *98*, 21409; doi:10.1029/93JA02080.

Chu, X. (2015), Configuration and Generation of Substorm Current Wedge, University of California, Los Angeles, ProQuest Dissertations Publishing, 2015. 3737851.

Chu, X., et al. (2014), Development and validation of inversion technique for substorm current wedge using ground magnetic field data, *J. Geophys. Res. Space Physics*, *119*, 1909–1924; doi:10.1002/2013JA019185.

Chu, X., R. L. McPherron, T.-S. Hsu, V. Angelopoulos, Z. Pu, Z. Yao, H. Zhang, and M. Connors (2015), Magnetic mapping effects of substorm currents leading to auroral poleward expansion and equatorward retreat, *J. Geophys. Res. Space Physics*, *120*, 253–265; doi:10.1002/2014JA020596.

Clauer, C. R., and R. L. McPherron (1974), Mapping the local time-universal time development of magnetospheric substorms using mid-latitude magnetic observations, *J. Geophys. Res.*, *79*, 2811; doi:10.1029/JA079i019p02811.

Dubyagin, S., V. Sergeev, S. Apatenkov, V. Angelopoulos, A. Runov, R. Nakamura, W. Baumjohann, J. McFadden, and D. Larson (2011), Can flow bursts penetrate into the inner magnetosphere?, *Geophys. Res. Lett.*, *38*, 8102; doi:10.1029/2011GL047016.

Eastwood, J. P., M. V. Goldman, H. Hietala, D. L. Newman, R. Mistry, and G. Lapenta (2015), Ion reflection and acceleration near magnetotail dipolarization fronts associated with magnetic reconnection, *J. Geophys. Res. Space Physics*, *120*(1), 511–525; doi:10.1002/2014ja020516.

Forsyth, C., et al. (2008), Observed tail current systems associated with bursty bulk flows and auroral streamers during a period of multiple substorms, *Annales Geophysicae*, *26*(1), 167–184; doi:10.5194/angeo-26-167-2008.

Forsyth, C., et al. (2014), In situ spatiotemporal measurements of the detailed azimuthal substructure of the substorm current wedge, *J. Geophys. Res. Space Physics*, *119*, 927–946; doi:10.1002/2013JA019302.

Fu, H. S., et al. (2013), Dipolarization fronts as a consequence of transient reconnection: In situ evidence, *Geophys. Res. Lett.*, *40*, 6023–6027; doi:10.1002/2013GL058620.

Fu, H. S., Y. V. Khotyaintsev, M. André, and A. Vaivads (2011), Fermi and betatron acceleration of suprathermal electrons behind dipolarization fronts, *Geophys. Res. Lett.*, *38*, 16104; doi:10.1029/2011GL048528.

Gabrielse, C., V. Angelopoulos, A. Runov, and D. L. Turner (2012), The effects of transient, localized electric fields on equatorial electron acceleration and transport toward the inner magnetosphere, *J. Geophys. Res.*, *117*, 10213; doi:10.1029/2012JA017873.

Ge, Y. S., J. Raeder, V. Angelopoulos, M. L. Gilson, and A. Runov (2011), Interaction of dipolarization fronts within multiple bursty bulk flows in global MHD simulations of a substorm on 27 February 2009, *J. Geophys. Res. Space Physics*, *116*(A5), A00I23; doi:10.1029/2010ja015758.

Greco, A., A. Artemyev, and G. Zimbardo (2014), Proton acceleration at two-dimensional dipolarization fronts in the magnetotail, *J. Geophys. Res. Space Physics*, *119*(11), 8929–8941; doi:10.1002/2014ja020421.

Guzdar, P. N., A. B. Hassam, M. Swisdak, and M. I. Sitnov (2010), A simple MHD model for the formation of multiple dipolarization fronts, *Geophys. Res. Lett.*, *37*, 20102; doi:10.1029/2010GL045017.

Hwang, K. J., M. L. Goldstein, E. Lee, and J. S. Pickett (2011), Cluster observations of multiple dipolarization fronts, *J. Geophys. Res. Space Physics*, *116*(A5), A00I32; doi:10.1029/2010ja015742.

Juusola, L., N. Østgaard, E. Tanskanen, N. Partamies, and K. Snekvik (2011), Earthward plasma sheet flows during substorm phases, *J. Geophys. Res. Space Physics*, *116*(A10), A10228; doi:10.1029/2011ja016852.

Keiling, A., et al. (2009), Substorm current wedge driven by plasma flow vortices: THEMIS observations, *J. Geophys. Res.*, *114*; doi:10.1029/2009JA014114.

Li, J.-Z., X.-Z. Zhou, V. Angelopoulos, J. Liu, A. Runov, D.-X. Pan, and Q.-G. Zong (2016), Contribution of ion reflection to the energy budgets of dipolarization fronts, *Geophys. Res. Lett.*, *43*(2), 493–500; doi:10.1002/2015gl067300.

Li, S.-S., V. Angelopoulos, A. Runov, X.-Z. Zhou, J. McFadden, D. Larson, J. Bonnell, and U. Auster (2011), On the force balance around dipolarization fronts within bursty bulk flows, *J. Geophys. Res.*, *116*; doi:10.1029/2010JA015884.

Liu, J., V. Angelopoulos, A. Runov, and X.-Z. Zhou (2013a), On the current sheets surrounding dipolarizing flux bundles in the magnetotail: The case for wedgelets, *J. Geophys. Res. Space Physics*, *118*, 2000–2020; doi:10.1002/jgra.50092.

Liu, J., V. Angelopoulos, X. Chu, X.-Z. Zhou, and C. Yue (2015), Substorm current wedge composition by wedgelets, *Geophys. Res. Lett.*, *42*, 1669–1676; doi:10.1002/2015GL063289.

Liu, J., V. Angelopoulos, X.-J. Zhang, D. L. Turner, C. Gabrielse, A. Runov, J. Li, H. O. Funsten, and H. E. Spence (2016), Dipolarizing flux bundles in the cis-geosynchronous magnetosphere: Relationship between electric fields and energetic particle injections, *J. Geophys. Res. Space Physics*, *121*(2), 1362–1376; doi:10.1002/2015ja021691.

Liu, J., V. Angelopoulos, X.-Z. Zhou, A. Runov, and Z. Yao (2013b), On the role of pressure and flow perturbations around dipolarizing flux bundles, *J. Geophys. Res. Space Physics*, *118*, 7104–7118; doi:10.1002/2013JA019256.

Liu, J., V. Angelopoulos, X.-Z. Zhou, and A. Runov (2014), Magnetic flux transport by dipolarizing flux bundles, *J. Geophys. Res. Space Physics*, *119*, 909–926; doi:10.1002/2013JA019395.

Lu, S., A. V. Artemyev, V. Angelopoulos, Q. Lu, and J. Liu (2016), On the current density reduction ahead of dipolarization fronts, *J. Geophys. Res. Space Physics*, *121*(5), 4269–4278; doi:10.1002/2016ja022754.

Lu, S., et al. (2015), Dipolarization fronts as earthward propagating flux ropes: A three-dimensional global hybrid simulation, *J. Geophys. Res. Space Physics*, *120*(8), 6286–6300; doi:10.1002/2015ja021213.

Lyons, L. R., T. Nagai, G. T. Blanchard, J. C. Samson, T. Yamamoto, T. Mukai, A. Nishida, and S. Kokubun (1999), Association between Geotail plasma flows and auroral poleward boundary intensifications observed by CANOPUS photometers, *J. Geophys. Res. Space Physics*, *104*(A3), 4485–4500; doi:10.1029/1998ja900140.

Lyons, L. R., Y. Nishimura, X. Xing, A. Runov, V. Angelopoulos, E. Donovan, and T. Kikuchi (2012), Coupling of dipolarization front flow bursts to substorm expansion phase phenomena within the magnetosphere and ionosphere, *J. Geophys. Res.*, *117*, 2212; doi:10.1029/2011JA017265.

Marklund, G. T., et al. (1998), Observations of the electric field fine structure associated with the westward traveling surge and large-scale auroral spirals, *J. Geophys. Res. Space Physics*, *103*(A3), 4125–4144; doi:10.1029/97ja00558.

McPherron, R. L., C. T. Russell, and M. P. Aubry (1973), Satellite studies of magnetospheric substorms on August 15, 1968. 9. Phenomenological model for substorms., *J. Geophys. Res.*, *78*, 3131–3149; doi:10.1029/JA078i016p03131.

Milan, S. E., A. Grocott, C. Forsyth, S. M. Imber, P. D. Boakes, and B. Hubert (2009), A superposed epoch analysis of auroral evolution during substorm growth, onset and recovery: open magnetic flux control of substorm intensity, *Annales Geophysicae*, *27*(2), 659–668; doi:10.5194/angeo-27-659-2009.

Nagai, T., M. Fujimoto, Y. Saito, S. Machida, T. Terasawa, R. Nakamura, T. Yamamoto, T. Mukai, A. Nishida, and S. Kokubun (1998), Structure and dynamics of magnetic reconnection for substorm onsets with Geotail observations, *J. Geophys. Res.*, *103*, 4419–4440; doi:10.1029/97JA02190.

Nakamura, R., D. N. Baker, T. Yamamoto, R. D. Belian, E. A. Bering, J. R. Benbrook, and J. R. Theall (1994), Particle and field signatures during pseudobreakup and major expansion onset, *J. Geophys. Res.*, *99*, 207–221; doi:10.1029/93JA02207.

Nakamura, R., et al. (2002), Motion of the dipolarization front during a flow burst event observed by Cluster, *Geophys. Res. Lett.*, *29*(20); doi:10.1029/2002GL015763.

Nakamura, R., et al. (2004), Spatial scale of high-speed flows in the plasma sheet observed by Cluster, *Geophys. Res. Lett.*, *31*, 9804; doi:10.1029/2004GL019558.

Nakamura, R., W. Baumjohann, R. Schödel, M. Brittnacher, V. A. Sergeev, M. Kubyshkina, T. Mukai, and K. Liou (2001), Earthward flow bursts, auroral streamers, and small expansions, *J. Geophys. Res.*, *106*, 10791–10802; doi:10.1029/2000JA000306.

Newell, P. T., and J. W. Gjerloev (2011), Substorm and magnetosphere characteristic scales inferred from the SuperMAG auroral electrojet indices, *J. Geophys. Res. Space Physics*, *116*(A12), A12232; doi:10.1029/2011ja016936.

Ohtani, S., H. J. Singer, and T. Mukai (2006), Effects of the fast plasma sheet flow on the geosynchronous magnetic configuration: Geotail and GOES coordinated study, *J. Geophys. Res.*, *111*, 1204; doi:10.1029/2005JA011383.

Ohtani, S., M. A. Shay, and T. Mukai (2004), Temporal structure of the fast convective flow in the plasma sheet: Comparison between observations and two-fluid simulations, *J. Geophys. Res.*, *109*, 3210; doi:10.1029/2003JA010002.

Pan, D.-X., X.-Z. Zhou, Q.-Q. Shi, J. Liu, V. Angelopoulos, A. Runov, Q.-G. Zong, and S.-Y. Fu (2015), On the generation of magnetic dips ahead of advancing dipolarization fronts, *Geophys. Res. Lett.*, *42*(11), 4256–4262; doi:10.1002/2015gl064369.

Panov, E. V., et al. (2010), Multiple overshoot and rebound of a bursty bulk flow, *Geophys. Res. Lett.*, *37*, 8103; doi:10.1029/2009GL041971.

Pontius, D. H., and R. A. Wolf (1990), Transient flux tubes in the terrestrial magnetosphere, *Geophys. Res. Lett.*, *17*, 49–52; doi:10.1029/GL017i001p00049.

Ritter, P., and H. Lühr (2008), Near-Earth magnetic signature of magnetospheric substorms and an improved substorm current model, *Ann. Geophys.*, *26*, 2781–2793; doi:10.5194/angeo-26-2781-2008.

Rostoker, G. (1991), Some observational constraints for substorm models, *Magnetospheric Substorms, Geophys. Monogr. Ser., AGU*, *64*, 61–72; doi:10.1029/GM064p0061.

Runov, A., et al. (2011a), Dipolarization fronts in the magnetotail plasma sheet, *Planet. Space Sci.*, *59*(7), 517–525; doi:10.1016/j.Planet. Space Sci.2010.06.006.

Runov, A., V. Angelopoulos, C. Gabrielse, J. Liu, D. L. Turner, and X. Z. Zhou (2015), Average thermodynamic and spectral properties of plasma in and around dipolarizing flux bundles, *J. Geophys. Res. Space Physics*, *120*(6), 4369–4383; doi:10.1002/2015ja021166.

Runov, A., V. Angelopoulos, M. I. Sitnov, V. A. Sergeev, J. Bonnell, J. P. McFadden, D. Larson, K.-H. Glassmeier, and U. Auster (2009), THEMIS observations of an earthward-propagating dipolarization front, *Geophys. Res. Lett.*, *36*, 14106; doi:10.1029/2009GL038980.

Runov, A., V. Angelopoulos, X.-Z. Zhou, X.-J. Zhang, S. Li, F. Plaschke, and J. Bonnell (2011b), A THEMIS multicase study of dipolarization fronts in the magnetotail plasma sheet, *J. Geophys. Res.*, *116*, 5216; doi:10.1029/2010JA016316.

Sato, T., and T. Iijima (1979), Primary sources of large-scale Birkeland currents, *Space Sci. Rev.*, *24*, 347–366; doi:10.1007/BF00212423.

Schmid, D., M. Volwerk, R. Nakamura, W. Baumjohann, and M. Heyn (2011), A statistical and event study of magnetotail dipolarization fronts, *Ann. Geophys.*, *29*, 1537–1547; doi:10.5194/angeo-29-1537-2011.

Semenov, V. S., T. Penz, V. V. Ivanova, V. A. Sergeev, H. K. Biernat, R. Nakamura, M. F. Heyn, I. V. Kubyshkin, and I. B. Ivanov (2005), Reconstruction of the reconnection rate from Cluster measurements: First results, *J. Geophys. Res.*, *110*, 11217; doi:10.1029/2005JA011181.

Sergeev, V. A., A. V. Nikolaev, N. A. Tsyganenko, V. Angelopoulos, A. V. Runov, H. J. Singer, and J. Yang (2014), Testing a two-loop pattern of the substorm current wedge (SCW2L), *J. Geophys. Res. Space Physics*, *119*(2), 947–963; doi:10.1002/2013ja019629.

Sergeev, V. A., K. Liou, P. T. Newell, S. I. Ohtani, M. R. Hairston, and F. Rich (2004), Auroral streamers: characteristics of associated precipitation,convection and field-aligned currents, *Annales Geophysicae*, *22*(2), 537–548; doi:10.5194/angeo-22-537-2004.

Sergeev, V. A., N. A. Tsyganenko, M. V. Smirnov, A. V. Nikolaev, H. J. Singer, and W. Baumjohann (2011), Magnetic effects of the substorm current wedge in a "spread-out wire model" and their comparison with ground, geosynchronous, and tail lobe data, *J. Geophys. Res.*, *116*, 7218; doi:10.1029/2011JA016471.

Sergeev, V. A., R. C. Elphic, F. S. Mozer, A. Saint-Marc, and J. A. Sauvaud (1992), A two satellite study of nightside flux transfer events in the plasma sheet, *Planet. Space Sci.*, *40*, 1551–1572; doi:10.1016/0032-0633(92)90052-P.

Sergeev, V. A., V. Angelopoulos, J. T. Gosling, C. A. Cattell, and C. T. Russell (1996), Detection of localized, plasma-depleted flux tubes or bubbles in the midtail plasma sheet, *J. Geophys. Res.*, *101*, 10817–10826; doi:10.1029/96JA00460.

Sitnov, M. I., M. Swisdak, and A. V. Divin (2009), Dipolarization fronts as a signature of transient reconnection in the magnetotail, *J. Geophys. Res.*, *114*, 4202; doi:10.1029/2008JA013980.

Sormakov, D. A., and V. A. Sergeev (2008), Topology of Magnetic Flux Ropes in the Magnetospheric Plasma Sheet as Measured by the Geotail Spacecraft, *Cosmic Res.*, *46*, 387–391.

Sun, W. J., S. Fu, G. K. Parks, Z. Pu, Q.-G. Zong, J. Liu, Z. Yao, H. Fu, and Q. Shi (2014a), Electric fields associated with dipolarization fronts, *J. Geophys. Res. Space Physics*, *119*(7), 5272–5278; doi:10.1002/2014ja020045.

Sun, W. J., S. Fu, Z. Pu, G. K. Parks, J. A. Slavin, Z. Yao, Q.-G. Zong, Q. Shi, D. Zhao, and Y. Cui (2014b), The current system associated with the boundary of plasma bubbles, *Geophys. Res. Lett.*, *41*(23), 8169–8175; doi:10.1002/2014gl062171.

Sun, W. J., S. Y. Fu, G. K. Parks, J. Liu, Z. H. Yao, Q. Q. Shi, Q.-G. Zong, S. Y. Huang, Z. Y. Pu, and T. Xiao (2013), Field-aligned currents associated with dipolarization fronts, *Geophys. Res. Lett.*, *40*, 4503–4508; doi:10.1002/grl.50902.

Vasyliunas, V. M. (1970), Mathematical models of magnetospheric convection and its coupling to the ionosphere, *Particles and Field in the Magnetosphere*, *17*, 60.

Walsh, A. P., et al. (2009), Cluster and Double Star multipoint observations of a plasma bubble, *Ann. Geophys.*, *27*(2), 725–743; doi:10.5194/angeo-27-725-2009.

Wiltberger, M., V. Merkin, J. G. Lyon, and S. Ohtani (2015), High-resolution global magnetohydrodynamic simulation of bursty bulk flows, *J. Geophys. Res. Space Physics*, *120*(6), 4555–4566; doi:10.1002/2015ja021080.

Xing, X., L. R. Lyons, V. Angelopoulos, D. Larson, J. McFadden, C. Carlson, A. Runov, and U. Auster (2009), Azimuthal plasma pressure gradient in quiet time plasma sheet, *Geophys. Res. Lett.*, *36*, 14105; doi:10.1029/2009GL038881.

Xing, X., L. R. Lyons, Y. Nishimura, V. Angelopoulos, E. Donovan, E. Spanswick, J. Liang, D. Larson, C. Carlson, and U. Auster (2011), Near-Earth plasma sheet azimuthal pressure gradient and associated auroral development soon before substorm onset, *J. Geophys. Res.*, *116*, 7204; doi:10.1029/2011JA016539.

Xing, X., L. R. Lyons, X.-Z. Zhou, V. Angelopoulos, E. Donovan, D. Larson, C. Carlson, and U. Auster (2012), On the formation of pre-onset azimuthal pressure gradient in the near-Earth plasma sheet, *J. Geophys. Res.*, *117*, 8224; doi:10.1029/2012JA017840.

Yang, J., F. R. Toffoletto, R. A. Wolf, and S. Sazykin (2011), RCM-E simulation of ion acceleration during an idealized plasma sheet bubble injection, *J. Geophys. Res.*, *116*, 5207; doi:10.1029/2010JA016346.

Yao, Z. H., et al. (2012), Mechanism of substorm current wedge formation: THEMIS observations, *Geophys. Res. Lett.*, *39*, 13102; doi:10.1029/2012GL052055.

Yao, Z. H., et al. (2013), Current structures associated with dipolarization fronts, *J. Geophys. Res. Space Physics*, *118*, 6980–6985; doi:10.1002/2013JA019290.

Yao, Z. H., et al. (2014), Current reduction in a pseudo-breakup event: THEMIS observations, *J. Geophys. Res. Space Physics*, *119*(10), 8178–8187; doi:10.1002/2014ja020186.

Yao, Z. H., et al. (2015), A physical explanation for the magnetic decrease ahead of dipolarization fronts, *Annales Geophysicae*, *33*(10), 1301–1309; doi:10.5194/angeo-33-1301-2015.

Zesta, E., L. Lyons, C.-P. Wang, E. Donovan, H. Frey, and T. Nagai (2006), Auroral poleward boundary intensifications (PBIs): Their two-dimensional structure and associated dynamics in the plasma sheet, *J. Geophys. Res.*, *111*, 5201; doi:10.1029/2004JA010640.

Zesta, E., L. R. Lyons, and E. Donovan (2000), The auroral signature of Earthward flow bursts observed in the magnetotail, *Geophys. Res. Lett.*, *27*, 3241–3244; doi:10.1029/2000GL000027.

Zhang, X.-J., V. Angelopoulos, A. Runov, X.-Z. Zhou, J. Bonnell, J. P. McFadden, D. Larson, and U. Auster (2011), Current carriers near dipolarization fronts in the magnetotail: A THEMIS event study, *J. Geophys. Res.*, *116*; doi:10.1029/2010JA015885.

Zhou, M., et al. (2013), Cluster observations of kinetic structures and electron acceleration within a dynamic plasma bubble, *J. Geophys. Res. Space Physics*, *118*, 674–684; doi:10.1029/2012JA018323.

Zhou, M., M. Ashour-Abdalla, X. Deng, D. Schriver, M. El-Alaoui, and Y. Pang (2009), THEMIS observation of multiple dipolarization fronts and associated wave characteristics in the near-Earth magnetotail, *Geophys. Res. Lett.*, *36*, 20107; doi:10.1029/2009GL040663.

Zhou, X.-Z., V. Angelopoulos, J. Liu, A. Runov, and D.-X. Pan (2014b), Asymmetric braking and dawnward deflection of dipolarization fronts: Effects of ion reflection, *Geophys. Res. Lett.*, *41*(20), 6994–7001; doi:10.1002/2014gl061794.

Zhou, X.-Z., V. Angelopoulos, J. Liu, A. Runov, and S.-S. Li (2014a), On the origin of pressure and magnetic perturbations ahead of dipolarization fronts, *J. Geophys. Res. Space Physics*, *119*, 211–220; doi:10.1002/2013JA019394.

Zhou, X.-Z., V. Angelopoulos, V. A. Sergeev, and A. Runov (2010), Accelerated ions ahead of earthward propagating dipolarization fronts, *J. Geophys. Res.*, *115*; doi:10.1029/2010JA015481.

Zhou, X.-Z., V. Angelopoulos, V. A. Sergeev, and A. Runov (2011), On the nature of precursor flows upstream of advancing dipolarization fronts, *J. Geophys. Res.*, *116*, 3222; doi:10.1029/2010JA016165.

20

Cusp Current System: An Energy Source View

M. Yamauchi[1] and R. Slapak[2]

ABSTRACT

Electric currents in the cusp region are reviewed from viewpoints of history and energy conversion. During late 1980s and early 1990s, there were debates on the cause of the cusp region current system (cusp Region 0 field-aligned current [FAC] and cusp Region 1 FAC, and relevant ionospheric current). The major debates were whether the cusp part Region 1 FAC is an extension of the non-cusp-part dayside Region 1 FAC or whether they are generated in different regions independently. The independency of the cusp current system, which was demonstrated by many observations, suggests additional extraction of energy from the magnetosheath-like flow (e.g., deceleration) inside the polar magnetosphere. An extra deceleration is theoretically possible by adding a substantial local obstacle such as the outflowing ionospheric ions through the mass-loading effect, which conserves momentum but not kinetic energy. Thus, two different dynamo ($\mathbf{J}\cdot\mathbf{E}<0$) mechanisms most likely exist between the dayside Region 1 and 2 FACs and cusp Region 1 and 0 FACs, forming "double openness," which was introduced by *Vasyliunas* [1995]. The other debates (e.g., roles of mesoscale FACs, mapping to high latitude, and current carriers problems) are also reviewed in the light of new observational knowledge after Cluster.

20.1. INTRODUCTION

Compared with the large-scale current system associated with the auroral region, including the Region 1 and Region 2 field-aligned currents (FACs) [*Birkeland*, 1908; *Axford and Hines*, 1961; *Akasofu*, 1964; *Nishida*, 1968; *Iijima and Potemra*, 1976a], the large-scale cusp current system has special importance because it flows in the region that is geomagnetically directly connected to the magnetosheath-like flow in the polar magnetosphere, and because the cusp current system is the most dominant current system in the dayside high-latitude region unless the interplanetary magnetic field (IMF) southward component (= $-Bz$) is greater than the dawn-dusk component $|By|$ [*Iijima and Potemra*, 1976b; *Yamauchi and Araki*, 1989].

[1] Swedish Institute of Space Physics (IRF), Kiruna, Sweden
[2] Division of Space Technology, Luleå University of Technology (LTU), Kiruna, Sweden

The direct link to the exterior cusp and plasma mantle through the open magnetic field lines made the mapping of this current system to its direct source (generator) region as one of the central focuses in the past discussions. The mapping destination was particularly important for the cusp Region 1 FAC that constitutes the equatorward part of the cusp current system, because the cusp field lines can be connected to very large regions of the magnetopause boundary layer: high-altitude dayside boundary (high-altitude cusp), high-latitude boundary (plasma mantle or polar magnetopause), low-latitude dayside, flank, and even to the lobe without reaching the magnetopause (e.g., Fig. 20.1a [*Tsyganenko*, 2002]) if we consider a slight offset of the current flow from the exact field-aligned direction, for example, by combination of the velocity filter and drift effects of current carriers, in the convecting magnetosphere. This fact caused confusion in the observational terminologies of particle regions and names of the relevant FACs as summarized in section 20.2.

Electric Currents in Geospace and Beyond, *Geophysical Monograph 235*, First Edition.
Edited by Andreas Keiling, Octav Marghitu, and Michael Wheatland.

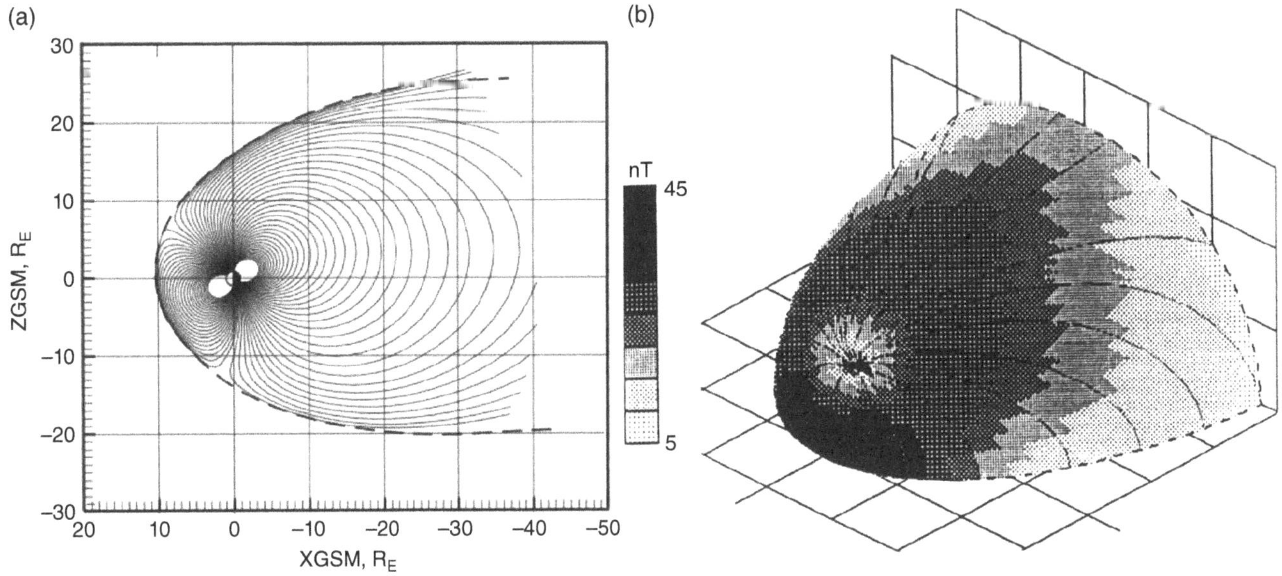

Figure 20.1 (a) A meridian cut of magnetic field lines of the Tsyganenko 2002 model. The entire region poleward part of the cusp is an open region and the mapping relation is more complicated than this model [*Tsyganenko*, 2002]. The density of the field lines is roughly proportional to the square root of the magnetic field strength (due to 2D cut) and is very weak near the cusp. (b) Strength of the magnetic field in Tsyganenko's 1987 model [*Stasiewicz*, 1991; *Tsyganenko* 1987]. The cusp is indeed an open hole in the magnetic obstacle as *Chapman and Ferraro* [1931] predicted.

The exterior cusp is originally postulated as a "hole" in the magnetic obstacle due to an extremely weak magnetic field, as shown in Figure 20.1b [*Chapman and Ferraro*, 1931; *Stasiewicz*, 1991]. Even in the antiparallel merging model [*Crooker*, 1979], which is a generalization of such a hole when the IMF is included (this is often called the reconnection site), the region of low magnetic pressure is very limited to high latitudes unless the IMF By is exactly zero. This is why the antiparallel merging model is considered to be different from the component reconnection model [e.g., *Lockwood*, 1994] that predicts the reconnection near the nose of the magnetopause due to the highest dynamic pressure.

Observations in the exterior cusp [e.g., *Sckopke et al.*, 1976; *Lundin*, 1985; *Eastman et al.*, 2000; *Lavraud et al.*, 2004] endorsed high plasma beta (ratio of plasma pressure to magnetic pressure) in this region with extremely small total magnetic field, forming a diamagnetic region. The extremely small total magnetic field is actually used to identify the high-altitude cusp [*Fung et al.*, 1997; *Zhou et al.*, 2000]. The situation is quite different from the Martian cusp, where the localized crust magnetic field is interconnected with the IMF due to the enhanced magnetic field at the ionopause [*Acuna et al.*, 1999; *Lundin et al.*, 2006]. Unfortunately, determination of FACs near the Martian cusp is not yet possible [cf. *Fillingim*, Chapter 26, this volume].

Since the open magnetic field region is involved, many studies on the FACs in the cusp region during the past three decades tried to identify the location of the reconnection site and resultant accelerated flow rather than how the energy of the field-aligned currents is provided from the ultimate source (solar wind). For example, the deflection of the narrow but intensified ionospheric convection "throat" near local noon at high latitude, to dawn or dusk direction depending on the IMF By polarity [*Heelis et al.*, 1976; *Burch*, 1979], has long been explained by the dawn-dusk acceleration of plasma by the magnetic tension force (a part of the $\mathbf{J} \times \mathbf{B}$ force that also includes the gradient of magnetic pressure, where $\mathbf{J}$ is the electric current and $\mathbf{B}$ is the magnetic field) after reconnection in the lobe magnetopause in magnetohydrodynamics (MHD) models [e.g., *Cowley et al.*, 1991; *Vennerstrom*, *et al.*, 2005].

However, deflection of plasma by the magnetic tension force requires sufficient magnetic field strength, that is, relatively low plasma beta where the magnetic field lines are most bent, whereas the magnetic pressure is too low to move "heavy" high-beta diamagnetic plasma in the exterior cusp. Observations of the high-altitude cusp never showed any magnetic pileup where the lobe reconnection is postulated.

Another problem with the view of using only the magnetic tension force (UB paradigm) is that the chain and locations of the energy conversion are not sufficiently considered [e.g., *Iijima*, 2001; *Tanaka et al.*, 2016]. The cusp current system is the strongest dayside current system unless IMF points southward within 45° clock angle [*Iijima et al.*, 1984; *Friis-Christensen et al.*, 1985; *Yamauchi and Araki*, 1989; *Anderson et al.*, 2008], and the response of the cusp current system and related ionospheric signatures to the changes in the solar wind condition are nearly immediate [*Sandholt et al.*, 1994; *Yamauchi et al.*, 1995; *Frey et al.*, 2002]. Therefore, the energy conversion from the solar wind to the cusp current system and the relevant ionospheric convection must be powered by a quick energy conversion mechanism involving both the deceleration of the inflow and powering the dynamo current in the shortest route, that is, on the open field line region just above the cusp. Note that this does not apply to the polar cap part of the NBZ FACs [*Iijima et al.*, 1984; *Potemra*, 1994] of which the response to the IMF is different from the cusp part, and might have a distant source [e.g., *Yamauchi et al.*, 1993b; *Tanaka et al.*, 2016]. The most intense part of the NBZ FAC is confined within the cusp and this part should be discussed separately from the rest of the NBZ FAC (Region 0 current), which most likely is related to the transpolar arcs rather than the cusp [*Yamauchi et al.*, 1993b, 1995].

One candidate location of the dynamo ($\mathbf{J} \cdot \mathbf{E} < 0$, where $\mathbf{J}$ and $\mathbf{E}$ are electric current and electric field) where the solar wind energy is extracted is the exterior cusp, where the magnetosheath-like flow is significantly decelerated as compared with immediate downstream of the bow shock according to observations [*Eastman et al.*, 1976; *Lundin*, 1985; *Sandahl et al.*, 1997; *Lavraud et al.*, 2004]. This localized mechanism in the cusp was not considered in the Dungey's simplified view for southward IMF conditions (in this chapter, Dungey's view means that the entire magnetosphere and ionosphere of the open field line regions downstream of the dayside reconnection site move according to the convection electric field of the connecting magnetosheath flow) [e.g., *Roederer*, 1977; *Cowley et al.*, 1991].

The deceleration means buildup of density and pressure (internal energy) to conserve the number and momentum fluxes unless the cross section of the flow increases (such rarefaction is predicted in the downstream of the exterior cusp in some models [*Yamauchi and Lundin*, 1994, 1997; *Tanaka*, 1995, 2003]). During or after such a pressure increase, charge separation against the convection electric field (including in the oblique direction) may take place, finally powering the cusp FACs. Inversely, one may regard such a compressive structure that extracts the kinetic energy as a dynamo as long as the related current satisfies $\mathbf{J} \cdot \mathbf{E} < 0$. While MHD simulations already predict such structure through fast/slow mode structure [*Yamauchi*, 1994; *Tanaka*, 1995, 2003], any non-MHD obstacles in the exterior cusp may also cause such a compressing structure [*Yamauchi and Lundin*, 2001]. Candidate obstacles include wave turbulences (their energy is higher than the magnetic energy according to observations) [*Savin et al.*, 2004] and outflowing O^+ (outflowing O^+ of density typically $0.1\,cm^{-3}$ is mixed with the incoming magnetosheath flow with density around $10\,cm^{-3}$) [*Nilsson et al.*, 2008; *Slapak et al.*, 2013, 2015]. As described in section 20.4, the mass-loading effect of such outflowing O^+ can convert a nonnegligible amount of kinetic energy to electrostatic energy.

Such an energy conversion viewpoint is somewhat missing in most past studies of the current system although sufficient observations already exist and some of the past debates are directly related to the dynamo location. Thus, synthesizing the past observations in terms of energy conversion is important and needed. We review past observations and arguments, including the terminology problem of the cusp current system, and discuss consequences of the observational knowledge from an energy conversion point of view.

20.2. EARLY HISTORY AND TERMINOLOGY UNTIL 1980S AND ITS RELEVANCE TO RECENT WORKS

The electric currents in and above the ionosphere, which consist of FACs and the connecting ionospheric Petersen and Hall currents (flowing along the ionospheric electric field and the electric equipotential lines, respectively), have long been known as causing the geomagnetic disturbances observed at ground [e.g., *Birkeland*, 1908]. Although the ground geomagnetic disturbances caused by FACs are canceled with those by the connecting Pedersen currents [*Fukushima* 1969, 1976], the closure of the FACs and the ionospheric Pedersen currents determines the ionospheric electric field that drives the Hall current, and hence this current circuit can be regarded as the ultimate cause of the ground geomagnetic activity.

Such a current system in the polar region is predominantly associated with nightside auroral activities such as the substorm-time auroral electrojet (DP1 current system [*Akasofu*, 1964]) and the dawn-to-dusk electric field over the polar cap during mainly presubstorm periods (DP2 current system [*Nishida*, 1968]), where DP is named after "(geomagnetic) disturbance in polar region." However, the current system associated with the cusp was not recognized until the mid 1970s [*Friis-Christensen and Wilhjelm*, 1975; *Maezawa*, 1976; *Iijima and Potemra*, 1976b; *Wilhjelm et al.*, 1978; *McDiarmid et al.*, 1978]. This is partly because the geomagnetic observatory network could not cover the cusp region effectively until the 1970s, partly because the current system becomes dominant over the Region 1 FAC only when IMF Bz is positive, and partly because the flow direction of the current system changes only its polarity depending on the IMF By polarity. The cusp current system is located in a limited local time range and at higher latitudes than the other major current systems [*Friis-Christensen and Wilhjelm*, 1975; *Iijima and Potemra*, 1976b], and could not be easily identified without a good observation network.

The cusp current system was first recognized from ground geomagnetic disturbances. After the geomagnetic observatory network extended to > 80° geomagnetic latitude (GMlat), it became possible to estimate the ionospheric current (and hence convection) over the cusp region [*Friis-Christensen and Wilhjelm*, 1975]. Figure 20.2a shows the estimated ionospheric electric potential from ground geomagnetic disturbances (data covered up to the pole). Since the ionospheric Hall current is a result of electron ExB drift with ions stopped by neutrals, the convection pattern is approximately along the electric equipotential contour lines. The figure demonstrates that the DP2 type antisunward two-cell convection pattern (given as equivalent potential contour shown in Fig. 20.2) is modified in the high-latitude midday sector when IMF $|By| > |Bz|$ or $Bz > |By|$. In the former case, the antisunward convection is strongly deflected to the east-west direction depending on the IMF By polarity [*Levitin et al.*, 1982]. In the latter case, that is, when IMF points northward, the antisunward convection completely disappears and even a slight sunward motion appears [*Maezawa*, 1976; *Levitin et al.*, 1982]. The relevant ionospheric currents are called the DPY current and the DPZ current, respectively [*Friis-Christensen and Wilhjelm*, 1975; *Friis-Christensen et al.*, 1985]. The DPY and DPZ currents are local, but yet the most intense large-scale current system in the dayside, as the dense counter lines near the high-latitude local noon indicate. This basic result and general pattern of the ionospheric electric potential for different IMF clock angles is reproduced in the modern dataset [e.g., *Weimer*, 2005] as shown in Figure 20.2b.

Ionospheric convection corresponding to the DPY and DPZ ionospheric currents is confirmed by the AE-C satellite convection measurements [*Heelis et al.*, 1976; *Burch*, 1979] and ground-based radar observations [*Clauer and Banks*, 1986]. Satellite observations also revealed that the DPY-type convection at around 80° latitude is enhanced in a throat-like narrowed channel (called "convection throat") compared to the upstream (lower latitude) and downstream (higher latitude) area, and that this convection throat is colocated with the cusp particle precipitation. Observations indicate that the DPY and DPZ currents are intense and related to the cusp. Note that the terminology "throat" was also used for the shape of the cusp in the vertical direction (filled with magnetosheath particles), causing some confusion on the terminology, but it originally and normally means the convection throat.

By assuming a Hall/Pedersen conductivity ratio distribution, it is possible to estimate the FACs [*Gurevich et al.*, 1976; *Kamide et al.*, 1981; *Friis-Christensen et al.*, 1985]. *Levitin et al.* [1982] extracted modes corresponding only to DPY and DPZ, respectively, to obtain relevant pairs of FACs as shown in Figure 20.3 [*Levitin et al.*, 1982]. This method is further generalized as the principal component method or orthogonal function method [*He et al.*, 2012; *Cousins et al.*, 2015; *Milan et al.*, 2015].

During the same decade when the DPY and DPZ ionospheric currents and the related convection were established, systematic low-altitude satellite observations of geomagnetic disturbances above the ionosphere became available, such as ISIS-2 [*Burrows et al.*, 1976; *McDiarmid et al.*, 1978, 1979] and Triad [*Iijima and Potemra*, 1976b, *Iijima et al.*, 1978], and later Magsat [*Araki et al.*, 1984] and DE2 [*Burch et al.*, 1985]. Among those, Triad and Magsat provided high-precision data suitable for obtaining FAC distributions for different IMF conditions, whereas ISIS-2 and DE2 provided simultaneous observations of FACs and precipitating magnetosheath-origin ions, suitable for identifying the location of the FACs relative to the particle cusp at low altitudes. In addition, a spot observation by the AE-C satellite (FACs were derived for a very short period) confirmed the relation between the extra current system with the convection throat, precipitation, and the FACs [*Klumpar*, 1979; *Bythrow et al.*, 1982].

From these various types of observations, the existence of large-scale FACs (Region 0 FAC) poleward of the Region 1 FAC at near local noon at high latitudes (>80°) was established in the middle 1980s for all IMF conditions. This Region 0 FAC, flowing poleward of the Region 1 FAC in the opposite direction, is originally defined for all local times [*Heikkila*, 1984]. However, the Region 0 FACs in the dawn, dusk and midnight sectors are very small compared to the dayside Region 0 FAC. To differentiate from them the Region 0 current near local noon is called the Cusp Region 0 (CR-0) FAC. We use this terminology in this chapter.

The CR-0 FAC was originally named "cusp region FAC" by *Iijima and Potemra* [1976b], but this terminology is often confused with the cusp part of the Region 1 current because the cusp current system should be closed within a limited region near the dayside cusp, that is, by the DPY currents and at least one pair of upward and downward FACs. Therefore, the terminology was soon revised to different names such as "poleward part of DPY-FAC," "NBZ FAC," and "mantle FAC" [*Friis-Christensen and Wilhjelm*, 1975; *Iijima et al.*, 1984; *Potemra et al.*, 1987; *Bythrow et al.*, 1988]. The Cusp Region 0 (CR-0) is the most neutral and general name because it is the same as the mantle FAC for southward IMF (mantle cusp disappears when the IMF points

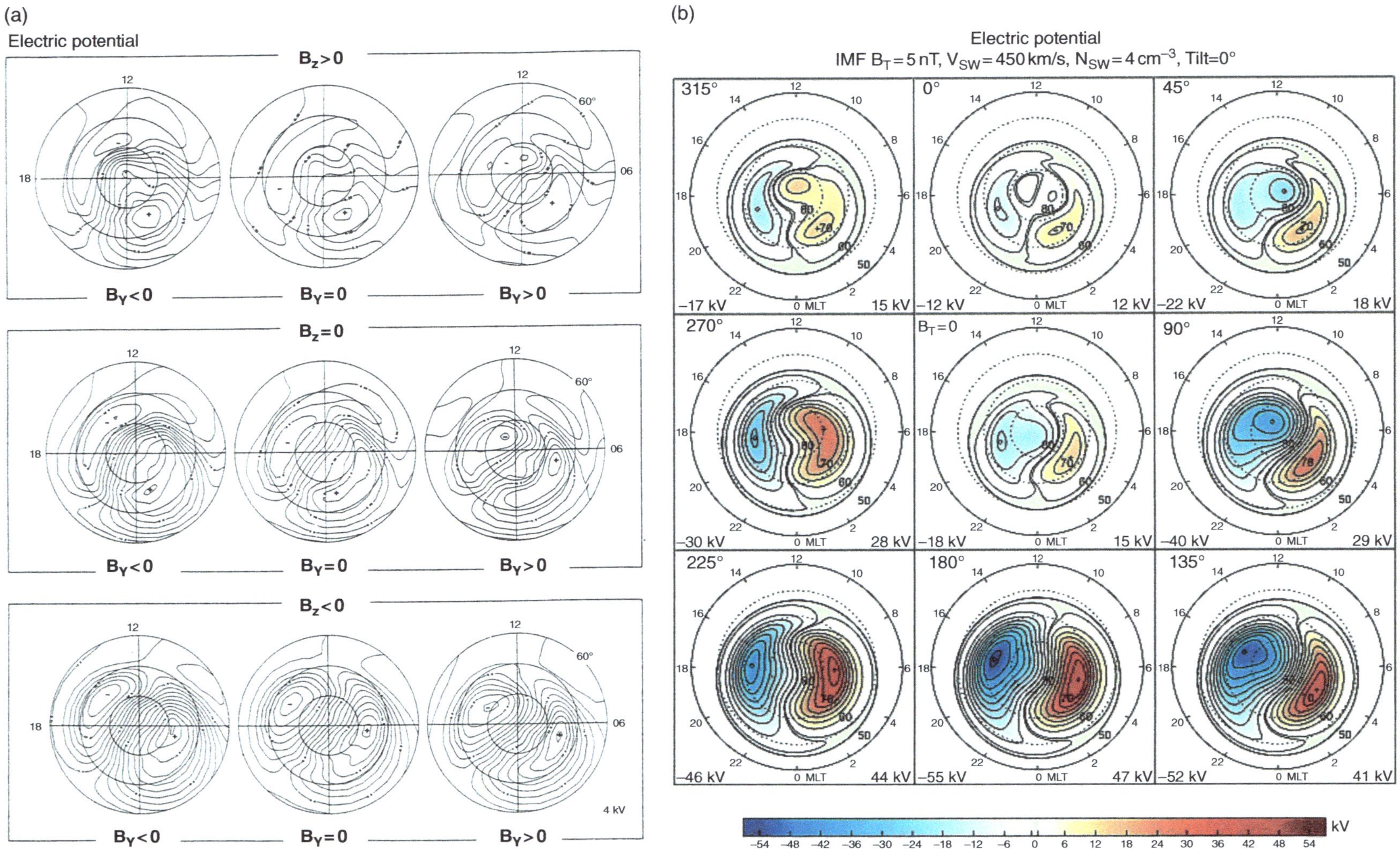

Figure 20.2 Ionospheric electric potential for different IMF directions: (a) obtained from ground magnetometer data [*Friis-Christensen et al.*, 1985]; (b) modeled from assimilated data including satellites [*Weimer*, 2005]. Over 20 yr difference, the qualitative nature has not changed very much, except that the old estimate is slightly tilted westward compared to the new model. For example, the contour lines are very dense at around 80° latitude near local noon, indicating that the convection is most intense near the cusp in both old and new estimates. The convection (Hall current) pattern returns back to a DP2-like two cell pattern already near the pole in both estimates.

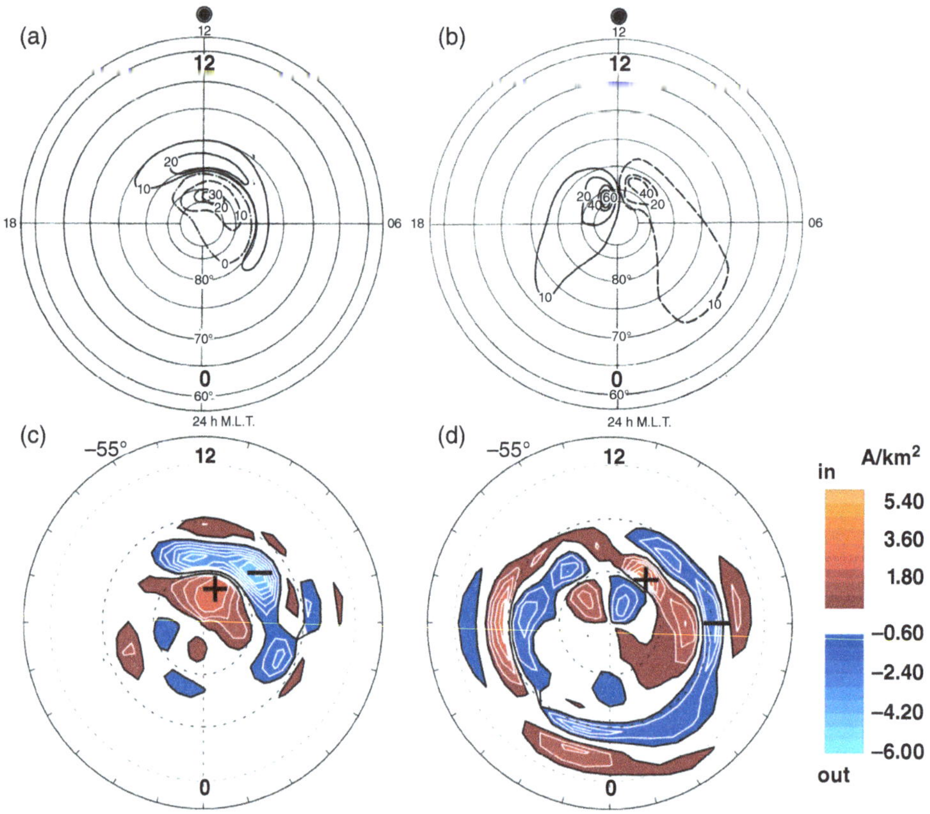

Figure 20.3 FACs modes corresponding to DPY and DPZ ionospheric currents by *Levitin et al.* [1982] (upper panels) and *Cousins et al.* [2015] (lower panels). The left panels show the FAC distribution that corresponds to the DPY current (IMF By > 0) and the right panels show the FAC distribution that corresponds to the DPZ current (IMF $Bz > 0$) near the cusp. (a) and (b) The solid and dashed lines correspond to FACs into and out of the northern hemisphere, respectively. The value (in μA/m²) is normalized for IMF $By = +1$ nT (left) and $Bz = +1$ nT (right) in the upper panels. (c) and (d) The second and third component in the southern hemisphere from one week data. The red and blue contours correspond to FACs into and out of the southern hemisphere, respectively.

northward), the (most intensive part of) NBZ FAC for northward IMF, and DPY-FAC for IMF $|By| > |Bz|$ with some rotated location [*Banks et al.*, 1984; *Yamauchi et al.*, 1993b; *Vennerstrom et al.*, 2002]. In fact, the intensity of CR-0 FAC for large IMF By component (|By| > |Bz|) is similar between during northward IMF and during southward IMF, as shown in Figure 20.4 [*Yamauchi and Araki*, 1989]. To avoid any prejudice in the interpretation, the name of the CR-0 FAC has been commonly used in the recent decade. Similarly, the name of the Cusp Region 1 (CR-1) is also commonly used for the low-latitude side counterpart.

The satellite studies of the FACs near the cusp mainly focused on their distributions and the dependence on the IMF polarities during 1980s. This confirmed the basic distribution patterns that are shown in Figure 20.3. The DPY current is located between the CR-1 and CR-0 FACs for strong IMF By, and the DPZ current is located between the prenoon-postnoon pairs of the CR-0 FACs for strong IMF

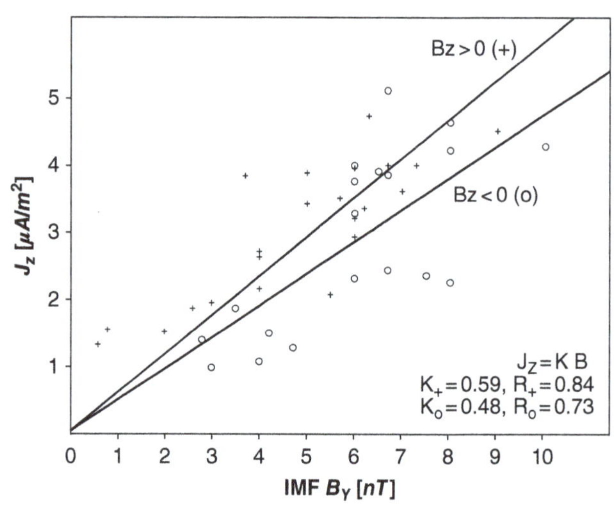

Figure 20.4 Peak intensity of the Cusp Region 0 FAC for IMF $|By| > |Bz|$ using Magsat data when the IMF is stable for more than half a day [*Yamauchi and Araki*, 1989].

$Bz > 0$. The CR-0 occupies an area rather than a band-like region such as the Region 1 FAC as in Figure 20.3, according to multitrajectory analysis covering a 2D area during the same IMF and solar wind conditions in the same season [*Iijima and Shibaji*, 1987; *Yamauchi and Araki*, 1989; *Papitashvili et al.*, 2002; *Korth et al.*, 2014]. This fact made it difficult to detect the CR-0 with satellites traversing the midday meridian, such as Triad [*Potemra*, 1994].

The seasonal effect turned out to be much more drastic for CR-1 and CR-0 than the dayside Region 1 FAC [*Fujii and Iijima*, 1987; *Yamauchi and Araki*, 1989; *Lu et al.*, 1994], even after the hemispheric difference is subtracted [*Papitashvili et al.*, 1994; *Green et al.*, 2006, 2009]. This is expected from the latitudinal difference between these two current systems because the ionospheric conductivity varies at higher latitudes (CR-0 and CR-1) more than at lower latitudes (dayside Region 1) [*Fujii and Iijima*, 1987]. This has been used as an observational indication that the current system is associated with a voltage driver (dynamo with $\mathbf{J} \cdot \mathbf{E} < 0$, e.g., by deceleration of flow as shown in Fig. 20.5c), rather than a current driver (divergence of perpendicular current without requirement of $\mathbf{J} \cdot \mathbf{E} < 0$). By examining the conductivity dependency for each current system in the dayside, *Haraguchi et al.* [2004] confirmed that the CR-1 and CR-0 currents are more associated with voltage drivers and the dayside Region 2 FAC is more associated with current drivers.

Traditional examples of current drivers are nonaligned gradients of pressure P and magnetic field strength $|B|$ (the $(\nabla|B| \times \nabla P) \cdot \mathbf{B}$ term in MHD) in the ring current [*Vasyliunas*, 1970] as illustrated in Figure 20.5a,b and the dayside magnetopause current (rotation of the magnetic field with a large tangential component across any layer means the current is flowing nearly along the magnetic field, as shown in Fig. 20.6b) [*Lee et al.*, 1985]. The former ($\nabla|B| \times \nabla P$ effect) is believed to be connected to the Region 2 FAC [*Wolf and Spiro*, 1985; *Iijima et al.*, 1990].

During the 1980s and 1990s, the cusp red aurora was also studied as a possible signature of magnetosheath plasma injection through the cusp [*Sandholt et al.*, 1986, 1994], because the red aurora is caused by low-energy electrons. If the solar wind electrons (temperature of 100 eV) precipitate to the ionosphere without acceleration, they may cause red aurora. If the ionospheric footpoint is different between electrons and ions due to the velocity filter effect [*Lockwood*, 1994], such electron motion may cause an electric current [*Newell et al.*, 2010]. However, the low-energy electron bursts in the cusp are most likely secondary electrons that have experienced minor potential structures (<300 V). Such a structure is colocated with the cusp precipitating protons (decelerated from the magnetosheath energy) [*Yamauchi et al.*, 2001; *Jacobsen et al.*, 2010]. Furthermore, the red aurora could be caused by accelerated electrons up to several hundred eV. Therefore, the relation between the red aurora and large-scale FACs is still unclear, although it might contribute to the mesoscale FACs.

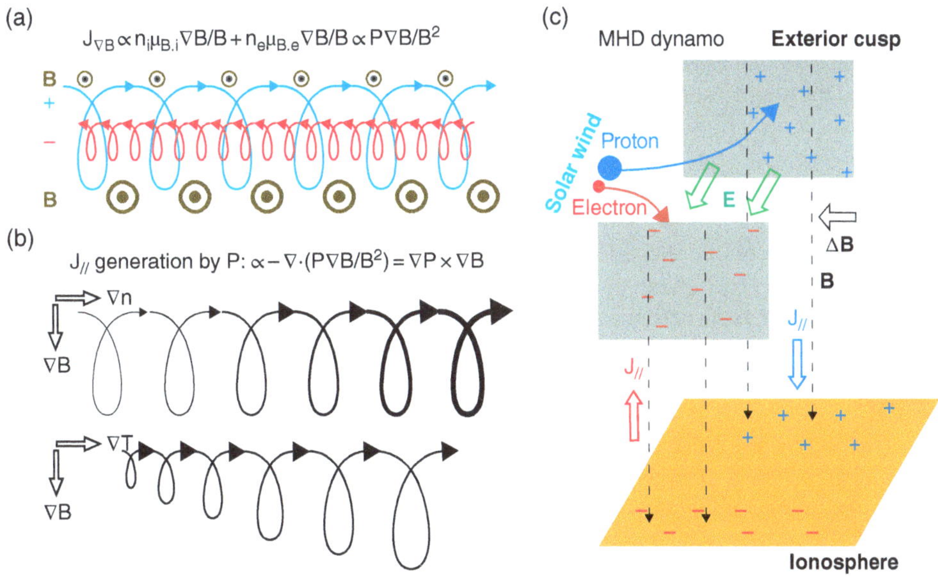

Figure 20.5 Illustration of electric current related to gradient B drift and traditional MHD dynamo [*Alfvén* and Fälthammar, 1963]. (a) Gradient B drift, (b) density effect and temperature effect of the divergence of the gradient B drift, (c) current related to change in $\mathbf{U} \times \mathbf{B}$.

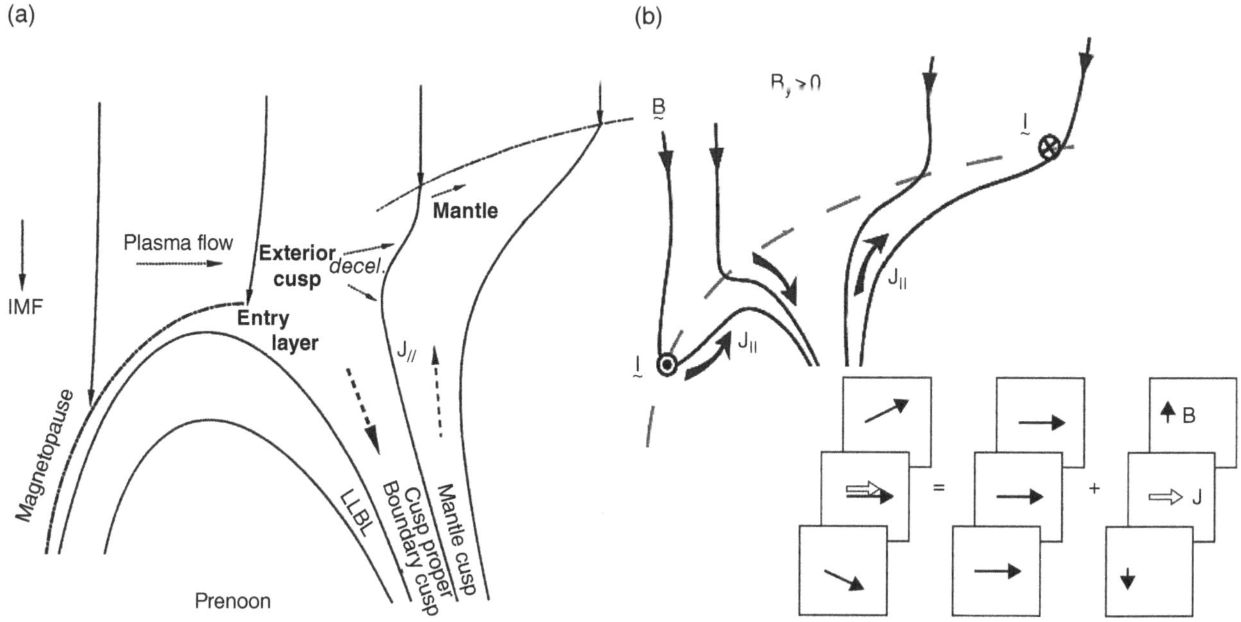

Figure 20.6 Two different scenarios on the generator location of the cusp current system: (a) the source is localized in the cusp; (b) the source of CR-0 and CR-1 are remote, for example, at the magnetopause [*Lee et al.*, 1985]. In the latter scenario, the CR-0 and CR-1 should appear as a part of DP2 current system.

20.3. MAIN DEBATES IN 1990S

20.3.1. Distribution

The observed FAC distribution raised two basic interpretations on the source: (1) Both CR-0 and CR-1 are generated near the exterior cusp including the high-latitude boundary (plasma mantle and polar magnetopause) as shown in Figure 20.6a [*Yamauchi and Lundin*, 1992], and they are independent from (noncusp) dayside Region 1 FAC as shown in Figure 20.7a, c, and d [*Erlandson et al.*, 1988; *Taguchi et al.*, 1993; *Yamauchi et al.*, 1993b; *Ohtani et al.*, 1995b; *Vennerstrom et al.*, 2002]. (2) The pair of CR-0 and CR-1 is a simple deformation of the dawn-dusk pair of the dayside Region 1 FACs as shown in Figure 20.7b [*Cowley et al.*, 1991; *Lockwood*, 1994], and this pair is directly connected to the magnetopause processes.

The first scenario requires an independent source location between the CR-0/CR-1 currents and the dayside Region 1 currents (related to the DP2 convection). In this scenario, the magnetosheath flow is significantly decelerated in the exterior cusp and plasma mantle at some distance from the magnetopause compared with the deceleration in the other open region. The second scenario considers that the current system is associated with DP2, and in most models is connected to the magnetopause Chapman-Ferraro current through the rotational discontinuity [*Lee et al.*, 1985; *Siscoe et al.*, 1991] as illustrated in Figure 20.6b. In this scenario, the energy comes from the nightside to the dayside.

Debate between these scenarios continued during 1990s for the following reasons:

1. Both scenarios (1) and (2) (or FAC patterns of Fig. 20.7a,b) produce the same ionospheric convection pattern. The convection pattern is essentially the same as the equipotential (cf. Fig. 20.2), which is mathematically equivalent to the integral of the FAC distribution. In such integration, the strongest current smears out the other FACs [*Yamauchi and Araki*, 1989]. Therefore, the convection pattern cannot solve this debate even after the SuperDarn network [*Greenwald et al.*, 1995] was developed.

2. Both scenarios (1) and (2) provide two to three FACs along a meridian, and satellite observations with a single trajectory cannot, in principle, solve this question.

3. Without knowing the relative location of the FACs compared with plasma regions (which give clues on mapping), it is difficult to judge the connectivity of the FACs between different altitudes.

4. A statistical distribution of FACs easily misleads the interrelationship between the dayside Region 1 FAC and CR-1/CR-0, because location of the cusp (and the cusp current system) is not solely determined by the IMF, but is also influenced by the solar wind dynamic pressure, seasons, and past history of magnetospheric conditions.

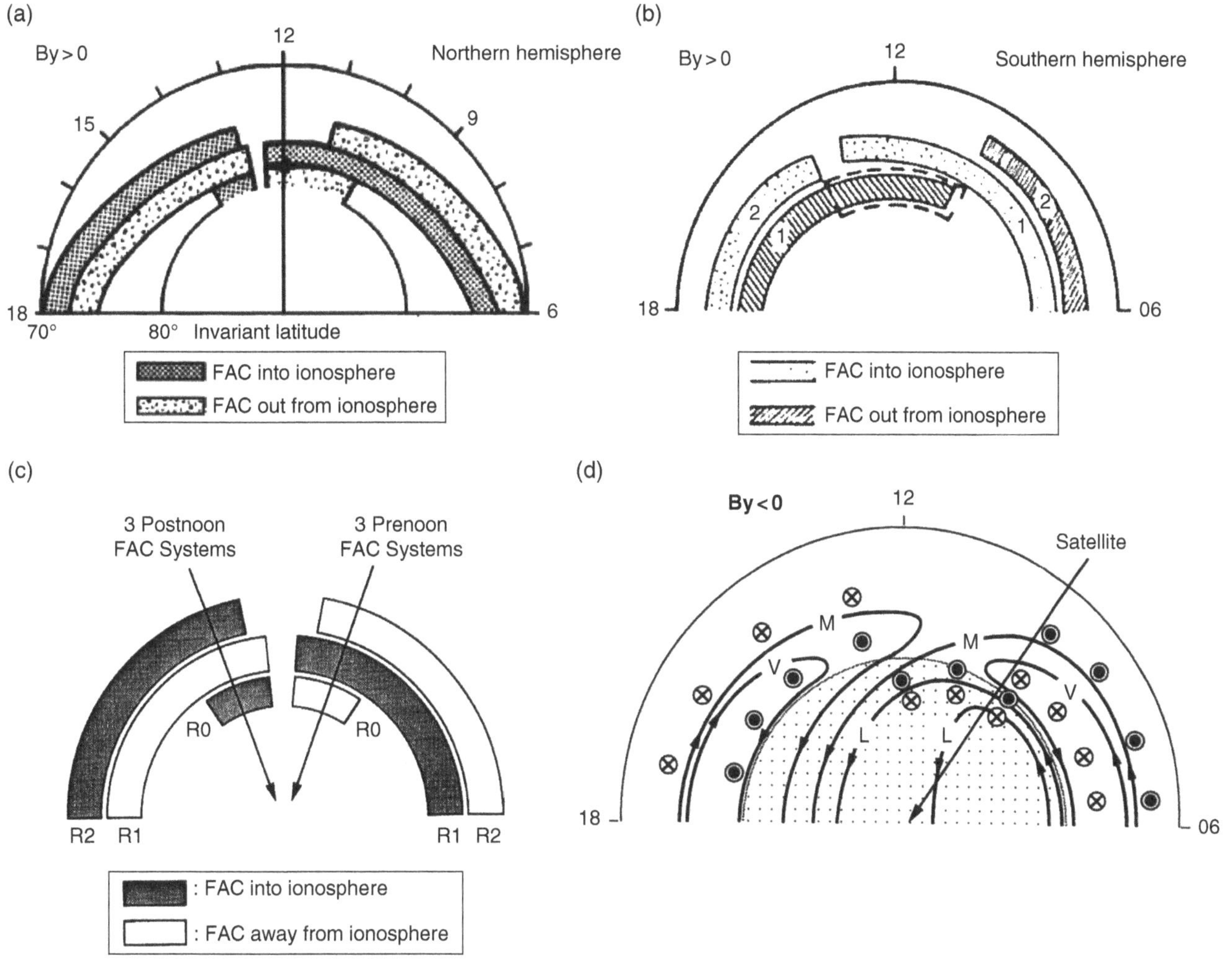

Figure 20.7 Connectivity of the CR-1 and CR-0 at the ionospheric height: (a) summary of satellite observations proposed by *Erlandson et al.* [1988], corresponding to 20.5a; (b) FACs predicted only by reconnection model [*Cowley et al.*, 1991; *Saunders*, 1992], corresponding to 20.5b (more discussion in section 20.4). To examine which is more likely, we need simultaneous satellites observation in the prenoon and postnoon sector like (c) [*Ohtani et al.*, 1995b]. Furthermore, satellite observation often found four currents like (d) [*Ohtani et al.*, 1995a].

Therefore, the gap between two distinct but closely located areas is smeared out in the statistical method, in the same manner as for the integration method.

To compensate the single trajectory weakness (second point above), a large dataset including both ground-based and satellite observations has been developed to obtain instantaneous 2D distributions of the FACs [*Lu et al.*, 1994; *Weimer*, 2005], and this effort is still going on. For example, the AMPERE project uses a dataset from 64 Iridium satellites [*Anderson et al.*, 2002, 2008; 2014; *Knipp et al.*, 2014]. However, such a huge dataset is still not sufficient to answer this question because the simultaneous satellite passes over the cusp region are still rare, and the particle data are missing in this method.

The most effective way to tackle this problem is to make systematic comparisons of the copopulating plasma with the magnetic field data, and further compare different spacecraft traversing both postnoon and prenoon simultaneously. *Ohtani et al.* [1995b] found such simultaneous traversals by DMSP and Viking with both particle and magnetic field data. Statistics of the relative positions of the particle region and FACs supported this independency [*Taguchi et al.*, 1993; *Yamauchi et al.*, 1993b, 2000; *Ohtani et al.*, 1995a; *Lukianova et al.*, 2001]. These observations as well as statistics on numbers of FACs (with confirmation by particle data) also showed that the Region 1 FAC and CR-0/CR-1 FACs are not connected, contrary to what simple statistics of distributions showed. Thus, any statistical connection (like Fig. 20.7b) is superficial. The most recent knowledge [*Wing et al.*, 2010; *Clausen et al.*, 2013; *Korth et al.*, 2014] is consistent with such an independency (different particle populations between the cusp part and noncusp dayside), while no solid counterevidence has been presented as far as the authors know.

20.3.2. Mapping of Plasma Regions

The method of distinguishing the dayside Region 1 FAC and CR-1 in terms of the cusp is not mature. This problem partly comes from the confusion in the definition of the cusp in the particle data, and partly from the assumptions when mapping from low altitudes to the magnetospheric boundaries. The observational terminologies of particle regions that are used worldwide was defined by DMSP observations [*Newell and Meng*, 1992] because the DMSP series provided a dataset from more than 30 yr of the observations with the same instrumentation. At 700 km altitude and without pitch angle information, DMSP's definition uses an assumption on mapping, even near the cusp where the field line reaches various regions of the magnetospheric boundary as shown in Figure 20.1a. In fact, a part of the plasma region that is marked as the low-latitude boundary layer (LLBL) in the DMSP definition is also called the "boundary cusp" in the mid-altitude Viking observations [*Woch and Lundin*, 1992], in which LLBL (Fig. 20.8a) and the boundary cusp (Fig. 20.8b) can be distinguished using the pitch-angle distribution information. Viking LLBL is rather similar to the poleward part of boundary plasma sheet (BPS) in the DMSP terminology [*Woch et al.*, 1993]. These two terminologies indicate that the mapping relation in the DMSP classification is ambiguous near the cusp.

In the Viking definition, Region 1 FACs are found in both the LLBL (Fig. 20.8a) and the boundary cusp (Fig. 20.8b). Thus, with pitch-angle information at mid-altitude, it is easy to distinguish the CR-1 (in the boundary cusp) from the dayside region 1 FAC (in the LLBL). In Figure 20.8b, the Region 1 FAC is found both inside the LLBL (closed region) and the cusp/boundary cusp (open region).

The discrepancy in the terminologies for the same population may mislead the mapping relation and even the physics involved. LLBL in the Viking terminology implies that it is mapped to the dayside/flank magnetopause and in a closed region. However, LLBL in the DMSP definition can partly be an open region because it might be mapped only up to the high-altitude boundary near the cusp without any guarantee of reaching farther to the low-latitude magnetopause boundary. Accordingly, FACs flowing in DMSP's LLBL [e.g., *Wing et al.*, 2010 and references therein] might come from the exterior cusp but not from low latitudes. Furthermore, Figure 20.8b

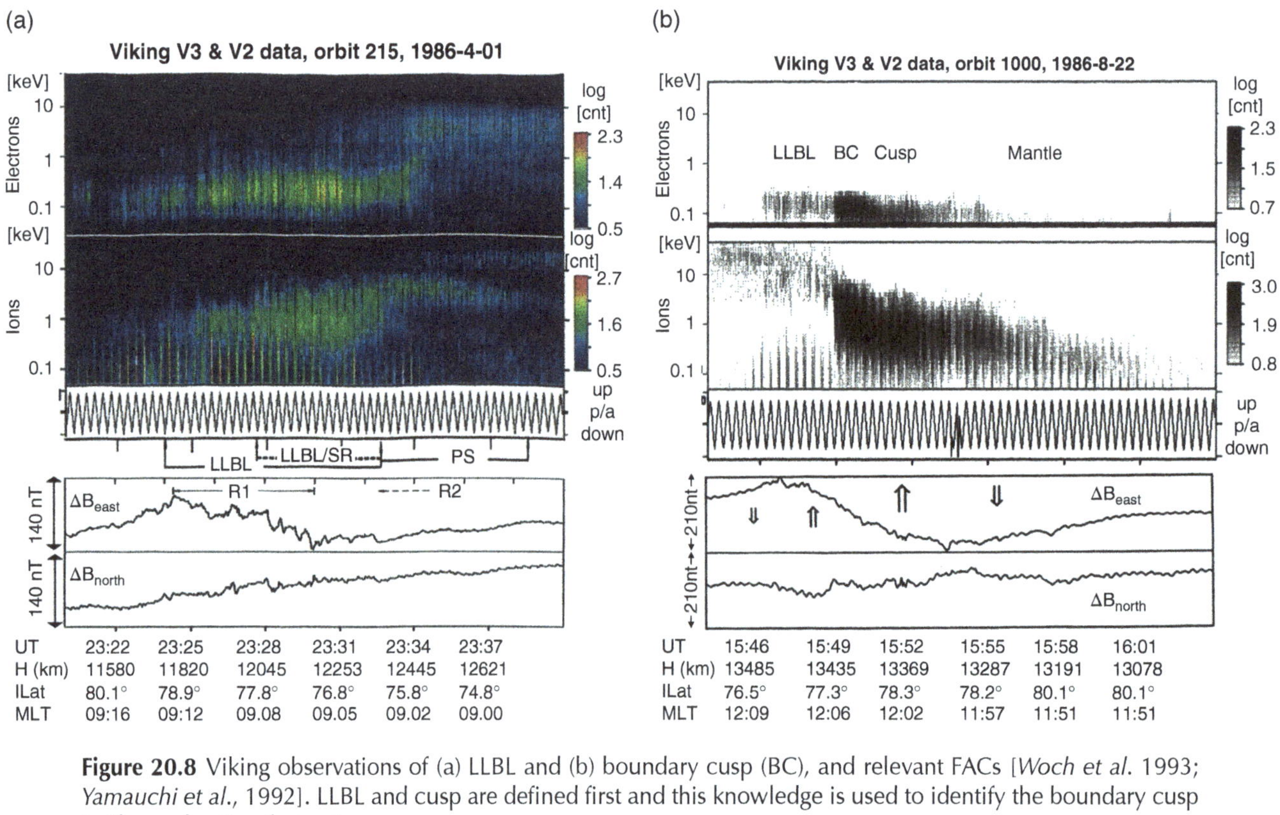

Figure 20.8 Viking observations of (a) LLBL and (b) boundary cusp (BC), and relevant FACs [*Woch et al.* 1993; *Yamauchi et al.*, 1992]. LLBL and cusp are defined first and this knowledge is used to identify the boundary cusp (with acceleration feature).

shows that the FAC intensity is stronger in the upstream region than in the downstream region as judged from particle data.

Thus, one may not simply map the low-altitude FACs beyond the high-altitude cusp. For precise definitions of the regions that are based on the mapping relation, high-altitude cusp observations are inevitable. However, the high-altitude observations, for example by HEOS-2 [e.g., *Haerendel et al.*, 1978], Haykeye [e.g., *Eastman et al.*, 2000], Prognoz-7 [e.g., *Lundin*, 1985], Interball [e.g., *Merka et al.*, 2002], and Polar [*Zhou et al.*, 2000] could not establish the mapping relation between the high-altitude cusp and low-altitude cusp except for the stagnant type cusp. Cluster was an ideal mission to study the exterior cusp [*Escoubet et al.*, 2001] by combining particle signatures [*Lavraud et al.*, 2004, 2005] with the 3D current system obtained from the Curlometer method [*Robert et al.*, 1998; *Dunlop et al.*, 2002]. However, not many studies were performed on the FACs except for mesoscale events [*Marchaudon et al.*, 2006, 2009].

Yet what was found so far is that the high-altitude observations of the cusp region FACs are similar to those at the mid-altitudes, and that the strong current (and large deviation from the Tsyganenko model [*Zhang et al.*, 2010]) is localized in the cusp at all altitudes, suggesting that the generator is located inside the (high-altitude) cusp and plasma mantle, no matter where the magnetic field line of the CR-1 is mapped beyond the cusp. In other words, the LLBL in the DMSP terminology might map to the boundary region of the exterior cusp. Note that the noncusp part of NBZ (Region 0) FAC, which is often related to the transpolar arcs [*Yamauchi and Lundin.*, 1993b], does not necessarily flow nearly along the geomagnetic field and might come from distant places such as the LLBL or plasma sheet [e.g., *Watanabe et al.*, 2014; *Tanaka et al.*, 2016].

20.3.3. Contribution From Mesoscale FAC Pairs

Smith et al. [1992] proposed that CR-0 and CR-1 can be summations of small pairs of FACs related to mesoscale injections in the cusp because particle injections in the cusp and the cusp red aurora (corresponding to low-energy electrons) often show an internal structure composed of many mesoscale injections [*Yamauchi and Lundin*, 1994; *Sandholt et al.*, 1986]. Such mesoscale injections are known to carry a pair of FACs [*Lundin et al.*, 1991; *Yamauchi et al.*, 1995, 1998; *Andersson et al.*, 1997], and therefore, the origin of such mesoscale particle signatures are again debated, whether (1) it is generated in the exterior cusp by, for example, natural turbulence relevant to the cusp outer boundary [*Yamauchi and Lundin*, 1997] or (2) it is generated in the magnetopause by, for example, flux transfer events [*Southwood*, 1987; Escoubet *et al.*, 1992] and plasma transfer events [*Woch and Lundin*, 1991; *Yamauchi et al.*, 2003].

According to mid-altitude observations, the mesoscale FACs only slightly modify semi-permanent CR-0 and CR-1 in most cases, as shown in Figure 20.9 [*Yamauchi et al.*, 1998]. This indicates that the mesoscale transient structure does not change the average dayside reconnection rate or the related cross-polar potential. Also, high-altitude observations showed that the internal mesoscale structure is also confined to the cusp [*Sandahl et al.*, 1997; *Fedorov et al.*, 2000; *Lauvaud et al.*, 2005], indicating that the mesoscale structure might be formed in the exterior cusp rather than at the magnetopause.

Marchaudon et al. [2009] applied the curlometer method to one mesoscale event when the Cluster was located near the postnoon edge of the cusp, and argued that the flux transfer event composes the major current in the cusp. However, their observation shows that the flux tube is actually filled with high-beta plasma inside the diamagnetic cavity, that is, opposite to the traditional low-beta magnetic flux tube in the magnetotail, in which strong magnetic field expels the plasma. The observation is rather consistent with a detached plasma cloud from the exterior cusp [*Yamauchi and Lundin*, 1997].

The observed high-beta diamagnetic flux tube (or slow-mode pressure pulse) accumulates positive charges in the front and negative charges in the back of the pressure pulse, as shown in Figure 20.10b. Such a longitudinal charge separation (polarization) does not agree with the prediction by *Southwood* [1987], but is predicted in MHD, when the compression (fast mode) and rarefaction (slow mode) are considered, as shown in Figure 20.10a [*Yamauchi et al.*, 1993a; *Yamauchi*, 1994; *Watanabe et al.*, 2014].

20.3.4. Current Carriers

While the framework of the CR-0 and CR-1 FACs is defined by the ion characteristic, which is quite different from that of the nightside FACs, the current carriers must be electrons [*Marklund et al.*, 1994; *Yamauchi et al.*, 1998]. The observations confirmed that majority of the FACs is carried by electrons, at least for the mesoscale cases [*Marchaudon et al.*, 2009]. The question is then how the dawn-dusk asymmetry (upward current and downward current) or changes in the flow direction in response to the IMF By polarity are accomplished.

Wing et al. [2015] showed that structured (polar arc-like) electrons are found in the CR-0 FAC region when the FAC flows upward, but not when the FAC flows downward. This relation is also found in the TIMED-Iridium comparison [*Korth et al.*, 2014] as shown in Figure 20.11. The energy flux for the downward CR-0 FAC is very low, and the energy flux is relatively low even for the upward CR-0 FAC except during strongly

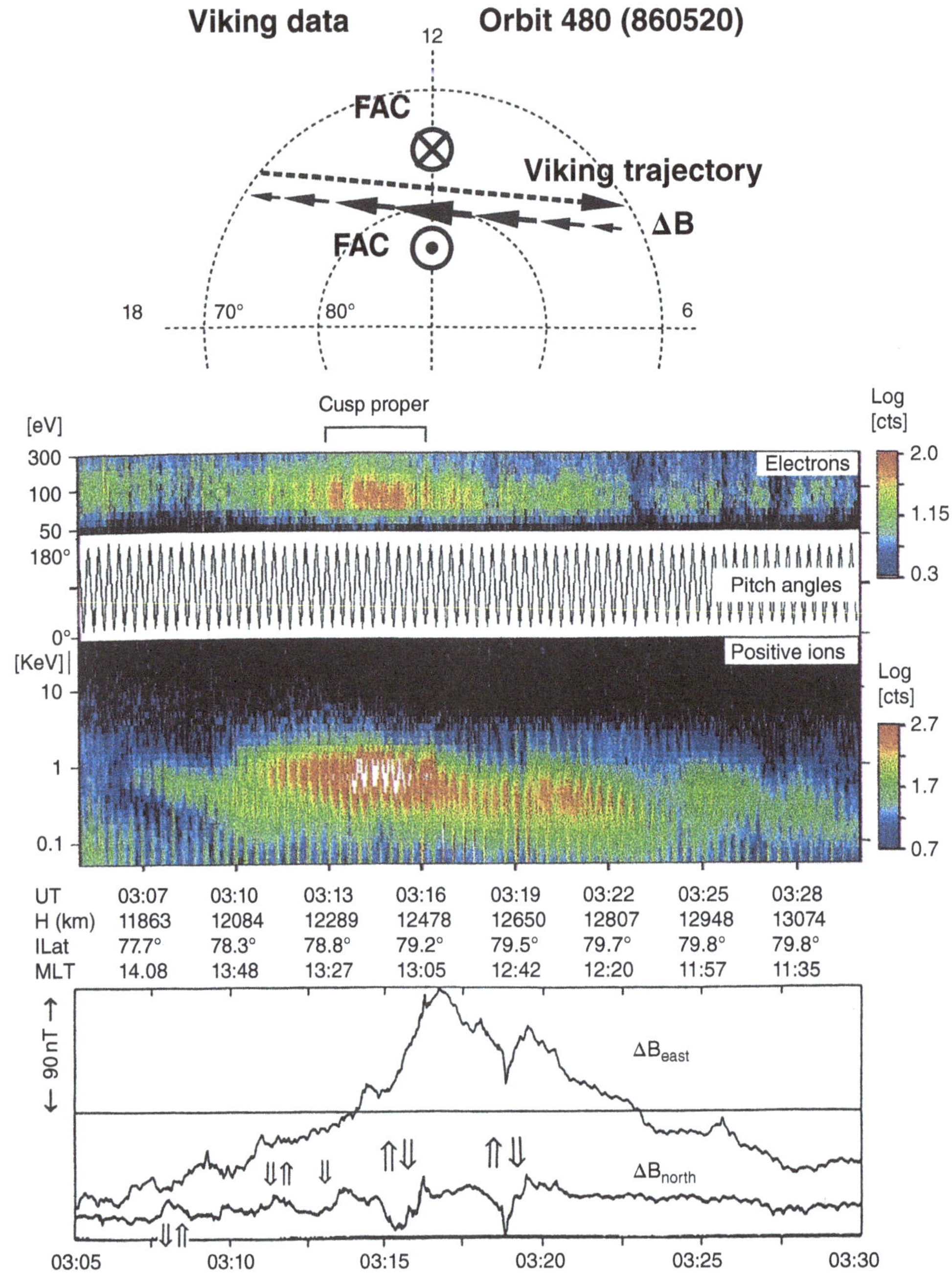

Figure 20.9 Particle and magnetic field data from Viking orbit 480 (20 May 1986). Positive (negative) slopes of B_{north} mean downward (upward) FACs. The positive deviation of B_{east} indicates upward FAC poleward of the satellite and/or downward FAC equatorward of the satellite [*Yamauchi et al.*, 1998].

northward IMF conditions. For southward IMF, the energy flux is low in the entire cusp, indicating that the current carrier could be thermal electrons for both upward and downward CR-1/CR-0. To answer this question, we need to measure thermal electrons, which is quite challenging.

20.4. MISSING DISCUSSION: ENERGY SOURCE

The source of the current system is often explained simply as "newly opened flux" under Dungey's simplified view, which only requires global-scale access of magnetosheath plasma to the magnetosphere across the magnetospheric

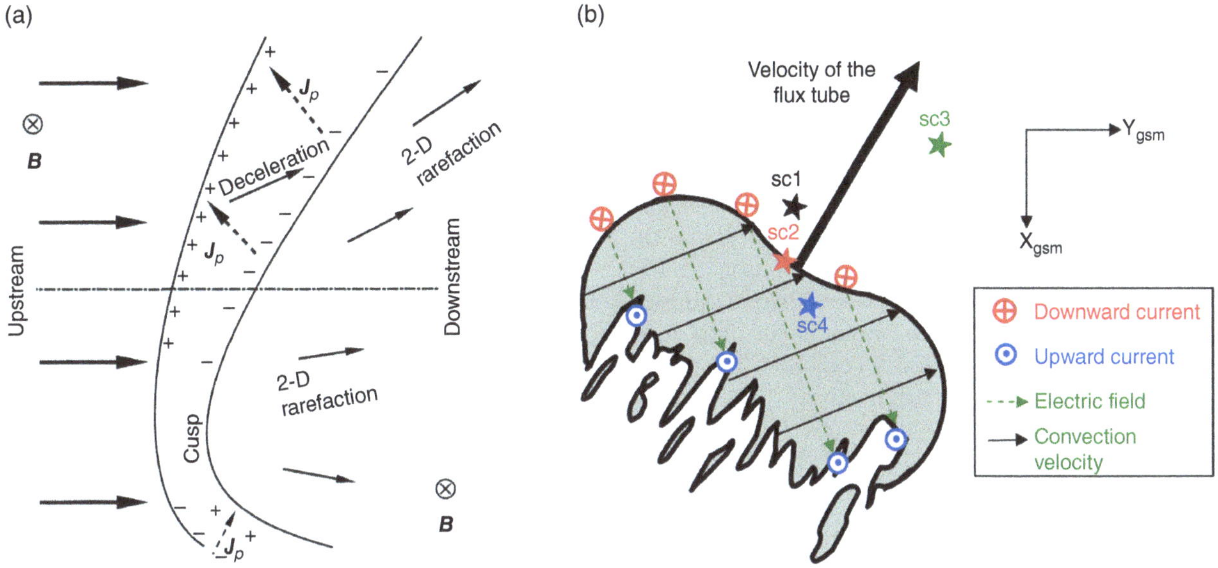

Figure 20.10 (a) Dynamic model of FAC generation in a compressional structure. The flow direction of the FAC (charge separation) is sensible to $\mathbf{U} \times \nabla P$, and can easily be asymmetric [*Yamauchi and Lundin*, 1992]. The sign of charges reverses for a diamagnetic (rarefaction) case. (b) Cluster observation of traveling high-beta tube [*Marchaudon et al.*, 2009].

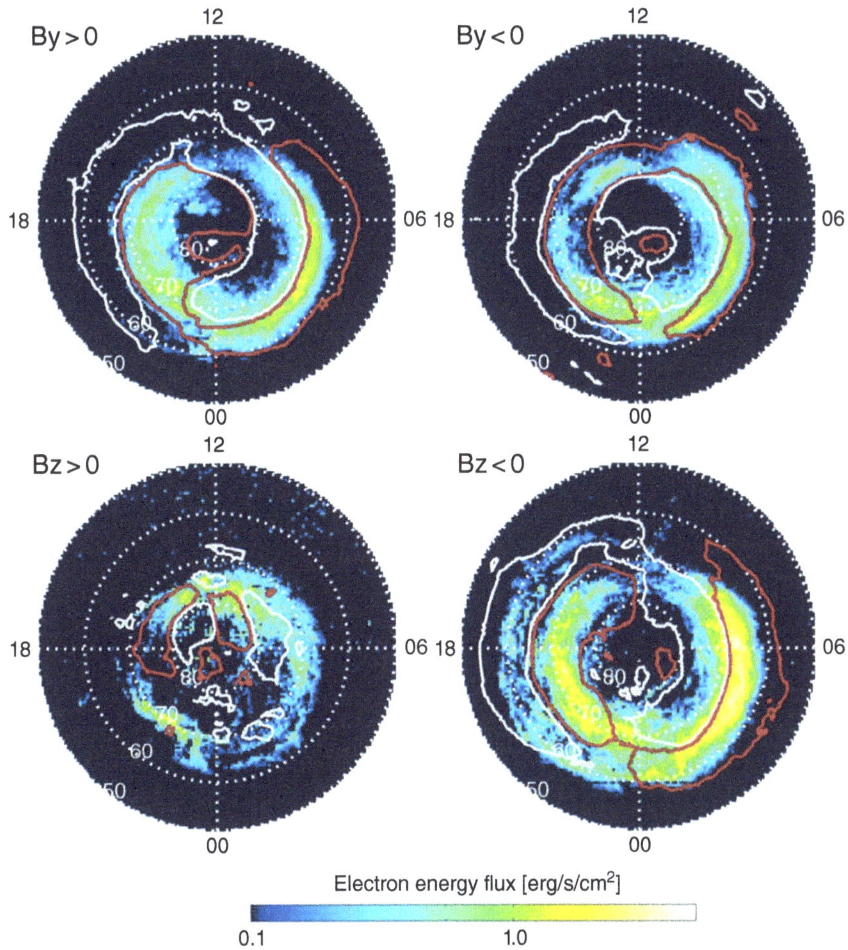

Figure 20.11 Statistical distributions of the electron energy flux observed in the northern hemisphere for different IMF conditions (with 45° angle) [*Korth et al.*, 2014]. The 2-sigma contours of the upward (red) and downward (grey) FACs are overlaid on each distribution.

boundary. Since positive and negative space charges are deflected by the geomagnetic field unless the electric field E cancels the Lorenz force, space charges are accumulated at both ends of an azimuthally limited flow until $\mathbf{E} = -\mathbf{U} \times \mathbf{B}$ is achieved, as illustrated in Figure 20.5c, where **U** is the flow velocity inside the magnetosphere [*Alfvén and Fälthammar*, 1963]. With 150 km/s access velocity and 10 nT field over the exterior cusp, this dynamo moves ions and electrons against the electric field directions, that is, powers charges against a potential drop across the exterior cusp of about 10 kV/R_E in the MHD regime, where R_E is Earth's radius.

Such a "traditional" MHD dynamo (cf. "traditional" means powering $\mathbf{J} \cdot \mathbf{E} < 0$ by the extraction of kinetic energy as illustrated in Fig. 20.5c) has been believed most effective at the mantle magnetopause where the Chapman-Ferraro current is flowing against the solar wind electric field [e.g., *Roederer*, 1977; *Siscoe et al.*, 2002; *Vennerstrom et al.*, 2005]. In fact, Interball and Cluster observations show that the mantle magnetopause is a rotational discontinuity [*Fedorov et al.*, 2000; *Lavraud et al.*, 2005], across which the flow is decelerated by the $\mathbf{J} \times \mathbf{B}$ force (the magnetic tension force). However, the exterior cusp and entire plasma mantle can also be the dynamo region because the traditional MHD dynamo requires only plasma motion across field lines that are connected to the load (ionosphere). Even the viscous-like interaction at the magnetopause can also produce the same type of charge separation with $\mathbf{J} \cdot \mathbf{E} < 0$ by the same principle. Furthermore, the dynamo mechanism is not limited to such transversal charge separation (current across the flow) by the Lorenz force shown in Figure 20.5c, but the pressure gradient after compression of the flow can also cause a charge separation with $\mathbf{J} \cdot \mathbf{E} < 0$ in the source region of the FACs [*Siscoe et al.*, 1991; *Yamauchi et al.*, 1993a; *Tanaka*, 1995, 2003].

The accumulated space charges flow toward the ionosphere as FACs, to compensate the ionospheric Pedersen current and to maintain the DP2-like antisunward convection. In steady states ($\nabla \times \mathbf{E} = 0$ according to Farady's law), the charge provided by this dynamo and charge consumed by the ionospheric load is balanced ($\nabla \cdot \mathbf{J} = 0$), such that the electric potential maintained by the dynamo can be mapped to the ionosphere. This is the microphysics of the frozen-in magnetic field, and of mapping the convection along the magnetic field [*Alfvén and Fälthammar*, 1963]. The total energy of the ionospheric convection and electric current system (Joule heating) is ultimately maintained by the deceleration of the newly accessed plasma.

In transit state, the space charges have inertia, requiring finite transit time of a few minutes along the magnetic field with the Alfvén velocity [*Sato and Iijima*, 1979] for such momentum/energy transfer along the geomagnetic field. Also, it takes tens of minutes for low-mass density plasma to move high mass density ionospheric plasma until ionospheric convection reaches steady state. In addition, the information of "system requires dynamo" propagates upward to the polar magnetopause.

In Dungey's simplified view, the FAC is concentrated on both ends of magnetosheath plasma access, that is, at locations corresponding to (noncusp) Region 1 FACs at dawn and dusk like in Figure 20.7b. No matter how the magnetic field geometry is deformed, one may not add independent FACs inside this antisunward convection like Figure 20.7 a, b, and d, because no more "newly opened" flux tubes can be added inside the already-open region in Dungey's simplified view. The extra FAC and enhanced localized convection in the cusp region means that the magnetosheath plasma flowing on field lines connected to the ionosphere (i.e., exterior cusp and plasma mantle) must experience extra deceleration, that is not included in Dungey's simplified view.

To avoid prejudice based on Dungey's simplified view (e.g., Fig. 20.7b, which is disqualified according to observations as mentioned in section 20.3), *Iijima* [2000] introduced a concept of stress balance, including pressure gradient force ($-\nabla P$) and inertia force ($-\rho d\mathbf{U}/dt$) in addition to the $\mathbf{J} \times \mathbf{B}$ force. In the traditional MHD dynamo as depicted in Figure 20.5c, loss of kinetic energy from the deceleration ($-\rho d\mathbf{U}/dt$) is directly converted to the Poynting flux of the FACs (and a pair of FAC causes a magnetic deviation $\Delta\mathbf{B}$ perpendicular with respect to both **E** and **B**, and $\mathbf{E} \times \Delta\mathbf{B}$ points in the field-aligned direction) and electric potential energy through $\mathbf{J} \cdot \mathbf{E} < 0$. However, the energy can also be temporally stored as plasma pressure (P) in the cusp due to its geometry [*Spreiter and Summers*, 1967; *Yamauchi and Lundin*, 1994; *Tanaka*, 2003; *Tanaka et al.*, 2016]. Also, physical characteristics of the cusp such as the high turbulence (Ponderomotive force replaces pressure gradient force) [*Savin et al.*, 2004; *Lundin and Guglielmi*, 2006] and the mass-loading effect of escaping ions [*Yamauchi and Lundin*, 1997] may work as an extra obstacle to increase the substantial pressure. *Yamauchi and Lundin* [1997] had to add such an extra obstacle downstream of the magnetopause in their original 1994 model because it is difficult to have a localized pressure increase (deceleration) that is independent of the entire flow (corresponding to DP2 ionospheric convection), and the ion escape is known to maximize in and around the cusp [*Yau and André*, 1997; *Norqvist et al.*, 1998].

The mass-loading effect has been overlooked because the amount of ion outflow in the exterior cusp and plasma mantle was not clear at that time [*Eklund et al.*, 1997]. The entire cusp at all altitudes is filled with intense wave activity, which is effective in energizing the ionospheric ions. This causes large amounts of O^+ to reach the high-altitude cusp and plasma mantle, with final destination either far downstream in the tail [*Nilsson et al.*, 2008] or directly into the high-latitude magnetosheath [*Slapak et al.*, 2013, 2015]. The outflowing O^+ density in the high-altitude cusp and mantle shows large variability but is typically about 0.1 cm^{-3}, mixing with the incoming magnetosheath H^+ flow of about 10 cm^{-3}.

The mass-loading effect by such amount of outflowing O^+ cannot be ignored due to 16 times heavier mass of O^+ than H^+ [*Yamauchi and Lundin*, 1997]. If the tailward velocity of O^+ becomes the same as that of the solar wind without significant local heating or turbulence, the conservation of momentum flux with zero initial tailward velocity for oxygen leads to

$$m_P n_P U_{IN}^2 = (m_P n_P + m_O n_O) U_{OUT}^2$$

or

$$(n_P + 16 \cdot n_O) U_{OUT}^2 = n_P U_{IN}^2 \qquad (20.1)$$

where m_P and $m_O = 16\ m_P$ are the proton an oxygen masses, n_P and n_O are the proton and oxygen density, and U_{IN} and U_{OUT} are tailward inflow and final velocities, respectively. The inflow velocity decreases by a factor of $\delta U / U_{IN} = 8 \cdot n_O / (n_P + 16 \cdot n_O)$.

The total kinetic energy flux after mass loading is

$$\begin{aligned} m_P (n_P + 16 \cdot n_O) \left[U_{IN}^2 n_P / (n_P + 16 \cdot n_O) \right]^{(3/2)} / 2 \\ = \left[n_P / (n_P + 16 \cdot n_O) \right]^{(1/2)} \cdot m_P n_P U_{IN}^3 / 2 \end{aligned} \qquad (20.2)$$

and hence the kinetic energy decreases by the same factor of $8 \cdot n_O / (n_P + 16 \cdot n_O)$. This energy is converted to the charge separation if no heating effect is involved.

A primitive estimate of the mass-loading of 0.1 cm^{-3} O^+ to 10 cm^{-3} magnetosheath flow results in about 7% (=0.8/11.6) loss of velocity. If the incoming flow velocity is 100 km/s, the loss of the kinetic energy (that is converted to the Poynting flux) reaches 2 W/km^2, which is comparable to the solar wind electromagnetic energy (that is used for the DP2 system), and is sufficient in powering the cusp current system although the access area is smaller than the magnetopause. Thus, it is quite possible that the magnetosheath flow in the exterior cusp, the location of which is determined by the antiparallel merging [*Crooker*, 1979; *Stasiewicz*, 1991], experiences extra deceleration (or pressure increase) downstream of the reconnection site by non-MHD effects, resulting in substantial dynamo power.

In solar wind rest frame, the inflow to the exterior cusp experiences information that "the field is connected to the ionosphere" twice: first by the loss of the charge that maintains the convection electric field to the ionosphere immediately after crossing the boundary, and then by meeting the ionospheric outflowing ions. In this sense, the cusp magnetosphere is "double open," in the similar sense as the double openness proposed by *Vasyliunas* [1995].

If we apply this result to the northward IMF case, the reconnection might well be a result of deceleration of the magnetosheath flow in the cusp where the escaping O^+ is found during such IMF conditions. The MHD theory does not determine any causality relation [*Vasyliunas*, 2001] whether the reconnection causes the deceleration of the solar wind or a deceleration in the cusp (that may include non-MHD effects) causes the reconnection downstream. However, in the latter case, sunward acceleration from the high-latitude reconnection site during northward IMF is expected only in a tiny region at the poleward edge of the cusp, which is what observations also indicate [*Woch and Lundin*, 1992; *Yamauchi and Lundin*, 1994].

20.5. SUMMARY

We have reviewed the historical development of observations of the cusp current system. While recent studies provide instantaneous and more precise 2D distributions (e.g., AMPERE) as well as 3D distributions (Cluster), qualitative knowledge on the distribution and IMF/seasonal dependencies was established in the 1990s. In addition, comparisons with particle data finally started to reveal the current carrier problem: it should not be the same for upward FACs and downward FACs. Further studies are needed in this direction.

We also summarized the main debate in 1990s, whether CR-0 and CR-1 are independent from the Region 1 current system related to the DP2 convection pattern (cf. Fig. 20.6). Unfortunately, results of such arguments are not well referred even in recent studies. The main reason for such a never-ending discussion is that the location of the dynamo is yet not clear for the cusp current system. However, if we include the mass-loading effect of escaping ionospheric ions, this problem might be solved quantitatively.

ACKNOWLEDGMENTS

The work is partly supported by Swedish National Space Board.

REFERENCES

Acuna M. H., J. E. P. Connerney, N. F. Ness, R. P. Lin, D. Mitchell, C. W. Carlson3, J. McFadden, et al. (1999), Global distribution of crustal magnetization discovered by the Mars Global Surveyor MAG/ER experiment, *Science*, *284*(5415), 790–793; doi:10.1126/science.284.5415.790.

Akasofu, S.-I. (1964), The development of the auroral substorm, *Planet. Space Sci.*, *12*, 273–282; doi: 10.1016/0032-0633(64)90151-5.

Alfvén, H., and C. G. Fälthammar (1963), *Cosmical Electrodynamics, Fundamental Principles*, Clarendon, Oxford.

Anderson, B. J., H., Korth, C. L. Waters, D. L. Green, and P. Stauning (2008) Statistical Birkeland current distributions from magnetic field observations by the Iridium constellation, *Ann. Geophys.*, *26*, 671-687, doi:10.5194/angeo-26-671-2008.

Anderson, B. J., H. Korth, C. L. Waters, D. L. Green, V. G. Merkin, R. J. Barnes, and L. P. Dyrud (2014), Development of large-scale Birkeland currents determined from the Active Magnetosphere and Planetary Electrodynamics Response Experiment, *Geophys. Res. Lett.*, *41*, 3017–3025; doi: 10.1002/2014GL059941.

Anderson, B. J., K. Takahashi, T. Kamei, C. L. Waters, and B. A. Toth (2002), Birkeland current system key parameters derived from Iridium observations: Method and initial validation results, *J. Geophys. Res.*, *107*(A6); doi: 10.1029/2001JA000080.

Andersson, L., H. Nilsson, and J.-E. Wahlund (1997), Mesoscale structure in the cusp/cleft region, *Phys. Chem. Earth*, *22*, 663–667.

Araki, T., T. Kamei, and T. Iyemori (1984), Polar cap vertical currents associated with northward interplanetary magnetic field, *Geophys. Res. Lett.*, *11*, 23–26; doi: 10.1029/GL011i001p00023.

Axford, W. I., and C.O. Hines (1961), A unifying theory of high-latitude geophysical phenomena and geomagnetic storms, *Can. J. Phys.*, *39*(10): 1433–1464; doi: 10.1139/p61-172.

Banks, P. M., T. Araki, C. R. Clauer, J. P. St. Maurice, and J. C. Foster (1984), The interplanetary electric field, cleft currents and plasma convection in the polar caps, planet. *Space Sci.*, *32*, 1551–1557.

Birkeland, K. (1908), *The Norwegian Aurora Polaris Expedition, 1902–1903*, 1, H. Aschehoug & Co., Christiania.

Burch, J. L. (1979), Effects of the interplanetary magnetic field on the auroral oval and plasmapause. *Space Sci. Rev.*, *23*(3), 449–464.

Burch, J. L., P. H. Reiff, J. D. Menietti, R. A. Heelis, W. B. Hanson, S. D. Shawhan, E. G. Shelley, M. Sugiura, D. R. Weimer, and J. D. Winningham (1985), IMF By - dependent plasma flow and Birkeland currents in the dayside magnetosphere: 1. Dynamics Explorer observations, *J. Geophys. Res.*, *90*(A2), 1577–1593; doi: 10.1029/JA090iA02p01577.

Burrows, J. R., M. D. Wilson, and, I. B. McDiarmid (1976), Simultaneous field aligned current and charged particle measurements in the cleft, in *Magnetospheric Particles and Fields*, 111–124; doi: 10.1007/978-94-010-1503-5_11.

Bythrow, P. F., T. A. Potemra, and R. A. Hoffman (1982), Observations of field-aligned currents, particles, and plasma drift in the polar cusps near solstice, *J. Geophys. Res.*, *87*(A7), 5131–5139; doi: 10.1029/JA087iA07p05131.

Bythrow, P. F., T. A. Potemra, R. E. Erlandson, L. J. Zanetti, and D. M. Klumpar (1988), Birkeland currents and charged particles in the high-latitude prenoon region: A new interpretation, *J. Geophys. Res.*, *93*, 9791–9803.

Chapman, S., and V. C. A. Ferraro (1931), A new theory of magnetic storms, Terr. Magn. *Atmos. Electr.*, *36*(2), 77–97; doi:10.1029/TE036i002p00077.

Clauer, C. R., and P. M. Banks (1986), Relationship of the interplanetary electric field to the high-latitude ionospheric electric field and currents: Observations and model simulation, *J. Geophys. Res.*, *91*, 6959–6971.

Clausen, L. B. N., J. B. H. Baker, J. M. Ruohoniemi, S. E. Milan, J. C. Coxon, S. Wing, S. Ohtani, and B. J. Anderson (2013), Temporal and spatial dynamics of the regions 1 and 2 Birkeland currents during substorms, *J. Geophys. Res. Space Physics*, *118*, 3007–3016; doi: 10.1002/jgra.50288.

Cousins, E. D. P., T. Matsuo, A. D. Richmond, and B. J. Anderson (2015), Dominant modes of variability in large-scale Birkeland currents, *J. Geophys. Res. Atmos.*, *120*, 6722–6735; doi: 10.1002/2014JA020462.

Cowley, S. W. H., J. P. Morelli, and M. Lockwood (1991), Dependence of convective flows and particle precipitation in the high-latitude dayside ionosphere on the X and Y components of the interplanetary magnetic field, *J. Geophys. Res.*, *96*, 5557–5564.

Crooker, N. U. (1979), Dayside merging and cusp geometry, *J. Geophys. Res.*, *84*(A3), 951–959; doi: 10.1029/JA084iA03p00951.

Dunlop, M. W., A. Balogh, K.-H. Glassmeier, and P. Robert (2002), Four-point Cluster application of magnetic field analysis tools: The Curlometer, *J. Geophys. Res.*, *107*(A11), 1384; doi: 10.1029/2001JA005088.

Eastman, T. E., E. W. Hones, S. J. Bame, and J. R. Asbridge (1976), The magnetospheric boundary layer: Site of plasma, momentum and energy transfer from the magnetosheath into the magnetosphere, *Geophys. Res. Lett.*, *3*, 685–688; doi:10.1029/GL003i011p00685.

Eastman, T. E., S. A. Boardsen, S.-H. Chen, S. F. Fung, and R. L. Kessel (2000), Configuration of high-latitude and high-altitude boundary layers, *J. Geophys. Res.*, *105*(A10), 23221–23238; doi: 10.1029/1999JA900269.

Eklund, U., R. Lundin, and I. Sandahl (1997), Measurements of O+ in the high latitude magnetosheath, *Phys. Chem. Earth*, *22*, 639–644.

Erlandson, R. E., L. J. Zanetti, T. A. Potemra, P. F. Bythrow, and R. Lundin (1988), IMF B_Y dependence of region 1 birkeland currents near noon, *J. Geophys. Res.*, *93*, 9804–9814.

Escoubet, C. P., M. F. Smith, S. F. Fung, P. C. Anderson, R. A. Hoffman, E. M. Basinska, and, J. M. Bosqued (1992), Staircase ion signature in the polar cusp: A case study, *Geophys. Res. Lett.*, *19*(17), 1735–1738; doi: 10.1029/92GL01806

Escoubet, C. P., M. Fehringer, and M. Goldstein (2001), Introduction: The Cluster mission, *Ann. Geophys.*, *19*, 1197–1200; doi: 10.5194/angeo-19-1197-2001.

Fedorov, A., E. Dubinin, P. Song, E. Budnick, P. Larson, and J.-A. Sauvaud (2000), Characteristics of the exterior cusp for steady southward interplanetary magnetic field: Interball observations, *J. Geophys. Res.*, *105*(A7), 15945–15957; doi: 10.1029/1999JA900482.

Frey, H. U., S. B. Mende, T. J. Immel, S. A. Fuselier, E. S. Claflin, J.-C. Gerard, and B. Hubert (2002), Proton aurora in the cusp, *J. Geophys. Res.*, *107*(A7); doi: 10.1029/2001JA900161.

Friis-Christensen, E., and J. Wilhjelm (1975), Polar cap currents for different directions of the interplanetary magnetic field in the YZ plane, *J. Geophys. Res.*, *80*, 1248–1260.

Friis-Christensen, E., Y. Kamide, A. D. Richmond, and S. Matsushita (1985), Interplanetary magnetic field control of high-latitude electric fields and currents determined from Greenland magnetometer data, *J. Geophys. Res.*, *90*, 1325–1338.

Fujii, R., and T. Iijima (1987), The control of the ionospheric conductivities on large-scale Birkeland current intensities under geomagnetic quiet conditions, *J. Geophys. Res.*, *92*, 4505–4513.

Fukushima, N. (1969), Equivalence in ground magnetic effect of Chapman-Vestine's and Birlekand-Alfven's current systems for polar magnetic storms, *Rep. Ionos. Space Res. Jpn.*, *22*, 219.

Fukushima, N. (1976), Generalized theorem for no ground magnetic effect of vertical currents connected with Pedersen currents in the uniform conducting ionosphere, *Rep. Ionos. Space Res. Jpn.*, *30*(1–2), 35–40.

Fung, S. F., T. E. Eastman, S. A. Boardsen, and S.-H. Chen (1997), High-altitude cusp positions samples by the Hawkewe satellite, Phys. *Chem. Earth*, *22*, 653–662.

Green, D. L., C. L. Waters, B. J. Anderson, H. Korth, and R. J. Barnes (Green) (2006), Comparison of large-scale Birkeland currents determined from Iridium and SuperDARN data, *Ann. Geophys.*, *24*, 941–959; doi: 10.5194/angeo-24-941-2006.

Green, D. L., C. L. Waters, B. J. Anderson, and H. Korth (2009), Seasonal and interplanetary magnetic field dependence of the field-aligned currents for both Northern and Southern Hemispheres, *Ann. Geophys.*, *27*, 1701–1715; doi: 10.5194/angeo-27-1701-2009.

Greenwald, R. A., K. B. Baker, J. R. Dudeney, M. Pinnock, T. B. Jones, E. C. Thomas, J.-P. Villain, et al. (1995), Darn/Superdarn: A global view of the dynamics of high-latitude convection, *Space Sci. Rev.*, *71*(1–4), 761–796; doi: 10.1007/BF00751350.

Gurevich, A. V., A. L. Krylov, and E. E. Tsedilina (1976), Electric fields in the Earth's magnetosphere and ionosphere, *Space Sci. Rev.*, *19*, 59–160; doi: 10.1007/BF00215629.

Haerendel, G., G. Paschmann, N. Sckopke, H. Rosenbauer, and P. C. Hedgecock (1978), The frontside boundary layer of the magnetosphere and the problem of reconnection, *J. Geophys. Res.*, *83*, 3195–3216.

Haraguchi K., H. Kawano, K. Yumoto, S. Ohtani, T. Higuchi, and G. Ueno (2004), Ionospheric conductivity dependence of dayside region-0, 1, and 2 field-aligned current systems: statistical study with DMSP-F7, *Ann. Geophys.*, *22*, 2775–2783.

He, M., J. Vogt, H. Lühr, E. Sorbalo, A. Blagau, G. Le, and G. Lu (2012), A high-resolution model of field-aligned currents through empirical orthogonal functions analysis (MFACE), *Geophys. Res. Lett.*, *39*, L18105; doi: 10.1029/2012GL053168.

Heelis, R. A., W. B. Hanson, and J. L. Burch (1976), Ion convection velocity reversals in the dayside cleft, *J. Geophys. Res.*, *81*(22), 3803–3809; doi: 10.1029/JA081i022p03803.

Heikkila, W. J. (1984), Magnetospheric topology of fields and currents, in *Magnetospheric Currents*, edited by T. A. Potemra, 208–222, American Geophysical Union, Washington, DC.

Iijima, T. (2000), Field-aligned currents in geospace: substance and significance, in *Magnetospheric Current System*, Geophysical Monograph, *118*, edited by S. Ohtani, R. Fujii, M. Hesse, and R. L. Lysak, 107–129, American Geophysical Union, Washington, DC.

Iijima, T., and T. A. Potemra (1976a), The amplitude distribution of field-aligned currents at northern high latitudes observed by Triad, *J. Geophys. Res.*, *81*, 2165.

Iijima, T., and T. A. Potemra (1976b), Field-aligned currents in the dayside cusp observed by Triad, *J. Geophys. Res.*, *81*, 5971–5979.

Iijima, T., and T. Shibaji (1987), Global characteristics of northward IMF-associated (NBZ) field-aligned currents, *J. Geophys. Res.*, *92*, 2408–2424.

Iijima, T., R. Fujii, T. A. Potemra, and N. A. Saflekos (1978), Field-aligned currents in the south polar cusp and their relationship to the Interplanetary magnetic field, *J. Geophys. Res.*, *83*, 5595–5603; doi: 10.1029/JA083iA12p05595.

Iijima, T., T. A. Potemra, and L. J. Zanetti (1990), Large-scale characteristics of magnetospheric equatorial currents, *J. Geophys. Res.*, *95*, 991–999.

Iijima, T., T. A. Potemra, L. J. Zanetti, and P. F. Bythrow (1984), Large-scale Birkeland currents in the dayside polar region during strongly northward IMF: A new Birkeland current system, *J. Geophys. Res.*, *89*, 7441–7452.

Jacobsen, K. S., J. I. Moen, and A. Pedersen (2010), Quasistatic electric field structures and field-aligned currents in the polar cusp region, *J. Geophys. Res.*, *115*, A10226; doi: 10.1029/2010JA015467.

Kamide, Y., A. D. Richmond, and S. Matsushita (1981), Estimation of ionospheric electric fields, ionospheric currents, and field-aligned currents from ground magnetic records, *J. Geophys. Res.*, *86*(A2), 801–813; doi: 10.1029/JA086iA02p00801.

Klupmar, D. M. (1979), Relationships between auroral particle distributions and magnetic field perturbations associated with field-aligned currents, *J. Geophys. Res.*, *84*, 6524–6532.

Knipp, D. J., T. Matsuo, L. Kilcommons, A. Richmond, B. Anderson, H. Korth, R. Redmon, B. Mero, and N. Parrish (2014), Comparison of magnetic perturbation data from LEO satellite constellations: Statistics of DMSP and AMPERE, *Space Weather*, *12*, 2–23; doi:10.1002/2013SW000987.

Korth, H., Y. Zhang, B. J. Anderson, T. Sotirelis, and C. L. Waters (2014), Statistical relationship between large-scale upward field-aligned currents and electron precipitation, *J. Geophys. Res. Space Physics*, *119*, 6715–6731; doi: 10.1002/2014JA019961.

Lavraud, B., A. Fedorov, E. Budnik, M. F. Thomsen, A. Grigoriev, P. J. Cargill, M. W. Dunlop, H. Rème, I. Dandouras, and A. Balogh (2005), High-altitude cusp flow dependence on IMF orientation: A 3-year Cluster statistical study, *J. Geophys. Res.*, *110*, A02209; doi: 10.1029/2004JA010804.

Lavraud, B., T. D. Phan, M. W. Dunlop, M. G. G. T. Taylor, P. J. Cargill, J. M. Bosqued, I. Dandouras, et al. (2004), The exterior cusp and its boundary with the magnetosheath: Cluster multi-event analysis, *Ann. Geophys.*, *22*, 3039–3054; doi: 10.5194/angeo-22-3039-2004.

Lee, L., J. R. Kan, and S.-I. Akasofu (1985), On the origin of cusp field aligned currents, *J. Geophys.*, *157*, 217–221.

Levitin, A. E., R. G. Afonina, B. A. Belov, and YA. I. Feldstein (1982), Geomagnetic variation and field-aligned currents at northern high-latitudes, and their relationship to the solar wind parameters, *Phil. Trans. R. Soc. Lond.*, *A304*, 253–301; doi: 10.1098/rsta.1982.0013.

Lockwood, M. (1994), Ionospheric signatures of pulsed magnetopause reconnection, in *Physical Signatures of Magnetospheric Boundary Layer Process*, edited by J. A. Holtet and A. Egeland, 229–243, Kluwer Academic Publishers, Dordrecht, Netherlands.

Lu, G., A. D. Richmond, B. A. Emery, P. H. Reiff, O. de la Beaujardiere, F. J. Rich, W. F. Denig, et al. (1994), Interhemispheric asymmetry of the high-latitude ionospheric convection pattern, *J. Geophys. Res.*, *99*, 6491–6510.

Lukianova, R., O. Troshichev, Y. Galperin, and N. Jorjio (2001), Determination of FAC patterns in the day-time cusp region with regard for edge effects of current sheets, Internat, *J. Geomag. Aeronomy*, *2*(3), 159–172.

Lundin, R. (1985), Plasma composition and flow characteristics in the magnetospheric boundary layers connected to the polar cusp, in *The Polar Cusp*, edited by J. A. Holtet and A. Egeland, 9–32, Reidel.

Lundin, R., and A. Guglielmi (2006), Ponderomotive forces in Cosmos, *Space Sci. Rev.*, *127*(1), 1–116; doi:10.1007/s11214-006-8314-8.

Lundin, R., D. Winningham, S. Barabash, R. Frahm, M. Holmström, J. A. Sauvaud, et al. (2006), Plasma acceleration above Martian magnetic anomalies, *Science*, *311*(5763), 980–983; dos:10.1126/science.1122071.

Lundin, R., J. Woch, and M. Yamauchi (1991), The present understanding of the cusp, *ESA SP*, *330*, 83–95.

Maezawa, K. (1976), Magnetospheric convection induced by the positive and negative z components of the interplanetary magnetic field: Quantitative analysis using polar cap magnetic records, *J. Geophys. Res.*, *81*, 2289–2303.

Marchaudon, A., J.-C. Cerisier, J.-M. Bosqued, C. J. Owen, A. N. Fazakerley, and A. D. Lahiff (2006), On the structure of field-aligned currents in the mid-altitude cusp, *Ann. Geophys.*, *24*, 3391–3401; doi:10.5194/angeo-24-3391-2006.

Marchaudon, A., J.-C. Cerisier, M. W. Dunlop, F. Pitout, J.-M. Bosqued, and A. N. Fazakerley (2009), Shape, size, velocity and field-aligned currents of dayside plasma injections: A multi-altitude study, *Ann. Geophys.*, *27*, 1251–1266; doi: 10.5194/angeo-27-1251-2009,.

Marklund, G., L. Blomberg, C.-G. Fälthammar, and, P.-A. Lindqvist (1994), On the diverging electric fields associated with black aurora, *Geophys. Res. Lett.*, *21*, 1859–1962.

McDiarmid, I. B., J. R. Burrows, and M. D. Wilson (1978), Magnetic field perturbations in the dayside cleft and their relationship to the IMF, *J. Geophys. Res.*, *83*, 5753–5756.

McDiarmid, I. B., J. R. Burrows, and M. D. Wilson (1979), Large-scale magnetic field perturbations and particle measurements at 1400 km on the dayside, *J. Geophys. Res.*, *84*, 1431–1441.

Merka, J., J. Safrankova, and Z. Nemecek (2002), Cusp-like plasma in high altitudes: A statistical study of the width and location of the cusp from Magion-4, *Ann. Geophys.*, *20*, 311–320; doi:10.5194/angeo-20-311-2002.

Milan, S. E., J. A. Carter, H. Korth, and B. J. Anderson (2015), Principal component analysis of Birkeland currents determined by the Active Magnetosphere and Planetary Electrodynamics Response Experiment, *J. Geophys. Res.*, *120*, 10415 10424; doi: 10.1002/2015JA021680.

Newell, P. T., and C.-I. Meng (1992), Mapping the dayside ionosphere to the magnetosphere according to particle precipitation characteristics, *Geophys. Res. Lett.*, *19*, 609–612.

Newell, P. T., A. R. Lee, K. Liou, S. Ohtani, S. Wing, and M. Hairston (2010), Multisatellite low-altitude observations of a magnetopause merging burst, *J. Geophys. Res.*, *115*, A11204, doi:10.1029/2010JA015438.

Nilsson, H., M. Waara, O. Marghitu, M. Yamauchi, R. Lundin, H. Rème, J.-A. Sauvaud, et al. (2008), Transients in oxygen outflow above the polar cap as observed by the Cluster spacecraft, *Ann. Geophys.*, *26*, 3365–3373.

Nishida, A. (1968), Coherence of geomagnetic DP 2 fluctuations with interplanetary magnetic variations, *J. Geophys. Res.*, *73*(17), 5549–5559; doi: 10.1029/JA073i017p05549.

Norqvist, P., M. André, and M. Tryland (1998), A statistical study of ion energization mechanisms in the auroral region, *J. Geophys. Res.*, *103*, 23459–23474.

Ohtani, S., T. A. Potemra, P. T. Newell, L. J. Zanetti, T. Iijima, M. Watanabe, L. G. Blomberg, et al. (1995a), Four large-scale field-aligned current systems in the dayside high-latitude region, *J. Geophys. Res.*, *100*, 137–153; doi: 10.1029/94JA01744.

Ohtani, S., T. A. Potemra, P. T. Newell, L. J. Zanetti, T. Iijima, M. Watanabe, M. Yamauchi, R. D. Elphinstone, O. de la Beaujardiere, and L. G. Blomberg (1995b), Simultaneous prenoon and postnoon observations of three field-aligned current systems from Viking and DMSP-F7, *J. Geophys. Res.*, *100*, 119–136; doi: 10.1029/94JA02073, 1995.

Papitashvili, V. O., B. A. Belov, D. S. Faermark, Y. I. Feldstein, S. A. Golyshev, L. I. Gromova, and A. E. Levitin (1994), Electric potential patterns in the northern and southern polar regions parameterized by the interplanetary magnetic field, *J. Geophys. Res.*, *99*(A7), 13251–13262; doi:10.1029/94JA00822.

Papitashvili, V. O., F. Christiansen, and T. Neubert (2002), A new model of field-aligned currents derived from high-precision satellite magnetic field data, *Geophys. Res. Lett.*, *29*(14); doi:10.1029/2001GL014207.

Potemra, T. A. (1994), Sources of large-scale Birkeland currents, in *Physical Signatures of Magnetospheric Boundary Layer Process*, edited by J. A. Holtet and A. Egeland, 3–27, Kluwer Academic Publishers, Dordrecht, Netherlands.

Potemra, T. A., L. J. Zanetti, R. E. Erlandson, P. F. Bythrow, G. Gustafsson, M. H. Acuna, and R. Lundin (1987), Observations of large-scale Birkeland currents with Viking, *Geophys. Res. Lett.*, 419–422.

Robert, P., M. W. Dunlop, A. Roux, and G. Chanteur (1998), Accuracy of current density determination, in *Analysis Methods for Multi-Spacecraft Data*, edited by G. Paschmann and P. W. Daly, 395–418, ISSI Sci. Rep., SR-001, Kluwer Acad., Norwell, Massachusetts.

Roederer, J. G. (1977), Global problems in magnetospheric plasma physics and prospects for their solution, *Space. Sci. Rev.*, *21*(1), 23–71; doi: 10.1007/BF00194998.

Sandahl, I., R. Lundin, M. Yamauchi, U. Eklund, J. Safrankova, Z. Nemecek, K. Kudela, et al. (1997), Cusp and boundary layer observations by INTERBALL, *Adv. Space Res.*, *20*, 823–832; doi:10.1016/S0273-1177(97)00515-2.

Sandholt, P. E., C. J. Farrugia, L. F. Burlaga, J. A. Holtet, J. Moen, B. Lybekk, B. Jacobsen, et al. (1994), Cusp/cleft auroral activity in relation to solar wind dynamic pressure, interplanetary magnetic field Bz and By, *J. Geophys. Res.*, *99*, 17323.

Sandholt, P. E., C. S. Deehr, A. Egeland, B. Lybekk, R. Viereck, and G. J. Romick (1986), Signatures in the dayside aurora of plasma transfer from the magnetosheath, *J. Geophys. Res.*, *91*, 10,063–10,079.

Sato, T., and T. Iijima (1979), Primary sources of large-scale Birkeland current, *Space Sci. Rev.*, *24*, 347–366.

Saunders, M.A. (1992), The morphology of dayside Birkeland currents, *J. Atmos Terr. Phys.*, *54*, 457–461.

Savin, S., L. Zelenyi, S. Romanov, I. Sandahl, J. Pickett, E. Amata, et al. (2004), Magnetosheath-cusp interface, *Ann. Geophys.*, *22*, 183–212; doi:10.5194/angeo-22-183-2004.

Sckopke, N., G. Paschmann, H. Rosenbauer, and D. H. Fairfield (1976), Influence of the interplanetary magnetic field on the occurrence and thickness of the plasma mantle, *J. Geophys. Res.*, *81*, 2687–2691.

Siscoe, G. L., W. Lotko, and B. U. Sonnerup (1991), A high-latitude, low-latitude boundary layer model of the convection current system, *J. Geophys. Res.*, *96*(A3), 3487–3495; doi: 10.1029/90JA02362.

Siscoe, G. L., G. M. Erickson, B. U. Ö. Sonnerup, N. C. Maynard, J. A. Schoendorf, K. D. Siebert, D. R. Weimer, W. W. White, and G. R. Wilson (2002), Hill model of transpolar potential saturation: Comparisons with MHD simulations, *J. Geophys. Res.*, *107*(A6), doi:10.1029/2001JA000109, 2002.

Slapak, R., H. Nilsson, and L.-G. Westerberg (2013), A statistical study on O+ flux in the dayside magnetosheath, *Ann. Geophys*, *31*, 1005–1010; doi: 10.5194/angeo-31-1005-2013.

Slapak, R., H. Nilsson, L. G. Westerberg, and R. Larsson (2015), O+ transport in the dayside magnetosheath and its dependence on the IMF direction, *Ann. Geophys.*, *33*, 301–307; doi:10.5194/angeo-33-301-2015.

Smith, M. F., M. Lockwood, and S. W. H. Cowley (1992), The statistical cusp: A flux transfer event model, *Planet. Space Sci.*, *40*(9), 1251–126; doi:10.1016/0032-0633(92)90082-Y.

Southwood, D. J. (1987) The ionospheric signature of flux transfer event, *J. Geophys. Res.*, *92*, 3207.

Spreiter, J. R., and A. L. Summers (1967), On conditions near the neutral points on the magnetosphere boundary, *Planet Space Sci.*, *15*, 787.

Stasiewicz, K. (1991), A global model of gyroviscous field lines merging at the magnetopause, *J. Geophys. Res.*, *96*, 77–86.

Taguchi, S., M. Sugiura, J. D. Winningham, and J. A. Slavin (1993), Characterization of the IMF By-dependent field-aligned currents in the cleft region based on DE 2 observations, *J. Geophys. Res.*, *98*, 1393–1407.

Tanaka, T. (1995), Generation mechanisms for magnetosphere-ionosphere current systems deduced from a three-dimensional MHD simulation of the solar wind-magnetosphere-ionosphere coupling processes, *J. Geophys. Res.*, *100*(A7), 12057–12074; doi:10.1029/95JA00419.

Tanaka, T. (2003), Formation of magnetospheric plasma population regimes coupled with the dynamo process in the convection system, *J. Geophys. Res.*, *108*(A8), 1315; doi: 10.1029/2002JA009668.

Tanaka, T., M. Watanabe, M. Den, S. Fujita, Y. Ebihara, T. Kikuchi, K. K. Hashimoto, and R. Kataoka (2016), Generation of field-aligned current (FAC) and convection through the formation of pressure regimes: Correction for the concept of Dungey's convection, *J. Geophys. Res. Space Physics*, *121*, 8695–8711; doi: 10.1002/2016JA022822.

Tsyganenko, N. A. (1987), Global quantitative models of the geomagnetic field in the cislunar magnetosphere for different disturbance levels, *Planet. Space Sci.*, *35*, 1347–1358; doi: 10.1016/0032-0633(87)90046-8.

Tsyganenko, N. A. (2002), A model of the magnetosphere with a dawn-dusk asymmetry, 1, Mathematical structure, *J. Geophys. Res.*, *107*(A8); doi: 10.1029/2001JA000219.

Vasyliunas, V. M. (1970), Mathematical models of magnetospheric convection and its coupling to the ionosphere, in *Particles and Fields in the Magnetosphere*, edited by B. McCormac, 60, D. Reidel, Hingham, Massachusetts.

Vasyliunas, V. M. (1995), Multiple-branch model of the open magnetopause, *Geophys. Res. Lets.*, *22*, 1145–1147; doi: 10.1029/95GL01068.

Vasyliunas, V. M. (2001), Electric field and plasma flow: What drives what?, *Geophys. Res. Lett.*, *28*(11), 2177–2180; doi: 10.1029/2001GL013014.

Vennerstrom, S., T. Moretto, L. Rastätter, and J. Raeder (2005), Field-aligned currents during northward interplanetary magnetic field: Morphology and causes, *J. Geophys. Res.*, *110*, A06205; doi: 10.1029/2004JA010802.

Vennerstrom, S., T. Moretto, N. Olsen, E. Friis-Christensen, A. M. Stampe, and J. F. Watermann (2002), Field-aligned currents in the dayside cusp and polar cap region during northward IMF, *J. Geophys. Res.*, *107*(A8); doi: 10.1029/2001JA009162.

Watanabe, M., S. Sakito, T. Tanaka, H. Shinagawa, and K. T. Murata (2014), Global MHD modeling of ionospheric convection and field-aligned currents associated with IMF By triggered theta auroras, *J. Geophys. Res. Space Physics*, *119*, 6145–6166; doi: 10.1002/2013JA019480.

Weimer, D. R. (2005), Improved ionospheric electrodynamic models and application to calculating Joule heating rates, *J. Geophys. Res.*, *110*, A05306; doi: 10.1029/2004JA010884.

Wilhjelm, J., E. Friis-Christensen, and T. A. Potemra (1978), The relationship between ionospheric and field-aligned currents in the dayside cusp, *J. Geophys. Res.*, *83*, 5586–5594.

Wing, S., D. H. Fairfield, J. R. Johnson, and Shin-I. Ohtani (2015), On the field-aligned electric field in the polar cap, *Geophys. Res. Lett.*, *42*, 5090–5099; doi:10.1002/2015GL064229.

Wing, S., S. Ohtani, P. T. Newell, T. Higuchi, G. Ueno, and J. M. Weygand (2010), Dayside field-aligned current source regions, *J. Geophys. Res.*, *115*, A12215; doi: 10.1029/2010JA015837.

Woch, J., and R. Lundin (1991), Temporal magnetosheath plasma injection observed with Viking: a case study, *Ann. Geophys.*, *9*, 133–142.

Woch, J., and R. Lundin (1992), Magnetosheath plasma precipitation in the polar cusp and its control by the interplanetary magnetic field, *J. Geophys. Res.*, *97*, 1421–1430.
Woch, J., M. Yamauchi, R. Lundin, T. A. Potemra, and L. J. Zanetti (1993), The low-latitude boundary layer at mid-altitudes: Relation to large-scale Birkeland currents, *Geophys. Res. Lett.*, *20*, 2251–2254.
Wolf, R. A., and R. W. Spiro (1985), Particle behavior in the magnetosphere, in *Computer Simulation of Space Plasma*, edited by H. Matsumoto and T. Sato, 227–254, D. Reidel, Hingham, Massachusetts.
Yamauchi, M. (1994), Numerical simulation of large-scale field-aligned current generation from finite-amplitude magnetosonic waves, *Geophys. Res. Lett. 21*, 851–854.
Yamauchi, M., and R. Lundin (1992), A new cusp model and its Viking observation (in Japanese with English abstract), in *3rd STEP Report*, edited by H. Oya, 666–672, Sendai, Japan.
Yamauchi, M., and R. Lundin (1994), Classification of large-scale and meso-scale ion dispersion patterns observed by Viking over the cusp-mantle region, in *Physical Signatures of Magnetospheric Boundary Layer Process*, edited by J. A. Holtet, and A. Egeland, 99–109, Kluwer Academic Publishers, Dordrecht, Netherlands; doi: 10.1007/978-94-011-1052-5_7.
Yamauchi, M., and R. Lundin (1997), The wave-assisted cusp model: Comparison to low-altitude observations, *Phys. Chem. Earth*, *22*, 729–734; doi: 10.1016/S0079-1946(97)00203-6.
Yamauchi, M., and R. Lundin (2001), Comparison of various cusp models with high- and low-resolution observations, *Space Sci., Rev.*, *95*, 457–468; doi: 10.1023/A:1005237630306.
Yamauchi, M., and T. Araki (1989), The interplanetary magnetic field BY-dependent field-aligned current in the dayside polar cap under quiet conditions, *J. Geophys. Res.*, *94*, 2684–2690; doi: 10.1029/JA094iA03p02684.
Yamauchi, M., L. Andersson, P.-A. Lindqvist, S. Ohtani, J. H. Clemmons, J.-E. Wahlund, L. Eliasson, and R. Lundin (2001), Acceleration signatures in the dayside boundary layer and the cusp, *Phys. Chem. Earth*, *26*, 195–200; doi: 10.1016/S1464-1917(00)00107-0.
Yamauchi, M., R. Lundin, and A. T. Y. Lui (1993a), Vorticity equation for MHD fast waves in geospace environment, *J. Geophys. Res.*, *98*, 13523–13528; doi: 10.1029/93JA00638.
Yamauchi, M., R. Lundin, and J. Woch (1993b), The interplanetary magnetic field By effects on large-scale field-aligned currents near local noon: Contributions from cusp part and noncusp part, *J. Geophys. Res.*, *98*, 5761–5767, doi: 10.1029/92JA02934.
Yamauchi, M., R. Lundin, and T. Potemra (1995), Dynamic response of the cusp morphology to the IMF changes: An example observed by Viking, *J. Geophys. Res.*, *100*, 7661–7670; doi: 10.1029/95JA00333.
Yamauchi, M., R. Lundin, L. Eliasson, S. Ohtani, and J. H. Clemmons (1998), Relationship between large-, meso, and small-scale field-aligned currents and their current carriers, in Polar Cap Boundary Phenomena, edited by J. Moens, et al., Kluwer Acad. Pub., 173–188; doi: 10.1007/978-94-011-5214-3_14.
Yamauchi, M., R. Lundin, L. Eliasson, S. Ohtani, P.-A. Lindqvist, and R. P. Lepping (2000), Independency of the dayside field-aligned current system: a restriction to cusp models, in *Magnetospheric Current System*, Geophysical Monograph, *118*, edited by S. Ohtani, R. Fujii, M. Hesse, and R. L. Lysak, 245–252, American Geophysical Union, Washington, DC; doi: 10.1029/GM118p0245.
Yamauchi, M., R. Lundin, O. Norberg, I. Sandahl., L. Eliasson, and D. Winningham (2003), Signature of direct magnetosheath plasma injections onto closed field-line regions based on observations at mid- and low-altitudes, in *Earth's Low-Latitude Boundary Layer*, Geophysical Monograph, *133*, edited by P. Newell and T. Onsager, 179–188, American Geophysical Union, Washington, DC; doi: 10.1029/133GM18.
Yau, A. W., and M. André (1997), Sources of ion outflow in the high latitude ionosphere, *Space Sci. Rev.*, *80*, 1–25.
Zhang, Q.-H., M. W. Dunlop, R. Holme, and E. E. Woodfield (2010), Comparison of eight years magnetic field data from Cluster with Tsyganenko models in the inner magnetosphere, *Ann. Geophys.*, *28*, 309–326; doi: 10.5194/angeo-28-309-2010.
Zhou, X. W., C. T. Russell, G. Le, S. A. Fuselier, and J. D. Scudder (2000), Solar wind control of the polar cusp at high altitude, *J. Geophys. Res.*, *105*, 245–251.

21

Magnetospheric and Atmospheric Controls of Giant Planet Auroral Currents

L. C. Ray

ABSTRACT

Planetary atmospheres and their surrounding plasma environments exchange energy and momentum through magnetosphere-ionosphere-thermosphere (MIT) coupling. This fundamental process is mediated by the planetary magnetic field, which connects the ionosphere, the ionized part of the atmosphere, to the magnetospheric plasma. The ionosphere, in turn, is collisionally coupled to the thermosphere, the upper portion of the neutral atmosphere. Momentum and energy are communicated through magnetic fields and electric fields generated by plasma flows, wave-particle interactions, varying gradients in the magnetic field. Associated currents are generated that flow along the magnetic field, diverging perpendicularly to the magnetic field in the magnetospheric equator and ionosphere to convey $\mathbf{J} \times \mathbf{B}$ forces in the local plasma populations. An observable consequence of MIT coupling is the aurora, which has been observed at every magnetized planet with an atmosphere. While the presence of aurora is ubiquitous, the phenomenology of the emissions is system dependent, detailing differences in the currents at work. We discuss auroral currents at the giant planets and how the planetary magnetospheres and atmospheres contribute to and control the current systems.

21.1. INTRODUCTION

A planet's thermosphere, ionosphere, and magnetosphere form a complex coupled system where energy and angular momentum is communicated between the planet and its surrounding plasma environment. Within the atmosphere, ion-neutral collisions couple the ionosphere and neutral thermosphere. Between the ionosphere and magnetosphere, energy and angular momentum are exchanged via electromagnetic forces that are associated with electric currents. The currents that result from this global system, along with magnetic field-aligned acceleration regions, result in particle precipitation into the underlying atmosphere, which generates bright auroral emissions. Planetary auroral emissions are a powerful tool that can be used as a diagnostic of the magnetospheric and atmospheric dynamics. In the absence of in situ spacecraft measurements, auroral observations provide much valued insight into the magnetospheric and atmosphere drivers present in a system. In conjunction with in situ data, auroral observations are evidence of the atmospheric consequences of processes measured in the magnetosphere and provide a global system view to complement local measurements. Therefore, it is critical to understand the physics behind auroral currents and their magnetospheric and atmospheric controls.

21.1.1. A Brief Review of Selected Solar System Auroral Emissions

Earth's aurora has been extensively investigated, both through in situ measurements, and remote and ground-based observations (see, e.g., *Akasofu* [2012]; *Henderson*

Department of Physics, Lancaster University, Lancaster, UK

Electric Currents in Geospace and Beyond, Geophysical Monograph 235, First Edition.
Edited by Andreas Keiling, Octav Marghitu, and Michael Wheatland.

[2012]). Ultimately, the emission is driven by the interaction of the solar wind and magnetosphere. Magnetic flux is opened through reconnection and transported from the dayside of the planet to the nightside [*Dungey*, 1961]. Flux is closed in the magnetotail, most dramatically during substorms, in which magnetic energy is rapidly converted into particle energy in the nightside magnetosphere [*Akasofu*, 1964; *McPherron et al.*, 1973]. The rapid dipolarization of the magnetic field during substorms generates currents that trigger bright auroral emissions.

The two dominant magnetosphere-ionosphere coupling current systems at Earth, Region 1 and Region 2, are fixed in the Sun-Earth inertial frame (see *Cowley* [2000] and *Khurana and Liu*, Chapter 2, this volume). Through these two current systems, information is conveyed between the atmosphere and magnetosphere. Region 1 currents flow into the atmosphere at the open-closed field line boundary on the dawn side of the planet, diverge in the ionosphere, with some current flowing toward low latitudes and the remainder crossing the polar cap to flow out along the field at the dusk open-closed field-line boundary and close through magnetopause currents. Region 2 currents sit at lower ionospheric latitudes, coupling to the partial ring current in the magnetosphere. Current travels outward along field lines at dawn, azimuthally in the magnetosphere to form a partial ring current, returns back toward the planet along field lines at dusk, and closes poleward through the auroral ionosphere. *Coxon et al.* [Chapter 16, this volume] review Earth's Birkeland currents in more detail.

Region 1 and Region 2 currents also exist at the outer planets, although they are oriented in the opposite sense due to the reverse in dipole field orientation at Jupiter and Saturn [*Khurana*, 2001; *Walker and Ogino*, 2003]. Fixed in the inertial frame, the Birkeland currents penetrate deep into the giant planet magnetospheres. Region 1 and 2 currents contribute to MIT coupling at the giant planets, in conjunction with corotation-driven radial currents. *Khurana* [2001] determined that the Region 2 currents have a similar magnitude to radial corotation currents. *Bonfond et al.* [2015] suggested that dawn-dusk auroral emission asymmetries could be attributed to a partial ring current in the magnetosphere. Modeling efforts to reproduce the auroral currents with corotation current alone fail to reproduce both the observed auroral intensity and magnetospheric plasma rotation profile, lending further support to the importance of a partial ring current in the magnetosphere [*Ray et al.*, 2014].

Jupiter's main emission is composed of three main components: satellite footprints, a main oval of emission, and a highly variable polar emission. Io's satellite-driven aurora is composed of an Alfvénic spot and a tail of emission that can extend as much as 120° around the planet (see *Bonfond* [2012]). Europa and Ganymede also generate Alfvénic spots. Transient auroral features driven by injections of plasma from the outer magnetosphere and/or magnetotail into the inner magnetosphere exist at low latitudes between the main emission and Io footprint [*Gray et al.*, 2016; *Kimura et al.*, 2015; *Mauk et al.*, 2002; *Radioti et al.*, 2009]. The main emission is poleward of the satellite aurora and transient emissions, and fixed in the frame rotating with the planet (see *Clarke* [2012]). While there is variability in Jupiter's main auroral emission with solar wind conditions (see *Delamere et al.* [2015a]), the dominant driving mechanisms are internal to the magnetosphere in contrast to the terrestrial system.

Saturn's auroral emissions respond both to the solar wind forcing and internal stresses. The location of the main emission shifts poleward and, morphologically, it changes from a ring to a spiral at times of high solar wind dynamic pressure [*Clarke et al.*, 2009]. Independent from solar wind effects, the dawn emission is typically brighter than the dusk emission (see *Badman et al.* [2015]). Unlike at Jupiter, Saturn's satellites do not consistently generate an auroral signature, with the Enceladus spot only observed three times over the course of the Cassini mission [*Pryor et al.*, 2011] despite numerous flybys. *Wannawichian et al.* [2008] determined that the currents associated with the Enceladus interaction are too weak to generate an auroral emission above the Hubble Space Telescope detectability threshold.

Both planets have polar auroral emissions, which are highly variable. The driving mechanism for polar emission remains an open question, although possible processes include internally driven tail reconnection [e.g., *Radioti et al.*, 2008], lobe reconnection [e.g, *Bunce et al.*, 2005], and magnetic flux circulation in the outer magnetosphere [*Delamere et al.*, 2015b].

Jupiter's and Saturn's auroral morphology indicate the solar wind driven current system has a secondary role to currents driven by rotational forces and dynamics triggered by internal plasma sources. Both giant planets have moons, which act as substantial plasma sources within the magnetosphere. Therefore, the dominant current system is related to the transfer angular momentum from the planet to the surrounding plasma disc. We note that other field-aligned current systems exist at Jupiter and Saturn, such as the Region 1 and 2 currents [*Khurana*, 2001; *Walker and Ogino*, 2003], Alfvénic currents driven by processes such as the motion of satellites through the magnetospheric plasma (see *Hess and Delamere* [2012]), currents associated with the loss of plasma down the planetary magnetotail [e.g., *Jia and Kivelson*, 2012], currents related to the radial transport of mass through the plasmadisc [e.g *Goldstein et al.*, 2014; *Liu et al.*, 2010] and plasma injections [e.g., *Mauk et al.*, 1997], and currents associated with the interaction of the solar wind with the magnetosphere [e.g., *Fukazawa et al.*, 2012; *Walker and*

Ogino, 2003]. A thorough discussion of these current systems is beyond the scope of this review chapter and the reader is referred to the listed references for more details.

Here, we focus on the quasi-static field-aligned currents associated with MIT coupling driven by the radial transport of angular momentum through giant planet magnetospheres. Despite differences in the systems, the same factors exert control over quasi-static auroral currents: the ionospheric Pedersen conductance, thermospheric angular velocity, high-latitude field-aligned potentials, and the planetary magnetic field (a more prominent control at Jupiter and Saturn). Below, we describe the components of giant planet MIT coupled systems, their auroral currents, and how the aforementioned four factors influence their development.

21.1.2. Giant Planet Plasma Discs

At Jupiter, the Galilean moon, Io, orbits at a distance of 5.9 R_J outgassing SO_2 based neutral material into the magnetosphere. While there is significant temporal variability in volcanic activity, *Delamere et al.* [2005] found a neutral source rate of 700–1200 kg/s to be consistent with measurements taken during the Cassini flyby of Jupiter. The neutral material forms a torus at the orbital radius of Io where it is ionized through a combination of electron impact ionization and charge exchange. Approximately half of the original outgassed neutral material is removed from the system by fast neutral escape (see *Thomas et al.* [2004], *Bagenal and Delamere* [2011], and *Achilleos et al.* [2015]). Thus, assuming the source rates from the Cassini era, 350–600 kg/s of ionized plasma remains to be transported outward through the magnetosphere. The transport time for plasma to move radially from 6 R_J to 10 R_J is ~14–60 days, beyond which point theoretical models suggest it rapidly spirals outward before exiting the system down the magnetotail [*Bagenal and Delamere*, 2011; *Copper et al.*, 2016; *Delamere and Bagenal*, 2010].

Within the Saturnian magnetosphere, Enceladus orbits at a radial distance of 4 R_S, ejecting from 3 to 1600 kg/s water-based neutral material into the magnetosphere [*Burger et al.*, 2007; *Hansen et al.*, 2006; *Saur et al.*, 2008; *Sittler et al.*, 2008; *Smith et al.*, 2010; *Tokar et al.*, 2006; *Waite et al.*, 2006]. Enceladus's location in the magnetosphere leads to a higher neutral to plasma ratio than in the Jovian system [*Delamere et al.*, 2007]. Unlike Jupiter, where charge exchange primarily occurs in the Io torus, ionization occurs over a broader radial range at Saturn. *Fleshman et al.* [2013] showed that charge exchange occurs from 4 R_S to 10 R_S, with the peak in the water-group ion profiles occurring between 4 and 6 R_S. Ionization through charge exchange dominates over photoionization and electron impact ionization [*Fleshman et al.*, 2010]. Plasma takes approximately 30 days to travel to 10 R_S, beyond which point ~60 kg/s [*Fleshman et al.*, 2013] is available for radial transport and eventual loss down the magnetotail.

In both systems, the presence of an internal plasma source acts to inflate the magnetosphere beyond the size predicted by simply balancing the planetary magnetic field pressure and solar wind ram pressure. In an "empty" system, the magnetic pressure from Jupiter's magnetic field would balance the solar wind ram pressure at ~43 R_J. With the Iogenic plasma source, the magnetopause standoff distance exhibits a bimodal distribution with peaks at 63 R_J and 92 R_J [*Joy et al.*, 2002]. A similar distribution exists in the Saturnian system, with observed subsolar distances of 22 or 27 R_S [*Achilleos et al.*, 2008], rather than the ~19 R_S predicted in the absence of an internal plasma source. The bimodality of the magnetopause standoff distance is not only dependent on the conditions of the local solar wind (compressions vs. expansions), but also on the plasma conditions within the magnetosphere [*Pilkington et al.*, 2015].

21.1.3. The Role of Currents

Jupiter's and Saturn's internal magnetospheric plasma populations are collisionless and therefore frozen in to the planetary magnetic field. Any deviations in the plasma angular velocity from corotation with the planet generate a current system, which acts to accelerate the plasma back toward corotation. Broadly speaking, there are two mechanisms that cause deviations between the plasma angular velocity and corotation velocity in the inner and middle magnetospheres of Jupiter and Saturn: local pickup [*Ergun et al.*, 2009; *Hill and Vasyliunas*, 2002; *Pontius and Hill*, 1982] and radial transport [*Cowley and Bunce*, 2001, 2003; *Hill*, 1979; *Nichols*, 2011; *Nichols and Cowley*, 2004; *Ray et al.*, 2010, 2012a, 2014; *Smith and Aylward*, 2009]. Local pickup can be described as momentum- and/or mass-loading. Momentum-loading occurs when neutrals orbiting the planet at the Keplerian velocities are ionized and plasma is created that lags the plasma rotation rate. Mass-loading occurs when neutrals are ionized generating additional plasma, thus increasing the plasma density of the magnetodisc and decreasing the local plasma rotation rate. Charge exchange, for example, momentum-loads the system, but does not mass-load it as the plasma density remains constant.

In both cases, the new ions must be accelerated from Keplerian velocity to that of the planetary rotation rate, $\Omega_{Jupiter} = 1.76 \times 10^{-4}$ rad s^{-1} or $\Omega_{Saturn} \approx 1.638 \times 10^{-4}$ rad s^{-1}, for Jupiter or Saturn, respectively. The relative velocity between corotation and Keplerian velocity is −57 km/s at Io's orbit and −26.4 km/s at Enceladus. Thus, new plasma, n_p, with the Keplerian velocity, v_{Kep}, decreases the bulk flow speed, v_{cor}, of the local plasma population, n.

This reduction in flow speed generates a motional electric field [e.g., *Vasyliunas*, 2016]

$$E = -\left(n_p v_{Kep} + n v_{corot.}\right) / \left(n_p + n\right) \times B \qquad (21.1)$$

which, in the corotating frame, points radially in toward the planet. Treating the magnetic field lines as equipotentials, the magnetospheric electric field maps to an equatorward pointing latitudinal ionospheric electric field. Via Ohm's law, the ionospheric electric field is related to latitudinal Pedersen currents, which diverge to generate upward field-aligned current, that is, electrons precipitate into atmosphere, which flows radially outward in the magnetosphere, and closes back along the magnetic field line. The radial magnetospheric current has an associated **J** × **B** force that acts to accelerate the subcorotating plasma from the Keplerian velocity to that of planetary corotation. The magnetospheric outer boundary of the pickup is determined by the location where ionization has decreased enough that the plasma is corotating with the planet. At Jupiter, the upward portion of this circuit generates the Io wake emission. Saturn's ionization is extended over a broader radial range, and hence a similar auroral signature does not exist.

Both Io and Enceladus sit in the inner magnetospheres of their respective systems. Plasma is transported radially outward through the magnetosphere by a rotationally driven centrifugal interchange instability [e.g., *Hill*, 2016; *Hill et al.*, 2005; *Mauk et al.*, 1997; *Thomsen et al.*, 2014] before eventually being lost down the magnetotail. In the absence of tangential forces, the angular velocity of a particle would decrease as it moves outward from the gravitational center due to conservation of angular momentum, $m\mathbf{v} \times \mathbf{r} = 0$. However, in the collisionless MHD limit, the magnetospheric plasma is frozen in to the magnetic field. Per Ampère's law, the bend-back in the magnetic field generated by the slowing plasma is linked to currents ($\nabla \times \mathbf{B} = \mu_0 \mathbf{J}$). Similar to the local pickup case, these currents run equatorward through the ionosphere, upward along magnetic field lines, radially outward through the magnetosphere, and close in the outer magnetosphere. Unlike in the local pickup case, the radial extent of the upward current region is not clearly bounded by a region of corotating flow and the location of the downward currents in the Jovian and Saturnian magnetospheres is not well defined. There is also a ring current [*Achilleos*, Chapter 8, this volume; *Sergis et al.*, Chapter 9, this volume] and current sheet [*Arridge and Martin*, Chapter 12, this volume] at each system and the interested reader is referred to the relevant review chapters within this volume.

Hill [1979] first described the current system associated with radial transport of magnetospheric plasma. His seminal model assumed a dipole magnetic field, uniform ionospheric Pedersen conductance, and equipotential field lines. Under these simplifying assumptions, the location of corotation breakdown within a planetary magnetosphere would occur at a distance of

$$L_0 = \left(\frac{\pi \Sigma_{Ped.} R_P^2 B_P^2}{M_{dot}}\right)^{\frac{1}{4}} \qquad (21.2)$$

where L_0 is in units of planetary radii, Σ_{Ped} is the ionospheric Pedersen conductance in mho, R_p is the radius of the planet, B_p is the planetary magnetic field strength, and M_{dot} is the radial mass transport rate. *Hill* [2001] and *Cowley and Bunce* [2001] linked the angular momentum currents to those driving the main auroral emission for the Jovian system. The simplified description of auroral currents, however, did not simultaneously match all of the measured constraints in the system: observed corotation breakdown location, width of auroral emission at the planet, estimated auroral precipitating auroral energy flux and precipitation energy, as in Table 21.1.

Subsequent descriptions of giant planet auroral currents have been modified to include variable ionospheric Pedersen conductance, feedback from changes in the thermospheric angular velocity, deviations in the magnetic field profile from a dipole, and the rotational decoupling from parallel electric fields. Figure 21.1 shows the net currents associated with radial transport. In addition to these broad net current structures, there are filamentary current structures. *Mauk and Saur* [2007] measured spatially and temporally structured upward and downward field-aligned electron beams in Jupiter's middle magnetosphere using Galileo EPD data. However, the exact nature of the fine-scale structure remains an open question.

Table 21.1 Summary of Jovian and Saturnian System Parameters Associated With the Auroral Current Systems.

	Saturn	Jupiter
Neutral chemistry base	H_2O	SO_2
Neutral source rate (kg s^{-1})	3–1600	700–3000
Mass-loading rates (kg s^{-1})	10–100	350–1500
Predicted corotation breakdown location, L_0 [*Hill*, 1979]	18–29 R_S	20–25 R_J
L0 / $R_{Magnetopause}$	0.72–1.16	0.24–0.30
Observed breakdown location	8–10 R_S*	~20 R_J†
Magnetospheric location of auroral currents	> ~12 R_S‡	15–50 R_J††**
Ionospheric Pedersen conductance	~2–20 mho	0.01–2 mho

* *Thomsen et al.* [2010];
† *Bagenal et al.* [2016];
‡ *Bunce et al.* [2008];
** *Vogt et al.* [2011];
†† *Bonfond et al.* [2012].

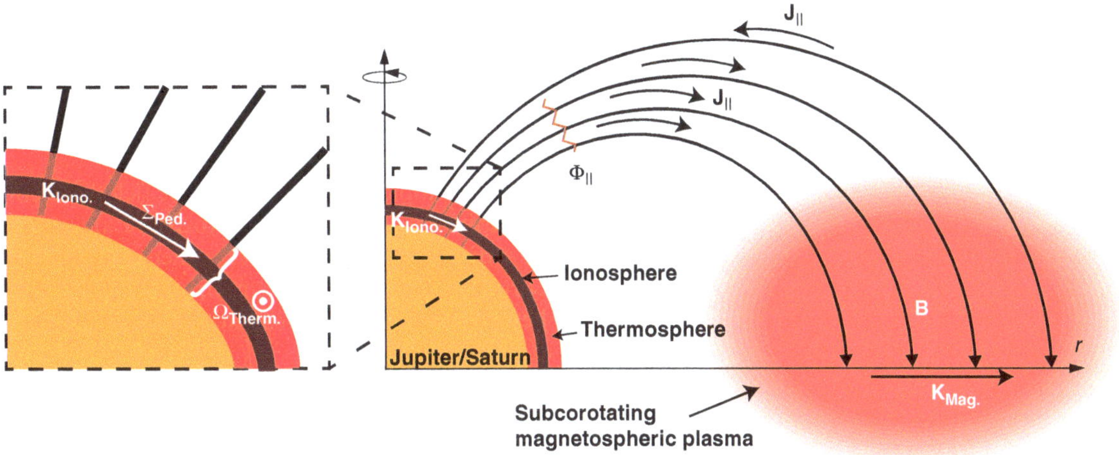

Figure 21.1 Current system associated with the radial transport of angular momentum. The solid black lines show the planetary magnetic field, **B**. The thermosphere is shown in red, with the coincident ionosphere in maroon. $K_{Iono.}$ and $K_{Mag.}$ denote the ionospheric and magnetospheric height-integrated currents, respectively, $\Sigma_{Ped.}$ is the ionospheric Pedersen conductance, $\Omega_{Therm.}$ is the thermospheric angular velocity, which is out of the page if subcorotating (in the corotating frame), $\Phi_{||}$ and the jagged orange line denote the field-aligned potentials. The curved arrows mark the direction of the field-aligned currents that couple the atmosphere and magnetosphere.

21.2. LIMITATIONS TO M-I COUPLING

21.2.1. Thermospheric Flows

Ultimately, angular momentum is conveyed from the deep interior of the planet to the magnetospheric plasma. Yet, the currents that transfer angular momentum between the planet and the magnetospheric plasma run through the ionosphere, which is confined to a narrow altitude range. The magnetospheric plasma "sees" the angular velocity of the thermospheric layers coincident with the ionosphere as the rotation rate of the planet. The auroral currents modify the dynamics of the ionosphere and thermosphere through ion drag forces and joule heating. Over time, if angular momentum cannot be transferred quickly enough from the lower atmosphere to the thermosphere, these effects can significantly alter the rotation of the dense, neutral thermosphere despite its having much more inertia than the ionosphere and magnetosphere. For example, the anticorotational **J** × **B** forces associated with the currents that generate the Io tail aurora are estimated to introduce deviations of up to 5% subcorotation in Jupiter's thermosphere [*Pontius*, 1995]. *Huang and Hill* [1989] suggested that angular momentum is transferred vertically through viscous forces. *Smith and Aylward* [2008] showed that meridional advection is a more effective means of transport.

Modifications to the neutral atmosphere's angular velocity affect the maximum angular momentum that can be conveyed. The magnitude of the ionospheric current is proportional to the difference in the bulk velocities of the ionospheric plasma and neutral thermosphere. Although any reference frame may be used to describe the ionospheric electric field, such as the plasma reference frame [e.g., *Song et al.*, 2001] or a weighted reference frame [e.g., *Vasyliunas and Song*, 2005], the large inertia of the giant planet thermosphere and long timescales necessary to modify the underlying atmospheric flows [*Pontius*, 1995] indicate that the rest frame of the neutral thermosphere is an appropriate choice at Jupiter and Saturn. Modifications to the thermospheric angular velocity affect the Pedersen conductance, which maximizes where the ion-neutral collision frequency is equal to the ion gyrofrequency. *Smith and Aylward* [2009] and *Tao et al.* [2009] showed that the self-consistently including thermospheric feedback in the Jovian MIT coupling system alters the magnitude and location of the auroral currents.

At Saturn, the thermospheric contribution to MIT coupling currents has been invoked as a possible explanation for Saturn's variable planetary-period magnetic field oscillations (PPO) [e.g., *Smith*, 2011]. Differences in the ionospheric rotation rate between the two planetary hemispheres may generate interhemispheric currents [e.g., *Jia et al.*, 2012; *Southwood and Kivelson*, 2007]. Subsequent seasonal or magnetospherically driven modifications to the ionospheric and thermospheric rotation rates would then alter the current system, changing the measured PPO signature.

21.2.2. Pedersen Conductance

The ionospheric Pedersen conductance, or height integrated Pedersen conductivity, controls the magnitude of the ionospheric currents for a given plasma-neutral velocity difference. Auroral electrons, the precipitating component of the upward current system, ionize the thermosphere. The degree of atmospheric ionization depends on the energy intensity profile and angle of the incoming population, but in general, auroral precipitation leads to increased densities of ionospheric H^+ and H_3^+ [e.g., *Galand et al.*, 2011; *Grodent et al.*, 2001] and an elevated Pedersen conductance [*Bougher et al.*, 2005; *Millward et al.*, 2002; *Mooire et al.*, 2010; *Ray et al.*, 2012b; *Tao et al.*, 2010]. The peak in the conductivity profile with altitude exists where the ion-neutral collision frequency is equal to the ion gyro-frequency. Therefore, the energy of the precipitating electrons strongly controls the enhancement. Large energy fluxes of low energy particles will not significantly modify Σ_{Ped}. Similarly, precipitating particles with high energies greater than ~60–80 keV [*Millward et al.*, 2002] will deposit their energy too low in the atmosphere, where newly generated ions quickly recombine into neutrals.

Increases in Σ_{Ped} enhance the field-aligned currents that couple the atmosphere and magnetosphere and enable more efficient angular momentum transfer. *Nichols and Cowley* [2004] first explored this feedback at Jupiter by modifying the Pedersen conductance with the field-aligned current density, $\Sigma_{Ped} = f(j_{\|})$. *Ray et al.* [2010] later derived a description for $\Sigma_{Ped} = f(EF, \Phi_{\|})$ that explored how enhancements with both energy flux and precipitation energy modified the auroral current system, capturing the decrease in Σ_{Ped} at large precipitation energies. Both studies used the *Millward et al.* [2002] Jovian ionospheric model to derive their Σ_{Ped} parameterizations and found that the Pedersen conductance enhancements lead to an increase in auroral currents and angular momentum transfer, with the magnetospheric plasma corotating out to larger equatorial distances.

At Saturn, the effect of the Pedersen conductance on the MIT coupling system has been primarily studied through "reverse" studies, which impose a magnetospheric angular velocity profile and then determine Σ_P [e.g., *Cowley and Bunce*, 2003; *Pontius and Hill*, 2009; *Saur et al.*, 2004]. These studies have predicted values of effective ionospheric conductance ranging from ~0.01 to 1 mho, which are scaled to simplify the reduction of the conductance in the inertial or corotating frames due to the presence of thermospheric subcorotation. Interestingly, this is smaller than the conductance predicted by ionospheric models, suggesting that the simplified description of radial currents does not adequately describe the Saturnian system. One possible culprit for this discrepancy is the large presence of dust within the Saturnian magnetosphere [*Sakai et al.*, 2013].

21.2.3. Field-Aligned Potentials

In the simplest case, the magnetic field lines connecting the atmosphere to the magnetosphere are considered to be equipotentials. However, when the magnitude of current necessary to communicate angular momentum between the two regions exceeds the electron thermal current density, $j_{\|_{therm}} = ne\sqrt{\left(\frac{T}{2\pi m_s}\right)}$, field-aligned potentials develop, shifting the electron distribution into the loss-cone. This increases the number of current carriers that can reach the ionosphere, allowing the current density to grow. If the magnitude of the field-aligned potential is such that the entire electron distribution is within the loss cone, then further increases in the potential magnitude will no longer modify the current density and the current density is saturated.

This current-voltage relationship between was first described by *Knight* [1973] and depends on the density and energy of electrons at the top of the acceleration region, and the magnetic mirror ratio, R_M between the top of the acceleration region and planet. In the limit that the ratio of the energy gained by the precipitating electrons to their initial energy is much smaller than the magnetic mirror ratio, $1 \ll \frac{e\Phi_{\|}}{k_B T} \ll R_M$, the current-voltage relation can be simplified to a linear relation dependent only on the density and energy of the electrons at the top of the acceleration region [*Lyons*, 1980].

At Jupiter and Saturn, the rapid rotation rates of the planets coupled with an internal, heavy plasma source lead to a field-aligned density profile confined to the magnetospheric centrifugal equator. Sulfur dioxide-based and water group ions are confined to the magnetospheric plane, while more mobile electrons can reach higher latitudes. The charge separation generates an ambipolar electric field that impedes the motion of the electrons toward higher latitudes. At the atmosphere, the ionospheric population is confined by strong gravitational forces. Therefore, there is a depletion in the plasma density at high magnetic latitudes, which is predicted to maximize at the minimum in the sum of the gravitational and centrifugal potentials [*Ray et al.*, 2009; *Su et al.*, 2003]. Thus, the full *Knight* [1973] current-voltage relation must be considered at the giant planet systems.

When a significant radial, or equivalently latitudinal, gradient exists in the field-aligned potential strength, the rotation rates of the ionosphere and magnetosphere are decoupled. *Ray et al.* [2010] showed that, at Jupiter, self-consistently including the rotational decoupling between the ionosphere and magnetosphere due to field-aligned potentials alters the auroral currents and magnetospheric plasma flow. Where the radial gradient in the field-aligned potential is positive, the magnitude of the ionospheric electric field exceeds that of the magnetospheric field, with the opposite scenario for a negative $\frac{d\Phi_{\parallel}}{dr}$. In addition to boosting the electron distribution into the loss cones, the field-aligned potentials also increase the energy of the precipitating auroral electrons, modifying Σ_{Ped} as discussed in section 21.2.2.

Rotational decoupling from field-aligned potentials is not expected to have as strong of an effect at Saturn where the relative abundance of protons is larger than that in the Jovian system. *Ray et al.* [2013] showed that cold magnetospheric electrons carry the majority of current for potential strengths less than 500 V, as opposed to the hot electron population required to carry current at Jupiter.

21.2.4. Planetary Magnetic Field

The final component that exerts control over the auroral currents is the planetary magnetic field strength. As the giant planet auroral current systems are driven by angular momentum transport, they're intimately related to the $\mathbf{J} \times \mathbf{B}$ force that accelerates the magnetospheric plasma toward corotation. In the inner magnetospheres, where the fields are relatively dipolar, the $\mathbf{J} \times \mathbf{B}$ force quickly brings the newly ionized plasma back to corotation, such that the auroral currents corresponding to pickup have a narrow latitudinal and azimuthal extent at the ionosphere.

The currents associated with the radial transport of angular momentum are quite different. The continuing demand for angular momentum as the plasma moves outward results in the auroral currents covering a broad radial extent within the magnetosphere. At the ionosphere, the narrow width of the main aurora reflects the distention of the planetary magnetic field by the current sheet, rather than a quickly reaccelerated magnetospheric plasma population. If we were to assume a constant current density with radial distance, the $\mathbf{J} \times \mathbf{B}$ force would diminish with distance due to the decrease in **B**. The radial current density is not constant within the Jovian and Saturnian magnetospheres, increasing with enhanced Pedersen conductance and the presence of field-aligned potentials. Yet it is the decreasing **B** that ultimately can no longer provide the necessary acceleration to the magnetospheric plasma. In addition to the radial variations in the planetary magnetic field strength, there are azimuthal trends in **B** [*Ray et al.*, 2014; *Vogt et al.*, 2011] fixed in local time and variations with plasma pressure [*Achilleos et al.*, 2010; *Nichols*, 2011; *Nichols et al.*, 2015], which can spatially and temporally modify the magnitude of the auroral currents. In addition to magnetospheric asymmetries, neither Jupiter nor Saturn host a perfect dipole magnetic field. Offsets, tilts, and higher order moments at the planet contribute to variations in the currents fixed in planetary longitude and asymmetries between the northern and southern hemispheres [*Gérard et al.*, 2013; *Nichols et al.*, 2009].

21.3. FUTURE PATHS

Ultimately, it is an interplay between all four factors than control the magnitude and behavior of the auroral currents at the outer planets. In the steady-state, *Ray et al.* [2015] compared three one-dimensional (1-D) models: one with thermospheric and Pedersen conductance variations [*Smith and Aylward*, 2009], one with rotational decoupling and Pedersen conductance variations [*Ray et al.*, 2010], and one with thermospheric and Pedersen conductance variations in a compressed magnetosphere [*Yates et al.*, 2012]. Their analysis showed that the Pedersen conductance variations dominated over thermospheric and rotational decoupling effects. However, this indicates that it is essential to properly understand how the Pedersen conductance varies with latitude and longitude, thermospheric angular velocity, and auroral precipitation. Additionally, none of the 1-D models self-consistently included variations from plasma pressure, nor explored how azimuthal variations in the system might feed back into the auroral currents. Initial time-variable investigations indicate that solar wind compressions strongly affect the coupled MIT system, heating and spinning up the thermosphere as well as increasing the angular velocity of the magnetosphere [*Yates et al.*, 2014].

As a community, we are approaching the limit of what 1-D, steady-state model can tell us about auroral current systems. Yet in order to appropriately treat the MIT coupled system we must self-consistently address the thermosphere, ionosphere, high-latitude magnetosphere, and equatorial magnetosphere: a feat beyond the capabilities of MHD or atmospheric models alone. This requires a full system approach that connects the physics, and model outputs, of different regions in a holistic manner. Answering the following open questions would contribute toward a full global description:

1. How does the location of the acceleration region vary with local time?
2. Under what conditions do currents associated with the radial transport of angular momentum drive auroral emission at Saturn?
3. How does Jupiter's tilted dipole, combined with the ion drag forces and joule heating from auroral current, alter the thermospheric angular velocity?
4. How do asymmetries induced by the solar wind interaction affect the internally driven currents at both planets?

REFERENCES

Achilleos, N., C. S. Arridge, C. Bertucci, C. M. Jackman, M. K. Dougherty, K. K. Khurana, and C. T. Russell (2008), Large-scale dynamics of Saturn's magnetopause: Observations by Cassini, *J. Geophys. Res.*, *113*, A11, 209; doi:10.1029/2008JA013265.

Achilleos, N., N. André, X. Blanco-Cano, P. C. Brandt, P. A. Delamere, and R. Winglee (2015), 1. Transport of mass, momentum and energy in planetary magnetodisc regions, *Space Sci. Rev.*, *187*, 229–299; doi:10.1007/s11214-014-0086-y.

Achilleos, N., P. Guio, and C. S. Arridge (2010), A model of force balance in Saturn's magnetodisc, *Monthly Not. R. Astron. Soc.*, *401*, 2349–2371; doi:10.1111/j.1365-2966.2009.15865.x.

Akasofu, S.-I. (1964), The development of the auroral substorm, *Planet. Space Sci.*, *12*, 273–282; doi:10.1016/0032-0633(64)90151-5.

Akasofu, S.-I. (2012), *Auroral Morphology: A Historical Account and Major Auroral Features During Auroral Substorms*, Geophysical Monograph, *197*, American Geophysical Union, Washington, DC; doi: 10.1029/2011GM001156.

Badman, S. V., G. Branduardi-Raymont, M. Galand, S. L. G. Hess, N. Krupp, L. Lamy, H. Melin, and C. Tao (2015), Auroral processes at the giant planets: Energy deposition, emission mechanisms, morphology and spectra, *Space Sci. Rev.*, *187*, 99–179; doi:10.1007/s11214-014-0042-x.

Bagenal, F., and P. A. Delamere (2011), Flow of mass and energy in the magnetospheres of Jupiter and Saturn, *J. Geophys. Res. Space Physics*, *116*, A05,209; doi:10.1029/2010JA016294.

Bagenal, F., R. J. Wilson, S. Siler, W. R. Paterson, and W. S. Kurth (2016), Survey of Galileo plasma observations in Jupiter's plasma sheet, *J. Geophys. Res. Planets*, *121*, 871–894; doi:10.1002/2016JE005009.

Bonfond, B. (2012), When moons create aurora: The satellite footprints on giant planets, in *Auroral Phenomenology and Magnetospheric Processes: Earth and Other Planets*, *197*; doi:10.1029/2011GM001169.

Bonfond, B., D. Grodent, J.-C. Gérard, T. Stallard, J. T. Clarke, M. Yoneda, A. Radioti, and J. Gustin (2012), Auroral evidence of Io's control over the magnetosphere of Jupiter, *Geophys. Res. Lett.*, *39*, L01105; doi:10.1029/2011GL050253.

Bonfond, B., J. Gustin, J.-C. Gérard, D. Grodent, A. Radioti, B. Palmaerts, S. V. Badman, K. K. Khurana, and C. Tao (2015), The far-ultraviolet main auroral emission at Jupiter, Part 1: Dawn-dusk brightness asymmetries, *Ann. Geophys.*, *33*, 1203–1209; doi:10.5194/angeo-33-1203-2015.

Bougher, S. W., J. H. Waite, T. Majeed, and G. R. Gladstone (2005), Jupiter thermospheric general circulation model (JTGCM): Global structure and dynamics driven by auroral and Joule heating, *J. Geophys. Res. Planets*, *110*, A04008; doi:10.1029/2003JE002230.

Bunce, E. J., C. S. Arridge, J. T. Clarke, A. J. Coates, S. W. H. Cowley, M. K. Dougherty, J. Gérard, et al. (2008), Origin of Saturn's aurora: Simultaneous observations by Cassini and the Hubble Space Telescope, *J. Geophys. Res. Space Physics*, *113*, A09209; doi:10.1029/2008JA013257.

Bunce, E. J., S. W. H. Cowley, and S. E. Milan (2005), Interplanetary magnetic field control of Saturn's polar cusp aurora, *Ann. Geophys.*, *23*, 1405–1431; doi:10.5194/angeo-23-1405-2005.

Burger, M. H., E. C. Sittler, R. E. Johnson, H. T. Smith, O. J. Tucker, and V. I. Shematovich (2007), Understanding the escape of water from Enceladus, *J. Geophys. Res. Space Physics*, *112*, A06219; doi:10.1029/2006JA012086.

Clarke, J. T. (2012), Auroral processes on Jupiter and Saturn, in *Auroral Phenomenology and Magnetospheric Processes: Earth and other Planets*, edited by A. Keiling, E. Donovan, F. Bagenal, and T. Karlsson, 113–122, American Geophysical Union, Washington, DC; doi:10.1029/2011GM001199.

Clarke, J. T., J. Nichols, J. Gérard, D. Grodent, K. C. Hansen, W. Kurth, G. R. Gladstone, et al. (2009), Response of Jupiter's and Saturn's auroral activity to the solar wind, *J. Geophys. Res. Space Physics*, *114*, A05,210; doi:10.1029/2008JA013694.

Copper, M., P. A. Delamere, and K. Overcast-Howe (2016), Modeling physical chemistry of the Io plasma torus in two dimensions, *J. Geophys. Res. Space Physics*, *121*, 6602–6619; doi:10.1002/2016JA022767.

Cowley, S. W. H. (2000), Magnetosphere-ionosphere interactions: A tutorial review, in *Magnetospheric Current Systems*, edited by S. Ohtani, R. Fujii, M. Hesse, and R. L. Lysak, 91–106, Geophysical Monograph 118, American Geophysical Union, Washington, DC.

Cowley, S. W. H., and E. J. Bunce (2001), Origin of the main auroral oval in Jupiter's coupled magnetosphere-ionosphere system, *Planet. Space Sci.*, *49*, 1067–1088.

Cowley, S. W. H., and E. J. Bunce (2003), Corotation-driven magnetosphere-ionosphere coupling currents in Saturn's magnetosphere and their relation to the auroras, *Ann. Geophys.*, *21*, 1691–1707; doi:10.5194/angeo-21-1691-2003.

Delamere, P. A., and F. Bagenal (2010), Solar wind interaction with Jupiter's magnetosphere, *J. Geophys. Res. Space Physics*, *115*, A10201; doi:10.1029/2010JA015347.

Delamere, P. A., A. Otto, X. Ma, F. Bagenal, and R. J. Wilson (2015b), Magnetic flux circulation in the rotationally driven giant magnetospheres, *J. Geophys. Res. Space Physics*, *120*, 4229–4245; doi:10.1002/2015JA021036.

Delamere, P. A., F. Bagenal, and A. Steffl (2005), Radial variations in the Io plasma torus during the Cassini era, *J. Geophys. Res. Space Physics*, *110*, A12223; doi:10.1029/2005JA011251.

Delamere, P. A., F. Bagenal, V. Dols, and L. C. Ray (2007), Saturn's neutral torus versus Jupiter's plasma torus, *Geophys. Res. Lett.*, *34*, L09105; doi:10.1029/2007GL029437.

Delamere, P. A., F. Bagenal, C. Paranicas, A. Masters, A. Radioti, B. Bonfond, L. Ray, X. Jia, J. Nichols, and C. Arridge (2015a), Solar wind and internally driven dynamics: Influences on magnetodiscs and auroral responses, *Space Sci. Rev.*, *187*, 51–97; doi:10.1007/s11214-014-0075-1.

Dungey, J. W. (1961), Interplanetary magnetic field and the auroral zones, *Phys. Rev. Lett.*, *6*, 47–48; doi:10.1103/PhysRevLett.6.47.

Ergun, R. E., L. Ray, P. A. Delamere, F. Bagenal, V. Dols, and Y. Su (2009), Generation of parallel electric fields in the Jupiter-Io torus wake region, *J. Geophys. Res. Space Physics*, *114*, A05201; doi:10.1029/2008JA013968.

Fleshman, B. L., P. A. Delamere, and F. Bagenal (2010), A sensitivity study of the Enceladus torus, *J. Geophys. Res. Space Physics*, *115*, E04,007; doi:10.1029/2009JE003372.

Fleshman, B. L., P. A. Delamere, F. Bagenal, and T. Cassidy (2013), A 1-D model of physical chemistry in Saturn's inner magnetosphere, *J. Geophys. Res. Planets*, *118*, 1567–1581; doi:10.1002/jgre.20106.

Fukazawa, K., T. Ogino, and R. J. Walker (2012), A magnetohydrodynamic simulation study of Kronian field-aligned currents and auroras, *J. Geophys. Res. Space Physics*, *117*, A02214; doi:10.1029/2011JA016945.

Galand, M., L. Moore, I. Mueller-Wodarg, M. Mendillo, and S. Miller (2011), Response of Saturn's auroral ionosphere to electron precipitation: Electron density, electron temperature, and electrical conductivity, *J. Geophys. Res. Space Physics*, *116*, A09,306; doi:10.1029/2010JA016412.

Gérard, J.-C., D. Grodent, A. Radioti, B. Bonfond, and J. T. Clarke (2013), Hubble observations of Jupiter's north-south conjugate ultraviolet aurora, *Icarus*, *226*, 1559–1567; doi:10.1016/j.icarus.2013.08.017.

Goldstein, J., T. W. Hill, J. H. Waite, and J. L. Burch (2014), Analytical model of rotating two-cell convection at Saturn, *J. Geophys. Res. Space Physics*, *119*, 1980–1993; doi:10.1002/2013JA019615.

Gray, R. L., S. V. Badman, B. Bonfond, T. Kimura, H. Misawa, J. D. Nichols, M. F. Vogt, and L. C. Ray (2016), Auroral evidence of radial transport at Jupiter during January 2014, *J. Geophys. Res. Space Physics*, *121*, 9972–9984; doi:10.1002/2016JA023007.

Grodent, D., J. H. Waite Jr., and J.-C. Gérard (2001), A self-consistent model of the Jovian auroral thermal structure, *J. Geophys. Res. Space Physics*, *106*, 12,933–12,952; doi:10.1029/2000JA900129.

Hansen, C. J., L. Esposito, A. I. F. Stewart, J. Colwell, A. Hendrix, W. Pryor, D. Shemansky, and R. West (2006), Enceladus' water vapor plume, *Science*, *311*, 1422–1425; doi:10.1126/science.1121254.

Henderson, M. G. (2012), *Auroral Substorms, Poleward Boundary Activations, Auroral Streamers, Omega Bands, and Onset Precursor Activity*, Geophysical Monograph 197, American Geophysical Union, Washington, DC; doi:10.1029/2011GM001165.

Hess, S. L. G., and P. A. Delamere (2012), Satellite-induced electron acceleration and related auroras, in *Auroral Phenomenology and Magnetospheric Processes: Earth and other Planets*, *197*; doi:10.1029/2011GM001175.

Hill, T. (2016), Penetration of external plasma into a rotation-driven magnetosphere, *J. Geophys. Res. Space Physics*, *121*(10), 10,032–10,036; doi:10.1002/2016JA023430, 2016JA023430.

Hill, T. W. (1979), Inertial limit on corotation, *J. Geophys. Res. Space Physics*, *84*, 6554–6558.

Hill, T. W. (2001), The Jovian auroral oval, *J. Geophys. Res. Space Physics*, *106*, 8101–8108; doi:10.1029/2000JA000302.

Hill, T. W., and V. M. Vasyliunas (2002), Jovian auroral signature of Io's corotational wake, *J. Geophys. Res. Space Phys.*, *107*, 1464; doi:10.1029/2002JA009514.

Hill, T. W., A. M. Rymer, J. L. Burch, F. J. Crary, D. T. Young, M. F. Thomsen, D. Delapp, N. André, A. J. Coates, and G. R. Lewis (2005), Evidence for rotationally driven plasma transport in Saturn's magnetosphere, *Geophys. Res. Lett.*, *32*, L14S10; doi:10.1029/2005GL022620.

Huang, T. S., and T. W. Hill (1989), Corotation lag of the Jovian atmosphere, ionosphere, and magnetosphere, *J. Geophys. Res. Space Physics*, *94*, 3761–3765.

Jia, X., and M. G. Kivelson (2012), Driving Saturn's magnetospheric periodicities from the upper atmosphere/ionosphere: Magnetotail response to dual sources, *J. Geophys. Res. Space Physics*, *117*, A11219; doi:10.1029/2012JA018183.

Jia, X., M. G. Kivelson, and T. I. Gombosi (2012), Driving Saturn's magnetospheric periodicities from the upper atmosphere/ionosphere, *J. Geophys. Res. Space Physics*, *117*(A4); doi:10.1029/2011JA017367, a04215.

Joy, S. P., M. G. Kivelson, R. J. Walker, K. K. Khurana, C. T. Russell, and T. Ogino (2002), Probabilistic models of the Jovian magnetopause and bow shock locations, *J. Geophys. Res. Space Physics*, *107*, 1309; doi:10.1029/2001JA009146.

Khurana, K. K. (2001), Influence of solar wind on Jupiter's magnetosphere deduced from currents in the equatorial plane, *J. Geophys. Res. Space Physics*, *106*, 25,999–26,016; doi:10.1029/2000JA000352.

Kimura, T., S. V. Badman, C. Tao, K. Yoshioka, G. Murakami, A. Yamazaki, F. Tsuchiya, et al. (2015), Transient internally driven aurora at Jupiter discovered by Hisaki and the Hubble Space Telescope, *Geophys. Res. Lett.*, *42*, 1662–1668; doi:10.1002/2015GL063272.

Knight, S. (1973), Parallel electric fields, *Planet. Space Sci.*, *21*, 741–750.

Liu, X., T. W. Hill, R. A. Wolf, S. Sazykin, R. W. Spiro, and H. Wu (2010), Numerical simulation of plasma transport in Saturn's inner magnetosphere using the Rice Convection Model, *J. Geophys. Res. Space Physics*, *115*, A12,254; doi:10.1029/2010JA015859.

Lyons, L. R. (1980), Generation of large-scale regions of auroral currents, electric potentials, and precipitation by the divergence of the convection electric field, *J. Geophys. Res. Space Physics*, *85*, 17–24.

Mauk, B. H., and J. Saur (2007), Equatorial electron beams and auroral structuring at Jupiter, *J. Geophys. Res. Space Physics*, *112*, A10221; doi:10.1029/2007JA012370.

Mauk, B. H., D. J. Williams, and R. W. McEntire (1997), Energy-time dispersed charged particle signatures of dynamic injections in Jupiter's inner magnetosphere, *Geophys. Res. Lett.*, *24*(23), 2949–2952; doi:10.1029/97GL03026.

Mauk, B. H., J. T. Clarke, D. Grodent, J. H. Waite, C. P. Paranicas, and D. J. Williams (2002), Transient aurora on Jupiter from injections of magnetospheric electrons, *Nature*, *415*, 1003–1005; doi:10.1038/4151003a.

McPherron, R. L., C. T. Russell, M. G. Kivelson, and P. J. Coleman Jr. (1973), Substorms in space: The correlation between ground and satellite observations of the magnetic field, *Radio Sci.*, *8*, 1059–1076; doi:10.1029/RS008i011p01059.

Millward, G., S. Miller, T. Stallard, A. D. Aylward, and N. Achilleos (2002), On the Dynamics of the Jovian ionosphere and thermosphere, III. The modeling of auroral conductivity, *Icarus*, *160*, 95–107; doi:10.1006/icar.2002.6951.

Moore, L., I. Mueller-Wodarg, M. Galand, A. Kliore, and M. Mendillo (2010), Latitudinal variations in Saturn's ionosphere: Cassini measurements and model comparisons, *J. Geophys. Res. Space Physics*, *115*, A11317; doi:10.1029/2010JA015692.

Nichols, J. D. (2011), Magnetosphere-ionosphere coupling currents in Jupiter's middle magnetosphere: Computations including a self-consistent current sheet magnetic field model, *J. Geophys. Res. Space Physics*, *116*, A10232; doi:10.1029/2011JA016922.

Nichols, J. D., and S. W. H. Cowley (2004), Magnetosphere-ionosphere coupling currents in Jupiter's middle magnetosphere: effect of precipitation-induced enhancement of the ionospheric Pedersen conductivity, *Ann. Geophys.*, *22*, 1799–1827; doi:10.5194/angeo-22-1799-2004.

Nichols, J. D., J. T. Clarke, J. C. Gérard, D. Grodent, and K. C. Hansen (2009), Variation of different components of Jupiter's auroral emission, *J. Geophys. Res. Space Physics*, *114*, A06210; doi:10.1029/2009JA014051.

Nichols, J. D., N. Achilleos, and S. W. H. Cowley (2015), A model of force balance in Jupiter's magnetodisc including hot plasma pressure anisotropy, *J. Geophys. Res. Space Physics*, *120*, 10; doi:10.1002/2015JA021807.

Pilkington, N. M., N. Achilleos, C. S. Arridge, P. Guio, A. Masters, L. C. Ray, N. Sergis, M. F. Thomsen, A. J. Coates, and M. K. Dougherty (2015), Internally driven large-scale changes in the size of Saturn's magnetosphere, *J. Geophys. Res. Space Physics*, *120*, 7289–7306; doi:10.1002/2015JA021290.

Pontius, D. H. (1995), Implications of variable mass loading in the Io torus: The Jovian flywheel, *J. Geophys. Res*, *100*, 19,531–19,540; doi:10.1029/95JA01554.

Pontius, D. H., and T. W. Hill (2009), Plasma mass loading from the extended neutral gas torus of Enceladus as inferred from the observed plasma corotation lag, *Geophys. Res. Lett.*, *36*, L23103; doi:10.1029/2009GL041030.

Pontius, D. H., Jr., and T. W. Hill (1982), Departure from corotation of the Io plasma torus - Local plasma production, *Geophys. Res. Lett.*, *9*, 1321–1324.

Pryor, W. R., A. M. Rymer, D. G. Mitchell, T. W. Hill, D. T. Young, J. Saur, G. H. Jones, et al. (2011), The auroral footprint of Enceladus on Saturn, *Nature*, *472*; doi:10.1038/nature09928.

Radioti, A., A. T. Tomás, D. Grodent, J.-C. Gérard, J. Gustin, B. Bonford, N. Krupp, J. Woch, and J. D. Menietti (2009), Equatorward diffuse auroral emissions at Jupiter: Simultaneous HST and Galileo observations, *Geophys. Res. Lett.*, *36*, L07101; doi:10.1029/2009GL037857.

Radioti, A., D. Grodent, J.-C. Gérard, B. Bonfond, and J. T. Clarke (2008), Auroral polar dawn spots: Signatures of internally driven reconnection processes at Jupiter's magnetotail, *Geophys. Res. Lett.*, *35*, L03104; doi:10.1029/2007GL032460.

Ray, L. C., M. Galand, L. E. Moore, and B. L. Fleshman (2012b), Characterizing the limitations to the coupling between Saturn's ionosphere and middle magnetosphere, *J. Geophys. Res. Space Physics*, *117*, A07210; doi:10.1029/2012JA017735.

Ray, L. C., M. Galand, P. A. Delamere, and B. L. Fleshman (2013), Current-voltage relation for the Saturnian system, *J. Geophys. Res. Space Physics*, *118*, 3214–3222; doi:10.1002/jgra.50330.

Ray, L. C., N. A. Achilleos, M. F. Vogt, and J. N. Yates (2014), Local time variations in Jupiter's magnetosphere-ionosphere coupling system, *J. Geophys. Res. Space Physics*, *119*, 4740–4751; doi:10.1002/2014JA019941.

Ray, L. C., N. A. Achilleos, and J. N. Yates (2015), The effect of including field-aligned potentials in the coupling between Jupiter's thermosphere, ionosphere, and magnetosphere, *J. Geophys. Res. Space Physics*, *120*, 6987–7005; doi:10.1002/2015JA021319.

Ray, L. C., R. E. Ergun, P. A. Delamere, and F. Bagenal (2010), Magnetosphere-ionosphere coupling at Jupiter: Effect of field-aligned potentials on angular momentum transport, *J. Geophys. Res. Space Physics*, *115*, A09,211; doi:10.1029/2010JA015423.

Ray, L. C., R. E. Ergun, P. A. Delamere, and F. Bagenal (2012a), Magnetosphere-ionosphere coupling at Jupiter: A parameter space study, *J. Geophys. Res. Space Physics*, *117*, A01,205; doi:10.1029/2011JA016899.

Ray, L. C., Y. Su, R. E. Ergun, P. A. Delamere, and F. Bagenal (2009), Current-voltage relation of a centrifugally confined plasma, *J. Geophys. Res. Space Physics*, *114*, A04214; doi:10.1029/2008JA013969.

Sakai, S., S. Watanabe, M. W. Morooka, M. K. G. Holmberg, J.-E. Wahlund, D. A. Gurnett, and W. S. Kurth (2013), Dust-plasma interaction through magnetosphere-ionosphere coupling in Saturn's plasma disk, *Planet. Space Sci.*, *75*, 11–16; doi:10.1016/j.pss.2012.11.003.

Saur, J., F. M. Neubauer, J. E. P. Connerney, P. Zarka, and M. G. Kivelson (2004), Plasma interaction of Io with its plasma torus, in *Jupiter: The Planet, Satellites and Magnetosphere*, edited by F. Bagenal, T. E. Dowling, and W. B. McKinnon, 537–560.

Saur, J., N. Schilling, F. M. Neubauer, D. F. Strobel, S. Simon, M. K. Dougherty, C. T. Russell, and R. T. Pappalardo (2008), Evidence for temporal variability of Enceladus' gas jets: Modeling of Cassini observations, *Geophys. Res. Lett.*, *35*, L20105; doi:10.1029/2008GL035811.

Sittler, E. C., N. Andre, M. Blanc, M. Burger, R. E. Johnson, A. Coates, A. Rymer, et al. (2008), Ion and neutral sources and sinks within Saturn's inner magnetosphere: Cassini results, *Planet. Space Sci.*, *56*, 3–18; doi:10.1016/j.pss.2007.06.006.

Smith, C. G. A. (2011), A saturnian cam current system driven by asymmetric thermospheric heating, *Monthly Not. R. Astron. Soc.*, *410*(4), 2315–2328; doi:10.1111/j.1365-2966.2010.17602.x.

Smith, C. G. A., and A. D. Aylward (2008), Coupled rotational dynamics of Saturn's thermosphere and magnetosphere: a thermospheric modelling study, *Ann. Geophys.*, *26*, 1007–1027; doi:10.5194/angeo-26-1007-2008.

Smith, C. G. A., and A. D. Aylward (2009), Coupled rotational dynamics of Jupiter's thermosphere and magnetosphere, *Ann. Geophys.*, *27*, 199–230.

Smith, H. T., R. E. Johnson, M. E. Perry, D. G. Mitchell, R. L. McNutt, and D. T. Young (2010), Enceladus plume variability and the neutral gas densities in Saturn's magnetosphere, *J. Geophys. Res. Space Physics*, *115*, A10252; doi:10.1029/2009JA015184.

Song, P., T. I. Gombosi, and A. J. Ridley (2001), Three-fluid Ohm's law, *J. Geophys. Res. Space Physics*, *106*, 8149–8156; doi:10.1029/2000JA000423.

Southwood, D. J., and M. G. Kivelson (2007), Saturnian magnetospheric dynamics: Elucidation of a camshaft model, *J. Geophys. Res. Space Physics*, *112*(A12), n/a–n/a; doi:10.1029/2007JA012254, a12222.

Su, Y., R. E. Ergun, F. Bagenal, and P. A. Delamere (2003), Io-related Jovian auroral arcs: Modeling parallel electric fields, *J. Geophys. Res. Space Physics*, *108*, 1094; doi:10.1029/2002JA009247.

Tao, C., H. Fujiwara, and Y. Kasaba (2009), Neutral wind control of the Jovian magnetosphere-ionosphere current system, *J. Geophys. Res. Space Physics*, *114*, A08307; doi:10.1029/2008JA013966.

Tao, C., H. Fujiwara, and Y. Kasaba (2010), Jovian magnetosphere-ionosphere current system characterized by diurnal variation of ionospheric conductance, *Planet. Space Sci.*, *58*, 351–364; doi:10.1016/j.pss.2009.10.005.

Thomas, N., F. Bagenal, T. W. Hill, and J. K. Wilson (2004), The Io neutral clouds and plasma torus, in *Jupiter. The Planet, Satellites and Magnetosphere*, edited by F. Bagenal, T. E. Dowling, and W. B. McKinnon, 561–591.

Thomsen, M. F., D. B. Reisenfeld, D. M. Delapp, R. L. Tokar, D. T. Young, F. J. Crary, E. C. Sittler, M. A. McGraw, and J. D. Williams (2010), Survey of ion plasma parameters in Saturn's magnetosphere, *J. Geophys. Res. Space Physics*, *115*, A10220; doi:10.1029/2010JA015267.

Thomsen, M. F., D. B. Reisenfeld, R. J. Wilson, M. Andriopoulou, F. J. Crary, G. B. Hospodarsky, C. M. Jackman, et al. (2014), Ion composition in interchange injection events in Saturn's magnetosphere, *J. Geophys. Res. Space Physics*, *119*(12), 9761–9772; doi:10.1002/2014JA020489, 2014JA020489.

Tokar, R. L., R. E. Johnson, T. W. Hill, D. H. Pontius, W. S. Kurth, F. J. Crary, D. T. Young, et al. (2006), The interaction of the atmosphere of Enceladus with Saturn's plasma, *Science*, *311*, 1409–1412; doi:10.1126/science.1121061.

Vasyliunas, V. M. (2016), Physical origin of pickup currents, *Ann. Geophys.*, *34*(1), 153–156; doi:10.5194/angeo-34-153-2016.

Vasyliunas, V. M., and P. Song (2005), Meaning of ionospheric joule heating, *J. Geophys. Res. Space Physics*, *110*(A2); doi:10.1029/2004JA010615, a02301.

Vogt, M. F., M. G. Kivelson, K. K. Khurana, R. J. Walker, B. Bonfond, D. Grodent, and A. Radioti (2011), Improved mapping of Jupiter's auroral features to magnetospheric sources, *J. Geophys. Res. Space Physics*, *116*, A03220; doi:10.1029/2010JA016148.

Waite, J. H., M. R. Combi, W.-H. Ip, T. E. Cravens, R. L. McNutt, W. Kasprzak, R. Yelle, et al. (2006), Cassini ion and neutral mass spectrometer: Enceladus plume composition and structure, *Science*, *311*, 1419–1422; doi:10.1126/science.1121290.

Walker, R. J., and T. Ogino (2003), A simulation study of currents in the Jovian magnetosphere, Planet. *Space Sci.*, *51*, 295–307; doi:10.1016/S0032-0633(03)00018-7.

Wannawichian, S., J. T. Clarke, and D. H. Pontius (2008), Interaction evidence between Enceladus' atmosphere and Saturn's magnetosphere, *J. Geophys. Res. Space Physics*, *113*, A07217; doi:10.1029/2007JA012899.

Yates, J. N., N. Achilleos, and P. Guio (2012), Influence of upstream solar wind on thermospheric flows at Jupiter, *Planet. Space Sci.*, *61*, 15–31; doi:10.1016/j.pss.2011.08.007.

Yates, J. N., N. Achilleos, and P. Guio (2014), Response of the Jovian thermosphere to a transient pulse in solar wind pressure, *Planet. Space Sci.*, *91*, 27–44; doi:10.1016/j.pss.2013.11.009.

22

The Ambivalent Role of Field-Aligned Electric Currents in the Solar Atmosphere

Manolis K. Georgoulis

ABSTRACT

The solar atmosphere above the photospheric interface is largely ruled by powerful magnetic fields. The presence of solar plasma nonetheless dictates electric currents flowing along and, in some cases, perpendicular to the magnetic lines of force. This brief review aims to discuss how important field-aligned currents are within the observed solar atmospheric stratification and under the severe limitation of incomplete solar magnetographic capabilities. Active regions, namely, areas with the strongest local-scale solar magnetic fields, are mostly highlighted. Discussion recounts a number of fundamental debates unfolding within the last 25 yr and reports on the latest developments that have arguably put some outstanding questions to rest. This said, scarce observations above the photosphere seem to point to an electric-currents situation that is far more complicated than what simplified models, motivated by the lack of actual measurements, have envisioned. The overall picture drawn is one of a highly variable (for different atmospheric heights) and ambivalent role of field-aligned currents, that may prove useful for specific purposes and misleading for others. This review foresees that field-aligned currents will continue to be used in the future, given observational and computational limitations, but one should always be mindful of their caveats and shortcomings.

22.1. INTRODUCTION

The early twentieth century brought a revolution in the relatively new field of solar physics, following the realization that sunspots, known and observed since the eleventh century AD, owed their existence to solar magnetic fields [*Hale*, 1908]. These fields are formed and evolve within the solar plasma, making it natural for electric currents to flow along interacting magnetic configurations, adding to the overall complexity that comprises the multiscale, dynamical collage of the solar magnetized atmosphere.

Detailed measurements of the actual magnetic field vector in the solar atmosphere, mostly in the photosphere and against the solar disk, took several more decades to achieve (e.g., *Severny* [1964], *Harvey* [1969], *Moreton and Severny* [1968]). Even before such vector maps were routinely available, however, indirect inferences and precise knowledge of the orientation of these fields were available from detailed Hα maps of the solar chromosphere (e.g., *Bruzek* [1967]; *Foukal* [1971]; see also a detailed account in Chapter 2 of *Zirin* [1988]). These maps (with an example given in Fig. 22.1) largely outline magnetic field lines, emphasizing mainly field-aligned electric currents (FACs) formed by the motion of plasma along the magnetic field. Early but detailed observational inferences of electric currents in the photosphere of solar active regions, in particular, were achieved by a number of authors [*Hagyard*, 1988; *Hofmann et al.*, 1988; *Wang et al.*, 1994a, *Leka et al.*, 1996], as soon as exploitable photospheric vector magnetograms became available. This volume contains accounts of both electric currents in the solar atmosphere [*Fleishman and Pevtsov*, Chapter 3,

RCAAM of the Academy of Athens, Athens, Greece

Electric Currents in Geospace and Beyond, Geophysical Monograph 235, First Edition.
Edited by Andreas Keiling, Octav Marghitu, and Michael Wheatland.

(a) (b)

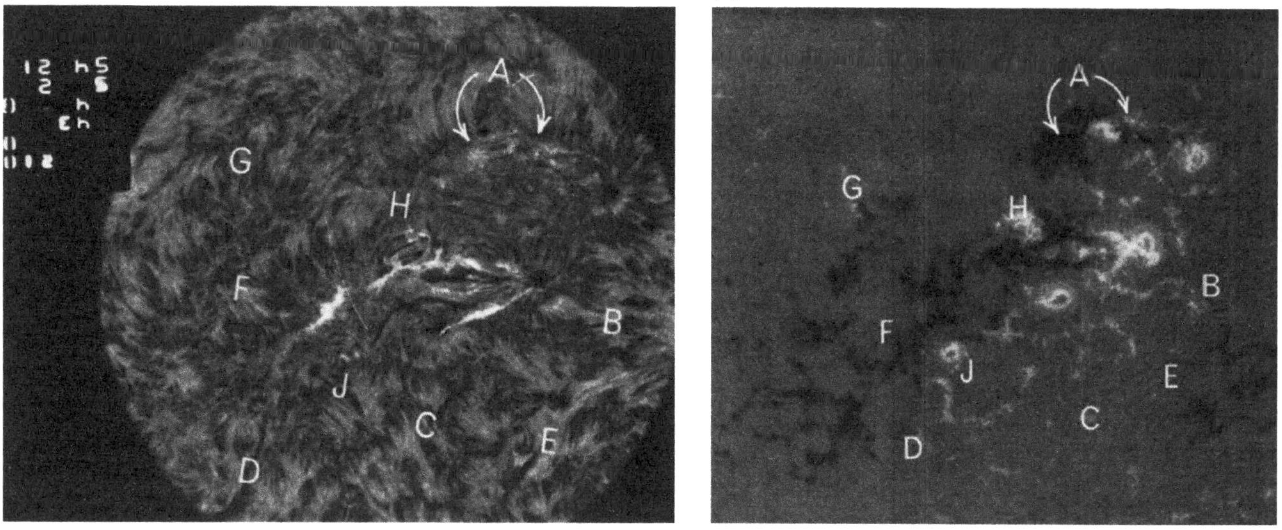

Figure 22.1 (a) A solar active region observed on 26 January 1970 in a Hα+0.4Å filtergram; (b) a Ca I (6103Å) photospheric line-of-sight magnetogram. Characters in both plates help guide the eye over the same areas, observed in different wavelengths. From *Foukal* [1971].

this volume] and the basic techniques employed to measure them [*Barnes and Leka*, Chapter 5, this volume].

Given a solar vector magnetogram, free from the otherwise inherent 180° azimuthal ambiguity in case of Zeeman-effect inference (e.g., *Harvey* [1969], *Metcalf et al.* [2006]), one may use Ampere's law to link the magnetic field **B** with the electric current density **J**:

$$\nabla \times \mathbf{B} = \frac{4\pi}{c}\mathbf{J}, \tag{22.1}$$

where the temporal variation of a cross-field induced electric field has been omitted. Equation (22.1) applies to the three-dimensional (3D) space occupied by **B** but, by necessity, it is evaluated only at the level of magnetic-field measurements, often approximated by a single Cartesian plane for areas small enough for solar curvature effects to be negligible. This information, however, is not sufficient to fully reveal FACs associated with the magnetic field vector **B**: for single-height measurements without additional or simplifying assumptions (such as, for example, the approaches of *Pevtsov and Peregud* [1990] and *Puschmann et al.* [2010]), only the normal component J_n of **J** can be inferred, by using the tangential derivatives of **B**. In Cartesian geometry, this implies knowledge of the vertical component J_z of ***J***, that is,

$$J_z = \frac{c}{4\pi}\left(\frac{\partial B_y}{\partial x} - \frac{\partial B_x}{\partial y}\right). \tag{22.2}$$

Equation (22.2) does not provide information on which part of J_z belongs to cross-field currents (CFCs) and which to FACs. Otherwise useful analyses, such as the one by *Zhang* [2001], in which **J** is written in term of twist and inhomogeneity terms, help us conceive the physical picture (section 22.2.3) but offer little in distinguishing FACs from CFCs.

Perhaps the only, indirect and hence qualitative, information in this direction stems from the plasma parameter β, namely the ratio between the plasma and the magnetic-field energy densities. This seems to be typically larger than unity ($\beta > 1$) even in the active-region solar photosphere, unless the photospheric magnetic field strength reaches ≈ 1 kG (see, e.g., *Georgoulis et al.* [2012a]). Otherwise, the dense photospheric material is able to shuffle the magnetic field anchored in it. A few Mm above the photosphere, the plasma parameter becomes smaller than unity ($\beta < 1$) [*Gary*, 2001], so the magnetic field expands to fill the available atmospheric space, contrary to the photosphere, and the plasma begins to "freeze" along the magnetic-field lines, unable to move in the cross-field direction. The only exception in this picture is, effectively, tangential discontinuities, also known as current sheets [*Parker*, 1972; *Parker*, 1979; *Priest*, 1982]. These rather exotic structures are characterized by the lack of magnetic field, that can either be zero on, or tangent to, their outer surface. In addition, moving from the photosphere to the corona, the temperature drops, up until the temperature-minimum region ($\approx$0.5 Mm above photosphere) and then begins to increase abruptly, strongly indicating that the plasma is heated due to the release of magnetic energy [*Withbroe and Noyes*, 1977] in a fashion whose details remain elusive to this day.

The above sketchy, qualitative understanding advocates for a generally plasma-dominated photosphere, allowing for both FACs and CFCs, and a magnetically dominated chromosphere and corona, allowing FACs only, particularly in the corona. In the following sections, we will recount developments of the past few decades to elaborate on and elucidate this fundamental physical picture of the debated, often equivocal FAC importance in the magnetized solar atmosphere.

22.2. PHYSICAL CONSIDERATIONS FOR THE STUDY OF ELECTRIC CURRENTS IN THE SOLAR ATMOSPHERE

22.2.1. Analysis Based on Single-Height Vector Magnetic Field Measurements

Assuming that one possesses only a single-height (photospheric, most commonly) vector magnetogram, one may use the vertical component of Ampere's law (equation [22.2]) to estimate the corresponding vertical current density J_z. Such an example is provided in Figure 22.2, where a high-resolution vector magnetogram of the Solar Optical Telescope/Spectropolarimeter (SOT/SP; *Lites et al.* [2008]) onboard the Hinode satellite has been used. While one notices the close correspondence between the B_z and J_z patterns, especially along the active region's magnetic polarity inversion line (PIL), one does not a priori know which part of these currents are FACs.

The electric current density **J** becomes fully known in the plane of magnetic field measurements if an assumption of FAC-only currents is adopted. In this case, one assumes

$$\nabla \times \mathbf{B} = \alpha(\mathbf{r})\mathbf{B}, \qquad (22.3)$$

where $\alpha(\mathbf{r})$ is a scalar force-free parameter. Complete knowledge of **J** is achieved for complete knowledge of $\alpha(\mathbf{r})$ that can either vary within the plane of the measurements ($\nabla\alpha(\mathbf{r}) \neq 0$), in which case a nonlinear force-free (NLFF) approximation is adopted, or can be kept constant in it ($\alpha(\mathbf{r}) \equiv \alpha$ for every **r**), in which case one imposes a linear force-free (LFF) approximation. Regardless, the force-free approximation generally dictates that $\alpha(\mathbf{r})$ is preserved along any given magnetic field line, for example, $\nabla\alpha \cdot \mathbf{B} = 0$, as can be readily shown from equation (22.3).

Force-freeness, in the case of FAC-only current densities, is guaranteed by the fact that such a current density **J** will give rise to zero Lorentz forces $\mathbf{F}_L = \left(\frac{1}{c}\right)\mathbf{J} \times \mathbf{B}$.

In the special case of the LFF field approximation, the value of the constant force-free parameter α is obtained by solving the Helmholtz equation, $\nabla^2\mathbf{B} + \alpha^2\mathbf{B} = 0$. Notice that the solution of this equation is gauge dependent and hence not unique, with an infinity of α-values possible, as one may define a new LFF field solution by means of B_z and an arbitrary choice of $|\alpha| < \left(\frac{2\pi}{L}\right)$, where L is the linear size of the boundary plane (e.g., *Alissandrakis* [1981]; *Pariat et al.* [2004]). For NLFF fields, the variable $\alpha(\mathbf{r})$ needs to be determined from equation (22.3), $\alpha(\mathbf{r}) = \frac{J_z}{B_z}$. However, one must solve consistency problems in this case, since $\alpha(\mathbf{r})$ has to be identical in the footpoints of a given field line. For an idea and an iterative method see, for example, *Cuperman et al.* [1997].

In the force-free approximation, equations (22.1) and (22.3) give rise to a FAC-only density

$$\mathbf{J}_{FAC}(\mathbf{r}) = \frac{c}{4\pi}\alpha(\mathbf{r})\mathbf{B}(\mathbf{r}). \qquad (22.4)$$

In addition, a number of works have been published on the calculation of the LFF constant-α value [*Leka and Skumanich*, 1999; *Leka*, 1999; *Georgoulis and LaBonte*, 2007]. However, making use of equation (22.4), one in principle does not know how far off the calculated FAC density is from the true current density **J**.

In more general settings, a number of authors used the Maxwell stress tensor to calculate the net Lorentz forces [*Metcalf et al.*, 1995; *Georgoulis and LaBonte*, 2004], or temporal variations thereof [*Hudson et al.*, 2012; *Fisher et al.*, 2012]. The idea facilitating much of this analysis came from *Low* [1984], who argued that the Lorentz force in a half space $z > 0$ above the plane $z = 0$, permeated by magnetic field, is simply the integral of the Maxwell stress tensor at $z = 0$. One may then infer the magnitude of the net Lorentz force above the photospheric plane by integrating the magnetic field components on this plane. In particular, the three Cartesian Lorentz-force components are found to be

$$\begin{aligned} F_x &= \frac{1}{4\pi}\int_{z=0} B_x B_z \, dS \\ F_y &= \frac{1}{4\pi}\int_{z=0} B_y B_z \, dS \\ F_z &= \frac{1}{8\pi}\int_{z=0} \left(B_z^2 - B_x^2 - B_y^2\right) dS, \end{aligned} \qquad (22.5)$$

where $dS = dx\,dy$. For the magnetic field to be sufficiently force-free, all above components should be much smaller than the characteristic amplitude F_0 of the Lorentz force, that is,

$$F_0 = \frac{1}{8\pi}\int_{z=0} \left(B_x^2 + B_y^2 + B_z^2\right) dS. \qquad (22.6)$$

Georgoulis and LaBonte [2004], on the other hand, decomposed the Lorentz force into pressure and tension components. This led to a compact inference of all Lorentz force components as

$$F_i = \frac{1}{4\pi}\mathbf{B} \cdot \nabla B_i - B\frac{\partial B}{\partial i}; \qquad i \equiv \{x, y, z\}. \qquad (22.7)$$

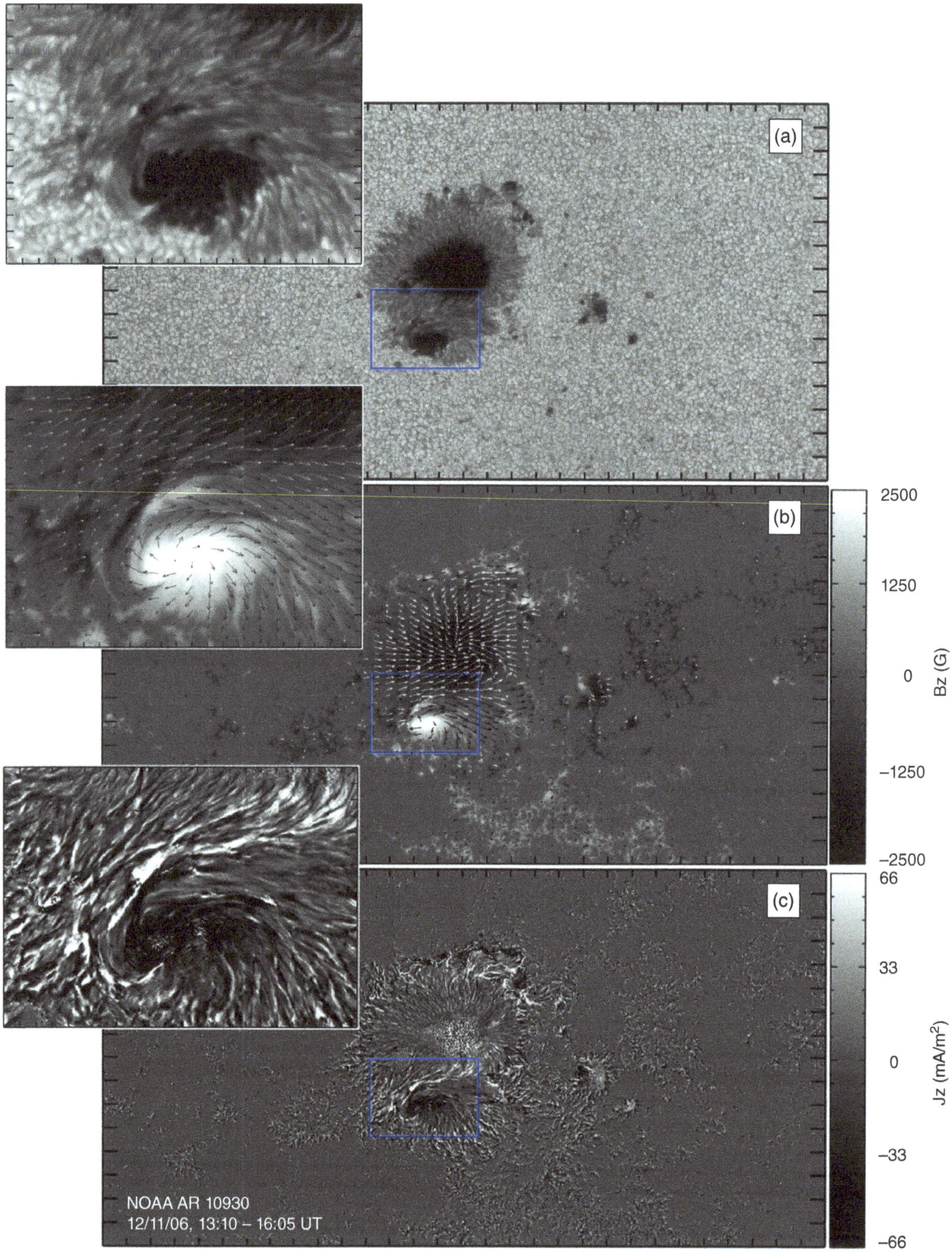

Figure 22.2 Vertical current density calculation (c) using a disambiguated high-resolution vector magnetogram (b) of NOAA AR 10930, as observed by Hinode's SOT/SP in December 2006. The co-temporal continuum image is also provided in (a). The insets in all frames magnify the area of the main magnetic polarity inversion line in the region, enclosed by the blue box. Tic mark separation is 10 arcsec (≈7250 km) in the full SOT/SP field of view and 2 arcsec (≈1550 km) in the insets, with a pixel size of 0.158 arcsec (≈114.6 km). North is up; west is to the right. From *Georgoulis et al.* [2012a].

In case of perfect force-freeness, therefore, $F_i = 0$ for all i, so the two terms in the rhs of equation (22.7) coincide. From equation (22.7) one can use the vertical component only, by writing

$$F_z = \frac{1}{4\pi} B_z \frac{\partial B_z}{\partial z} - B\frac{\partial B}{\partial z}, \qquad (22.8)$$

of which the only unknown is the height derivative $\partial B/\partial z$, as the height derivative $\partial B_z/\partial z$ can be inferred by the divergence-free, solenoidal condition $\nabla \cdot \mathbf{B} = 0$ of the magnetic field vector **B**. The unknown height derivative was approximated by the so-called "minimum-structure" expression, namely

$$\frac{\partial B}{\partial z} = \frac{b_z^2}{b_x^2 + b_y^2}\left(b_x \frac{\partial B}{\partial x} + b_y \frac{\partial B}{\partial y}\right), \qquad (22.9)$$

where $\mathbf{B} = B\hat{\mathbf{b}}$ denotes a decomposition of **B** into its amplitude (strength) B and the unit vector $\hat{\mathbf{b}} \equiv (b_x, b_y, b_z)$. equation (22.9) was deemed to favor situations in which the magnetic field **B** becomes space filling (i.e., force-free).

22.2.2. Analysis Based on Multiple-Height Vector Magnetic Field Measurements

In case one possesses coaligned, nearly simultaneous magnetograms of the same solar magnetic structure at different heights, then the analysis of the previous section may become largely obsolete in the ideal case that the height variation of the magnetic field components can be inferred from the multi-height-field measurements. If so, one can define an intermediate height for the calculation of vertical derivatives and then simply compare $\nabla \times \mathbf{B}$ with **B** at this height.

While such multi-height measurements at science-grade quality are still a rare commodity, working with them should also ensure that they satisfy the divergence-free condition. This task has been proven less than straightforward in several cases (e.g., *Crouch* [2013, 2015]; *Bommier* [2013]). Indeed, a number of issues, such as imperfections in the measurements, the variable atmospheric seeing (all multi-height magnetic-field measurements are, essentially, ground based [*Kotov*, 1971; *Mickey et al.*, 1996; *López Ariste et al.*, 2000]), or even the highly inhomogeneous expansion of magnetic fields with height, depending on the photospheric structure and exacerbated by the finite, imperfect spatial resolution, give rise to discrepancies that are either very difficult or untenable to handle.

22.2.3. On the Relation Between Force-Freeness and the Neutralization of Electric Currents

A distinct, but equally meaningful, line of attack in the understanding of FACs and CFCs refers to how these currents relate to the neutralization of electric currents in the solar atmosphere. Ampere's law, equation (22.1), directly implies that due to the solenoidality of the magnetic field the electric current density **J** is also divergence free (i.e., $\nabla \cdot \mathbf{J} = 0$). Hence, any isolated magnetic structure embedded in a field-free space must have electric currents closing within the structure. If this structure permeates a given surface S, which is the case of solar magnetic structures permeating the photosphere and extending into the corona, current balance, an immediate consequence of current solenoidality, dictates that the total current I_{tot} of the structure in the photosphere is zero, that is,

$$I_{tot} = \int_S \mathbf{J} \cdot d\mathbf{S} = \int_S J_z \, dS = 0, \qquad (22.10)$$

where S corresponds to the total photospheric area occupied by the structure. Here, again, we have assumed Cartesian geometry so the normal component of **J** translates directly to J_z.

While equation (22.10) and its rationale are unchallenged, debate pertains over whether electric currents for a *given* magnetic polarity are balanced. In this case, the structure is said to be "current neutralized," namely

$$\int_{S_+} J_z \, dS_+ = \int_{S_-} J_z \, dS_- = 0, \qquad (22.11)$$

where S_+, S_- correspond to the total areas occupied by the two polarities of the magnetic structure. The algebraic sum of both terms of equation (22.11) provides I_{tot} of equation (22.10) ($I_{tot} = I_+ + I_-$, with I_+ and I_- corresponding to the first and second term of equation [22.11], respectively). Hence, in situations of current neutrality, both I_+ and I_- are zero. In situations of nonneutrality, however, $I_+ = -I_-$, leading to equation (22.10).

What does neutrality of electric currents physically mean, however? Fundamentally, it means that emergence of solar magnetic flux does not imply the deposition of *net* electric currents in the corona. In the opposite case of nonneutrality, net currents are formed in the solar atmosphere and can only be neutralized by dissipation via magnetic reconnection. The latter school of thought became prevalent in solar physics, after science-grade observations of current-carrying magnetic flux became available [*Leka et al.*, 1996], although the former school of thought advocating for no net currents continued to appeal to a significant part of the solar physics community.

What is the relation between current neutralization and FACs? First, it seems clear that most of the photosphere is fairly unmagnetized at observable scales: this is where the concept of magnetic carpet, thought as a "forest" of mixed-polarity flux tubes embedded in a field-free space, was based [*Schrijver et al.*, 1998]. Active regions at photospheric levels are also thought of as relatively isolated structures embedded in much larger areas of quiet Sun.

This is basically the idea behind "sheath" currents that ought to outline the boundaries of the photospheric footprints of flux tubes, bounding them from the surrounding unmagnetized medium [*Pizzo*, 1986; *Ding et al.*, 1987]. Indeed, observations show a fairly sharp interface between strong magnetic-flux areas and surrounding quiet Sun, which is not the case if a force-free photospheric model is adopted. While complete knowledge of a 3D magnetic field can quantify FACs and CFCs, including sheath currents, existence of the latter can be exemplified by the analysis of *Zhang* [2001] where, substituting again $\mathbf{B} = B\hat{\mathbf{b}}$, Ampere's law of equation (22.1) gives

$$\mathbf{J} = \frac{cB}{4\pi}\nabla\times\hat{\mathbf{b}} + \frac{c}{4\pi}\nabla\mathbf{B}\times\hat{\mathbf{b}}. \tag{22.12}$$

Nonnegligible field inhomogeneities ∇B indeed exist within a flux-tube footprint. However, ∇B is clearly prevalent at the boundaries of photospheric flux tubes, where the transition between magnetized and unmagnetized areas takes place. This relates sheath currents with the second term of the rhs of equation (22.12). Therefore, sheath currents are mainly CFCs. On the other hand, the first term of the rhs of equation (22.12) includes both the entire contribution of FACs and, generally, some contribution from CFCs.

In the most general force-free approximation, one clearly has

$$B\left(\nabla\times\hat{\mathbf{b}}\right)_\perp = -\nabla B\times\hat{\mathbf{b}} \tag{22.13}$$

so, by virtue of equation (22.12), $\mathbf{J} = \frac{cB}{4\pi}(\nabla\times\hat{\mathbf{b}})_\parallel$, where $\parallel$ and $\perp$ denote components parallel (field-aligned) and perpendicular (cross-field) to **B**, respectively. As a special case, one could also have

$$B(\nabla\times\hat{\mathbf{b}})_\perp = \nabla B\times\hat{\mathbf{b}} = 0 \tag{22.14}$$

with the same end result. This could be the case for the corona, where the pressure component of the Lorentz force (second term in the rhs of equation [22.7]) acts so that the magnetic field fills the entire space, except for current sheets. Above the photosphere, surface sheath currents must gradually disappear to give way to volume currents, as exemplified by *Longcope and Welsch* [2000], reproduced here in Figure 22.3. The (volume) current density in the corona corresponds to FACs, then, with (surface) sheath currents corresponding to CFCs.

The above said, single-height photospheric field measurements again preclude knowledge of the relative significance between the two rhs terms in equation (22.12) and, of course, the cross-field and field-aligned component of $\nabla\times\hat{\mathbf{b}}$. In addition, in the situation of space-filling

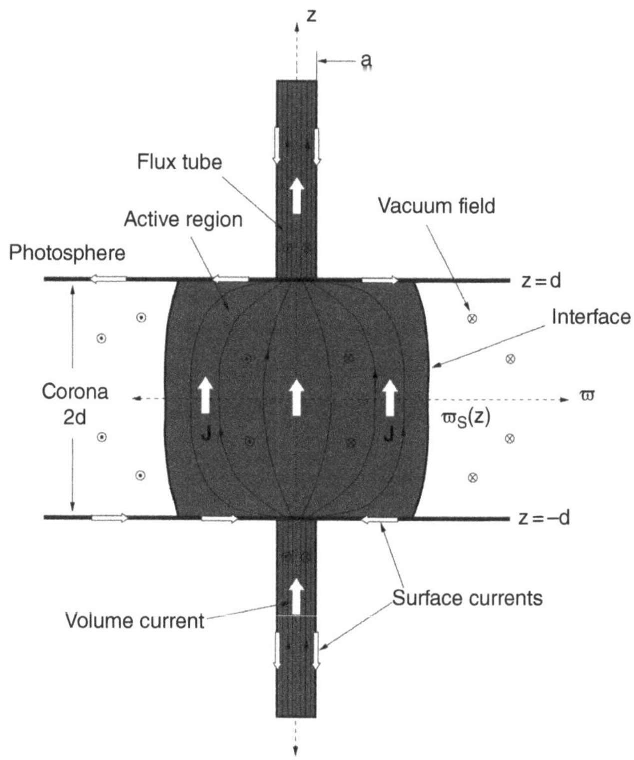

Figure 22.3 Two-dimensional schematic showing the behavior of a twisted magnetic flux tube that has emerged into the corona through the photospheric boundary. A balance between volume and surface (sheath) currents maintains the flux-tube integrity in the subphotosphere. The action of surface currents (CFCs) continues to the photosphere but vanishes above it, with the flux tube expanding significantly within an ambient vacuum field, letting the volume currents (FACs) dominate in the corona. From *Longcope and Welsch* [2000].

coronal FACs of Figure 22.3, any coronal cross section will include a net current. Hence, coronal FACs are non-neutralized. The divergence-free condition $\nabla\cdot\mathbf{J} = 0$ is then satisfied on, or slightly above, the photosphere in the case of neutralized photospheric currents. In the opposite case of photospheric current nonneutrality, currents must close below the photosphere (see section 22.4 for a brief account of this debate).

Photospheric current neutralization (or absence, thereof) continues to be a subject of debate. Recent developments, summarized in section 22.3.2, may help bring this problem to a closure. At this point, it suffices to say that nonneutralized currents pertain to a situation of space-filling FACs or, equivalently, that force-freeness implies nonneutralized currents. For CFCs in a non-force-free situation, current neutralization is still under investigation, with current understanding discussed in section 22.3.2.

22.3. APPLICATION TO OBSERVATIONS: MAIN FINDINGS

22.3.1. Analysis of Photospheric and Chromospheric Vector Magnetograms

Applying the analysis of section 22.2 to observed solar vector magnetograms, one must first deal with two key problems besides the inevitable azimuth disambiguation of the magnetograms: first, significant, sometimes substantial, uncertainties and, second, observations with flux-imbalanced, sometimes severely imbalanced, active regions. Noise and uncertainties in the inference of magnetic field components affect the calculation of every quantity discussed in section 22.2 (for a propagation of these uncertainties in the vertical current, force-free parameter and Lorentz-force components, see the appendix of *Georgoulis and LaBonte* [2004]). Magnetic flux imbalance further affects all photospheric surface integrals, including those for the estimation of Lorentz forces (equation [22.5], [22.6]), total currents (equation [22.10]), and their polarity terms (equation [22.11]). Keeping such limitations in mind, we attempt to describe some key findings below:

In the middle 1990s, *Metcalf et al.* [1995] independently applied the Lorentz-force components of equations (22.5) and (22.6) to both photospheric and chromospheric vector magnetograms at six different wavelengths obtained by the Haleakala Stokes Polarimeter (HSP; *Mickey* [1985]) to find that individual ratios F_x/F_0, F_y/F_0, F_z/F_0 become sufficiently small only above the photosphere, independent of a filling factor accounting for unresolved magnetic fields. In addition, photospheric fields were found incompatible with the magnetic energy from the Virial theorem, which applies to force-free fields [*Aly*, 1989]. They concluded that solar magnetic fields start becoming force-free at ≈400 km above the photosphere ($\tau = 1$), close to the temperature-minimum region, with the photosphere dominated by CFCs.

A decade later, *Georgoulis and LaBonte* [2004] estimated values of the vertical Lorentz force F_z (equations [22.8] and [22.9]), from which a magnitude of the horizontal CFC density could be found. Comparing this magnitude with the inferred J_z, they found it to be significantly larger than J_z, both for an eruptive and for a quiescent active region. Hence, provided that the minimum-structure assumption of equation (22.8) was correct, they also concluded that the photospheric magnetic field cannot be force-free. Their data came also from the ground-based Imaging Vector Magnetograph (IVM; *Mickey et al.* [1996]). Later on, *Petrie* [2012, 2013] and *Sun* [2014], among others, used the time derivatives of the Lorentz force derived by *Fisher et al.* [2012] to estimate changes before and after solar eruptions on eruption-prolific active regions. These data were of relatively high resolution (≈1 arcsec; ≈ 0.5 arcsec per pixel) and of high cadence (12 min), provided by the Helioseismic and Magnetic Imager (HMI; *Scherrer et al.* [2012]) on board the Solar Dynamics Observatory mission. Using much improved data did not change the overall picture: Lorentz forces were found to be very significant, particularly along magnetic PILs.

This overall picture of a CFC-dominated active-region photosphere was corroborated by photospheric and three-dimensional magnetostatic and magnetohydrodynamic models (e.g., *Manchester* et al. [2004]; *Rempel* [2011]). In a related study, *Puschmann et al.* [2010] used a geometrical three-dimensional model of a sunspot to estimate the plasma β-parameter on the photosphere and above. They reported that studies focusing on J_z strongly underestimate the magnitude J of the current density. In addition, in topologically important parts of the photosphere, such as intraspines, they found an overall angular difference between **B** and **J** of the order 20°. As such, photospheric fields were also found to be generally non-force-free. The reader is referred also to *Balthasar* [2006], for detailed calculations of J_z and to *Wheatland* [2015] for efforts to calculate J_z in the presence of significant noise.

In a comprehensive review, *Borrero and Ichimoto* [2011] examined the three-dimensional magnetic structures of active-regions (sunspots, in particular), where the Lorentz-force dominated, CFC-prevailing photospheric picture seems established. In this important direction, *Socas-Navarro* [2005] used vector magnetogram data from the Spectro-Polarimeter for Infrared and Optical Regions (SPINOR; *Socas-Navarro et al.* [2006]) from 0 to 1.6 Mm above the photosphere and was able to deduce $\nabla \times \mathbf{B}$ to be compared with **B**. The result is reproduced in Figure 22.4, showing substantial angular differences at three different altitudes (0.2 Mm, 0.65 Mm, and 1.6 Mm) for most of an observed sunspot. This behavior seems to mollify at higher altitudes in favor of FACs but, even at the chromosphere of this sunspot, it is probably safe to say that the magnetic field was generally forced.

A notable (perhaps the only) exception in the above mounting evidence for a CFC-dominated magnetized photosphere stems from *Moon et al.* [2002]. They also used a number of HSP vector magnetograms and calculated the ratios F_x/F_0, F_y/F_0, F_z/F_0 along the lines of *Metcalf et al.* [1995], reporting much lower values, hence photospheric fields closer to force-freeness. They even found that one, particularly eruptive, active region can be reasonably well reproduced by a single α-value, prompting them to report that even a LFF approximation would work for this particular case. They attributed the difference between their results and those of *Metcalf et al.* [1995] to magnetic-field inversion issues and the calibration of the data.

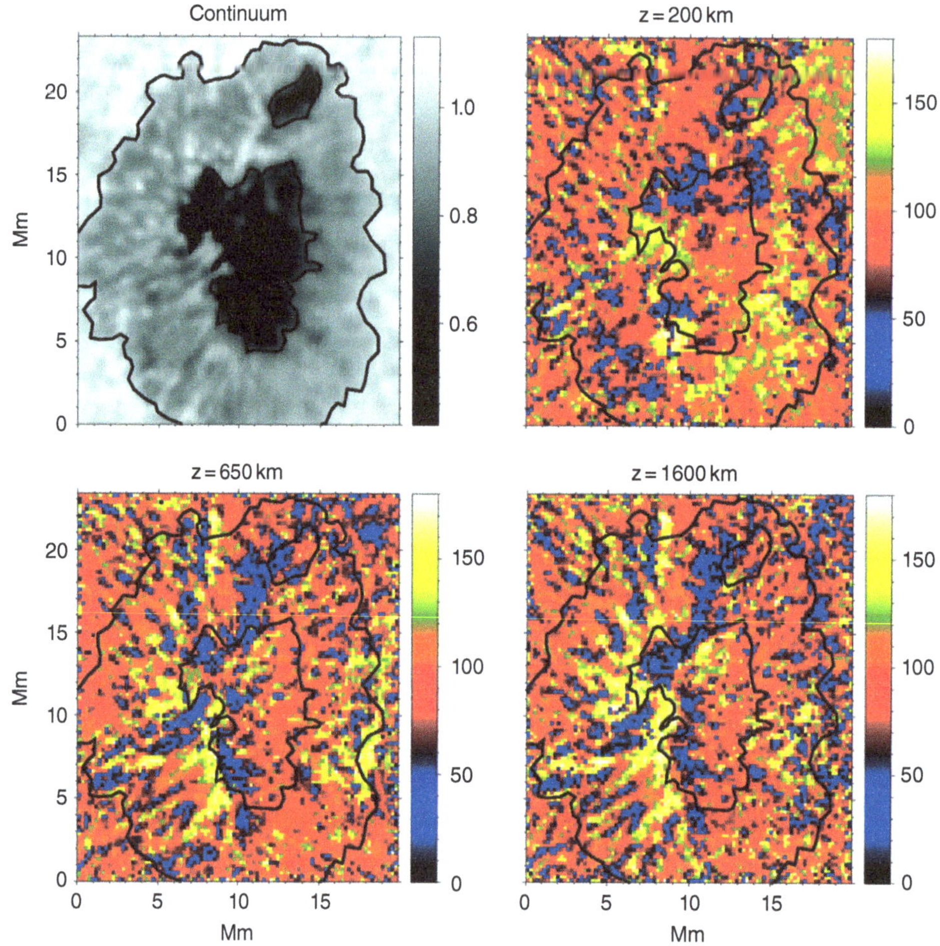

Figure 22.4 Multi-height observations of a sunspot, where the inferred three-dimensional magnetic fields **B** allow for the calculation of $\nabla\times\mathbf{B}$ and its comparison with the field. Color images (top right; bottom) show the angular difference (deg) between the two vectors at different altitudes above the photosphere. Black/dark blue and yellowish/white areas indicate roughly parallel and antiparallel, respectively, current densities, hence FACs. All other areas indicate mainly CFCs. From *Socas-Navarro* [2005].

This differing view notwithstanding, FACs seem to be less significant than CFCs in the magnetized, active-region photosphere. Yet, due to their easier applicability, FACs have received the lion's share of attention in efforts to model the unknown coronal magnetic field. This apparent inconsistency and its reasoning are discussed in section 22.5.

22.3.2. Current Neutralization of Photospheric Magnetic Fields

A practical application of calculating the total electric current I_{tot} equivalently to equation (22.10) over a photospheric patch of area S is by utilizing the Stokes theorem, that is,

$$I_{\text{tot}} = \frac{c}{4\pi}\int_S \nabla\times\mathbf{B}\cdot d\mathbf{S} = \frac{c}{4\pi}\oint_C \mathbf{B}\cdot d\ell, \qquad (22.15)$$

where C is a suitably chosen contour encompassing S. Notice that any choice of the closed path C will return, by virtue of equation (22.15), the total current enclosed by this contour.

Applying equation (22.15) to different contours gradually covering a sunspot, *Wilkinson et al.* [1992] found that including the entire sunspot but no patches of opposite polarity leads to $I_{\text{tot}} \approx 0$. They then concluded that either the photospheric currents of the sunspot are neutralized, indeed, or that observational artifacts, such as the Faraday rotation, give rise to a fictitious current that appears neutralized. A barrage of later works [*Leka et al.*, 1996; *Semel and Skumanich*, 1998; *Wheatland*, 2000; *Falconer*, 2001], however, concluded that photospheric total currents over a given polarity are *not* neutralized, without challenging the general notion of a balanced current in a flux-balanced active region.

The nonneutralized opposite polarity currents, therefore, cancel within the two active-region polarities. These studies, however, including that of *Wilkinson et al.* [1992], invariably suggested that the analysis should be repeated with higher-resolution observations featuring high-quality vector magnetograms.

This was the motivation behind the study of *Georgoulis et al.* [2012a] that utilized ultra-high-resolution data (≈0.32 arcsec, or ≈0.16 arcsec per pixel) from Hinode's SOT/SP. These authors partitioned the photospheric magnetic flux, identifying unique, nonoverlapping contours C that outline a given flux partition. In this sense, the total current of each partition could be calculated, with the total current for each magnetic polarity being the algebraic sum of all partition total currents belonging to this polarity. In the limit of the total current imbalance being much smaller than the magnetic flux imbalance, *Georgoulis et al.* [2012a] also devised the overall current non-neutrality factor I_{nn}, where

$$I_{nn} = \frac{1}{2}\left(\frac{|I_+|}{\sum_{i=1}^{N}|s_i^+ I_i|} + \frac{|I_-|}{\sum_{i=1}^{N}|s_i^- I_i|} \right). \qquad (22.16)$$

Here I_i is the total current for partition i and $s_i^{\pm}$ $(i \equiv \{1,\ldots,N\})$ corresponds to 0 or 1, making sure that only total currents of the correct polarity take part in the respective algebraic sum. The nonneutrality factor can also be called a "coherence factor": if all partitions of a given polarity have a certain sense (sign) of total current, then $I_{nn} = 1$, implying a perfect current coherence. If $I_{nn} = 0$, total currents have a random sense for a given polarity, implying a totally incoherent, and current neutralized, structure.

The findings of *Georgoulis et al.* [2012a] over two active regions, an eruptive and a quiescent one, were the following: currents over entire active regions are balanced within uncertainties, as expected, but nonneutralized currents exist and become more intense for more compact active regions, that is, active regions with stronger PILs. The current coherence of equation (22.16) is much higher for compact active regions than for dispersed ones. Invariably, contribution to nonneutralized currents stems exclusively from the PIL area of an active region (e.g., the blue box in Fig. 22.2). For every other flux patch of a given polarity, the total current is practically zero. The coherence of currents around PILs was interpreted as being of subphotospheric origin, implying that these flux tubes inject net currents in the solar atmosphere. Further, the existence of net currents along PILs only was interpreted as the result of a Lorentz tension force generating magnetic shear along strong PILs, in agreement with existing three-dimensional MHD models (e.g., *Manchester et al.* [2004]). It is worth noticing that *Georgoulis et al.* [2012a] tested the effects of deteriorating spatial resolution on their analysis and concluded that the qualitative results and conclusions did not change. Conversely, in the absence of PILs and shear, no net currents are found, even using similarly high-resolution data from Hinode SOT/SP [*Venkatakrishnan and Tiwari*, 2009].

Strong PILs imply stronger, more widespread shear, which is a long-standing observational fact (e.g., *Hagyard et al.* [1984]). PIL-aligned Lorentz-force tension according to *Georgoulis et al.* [2012a] occurs as a result of the loss of azimuthal symmetry along PILs, with strong B_z-amplitudes along both sides of a PIL (but not on it, obviously) giving rise to opposite shear patterns in the two PIL sides. In addition, loss of symmetry is spawned by distorted sheath-current patterns that are severely perturbed by the approach and interaction of two (or more) opposite-polarity magnetic footprints, forming the PIL area.

A similar picture with strong net currents related to shear in the same active region, also attributed to Lorentz force but in a different way, was reported by *Ravindra et al.* [2011]. In this case, Lorentz force facilitated spinning and braiding of emerging flux tubes giving rise to the twist observed at photospheric heights and the ensuing complex behavior in NOAA AR 10930. In spite of different interpretations of the Lorentz-force action, notice the common underlying origin of the shear flows between the two studies.

Some interest within the community was spurred by the above results: in the recent three-dimensional MHD model of *Török et al.* [2014], describing both the emergence of magnetic flux and its expansion in the solar atmosphere, a significant departure from current neutrality was obtained as soon as the magnetic structure emerged and shear developed along its main PIL. This situation prevailed for as long as flux emergence lasted in the modeled active region. The situation reversed thereafter, with volume currents clearly exceeding sheath currents. However, *Török et al.* [2014] did not go as far in their interpretation as involving the Lorentz force in the generation of shear. Further on, *Dalmasse et al.* [2015a] were able to prescribe twist-generating rotation and shearing flows in a model photosphere in the course of the emergence of a NLFF flux tube. They achieved a net current in the absence of a Lorentz force, thus offering an alternative possibility to the Lorentz-force interpretations of *Ravindra et al.* [2011] and *Georgoulis et al.* [2012a]. In this scenario, however, photospheric shearing flows were externally imposed, leaving the question of their origin unaddressed.

In conclusion, the question of nonneutralized electric currents in active regions observed by means of high-resolution photospheric vector magnetic field data seems reasonably tackled: all recent works, observational and modeling, report significant, CFC-stemming net currents flowing exclusively along shear-dominated PILs. The origin of shear and the role of the Lorentz force in this fundamental physical picture are still in debate. FACs perhaps dominate in the photosphere after flux emergence (and shear) has ceased, while they clearly dominate in the overlaying corona.

22.4. IMPLICATIONS OF CURRENT (NON) NEUTRALITY AND THE VERY REALISM OF PHOTOSPHERIC CURRENT DENSITY

Up until the early 1990s, the common perception was that electric currents in the active-region photosphere are neutralized, with gradual storage and impulsive release of magnetic energy achieved via photospheric flows and shuffling (e.g., *Parker* [1972, 1979]; *Rosner and Vaiana* [1978]). This picture directly supports divergence-freeness of the electric current density, with currents closing around the photosphere. An explanation of this behavior via a closed-circuit analog was proposed by *Spicer* [1982]. However, *Melrose* [1991] argued that photospheric current neutralization via sheath currents should imply a current neutralization in the corona, as well, where CFC sheath currents do not exist. Sensing an inconsistency, he argued that either sheath currents evade detection in the corona, so the coronal field is not nearly force-free, or photospheric currents should be current neutralized, as well. He offered two possible explanations for photospheric current neutralization: nearly concentric opposite-polarity patches in every flux-tube footprint, in case of isolated flux tubes (i.e., in the absence of a strong PIL), or, in case of shearing motions along the photospheric PIL, two "zebra-type" current patterns alongside the PIL that would act to neutralize currents. Further arguing that none of these scenarios seems to enjoy observational support, he offered the possibility of nonneutralized photospheric currents closing much below the photosphere, at the base of the convection zone where active-region flux tubes are necessarily rooted (also known as tachocline). Somewhat later, *Wang et al.* [1994b] reported an intriguing case of a shear increase following a major flare. Citing this observation, *Melrose* [1995] further supported an electric circuit scenario closing at the tachocline that would act to replenish the energy released during the flare at observed timescales much faster than what photospheric line-tying would allow (Fig. 22.5a). This "current-path" interpretation facilitated what would be later labeled as the **(E, J)** paradigm of solar magnetic structures, challenging the prevalent in situ (i.e., photosphere and slightly above) storage-and-release notion of solar flare occurrence, that would be justified by a current closure such as the one shown in Figure 22.5b.

This picture was intensely criticized by *Parker* [1996a], who argued that the action of the photospheric velocity field on the magnetic field (counterposed as the **[B, u]** paradigm), forms the basis of the generally accepted MHD interpretation of magnetic-field evolution in the solar atmosphere. The **(B, u)** paradigm does not allow

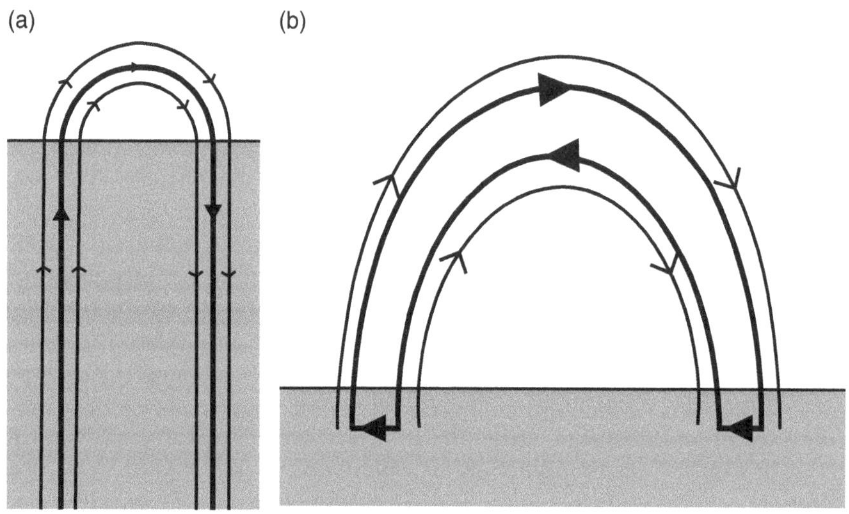

Figure 22.5 Current path models in emerged magnetic flux tubes. (a) Model proposed to interpret observations of a fast shear increase following a major solar flare that, as per the analysis, could not be explained by photospheric convection. Divergence-freeness of the electric current density is guaranteed by currents essentially closing at the tachocline. (b) Model with electric currents closing slightly below the photosphere, adhering to photospheric current neutrality in case shearing and twisting motions enable in situ magnetic energy storage and release. From *Melrose* [1995].

alternative interpretations or modifications. In a dimensional calculation, *Parker* [1996a] provided evidence that solar photospheric magnetic fields are "overwhelmed" by convection, with an effective line tying essentially in place at ≈5 Mm below the photosphere. Later, *Parker* [2007] further denounced the (**E**, **J**) paradigm in the Sun and other MHD-described natural systems, such as the terrestrial magnetosphere (see, e.g., *Parker* [1997]), as well as alternative approaches effectively rejecting MHD (e.g., *Parks* [2004]), assessing that analysis along such lines soon leads to both unphysical and mathematically intractable sets of equations.

Parker [1996b], possibly aiming to abolish the notion of nonneutralized photospheric electric currents altogether, argued that inferring J_z (the only current-density component directly accessible by photospheric vector magnetograms) is meaningless in situations of unresolved magnetic fibrils, where magnetic structures are imperfectly observed. In this case, one can only obtain a spatially averaged J_z over a given resolution element. However, for a given fibril composed of a set of concentric annuli, each carrying no net current, the spatially averaged nonzero J_z-map would be fictitious and depending totally on the resolution, converging to zero for progressively better resolved magnetic structures. The typical situation of twisted fibrils will result in spiraling annuli (Fig. 22.6) without, however, changing the final result or conclusion. For different reasons, this general picture applies to both space-filling force-free fields and non-force-free situations, featuring field-free interfaces (i.e., isolated flux tubes).

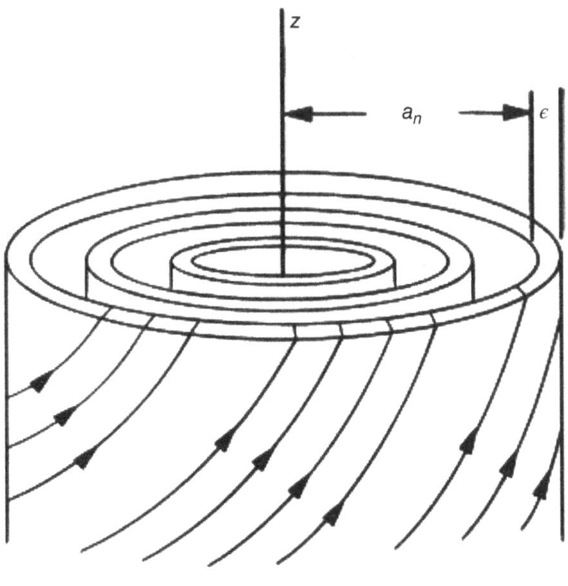

Figure 22.6 An envisioned fully resolved twisted magnetic fibril of radius a_n composed of spiraling, tightly packed and concentric annuli of thickness ε. Each annulus carries no net current, hence the entire structure carries no net current, although FACs are flowing along its axis. From *Parker* [1996b].

Clearly, if this picture were adopted by the community, it would be of little meaning to infer any electric current densities in the solar atmosphere and to pursue any physical analysis for the understanding of either FACs or CFCs through them. What is more unsettling is that partially resolved potential-field vector magnetograms give rise to nonzero J_z-maps (see, e.g., the two variations of the "flowers" synthetic magnetogram produced by *Leka et al.* [2009] and *Georgoulis* [2012]). This said, while not necessarily adopting the (**E**, **J**) paradigm of an electric circuit closing in the tachocline, several nearly cotemporal observational works refuted the *Parker* [1996b] arguments of totally fictitious J_z-maps. Typically, these studies largely overlap with the works advocating for nonneutralized photospheric current patterns (*Leka et al.* [1996]; *McClymont et al.* [1997]; *Semel and Skumanich* [1998]; *Wheatland* [2000]; *Falconer* [2001], among others). Of them, only *Wheatland* [2000] appeared inclined to support the (**E**, **J**) paradigm. Interesting arguments were voiced in these works such as, for example, that since stronger J_z patterns are inferred in sunspots where the fibril state of magnetic structures seems to break down, inclined flux tubes need a real FAC in order to maintain their stability, or that J_z patterns along sheared PILs assume ribbon-like structures aligning with the shear flow. In addition, it was argued that the increase in the current density in cases of emerging flux cannot be fully accounted for by photospheric motions alone.

In what later proved to be an important comment, voiced by *Semel and Skumanich* [1998], but also alluded to by *Melrose* [1995], current nonneutrality appears in the case of interacting flux tubes, that were not considered by *Parker* [1996b]. This agrees with recent works cited in section 22.3.2 about net currents appearing exclusively along sheared PILs that necessarily involve distorted, due to interaction, footprints of photospheric flux tubes. Conversely, there are cases of apparently noninteracting, nonpotential, nearly vertical flux tubes whose J_z patterns are nearly vanishing, even with imperfectly resolved vector magnetograms. These are plage fields in active regions (e.g., *Martinez Pillet et al.* [1997]). Extended plage areas and their corresponding J_z-maps can be seen in Figure 22.2, south and west of the main sunspot complex in NOAA AR 10930. Plage areas and their vanishing vertical current densities were quoted as "Parker's gardens" by *Semel and Skumanich* [1998].

22.5. APPLICABILITY AND PHYSICAL SIGNIFICANCE OF FIELD-ALIGNED CURRENTS

Discussion in the previous sections apparently justifies the ambivalence of the FAC role in the solar atmosphere: FAC dominance is at the very least debated, if not ruled out, in comparison to that of CFCs at photospheric heights, while it is almost unanimously acknowledged just above the photosphere and in the entire solar corona, until the source surface of the inner heliospheric magnetic field (typically 2.5 solar radii, *Hoeksema* [1984]; *Wang and Sheeley* [1990]). However, reliable measurements of the magnetic field vector, giving rise to a routine inference of one component of the electric current density, have been achieved for the photospheric layer alone in, by far, the most cases. This incomplete information has led to a substantial body of literature aiming to understand the evolutionary behavior of the solar magnetized atmosphere. In the following, we briefly review part of this literature that, in spite of considerable ambiguity, has managed to arguably show that one would be better off using the information of photospheric FACs than refrain from using it at all.

A strong motivation for using photospheric FACs is for extrapolating to the coronal magnetic field above the photosphere, using the photospheric magnetic field vector and/or J_z-map as the necessary boundary condition. While LFF field extrapolation that uses primarily the photospheric B_z and, in some cases, the photospheric horizontal field to infer a unique α-value (section 22.2.1), is a well-defined problem in most cases, NLFF field extrapolation is clearly ill posed: typically, one would need the vector field on all boundaries of the extrapolation volume, bottom, top, and lateral. As this is obviously infeasible, different assumptions give rise to different solutions that do not always agree satisfactorily. The works of *Schrijver et al.* [2006], *Metcalf et al.* [2008], *DeRosa et al.* [2009; 2015], and the comprehensive review of *Wiegelmann and Sakurai* [2012] discuss extensively several facets of the problem and compare the results of numerous NLFF field extrapolation methods.

Besides the ambiguity caused by the lack of information needed for a NLFF field extrapolation, which is the most general and realistic force-free extrapolation approach, photospheric non-force-freeness creates further problems that impact different NLFF field extrapolation methods in different ways. Photospheric CFCs cannot be used in force-free extrapolations but can be used in more general magnetostatic, magnetohydrostatic, and MHD modeling efforts, labeled accordingly to the terms ignored in the general MHD equation of motion. NLFF equilibrium is a special case of magnetostatic equilibria assuming force-freeness on top of immovability ($D/Dt = 0$, where D denotes the convective derivative) in a $\beta = 0$ plasma. A $\beta \neq 0$ triggered by a nonnegligible plasma pressure

Figure 22.7 Two different extrapolation approaches on a three-dimensional MHD model of an emerging active region in the solar atmosphere (right column). The top row shows the magnetic field configuration using the model's photosphere as lower boundary, while the bottom row shows the same using the modeled chromospheric boundary. A potential field extrapolation (left column) uses the photospheric and chromospheric magnetograms as the required boundary condition, while a NLFF field extrapolation is also included for comparison (middle column). From *Abbett et al.* [2004].

gradient balancing a nonzero Lorentz force composes the more general magnetostatic equilibrium case, while the consideration of the gravity force adds the "hydro" characterization. Hence, NLFF field extrapolation results can only be directly compared with magnetostatic solutions (e.g., *Hu et al.* [2008]; *Gilchrist and Wheatland* [2013]). Comparison with MHD models could be done if a velocity flow prescription is available on the lower boundary (i.e., photosphere; see, for example, *Abbett et al.* [2004]), to allow application of these models. Alternatively, one sometimes creates a tailored MHD model reproducing only abstractly the observed photospheric boundary condition (e.g., *Savcheva et al.* [2012]).

Comparisons between NLFF and more sophisticated extrapolations and models show nonnegligible, sometimes significant, differences that underline the existence of CFCs in the photospheric boundary (see, e.g., Fig. 22.7 for an example). At the same time, they conceivably show a general qualitative agreement that might convince one to use these results for specific purposes, such as the overall coronal magnetic structure and topological features of it, including null points, separatrices, and quasi-separatrix layers [*Priest and Titov*, 1996; *Demoulin et al., 1997*; *Titov*, 2007]. Prominently among the objectives for using FACs is to address the question of preeruption and posteruption magnetic field and current patterns discussed by *Schmieder and Aulanier* [Chapter 23, this volume], among many others.

In multiple other situations, one gathers this mixed impression of, first, NLFF field extrapolations leading to discrepancies from observations or to paradoxes altogether and, second, results of NLFF field extrapolations being used for state-of-the-art interpretation, with varying, sometimes high, levels of success. Deviations in the first category should be attributed, at least to some degree, to the photospheric existence and action of CFCs. In particular, one sees the discrepancies between traced coronal magnetic loops in the extreme ultraviolet and NLFF field extrapolations (e.g., *Aschwanden et al.* [2014], among others), reproduced here in Figure 22.8, or the calculation of significantly different free magnetic energies from two energy volume integrals that should otherwise be equivalent for fully force-free and solenoidal extrapolation results [*Moraitis et al.*, 2014]. Further on, using the photospheric J_z-map (or α-map) of one polarity for the NLFF extrapolation, one is led to nonequivalent results depending on the choice of polarity that, however, could be reasonably tackled by enforcing a kind of self-consistency check [*Wheatland and Regnier*, 2009; *Wheatland and Leka*, 2011] or the inclusion of possible noise in the boundary condition [*Wiegelmann and Inhester*, 2010; *Wheatland*, 2015]. The task of "preprocessing" photospheric vector magnetograms to bring them closer to a NLFF configuration (e.g., *Wiegelmann et al.* [2006]), that inevitably alters the original magnetogram, also attests to the difficulties faced by (some, at least) NLFF field extrapolation methods.

The above taken, NLFF field extrapolations, or the NLFF approximation in general, have greatly helped

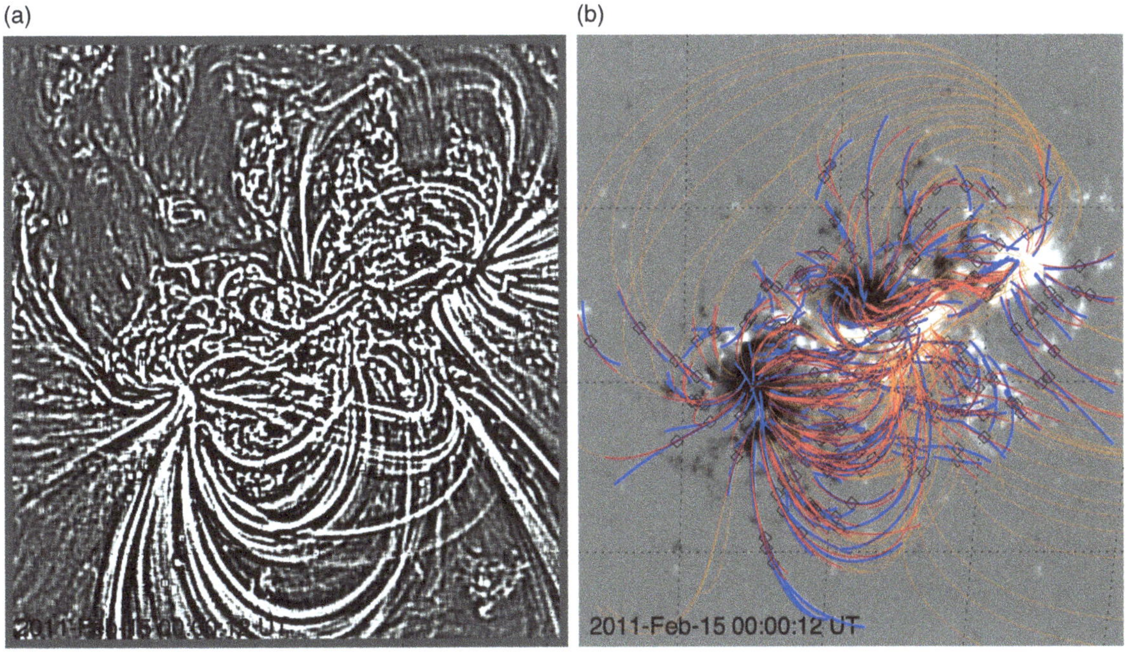

Figure 22.8 Comparison between traced loops in the extreme ultraviolet corona (171 Å in [a] and blue/red curves in [b]) and forward-fitting NLFF field extrapolation results (orange lines in [b]). The grayscale background in [b] corresponds to the photospheric normal magnetic field of NOAA AR 11158, an eruptive active region observed in February 2011. From *Aschwanden et al.* [2014].

infer reasonable values of the free magnetic energy and the relative magnetic helicity, including their temporal variations, in active regions (for example, *Kusano et al.* [2002]; *Nindos et al.* [2003]; *LaBonte et al.* [2007]; *Georgoulis et al.* [2009]; *Sun et al.* [2012]; *Vemareddy et al.* [2012]; *Tziotziou et al.* [2013], among many others). Free energy and relative helicity are fundamental nonpotentiality parameters that can explain the powering of solar eruptions (see, e.g., *Tziotziou et al.* [2012]; *Valori et al.* [2016] and references therein). In fact, some of the studies above have even identified possible eruption-related variations to free-energy and relative-helicity time series. Moreover, NLFF field extrapolations have achieved plausible interpretations of particular eruption cases (e.g., *Schrijver et al.* [2008]; *He et al.* [2011]; *Tadesse et al.* [2012]; *Chintzoglou et al.* [2015]) or even frustrated eruptions (e.g., *Sun et al.* [2015]). In an interesting case, *Dalmasse et al.* [2015b] managed to interpret an "atypical" flare exhibiting both eruptive and confined characteristics by means of a topological analysis based on a LFF field extrapolation(!).

Last but not least, FACs appear instrumental for at least some ideal (i.e., in the absence of magnetic reconnection) MHD solar eruption mechanisms and are important for reconnection-related particle acceleration and heating in the solar atmosphere.

In terms of eruption mechanisms, the helical kink instability [*Baty and Heyvaerts*, 1996; *Török and Kliem*, 2005; *Rust and LaBonte*, 2005], as well as the torus instability [*Kliem and Török*, 2006; *Demoulin and Aulanier*, 2010], appear as plausible eruption-initiation mechanisms not intimately related to magnetic reconnection, although their action can lead to the latter, resulting in major flares. The helical kink instability, for instance, depends on the internal twist of so-called magnetic flux "ropes", namely, twisted flux tubes with a relatively untwisted core and a much more twisted periphery, and imposes a critical twist threshold for triggering. This may or may not imply CFCs, hence a Lorentz force, but twist certainly implies FACs. The torus instability, on the other hand, explicitly involves the Lorentz self (or "hoop") force that acts to an expanding flux rope and can totally destabilize it leading to an uncontained expansion if the overlaying, "strapping" global magnetic field does not exert enough magnetic pressure to contain it. In this physical picture of the torus instability, it is harder to see a fundamental FAC role than in the helical kink instability but FACs do exist in the formation of the flux rope per se as, again, twist essentially involves FACs (see, e.g., *Malanushenko et al.* [2011]; *Inoue et al.* [2011]). In case of resistive MHD eruption mechanisms (i.e., relying on magnetic reconnection; see, e.g., *Forbes et al.* [2006]; *Mikic and Lee* [2006] for reviews) it is probably harder to directly link FACs to the eruption.

In terms of particle acceleration and subsequent heating of the coronal plasma, mechanisms requiring magnetic reconnection as the source of acceleration and heating (traditionally known as "DC" mechanisms [*Heyvaerts and Priest*, 1984; *Milano et al.*, 1997]) need an overall knowledge of field lines and FACs in order to understand and interpret propagation of accelerated particles and heating-related energy losses due to collisions within coronal loops (see, e.g., *Vlahos* [2007], *Klimtchuk et al.* [2008], *Cargill et al.* [2012]). Indeed, although the theory for cross-field particle acceleration, resulting in runaway particles, is long existing [*Dreicer*, 1959], the apparent lack of such strong electric fields in the corona due to the absence of cross-field flows (the plasma "frozen-in" condition) leaves this scenario without adequate support in the solar atmosphere. On the other hand, there are models that exclusively rely on FACs to study heating and acceleration processes with apparent success: for instance, one-dimensional coronal loop models have extensively studied DC loop heating (e.g., *Patsourakos and Klimchuk* [2006]), while 3D NLFF field extrapolation models have been used to study DC heating and particle acceleration in tandem (e.g., *Moraitis et al.* [2016]).

22.6. SUMMARY AND CONCLUSION

This brief review attempts to summarize the copious efforts, main results, debates, and controversies accompanying the quantitative exploitation of solar magnetic field measurements, particularly highlighting the implications for the electric currents in the solar atmosphere and additionally focusing on currents aligned (parallel or antiparallel) to the local magnetic field vector (FACs). An overwhelming majority of magnetic field measurements are photospheric, with some rare chromospheric data extending mostly up to ≈ 2 Mm from the $\tau = 1$ interface. Solar vector magnetograms discussed here invariably stem from the well-known Zeeman effect giving rise to circularly and linearly polarized Stokes images via which magnetic field components are inverted. However, the fundamental problems of magnetic field inversion and 180° azimuth disambiguation, a degeneracy of the Zeeman effect properties, are omitted here to allow better focus on electric currents.

A key issue inhibiting much of the analysis and causing much of the existing controversy is the incomplete inference of electric currents in the solar atmosphere. For single-height, photospheric-only vector magnetograms, only the vertical (i.e., normal to the photosphere) component of the electric current density can be inferred without further assumptions. Only in less than a handful of cases has the combined quality and correspondence of photospheric and chromospheric vector magnetograms been high enough to allow a two-dimensional tomographic analysis of fields and currents as functions of height. Therefore, key questions pertain to the a priori unknown role of FACs in the solar atmosphere and how FACs influence the observed eruptive and noneruptive manifestations of solar magnetism.

To this author's understanding, based also on the literature reviewed here, FAC-based results and the role of FACs should be taken *cum grano salis*: the relevance and significance of FACs directly reflect on the question one is asking. One should also be aware of the FAC limitations. Lack or gaps of knowledge and understanding are the core reasons for the ambivalence noticed in this discussion.

Observations and models seem to converge on a widely variable significance of FACs in the solar atmosphere: from the photosphere, where FACs succumb to the dominance of CFCs flowing perpendicular to the local magnetic field vector, to slightly higher (up to 1–2 Mm above, toward the corona), where the situation is largely mixed (see, e.g., Fig. 22.4), to the overlaying corona, where FACs must overwhelmingly dominate as per the observed plasma properties. FACs seem to defy the notion of neutralized currents in the solar atmosphere, a more-or-less generally accepted principle until the first high-resolution solar vector magnetograms appeared: in fact, one seems to require CFCs to achieve current neutrality and it is only due to perturbed CFC configurations and CFC inhomogeneities and asymmetries appearing exclusively along sheared magnetic PILs that current nonneutralization appears in the photosphere (sections 22.3.2 and 22.4). One could understand in a simplified way why this is the case by recalling that the Stokes theorem of equation (22.15) implies complete liberty in selecting the contour within which a footprint is enclosed. For isolated magnetic structures embedded in mostly field-free areas, a contour selection within the quiet Sun will render a zero total current, within uncertainties. In PILs, however, there is only one path that a given contour can follow to fully enclose the involved footprints: along the PIL itself, where CFCs are perturbed and distorted. This causes nonneutralization and net currents injected in the solar atmosphere. The physical origin of shear along photospheric PILs remains debated, with only one proposed, yet to be confirmed, origin, namely the action of the tension component of the Lorentz force.

Although active-region photospheric fields are more or less nonneutralized along their PILs, the scenario favoring electric circuits and the (**E**, **J**) paradigm, with a closure of electric currents in the tachocline (Fig. 22.5a), seems rather unsupported by observations and models (section 22.4). Instead, the MHD (**B**, **u**) paradigm, with current closure around the photosphere (on, slightly above or slightly below) appears more likely (Fig. 22.5b). This said, subphotospheric magnetic field measurements are still (and likely will continue to be) lacking, so physical models will continue to be our best means for future discussion.

Also not fully founded is the notion that the inferred photospheric vertical current density is meaningless due to the existing, partly unresolved, photospheric observations (section 22.4). There is considerable evidence that at least some of the observed J_z-patterns are real, particularly along PILs. Horizontal current densities, estimated following certain assumptions, are also significant in the photosphere, even higher than J_z. However, in unipolar plage areas consisting of nearly vertical fields, $J_z \approx 0$. Even if one (partly) trusts J_z, however, the question of how is this current density distributed between FACs and CFCs remains outstanding for single-height measurements: only magnetostatic extrapolations and MHD models of active regions are in position to address this, subject to their own shortcomings and limitations. Alternatively, detailed multi-height measurements may determine this, provided that various practical issues are resolved (section 22.2.2).

Why are FACs still important to study and exploit, then, in spite of the above caveats? Arguably the best use of FACs is when one extrapolates to the unknown coronal magnetic field by using the observed photospheric/low-chromospheric field as boundary condition. FACs are necessary in this case, with chromospheric vector magnetograms certainly making a more intuitive boundary condition than photospheric ones. Force-free extrapolations are often necessary to achieve an overall, imperfect but still useful, quantitative understanding of the coronal magnetic configuration above a set of photospheric flux footprints (section 22.5 and Figs. 22.7, 22.8). This understanding would be lacking otherwise and CFCs cannot help in achieving it. This information is more valuable in case an abstract, topological examination of the coronal magnetic configuration is attempted, particularly for features such as null points and separatrices, that are considered topological invariants. In this basis, discrepancies between current-free, LFF and NLFF extrapolation results are secondary to the knowledge of the location of these features, that can help interpret the active-region evolution, as well as the triggering of eruptions via a well-defined topological skeleton. Remarkably, FACs seem to favor popular ideal-MHD eruption mechanisms, such as the kink and torus instabilities.

FAC-based, force-free extrapolations (or approaches, in general) offer a practical way to infer budgets and time variations of magnetic free energy and helicity in the active-region atmosphere, that are instrumental for interpreting solar eruptions. Three-dimensional magnetostatic or MHD models also allow this but their routine application to long time series of high-resolution vector magnetograms seems out of reach at the moment. Perhaps most importantly, FAC-based methods allow one to fundamentally realize extremes in free magnetic energy and relative helicity: the Woltjer-Taylor theorem [*Woltjer*, 1958; *Taylor*, 1974, 1986] implies that a LFF field is a minimum free-energy field for a given relative helicity. Hence, a fully relaxed, confined, helical magnetic configuration is essentially a LFF field configuration. This hypothesis has also been debated in the literature (e.g., *Antiochos et al.* [2002]; *Kusano* [2005]; *Georgoulis et al.* [2012b]) but only on the physical basis of whether a nonpotential solar magnetic structure is left to relax to this simple state. This does not diminish the, theoretical, at least, significance of FACs in constraining magnetic energy and helicity budgets.

FACs are also used for the correspondence between solar radio emission and coronal magnetic fields (e.g., *Yan et al.* [2006]; *Tun et al.* [2011]), including the understanding of Type III bursts as FAC-related electron beams, and even for advancing the discrimination between flaring and nonflaring active regions (e.g., *Bobra and Couvidat* [2015]; *Barnes et al*, [2016]). It is, therefore, natural to assume that they will continue to be used in the future, only to become a secondary preference when computing capabilities on high-resolution, high-cadence magnetograms amply exceed today's norm, allowing for detailed, routine magnetostatic or MHD modeling.

Concluding, this review is evidently oriented toward solar active-region magnetic fields. This is because active regions have arguably attracted much of the interest in local-scale solar magnetism. Notwithstanding, there are important studies on quiet-Sun magnetic fields, most recently by the SUNRISE Balloon-Borne mission [*Solanki et al.*, 2010]. In one of these works, *Wiegelmann et al.* [2015] reported a nonnegligible CFC presence in the photosphere, giving way to FACs ≈ 2 Mm above. Very weak, if any, nonneutralized currents should be expected in the quiet-Sun photosphere, that should be approached either by current-free or by NLFF field extrapolations. Indeed, the randomness in the convective formation of these flux elements renders a coherent, homogeneous LFF approach meaningless. Moreover, a quiet-Sun situation mostly resembles the *Parker* [1996b] analysis, so the presence of FACs (or electric currents altogether) in the quiet Sun's low atmosphere, from photosphere to the outskirts of the corona, should be scrutinized beyond artifacts and applicable uncertainties.

ACKNOWLEDGMENTS

The author sincerely thanks the organizers of the AGU Chapman Conference on Currents in Geospace in Beyond for their kind invitation to deliver a talk and to contribute this chapter to the present volume. Partial financial support by the AGU to attend the conference is also gratefully acknowledged. Part of the author's work presented here was supported by a Marie Curie grant of the European Commission's Seventh Framework Programme entitled "Solar Magnetic Eruptions: Understanding and Forecasting" (grant agreement no. PIRG07 GA 2010 268245), while meeting participation was partially supported by a research grant of the Academy of Athens Research Committee entitled "Dissemination of the A-EFFort Project Results" (grant no. 200/857). The contribution of three anonymous reviewers to the final shaping of the review is gratefully acknowledged.

REFERENCES

Abbett, W. P., Z. Mikic, J. A. Linker, J. M. McTiernan, T. Magara, and, G. A. Fisher (2004), The photospheric boundary of sun-to-earth coupled models, *J. Atmos. Solar Terr. Phys.*, *66*, 1257.

Alissandrakis, C. (1981), On the computation of constant alpha force-free magnetic field, *Astron. Astrophys.*, *100*, 197.

Aly, J. J. (1989), On the reconstruction of the nonlinear force-free coronal magnetic field from boundary data, *Solar Phys.*, *120*, 19.

Antiochos, S. K., J. T. Karpen, and C. R. DeVore (2002), Coronal magnetic field relaxation by null-point reconnection, *Astrophys. J.*, *575*, 587.

Aschwanden, M. J., X. Sun, and L. Yang (2014), The magnetic field of active region 11158 during the 2011 February 12–17 flares: Differences between photospheric extrapolation and coronal forward-fitting methods, *Astrophys. J.*, *785*, 34.

Balthasar, A. (2006), Vertical current densities and magnetic gradients in sunspots, *Astron. Astrophys.*, *449*, 1169.

Barnes, G., K. D. Leka, C. J. Schrijver, et al. (2016), A comparison of flare forecasting methods, I. Results from the "all-clear" workshop, *Astrophys. J.*, *829*, 89.

Baty, H., and J. Heyvaerts (1996), Electric current concentration and kink instability in line-tied coronal loops, *Astron. Astrophys.*, *308*, 935.

Bobra, M. G., and S. Couvidat (2015), Solar flare prediction using SDO/HMI vector magnetic field data with a machine-learning algorithm, *Astrophys. J.*, *798*, 135.

Bommier, V. (2013), Reconciliating the vertical and horizonal gradients of the sunspot magnetic field, *Phys. Res. Int.*, article id. 195403.

Borrero, J. M., and K. Ichimoto (2011), Magnetic structure of sunspots, *Living Rev. Solar Phys.*, *8*, article id. 4.

Bruzek, A.(1967), On arch-filament systems in spotgroups, *Solar Phys.*, *2*, 451.

Cargill, P. J., L. Vlahos, G. Baumann, J. F. Drake, and A. Nordlund (2012), Current fragmentation and particle acceleration in solar flares, *Space Sci. Rev.*, *173*, 223.

Chintzoglou, G., S. Patsourakos, and A. Vourlidas (2015), Formation of magnetic flux ropes during a confined flaring well before the onset of a pair of major coronal mass ejections, *Astrophys. J.*, *809*, 34.

Crouch, A. D. (2013), Resolving the azimuthal ambiguity in vector magnetogram data with the divergence-free condition: The effects of noise and limited spatial resolution, *Solar Phys.*, *282*, 107.

Crouch, A. D. (2015), Resolving the azimuthal ambiguity in vector magnetogram data with the divergence-free condition: implementations for disambiguating each height independently, *Solar Phys.*, *290*, 2677.

Cuperman, S., C. Bruma, and D. Heristchi (1997), Three-dimensional chromospheric magnetic field configurations based on photospheric-vector and chromospheric-multi-level longitudinal-magnetic field observations, *Astron. Astrophys. Suppl. Series*, *122*, 381.

Dalmasse, K., G. Aulanier, P. Demoulin, B. Kliem, T. Torok, and E. Pariat (2015a), The origin of net electric currents in solar active regions, *Astrophys. J.*, *810*, article id. 17.

Dalmasse, K., R. Chandra, B. Schmieder, and G. Aulanier (2015b), Can we explain atypical solar flares?, *Astron. Astrophys.*, *574*, id. A37.

Demoulin, P., and Aulanier, G., (2010), *Astrophys. J.*, 718, 1388–1399; doi:10.1088/0004-637X/718/2/1388.

Demoulin, P., L. G. Bagala, C. H. Mandrini, J. C. Henoux, and M. G. Rovira (1997), Quasi-separatrix layers in solar flares, II. Observed magnetic configurations, *Astron. Astrophys.*, *325*, 305.

DeRosa, M. L., C. J. Schrijver, G. Barnes, et al. (2009), A critical assessment of nonlinear force-free field modeling of the solar corona for active region 10953, *Astrophys. J.*, *696*, 1780.

DeRosa, M. L., M. S. Wheatland, K. D. Leka, et al. (2015), The influence of spatial resolution on nonlinear force-free modeling, *Astrophys. J.*, *811*, article id. 107.

Ding, Y. J., Q. F. Hong, M. J. Hagyard, A. C. Deloach, and X. P. Liu (1987), Explorations of electric current system in solar active regions, I. Empirical inferences of the current flows, *Solar Phys.*, *109*, 307.

Dreicer, H. (1959), Electron and ion runaway in a fully ionized gas, I, *Phys. Rev.*, *115*, 238.

Falconer, D. A.(2001), A prospective method for predicting coronal mass ejections from vector magnetograms, *J. Geophys. Res.*, *106* (A11), 25185.

Fisher, G. H., D. J. Bercik, B. T. Welsch, and H. S. Hudson (2012), Global forces in eruptive solar flares: The Lorentz force acting on the solar atmosphere and the solar interior, *Solar Phys.*, *277*, 59.

Forbes, T., J. A. Linker, J. Chen, et al. (2006), CME theory and models, *Space Sci. Rev.*, *123*, 251.

Foukal, P. (1971), Hα fine structure and the chromospheric field, *Solar Phys.*, *20*, 298.

Gary, G. A. (2001), Plasma beta above a solar active region: Rethinking the paradigm, *Solar Phys.*, *203*, 71.

Georgoulis, M. K. (2012), Comment on "Resolving the 180° ambiguity in solar vector magnetic field data: Evaluating the effects of noise, spatial resolution, and method assumptions," *Solar Phys.*, *276*, 423.

Georgoulis, M. K., and B. J. LaBonte (2004), Vertical Lorentz force and cross-field currents in the photospheric magnetic fields of solar active regions, *Astrophys. J.*, *615*, 1029.

Georgoulis, M. K., and B. J. LaBonte (2007), Magnetic energy and helicity budgets in the active region solar corona, I. Linear force-free approximation, *Astrophys. J.*, *671*, 1034.

Georgoulis, M. K., D. M. Rust, A. A. Pevtsov, P. N. Bernasconi, and K. M. Kuzanyan (2009), Solar magnetic helicity injected into the heliosphere: magnitude, balance, and periodicities over solar cycle 23, *Astrophys. J.*, *705*, L48.

Georgoulis, M. K., V. S. Titov, and Z. Mikic (2012a), Non-neutralized electric current patterns in solar active regions: Origin of the shear-generating Lorentz force, *Astrophys. J.*, *761*, article id. 61.

Georgoulis, M. K., K. Tziotziou, and, N.-E. Raouafi (2012b), Magnetic energy and helicity budgets in the active-region solar corona, II. Nonlinear force-free approximation, *Astrophys. J.*, *759*, article id. 1.

Gilchrist, S. A., and M. S. Wheatland (2013), A magnetostatic grad-rubin code for coronal magnetic field extrapolations, *Solar Phys.*, *282*, 283.

Hagyard, M. J. (1988), Observed nonpotential magnetic fields and the inferred flow of electric currents at a location of repeated flaring, *Solar Phys.*, *115*, 107.

Hagyard, M. J., D. Teuber, E. A. West, and, J. B. Smith (1984), A quantitative study relating observed shear in photospheric magnetic fields to repeated flaring, *Solar Phys.*, *91*, 115.

Hale, G. E.(1908), On the probable existence of a magnetic field in sun-spots, *Astrophys. J.*, *28*, 315.

Harvey, J. W. (1969), Magnetic Fields Associated with Solar Active-Region Prominences, PhD Thesis, University of Colorado, Boulder, Colorado.

He, H., H. Wang, and, Y. Yan (2011), Nonlinear force-free field extrapolation of the coronal magnetic field using the data obtained by the Hinode satellite, *J. Geophys. Res.*, *116* (A11), ID A01101.

Heyvaerts, J., and E. R. Priest (1984), Coronal heating by reconnection in DC current systems: A theory based on Taylor's hypothesis, *Astron. Astrophys.*, *137*, 63.

Hoeksema, J. T. (1984), *Structure and Evolution of the Large Scale Solar and Heliospheric Magnetic Fields*, PhD Thesis, Stanford University, California.

Hofmann, A., V. M. Grigorev, and V. L. Selivanov (1988), Analysis and results of cooperative magnetographic measurements, II. Comparison and interpretation of the current densities, *Astron. Nachr.*, *309*, 373.

Hu, Q., B. Dasgupta, D. P. Choudhary, and J. Buchner (2008), A practical approach to coronal magnetic field extrapolation based on the principle of minimum dissipation rate, *Astrophys. J.*, *679*, 848.

Hudson, H. S., L. Fletcher, G. H. Fisher, W. P. Abbett, and A. Russell (2012), Momentum distribution in solar flare processes, *Solar Phys.*, *277*, 77.

Inoue, S., K. Kusano, T. Magara, D. Shiota, and T. T. Yamamoto (2011), Twist and connectivity of magnetic field lines in the solar active region NOAA 10930, *Astrophys. J.*, *738*, 161.

Kliem B., and Török, T., (2006), *Phys. Rev. Let.*, *96*, id. 255002; doi:10.1103/PhysRevLett.96.255002.

Klimtchuk, J. A., S. Patsourakos, and J. P. Cargill (2008), Highly efficient modeling of dynamic coronal loops, *Astrophys. J.*, *682*, 1351.

Kotov, V. A. (1971), On the structure of magnetic field and electric currents of a unipolar sunspot (presented by A. Severny), in *Solar Magnetic Fields*, edited by R. Howard, IAU Symp. *43*, 212.

Kusano, K. (2005), Simulation study of the formation mechanism of sigmoidal structure in the solar corona, *Astrophy. J.*, *631*, 1260.

Kusano, K., T. Maeshiro, T. Yokoyama, and T. Sakurai (2002), Measurement of magnetic helicity injection and free energy loading into the solar corona, *Astrophys. J.*, *577*, 501.

LaBonte, B. J., M. K. Georgoulis, and D. M. Rust (2007), Survey of mMagnetic helicity injection in regions producing X-class flares, *Astrophys. J.*, *671*, 955.

Leka, K. D. (1999), On the value of "αAR" from vector magnetograph data, II. Spatial resolution, field of view, and validity, *Solar Phys.*, *188*, 21.

Leka, K. D., G. Barnes, A. D. Crouch, T. R. Metcalf, G. A. Gary, J. Jing, and Y. Liu (2009), Resolving the 180° ambiguity in solar vector magnetic field data: Evaluating the effects of noise, spatial resolution, and method assumptions, *Solar Phys.*, *260*, 83.

Leka, K. D., R. C. Canfield, A. N. McClymont, and, L. van Driel-Gesztelyi (1996), Evidence for current-carrying emerging flux, *Astrophys. J.*, *462*, 547.

Leka, K. D., and A. Skumanich (1999), On the Value of "αAR" from vector magnetograph data, I. Methods and caveats, *Solar Phys.*, *188*, 3.

Lites, B. W., M. Kubo, H. Socas-Navarro, et al. (2008), The horizontal magnetic flux of the quiet-sun internetwork as observed with the Hinode spectro-polarimeter, *Astrophys. J.*, *672*, article id. 1237.

Longcope, D. W., and, B. T. Welsch (2000), A model for the emergence of a twisted magnetic flux tube, *Astrophys. J.*, *545*, 1089.

López Ariste, A., J. Rayrole, and M. Semel (2000), First results from THEMIS spectropolarimetric mode, *Astron. Astrophys. Suppl.*, *142*, 137.

Low, B. C. (1984), On the large-scale magnetostatic coronal structures and their stability, *Astrophys. J.*, *286*, 772.

Malanushenko, A., M. H. Yusuf, and D. W. Longcope (2011), Direct measurements of magnetic twist in the solar corona, *Astrophys. J.*, *736*, article id. 97.

Manchester, W., IV, T. Gombosi, D. DeZeeuw, and Y. Fan (2004), Eruption of a buoyantly emerging magnetic flux rope, *Astrophys. J.*, *610*, 588.

Martinez Pillet, V., B. W. Lites, and A. Skumanich (1997), Active region magnetic fields, I. Plage fields, *Astrophys. J.*, *474*, 810.

McClymont, A. N., L. Jiao, and Z. Mikic (1997), Problems and progress in computing three- dimensional coronal active region magnetic fields from boundary data, *Solar Phys.*, *174*, 191.

Melrose, D. B. (1991), Neutralized and unneutralized current patterns in the solar corona, *Astrophys. J.*, *381*, 306.

Melrose, D. B. (1995), Current paths in the corona and energy release in solar flares, *Astrophys. J.*, *451*, 391.

Metcalf, T. R., K. D. Leka, G. Barnes, et al. (2006), An overview of existing algorithms for resolving the 180° ambiguity in vector magnetic fields: Quantitative tests with synthetic data, *Solar Phys.*, *237*, 267.

Metcalf, T. R., L. Jiao, A. N. McClymont, R. C. Canfield, and H. Uitenbroek (1995), Is the solar chromospheric magnetic field force-free?, *Astrophys. J.*, *439*, 474.

Metcalf, T. R., M. L. DeRosa, C. J. Schrijver, G. Barnes, A. A. van Ballegooijen, T. Wiegelmann, M. S. Wheatland, G. Valori, and, J. M. McTiernan (2008), Nonlinear force-free modeling of coronal magnetic fields, II. Modeling a filament arcade and simulated chromospheric and photospheric vector fields, *Solar Phys.*, *247*, 269.

Mickey, D. L. (1985), The Haleakala stokes polarimeter, *Solar Phys.*, *97*, 223.

Mickey, D. L., R. C. Canfield, B. J. LeBonte, K. D. Leka, M. F. Waterson, and H. M. Weber (1996), The imaging vector magnetograph at Haleakala, *Solar Phys.*, *168*, 229.

Mikic, Z., and M. A. Lee (2006), An introduction to theory and models of CMEs, shocks, and solar energetic particles, *Space Sci. Rev.*, *123*, 57.

Milano, L. J., D. O. Gomez, and P. C. H. Martens (1997), Solar coronal heating: AC versus DC, *Astrophys. J.*, *490*, 442.

Moon, Y.-J., G. S. Choe, H. S. Yun, Y. D. Park, and, D. L. Mickey (2002), Force-freeness of solar magnetic fields in the photosphere, *Astrophys. J.*, *568*, 422.

Moraitis, K., A. Toutountzi, H. Isliker, M. K. Georgoulis, L. Vlahos, and G. Chintzoglou (2016), An observationally driven kinetic approach to coronal heating, *Astron. Astrophys.*, *596*, id. A56.

Moraitis, K., K. Tziotziou, M. K. Georgoulis, and V. Archontis (2014), Validation and benchmarking of a practical free magnetic energy and relative magnetic helicity budget calculation in solar magnetic structures, *Solar Phys.*, *289*, 4453.

Moreton, G. E., and A. B. Severny (1968), Magnetic fields and flares in the region CMP 20 September 1963, *Solar Phys.*, *3*, 282.

Nindos, A., J. Zhang, and H. Zhang (2003), The magnetic helicity budget of solar active regions and coronal mass ejections, *Astrophys. J.*, *594*, 1033.

Pariat, E., G. Aulanier, B. Schmieder, M. K. Georgoulis, D. M. Rust, and P. N. Bernasconi (2004), Resistive emergence of undulatory flux tubes, *Astrophys. J.*, *614*, 1099.

Parker, E. N. (1972), Topological dissipation and the small-scale fields in turbulent gases, *Astrophys. J.*, *174*, 499.

Parker, E. N. (1979), *Cosmical Magnetic Fields: Their Origin and Their Activity*, Oxford University Press, New York.

Parker, E. N. (1996a), Comment on "Current Paths in the Corona and Energy Release in Solar Flares," *Astrophys. J.*, *471*, 489.

Parker, E. N. (1996b), Inferring mean electric currents in unresolved fibril magnetic fields, *Astrophys. J.*, *471*, 485.

Parker, E. N. (1997), Reply, *J. Geophys. Res.*, *102*(A5), 9657.

Parker, E. N. (2007), *Conversations on Electric and Magnetic Fields in the Cosmos*, Princeton University Press, Princeton.

Parks, G. K. (2004), Why space physics needs to go beyond the MHD box, *Space Sci. Rev.*, *113*, 97.

Patsourakos, S., and J. A. Klimchuk (2006), Nonthermal spectral line broadening and the nanoflare model, *Astrophys. J.*, *647*, 1452.

Petrie, G. J. D. (2012), The abrupt changes in the photospheric magnetic and Lorentz force vectors during six major neutral-line flares, *Astrophys. J.*, *759*, article id. 50.

Petrie, G. J. D. (2013), A spatio-temporal description of the abrupt changes in the photospheric magnetic and Lorentz-force vectors during the 15 February 2011 X2.2 flare, *Solar Phys.*, *287*, 415.

Pevtsov, A. A., and, N. L. Peregud (1990), *Physics of Magnetic Flux Ropes*, edited by C. T. Russel, E. R. Priest, and C. Lee, 161, Geophysical Monograph, 58, American Geophysical Union, Washington, DC.

Pizzo, V. J. (1986), Numerical solution of the magnetostatic equations for thick flux tubes, with application to sunspots, pores, and related structures, *Astrophys. J.*, *302*, 785.

Priest, E. R. (1982), *Solar Magneto-hydrodynamics*, D. Reidel Pub. Co., Hingham.

Priest, E. R., and V. S. Titov (1996), Magnetic reconnection at three-dimensional null points, *Phil. Trans. Math. Phys. Eng. Sci.*, *354*, 2951.

Puschmann, K. G., B. Ruiz Cobo, and V. Martinez Pillet (2010), The electrical current density vector in the inner penumbra of a sunspot, *Astrophys. J.*, *721*, L58.

Ravindra, B., P. Venkatakrishnan, S. K. Tiwari, and R. Bhattacharyya (2011), Evolution of currents of opposite signs in the flare-productive solar active region NOAA 10930, *Astrophys. J.*, *740*, article id. 19.

Rempel, M. (2011), Penumbral fine structure and driving mechanisms of large-scale flows in simulated sunspots, *Astrophys. J.*, *729*, article id. 5.

Rosner, R., and G. S. Vaiana (1978), Cosmic flare transients: Constraints upon models for energy storage and release derived from the event frequency distribution, *Astrophys. J.*, *222*, 1104.

Rust, D. M., and B. J. LaBonte (2005), Observational evidence of the kink instability in solar filament eruptions and sigmoids, *Astrophys. J.*, *622*, L69.

Savcheva, A., E. Pariat, A. A. van Ballegooijen, G. Aulanier, and E. DeLuca (2012), Sigmoidal active region on the sun: Comparison of a magnetohydrodynamical simulation and a nonlinear force-free field model, *Astrophys. J.*, *750*, article id. 15.

Scherrer, P. H., J. Schou, R. I. Bush, et al.(2012), The helioseismic and magnetic imager (HMI), Investigation for the Solar Dynamics Observatory (SDO), *Solar Phys.*, *275*, 207.

Schrijver, C. J., A. M. Title, K. L. Harvey, N. R. Sheeley, Y.-M. Wang, G. H. J. van der Oord, R. A. Shine, T. D. Tarbell, and, N. E. Hurlburt (1998), Large-scale coronal heating by the small-scale magnetic field of the Sun, *Nature*, *394*, 152.

Schrijver, C. J., M. L. De Rosa, T. R. Metcalf, Y. Liu, J. McTiernan, S. Regnier, G. Valori, M. S. Wheatland, and T. Wiegelmann (2006), Nonlinear force-free modeling of coronal magnetic fields, Part I: A quantitative comparison of methods, *Solar Phys.*, *235*, 161.

Schrijver, C. J., M. L. De Rosa, T. R. Metcalf, et al. (2008), Nonlinear force-free field modeling of a solar active region around the time of a major flare and coronal mass ejection, *Astrophys. J.*, *675*, 1637.

Semel, M., and A. Skumanich (1998), An ambiguity-free determination of J_z in solar active regions, *Astron. Astrophys.*, *331*, 383.

Severny, A. B. (1964,), Solar flares, *Ann. Rev. Astron. Astrophys.*, *2*, 363.

Socas-Navarro, H. (2005), The three-dimensional structure of a sunspot magnetic field, *Astrophys. J.*, *631*, L167.

Socas-Navarro, H., D. Elmore, A. Pietarila, A. Darnell, B. W. Lites, S. Tomczyk, and S. Hegwer (2006), Spinor: Visible and infrared spectro-polarimetry at the National Solar Observatory, *Solar Phys.*, *235*, 55.

Solanki, S. K., P. Barthol, S. Danilovic, et al.(2010), SUNRISE: Instrument, mission, data, and first results, *Astrophys. J.*, *723*, L127.

Spicer, D. S. (1982), Magnetic energy storage and conversion in the solar atmosphere, *Space Sci. Rev.*, *31*, 351.

Sun, X. (2014), The CGEM Lorentz force data from HMI vector magnetograms, arXiv, *1405*, 7353.

Sun, X., J. T. Hoeksema, Y. Liu, T. Wiegelmann, K. Hayashi, Q. Chen, and J. Thalmann (2012), Evolution of magnetic field and energy in a major eruptive active region based on SDO/HMI observation, *Astrophys. J.*, 748, article id. 77.

Sun, X., M. G. Bobra, J. T. Hoeksema, Y. Liu, Y. Li, C. Shen, S. Couvidat, A. A. Norton, and G. Fisher (2015), Why is the great solar active region 12192 flare-rich but CME-poor?, *Astrophys. J.*, *804*, L28.

Tadesse, T., T. Wiegelmann, B. Inhester, and A. Pevtsov (2012), Coronal magnetic field structure and evolution for flaring AR 11117 and its surroundings, *Solar Phys.*, *281*, 53.

Taylor, J. B. (1974), Relaxation of toroidal plasma and generation of reverse magnetic fields, *Phys. Rev. Lett.*, *33*, 1139.

Taylor, J. B. (1986), Relaxation and magnetic reconnection in plasmas, *Rev. Mod. Phys.*, *58*, 741.

Titov, V. S. (2007), Generalized squashing factors for covariant description of magnetic connectivity in the solar corona, *Astrophys. J.*, *660*, 863

Török, T., and B. Kliem (2005), Confined and ejective eruptions of kink-unstable flux ropes, *Astrophys. J.*, *630*, L97.

Török, T., J. E. Leake, V. S. Titov, V. Archontis, Z. Mikic, M. G. Linton, K. Dalmasse, G. Aulanier, and B. Kliem (2014), Distribution of electric currents in solar active regions, *Astrophys. J.*, *782*, article id. 10.

Tun, S. D., D. E. Gary, and M. K. Georgoulis (2011), Three-dimensional structure of a solar active region from spatially and spectrally resolved microwave observations, *Astrophys. J.*, *728*, article id. 1.

Tziotziou, K., M. K. Georgoulis, and N.-E. Raouafi (2012), The magnetic energy-helicity diagram of solar active regions, *Astrophys. J.*, *759*, article id. L4.

Tziotziou, K., M. K. Georgoulis, and Y. Liu (2013), Interpreting eruptive behavior in NOAA AR 11158 via the region's magnetic energy and relative-helicity budgets, *Astrophys. J.*, *772*, article id. 115.

Valori, G., E. Pariat, S. Anfinogentov, F. Chen, M. K. Georgoulis, Y. Guo, Y. Liu, K. Moraitis, J. K. Thalmann, and, S. Yang (2016), Magnetic helicity estimations in models and observations of the solar magnetic field, Part I: Finite volume methods, *Space Sci. Rev.*, *201*, 147.

Vemareddy, P., A. Ambastha, and R. A. Maurya (2012), On the role of rotating sunspots in the activity of solar active region NOAA 11158, *Astrophys. J.*, *761*, article id. 60.

Venkatakrishnan, P., and S. K. Tiwari (2009), On the absence of photospheric net currents in vector magnetograms of sunspots obtained from Hinode (Solar Optical Telescope/Spectro-Polarimeter), *Astrophys. J.*, *706*, L114.

Vlahos, L. (2007), Magnetic complexity, fragmentation, particle acceleration and radio emission from the Sun, in *The High Energy Solar Corona: Waves, Eruptions, Particles*, Lect. Notes Phys., *725*, 15.

Wang, H., M. W. Ewell Jr., H. Zirin, and G. Ai (1994b), Vector magnetic field changes associated with X-class flares, *Astrophys. J.*, *424*, 436.

Wang, T., A. Xu, and H. Zhang (1994a), Evolution of vector magnetic fields and vertical currents and their relationship with solar flares in AR 5747, *Solar Phys.*, *155*, 99.

Wang, Y.-M., and N. R. Sheeley Jr. (1990), Solar wind speed and coronal flux-tube expansion, *Astrophys. J.*, *355*, 726.

Wheatland, M. S. (2000), Are electric currents in solar active regions neutralized?, *Astrophys. J.*, *532*, 616.

Wheatland, M. S.(2015), Estimating electric current densities in solar active regions, *Solar Phys.*, *290*, 1147.

Wheatland, M. S., and K. D. Leka (2011), Achieving self-consistent nonlinear force-free modeling of solar active regions, *Astrophys. J.*, *728*, article id. 112.

Wheatland, M. S., and S. Regnier (2009), A self-consistent nonlinear force-free solution for a solar active region magnetic field, *Astrophys. J.*, *700*, L88.

Wiegelmann, T., and B. Inhester (2010), How to deal with measurement errors and lacking data in nonlinear force-free coronal magnetic field modelling?, *Astron. Astrophys.*, *516*, id. A107.

Wiegelmann, T., B. Inhester, and T. Sakurai (2006), Preprocessing of vector magnetograph data for a nonlinear force-free magnetic field reconstruction, *Solar Phys.*, *233*, 215.

Wiegelmann, T., and T. Sakurai (2012), Solar force-free magnetic fields, *Liv. Rev. Solar Phys.*, *9*, article id. 5.

Wiegelmann, T., T. Neukirch, D. H. Nicheler, S. K. Solanki, V. Martinez Pillet, and, J. M. Borrero (2015), Magneto-static modeling of the mixed plasma beta solar atmosphere based on sunrise/IMaX data, *Astrophys. J.*, *815*, article id. 10.

Wilkinson, L. K., A. G. Emslie, and G. A. Gary (1992), On neutralized currents in the solar corona, *Astrophys. J.*, *392*, L39.

Withbroe, G. L., and R. W. Noyes (1977), Mass and energy flow in the solar chromosphere and corona, *Ann. Rev. Astron. Astrophys.*, *15*, 363.

Woltjer, L. (1958), A theorem on force-free magnetic fields, *Proc. NAS*, *44*, 489.

Yan, Y., M. Pick, M. Wang, S. Krucker, and A. Vourlidas (2006), A radio burst and its associated CME on March 17, 2002, *Solar Phys.*, *239*, 277.

Zhang, H. (2001), Electric current and magnetic shear in solar active regions, *Astrophys. J.*, *557*, L71.

Zirin, H. (1988), *Astrophysics of the Sun*, Cambridge University Press, Cambridge and New York.

23

Solar Active Region Electric Currents Before and During Eruptive Flares

B. Schmieder and G. Aulanier

ABSTRACT

This chapter discusses electric currents in the preeruption state and in the course of eruptions of solar magnetic structures, using information from solar observations, nonlinear force-free field extrapolations relying on these observations, and three-dimensional magnetohydrodynamic (MHD) models. The discussion addresses the issue of neutralized versus nonneutralized currents in active regions and concludes that MHD models are able to explain nonneutralized currents in active regions by the existence of strong magnetic shear along the polarity inversion lines, thus confirming previous observations that already contained this result. The models have also captured the essence of the behavior of electric currents in active regions during solar eruptions, predicting current-density increases and decreases inside flare ribbons and in the interior of expanding flux ropes, respectively. The observed photospheric current-density maps, inferred from vector magnetic field observations, exhibit whirling ribbon patterns similar to the MHD model results, which are interpreted as the signatures of flux ropes and of quasi-separatrix layers (QSLs) between the magnetic systems in active regions. Enhancement of the total current in these QSLs during the eruptions and decreasing current densities at the footpoint of erupting flux ropes, has been confirmed in the observations.

23.1. INTRODUCTION

The occurrence of finite electric currents in the Sun's corona is a fundamental requirement for the existence of an active phenomena, such as solar flares and CMEs. Several reasons can be put forward. In the dilute corona, the plasma beta (β) is much lower than unity. So the structure of the coronal plasma is dominated by the magnetic field, and the magnetic energy is the main reservoir for all dynamics. Also, active phenomena occur on timescales far shorter than those on which the dense photosphere evolves. So the distribution of magnetic flux crossing the photosphere hardly changes during solar transients. Given the solenoidal nature of the magnetic field, the potential (i.e., current-free) field corresponds to the lowest magnetic energy state for a fixed photospheric flux distribution. However, slow photospheric motions can induce twist and stress of the magnetic field lines and lead to a nonpotential magnetic field, which carries electric currents. So current-carrying magnetic fields constitute the source of energy for active phenomena. Also beyond this global property, electric currents play several key local roles in magnetohydrodynamics (MHD). The most significant example is their presence in the Lorentz force, as well as in several terms of the generalized Ohm's law, including the usual resistive term and the Hall electric field, which are both involved in magnetic reconnection within current sheets.

In spite of their importance, few studies have been realized on solar electric currents over several decades and up to very recently. The reason is probably theoretical and observational (See also Chapters 03 and 22).

LESIA, Observatoire de Paris, PSL Research University, CNRS, Sorbonne Universités, UPMC University Paris 06, University Paris Diderot, Sorbonne Paris Cité, Meudon, France

Electric Currents in Geospace and Beyond, Geophysical Monograph 235, First Edition.
Edited by Andreas Keiling, Octav Marghitu, and Michael Wheatland.

Theoretically, several historical models had initially considered the electric current to be an input parameter. For example, "circuit models" for solar flares prescribed the total current I as it is in laboratory experiments [see e.g., *Alfvén and Carlqvist*, 1967; *Spicer*, 1982]. More generally, one approach to plasma physics that is often used in the magnetospheric community (and is advocated by, e.g., *Heikkila* [1997] and *Melrose* [1995]) is to make every calculation with the equations that use the electric field **E** and the current density **J** as variables. This is the so-called **E; J** paradigm. But both circuit models and the **E; J** paradigm have been criticized throughout the years (see *Parker* [1996a,b; 2001]). In this line, E. N. Parker argued that they should be substituted by the **B; v** paradigm in MHD. In the latter, the current densities **J** merely result from the generation of a finite curl of the magnetic field **B** through Ampère's equation, as a direct result of plasma flows **v** through the ideal induction equation. So, the current is not a prime variable, since it does not even appear in the governing equations of MHD. For this reason, electric currents have been underlooked for years.

Observationally, the photospheric magnetic field vector in solar active regions provides the vertical component of curl **B**. However the vector magnetic field can be measured only in the photosphere, and thus only the vertical component of the current density J_z passing through the photosphere to the corona can be identified. An idealized active region (AR) consists of two sunspots represented by a bipole with two opposite polarities linked by a flux tube (called a flux rope when it is twisted) in the corona. Since $\beta < 1$ in sunspots, the AR-scale magnetic-field should be close to force free not only in the corona but also in the photosphere. Thus, electric current density measured in sunspots indicates the flow of currents in the large-scale coronal flux tube. Most (if not all) vector magnetic field measurements used are quite sparse as well as relatively unreliable outside sunspots and their close vicinity. So photospheric electric-current densities are rather difficult to measure using the transverse magnetic field, which was and still is, commonly very noisy. In past observations, only one sign of electric current was detected in each polarity. A full neutralization of currents in an active region implies that the total net current is zero in each polarity of a bipole. Therefore, results such as net currents in active regions were regarded as uncertain (as reported by, e.g., *Gary et al.* [1987], *Hagyard* [1988], *Leka et al.* [1996], *Leka and Skumanich*, [1999], and *Wilkinson et al.*, [1992]). With the development of new-generation and dedicated ground-based telescopes (e.g., THEMIS in Canary Islands) and space-borne (e.g., Hinode, Solar Dynamic Observatory [SDO]) missions, more complicated electric current patterns have been detected in each polarity of ARs.

Table 23.1 Electric Current Density Signs for Positive and Negative Magnetic Helicity at Active Region Scale.

Magnetic helicity	Direct current	Return current
$H_B>0$	$J_Z\,B_Z>0$	$J_Z\,B_Z<0$
$H_B<0$	$J_Z\,B_Z<0$	$J_Z\,B_Z>0$

Thus, new MHD simulations have been developed during recent years. Some correspondence has been found with the observations. MHD simulations based on the existence of twisted flux ropes show that current densities of both signs exist. A twist of finite radius theoretically implies the occurrence of a sheath of currents at the edge of the twisted flux tube, which flows in an opposite direction to the direct currents. These are called the return currents. A twisted flux tube of left-handed twist has a negative helicity and a direct electric current that flows antiparallel to the magnetic field (and vice versa) [*Démoulin*, 2007; *Démoulin and Pariat*, 2009]. The modern conventional definition in MHD simulations, which holds for AR large-scale quasi-force-free coronal field is summarized in Table 23.1. Actually, this definition is consistent with the inference of direct current (i.e., with their dominant sign, as mentioned above) in past observations.

Indeed, observational studies by *Falconer et al.* [2002], *Georgoulis et al.* [2012], *Ravindra et al.* [2011], *Schmieder and Aulanier* [2012], and *Wheatland* [2000] showed new and frequent evidences of net current in solar active regions. These relatively recent observations have revived the old debate from the 1990s about the neutralization, or not, of electric currents in active regions [*Melrose*, 1991; Parker, 1996a]. In this perspective, the MHD simulations developed by *Dalmasse et al.* [2015] and *Török et al.* [2014] have explained how and why unneutralized current densities occur in active regions when magnetic shear is present along polarity inversion lines (PIL).

In parallel, progress on the description of the magnetic topology of active regions has contributed significantly to the interpretation of the current-density patterns observed before and during eruptions. The field-aligned currents in line-tied force-free coronal structures imply a direct continuity between the coronal and the photospheric currents, especially above sunspots where $\beta < 1$. Because of magnetic flux conservation along the flux tube, one can expect to see the footprint of the coronal current densities in the photosphere. In other words, the photospheric current pattern in sunspots should show a cross section of the electric current density distribution inside the flux tube. In addition to twisted flux tubes with their volume currents, other magnetic structures also exist in the corona and connect down to the photosphere. Among such structures, one can find open-closed boundaries that separate closed magnetic field above active regions from open field in their surroundings, and separatrices between different flux systems, or quasi-separatrix layers (QSLs), which are very narrow volumes across which field lines quickly change connectivity. In these thin volumes, electric current sheet can develop along their length and eventually trigger reconnection of the magnetic field lines, leading to flares [*Démoulin et al.*, 1996a]. Therefore, during eruptions, QSLs become very important structures. The footprints in the photosphere of these structures are thin lanes of QSLs, which appear as long and narrow hooks. These structures reveal the evolution of the electric current densities before and during eruptions, observationally and in MHD models [*Janvier et al.*, 2014].

This chapter addresses some of these issues, and describes new publications with recent findings and interpretations resulting from coupling current measurements and MHD models. The next section is focused on the evidence of preeruption currents in active regions (section 23.2), and the following section concerns the evolution of the currents during eruptions (section 23.3). We conclude in section 23.4.

23.2. PRE-ERUPTION CURRENTS IN ACTIVE REGION

23.2.1. Observations of Direct and Return Current

The photospheric current distribution is derived from measurements of the vector magnetic field using Ampère's law. The vector magnetic field is measured by using the Zeeman effect, which is sensitive mainly in photospheric lines. Hence, only the vertical component of electric current density J_z can be readily calculated in the photosphere. The first published calculations of J_z were presented by *Severny* [1964] and *Moreton and Severny* [1968] (see the thesis of *Harvey* [1969]) and later by *Hagyard* [1988], *Wilkinson et al.* [1992], and *Canfield et al.* [1993]. The J_z observations showed principally two areas of opposite sign electric current density partly overlying the leading and following polarities of active regions (AR). Electric currents in magnetically isolated regions flow from one polarity to the other polarity, and correspond to direct currents. This was the basis of flare models with unneutralized active regions.

Due to recent improvements in vector magnetographs, the measurements of electric current density are more reliable and now direct and return electric currents are detectable. Recent measurements indicate commonly that the return currents in total are much smaller than the direct currents and a net current still exists. From recent observations using high spatial resolution vector magnetograms (the Helioseismic and Magnetic Imager [HMI] aboard SDO, the Spectro Polarimeter [SP] aboard Hinode) and ground-based telescopes (e.g., THEMIS) active region current maps are routinely obtained. ARs with neutralized currents and ARs with unneutralized currents have both been observed [*Cheng and Ding*, 2016; *Georgoulis et al.*, 2012; *Gosain et al.*, 2014; *Metcalf et al.*, 2006; *Ravindra et al.*, 2011; *Schmieder et al.*, 2015; *Wheatland*, 2000; *Zhao et al.*, 2016].

We select one example of an AR with electric current density observations obtained with THEMIS. From the retrieved Stokes parameters *I, Q, U, V* of the vector magnetogram, the three magnetic field components were obtained by using the Milne-Eddington inversion code UNNOFIT [*Bommier et al.*, 2007]. The active region was a newborn active region (AR) [*Bommier*, 2013; *Canou et al.*, 2009; *Li et al.*, 2007; *Schmieder and Aulanier*, 2012; *Schmieder et al.*, 2015]. The photospheric magnetic field of the active region presented two elongated tongues of opposite polarities (Fig. 23.1 a). The photospheric magnetic field vector direction (see the arrows in Fig. 23.1a) suggests that the twist of the emerging flux tube is right handed, that is, has a positive helicity. It has been shown that such a tongue-shape pattern in the photospheric magnetograms indicates that the emergence of the flux tube is not completed [*Chandra et al.*, 2009; *Luoni et al.*, 2011]. In fact, the tongue elongation pattern indicates the existence of an azimuthal component of the magnetic field around the axis of the omega-shaped coronal flux tube, and therefore the sign of the twist of the emerging flux tube can be deduced (see section 23.2.2). The current pattern shows J-shaped current ribbons (Fig. 23.1b). The direct current occupies two J-shaped areas with two elongated lanes in the AR center along the PIL. The direct current density J_z has a positive sign over the positive magnetic polarity and vice versa according to the positive helicity of the active region (Table 23.1). At low resolution, return currents are observed at the periphery of each sunspot as much weaker current densities and narrower current lanes. At high spatial resolution, it should be still noted that the fibril nature of sunspot penumbrae there leads to mixed strong direct and return currents at small scale [*Venkatakrishnan and Tiwari*, 2009]. The observations of such return currents are at the limit of what can be identified on the edge of and away from sunspots because the photosphere is there far from force free, because at small scale, the magnetic field is subject to granular motions. So, we are only concentrating here on current in sunspots where **B** is sufficiently strong ($\beta < 1$).

These observations have been compared with theoretical results obtained by a MHD model created from a bipolar potential field by photospheric vortex flows using the OHM code [*Aulanier et al.*, 2012] (see section 23.2.3). The $\beta = 0$ simulation performed with the OHM code reproduced the evolution of an initially torus-unstable flux rope (Fig. 23.1 right column).

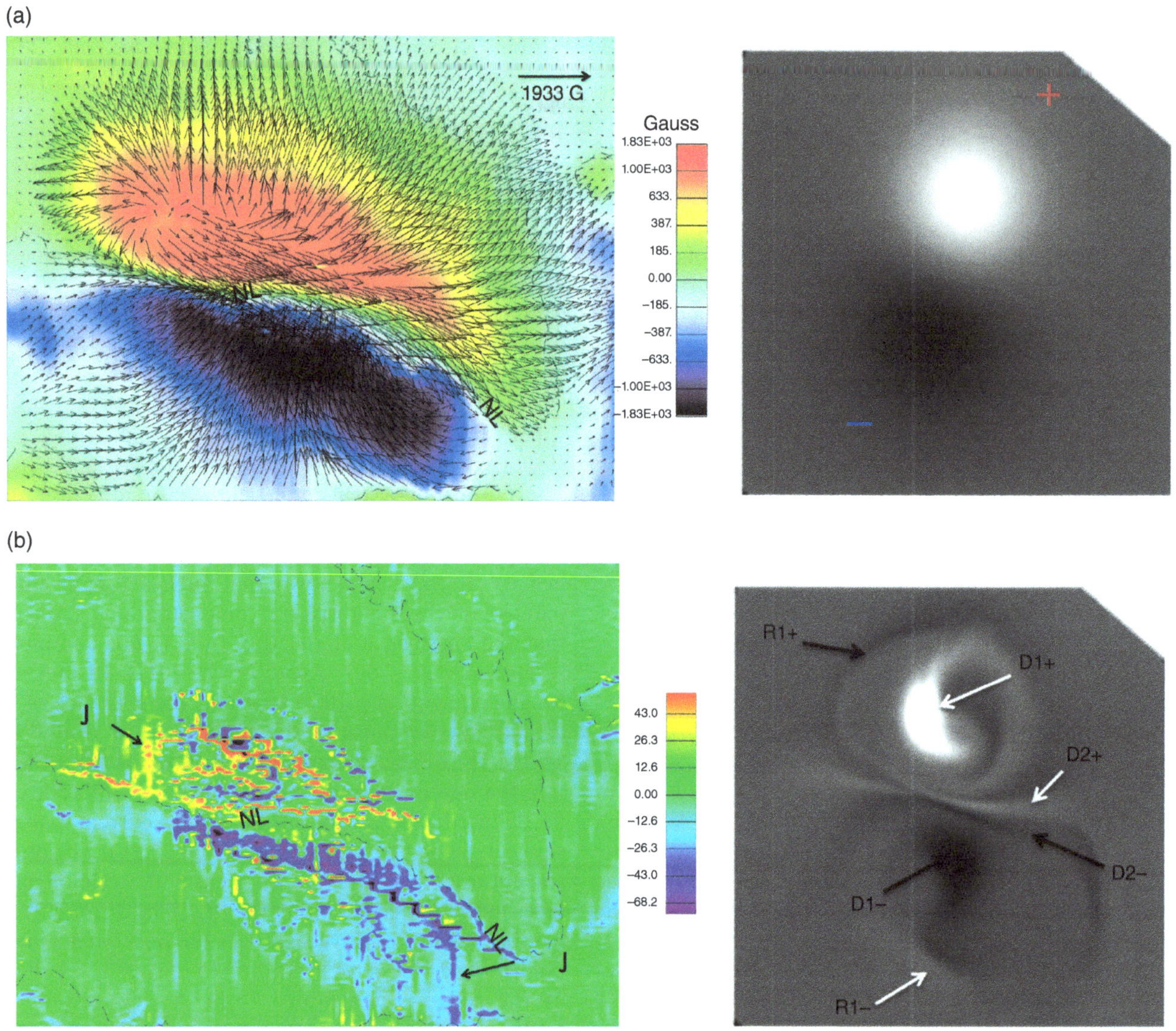

Figure 23.1 Vector magnetogram and vertical component of the electric current density map in the emerging active region AR 10808 with positive helicity observed with the THEMIS vector magnetograph on 13 September 2005 (courtesy of Bommier). (a) B_z map with two elongated tongues symmetrical to the neutral line (NL or PIL) where the arrows represent the horizontal field (red-yellow represents positive polarity, blue-violet represents negative polarity between +/– 1800 gauss). (b) Vertical component of the electric current density in units of mA m^{-2}; yellow represents positive current; violet represents negative current. The letters J and arrows point out the J-shaped current ribbons. (Right panels) Initial B_z in a bipole and vertical electric current density after applying a positive twist in the OHM simulation (from *Schmieder and Aulanier* [2012], section 2.2). White areas correspond to positive B_z and J_z, dark-grey areas correspond to negative B_z and J_z. D indicates direct currents, and R shows return currents. Their sign is defined by the magnetic helicity convention shown in Table 23.1.

The global pattern of the photospheric electric current density shows intense direct currents in the polarities surrounded by return currents with a J-shape, similar to the observations. The dominant direct current in the strong fields (encircled by hooks) are D1+ and D1–. The hooks primarily have direct currents, surrounded by weaker return current densities (R1+ and R1–). D2+ and D2– are direct current-density lanes in the central part of the active region.

The presence of these relatively narrow current-density ribbons in the photosphere can be explained as follows. The connectivity domains are bordered by the quasi-separatrix layers (QSLs) where large magnetic field distortion could exist. Between the flux rope and the environment, coronal current-density layers can be formed during the preeruptive phase. Along the QSLs, any small perturbations can induce an increase of the currents in these narrow layers, which are rooted in the photosphere [*Démoulin et al.*, 1996a]. Therefore, analyzing the electric current-density maps in the photosphere can inform on the existence and the geometry of QSLs, and hence on the location of a flux rope. In particular, the hook-shape extremities of the current ribbons in the current-density maps are the signatures of the existence of a twisted flux rope. The curvature of the hook part of the J depends on the twist of the flux rope [*Démoulin et al.*, 1996b]. The review of *Gibson et al.* [2006] also provided the evolutionary picture accompanying the flux-rope formation with hook structures.

23.2.2. Current Neutralization in Active Regions: History

In the previous subsection, we reported on how observations, as well as one numerical simulation, showed the coexistence of direct and return currents at active region scales. Let us now discuss the issue of current neutralization, that is, of when, how, and why, direct and current currents can become equal. Before describing the most recent joint observational and MHD results, we start by presenting an historical view of a long-lasting debate about whether or not a net current can exist in an active region [*Melrose*, 1991; *Parker*, 1996a].

According to coronal models for solar flares [*Sturrock and Stern*, 1980], the energy storage involves a nonpotential component of the magnetic field, which can be achieved by a twist or shear of the coronal magnetic field. *Melrose* [1991] proposed a reductio ad absurdum argument. As a first point in his discussion, he argued that in an idealized case, if the currents were generated by photospheric or subphotospheric stresses after an isolated magnetic flux tube had emerged, then the currents (directed along the flux tube axis) should be neutralized. He pursued his reasoning by saying that if electric currents were neutralized in the corona, the sum of direct currents should equal the sum of return currents. In the case of isolated flux ropes, each direct current should be surrounded by return current on each side of the polarity inversion line (PIL). In the case of coherent shear, two elongated lanes of opposite currents along each side of the PIL should be observed. It is interesting to see that he predicted correctly the shape of the current-density pattern obtained more than a decade later by MHD simulations (see Fig. 23.2e here and Fig. 2 in *Melrose* [1991]). But he did not correctly predict the neutralization. The reason is that in his argument, he considered a cylindrical and vertical flux tube anchored in the photosphere, which has emerged completely. Treating the legs of active-region-scale coronal-flux rope as being cylindrical and orthogonal to the photosphere was an oversimplification. Indeed, new MHD models show that these legs are inclined and partially emerged from the solar interior and the curvature of the flux tube is important [*Dalmasse et al.*, 2015; *Leake et al.*, 2013; *Török et al.*, 2014].

In a second point of his reasoning, Melrose argued that neutralized currents were not measured (see *Hagyard* [1988] and section 23.2.3). With these observational considerations, *Melrose* [1991] concluded that his basic assumption of neutralization of currents should not be valid, and he declared that the currents are unneutralized. This was a good conclusion, however this raised the question about where do the currents close. *Spicer* [1982] proposed an idea in the frame of a flare model based on an electric circuit. In that case the electric circuit was closed in the photosphere by cross-field horizontal currents. Nevertheless, this was not the solution proposed by *Melrose* [1991], who proposed instead that the electric current should close under the solar surface, deep in the solar dynamo region. This idea allowed the reconciliation of the observational results from this time with the theoretical approach. But then, one problem was the occurrence of elongated paths of net/unneutralized currents in the Sun's interior. Such net currents implied the existence of nonisolated flux tubes, with structured azimuthal fields existing around the current paths everywhere inside the Sun. The problem was that these properties contradicted intuitive high-β behavior that isolates in principle subphotospheric flux ropes from their nonneutralized environment.

Melrose's physical thinking of unneutralized currents was correct. But it was impossible to see the role of magnetic shear at the PIL with a cylindrical geometry, since PILs simply do not exist in such a geometry. Also the poor observations available at this time were not granted since their noise hardly allowed to measure return current densities, which could have been weaker than the direct current densities.

All these ideas were debated in a series of papers (e.g., *Melrose* [1995]; *Parker* [1996a,b]). Instead, *Parker* [1996b] argued that the net current must be zero, but that, because of a failure to resolve magnetic fibril structure, measurements of the field would incorrectly infer a nonzero net current.

The lack of return current in the past observations was due either to the return current strength being below the threshold of the measurements, or to the low spatial resolution of the former vector magnetographs [*Leka et al.*, 1996], or to artifacts (e.g., Faraday rotation) as suggested by *Wilkinson et al.* [1992]. The MHD simulations now consider 3D omega-shaped loops that have different properties than the cylindrical flux tubes that were considered in the 1990s for the discussion about electric currents.

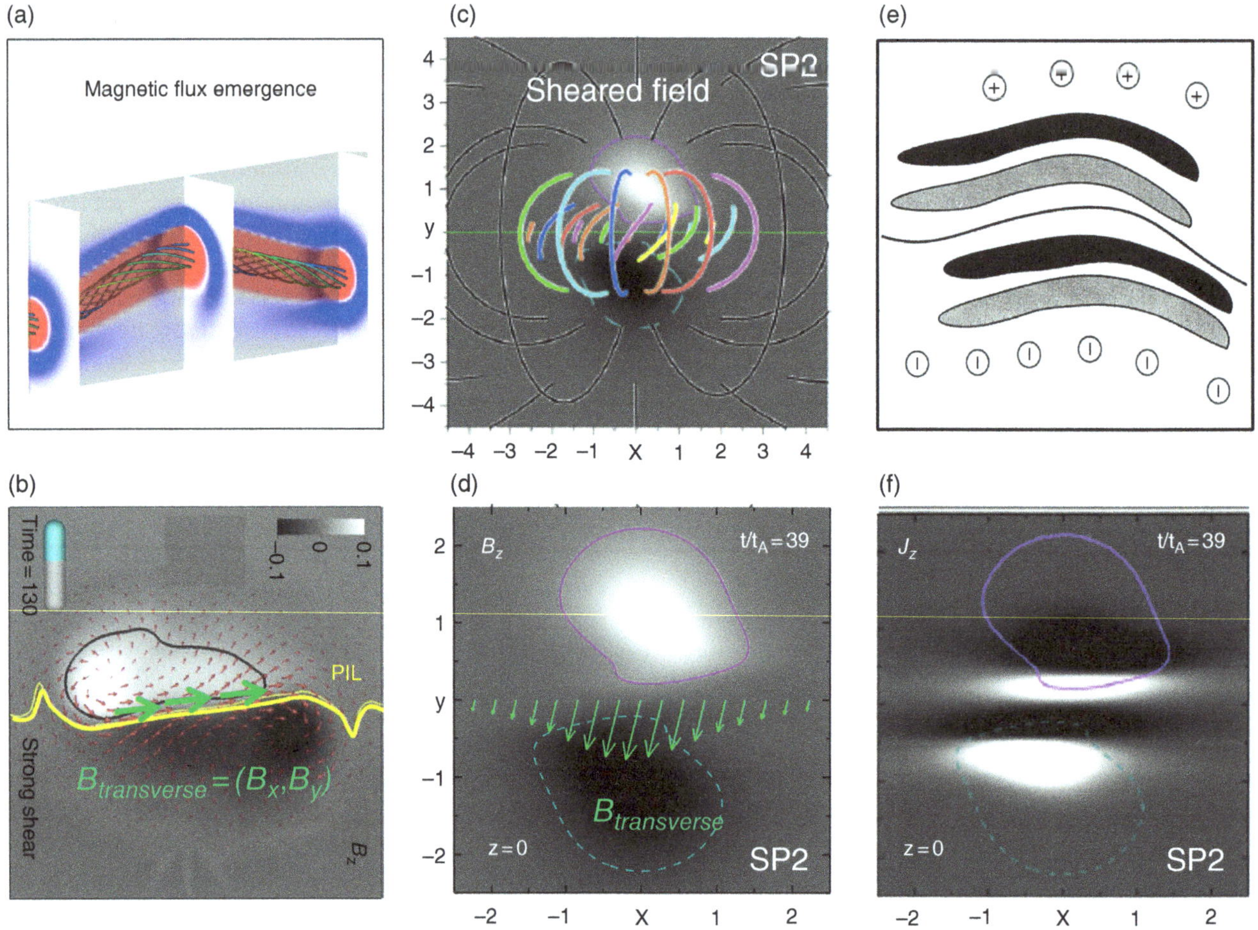

Figure 23.2 MHD simulations of magnetic flux emergence. (a) A model of emerging flux rope from *Leake et al.* [2013]. (b), (c), (d), and (f) Results from the simulations of *Dalmasse et al.* [2015] showing in (b), (c), and (d) the magnetic field B_z for different configurations of shear, the green arrows represent the transverse field between the two polarities. In (f), the photospheric current density map J_z in the configuration shown in (c) and (d) with partially unneutralized currents to be compared with (e) showing the predicted sketch of *Melrose* [1991] for the idealized case of neutralization. Black areas display direct current, and white areas show return current on the positive polarity and the opposite on the negative polarity according to the negative sign of the helicity (left-hand side twist) (see Table 23.1 and section 23.2.2).

Both modern observational and theoretical improvements now permit us to resolve some of the problems debated between Parker and Melrose (see next section).

23.2.3. Current Neutralization in Active Regions: Recent MHD Models

In the eruptive flare models of torus-type, only a net current is involved [*Forbes and Isenberg*, 1991; *Forbes and Priest*, 1995; *Kliem and Török*, 2006; *Lin and Forbes*, 2000; *Lin et al.*, 2001; *Martens and Kuin*, 1989; *Molodenskii and Filippov*, 1987; *van Tend and Kuperus*, 1978]. *Forbes* [2010] showed that if return currents were introduced, they could stop the MHD instability. Since the Melrose and Parker discussions, many theoretical models and MHD simulations have been proposed for flares and CMEs [*Amari et al.*, 2014, 2003a,b; *Aulanier et al.*, 2010; *Fan*, 2010; *Fan and Gibson*, 2003, 2007; *Inoue*, 2016].

The recent flare and CME MHD models have been based on two mechanisms: emergence of current-carrying magnetic flux tubes through the photosphere [*Cheung and Isobe*, 2014; *Leka et al.*, 1996] or shearing of coronal field by photospheric horizontal flows [*Aulanier et al.*, 2005, 2010; *Klimchuk and Sturrock*, 1992; *Török and Kliem*, 2003].

The question of whether net currents exist or not have been the motivation for the development of recent MHD models, which were dedicated to this question [*Dalmasse et al.*, 2015; *Török et al.*, 2014]. Let us discuss in particular these two sets of theoretical papers based on mechanisms of injection of stress in the corona.

Török and Kliem [2003] studied the stressing of coronal magnetic field by photospheric motions of a bipole and noticed that net currents develop if the vortices are close enough to the PIL. Later *Török et al.* [2014] and *Dalmasse et al.* [2015] used respectively the Lare3D code [*Leake et al.*, 2013] and the Observationally driven High-order scheme Magnetohydrodynamic code (OHM) [*Aulanier et al.*, 2005] to quantify net currents and to explain why they exist. In the simulation of *Török et al.* [2014], a subphotospheric magnetic flux rope containing neutralized current was considered (Fig. 23.2a). In the case of emergence of a flux tube, it has been shown that the flux tube looks like an omega loop, starts to flatten below the solar surface, and emerges progressively [*Archontis and Török*, 2008; *Schmieder et al.*, 2014]. Unneutralized (i.e., net) currents appear in the modeled photosphere when the flux tube has not completely emerged. The top part with direct currents has emerged while a nonnegligible part of return currents surrounding the flux tube is still below the photosphere.

Dalmasse et al. [2015] revisited the case of photospheric motions inducing stressed magnetic field by rotating and shearing the polarities of a bipole. In the case of twist, the photospheric vertical current-density maps display direct and strong currents in the core of each polarity surrounded by a shell of return currents with a swirling pattern (see Fig. 3 in *Dalmasse et al.* [2015]). This asymmetry is due to the effect of field line length resulting from the flux tube curvature. The stronger currents develop at the footpoints of the shorter field lines. This pattern can exist only in a 3D configuration and not in 2.5 D cylindrical geometry. It explains on its own the whirling pattern of the current density but not the unneutralized currents.

Figure 23.2 is a composite figure with panels of different models of twist and shear, all showing an important shear along the PIL, which is the clue to obtaining a net current. The left column concerns an emerging flux model and shows the magnetic field vectors parallel to the PIL (from *Török et al.* [2014]). In such a magnetic configuration, there is a strong shear along the PIL. The two right columns concern the sheared field model of *Dalmasse et al.* [2015] with, at the top right (e) for comparison, the sketch by *Melrose* [1991] on neutralized currents with lanes of positive alternated with negative currents. This pattern is very similar, in fact, to the current density pattern (f) found in *Dalmasse et al.* [2015] for a weak shear, in which the currents are not neutralized. When the PIL is sheared, the magnetic field does not have, anymore, a current-free region around the PIL.

Dalmasse et al. [2015] concluded that magnetic shear along the PIL caused by the motions imposed in the photosphere is responsible for the unneutralized (i.e., net current) current observed in ARs and shown in current-density maps. The magnetic shear generates a force-free net current. According to the different models, unneutralized currents occur when the twist/shear motions (i.e., flows) reach the PIL [*Dalmasse et al.*, 2015]. In that case, there is no way to have a potential field at the PIL, that is, a flux rope with direct current in its core surrounded by return currents. The shear inhibits the return current along to the PIL and only direct current remains there.

In conclusion, whatever the magnetic field geometry is, the net current depends on the length of the part of the PIL above which the magnetic field is sheared, and also depends on how strong and how sheared the field is. This idea is consistent with past observations [*Falconer et al.*, 2002; *Ravindra et al.*, 2011; *Wheatland*, 2000], and with the magnetic conditions found for eruptions (*Falconer et al.* [2002] and *Guennou et al.* [2017]).

Several other attempts have been made to interpret the observed current-density pattern. In particular, *Georgoulis et al.* [2012] conjectured a mechanism for producing non-neutralized currents based on a dynamical compression, which would generate a Lorentz force along the PIL.

23.2.4. Link Between Coronal Structures and Photospheric Electric Currents

In addition to the question of current neutralization, another question is, can we detect flux ropes in the corona before eruptions? In fact, the observations of the corona in multiwavelengths exhibit many structures that look like flux ropes. Many structures in the corona are considered as direct signatures of current-carrying magnetic fields such as the following:

1. Forward or reversed sigmoids are bright structures with a S-shape or reverse S-shape, observed in EUV with the Atmospheric Imaging Assembly (AIA) imager aboard the Solar Dynamics Observatory (SDO) with the 131 Å and 94 Å filters [*Gibson et al.*, 2006; *Savcheva et al.*, 2012a,b], and in X-ray with Yohkoh [*Canfield et al.*, 1999]. Sigmoids are frequently inferred to be bundles of whirling field lines forming a flux rope [*Cheng and Ding*, 2016; *Cheng et al.*, 2011; *Dudík et al.*, 2014; *Janvier et al.*, 2013; *Zhao et al.*, 2016, 2014]. They are considered to be the surrounding of the erupting flux rope.

2. The prominences, called filaments when observed on the disk, are formed by cool plasma suspended in magnetic structures with **J** parallel to **B** [*Aulanier and Schmieder*, 2002; *Guo et al.*, 2010; *Mackay et al.*, 2010]. Dark cavities with spinning motions observed at the limb in coronal lines (i.e., 193 Å and 171 Å with SDO/AIA) are also interpreted as being the signatures of flux ropes [*Gibson et al.*, 2010; *Parenti et al.*, 2012]. They are formed by persistent shear flows and flux cancellation at PILs [*van Ballegooijen and Martens*, 1989].

The intrinsic relationship between filaments, sigmoids, and magnetic flux ropes is not obvious. It is commonly very difficult to prove that the sigmoid forms a flux rope [*Zhao et al.*, 2016]. Many studies interpret sigmoids and filaments as indicating the presence of a flux rope because it is a direct interpretation for eruptions and coronal mass ejections. In the CME models, filaments are identified as a tracer of the flux rope, which can erupt due to the torus or breakout instabilities [*Kliem and Török*, 2006; *Schmieder et al.*, 2015]. Flux ropes are important because they indicate the presence of strong electric currents. Their footpoints in the photosphere should be a region with intense currents and a possible signature of their existence.

However, the detection of flux ropes in the corona and their footpoints in the photosphere is not trivial. In many observations, it is not clear that the filament channel or the sigmoid plays a role in the eruption and the eruptive sigmoid can be quite different from it. For example, on 12 July 2012, the filament in AR 11520 remained unperturbed during the entire flare. The flare started in a pre-existing coronal sigmoid, which did not take part of the eruption. The eruption was occurring via a second sigmoid formed by continuous slipping magnetic reconnection of the overlying arcades [*Dudík et al.*, 2014]. In this complicated case, it is very difficult to define the flux rope and even more its footpoints, which are moving during the entire period of the eruption. However in many cases, sigmoid magnetic field lines can be considered as the envelope of the flux rope. Commonly, the field lines are anchored in a strong magnetic field and cannot shift easily. Therefore, sigmoids can be used as a proxy of a flux rope (See section 23.3.2).

Nonlinear force-free field (NLFFF) extrapolations are also in principle a good tool for testing the existence of a flux rope. However, the magnetic field in the boundary (photosphere) has to be modified (e.g., preprocessed), which may lead to photospheric magnetic field with less free energy [*Wiegelmann et al.*, 2010]. This is the limitation of the NLFF field extrapolation studies. If the flux rope can be detected by a NLFFF extrapolation, the footpoints of the flux rope would be nevertheless better defined. We have to keep in mind that many different NLFFF methods exist and could lead to different locations of flux rope footprints. Thus, the solution is also not unique.

We present in the following a case of a flux rope anchored in a sunspot with strong magnetic field ($\beta < 1$) in which case the problem of the non-force-free photosphere has been somewhat avoided [*Zhao et al.*, 2016]. Figure 23.3 illustrates the magnetic reconstruction of the UV sigmoid observed on 10 September 2014 with AIA 94 Å using a NLFFF extrapolation based on the Grad-Rubin method [*Gilchrist and Wheatland*, 2014; *Wheatland and Gilchrist*, 2013; *Zhao et al.*, 2016]. One footpoint of the flux rope is surrounded by whirling bundles of extrapolated field lines anchored in the leading polarity. The second footpoint corresponds to the other end of the field bundles, which is localized in the weak magnetic field negative polarity. The S shape of the sigmoid indicates that the twist of the flux rope has a negative helicity. It is a left-handed twist with currents antiparallel to the magnetic field. The right panel in Figure 23.3 presents the smoothed electric current map computed from the data of HMI, the vector magnetograph on board SDO. The current density J_z in the positive leading polarity has therefore a negative sign. A ring of positive currents are well identified at the edge of this positive polarity. They correspond to the return currents. The J_z values in the negative following polarity have mainly a positive sign and represent the direct current density. The return current in that polarity is not detectable, presumably because of unresolved small flux tubes of the network (following Parker, see section 23.2.1).

23.3. ELECTRIC CURRENTS DURING ERUPTIONS

23.3.1. Current Ribbons

In the previous sections, we discussed how pre-eruptive ARs are often associated with flux rope signatures. We have shown an example of direct/return currents in current-density maps in the center of an active region formed by emerging magnetic flux before an eruption. We have seen the existence of two parallel elongated current-density lanes symmetrically (each ending with one hook) located on each side of the PIL (Fig. 23.1).

Recent studies only have also started to address the properties of active currents during solar flares. Flares are characterized by a fast increase of the light emission in a wide range of the electromagnetic spectrum. During the impulsive phase, the sudden increase of brightening of two elongated ribbon-like structures, which could already exist in the prephase as it has been shown previously is observed in visible radiation (Hα) as well as in UV wavelengths (in C IV with SMM/UVSP in *Schmieder et al.* [1987], in 304 Å with TRACE in *Chandra et al.* [2009], in different filters of AIA in *Dudík et al.* [2014]; *Janvier et al.* [2014]), and sometimes in hard X-ray sources [*Krucker et al.*, 2007; *Musset et al.*, 2015]. The two-ribbon flare models explained the bright ribbons as due to the impact in the chromosphere of the accelerated particles from the reconnection site [*Forbes et al.*, 1989]. MHD simulations obtained with the OHM code allow a 3D view of the phenomena. As described in the prephase of the flare, the elongated Hα and UV brightenings coincide with the photospheric footprints of the QSLs, manifested by lanes of high current density that enclose the flux rope

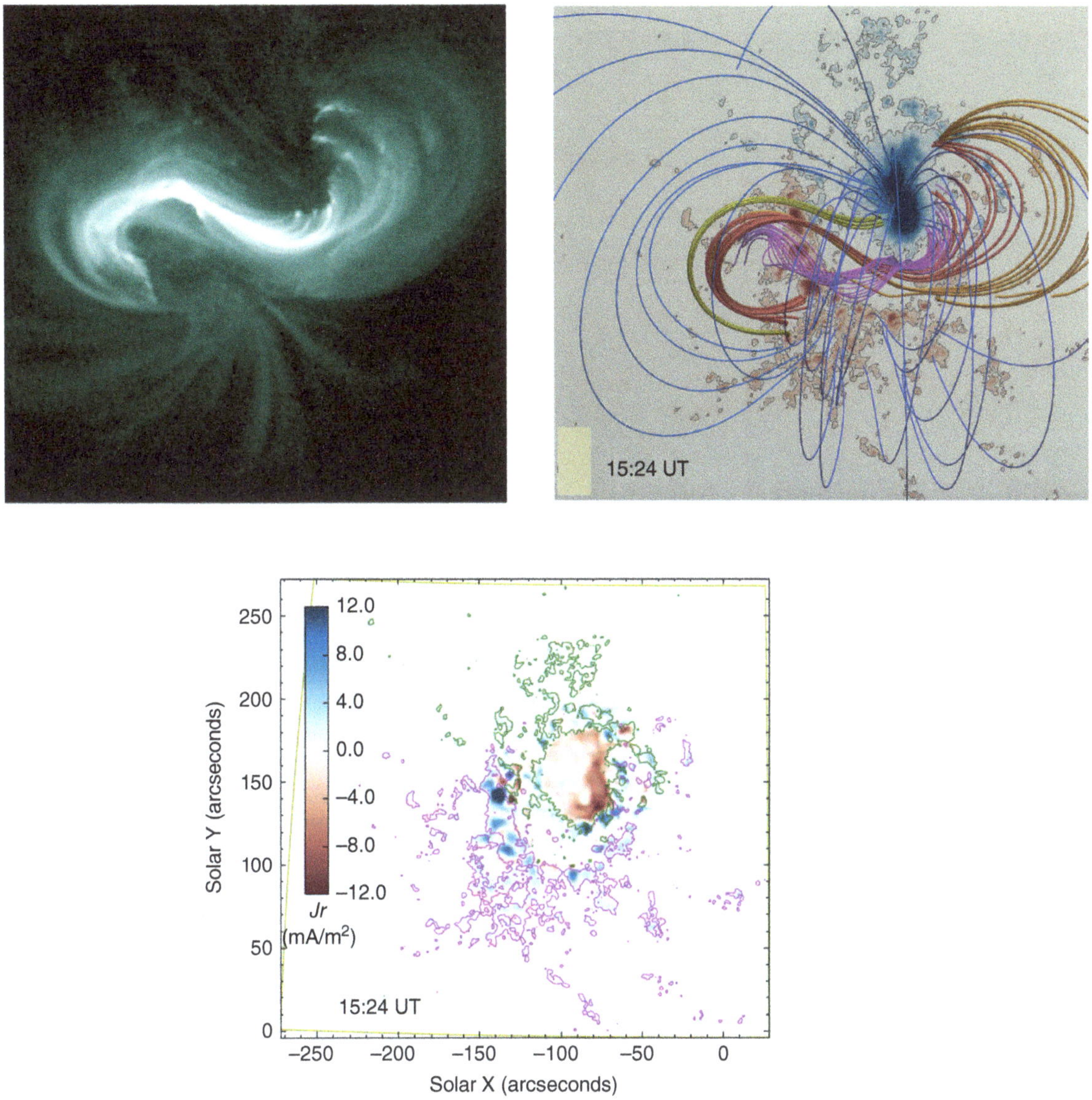

Figure 23.3 Sigmoidal field lines wrapping around a flux rope in the AR 12158, which has a negative helicity (reverse S, left-hand twist) on 10 September 2014, 2 hr before a flare. The top left panel shows an observation in AIA 94 Å; the top right panel shows sigmoidal magnetic field lines using a NLFFF extrapolation based on the Grad-Rubin method [*Gilchrist and Wheatland*, 2014], overlying a B_z map saturated at 2000 gauss; and the bottom panel shows direct and return electric currents computed from HMI data after smoothing the transverse component of the observed photospheric magnetic field. The green contour outlines leading/positive magnetic field, while the purple contour outlines trailing/negative magnetic field at a field strength of +/− 500 gauss. Under the force-free condition, the direct currents are red and blue in the leading positive and following negative polarity, respectively, according to Table 23.1. The return currents are blue on the right side of the main positive polarity in the green contours. The return currents (red) corresponding to the following negative polarity are more difficult to visualize. From *Zhao et al.* [2016].

[*Janvier et al.*, 2013]. Reconnection occurs in current sheets in the quasi-separatrix layers (QSLs), and in particular in the thin high region underneath the flux rope. The simulations predict the increase of current density in the QSLs.

We have to point out that there are very few examples of observations of current-density maps during eruptions, up to now, because it requires a rapid acquisition of the *I, Q, U, V* (Stokes parameters) in a wide region. However, the time to scan an active region requires one hour, half

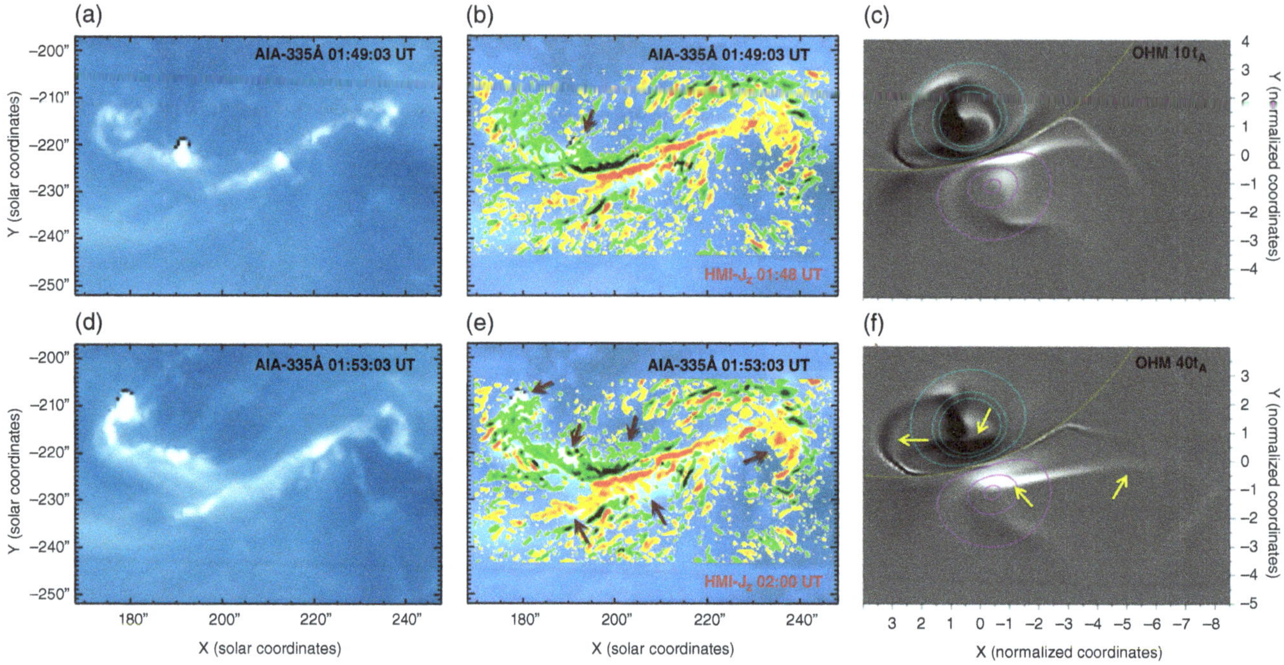

Figure 23.4 (Left column) Flare ribbons at the onset (a) and during (d) the peak of the flare observed in 335 Å with SDO/AIA in AR 11158 on 15 February 2011. (Middle column) Vertical current density J_z. maps (red/green are positive/negative currents saturated to +/− 0.2 A.m^{-2}) from SDO/HMI before (b) and after (e) the impulsive phase superposed over a 335 Å image showing the ribbons. The flare ribbons are very well coaligned with the current ribbons. The brown arrows show similar structures found for J_z and the ribbon maps. (Right column) J_z photospheric maps from OHM 3D simulations of an eruptive flare. The yellow arrows point out the differences seen around the end of the simulation compared to the beginning. From *Janvier et al.* [2014].

an hour, and 12 min with the THEMIS, Hinode, and HMI vector magnetographs respectively. Therefore, only HMI, with its relatively high cadence, is able to follow the evolution of the current density during an eruption with a high spatial resolution. The low spectral resolution of HMI (five points along the line profile only) apparently does not strongly influence the determination of the currents because for the inversion, the code has to fit simultaneously the four Stokes parameter profiles for one pixel. So in fact it is 20 points and not five, that are fitted simultaneously. This is still a relatively small number, but it brings a lot of information. What is also important in the observations of the locations of strong current densities is the spatial continuity in the 2D maps. For isolated current concentrations that only cover one (or a few pixels), the signal could disappear in the noise.

We focus our review on the pioneering paper of *Janvier et al.* [2014], and on the second paper [*Janvier et al.*, 2016] confirming the results found in the first paper for a different active region. The first paper concerns the active region AR 11158 of 15 February 2011, where a X2.1 class flare occurred. This active region was located in the Northern Hemisphere. This region was intensively studied [*Inoue*, 2016; *Inoue et al.*, 2015; *Zhao et al.*, 2014, *Musset et al.*, 2015]. The region consists of the central portion of two emerging active regions, which, a few days earlier, joined each other (see Fig 3.4 in Chapter 03). Therefore a strong shear was developing between the two central polarities (leading and following) and led to X-class flares [*Schrijver et al.*, 2011]. The sigmoid observed in the AIA filter by SDO, not shown here, indicates the presence of a flux rope of positive helicity before the eruption.

Figure 23.4 focuses on the two polarities in the AR center. In the left panels are displayed the ribbons, in the middle panels, the current-density J_z maps obtained from HMI. The two bright ribbons observed in AIA 335 Å and their corresponding electric current density ribbons, J_z (negative/positive current over the negative/positive polarity) appear at the onset of the flare. The ribbons were thin and overlapped the elongated J-shaped current-density lanes observed before the flare, and during the impulsive phase. They very rapidly became thicker, because the post-flare loops were already developing below. The hooks became more prominent with an increase of the current density during the impulsive phase of the flare. During the eruption, the hooks increased in size. Quantitative results for the total current I integrated in the surface of boxes covering the hooks showed an

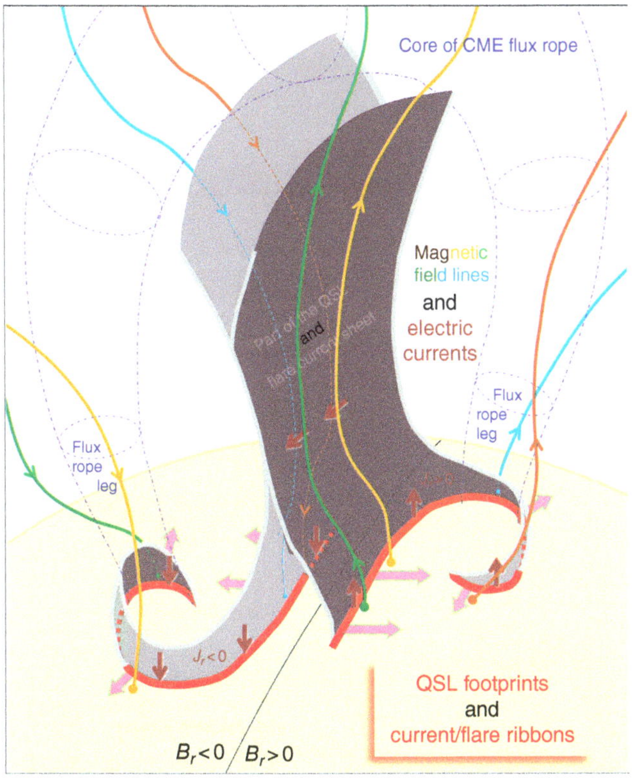

Figure 23.5 Cartoon for the standard model in 3D with four prereconnected magnetic field lines (yellow, green). The pink-grey area represents parts of the 3D volume of the QSLs and the current layers. The red lines with hooks represent the footprints of the QSLs in the photosphere and are similar to two ribbon flares. The dashed lines with oval sections in the overlaying area of the QSLs represent the flux rope with its hook footprints in the photosphere. From *Janvier et al.* [2014].

increase by a factor reaching 2 [*Janvier et al.*, 2014]. In the second paper, the increase of the total current reached an increase of 2.8 [*Janvier et al.*, 2016].

These results have been discussed in the frame of two theoretical MHD models [*Janvier et al.*, 2016]: one using OHM [*Aulanier et al.*, 2010] for the first paper and the other paper the model of *Savcheva and van Ballegooijen* [2009]. Figure 23.4 (right panels) shows the evolution of current-density maps resulting from the OHM simulation of an eruptive flare [*Aulanier et al.*, 2010; *Janvier et al.*, 2014].

These main features are summarized in a sketch with the QSLs (grey-pink areas) and magnetic field lines (four lines) enveloping a flux rope (Fig. 23.5). The QSLs are coronal current layers that extend to the photosphere and are the locations of high electric current densities. The magnetic field lines are distorted at the QSLs, which are a possible site of reconnection, as indicated by red arrows. The footprints of the QSLs and current layers are J-shaped (Fig. 23.4, right panels) and the hooks surround the legs of the flux rope. The magnetic topology analysis of the 15 February 2011 active region confirmed that the bright ribbons of the flare (observed in AIA/304 Å) overlaid the QSL footprints in the photosphere (see Fig. 4 in *Zhao et al.* [2014]). The QSL footprints wrapped around the flux rope footpoints in the photosphere.

The numerical simulations show that, in torus-instability models, the photospheric current ribbons can be interpreted qualitatively as the footprints of the 3D coronal current following the QSLs [*Janvier et al.*, 2014, 2016]. The simulations being running dimensionless, no quantitative analysis has been performed on these simulations yet. As the torus instability develops, stronger electric currents are formed along the QSLs and reconnection is not fast enough to destroy them, as it is limited by the Alfvén speed [*Lin et al.*, 2001]. The first observations have confirmed that the total current increased by a factor around 2 during the flare.

23.3.2. Current Decrease at the Footpoints of the Erupting Flux Tube During and After the Eruption

As explained in the previous section, the prediction of MHD models is that the localized electric currents increase inside the flare ribbons, which are the intersections of the QSLs and the photosphere [*Aulanier et al.*, 2012]. On the other hand, the MHD model predicts also that the electric current inside the flux rope should decrease due to the expansion of the flux rope [*Aulanier et al.*, 2005]. This may be deduced from flux conservation. It was demonstrated that for a cylindrical flux rope with a constant end-to-end twist, the current J_z is proportional to L^{-1}, L being its length (see equations (46) and (47) in *Aulanier et al.* [2005]). During the eruption, the flux rope is stretched so its length increases, and the footprint currents in the photosphere should decrease. The current decreases also in catastrophic models [*Forbes and Lin*, 2000].

However the decrease of the electric current inside the flux rope is difficult to test because of the uncertainty of the identification of the flux rope ends (see section 23.2.2). However, combining two recent papers studying the same active region AR 12158 on 10 September 2014, we can conjecture that the ends of the sigmoidal loops wrapping around the flux rope found by extrapolation by *Zhao et al.* [2016] could correspond to the ends of the sigmoid observed in the hot channel of AIA (94 Å) and used as proxy of the flux rope ends in the *Cheng and Ding* [2016] paper (see Figs. 23.3 and 23.6). *Cheng and Ding* [2016] studied the evolution of the magnetic field and the currents in two boxes contained the footpoints of the sigmoid. One flux rope footpoint is located in a strong field positive polarity. The second one, being in a weaker negative field, corresponds to weak currents, which could

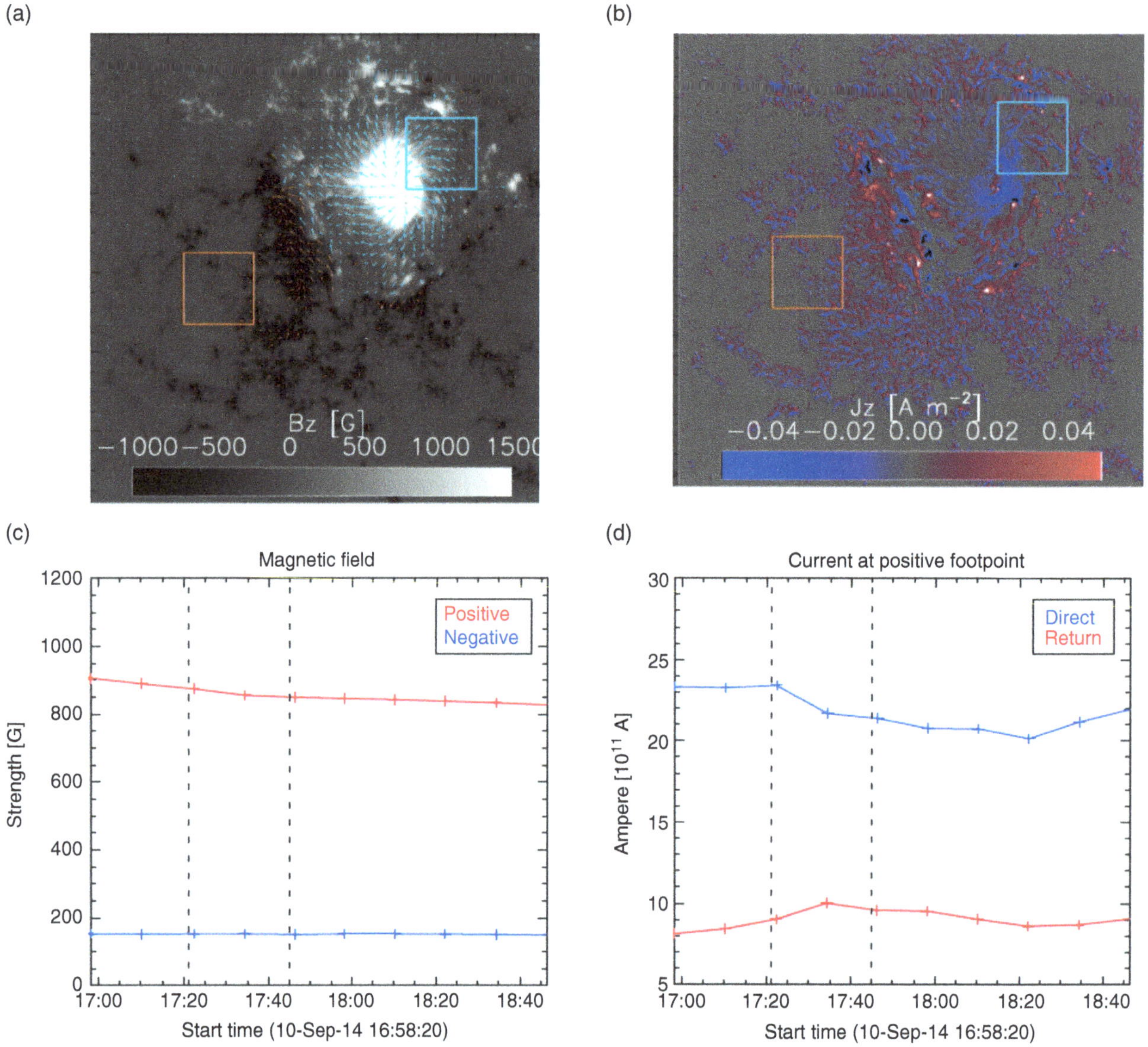

Figure 23.6 Photospheric electric currents in the footpoints of the sigmoid/flux rope in AR 12158 on 10 September 2014 (the same AR as in Fig. 23.3). (a) HMI vector magnetogram with two boxes surrounding the deduced two flux rope footpoints; (b) HMI electric current maps; (c) the mean strength of the magnetic field measured in the two boxes of panels (a) and (b); (d) the direct and return current in the positive magnetic polarity box. The two vertical lines indicate the onset and peak of the flare. From *Cheng and Ding* [2016].

be at the limit of the measurements. A decrease of the direct current is measured in the strong positive polarity (Fig. 23.6d). These results agree with the qualitative behavior of numerical simulations, which treat asymmetrical configurations, similar to the observed configuration of this AR [*Aulanier et al.*, 2010].

The decrease of the currents is predicted by MHD models because of end-to-end twist conservation induced by line tying [*Aulanier et al.*, 2005] and catastrophic models because of flux conservation below an erupting flux rope [*Démoulin and Aulanier*, 2010]. This is opposite to the behavior that is built in circuit models [*Melrose*, 1991] where the total current is prescribed. Direct and quantitative comparisons between models and observations are therefore needed to better characterized the time-evolution of currents during solar flares.

23.4. CONCLUSION

This chapter discusses electric currents in the preeruption state and in the course of eruptions of solar magnetic structures, using information from solar observations, nonlinear force-free (NLFF) field extrapolations relying on these observations, and three-dimensional magnetohydrodynamical (MHD) models. With the new generation of vector magnetographs, new current-density maps have been obtained showing in each polarity a complex whirling pattern of current density [*Schmieder and Aulanier*, 2012].

The main topic of this chapter is to show how the new current-density maps in active regions combined with the development of new 3D MHD models allows us to progressively interpret the observations and better fine-tune the understanding of MHD mechanisms, in the context of eruptions. MHD simulations of preeruptive active regions, based on the coronal flux rope concept, show a whirling electric current-density pattern, similar to the observations in the photosphere with direct and return current in each polarity of the initial bipole [*Aulanier et al.*, 2005, 2010; *Janvier et al.*, 2014]. The current-density occurrence in bright flare ribbons in the footprints of the QSLs that separate the flux rope from the overlying arcades was an important discovery.

Before discussing eruptions, we have reported how the old problem of unneutralized currents has been resolved. The new observations of electric current density of preeruptive active regions have reopened the long debate on the neutralization of the currents in the 1990s by *Melrose* [1991] and *Parker* [1996a]. The key parameter determining the existence of unneutralized (i.e., net current) active regions in the preeruptive phase has been proven to be the shear along the PIL of a bipole [*Dalmasse et al.*, 2015; *Török et al.*, 2014]. Unneutralized photospheric current-density patterns appear whenever magnetic shear is present along polarity inversion lines (which is typical of observed intense PIL, e.g., in delta spots). In the models of *Dalmasse et al.* [2015], shear and twist were created in the corona by artificial line-tied surface motions that were primarily aimed at building a series of differently sheared quasi-force-free configurations. In reality, shear could be due to the coupling of the dynamics of the emergence of a flux rope and the convection in the subphotosphere. This shear creates the net current when the flux rope does not emerge completely. Shear along the PIL and net current are necessary for the occurrence of flares and eruptions [*Falconer et al.*, 2002; *Kliem and Török*, 2006].

A whirling current-density pattern, observed in some active regions before eruptions, has been interpreted as the photospheric footprint of flux ropes in the corona. The currents have a hook shape, which, according to the models, encircles the flux rope footprint. Inside the active region, two elongated current-density lanes are detected, symmetrically located along the PIL when there is a strong shear before the flare. All regions do not show such a structure before the eruption but already a few cases have been published [*Georgoulis et al.*, 2012; *Janvier et al.*, 2014; *Schmieder and Aulanier*, 2012]. In the observed current-density pattern with J-shaped hooks, the high spatial resolution allow us to distinguish the currents in the hooks and the surrounding mixed sign currents. Considering one magnetic polarity, that is, above a sunspot, it is clear that both direct and return currents are present. The direct current is generally dominant. During the eruptions, more intense current ribbons are observed and an increase of the current density in the straight part of the ribbons has been measured. The total current during an eruption in the current ribbons including their hooks has been quantified in two observational cases up to now, and shows an increase by a factor of 2 [*Janvier et al.*, 2014, 2016]. This can be explained qualitatively. When a flare begins, new narrow J-shaped current structures develop on top of the preexisting relatively broad elongated current-density lanes, also in J-shape. These new structures appear to match bright flare ribbons, visible from the visible to EUV wavelengths. On the contrary, in regions encircled by the hooks, a decrease has been detected [*Cheng and Ding*, 2016]. This can be explained by the expansion of the flux rope.

Several authors analyzed MHD simulations before and during an eruption from a qualitatively point of view [*Dalmasse et al.*, 2015; *Janvier et al.*, 2014; *Török et al.*, 2014]. The simulations use dimensionless units and have neither been yet analyzed to give quantified results on the values of the current increase during an eruption, nor to quantify how much of a decrease in the current density in the flux rope we can expect. No attempt of distinguishing direct and return current in ARs during eruptions has been carried out yet. The work still remains to be done. With the increase of computer facilities and the parallelization of codes, it should be possible to estimate the net current in each polarity and to compute their variation during eruptions.

Today we have vector magnetograms with high spectral resolution (Hinode/SOT, THEMIS) in a small field of view with low time resolution. We have HMI full disk vector magnetograms with high cadence but relatively low sensitivity. In the future, we may expect to have sensitive vector magnetograms with higher spatial resolution to compute electric current density with accuracy and with higher cadence to study the evolution during eruptions.

DKIST and EST (the future 4 m telescopes) and Solar Orbiter/PHI (the Polarimetric and Helioseismic Imager), which will fly close to the Sun, will certainly bring very surprising results in this domain, on small fields of view with DKIST and EST and on the full disk with PHI.

REFERENCES

Alfvén, H., and P. Carlqvist (1967), Currents in the solar atmosphere and a theory of solar flares, *Solar Phys.*, *1*, 220–228; doi: 10.1007/BF00150857.

Amari, T., A. Canou, and J.-J. Aly (2014), Characterizing and predicting the magnetic environment leading to solar eruptions, *Nature*, *511*, 165–169.

Amari, T., J. F. Luciani, J. J. Aly, Z. Mikic, and J. Linker (2003a), Coronal mass ejection: Initiation, magnetic helicity, and flux ropes, I. Boundary motion-driven evolution, *Astrophys. J*, *585*, 1073.

Amari, T., J. F. Luciani, J. J. Aly, Z. Mikic, and J. Linker (2003b), Coronal mass ejection: Initiation, magnetic helicity, and flux ropes, II. Turbulent diffusion-driven evolution, *Astrophys. J.*, *595*, 1231.

Archontis, V., and T. Török, (2008), Eruption of magnetic flux ropes during flux emergence, *Astron. Astrophys.*, *492*, L35.

Aulanier, G., and B. Schmieder (2002), The magnetic nature of wide EUV filament channels and their role in the mass loading of CMEs, *Astron. Astrophys.*, *386*, 1106.

Aulanier, G., M. Janvier, and B. Schmieder (2012), The standard flare model in three dimensions, I. Strong-to-weak shear transition in post-flare loops, *Astron. Astrophys.*, *543*, A110.

Aulanier, G., E. Pariat, and P. Démoulin (2005), Current sheet formation in quasi-separatrix layers and hyperbolic flux tubes, *Astron. Astrophys.*, *444*, 961.

Aulanier, G., T. Török, P. Démoulin, and E. E. DeLuca (2010), Formation of torus-unstable flux ropes and electric currents in erupting sigmoids, *Astrophys. J.*, *708*, 314

Bommier, V. (2013), Reconciliating the vertical and horizontal gradients of the sunspot magnetic field, *Phys. Res. Int.*, *2013*, Article ID 195403.

Bommier, V., E. Landi Degl'Innocenti, M. Landolfi, and G. Molodij (2007), UNNOFIT inversion of spectro-polarimetric maps observed with THEMIS, *Astron. Astrophys.*, *464*, 323.

Canfield, R. C., H. S. Hudson, and D. E. McKenzie (1999), Sigmoidal morphology and eruptive solar activity, *Geophys. Res. Lett.*, *26*, 627.

Canfield, R. C., J.-F. de La Beaujardiere, Y. Fan, et al. (1993), The morphology of flare phenomena, magnetic fields, and electric currents in active regions, I. Introduction and methods, *Astrophys. J.*, *411*, 362.

Canou, A., T. Amari, V. Bommier, B. Schmieder, G. Aulanier, and H. Li (2009), Evidence for a pre-eruptive twisted flux rope using the THEMIS vector magnetograph, *Astrophys. J. Lett.*, *693*, L27.

Chandra, R., B. Schmieder, G. Aulanier, and J. M. Malherbe (2009), Evidence of magnetic helicity in emerging flux and associated flare, *Solar Phys.*, *258*, 53.

Cheng, X., and M. D. Ding (2016), The characteristics of the footpoints of solar magnetic flux ropes during eruptions, *Astrophys. J. Supp.*, *225*, 16.

Cheng, X., J. Zhang, Y. Liu, and M. D. Ding (2011), Observing flux rope formation during the impulsive phase of a solar eruption, *Astrophys. J.Lett.*, *732*, L25.

Cheung, M. C. M., and H. Isobe (2014), Flux emergence (theory), *Living Rev. Sol. Phys.*, *11* (1), article id. 3.

Dalmasse, K., G. Aulanier, P. Démoulin, B. Kliem, T. Török, and E. Pariat (2015), The origin of net electric currents in solar active regions, *Astrophys. J.*, *810*, 17.

Démoulin, P. (2007), Recent theoretical and observational developments in magnetic helicity studies, *Adv. Space Res.*, *39*, 1674.

Démoulin, P., and E. Pariat (2009), Modelling and observations of photospheric magnetic helicity, *Adv. Space Res.*, *43*, 1013.

Démoulin, P., and G. Aulanier (2010), Criteria for flux rope eruption: Non-equilibrium versus torus instability, *Astrophys. J.*, *718*, 1388.

Démoulin, P., E. R. Priest, and D. P. Lonie (1996b), Three-dimensional magnetic reconnection without null points 2. Application to twisted flux tubes, *J. Geophys. Res.*, *101*, 7631.

Démoulin, P., J. C. Hénoux, E. R. Priest, and C. H. Mandrini (1996a), Quasi-separatrix layers in solar flares, I. Method, *Astron. Astrophys.*, *308*, 643.

Dudík, J., M. Janvier, G. Aulanier, G. Del Zanna, M. Karlický, H. E. Mason, and B. Schmieder (2014), Slipping magnetic reconnection during an X-class solar flare observed by SDO/AIA, *Astrophys. J.*, *784*, 144.

Falconer, D. A., R. L. Moore, and G. A. Gary (2002), Correlation of the coronal mass ejections productivity of solar active regions with measures of their global nonpotentiality from vector magnetograms: Baseline results, *Astrophys. J.*, *569*, 1016.

Fan, Y. (2010), On the eruption of coronal flux ropes, *Astrophys. J.*, *719*, 728.

Fan, Y., and S. E. Gibson (2003), The emergence of a twisted magnetic flux tube into a preexisting coronal arcade, *Astrophys. J.*, *589*, L105.

Fan, Y., and S. E. Gibson (2007), Onset of coronal mass ejections due to loss of confinement of coronal flux ropes, *Astrophys. J.*, *668*, 1232.

Forbes, T. (2010), Models of coronal mass ejections and flares, edited by C. J. Schrijver and G. L. Siscoe, Cambridge University Press, 159.

Forbes, T. G., and E. R. Priest (1995), Photospheric magnetic field evolution and eruptive flares, *Astrophys. J.*, *446*, 377.

Forbes, T. G., and, P. A. Isenberg (1991), A catastrophe mechanism for coronal mass ejections, *Astrophys. J.*, *373*, 294.

Forbes, T.G. and Lin J. (2000), What can we learn about reconnection from coronal mass ejections?, *Journal of Atmospheric and Solar-Terrestrial Physics*, *62* (16), 1499–1507.

Forbes, T. G., J. M. Malherbe, and E. R. Priest (1989), The formation flare loops by magnetic reconnection and chromospheric ablation, *Solar Phys.*, *120*, 285.

Gary, G. A., R. L. Moore, M. J. Hagyard, and B. M. Haisch (1987), Nonpotential features observed in the magnetic field of an active region, *Astrophys. J.*, *314*, 782.

Georgoulis, M. K., V. S. Titov, and, Z. Mikić (2012), Non-neutralized electric current patterns in solar active regions: Origin of the shear-generating Lorentz force, *Astrophys. J.*, *761*, 61.

Gibson, S. E., T. A. Kucera, D. Rastawicki, J. Dove, G. de Toma, J. Hao, S. Hill, et. al. 2010), Three-dimensional morphology of a coronal prominence cavity, *Astrophys. J.*, *724*, 1133.

Gibson, S. E., Y. Fan, T. Török, and B. Kliem (2006), The evolving sigmoid: Evidence for magnetic flux ropes in the corona before, during, and after CMEs, *Space Sci. Rev.*, *124*, 131.

Gilchrist, S. A., and M. S. Wheatland (2014), Force-free modeling of the corona in spherical coordinates, *Solar Phys.*, *289*, 1153.

Gosain, S., P. Démoulin, and, M. López Fuentes (2014), Distribution of electric currents in sunspots from photosphere to corona, *Astrophys. J.*, *793*, 15.

Guennou, C., E. Pariat, J.E. Leake, N. Vilmer, (2017) , Testing predictors of eruptivity using parametric flux emergence simulations, *Journal of Space Weather and Space Climate*, *7*, A17.

Guo, Y., M. D. Ding, B. Schmieder, H. Li, T. Török, and T. Wiegelmann (2010), Driving mechanism and onset condition of a confined eruption, *Astrophys. J. Lett.*, *725*, L38.

Harvey, J.W. (1969) Magnetic Fields Associated with Solar Active-Region Prominences, PhD Thesis, *Dissertation Abstracts International*, *30-06* (B), 2519, University of Colorado, Boulder, Colorado.

Hagyard, M. J. (1988), Observed nonpotential magnetic fields and the inferred flow of electric currents at a location of repeated flaring, *Solar Phys.*, *115*, 107.

Heikkila, W. J. (1997), Comment on "The alternative paradigm for magnetospheric physics" by E. N. Parker, *J. Geophys. Res.*, *102*, 9651.

Inoue, S. (2016), Magnetohydrodynamics modeling of coronal magnetic field and solar eruptions based on the photospheric magnetic field, *Prog. Earth Planet. Sci.*, *3*, 19.

Inoue, S., K. Hayashi, T. Magara, G. S. Choe, and Y. D. Park (2015), Magnetohydrodynamic simulation of the X2.2 solar flare on 2011 February 15, II. Dynamics connecting the solar flare and the coronal mass ejection, *Astrophys. J.*, *803*, 73.

Janvier, M., A. Savcheva, E. Pariat, et al. (2016), Evolution of flare ribbons, electric currents, and quasi-separatrix layers during an X-class flare, *Astron. Astrophys.*, *591*, A141.

Janvier, M., G. Aulanier, E. Pariat, and P. Démoulin (2013), The standard flare model in three dimensions, III. Slip-running reconnection properties, *Astron. Astrophys.*, *555*, A77.

Janvier, M., G. Aulanier, V. Bommier, et al. (2014), Electric currents in flare ribbons: Observations and three-dimensional standard model, *Astrophys. J.*, *788*, 60

Kliem, B., and T. Török (2006), Torus instability, *Phys. Rev. Lett.*, *96*, 255002.

Klimchuk, J. A., and, P. A. Sturrock (1992), Three-dimensional force-free magnetic fields and flare energy buildup, *Astrophys. J.*, *385*, 344.

Krucker, S., S. M. White, and R. P. Lin (2007), Solar flare hard x-ray emission from the high corona, *Astrophys. J.*, *669*, L49.

Leake, J. E., M. G. Linton, and T. Török (2013), Simulations of Emerging Magnetic Flux, I. The formation of stable coronal flux ropes, *Astrophys. J.*, *778*, 99.

Leka, K. D., and A. Skumanich (1999), On the value of "αAR" from vector magnetograph data, I. Methods and caveats, *Solar Phys.*, *188*, 3.

Leka, K. D., R. C. Canfield, A. N. McClymont, and L. Van Driel-Gesztelyi (1996), Evidence for current-carrying emerging flux, *Astrophys. J.*, *462*, 547.

Li, H., B. Schmieder, M. T. Song, and V. Bommier (2007), Interaction of magnetic field systems leading to an X1.7 flare due to large-scale flux tube emergence, *Astron. Astrophys.*, *475*, 1081.

Lin, J., and T. G. Forbes (2000), Effects of reconnection on the coronal mass ejection process, *J. Geophys. Res.*, *105*, 2375.

Lin, J., T. G. Forbes, and P. A. Isenberg (2001), Prominence eruptions and coronal mass ejections triggered by newly emerging flux, *J. Geophys. Res.*, *106*, 25053.

Luoni, M. L., P. Démoulin, C. H. Mandrini, and L. van Driel-Gesztelyi (2011), Twisted flux tube emergence evidenced in longitudinal magnetograms: Magnetic tongues, *Solar Phys.*, *270*, 45.

Mackay, D. H., J. T. Karpen, J. L. Ballester, B. Schmieder, and G. Aulanier (2010), Physics of solar prominences, II. Magnetic structure and dynamics, *Space Sci. Rev.*, *151*, 333.

Martens, P. C. H., and N. P. M. Kuin (1989), A circuit model for filament eruptions and two-ribbon flares, *Solar Phys.*, *122*, 263.

Melrose, D. B. (1991), Neutralized and unneutralized current patterns in the solar corona, *Astrophys. J.*, *381*, 306.

Melrose, D. B. (1995), Current paths in the corona and energy release in solar flares, *Astrophys. J.*, *451*, 391; doi:10.1086/176228.

Metcalf, T. R., K. D. Leka, G. Barnes, et al. (2006), An overview of existing algorithms for resolving the 180° ambiguity in vector magnetic fields: Quantitative tests with synthetic data, *Solar Phys.*, *237*, 267.

Molodenskii, M. M., and B. P. Filippov (1987), Rapid motion of filaments in solar active regions, *Sov. Astron.*, *31*, 430.

Moreton, G. E., and, A. B. Severny (1968), Magnetic fields and flares in the region CMP 20 September 1963, *Solar Phys.*, *3*, 282.

Musset, S., N. Vilmer, and V. Bommier (2015), Hard X-ray emitting energetic electrons and photospheric electric currents, *Astron. Astrophys.*, *580*, A106.

Parenti, S., B. Schmieder, P. Heinzel, and L. Golub (2012), On the nature of prominence emission observed by SDO/AIA, *Astrophys. J.*, *754*, 66.

Parker, E. N. (1996a), Comment on "Current paths in the corona and energy release in solar flares," *Astrophys. J.*, *471*, 489.

Parker, E. N. (1996b), The alternative paradigm for magnetospheric physics, *J. Geophys. Res.*, *101*, 10587.

Parker, E. N. (2001), A critical review of sun-space physics, *Astrophys. Space Sci.*, *277*, 1.

Ravindra, B., P. Venkatakrishnan, S. K. Tiwari, and R. Bhattacharyya (2011), Evolution of currents of opposite signs in the flare-productive solar active region NOAA 10930, *Astrophys. J.*, *740*, 19.

Savcheva, A., and A. van Ballegooijen (2009), Nonlinear force-free modeling of a long-lasting coronal sigmoid, *Astrophys. J.*, *703*, 1766.

Savcheva, A. S., A. A. van Ballegooijen, and E. E. DeLuca (2012b), Field topology analysis of a long-lasting coronal sigmoid, *Astrophys. J.*, *744*, 78.

Savcheva, A. S., L. M. Green, A. A. van Ballegooijen, and, E. E. DeLuca (2012a), Photospheric flux cancellation and the build-up of sigmoidal flux ropes on the Sun, *Astrophys. J.*, *759*, 105.

Schmieder, B., V. Archontis, and E. Pariat (2014), Magnetic flux emergence along the solar cycle, *Space Sci. Rev.*, *186*, 227.

Schmieder, B., and G. Aulanier (2012), What are the physical mechanisms of eruptions and CMEs?, *Adv. Space Res.*, *49*, 1598.

Schmieder, B., G. Aulanier, and B. Vršnak (2015), Flare-CME models: An observational perspective (invited review), *Solar Phys.*, *290*, 3457.

Schmieder, B., T. G. Forbes, J. M. Malherbe, and M. E. Machado (1987), Evidence for gentle chromospheric evaporation during the gradual phase of large solar flares, *Astrophys. J.*, *317*, 956.

Schrijver, C. J., G. Aulanier, A. M. Title, E. Pariat, and C. Delannée (2011), The 2011 February 15 X2 flare, ribbons, coronal front, and mass ejection: Interpreting the three-dimensional views from the Solar Dynamics Observatory and STEREO guided by magnetohydrodynamic flux-rope modeling, *Astrophys. J.*, *738*, 167.

Severny, A. B. (1964), Solar flares, *Ann. Rev. Astron. Astrophys*, *2*, 363.

Spicer, D. S. (1982), Magnetic energy storage and conversion in the solar atmosphere, *Space Sci. Rev.*, *31*, 351.

Sturrock, P. A., and R. Stern (1980), Is the galactic corona produced by galactic flares, *Astrophys. J.*, *238*, 98.

Török, T., and B. Kliem (2003), The evolution of twisting coronal magnetic flux tubes, *Astron. Astrophys.*, *406*, 1043.

Török, T., J. E. Leake, V. S. Titov, V. Archontis, Z. Mikić, M. G. Linton, K. Dalmasse, G. Aulanier, and B. Kliem (2014), Distribution of electric currents in solar active regions, *Astrophys. J. Lett.*, *782*, L10.

van Ballegooijen, A. A., and P. C. H. Martens (1989), Formation and eruption of solar prominences, *Astrophys. J.*, *343*, 971.

van Tend, W., and, M. Kuperus (1978), The development of coronal electric current systems in active regions and their relation to filaments and flares, *Solar Phys.*, *59*, 115.

Venkatakrishnan, P., and S. K. Tiwari (2009), On the absence of photospheric net currents in vector magnetograms of sunspots obtained from Hinode (Solar Optical Telescope/Spectro-Polarimeter), *Astrophys. J.*, *706*, L114.

Wheatland, M. S. (2000), Flare frequency-size distributions for individual active regions, *Astrophys. J.*, *532*, 1209.

Wheatland, M. S., and S. A. Gilchrist (2013), The state of non-linear force-free magnetic field extrapolation, *J. Phys. Conference Series*, *440*, 012037.

Wiegelmann, T., L. Yelles Chaouche, S. K. Solanki, and A. Lagg (2010), Nonlinear force-free modelling: Influence of inaccuracies in the measured magnetic vector, *Astron. Astrophys.*, *511*, A4.

Wilkinson, L. K., A. G. Emslie, and G. A. Gary (1992), On neutralized currents in the solar corona, *Astrophys. J.*, *392*, L39.

Zhao, J., H. Li, E. Pariat, B. Schmieder, Y. Guo, and T. Wiegelmann (2014), Temporal evolution of the magnetic topology of the NOAA active region 11158, *Astrophys. J.*, *787*, 88.

Zhao, J., S. A. Gilchrist, G. Aulanier, B. Schmieder, E. Pariat, and H. Li (2016), Hooked flare ribbons and flux-rope-related QSL footprints, *Astrophys. J.*, *823*, 62.

Part V: Ionospheric Currents

24

Review of Data Analysis Techniques for Estimating Ionospheric Currents Based on MIRACLE and Satellite Observations

H. Vanhamäki[1,2] and L. Juusola[2]

ABSTRACT

We present an updated review of selected data-analysis tools that have been developed for studying ionospheric current systems at high latitudes. As we concentrate on electric currents, we will ignore ionization/recombination processes, plasma transport, kinetic effects and many other interesting phenomena. Our emphasis is on tools developed for the MIRACLE network and selected low Earth-orbiting satellites, but will include some discussion about tools used in global or hemispherical studies. We will mostly limit the discussion to "data driven" tools, meaning that they should not rely on statistical background models. For our purposes, the ionospheric electrodynamic state is described by the electric field **E**, horizontal sheet current density **J**, field-aligned current density $j_{\parallel}$, and the height-integrated Pedersen and Hall conductances $\Sigma_{P,H}$, respectively. We demonstrate that there exist some tools for analyzing most combinations of measured data, so that the other parameters can be calculated. In many cases, we make a parallel discussion using either scalar potentials or Spherical Elementary Current Systems (SECS) to represent the ionospheric current or electric field. This may complicate the discussion a bit, but it shows that in many situations either of these mathematical approaches can be used.

24.1. INTRODUCTION

We present a concise review of some selected data-analysis tools that have been developed for studying ionospheric electrodynamics and especially ionospheric current systems at high latitudes. This is an updated version of the review presented by *Vanhamäki and Amm* [2011]. Like before, we will emphasize data-analysis tools developed for the MIRACLE network (http://space.fmi.fi/MIRACLE/) and selected low Earth-orbiting satellites, but will include a bit more discussion about tools used in global or hemispherical studies. Naturally, the same methods can also be applied to other sources that provide similar data.

[1] *Ionospheric Research Unit, University of Oulu, Oulu, Finland*
[2] *Finnish Meteorological Institute, Helsinki, Finland*

The MIRACLE network is located in northern Europe (see Fig. 24.1) and consists of magnetometers, all-sky cameras, and tomographic beacon receivers. The network used to contain also the STARE coherent scatter radar, which was decommissioned in 2005. This loss has been partly compensated by using ionospheric plasma convection and electric field data provided by the SuperDARN radar network (http://vt.superdarn.org/) [e.g., *Ruohoniemi and Baker*, 1996], although with reduced spatial resolution and temporal coverage. Additionally, there are several other instruments in the same area, perhaps most notably the EISCAT (https://www.eiscat.se/) incoherent scatter radars as well as several different all-sky cameras and riometers.

We will mostly concentrate on analysis tools that are suitable for event studies. This means that the tools should be data driven and not contain any statistical background model that is updated with the measurements or fitted to

Electric Currents in Geospace and Beyond, Geophysical Monograph 235, First Edition.
Edited by Andreas Keiling, Octav Marghitu, and Michael Wheatland.

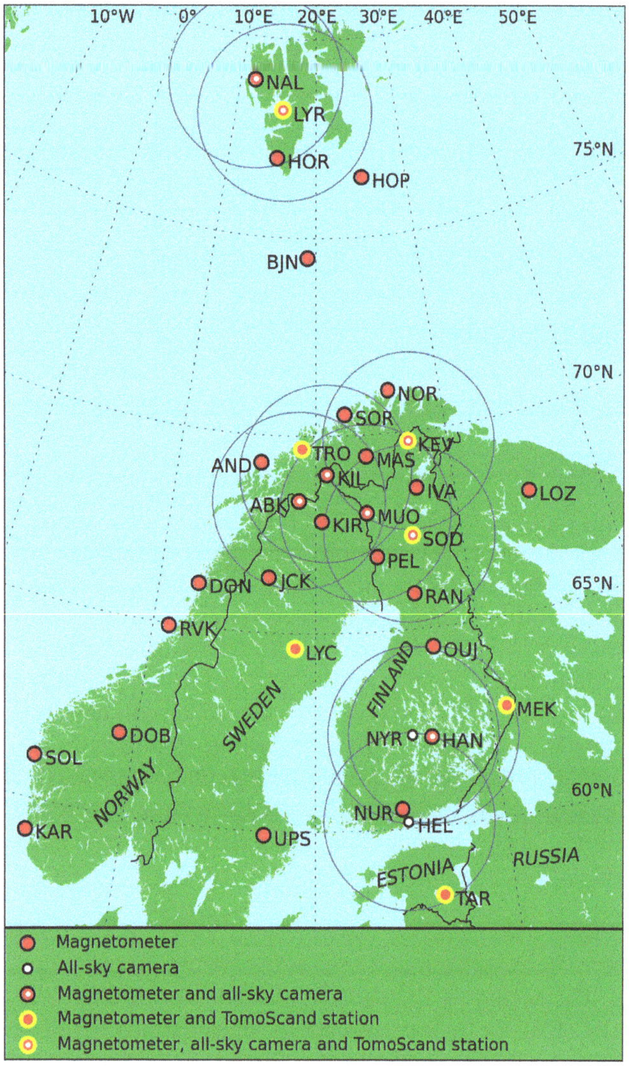

Figure 24.1 The MIRACLE network in September 2016.

the data. This rules out some very useful tools, such as the AMIE procedure for estimating the ionospheric electrodynamic state via data assimilation [e.g. *Cousins et al.*, 2013; *Richmond et al.*, 1998], or the partly statistical method used by *Waters et al.* [2015] to derive global current maps from SuperMAG (http://supermag.jhuapl.edu/) data.

We will start with a very short review of ionospheric electrodynamics, as understood in the context of this review. Then we will discuss various examples of data-analysis tools as listed in Table 24.1.

24.2. BACKGROUND

In this review we concentrate on ionospheric horizontal and field-aligned currents, including the electric fields that drive them and the magnetic disturbances they create. Thus, we will largely ignore ionization/recombination processes, plasma transport, kinetic effects, and many other interesting phenomena. Below we present a minimal summary of ionospheric electrodynamics. A much more comprehensive review has been given, for example, by *Richmond and Thayer* [2000].

24.2.1. Overview of Ionospheric Electrodynamics

For our purposes, the ionospheric electrodynamic state can be described by specifying the electric field **E**, horizontal sheet current density **J**, field-aligned current density $j_{\parallel}$, and the height-integrated Pedersen and Hall conductances $\Sigma_{P,H}$, respectively. These parameters are connected through the Maxwell's equations, current continuity, and ionospheric Ohm's law,

$$(\nabla \times \mathbf{E})_r = -\frac{\partial B_r}{\partial t} \tag{24.1}$$

Table 24.1 Selected Analysis Methods Discussed in This Review.

Input	Assumptions	Output	Method	Section
$\mathbf{B}_G$	–	$\mathbf{J}_{eq,ion}, \mathbf{J}_{eq,int}$	Field continuation and separation	3
$\mathbf{B}_{sat}$	–	$j_{\parallel}, \mathbf{J}$	–	4
$\mathbf{V}_{radar}$	Usually $\mathbf{E} = -\nabla\Phi_E$	$\mathbf{E}$	Potential mapping	5
$\mathbf{B}_G, j_{\parallel}, \mathbf{E}$	–	$\mathbf{J}, \Sigma_P, \Sigma_H$	–	6
$\mathbf{B}_G, \Sigma_P, \Sigma_H$	Usually $\mathbf{E} = -\nabla\Phi_E$	$\mathbf{E}, \mathbf{J}$	KRM	7
$j_{\parallel}, \Sigma_P, \Sigma_H$	Usually $\mathbf{E} = -\nabla\Phi_E$	$\mathbf{E}, \mathbf{J}$	Potential solver	7
$\mathbf{B}_G, \mathbf{E}$	$\alpha = \Sigma_H/\Sigma_P$	$\Sigma_H, \Sigma_P, \mathbf{J}$	$\mathbf{J}_{eq}$-based Method of characteristics	8
$j_{\parallel}, \mathbf{E}$	$\alpha = \Sigma_H/\Sigma_P$	$\Sigma_H, \Sigma_P, \mathbf{J}$	$j_{\parallel}$-based Method of characteristics	8

$$\frac{1}{\mu_o}\nabla\times\mathbf{B}=\mathbf{j}=\mathbf{J}\delta(r-R_I)-j_{\parallel}\hat{\mathbf{e}}_r \quad (24.2)$$

$$\mathbf{J}=\Sigma_P\mathbf{E}-\Sigma_H\hat{\mathbf{e}}_r\times\mathbf{E} \quad (24.3)$$

$$j_{\parallel}=\nabla\cdot\mathbf{J} \quad (24.4)$$

In the above equations, we have made several simplifying assumptions. First of all, we neglect the three-dimensional structure of the ionospheric current systems by assuming altitude-independent electric field and height integrating the conductances and the horizontal current density into a thin sheet at radius R_I. This can be motivated by the fact that the altitude range where the Hall and Pedersen conductivities peak in the ionospheric E-region is rather thin compared to the spatial extent of many typical current systems. However, 3D modeling of ionospheric currents may be important in some cases, as discussed, for example, by *Amm et al.* [2008] and *Fujii et al.* [2011]. Also, our assumption of altitude-independent **E** may break down in small scales, such as in thin auroral arcs [see, e.g., *Karlsson and Marklund*, 1998; *Vanhamäki et al.*, 2015].

Moreover, in equations (24.1)–(24.4), we have assumed radial background magnetic field and FAC, so that $\hat{e}_{\parallel}=\mathbf{B}/|\mathbf{B}|=-\hat{\mathbf{e}}_r$ in the Northern Hemisphere. This is a reasonable approximation at high magnetic latitudes, where the Earth's magnetic field is almost vertical, but it breaks down at midlatitudes where the inclination becomes significant. *Untiedt and Baumjohann* [1993] and *Amm* [1998] estimate that the radial field assumption is reasonable with inclinations $\chi>75°$, which covers the auroral regions. In the above equations, tilt of the magnetic field could be taken into account by writing equation (24.4) as $j_{\parallel}=\nabla\cdot\mathbf{J}/\sin\chi$ and by modifying the conductances in equation (24.3) as described by *Brekke* [1997] (see Chapter 7.12). However, this would complicate many of the analysis tools discussed below.

We have neglected the neutral wind and its effect on the ionospheric currents. We do it mainly because the neutral wind is difficult to measure, so suitable data are seldom available. However, if the neutral wind velocity **u** is known, its effect on ionospheric currents can be included by using the effective electric field $\mathbf{E}'=\mathbf{E}+\mathbf{u}\times\mathbf{B}$ in equation (24.3). Finally, we ignore the gravity and pressure-gradient driven currents, as their effect is rather small and probably important only at low latitudes [*Alken et al.*, 2016].

24.2.2. 2D Vector Fields

In the thin sheet approximation, the ionospheric electric field and horizontal current are two-dimensional vector fields on a sphere. These can be represented in terms of potentials, such as

$$\mathbf{E}(\theta,\phi)=-\nabla\Phi_E-\hat{\mathbf{e}}_r\times\nabla\psi_E \quad (24.5)$$

$$\mathbf{J}(\theta,\phi)=-\nabla\Phi_J-\hat{\mathbf{e}}_r\times\nabla\psi_J \quad (24.6)$$

Here Φ_E is the electrostatic potential, which we assume to be height independent, while ψ_E represents the rotational part. Usually it is assumed that $\psi_E=0$, but in principle this inductive part could be calculated from Faraday's law in equation (24.1). The electrostatic approximation is usually well founded and simplifies analysis considerably, but in some cases, inductive phenomena can be important. For example, *Yoshikawa and Itonaga* [1996] showed that induction affects the reflection of shear Alfén waves at the ionospheric boundary, while the model calculations done by *Vanhamäki et al.* [2007] indicate that the rotational electric field can drive significant FAC in very dynamical ionospheric systems, such as the westward traveling surge. However, for the rest of this review, we will ignore inductive effects and set $\psi_E=0$.

Similarly, the current potential Φ_J represent the irrotational or curl-free part of the current, which is connected to the FAC via equation (24.4). The other potential ψ_J describes a rotational or divergence-free current, which closes within the ionospheric sheet. In section 24.3, we'll see that it's closely connected to the ground magnetic disturbances.

The representations in equations (24.5) and (24.6) can be further developed by expanding the potentials in terms of suitable basis functions, such as spherical harmonics or spherical cap harmonics. This approach is used in many analysis methods. However, instead of the potentials, we may equally well represent the vector fields **E** and **J** in terms of their divergences and rotations. This is always possible in accordance with the Helmholtz's theorem. In analogy with the potential representation, the divergences and rotations of the vector field can then be composed of simple basis functions. One possible choice for the basis functions are the Spherical Elementary Current Systems (SECS) defined by *Amm* [1997]. Although for historical reasons their name refers to currents, they can be used to represent any 2D vector field on a sphere.

The SECS consist of two different sets of basis functions, one being curl-free (CF) and the other divergence-free (DF). These are illustrated in Figure 24.2. Use of the SECS has been covered in several previous works by *Amm and Viljanen* [1999], *Pulkkinen et al.* [2003a,b], and *Weygand et al.* [2011]. It will also be described in detail by *Vanhamäki and Juusola* [2018] in a forthcoming book. Therefore, we do not go into the details here.

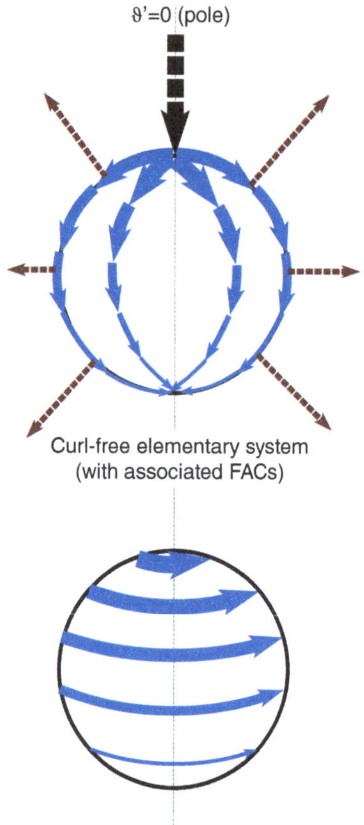

Figure 24.2 Spherical elementary current systems (SECS).

We just note that each individual SECS represents a localized curl or divergence of the vector field at the location of the elementary system's center point or pole. By placing several CF and/or DF SECS of different amplitudes at a suitable grid in the ionosphere, one can represent any 2D vector field in terms of its sources. This way many integral and differential equations involving the vector field can be reduced to linear algebraic equations involving the elementary systems' amplitudes, in analogy with the way, for example, spherical harmonic expansion can be used.

24.3. ANALYZING GROUND MAGNETIC FIELD

Ground magnetic measurements are an important data source for studying ionospheric current systems. *Gjerloev* [2012] give a short summary of some important findings, while *Untiedt and Baumjohann* [1993] give a detailed description of results obtained using the IMS Scandinavian magnetometer array (predecessor of MIRACLE). Estimation of ionospheric currents based on ground magnetic measurements is thus an important problem in geophysics, and many methods have been developed to accomplish this task.

As we are not interested in modeling the Earth's main field or crustal fields, the ground magnetic variation field $\mathbf{B}_G$ must first be estimated by removing unwanted sources,

$$\mathbf{B}_G = \mathbf{B}_{measured} - \mathbf{B}_{main} - \mathbf{B}_{crustal} - \ldots \tag{24.7}$$

Other possible sources include, for example, magnetospheric current systems, but also instrumental effects. The weaker the studied current systems are, the more careful processing is needed for determining the correct baseline to be removed from the data. We refer to *Gjerloev* [2012] and *Van de Kamp* [2013] for further discussion of this step.

24.3.1. The Concept of Equivalent Currents

As we see in Table 24.1, many analysis methods use the ground magnetic field as input data. In practice, this is always done by first converting the ground magnetic measurements into equivalent currents, which are then used in further analysis.

By definition, ionospheric equivalent current $\mathbf{J}_{eq,ion}$ is a divergence-free sheet current that produces the same magnetic field below the ionosphere as the real ionospheric current and associated FAC. As we try to allude in section 24.3.2, in global scales such an equivalent current representation always exists and is in principle uniquely defined, assuming perfect data coverage. In local studies, unknown boundary conditions may cause further uncertainty in the equivalent current solution.

The relationship between the equivalent current determined from $\mathbf{B}_G$ and the real ionospheric sheet current $\mathbf{J}$ is naturally a very important question. At high magnetic latitudes, FAC are almost radial. In that case, it can be shown that the curl-free part of the ionospheric sheet current together with the associated FAC does not create any magnetic field below the ionosphere [*Fukushima*, 1976, 1994]. Fukushima derived his result assuming uniform ionospheric conductances, but as *Amm* [1997] showed, the result is valid independent of the conductances. Therefore, if we take for a moment granted that $\mathbf{J}_{eq,ion}$ can be uniquely defined as a divergence-free sheet current, we can conclude that $\mathbf{J}_{eq,ion}$ is equal to the divergence-free part of $\mathbf{J}$. In the potential representation of equation (24.6), this means

$$\mathbf{J}_{eq,ion} = -\hat{\mathbf{e}}_r \times \nabla\psi_J \tag{24.8}$$

In general, this is all that we can accurately say about the ionospheric current based on ground magnetic data alone. Estimating the total ionospheric current from its divergence-free part requires some additional measurements or assumptions about the ionospheric conductances

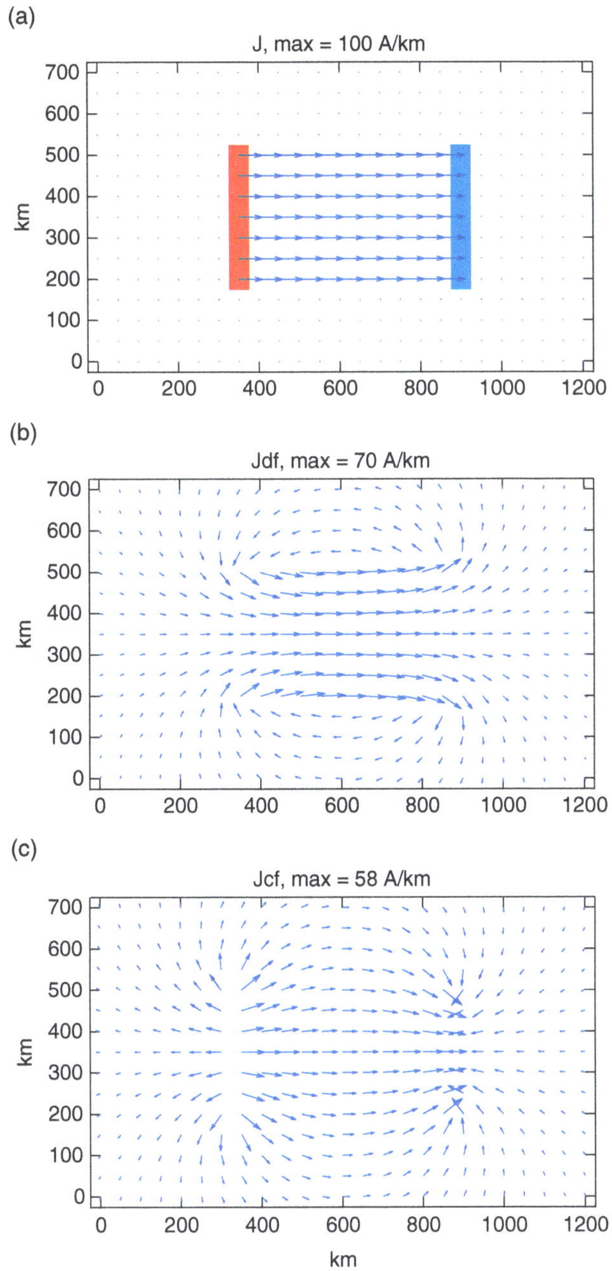

Figure 24.3 Example of an ionospheric current distribution (a) divided into divergence-free (b) and curl-free (c) parts. The divergence-free current produces the same magnetic field below the ionosphere as the total current system. On the top panel upward FAC are marked in blue and downward FAC in red.

or electric field, as indicated in Table 24.1 and discussed in sections 24.6–24.8 below.

It should also be kept in mind that equation (24.8) is valid only at high magnetic latitudes, where the FAC are approximately radial. Even there one must carefully interpret the meaning of $\mathbf{J}_{\text{eq,ion}}$. Figure 24.3 shows an extreme example, where the ionospheric current is confined to a high-conductivity channel and vanishes outside. The equivalent current calculated from ground magnetic measurements would correspond to the divergence-free current shown in the middle panel. Inside the high-conductivity band the curl-free and divergence-free currents enhance each other, while outside they completely cancel. For further discussion, we refer to *Laundal et al.* [2016], who studied how the relationship between horizontal and equivalent currents depends on sunlight conditions, as it affects the ionospheric conductances and total current pattern.

At lower latitudes, the curl-free sheet current and FAC do create some ground magnetic signatures, thus distorting the above relationship. We can, of course, derive the equivalent current solution as before, but equation (24.8) is no longer valid and interpretation of $\mathbf{J}_{\text{eq,ion}}$ is more difficult.

An additional complication is that temporally varying ionospheric currents create an induced electric field, which will drive currents in the oceans, crust, and even deeper in the lithosphere. Thus, the measured magnetic disturbance has both external (ionospheric) and internal (induced telluric) sources. The internal part becomes more significant during active geomagnetic conditions, when time derivatives are large [e.g., *Tanskanen et al.*, 2001]. As we will see, the internal and external sources can be separated, and both can be represented in terms of corresponding equivalent currents.

24.3.2. Harmonic Analysis

In section 24.3.3 we will discuss the SECS method, which is nowadays the standard tool in MIRACLE magnetic data analysis. However, we want to first briefly discuss harmonic analysis, as that may be more familiar to most readers and is widely applied in analysis of magnetic data.

Electric currents are negligible in the neutral atmosphere, so according to equation (24.2), the magnetic field there is curl-free and can be presented in terms of a potential,

$$\mathbf{B} = -\nabla\Phi_B \tag{24.9}$$

In harmonic analysis, the potential is expanded as a sum of basis functions, and the coefficients are fitted to the measured variation field. Spherical Harmonic Analysis (SHA) can be applied in global scales, while Spherical Cap Harmonic Analysis (SCHA) introduced by *Haines* [1985] works over a spherical cap of tunable half-width λ_c. In the limit $\lambda_c \to 180^\circ$ SCHA approaches SHA.

Using spherical cap harmonics, the magnetic potential in equation (24.9) is written as

$$\Phi_B(r,\theta',\phi') = R_E \sum_{k=0}^{K^i}\sum_{m=0}^{k}\left(\frac{R_E}{r}\right)^{n_k(m)+1}\left[g_{km}^i\cos(m\phi') + h_{km}^i\sin(m\phi')\right]P_{n_k(m)}^m(\cos\theta') + R_E\sum_{k=0}^{K^e}\sum_{m=0}^{k}\left(\frac{r}{R_E}\right)^{n_k(m)}\left[g_{km}^e\cos(m\phi') + h_{km}^e\sin(m\phi')\right]P_{n_k(m)}^m(\cos\theta') \tag{24.10}$$

Here R_E is Earth's radius, the colatitude θ' and longitude φ' are given in a coordinate system centered at the chosen spherical cap, $K^{i,e}$ give the truncation points of the series expansion and $P_{n_k(m)}^m(\cos\theta')$ are the associated Legendre functions of the first kind with noninteger degree $n_k(m)$ and order m. The superscripts i and e denote the internal (decreasing with r) and external (increasing with r) parts of the magnetic field, respectively. The only difference to SHA is that in SCHA the degree $n_k(m)$ are not integers. The details of calculating the polynomials $P_{n_k(m)}^m(\cos\theta')$ and fitting the coefficients to the measured data are described by *Haines* [1985] and *Haines and Torta* [1994], and also will be covered by *Fiori et al.* [2018] in a forthcoming book.

When we calculate the magnetic field components from $-\nabla\Phi_B$ and examine their form, it becomes clear that the horizontal magnetic field can be explained using either the internal part (coefficients g^i, h^i) or the external part (g^e, h^e) alone. The same is true for the vertical field. Only by fitting both vertical and horizontal components at the same time can we separate the ground magnetic perturbation into internal and external parts. In global SHA, this separation can be done unambiguously, at least in principle. Unfortunately this is not true in localized SCHA. *Thébault et al.* [2006] showed that for cap sizes $\lambda_c < 180°$ the internal and external parts get mixed up to some extent. Despite this ambiguity, the results can still be useful [e.g. *Gaya-Piqué et al.*, 2008].

After fitting the coefficients, we can calculate the magnetic field at different altitudes, but only in the region where $\nabla\times\mathbf{B}=0$. We can determine $\mathbf{J}_{\text{eq,ion}}$ by upward continuing the magnetic field to the ionosphere, and calculating the sheet current from the jump condition $\Delta\mathbf{B}_\perp = -\mu_0\hat{\mathbf{e}}_r\times\mathbf{J}_{\text{eq,ion}}$ by assuming that all external currents are concentrated in the sheet. This procedure is explained in detail by *Haines and Torta* [1994]. Similarly, we can calculate the internal equivalent current $\mathbf{J}_{\text{eq,int}}$ by downward continuing the field and introducing another fictional sheet current layer at some depth inside the Earth.

Although not shown in detail here, from the completeness and orthogonality of the global spherical harmonic expansions representing the magnetic potential and equivalent sheet current, we can conclude that any ground magnetic perturbation, if measured with sufficient spatial coverage and resolution, can indeed be uniquely represented in terms of divergence-free internal and external sheet currents.

24.3.3. SECS Analysis

When applied to regional magnetometer networks, such as the MIRACLE, the SCHA has some drawbacks. The geometry is limited to that of a spherical cap, and we must select a rigid wavelength cutoff by specifying the constants $K^{i,e}$ in equation (24.10). Typically our magnetometer network has a more general shape, and it may contain denser parts where we would like to use enhanced spatial resolution. The SECS method can meet these requirements, as demonstrated by *Amm and Viljanen* [1999].

In the SECS method we make direct use of the result that the ground magnetic variation field can be represented in terms of internal and external divergence-free sheet currents. So instead of representing the magnetic field with a potential, we represent the equivalent currents in terms of DF SECS.

The basic idea in the SECS method is to calculate the ionospheric current and magnetic field produced by a single DF SECS, place several DF SECS over and below the magnetometer network, and then fit their amplitudes to the measured data. This procedure has been described in detail by *Amm and Viljanen* [1999], who also derived formula for the magnetic field of a single DF SECS. In practice, we end up with a matrix equation

$$\mathcal{B}_G = \mathbf{M}\cdot\mathcal{S}^{df} \tag{24.11}$$

where the vector $\mathcal{B}_G$ contains the measured magnetic field components at different stations (locations $\mathbf{r}_n = (R_E, \theta_n, \phi_n), n = 1\ldots L$),

$$\mathcal{B}_G = \left[B_x(\mathbf{r}_1)\; B_y(\mathbf{r}_1)\; B_z(\mathbf{r}_1)\; B_x(\mathbf{r}_2)\ldots\right]^T \tag{24.12}$$

while the vector $\mathcal{S}^{df}$ contains the amplitudes or scaling factors of the elementary systems placed at locations $\mathbf{r}_k^{el} = (r_k, \theta_k, \phi_k), k = 1\ldots K$,

$$\mathcal{S}^{df} = \left[S^{df}(\mathbf{r}_1^{el})\; S^{df}(\mathbf{r}_2^{el})\; S^{df}(\mathbf{r}_3^{el})\ldots\right]^T \tag{24.13}$$

Components of the system matrix M tell how individual DF SECS affect the magnetic field at different stations. For example, $M_{2,3}$ gives the y-component of the magnetic field at station $\mathbf{r}_1$ created by the unit SECS located at $\mathbf{r}_3^{el}$. Equation (24.11) must be inverted for the unknown scaling factors $\mathcal{S}^{df}$, after which the equivalent current can be calculated.

We note that also in the SECS analysis we can choose to use all three components of the magnetic field in equations (24.11) and (24.12) and separate the field into internal and external parts, or we can use only the horizontal components in estimating just the external equivalent current. In the former case, we place the DF SECS both above (in the ionosphere) and below the magnetometer array, whereas, in the latter case, we use only ionospheric DF SECS.

Amm and Viljanen [1999] describe how to form the system matrix M, how to invert it using truncated singular value decomposition (SVD) and how to calculate the resulting equivalent current. They also present a comparison between the SCHA and SECS methods using synthetic model data, demonstrating that the SECS method gives more accurate results, and is able to incorporate variable spatial resolution. *Pulkkinen et al.* [2003a,b] carried out more thorough testing and applications to real data, including some error analysis and separation of the variation field into internal and external sources. *Weygand et al.* [2011] applied the method to magnetometers in North America and Greenland, which required use of an irregularly shaped analysis grid. They also presented a detailed optimization of the truncation parameter in the matrix inversion. However, we emphasize that even though truncated SVD has traditionally been used in inverting equation (24.11), any other regularization method could be used equally well.

A detailed description of the SECS method with further discussion of applications will be given by *Vanhamäki and Juusola* [2018] in a forthcoming book. Here we just note that the SECS method is routinely used to derive maps of ionospheric equivalent currents based on data from the MIRACLE network or the magnetometers in North America and Greenland. The latter application is demonstrated in Figure 24.4 reproduced from *Weygand et al.* [2011]. This snapshot calculated using data from the THEMIS magnetometer network illustrates the equivalent current system around the local midnight and shows the presence of a Harang discontinuity between the eastward and westward electrojets. These maps are freely available for quick-look purposes (see http://space.fmi.fi/MIRACLE/iono_2D.php and http://www.igpp.ucla.edu/jweygand/htmls/EICS_NA_Greenland.html).

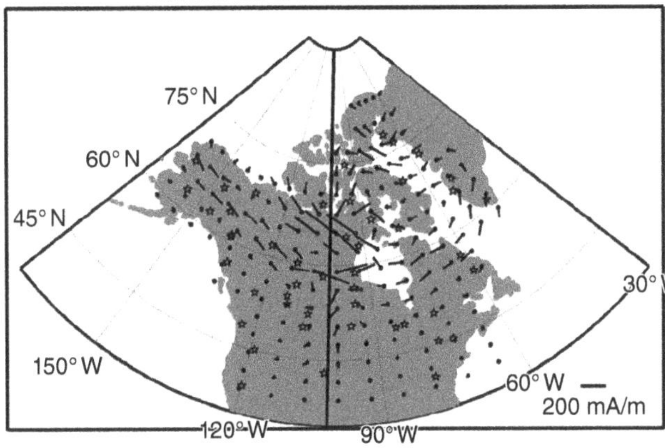

Figure 24.4 Equivalent ionospheric currents (EIC) calculated from the THEMIS magnetometer network showing a Harang discontinuity. Scale arrow is in the lower right corner, while the black line at about 100° W indicates local midnight. The stars mark the location of the ground magnetometers used in the analysis. From *Weygand et al.* [2011].

24.4. ANALYZING SATELLITE MAGNETIC FIELD

In the previous section, we saw that ground magnetic measurements reveal only part of the ionospheric current system, namely its divergence-free part. Thus, we get very little or no information about FAC. The situation is completely different when we have access to magnetic field data recorded above the ionospheric current layer.

Low-orbit satellites move in the ionospheric F-region, far above the E-region current layer and far below the magnetospheric currents. In this region at high latitudes, the magnetic perturbation field is dominated by the FACs flowing between the magnetosphere and ionosphere, so satellite magnetic measurements give us direct information about FAC and associated curl-free part of the height-integrated ionospheric current. As the divergence-free part of the ionospheric current is confined to the E-region current sheet, its contribution to the satellite magnetic field decreases steadily with altitude. This limits our ability to resolve the divergence-free currents from satellite data, although the large-scale electrojets are often easily detectable.

Above the E-region ionosphere, the divergence-free part of the ionospheric sheet current ($\hat{\mathbf{e}}_r \times \nabla\psi_J$ in equation [24.6]) causes a magnetic field that can be presented as

$$\mathbf{B}_\psi(r, \theta, \phi) = -\nabla V \tag{24.14}$$

The potential V can further be expanded in terms of spherical harmonic functions as the internal part of equation (24.10), showing that small-scale features are quickly suppressed with increasing altitude. On the other hand, the magnetic field due to the curl-free sheet current ($\nabla\Phi_J$ in equation [24.6]) and associated radial FAC is

$$\mathbf{B}_\Phi(r, \theta, \phi) = \begin{cases} \dfrac{-\mu_0 R_I}{r}\hat{\mathbf{e}}_r \times \left[\dfrac{r}{R_I}\nabla\Phi_J(\theta, \phi)\right], & r > R_I \\ 0, & r < R_I \end{cases} \tag{24.15}$$

This result can be easily verified by checking that $\nabla \cdot \mathbf{B}_\Phi = 0$, curl of $\mathbf{B}_\Phi$ gives the correct FAC, which scales as $(R_I/r)^2$ assuming radial main field, and the jump condition at $r = R_I$ is satisfied. However, this is in fact a special case of the poloidal-toroidal decomposition [e.g., *Backus*, 1986; *Laundal et al.*, 2016], which can be applied to any divergence-free vector field.

In many statistical studies, satellite magnetic measurements from several orbits, or typically from several years, are merged together and used to determine the FAC and associated curl-free ionospheric current via equation (24.15), or some variation of it. Usually the current $\nabla\Phi_J$ or the current potential is expanded in terms of some basis functions, such as spherical (cap) harmonics or CF SECS, which are then fitted to the data. These studies have given us invaluable insight to the high-latitude current system and its variations during different solar wind and geomagnetic conditions. However, in this review, we are primarily interested in methods suitable for event studies, so we simply refer to *Weimer* [2005], *Stauning* [2002], *Juusola et al.* [2014], and *Laundal et al.* [2016], and to the studies cited by them, for further details and results of such statistical investigations.

24.4.1. Data From Single Satellite

If we want to determine the FAC or ionospheric horizontal current in a specific event and have data from only a single satellite, we must make some additional assumptions about the current geometry [*Lühr et al.*, 1996]. For example, if we assume that the FAC forms longitudinally uniform sheets, we can estimate the current density from Faraday's law by setting the derivatives in eastward direction to zero. The result is

$$j_{\parallel} = \frac{1}{\mu_0}\frac{\partial B_y}{\partial x} \qquad (24.16)$$

where x is northward and B_y is eastward component of the variational magnetic field. Also in this case, we must carefully extract the variational field from the total measured field as in equation (24.7). This can be done by removing a suitable main field model, such as CHAOS (http://www.space.dtu.dk/MagneticFieldModels)[*Finlay et al.*, 2016], from the data.

This method, with some variations, has been applied very successfully to magnetic measurements from various satellites beginning from the seminal work by *Iijima and Potemra* [1976]. When applying equation (24.16), we can use the Maximum Variance Analysis (MVA, e.g., *Sonnerup and Scheible* [1998]) or similar techniques for checking the validity of assuming elongated current sheets, and also to determine the most likely orientation of the sheet.

The DF part of the current can be estimated from single satellite data using the method developed by *Olsen* [1996]. In the method, the DF current is expanded in terms of infinitely long, straight, horizontal line currents placed at the altitude of the E-region and oriented perpendicular to the satellite track. The strengths of the line currents are determined by fitting the measured residual magnetic field to the superposed field of the line currents. This involves a matrix inversion, similar to section 24.3.3. In order to minimize the contribution from the FAC, the component of the residual field along the main field is used in the fitting. Telluric currents can be taken into account by assuming an infinite conductor at depth d, implemented as mirror line currents at depth $2d$ [see also *Juusola et al.*, 2016a].

Alken et al. [2013] applied the line current method to estimating the equatorial electrojet from satellite magnetic data. They also modified the method so that instead of infinite straight lines, the current is constructed from short, straight segments. This way, by placing the segments carefully, the electrojet current can flow around the world along curves of constant magnetic latitude.

24.4.1.1. 1D SECS

One advantage of the SECS method is that all three current components (DF, CF, and FAC) can be estimated simultaneously and with a compatible spatial resolution. This makes it possible to combine the DF and CF currents as the true horizontal current **J**. With a single satellite, though, we must make the assumption that the currents vary only in one direction (usually geographical or geomagnetic latitude). A 1D variant of the SECS method has been developed for this purpose by *Vanhamäki et al.* [2003] and *Juusola et al.* [2006]. The 1D CF and DF SECS were constructed by placing infinitely many 2D CF or DF SECS poles at a constant latitude ($\theta = \theta_0$). The resulting 1D SECS are independent of longitude in the chosen coordinate system. Unlike the 2D SECS equations that are given with respect to a local pole ($\theta' = 0$) and need transformation to a global coordinate system (e.g., geographic or geomagnetic), the 1D SECS are already defined in a global system. While the 2D SECS have a localized curl or divergence at its pole, the 1D SECS have a ring of curl or divergence at a constant latitude.

The 1D SECS analysis principle is the same as with the methods of *Lühr et al.* [1996] and *Olsen* [1996] above, except that the basis functions are different. The DF current at 110 km altitude is expanded in terms of DF 1D SECS and the amplitudes are determined by fitting the field-aligned component of the residual magnetic field. The telluric currents can be taken into account as mirror currents [*Juusola et al.*, 2016a]. The CF currents at 110 km altitude and the associated radial FAC are expanded in terms of CF 1D SECS. The amplitudes are determined by fitting the eastward component of the residual magnetic field. Spatial resolutions as well as other practical details of applying the method have been discussed by *Juusola et al.* [2016a]. Applications of the method include derivation of auroral ionospheric currents from CHAMP [*Juusola et al.*, 2007] and Swarm [*Juusola et al.*, 2016a,b].

24.4.2. Swarm

The multisatellite Swarm mission [*Olsen et al.*, 2013] provides information not only along the orbit but also in the cross-track direction. This allows relaxation of the 1D assumption. *Amm et al.* [2015] developed a method based on a combination of the 1D and 2D SECS analysis, suitable for deriving latitude-longitude maps of the FAC, DF, and CF currents around the E-region footprints of the side-by-side flying Swarm satellite pair. *Amm et al.* [2015] reasoned that as there is much more information on the latitudinal than longitudinal variation, it is better to first use the 1D SECS to describe large-scale electrojet-type systems, and only try to fit the remnant field using the more localized 2D SECS. An example of the current maps derived using this method is shown in Figure 24.5.

Juusola et al. [2016a] compared the resulting DF currents with those derived from MIRACLE using the 2D SECS method (described in section 24.3.3). They found that the current longitudinal separation of the side-by-side flying Swarm pair (~60 km above MIRACLE) may be too small compared with the distance to E-region (~350 km) to properly see the longitudinal variation in the DF current. However, in the eastward current component, mostly dominated by electrojet currents, the agreement between satellite and ground-based currents is excellent. In addition to magnetic field analysis and currents, the method of *Amm et al.* [2015] includes also analysis of Swarm's electric field measurements and estimation of ionospheric Hall and Pedersen conductances, as discussed in section 24.6 below.

Besides *Amm et al.* [2015], also various other multispacecraft analysis techniques, many of them originally developed for the Cluster mission, have been applied to the unique configuration of Swarm. *Ritter et al.* [2013] used the integral form of Ampere's law for computing FAC from a quad of measurement points given by the two side-by-side flying Swarm satellites. This quad

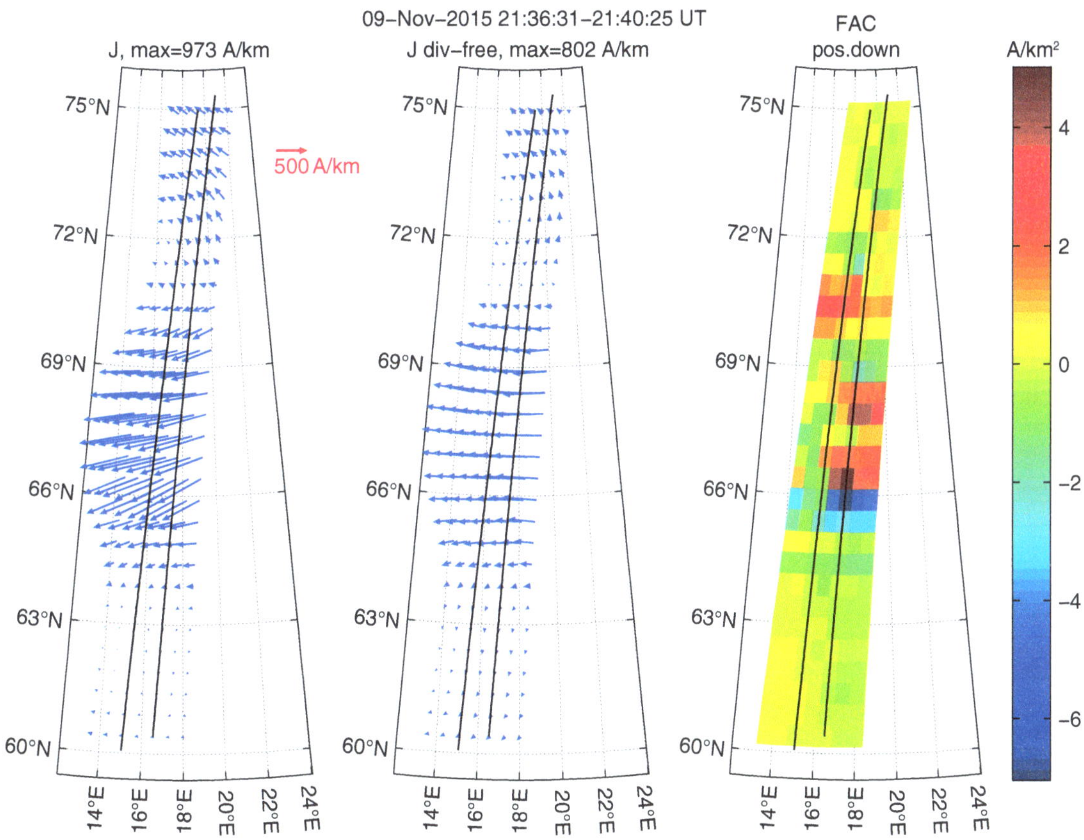

Figure 24.5 An example of ionospheric current maps derived from Swarm satellite's magnetic measurements using the method by *Amm et al.* [2015]. Footprints of the Alpha and Charlie satellites, projected to 110 km altitude along the magnetic field lines, are shown as black lines. The divergence-free current shown in the middle panel can be directly compared with the equivalent current derived from ground magnetic data.

estimate, together with traditional single-satellite estimates, is available as Swarm level-2 data product (https://earth.esa.int/swarm/data-access). *Dunlop et al.* [2015] have extended the quad method by applying various curlometer configurations to data from two or three Swarm spacecraft, whereas *Vogt et al.* [2013] have investigated the general problem of gradient estimation, including FAC estimation, using generalized reciprocal vectors and data from two or three spacecraft. For a general discussion about single- and multisatellite techniques, we refer to *Vogt* [2014], while in Chapter 4 of this book, *Dunlop et al.* discuss analysis of Cluster measurements.

24.4.3. AMPERE

As stated above, the two side-by-side flying Swarm satellites provide some local information about the cross-track gradients in the magnetic field. In contrast, the AMPERE project (http://ampere.jhuapl.edu/), based on the 66 Iridium communication satellites, provides an instantaneous global view of the magnetic variations, albeit at reduced spatial resolution. These data can be used to derive hemispherical maps of the FAC by employing equation (24.15) and spherical (cap) harmonic expansion of the magnetic potential, as described by *Waters et al.* [2001] and *Green et al.* [2006]. However, due to the satellites altitude (~780 km) and the limited resolution of the magnetometers, the divergence-free current can not be determined from the data. The AMPERE project, analysis method, and results are discussed in detail by *Coxon et al.* in Chapter 16 of this book.

24.5. IONOSPHERIC ELECTRIC FIELD: FITTING LINE-OF-SIGHT MEASUREMENTS

In this review, we are mostly concerned with ionospheric currents, but they are intimately related to the electric field and plasma convection. The main instruments for measuring ionospheric electric fields in mesoscales are various coherent scatter radars, the prime example being the SuperDARN radar network.

The radars measure the line-of-sight (LOS) plasma velocity, usually at F-region altitudes. When two or more radars measure the velocity from different directions, we can reconstruct the horizontal plasma velocity vector from the LOS components, and convert it to electric field by assuming collisionless $\mathbf{E}\times\mathbf{B}$ drift. Simultaneous and overlapping measurements from multiple radars are not always available, but we can still try to combine regional or global LOS measurements by making some additional assumptions of the velocity field.

24.5.1. Global SuperDARN Potential Mapping

For the SuperDARN radars, the standard method of estimating the global (or hemispherical) convection pattern is the potential mapping technique introduced by *Ruohoniemi and Baker* [1996]. The electric field is assumed to be given by a potential, $\mathbf{E} = -\nabla\Phi_E$, which corresponds to incompressible plasma flow. The electric potential is expanded in terms of spherical harmonics, whose coefficients are determined by least-squares fit to the measured LOS velocities. Unfortunately, the amount and spatial coverage of the radar data are rarely sufficient for determining the global convection pattern. Therefore, extra "data points" are taken from a statistical model, which is based on previous radar measurements and is parameterized with solar wind data.

As the SuperDARN potential mapping technique is global and relies on a statistical background model, it does not quite fit in our criterion of regional and data-driven analysis methods. Nevertheless, we wish to mention it here, as it is a very commonly used tool in studies of ionospheric electrodynamics. Recent developments, like the SuperDARN assimilative mapping (SAM) developed by *Cousins et al.* [2013], may use more advanced fitting techniques and statistical models, but do not change the fundamentals.

24.5.2. Regional Analysis of SuperDARN Data

If we forget the global convection pattern and concentrate on regional studies, we may be able to find an area where the radars have very good data coverage. In those regions, we can try to reconstruct the electric field without any statistical assumptions. At least two methods have been developed for this task, using either spherical cap harmonics [*Fiori et al.*, 2010] or curl-free SECS [*Amm et al.*, 2010] as the basis functions for representing the electric field.

The SCHA method suffers from the same limitations as mentioned in section 24.3 regarding ground magnetic field analysis, namely that the geometry is limited to a spherical cap and we have to select the same wavelength cutoff for the whole area. Also the unknown boundary conditions at the edges of the cap may affect the solution.

The SECS-approach used by *Amm et al.* [2010] does not have similar limitations. Spatial resolution and the shape of the analysis area can be selected freely by placing the elementary systems at suitable locations. Additionally, there are no explicit boundary conditions, although it is implicitly assumed that all the sources of the electric field are located inside the analysis area. Naturally, other boundary conditions could be included by adding virtual measurement points at the edges of the analysis area.

Details of the SECS technique and various test cases are presented by *Amm et al.* [2010]. Essentially, the ionospheric plasma velocity field is composed of DF SECS placed at a discrete grid covering the radars' combined field of view. This results in a matrix equation between the measured LOS velocities and the unknown amplitudes of the elementary systems,

$$\mathcal{V}_{\text{los}} = \mathbf{M} \cdot \mathcal{S}^{df} \tag{24.17}$$

The vector $\mathcal{V}_{\text{los}}$ contains the measured LOS components at locations $\mathbf{r}_n = (R_E, \theta_n, \phi_n), n = 1 \ldots L$,

$$\mathcal{V}_{\text{los}} = \left[V(\mathbf{r}_1)\ V(\mathbf{r}_2)\ V(\mathbf{r}_3)\ldots\right]^T \tag{24.18}$$

while the vector $\mathcal{S}^{df}$ contains the SECS amplitudes, as in equation (24.13). The geometrical system matrix M can be calculated once we specify the SECS locations and know the LOS directions and locations. Equation (24.17) must be inverted for the unknown amplitudes, after which the plasma velocity and thereof the convection electric field can be estimated at any location within the analysis area. The test cases studied by *Amm et al.* [2010] show that the method works reliably with surprisingly few measurement points.

24.6. ESTIMATING CONDUCTANCES FROM B_G, B_{sat}, AND E

In the previous sections, we have seen how to estimate the ionospheric equivalent current (or divergence-free current) from ground magnetic data, the FAC and associated curl-free current from satellite data, and the electric field from radar LOS velocity measurements. If we have both ground- and satellite-based magnetic measurements, we can estimate the total ionospheric current in a very straightforward manner by combining the curl-free and divergence-free parts. As discussed in section 24.4, in principle satellite measurements alone are enough to determine the whole current system, but in practice it may be difficult to reliably determine the divergence-free currents even from low-altitude satellite data.

It should be mentioned that even though the ionospheric sheet current (or any other horizontal vector field) may be globally obtained as a unique sum of divergence- and curl-free parts, this is not necessarily true locally. In regional studies we should also specify the boundary conditions, as they may manifest themselves as an additional Laplacian vector field, which is both curl-free and divergence-free inside our analysis area. In global scales such a Laplacian field must vanish, or be physically unreasonable.

If we know also the ionospheric electric field, in addition to the CF and DF currents, we can try to estimate the ionospheric Pedersen and Hall conductances from Ohm's law in equation (24.3). The explicit formulas are

$$\Sigma_P = \frac{\mathbf{J} \cdot \mathbf{E}}{|\mathbf{E}|^2}, \qquad \Sigma_H = \frac{(\mathbf{J} \times \mathbf{E})_r}{|\mathbf{E}|^2} \tag{24.19}$$

This approach has been proposed, for example, by *Amm* [2001], who considered combining MIRACLE ground magnetic measurements with FAC estimates from the Cluster spacecraft in order to obtain a regional map of the ionospheric sheet current. When further combined with electric field measured by the STARE radar, he could estimate the Hall and Pedersen conductances in a reasonably large area over the northern part of the MIRACLE network.

However, in practice it may be difficult to combine separate measurements, with often very different spatial and temporal resolutions, in this simple way. For example, *Green et al.* [2007] estimated ionospheric conductances from ground- and satellite-based magnetic measurements, with **E** taken from SuperDARN. Their results for an event study agree with statistical models, but sparse data coverage resulted in significant uncertainties. *Juusola et al.* [2014] published statistical maps of ionospheric currents calculated from CHAMP satellite's magnetic data and ionospheric electric field assembled from SuperDARN measurements. Nevertheless, they did not combine these two maps to estimate the Hall and Pedersen conductances, as this would have resulted in areas of negative conductances, especially in the nightside around the Harang discontinuity.

The SECS-based technique developed by *Amm et al.* [2015] for analyzing electric and magnetic field measurements provided by the Swarm satellites is a continuation of the approach proposed by *Amm* [2001]. As described in section 24.4.2, *Amm et al.* [2015] estimate both the CF and DF parts of the ionospheric current from the Swarm magnetic data. Additionally, the electric field measurements from the side-by-side flying satellite pair are analyzed independently, using the elementary system method described in section 24.5.2. This provides a spatially extended 2D (latitude-longitude) map of the electric field around the satellite tracks, similar to the current maps shown in Figure 24.5.

As discussed by *Juusola et al.* [2016a], the spatial resolution in determining the DF current is limited both by the separation between the satellite pair, and (more severely) by the altitude of the satellites. Even though the FAC and associated CF ionospheric current, as well as the electric field, could in principle be obtained with better spatial resolution, for conductance estimates, we must use the same resolution for all analysis, in this case limited by the DF current determination.

In the synthetic test cases presented by *Amm et al.* [2015], both the magnetic field and electric field analysis performed very well. Uncertainties naturally increase when the independently derived electric field and current maps are used to calculate the conductances using equation (24.19). Nevertheless, in the test cases also the conductances were reproduced well, giving a good qualitative and even quantitative view of the situation.

In practice, due to some limitations and calibration problems with Swarm electric field instruments, only the magnetic field analysis has been extensively applied to real data by *Juusola et al.* [2016a], as described in section 24.4.2. However, *Juusola et al.* [2016b] presented an event study of a morning sector auroral arc, where they were ably to assume vanishing longitudinal gradients. This made it possible to use a limited version of the analysis, with electric and magnetic field data from only one satellite. Their results demonstrate the usability of the method by *Amm et al.* [2015] and the potential of Swarm satellite data in detailed studies of auroral current systems, including conductance distributions.

Although we emphasize regional studies, we should mention that the combination of SuperMAG ground magnetic data, SuperDARN convection maps and AMPERE FAC maps give a unique opportunity to estimate the ionospheric electrodynamic state in hemispherical scales. This approach was already studied by *Green et al.* [2007], but direct calculation of the Hall and Pedersen conductances based on these data is hampered by the aforementioned differences in spatial and temporal resolutions, as well as by the effects of statistical background models used at least in the SuperDARN analysis. However, progress is being made toward resolving these issues [*Gjerloev and Waters*, private communication].

24.7. ASSUMING KNOWN CONDUCTANCES

If we happen to know the ionospheric conductances from some other measurements, we may combine them with $\mathbf{J}_{\text{eq,ion}}$ or FAC determined from magnetic data, and solve the ionospheric electric field and total current. When we calculate the curl and divergence of Ohm's law (equation [24.3]), and further assume that the ionospheric electric field is given by a potential, $\mathbf{E} = -\nabla\Phi_E$, we get two scalar equations

$$\Sigma_H \nabla^2 \Phi_E + \nabla\Sigma_H \cdot \nabla\Phi_E - \left(\nabla\Sigma_P \times \nabla\Phi_E\right)_r = \left(\nabla \times \mathbf{J}_{\text{eq,ion}}\right)_r \tag{24.20}$$

$$\Sigma_P \nabla^2 \Phi_E + \nabla\Sigma_P \cdot \nabla\Phi_E - \left(\nabla\Sigma_H \times \nabla\phi_E\right)_r = -j_{\parallel} \tag{24.21}$$

These are elliptic partial differential equations, which can be solved for the electric potential. The required input data consists of the Hall and Pedersen conductances, and either FAC or $\mathbf{J}_{\text{eq,ion}}$. If the electric field is solved in a limited region, we also need boundary condition for Φ_E. Incidentally, we note that by making the further assumptions

$$\left(\nabla\Sigma_P \times \mathbf{E}\right)_r = \left(\nabla\Sigma_H \times \mathbf{E}\right)_r = 0, \qquad \alpha = \Sigma_H / \Sigma_P = constant \tag{24.22}$$

the above equations imply a direct proportionality between the curl of equivalent current and FAC:

$$\left(\nabla \times \mathbf{J}_{\text{eq,ion}}\right)_r = -\alpha j_{\parallel} \tag{24.23}$$

This relationship has been used, for example, by *Amm et al.* [2002] and *Weygand and Wing* [2016] for estimating FAC from ground-magnetic data. However, due to the assumptions needed in deriving the relationship, the result should be considered only as a proxy for the FAC. In individual case studies, especially during active conditions, there may be large conductance variations that do not satisfy the conditions in equation (24.22).

24.7.1. Combining Equivalent Current and Conductances: KRM

Solving equation (24.20) for the electric field is called the KRM method, named after *Kamide et al.* [1981]. The equivalent current needed as input data can be obtained quite straightforwardly from ground magnetic measurements, as discussed in section 24.3, but the required conductance estimates are much more difficult to obtain. Once the input data are known, we can solve the electric field and thereof the ionospheric current and FAC.

As discussed in more detail by *Vanhamäki and Amm* [2011] and *Murison et al.* [1985], another problem in applying equation (24.20) to regional studies is the unknown boundary conditions for the electric potential Φ_E, and their effect on the solution. *Kamide et al.* [2003] limited solution of equation (24.20) to areas with good magnetometer data coverage, and took the necessary boundary conditions from the data-assimilation code AMIE [*Richmond et al.*, 1998]. However, when the global data coverage is poor, the boundary conditions provided by AMIE may depend heavily on the statistical background models used in the data assimilation.

Vanhamäki and Amm [2007] took a different approach and rewrote the curl of Ohm's law, the basis of equation (24.20), in terms of CF and DF SECS. This converts the differential equation into matrix algebra, and has the side effect of removing all explicit boundary conditions. The reason is the same as mentioned in section 24.5.2. In the SECS analysis, we implicitly assume that all the sources of the electric field are located inside the analysis area, and outside the area it continues as a Laplacian field with zero curl and divergence. This is not the correct boundary condition either, but in practice it seems to work rather well.

The details of the local SECS-based KRM method are described by *Vanhamäki and Amm* [2011] and the references cited therein. Here we just note that the method has been applied to study auroral omega bands [*Vanhamäki et al.*, 2009] and an ionospheric trough [*Vanhamäki et al.*, 2016] in the MIRACLE area. In the former case, the conductances were estimated from satellite-based UV images and by combining ground-based all-sky images with riometer measurements, while in the latter case they were taken from EISCAT incoherent scatter radar data.

24.7.2. Combining FAC and Conductances

In analogy with the KRM method, equation (24.21) can be solved for the electric field. This is typically used as the ionospheric part of coupled magnetohydrodynamic magnetosphere-ionosphere simulations [e.g., *Janhunen et al.*, 2014], or in statistically motivated model calculations such as performed by *Nakamizo et al.* [2012]. To our knowledge, it has not been applied to any case studies, possibly due to the fact that regional FAC distribution is more difficult to estimate than the equivalent current.

Also boundary conditions may be more problematic in this case. Using SECS and synthetic input data, *Vanhamäki* [2011] solved a time-dependent form of equation (24.21), where induction effects are included and the electric field is no longer given by a potential. They noted that a Cowling channel-type structure (i.e., a current system where FAC is locally zero) could be added to the solution, without affecting the input data. This can be a serious limitation, as, for example, the auroral electrojet may be a partial Cowling channel (see Chapter 25 in this book, by *Yoshikawa and Fujii*).

24.7.3. The ALADYN Method

A rather different approach is taken in the Auroral Arc Electrodynamics or ALADYN method developed by *Marghitu et al.* [2004] and *Marghitu et al.* [2011]. This is a single-satellite analysis tool, but we choose to discuss it here, as the method needs ionospheric conductances as the input data. ALADYN has mostly been applied to FAST satellite observation, in which case the conductances can be estimated from the particle data [see *Marghitu et al.*, 2004].

Basically, the ALADYN method can be used to estimate the 1D cross-arc profiles of the electric currents and electric field, and also to estimate how well the 1D assumption of vanishing along-arc gradients holds. The current continuity equation is written in the cross-arc direction (x) as

$$j_{\|}(x) = \frac{\partial J_x(x)}{\partial x} + c_1 \tag{24.24}$$

Here deviations from a perfect 1D situation are modeled by the constant c_1, which describes feeding or draining of the along-arc current. Using equation (24.16), this can be integrated to

$$\mu_0^{-1} B_y(x) = J_x(x) + c_0 + x c_1 \tag{24.25}$$

where c_0 represents a constant background current in the cross-track direction. The along-arc electric field is assumed to be constant, denoted b_0, while the cross-arc electric field is written as

$$E_x(x) = \Phi_0 / L + \delta E_x(x) \tag{24.26}$$

The average electric field is Φ_0/L, that is, the total potential drop divided by the arc width, while $\delta E_x(x)$ represent a polarization field localized around the arc. The polarization field can be further expanded, for example, in terms of Legendre polynomials, as done by *Marghitu et al.* [2004].

Equations (24.25) and (24.26) can be rearranged as

$$\Sigma_P \delta E_x(x) - \Sigma_H(x) b_0 + c_0 + x c_1 = \mu_0^{-1} B_y(x) - \Sigma_P(x) \Phi_0 / L \tag{24.27}$$

The input data to the ALADYN method consist of the magnetic variation B_y and the potential difference Φ_0 measured by a low-orbiting satellite, as well as the conductances $\Sigma_{P,H}$ estimated, for example, from energetic particle measurements. Thus, the right-hand side of equation (24.27) is known. When the parameter b_0 is fixed, the other parameters can be estimated by fitting the left-hand side to the measured data at several points $x = x_n$. Repeating the fitting with several different b_0, one can determine the most likely values by comparing the estimated electric field $\Phi_0/L + \delta E_x(x)$ against the field measured by the satellite (including possible parallel potential drops), as well as by checking how constant the parameters c_0 and c_1 actually are when the fitting is done separately over several sections across the arc. A detailed description of the method, including generalization to tilted arcs, and application examples are given by *Marghitu et al.* [2004] and *Marghitu et al.* [2011].

24.8. METHOD OF CHARACTERISTICS

Yet other combinations of input data can be analyzed with the Method of Characteristics (MoC). There are two variants of the MoC, $\mathbf{J}_{eq}$-based and FAC-based, which refer to the part of current that is assumed to be known (see Table 24.1). Additionally, in both methods, we need to know the ionospheric electric field and the conductance ratio $\alpha = \Sigma_H/\Sigma_P$ from measurements. Then the ionospheric conductances can be solved by integrating a first-order partial differential equation along its characteristic curves (see, e.g., https://en.wikipedia.org/wiki/Method_of_characteristics), giving the method its name. One advantage of the MoC over the KRM method discussed in the previous section is that the conductance ratio α is less variable and easier to estimate e.g. from auroral images than the conductances themselves.

The original development and possible extensions of the MoC has been discussed by *Vanhamäki and Amm* [2011] and references cited there, so here we give only a very brief outline, concentrating on the implementation in spherical geometry, as presented by *Amm* [1998] and *Amm* [2002]. The starting point is again Ohm's law. If we define a vector field **V** as

$$\mathbf{V} = \mathbf{E} - (\mathbf{E} \times \hat{\mathbf{e}}_r)/\alpha \quad (24.28)$$

and two scalar fields C and D as

$$C = \nabla \cdot \mathbf{V} \quad (24.29)$$

$$D = -(\nabla \times \mathbf{J}_{\text{eq,ion}})_r \quad (24.30)$$

we can write the curl of Ohm's law as

$$\frac{1}{R_I}\frac{\partial \Sigma_H}{\partial \theta}V_\theta + \frac{1}{R_I \sin\theta}\frac{\partial \Sigma_H}{\partial \phi}V_\phi + C\Sigma_H = D \quad (24.31)$$

This is a first-order partial differential equation for the unknown Hall conductance Σ_H. It should be noted that in deriving this equation, we did not assume a potential electric field. Furthermore, the new variables **V**, C, and D are known from the input data.

As described in detail by *Amm* [1998], equation (24.31) can be divided into two ordinary differential equations. The first defines the characteristic curves $\mathbf{r}(l)$ as

$$\frac{\mathrm{d}}{\mathrm{d}l}\mathbf{r}(l) = \frac{V_\theta(\mathbf{r}(l))\sin\theta\hat{\mathbf{e}}_\theta + V_\phi(\mathbf{r}(l))\hat{\mathbf{e}}_\phi}{R_I \sin\theta|\mathbf{V}(\mathbf{r}(l))|} \quad (24.32)$$

where l is the geometric path length along the characteristic. Then the second equation can be integrated along the curve $\mathbf{r}(l)$, giving the Hall conductance as

$$\Sigma_H(\mathbf{r}(l)) = \Sigma_H(\mathbf{r}_0)e^{-I(0,l)} + \int_0^l \frac{D(\mathbf{r}(l'))e^{-I(l',l)}}{|\mathbf{V}(\mathbf{r}(l'))|}\mathrm{d}l' \quad (24.33)$$

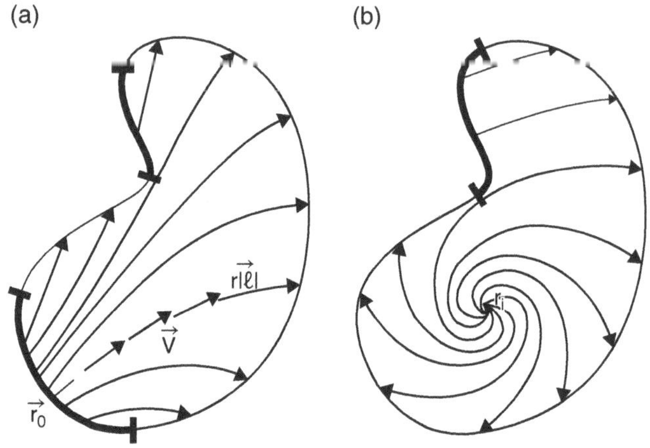

Figure 24.6 (a) Sketch of the vector field $\vec{V}$ and the characteristic curves $\mathbf{r}(l)$. The boundary curve is drawn as a heavy line where it influences the solution. In (b) there is a singular point where the characteristics start. From *Inhester et al.* [1992].

with

$$I(l', l) = \int_{l'}^{l} \frac{C(\mathbf{r}(l''))}{|\mathbf{V}(\mathbf{r}(l''))|}\mathrm{d}l'' \quad (24.34)$$

Examples of the characteristic curves are shown in Figure 24.6.

In equation (24.33), $\Sigma_H(\mathbf{r}_0)$ is the initial value of the Hall conductance, given at the boundary of the analysis domain where the integration is started. However, influence of the boundary value is guaranteed to decrease exponentially, if the direction of integration is suitably chosen. Moreover, due to the geometry of the characteristic curves, at most 50% of the boundary influences the solution, as illustrated in Figure 24.6. There can be additional singular points, where $\mathbf{E} = 0$ but $\nabla \cdot \mathbf{E} \neq 0$. At these points $\Sigma_H = D/C$, and they also serve as starting points of several characteristic curves. In global (and probably also hemispherical) studies, all characteristics are guaranteed to start at a singular point, removing the need for any boundary conditions.

The FAC-based MoC can be derived from the divergence of Ohm's law in a similar way. In fact, with substitutions

$$\mathbf{V} \quad \rightarrow \quad \mathbf{W} = \mathbf{E}/\alpha + \mathbf{E} \times \hat{\mathbf{e}}_r \quad (24.35)$$

$$C \quad \rightarrow \quad Q = \nabla \cdot \mathbf{W} \quad (24.36)$$

$$D \quad \rightarrow \quad j_\parallel = \nabla \cdot \mathbf{J} \quad (24.37)$$

the equations are completely analogous.

24.9. DISCUSSION AND CONCLUSIONS

We have presented a review of selected data analysis techniques applied in studies of ionospheric electrodynamics. This can be viewed as an update of the previous review by *Vanhamäki and Amm* [2011], with more discussions about tools used in global or hemispherical studies. Our emphasis was still on auroral regions and data-driven methods, excluding statistical or data-assimilation techniques. We neglected some topics covered by *Vanhamäki and Amm* [2011], most notably ionospheric induction effects, which may locally have an important role during very active conditions, such as substorm onsets.

Our selection of data analysis methods is naturally somewhat biased, but it demonstrates that there exist several tools for studying various sets of measured data. In many sections, we have made a parallel discussion using both potential and SECS representations of the fields. This may have complicated the discussion a bit but shows that, in many situations, either of these mathematical tools can be used. One naturally wants to choose the best tool for each job, and we believe that in most reviewed cases, it should be the SECS method. Others may have different preferences.

Naturally, other sets of basis functions apart from spherical (cap) harmonics and SECS may be used in analyzing magnetic and electric field measurements. In recent years, there has been a lot of effort in applying various forms of empirical orthogonal function (EOF) analysis and related methods to derive new and improved sets of basis functions for data analysis needs. For example, *Waters et al.* [2015] have applied a combination of EOF-based interpolation functions and SCHA in deriving global maps of ionospheric equivalents currents from SuperMAG data, while *Cousins et al.* [2013] have used EOF to derive a set of statistical convection patterns and applied those in assimilative mapping of SuperDARN data. *Alken et al.* [2017] may have taken this development one step farther, when they derived statistical basis functions of low-latitude current systems from a large set of simulation data, instead of relying on measured data. In our opinion, these methods rely on statistical background models, so they do not quite meet our criterion of purely data-driven analysis tools. Therefore, we have not included them in this review, but in their own class, they perform very well and are certainly worth becoming acquainted with.

In their review, *Vanhamäki and Amm* [2011] noted that "at present, there is no single measurement device that can measure all ionospheric electrodynamic parameters directly and simultaneously, with good spatial and temporal resolution and coverage." This is still true today, so there is a continuous need to create new methods and update old ones, especially when new types of measurements become available. A prime example for such development is the three-satellite Swarm mission, whose full potential for ionospheric research can not be capitalized using only traditional single-satellite analysis methods.

Nevertheless, we want to point out that in the near future the proposed EISCAT_3D radar system (https://eiscat3d.se/)[*McCrea et al.*, 2015] may be able to measure all the electrodynamic parameters discussed in this review within an extended volume of space. The multistatic, phased array radar system will give direct measurement of the ionospheric plasma motion and associated electric field in a mesoscale volume over northern Scandinavia. The measured plasma density and temperature, together with a neutral atmosphere model, can be used to estimate the 3D distribution of ionospheric conductances and thereof the 3D current distribution. Last but not least, simultaneous multistatic measurements in the ionospheric E-region and F-region can be used also to estimate the neutral wind.

ACKNOWLEDGMENTS

The European Space Agency (ESA) is acknowledged for providing the Swarm data.

REFERENCES

Alken, P., A. Maute, and A. D. Richmond (2016), The F-region gravity and pressure gradient current systems: A review, *Space Sci. Rev.*; doi:10.1007/s11214-016-0266-z.

Alken, P., A. Maute, A. D. Richmond, H. Vanhamäki, and G. D. Egbert (2017), An application of principal component analysis to the interpretation of ionospheric current systems, *J. Geophys. Res. Space Physics*, *122*, doi:10.1002/2017JA024051.

Alken, P., P. Maus, S. Vigneron, O. Sirol, and, G. Hulot (2013), Swarm SCARF equatorial electric field inversion chain, *Earth Planet Space*, *65*, 1309–1317; doi:10.5047/eps.2013.09.008.

Amm, O. (1997), Ionospheric elementary current systems in spherical coordinates and their application, *J. Geomag. Geoelectr.*, *49*, 947–955.

Amm, O. (1998), Method of characteristics in spherical geometry applied to a Harang-discontinuity situation, *Ann. Geophys.*, *16*, 413–424, 1998.

Amm, O. (2001), The elementary current method for calculating ionospheric current systems from multisatellite and ground magnetometer data, *J. Geophys. Res.*, *106*, 24843–24855.

Amm, O. (2002), The method of characteristics for calculating ionospheric electrodynamics from multi-satellite and ground-based radar data, *J. Geophys. Res.*, *107*(A10), 1270; doi:10.1029/2001JA005077.

Amm, O., and A. Viljanen (1999), Ionospheric disturbance magnetic field continuation from the ground to the ionosphere using spherical elementary current systems, *Earth Planet Space*, *51*, 431–440.

Amm, O., A. Aruliah, S. C. Buchert, R. Fujii, J. W. Gjerloev, A. Ieda, T. Matsuo, C. Stolle, H. Vanhamäki, and A. Yoshikawa (2008) Towards understanding the electrodynamics of the 3-dimensional high-latitude ionosphere: present and future, *Ann. Geophys.*, *26*, 3913–3932; doi:10.5194/angeo-26-3913-2008.

Amm, O., A. Grocott, M. Lester, and T. K. Yeoman (2010), Local determination of ionospheric plasma convection from coherent scatter radar data using the SECS technique, *J. Geophys. Res.*, *115*, A03304; doi:10.1029/2009JA014832.

Amm O., H. Vanhamäki, K. Kauristie, C. Stolle, F. Christiansen, R. Haagmans, A. Masson, M. G. G. T. Taylor, R. Floberghagen, and C. P. Escoubet (2015), A method to derive maps of ionospheric conductances, currents, and convection from the Swarm multisatellite mission, *J. Geophys. Res.*, *120*, 3263–3282; doi:10.1002/2014JA020154.

Amm, O., M. Engebretson, T. Hughes, L. Newitt, A. Viljanen, and J. Watermann (2002), A traveling convection vortex event study: Instantaneous ionospheric equivalent currents, estimation of field-aligned currents, and the role of induced currents, *J. Geophys. Res.*, *107* (A11), 1334; doi:10.1029/2002JA009472.

Backus, G. (1986), Poloidal and toroidal fields in geomagnetic field modeling, *Rev. Geophys. 24*, 75–109.

Brekke, A. (1997), *Physics of the Upper Polar Atmosphere*, John Wiley & Sons.

Cousins, E., T. Matsuo, and A. Richmond (2013), SuperDARN assimilative mapping, *J. Geophys. Res.*, *118*, 7954–7962; doi:10.1002/2013JA019321.

Dunlop, M. W., et al. (2015), Multispacecraft current estimates at swarm, *J. Geophys. Res.*, *120*, 8307–8316; doi:10.1002/2015JA021707.

Finlay C. C., N. Olsen, S. Kotsiaros, N. Gillet, and L. Tøffner-Clausen (2016), Recent geomagnetic secular variation from Swarm and ground observatories as estimated in the CHAOS-6 geomagnetic field model, *Earth Planet Space*, *68*, 112; doi:10.1186/s40623-016-0486-1.

Fiori, R., D. H. Boteler, A. V. Koustov, G. V. Haines, and J. M. Ruohoniemi (2010), Spherical cap harmonic analysis of Super Dual Auroral Radar Network (SuperDARN) observations for generating maps of ionospheric convection, *J. Geophys. Res.*, *115*, A07307; doi:10.1029/2009JA015055.

Fiori, R., et al. (2018), Review of spherical cap harmonic modelling techniques, *Ionospheric Multi-Spacecraft Analysis Tools*, in preparation.

Fujii, R., O. Amm, A. Yoshikawa, A. Ieda, and H. Vanhamäki (2011), Reformulation and energy flow of the Cowling channel, *J. Geophys. Res.*, *116*, A02305; doi:10.1029/2010JA015989.

Fukushima, N. (1976), Generalized theorem for no ground magnetic effect of vertical currents connected with Pedersen currents in the uniform-conductivity ionosphere, *Rep. Ionos. Space. Res. Japan*, *30*, 35–40.

Fukushima, N. (1994), Some topics and historical episodes in geomagnetism and aeronomy, *J. Geophys. Res.*, *99*, 19113–19142; doi:10.1029/94JA00102.

Gaya-Piqué, L. R., J. J. Curto, J. M. Torta, and A. Chulliat (2008), Equivalent ionospheric currents for the 5 December 2006 solar flare effect determined from spherical cap harmonic analysis, *J. Geophys. Res.*, *113*, A07304; doi:10.1029/2007JA012934.

Gjerloev, J. W. (2012), The SuperMAG data processing technique, *J. Geophys. Res.*, *117*, A09213; doi:10.1029/2012JA017683.

Green, D. L., C. L. Waters, B. J. Anderson, H. Korth, and R. J. Barnes (2006), Comparison of large-scale Birkeland currents determined from Iridium and SuperDARN data, *Ann. Geophys.*, *24*, 941–959.

Green, D. L., C. L. Waters, H. Korth, B. J. Anderson, A. J. Ridley, and, R. J. Barnes (2007), Technique: Large-scale ionospheric conductance estimated from combined satellite and ground-based electromagnetic data, *J. Geophys. Res.*, *112*, A05303; doi:10.1029/2006JA012069.

Haines, G. V. (1985), Spherical cap harmonic analysis, *J. Geophys. Res.*, *90*, 2583–2591.

Haines, G. V., and, J. M. Torta (1994), Determination of equivalent current sources from spherical cap harmonic models of geomagnetic field variations, *Geophys. J. Int.*, *118*, 499–514.

Iijima, T., and, T. A. Potemra (1976), The amplitude distribution of field-aligned currents at northern high latitudes observed by Triad, *J. Geophys. Res.*, *81*(13), 21652174; doi:10.1029/JA081i013p02165.

Inhester, B., J. Untiedt, M. Segatz, and M. Kürschner (1992), Direct determination of the local ionospheric Hall conductance distribution from two-dimensional electric and magnetic field data, *J. Geophys. Res.*, *97*, 4073–4083.

Janhunen, P., M. Palmroth, T. Laitinen, I. Honkonen, L. Juusola, G. Facsko, and T. I. Pulkkinen (2014), The GUMICS-4 global MHD magnetosphereionosphere coupling simulation, *J. Atm. Solar Terr. Phys.*, *80*, 48–59; doi:10.1016/j.jastp.2012.03.006.

Juusola L., K. Kauristie, H. Vanhamäki, A. Aikio, and M. van de Kamp (2016a), Comparison of auroral ionospheric and field-aligned currents derived from Swarm and ground magnetic field measurements, *J. Geophys. Res. 121*, 9256–9283; doi:10.1002/2016JA022961.

Juusola, L., O. Amm, and A. Viljanen (2006), One-dimensional spherical elementary current systems and their use for determining ionospheric currents from satellite measurements, *Earth Planet Space*, *58*, 667–678.

Juusola, L., O. Amm, K. Kauristie, and A. Viljanen (2007), A model for estimating the relation between the Hall to Pedersen conductance ratio and ground magnetic data derived from CHAMP satellite statistics, *Ann. Geophys.*, *25*, 721–736.

Juusola, L., S. E. Milan, M. Lester, A. Grocott, and S. M. Imber (2014), Interplanetary magnetic field control of the ionospheric field-aligned current and convection distributions, *J. Geophys. Res.*, *119*, 3130–3149; doi:10.1002/2013JA019455.

Juusola L., W. Archer, K. Kauristie, H. Vanhamäki, and A. Aikio (2016b), Ionospheric conductances of a morning-sector auroral arc from Swarm measurements, *Geophys. Res. Lett.*, *43*; doi:10.1002/2016GL070248.

Kamide, Y., A. Richmond, and S. Matsushita (1981), Estimation of ionospheric electric fields, ionospheric currents, and field-aligned currents from ground magnetic records, *J. Geophys. Res.*, *86*, 801–813.

Kamide, Y., E. A. Kihn, A. J. Ridley, E. W. Cliver, and Y. Kadowaki (2003), Real-time specifications of the geospace environment, *Space Sci. Rev.*, *107*, 307–316.

Karlsson, T., and G. Marklund (1998), Simulations of effects of small-scale auroral current closure in the return current region, *Phys. Space Plasmas*, *15*, 401–406.

Laundal, K. M., C. Finlay, and N. Olsen (2016), Sunlight effects on the 3D polar current system determined from low Earth orbit measurements, *Earth, Planet Space*, *68*, 142; doi:10.1186/s40623-016-0518-x

Lühr, H., J. J. Warnecke, and M. Rother (1996), An algorithm for estimating field-aligned currents from single-spacecraft magnetic field measurements: A diagnostic tool applied to Freja satellite data, *IEEE Trans. Geosci. Remote Sens.*, *34*, 1369–1376.

Marghitu, O., C. Bunescu, T. Karlsson, B. Klecker, and H. C. Stenbaek-Nielsen (2011), On the divergence of the auroral electrojets, *J. Geophys. Res.*, *116*, A00K17; doi:10.1029/2011JA016789.

Marghitu, O., N. Klecker, H. Haerendel, and J. McFadden (2004), ALADYN: A method to investigate auroral arc electrodynamics from satellite data, *J. Geophys. Res.*, *109*, A11305; doi:10.1029/2004JA010474.

McCrea, I., et al. (2015), The science case for the EISCAT_3D radar, *Prog. Earth Planet. Sci.*, *2*, 21; doi:10.1186/s40645-015-0051-8.

Murison, M., A. Richmond, S. Matsushita, and W. Baumjohann (1985), Estimation of ionospheric electric fields and currents from a regional magnetometer array, *J. Geophys. Res.*, *90*, 3525–3530.

Nakamizo, A., et al. (2012), Effect of R2-FAC development on the ionospheric electric field pattern deduced by a global ionospheric potential solver, *J. Geophys. Res.*, *117*, A09231; doi:10.1029/2012JA017669.

Olsen, N. (1996), A new tool for determining ionospheric currents from magnetic satellite data, *Geophys. Res. Lett.*, *23*, 3635–3638.

Olsen, N., et al. (2013), The Swarm satellite constellation application and research facility (SCARF) and Swarm data products, *Earth Planet Space*, *64*(11), 1189–1200; doi:10.5047/eps.2013.07.001.

Pulkkinen, A., O. Amm, A. Viljanen, and BEAR Working Group (2003a), Ionospheric equivalent current distributions determined with the method of spherical elementary current systems, *J. Geophys. Res.*, *108*(A2), 1053; doi:10.1029/2001JA005085.

Pulkkinen, A., O. Amm, A. Viljanen, and BEAR Working Group (2003b), Separation of the geomagnetic variation field on the ground into external and internal parts using the spherical elementary current system method, *Earth Planet Space*, *55*, 117–129.

Richmond, A. D., and J. P. Thayer (2000), Ionospheric electrodynamics: A tutorial, in *Magnetospheric Current Systems*, Geophysical Monograph Series 118, edited by S. Ohtani et al., 131–146, American Geophysical Union, Washington, DC.

Richmond, A. D., G. Lu, B. A. Emery, and D. J. Knipp (1998), The AMIE procedure: Prospects for space weather specification and prediction, *Adv. Space Res.*, *22*, 103–112.

Ritter, P., H. Lühr, and J. Rauberg (2013), Determining field-aligned currents with the Swarm constellation mission, *Earth Planet Space*, *65*, 1285–1294; doi:10.5047/eps.2013.09.006

Ruohoniemi, J. M., and K. B. Baker (1998), Large-scale imaging of high-latitude convection with Super Dual Auroral Radar Network HF radar observations, *J. Geophys. Res.*, *103*(A9), 20797–20811; doi:10.1029/98JA01288.

Sonnerup, B., and M. Scheible (1998), *Minimum and Maximum Variance Analysis*, ISSI Sci. Rep. Ser., *1*, 185–220.

Stauning, P. (2002), Field-aligned ionospheric current systems observed from Magsat and Oersted satellites during northward IMF, *Geophys. Res. Lett.*, *29*(15); doi:10.1029/2001GL013961.

Tanskanen, E., A. Viljanen, T. Pulkkinen, R. Pirjola, L. Häkkinen, A. Pulkkinen, and O. Amm (2001), At substorm onset 40% of AL comes from underground, *J. Geophys. Res.*, *106* (A7), 13119–13134.

Thébault, E., J. J. Schott, and, M. Mandea (2006), Revised spherical cap harmonic analysis (R-SCHA): Validation and properties, *J. Geophys. Res.*, *111*, B01102; doi:10.1029/2005JB003836.

Untiedt, J., and, W. Baumjohann (1993), Studies of polar current systems using the IMS Scandinavian magnetometer array, *Space Sci. Rev.*, *63*, 245–390; doi:10.1007/BF00750770.

van de Kamp, M. (2013), Harmonic quiet-day curves as magnetometer baselines for ionospheric current analyses, *Geosci. Instrum. Method. Data Syst.*, *2*, 289–304; doi:10.5194/gi-2-289-2013.

Vanhamäki, H. (2011), Inductive ionospheric solver for magnetospheric MHD simulations, *Ann. Geophys.*, *29*, 97–108; doi:10.5194/angeo-29-97-2011.

Vanhamäki, H., and L. Juusola (2018), Review of SECS modelling techniques, in *Ionospheric Multi-Spacecraft Analysis Tools*, in preparation.

Vanhamäki, H., and O. Amm (2007), A new method to estimate ionospheric electric fields and currents using data from a local ground magnetometer network, *Ann. Geophys.*, *25*, 1141–1156.

Vanhamäki, H., and O. Amm (2011), Analysis of ionospheric electrodynamic parameters on mesoscales: A review of selected techniques using data from ground-based observation networks and satellites, *Ann. Geophys.*, *29*, 467–491; doi:10.5194/angeo-29-467-2011.

Vanhamäki, H., A. Aikio, M. Voiculescu, L. Juusola, T. Nygrén, and R. Kuula (2016), Electrodynamic structure of the morning high-latitude trough region, *J. Geophys. Res.*, *121*, 2669–2682; doi:10.1002/2015JA022021.

Vanhamäki, H., et al. (2015), Altitude dependent polarization electric field in the ionospheric E-region, 26th IUGG General Assembly, 22 Jun–2 July 2015, Prague, Czech Republic, session A25.

Vanhamäki, H., K. Kauristie, O. Amm, A. Senior, D. Lummerzheim, and S. Milan (2009), Electrodynamics of an omega-band as deduced from optical and magnetometer data, *Ann. Geophys.*, *27*, 3367–3385; doi:10.5194/angeo-27-3367-2009.

Vanhamäki, H., O. Amm, and A. Viljanen (2003), One-dimensional upward continuation of the ground magnetic field disturbance using elementary current systems, *Earth Planet Space*, *55*, 613–625.

Vanhamäki, H., O. Amm, and A. Viljanen (2007), Role of inductive electric fields and currents in dynamical ionospheric situations, *Ann. Geophys.*, *25*, 437–455.

Vogt, J. (2014), Analysis of data from multi-satellite geospace missions, in *Handbook of Geomathematics*, edited by F. Willi, M. Z. Nashed, and T. Sonar, Springer, Berlin, 1–28.

Vogt, J., E. Sorbalo, M. He, and A. Blagau (2013), Gradient estimation using configurations of two or three spacecraft, *Ann. Geophys.*, *31*, 1913–1927; doi:10.5194/angeo-31-1913-2013.

Waters, C. L., B. J. Anderson, and, K. Liou (2001), Estimation of global field aligned currents using the iridium® System magnetometer data, *Geophys. Res. Lett.*, *28*, 2165–2168; doi:10.1029/2000GL012725.

Waters, C. L., J. W. Gjerloev, M. Dupont, and R. J. Barnes (2015), Global maps of ground magnetometer data, *J. Geophys. Res.*, *120*, 9651–9660; doi:10.1002/2015JA021596.

Weimer, D. R. (2005), Improved ionospheric electrodynamic models and application to calculating Joule heating rates, *J. Geophys. Res.*, *110*, A05306; doi:10.1029/2004JA010884.

Weygand, J. M., and S. Wing (2016), Comparison of DMSP and SECS region-1 and region-2 ionospheric current boundary, *J. Atmos. Solar Terr. Phys.*, *143–144*, 8–13; doi:10.1016/j.jastp.2016.03.002.

Weygand, J. M., O. Amm, A. Viljanen, V. Angelopoulos, D. Murr, M. J. Engebretson, H. Gleisner, and I. Mann (2011), Application and validation of the spherical elementary currents systems technique for deriving ionospheric equivalent currents with the North American and Greenland ground magnetometer arrays, *J. Geophys. Res.*, *116*, A03305; doi:10.1029/2010JA016177.

Yoshikawa, A., and M. Itonaga (1996), Reflection of shear Alfvén waves at the ionosphere and the divergent Hall current, *Geophys. Res. Lett.*, *23*, 101–104.

25

Earth's Ionosphere: Theory and Phenomenology of Cowling Channels

Akimasa Yoshikawa[1] and Ryoichi Fujii[2]

ABSTRACT

The Cowling channel is a generic name of a current system forming inside a high conductivity band in which a secondary polarization electric field modifies the current flow. The polarization field is excited when a divergent part of Hall current driven by the primary electric field is prevented from flowing out to the magnetosphere as the field-aligned current. The purpose of this chapter is to review the recent development of the Cowling channel model. Recent work provides an extension of the theoretical description of the classical Cowling channel with respect to the following aspects: (1) taking into account the 3D nature of the ionosphere by introducing two current layers at different altitudes, and (2) considering finite length of the Cowling channel by introducing a conductance boundary not only at the meridional borders of the Cowling channel, but also at its zonal boundaries. Using this improved model, we discuss current closure and the energy principle for evolution of the Cowling channel. Energy flow inside the Cowling channel and impact of the polarization effect on Joule dissipation in the more general M-I coupling scheme are also provided. We also clarify how shear Alfvén waves interact to the Cowling channel and their application to the global magnetosphere-ionosphere coupling simulations.

25.1. INTRODUCTION

The Cowling channel is a generic name given to a current system that forms inside a high-conductivity band, in which a secondary polarization electric field modifies the current flow. The polarization field is excited when a divergent part of the Hall current driven by the primary electric field is prevented from flowing out to the magnetosphere as the field-aligned current (FAC).

The Cowling effect is now well known as the enhancement of current flow in the direction of the primary electric field by the secondary Hall current [*Chapman*, 1956]. It was first investigated by *Cowling* [1932] in connection with the solar atmosphere. The generation mechanism [*Cowling and Borger*, 1948] was adopted to account for the equatorial electrojet [*Hirono*, 1950; *Untiedt*, 1967] and auroral electrojet [*Boström*, 1964]. The Cowling effect has been investigated theoretically and observationally [e.g., *Baujohann*, 1983; *Yasuhara et al.*, 1985; *Haerendel*, 2008; *Amm et al.*, 2008; *Amm and Fujii*, 2008; *Marghitu et al.*, 2011].

Figure 25.1 shows a conventional view of the two-dimensional Cowling channel model elongated along the east-west direction [e.g., *Baumjohann*, 1983], in which the ionospheric Hall and Pedersen conductivities are height-integrated. The primary westward electric field ($\mathbf{E}_1$) drives the northward Hall current and the westward Pedersen current. The southward secondary field ($\mathbf{E}_2$) is generated such that the Pedersen current cancels the

[1] Department of Earth and Planetary Sciences, Kyushu University; International Center for Space Weather Science and Education, Kyushu University

[2] Institute for Space-Earth Environmental Research, Nagoya University

Electric Currents in Geospace and Beyond, Geophysical Monograph 235, First Edition.
Edited by Andreas Keiling, Octav Marghitu, and Michael Wheatland.

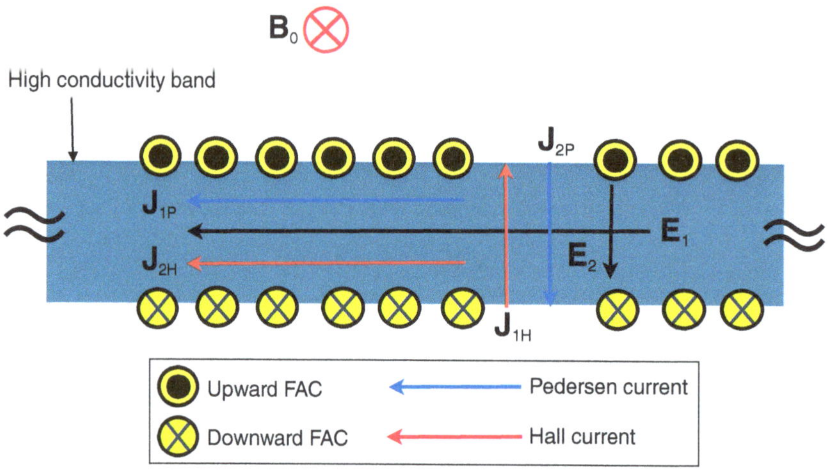

Figure 25.1 Traditional picture of two-dimensional Cowling channel model elongated along east-west direction.

primary Hall current between the conductivity gradients. The secondary Hall current flows in the same direction as the primary Pedersen current and forms the electrojet system.

In general, it is difficult to specify the polarization effects in the ionosphere from ground-based data alone. These data only facilitate inference of the resultant total electrodynamic fields; they cannot trace the chain of cause and consequence that leads to the physical situation resulting in these observed total fields. Thus, by using ground-based data alone, in most cases, we can only state whether an observed situation is consistent with the expectations from an active polarization effect.

To quantify the Cowling effect, we need to determine the relative strength of the polarization electric field compared to the total electric field as well as the extent to which it cancels the primary Hall current. This problem is complementary to the following question: How much curl-free Hall current flows out to the magnetosphere as the FAC?

To solve this problem posed by *Amm et al.* [2008], modeling of the Cowling channel has been further developed.

To describe the Cowling channel, *Amm et al.* [2011] and *Fujii et al.* [2011] introduced a parameter called the Cowling efficiency. It is defined as the proportion of the Hall current that is confined inside the ionosphere by the secondary Pedersen current excited by the polarization electric field. The definition of the Cowling efficiency is practically significant. It provides a general way to quantitatively calculate the polarization electric field if the Cowling efficiency, the conductance, and either the primary electric field or the total electric field are known [*Amm et al.*, 2013].

It has been suggested that to identify the Cowling efficiency for a specific phenomenon, one needs to determine the impedance of the magnetospheric circuit that completes the current circuit in the M-I system via the FAC [e.g., *Fujii et al.*, 2011]. However, assigning a magnetospheric impedance for the steady state is debatable because the M-I system is dynamic.

The M-I coupling process via shear Alfvén waves has been used to investigate nonstationary FAC closure by the ionospheric conducting current [e.g., *Scholer*, 1970]. By assuming specific electric field configurations of an incident wave, *Glassmeier* [1983] and *Itonaga and Kitamura* [1988] have shown that a secondary polarization field due to gradients of the Hall conductance can appear in the reflected wave. Actually, the Alfvén wave approach can be used to describe not only local and dynamical phenomena but also more general global quasi-static M-I coupling processes [*Yoshikawa et al.*, 2010]. Therefore, it is very important to understand how the shear Alfvén wave interacts with the Cowling channel.

Yoshikawa et al. [2011a] proposed a general theory about M-I coupling, independent of specific geometries or situations. This theory, which is based on the use of Alfvén waves as a basis function for the M-I coupling process, was later applied by *Yoshikawa et al.* [2013a, 2013b] specifically to a Cowling channel situation; however, it can be applied to any general case.

Most of the Cowling channel models introduced thus far rely on a thin-sheet ionosphere [e.g., *Baujohann*, 1983]. However, in a realistic ionospheric E layer, vertical distributions of the Pedersen conductivity and Hall conductivity have maximum peaks around altitudes of 125 km and 110 km, respectively [*e.g., Richmond and Thayer*, 2000]. To properly consider the three-dimensional ionospheric current closures, one should also take the ionospheric thickness into account [*Amm et al.*, 2008]. One step in this direction is to assume that the Pedersen

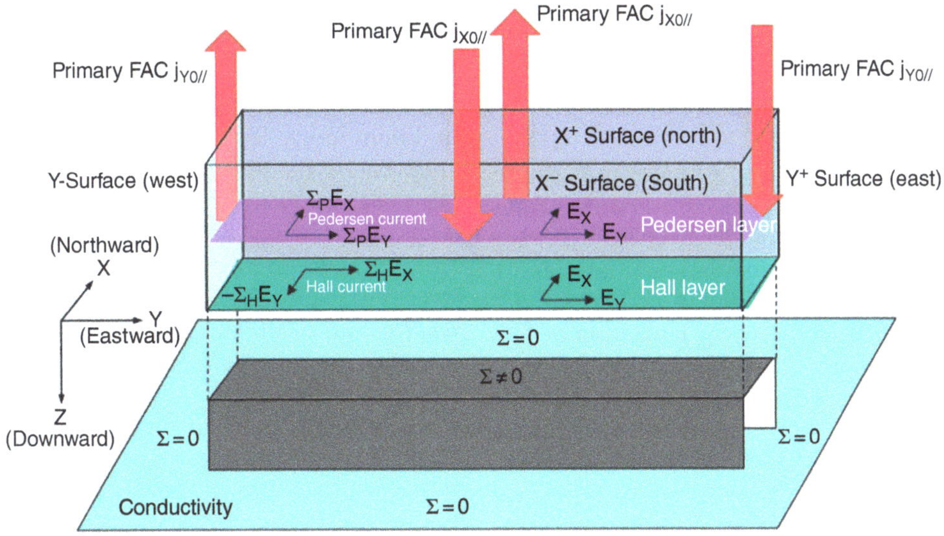

Figure 25.2 Two-layer current model for finite Cowling channel. After *Fujii et al.* [2011].

and Hall currents flow through thin layers at different altitudes [*Fujii et al.*, 2011; *Amm et al.*, 2011; *Yoshikawa et al.*, 2011b].

The classical view shown in Figure 25.1 describes divergence-free approximation of the auroral electrojet. However, a longitudinal border of the Cowling channel is also important for considering finite auroral arc formation and the Harang reversal region [*Harang*, 1946; *Heppner*, 1972; *Marghitu et al.*, 2011], where the auroral electrojet diverges.

Amm et al. [2008] presented a review of the work available in the literature until 2008 with regard to the following aspects of ionospheric electrodynamics and magnetosphere-ionosphere (M-I) coupling:

1. Polarization effect in the ionosphere (often referred to as Cowling effect)
2. Inductive effect in the ionosphere
3. Effect of the three-dimensional (3D) nature of the ionosphere for ionospheric electrodynamics
4. Consequences of the above-mentioned aspects for M-I coupling

Marghitu [2012] provided an excellent review of auroral arc electrodynamics by considering the 1D thin uniform arc, 2D thick uniform arc, and nonuniform arc. The various arc features are assembled together in a tentative 3D arc model.

The purpose of this chapter is to review the recent development of the Cowling channel model subsequent to the work of *Amm et al.* [2008] and *Marghitu* [2012]. Recent studies have extended the theoretical description of the classical Cowling channel in the following aspects: (1) taking the 3D nature of the ionosphere into account by introducing two current layers at different altitudes, and (2) considering finite length of the Cowling channel by introducing a conductance boundary not only at the meridional borders of the Cowling channel but also at its zonal boundaries. Separation of the Hall-Pedersen current layer would have no practical effect as seen from the magnetosphere or below from the ground if the spatial scale of the phenomena is much larger than the E-layer thickness. The most important issues are performing the first step for investigating 3D current closures and determining how and where the polarization process occurs. Hence, the theory laid out in this chapter relies heavily on height-integrated parameters.

Using this improved model, which is schematically illustrated in Figure 25.2 with the description of the Cowling efficiency, we discuss current closure and the energy principle for the evolution of the Cowling channel. The energy flow inside the Cowling channel and the impact of the polarization effect on the Joule dissipation in more general M-I coupling schemes are also investigated. In addition, we clarify how the shear Alfvén wave interacts with the Cowling channel and examine their application to global magnetosphere-ionosphere coupling simulations.

25.2. GENERATION OF IONOSPHERIC POLARIZATION FIELD IN THE M-I COUPLING PROCESS

By using Ampere's law ($\mu_0 \mathbf{j} = \nabla \times \mathbf{B}$) and Faraday's induction law ($-\partial \mathbf{B}/\partial t = (\nabla \times \mathbf{E})$) while assuming no parallel electric field ($\mathbf{E} = \mathbf{E}_\perp$), we can derive a general relation between the evolution of the FAC density and the gradient of the electric field divergence density along the background magnetic field $\mathbf{B}_0$:

$$\mu_0 = \frac{\partial}{\partial t} j_{\|} = \frac{\partial}{\partial s}(\nabla \cdot \mathbf{E}) \tag{25.1}$$

where μ_0 is the magnetic permeability of space, $j_{\|} = \mu_0 \hat{\mathbf{e}}_B \cdot (\nabla \times \mathbf{B})$ is the FAC density, $\hat{\mathbf{e}}_B = \mathbf{B}_0 / |\mathbf{B}_0|$ is the unit vector along the background magnetic field $\mathbf{B}_0$, s is a coordinate measured along $\mathbf{B}_0$, and the subscripts $\|$ and $\perp$ denote quantities or operators along $\mathbf{B}_0$ and on the surface perpendicular to $\mathbf{B}_0$, respectively. Excitation of $\nabla \cdot \mathbf{E}$ proportional to the space charge density in the ionosphere is very important. If $\nabla \cdot \mathbf{E}$ is generated in the ionospheric conductivity gradient region and there are is corresponding $\nabla \cdot \mathbf{E}$ along $\mathbf{B}_0$ on the magnetospheric side, the FAC of ionospheric origin will have to be excited until $\nabla \cdot \mathbf{E}$ is distributed along $\mathbf{B}_0$. Another important factor is the current continuity condition

$$\nabla_{\|} \cdot \mathbf{J}_{\|} = -\nabla_{\perp} \cdot \mathbf{j}_{\perp} \tag{25.2}$$

which describes current closure between the FAC and the perpendicular current. In general, the FAC is converted into curl-free perpendicular current in the region of $\nabla_{\|} \cdot \mathbf{j} \neq 0$ and the torsional-type Ampere force acting on a plasma body produces plasma vorticity. In the ionospheric E layer, this type of Ampere force nearly balances the ion friction force caused by momentum exchange between ions and neutrals via collisions that lead to shear of the ion momentum flow [e.g., *Strangeway*, 2012]. Thus, it is very important to understand the formation process of the Cowling channel from the viewpoint of weakly ionized plasma dynamics. However, in this section, we continue to focus on the generation of a polarization electric field accompanied by Hall current divergence on the basis of the conventional M-I coupling theory and current closure between the FAC and the ionospheric current.

Our formulation starts with basic equations for the height-integrated currents and their closure of the FAC in the northern ionosphere:

$$\mathbf{J}_{\perp} = \Sigma_P \mathbf{E}_{\perp} + \Sigma_H \hat{\mathbf{e}}_B \times \mathbf{E}_{\perp} \tag{25.3}$$

$$j_{\|} = \nabla \cdot \mathbf{J}_{\perp} \tag{25.4}$$

where the variables $\mathbf{J}_{\perp}$, $\mathbf{E}$, and $j_{\|}$, are the height-integrated ionospheric current, electric field, and FAC density, while Σ_P and Σ_H are the height-integrated Pedersen and Hall conductivities, respectively.

From equations (25.3) and (25.4), the divergence of the electric field proportional to the space charge densities excited by the M-I current closure process is given by

$$\nabla \cdot \mathbf{E} = \frac{j_{\|} - (\nabla \Sigma_P - \hat{\mathbf{e}}_B \times \nabla \Sigma_H) \cdot \mathbf{E} + \Sigma_H (\nabla \times \mathbf{E})_{\|}}{\Sigma_P} \tag{25.5}$$

The first term on the right-hand side (RHS) of (25.5) denotes the charging or discharging effect of the ionosphere when the FAC is flowing into or out of the ionosphere, respectively [e.g., *Sato et al.*, 1978]. The second and third terms denote the resultant contributions to the space charge density when the Pedersen and Hall currents cross the conductivity gradient region, respectively [e.g., *Glassmeier*, 1983]. The last term denotes the contribution from the Hall current divergence excited by nonzero curl of the electric field [*Yoshikawa and Itonaga*, 1996; *Yoshikawa and Itonaga*, 2000]. This inductive divergent Hall current is very important not only for the evolution of the curl-free Hall current system [*Yoshikawa*, 2002a, 2002b] but also at the head of the auroral surge [*Vanhamäki et al.*, 2007]. However, in this chapter, we mainly focus on the polarization effect caused by the conductivity gradient region.

25.2.1. Separation of Polarization Field Into Primary and Secondary Components

To focus on the polarization by the Hall conductivity gradient region, we consider the following separations of the ionospheric current into the primary and secondary current systems [*Yoshikawa et al.*, 2013a]:

$$\mathbf{J}_1 = \Sigma_P \mathbf{E}_1, \qquad j_{\|,1} = \nabla \cdot \mathbf{J}_1 \tag{25.6}$$

$$\mathbf{J}_2 = \Sigma_P \mathbf{E}_2 + \Sigma_H (\hat{\mathbf{e}}_B \times \mathbf{E}), \qquad j_{\|,2} = \nabla \cdot \mathbf{J}_2 \tag{25.7}$$

where $\mathbf{E}_1$ is the primary electric field, $\mathbf{J}_1$ is the Pedersen current driven by $\mathbf{E}_1$, and equation (25.6) describes the current closure process between the magnetospheric FAC and the Pedersen current system when there is no Hall current divergence. Furthermore, $\mathbf{E}_2$ is the secondary electric field induced by the existence of Hall current divergence, and $\mathbf{J}_2$ consists of the Hall current and Pedersen current driven by $\mathbf{E}_2$.

In the two-layer model, $\mathbf{J}_1$ is confined within the Pedersen layer, whose curl-free part only closes the magnetospheric FAC of $j_{\|,1}$. On the other hand, in the secondary current system $\mathbf{J}_2$, the curl-free part of the height-integrated Hall current in the Hall layer and the curl-free part of the secondary Pedersen current in the Pedersen layer are connected via the FAC between the Pedersen and Hall layers.

The divergence densities of the polarization electric fields produced in the primary current system and secondary current system become

$$\nabla \cdot \mathbf{E}_1 = \frac{j_{\|,1} - \nabla \Sigma_P \cdot \mathbf{E}_1}{\Sigma_P} \tag{25.8}$$

$$\nabla \cdot \mathbf{E}_2 = \frac{j_{\|,2} - \nabla \Sigma_P \cdot \mathbf{E}_2 + (\hat{\mathbf{e}}_B \times \nabla \Sigma_H) \cdot (\mathbf{E}_1 + \mathbf{E}_2) + \Sigma_H (\nabla \times \mathbf{E})_{\|}}{\Sigma_P} \tag{25.9}$$

respectively. We assume that the primary current system describes the background current closure between the ionosphere and the magnetosphere. Therefore, the distribution of $\mathbf{E}_1$ essentially reflects a priori distributed magnetospheric FACs and the polarization effect by the Pedersen conductivity gradient. By contrast, the secondary current system is excited by the Hall effect. Excitation of $\mathbf{E}_2$ is controlled by the amount of curl-free Hall current that is confined within the ionosphere by closure to the secondary Pedersen current and/or the amount that flows out to the magnetosphere as secondary FACs. If the there is no Hall conductivity gradient, the secondary electric field is not excited and $\mathbf{J}_2$ simply represents the divergence-free Hall current.

25.2.2. Cowling Efficiency Description

From equations (25.6)–(25.9), it is seen that we cannot determine $\mathbf{E}_1$ and $\mathbf{E}_2$ until we find a way to identify how $j_{\parallel,1}$ and $j_{\parallel,2}$ are distributed. To avoid this problem, *Amm et al.* [2011] and *Fujii et al.* [2011] introduced a free parameter, that is, the Cowling efficiency α_C, as follows:

$$\nabla\cdot(\Sigma_P\mathbf{E}_2)+\alpha_C\nabla\cdot(\Sigma_H\hat{\mathbf{e}}_B\times\mathbf{E})=0 \tag{25.10}$$

It indicates how much Hall current divergence is confined within the ionosphere by excitation of a secondary Pedersen current divergence; thus, the residual FAC is

$$\begin{aligned} j_{\parallel,2} &= (1-\alpha_C)\nabla\cdot(\Sigma_H\hat{\mathbf{e}}_B\times\mathbf{E}) \\ &= \left(\frac{\alpha_C-1}{\alpha_C}\right)\nabla\cdot(\Sigma_P\mathbf{E}_2) \end{aligned} \tag{25.11}$$

From (25.10) and (25.11) we get

$$j_{\parallel,2}=\nabla\cdot(\Sigma_P\mathbf{E}_2+\Sigma_H\hat{\mathbf{e}}_B\times\mathbf{E}) \tag{25.12}$$

which is equivalent to (25.7).

The Cowling efficiency is a free parameter that takes values in the range of $0\le\alpha_C\le 1$. To concretely determine the Cowling efficiency, we need to know the current-voltage relation of the secondary FAC ($j_{\parallel,2}$). In a transient process of M-I coupling, $j_{\parallel,2}$ should be accompanied by the shear Alfvén wave. This specific case will be discussed in section 25.3. If $j_{\parallel,2}$ is closed by the static magnetospheric current and/or ionospheric current system in the opposite hemisphere, we will need to specify the magnetospheric and/or ionospheric impedances and how they form the total current system. Hence, determination of the Cowling efficiency is beyond the scope of this study, except for the case of the M-I coupling process via the shear Alfvén wave.

Further, $\alpha_C=0$ implies that all of the curl-free Hall current flows out to the magnetosphere without producing a secondary electric field. On the other hand, $\alpha_C=1$ implies that all of the curl-free Hall current is confined within the ionosphere by connection to the curl-free part of the secondary Pedersen current. Hence, the effect of the total electric field variation from the primary electric field becomes maximum in the case of $\alpha_C=1$.

A general solution for calculating polarization electric fields in the auroral ionosphere by using the Cowling efficiency was investigated by *Amm et al.* [2013]. This method is practically significant. In contrast to most previous studies concerning the Cowling channel, where the secondary polarization field has been discussed only qualitatively in an approximation that holds for a specific geometry, such as an auroral arc, the application of (25.10) provides a general way to quantitatively calculate the polarization electric field in any geometry. However, we still need to determine the Cowling efficiency, the conductance, and either the primary electric field or the total electric field.

25.2.3. Current Closure in the Finite Cowling Channel

In general, (25.10) can be applied to any geometry of conductivity distribution. However, for step-by-step understanding of the Cowling channel, we consider current closure inside a finite, rectangular, high-conductivity band. By using this model, we will examine two special cases of Cowling channel formation excited by north-south (meridional) closure of the primary current across the auroral arc and east-west (azimuthal) closure of the primary current at the edge of the auroral arc. Of course, these special cases are not realistic. However, insights from this analysis might be useful for understanding the basic nature of the Cowling channel.

The primary electric field $\mathbf{E}_1$ is assumed to be aligned with the upper and lower borders of conductivity band A, where the Cowling efficiency is given by α_C^A and produces a secondary electric field $\mathbf{E}_2^A$. At the right and left borders of conductivity band B, the Hall current excited by $\mathbf{E}_2^A$ produces the secondary electric field $\mathbf{E}_2^B$ with the Cowling efficiency α_C^B. This situation is expressed as

$$j_{\parallel,1}=\nabla\cdot(\Sigma_P\mathbf{E}_1) \tag{25.13}$$

$$\nabla\cdot\left(\Sigma_P\mathbf{E}_2^A\right)+\alpha_C^A\nabla\cdot\left\{\Sigma_H\hat{\mathbf{e}}_B\times\left(\mathbf{E}_1+\mathbf{E}_2^B\right)\right\}=0 \tag{25.14}$$

$$\nabla\cdot\left(\Sigma_P\mathbf{E}_2^B\right)+\alpha_C^B\nabla\cdot\left(\Sigma_H\hat{\mathbf{e}}_B\times\mathbf{E}_2^A\right)=0 \tag{25.15}$$

Further, we assume that the Cowling efficiency, the ratio of the height-integrated Hall to Pedersen conductivities $\alpha_H=\Sigma_H/\Sigma_P$, and the conductivity distribution inside the conductivity band are uniform throughout the system. Thus, equations (25.14) and (25.15) can be reduced to

$$\mathbf{E}_2^A \simeq -\alpha_C^A \alpha_H \hat{\mathbf{e}}_\mathrm{B} \times \left(\mathbf{E}_1 + \mathbf{E}_2^B\right) \tag{25.16}$$

$$\mathbf{E}_2^B \simeq -\alpha_C^B \alpha_H \hat{\mathbf{e}}_\mathrm{B} \times \mathbf{E}_2^A \tag{25.17}$$

These lead to

$$\mathbf{E}_2^B \simeq -\left(\frac{\bar{\alpha}_C^2 \alpha_H^2}{1+\bar{\alpha}_C^2 \alpha_H^2}\right)\mathbf{E}_1$$

where $\bar{\alpha}_C^2 = \alpha_C^A \alpha_C^B$, and the total electric field and total perpendicular current are given by

$$\mathbf{E} \simeq \left(1 - \frac{\bar{\alpha}_C^2 \alpha_H^2}{1+\bar{\alpha}_C^2 \alpha_H^2}\right)\mathbf{E}_1 - \left(\frac{\alpha_C^A \alpha_H}{1+\bar{\alpha}_C^2 \alpha_H^2}\right)(\hat{\mathbf{e}}_\mathrm{B} \times \mathbf{E}_1) \tag{25.18}$$

$$\mathbf{J} = \Sigma_P \left\{1 + \alpha_C^A \alpha_H^2 \left(\frac{1-\alpha_C^B}{1+\bar{\alpha}_C^2 \alpha_H^2}\right)\right\}\mathbf{E}_1 + \Sigma_H \left\{1 - \alpha_C^A \left(\frac{1+\alpha_C^B \alpha_H^2}{1+\bar{\alpha}_C^2 \alpha_H^2}\right)\right\}(\hat{\mathbf{e}}_\mathrm{B} \times \mathbf{E}_1) \tag{25.19}$$

By substituting (25.19) into (25.13)–(25.15), we obtain the FAC densities at conductivity borders A and B, that is,

$$\begin{aligned} j_{\parallel,A} &= \left(1-\alpha_C^A\right)\nabla\left\{\Sigma_H \hat{\mathbf{e}}_\mathrm{B} \times \left(\mathbf{E}_1 + \mathbf{E}_2^B\right)\right\} \\ &= \left\{1 - \alpha_C^A \left(\frac{1+\alpha_C^B \alpha_H^2}{1+\bar{\alpha}_C^2 \alpha_H^2}\right)\right\}\nabla \cdot (\Sigma_H \hat{\mathbf{e}}_\mathrm{B} \times \mathbf{E}_1) \end{aligned} \tag{25.20}$$

$$\begin{aligned} j_{\parallel,B} &= \nabla \cdot (\Sigma_P \mathbf{E}_1) + \left(1-\alpha_C^B\right)\nabla \cdot \left(\Sigma_H \hat{\mathbf{e}}_\mathbf{B} \times \mathbf{E}_2^A\right) \\ &= \left\{1 + \alpha_C^A \alpha_H^2 \left(\frac{1-\alpha_C^B}{1+\bar{\alpha}_C^2 \alpha_H^2}\right)\right\} j_{\parallel,1} \end{aligned} \tag{25.21}$$

respectively.

In (25.18) and (25.19), the first and second terms on the RHS represent the components parallel to conductivity border A ($\mathbf{E}_\mathrm{A}$, $\mathbf{J}_\mathrm{A}$ parallel to $\mathbf{E}_1$) and conductivity border B($\mathbf{E}_\mathrm{B}$, $\mathbf{J}_\mathrm{B}$ perpendicular to $\mathbf{E}_1$), respectively.

Further, (25.18) implies that the generation of $\mathbf{E}_2^A$ causes rotation of the electric field from $\mathbf{E}_1$; the rotation angle θ is given by

$$\tan\theta = \alpha_C^A \alpha_H \tag{25.22}$$

and the generation of $\mathbf{E}_2^B$ leads to the shielding effect of the electric field in the direction of $\mathbf{E}_1$. The second line of (25.19) consists of the Pedersen and Hall currents excited by $\mathbf{E}_1$ and the residual parts excited by $\mathbf{E}_2$. In $\mathbf{J}_\mathrm{A}$, the secondary excited part is always positive, and, therefore, this term enhances the current in the direction of $\mathbf{E}_1$. On the other hand, in $\mathbf{J}_\mathrm{B}$, the secondary excited part is always negative; hence, this term causes the depression effect from the Hall current excited by $\mathbf{E}_1$. In other words, a Hall-Pedersen antiparallel channel is established in $\mathbf{J}_\mathrm{B}$ and a Hall-Pedersen parallel channel is produced in $\mathbf{J}_\mathrm{A}$. However, the total current in $\mathbf{J}_\mathrm{A}$ will be reduced compared with the case of no $\mathbf{E}_2^B$ ($\alpha_C^B = 0$).

The total Joule dissipation is given by

$$\mathbf{J} \cdot \mathbf{E} = \Sigma_P \left|E_1^2\right| \frac{1+\left(\alpha_C^A \alpha_H\right)^2}{\left(1+\bar{\alpha}_C^2 \alpha_H^2\right)^2} \tag{25.23}$$

which is equal to all of the Poynting flux absorbed into the ionosphere. The total Joule dissipation can be separated into

$$\mathbf{J}_\mathrm{A} \cdot \mathbf{E}_\mathrm{A} = \Sigma_P \left|E_\mathrm{A}^2\right| + \Sigma_H \left|E_\mathrm{A}\right|\left|E_\mathrm{B}\right| = \frac{1+\alpha_C^A \alpha_H^2}{\left(1+\bar{\alpha}_C^2 \alpha_H^2\right)^2} > 0 \tag{25.24}$$

$$\begin{aligned} \mathbf{J}_\mathrm{B} \cdot \mathbf{E}_\mathrm{B} &= \Sigma_P \left|E_\mathrm{B}^2\right| - \Sigma_H \left|E_\mathrm{A}\right|\left|E_\mathrm{B}\right| \\ &= -\left(\frac{\alpha_C^A \alpha_H^2}{1+\alpha_C^A \alpha_H^2}\right)\left(1-\alpha_C^A\right)\Sigma_P \left|E_1\right|^2 < 0 \end{aligned} \tag{25.25}$$

where $\mathbf{J}_\mathrm{A} \cdot \mathbf{E}_\mathrm{A}$ is equal to the Poynting flux absorbed into the ionosphere (see equation [25.34]) through current channel $\mathbf{J}_\mathrm{A}$, while $\mathbf{J}_\mathrm{B} \cdot \mathbf{E}_\mathrm{B}$ is equal to the Poynting flux absorbed into the ionosphere through current channel $\mathbf{J}_\mathrm{B}$. It is clear that a part of the Poynting flux absorbed into the ionosphere through $\mathbf{J}_\mathrm{A}$ is used not only for the Joule dissipation but also for the term of $+\Sigma_H \left|E_\mathrm{A}\right|\left|E_\mathrm{B}\right| > 0$ in (25.24). This energy is converted into a dynamo for $\mathbf{J}_\mathrm{B}$ through the term $-\Sigma_H \left|E_\mathrm{A}\right|\left|E_\mathrm{B}\right| < 0$ in (25.25).

By definition, the Poynting flux supplied from the magnetosphere through the primary current system balances the Joule dissipation in the primary current system:

$$S_1 = \mathbf{J}_1 \cdot \mathbf{E}_1 = \Sigma_P \left|E_1\right|^2 \tag{25.26}$$

As shown by (25.23), in the case of $\alpha_C^B = 0$ (which also means $\bar{\alpha}_C = 0$), $\mathbf{J} \cdot \mathbf{E}$ is always larger than $\Sigma_P \left|E_1\right|^2$. Thus, an infinite Cowling channel can indeed exist; however, in this case, the secondary current system also requires additional Poynting flux input for a steady-state solution to exist.

If $\mathbf{J} \cdot \mathbf{E} < \Sigma_P \left|E_1\right|^2$ is satisfied, S_1 can account for the total Joule dissipation ($S_1 > \mathbf{J} \cdot \mathbf{E}$) and Poynting flux flow out of the ionosphere through the formation of the secondary current system that can drive the magnetosphere. This condition is given by

$$\frac{\sqrt{1+\left(\alpha_C^A \alpha_H\right)^2} - 1}{\left(\alpha_C^A \alpha_H\right)^2} < \left(\alpha_C^B\right)^2 < 1 \tag{25.27}$$

for $0 < \alpha_C^B < 1$. This condition is always satisfied in the case of $\alpha_C^A = \alpha_C^B$. Thus, we may say that for generating the Cowling channel, the Hall current excited by the polarization electric field needs to be connected to the primary current with the production of an additionally excited polarization electric field, which leads to

$$\mathbf{J}_1 \cdot \mathbf{E}_1 > \mathbf{J}_A \cdot \mathbf{E}_A > \mathbf{J} \cdot \mathbf{E} \text{ with } \mathbf{J}_B \cdot \mathbf{E}_B < 0$$

The relation $\mathbf{J}_1 \cdot \mathbf{E}_1 > \mathbf{J}_A \cdot \mathbf{E}_A$ suggests that the Poynting flux is supplied from the magnetosphere. The primary current system in $\mathbf{J}_A$ is not only absorbed into the ionosphere through $\mathbf{J}_A \cdot \mathbf{E}_A$ but also reflected back to the magnetosphere with Poynting flux ΔS_1 satisfying

$$S_1 + \Delta S_1 = \mathbf{J}_A \cdot \mathbf{E}_A > 0(\Delta S_1 < 0)$$

The current channel $\mathbf{J}_B$ does not connect to the primary current system, and the Poynting flux only flows out of this current system

$$\Delta S_2 = \mathbf{J}_B \cdot \mathbf{E}_B < 0$$

25.2.3.1. Example 1: North-South (Meridional) Closure of Primary Current

If the primary current is purely northward, conductivity band A corresponds to the northward and southward borders of the auroral arc, and conductivity band B corresponds to the westward and eastward borders of the auroral arc (Boström Type II current that generator force parallel to arc current [*Boström*,1964]). This situation has been investigated in detail by *Fujii et al.* [2012]. In this case, the primary Hall current is directed eastward and will negatively charge up at the westward border of the auroral arc. This produces a strong westward electric field at the western edge of the conductivity border. According to equation (25.18), the ratio of the total electric field at the westward border to that at the northward border is given by $\alpha_C^A \alpha_H^2$. We expect that with increasing Cowling efficiency, the westward $\mathbf{E} \times \mathbf{B}$ drift caused by the northward primary electric field $\mathbf{E}_1$ is efficiently reduced and the equatorward $\mathbf{E} \times \mathbf{B}$ drift caused by the westward secondary polarization is strongly enhanced. The FAC ($j_{\parallel,A}$) at the westward edge is downward ($\nabla \cdot (\Sigma_H \hat{\mathbf{e}}_B \times \mathbf{E}_1) > 0$). However, as suggested by (25.20), $j_{\parallel,A}$ is always small compared with $\nabla \cdot (\Sigma_H \hat{\mathbf{e}}_B \times \mathbf{E}_1)$. On the other hand, the FAC caused by divergence of the total meridional current ($j_{\parallel,B}$) may be enhanced compared with $j_{\parallel,1}$.

The northward primary electric field is generally accompanied by the westward plasma flow on the magnetospheric side. Reduction of the northward electric field by the Cowling effect implies deceleration of this westward plasma flow as a result of M-I coupling. The electromagnetic energy for this braking of plasma flow is supplied from the ionosphere as a reflected field (with upward Poynting flux).

The polarization process at the westward edge induces strong negative charge up and small downward FAC. However, once negative charge density ($\nabla_\perp \cdot \mathbf{E}_B < 0$) is produced at the westward border, the induction process ($(\nabla \times \mathbf{E}_B)_\perp \approx \partial E/\partial s \neq 0$) at the conductivity boundary induces shear of toroidal magnetic field disturbances that form strong upward FAC. Therefore, elimination of this apparent inconsistency depends on the detailed physics of M-I coupling. However, this topic is beyond the scope of this chapter.

25.2.3.2. Example 2: East-West (Azimuthal) Closure of Primary Current

If the primary current is purely eastward, conductivity band A corresponds to the westward and eastward borders of the auroral arc, and conductivity band B corresponds to the northward and southward borders of the auroral arc (Boström Type I current that generator force perpendicular to arc current [*Boström*,1964]). In this case, the primary Pedersen current is directed eastward and will negatively charge up at the westward border of the auroral arc.

Consequently, the Hall polarization effect would cause depression of the total westward electric field, enhancement of the southward electric field (small equatorward $\mathbf{E} \times \mathbf{B}$ drift, large eastward $\mathbf{E} \times \mathbf{B}$ drift), and negative charge up with the upward electric field at the western edge. However, the total westward current becomes fairly small in the case of relatively large α_C^B compared with the westward current expected from the Cowling channel model infinitely elongated along the east-west direction.

25.2.4. Effect of the 3D Electron Flux Divergence on the Current Closure in the Cowling Channel

An electron continuity equation in the ionospheric E region strongly controls the spatiotemporal development of the ionospheric current system through conductivity evolution [e.g., *Atkinson*, 1970; *Sato*, 1978]. The continuity equation can be written as

$$\frac{\partial}{\partial t} n + \nabla \cdot (n \mathbf{v}_e) = S - L \tag{25.28}$$

where S is the source term for plasma ionization by energetic electron precipitation and disassociation processes, such as solar UV radiation, and L is the loss term due to recombination. The 3D electron flux divergence density $\nabla \cdot (n\mathbf{v}_e)$ will provide the source or loss effect for electron density by advection transportation and flow divergence. By assuming that the ion velocity has no parallel component, we can express the current density as $\mathbf{j} = en(-\mathbf{v}_{e\|} - \mathbf{v}_{e\perp} + \mathbf{v}_{i\perp})$; then, the current continuity condition $\nabla \cdot \mathbf{j} = 0$ gives

$$\nabla \cdot (n\mathbf{v}_e) = \nabla \cdot (n\mathbf{v}_{i,\perp}) \quad (25.29)$$

The finite electron flux divergence implies that the FAC carried by electrons is not only closed by the perpendicular current carried by the electrons but also by the ionic current. In other words, the current carrier exchange process from electron to ion has to occur in the region where the FAC closes the perpendicular current [*Yoshikawa et al.*, 2011a].

In the ionospheric E layer, we regard the electrons as the main carriers of the Hall current and the ions as the main carriers of the Pedersen current [*Sato*, 1978].

$$\nabla \cdot (n\mathbf{v}_e) \simeq -e^{-1}(\nabla_\| \cdot \mathbf{j}_\| + \partial_\perp j_H) = e^{-1}\partial_\perp j_P \quad (25.30)$$

where $\partial_\perp j_H$ and $\partial_\perp j_P$ denote the Hall current divergence and Pedersen current divergence, respectively.

In a special case, where there is no Hall current divergence in the system and the FAC only closes the Pedersen current, equation (25.30) reduces to

$$\nabla \cdot (n\mathbf{v}_e) \simeq (el)^{-1} j_{\|,1} = (el)^{-1} \nabla \cdot (\Sigma_P \mathbf{E})$$

where l is the effective thickness of the ionospheric E layer. Therefore, in the primary current system, the plasma density is enhanced ($\nabla \cdot (n\mathbf{v}_e) < 0$ in [25.28]) in the upward FAC region ($j_{\|,1} < 0$) and depleted ($\nabla \cdot (n\mathbf{v}_e) > 0$ in [25.28]) in the downward current region ($j_{\|,1} > 0$), which causes increase and decrease in conductivity, respectively [e.g., *Doe et al.*, 1995; *Blixt and Brekke*, 1996; *Karlsson and Marklund*, 1998; *Karlsson et al.*, 2005; *Russell et al.*, 2010]. Thus, concentration should occur in the upward FAC region and current spreading should occur in the downward FAC region, so that the current continuity can be maintained [e.g., *Cran-McGreehin et al.*, 2007; *Streltsov*, 2007].

However, if we consider the Hall current divergences that induces the Cowling channel, we need to carefully consider the relation between FAC density, electron-flux divergence, and plasma density evolutions. The condition $\nabla \cdot (n\mathbf{v}_e) = 0$ corresponds to the special case in which the Pedersen current has no divergence and all of the FAC is connected to the curl-free part of the Hall current. In this case, the FAC contributes to the evolution of the plasma density. Further, $\nabla \cdot (n\mathbf{v}_e) \neq 0$ implies that the FAC carried by the electron is partly connected to the Pedersen current ($\alpha_C \neq 0$) and $\partial_\perp j_P$ directly controls the evolution of the plasma density according to its sign.

More specifically, we consider the electron-flux divergence in the finite Cowling channel discussed in equations (25.13)–(25.21). By applying primary and secondary current separation to (25.30) and considering the electron flux divergence at the east-west and north-south edges of the Cowling channel, we get

$$\nabla \cdot (n\mathbf{v}_e)|^B \simeq (el)^{-1}\left(j_{\|,A} - \nabla.\{\Sigma_H \hat{\mathbf{e}}_B \times \mathbf{E}_A\}\right) = (el)^{-1}\left(\frac{1}{1+\bar{\alpha}_C^2\alpha_H^2}\right) j_{\|,1} = (el)^{-1}\nabla \cdot (\Sigma_P \mathbf{E}_B) \quad (25.31)$$

$$\nabla \cdot (n\mathbf{v}_e)|^A \simeq (el)^{-1}\left(j_{\|,B} - \nabla.\{\Sigma_H \hat{\mathbf{e}}_B \times \mathbf{E}_B\}\right) = -(el)^{-1}\left(\frac{\alpha_C^A}{1-\alpha_C^A}\right) j_{\|,A} = (el)^{-1}\nabla \cdot (\Sigma_P \mathbf{E}_A) \quad (25.32)$$

Figure 25.3 shows schematic views of the effect of the 3D electron flux density divergence on the current closure process excited by the westward primary electric field, when taking the perpendicular electron flux density divergence (the Cowling effect) into account.

As indicated by equation (25.31), in the region of the downward current of current channel $\mathbf{J}_A$, $j_{\|,B} < 0$ (upward current of $j_{\|,B} > 0$), the electron flux divergence term becomes positive (negative) and it will contribute to the depression (enhancement) of plasma density. In the current channel $\mathbf{J}_A$, the density enhancement and depletion occur in the upward FAC region and downward FAC, respectively. Therefore, combined with the density enhancement effect by energetic particle precipitation in the upward FAC region of the primary current system, it is natural to consider the primary channel producing the conductivity border of the Cowling channel.

Similarly, equation (25.32) shows that in the region of downward current of current channel $\mathbf{J}_B$, $j_{\|,A} < 0$ (upward current of $j_{\|,A} > 0$); the electron flux divergence term becomes negative (positive) and it will contribute to the enhancement (depression) of plasma density. Thus, in the current channel $\mathbf{J}_B$, the density enhancement and depletion occur in the downward FAC and upward FAC regions, respectively. This is a specific property of the Cowling channel.

25.2.5. Geometrical Effect of Ionospheric Polarization on the Total Electric Field Distribution and Plasma Flow

According to equation (25.9), when the primary Hall current flows from lower to higher conductivity $\nabla\Sigma_H \cdot (\hat{\mathbf{e}}_B \times \mathbf{E}) < 0$, a positive polarization charge is induced in the conductivity gradient region. This causes clockwise turning of the electric field along $\mathbf{B}_0$ in the region with low conductivity $\nabla\Sigma_H \cdot (\hat{\mathbf{e}}_B \times \mathbf{E}) > 0$ and anticlockwise turning along $\mathbf{B}_0$ in the region with high

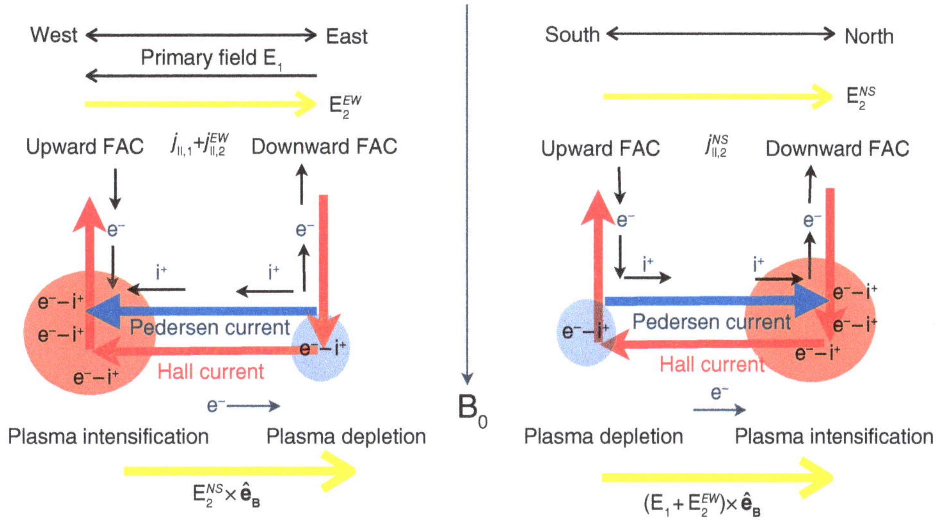

Figure 25.3 Schematic pictures for effect of the three-dimensional electron flux density divergence on the current closure process excited by the westward primary electric field, when taking into account the perpendicular electron flux density divergence (the Cowling effect). After *Yoshikawa et al.* [2011b].

conductivity [*Yoshikawa et al.*, 2013b]. We can say that the polarization charge produced by Hall current divergence causes meandering of the convection flow produced by the primary electric field across the conductivity gradient region. The streamlines of the convection flow will be deformed to be more tangential to the conductivity gradients (see numerical demonstration by *Yoshikawa et al.* [2013b]). The deformation of the ionospheric convection by excitation of the polarization charge along the conductivity gradient region is expected to occur at all the edges of the auroral shapes, at the dawn-dusk conductivity terminator line, and in the equatorial electrojet regions.

A rotationally symmetric vortex where the electric field and the ionospheric conductivity gradients can reasonably be assumed to point in the radial direction with respect to the vortex center, when it is possible to separate the total observed current system into the parts caused by the polarization effect and the remaining parts. Such a case was analyzed by *Amm and Fujii* [2008] for a substorm break up spiral; they followed up on a long-standing discussion about whether the upward FAC in such a spiral is provided by the local downward FAC in the vicinity of the spiral or carried remotely via a Cowling channel through the evolving substorm bulge toward the spiral. Further, they showed that around two-thirds of the upward FAC were provided by local current closure, and one-third was provided by remote (Cowling channel) current closure. However, in a certain region at the western edge of the spiral, the contribution of the remote closure was more than 80%. This region coincided with steep zonal conductance gradients and was also clearly discriminated in the auroral display.

A physical mechanism of latitudinal motion of the auroral arc can be explained by a four-side-bounded Cowling channel model [*Fujii et al.*, 2012]. In this case, even with a purely meridionally directed primary electric field (e.g., northward in the afternoon sector), a zonal secondary electric field component is generated owing to the divergence of the Hall current at the zonal boundary of the arc. In the afternoon sector, this secondary electric field is directed westward, thus corresponding to an equatorward plasma drift. This mechanism may, in some cases, explain the observed equatorward drift of auroral arcs in the afternoon sector [e.g., *Pellinen at al.*, 1982; *Opgenoorth et al.*, 1990; *Haerendel et al.*, 1993; *Williams et al.*, 1998]. In the morning sector, an oppositely directed eastward zonal electric field component is expected to be generated by the polarization mechanism.

25.2.6. Energy Balance in the Cowling Channel

Here, we discuss the energy balance in the Cowling channel in a more general manner to facilitate application to any geometry of conductivity distributions.

The height-integrated ionospheric sheet current satisfies

$$\mathbf{J} = \mu_0^{-1} e \hat{\mathbf{n}} \times \delta \mathbf{b} \tag{25.33}$$

where $\delta\mathbf{b}$ is the jump of the horizontal magnetic field perturbation across the ionosphere produced by $\mathbf{J}$, $\hat{\mathbf{n}} = -\hat{\mathbf{e}}_B$ (in the northern ionosphere) is a unit vector directed vertically upward from the ionosphere, and μ_0 is the magnetic permeability in space.

By taking the inner product between **J** and **E**, we get the energy balance equation

$$\mathbf{J}\cdot\mathbf{E} = -\mu_0^{-1}\hat{\mathbf{n}}\cdot(\mathbf{E}\times\delta\mathbf{b}) = \Sigma_P E^2 \quad (25.34)$$

where $-\mu_0^{-1}\hat{\mathbf{n}}\cdot(\mathbf{E}\times\delta\mathbf{b})$ is the Poynting flux absorbed into the ionosphere, which balances the Joule dissipation of $\Sigma_P E^2$.

By taking the inner product between **J** and $\mathbf{E}_1$ and between **J** and $\mathbf{E}_2$, the energy balance equations in the primary and secondary components of the Cowling channel become

$$\begin{aligned} S_1 &\equiv \mu_0^{-1}\hat{\mathbf{n}}\cdot(\mathbf{E}_1\times\delta\mathbf{b}) \\ &= \Sigma_P|E_1|^2 + \Sigma_P\mathbf{E}_2\cdot\mathbf{E}_1 + \Sigma_H(\hat{\mathbf{e}}_B\times\mathbf{E}_2)\cdot\mathbf{E}_1 \end{aligned} \quad (25.35)$$

$$\begin{aligned} S_2 &\equiv -\mu_0^{-1}\hat{\mathbf{n}}\cdot(\mathbf{E}_2\times\delta\mathbf{b}) \\ &= \Sigma_P|E_2|^2 + \Sigma_P\mathbf{E}_1\cdot\mathbf{E}_2 + \Sigma_H(\hat{\mathbf{e}}_B\times\mathbf{E}_1)\cdot\mathbf{E}_2 \end{aligned} \quad (25.36)$$

respectively.

Here, S_1 is the Poynting flux absorbed into the ionosphere through the primary component of the Cowling channel ($S_1>0$), which includes energy flow from the magnetosphere to the ionosphere for driving the ionospheric current system. The first two terms on the RHS of equation (25.35) are the Joule dissipation in $\mathbf{J}\cdot\mathbf{E}_1$ and the last term is the power produced by applying $\mathbf{E}_1$ to the total Hall current. On the other hand, S_2 is the Poynting flux flowing out from the ionosphere ($S_2<0$) because $\mathbf{E}_2$ is generated as a result of Hall current divergence in the ionosphere. The first two terms on the RHS of equation (25.36) are the Joule dissipation in $\mathbf{J}\cdot\mathbf{E}_2$ and the last term is the power generated by applying $\mathbf{E}_2$ to the Hall current. Cleary, the relation

$$\Sigma_H(\hat{\mathbf{e}}_B\times\mathbf{E}_2)\cdot\mathbf{E}_1 = -\Sigma_H(\hat{\mathbf{e}}_B\times\mathbf{E}_1)\cdot\mathbf{E}_2 > 0 \quad (25.37)$$

is satisfied, because the total Hall current never does work on an external system. However, (25.35) shows that the net energy absorbed into the ionosphere through the primary component does not simply balance the Joule dissipation in $\mathbf{J}\cdot\mathbf{E}_1$; some energy is deposited in the external system by the Hall effect, because the secondary Hall current generally has a component parallel to the primary electric field. Similarly, (25.36) shows that the free energy for the outward Poynting flux and the Joule dissipation in the secondary component are supplied through the Hall effect because the primary Hall current generally has an antiparallel component to the secondary electric field. Hence, we can regard the term in (25.37) as the energy transfer function by the polarization Hall effect in the Cowling channel, which transmits the free energy absorbed through the primary component in the Cowling channel to the secondary component and is used for excitation of the radiant Poynting flux from the secondary channel and Joule dissipation in it.

25.2.7. Impact of the Polarization Effect to the Joule Heating

Finally, we inspect the impact of the polarization effect on Joule heating under electrostatic approximation of the M-I coupled system. As discussed by *Amm et al.* [2013], if the Cowling efficiency α_C can be assumed to be uniform, we can integrate the secondary Pedersen current from equation (25.10):

$$\mathbf{J}_{2,P} = \Sigma_P\mathbf{E}_2 = \mathbf{A}^{df} - \alpha_C\Sigma_H(\hat{\mathbf{e}}_B\times\mathbf{E}) \quad (25.38)$$

where $\mathbf{A}^{df}$ is an unknown divergence-free vector field. Thus, the Joule dissipation by the secondary Pedersen current is

$$\mathbf{J}_{2,P}\cdot\mathbf{E} = \Sigma_P|\mathbf{E}_2|^2 + \Sigma_P\mathbf{E}_2\cdot\mathbf{E}_1 = \mathbf{A}^{df}\cdot\mathbf{E} \quad (25.39)$$

Richmond [2010] derived the Equipotential Boundary Poynting Flux theorem, according to which the Poynting flux within a flux tube whose boundary is an equipotential curve is dissipated inside the ionospheric foot point of the flux tube. By applying this theorem, *Vanhamäki et al.* [2012] showed that Joule heating of a divergence-free current system integrated over the entire ionosphere or any equipotential curve (area D in the following integrals) is zero. Thus, we see that the net secondary Pedersen current is dissipationless.

$$\int_D \mathbf{J}_{2,P}\cdot\mathbf{E}da = \int_D A^{df}\cdot\mathbf{E}da = 0 \quad (25.40)$$

This result can be rewritten as

$$\int_D \Sigma_P|E_2|^2 da = -\int_D \Sigma_P\mathbf{E}_2\cdot\mathbf{E}_1 da \quad (25.41)$$

Thus, the total Joule heating in the Cowling channel becomes

$$\begin{aligned} \int_D (S_1+S_2)da &= \int_D \mathbf{J}\cdot\mathbf{E}da \\ &= \int_D \Sigma_P\left(|E_1|^2 + 2\mathbf{E}_1\cdot\mathbf{E}_2 + |E_2|^2\right)da \\ &= \int_D \Sigma_P\left(|E_1|^2 - |E_2|^2\right)da \\ &\le \int_D \Sigma_P|E_1|^2 da = -\int_D \mu_0\hat{\mathbf{n}}\cdot(\mathbf{E}_1\times\delta\mathbf{b}_1)da \end{aligned} \quad (25.42)$$

where $\delta \mathbf{b}_1$ satisfies $\mathbf{J}_1 = \mu_0^{-1}\hat{\mathbf{n}} \times \delta \mathbf{b}_1$. Hence, we see that at least for the case where α_C is uniform, the Hall polarization effect always decreases the total ionospheric Joule dissipation compared with the case of no Cowling effect ($\mathbf{E}_2 = 0$).

From equations (25.35), (25.36), and (25.41), we get

$$\int_D S_1 da = \int_D \Sigma_P \left(|E_1|^2 - |E_2|^2\right) da - \int_D S_2 da \tag{25.43}$$

$$\int_D S_2 da = \int_D \Sigma_H \left(\hat{\mathbf{e}}_B \times \mathbf{E}_1\right) \cdot \mathbf{E}_2 \, da < 0 \tag{25.44}$$

Thus, the net Poynting flux absorbed through the primary system (the LHS of [25.43]) is used for total Joule dissipation and Poynting flux flow out of the ionosphere by the Cowling effect through $\int_D S_2 da$.

25.2.8. Energy Conversion in the Inductive Cowling Channel

Yoshikawa [2002a, 2002b] discussed energy balance in the case of uniform conductance but included the induction (rotational) electric field that satisfies

$$-\partial B_{\parallel}/\partial t = (\nabla \times \mathbf{E})_{\parallel}$$

It is well known that FAC systems lose their energy in the ionosphere through Joule dissipation that is caused by their closure via the ionospheric Pedersen current, and that the ionospheric Hall current cannot contribute to the total energy dissipation. However, it is also true that the divergence-free Hall current is excited by the incident FAC, and it produces a Poynting flux that includes a poloidal-type magnetic field. Even if the Hall effect cannot do any work on an external system, what is the significance of its contribution to the accumulation of poloidal magnetic energy? *Yoshikawa* [2002a, 2002b] clarified that the divergent Hall current, excited during the transient phase of M-I coupling, closes via the FAC and produces a Hall current generator, which pumps up the energy of the FAC system to increase the ionospheric rotational Hall current (together with its associated poloidal magnetic field). This inductive Hall effect also causes changes in the eigenfrequency of the field line oscillation [*Yoshikawa et al.*, 1999]

In the reflection process of the shear Alfvén waves, including the ionospheric induction effect, the divergent Hall current driven by the rotational electric field (corresponding to the last term on the RHS of [25.5]) is closed by the FAC [*Yoshikawa and Itonaga*, 1996, 2000; *Buchert and Budnik*, 1997]. As shown by Figure 25.4, in the homogeneous but anisotropically conducting ionosphere,

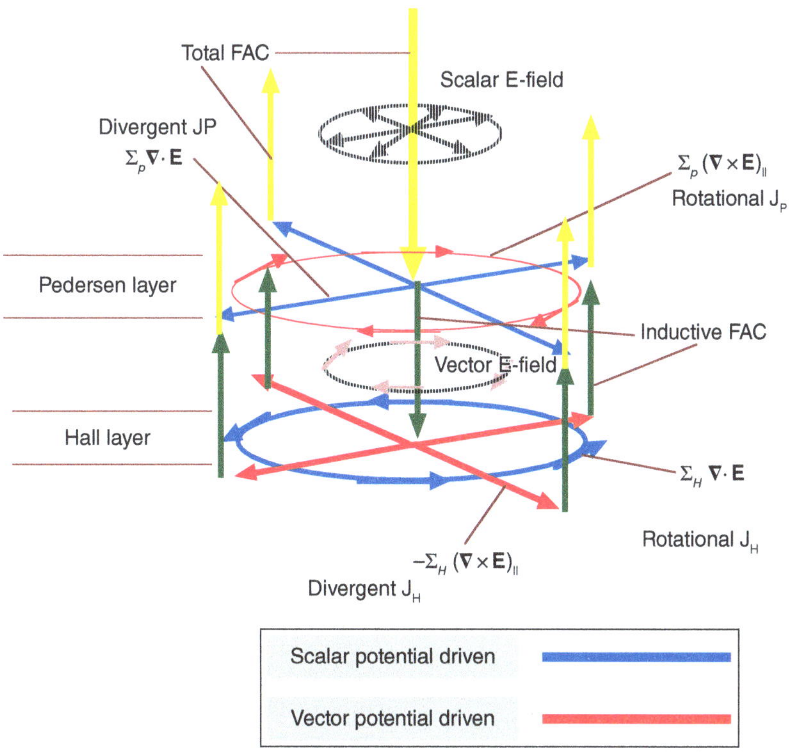

Figure 25.4 Schematic picture for Cowling channel excited by inductive effect. After *Yoshikawa and Itonaga* [2000]).

the inductive rotational electric field accompanied by the time-dependent divergence-free Hall current produces the divergence-free Pedersen current. Thus, they form the Hall-Pedersen antiparallel circuit (when the incident FAC increases) [*Yoshikawa and Itonaga*, 2000]. The inductive divergent Hall current excited by the Hall effect for the rotational electric field is in the same direction as the divergent Pedersen current. Thus, they form the Hall-Pedersen parallel circuit. This inductive divergent Hall current provides the free energy from the magnetosphere for increasing the magnetic energy held by the divergence-free current system [*Yoshikawa*, 2002a, 2002b].

This type of energy conversion in the ionosphere, caused by the inductive Hall effect, is similar to the energy flow in the Cowling channel discussed in this section. The difference is there is no conductivity gradient in equation (25.9), that is,

$$\nabla \cdot \mathbf{E}_2 = \frac{j_{\parallel,2} + \Sigma_H (\nabla \times \mathbf{E})_{\parallel}}{\Sigma_P} \tag{25.45}$$

and the divergence-free component of $\mathbf{E}_2$ is also generated in the above-mentioned induction process. The divergence-free component is perpendicular to $\mathbf{E}_1$. Thus, S_2 consists of the radiated Poynting flux of the compressional wave to the magnetosphere and poloidal magnetic disturbances to the atmosphere. This $\mathbf{E}_2$ becomes effective when the primary field $\mathbf{E}_1$ is time varying (induction phase). Therefore, clearly, the magnetospheric energy carried by the Poynting flux of the shear Alfvén wave through the primary current is used for generation of the Hall current through the Hall current generator [*Yoshikawa*, 2002b].

25.3. COWLING CHANNEL COUPLING TO THE SHEAR ALFVÉN WAVE

The Cowling efficiency corresponds to the ratio of space charge production, that is, charging by Hall current divergence and discharging by secondary FAC excitation. As described by equation (25.1), excitation of div($\mathbf{E}_2$) should induce evolution of FAC. Therefore, to obtain the Cowling efficiency, we need to determine the relationship between the polarization field and the FAC densities.

In the M-I coupling process via shear Alfvén waves, the ratio between the FAC density and the divergence of the wave electric field in a propagating fields is given by the Alfvén wave conductance $\Sigma_A = (\mu_0 V_A)^{-1}$ [*Walén*, 1944], where V_A is the Alfvén velocity.

The incident electric field to the ionosphere becomes a driver of the ionospheric current system and produces a reflected electric field to totally satisfy the current continuity condition. Hence, it is natural to combine this incident electric field with the primary electric field of the Cowling channel and to consider the reflected wave field to be composed of a reflected field from the primary current system plus the secondary current system of the Cowling channel [*Yoshikawa et al.*, 2013a]. Actually, the Alfvén wave approach can be used to describe not only local and dynamical phenomena but also more general global quasi-static M-I coupling processes [*Yoshikawa et al.*, 2010].

The total electric field and FAC density of shear Alfvén wave just above the ionosphere can be separated into

$$\begin{aligned} \mathbf{E} &= \mathbf{E}^{(in)} + \mathbf{E}^{\mathrm{MI}} \\ j_{\parallel} &= \Sigma_A \left(\nabla \cdot \mathbf{E}^{(in)} - \nabla \cdot \mathbf{E}^{\mathrm{MI}} \right) \end{aligned} \tag{25.46}$$

[e.g., *Alfvén and Fälthammar*, 1963; *Mallinckrodt and Carlson*, 1978; *Lysak*, 1990]. Here, $\mathbf{E}^{(in)}$ is the electric field accompanied by the incident component, which brings about an electromotive force for driving the M-I coupled current system from the magnetosphere, and $\mathbf{E}^{\mathrm{MI}}$ is the electric field generated as a result of the M-I coupling process.

By substituting equation (25.46) into (25.3) and (25.4), we get the total current closure equation [e.g., *Glassmeier*, 1983, 1984; *Zhu et al.*, 1993; *Vogt*, 2002]

$$\begin{aligned} &\Sigma_A \nabla_\perp \cdot \mathbf{E}^{(in)} - \nabla_\perp \cdot \left(\Sigma_P \mathbf{E}^{(in)} + \Sigma_H \hat{\mathbf{e}}_{\mathrm{B}} \times \mathbf{E}^{(in)} \right) \\ &= \Sigma_A \nabla_\perp \cdot \mathbf{E}^{\mathrm{MI}} + \nabla_\perp \cdot \left(\Sigma_P \mathbf{E}^{\mathrm{MI}} + \Sigma_H \hat{\mathbf{e}}_{\mathrm{B}} \times \mathbf{E}^{\mathrm{MI}} \right) \end{aligned} \tag{25.47}$$

This can be solved for the total feedback electric field $\mathbf{E}^{\mathrm{MI}} = -\nabla_\perp \Phi^{\mathrm{MI}}$ when the driving fields $\mathbf{E}^{(in)}$ and distribution of (Σ_A, Σ_P, Σ_H) are known.

The primary current system defined in equation (25.6) describes the current closure process (1) between the Alfvénic FAC (composed of incident and reflected components of the Alfvén wave) and the homogeneous Pedersen current and (2) between the reflected Alfvénic FAC and the Pedersen current excited by the polarization process. The primary electric field and Alfvénic FAC in the primary current system are given by

$$\begin{aligned} \mathbf{E}_1 &= \mathbf{E}^{(in)} + \mathbf{E}_1^{(ref)} + \mathbf{E}_1^{(pol)} \\ j_{\parallel,1} &= \Sigma_A \nabla \cdot \mathbf{E}^{(in)} - \Sigma_A \nabla \cdot \left(\mathbf{E}_1^{(ref)} + \mathbf{E}_1^{(pol)} \right) \end{aligned} \tag{25.48}$$

where the first, second, and third terms on the RHS of (25.48) denote the incident, reflected, and polarization components, respectively. On the other hand, the secondary current system describes the current closure process among the reflected Alfvénic FAC, secondary Pedersen current, and curl-free component of the Hall current. The secondary electric field and Alfvénic FAC in the secondary current system are given by

$$\mathbf{E}_2 = \mathbf{E}_2^{(pol)}, \qquad j_{\parallel,2} = -\Sigma_A \nabla \cdot \mathbf{E}_2^{(pol)} \tag{25.49}$$

By substituting these relations into equations (25.6) and (25.7), the primary FAC can be decomposed into an incident and reflection process, given by

$$\Sigma_A \nabla\cdot\left(\mathbf{E}^{(in)} - \mathbf{E}_1^{(ref)}\right) = \Sigma_P \nabla\cdot\left(\mathbf{E}^{(in)} + \mathbf{E}_1^{(ref)}\right) \tag{25.50}$$

and a polarization process by the Pedersen conductivity gradient, given by

$$-\Sigma_A \nabla\cdot\mathbf{E}_1^{(pol)} = \nabla\cdot\left(\Sigma_P \mathbf{E}_1^{(pol)}\right) + \nabla\Sigma_P\cdot\left(\mathbf{E}^{(in)} + \mathbf{E}_1^{(ref)}\right) \tag{25.51}$$

At the same time, the secondary FAC satisfies

$$-\Sigma_A \nabla\cdot\mathbf{E}_2^{(pol)} = \nabla\cdot\left(\Sigma_P \mathbf{E}_2^{(pol)}\right) + \left(\hat{\mathbf{e}}_B \times \nabla\Sigma_H\right)\cdot\mathbf{E} \tag{25.52}$$

From equations (25.50), (25.51), and (25.52), the divergence densities of the reflection field, the polarization field by the Pedersen conductivity gradient, and the polarization effect by the Hall conductivity gradient are given by

$$\begin{aligned}\nabla\cdot\mathbf{E}_1^{(ref)} &= R_A^{(ref)} \nabla\cdot\mathbf{E}^{(in)}, \nabla\cdot\mathbf{E}_1^{(pol)} = \vec{l}_P\cdot\mathbf{E}_1 \\ \nabla\cdot\mathbf{E}_2^{(pol)} &= \vec{l}_P\cdot\mathbf{E}_2 + \vec{l}_H\cdot\mathbf{E}\end{aligned} \tag{25.53}$$

respectively. Here,

$$R_A^{(ref)} = \left(\frac{\Sigma_A - \Sigma_P}{\Sigma_A + \Sigma_P}\right) \tag{25.54}$$

is the reflection coefficient of a shear Alfvén wave from the homogeneous ionosphere [*Scholer*, 1970],

$$\vec{l}_P = -\left(\frac{\nabla\Sigma_P}{\Sigma_A + \Sigma_P}\right) \tag{25.55}$$

is the production operator of the polarization field when the Pedersen current crosses the conductivity gradient region, and

$$\vec{l}_H \equiv -\left(\frac{\hat{\mathbf{e}}_B \times \nabla\Sigma_H}{\Sigma_A + \Sigma_P}\right) \tag{25.56}$$

is the production operator of the polarization field when the Hall current crosses the conductivity gradient region. Based on these relations, the FAC and divergence density of the total electric field can be written as

$$\begin{aligned} j_\parallel = \Sigma_P\left(1 + R_A^{(ref)}\right)\nabla\cdot\mathbf{E}^{(in)} + \left(1 - \alpha_C^{(Alfven)}\right) \\ \left(\nabla\Sigma_P + \hat{\mathbf{e}}_B \times \nabla\Sigma_H\right)\cdot\mathbf{E}\end{aligned} \tag{25.57}$$

$$\nabla\cdot\mathbf{E} = \left(1 + R_A^{(ref)}\right)\nabla\cdot\mathbf{E}^{(in)} + \left(\vec{l}_P + \vec{l}_H\right)\cdot\mathbf{E} \tag{25.58}$$

where

$$\alpha_C^{(Alfven)} = \left(\frac{\Sigma_P}{\Sigma_A + \Sigma_P}\right) \tag{25.59}$$

is the Cowling efficiency defined in the form of equation (25.10) but naturally extended to include the effect of the Pedersen gradients

$$\Sigma_P \nabla\cdot\mathbf{E}_2 + \alpha_C^{Alfven}\left(\nabla\Sigma_P\cdot\mathbf{E} + \nabla\cdot\left(\Sigma_H \hat{\mathbf{e}}_B \times \mathbf{E}\right)\right) = 0 \tag{25.60}$$

The first term on the RHS of equation (25.57) denotes the Alfvénic FAC accompanied by the incident and reflection process of the shear Alfvén wave and the second term denotes the Alfvénic FAC excited by the polarization process with discharging ratio (1-$\alpha_C^{(Alfvén)}$). The first term on the RHS of (25.58) shows the polarization charge density accompanied by the incident and reflection process of the shear Alfvén wave, and the second term corresponds to the total polarization charge densities $\nabla\cdot\mathbf{E}^{(pol)}$. Accordingly, the Alfvén wave conductance Σ_A can be obtained as

$$\Sigma_A = -\frac{\left(\nabla\Sigma_P - \hat{\mathbf{e}}_B \times \nabla\Sigma_H\right)\cdot\mathbf{E}}{\nabla\cdot\mathbf{E}^{(pol)}} - \Sigma_P \tag{25.61}$$

if we somehow determine $\nabla\cdot\mathbf{E}^{(pol)}$ from the observational data.

From the Waléen relation of (25.46), the incoming wave field $\mathbf{E}^{(in)}$ satisfies

$$\nabla\cdot\mathbf{E}^{(in)} = \frac{\Sigma_A^{-1} j_\parallel + \nabla\cdot\mathbf{E}}{2} \tag{25.62}$$

which can be extracted from the total observed wave field. Thus, the primary and secondary fields can be calculated for any distribution of the electrodynamics parameters, not only for special cases such as the classical Cowling channel. Although the use of the Cowling efficiency parameter is not required theoretically, it can be quantitatively estimated from the results a posteriori.

Numerical demonstration of the above-mentioned M-I coupling-based separation of the ionospheric fields into primary and secondary (polarization) electrodynamic systems was provided by *Yoshikawa et al.* [2013b]. They found that, as a principle response, the polarization effect caused by the Pedersen current divergence cancels (intensifies) the ambient background electric field inside (outside) the high-conductivity band. By contrast, the polarization effect caused by the Hall current divergence rotates the electric field from the background one, which causes a meandering of the convection flow along the conductivity gradient region. They concluded that while the Hall and Pedersen currents never cancel each other when the conductances are homogeneous, they can do so in the conductivity gradient region. This is the reason why a Hall polarization charge is induced and a resultant Cowling channel is formed.

In this section, we discuss the scale-independent reflection process of the shear Alfvén wave by assuming no induction fields. The scale-dependent M-I coupling process caused by parallel potential drops between the auroral acceleration region and the ionosphere, and the Pedersen and Hall current layer inside the ionospheric E layer have been discussed in detail by *Yoshikawa et al.* [2011a].

25.3.1. Application to the Global Magnetosphere-Ionosphere Coupling Simulation

If equation (25.4) is solved for the total electric field $\mathbf{E}=-\nabla_{\perp}\Phi$ by assuming a fixed primary FAC distribution, as is the case in most global M-I coupling simulations [e.g., *Tanaka*, 1995; *Raeder et al.*, 2001; *Lyon et al.*, 2004], a nonzero polarization electric field can be produced even in areas with zero primary FAC, provided that a finite primary electric field ($\mathbf{E}_1=-\nabla_{\perp}\Phi_1$) and conductivity gradient exist. In this case, the local current continuity relation becomes $\nabla\cdot\left(\Sigma_P\left(\mathbf{E}_1+\mathbf{E}_2\right)+\Sigma_H\hat{\mathbf{e}}_B\times\mathbf{E}\right)=0$. This satisfies $j_{\parallel,2}=0$, but there is finite $\mathbf{E}_2=-\nabla_{\perp}\Phi_2$, which implies that the Hall current divergence is perfectly canceled out by the Pedersen current divergence produced by the secondary electric field ($\alpha_C=1$). In other words, a perfect Cowling channel is always assumed in ionospheric potential solvers whenever a fixed FAC distribution is prescribed.

The conventional potential solver has been used so that the FAC densities produced by the MHD simulator are taken as the input for the solver, which produces ionospheric electrostatic potential as the output. This is used to update the electrostatic electric field (or plasma flow) at the inner edge of the magnetospheric simulation (magnetospheric inner boundary condition). However, as discussed by *Yoshikawa et al.* [2010], this algorithm cannot give a correct reflected field in a temporal M-I coupling process. In reality, the perturbation components of the ionospheric potential act not only as a mere electrostatic field but also as an inductive field of Alfvénic FACs in the sense of equation (25.1), which preserves the conservation law of energy and momentum in the temporal M-I coupling. On the other hand, the M-I coupling theory through shear Alfvén waves presented in this chapter, which can also be used in association with an ionospheric solver of a magnetospheric MHD simulation, allows the ionospheric and magnetospheric boundary conditions for the currents (magnetic field) and electric potential (or plasma flow) to be simultaneously satisfied [*Yoshikawa et al.*, 2010]. This operation is achieved by Walén separation of the magnetic field and the plasma flow field at the inner edge of the magnetosphere into inward and outward propagating waves. The inward propagating wave is reflected at the ionosphere, and the reflected wave is used to update the boundary condition at the inner edge of the simulation.

Furthermore, *Vanhamäki* [2011] explicitly considered the effect of induction in the ionosphere on M-I coupling simulation. He presented an ionospheric solver for an MHD simulation that simultaneously solves Faraday's law in the ionosphere in addition to Ohm's law and current continuity. When compared with an MHD solver that ignores these inductive effects, the solution for the ionospheric electric field with the same input conditions is found to differ by up to 10 mV/m, or 20%–40% of the total electric field. These maximum values are obtained from the analysis of just one time step forward in the ionospheric solver of an MHD simulation. Therefore, when using this solver for MHD runs with longer intervals, the discrepancy caused by ionospheric inductive effects can be expected to become even larger. Inductive effects are important issues in the dynamical evolution of plasma flow inside the ionospheric E layer [*Vanhamaki et al.*, 2006, 2007, *Vanhamaki*, 2011], such as in the westward traveling surge and auroral streamers.

Furthermore, recent advancements in computing power enable us to simulate the electrodynamics of the current closure process in the ionosphere, including chemical reaction, precipitation, and nonlinear advection effects on the evolution of ionospheric conductivities [e.g., *Doe et al.*, 1995; *de Boer et al.*, 2010]. However, at present, these simulations are two-dimensional (field-aligned and one perpendicular direction) and the ionospheric electric field is only produced by the FAC-Pedersen current closure equation (no Hall current divergence). Moreover, as discussed by *Yoshikawa et al.* [2011a], the FAC and ionospheric current closure and their evolutions with ionospheric conductivity in the Cowling channel are significantly modified by the existence of the Hall polarization field.

Therefore, the Hall-Pedersen closure circuit in the Cowling channel should be understood in the context of 3D inductive M-I coupling.

25.4. SUMMARY AND FUTURE WORK

In this chapter, we summarized recent developments with regard to the Cowling channel model and M-I coupling theory. We made significant contributions in terms of providing a new perspective on M-I coupling by including the polarization effect, induction effect, and the effect of the 3D nature of the ionosphere. However, many questions remain unanswered and much work needs to be done before we can fully understand the very complex nature of the M-I coupling process, at least at the mesoscale level, with which we are mainly concerned. Some directions for future research are listed below.

1. Combining new M-I coupling theory with magnetospheric MHD (or other) simulations
2. Full inclusion of the inductive electric field effect in the M-I coupling theory (thus far, the two have been treated somewhat separately)
3. Effect of conductivity evolution on the Cowling channel evolution
4. Applications of the new M-I coupling theory with ground-based and satellite data sets, and determination of polarization electric fields, currents, and the Cowling efficiency from the data
5. Detailed analysis of specific ionospheric structures with emphasis on M-I coupling, such as substorm current wedge, westward travelling surge, and omega bands, using the above-mentioned approach
6. A truly 3D treatment of the ionosphere (not only a two-layer approach as presently implemented)
7. A particle simulation for inspecting the 3D-kinetic and electromagnetic aspects of polarization charge in the ionosphere.

The combination of the inductive and polarization-Hall effects is a very important albeit relatively unexplored aspect of the M-I coupling process.

ACKNOWLEDGMENTS

We thank members of the International Cowling channel team (Dr. Amm, Dr. Vanhamäki, Dr. Ieda, Dr. Nakamizo, and Dr. Ohtani), who contributed together to development of the theory of Cowling channel since 2008. We acknowledge Dr. Amm for his dynamic leadership of this international Cowling channel team. Work at Kyushu University was supported by MEXT/JSPS KAKENHI Grant Number 15H05815.

REFERENCES

Alfvén, H., and C.-G. Fälthammar (1963), *Cosmical Electrodynamics: Fundamental Principles*, Oxford University Press, London.

Amm O., and R. Fujii (2008), Separation of Cowling channel and local closure currents in the vicinity of a substorm breakup spiral, *J. Geophys. Res.*, *113*, A06304; doi:10.1029/2008JA013021.

Amm, O., A. Aruliah, S. C. Buchert, R. Fujii, J. W. Gjerloev, A. Ieda, T. Matsuo, C. Stolle, H. Vanhamäki, and A. Yoshikawa (2008), Towards understanding the electrodynamics of the 3-dimensional high-latitude ionosphere: Present and future, *Ann. Geophys.*, *26*, 3913–3932.

Amm, O., R. Fujii, H. Vanhamäki, A. Yoshikawa, and A. Ieda (2013), General solution for calculating polarization electric fields in the auroral ionosphere and application examples, *J. Geophys. Res. Space Phys.*, *118*, 2428–2437; doi:10.1002/jgra.50254.

Amm, O., R. Fujii, K. Kauristie, A. Aikio, A. Yoshikawa, A. Ieda, and H. Vanhamäki (2011), A statistical investigation of the Cowling channel efficiency in the auroral zone, *J. Geophys. Res.*, *116*, A02304; doi:10.1029/2010JA015988.

Atkinson, G. (1970), Auroral arcs: Results of the interaction of a dynamic magnetosphere with the ionosphere, *J. Geophys. Res.*, *75*, 4746–4755; doi:10.1029/JA075i025p04746.

Baumjohann, W. (1983), Ionospheric and field-aligned current systems in the auroral zone: A concise review, *Adv. Space Res.*, *2*(10), 55–62.

Blixt, E. M., and A. Brekke (1996), A model of currents and electric fields in a discrete auroral arc, *Geophys. Res. Lett.*, *23*(18), 2553–2556; doi:10.1029/96GL02378.

Boström, R. (1964), A model of the auroral electrojets, *J. Geophys. Res.*, *69*, 4983–4999; doi:10.1029/JZ069i023p04983.

Buchert, S. C., and F. Budnik (1997), Field-aligned current distributions gen-erated by a divergent Hall current, *Geophys. Res. Lett.*, *24*; doi:10.1029/ 97GL00064.

Chapman, S. (1956), The electrical conductivity of the ionosphere: A review, *Nuovo Cim.*, *4*(Suppl. 4), 1385–1412; doi:10.1007/BF02746310.

Cowling, T. G. (1932), The electrical conductivity of an ionised gas in the presence of a magnetic field, *Mon. Not. R. Astron. Soc.*, *93*, 90–98.

Cowling, T. G., and R. Borger (1948), Electric conductivity of the ionospheric D-region, *Nature*, *162*, 142–143; doi:10.1038/162143a0.

Cran-McGreehin, A. P., A. N. Wright, and A. W. Hood (2007), Ionospheric depletion in auroral downward currents, *J. Geophys. Res.*, *112*, A10309; doi:10.1029/2007JA012350.

de Boer, J. D., J. M. A. Noel, and J. P. St. Maurice (2010), The effect of mesoscale regions of precipitation on the ionospheric dynamics, electro- dynamics and electron density in the presence of strong ambient electric fields, *Ann. Geophys.*, *28*, 1345–1360; doi:10.5194/angeo-28-1345-2010.

Doe, R., J. Vickrey, and M. Mendillo (1995), Electrodynamic model for the formation of auroral ionospheric cavities, *J. Geophys. Res.*, *100*, 9683–9696; doi:10.1029/95JA00001.

Fujii, R., O. Amm, A. Yoshikawa, A. Ieda, and H. Vanhamäki (2011), Reformulation and energy flow of the Cowling channel, *J. Geophys. Res.*, *116*, A02305; doi:10.1029/2010JA015989.

Fujii, R., O. Amm, H. Vanhamäki, A. Yoshikawa, and A. Ieda (2012), An application of the finite length Cowling channel model to auroral arcs with longitudinal variations, *J. Geophys. Res.*, *117*, A11217; doi:10.1029/2012JA017953.

Glassmeier, K. H. (1983), Reflection of MHD-waves in the Pc4–5 period range at ionospheres at non-uniform conductivity distributions, *Geophys. Res. Lett.*, *10*, 678–681; doi:10.1029/GL010i008p00678.

Glassmeier, K. H. (1984), On the influence ionospheres with non-uniform con- ductivity distribution on hydromagnetic wave, *J. Geophys.*, *54*, 125–137.

Haerendel, G. (2008), Auroral arcs as current transformers, *J. Geophys. Res.*, *113*, A07205; doi:10.1029/2007JA012947.

Haerendel, G., S. Buchert, C. La Hoz, B. Raaf, and E. Rieger (1993), On the proper motion of auroral arcs, *J. Geophys. Res.*, *98*, 6087–6099; doi:10.1029/ 92JA02701.

Harang, L. (1946), The mean field of disturbance of polar geomagnetic storms, *Terrest. Mang. Atm. Electr.*, *51*, 353–380.

Heppner, J. (1972), The Harang discontinuity in auroral belt ionospheric currents, *Geophys. Publ.*, *29*, 105.

Hirono, M. (1950), On the influence of the Hall current to the electrical conductivity of the ionosphere II, *J. Geomagn. Geoelectr.*, *2*, 113–120.

Itonaga, M., and T. Kitamura (1988), Effect of non-uniform ionospheric conductivity distribution on PC3-5 magnetic pulsations: Alfvén wave incidence, *J. Geomagn. Geoelectr.*, *40*, 1413–1435.

Karlsson, T., and G. T. Marklund (1998), Simulations of effects of small- scale auroral current closure in the return current region, *Phys. Space Plasmas*, *15*, 401–406.

Karlsson, T., G. Marklund, N. Brenning, and I. Axnäs (2005), On enhanced aurora and low-altitude parallel electric fields, *Phys. Scr.*, *72*, 419–422; doi:10.1238/Physica.Regular.072a00419.

Lyon, J. G., J. A. Fedder, and C. M. Mobarry (2004), The Lyon-Fedder Mobarry (LFM) global MHD magnetospheric simulation code, *J. Atmos. Sol. Terr. Phys.*, *66*, 1333–1350, doi:10.1016/j.jastp.2004.03.020.

Lysak, R. L. (1990), Electrodynamic coupling of the magnetosphere and ionosphere, *Space Sci. Rev.*, *52*, 33–87.

Mallinckrodt, A. J., and C. W. Carlson (1978), Relations between transverse electric fields and field-aligned currents, *J. Geophys. Res.*, *83*, 1426.

Marghitu, O. (2012) Auroral Arc Electrodynamics: Review and Outlook, in *Auroral Phenomenology and Magnetospheric Processes: Earth And Other Planets*, edited by A. Keiling, E. Donovan, F. Bagenal and T. Karlsson), American Geophysical Union, Washington, D. C. doi: 10.1029/2011GM001189

Marghitu, O., C. Bunescu, T. Karlsson, B. Klecker, and H. C. Stenbaek-Nielsen (2011), On the divergence of the auroral electrojets, *J. Geophys. Res.*, *116*, A00K17; doi:10.1029/2011JA016789.

Opgenoorth, H. J., I. Häggström, P. J. S. Williams, and G. O. L. Jones (1990), Regions of strongly enhanced perpendicular electric fields adjacent to auroral arcs, *J. Atmos. Terr. Phys.*, *52*, 449–458; doi:10.1016/ 0021-9169(90)90044-N.

Pellinen, R. J., W. Baumjohann, W. J. Heikkila, V. A. Sergeev, A. G. Ianin, A. O. Melnikov, and G. Marklund (1982), Event study on pre-substorm phases and their relation to the energy coupling between solar wind and magnetosphere, *Planet. Space Sci.*, *30*, 371–388; doi:10.1016/0032- 0633(82)90043-5.

Raeder, J., R.L. McPherron, L.A. Frank, S.Kokubun, G. Lu, T. Mukai, W.R.Paterson, J.B. Sigwarth, H.J. Singer, and J.A. Slavin, (2001) Global simulation of the Geospace Environment Modeling substorm challenge event, *J. Geophys. Res.*, *106*, 381–395.

Richmond, A. D (2010), On the ionospheric application of Poynting's theorem, *J. Geophys. Res.*, *115*, A10311; doi:10.1029/2010JA015768.

Richmond, A. D., and J. P. Thayer (2000), Ionospheric electrodynamics: A tutorial, in *Magnetospheric Current Systems*, edited by S. Ohtani et al., 131–146, *Geophysical. Monograph 118*, American Geophysical Union, Washington, DC; doi:10.1029/JA089iA10p08941.

Russell, A. J. B., A. N. Wright, and A. W. Hood (2010), Self-consistent ionospheric plasma density modifications by field-aligned currents: Steady state solutions, *J. Geophys. Res.*, *115*, A04216; doi:10.1029/ 2009JA014836.

Sato, T. (1978), A theory of quiet auroral arc, *J. Geophys. Res.*, *83*(A3), 1042–1048; doi:10.1029/JA083iA03p01042.

Scholer, M. (1970), On the motion of artificial ion clouds in the magnetosphere, *Planet. Space Sci.*, *18*, 977–1004; doi:10.1016/0032-0633(70) 90101-7.

Strangeway, R. J. (2012), The equivalence of Joule dissipation and frictional heating in the collisional ionosphere, *J. Geophys. Res.*, *117*, A02310; doi:10.1029/2011JA017302.

Streltsov, A. V. (2007), Narrowing of the discrete auroral arc by the ionosphere, *J. Geophys. Res.*, *112*, A10218; doi:10.1029/2007JA012402.

Tanaka, T. (1995), Generation mechanisms for magnetosphere-ionosphere current systems deduced from a three-dimensional MHD simulation of the solar wind-magnetosphere-ionosphere coupling processes, *J. Geophys. Res.*, *100*, 12,057–12,074; doi:10.1029/95JA00419.

Untiedt, J. (1967), A model of the equatorial electrojet involving meridional currents, *J. Geophys., Res.*, *72*, 5799–5810.

Vanhamäki, H. (2011), Inductive ionospheric solver for magnetospheric MHD simulations, *Ann. Geophys.*, *29*, 97–108.

Vanhamäki, H., A. Yoshikawa, O. Amm, and R. Fujii (2012), Ionospheric Joule heating and Poynting flux in quasi-static approximation, *J. Geophys. Res.*, *117*, A08327; doi:10.1029/2012JA017841.

Vanhamäki, H., O. Amm, and A. Viljanen (2006), New method for solving inductive electric fields in the nonuniform conducting ionosphere, *Ann. Geophys.*, *24*, 2573–2582; doi:10.5194/angeo-24-2573-2006.

Vanhamäki, H., O. Amm, and A. Viljanen (2007), Role of inductive electric fields and currents in dynamical ionospheric situations, *Ann. Geophys.*, *25*, 437–455.

Vogt, J. (2002), Alfvén wave coupling in the auroral current circuit, *Surv. Geophys.*, *23*, 335–377.

Walén, C. (1944), On the theory of sunspots, *Arc. Mat. Astron. Fys.*, *31*(3), 1–87.

Williams, P. J. S., C. F. del Pozo, I. Hiscock, and R. Fallows (1998), Velocity of auroral arcs drifting equatorward from the polar cap, *Ann. Geophys.*, *16*, 1322–1331; doi:10.1007/s00585-998-1322-0.

Yasuhara, F., Y. Kamide, and J. F. Vickrey (1985), On the efficiency of the Cowling mechanism in the auroral electrojet, *Geophys. Res. Lett.*, *12*, 389–392.

Yoshikawa, A. (2002a), How does the ionospheric rotational Hall current absorb the increasing energy from the field-aligned current system?, *Geophys. Res. Lett.*, *29*(7), 1133; doi:10.1029/2001GL014125.

Yoshikawa, A. (2002b), Excitation of a Hall-current generator by field- aligned current closure, via an ionospheric, divergent Hall-current, during the transient phase of magnetosphere–ionosphere coupling, *J. Geophys. Res.*, *107*, 1445; doi:10.1029/2001JA009170.

Yoshikawa, A., and M. Itonaga (1996), Reflection of shear Alfvén waves at the ionosphere and the divergent Hall current, *Geophys. Res. Lett.*, *23*, 101–104; doi:10.1029/95GL03580.

Yoshikawa, A., and M. Itonaga (2000), The nature of reflection and mode conversion of MHD waves in the inductive ionosphere: Multistep mode conversion between divergent and rotational electric fields, *J. Geophys. Res.*, *105*(A5), 10,565–10,584; doi:10.1029/1999JA000159.

Yoshikawa, A., A. Nakamizo, O. Amm, H. Vanhamäki, R. Fujii, Y.-M. Tanaka, T. Uozumi, K. Yumoto, and S. Ohtani (2011b), Self-consistent formulation for the evolution of ionospheric conductances at the ionospheric E region within the M-I coupling scheme, *J. Geophys. Res.*, *116*, A09223; doi:10.1029/2011JA016449.

Yoshikawa, A., H. Nakata, A. Nakamizo, T. Uozumi, M. Itonaga, S. Fujita, K. Yumoto, and T. Tanaka (2010), Alfvénic-coupling algorithm for global and dynamical magnetosphere-ionosphere coupled system, *J. Geophys. Res.*, *115*, A04211; doi:10.1029/2009JA014924.

Yoshikawa, A., M. Itonaga, S. Fujita, H. Nakata, and K. Yumoto (1999), Eigenmode analysis of field line oscillations interacting with the ionosphere-atmosphere-solid earth electromagnetic coupled system, *J. Geophys. Res.*, *104*(A12), 28437–28457; doi:10.1029/1999JA900287.

Yoshikawa, A., O. Amm, H. Vanhamäki, A. Nakamizo, and R. Fujii (2013a), Theory of Cowling channel formation by reflection of shear Alfvén waves from the auroral ionosphere, *J. Geophys. Res. Space Phys.*, *118*, 6416–6425; doi:10.1002/jgra.50514.

Yoshikawa, A., O. Amm, H. Vanhamäki, and R. Fujii (2013b), Illustration of Cowling channel coupling to the shear Alfvén wave, *J. Geophys. Res. Space Phys.*, *118*; doi:10.1002/jgra.50513.

Zhu, L., J. J. Sojka, R. W. Schunk, and D. J. Crain (1993), Influence of horizontal inhomogeneity in the ionosphere on the reflection of Alfvén waves, *Geophys. Res. Lett.*, *20*(4), 313–316; doi:10.1029/93GL00079.

26

Ionospheric Currents at Mars and Their Electrodynamic Effects

Matthew Fillingim

ABSTRACT

Mars has a complex magnetic topology which can lead to a variety of electrodynamic plasma processes in the ionosphere. Several previous authors have attempted to quantitatively estimate the magnitude of currents in the ionospheric dynamo region and their magnetic perturbations to the ambient magnetic field. Static currents driven by pressure gradients have been estimated to be ~10^{-8} A m^{-2} while dynamic, neutral wind driven currents have been estimated to be on the order of one to a several 10^{-6} A m^{-2}. To date, there have been very few quantitative measurements of currents in the ionosphere of Mars. Although flux ropes have been observed in the ionosphere by many authors, there are few reports of the currents in the flux ropes. Those reports that do exist suggest that the average currents in flux ropes are ~10^{-6} A m^{-2}. Also, in regions of magnetic cusps, vertical field-aligned currents with magnitudes of ~10^{-6} A m^{-2} have been reported in general agreement with previous predictions. We outline a method for quantitatively calculating ionospheric currents given appropriate, somewhat restrictive, assumptions. We find ionospheric currents on the order of 10^{-6} A m^{-2}, in line with model predictions.

26.1. INTRODUCTION

Interactions of solar plasma and magnetic fields with plasma and magnetic fields of planetary origin create currents in near-planetary space. The variety of currents that result from these interactions depend upon, among other things, the driving forces and current carriers. A recent review by *Baumjohann et al.* [2010] summarized the main current systems that arise in and around the planetary objects in the solar system. Currents occurring at various altitudes can couple different plasma regions electrodynamically. Currents in near-planetary space at high altitude can close through resistive ionospheres coupling different altitude regions. Therefore, ionospheric currents provide a conduit for mass, energy, and momentum transfer from the near-planetary space environment to planetary atmospheres. One example of such a transfer is Joule heating of the neutral atmosphere as a result of dissipative ionospheric currents. This review focuses on currents in the ionosphere of Mars, both model predictions and observations, and considers some of the possible electrodynamic effects of these currents.

Mars lacks a global magnetic field. The interplanetary magnetic field (IMF) drapes over a large portion of the dayside of the planet. Currents induced by this draping create an induced magnetosphere. However, regions of significant crustal remanent magnetization also exist at Mars, primarily in the southern hemisphere. Figure 26.1, from *Acuña et al.* [2001], shows a map of the radial component of the magnetic field at an altitude of 400 km as measured by Mars Global Surveyor (MGS). In the region between 150° and 210° longitude and poleward of –30° latitude, there are several "stripes" where the radial component changes sign over a short latitudinal distance. These are magnetic arcade structures, magnetic loops extending from the surface to well above 400 km altitude, with lengths

Space Sciences Laboratory, University of California, Berkeley, California, USA

Electric Currents in Geospace and Beyond, Geophysical Monograph 235, First Edition.
Edited by Andreas Keiling, Octav Marghitu, and Michael Wheatland.

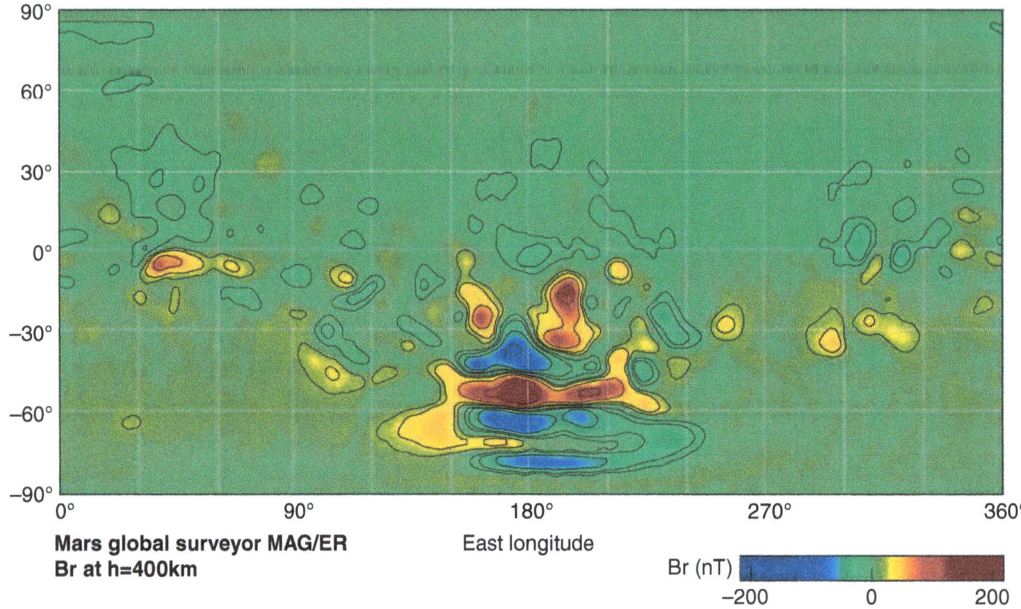

Figure 26.1 Map of the magnetic field of Mars at the MGS mapping altitude of 400 (±20) km. The radial component of the magnetic field (in planetocentric coordinates) associated with crustal sources is illustrated using the color scale shown. From *Acuña et al.* [2001].

approaching 2000 km and widths of approximately one half to one third of their length. The presence of these localized magnetic fields (along with other small scale crustal magnetization seen in Fig. 26.1) creates a very complex magnetic topology in the southern hemisphere.

Brain et al. [2007], using electron loss cones (or lack thereof) as tracers of magnetic field topology, showed that regions of strong radial crustal field are typically open, allowing solar wind or magnetosheath electrons access to the atmosphere. They also noted that the topology was time variable and likely influenced by external conditions. It was in this same region (containing strong radial crustal fields) where auroral emission was first observed at Mars [*Bertaux et al.*, 2005]. Additionally, accelerated "auroral-like" electrons and ions have been observed collocated with strong radial crustal fields [*Brain et al.*, 2006; *Lundin et al.*, 2006a]. Therefore, taken together, the data suggest that between the magnetic arcades are magnetic "cusps," which are open to external sources of plasma. Acceleration processes on these open field lines can accelerate the plasma to sufficient energies to produce auroral emission as the plasma interacts with the atmosphere analogous to processes occurring in Earth's auroral region.

The strength and dynamics of ionospheric currents are governed in large part by the relative importance of charged particle collisions with neutrals (primarily determined by the number density of the neutrals) and charge particle gyro motion in the presence of magnetic fields (primarily determined by the strength of the magnetic field, B, since the gyrofrequncy, ω, is given by qB/m). Since currents are the result of differential motion of positive and negative charges (primarily ionospheric ions and electrons), where this difference in motion is maximized is where the strongest ionospheric currents may be expected. This difference is maximized in the ionospheric dynamo region. In the altitude range of the dynamo region, the ion gyrofrequency, ω_i, is less than the ion-neutral collision frequency, ν_{in}, ($\omega_i/\nu_{in} < 1$) while the electron gyrofrequency, ω_e, is greater than the electron-neutral collision frequency, ν_{en}, ($\omega_e/\nu_{en} > 1$); that is, the ions are collisional while the electrons are magnetized and their motions are decoupled. The dynamo region is bounded at high altitude by the condition that $\omega_i/\nu_{in} = 1$ and at low altitude by $\omega_e/\nu_{en} = 1$. At Mars, this region typically bounds the peak in ionospheric density. As shown by, for example, *Withers et al.* [2005] and *Opgenoorth et al.* [2010], the dynamo region can extend from ~120 km to ~200 km on the dayside while the dayside ionospheric peak is typically observed between 120 km and 140 km (e.g., *Hantsch and Bauer*, 1990; *Zou et al.*, 2005). Since the current density can be defined as $\mathbf{j} = nq\,(\mathbf{v}_i - \mathbf{v}_e)$, the strongest ionospheric currents would be expected to exist where the maximum in plasma density (the ionospheric peak) overlaps with the region of ion-electron velocity decoupling (the dynamo region).

The ionospheric magnetic field at Earth is about 100 to 1000 stronger than it is at Mars. This stronger magnetic field yields a larger gyrofrequency and lowers the altitude of the dynamo region. The terrestrial dynamo region as defined above (between the altitudes where $\omega_i/\nu_{in} = 1$ and $\omega_e/\nu_{en} = 1$) extends from about 60 or 70 km to ~125 km encompassing most of the D and E layers of the terrestrial ionosphere. At the bottom of the terrestrial dynamo region, the plasma density can be quite low. The peak ionospheric density occurs in the F region, typically a hundred km or more above the top of the dynamo region. However, there is significant ionization in the E region (90–150 km), which overlaps the higher altitude portion of the dynamo region where the largest terrestrial ionospheric currents are predominantly found.

One of the most persistent ionospheric currents present in the terrestrial E region ionosphere at low-latitudes and midlatitudes is the so called "solar quiet" (Sq) current system. The Sq current system is a result of solar heating, which drives neutral atmospheric winds. In the dynamo region (E region), the collisional ions are coupled to the neutral winds while the magnetized electrons drift primarily perpendicular to the neutral winds resulting in the generation of ionospheric currents. Typical magnitudes of the current densities are on the order of 10^{-6} A m^{-2}. For additional details on the variability of the terrestrial Sq current system, see the recent review by *Yamazaki and Maute* [2017] and references therein. In addition, magnetospheric processes drive E region currents in the high-latitude (auroral) ionosphere. Magnetospheric plasma pressure gradients, magnetospheric convection, and auroral acceleration processes drive field-aligned currents along vertical magnetic fields. These parallel currents close through horizontal currents in the E region ionosphere.

Similarly, ionospheric currents at Mars can be driven by a variety of both external (electric fields) and internal (neutral winds) forcing. One may expect Sq-type currents in the dayside ionosphere driven by solar forcing, particularly in the northern hemisphere away from crustal magnetic fields where the IMF is draped across the ionosphere (a similar geometry to the terrestrial low-latitudes and midlatitudes). Additionally, one may expect auroral-type current systems (vertical field-aligned currents closing through horizontal currents) in the vicinity of magnetic cusps. These types of currents and their interactions, in the presence of a magnetic configuration and topology that changes over short (~100 km) spatial scales, may lead to a rich variety of electrodynamic processes occurring in the ionosphere of Mars.

Since there are very few published observations of ionospheric currents, this review focuses on analytic estimates and simulations of ionospheric currents at Mars in section 26.2. The relatively few observations that have been reported are briefly reviewed in section 26.3. In section 26.4, new quantitative determinations of horizontal ionospheric currents and their electrodynamic effects are shown.

26.2. ANALYTIC ESTIMATES AND SIMULATIONS

Withers et al. [2005] were among the first authors to make quantitative estimates of currents in the ionosphere of Mars and their electrodynamic effects. Assuming a crustal magnetic field strength of 100 nT at ionospheric altitudes, they considered the effect of a 100 m s^{-1} neutral wind. They estimated a height-integrated current of ~30 A km^{-1} in the dynamo region. Given a 70 km dynamo region thickness, this gives an average current density of <~ 10^{-6} A m^{-2} or 1 μA m^{-2}. Also, *Withers et al.* [2005] estimated that the current would in turn induce a magnetic field of ~40 nT. This is a significant fraction of their assumed 100 nT ionospheric field, demonstrating that ionospheric currents at Mars may significantly modify the ambient magnetic field.

Later, *Withers* [2008] calculated ionospheric currents in a non-crustal-field region. Using a more rigorous, self-consistent model, which included pressure gradients and gravitational forces as well as ambipolar diffusion, *Withers* [2008] calculated ionospheric currents of ~10^{-8} A m^{-2} in the horizontal direction (the vertical current was zero in their one-dimensional simulation) for a Mars-like ionosphere. These currents could be considered "background" ionospheric currents in that they were supported only by pressure gradients, gravity, and ambipolar diffusion in the vertical direction; no driving neutral winds were assumed as was done by *Withers et al.* [2005].

Using Earth's auroral region as an analogy, *Dubinin et al.* [2008] computed field-aligned currents in the vicinity of magnetic cusps using a Knight-type relation [*Knight*, 1973]. Using values appropriate for Mars, electron densities and temperatures and magnetic field strengths, they estimated field-aligned currents in the range of 0.5–2 μA m^{-2} near the terminator and on the nightside. These values are in the same ranges as the field-aligned currents observed in the terrestrial cusp (cf. *Iijima and Potemra* [1976]). Additionally, they numerically calculated the spatial extent of field-aligned currents and potentials. They found that relatively narrow regions (few 100 km or less) of field-aligned currents could be supported in low conductivity regions (e.g., on the nightside). The width of these currents was comparable to or smaller than the width of regions of strong radial magnetic field, implying that narrow currents sheets could be supported in the cusps under their assumed conditions.

Fillingim et al. [2010] analytically computed horizontal currents in the vicinity of magnetic cusps in the nightside ionosphere. Using Mars Global Surveyor (MGS) observations of downgoing (precipitating) electrons as input to an electron transport model, they calculated the ionospheric density resulting from particle precipitation and its spatial variability. During the time of observation, MGS was moving southward in its polar orbit from about −10° to −70° latitude along the 210° east longitude meridian at 02 local time at a nearly constant altitude of 400 km. Figure 26.2a shows the observed electron energy spectrogram. Strong radial magnetic fields (cusps) and large fluxes of energetic electrons were encountered at about 02:24 and 02:28 corresponding to spacecraft latitudes of about −52° and −62°, respectively (compare with Fig. 26.1). Figure 26.2b shows the modeled ionospheric density resulting from the incident electron flux. Since the largest energy flux of electrons occurs along open field lines in cusps, the largest nighttime ionospheric densities were calculated to be present in cusps. The black curves bound the altitude range of the ionospheric dynamo region. Variations in the height of the dynamo region are due to variations in the strength of the crustal magnetic field.

From the ion and electron equations of motion, *Fillingim et al.* [2010] computed the ionospheric currents due to different forces acting on the plasma. They first considered the horizontal plasma pressure gradient force resulting from the spatial varying plasma density. They computed ionospheric currents of $\sim 10^{-8}$ A m^{-2} in the horizontal direction, similar to that predicted by *Withers* [2008]. However, while *Withers* [2008] considered vertical pressure gradients (along with gravity and ambipolar diffusion), the currents computed by *Fillingim et al.* [2010] were in the horizontal (latitudinal) direction.

Fillingim et al. [2010] also considered the currents driven by horizontal neutral winds in the vicinity of magnetic cusps in the presence of nonuniform plasma density. They considered both meridional wind and zonal winds. For simplicity, they assumed the winds were uniform in latitude and longitude near the MGS trajectory and constant with altitude at 100 m/s. These winds are consistent with the postmidnight, southern hemisphere thermospheric winds computed by *Bougher et al.* [1999; 2000]; thermospheric winds are primarily meridional at equinox and zonal at the solstices in this region. As expected from the above discussion, the largest currents were calculated to be in the dynamo region where the ionospheric densities were the largest, that is, in magnetic cusps. Their maximum calculated current densities were $\sim 1\,\mu$A m^{-2}, comparable to the estimates of wind-driven currents given by *Withers et al.* [2005]. *Fillingim et al.* [2010] found that currents in the direction of the neutral wind were produced primarily by ion drag (Fig. 26.2c). Horizontal currents perpendicular to both the neutral wind direction and the vertical magnetic field were produced by drifting magnetized particles (Fig. 26.2d). At adjacent magnetic cusps on opposite sides of magnetic arcades, the direction of the Hall-like current reversed as the radial component of the magnetic field changed sign (from vertically upward to vertically downward). Panels (c) and (d) of Figure 26.2 illustrate the currents resulting from calculations assuming a uniform northward (+ X) neutral wind in the northward (+ X) direction and the westward (+ Y) direction, respectively.

Fillingim et al. [2010] noted that gradients in the plasma density, as well as variations in the strength of the magnetic field (through the gyrofrequency), led to horizontal gradients in the conductivity and horizontal gradients in the horizontal currents. *Fillingim et al.* [2012] considered the possible electrodynamic effects of such gradients in the horizontal currents. Under the assumption of no field-aligned currents (*Fillingim et al.* [2012] acknowledged that this may be an unreasonable assumption, but it was presented as a limiting case, which also simplified the analytic estimates), the horizontal current must be continuous across conductivity gradients in steady state. Otherwise, charges will accumulate at the gradients of the current, and polarization electric fields will arise to attempt to make the horizontal current continuous across conductivity gradients (as shown schematically in Fig. 26.2c). Such polarization electric fields will drive secondary Hall currents perpendicular to the conductivity gradients (i.e., along the axes of the cusps, as shown schematically in Fig. 26.2d). *Fillingim et al.* [2012] noted that if the driving neutral winds were in the north-south direction, perpendicular to the axes of the cusps, the secondary Hall currents act in the same direction as the initial Hall-like currents, enhancing them and creating "electrojets" with maximum magnitudes of several μA m^{-2}. Figure 26.2e shows the secondary Hall currents in the + Y direction, and Figure 26.2f shows the total current (initial plus secondary) in the + Y direction. (The total current in the + X direction is small and constant, i.e., continuous.) If the driving winds were in the east-west direction, along the axes of the cusps, the secondary currents were opposite to the initial currents resulting in weaker overall currents. *Boström* [1964] came to a similar conclusion when considering the source of Earth's auroral electrojet. (As an aside, unpublished work by the author, presented at the EPSC-DPS Joint Meeting in 2011, suggests that in the other limiting case, where the divergence of the horizontal current is equal to a field-aligned current, field-aligned current densities of a few tenths of μA m^{-2} may be achieved.)

Finally, *Fillingim et al.* [2012] calculated the magnetic perturbation due to their proposed electrojets. The strength and direction of the magnetic perturbations are shown in Figure 26.2f. In the ionosphere, the magnetic perturbation was predicted to be ~50 nT, comparable to the estimates made by *Withers et al.* [2005]; however, the ambient magnetic field used by *Fillingim et al.* [2012] came from the model of *Cain et al.* [2003], which gave a field magnitude of ~500 nT in the region of interest for a perturbation of about 10% rather than 40% as quoted by

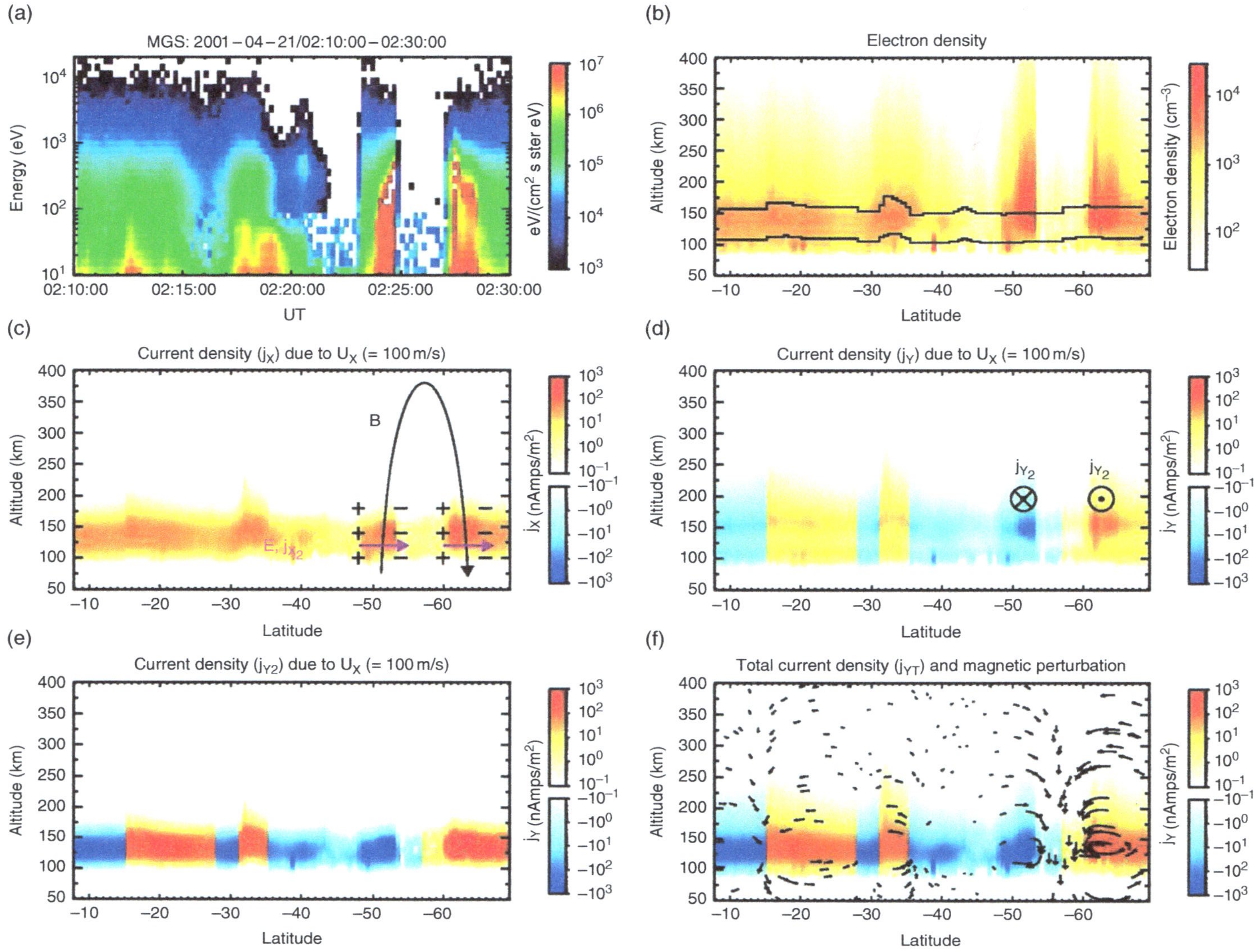

Figure 26.2 (a) Electron energy spectrogram observed by MGS at an altitude of 400 km. During the time of observation, MGS was at an altitude of 400 km and moving southward from −10° to −70° latitude along the 210° east longitude meridian at 02 local time. (b) Calculated electron density as a function of altitude and spacecraft latitude resulting from the observed downgoing electrons. The black curves denote the boundaries of the ionospheric dynamo region. (c) Calculated northward (+ X) current due to a uniform northward neutral wind of 100 m/s. Positive currents are to the north (left). The black curve represents an arcade magnetic field line, the plus and minus signs indicate where charges accumulate, and the purple arrows represent the direction of polarization electric fields and secondary Pedersen currents. (d) Calculated westward (+ Y) current due to a uniform northward neutral wind of 100 m/s. Positive currents are to the west (out of the page). The symbols represent the direction of the secondary Hall currents. (e) Calculated secondary westward current due to polarization electric fields acting to make the northward current continuous. (f) Total (initial plus secondary) westward current and the direction of the associated magnetic perturbations (black curves). From *Fillingim et al.* [2012].

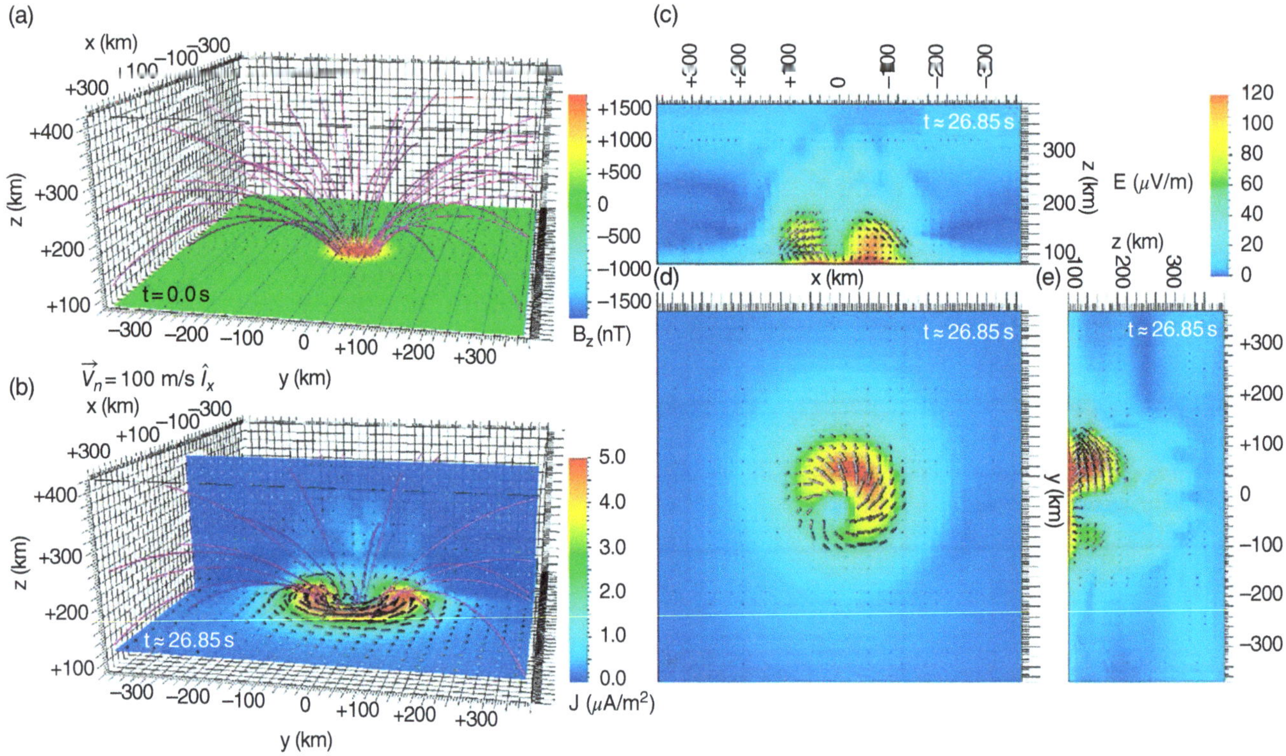

Figure 26.3 Results for a magnetic cusp created by a single dipole placed 20 km below the Martian surface. (a) Magnitude of the vertical magnetic field B_z at t=0 s (color map) and magnetic field lines (solid magenta lines). (b) Magnitude (color map) and direction (solid black arrows) of the current density in the planes X=0 km and Z=112 km. The dynamo current is confined in the altitude range 110 $\lesssim z \lesssim$ 150 km. Magnitude (color map) and projected direction (solid black arrows) of the electric field E in three cut planes: (c) at Y=0 km, (d) at Z=112 km, in the dynamo region, and (e) at X=0 km. From *Riousset et al.* [2014].

Withers et al. [2005]. The magnetic perturbation measured on the surface of Mars was predicted to be ~1% of the ambient field in the vicinity of magnetic cusps. Interestingly, this value is comparable to that measured at Earth during moderate auroral activity (perturbations of ~500 nT in an ambient field of ~50,000 nT).

Riousset et al. [2013; 2014] were the first to investigate the three-dimensional behavior of ionospheric currents using a self-consistent, multifluid magnetohydrodynamic model. Initially, *Riousset et al.* [2013] considered wind-driven ionospheric currents in the presence of a uniform background ionospheric magnetic field away from crustal magnetic sources. They also assumed a constant (in time), uniform (in altitude and horizontal extent) wind field in the +X direction similar to that considered by *Fillingim et al.* [2010; 2012]. In their simulations, the wind-driven currents maximized in the dynamo region; the maximum magnitude of the current density was between 0.01 and 0.1 μA m^{-2} depending upon the strength of the background ionospheric magnetic field. A stronger magnetic field in the ionosphere lowered the altitude of the dynamo region so that it overlapped more with the peak of the ionospheric density leading to an increase in the current density. Interestingly, these currents were one to two orders of magnitude smaller than those estimated by *Withers et al.* [2005] and *Fillingim et al.* [2010]. These modeled currents produced relatively weak perturbations to the ambient magnetic field. Currents in the presence of a weak background field (20 nT) led to a 5% to 10% perturbation while currents in a strong background field (2000 nT) led to a perturbation of only 0.2%; the increase in the current is not proportional to the increase in the ionospheric magnetic field.

Later, *Riousset et al.* [2014] simulated wind-driven currents in the presence of strongly inhomogeneous ionospheric magnetic fields representing crustal magnetic field sources. Two configurations were considered: an isolated magnetic cusp, represented by a single vertical dipole buried 20 km below the surface (shown in Fig. 26.3a), and a magnetic arcade structure, represented by nine buried dipoles (three vertically upward, three vertically downward, and three vertically upward with the rows separated by 250 km, shown in Fig. 26.4a).

In the isolated cusp geometry, a generally counterclockwise current around the axis of the dipole with a

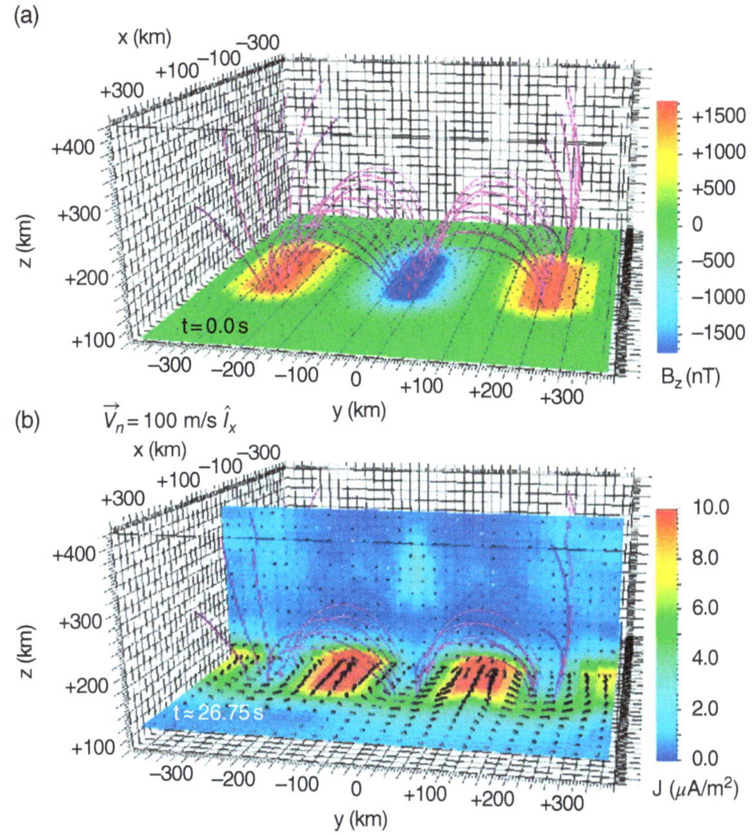

Figure 26.4 Results for magnetic arcades created by nine dipoles. (a) Magnitude of the vertical magnetic field B_z at t = 0 s (color map) and magnetic field lines (solid magenta lines). (b) Magnitude (color map) and direction (solid black arrows) of the current density in the planes X = 0 km and Z = 112 km. From *Riousset et al.* [2014].

maximum magnitude of ~5 μA m^{-2} was found (shown in Fig. 26.3b). They interpreted this counterclockwise current as being predominantly carried by electrons drifting in the presence of electric fields, particularly Hall electric fields, which were derived from a generalized form of Ohm's law. The modeled electric field structure is shown in panels (c) to (e) of Figure 26.3.

For the magnetic arcade geometry, *Riousset et al.* [2014] found strong horizontal currents flowing along the axes of the arcades. The direction of the current reversed, from + X to − X, in adjacent arcades. Along the axes of the arcades, the maximum current density was ~10 μA m^{-2}, about double the maximum current in the isolated cusp case and about two orders of magnitude larger than the maximum current in the strong uniform background magnetic field case. These simulated currents are shown in Figure 26.4b. Though not addressed by the authors, it appears as though the inclusion of inhomogeneous magnetic fields led to a substantial increase in the simulated ionospheric currents: several μA m^{-2} for the inhomogeneous cases versus several hundredths of μA m^{-2} for the uniform cases. By looking at the upward ion flux through the z = 300 km plane, *Riousset et al.* [2014] suggested that the magnetic arcade structure may act to shield the local atmosphere and ionosphere by decreasing the upward ion flux as compared to the isolated cusp case. In this way, the magnetic field geometry may affect the rate of atmospheric loss.

In summary, most of the previous estimates and calculations of the different types of horizontal ionospheric currents are in general agreement with each other even though they have been derived by a variety of methods under different assumptions. *Withers* [2008] and *Fillingim et al.* [2010] considered currents under nondynamic conditions, that is, currents due to pressure gradients (and ambipolar diffusion in the case of *Withers* [2008]). They both estimated horizontal currents on the order of 10^{-8} A m^{-2}. Additionally, *Withers et al.* [2005], *Fillingim et al.* [2010, 2012], and *Riousset et al.* [2014] all considered wind-driven ionospheric currents, and all reported maximum current densities on the order of 10^{-6} A m^{-2}, about 2 orders of magnitude larger than the nondynamic case. The only quantitative calculations of vertical field-aligned currents in the ionosphere have been reported by *Dubinin et al.* [2008] who reported current densities on the order of 10^{-6} A m^{-2}. In the next section, these model calculations will be compared to observations of ionospheric currents at Mars.

26.3. OBSERVATIONS

While a wide variety of magnetic field structures have been observed in the ionosphere of Mars, there have been very few attempts to quantitatively estimate currents present there. Flux ropes have been frequently observed in the ionosphere of Mars [e.g., *Cloutier et al.*, 1999; *Vignes et al.*, 2004; *Briggs et al.*, 2011]. Flux ropes are helical magnetic field structures with strong axial magnetic fields near the center of the structure and increasingly weaker and more azimuthal fields away from the center [*Russell and Elphic*, 1979]. A strong axial current must be present to support these structures. About half of the flux ropes reported by *Vignes et al.* [2004] were at altitudes below 200 km, within the nominal dynamo region of the ionosphere. They also found that such flux ropes typically had widths of a few tens of kilometers up to ~100 km and typical magnetic field amplitudes of just over 10 nT. *Briggs et al.* [2011] found similar sizes and amplitudes for flux ropes observed at MGS's 400 km mapping orbit altitude, well above the ionospheric dynamo region. Although the exact generation mechanism is unknown, it is thought that these ionospheric flux ropes may form as a result of a shear instability between the magnetosheath flow and the upper ionosphere [*Briggs et al.*, 2011].

Briggs et al. [2011] also noted another class of flux ropes that are larger in scale (hundreds of kilometers in size) and amplitude (>100 nT) and are primarily seen downstream of strong crustal magnetic fields in the southern hemisphere [*Brain et al.*, 2010; *Morgan et al.*, 2011; *Beharrell and Wild*, 2012; *Hara et al.*, 2014a,b]. These larger-scale flux ropes are typically observed at higher altitude (~400 km) and are thought to form by reconnection within stretched magnetic fields of crustal origin. It is interesting that these larger-scale flux ropes often contain ionospheric plasma suggesting they form at lower altitude and may be an effective mechanism for planetary plasma escape [*Brain et al.*, 2010; *Morgan et al.*, 2011; *Hara et al.*, 2014a, b]. Only *Beharrell and Wild* [2012] have reported estimates of currents associated with flux ropes. They estimated the total current within flux ropes to be between 10 and 100 kA with larger total currents being associated with larger flux ropes. By simply dividing the total current by the estimated cross-sectional area, the average current densities are found to be ~1 μA m^{-2} within the flux rope.

Analyzing MGS low-altitude (aerobraking) magnetic field data, *Dubinin et al.* [2014a] found strong magnetic field signatures with amplitudes between 50 and 100 nT in the northern hemisphere far from crustal magnetic fields. These magnetic field peaks were seen in the Y MSO direction (toward dusk) at altitudes of 150–230 km, within or just above the ionospheric dynamo region. *Dubinin et al.* [2014a] attributed these magnetic field peaks to strong Hall currents flowing in the ionospheric dynamo region. In their interpretation, downward electron drifts in the presence of vertical gradients in the vertical plasma density (thus vertical gradients in the vertical currents) lead to polarization electric fields, which in turn drive secondary currents. (They noted that a similar situation occurs in the equatorial electrojet in Earth's ionosphere.) *Dubinin et al.* [2014a] estimated that the magnetic perturbation due to these secondary currents to be ~50 nT in an assumed background field of ~20 nT. Based on their assumed ionospheric parameters (conductivity and plasma velocity), this perturbation could be generated by current densities of ~10^{-6} A m^{-2} (or ~1 μA m^{-2}). Their estimated magnetic field perturbations were in general agreement with previous estimates of magnetic field perturbations given by *Withers et al.* [2005] and *Fillingim et al.* [2012]. It is interesting that *Dubinin et al.* [2014b] reported similar observations of low-altitude magnetic field peaks at Venus and offered a similar mechanism for their formation.

Brain et al. [2006], when reporting observations of accelerated "auroral-like" electrons near magnetic cusps, noted that perturbations in the magnetic field implied an upward field-aligned current of ~1 μA m^{-2}. The observed downgoing electrons were calculated to carry about half of the current. Presumably, upgoing ions (not observed by MGS) carried the other half of the current. *Lundin et al.* [2006b] proposed a simple electric circuit analogy suggesting that such field-aligned currents occur in pairs, are driven by solar wind–magnetosphere interactions (the dynamo/generator), and close through the resistive ionosphere (the load), similar to what occurs in Earth's auroral region. To date, no observations of ionospheric currents have been made in the presumed closure region in magnetic cusps.

26.4. NEW OBSERVATIONS

Given the relative dearth of observations of ionospheric currents at Mars, we present a new method for determining ionospheric currents at Mars. The elliptical orbits of MGS during aerobraking allowed the spacecraft to sample vertical profiles of the magnetic field through the ionosphere (under the rather restrictive assumption of stationarity and uniformity along the spacecraft trajectory). Under the appropriate conditions, variations in the magnetic field as a function of altitude can be used to calculate currents from a simplified form of Ampére's law. In the following section, we follow the method of *Ågren et al.* [2011], who determined currents in the ionosphere of Titan from Cassini observations.

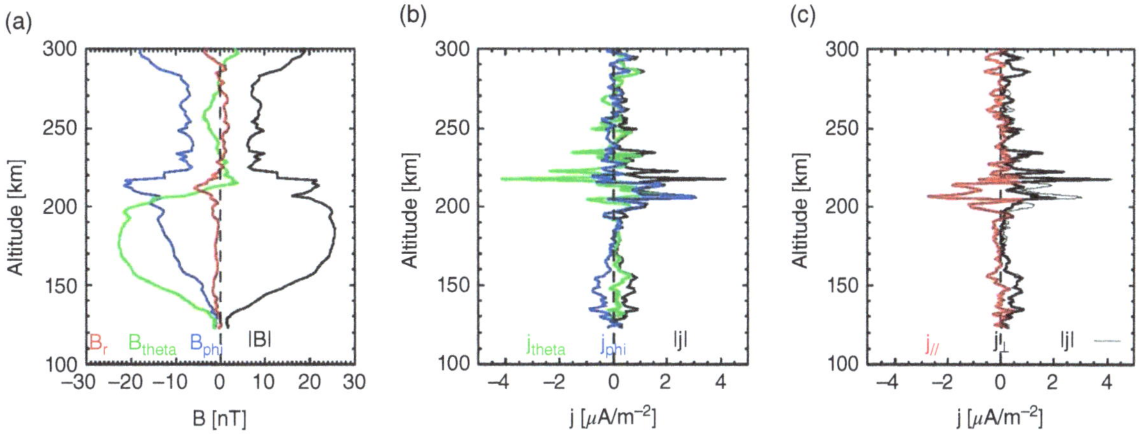

Figure 26.5 (a) Magnetic field profile for the outbound pass on 8 December 1998 observed by MGS in planetocentric coordinates. The data have been resampled to a uniform 1 km vertical resolution. (b) Current density calculated from the magnetic field profile in planetocentric coordinates. (c) Components of the current density parallel (red) and perpendicular (black) to the magnetic field.

An outbound magnetic field profile observed by MGS on 8 December 1998 is shown in Figure 26.5a. During this pass, the spacecraft reached a periapsis of ~120 km altitude over the Tharsis province (an area with the weakest crustal fields on the planet [Lillis *et al.*, 2009]). The data in Figure 26.5a span from periapsis to 300 km altitude, a time interval of about 8 min. As the spacecraft moved up in altitude, it moved from 15° to nearly 50° latitude while longitude and solar zenith angle were nearly constant at 255° and 45°, respectively. Shown are the radial (positive outward, in red), theta (positive southward, in green), and phi (positive eastward, in blue) components of the magnetic field. The magnitude of the magnetic field is shown in black. The data have been resampled so that the vertical resolution is a uniform 1 km. At high altitude, the data are consistent with a draped, compressed solar wind magnetic field: predominantly westward (negative phi), with a magnitude between 20 and 30 nT. Between 280 and 220 km altitude, the magnetic field decreases to less than 10 nT. Starting at 230 km and down to 185 km, the magnetic field increases again to over 25 nT. As the field increases, the direction of the field rotates from predominantly westward to predominantly northward (negative theta). Below 185 km, the magnetic field magnitude steadily decreases until it is nearly zero (< ~ 1 nT) below 130 km.

Throughout this interval, the radial component of the magnetic field, B_r, is relatively small, less than 5 nT. The theta and phi components, B_θ and B_φ, respectively, dominate and range from a few to > 25 nT; that is, the magnetic field is predominantly horizontal. Since B_r is small and since the dominant variations are in B_θ and B_φ, we assume that the gradient in the magnetic field is primarily in the radial direction. We can use this essential initial assumption to derive ionospheric currents from single spacecraft data.

In general, in the absence of time-dependent displacement currents, Ampère's law states

$$\nabla \times \mathbf{B} = \mu_0 \mathbf{j} \tag{26.1}$$

where **B** is the magnetic field, μ_o is the permittivity of free space ($4\pi \times 10^{-7}$ H m^{-1}), and **j** is the current density. In spherical coordinates, the current density is given by

$$\mathbf{j} = \frac{1}{\mu_0}\left[\frac{1}{r\sin\theta}\left(\frac{\partial}{\partial\theta}(B_\varphi \sin\theta) - \frac{\partial B_\theta}{\partial\varphi}\right)\hat{r} + \frac{1}{r}\left(\frac{1}{\sin\theta}\frac{\partial B_r}{\partial\varphi} - \frac{\partial}{\partial r}(rB_\varphi)\right)\hat{\theta} + \frac{1}{r}\left(\frac{\partial}{\partial r}(rB_\theta) - \frac{\partial B_r}{\partial\theta}\right)\hat{\varphi}\right] \tag{26.2}$$

However, since we only have one spacecraft, we must make assumptions about the uniformity of the magnetic field and the direction of magnetic field gradients. If we are able to assume that the principal gradient in the magnetic field is in the radial direction and that gradients in the theta and phi (horizontal) directions are negligible, above equation reduces to

$$\mathbf{j} \approx \frac{1}{\mu_0}\left[\frac{1}{r}\left(-\frac{\partial}{\partial r}(rB_\varphi)\right)\hat{\theta} + \frac{1}{r}\left(\frac{\partial}{\partial r}(rB_\theta)\right)\hat{\varphi}\right] \tag{26.3}$$

with the components of the current only in the horizontal (theta and phi) directions. This assumption is most likely satisfied when the magnitude and variation of radial component of the magnetic field are small compared to the magnitude and variation of the theta and phi components; that is, when the magnetic field is predominantly horizontal. In general, uniform, horizontal magnetic fields are most likely to be encountered in regions away from crustal magnetic fields (as in this example).

Once the theta and phi components of the current are calculated, the parallel and perpendicular components of the current can be determined from knowledge of the magnetic field. The parallel component is given by

$$\mathbf{j}_{\parallel} = \left(\mathbf{j}\cdot\hat{b}\right)\hat{b} \tag{26.4}$$

where $\hat{b}$ is the unit vector along **B**, and the perpendicular component is

$$\mathbf{j}_{\perp} = \mathbf{j} - \mathbf{j}_{\parallel}. \tag{26.5}$$

Figure 26.5b,c shows the currents caclulated from the magnetic field in spherical planetocentric and magnetic coordinates, respectively. Since gradients in the theta and phi directions are neglected, the radial component of the current is zero; currents are only in the theta and phi directions.

At altitudes above 230 km, where the magnetic field varies slowly, currents are relatively small, < ~1 μA m^{-2}, and variable. Between 230 and 200 km, the currents are stronger, up to several μA m^{-2}, but highly variable. Changes in the magnitude and direction of the magnetic field occur over altitude spans of a few kilometers giving rise to strong currents. From 230 to 215 km, the excursions are predominantly northward; there is a signficant eastward component between 215 and 200 km. Between 215 and 200 km, there is also a significant component of the current antiparallel to the magnetic field. In this altitude range, both the horizontal magnetic field and the horizontal current are rotating; the magnetic field is rotating from west to north while the current is rotating from north to east. During this rotation, a component of the current briefly becomes antiparallel to the magnetic field. The currents are extremely weak, ~0.1 μA m^{-2}, from 200 km down to 160 km as the magnitude and direction of the magnetic field change very slowly with altitude. Below 160 km to about 130 km, currents are relatively moderate in strength, ~1 μA m^{-2}, less variable, and primarily westward and perpendicular to the magnetic field. These currents arise from the relatively rapid decrease in the magnitude of the magnetic field. In this lowermost region, below 200 km, the spacecraft is in the ionospheric dynamo region where currents are driven by differential motion between the collisional ions and magnetized electrons.

If we make the common assumption that electric forces dominate over gravitational and pressure gradient forces in the dynamo region (cf. *Schunk and Nagy* [2000], Chapter 5), we can relate the ionospheric electric field to the ionospheric currents through a simplified form of Ohm's law

$$\mathbf{j} = \sigma_{\parallel}\mathbf{E}_{\parallel} + \sigma_P\mathbf{E}_{\perp} - \sigma_H\mathbf{E}_{\perp}\times\hat{b} \tag{26.6}$$

where $\mathbf{E}_{\parallel}$ and $\mathbf{E}_{\perp}$ are the components of the electric field parallel and perpendicular to the magnetic field in the rest frame of the neutrals and $\sigma_{\parallel}$, σ_P, and σ_H are the parallel (or direct), Pedersen, and Hall conductivities. This equation can be solved for the perpendicular and parallel electric fields:

$$\mathbf{E}_{\perp} = \frac{\sigma_P}{\sigma_P^2+\sigma_H^2}\left(\mathbf{j}_{\perp} + \frac{\sigma_H}{\sigma_P}\mathbf{j}_{\perp}\times\hat{b}\right) \tag{26.7}$$

$$\mathbf{E}_{\parallel} = \frac{\mathbf{j}_{\parallel}}{\sigma_{\parallel}}$$

It is clear from these equations that to determine the ionospheric electric field, we must first have knowldege of the ionospheric conductivities. The parallel, Pedersen, and Hall conductivities are given by

$$\sigma_{\parallel} = \frac{q^2 n_e}{m_e \nu_e} + \sum \frac{q^2 n_i}{m_i \nu_i} \tag{26.8}$$

$$\sigma_P = \sigma_{\parallel e}\frac{\nu_e^2}{\left(\omega_e^2+\nu_e^2\right)} + \sum \sigma_{\parallel i}\frac{\nu_i^2}{\left(\omega_i^2+\nu_i^2\right)} \tag{26.9}$$

$$\sigma_H = \sigma_{\parallel e}\frac{\omega_e \nu_e}{\left(\omega_e^2+\nu_e^2\right)} - \sum \sigma_{\parallel i}\frac{\omega_i \nu_i}{\left(\omega_i^2+\nu_i^2\right)} \tag{26.10}$$

where the summation is over the different ion species.

Computation of the conductivities, in turn, requires knowledge of the plasma and neutral composition and density, the ion-neutral and electron-neutral collision frequencies, and the ion and electron gyrofrequencies. In order to calculate the collision frequencies, $\nu_{e,i}$, we need knowledge of the neutral composition and density along with the elastic and inelastic collision frequency coefficients and expressions (given in, for example, *Schunk and Nagy* [2000]). The total electron (or ion) collision frequency, ν_e (or ν_i), is then the sum of the electron (or ion) collision frequency with each neutral species over all neutral species. That is,

$$\nu_e = \sum_n \nu_{en} \text{ and } \nu_i = \sum_n \nu_{in} \tag{26.11}$$

Finally, determination of the electron-neutral collision frequencies, ν_{en}, requires knowledge of the electron temperature, T_e. It should be noted that the electron-ion collision frequencies have been neglected in this brief description but can be included in the more general case.

Since MGS did not measure the neutral or plasma composition and density, we employ a model of the neutral atmosphere and ionosphere. We use neutral densities from the Mars Climate Database version 4.3 [*Millour et al.*, 2008], which comes directly from a large ensemble of simulation runs of the Laboratoire de

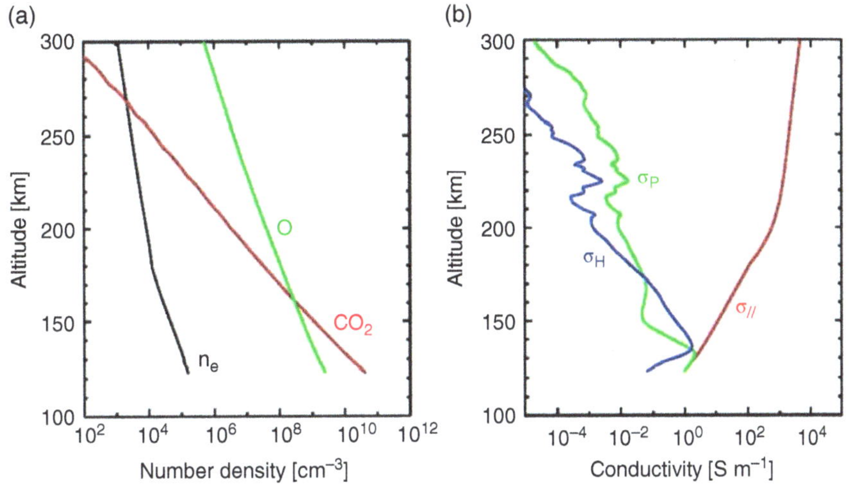

Figure 26.6 (a) Major neutral atmospheric species (CO_2 in red and O in green) from the LMD MGCM taken along the spacecraft trajectory and the computed ionospheric density (black) resulting from photoionization of the neutrals. (b) Calculated ionospheric conductivities. The parallel (or direct) conductivity, $\sigma_{||}$, is in red; the Pedersen conductivity, σ_P, is in green; and the Hall conductivity, σ_H, is in blue.

Météorologie et Dynamique (LMD) Mars Global Circulation Model (MGCM). We extract the neutral densities (only CO_2, and atomic oxygen for simplicity) and neutral temperature along the spacecraft track from periapsis to the upper boundary of the model, which is ~300 km. The neutral densities are shown in Figure 26.6a (red and green curves).

The ionospheric plasma density is computed by calculating the photoionization rate of the neutral atmosphere and assuming photochemical equilibrium. Following *Lillis et al.* [2012], photoionization rates are calculated using standard equations (e.g., *Schunk and Nagy* [2000] equations 9.17 and 9.26) and standard photoabsorption and photoionization cross sections of CO_2 and atomic oxygen (e.g., *Avakyan et al.* [1998]), including photoelectron impact ionization (following the parameterization of *Mendillo et al.* [2011]). The solar spectrum used to compute the photoionization rate is taken from TIMED SEE data at Earth and scaled to the Mars-Sun distance. Since there are no TIMED SEE data from 1998, the average solar spectrum from December 2009, one solar cycle later, is used as input. Electron and ion densities are computed from the total ionization rate using the electron temperature-dependent dissociative recombination rate of O_2^+ [*Mehr and Biondi*, 1969; *Sheehan and St.-Maurice*, 2004] and the dayside electron temperature profiles of *Chen et al.* [1978], implicitly assuming all ions eventually become O_2^+, following *Fillingim et al.* [2007, 2010, 2012]. The resulting ionospheric density is also shown in Figure 26.6a (black curve). The altitude of the ionospheric peak is computed to be ~115 km, just below the MGS periapsis of ~120 km.

Figure 26.6b shows the conductivities determined from the above models and assumptions. As the neutral density drops off rapidly with altitude (more rapidly than the plasma density), the parallel conductivity rapidly increases. Through the collision frequency terms, the parallel conductivity is inversely proportional to the neutral density. The Pedersen and Hall conductivities, in contrast, decrease with altitude. As the collision frequencies become vanishingly small, the relatively stable gyrofrequency terms in the denominator prevent the conductivities from rapidly increasing with altitude. The conductivities become roughly proportional to the product of the plasma density times the neutral density (divided by the magentic field), which similarly decreases with altitude. Additional structure in the Pedersen and Hall conductivities is due to variations in the magnetic field magnitude with altitude (cf. *Opgenoorth et al.* [2010]).

The Pedersen conductivity reaches its maximum value of ~1 S m^{-1} near 130 km and becomes approximately equal to the parallel (direct) conductivity as both the electrons and ions become collision dominated (at the bottom of the dynamo region). The Hall conductivity reaches a similar maximum (~1 S m^{-1}) at a slightly higher altitude (~135 km).The Pedersen and Hall conductivities are comparable to those reported by *Opgenoorth et al.* [2010] for Mars and roughly three to four orders of magnitude larger than typical values found in the terrestrial ionosphere. Also, at Earth the altitude of the peak of the Pedersen conductivity is higher than the altitude of the peak of the Hall condcutivity. We find the opposite at Mars (as did *Opgenoorth et al.* [2010]). *Opgenoorth et al.* [2010] noted the double peaked nature of the Pedersen

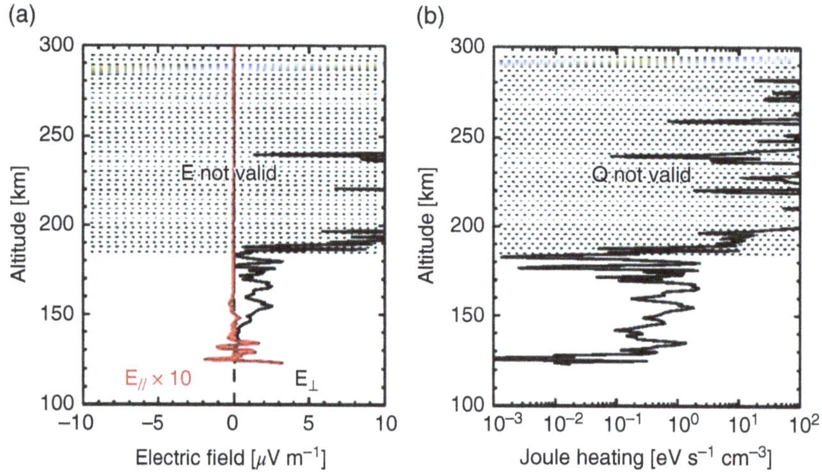

Figure 26.7 (a) The perpendicular (black) and parallel (red) components of the ionosphere electric field computed from the current density and conductivities. The parallel component of the electric field has been multiplied by a factor of 10 for clarity. (b) Joule heating rates estimated from the ionospheric current density and electric field.

conductivity (*Rosenqvist et al.* [2009] noted a similar structure at Titan as well). The higher altitude peak in the Pedersen conductivity around 175 km, above the peak in the Hall conductivity, is analogous to the peak in the Pederson condcutivity seen at Earth. The lower altitude peak in the Pedersen conductivity is not seen at Earth due to a lack of sufficient ionization at lower altitudes.

Between about 175 and 135 km, the Hall conductivity is larger than the Pedersen conductivity. This is also the altitude range of the dynamo region. This suggests that the low-altitude currents (those seen below 160 km) in the dynamo region are Hall currents carried by magnetized electrons.

Figure 26.7a shows the perpendicular (black) and parallel (red) components of the computed ionospheric electric field. The parallel electric field has been multiplied by a factor of 10 for clarity. Since the magnetic field and parallel current are both horizontal, the parallel electric field is likewise horizontal (similar to the terrestrial equitorial ionosphere and in contrast to the terrestrial polar ionosphere). The very large parallel conductivity at altitudes above periapsis requires a very small parallel electric field. At all altitudes, the perpendicular component of the electric field is dominant and reaches a maximum of several of $\mu V\ m^{-1}$ below ~175 km altitude. At altitudes above the dynamo region, above ~175 km, the Pedersen and Hall conductivities become very small and the electric field becomes very large. Such large electric fields are unphysical. The problem lies in our use of an incomplete Ohm's law to compute the electric field. The above version of Ohm's law assumes only collisions relate the current density to the electric field. Above ~175 km, the ions become noncollisional (the electrons are noncollisional above ~135 km), transport becomes important, and our assumed simplified Ohm's law breaks down. As shown by *Withers* [2008], pressure gradients, gravitational forces, and ambipolar diffusion become important in driving and supporting currents at high altitudes.

Currents passing through a conductive medium can dissipate energy in the form of Joule heating but only if the current is parallel to the electric field. Once we have calculated both the current density and the electric field, the Joule heating rate in the upper atmosphere can be estimated from **j·E**. Figure 26.7b shows our estimation of Joule heating. Joule heating rates are ~0.1 to 1 $eV\ s^{-1}\ cm^{-3}$ (or ~10^{-14} to 10^{-13} $J\ s^{-1}\ m^{-3}$) at altitudes below 175 km. Again, at altitudes higher than ~175 km, the electric field is unphysical, leading to unphysical estimates of Joule heating rates.

As mentioned above, *Ågren et al.* [2011] computed currents and electric fields in the ionosphere of Titan using a similar methodology. Under the assumption of principally radial gradients in the magnetic field, they computed ionospheric currents from Cassini measurements. They found currents from a few tens to ~100 $nA\ m^{-2}$, about 10 to 100 times weaker than reported here for Mars. Rather than relying on a modeled atmosphere and ionosphere, *Ågren et al.* [2011] were able to use conductivities from simultaneous particle measurements reported by *Rosenqvist et al.* [2009]. The Pedersen and Hall conductivity maxima were ~10^{-2} and ~10^{-1} $S\ m^{-1}$, respectively. The computed electric fields were a few $\mu V\ m^{-1}$, similar to the values derived here for Mars. It is interesting that the dynamo region and associated currents and electric fields occur at a much higher altitude on Titan due to its extended atmosphere. As shown by *Rosenqvist et al.* [2009], the dynamo region at Titan is between ~1000 km and 1400 km altitude.

26.5. SUMMARY

Mars has a complex magnetic topology, which can lead to a variety of electrodynamic plasma processes in the ionosphere. Several previous authors have attempted to quantitatively estimate the magnitude of currents in the ionospheric dynamo region and their magnetic perturbations to the ambient magnetic field. Static currents driven by pressure gradients have been estimated to be ~10^{-8} A m^{-2} while dynamic, neutral wind-driven currents have been estimated to be on the order of one to a several 10^{-6} A m^{-2}.

To date, there have been very few quantitative measurements of currents in the ionosphere of Mars. Although flux ropes have been observed in the ionosphere by many authors, there are few reports of the currents in the flux ropes. Those reports that do exist suggest that the average currents in flux ropes are ~10^{-6} A m^{-2}. Also, in regions of magnetic cusps, vertical field-aligned currents with magnitudes of ~10^{-6} A m^{-2} have been reported in general agreement with previous predictions. We outline a method for quantitatively calculating horizontal ionospheric currents given appropriate, somewhat restrictive, assumptions. We find ionospheric currents on the order of 10^{-6} A m^{-2}, in line with model predictions. Coincidentally, 10^{-6} A m^{-2} is also of the same order of magnitude for currents typically seen in the terrestrial ionosphere at both low latitudes to midlatitudes (Sq currents) and at high latitude (field-aligned currents).

During the aerobraking phase, MGS passed through the ionosphere roughly a thousand times. Additionally, the Mars Atmosphere and Volatile EvolutioN (MAVEN) mission currently orbiting Mars has passed through the ionosphere during every 4.5 hr orbit since late 2014. Instruments on MAVEN measure the magnetic field and the neutral and plasma composition and densities in situ obviating the need for atmospheric and ionospheric models, which were used here. Future work utilizing both MGS and MAVEN data has the potential to greatly enhance our understanding of currents in the ionosphere of Mars and their electrodynamic effects.

REFERENCES

Acuña, M. H., J. E. P. Connerney, P. Wasilewski, R. P. Lin, D. Mitchell, K. A. Anderson, C. W. Carlson, et al. (2001), Magnetic field of Mars: Summary of results from the aerobraking and mapping orbits, *J. Geophys. Res.*, *106*, 23403–23417; doi:10.1029/2000JE001404.

Ågren, K., D. J. Andrews, S. C. Buchert, A. J. Coates, S. W. H. Cowley, M. K. Dougherty, N. J. T. Edberg, et al. (2011), Detection of currents and associated electric fields in Titan's ionosphere from Cassini data, *J. Geophys. Res.*, *116*, A04313; doi:10.1029/2010JA016100.

Avakyan, S. V., R. N. Ii'In, V. M. Lavrov, and G. N. Ogurtsov (1998), *Collision Processes and Excitation of the Ultraviolet Emission from Planetary Atmospheric Gases: A Handbook of Cross Sections*, Gordon and Breach, London.

Baumjohann, W., M. Blanc, A. Federov, and K.-H. Glassmeier (2010), Current systems in planetary magnetospheres and ionospheres, *Space Sci. Rev.*, *152*, 99–134; doi:10.1007/s11214-010-9629-z.

Beharrell, M. J., and J. A. Wild (2012), Stationary flux ropes at the southern terminator of Mars, *J. Geophys. Res.*, *117*, A12212; doi:10.1029/2012JA017738.

Bertaux, J.-L., F. Leblanc, O. Witasse, E. Quémerais, J. Lilensten, A. S. Stern, B. Sandel, and O. Korablev (2005), Discovery of aurora on Mars, *Nature*, *435*, 790–794; doi:10.1038/nature03603.

Boström, R. (1964), A model of the auroral electrojets, *J. Geophys. Res.*, *69*, 4983–4999; doi:10.1029/JZ069i023p04983.

Bougher, S. W., S. Engel, R. G. Roble, and B. Foster (1999), Comparative terrestrial planet thermospheres, 2. Solar cycle variation of global structure and winds at equinox, *J. Geophys. Res.*, *104*, 16591–16611; doi:10.1029/1998JE001019.

Bougher, S. W., S. Engel, R. G. Roble, and B. Foster (2000), Comparative terrestrial planet thermospheres: 3. Solar cycle variation of global structure and winds at solstices, *J. Geophys. Res.*, *105*, 17669–17692; doi:10.1029/1999JE001232.

Brain, D. A., A. H. Baker, J. Briggs, J. P. Eastwood, J. S. Halekas, and T.-D. Phan (2010), Episodic detachment of Martian crustal magnetic fields leading to bulk atmospheric plasma escape, *Geophys. Res. Lett.*, *37*, L14108; doi:10.1029/2010GL043916.

Brain, D. A., J. S. Halekas, L. M. Peticolas, R. P. Lin, J. G. Luhmann, D. L. Mitchell, G. T. Delory, S. W. Bougher, M. H. Acuña, and H. Rème (2006), On the origin of aurorae on Mars, *Geophys. Res. Lett.*, *33*, L01201; doi:10.1029/2005GL024782.

Brain, D. A., R. J. Lillis, D. L. Mitchell, J. S. Halekas, and R. P. Lin (2007), Electron pitch angle distributions as indicators of magnetic field topology near Mars, *J. Geophys. Res.*, *112*, A09201; doi:10.1029/2007JA012435.

Briggs, J. A., D. A. Brain, M. L. Cartwright, J. P. Eastwood, and J. S. Halekas (2011), A statistical study of flux ropes in the Martian magnetosphere, *Planet. Space Sci.*, *59*, 1498–1505; doi:10/1016.j.pss.2011.06.010.

Cain, J. C., B. B. Ferguson, and D. Mozzoni (2003), An n = 90 internal potential function of the Martian crustal magnetic field, *J. Geophys. Res.*, *108*, 5008; doi:10.1029/2000JE001487.

Chen, R. H., T. E. Cravens, and A. F. Nagy (1978), The Martian ionosphere in light of the Viking observations, *J. Geophys. Res.*, *83*, 3871–3876, doi:10.1029/JA083iA08p03871.

Cloutier, P. A., C. C. Law, D. H. Crider, P. W. Walker, Y. Chen, M. H. Acuña, J. E. P. Connemey, et al. (1999), Venus-like interaction of the solar wind with Mars, *Geophys. Res. Lett.*, *26*, 2685–2688; doi:10.1029/1999GL900591.

Dubinin, E., G. Chanteur, M. Fraenz, and J. Woch (2008), Field-aligned currents and parallel electric field potential drops at Mars: Scaling from the Earth's aurora, *Planet. Space Sci.*, *56*, 868–872; doi:10.1016/j.pss.2007.01.019.

Dubinin, E., M. Fraenz, T. L. Zhang, J. Woch, and Y. Wei (2014a), Magnetic fields in the Mars ionosphere of a noncrustal origin: Magnetization features, *Geophys. Res. Lett.*, *41*, 6329–6334; doi:10.1002/2014GL061453.

Dubinin, E., M. Fraenz, T. L. Zhang, J. Woch, and Y. Wei (2014b), Magnetic fields in the Venus ionosphere: Dependence on the IMF direction: Venus express observations, *J. Geophys. Res.*, *119*, 7587–7600; doi:10.1002/2014JA020195.

Fillingim, M. O., L. M. Peticolas, R. J. Lillis, D. A. Brain, J. S. Halekas, D. L. Mitchell, R. P. Lin, D. Lummerzheim, S. W. Bougher, and D. L. Kirchner (2007), Model calculations of electron precipitation induced ionization patches on the nightside of Mars, *Geophys. Res. Lett.*, *34*, L12101; doi:10.1029/2007GL029986.

Fillingim, M. O., L. M. Peticolas, R. J. Lillis, D. A. Brain, J. S. Halekas, D. Lummerzheim, and S. W. Bougher (2010), Localized ionization patches in the nighttime ionosphere of Mars and their electrodynamic consequences, *Icarus*, *206*, 112–119; doi:10.1016/j.icarus.2009.03.005.

Fillingim, M. O., R. J. Lillis, S. L. England, L. M. Peticolas, D. A. Brain, J. S. Halekas, C. Paty, D. Lummerzheim, and S. W. Bougher (2012), On wind-driven electrojets at magnetic cusps in the nightside ionosphere of Mars, *Earth Planets Space*, *64*, 93–103; doi:10.5047/eps.2011.04.010.

Hantsch, M. H., and S. J. Bauer (1990), Solar control of the Mars ionosphere, *Planet. Space Sci.*, *38*, 539–542; doi:10.1016/0032-0633(90)90146-H.

Hara, T., K. Seki, H. Hasegawa, D. A. Brain, K. Matsunaga, and M. H. Saito (2014a), The spatial structure of Martian magnetic flux ropes recovered by the Grad-Shafranov reconstruction technique, *J. Geophys. Res. Space Physics*, *119*, 1262–1271; doi:10.1002/2013JA019414.

Hara, T., K. Seki, H. Hasegawa, D. A. Brain, K. Matsunaga, M. H. Saito, and D. Shiota (2014b), Formation processes of flux ropes downstream from Martian crustal magnetic fields inferred from Grad-Shafranov reconstruction, *J. Geophys. Res.*, *119*, 7947–7962; doi:10.1002/2014JA019943.

Iijima, T., and T. A. Potemra (1976), Field-aligned currents in the dayside cusp observed by Triad, *J. Geophys. Res.*, *81*, 5971–5979; doi:10.1029/JA081i034p05971.

Knight, S. (1973), Parallel electric fields, *Planet. Space Sci.*, *21*, 741–750; doi:10.1016/0032-0633(73)90093-7.

Lillis, R. J., D. A. Brain, G. T. Delory, D. L. Mitchell, J. G. Luhmann, and R. P. Lin (2012), Evidence for superthermal secondary electrons produced by SEP ionization in the Martian atmosphere, *J. Geophys. Res.*, *117*, E03004; doi:10.1029/2011JE003932.

Lillis, R. J., J. Dufek, J. E. Bleacher, and M. Manga (2009), Demagnetization of crust by magmatic intrusion near the Arsia Mons volcano: magnetic and thermal implications for the development of the Tharsis province, Mars, *J. Volcan. Geotherm. Res.*, *185*, 123–138; doi:10.1016/j.jvolgeores.2008.12.007.

Lundin, R., D. Winningham, S. Barabash, R. Frahm, D. Brain, H. Nilsson, M. Holmström, et al. (2006b), Auroral plasma acceleration above martian magnetic anomalies, *Space Sci. Rev.*, *126*, 333–354; doi:10.1007/s11214-006-9086-x.

Lundin, R., D. Winningham, S. Barabash, R. Frahm, M. Holmström, J.-A. Sauvaud, A. Fedorov, et al. (2006a), Plasma acceleration above Martian magnetic anomalies, *Science*, *311*, 980–983; doi: 10.1126/science.1122071.

Mehr, F. J., and M. A. Biondi (1969), Electron temperature dependence of recombination of O_2^+ and N_2^+ ions with electrons, *Phys. Rev.*, *181*, 264–271.

Mendillo, M., A. Lollo, P. Withers, M. Matta, M. Pätzold, and S. Tellmann (2011), Modeling Mars' ionosphere with constraints from same-day observations by Mars Global Surveyor and Mars Express, *J. Geophys. Res.*, *116*, A11303; doi:10.1029/2011JA016865.

Millour, E., F. Forget, F. González-Galindo, A. Spiga, S. Lebonnois, L. Montabone, S. R. Lewis, et al. (2008), The latest (version 4.3) Mars climate database, *LPI Contrib.*, *1447*, 9029.

Morgan, D. D., D. A. Gurnett, F. Akalin, D. A. Brain, J. S. Leisner, F. Duru, R. A. Frahm, and J. D. Winningham (2011), Dual-spacecraft observation of large-scale magnetic flux ropes in the Martian ionosphere, *J. Geophys. Res.*, *116*, A02319; doi:10.1029/2010JA016134.

Opgenoorth, H. J., R. S. Dhillon, L. Rosenqvist, M. Lester, N. J. T. Edberg, S. E. Milan, P. Withers, and D. Brain (2010), Day-side ionospheric conductivities at Mars, *Planet. Space Sci.*, *58*, 1139–1151; doi:10.1016/j.pss.2010.04.004.

Riousset, J. A., C. S. Paty, R. J. Lillis, M. O. Fillingim, S. L. England, P. G. Withers, and J. P. M. Hale (2013), Three-dimensional multifluid modeling of atmospheric electrodynamics in Mars' dynamo region, *J. Geophys. Res.*, *118*, 3647–3659; doi:10.1002/jgra.50328.

Riousset, J. A., C. S. Paty, R. J. Lillis, M. O. Fillingim, S. L. England, P. G. Withers, and J. P. M. Hale (2014), Electrodynamics of the Martian dynamo region near magnetic cusps and loops, *Geophys. Res. Lett.*, *41*, 1119–1125; doi:10.1002/2013GL059130.

Rosenqvist, L., J.-E. Wahlund, K. Ågren, R. Modolo, H. J. Opgenoorth, D. Strobel, I. Müller-Wodarg, P. Garnier, and C. Bertucci (2009), Titan ionospheric conductivities from Cassini measurements, *Planet. Space Sci.*, *57*, 1828–1833; doi:10.1016/j.pss.2009.01.007.

Russell, C. T., and R. C. Elphic (1979), Observation of magnetic flux ropes in the Venus ionosphere, *Nature*, *279*, 616–618; doi:10.1038/279616a0.

Schunk, R. W., and A. F. Nagy (2000), *Ionospheres: Physics, Plasma Physics, and Chemistry*, Cambridge University Press, New York.

Sheehan, C. H., and J. P. St.-Maurice (2004), Dissociative recombination of N_2^+, O_2^+, and NO^+: Rate coefficients for ground state and vibrationally excited ions, *J. Geophys. Res.*, *109*, A03302; doi:10.1029/2003JA010132.

Vignes, D., M. H. Acuña, J. E. P. Connerney, D. H. Crider, H. Réme, and C. Mazelle (2004), *Space Sci. Rev.*, *111*, 223–231; doi:10.1023/B:SPAC.0000032716.21619.f2.

Withers, P. (2008), Theoretical models of ionospheric electrodynamics and plasma transport, *J. Geophys. Res.*, *113*, A07301; doi:10.1029/2007JA012918.

Withers, P., M. J. Mendillo, H. Rishbeth, D. P. Hinson, and J. Arkani-Hamed (2005), Ionospheric characteristics above Martian crustal magnetic anomalies, *Geophys. Res. Lett.*, *32*, L16204; doi:10.1029/2005GL023483.

Yamazaki, Y., and A. Maute (2017), Sq and EEJ:A review on the daily variation of the geomagnetic field caused by ionospheric dynamic currents, *Space Sci. Rev.*, *206*, 299–405; doi:10.1007/s11214-016-0282-z.

Zou, H., J.-S. Wang, and E. Nielsen (2005), Effect of the seasonal variations in the lower atmosphere on the altitude of the ionospheric main peak at Mars, *J. Geophys. Res.*, *110*, A09311; doi:10.1029/2004JA010963.

27

Ionospheric Currents due to Ionosphere-Magnetosphere Coupling at Jupiter and Saturn

G. J. Hunt[1,2], S. W. H. Cowley[1], and J. D. Nichols[1]

ABSTRACT

In this chapter, we review the high-latitude ionospheric currents due to ionosphere-magnetosphere coupling at Jupiter and Saturn. We begin with a discussion on the theoretical framework, which relates the ionospheric meridional Pedersen current and effective Pedersen conductivity with the angular velocity of plasma in a rotation-dominated magnetosphere. We then review modeling results of this framework when applied to Jupiter in terms of the physical parameters projected to the ionosphere. Lastly, we review recent estimations of Saturn's horizontal meridional ionospheric currents; these are determined from magnetic field perturbations, in the azimuthal direction, associated with field-aligned currents as observed by the Cassini spacecraft. These ionospheric currents are then discussed with regard to the physics relating to ionosphere-magnetosphere coupling and the ubiquitous near planetary period oscillations observed in Saturn's magnetosphere.

27.1. INTRODUCTION

As for all the magnetized planets, the principal electric currents that flow in the ionospheres of the gas giants Jupiter and Saturn do so as a result of the transfer of flow-related stress via the magnetic field between the upper planetary atmosphere and the external plasma environment. Two principal sources of flow momentum are generally present. The first is the flow of the solar wind away from the Sun, transferred via reconnection and/or Kelvin-Helmholtz coupling at the magnetopause boundary, which is communicated along field lines into the interior to produce a twin-vortex flow within the magnetosphere and ionosphere [*Dungey*, 1961; *Axford and Hines*, 1961]. The second is the rotation of the planetary body, coupled through the atmosphere to the ionosphere, and thence outward along planetary field lines throughout the magnetosphere [*Nishida*, 1966; *Dungey*, 1967]. Stresses are applied where the associated electric current flows transverse to the magnetic field via the $\mathbf{j} \times \mathbf{B}$ force per unit volume on the plasma, in the magnetosphere at one end of the field lines and in the ionosphere at the other where the force balances that due to ion-neutral collisions, with the current circuit then being closed via force-free field-aligned currents that flow between these regions. In the case of the terrestrial system, the currents are dominated by the solar wind-driven flow, involving principal Region 1 field-aligned currents of a few MA magnitude that flow into the high-latitude ionosphere near the open-closed field boundary on the dawn side and out of the ionosphere on the dusk side, together with Region 2 currents of reversed sense that flow in the equatorward region of the twin-vortex flow system [*Iijima and Potemra*, 1976, 1978]. These currents close within the ionosphere principally via transpolar and auroral zone Pedersen currents, which transmit the solar-wind stress to the ionospheric plasma and upper atmospheric neutral gas (e.g., *Cowley* [2000]). Ionospheric Hall currents of comparable magnitude also flow around the twin-vortex

[1] *Department of Physics and Astronomy, University of Leicester, Leicester, UK*
[2] *Blackett Laboratory, Imperial College London, London, UK*

Electric Currents in Geospace and Beyond, Geophysical Monograph 235, First Edition.
Edited by Andreas Keiling, Octav Marghitu, and Michael Wheatland.

plasma flow streamlines, but close largely within the polar ionosphere, except in regions such as the auroral electrojets where the conductivity varies significantly along the streamlines. The overall terrestrial magnetic flux transport in this principally reconnection-driven flow, the Dungey cycle, lies typically between ~20 and ~100 kWb s^{-1}, equivalent to convection electric field voltages of ~20 to ~100 kV via Faraday's law, depending on the direction of the interplanetary magnetic field [e.g., *Milan et al.*, 2007]. In addition, Earth's atmosphere drives a rotational flow in the system with a total flux transport of ~90 kV, which is the principal flow component in the near-planet quasi-dipolar magnetosphere, while the solar wind-driven Dungey cycle dominates the outer part of the system and magnetotail. The currents associated with the rotational system are negligibly small.

In the case of the gas giants, however, the relative roles of solar-wind driving and planetary rotation are reversed, due to the large internal fields and rapid rotation of these planets. At Jupiter, the flux transport driven by planetary rotation is ~400 MV compared with estimates of ~0.1–1 MV for the Dungey cycle depending on solar-wind conditions, while at Saturn the rotational flux transport is ~10 MV compared with ~20–200 kV for the Dungey cycle [*Jackman et al.*, 2004; *Nichols et al.*, 2006; *Badman and Cowley*, 2007]. A direct comparison of these quantities overemphasizes the point, however, since, like the Earth, the majority of the equatorial rotational transport takes place close to the planet where the magnetic field is strong, while the rotational flux transport in the middle and outer magnetospheres are a few MV in the case of Jupiter and a few hundred kV in the case of Saturn, more comparable with the Dungey cycle estimates. It thus remains possible for solar-wind driving to sometimes make a sensible contribution to flux transport in these outer magnetospheric regimes. These considerations illustrate, however, why discussion of transport processes and related current systems focuses primarily on planetary-driven rotation at the gas giants but on solar-wind driving at Earth.

A second major factor at both gas giants is that the primary plasma source within the rotation-driven flow is not the planetary ionosphere as at Earth, forming the plasmasphere in that case, but moons that orbit deep within the magnetosphere that exhibit volcanism and mass loss as a consequence of external tidal forcing. Specifically, the moon Io at Jupiter, orbiting at a distance of ~6 R_J (compared with a typical subsolar magnetopause at ~50 R_J) is the source of ~1000 kg s^{-1} of sulfur and oxygen plasma [e.g., *Thomas et al.*, 2004], while the moon Enceladus at Saturn, orbiting at a distance of ~4 R_S (compared with a typical subsolar magnetopause at ~25 R_S) is the source of ~100 kg s^{-1} of hydrogen and oxygen (water product) plasma [*Bagenal and Delamere*, 2011]. (R_J is the 1 bar equatorial radius of Jupiter equal to 71,492 km, and R_S the 1 bar equatorial radius of Saturn equal to 60,268 km.) The plasma from these sources forms a near-corotating cool dense torus near the moon's orbit, which is then transported outward by centrifugally driven small-scale flux tube interchange motions, being replaced by hotter tenuous plasma from the outer magnetosphere [e.g., *Krupp et al.*, 2004; *Chen et al.*, 2010]. The moon plasma is eventually lost by plasmoid release down-tail, a process that may be either large-scale and continuous, giving rise to the Vasyliunas flow cycle, or small-scale and time dependent [*Vasyliunas*, 1983; *Kivelson and Southwood*, 2005; *Cowley et al.*, 2015]. Dungey cycle flow may continue simultaneously, though in this case restricted by planetary rotation to a single vortex returning sunward via dawn [*Cowley et al.*, 2003a]. These flows may be distinguished by the dominance of light ions from the solar wind and planetary ionosphere, rather than heavy ions from the moon sources that dominate the near-corotation and Vasyliunas cycle flows [*Badman and Cowley*, 2007; *Thomsen et al.*, 2015].

The principal dynamics of the gas giant magnetospheres thus center on the outward transport of moon-related plasma in a system dominated by planetary rotation [*Hill*, 1979; *Pontius*, 1997]. Although this plasma will near-rigidly rotate with the planet in the inner system near the sources where the planetary field is very strong (though pickup of ions and electrons newly created from the injected neutral gas will produce a small local departure from rigid corotation), outward transport causes the angular velocity to drop by conservation of angular momentum, as $\sim 1/r^2$ if no torques act. Any such departure from corotation, however, which under steady conditions is communicated along field lines down to the ionosphere, will result in a net force being imposed in the sense of planetary rotation by ion-neutral collisions in the Pedersen layer of the ionosphere. The associated torque causes the magnetospheric field lines to bend back out of meridian planes into the "lagging" configuration illustrated in Figure 27.1a, where we view the northern pole of the planet whose anticlockwise rotation is indicated. The field line arrows indicate the field direction north of the equator given the sense of the planetary field at both Jupiter and Saturn (opposite to Earth), with a negative azimuthal (clockwise) field directed around the planet. The sense of the azimuthal field reverses to positive (anticlockwise) south of the equator, with the $\mathbf{j} \times \mathbf{B}$ field tension force associated with the field bend producing a torque on the equatorial plasma in the sense of planetary rotation. The overall current system is shown in a view looking down on the north pole in Figure 27.1b, and in a magnetic meridian in Figure 27.1c, where the black arrowed lines indicate the near-axisymmetric background field. The green lines and symbols indicate the current flow, the blue lines and symbols the resulting perturbation field, while the circled dots and crosses

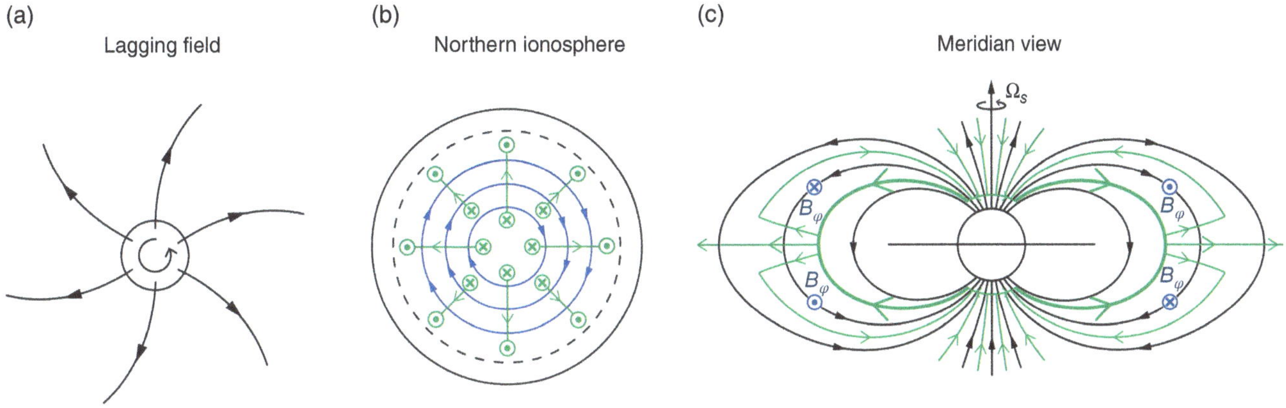

Figure 27.1 The magnetic perturbations and electric currents associated with plasma subcorotation in a rotation-dominated magnetosphere. Panel (a) shows the "lagging" field produced by the magnetosphere-ionosphere coupling currents in a view looking down on the north pole of the planet, with the direction of rotation indicated by the arrow. The curved lines indicate the field lines, with the arrows showing the direction in the northern hemisphere, reversing sense across the equator in the southern hemisphere. Panels (b) and (c) show the currents (green arrowed lines and symbols) and perturbation fields (blue arrowed lines and symbols) in both the northern ionosphere and in a magnetic meridian plane, respectively. Circled crosses and dots indicate vectors pointing into and out of the plane of the diagrams, respectively. The arrowed black lines indicate the near-axisymmetric background magnetospheric field. From Figure 1 of *Hunt et al.* [2014].

indicate vectors pointing out of and into the plane of the diagrams, respectively. Meridional Pedersen currents flow from pole to equator in both hemispheres, the $\mathbf{j}\times\mathbf{B}$-related torque of which balances the ion-neutral collisional torque on the ionospheric plasma, and close in radially outward currents in the equatorial magnetosphere, which impose an equal and opposite torque on the equatorial plasma in the direction of planetary rotation. The angular velocity of the plasma then declines with radial distance more slowly than indicated above. The distributed field-aligned closure currents flow downward into the polar ionosphere at highest latitudes and return upward out of the ionosphere at lower latitudes. The upward currents may then be associated with auroral emissions if the current density is sufficiently intense to require downward acceleration of magnetospheric electrons [e.g., *Cowley and Bunce*, 2001; *Cowley et al.*, 2003b]. Of course, the current system may become more structured than shown here if for any reason the flow or ionospheric conductance becomes more structured. As we will see in sections 27.3 and 27.4, the overall total current flowing, northern and southern hemispheres combined, is ~100 MA for Jupiter [*Khurana*, 2001] and ~15 MA for Saturn [*Hunt et al.*, 2014, 2015]. Ionospheric Hall currents of roughly comparable strength will also be present [*Millward et al.*, 2002; *Galand et al.*, 2011], flowing eastward for subcorotation, but will again close within the ionosphere under conditions of near axisymmetry. As at Earth, however, Hall currents may play a greater role when axisymmetry is broken, such as in the "planetary period oscillation" (PPO) phenomenon at Saturn [e.g., *Smith*, 2014]. Axisymmetry will also be broken by the simultaneous presence of solar wind-driven flows, which cause Region 1 and 2 currents to be superposed, opposite in sense to Earth due to the oppositely directed planetary field [e.g., *Jackman and Cowley*, 2006]. However, no such solar-wind effect has been unequivocally identified to date in current signatures observed by the Cassini spacecraft at Saturn [e.g., *Hunt et al.*, 2016].

In this chapter, we thus focus primarily on the current system associated with the transfer of angular momentum between the planetary atmosphere and the magnetosphere sketched in Figure 27.1, starting with related theory in section 27.2. In sections 27.3 and 27.4, we consider applications to Jupiter and Saturn, while in section 27.5 we review the PPO phenomenon at Saturn, and its associated magnetosphere-ionosphere current system, which has no evident counterpart at Jupiter. We do not cover related phenomena at either Uranus or Neptune, despite the recent observation of weak auroras at the former planet [*Lamy et al.*, 2012], due both to the paucity of in situ data, having been visited to date only by Voyager 2, and to the major differences of spin and magnetic axis orientation that make consideration of plasma flows and coupling currents very different from those of Jupiter and Saturn [e.g., *Cowley*, 2013; *Mejnertsen et al.*, 2016].

27.2. THEORETICAL DEVELOPMENT

27.2.1. Aligned Axial Rotator

Given the physical picture for a rotation-dominated system sketched in Figure 27.1, we now develop theory specifically for an aligned axial rotator, which applies accurately at Saturn where the magnetic and spin axes are aligned to better than ~0.1° [*Burton et al.*, 2010]. In section 27.2.5, we indicate how the theory can also be applied approximately to Jupiter, where the magnetic dipole is tilted to the spin axis through ~9.5° [*Connerney et al.*, 1998]. In the aligned case, the ionospheric parameters may be taken simply to be functions of colatitude from the northern pole, θ_i, including the radial distance of the Pedersen layer of the ionosphere R_i, the planetary poloidal ionospheric field $\mathbf{B}_i$, the height-integrated Pedersen conductivity Σ_P, and the plasma angular velocity ω (constant on each closed flux shell). The planetary angular velocity is given by the fixed value Ω. With regard to the ionospheric radial distance R_i, the 1 bar atmospheric layer is taken to form a spheroid of revolution about the spin axis with the polar radius flattened relative to the equator by ~10% due to the rapid planetary rotation, and the Pedersen layer lying ~500 km above for Jupiter and ~1000 km above for Saturn [*Millward et al.*, 2002; *Galand et al.*, 2011]. The flattening has a significant effect on the strength of the polar ionospheric field, which at Saturn has the value of ~62 μT at the south pole but ~75 μT at the north pole due also to the presence of the axial quadrupole [*Burton et al.*, 2010]. The polar fields at Jupiter are more than an order larger at ~1 mT [*Connerney et al.*, 1998]. The Pedersen conductivity may also vary with colatitude due to particle precipitation from the magnetosphere. Typical Pedersen conductivities of order ~0.1 mho at Jupiter and ~1 mho at Saturn could be increased by up to an order of magnitude by such effects [*Millward et al.*, 2002; *Nichols and Cowley*, 2004; *Galand et al.*, 2011]. Similar comments apply also to the Hall conductivity, centered a little lower in altitude than the Pedersen conductivity, whose height-integrated values are expected to be comparable to the Pedersen conductivity at Saturn [*Galand et al.*, 2011], but generally rather less than the Pedersen conductivity at Jupiter [*Millward et al.*, 2002].

27.2.2. Magnetosphere-Ionosphere Coupling Currents

In terms of these parameters, the height-integrated Pedersen current per radian of azimuth, directed equatorward (for subcorotation) transverse to the ionospheric magnetic field of magnitude B_i, is given from Ohm's law by

$$I_P(\theta_i) = \rho_i \Sigma_P E_i = \Sigma_P \rho_i^2 B_i(\Omega^* - \omega) \qquad (27.1)$$

where $\rho_i = R_i \sin\theta_i$ is the perpendicular distance from the rotation axis, and the meridional convection electric field E_i is that in the rest frame of the atmospheric neutral gas, whose angular velocity is Ω^* [e.g., *Cowley and Bunce*, 2003]. This angular velocity may differ from that of the planet Ω due to ion-neutral frictional drag, such that for some value of constant k we may put

$$(\Omega - \Omega^*) = k(\Omega - \omega) \qquad (27.2)$$

so that $(\Omega^* - \omega) = (1-k)(\Omega - \omega)$. Equation (27.1) then becomes

$$I_P(\theta_i) = (1-k)\Sigma_P \rho_i^2 B_i(\Omega - \omega) = \Sigma_P^* \rho_i^2 B_i(\Omega - \omega) \qquad (27.3)$$

where $\Sigma_P^* = (1-k)\Sigma_P$ is the effective height-integrated Pedersen conductivity, reduced from the true value by slippage of the neutral atmospheric flow. Although the coupling between thermosphere, ionosphere, and magnetosphere is in general complex, modeling studies suggest that $k \approx 0.5$ may be a simple reasonable value at both Jupiter and Saturn [*Millward et al.*, 2005; *Smith and Aylward*, 2009; *Galand et al.*, 2011]. We note that the power per unit area of ionosphere extracted from planetary rotation, p, is equal to Ω times the torque per unit area $p = \Omega I_P B_i$, of which the fraction going to the magnetosphere is $p_M = \omega I_P B_i$, while the remainder $p_A = (\Omega - \omega) I_P B_i$ goes to atmospheric heating as the sum of Joule heating and ion drag [e.g., *Hill*, 2001; *Smith et al.*, 2005].

The current given by equations (27.1) and (27.3) is directed transverse to the magnetic field, which is in general tilted in the meridian through angle α_i relative to the ionospheric normal. We then suppose that polarization of the ionosphere produces a small field-aligned current within the layer so that the total meridional current becomes exactly "horizontal" [e.g., *Cowley and Bunce*, 2003], given by

$$I_m(\theta_i) = \frac{I_P(\theta_i)}{\cos\alpha_i} = \frac{\Sigma_P^* \rho_i^2 B_i(\Omega - \omega)}{\cos\alpha_i} \qquad (27.4)$$

From current continuity, the field-aligned current just above the ionosphere is given by the colatitudinal gradient of I_m. In the northern hemisphere, the total parallel current per radian of azimuth flowing between θ_{i1} and θ_{i2} (with $\theta_{i2} > \theta_{i1}$) is given simply by

$$I_\parallel = -\left(I_m(\theta_{i2}) - I_m(\theta_{i1})\right) \qquad (27.5)$$

where negative values indicate downward current and positive values upward current. The same equation applies in the southern hemisphere with the same meaning if I_m is treated as positive equatorward as indicated above and θ_i is colatitude from the southern pole. The field-aligned current density is then given by

$$j_{\parallel i}(\theta_i) = -\frac{1}{R_i^2 \sin\theta_i} \frac{\cos\alpha_n}{\cos\alpha_i} \frac{dI_m}{d\theta_i} \qquad (27.6)$$

with the same meaning of the sign as in equation (27.5), where α_n is the angle between the ionospheric normal and the radial vector. The parallel current then flows along magnetospheric field lines with $(j_\parallel/B)$ constant until closing in the equatorial radial current as in Figure 27.1.

27.2.3. Azimuthal Perturbation Field

The azimuthal perturbation field just above the ionosphere associated with the above current system (see Fig. 27.1a) is given by Ampère's law as $B_{\varphi i}(\theta_i) = (\mu_o I_m(\theta_i)/\rho_i)$. Ampère's law also shows that ρB_φ is constant along each flux shell between the ionosphere and the equatorial closure currents [e.g., *Hunt et al.*, 2014], where ρ is perpendicular distance from the axis, such that the meridional ionospheric current flowing at the feet of the field lines can be determined from measurement of B_φ at any point between these regions as

$$I_m = \mp \rho_i B_{\varphi i}/\mu_o = \mp \rho B_\varphi/\mu_o \qquad (27.7)$$

where upper and lower signs correspond to measurement and ionospheric points in the northern and southern hemispheres, respectively, and I_m is again taken positive directed equatorward in both cases. Measurement of the azimuthal perturbation field along the trajectory of a near-polar orbiting spacecraft thus allows determination of the meridional current profile in the corresponding hemisphere through equation (27.7), and thus also of the field-aligned current from equation (27.6), as will be applied to Cassini data from Saturn in section 27.4 [*Hunt et al.*, 2014, 2015].

27.2.4. Plasma Angular Velocity

The principal dynamical parameter in the above equations is the plasma angular velocity ω, specifically its deviation from the planetary angular velocity Ω. In the outer part of the magnetospheric system, this may be influenced by complex factors relating to the Dungey and Vasyliunas cycles, which may be accessible only through empirical investigation or global modeling. As indicated in section 27.1, however, the quasi-axisymmetric inner and middle magnetosphere regions will instead be dominated by the outward transport of plasma from the inner moon sources. Application of Newton's second law then yields

$$\frac{d}{d\rho_e}\left(\dot{M}\rho_e^2\omega\right) = 2\pi\rho_e B_e I_\rho \qquad (27.8)$$

where ρ_e is radial distance in the equatorial plane, B_e is the equatorial field strength, $\dot{M}$ is the radial mass flux equal in the steady state to the source production rate, and I_ρ is the equatorial radial current per radian of azimuth [*Hill*, 1979; *Pontius*, 1997]. The left side of this expression is the radial gradient of the outward flux of plasma angular momentum, while the right side is the azimuth-integrated electromagnetic torque on the plasma, assumed confined to the equatorial plane, per unit radial distance. Current continuity requires $I_\rho(\rho_e) = 2I_m(\theta_i)$ assuming near-equal current contributions from the two hemispheres, where $I_m(\theta_i)$ is given by equation (27.4) and θ_i is the ionospheric colatitude of the field line that passes through the equator at radial distance ρ_e. In the simple case where the system parameters are taken to be constants, equation (27.8) represents a first order linear equation for $\omega(\rho_e)$ (or equivalently $\omega(\theta_i)$), which can be solved subject to one initial or boundary condition. There is only one solution, however, that does not diverge to either large positive or negative values at small ρ_e, this being the solution in which the plasma near-rigidly corotates at small distances. Setting $\omega \approx \Omega$ in the left side of equation (27.8), the meridional current required to maintain near-rigid corotation at small distances (large θ_i) is given by

$$I_m \approx \frac{\dot{M}\Omega}{2\pi B_e} \qquad (27.9)$$

which depends only on $\dot{M}$, while from equation (27.4), the associated plasma departure from rigid corotation, assumed small, is given by

$$\frac{\Delta\omega}{\Omega} = \left(1 - \frac{\omega}{\Omega}\right) \approx \frac{\dot{M}\cos\alpha_i}{2\pi\Sigma_P^*\rho_i^2 B_i B_e} \qquad (27.10)$$

depending on both $\dot{M}$ and Σ_P^* [*Nichols and Cowley*, 2003]. A large radial distance (small θ_i) approximation may also be obtained by putting $\omega \ll \Omega$ in equation (27.4), showing that I_m then depends only on Σ_P^* and not on $\dot{M}$.

27.2.5. Nonaligned Rotator

We now comment on the applicability of this theory to Jupiter, for which the spin and magnetic axes are not coaligned. In this case, colatitude θ_i in the above formulae must be defined with respect to the northern magnetic dipole axis, tilted at ~9.5° relative to the spin axis, while the angular velocity difference $(\Omega - \omega) = \Delta\omega(\theta_i)$ is correspondingly the angular velocity of the plasma around the magnetic axis in the planetary rest frame (as in equation [27.10]), positive clockwise as viewed from above the northern pole for plasma subcorotation [*Cowley et al.*, 2008]. Then $\Delta\omega = 0$ corresponds to rigid corotation, while $\Delta\omega = \Omega$ corresponds to nonrotation of the plasma in the inertial frame (the field lines simply rock up and down with the rotation of the dipole axis around the planet). However, the assertion above that the parameters are simple functions of colatitude will no longer be strictly valid in some cases. The ionospheric radius R_i, for example, is instead a function of colatitude from the spin axis, but since we are generally interested only in jovigraphic colatitudes between the spin pole ($R_i \approx 67,354$ km) and ~25° (~68,117 km), a radial difference of only ~1%, use of a fixed polar value represents a good approximation. Neglect of nonsymmetric field terms and neutral atmosphere flows (due, e.g., to the Coriolis force), represent more serious approximations, but can be incorporated only at the expense of far greater model complexity.

27.3. MODELING OF JUPITER'S COUPLING CURRENT SYSTEM

At the time of writing, late 2016, knowledge of the coupling currents at Jupiter are on the verge of transformation through the acquisition of the first close polar data by the Juno spacecraft. Here, therefore, we briefly review and illustrate previous modeling studies informed by near-equatorial and auroral data, particularly as they aid understanding of polar data obtained at Saturn by the Cassini spacecraft, the subject of sections 27.4 and 27.5. Equation (27.8) was first derived by *Hill* [1979] and solved analytically for the case of a dipole planetary magnetic field. As indicated above, the only nondivergent solution is one in which near-rigid corotation is maintained close to the planet, breaks down at an equatorial distance that depends on $(\Sigma_P^*/\dot{M})^{1/4}$ as the ionospheric torque becomes insufficient to maintain that condition, with the angular velocity then decreasing as $\sim 1/\rho_e^2$ (angular momentum per unit mass conserved) at large distances. *Pontius* [1997] subsequently showed that similar angular velocity profiles also apply to the more realistic case where the equatorial field is stretched out by plasma stresses into a current sheet configuration. While neither of these papers considered the related current system explicitly, *Cowley and Bunce* [2001] and *Cowley et al.* [2002] showed that the inclusion of a current sheet field is crucial in correctly determining the mapping and strength of the field-aligned currents in the ionosphere, and hence the location and intensity of the consequent auroras. In brief, they showed that the upward field-aligned currents associated with the breakdown of rigid corotation at ~20–30 R_J in Jupiter's middle magnetosphere, map into and are of sufficient strength to produce the "main emission" auroras observed at ~15° colatitude with respect to Jupiter's magnetic pole. Subsequently, *Nichols and Cowley* [2003] examined the nature of the solutions over wide ranges of Σ_P^* and $\dot{M}$, assumed constants, for both a dipole field (following *Hill* [2001]) and an empirical current sheet field.

Nichols and Cowley [2004] further investigated the effect of variations in ionospheric Pedersen conductivity due to auroral electron precipitation in regions of upward field-aligned current, based on the ionospheric modeling results of *Millward et al.* [2002], which allow the variation in conductivity with field-aligned current density $\Sigma_P^* = \Sigma_P^*(j_{\|i}, \theta_i)$ to be estimated. In this case, equations (27.8) and (27.6) form two coupled first-order differential equations for ω and $j_{\|i}$ that can be solved with the use of two initial values, conveniently chosen at the outer boundary of the empirical equatorial field model at $\rho_e = 100\ R_J$. Choice of ω and $j_{\|i}$ (hence Σ_P^*) at the outer boundary also sets the value of I_m at the boundary through equation (27.4), such that we may alternatively set I_m and one or other of ω or $j_{\|i}$, where I_m can be chosen to agree with empirical data. Either $j_{\|i}$ or ω is then iterated at fixed I_m to find the nondivergent solution, which near-rigidly corotates close to the planet. Illustrative results are provided in Figure 27.2 for a mass outflow $\dot{M}$ of 1000 kg s^{-1}, a conductivity such that Σ_P^* increases near-linearly from 0.0275 mho when $j_{\|i}$ is near-zero (based on the results of *Hill* [1980]) to, for example, ~0.7 mho when $j_{\|i}$ increases to ~0.25 μA m^{-2}, and a total azimuth-integrated equatorial radial current at the outer boundary of ~100 MA, consistent with the nightside Galileo results of *Khurana* [2001]. Figure 27.2 shows physical parameters of the solution projected along field lines into Jupiter's ionosphere and plotted versus colatitude from the magnetic pole, showing (a) the normalized plasma angular velocity (ω/Ω), (b) the meridional Pedersen current I_m (MA rad^{-1}), (c) the field-aligned current density just above the ionosphere $j_{\|i}$ (μA m^{-2}), and (d) the effective Pedersen conductivity Σ_P^* (mho). The plots extend from ~15.7° colatitude mapping to the outer limit of the model equatorial field at 100 R_J, to 19° mapping to ~12.1 R_J, where we note that the Io source at ~5.9 R_J maps in the ionosphere to ~24°. The solid lines in each panel show the full numerically integrated solution, the dashed lines the solution for fixed Σ_P^* equal to 0.0275 mho (the $j_{\|i} \approx 0$ value), and the dot-dashed lines the near-rigid corotation approximation for the inner magnetosphere given by equations (27.9) and (27.10). Initially, the solution closely follows the latter approximation as the colatitude decreases, the meridional current increasing as the equatorial magnetic field decreases (equation [27.9]) so as to maintain $\omega \approx \Omega$, with corresponding major increases occurring in $j_{\|i}$ and hence Σ_P^*. This regime is found to be maintained to increasingly large equatorial distances as the boundary value of I_m is increased. When I_m has reached roughly half of its peak (outer boundary) value, however, it rapidly diverges from this condition to values much lower than those required to maintain

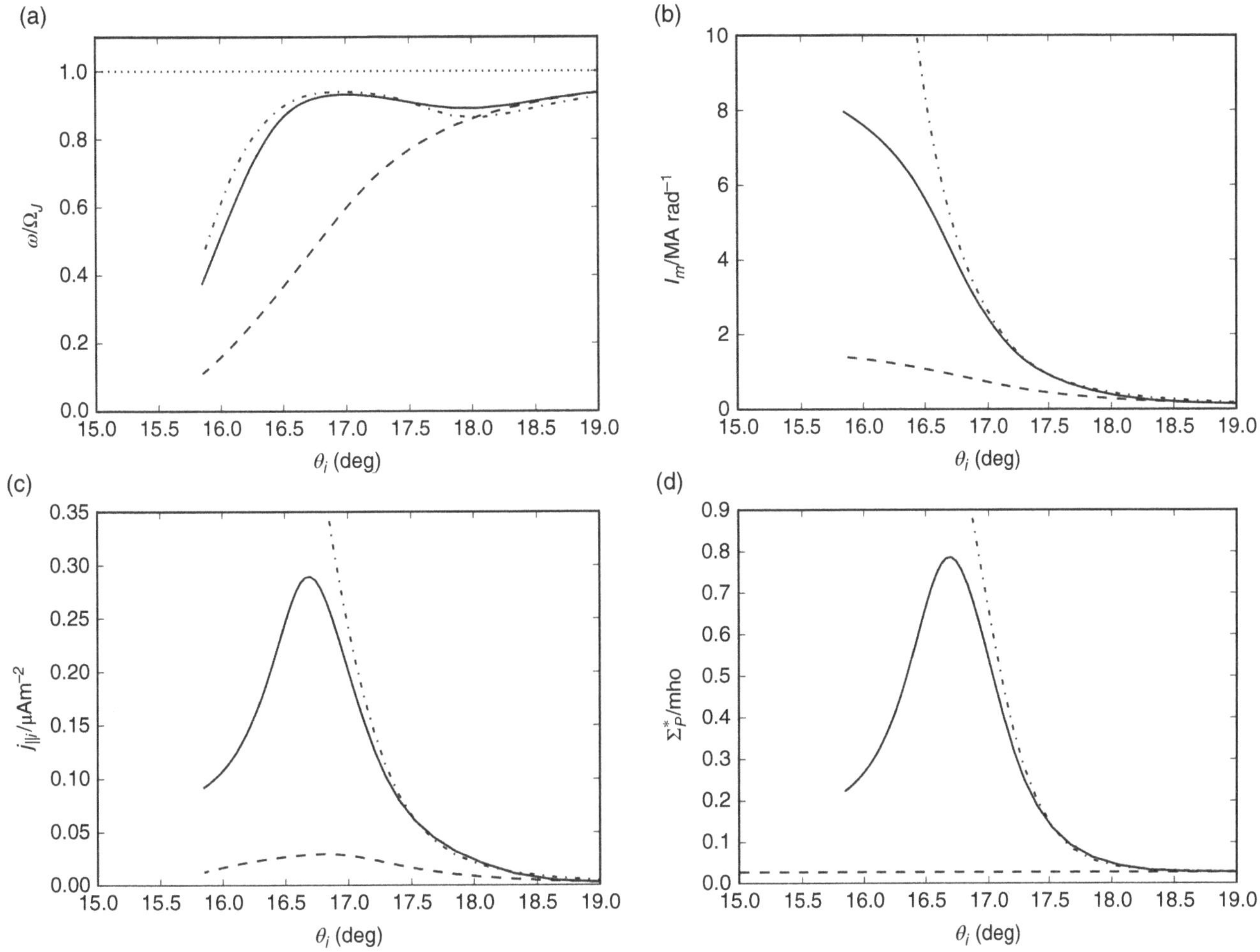

Figure 27.2 Modeled plasma and current parameters for Jupiter's magnetosphere-ionosphere coupling current system, shown versus ionospheric colatitude between 15° and 19°. The model assumes a constant mass outflow rate of $\dot{M}$ =1000 kg s^{-1}, an ionospheric conductivity that varies near-linearly with the field-aligned current density in line with the modeling results of *Millward et al.* [2002], and a total current flowing in the system that is in line with the Galileo results of *Khurana* [2001]. The panels show (a) the plasma angular velocity ω normalized to the planetary value Ω; (b) the meridional Pedersen current I_m per radian of azimuth (MA rad^{-1}), positive equatorward; (c) the field-aligned current density just above the ionosphere $j_{\parallel i}$ (μA m^{-2}), positive upward; and (d) the effective height-integrated Pedersen conductivity Σ_P^* (mho). The solid lines show the full solution, the dashed lines show the solution for a constant conductivity of 0.0275 mho (the $j_{\parallel i} \approx 0$ value), and the dot-dashed lines show the near-rigid corotation approximation given by equations (27.9) and (27.10). From Figure 16 of *Nichols and Cowley* [2004].

near-rigid corotation, at which point the plasma angular velocity decreases strongly with decreasing colatitude (i.e., increasing radial distance). In this regime, I_m increases less rapidly with decreasing colatitude despite the decrease in the plasma angular velocity (equation [27.4]), due to the strongly falling value of Σ_P^* associated with falling $j_{\parallel i}$. This discussion will be employed in section 27.4 to illuminate Cassini observations at Saturn.

In further development of Jupiter magnetosphere-ionosphere coupling models, *Cowley et al.* [2005b] proposed a global axisymmetric model of the plasma angular velocity in which the behavior in the middle magnetosphere follows Figure 27.2, while *Cowley et al.* [2007] considered the consequences of solar-wind-induced modulations in system size, which affects both the plasma angular velocity and the magnetospheric field. A particular goal was to quantify the energy deposited in the polar atmosphere by Joule heating, ion drag, and particle precipitation, to help address the unexplained high temperatures in Jupiter's thermosphere. These models were subsequently coupled to detailed thermospheric models to investigate this issue, but while it was found that the input has a major effect on the polar thermosphere, the energy is principally transported poleward rather than equatorward, thus leaving unresolved the issue of the high thermospheric temperatures in the lower-latitude region [*Tao et al.*, 2009; *Smith and Aylward*, 2009; *Yates et al.*, 2012; *Ray et al.*, 2015]. The above models have also been employed to derive predictions for the Juno mission, specifically of the azimuthal field associated with magnetosphere-ionosphere coupling currents (equation [27.7]) [*Cowley et al.*, 2008]. *Nichols* [2011b] has further incorporated a self-consistent axisymmetric model of the middle magnetosphere current sheet, rather than an empirical model as employed in the above studies, while *Nichols* [2011a] and *Nichols et al.* [2012] have applied the above magnetosphere-ionosphere coupling theory to both exoplanets and ultracool dwarf stars with the aim of estimating the detectability of related auroral radio emissions. Initial attempts have also been made by *Tao et al.* [2010] and *Ray et al.* [2014] to relax aspects of the axisymmetric assumption, while *Chané et al.* [2017] have recently presented the results of a time-dependent global magnetosphere model, which shows that while significant magnetospheric dawn-dusk asymmetries are generated by the confinement of the magnetosphere by the solar-wind flow, the essential physics of the magnetosphere-ionosphere coupling currents remains as described above.

27.4. OBSERVATIONS OF SATURN'S CURRENT SYSTEM

The magnetosphere-ionosphere coupling current system can conveniently be studied by measurement of the azimuthal field generated in the region between the ionospheric and equatorial transverse currents, as discussed in section 27.2.3. In particular, B_φ profiles obtained on north-south polar passes at modest radial distances above the ionosphere provide means to determine the colatitude profile of the meridional ionospheric current I_m through equation (27.7). The most suitable and most studied such sequence of passes from the Cassini mission at Saturn occurred during late southern summer in 2008 (prior to vernal equinox in mid-2009), when 37 passes took place at radial distances ~3–5 R_S over the northern and southern auroral zones in the midnight sector. Initial examination by *Talboys et al.* [2009, 2011] demonstrated a broader variety of signatures than expected from simple Saturn models constructed on a basis similar to the Jupiter models described above [*Cowley and Bunce* 2003; *Cowley et al.*, 2004]. Examples of data from southern hemisphere passes are shown in Figure 27.3, where each panel shows (a) an electron count rate spectrogram from the Cassini Plasma Spectrometer-Electron Spectrometer (CAPS-ELS) covering energies ~0.6 eV–28 keV, (b) the I_m (MA rad^{-1}) profile derived from the B_φ field component (equation [27.7]), and (c) the phase of the southern hemisphere PPO oscillation (to be discussed in detail in section 27.5), all plotted versus ionospheric colatitude from the southern magnetic/spin pole θ_{iS} (mapped along model field lines from the spacecraft). Overall, these profiles follow expectations based on Figures 27.1 and 27.2, that is, I_m increases with colatitude in the polar region (defined by lack of hot electrons) indicative of a distributed downward current into ionosphere, followed by a fall to smaller values near ~18°–19° colatitude (mapping to ~10–15 R_S in the equatorial plane) associated with strong upward currents colocated with Saturn's auroral zone. However, significant variations are clearly present depending on the southern PPO phase Ψ_S, associated with additional superposed downward currents for $\Psi_S \sim 90°$ and upward currents for $\Psi_S \sim 270°$. These are made evident in Figure 27.3 by arranging the profiles in order of PPO phase (indicated specifically at $\theta_{iS} = 19°$ at the top of each panel) around a phase dial showing a set of Ψ_S phase arrows, which point approximately toward the corresponding panel.

Hunt et al. [2014] showed that the PPO-independent component of the current associated with plasma subcorotation and the PPO-dependent current can be separated by exploiting the opposite azimuthal symmetries of these systems. Specifically, the southern hemisphere data such as those illustrated in Figure 27.3 were separated into eight 45° sectors of southern PPO phase Ψ_S, with "opposite" sectors (differing by 180°) first being added and divided by two to estimate the PPO-independent current profile, and then subtracted and divided by two to estimate the PPO-dependent current profile. This procedure thus resulted in four (very similar) estimates of the subcorotation I_m profile, which are averaged together and shown by the red line in Figure 27.4a. In accordance with Figure 27.1, this profile exhibits a slow rise with colatitude over the polar region indicative of distributed downward currents, followed by a rapid fall to near-zero values at auroral latitudes indicative of more intense upward currents, the latter currents beginning ~1.5° equatorward of the averaged southern open-closed field line boundary (OCB) determined by *Jinks et al.* [2014], marked by a vertical dashed line. The unexpected small negative values at largest colatitudes are likely due to the lack of exact antisymmetry in the PPO-dependent currents assumed in the analysis. The corresponding PPO-dependent current profiles will be discussed in section 27.5.

For the northern data, the procedure is complicated by the finding by *Hunt et al.* [2015] that these depend on both the northern and southern PPO phases, such that in this case the PPO-independent current was estimated by averaging the data obtained in intervals in which both northern and southern phases were within ± 45° of 0° or 180°, for which the PPO-dependent currents are smallest

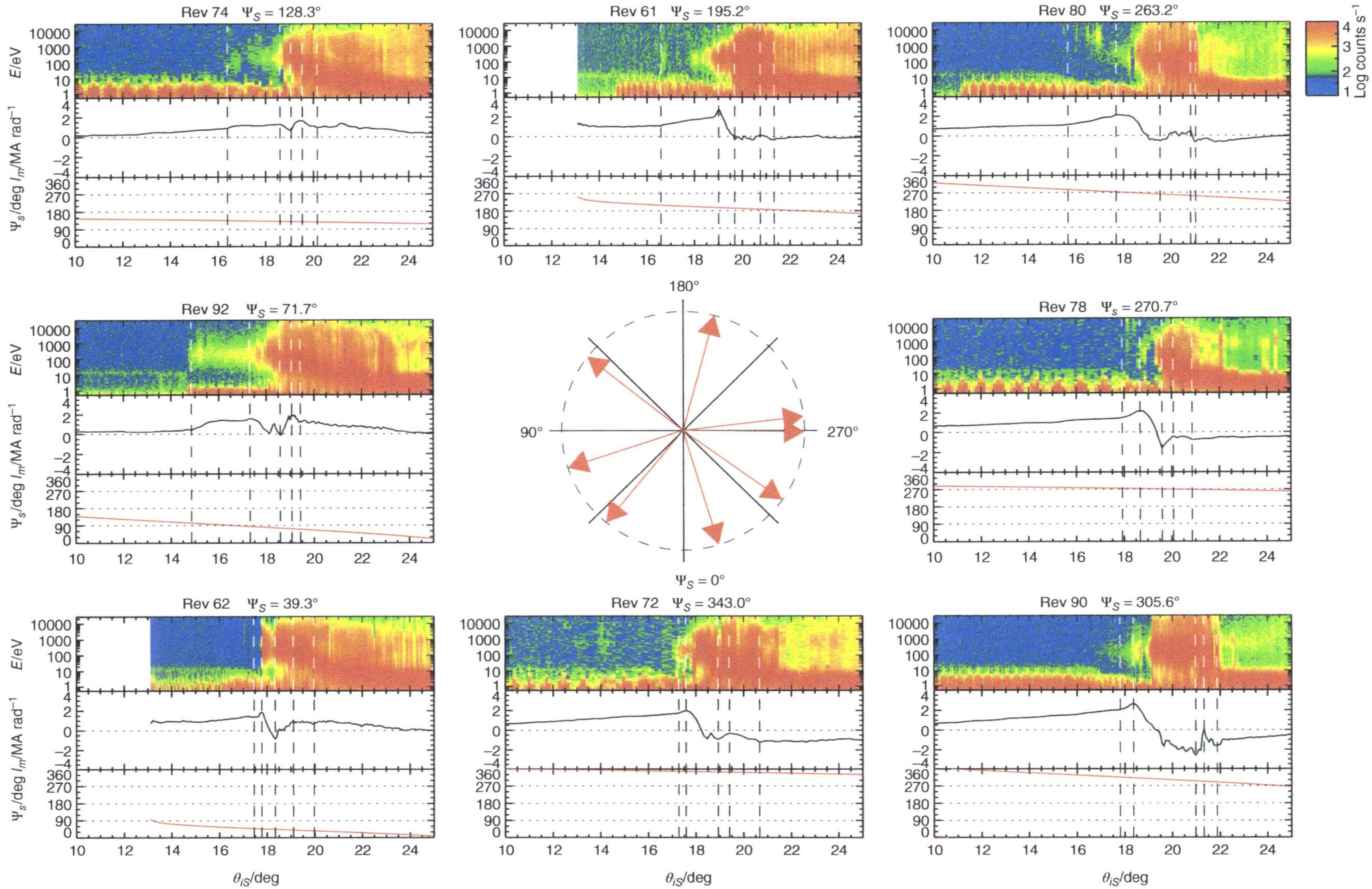

Figure 27.3 Sequence of eight data plots exemplifying the dependence of Saturn's southern hemisphere currents on the southern PPO phase Ψ_S. The panels in each plot show from top to bottom (a) a CAPS/ELS electron count rate spectrogram covering the energy range ~0.6 eV–28 keV color-coded as shown at upper right, (b) the meridional ionospheric current I_m (MA rad^{-1}), and (c) the southern PPO phase Ψ_S (deg), all plotted versus southern ionospheric colatitude θ_{iS} (deg) over the range 10°–25°. The vertical dashed lines indicate four main field-aligned current sheets in each plot drawn between extrema in the I_m values, directed with increasing colatitude down, up, down, and up with respect to the southern ionosphere. The plots are arranged clockwise in order of increasing PPO phase, with the Ψ_S value at the point where $\theta_{iS} = 19°$ being given at the top of each plot together with the Cassini Rev number. Phases ~0°/360° are shown at the bottom of the figure, ~90° to the left, ~180° at the top, and ~270° to the right. A Ψ_S phase dial is shown in the center of the figure displaying these phases with red arrows, pointing approximately toward the corresponding data plot.

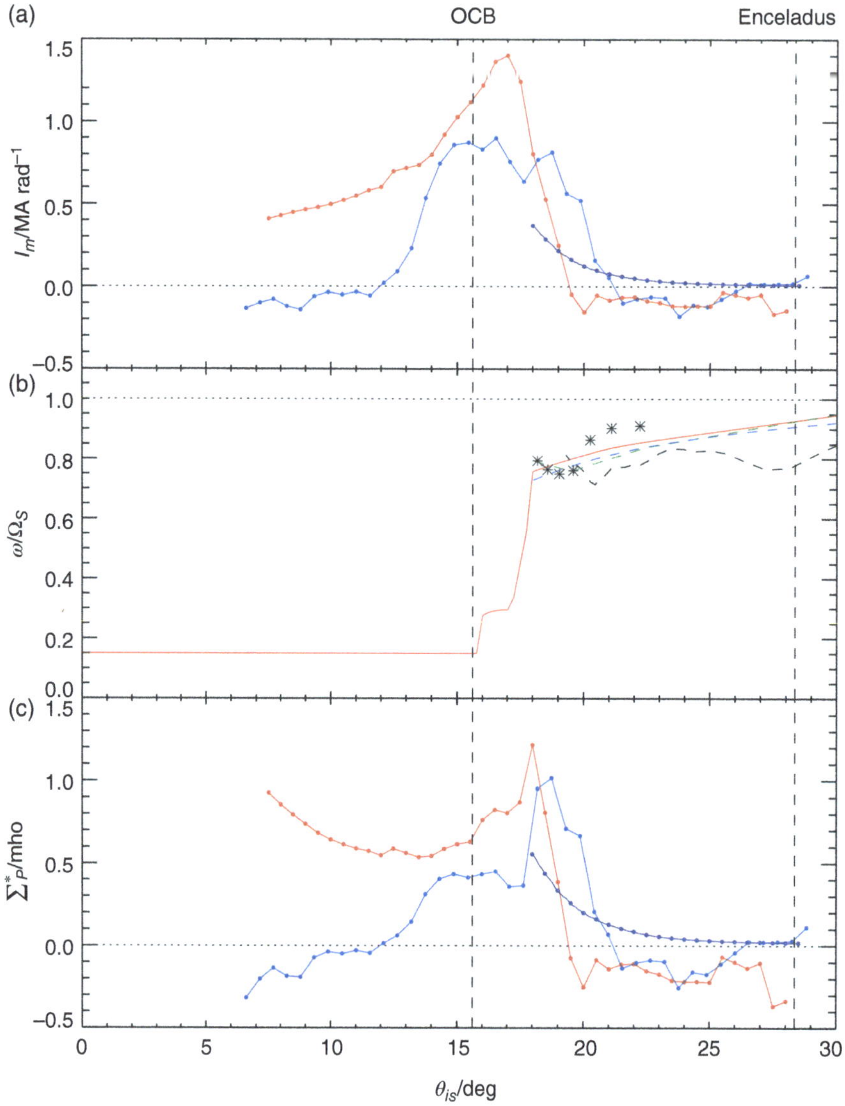

Figure 27.4 Profiles of the PPO-independent ionospheric meridional current I_m, the normalized plasma angular velocity (ω/Ω_S), and the effective Pedersen conductivity Σ^*_P, all plotted versus colatitude from the southern pole θ_{iS}. (a) The red line shows the southern current profile (MA rad^{-1}); the blue line shows the northern profile mapped along field lines into the southern hemisphere. The purple line shows the theoretical current profile calculated from equation (27.9) for $M = 100\,\text{kg s}^{-1}$. (b) The red line shows a model normalized angular velocity profile, based for $\theta_{iS} \geq 18°$ on the empirical data shown by the stars and dashed lines. (c) The red and blue lines show the Σ^*_P profiles (mho) for the southern and northern hemispheres, respectively, derived from equation (27.4); the purple line shows values determined from equation (27.10) using the theoretical current profile (purple line) in (a) and the angular velocity model in (b). The southern OCB of *Jinks et al.* [2014] and the ionospheric footprint of Enceladus are shown by the labeled vertical dashed lines. From Figure 16 of *Hunt et al.* [2014] and Figure 13 of *Hunt et al.* [2015].

(Fig. 27.3) and should average near zero. This profile is shown by the blue line in Figure 27.4a mapped along field lines to the southern hemisphere for direct comparison with the southern profile. An important difference occurs at small colatitudes where significant currents are present in the south but not in the north, this presumably reflecting the late southern summer conditions prevailing at Saturn, during which the northern pole was tilted away from the Sun between ~8° and ~4° over the interval concerned. The unexpected difference in colatitude at which the two profiles fall to near zero at larger colatitudes may also reflect imperfect removal of the PPO-dependent current from the more problematical northern data.

To relate these results to the theory in section 27.2, the red line in Figure 27.4b shows an empirical model of the plasma angular velocity normalized to that of the planet (corresponding specifically to a System III rotation period of 10.65622h). For colatitudes beyond 18° (mapping inside ~13 R_S in the equatorial plane), the profile is based on a variety of empirical sources detailed by *Hunt et al.* [2014], which indicate that the plasma angular velocity increases from ~75% of rigid corotation at ~18° nearer to rigid corotation in the inner magnetosphere. The principal plasma source near the moon Enceladus, located at 3.95 R_S in the equatorial plane, maps to ~28.3° southern colatitude as also marked by a vertical dashed line. In the more distant tail plasma sheet mapping closer to the pole the angular velocity is then observed to fall further to ~30% of rigid corotation [*Thomsen et al.*, 2014], but the ionospheric mapping is then unclear due to the lack of suitable magnetic models at such distances, save for the fact that it must map equatorward of the OCB. In agreement, for example with Figure 27.2, we have assumed in Figure 27.4b that the plasma angular velocity falls precipitately to ~30% of rigid corotation poleward of ~18°, and further to ~15% of rigid corotation across the OCB on open field lines, the latter value being in agreement with the angular velocity of reconnection-related polar auroral structures reported by *Radioti et al.* [2013].

Given the I_m and (ω/Ω_S) profiles in Figures 27.4a and 27.4b, respectively, we can then invert equation (27.4) to determine the effective ionospheric Pedersen conductivity profile, Σ_P^*, shown for the northern and southern hemispheres by the blue and red lines in Figure 27.4c. In each case, the conductivity peaks at ~1 mho within the upward current layer, where it is presumably augmented by auroral electron precipitation similar to Figure 27.2c, and falls to ~0.5 mho in both hemispheres as the field-aligned current reverses to downward in the poleward region, and then to near-zero (and unphysical negative) values in the northern hemisphere. We note that while the model value $(\omega/\Omega_S) = 0.15$ taken for the polar region rests on a rather small evidence base, significantly larger values would result in larger deduced conductivities via equation (27.4), comparable with and exceeding those deduced for the auroral region. If instead we chose $(\omega/\Omega_S) = 0.9$, for example, the deduced conductivities in the polar region would be a factor of 8.5 larger, of order ~5 mho rather than ~0.5 mho deduced here, which seems physically highly unlikely.

The region equatorward of the main ionospheric currents is potentially more problematic, however, where despite the observed plasma subcorotation, small negative (apparently super-corotational) currents are deduced in both hemispheres, likely due to imperfect removal of the PPO current as indicated above. We then conclude with *Hunt et al.* [2014, 2015] that the actual ionospheric currents in this region are small, at a level of ~0.1 MA rad^{-1} or less, and essentially unmeasurable via this analysis. To investigate whether this conclusion is reasonable despite the expected outward transport of Enceladus plasma in this region, we can calculate the current values expected in the near-rigid corotation approximation given by equation (27.9), which should apply in this region given the angular velocities in Figure 27.4b (see, for example, the dot-dashed lines in Fig. 27.2). We note that the angular velocity of the near-rigidly corotating plasma at the Enceladus source (at ~4 R_S in the equatorial plane) would fall to less than $(\omega/\Omega_S) \approx 0.1$ at ~18° colatitude (~13 R_S) in the absence of an ionospheric torque, rather than the $(\omega/\Omega_S) \approx 0.75$ observed. These currents are shown by the purple line in Figure 27.4a, where we have taken $\dot{M} = 100$ kg s^{-1} in equation (27.9) together with an equatorial field model consisting of a planetary plus empirical ring current field as employed in the above studies. The values are seen to be very small, less than ~0.1 MA rad^{-1}, equatorward of ~20° southern colatitude, and only increase to larger (measureable) values where the observed values also become significantly large. The occurrence of small, essentially unmeasurable, current values equatorward of ~20° is thus not inconsistent with the maintenance of near-rigid corotation in this region with an outflow rate of ~100 kg s^{-1}.

Using the theoretical current value, however, we can use the empirical (ω/Ω_S) profile in Figure 27.4b to estimate Σ_P^* from equation (27.10), shown by the purple profile in Figure 27.4c. The derived values are comparable with those deduced from the empirical current profile in the equatorward part of the upward current region, but then fall to very small values, less than ~0.1 mho beyond ~20° colatitude. Given the theoretical currents in Figure 27.4a required for a near-rigidly corotating outflow at ~100 kg s^{-1}, this implies that the departure of the plasma from rigid corotation would be expected to be much smaller than that indicated in Figure 27.4b in this region if the effective conductivity were higher, comparable, for example, to that in the polar region. Since the actual Pedersen conductivity of the ionosphere seems unlikely to be larger in the polar region than at subauroral latitudes, this result suggests that upper atmospheric slippage due to ion-neutral coupling, measured by the k parameter in equation (27.2), is considerably more significant at lower than at higher latitudes in both hemispheres. To put this another way, these results suggest that the subcorotation in the subauroral region shown in Figure 27.4b is much more a property of the thermosphere relative to the System III rotation period, rather than of the magnetosphere and ionosphere relative to the thermosphere.

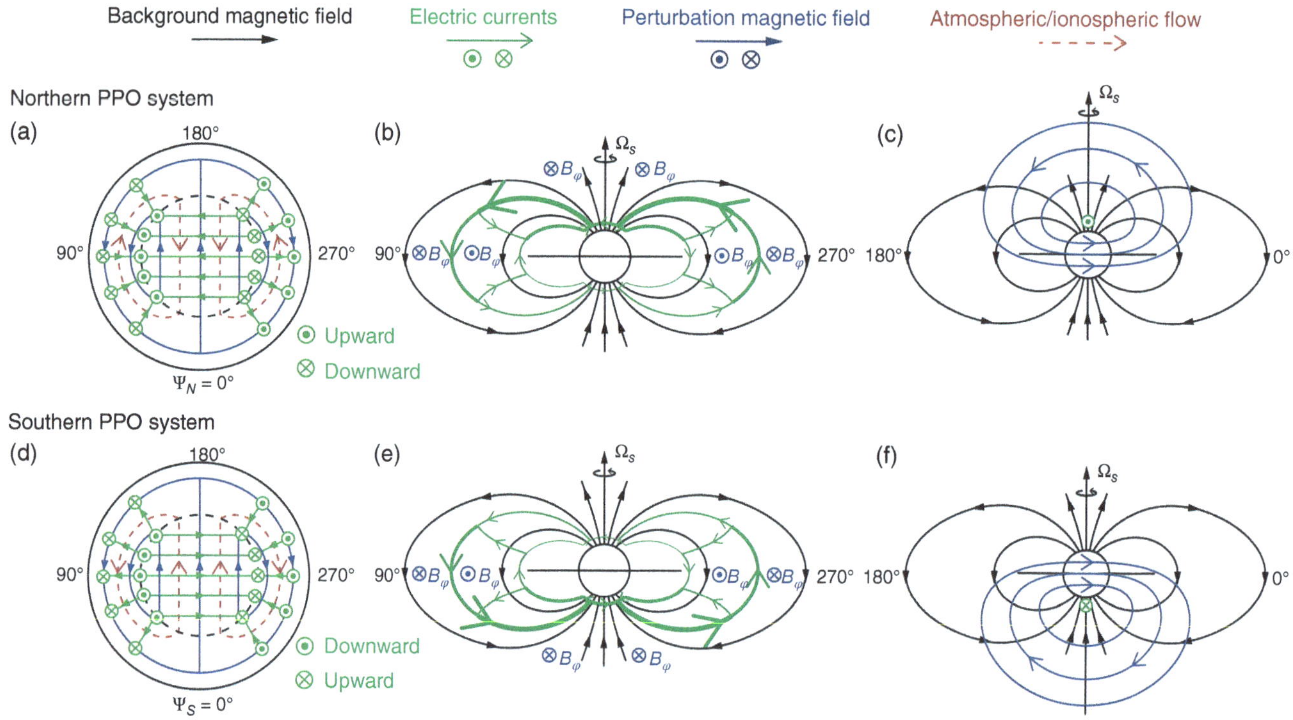

Figure 27.5 The northern (a)–(c) and southern (d)–(f) PPO current systems and field perturbations, where arrowed black lines show the near axisymmetric background field, green lines and symbols the currents, and blue lines and symbols the field perturbations. Circled dots and crosses indicate vectors pointing out of and into the plane of the figure, respectively. Panels (a) and (d) show views of the northern and southern polar ionospheres, both viewed from the north, where the red dashed lines also show the directions of the driving twin-vortex atmosphere-ionosphere flow. Panels (b) and (e) show the field and currents in the $\Psi_{N,S} = 90° - 270°$ meridian (see also [a] and [d]); panels (c) and (f) similarly show the orthogonal $\Psi_{N,S} = 0° - 180°$ meridian. These systems then rotate about the planetary spin-magnetic axis with their respective northern and southern periods, close to the planetary rotation period. From Figure 1 of *Hunt et al.* [2015].

27.5. SATURN'S PLANETARY PERIOD OSCILLATIONS

One of the major but less anticipated study topics at Saturn during the Cassini mission has been the ubiquitous PPOs that are observed to modulate essentially all magnetospheric plasma, field, and wave observations. Initially detected as ~10.7 h variations in auroral radio emissions observed during the Voyager flybys, Cassini field and radio data have shown that two separate periodicities are usually present, one associated with the northern hemisphere and the other with the southern, whose periods vary seasonally by ± 1% about the above value (see, e.g., the review by *Carbary and Mitchell* [2013] and *Provan et al.* [2016]). Like the corotation system, the PPOs are believed to be driven from the atmosphere outward into the magnetosphere, but now by variable twin-vortical flows rotating individually near the planetary period in the two polar upper stratospheres and thermospheres [*Jia et al.*, 2012; *Hunt et al.*, 2014; *Smith*, 2014]. The form of the flow and current systems determined from magnetic field observations is illustrated in Figure 27.5 in a similar format to Figure 27.1, based on the discussions of *Southwood and Kivelson* [2007], *Andrews et al.* [2010b], and *Southwood and Cowley* [2014], and adapted from Figure 1 of *Hunt et al.* [2015]. Figure 27.5a–c pertains to the northern PPO system and 27.5d–f to the southern, where Figure 27.5a and d shows views looking down on the polar region of the planet from the north ("through" the planet in Fig. 27.5d), Figure 27.5b and e shows a principal meridian plane illustrating the principal currents involved (green lines and symbols), and Figure 27.5c and d shows the orthogonal meridian plane (viewing Fig. 27.5b and e from the left) illustrating the associated magnetic field perturbations (blue lines and symbols). The principal currents flow into the ionosphere on one side of the planet and out on the other, thus to a first approximation opposite on opposite sides of the planet as assumed in the "separation" analysis outlined in section 27.4, and close partly through the equatorial plasma and partly through the opposite ionosphere. Each system creates a perturbation field, which is quasi-uniform in the equatorial region, and quasi-dipolar over the corresponding pole. The associated atmospheric-ionospheric twin-vortex flow is shown by the red dashed lines in Figure 27.5a and d. Each system rotates with time at its particular period near that of the planetary period, producing oscillations in the field and plasma at the corresponding PPO period at given point. Azimuthal location with respect to these systems is given by the phase functions illustrated in Fig. 27.5a and d, Ψ_N for the northern system and Ψ_S for the southern system, where we note that Figure 27.5b and e shows the $\Psi_{N,S} = 90° - 270°$ meridians, and Figure 27.5c and f shows the orthogonal $\Psi_{N,S} = 0° - 180°$ meridians, as indicated. Empirically, these phase functions are obtained by using magnetic data from the equatorial or polar regions to determine the orientation of the $\Psi_{N,S} = 0°$ meridians with respect to noon at any time, $\Phi_{N,S}(t)$, taken positive in the sense of planetary rotation, from which the PPO phase at any azimuth φ with respect to noon is given by

$$\Psi_{N,S}(\varphi, t) = \Phi_{N,S}(t) - \varphi \quad (27.11)$$

The $\Phi_{N,S}(t)$ magnetic field phases have been determined near-continuously throughout the Cassini mission by *Andrews et al.* [2012] for the southern summer interval whose data are discussed in section 27.4, and by *Provan et al.* [2013, 2016] post-equinox.

Colatitude profiles of the meridional ionospheric current associated with the northern and southern PPOs are shown in Figure 27.6, obtained from the "sum and difference" analysis of the 2008 dataset discussed in section 27.4. Specifically, Figure 27.6b shows the equal and opposite southern hemisphere PPO profiles for the $\Psi_S = 90°$ (blue) and $\Psi_S = 270°$ (red) meridians where the current should maximize (Fig. 27.5), plotted versus colatitude from the southern pole, while in Figure 27.6a, we show the southern PPO-independent (subcorotation) current profile repeated from Figure 27.4a for comparison. The main upward and downward PPO currents indicated by the gradients in I_m flow between ~16° and ~20° southern colatitude, from near the OCB indicated by the vertical dashed line to near the equatorward boundary of the auroral upward current region of the subcorotation system (mapping to ~10 R_S in the equatorial plane), thus directly overlapping the latter current. The magnitudes of the field-aligned currents in these two systems are also comparable at ~1.5 MA rad^{-1}, such that when the main PPO current is directed upward, it approximately doubles the subcorotation current, and when it is directed downward, it approximately cancels the subcorotation current. A smaller reversed current also flows at larger colatitudes to ~24° (mapping to ~5.5 R_S in the equatorial plane) in the region equatorward of the main subcorotation current (see Fig. 27.5), which attenuates the field perturbations in the inner magnetosphere. This agrees with the behavior of the PPO-related B_φ field perturbations found in southern summer equatorial data by *Andrews et al.* [2010a], which grow in amplitude between ~5 and 10 R_S on the nightside and then fall at larger distances.

Figure 27.6c–e similarly shows the northern meridional current profiles associated with subcorotation (repeated from Fig. 27.4a), the northern PPO system, and the southern PPO system, respectively, now plotted versus northern colatitude (see *Hunt et al.* [2015] for analysis details). It should be noted that no corresponding modulation of the southern currents by the northern PPOs could be detected in the southern hemisphere data. Again, the principal PPO field-aligned currents flow from the vicinity of the OCB to the equatorward boundary of the main subcorotation current, with all three components having similar ~1 MA rad^{-1} magnitudes. Weaker reversed currents also flow in the equatorward region in both PPO systems, reducing the magnitude of the field perturbations in the inner magnetosphere. Although it can thus be seen that the subcorotation and PPO currents flow in the same region in both hemispheres, and have similar magnitudes, the dynamic connection between these current systems has yet to be elucidated in detail.

27.6. SUMMARY

Magnetosphere-ionosphere coupling at the gas giant planets Jupiter and Saturn is dominated by currents associated with the transfer of angular momentum of planetary rotation from the upper atmosphere to the magnetosphere. However, this does not preclude physically significant flows and currents also being driven by the solar-wind interaction in the outer part of the magnetosphere that would lead to dawn-dusk asymmetries fixed in local time. Indeed, at Saturn the signatures of open flux production at the dayside magnetopause and closure in the tail have both been reported [e.g., *Radioti et al.*, 2013; *Thomsen et al.*, 2015], it also being suggested that episodic "auroral storms" are produced by intervals of strong tail reconnection induced by the solar-wind compression events with which they are associated [*Cowley et al.*, 2005a]. Open flux at Saturn may usually be slowly accumulated in the system over intervals of days, and intermittently released in such events [e.g., *Jackman et al.*, 2004]. The role of the Dungey cycle at Jupiter remains unclear and subject to further study at the present time.

On the other hand, extensive analysis of magnetic data from the Cassini mission has shown the continual presence of currents associated with planetary angular momentum transfer at Saturn, in the form of distributed downward currents over the polar open field region and

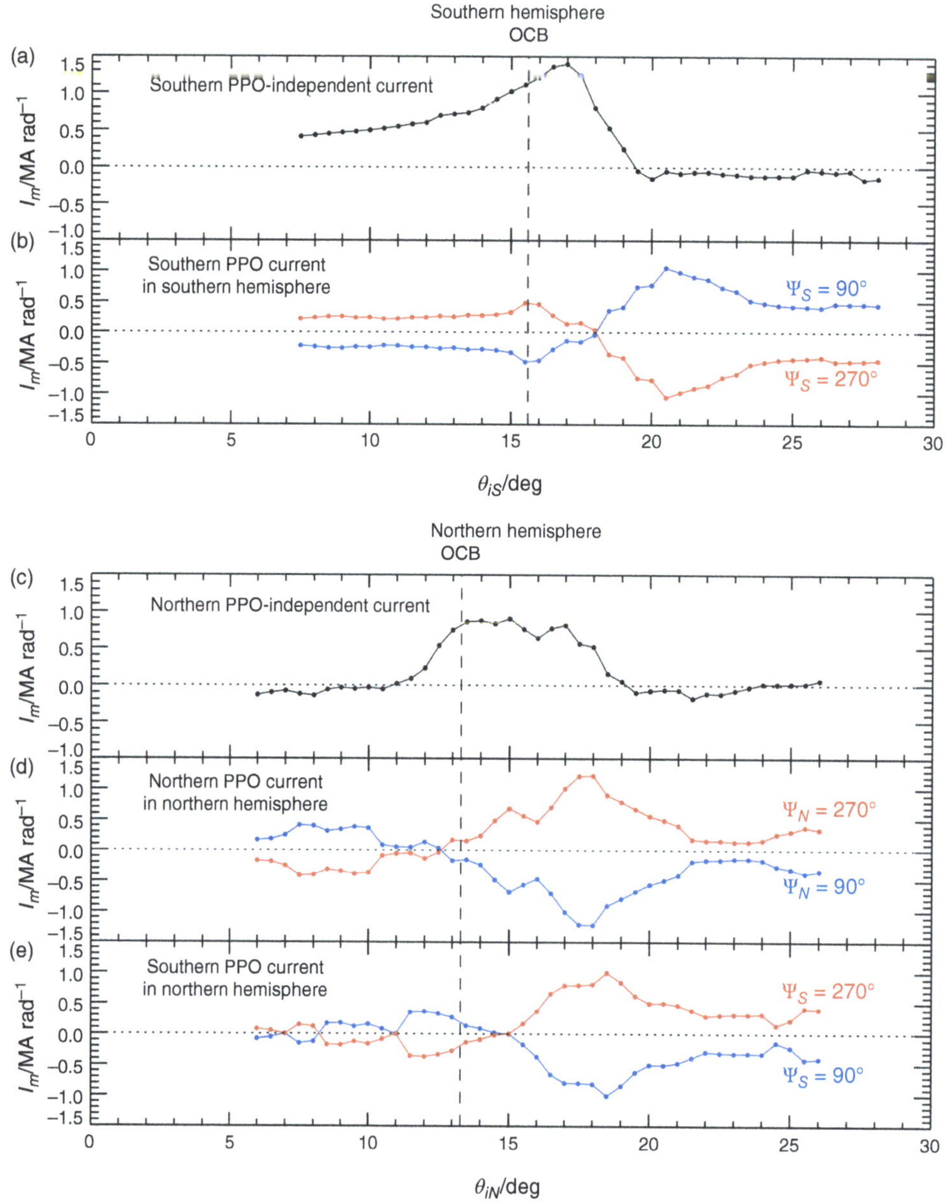

Figure 27.6 Subcorotation and PPO meridional current profiles determined from the "sum and difference" analysis of Cassini data during a sequence of high-latitude passes over the midnight sector northern and southern auroral zones during 2008 (late southern summer). (a) The southern hemisphere subcorotation current profile plotted versus colatitude from the southern pole (as in Fig. 27.4a); (b) this panel similarly shows the southern PPO profiles for the $\Psi_S = 90°$ (blue) and $\Psi_S = 270°$ (red) meridians where the PPO currents maximize. (c) This panel shows northern hemisphere subcorotation current profile plotted versus colatitude from the northern pole (Fig. 27.4a), while (d) and (e) show the PPO currents associated with the northern and southern systems, respectively. The vertical dashed lines mark the averaged OCBs in both hemispheres as determined by *Jinks et al.* [2014]. From Figure 15 of *Hunt et al.* [2014] and Figure 7 of *Hunt et al.* [2015].

outermost closed field magnetosphere, followed by a ring of upward current corresponding to the auroral zone mapping inward to ~10 R_S in the equatorial plane. The upward current is associated with plasma angular velocities increasing rapidly toward the planetary angular velocity with increasing colatitude from the pole, together with enhanced ionospheric Pedersen conductivities associated with auroral electron precipitation. At smaller equatorial radial distances, mapping beyond ~20° colatitude in the southern hemisphere and ~18° in northern hemisphere, the expected ionospheric currents associated with the maintenance of near-rigid corotation in the presence of ~100 kg s^{-1} outward transport from Enceladus are too weak to measure reliably (less than ~0.1 MA rad^{-1}), while the observed degree of departure from rigid corotation in this region, typically ~75%–85%, indicates very weak effective ionospheric Pedersen conductivities typically below ~0.1 mho, compared with peaks of ~1 mho in the auroral region and ~0.5 mho in the southern summer polar region. This suggests the presence of stronger slippage of the upper atmospheric flow from rigid corotation at these lower latitudes, such that the observed degree of subcorotation of the plasma is more related to subcorotation of the thermosphere (relative to System III) than to subcorotation of the magnetospheric-ionospheric plasma relative to the thermosphere. Overall, the total current flowing in the subcorotation current system at Saturn is ~15 MA for both hemispheres combined. PPO-related currents are then superposed, driven by twin-vortex atmospheric flows rotating near the planetary period in the two polar thermospheres. The total principal up and down currents on either side of the pole, essentially collocated with the upward auroral currents of the subcorotation current system, are ~3.5 MA in each hemisphere, this being directly comparable with half of the subcorotation current in each hemisphere.

At the time of writing, Jupiter studies are about to be transformed by the acquisition of the first polar data by the Juno space mission. Here we have therefore concentrated on prior modeling results informed by equatorial field data and studies of related UV auroral emissions that form a baseline for comparison with the Juno data. However, the extent to which these expectations will be borne out in practice, similar to the Saturn system, remains to be determined.

ACKNOWLEDGMENTS

Work at Leicester and Imperial College was supported by STFC Consolidated Grants ST/N000749/1 and ST/N000692/1, respectively. G. J. Hunt was supported by STFC Quota Studentship ST/K502121/1. CAPS-ELS operations were supported by ESA via the UK Space Agency. We thank Steve Kellock and the Cassini Mag team at Imperial College for access to processed magnetometer data, and Andrew Coates at University College London for access to CAPS-ELS data. Calibrated data from the Cassini mission are available from the NASA Planetary Data System (PDS) at the Jet Propulsion Laboratory (https://pds.jpl.nasa.gov/).

REFERENCES

Andrews, D. J., A. J. Coates, S. W. H. Cowley, M. K. Dougherty, L. Lamy, G. Provan, and P. Zarka (2010b), Magnetospheric period oscillations at Saturn: Comparison of equatorial and high-latitude magnetic field periods with north and south Saturn kilometric radiation periods, *J. Geophys. Res.*, *115*, A12252; doi: 10.1029/2010JA015666.

Andrews, D. J., S. W. H. Cowley, M. K. Dougherty, and G. Provan (2010a), Magnetic field oscillations near the planetary period in Saturn's equatorial magnetosphere: Variation of amplitude and phase with radial distance and local time, *J. Geophys. Res.*, *115*, A04212; doi: 10.1029/2009JA014729.

Andrews, D. J., S. W. H. Cowley, M. K. Dougherty, L. Lamy, G. Provan, and D. J. Southwood (2012), Planetary period oscillations in Saturn's magnetosphere: Evolution of magnetic oscillation properties from southern summer to post-equinox, *J. Geophys. Res.*, *117*, A04224; doi: 10.1029/2011JA017444.

Axford, W. I., and C. Hines (1961), A unifying theory of high-latitude geophysical phenomena and geomagnetic storms, *Can. J. Phys.*, *39*, 1433–1464.

Badman, S. V., and S. W. H. Cowley (2007), Significance of Dungey-cycle flows in Jupiter's and Saturn's magnetospheres, and their identification on closed equatorial field lines, *Ann. Geophys.*, *25*, 941–951.

Bagenal, F., and P. A. Delamere (2011), Flow of mass and energy in the magnetospheres of Jupiter and Saturn, *J. Geophys. Res.*, *116*, A05209; doi:10.1029/2010JA016294.

Burton, M. E., M. K. Dougherty, and C. T. Russell (2010), Saturn's internal planetary magnetic field, *Geophys. Res. Lett.*, *37*, L24105; doi:10.1029/2010GL045148.

Carbary, J. F., and D. G. Mitchell (2013), Periodicities in Saturn's magnetosphere, *Rev. Geophys.*, *51*, 1–30; doi: 10.1002/rog.20006.

Chané, E., J. Saur, R. Keppens, and S. Poedts (2017), How is the Jovian main auroral emission affected by the solar wind? *J. Geophys. Res.*, *122*, in press; doi:10.1002/2016JA023318.

Chen, Y., T. W. Hill, A. M. Rymer, and R. J. Wilson (2010), Rate of radial transport of plasma in Saturn's inner magnetosphere, *J. Geophys. Res.*, *115*, A10211; doi:10.1029/2010JA015412.

Connerney, J. E. P., M. H. Acuña, N. F. Ness, and T. Satoh (1998), New models of Jupiter's magnetic field constrained by the Io flux tube footprint, *J. Geophys. Res.*, *103*, 11929–11939.

Cowley, S. W. H. (2000), Magnetosphere-ionosphere interactions: A tutorial review, in *Magnetospheric Current Systems*, edited by S. Ohtani, R. Fujii, M. Hesse, and R. L. Lysak, 91–106, Geophysical Monograph 118, American Geophysical Union, Washington, DC.

Cowley, S. W. H. (2013), Response of Uranus' auroras to solar wind compressions at equinox, *J. Geophys. Res.*, *118*, 2897–2902; doi: 10.1002/jgra.50323.

Cowley, S. W. H., and E. J. Bunce (2001), Origin of the main auroral oval in Jupiter's coupled magnetosphere-ionosphere system, *Planet. Space Sci.*, *49*, 1067–1088.

Cowley, S. W. H., and E. J. Bunce (2003), Corotation-driven magnetosphere-ionosphere coupling currents in Saturn's magnetosphere and their relation to the auroras, *Ann. Geophys.*, *21*, 1691–1707.

Cowley, S. W. H., A. S. Deason, and E. J. Bunce (2008), Axi-symmetric models of auroral current systems in Jupiter's magnetosphere with predictions for the Juno mission, *Ann. Geophys.*, *26*, 4051–4074.

Cowley, S. W. H., E. J. Bunce, and J. D. Nichols (2003b), Origins of Jupiter's main oval auroral emissions, *J. Geophys. Res.*, *108* (A4), 8002; doi:10.1029/2002JA009329.

Cowley, S. W. H., E. J. Bunce, and J. M. O'Rourke (2004), A simple quantitative model of plasma flows and currents in Saturn's polar ionosphere, *J. Geophys. Res.*, *109*, A05212; doi:10.1029/2003JA010375.

Cowley, S. W. H., E. J. Bunce, T. S. Stallard, and S. Miller (2003a), Jupiter's polar ionospheric flows: Theoretical interpretation, *Geophys. Res. Lett.*, *30* (5), 1220; doi:10.1029/2002GL016030.

Cowley, S. W. H., I. I. Alexeev, E. S. Belenkaya, E. J. Bunce, C. E. Cottis, V. V. Kalegaev, J. D. Nichols, R. Prangé, and F. J. Wilson (2005b), A simple axisymmetric model of magnetosphere-ionosphere coupling currents in Jupiter's polar ionosphere, *J. Geophys. Res.*, *110*, A11209; doi:10.1029/2005JA011237.

Cowley, S. W. H., J. D. Nichols, and C. M. Jackman (2015), Down-tail mass loss by plasmoids in Jupiter's and Saturn's magnetospheres, *J. Geophys. Res.*, *120*, 6347–6356; doi:10.1002/2015JA021500.

Cowley, S. W. H., J. D. Nichols, and D. J. Andrews (2007), Modulation of Jupiter's plasma flow, polar currents, and auroral precipitation by solar wind-induced compressions and expansions of the magnetosphere: A simple theoretical model, *Ann. Geophys.*, *25*, 1433–1463.

Cowley, S. W. H., J. D. Nichols, and E. J. Bunce (2002), Distributions of current and auroral precipitation in Jupiter's middle magnetosphere computed from steady-state Hill-Pontius angular velocity profiles: solutions for current sheet and dipole magnetic field models, *Planet. Space Sci.*, *50*, 717–734.

Cowley, S. W. H., S. V. Badman, E. J. Bunce, J. T. Clarke, J.-C. Gérard, D. Grodent, C. M. Jackman, S. E. Milan, and T. K. Yeoman (2005a), Reconnection in a rotation-dominated magnetosphere and its relation to Saturn's auroral dynamics, *J. Geophys. Res.*, *110*, A02201; doi:10.1029/2004JA010796.

Dungey, J. W. (1961), Interplanetary magnetic field and the auroral zones, *Phys. Rev. Lett.*, *6*, 47–48.

Dungey, J. W. (1967), The theory of the quiet magnetosphere, in *Solar-Terrestrial Physics*, edited by J. W. King and W. S. Newman, 91–106, Academic Press, London.

Galand, M., L. Moore, I. Mueller-Wodarg, M. Mendillo, and S. Miller (2011), Response of Saturn's auroral ionosphere to electron precipitation: Electron density, electron temperature, and electrical conductivity, *J. Geophys. Res.*, *116*, A09306; doi:10.1029/2010JA016412.

Hill, T. W. (1979), Inertial limit on corotation, *J. Geophys. Res.*, *84*, 6554–6558.

Hill, T. W. (1980), Corotation lag in Jupiter's magnetosphere: Comparison of observation and theory, *Science*, *207*, 301–302.

Hill, T. W. (2001), The Jovian auroral oval, *J. Geophys. Res.*, *106*, 8101–8107.

Hunt, G. J., S. W. H. Cowley, G. Provan, E. J. Bunce, I. I. Alexeev, E. S. Belenkaya, V. V. Kalegaev, M. K. Dougherty, and A. J. Coates (2014), Field-aligned currents in Saturn's southern nightside magnetosphere: Subcorotation and planetary period oscillation components, *J. Geophys. Res.*, *119*, 9847–9899; doi:10.1002/2014JA020506.

Hunt, G. J., S. W. H. Cowley, G. Provan, E. J. Bunce, I. I. Alexeev, E. S. Belenkaya, V. V. Kalegaev, M. K. Dougherty, and A. J. Coates (2015), Field-aligned currents in Saturn's northern nightside magnetosphere: Evidence for interhemispheric current flow associated with planetary period oscillations, *J. Geophys. Res.*, *120*, 7552–7584; doi:10.1002/2015JA021454.

Hunt, G. J., S. W. H. Cowley, G. Provan, E. J. Bunce, I. I. Alexeev, E. S. Belenkaya, V. V. Kalegaev, M. K. Dougherty, and A. J. Coates (2016), Field-aligned currents in Saturn's magnetosphere: Local time dependence of southern summer currents in the dawn sector between midnight and noon, *J. Geophys. Res.*, *121*, 7785–7804; doi:10.1002/2016JA022712.

Iijima, T., and T. A. Potemra (1976), The amplitude distribution of field-aligned currents at northern high latitudes observed by Triad, *J. Geophys. Res.*, *81*, 2165–2174.

Iijima, T., and T. A. Potemra (1978), Large-scale characteristics of field-aligned currents associated with substorms, *J. Geophys. Res.*, *83*, 599–615.

Jackman, C. M., and S. W. H. Cowley (2006), A model of the plasma flow and current in Saturn's polar ionosphere under conditions of strong Dungey cycle driving, *Ann. Geophys.*, *24*, 1029–1055.

Jackman, C. M., N. Achilleos, E. J. Bunce, S. W. H. Cowley, M. K. Dougherty, G. H. Jones, S. E. Milan, and E. J. Smith (2004), Interplanetary magnetic field at ~9 AU during the declining phase of the solar cycle and its implications for Saturn's magnetospheric dynamics, *J. Geophys. Res.*, *109*, A11203; doi:10.1029/2004JA010614.

Jia, X., M. G. Kivelson, and T. I. Gombosi (2012), Driving Saturn's magnetospheric periodicities from the upper atmosphere/ionosphere, *J. Geophys. Res.*, *117*, A04215; doi:10.1029/2011JA017367.

Jinks, S. L., et al. (2014), Cassini multi-instrument assessment of Saturn's polar cap boundary, *J. Geophys, Res.*, *119*, 8161–8177; doi:10.1002/2014JA020367.

Khurana, K. K. (2001), Influence of solar wind on Jupiter's magnetosphere deduced from currents in the equatorial plane, *J. Geophys. Res.*, *106*, 25999–26016.

Kivelson, M. G., and D. J. Southwood (2005), Dynamical consequences of two modes of centrifugal instability in Jupiter's outer magnetosphere, *J. Geophys. Res.*, *110*, A12209; doi:10.1029/2005JA011176.

Krupp, N., V. M. Vasyliunas, J. Woch, A. Lagg, K. K. Khurana, B. H. Mauk, D. J. Williams, et al. (2004), Dynamics of the Jovian Magnetosphere, in *Jupiter: The Planet, Satellites, and Magnetosphere*, edited by F. Bagenal et al., Cambridge University Press, New York, 617–638.

Lamy, L., et al. (2012), Earth-based detection of Uranus' aurorae, *Geophys. Res. Lett.*, *39*, L07105; doi: 10.1029/2012Gl051312.

Mejnertsen, L., J. P. Eastwood, J. Chittenden, and A. Masters (2016), Global MHD simulations of Neptune's magnetosphere, *J. Geophys. Res.*, *121*, 7497–7513; doi:10.1002/2015JA022272.

Milan, S. E., G. Provan, and B. Hubert (2007), Magnetic flux transport in the Dungey cycle: A survey of dayside and nightside reconnection rates, *J. Geophys. Res.*, *112*, A01209; doi:10.1029/2006JA011642.

Millward, G., S. Miller, T. Stallard, A. D. Aylward, and N. Achilleos (2002), On the dynamics of the Jovian ionosphere and thermosphere, III. The modelling of auroral conductivity, *Icarus*, *160*, 95–107.

Millward, G., S. Miller, T. Stallard, N. Achilleos, and A. D. Aylward (2005), On the dynamics of the Jovian ionosphere and thermosphere, IV. Ion-neutral coupling, *Icarus*, *173*, 200–211.

Nichols, J. D. (2011a), Magnetosphere-ionosphere coupling at Jupiter-like exoplanets with internal plasma sources: implications for detectability of auroral radio emissions, *Monthly Not. R. Astron. Soc.*, *414*, 2125–2138; doi:10.1111/j.1365-2966.2011.18528.x.

Nichols, J. D. (2011b), Magnetosphere-ionosphere coupling in Jupiter's middle magnetosphere: Computations including a self-consistent current sheet magnetic field model, *J. Geophys. Res.*, *116*, A10232; doi:10.1029/2011JA016922.

Nichols, J. D., and S. W. H. Cowley (2003), Magnetosphere-ionosphere coupling currents in Jupiter's middle magnetosphere: dependence on the effective ionospheric Pedersen conductivity and iogenic plasma mass outflow rate, *Ann. Geophys.*, *21*, 1419–1441.

Nichols, J. D., and S. W. H. Cowley (2004), Magnetosphere-ionosphere coupling currents in Jupiter's middle magnetosphere: effect of precipitation-induced enhancement of the ionospheric Pedersen conductivity, *Ann. Geophys.*, *22*, 1799–1827.

Nichols, J. D., et al. (2012), Origin of electron cyclotron maser-induced radio emissions at ultra-cool dwarfs: magnetosphere-ionosphere coupling currents, *Astrophys. J.*, *760*, 59; doi:10.1088/0004-637X/760/1/59.

Nichols, J. D., S. W. H. Cowley, and D. J. McComas (2006), Magnetopause reconnection rate estimates for Jupiter's magnetosphere based on interplanetary measurements at ~5 AU, *Ann. Geophys.*, *24*, 393–406.

Nishida, A. (1966), Formation of plasmapause, or magnetospheric plasma knee, by the combined action of magnetospheric convection and plasma escape from the tail, *J. Geophys. Res.*, *71*, 5669–5679.

Pontius, D. H., Jr. (1997), Radial mass transport and rotational dynamics, *J. Geophys. Res.*, *102*, 7137–7150.

Provan, G., S. W. H. Cowley, J. Sandhu, D. J. Andrews, and M. K. Dougherty (2013), Planetary period magnetic field oscillations in Saturn's magnetosphere: Postequinox abrupt nonmonotonic transitions to northern system dominance, *J. Geophys. Res. Space Physics*, *118*, 3243–3264; doi:10.1002/jgra.50186.

Provan, G., S. W. H. Cowley, L. Lamy, E. J. Bunce, G. J. Hunt, P. Zarka, and M. K. Dougherty (2016), Planetary period oscillations in Saturn's magnetosphere: Coalescence and reversal of northern and southern periods in late northern spring, *J. Geophys. Res.*, *121*, 9829–9862; doi:10.1002/2016JA023056.

Radioti, A., D. Grodent, J.-C. Gérard, B. Bonford, J. Gustin, W. Pryor, J. M. Jasinski, and C. S. Arridge (2013), Auroral signatures of multiple magnetopause reconnection at Saturn, *Geophys. Res. Lett.*, *40*, 4498–4502; doi:10.1002/grl.50889.

Ray, L. C., N. A. Achilleos, and J. N. Yates (2015), The effect of including field-aligned potentials in the coupling between Jupiter's thermosphere, ionosphere, and magnetosphere, *J. Geophys. Res. Space Physics*, *120*, 6987–7005; doi:10.1002/2015JA021319.

Ray, L. C., N. A. Achilleos, M. F. Vogt, and J. N. Yates (2014), Local time variations in Jupiter's magnetosphere-ionosphere coupling system, *J. Geophys. Res.*, *119*, 4740–4751; doi:10.1002/2014JA019941.

Smith, C. G. A. (2014), On the nature and location of the proposed twin vortex systems in Saturn's polar upper atmosphere, *J. Geophys. Res.*, *119*, 5964–5977; doi:10.1002/2014JA019934.

Smith, C. G. A., and A. D. Aylward (2009), Coupled rotational dynamics of Jupiter's thermosphere and magnetosphere, *Ann. Geophys.*, *27*, 199–230.

Smith, C. G. A., S. Miller, and A. D. Aylward (2005), Magnetospheric energy inputs into the upper atmospheres of the giant planets, *Ann. Geophys.*, *23*, 1943–1947.

Southwood, D. J., and M. G. Kivelson (2007), Saturnian magnetospheric dynamics: Elucidation of a camshaft model, *J. Geophys. Res.*, *112*, A12222, doi:10.1029/2007JA012254.

Southwood, D. J., and S. W. H. Cowley (2014), The origin of Saturn's magnetic periodicities: Northern and southern current systems, *J. Geophys. Res.*, *119*, 1563–1571; doi:10.1002/2013JA019632.

Talboys, D. L., C. S. Arridge, E. J. Bunce, A. J. Coates, S. W. H. Cowley, M. K. Dougherty, and K. K. Khurana (2009), Signatures of field-aligned currents in Saturn's nightside magnetosphere, *Geophys. Res. Lett.*, *36*, L19107; doi:10.1029/2009GL039867.

Talboys, D. L., E. J. Bunce, S. W. H. Cowley, C. S. Arridge, A. J. Coates, and M. K. Dougherty (2011), Statistical characteristics of field-aligned currents in Saturn's nightside magnetosphere, *J. Geophys. Res.*, *116*, A04213; doi:10.1029/2009JA016102.

Tao, C., H. Fujiwara, and Y. Kasaba (2009), Neutral wind control of the Jovian magnetosphere-ionosphere current system, *J. Geophys. Res.*, *114*, A08307; doi:10.1029/2008JA013966.

Tao, C., H. Fujiwara, and Y. Kasaba (2010), Jovian magnetosphere-ionosphere current system characterized by diurnal variation of ionospheric conductance, *Planet. Space Sci. 58*, 351–364; doi:10.1016/j.pss.2009.10.005.

Thomas, N., F. Bagenal, T. W. Hill, and J. K. Wilson (2004), The Io neutral clouds and plasma torus, in *Jupiter: The Planet, Satellites, and Magnetosphere*, edited by F. Bagenal et al., Cambridge University Press, New York, 561–591.

Thomsen, M. F., C. M. Jackman, D. G. Mitchell, G. Hospodarsky, W. S. Kurth, and K. C. Hansen (2015), Sustained lobe reconnection in Saturn's magnetotail, *J. Geophys. Res. Space Physics*, *120*, 10,257–10,274; doi:10.1002/2015JA021768.

Thomsen, M. F., C. M. Jackman, R. L. Tokar, and R. J. Wilson (2014), Plasma flows in Saturn's nightside magnetosphere, *J. Geophys. Res. Space Physics*, *119*, 4521–4535; *doi*:10.1002/2104JA019912.

Vasyliunas, V. M. (1983), Plasma distribution and flow, in *Physics of the Jovian Magnetosphere*, edited by A. J. Dessler, Cambridge University Press, New York, 395–453.

Yates, J. N., N. Achilleos, and P. Guio (2012), Influence of upstream solar wind on thermospheric flows at Jupiter, *Planet. Space Sci.*, *61*, 15–31; doi:10.1016/j.pss.2011.08.007.

Part VI: Other Current Systems

28

The Bow Shock Current System

Ramon E. Lopez

ABSTRACT

The solar wind is a supermagnetosonic flow, thus a shock wave forms in front of Earth to slow the solar wind. The shock transfers solar-wind flow energy into plasma thermal energy and electromagnetic energy as the plasma is heated and compressed across the shock. The compression of the magnetic field transverse to the shock means that there is a dynamo current that flows on the shock, and this current exists at all times. When the solar-wind magnetosonic Mach number is below about 3, the conversion of solar-wind flow energy into electromagnetic energy at the bow shock is the dominant form of energy conversion and the force from the bow shock current is the largest force on the solar wind in the system. The dynamo current flowing on the bow shock deposits its energy mostly in the magnetosheath plasma, accelerating the flow around the magnetospheric obstacle. However, some current may close through Earth's ionosphere, directly depositing energy into the ionospheric load. In this chapter, we consider these issues, discuss the closure of the bow shock current for various orientations of the interplanetary magnetic field, and briefly consider the broader question of shock dynamo and load currents in space plasmas.

28.1. INTRODUCTION

The interplay between plasmas and magnetic fields in space on the large scale is well described by magnetohydrodynamics (MHD). The typical form of the MHD equations, particularly the momentum equation, is grounded in the language of forces and plasma flows. From this perspective, the forces on the plasma from pressure gradients and $\mathbf{J}\times\mathbf{B}$ are equal to the inertial term. The self-consistent current is given from the curl of the magnetic field, and Ampère's law is a diagnostic equation, not an evolution equation. The electric field in any given frame of reference is given by the plasma velocity in that frame [*Vasyliūnas*, 2001]. So, from this perspective, the forces on the plasma make the plasma move, dragging the magnetic field with it, creating an electric field in any given frame of reference, and electric currents flow self consistently according to the curl of the magnetic field.

Depending on the direction of the forces relative to the flow and of the electric field relative to the current, the electric currents are associated with the transfer of energy from the plasma flow to the magnetic field (dynamo currents), or with the transfer of energy from the magnetic field to the plasma (load currents). Since the solar wind is always supersonic, the energy in the solar-wind flow is always larger than the energy in the interplanetary magnetic field (IMF). Understanding the extraction and transfer of solar-wind flow energy to power magnetospheric processes is one of the most fundamental issues in magnetospheric physics.

The textbook approach for calculating the position of the magnetopause is to determine the point where the dynamic pressure of the solar wind is balanced by the magnetic pressure of Earth's dipole [e.g., *Kivelson and Russell*, 1995]. The Chapman-Ferraro currents then flow along the boundary, containing the geomagnetic field

Department of Physics, The University of Texas at Arlington, Arlington, Texas, USA

Electric Currents in Geospace and Beyond, Geophysical Monograph 235, First Edition.
Edited by Andreas Keiling, Octav Marghitu, and Michael Wheatland.

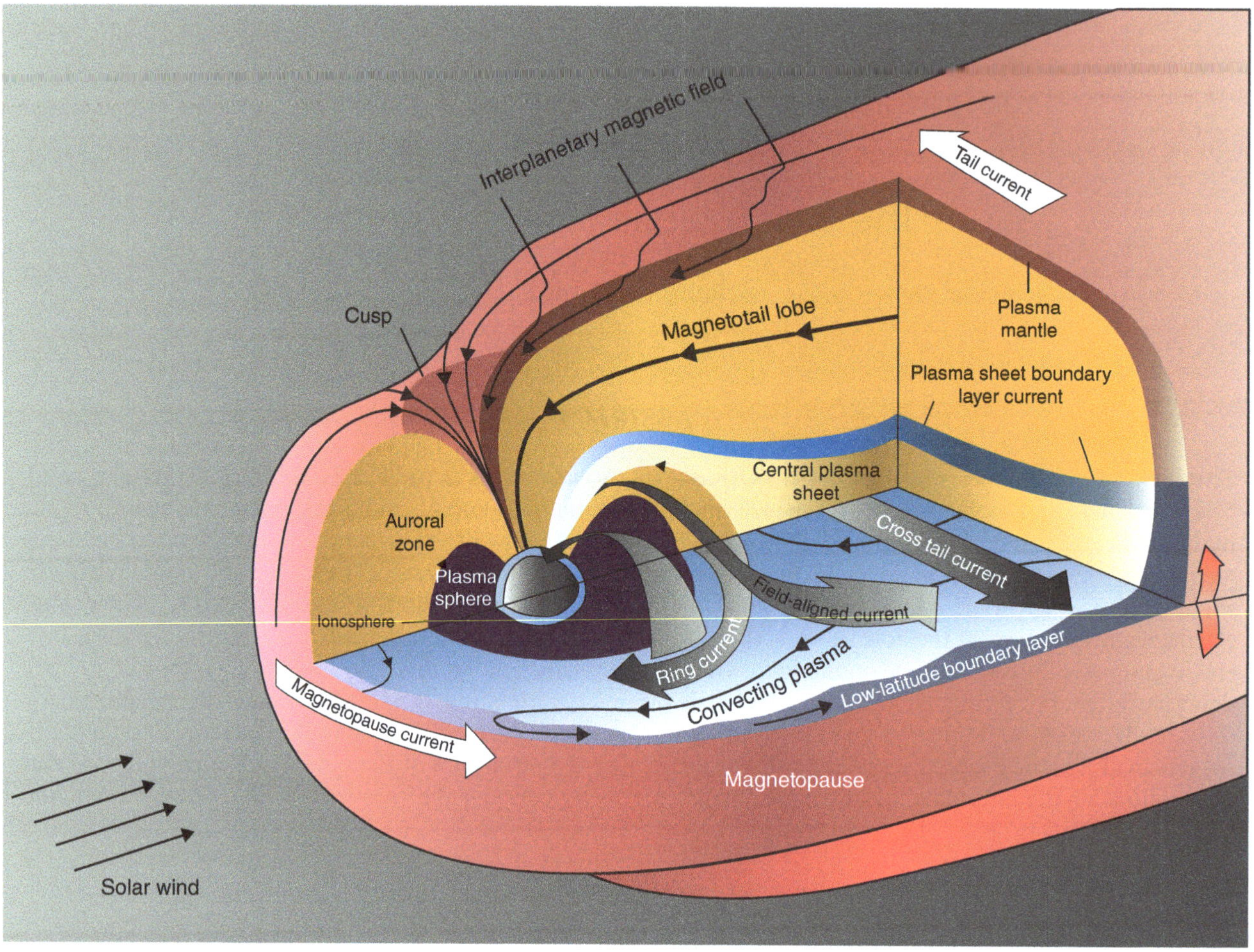

Figure 28.1 A schematic of the magnetosphere and magnetospheric currents.

inside the magnetosphere. This "zero-IMF" approach results in the Chapman-Ferraro current on the dayside acting as a load, and at high latitudes poleward of the cusp (and further down the tail), the current acts as a generator. An illustration of these boundary currents, along with other currents in the magnetosphere, can be seen in Figure 28.1, for a situation where the IMF is southward or zero. In that case, the dayside boundary current runs from dawn to dusk, consistent with the northward geomagnetic field contained inside of the magnetosphere.

Figure 28.1 is just an artist's conception of the current system, but much more illustrative results can be obtained using global magnetohydrodynamic models [e.g., *Siscoe et al.*, 2000]. Figure 28.2 shows the quantity **J•E** in the meridional (*X-Z*) plane calculated using the University of Michigan BATS-R-US code [*Powell et al.*, 1999]. Where **J•E** is negative, mechanical energy is being extracted from the solar-wind flow and converted into the electromagnetic energy (dynamo), and where it is positive, electromagnetic energy is being transferred into the plasma (load).

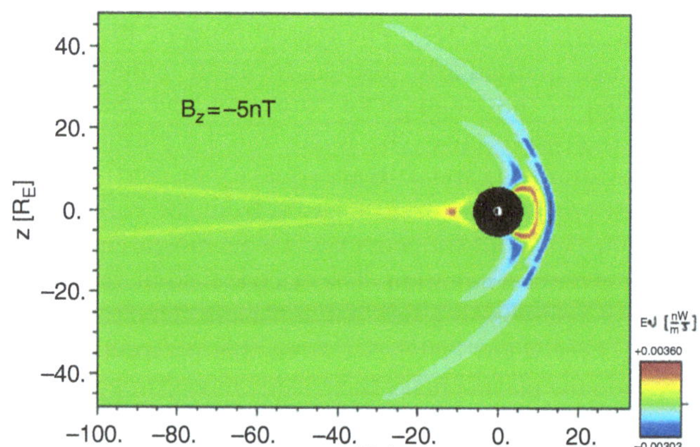

Figure 28.2 BATS-R-US MHD model results showing dynamo (blue) and load (red) regions.

As expected, the dayside region at latitudes lower than that of the cusp are loads while the area at latitudes higher than the cusp are dynamos. However, another feature is clearly visible. All along the bow shock, energy is being taken out of the solar-wind flow and being put into the magnetic field (as well as thermal energy). While this figure was made from a specific run archived at the Community Coordinated Modeling Center with steady solar-wind conditions (V_x=–400 km/s, n=5 cm^{-3}, and B_z=–5 nT), the basic result regarding the bow shock current does not depend on the details of the solar wind or the specific MHD code used. The bow shock current is always a dynamo current extracting mechanical energy from the solar wind, and it is to that dynamo that we now turn.

28.2. THE BOW SHOCK DYNAMO

The large-scale features of MHD shocks are well understood, the subject has been a textbook topic for a long time [e.g., *Boyd and Sanderson*, 1969], and the properties of Earth's bow shock were documented by early spacecraft missions [e.g., *Formisano et al.*, 1973]. The conservation of energy and momentum across the shock lead to the Rankine-Hugoniot jump conditions. On crossing the shock, the plasma slows, and is compressed and heated. The magnetic field is also compressed by the same factor as the plasma density, and the theoretical maximum compression ratio for a very high Mach number shock is four.

The situation is illustrated in Figure 28.3 for southward IMF (negative B_z). Since the plasma is heated and compressed, the plasma pressure downstream is much larger than upstream, and the pressure gradient force points upstream. This force does work against the solar-wind plasma as it flows across the shock, converting ordered flow energy into disordered thermal energy. The magnetic field compression across the shock means that there is a curl of the magnetic field, that is, a current, which flows on the face of the shock. The resulting **J**×**B** force, like the pressure gradient force, does work against the plasma flow, converting solar-wind mechanical energy into magnetic energy.

The orientation of the magnetic field in Figure 28.3 is not a factor in any of the preceding discussion. The field could have been draw in the opposite direction (northward IMF) and the only thing that would change would be the direction of the current on the bow shock, which would reverse. However, the bow shock current would still be a dynamo current. Moreover, were the field in the diagram to be labeled B_y, there would still be no change in the dynamics of the situation.

Under normal solar-wind conditions, the magnetosonic Mach number of the solar wind is large enough that the plasma pressure downstream of the shock is larger than the magnetic pressure. Under these conditions, the plasma pressure gradient is the largest force on the solar wind and solar-wind kinetic energy is converted preferentially into plasma thermal energy. The **J**×**B** force at the shock is relatively small, as is the fraction of solar-wind kinetic energy that is converted into magnetic energy (although a current always exists, and it is a dynamo current, no matter how small).

However, under conditions of low magnetosonic Mach number (around 3 or less), the flow downstream of the shock can be magnetically dominated and the bow shock current can exert the majority of the force on the solar wind [*Siscoe et al.*, 2004; *Lopez et al.*, 2010]. Moreover, when the solar-wind magnetosonic Mach number is low, the high-latitude magnetopause dynamo region essentially disappears and the bow shock is the only place in the system where substantial energy is extracted from the solar-wind flow [*Lopez et al.*, 2011]. Such conditions are relatively rare. From 1 January 2013 to 1 January 2016, only 1.3% of the 5 min averaged OMNI data intervals [*King and Papitashvili*, 2005] had a magnetosonic Mach number of 3.1 or less. However, a low magnetosonic Mach number is generally an indicator of large IMF magnitudes, and such periods are of disproportionate interest since they drive space weather events.

Figures 28.2 and 28.3 are both slices in the meridional plane showing the currents going into or out of the page. However, the currents must close in some three-dimensional configuration, and the solar-wind kinetic energy extracted at the bow shock must be delivered to a load somewhere. While the fact that the **J**×**B** force arising from the bow shock current does work against the solar wind does not depend on the IMF direction, the direction of the current and its subsequent closure does depend on the IMF direction. Therefore, we will examine various IMF orientations separately, beginning with southward IMF.

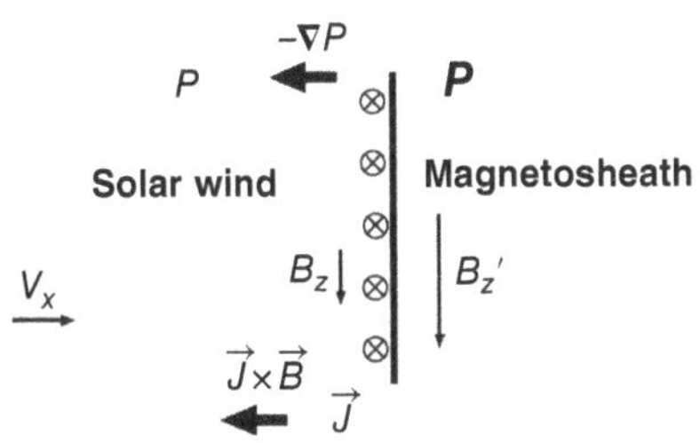

Figure 28.3 Forces at the bow shock.

28.3. BOW SHOCK CURRENT CLOSURE FOR IMF B_z

28.3.1. Southward IMF

Figure 28.4 shows current streamlines in the Lyon-Fedder-Mobarry, or LFM, global MHD code [*Lyon et al.*, 2004] under nominal, southward IMF solar-wind conditions (V_x=–400 km/s, n=5 cm^{-3}, and B_z=–5 nT).

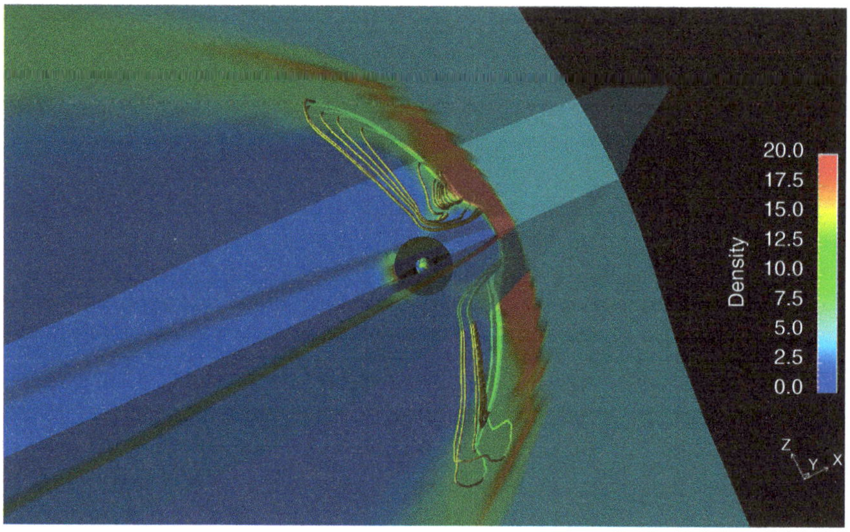

Figure 28.4 Current closure for moderate southward IMF in the LFM global MHD code. Current streamlines (yellow) are started at the bow shock (Southern Hemisphere) or on the magnetopause (Northern Hemisphere).

The current streamlines in the Northern Hemisphere show the Chapman-Ferraro current circulating around the cusp, producing a load at lower latitudes as the current flows from dawn to dusk, and a dynamo poleward of the cusp where the current reverses direction. The Southern Hemisphere current streamlines were initiated at the bow shock, and we can see that they cross the magnetosheath and close on the magnetopause. *Siebert and Siscoe* [2002] discussed this closure in detail, and pointed out that dayside reconnection is a load and the only dynamo current that connects to the Chapman-Ferraro current in the reconnection region is the bow shock current. Thus, they argue that the bow shock current dynamo powers the dayside reconnection region. It also provides power to the nightside reconnection region since the bow shock current is partially closed by the cross-tail current in the magnetotail [*Tang et al.*, 2009].

The bow shock current also connects directly to the Region 1 Birkeland currents [e.g., *Iijima and Potemra*, 1976], a situation that is most readily seen when the IMF is large [e.g., *Siscoe et al.*, 2002; *Lopez et al.*, 2008]. Figure 28.5 shows a case with $V_x = -400$ km/s, $n = 5\,\text{cm}^{-3}$, and $B_z = -20$ nT, with several current streamlines started at the bow shock. We see that they connect to the currents going into and out of the inner boundary of the MHD calculation. Those currents are then projected into the ionosphere where they are the source terms for the ionospheric potential.

The realization that the bow shock currents connect directly to Region 1 currents is not a new one. It was first proposed by *Fedder et al.* [1997] based on a comparison of LFM results and Polar observations, and it is an essential feature for understanding the interaction of the solar wind with the magnetosphere during periods of low solar-wind magnetosonic Mach number [*Siscoe et al.*, 2002, 2004; *Borovsky et al.*, 2009; Lopez *et al.*, 2010, 2011]. Under conditions of moderate driving, bow shock current contribution to the total Region 1 current system is modest and the majority of the Region 1 current is generated by dayside merging [e.g., *Fedder and Lyon*, 1987], the high-latitude dynamo current poleward of the cusp [*Tanaka et al.*, 2016], and pressure gradients at magnetospheric boundaries [*Yang et al.*, 1994; Tanaka *et al.*, 2016]. However, during strong driving conditions, the contribution to the Region 1 current system from currents flowing on open field lines that connect to the bow shock dynamo can be a large fraction of the total Region 1 Birkeland current [*Lopez et al.*, 2008].

Figure 28.6 shows a schematic of the bow shock current system for southward IMF. The bow shock current is a maximum near the nose where the flow is perpendicular to the shock and the compression of the solar-wind flow is the greatest and smaller farther from noon. Since the magnitude of the dusk-to-dawn shock current depends on local time dependence, there must be a divergence of the current as indicated in Figure 28.6. MHD simulations show that there is a volume current across the magnetosheath that has Region 1 polarity (see Fig. 11 of *Lopez et al.* [2010]). This produces a **J**×**B** force in the magnetosheath that diverts the flow around the magnetospheric obstacle and accelerates the magnetosheath plasma so that it rejoins the solar wind some distance down the magnetotail. The plasma pressure gradient points toward noon so that pressure force also works to divert the magnetosheath flow around the magnetosphere and accelerate the magnetosheath flow in an antisunward direction.

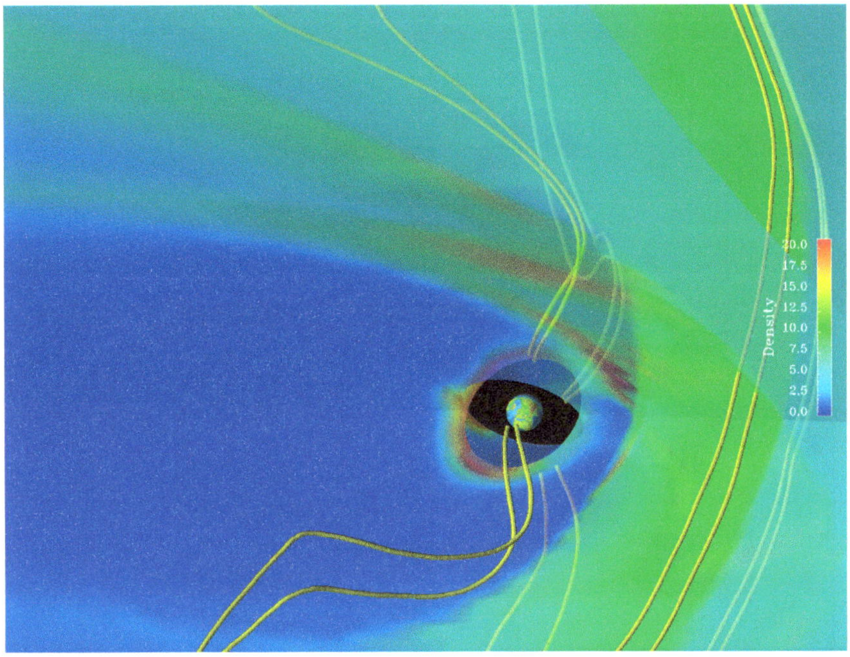

Figure 28.5 Current closure for large IMF with bow shock current closing as Region 1 current through the ionosphere.

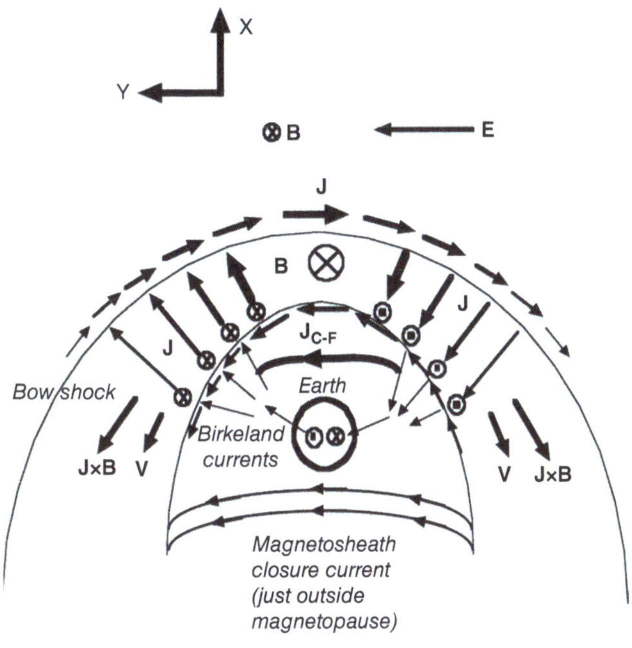

Figure 28.6 Schematic of bow shock current closure for southward IMF.

Those forces determine what fraction of the incident solar-wind flow is able to reach the dayside merging line, thus establishing the geoeffective length in the solar wind [*Lopez et al.*, 2010; *Lopez*, 2016].

Under conditions of low solar-wind magnetosonic Mach number, the magnetosheath is magnetically dominated and the **J**×**B** force is the largest force on the magnetosheath plasma. Under those conditions, just as most of the solar-wind kinetic energy is converted into magnetic energy, most of that energy is delivered back to the plasma by the **J**×**B** force in the magnetosheath. In fact, under all conditions, the acceleration of the magnetosheath plasma is the primary load for the bow shock dynamo. The majority of the bow shock current closes as current on or just outside the magnetospheric boundary, as well as through the cross-tail current in the magnetotail [*Tang et al.*, 2009]. When the bow shock current is very large, this overwhelms the Chapman-Ferraro current and produces a load along the region poleward of the cusp since the direction of the bow shock closure current is opposite to the direction of the Chapman-Ferraro current.

As noted above, a portion of the bow shock current closes directly through the ionosphere in the polar cap as part of Region 1, but on open field lines. There the Poynting flux associated with the current can directly deposit energy extracted at the bow shock without the intermediary process of magnetic reconnection. This direct coupling of the ionosphere to the bow shock dynamo represents a qualitatively different mode of coupling between the ionosphere and the solar wind than either the merging interaction [*Dungey*, 1961] or the viscous interaction [*Axford and Hines*, 1961].

Figure 28.7 shows OMNI data on 1 June 2013. After 0200 UT, the solar wind was in a low Mach number state during which time the bow shock current would

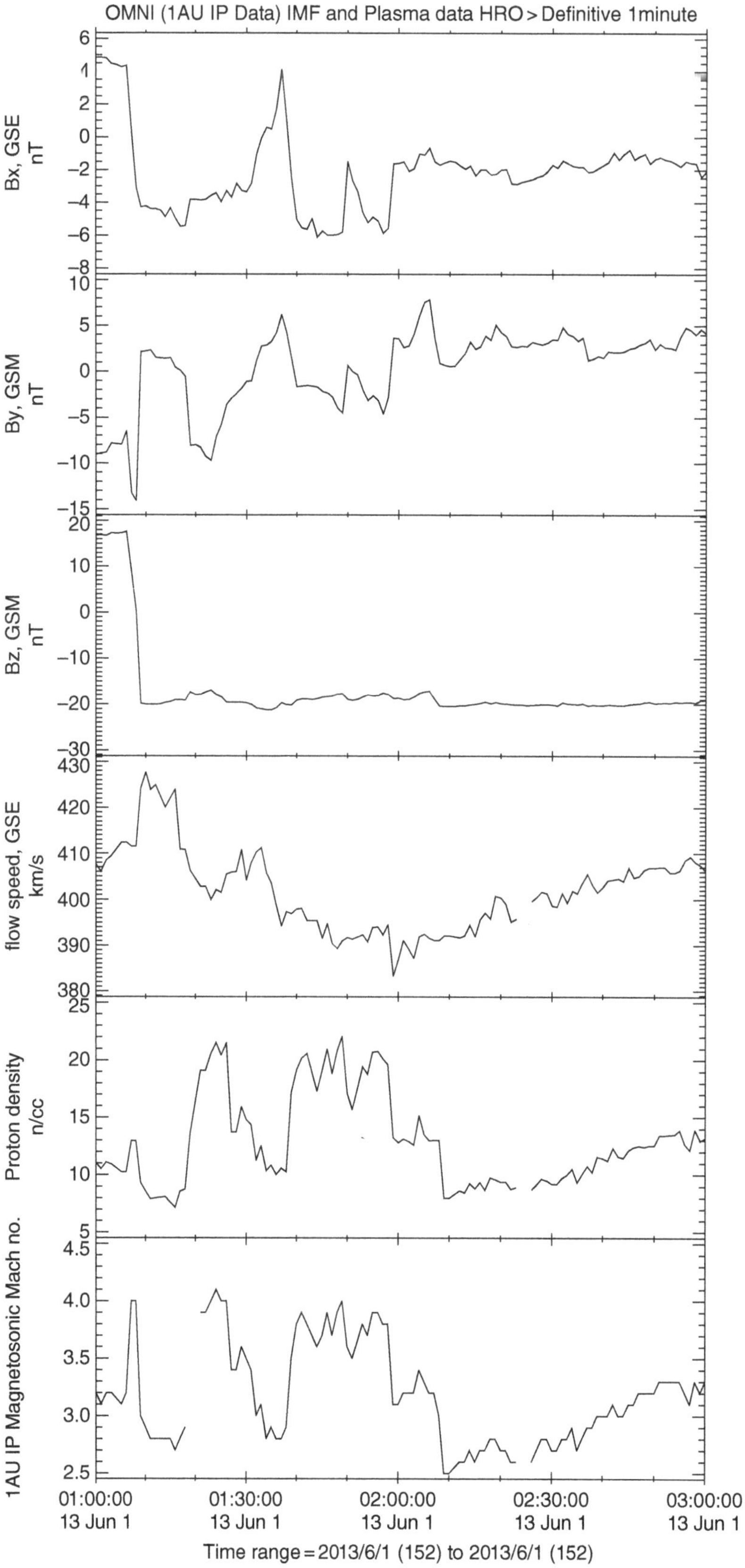

Figure 28.7 OMNI data for 1 June 2013.

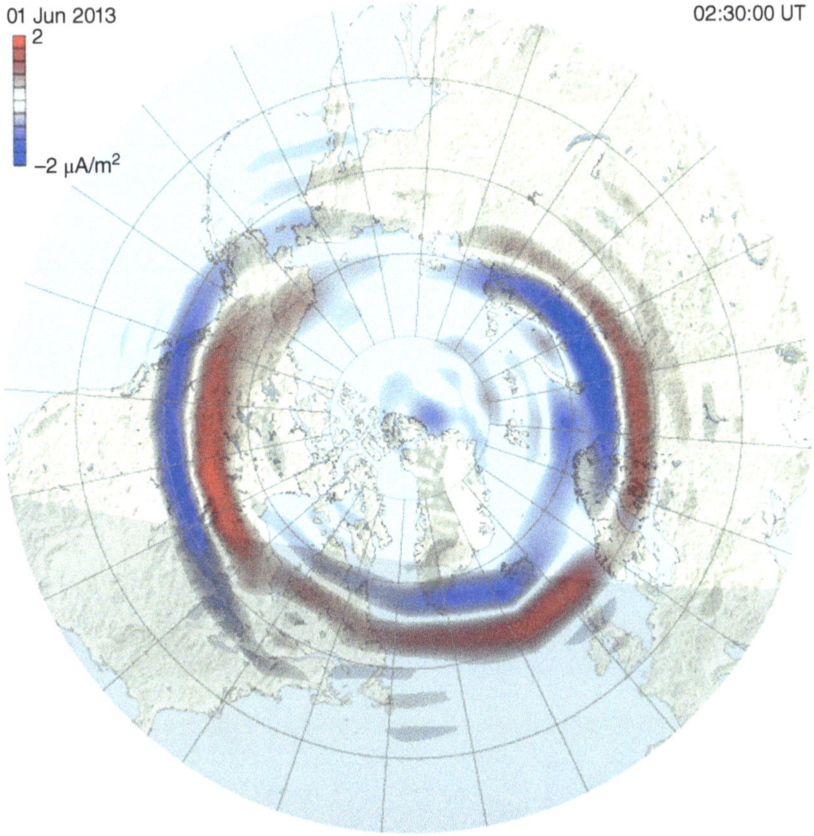

Figure 28.8 AMPERE data for 1 June 2013. The grid is for geomagnetic coordinates.

have been dynamically important. Figure 28.8 presents observations on that day from the Active Magnetosphere and Planetary Electrodynamics Response Experiment (AMPERE), which provides global snapshots of the ionospheric Birkeland currents. At 0230 UT, a classically southward IMF Birkeland current system was evident with very large Region 1 (downward on the dawn and upward on the dusk) and Region 2 (downward on the dusk and upward on the dawn) currents. There were strong currents in the nightside since at 0220 UT, the AE index showed the onset of a period of activity reaching 1200 nT.

The DMSP spacecraft are in low-altitude, sun-synchronous orbits, and they carry a variety of instrumentation, in particular precipitating particle sensors. Those data provide a robust determination of the open-closed field line boundary [*Gussenhoven and Mullen*, 1989] from which one can determine if auroral current systems are on open or closed field lines [e.g., *Lopez et al.* 1991, 2008]. The boundary between open and closed field lines is marked by a cutoff of the precipitation of the keV ions trapped in the magnetosphere. Figure 28.9, which presents data from DMSP F16, is a good example. The spacecraft crossed the polar cap around 0230 UT on 1 June 2013. At 02:33:43, DMSP F16 crossed from open to closed field lines at MLAT = 66.10° and MLT = 0723, exiting from the polar rain precipitation (unstructured, low-energy electrons) typical of the polar cap to the keV ions and structured electron precipitation associated with auroral arcs. Inspecting Figure 28.8, we see that this point was well within the distribution of downward Region 1 Birkeland current. In fact, it appears that about half or more of the Region 1 Birkeland current was on open field lines at that MLT. The current on open fields lines must have closed at least in part on the bow shock as illustrated in Figure 28.6.

28.3.2. Northward IMF

For northward IMF, much of Figure 28.6 still applies, except that the directions of all of the currents are reversed so that the bow shock current flows from dawn to dusk and the closure currents flow from dusk to dawn. However, the locations of dynamo and load for the bow shock currents remain the same, and the electromagnetic energy extracted at the shock is delivered back to the plasma in the magnetosheath. The bow shock current still will close primarily along or just outside the magnetospheric boundary, and none will close through the cross-tail current, which is directed from dawn to dusk.

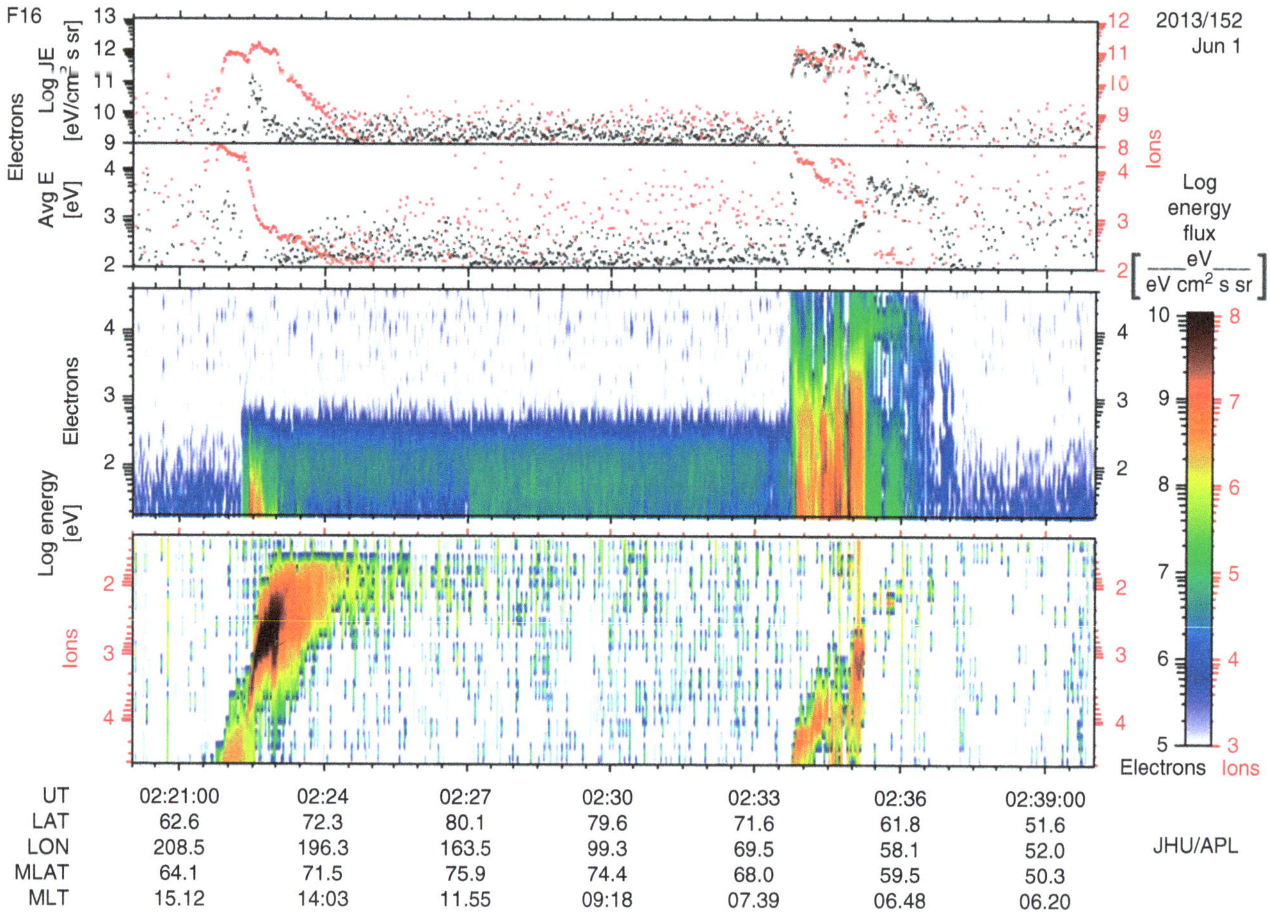

Figure 28.9 DMSP F16 particle precipitation data for 1 June 2013.

The major difference with the southward IMF case is that magnetic merging is restricted to a high-latitude region poleward of the cusp [*Crooker*, 1992; *Fedder and Lyon*, 1995; *Bhattarai et al.*, 2012]. As a consequence, the overall merging interaction is limited because of the restricted geoeffective length [*Lopez*, 2016] and the reverse convection NBZ current system is much weaker than Birkeland current system produced by a southward IMF of the same magnitude. Moreover, since merging and reconnection for purely northward IMF occur at a single point [*Song et al.*, 2000; *Bhatterai and Lopez*, 2013], in reality there will be very little open flux on which current can flow. Thus, the amount of bow shock current that closes in the ionosphere is very small.

We can see that this is the case by considering data from 25 October 2011. Figure 28.10 presents the OMNI data for this day, and we can see that the IMF was dominated by a large northward B_z component, and around 1300 UT, B_y was only 2 nT. Figure 28.11 shows the AMPERE data for this period, and the NBZ system (downward on the dusk and upward on the dawn) was present at high latitudes along the terminator. Note that the solar-wind values and the magnitudes of the components of the IMF were quite similar on 25 October 2011 and 1 June 2013. This means that the bow shock current in the two cases would have been of similar magnitude (and reversed direction), but the amount of Birkeland current into and out of the ionosphere was very different, with much more Birkeland current during southward IMF.

DMSP F17 flew through the approximate poleward edge of the postnoon portion of the NBZ current system from 1253 UT to 1254 UT (data not shown), and at that time encountered precipitating keV ions. We surmise that these ions were formerly magnetosheath plasma on closed field lines on the dayside of the magnetosphere that was trapped there by the northward IMF reconnection cycle, and which is very evident in MHD simulations with northward IMF [*Bhatterai and Lopez*, 2013]. There was no evidence of an open polar cap, though some small sections of field lines may have been open due to the small IMF B_y component. However, there really was little access for any current on open field lines to reach the ionosphere in this case. Thus, even in cases where there is a very strong NBZ current and a northward IMF merging

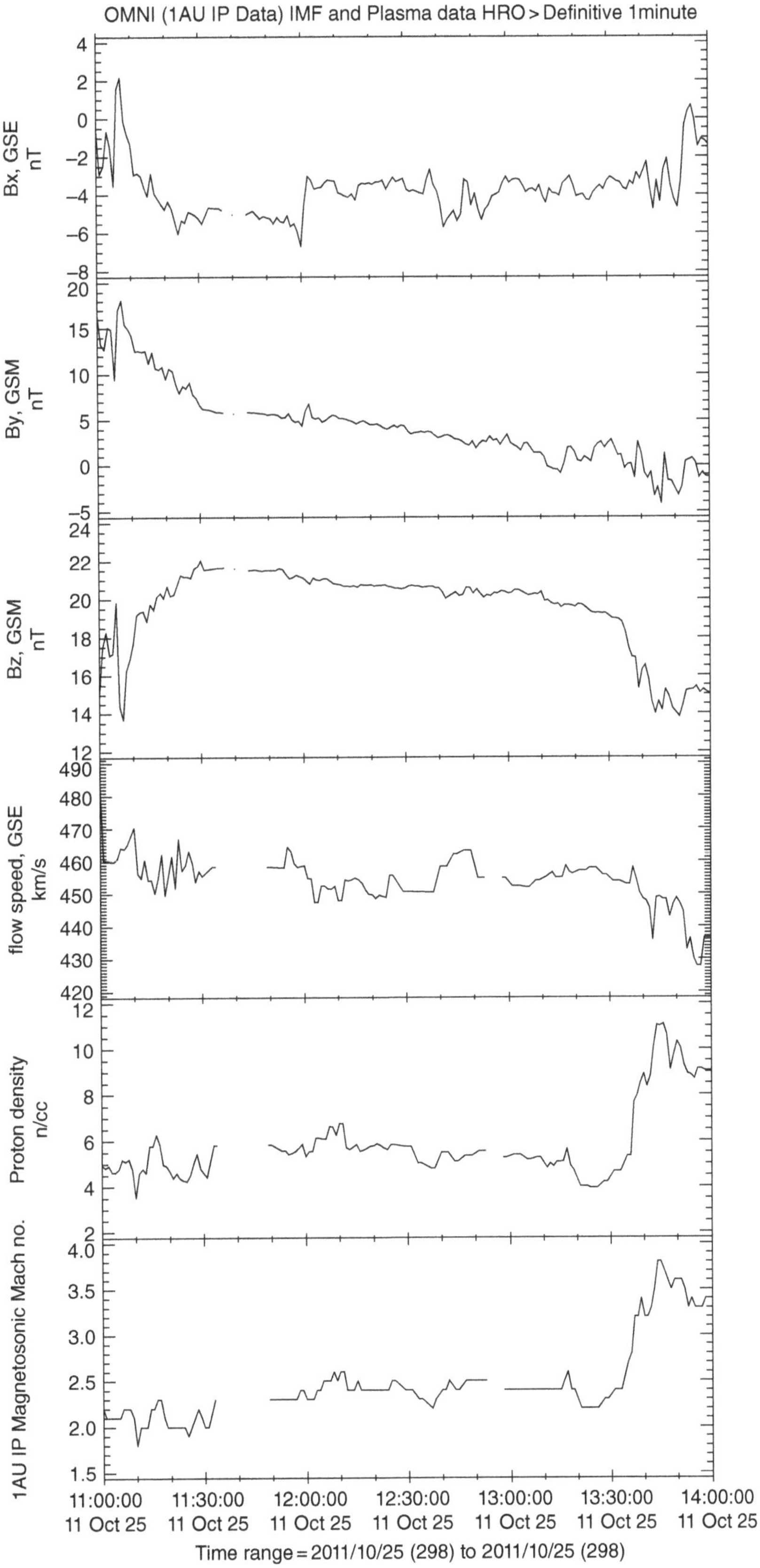

Figure 28.10 OMNI data for 25 October 2011.

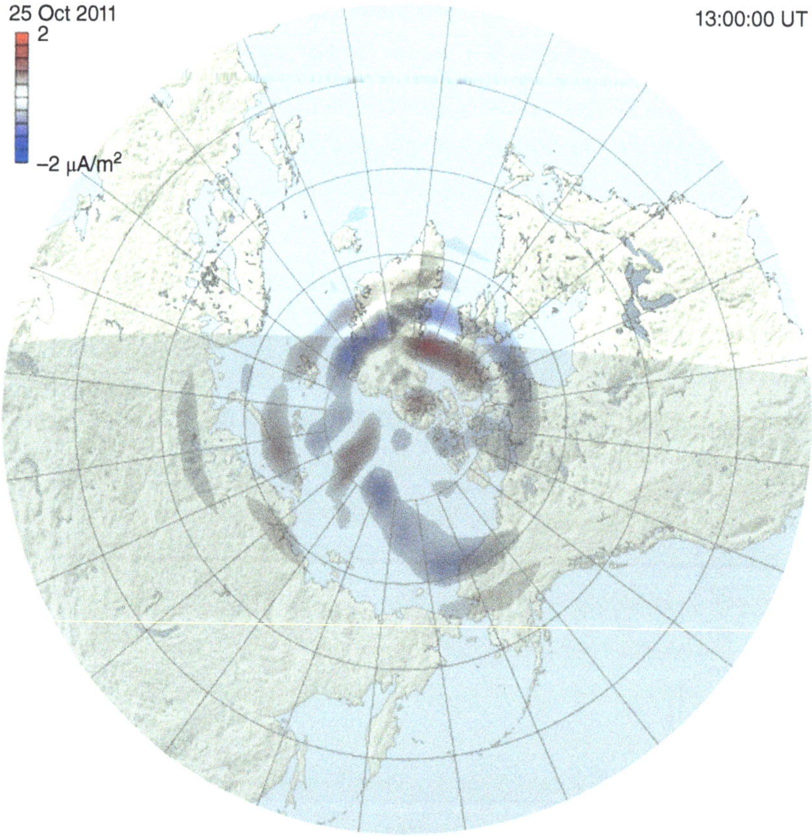

Figure 28.11 AMPERE data for 25 October 2013. The grid is for geomagnetic coordinates.

cycle [e.g., *Clauer et al.*, 2016], we would not expect the direct ionospheric impact of the bow shock current to be large, even if that current dominates the overall interaction with the solar wind.

28.4. BOW SHOCK CURRENT CLOSURE FOR EAST-WEST IMF

Figure 28.12 shows current streamlines originating in the bow shock in the LFM global MHD code for a situation with a purely negative IMF B_y (V_x=–400 km/s, n=5 cm^{-3}, and B_y=–10 nT). The current streamlines cross the magnetosheath, loop along the magnetopause, and some of the current flows on open field lines to the inner boundary of the MHD simulation, when it is projected into the ionosphere. The generalized bow shock current system for this situation is illustrated in Figure 28.13. The current on the bow shock now flows from south (north) to north (south) for negative (positive) IMF B_y. The Chapman-Ferraro current drawn in the diagram is actually just the north-south component of the current. The total Chapman-Ferraro current still flows from dawn to dusk at low latitudes and from dusk to dawn at high latitudes as in the case of southward (or zero) IMF. This total current is consistent with the magnetic shear across the magnetopause.

The current flowing on open field lines in this case is a unipolar current, shown in Figure 28.13 as going into the Northern Hemisphere and coming out of the Southern Hemisphere. In the ionosphere, this current must be closed by Pedersen currents that connect to the Region 2 current system driven by pressure gradients inside the magnetosphere [e.g., *Tanaka et al.*, 2016]. The magnetic stress delivered to the ionosphere by this unipolar current produces a circulation of ionospheric plasma that is part of the lobe cell convection described by *Crooker et al.* [1998].

As discussed above, the amount of current in the bow shock depends on the solar- wind magnetosonic Mach number. When the magnetosonic Mach number is low, the bow shock current will be large and dynamically important to the system. Figure 28.14 presents OMNI data from 17 September 2011 for such a period during which a large positive IMF B_y component dominated the solar wind–magnetosphere interaction. The IMF B_z was negative, but not large, and B_x was also small. This would produce a bow shock current that would flow mostly from north to south. At 1900 UT, the magnetosonic Mach number dropped to about 2, and stayed at that level for most of the next hour.

Figure 28.15 presents AMPERE observations for this event. At 1910 UT, the system had just entered into the

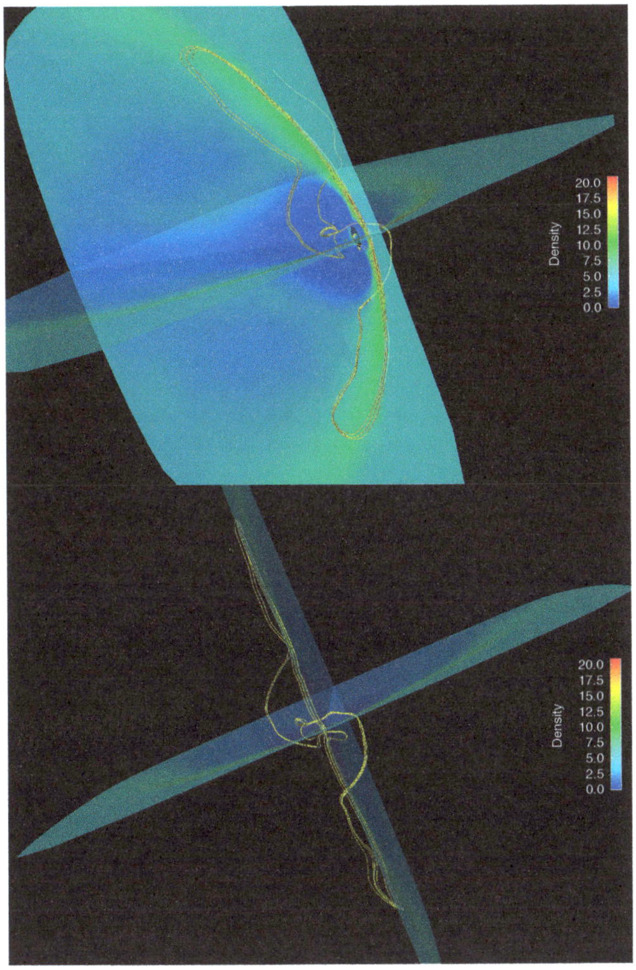

Figure 28.12 Bow shock current streamlines in the LFM code for IMF B_y.

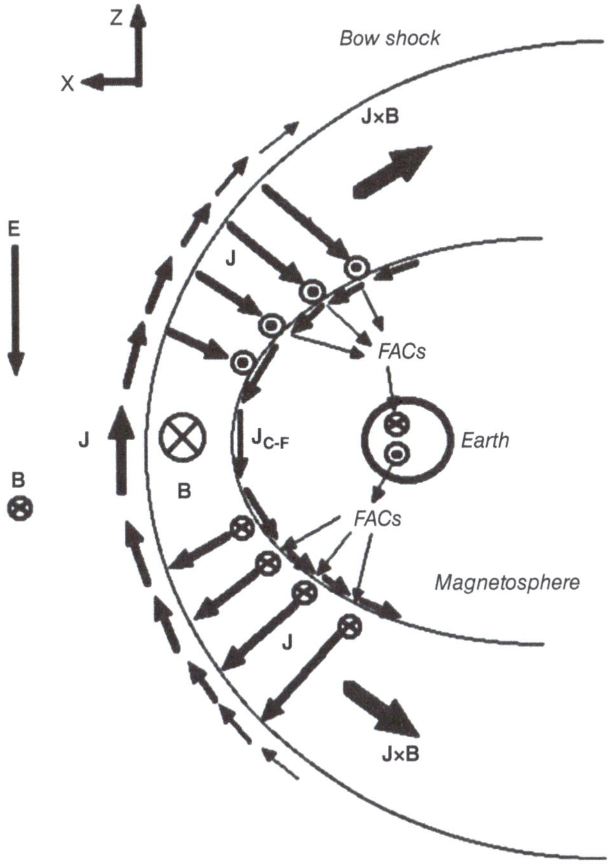

Figure 28.13 A schematic of the bow shock current closure for negative IMF B_y.

low Mach number state. There was a distorted two-cell pattern with Region 1 current that is asymmetric with a larger dawn current relative to the dusk current. The pattern was reversed in the Southern Hemisphere, as expected (not shown). At lower latitudes, a Region 2 current was evident. In addition, an upward Birkeland current at high latitudes overlapping the downward current near noon was evident, part of which must have been the current that connected to the bow shock current. If that is correct, then this current must have been on open field lines. AMPERE data later in the hour (not presented here) show this high-latitude current persisting until 2000 UT, when the solar wind changed.

Data (not shown) from DMSP F16 and F17 show an open polar cap with clear boundaries marking the transition from open to closed field lines. The auroral boundaries were at about 74° magnetic latitude. Examining Figure 28.15, we see that most of the high-latitude current near noon that we identified as the Birkeland current that connected to the bow shock was at or above 74° magnetic latitude. Unfortunately, DMSP F16 and F17 have orbital planes aligned roughly on the dawn to dusk direction, so we do not have direct observations of the particle precipitation characteristics of the field lines threading the current in question. Therefore, we cannot unambiguously determine if the current was on open field lines or not. However, *Sotirelis et al.* [1998] found that a circle fit the polar cap boundary quite well in half the cases examined with a standard deviation of 1.8°. Moreover, the boundaries determined from the DMSP F16 and F17 data were stable from 1900 UT to 1950 UT. The Birkeland current pattern seen in Figure 28.15 is also stable over the same time period. Therefore, the most likely interpretation is that a substantial portion of the unipolar knot of Birkeland current [e.g., *Mitchell et al.*, 2011] seen in the AMPERE data at high latitudes overlapping the Region 1 current was flowing on open field lines and therefore must have closed on the bow shock.

The bow shock current is at its most dynamically important when the shock is a low Mach number shock, however, the bow shock current is always closing somewhere, and to the extent that some ionospheric field lines are open, some of the shock current will close in the ionosphere. Figure 28.16 presents OMNI data from the end of 30 July and the beginning of 31 July 2012, when there

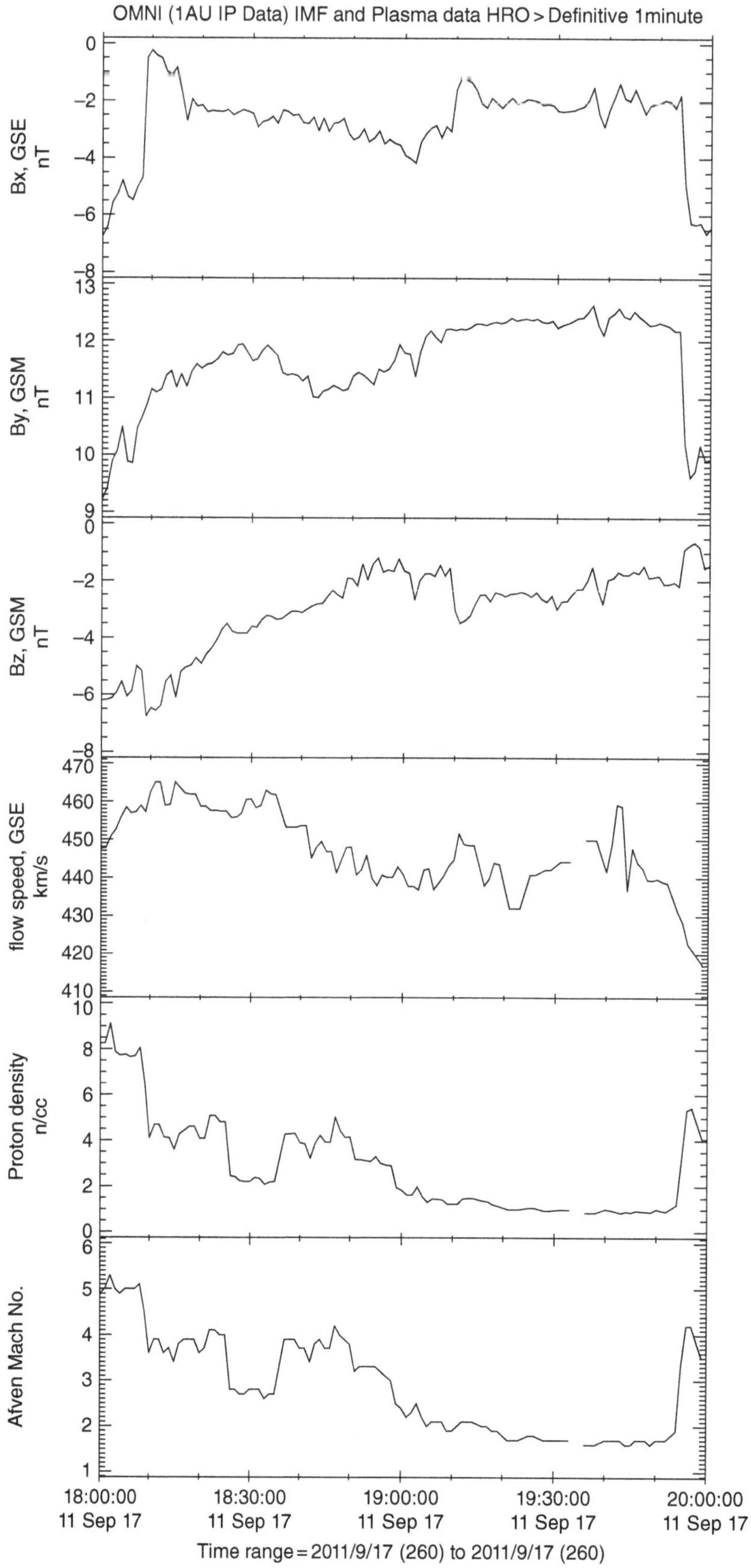

Figure 28.14 OMNI data for 17 September 2011.

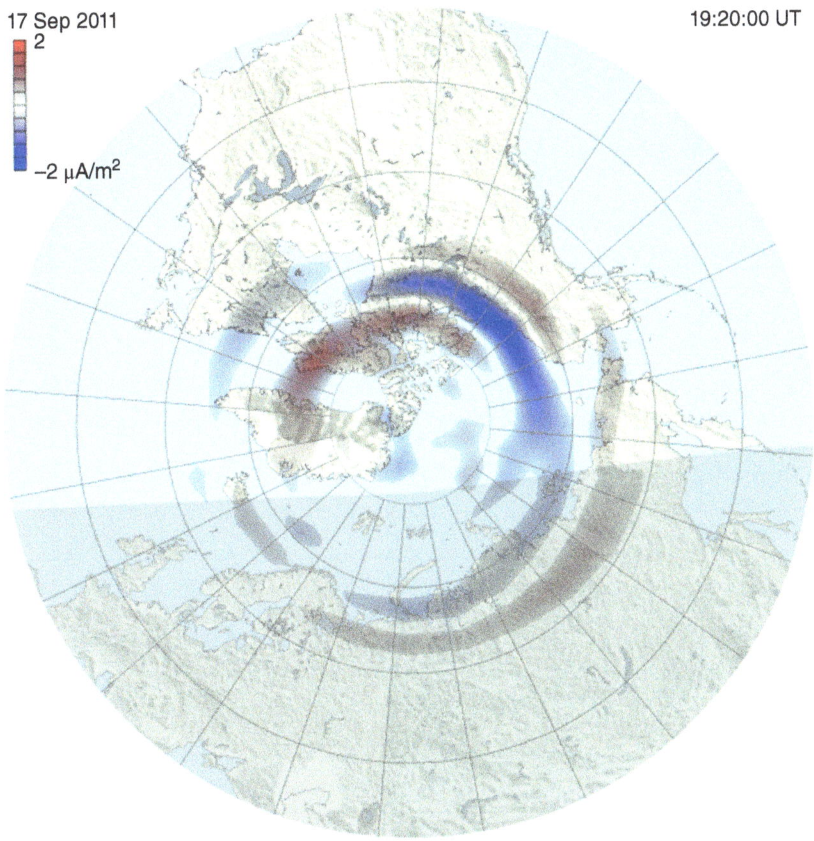

Figure 28.15 AMPERE data for 17 September 2011. The grid is for geomagnetic coordinates.

was a substantial east-west IMF orientation. The B_y component of the IMF dominated merging, but the solar wind was not in a low Mach number state and the magnetosheath was not magnetically dominated. This period has been simulated using the solar-wind input to drive the LFM code, and a snapshot of the ionospheric Birkeland currents at 0010 UT is presented in Figure 28.17. The basic pattern is a four-cell system with a very weak NBZ current system, as we would expect for this parameter range [e.g., *Bhattarai et al.*, 2012; *Bhatterai and Lopez*, 2013], with the B_y current superposed [e.g., *Lopez et al.*, 2014]. The AMPERE Birkeland current pattern for the same time is presented in Figure 28.18. The spatial correspondences are clear, although the magnitude of the simulated currents are about one half of the observed values, which is probably related to the low ionospheric conductivities the model develops as discussed by *Wiltberger et al.* [2012].

Figure 28.19 presents the data from DMSP F16 for this event. At 0012 UT, DMSP F16 exited a region of no ion precipitation, and we identify that point (78.5° MLAT, 1025 MLT) as the boundary between open and closed field lines. Poleward of that boundary, extending up to 79.3° MLAT and 1239 MLT, there was a region of weak, structured electron precipitation with average energies of a few hundred eV. This region just poleward of the open-closed field line boundary corresponds to the poleward edge of dawn side of the high-latitude downward Birkeland current associated with the IMF B_y that is evident both in the simulation and the AMPERE data. Therefore, we argue that at least part of that current was on open field lines and connected to the bow shock. Since this was not a period when one would expect the magnetic forces to play the leading role at the bow shock or in the magnetosheath, the observations support the idea that at all times there is at least some current that directly connects the ionosphere to the bow shock current along open field lines. It is also worthwhile to point out that for east-west IMF, the ionospheric convection pattern has a thick cell flanked by a banana cell [*Heelis*, 1984] and that the potentials in the two cells are not equal. This is what one would expect if a unipolar current were superposed onto the ionospheric convection pattern.

28.5. SOME ASTROPHYSICAL IMPLICATIONS

Shocks in space plasmas are ubiquitous, and shocks are critical for the acceleration of cosmic rays in interstellar space [e.g., *Lagage and Cesarsky*, 1983] as well as for the acceleration of energetic protons in the heliosphere

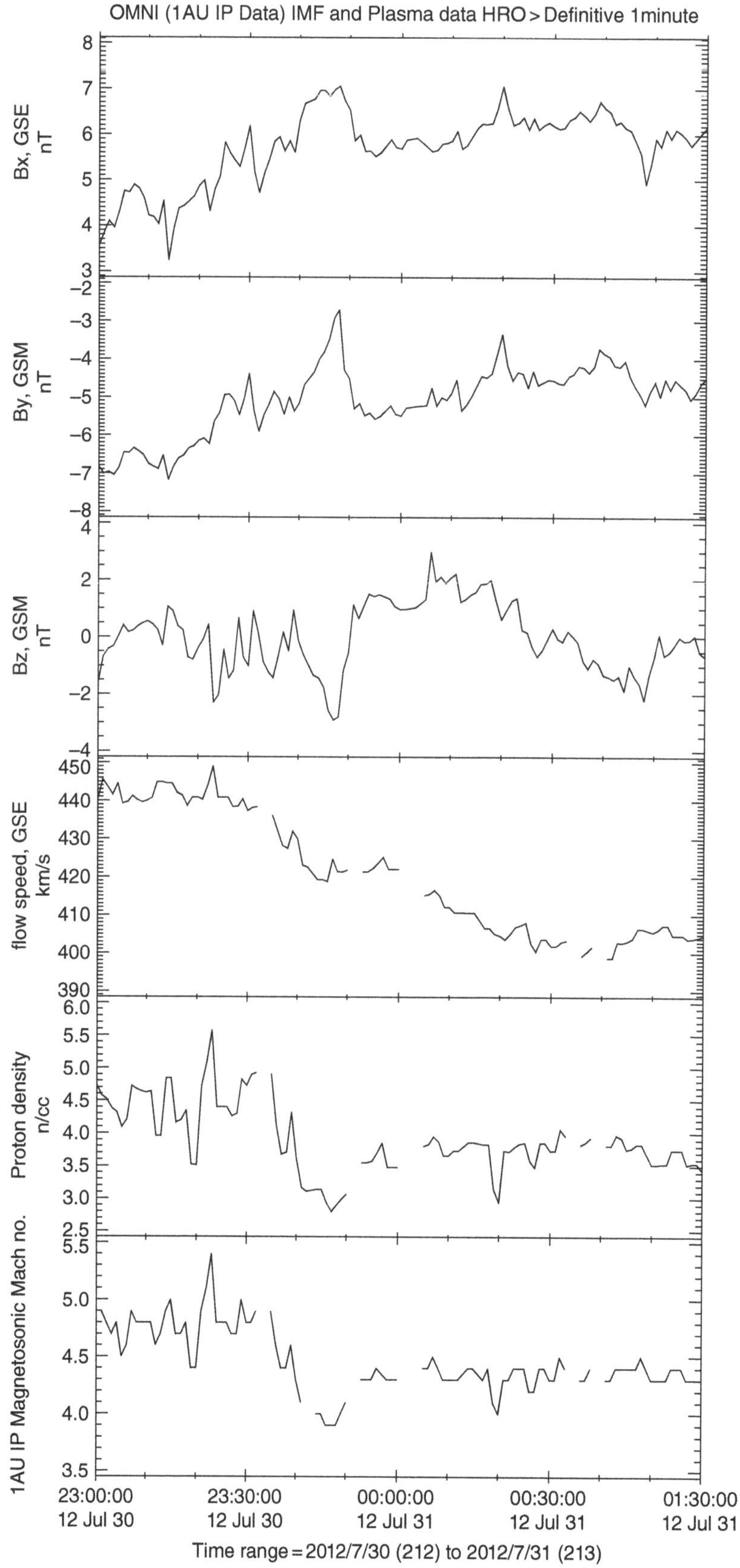

Figure 28.16 OMNI data for 31 July 2012.

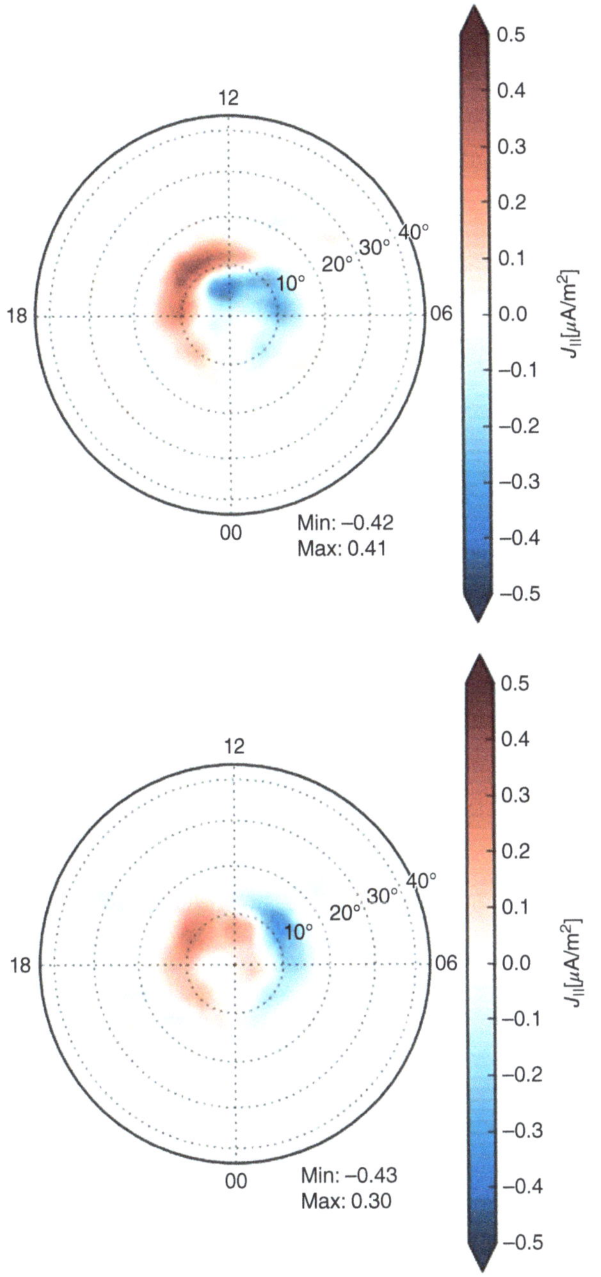

Figure 28.17 LFM-MIX simulation of ionospheric Birkeland currents at 0010 UT on 31 July 2012.

[e.g., *Fisk and Lee*, 1980]. However, the only extended discussion one finds in the literature concerning energy transfer and shocks involves the acceleration of high-energy particles. The issue discussed in this paper, the closure of currents on the shock, is essentially absent from the literature. For example, while *Washimi and Tanaka* [1996] discuss the large-scale current closure around the heliosphere, including the termination shock current, they never address the issue of what parts of the system are dynamos, what parts are load, and how this plays into the dynamics of the heliosphere as a whole. Moreover, the question of the closure of currents on shocks in the heliosphere, either around Coronal Mass Ejections (CMEs) or Corotating Interaction Regions (CIRs), seems not to be addressed at all.

However, in all cases of shocks in space plasmas, there are regions where energy is taken out of plasma motion and converted to electromagnetic energy in a dynamo. That energy is transported somewhere and deposited in a load. In the case of Earth, the loads are the magnetosheath plasma, reconnection regions, and the ionosphere. But in the case of CMEs and CIRs, where is the load? How and where does the energy taken out of the flow by the current on the shock find its way back into the plasma? The same question can be asked about interstellar shocks. While the shock around a supernova will initially be a very high Mach number shock, eventually that shock wave will slow and dissipate. At some point it will become a low Mach number shock and the remaining energy in the expansion will be converted primarily into electromagnetic energy. Where does that energy go and how is the shock current closed? Perhaps addressing this question could provide some useful perspectives in considering the energetics of astrophysical systems.

28.6. CONCLUSIONS

The solar wind is slowed, heated, and compressed by the bow shock in front of Earth's magnetosphere. The compression of the plasma also compresses the magnetic field, which means that there is a current that flows on the shock. This current does work against the solar-wind flow, extracting mechanical energy and converting it into electromagnetic energy. Under low magnetosonic Mach number conditions, the majority of the force on the solar wind is exerted by the bow shock current, and the majority of the energy extracted by the shock goes into electromagnetic energy.

The dynamo current of the shock deposits energy into the magnetosheath flow, and this is the primary load for the system. Some energy is also transferred to the merging regions on the magnetopause and the magnetotail, as well as directly into the ionosphere, all of which are loads. This paper has presented both MHD simulation results and schematic diagrams of the bow shock current closure for southward, northward, and east-west IMF. The ionospheric deposition of energy occurs on open field lines, as illustrated in the cases presented. This direct transfer of energy from the bow shock to the ionosphere represents a form of energy transfer from the solar wind that is qualitatively different from either merging or the viscous interaction. For northward IMF, there is relatively little closure of bow shock current through the ionosphere due to the relatively closed magnetospheric topology, but for

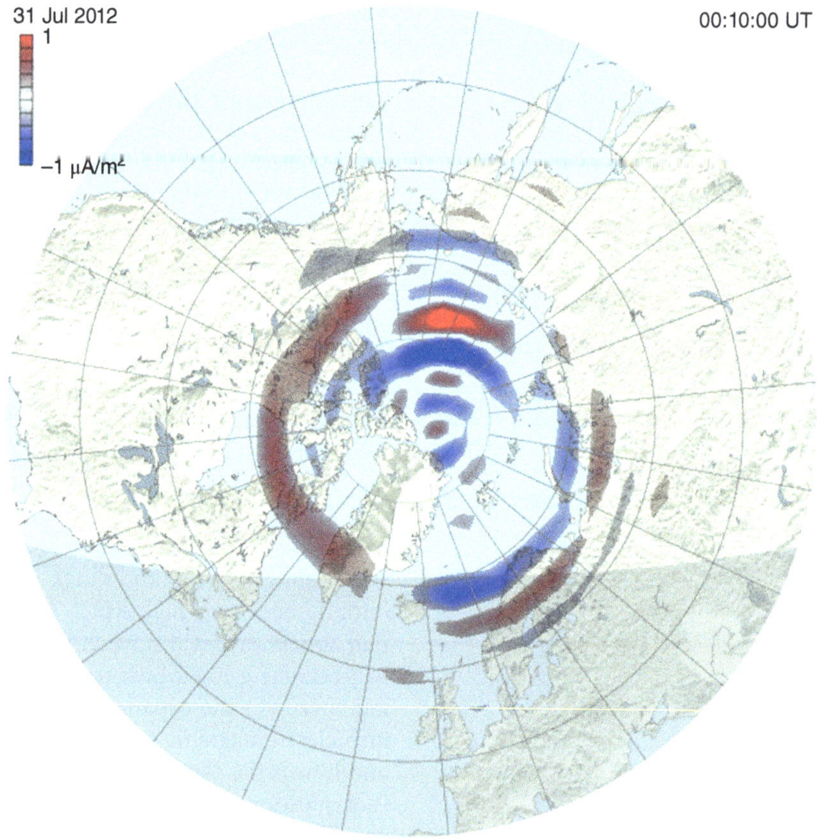

Figure 28.18 AMPERE data in the Northern Hemisphere at 0010 UT on 31 July 2012.

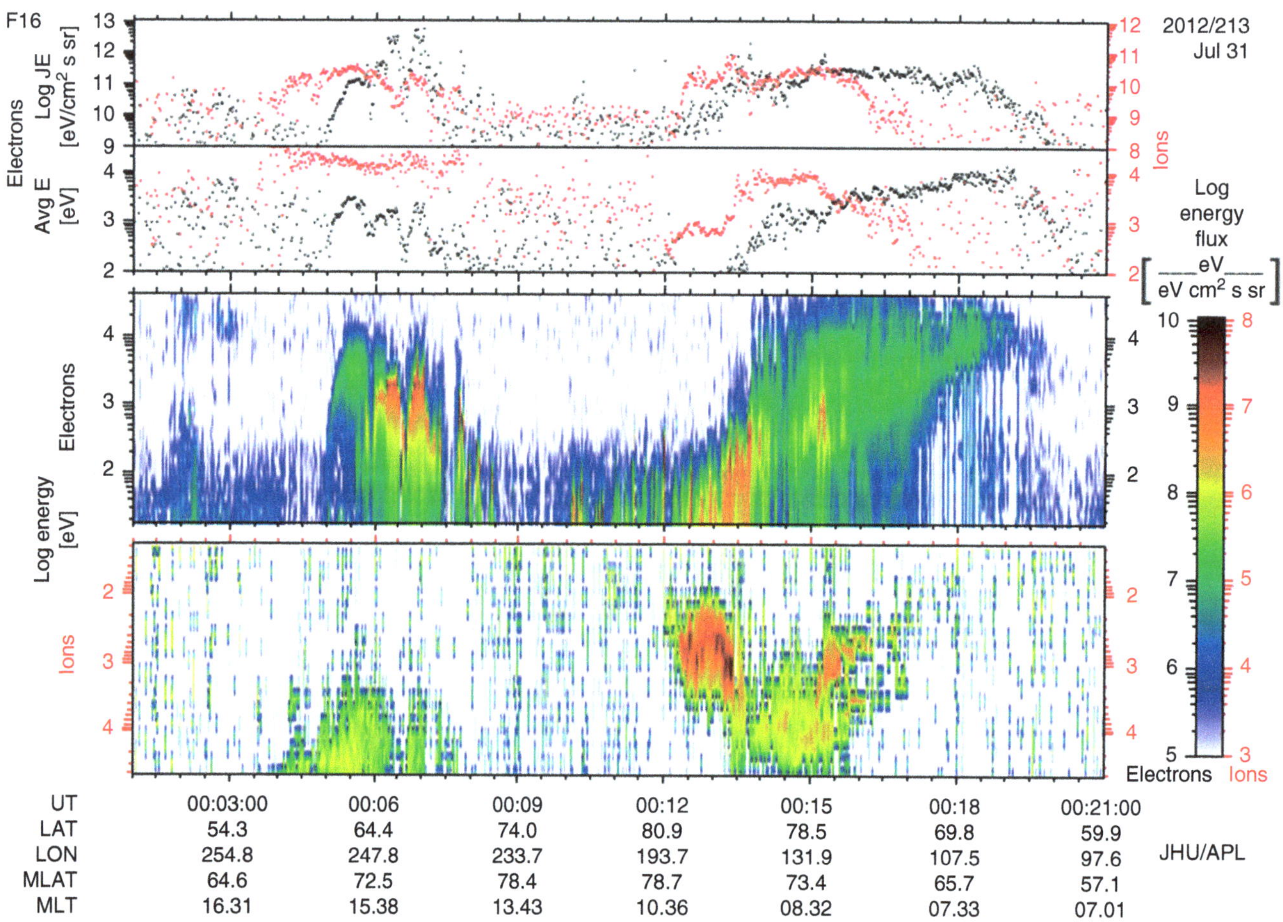

Figure 28.19 DMSP F16 particle precipitation data on 31 July 2012.

east-west and especially southward IMF, a considerable amount of Birkeland current can connect to the bow shock current on open field lines. This is true even for cases where the solar-wind flow has a relatively large magnetosonic Mach number and the magnetosheath is not magnetically dominated.

Finally, we briefly discussed the issue of the closure of shock currents in other plasma astrophysical settings. These currents must exist throughout space, and the currents generated must close and deposit energy somewhere. By considering this question, it is possible that new insights into astrophysical systems may be obtained.

ACKNOWLEDGMENTS

We acknowledge use of NASA/GSFC's Space Physics Data Facility's OMNIWeb and CDAWeb service, and OMNI data. Support for AMPERE has been provided under NSF sponsorship under grants ATM-0739864 and AGS-1420184. We are greatly indebted to the contribution of Iridium Communications for providing the data for AMPERE. The AMPERE data used in paper are available at http://ampere.jhuapl.edu/. The DMSP particle detectors were designed by Dave Hardy of AFRL, and data obtained from JHU/APL. Simulation results have been provided by the Community Coordinated Modeling Center at Goddard Space Flight Center through its public Runs on Request system (http://ccmc.gsfc.nasa.gov). The CCMC is a multiagency partnership between NASA, AFMC, AFOSR, AFRL, AFWA, NOAA, NSF, and ONR. We acknowledge useful conversation on this topic with M. Wiltberger, J. Lyon, and V. Merkin. This work was supported by NASA grants NNX09AI63G and NNX15AJ03G.

REFERENCES

Axford, W.I., Hines, C.O (1961), A unifying theory of high-latitude geophysical phenomena and geomagnetic storms. *Canadian Journal of Physics*, *39*, 1433–1464.

Bhattarai, S. K., and R. E. Lopez (2013), Reduction of viscous potential for northward interplanetary magnetic field as seen in the LFM simulation, *J. Geophys. Res. Space Physics*, *118*; doi:10.1002/jgra.50368.

Bhattarai, S. K., R. E. Lopez, R. Bruntz, J. G. Lyon, and M. Wiltberger (2012), Simulation of the polar cap potential during periods with northward interplanetary magnetic field, *J. Geophys. Res.*, *117*, A04219; doi:10.1029/2011JA017143.

Borovsky, J. E., B. Lavraud, and M. M. Kuznetsova (2009), Polar cap potential saturation, dayside reconnection, and changes to the magnetosphere, *J. Geophys. Res.*, *114*, A03224; doi:10.1029/2009JA014058.

Boyd, T. J. M., and J. J. Sanderson (1969), *Plasma Dynamics*, Barnes and Noble, Inc, New York.

Clauer, C. R., Z. Xu, M. Maimaiti, J. M. Ruohoneimi, W. Scales, M. D. Hartinger, M. J. Nicolls, S. Kaeppler, F. D. Wilder, and R. E. Lopez (2016), Investigation of a rare event where the polar ionospheric reverse convection potential does not saturate during a period of extreme northward IMF solar wind driving, *J. Geophys. Res. Space Physics*, *121*; doi:10.1002/2016JA022557.

Crooker, N. U. (1992), Reverse convection, *J. Geophys. Res.*, *97*(A12), 19,363–19,372; doi:10.1029/92JA01532.

Crooker, N. U., J. G. Lyon, and J. A. Fedder (1998), MHD model merging with IMF By: Lobe cells, sunward polar cap convection, and overdraped lobes, *J. Geophys. Res.*, *103*, 9143–9151; doi:10.1029/97JA03393.

Dungey, J. W. (1961), Interplanetary magnetic field and the auroral zones, *Phys. Rev. Lett.*, *6*, 47–48, doi:10.1103/PhysRevLett.6.47.

Fedder, J. A., and J. G. Lyon (1987), The solar wind-magnetosphere-ionosphere current-voltage relationship, *Geophys. Res. Lett.*, *14*(8), 880–883.

Fedder, J. A., and J. G. Lyon (1995), The Earth's magnetosphere is 165 RE long: Self-consistent currents, convection, magnetospheric structure, and processes for northward interplanetary magnetic field, *J. Geophys. Res.*, *100*(A3), 3623–3635; doi:10.1029/94JA02633.

Fedder, J. A., S. P. Slinker, J. G. Lyon, C. T. Russell, F. R. Fenrich, and J. G. Luhmann (1997), A first comparison of POLAR magnetic field measure- ments and magnetohydrodynamic simulation results for field aligned currents, *Geophys. Res. Lett.*, *24*, 2491–2494.

Fisk, L. A., and M. A. Lee (1980). Shock acceleration of energetic particles in corotating interaction regions in the solar wind, *Astrophys. J.*, *237*, 620–626.

Formisano, V., P. C. Hedgecock, G. Moreno, F. Palmiotto, and J. K. Chao (1973), Solar wind interaction with the Earth's magnetic field, 2. Magnetohydrodynamic bow shock, *J. Geophys. Res.*, *78*(19), 3731–3744; doi:10.1029/JA078i019p03731.

Gussenhoven, M. S., and E. G. Mullen, (1989), Simultaneous relativistic-electron and auroral-particle access to the polar caps during interplanetary magnetic field Bz northward: A scenario for an open-field line source of auroral particles, *J. Geophys. Res.*, *94*, 17121–17132.

Heelis, R. A. (1984). The effects of interplanetary magnetic field orientation on dayside high-latitude ionospheric convection, *J. Geophys. Res. Space Physics*, *89*(A5), 2873–2880.

Iijima, T., and T. A. Potemra (1976), The amplitude distribution of field-aligned currents at northern high latitudes observed by Triad, *J. Geophys. Res.*, *81*(13), 2165–2174; doi:10.1029/JA081i013p02165.

King, J. H., and N.E. Papitashvili (2005), Solar wind spatial scales in and comparisons of hourly Wind and ACE plasma and magnetic field data, *J. Geophys. Res.*, *110*, A02104.

Kivelson, M. G., and C. T. Russell (1995), *Introduction to Space Physics*, University Press, Cambridge.

Lagage, P. O., and, C. J. Cesarsky (1983). The maximum energy of cosmic rays accelerated by supernova shocks, *Astron. Astrophys.*, *125*, 249–257.

Lopez, R. E. (2016), The integrated dayside merging rate is controlled primarily by the solar wind, *J. Geophys. Res. Space Physics*, *121*; doi:10.1002/ 2016JA022556.

Lopez, R. E., H. Spence, and C.-I. Meng (1991), DMSP F7 observations of a substorm field-aligned current, *J. Geophys. Res.*, *96*, 19409–19415.

Lopez, R. E., R. Bruntz, and K. Pham (2014), Linear separation of orthogonal merging component and viscous interactions in solar wind-geospace coupling, *J. Geophys. Res. Space Physics*, *119*, 7566–7576; doi:10.1002/2014JA020153.

Lopez, R. E., R. Bruntz, E. J. Mitchell, M. Wiltberger, J. G. Lyon, and V. G. Merkin (2010), The role of magnetosheath force balance in regulating the dayside reconnection potential, *J. Geophys. Res.*, *115*, A12216; doi:10.1029/2009JA014597.

Lopez, R. E., S. Hernandez, K. Hallman, R. Valenzuela, J. Seiler, P. Anderson, and M. Hairston (2008), Field-aligned currents in the polar cap during saturation of the polar cap potential, *J. Atmos. Solar Terr. Phys.*, *70*, 555–563; doi:10.1016/j.jastp.2007.08.072.

Lopez, R. E., V. G. Merkin, and J. G. Lyon (2011), The role of the bow shock in solar wind-magnetosphere coupling, *Ann. Geophys.*, *29*, 1129–1135; doi:10.5194/angeo-29-1129-2011.

Lyon, J. G., J. A. Fedder, and C. M. Mobarry (2004), The Lyon-Fedder-Mobarry (LFM) global MHD magnetospheric simulation code, *J. Atmos. Solar Terr. Phys.*, *66*, 1333–1350; doi:10.1016/j.jastp.2004.03.020.

Mitchell, E. J., R. E. Lopez, R. Bruntz, M. Wiltberger, J. G. Lyon, R. C. Allen, S. J. Cockrell, and P. L. Whittlesey (2010), Saturation of transpolar potential for large Y-component interplanetary magnetic field, *J. Geophys. Res.*, *115*, A06201; doi:10.1029/2009JA015119.

Powell, K., P. Roe, T. Linde, T. Gombosi, and D. L. De Zeeuw (1999), A solution-adaptive upwind scheme for ideal magnetohydrodynamics, *J. Comp. Phys.*, *154*, 284–309.

Siebert, K. D., and G. L. Siscoe (2002), Dynamo circuits for magnetopause reconnection, *J. Geophys. Res.*, *107*(A7), 1095, 10.1029/2001JA000237.

Siscoe, G., J. Raeder, and A. J. Ridley (2004), Transpolar potential saturation models compared, *J. Geophys. Res.*, *109*, A09203; doi:10.1029/2003JA010318.

Siscoe, G. L., N. U. Crooker, and K. D. Siebert (2002), Transpolar potential saturation: Roles of Region 1 current system and solar wind ram pressure, *J. Geophys. Res.*, *107*(A10), 1321; doi:10.1029/2001JA009176.

Song, P., T. I. Gombosi, D. L. DeZeeuw, K. G. Powell, and C. P. T. Groth (2000), A model of solar wind-magnetosphere-ionosphere coupling for due northward IMF, *Planet Space Sci.*, *48*(1), 29–39; doi:10.1016/S0032- 0633(99)00065-3.

Sotirelis, T., Newell, P. T., & Meng, C. I. (1998), Shape of the open–closed boundary of the polar cap as determined from observations of precipitating particles by up to four DMSP satellites. *Journal of Geophysical Research: Space Physics*, *103*(A1), 399–406.

Tanaka, T., M. Watanabe, M. Den, S. Fujita, Y. Ebihara, T. Kikuchi, K. K. Hashimoto, and R. Kataoka (2016), Generation of field-aligned current (FAC) and convection through the formation of pressure regimes: Correction for the concept of Dungey's convection, *J. Geophys. Res. Space Physics*, *121*, 8695–8711; doi:10.1002/2016JA022822.

Tang, B. B., X. C. Guo, C. Wang, Y. Q. Hu, and J. R. Kan (2009), Bow shock and magnetopause contributions to the cross-tail current from global MHD simulations, *J. Geophys. Res.*, *114*, A08203; doi:10.1029/2009JA014325.

Vasyliūnas, V. M. (2001). Electric field and plasma flow: What drives what?, *Geophys. Res. Lett.*, *28*(11), 2177–2180.

Washimi, H., and T. Tanaka (1996), 3-D magnetic field and current system in the heliosphere, *Space Sci. Rev.*, *78*(1–2), 85–94.

Wiltberger, M., L. Qain, C.-L. Huang, W. Wang, R. E. Lopez, A. G. Burns, S. C. Solomon, Y. Deng, and Y. Huang (2012), CMIT study of CR2060 and 2068 comparing L1 and MAS solar wind drivers, *J. Atmos. Solar Terr. Phys.*, *83*, 39–50; doi:10.1016/j.jastp.2012.01.005.

Yang, Y. S., R. W. Spiro, and R. A. Wolf (1994), Generation of region 1 current by magnetospheric pressure gradients, *J. Geophys. Res. Space Physics*, *99*(A1), 223–234.

29

Current Systems of Inert Moons

M. Holmström[1] and S. Fatemi[2]

ABSTRACT

We present the different current systems associated with inert moons (nonconducting and lacking ionosphere), along with examples for Earth's Moon, and Saturn's moons Tethys and Rhea. The interaction of these moons with the solar wind or magnetospheric plasma is at first sight simple. They are obstacles to the plasma flow, leaving a wake behind. However, there are many processes occurring, from surface interactions to complex plasma physics governing the refill of the wake, generating a wealth of different current systems. Still, for all the complexity of the interactions, we find that the plasma interactions of Earth's Moon and its associated current systems are very similar to other inert moons. This is a view that has emerged lately from observations and modeling.

29.1. INTRODUCTION

What is different with moons compared to planets from a plasma point of view? One difference is that moons can encounter plasma regimes different from the solar wind as they orbit in the magnetosphere of a planet. Another difference is that most of the moons are smaller than any planet. The exceptions are Ganymede and Titan, which are slightly larger than Mercury. The smaller size of the moons implies that kinetic effects (e.g., from the finite gyroradius of ions) are more important for moons than for planets. The interaction of moons and planets with the surrounding plasma is often discussed in terms of the magnetic field, instead of the currents that are responsible for the field. One obvious reason is that the magnetic field is much easier to directly observe, compared with currents [*Balogh*, 2010]. Current systems related to the moons have not been studied as extensively as for the planets. These current systems can be divided into three broad classes.

1. *Currents in the interiors of moons.* Currents can be induced in the interior of moons due to changes in the surrounding field and plasma environment if the interior has high conductivity. If a moon has a resistive crust, there can be a conducting core (as for the Moon [*Weber et al.*, 2011]), or conductive layers, for example, an ocean (as is probably the case for Europa [*Kivelson et al.*, 1999, 2000] and Ganymede [*Kivelson et al.*, 2002]). Analysis of Galileo mission data also suggests that Io [*Khurana et al.*, 2011] and Callisto [*Kivelson et al.*, 1999] generate induced magnetic fields, likely by subsurface conducting layers. There can also be crustal magnetizations (e.g., the Moon [*Mitchell et al.*, 2008]), and permanent fields generated by an interior dynamo (e.g., Ganymede).

2. *Currents in the surrounding plasma.* There can be currents in the solar wind or magnetospheric plasma that surrounds the moon. These currents could be preexisting, convected toward the moon, or they could be a result of the interaction between the moon and the surrounding plasma, for example, in the wake behind the moon.

[1] *Swedish Institute of Space Physics, Kiruna, Sweden*

[2] *Space Sciences Laboratory, University of California, Berkeley, California, USA; Solar System Exploration Research Virtual Institute, NASA Ames Research Center, Moffett Field, California, USA*

Electric Currents in Geospace and Beyond, Geophysical Monograph 235, First Edition.
Edited by Andreas Keiling, Octav Marghitu, and Michael Wheatland.

3. *Currents in the ionospheres of moons*. If a moon has a sufficiently dense atmosphere (e.g., Io) to allow the generation of a significant ionosphere, there will be current systems in the ionosphere. There can also be localized regions of high neutral density that will generate a significant charge density, as is the case for the neutral plumes observed at Enceladus and Europa. Some of the moons have too tenuous atmospheres to generate any significant ionosphere and associated currents.

These current systems will depend on each other through the action of the magnetic fields generated by the currents. So, in the general case, one has to account for all the different currents in the system, making any analysis complicated.

Here, we will focus on the current systems associated with inert moons. Following *Jia* [2015], we define an inert moon as a moon lacking any significant atmosphere or ionosphere. Also, the exosphere (collisionless upper part of the atmosphere) reaches all the way to the surface of the moon (the surface is the exobase). Thus, the incoming plasma will directly impinge onto the surface of the moon, creating a wake in the plasma downstream of the moon. The existence, structure, and extent of the wake will depend on the ambient plasma and magnetic field conditions. An inert moon also has an interior that is to a large extent resistive (low conductivity) and also does not possess any significant permanent magnetic field from any dynamo action. So, for an inert moon, the major currents will be currents in the surrounding plasma (item 2 in the list above). In the following sections, we discuss some selected inert moons for which the associated current systems have been studied to some extent: Earth's Moon and the Saturnian moons Tethys and Rhea.

Even though we restrict the discussion to the Moon, Tethys, and Rhea, much will be applicable to other inert moons in the solar system, for example, the Jovian moon Callisto and the moons of Uranus and Neptune.

First, in order to characterize the plasma interaction with an object, we define some basic dimensionless parameters for the plasma flow that indicate the importance of various physical processes for the interaction. These parameters are the Alfvén Mach number, $M_A = v_p/v_A$, the sonic Mach number, $M_s = v_p/c_s$, the magnetosonic Mach number, $M_{ms} = v_p/\sqrt{v_A^2 + c_s^2}$, and the plasma beta, $\beta = P_{th}/P_m$, where v_p is the plasma flow speed, v_A is the Alfvén speed, c_s is the sound speed, P_{th} is the plasma thermal pressure, and P_m is the magnetic pressure. The Mach numbers indicate whether the plasma flow speed or the wave speed dominates, whereas the plasma beta is an indication of the dominant pressure. While the solar wind plasma near the Moon is supersonic ($M_s > 1.0$) and super-Alfvénic ($M_A > 1.0$) with plasma $\beta \simeq 1.0$, the magnetospheric plasma near the orbit of Saturn's inert moons is mainly transonic ($M_s \simeq 1.0$), sub-Alfvénic ($M_A < 1.0$), and the plasma β is lower than unity. We summarize typical parameters for the Moon and for the Saturnian inert moons Mimas, Tethys, Dione, and Rhea, in Table 29.1. The parameters are averages based on limited flyby observations at the orbit of the Saturnian moons. Near Saturn, the thermal particle pressure dominates, while outside Rhea's orbital distance, the pressure from suprathermal particles becomes comparable to the thermal pressure. At Rhea's distance the plasma beta can become larger than unity, due to this suprathermal particle pressure [*Sergis et al.*, 2010].

Regarding coordinate systems, in all the results presented, the origin of the coordinate system is centered at the moon, with the $+x$-axis directed along the plasma flow direction ($\mathbf{v}_p = |v_0|\hat{\mathbf{x}}$), the $+z$-axis pointing opposite to the magnetic field ($\mathbf{B} = -|B_0|\hat{\mathbf{z}}$), and the $+y$-axis completing the right-handed system. For the Saturnian moons discussed later, the $+y$-axis is pointing toward Saturn. The convective electric field, $\mathbf{E} = -\mathbf{v} \times \mathbf{B} = -|v_0||B_0|\hat{\mathbf{y}}$, is then pointing opposite to the y-axis.

The general wake structure for an inert moon is shown in Figure 29.1. As the plasma is absorbed by the surface of the moon on the hemisphere facing the plasma flow, a region of depleted density (refilling region) will develop, starting at the terminator and continuing downstream. The surrounding plasma expands into this region. Depending on the plasma flow velocity and the expansion velocity, a plasma void region may develop. The refill of the wake is preferentially along the magnetic field lines. Perpendicular to the field lines the refill will stop when there is a balance between the plasma and magnetic field pressures, and an expansion region is formed. The effect is that the magnetic field lines are pinched in the wake.

29.2. THE MOON

The exosphere of the Moon is too tenuous to generate any significant ionosphere, although there are observations of ionospheric charge densities, possibly due to charged dust [*Stubbs et al.*, 2011]. Ionization of neutrals in the exosphere will create pickup ions that are accelerated by the solar wind convective electric field [*Halekas et al.*, 2015b; *Lipatov et al.*, 2012; *Manka and Michel*, 1971; *Poppe et al.*, 2016]. These, mostly dayside, currents from pickup ions are especially prominent when the Moon is in the lobes of Earth's magnetosphere, since the surrounding plasma density is low [*Poppe et al.*, 2012; *Zhou et al.*, 2014].

The Moon has a surface layer of highly resistive regolith, and is thought to have a conductive interior [*Weber et al.*, 2011]. So the conductivity of the Moon will probably increase with depth below the surface [*Grimm et al.*, 2012]. Since the upper part of the lunar interior

Table 29.1 Inert Moons.

Satellite name	Moon*	Mimas	Tethys	Dione	Rhea
Characteristics					
Radius (km)	1736	195[b]	533[a]	562[a]	764[a]
Radial distance	$60.26R_E$	$3.1R_S$ [b]	$4.89R_S$ [a]	$6.26R_S$ [a]	$8.74R_S$ [a]
Mass ($\times 10^{20}$ kg)	734.2	0.46[b]	7.6[b]	10.5[b]	24.9[b]
Orbital period (h)	708.7	22.6[b]	45.3[b]	65.7[b]	108.4[b]
Parameters					
V relative (km/s)	350	15.6[b]	33.6[b]	39.9[b]	56.6[b]
$\|B_0\|$ (nT)	5.0	722[b]	167[a], 183[b]	75[a], 86[b]	25[a], 32[b]
n_i thermal (cm^{-3})	6.0	90[b]	40[b]	20[b]	3.5[b]
n_e (cm^{-3})	6.2	100[c]	30[a]	13[a]	2[a]
m_i (amu)	1.0	17 (?)	17[a]	17[a]	17[a]
T_i for H^+ (eV)	8.0	50[b]	50[a]	90[a]	100[a]
T_i for water ions (eV)	n/a	850[c]	850[c]	1530[c]	1700[c]
T_e (eV)	10.0	8[c]	20[a]	60[a]	100[a]
Convective E (mV/m)	1.75	11.3	5.61	2.99	1.42
Relative length scales (L/Radius)[d]					
Ion thermal gyro radius	0.05	0.03	0.047	0.13	0.31
Electron gyro radius	0.0012	5.12×10^{-5}	0.00017	0.00062	0.0017
Ion skin depth	0.06	3.02	2.02	2.91	5.45
Electron skin depth	0.0015	0.017	0.011	0.016	0.03
Debye length	0.4×10^{-5}	1.07×10^{-5}	1.13×10^{-5}	2.84×10^{-5}	6.87×10^{-5}
Frequencies (Hz)[d]					
Ion gyro freq.	0.07	0.65	0.15	0.067	0.022
Electron gyro freq.	139.9	20210.5	4674.7	2099.4	700
Ion plasma freq.	513	508.3	278.6	183.4	71.9
Elect. plasma freq.	2.2×10^4	8.98×10^4	4.92×10^4	3.24×10^4	1.27×10^4
Speeds (km/s)[d]					
Sound speed	53.6	23.3	25.6	37.5	43.3
Alfvén speed	44.5	381.9	161.3	110	93.5
Fast magnetosonic	69.7	382.6	163.3	116.2	103
Ion thermal speed	39.2	23.7	23.7	31.8	33.5
Electron thermal speed	1875.5	1677.2	2651.9	4593.2	5929.8
Escape velocity	2.38	0.18	0.44	0.5	0.66
Pressures (nPa)[d]					
Ion thermal pressure	7.7×10^{-3}	0.8	0.24	0.19	0.032
Ion dynamic pressure	1.23	0.69	0.96	0.59	0.182
Electron thermal pressure	9.6×10^{-3}	0.13	0.1	0.12	0.032
Magnetic pressure	0.01	207.4	11.1	2.24	0.249
Miscellaneous[d]					
Plasma beta[e]	1.3	0.0045	0.03	0.14	0.25
Sonic Mach	6.5	0.66	1.31	1.06	1.3
Alfven Mach	7.8	0.04	0.21	0.36	0.6
Magnetosonic Mach	5.1	0.04	0.2	0.34	0.54

*The Moon in the solar wind.
[a]*Khurana et al.* [2008].
[b]*Saur and Strobel* [2005].
[c]*Sittler et al.* [2006].
[d]Calculated based on observed parameters.
[e]Only plasma thermal pressure is considered.

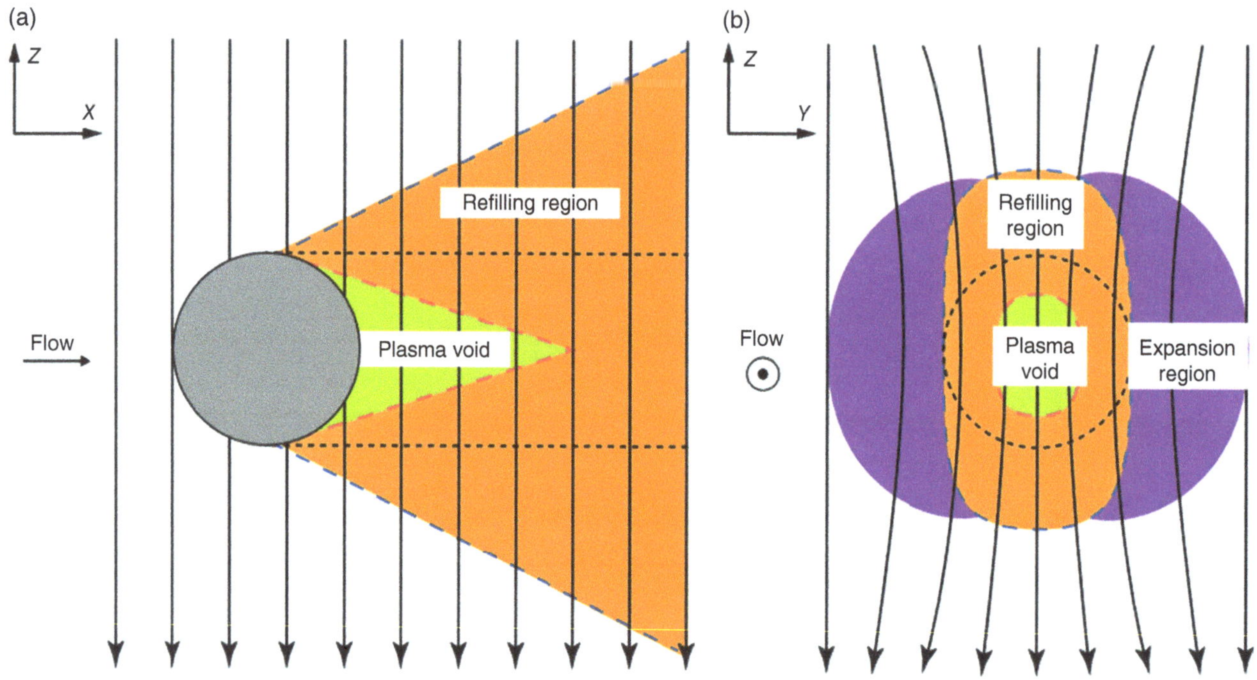

Figure 29.1 Illustration of plasma interaction with inert moons and structure of plasma wake in (a) a parallel plane to the background magnetic field and (b) a perpendicular plane to plasma flow. From *Jia* [2015].

(the lunar crust) is highly resistive, and the bulk interior conductivity of the moon is not large enough to generate strong induced currents [*Dyal et al.*, 1974; *Sonett*, 1982], the lunar induced magnetic fields are confined into the lunar dayside by the solar wind pressure, and they only extend downstream in the lunar wake. This view that the lunar induced magnetic fields are confined to the interior of the Moon and to the space of the cavity behind the Moon, has been denoted the lunar magnetosphere by *Schubert et al.* [1973]. It has, however, been predicted from recent hybrid modeling by *Fatemi et al.* [2015] that the lunar induced magnetic fields are not necessarily confined into the plasma wake and can extend outside the wake.

Since the Moon absorbs most of the supersonic solar wind plasma that precipitates on the dayside, without any prior slowdown, there is no global bow shock formed upstream in the solar wind. However, over magnetic anomalies, a large fraction of the solar wind that passes by is deflected [*Lue et al.*, 2011] and this is predicted to give rise to disturbances in the charge density and in the magnetic field and to generate associated currents [*Fatemi et al.*, 2014]. Strong magnetic anomalies may even be able to completely stand off the solar wind, generating a collisionless shock [*Xie et al.*, 2015], as may have been observed [*Halekas et al.*, 2014]. The nature of the interaction between a magnetic anomaly and the solar wind will however strongly depend on the geometry. As the Moon rotates, the solar wind will come from different directions for a given magnetic anomaly.

Close to the surface of the Moon there will also be currents associated with the plasma-surface interaction. On the dayside, photoelectrons are emitted and the surface charges positive, creating a photoelectron sheath. On the nightside, since electrons have higher thermal velocity than ions, the surface will charge negative, creating a plasma sheath above the surface. These charge imbalances will drive radial currents. However, the scale of these sheaths is on the order of the Debye length, so they will be restricted to at most a few hundreds of meters close to the surface [*Nitter and Havnes*, 1992]. The largest extent will be when the Moon is in the lobes of Earth's magnetosphere, and at the nightside, due to the low plasma densities [*Poppe et al.*, 2011].

There is not a wealth of literature on the morphology, in general, of the currents in the wake of the Moon, and on the closure of the currents, in particular. On the dayside of the Moon, the solar wind plasma precipitates onto the surface. As the plasma passes the terminator, it will be able to expand into the wake behind the Moon. The expansion is driven by the thermal motion of the plasma, but also from the fact that the electrons have higher velocity than the ions. The electrons will expand into the wake ahead of the ions, setting up an electrostatic ambipolar electric field that drags the ions into the wake faster than the thermal expansion. If the expansion is at an angle to the interplanetary magnetic field (IMF), the magnetic field lines will also be pushed into the wake, enhancing the field in the wake until a pressure balance is reached between the plasma pressure and the magnetic field pressure. As the plasma is expanding into the wake behind the terminator, the density will be decreased. This rarefaction wave will propagate into the downstream plasma, starting at the terminator, creating a cone. At the point where a balance between the plasma and magnetic pressure is reached, the density will increase again, forming the downstream recompression boundary of the rarefaction wave and a second cone boundary. At both boundaries of the rarefaction wave, the magnetic field is bent and there will be currents formed. In the wake where the plasma expands into vacuum, the density gradient will give rise to diamagnetic currents [*Colburn et al.*, 1967]. These currents increase the magnetic field magnitude in the wake, and decrease the field in a region outside the wake. This general picture of the Moon–solar wind interaction was put forward already in the Apollo times [*Ness et al.*, 1968; *Schubert et al.*, 1973], but the details have emerged more recently through observations and modeling, as summarized by *Halekas et al.* [2015a]. Confirmation has come with observations by the WIND spacecraft [*Ogilvie et al.*, 1996], and later by ARTEMIS [*Zhang et al.*, 2014]. The wake current systems can also be inferred from hybrid simulations [*Holmström et al.*, 2012; *Wiehle et al.*, 2011]. Later hybrid modeling work by *Vernisse et al.* [2013] and by *Fatemi et al.* [2013] have studied these current systems in more detail. The exact topology of these currents can be very complex, depending on the direction of the interplanetary magnetic field (IMF). The extreme cases when the IMF is perpendicular or parallel to the solar wind will lead to simpler current systems, while the current system for other IMF directions will be a mix of these simpler current topologies [*Fatemi et al.*, 2013]. It is often difficult to use model results to directly plot the currents, since they are noisy due to the fact that the currents are computed from derivatives of the magnetic field, $\mu_0\mathbf{J} = \nabla\times\mathbf{B}$. The current systems in the lunar wake when the solar wind is antiparallel and perpendicular to the IMF are illustrated in Figure 29.2. Currents are formed in the plasma behind the Moon at the Mach cone boundary, at the recompression boundary, and at the void boundary. We note that an earlier illustration of the wake current system by *Owen et al.* [1996] requires closure on the dayside of the Moon, and also cross the wake.

Apart from the currents at the wake boundary, and in the rarefaction waves emanating from the lunar terminator, there are also field-aligned currents in the wake, from the pressure gradient along the wake, resulting in a bending of the field lines in the wake toward the Moon. Although the field line bending had been observed in simulations [*Holmström et al.*, 2012], the explanation that such Alfvén wing structures form for inert moons in the super-Alfvénic solar wind flow was provided by *Zhang et al.* [2016] from ARTEMIS observations.

29.3. INERT ICY MOONS OF SATURN

Saturn possesses a diverse collection of planetary moons, several of which are inert icy moons that directly interact with Saturnian plasma and electromagnetic energy (e.g., Rhea, Tethys, and Dione). Their interaction is in some ways analogous to the Earth's Moon interaction with the solar wind. The inert icy moons, similar to the Earth's Moon, have neither a strong intrinsic magnetic field, nor a substantial source of internal conductivity, nor a considerable source of neutral atmosphere. The major difference between the plasma interaction of these satellites and the Moon, besides their shape and size, is the plasma characteristics and electromagnetic environments near these objects.

Since 2004, the Cassini spacecraft has made four close flybys of Rhea, five close flybys of Dione, and one at each of Tethys, Hyperion, Iapetus, and Phoebe with some distant flybys of Mimas (for a complete list of Cassini's targeted flybys see appendix tables in *Simon et al.* [2015]). Since the magnetic field data for Hyperion, Iapetus, and Phoebe either did not indicate any clear signature associated with these satellites (e.g., Hyperion [*Nordheim et al.*, 2014]) or has not been published in peer-reviewed articles (e.g., Iapetus [*Leisner et al.*, 2008]), we do not include them in this review study. Thus, we mainly focus on Rhea and Tethys and compare the current systems around them with those around the Earth's Moon.

Within Saturn's magnetosphere, plasma corotates with Saturn at a speed faster than the Keplerian orbital speed of the moons and impacts the trailing hemisphere of the moons. In the case of the inert moons, magnetospheric plasma directly impacts the moons and is absorbed by their surface, leaving a plasma wake downstream on the leading side of the moons. In the sub/transonic magnetized plasma, the large thermal speed of the magnetospheric ions allows the absorption of a significant fraction of the ions by the surfaces of the inert moons. This process forms an extended plasma wake in the plane of the magnetic field downstream of the moons and the signatures are observable at far distances [e.g., *Khurana et al.*, 2008; *Roussos et al.*, 2008]. Examples include Cassini observations of magnetic field compression as far as 7–8 R_{Rh} downstream in Rhea's wake during the untargeted R1.5 flyby [*Roussos et al.*, 2012] and Alfvén wing observations associated with the plasma wake of Rhea at 100 R_{Rh} on 3 June 2010 [*Khurana et al.*, 2017]. However, these large extensions of the plasma wakes behind the inert moons of Saturn (particularly for Rhea) have not yet been captured by simulations. The hybrid simulations of

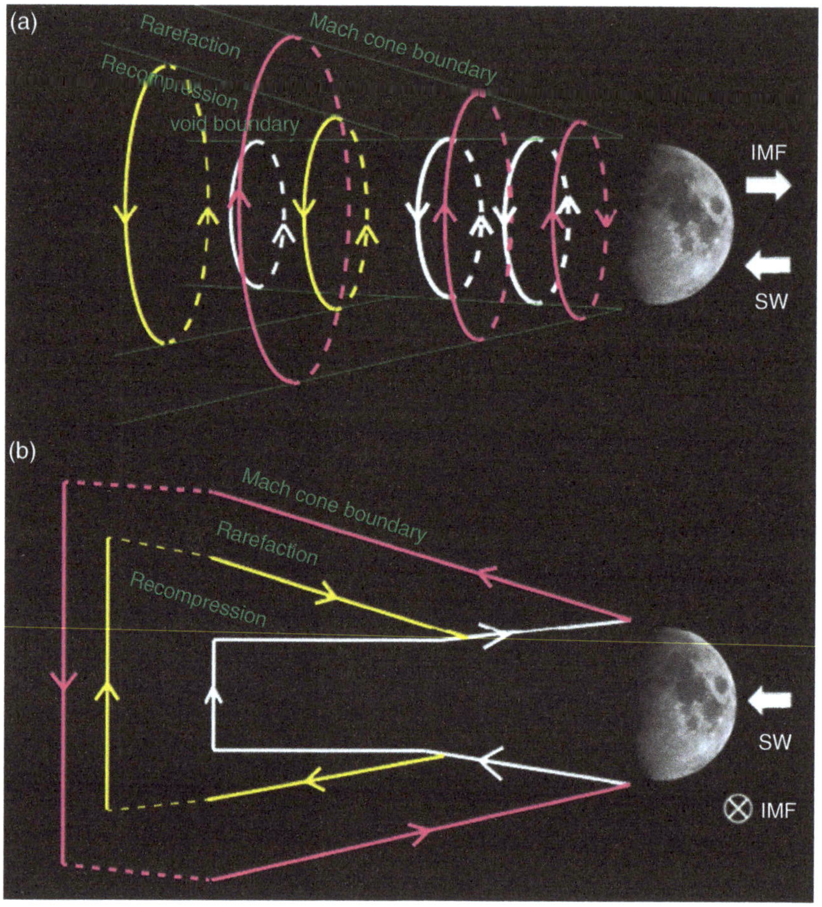

Figure 29.2 Current systems in the lunar wake for the IMF (a) antiparallel to the solar wind, and (b) perpendicular to the solar wind. Currents are formed at the Mach cone boundary (red), at the recompression boundary (yellow), and at the vacuum boundary (white). From *Fatemi et al.* [2013].

Rhea show that the wake disappears already at 4 R_{Rh} [e.g., *Khurana et al.*, 2017; *Roussos et al.*, 2008; *Simon et al.*, 2012; *Teolis et al.*, 2014], and they have never shown magnetic structures related to Rhea's wake at distances larger than 4–6 R_{Rh}. From this we can conclude that hybrid simulations of Saturn's inert moons are validated only for the immediate vicinity of the moons. The reasons for this are not fully understood, but it may be related to the energetic particle population. *Roussos et al.* [2012] state that the energetic particle observations show a variety of unexpected and complex interaction features that do not conform with our current understanding about plasma-absorbing interactions.

Figure 29.3 shows different examples of the magnetic field perturbations observed downstream of three inert moons. Figure 29.3a shows magnetic field disturbances observed by the ARTEMIS spacecraft at the closest approach of nearly 1.8 R_L downstream of the Moon, where R_L =1730 km is the radius of the Moon. Figure 29.3b shows similar magnetic field perturbations observed by Cassini downstream of Tethys on 24 September 2005 at the closest approach distance of 3.83 R_{Th}, and nearly 2.5 R_{Th} south of Tethys, where $R_{Th} = 533$ km is the mean radius of Tethys [*Khurana et al.*, 2008]. Please note that the direction of the solar wind magnetic field during the ARTEMIS event presented in Figure 29.3a is quasi-parallel to the solar wind flow (mainly along the x-axis), while the Saturnian magnetic field is always perpendicular to the corotating magnetospheric plasma (along the z-axis). Similar perturbations have also been observed in the wake of Rhea. An example is shown in Figure 29.3c from a Cassini flyby on 26 November 2005 at a closest approach distance of 1.67 R_R and nearly 0.3 R_R south of Rhea, where $R_R = 764$ km is the mean radius of Rhea. These perturbations, which are typical indications of plasma wakes at inert moons (see illustration in Fig. 29.1), have the following general signatures: (1) the magnetic fields are compressed near the closest approach and depressed in the surrounding expansion region (e.g., see B_z and $|B|$ in Fig. 29.3b at closest approach ≈ 02:43 and compare them with the gray-colored expansion region); (2) the magnetic field lines are bending and compressing

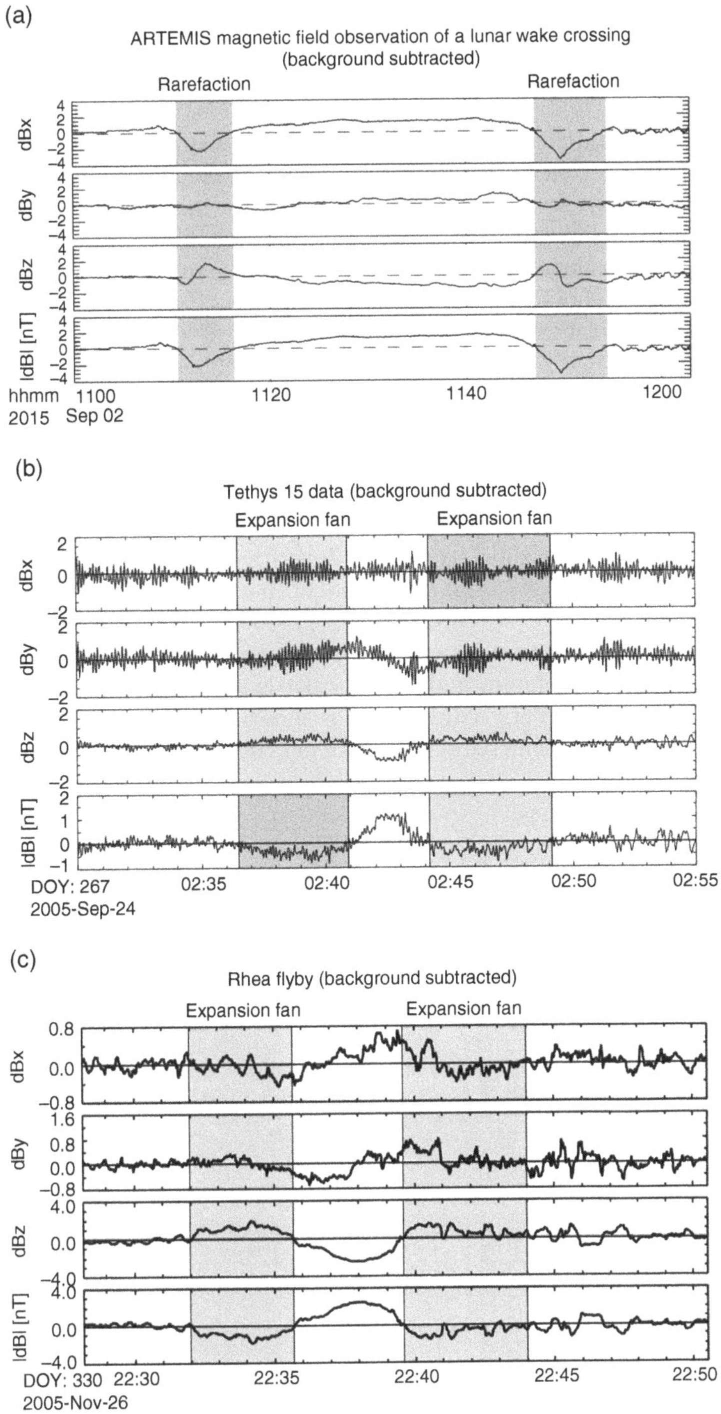

Figure 29.3 Magnetic field perturbations observed by (a) ARTEMIS spacecraft during a lunar wake crossing on 2 September 2015; and (b) and (c) Cassini spacecraft during a Tethys flyby on 24 September 2005 and a Rhea flyby on 26 November 2005, respectively. The average background magnetic field B_0 is subtracted from the perturbations. Note that the horizontal axis is observation time and is not scaled to the object's size. The direction of the solar wind magnetic field in (a) is quasi-parallel to the solar wind flow whereas the Saturnian background magnetic field is always perpendicular to plasma flow in (b) and (c). Cassini observations from *Khurana et al.* [2008]; ARTEMIS observation courtesy of Andrew R. Poppe.

toward the center of the wake indicated by the bipolar signature in the transverse component of the magnetic field (B_y). Please note that the horizontal axis in Figure 29.3 shows observation time and is not scaled to the radius of the moons. Therefore, the spatial scale of the wake and its associated magnetic field disturbances should not be compared between the three panels. However, comparing the global structure of magnetic field perturbations across outer planetary inert moons wakes (Fig. 29.3b,c) with those of a lunar wake crossing (e.g., Fig. 29.3a), we see that the perturbations are qualitatively similar.

Figures 29.4 and 29.5 compare the lunar wake structure (top panels) with the wake of Rhea (bottom panels). These simulation results were obtained from a self-consistent three-dimensional hybrid model of plasma that has previously been applied to both of these objects and successfully validated against observations [e.g., *Fatemi et al.*, 2012; *Holmström et al.*, 2012; *Khurana et al.*, 2017; *Poppe et al.*, 2014; *Zhang et al.*, 2016]. We assumed that both the Moon and Rhea are resistive obstacles to the magnetic field with resistivity 0.5×10^7 ohm·m and that their surfaces are perfect plasma absorbers without a neutral atmosphere or exosphere. The plasma parameters applied upstream in our simulations are taken from Table 29.1. It is worth noting that the Rhea simulation results presented in Figures 29.4 and 29.5 are similar to the hybrid simulations of Rhea by *Roussos et al.* [2008] and *Simon et al.* [2012]. Since the details of the simulations are not identical, in terms of parameters and implementation, this gives some confidence that the essentials of the interaction are captured by the models.

As shown in Figure 29.4a and b, the supersonic flow of the solar wind plasma near the Moon results in the extension of the vacuum region ($n/n_0 < 0.1$) behind the moon far downstream. This is while the transonic flow of magnetospheric plasma around Rhea fills in the vacuum region downstream of Rhea at closer distances (Fig. 29.4c,d). Moreover, the transonic plasma flow near Rhea, in contrast to the supersonic plasma around the Moon, allows a further extension of Rhea's wake downstream in a plane parallel to the magnetospheric field lines (Fig. 29.4d) compared to that of the Moon (Fig. 29.4b) and forms a wider angle Mach cone. In a plane perpendicular to the magnetic fields, however, the wake extension is constrained by the magnetic fields, as shown in Figure 29.4c and also discussed by *Roussos et al.* [2008] and *Khurana et al.* [2008].

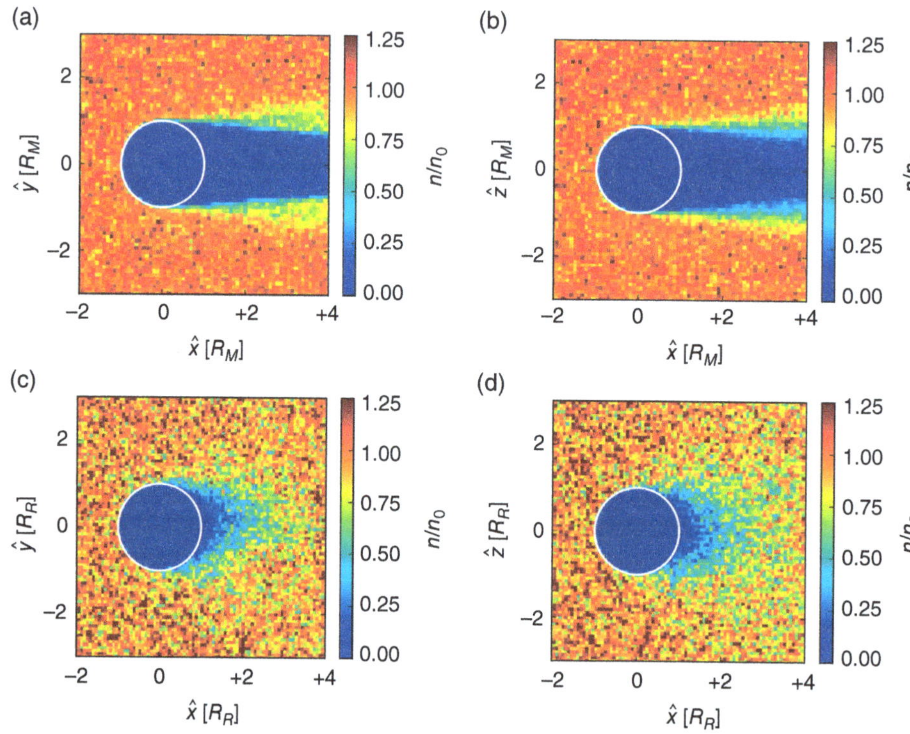

Figure 29.4 Plasma density obtained from hybrid simulations for (a) and (b) solar wind plasma interaction with the Moon, and (c) and (d) magnetospheric plasma interaction with Rhea, normalized to the background plasma density shown in Table 29.1. Panels (a) and (c) are cuts in the *xy*-plane at $z = 0$; panels (b) and (d) are cuts in the *xz*-plane at $y = 0$. In all panels, plasma flows along the *x*-axis and magnetic field is along the $-z$-axis. The white circles in each plane represent the moons.

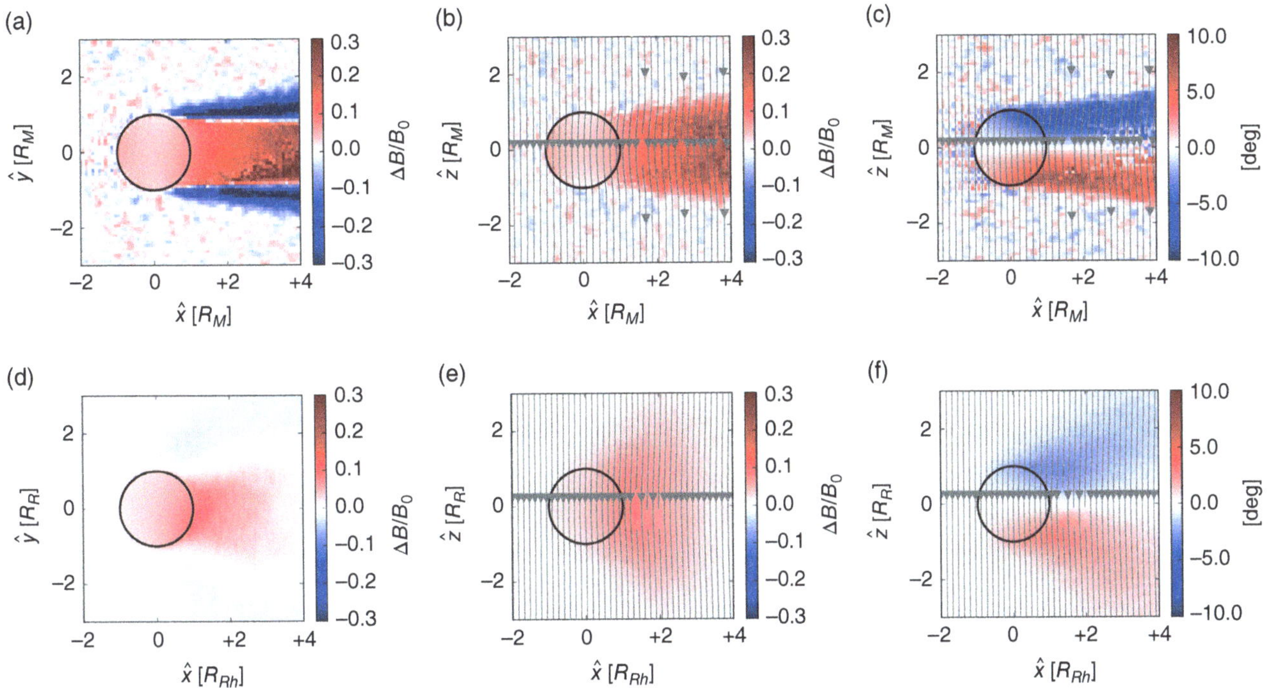

Figure 29.5 Magnetic field perturbations ΔB obtained from hybrid simulations for (a) and (b) solar wind plasma interaction with the Moon and (d) and (e) magnetospheric plasma interaction with Rhea, normalized to the background magnetic field from Table 29.1. Panels (c) and (f) show the bending of the magnetic field lines in degrees. The geometry of the cuts is the same as those in Figure 29.4.

The maximum compression of the magnetic fields in the central wake ($\Delta B = |\mathbf{B} - \mathbf{B}_0|$) is theoretically shown to occur when the magnetic field and the plasma flow direction are (anti-)aligned and is a function of plasma beta [*Fatemi*, 2014; *Simon et al.*, 2012],

$$\frac{\Delta B}{B_0} = \sqrt{1+\beta} - 1. \qquad (29.1)$$

For plasma $\beta = 0.03$ at the orbit of Tethys, 0.25 at the orbit of Rhea, and around 1.3 near the Earth's Moon (see Table 29.1) the maximum $\Delta B/B_0$ is around 0.015, 0.12, and 0.51, respectively. For the case of Saturn's inert moons, the magnetic field and corotating plasma are roughly perpendicular to each other, thus the magnetic field compression would be smaller than that predicted by the theoretical expression in equation (29.1). This is consistent with Cassini observations shown in Figure 29.3 (note that only ΔB is shown in Fig. 29.3). For the Earth's Moon, however, alignment of the interplanetary magnetic field and solar wind plasma is possible, thus the fields in the central wake of the Moon can enhance to the maximum compression predicted by equation (29.1).

Figure 29.5 compares the magnetic field compression and bending of the magnetic field lines obtained from our hybrid simulations between the Moon (top panels) and Rhea (bottom panels) when the magnetic fields are perpendicular to the plasma flow. Figure 29.5b and e shows that in the plane of the magnetic fields, the fields are compressed in the wake and bent toward the moons in the central wake. In the plane perpendicular to the magnetic fields (Fig. 29.5a, d), the magnetic field compression in the central wake is accompanied by the field reduction in the surrounding to ensure $\nabla \cdot \mathbf{B} = 0$. Figure 29.5c and f shows the bending angle of the field lines and the associated Alfvén wing structure.

Changes in the magnetic field strength and direction in the wake of Saturn's inert moons are indications for the existence of current systems flowing in their wake with qualitative similarities to those in the wake of Earth's Moon. *Roussos et al.* [2008] using a 3D hybrid model of plasma found a diamagnetic current system flowing perpendicular to the magnetic fields at the boundary of the plasma wake downstream of Rhea, shown in Figure 29.6a. This diamagnetic current flows either parallel or antiparallel to the plasma flow, similar to the case of the Moon (the white current in Fig. 29.2b). These currents perturb the magnetic fields in the wake of inert moons and the resulting $\mathbf{J} \times \mathbf{B}$ force makes an asymmetric velocity distribution on the two sides of the wake flanks (Fig. 29.6b).

Simon et al. [2012] examined the diamagnetic currents in the wake of Rhea in a plane perpendicular to that

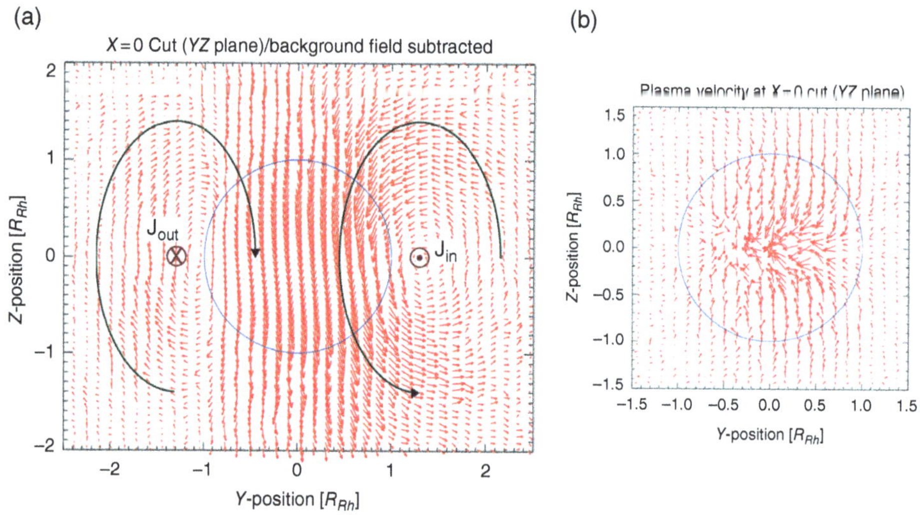

Figure 29.6 (a) Magnetic field perturbations in the yz-plane at $x = 0$ obtained from hybrid simulations of plasma interaction with Rhea (blue circle) by *Roussos et al.* [2008]. The background field has been subtracted to better present the field perturbations. The peak amplitude of the residual B_y in this simulation is about 1.5 nT. Direction of diamagnetic currents associated with plasma absorption by Rhea is shown by direction arrows. (b) Associated plasma velocity perturbations behind Rhea. The peak amplitude of the velocity components in this plane is about 30 km/s. The geometry of the cut is the same as in panel (a). From *Roussos et al.* [2008].

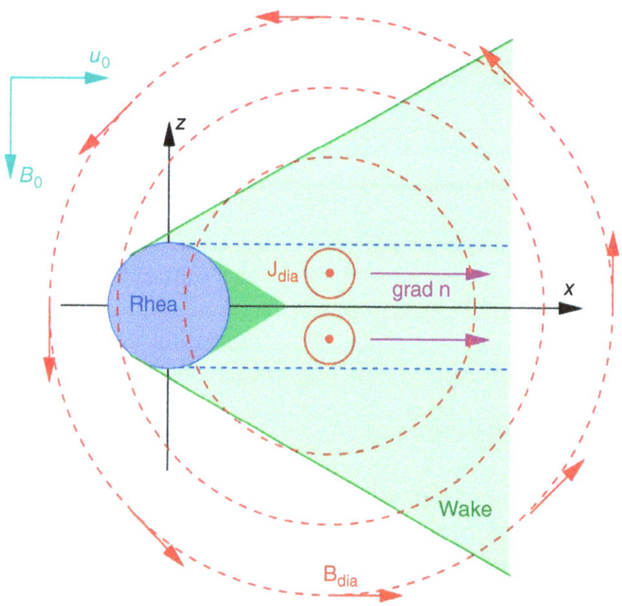

Figure 29.7 An illustration of diamagnetic currents $\mathbf{J}_{dia}$ in the wake of Rhea (green area). The magnetic field generated by this current $\mathbf{B}_{dia}$ is shown by dashed circles. From *Simon et al.* [2012].

investigated by *Roussos et al.* [2008]. Figure 29.7 shows a schematic illustration of Rhea's diamagnetic current and its associated magnetic field perturbations suggested by *Simon et al.* [2012]. In this figure, the dark green area is the density cavity and the light green area is the expansion cone. *Simon et al.* [2012] assumed that the vacuum region downstream of Rhea forms a pressure gradient along the plasma flow direction in the wake, shown in Figure 29.7. Since Saturn's magnetic field is perpendicular to the plasma flow direction, the suggested diamagnetic current should be generated along the convective electric field and perpendicular to both the magnetic field and plasma flow direction. *Simon et al.* [2012] showed that the magnetic fields generated by this diamagnetic current

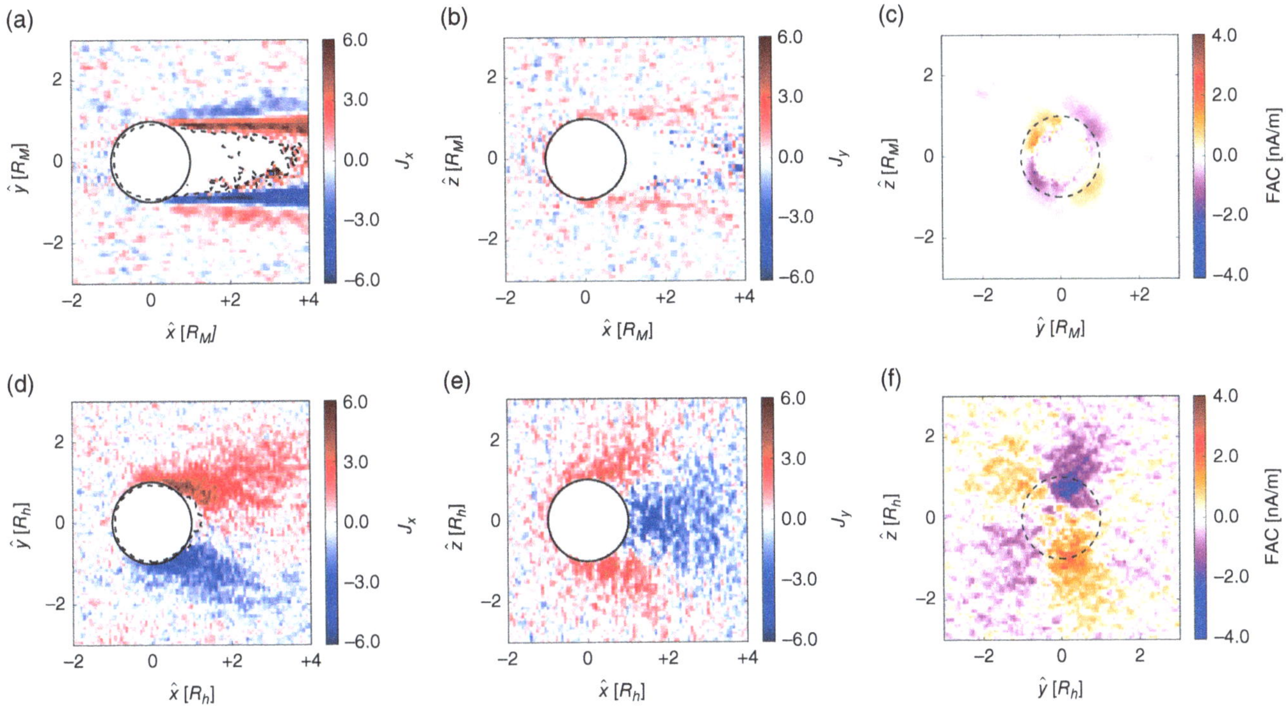

Figure 29.8 Components of different currents obtained from hybrid simulations for (a), (b), and (c), solar wind plasma interaction with the Moon; and (d), (e), and (f), magnetospheric plasma interaction with Rhea. (a) and (b) Diamagnetic and confinement currents in the lunar wake; (c) field-aligned currents in the lunar wake; (d) and (e) diamagnetic currents in Rhea's wake; (f) field-aligned currents in Rhea's wake. The geometry of the cuts in (a), (b), (d), and (e) are the same as those in Figure 29.4. The cuts in (c) and (f) are in the *yz*-plane at $x = +1.7$ moon radii. The boundaries of plasma cavity ($n/n_0 < 0.1$) are shown by dashed green contour lines in (a) and (d).

(dashed red curves in Fig. 29.7) add an antiflow component to the magnetic field at the north pole of Rhea and a component along the flow at the south pole. They suggested that this additional component results in the bending of the magnetic fields and forms the Alfvén wings of Rhea. We see from Figure 29.4c and d, which is consistent with simulation results presented by Simon *et al.* [2012], that the pressure gradient is perpendicular to the magnetic fields at all directions across the plasma wake and not only along the flow direction. Thus, the direction of diamagnetic currents in the wake of Rhea, as well as for the other inert moons, should not be only along the direction of the convective electric field.

Figure 29.8 compares the currents of the Moon (top panels) and of an inert moon of Saturn, Rhea (bottom panels), obtained from our hybrid plasma simulations, presented in Figures 29.4 and 29.5. Using the general Ampère's law, we calculate the current density $\mathbf{J} = \mu_0^{-1}\nabla\times\mathbf{B}$, where μ_0 is the vacuum permeability. Figure 29.8d shows that the diamagnetic currents in the wake of the inert moons flow around the wake boundaries (anti-)parallel to the plasma flow, consistent with the *Roussos et al.* [2008] simulations shown in Figure 29.6a. Figure 29.8e, on the other hand, shows that the diamagnetic currents flow across the wake and along the convective electric field in the central wake, consistent with the illustration by *Simon et al.* [2012] (shown in Fig. 29.7). Thus, the diamagnetic current system around the inert moons of Saturn is not only the one shown by *Roussos et al.* [2008] and not only the one shown by *Simon et al.* [2012], but is a combination of both. These diamagnetic currents and their flow direction are qualitatively similar to the diamagnetic currents around the Earth's Moon, shown in Figure 29.8a and b and previously shown in Figure 29.2. Figure 29.8c and f also shows the field-aligned currents (FACs) downstream of the Moon and Rhea, respectively. We see the FACs, which are responsible for formation of Alfvén wings, are qualitatively similar for all the inert moons, regardless of their plasma environment. However, as we discuss below, the closure of the currents in the supersonic, super-Alfvénic, plasma environment of the Moon is different compared to that at the sub/transonic sub-Alfvénic environment of the inert moons.

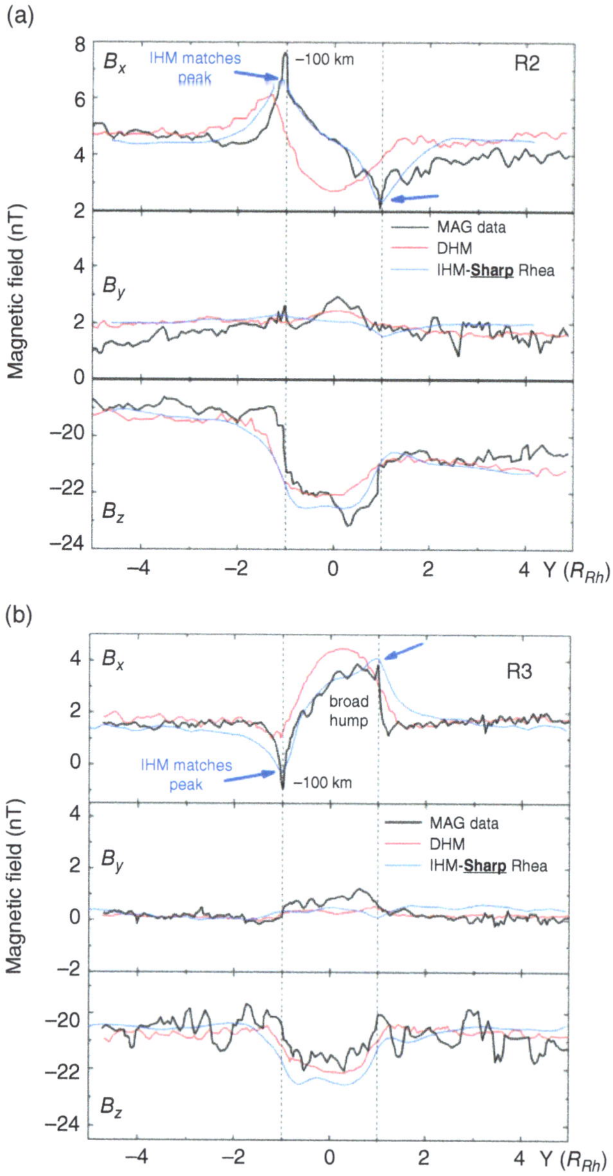

Figure 29.9 Cassini magnetic field observations (black lines) during (a) the R2 and (b) R3 Rhea flybys. Superimposed are hybrid simulations (red lines) from *Simon et al.* [2012] and (blue lines) from *Teolis et al.* [2014]. From *Teolis et al.* [2014].

During the only two polar flybys of Rhea, R2 and R3, Cassini observed nonzero magnetic field perturbations along the plasma flow (B_x) below Rhea's south pole and opposite to the plasma flow direction above Rhea's north pole, as shown in Figure 29.9. These signatures are associated with an Alfvén wing structure in the wake of Rhea [*Simon et al.*, 2012]. The observation of Rhea's Alfvén wings was one of the surprising aspects of Cassini's observations, because Rhea has neither a highly conductive interior, nor a tenuous atmosphere to slow down the corotating plasma upstream and bend the magnetic fields to form the Alfvén wings [*Anderson and Schubert*, 2007; *Teolis et al.*, 2010]. *Simon et al.* [2012], as explained previously, suggested that the B_x perturbations of the magnetic fields observed by Cassini during R2 and R3, and consequently the formation of Rhea's Alfvén wings, are due to the diamagnetic currents, shown in Figure 29.7. They also discussed that the Alfvén wing currents close through the diamagnetic currents across the wake [*Simon et al.*, 2012, 2015]. Recently, *Khurana et al.* [2017], using hybrid plasma simulations, suggested that the formation of Rhea's Alfvén wings is because of the pressure gradient associated with plasma absorption on the trailing side of Rhea. They showed that the pressure gradient slows down the plasma moving into the wake along the magnetic field lines, while the magnetic flux tubes connected to the wake at farther distances are moving with corotating magnetospheric plasma. This differentiated motion of plasma bends the magnetic field lines and forms the Alfvén wings [*Khurana et al.*, 2017].

In addition to the non-zero B_x component of the magnetic field during the R2 and R3 flybys, there are noticeable sharp magnetic field perturbations in the B_x component over the poles at close distances to Rhea near the surface of Rhea's flux tube (pointed to by blue arrows in Fig. 29.9) [*Simon et al.*, 2012; *Teolis et al.*, 2014]. These perturbations occurred over short distances, much smaller than the ion inertial length, and thus cannot be generated by Alfvénic field-aligned currents (FACs) [*Teolis et al.*, 2014]. However, since these perturbations were observed during the two Cassini close polar encounters in opposite hemispheres at Rhea's flux tube boundaries, they seem to be a consistent feature of the plasma interaction with Rhea [*Simon et al.*, 2015]. *Teolis et al.* [2014] showed that the ions are approximately uniformly absorbed by the surface of Rhea because of their large gyroradius, while the small gyroradius of electrons confines their motion along the magnetic field lines. This differentiated plasma absorption requires formation of additional FACs generated near the surface of Rhea and extending to large distances. These flux tube currents satisfy surface current balance and they are expected to be a general current structure for any inert object interacting with plasma [*Teolis et al.*, 2014].

Similar to the Earth's Moon, the surface of an inert object immersed in plasma and solar ultraviolet irradiation is charged to an electrostatic potential. For example, *Nordheim et al.* [2014] using Cassini observations remotely found evidences of a strongly negative (≈ −200 V) surface potential on Hyperion. Unlike the Moon, the corotating magnetospheric plasma flow direction near moons in outer planets' magnetospheres is not always coaligned with the solar UV [*Roussos et al.*, 2010]. Depending on the moon's local time, the magnetospheric plasma forms any angle with the solar UV direction, from perfectly aligned and antialigned directions to perfectly perpendicular directions. This forms a wide range of scenarios for surface charging as a function of the relative plasma and solar UV incidence angle [*Poppe et al.*, 2015; *Roussos et al.*, 2010].

Poppe et al. [2015] using electrostatic particle simulations showed the complex structure of electrostatic potential building up at the surface and near the inert outer planetary moons for aligned, perpendicular, and antialigned plasma flow directions relative to the solar UV (Fig. 29.10). They suggested that the plasma wake refilling may become highly asymmetric when the plasma flow and solar UV are not aligned. However, their electrostatic simulations did not include the magnetic fields, which adds another direction to the wake-refilling geometry and influences the current systems in the moon's plasma environment. They also did not include secondary emissions, which could also play an important role for surface charging of the outer planetary moons [*Poppe et al.*, 2015].

Rhea does not possess noticeable geological activity and does not have any active plumes [e.g., *Scipioni et al.*, 2014; *Wagner et al.*, 2007, 2010]. It also has a tenuous exosphere mainly composed of O_2 and CO_2, but the neutral gas and plasma production rates are too low to mass-load the corotating magnetospheric plasma [*Simon et al.*, 2012; *Teolis et al.*, 2010; *Teolis and Waite*, 2016]. *Simon et al.* [2012] theoretically estimated that the Pederson conductance of Rhea is nearly two orders of magnitude smaller than the Alfvén conductance of plasma. Therefore, Rhea is neither a mass-loading object nor a conducting obstacle. Dione, although smaller than Rhea, has a relatively denser exosphere with similar composition to Rhea, but it is still too weak to slow down the magnetospheric plasma upstream and considerably influence the plasma interaction [*Burch et al.*, 2007; *Krupp et al.*, 2013; *Teolis and Waite*, 2016]. In essence, the exosphere and the internal conductivity of the inert moons of Saturn, similar to the Earth's Moon, do not considerably influence the plasma interaction, nor alter the wake structure and the current systems associated with plasma absorption by the moons. However, *Simon et al.* [2011] argued for a mass-loading signature during the Dione D1 upstream flyby, driven by the presence of an exosphere that may have been generated due to the occurrence of a transient radiation belt.

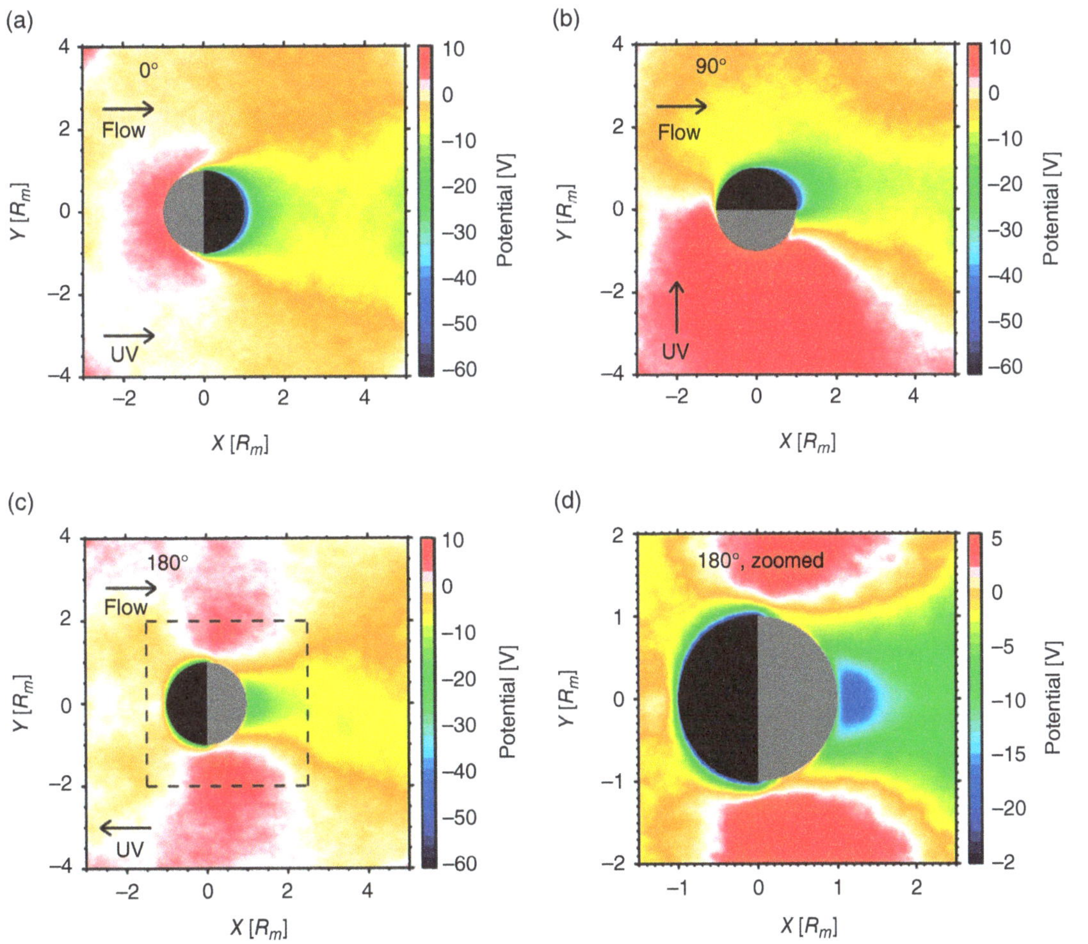

Figure 29.10 The electrostatic potential environment around an inert moon obtained from electrostatic particle simulations for UV vs. plasma flow angles (a) 0°, (b) 90°, (c) 180°, (d) zoomed in from the dashed box in (c). From *Poppe et al.* [2015].

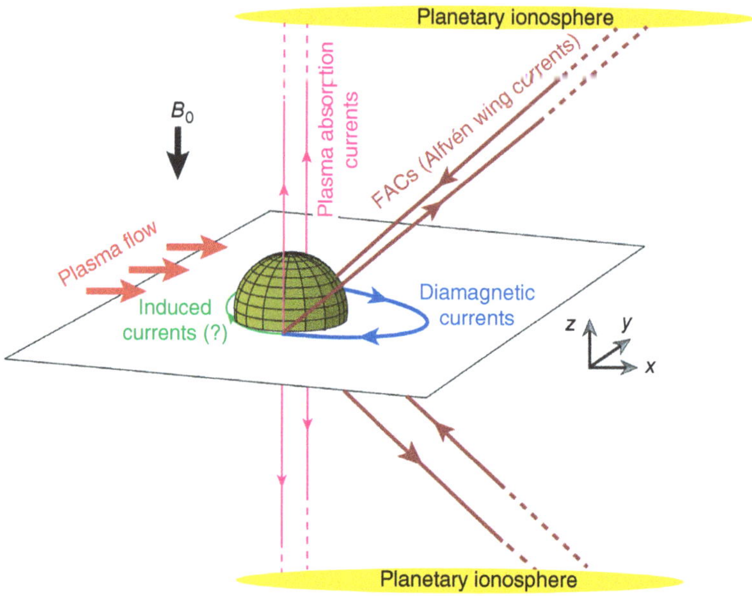

Figure 29.11 An illustration of current systems around inert moons in a trans/subsonic and sub-Alfvénic plasma.

Figure 29.11 shows a simplified illustration of our current understanding of the current systems around the inert moons of Saturn. Diamagnetic currents, shown in blue, form around the plasma cavity and compress the fields in the cavity to maintain pressure balance. FACs or Alfvén wing currents form due to the pressure gradient associated with plasma absorption by the moons, resulting in bending of the field lines close to the planetary ionosphere and resulting in diamagnetic currents. Plasma absorption currents also form due to absorption of plasma by the surface of the inert moons of Saturn. If the moons have a conductive interior and/or an ionosphere, an induced current would flow upstream of the moons and would close the wake current systems. In addition, since ions are absorbed ahead of the electrons, due to their larger gyroradius, currents should form near the surface on the upstream hemisphere, as suggested by *Owen et al.* [1996]. This simplified image of current systems would be valid for any inert moon located in the sub/transonic and sub-Alfvénic plasma environment.

However, at the satellites close to Saturn, that is, Dione and Tethys, the plasma beta is very small and the plasma pressure cannot compress the strong field lines in the wake of these objects, so we anticipate that these currents are not significant.

29.4. SUMMARY

We have presented the different current systems associated with inert moons (nonconducting and lacking ionosphere), along with examples for Earth's Moon, and Saturn's moons Tethys and Rhea. The interaction of these moons with the solar wind or magnetospheric plasma is at first sight simple. They are obstacles to the plasma flow, leaving a wake behind. However, there is a wealth of processes occurring, from surface interactions to complex plasma physics governing the refill of the wake, generating a wealth of different current systems. Still, for all the complexity of the interactions, we find that the plasma interactions of Earth's Moon and its associated current systems are very similar to other inert moons. This is a view that has emerged lately, from observations and modeling.

ACKNOWLEDGMENTS

We thank Andrew R. Poppe for the plot of ARTEMIS observations in Figure 29.3. S. Fatemi acknowledges support from NASA's Solar System Exploration Research Virtual Institute (SSERVI), grant NNX14AG16A, and partial support from NASA's Cassini Data Analysis program, grant NNX15AL20G.

REFERENCES

Anderson, J. D., and G. Schubert (2007), Saturn's satellite Rhea is a homogeneous mix of rock and ice, *Geophys. Res. Lett.*, *34* (2).

Balogh, A. (2010), Planetary magnetic field measurements: Missions and instrumentation, *Space Sci. Rev.*, *152*, 23–97.

Burch, J. L., J. Goldstein, W. S. Lewis, D. T. Young, A. J. Coates, M. K. Dougherty, and N. André (2007), Tethys and Dione as sources of outward-flowing plasma in Saturn's magnetosphere, *Nature*, *447*, 833–835.

Colburn, D. S., R. G. Currie, J. D. Mihalov, and C. P. Sonett (1967), Diamagnetic solar-wind cavity discovered behind Moon, *Science*, *158*, 1040–1042.

Dyal, P., C. W. Parkin, and W. D. Daily (1974), Magnetism and the interior of the Moon, *Rev. Geophys.*, *12*(4), 568–591.

Fatemi, S. (2014), *Kinetic Modeling of the Solar Wind Plasma Interaction With the Moon*, PhD Thesis, Luleå University of Technology.

Fatemi, S., H. A. Fuqua, A. R. Poppe, G. T. Delory, J. S. Halekas, W. M. Farrell, and M. Holmström (2015), On the confinement of lunar induced magnetic fields, *Geophys. Res. Lett.*, *42*, 6931–6938.

Fatemi, S., M. Holmström, and Y. Futaana (2012), The effects of lunar surface plasma absorption and solar wind temperature anisotropies on the solar wind proton velocity space distributions in the low-altitude lunar plasma wake, *J. Geophys. Res.*, *117* (A10105).

Fatemi, S., M. Holmström, Y. Futaana, C. Lue, M. R. Collier, S. Barabash, and G. Stenberg (2014), Effects of protons reflected by lunar crustal magnetic fields on the global lunar plasma environment, *J. Geophys. Res.*, *119*, 6095–6105.

Fatemi, S., M. Holmström, Y. Futaana, S. Barabash, and C. Lue (2013), The lunar wake current systems, *Geophys. Res. Lett.*, *40* (1), 17–21.

Grimm, R. E., and G. T. Delory (2012), Next-generation electromagnetic sounding of the Moon, *Adv. Space Res.*, *50* (12), 1687–1701.

Halekas, J. S., A. R. Poppe, J. P. McFadden, V. Angelopoulos, K.-H. Glassmeier, and D. A. Brain (2014), Evidence for small-scale collisionless shocks at the Moon from ARTEMIS, *Geophys. Res. Lett.*, *41*, 7436–7443.

Halekas, J. S., D. A. Brain, and M. Holmström (2015a), The Moon's plasma wake, in *Magnetotails in the Solar System*, edited by A. Keiling, C. M. Jackman, and P. A. Delamere, 149–167, Geophysical Monograph 207, American Geophysical Union, Washington, DC.

Halekas, J. S., M. Benna, P. R. Mahaffy, R. C. Elphic, A. R. Poppe, and G. T. Delory (2015b), Detections of lunar exospheric ions by the LADEE neutral mass spectrometer, *Geophys. Res. Lett.*, *42*, 5162–5169.

Holmström, M., S. Fatemi, Y. Futaana, and H. Nilsson (2012), The interaction between the Moon and the solar wind, *Earth Planets Space*, *64*, 237–245.

Jia, X. (2015), Satellite magnetotails, in *Magnetotails in the Solar System*, edited by A. Keiling, C. A. Jackman, and P. A. Delamere, 135–148, Geophysical Monograph, American Geophysical Union, Washington DC.

Khurana, K. K., C. T. Russell, and M. K. Dougherty (2008), Magnetic portraits of Tethys and Rhea, *Icarus*, *193* (2), 465–474.

Khurana, K. K., S. Fatemi, J. Lindkvist, E. Roussos, N. Krupp, M. Holmström, C. T. Russell, and M. Dougherty (2017), The role of plasma slowdown in the generation of Rhea's Alfvén wings, *J. Geophys. Res.*, *122*, 1778–1788.

Khurana, K. K., X. Jia, M. G. Kivelson, F. Nimmo, G. Schubert, and C. T. Russell (2011) Evidence of a global magma ocean in Io's interior, *Science*, *332*, 1186–1189.

Kivelson, M. G., K. K. Khurana, and M. Volwerk (2002), The permanent and inductive magnetic moments of Ganymede, *Icarus*, *157*, 507–522.

Kivelson, M. G., K. K. Khurana, C. T. Russell, M. Volwerk, R. J. Walker, and C. Zimmer (2000), Galileo magnetometer measurements: A stronger case for a subsurface ocean at Europa, *Science*, *289*, 1340–1343.

Kivelson, M. G., K. K. Khurana, D. J. Stevenson, L. Bennett, S. Joy, C. T. Russell, R. J. Walker, C. Zimmer, and C. Polanskey (1999), Europa and Callisto: Induced or intrinsic fields in a periodically varying plasma environment, *J. Geophys. Res.*, *104*, A3, 4609–4625.

Krupp, N., E. Roussos, H. Kriegel, P. Kollmann, M. G. Kivelson, A. Kotova, C. Paranicas, D. G. Mitchell, S. M. Krimigis, and K. K. Khurana (2013), Energetic particle measurements in the vicinity of Dione during the three Cassini encounters 2005–2011, *Icarus*, *226*, 617–628.

Leisner, J. S., C. T. Russell, R. J. Strangeway, N. Omidi, M. K. Dougherty, and W. S. Kurth (2008), *The Interior of Iapetus: Constraints Provided by the Solar Wind Interaction*, AGU Fall Meeting Abstracts (2008), C8.

Lipatov, A. S., J. F. Cooper, E. C. Sittler Jr., and R. E. Hartle (2012), Effects of Na^+ and He^+ pickup ions on the lunar-like plasma environment: 3D hybrid modeling, *Adv. Space Res.*, *50* (12), 1583–1591.

Lue, C., Y. Futaana, S. Barabash, M. Wieser, M. Holmström, A. Bhardwaj, M. B. Dhanya, and P. Wurz (2011), Strong influence of lunar crustal fields on the solar wind flow, *Geophys. Res. Lett.*, *38*, L03202.

Manka, R. H., and F. C. Michel (1971), Lunar atmosphere as a source of lunar surface elements, *Proceedings of the Second Lunar Science Conference*, The M.I.T. Press, 2, 1717–1728.

Mitchell, D. L., J. S. Halekas, R. P. Lin, S. Frey, L. L. Hood, M. H. Acuña, and A. Binder (2008), Global mapping of lunar crustal magnetic fields by Lunar Prospector, *Icarus*, *194* (2), 401–409.

Ness, N. F., K. W. Behannon, H. E. Taylor, and Y. C. Whang (1968), Perturbations of the interplanetary magnetic field by the lunar wake, *J. Geophys. Res.*, *73* (11), 3421–3440.

Nitter, T., and O. Havnes (1992), Dynamics of dust in a plasma sheath and injection of dust into the plasma sheath above Moon and asteroidal surfaces, *Earth Moon Planet*, *56* (7), 7–34.

Nordheim, T. A., G. H. Jones, E. Roussos, J. S. Leisner, A. J. Coates, W. S. Kurth, K. K. Khurana, N. Krupp, M. K. Dougherty, and J. H. Waite (2014), Detection of a strongly negative surface potential at Saturn's moon Hyperion, *Geophys. Res. Lett.*, *41* (20), 7011–7018.

Ogilvie, K. W., J. T. Steinberg, R. J. Fitzenreiter, C. J. Owen, A. J. Lazarus, W. M. Farrell, and R. B. Torbert (1996), Observations of the lunar plasma wake from the WIND spacecraft on December 27, 1994, *Geophys. Res. Lett.*, *23* (10), 1255–1258.

Owen, C. J., R. P. Lepping, K. W. Ogilvie, J. A. Slavin, W. M. Farrell, and J. B. Byrnes (1996), The lunar wake at 6.8 R_L: WIND magnetic field observations, *Geophys. Res. Lett.*, *23* (10), 1263–1266.

Poppe, A. R., J. S. Halekas, and M. Horányi (2011), Negative potentials above the day-side lunar surface in the terrestrial plasma sheet: Evidence of non-monotonic potentials, *Geophys. Res. Lett.*, *38* (2), L02103.

Poppe, A. R., J. S. Halekas, J. R. Szalay, M. Horányi, Z. Levin, and S. Kempf (2016), LADEE/LDEX observations of lunar pickup ion distribution and variability, *Geophys. Res. Lett.*, *43*, 3069–3077.

Poppe, A. R., M. I. Zimmerman, J. S. Halekas, and W. M. Farrell (2015), The electrostatic plasma environment of a small airless body under non-aligned plasma flow and UV conditions, *Planet. Space Sci.*, *119*, 111–120.

Poppe, A. R., R. Samad, J. S. Halekas, M. Sarantos, G. T. Delory, W. M. Farrell, V. Angelopoulos, and J. P. McFadden (2012), ARTEMIS observations of lunar pick-up ions in the terrestrial magnetotail lobes, *Geophys. Res. Lett.*, *39*, L17104.

Poppe, A. R., S. Fatemi, J. S. Halekas, M. Holmström, and G. T. Delory (2014), ARTEMIS observations of extreme diamagnetic fields in the lunar wake, *Geophys. Res. Lett.*, *41* (11), 3766–3773.

Roussos, E., J. Müller, S. Simon, A. Bößwetter, U. Motschmann, N. Krupp, M. Fränz, J. Woch, K. K. Khurana, and M. K. Dougherty (2008), Plasma and fields in the wake of Rhea: 3-D hybrid simulation and comparison with Cassini data, *Ann. Geophys.*, *26* (3), 619–637.

Roussos, E., N. Krupp, H. Krüger, and G. H. Jones (2010), Surface charging of Saturn's plasma-absorbing moons, *J. Geophys. Res.*, *115*, A08225.

Roussos, E., P. Kollmann, N. Krupp, C. Paranicas, S. M. Krimigis, D. G. Mitchell, A. M. Persoon, et al. (2012), Energetic electron observations of Rhea's magnetospheric interaction, *Icarus*, *221*, 116–134.

Saur, J., and D. F. Strobel (2005), Atmospheres and plasma interactions at Saturn's largest inner icy satellites, *Astrophys. J. Lett.*, *620* (2), L115.

Schubert, G., C. P. Sonett, K. Schwartz, and H. J. Lee (1973), Induced magnetosphere of the Moon, 1. Theory, *J. Geophys. Res.*, *78* (13), 2094–2110.

Scipioni, F., F. Tosi, K. Stephan, G. Filacchione, M. Ciarniello, F. Capaccioni, and P. Cerroni (2014), Spectroscopic classification of icy satellites of Saturn II: Identification of terrain units on Rhea, *Icarus*, *234*, 1–16.

Sergis, N., S. M. Krimigis, E. C. Roelof, C. S. Arridge, A. M. Rymer, D. G. Mitchell, D. C. Hamilton, et al. (2010), Particle pressure, inertial force, and ring current density profiles in the magnetosphere of Saturn, based on Cassini measurements, *Geophys. Res. Lett.*, *37*, L02102.

Simon, S., E. Roussos, and C. S. Paty (2015), The interaction between Saturn's moons and their plasma environments, *Phys. Rep.*, *602*, 1–65.

Simon, S., J. Saur, F. M. Neubauer, A. Wennmacher, and M. K. Dougherty (2011), Magnetic signatures of a tenuous atmosphere at Dione, *Geophys. Res. Lett.*, *38* (15), 1–5.

Simon, S., H. Kriegel, J. Saur, A. Wennmacher, F. M. Neubauer, E. Roussos, U. Motschmann, and M. K. Dougherty (2012), Analysis of Cassini magnetic field observations over the poles of Rhea, *J. Geophys. Res.*, *117*, A07211.

Sittler, E. C., Jr., M. Thomsen, R. E. Johnson, R. E. Hartle, M. Burger, D. Chornay, M. D. Shappirio, et al. (2006), Cassini observations of Saturn's inner plasmasphere: Saturn orbit insertion results, *Planet. Space Sci.*, *54* (12), 1197–1210.

Sonett, C. P. (1982), Electromagnetic induction in the Moon, *Rev. Geophys.*, *20* (3), 411–455.

Stubbs, T. J., D. A. Glenar, W. M. Farrell, R. R. Vondrak, M. R. Collier, J. S. Halekas, and G. T. Delory (2011), On the role of dust in the lunar ionosphere, *Planet. Space Sci.*, *59*, 1659–1664.

Teolis, B. D., and J. H. Waite (2016), Dione and Rhea seasonal exospheres revealed by Cassini CAPS and INMS, *Icarus*, *272*, 277–289.

Teolis, B. D., G. H. Jones, P. F. Miles, R. L. Tokar, B. A. Magee, J. H. Waite, E. Roussos, et al. (2010), Cassini finds an oxygen-carbon dioxide atmosphere at Saturn's icy moon Rhea, *Science*, *330* (6012), 1813–1815.

Teolis, B. D., I. Sillanpää, J. H. Waite, and K. K. Khurana (2014), Surface current balance and thermoelectric whistler wings at airless astrophysical bodies: Cassini at Rhea, *J. Geophys. Res.*, *119* (11), 8881–8901.

Vernisse, Y., H. Kriegel, S. Wiehle, U. Motschmann, and K.-H. Glassmeier (2013), Stellar winds and planetary bodies simulations: Lunar type interaction in super-Alfvénic and sub-Alfvénic flows, *Planet. Space Sci.*, *84*, 37–47.

Wagner, R. J., G. Neukum, B. Giese, T. Roatsch, and U. Wolf (2007), *The Global Geology of Rhea: Preliminary Implications from the Cassini ISS Data*, Lunar and Planetary Institute Conference, Abstracts 38, abstract 1958.

Wagner, R. J., G. Neukum, B. Giese, T. Roatsch, T. Denk, U. Wolf, and C. C. Porco (2010), *The Geology of Rhea: A First Look at the ISS Camera Data From Orbit 12 (Nov. 21, 2009) in Cassini's Extended Mission*, Lunar and Planetary Institute Conference, Abstracts 41, abstract 1672.

Weber, R. C., P. Lin, E. J. Garnero, Q. Williams, and P. Lognonne (2011), Seismic detection of the lunar core, *Science*, *331* (6015), 309–312.

Wiehle, S., F. Plaschke, U. Motschmann, K.-H. Glassmeier, H. U. Auster, V. Angelopoulos, J. Mueller, et al. (2011), First lunar wake passage of ARTEMIS: Discrimination of wake effects and solar wind fluctuations by 3D hybrid simulations, *Planet. Space Sci.*, *59*, 661–671.

Xie, L., L. Li, Y. Zhang, Y. Feng, X. Wang, A. Zhang, and L. Kong (2015), Three-dimensional Hall MHD simulation of lunar minimagnetosphere: General characteristics and comparison with Chang'E-2 observations, *J. Geophys. Res.*, *120*, 6559–6568.

Zhang, H., K. K. Khurana, M. G. Kivelson, S. Fatemi, M. Holmström, V. Angelopoulos, Y. D. Jia, et al. (2016), Alfvén Wings in the Lunar Wake: The role of pressure gradients, *J. Geophys. Res.*, *121*, 10698–10711.

Zhang, H., K. K. Khurana, M. G. Kivelson, V. Angelopoulos, W. X. Wan, L. B. Liu, Q.-G. Zong, Z. Y. Pu, Q. Q. Shi, and W. L. Liu (2014), Three-dimensional lunar wake reconstructed from ARTEMIS data, *J. Geophys. Res.*, *119*, 5220–5243.

Zhou, X.-Z., V. Angelopoulos, A. R. Poppe, J. S. Halekas, K. K. Khurana, M. G. Kivelson, S. Fatemi, and M. Holmström (2014), Lunar dayside current in the terrestrial lobe: ARTEMIS observations, *J. Geophys. Res.*, *5* (119), 3381–3391.

30
Currents in Cometary Comae

Martin Volwerk

ABSTRACT

In this chapter, a review is given about currents in cometary comae. First, from a historical viewpoint, developments about the knowledge of comets are given. The first space missions to comets are discussed, in particular the missions to comet 1P/Halley, which gave many in situ observations about the plasma environment around the nucleus and the interaction with the solar wind. Solar-wind draping around the nucleus results in current sheets. Possible reconnection in the tail or in the nested draped field regions also generates currents. After the historic view, the focus moves to the Rosetta mission around comet 67P/Churyumov-Gerasimenko and the latest results on the observations of electric currents in the coma.

30.1. INTRODUCTION

The appearance of a comet in the sky was usually accepted as a sign from heaven, for good or for evil (usually the latter). The most famous of appearances is probably the comet in 1066, well known because of its portrayal on the Bayeux tapestry (see Fig. 30.1). Just after Harold II was crowned king of England on 6 January 1066, a "comet-star" appeared on the firmament for a week. Later in October, Harold II was defeated by William II in the Battle of Senlac Hill (usually referred to as the Battle of Hastings; for the full story see, e.g., *Howarth*, [1977]). Depending on whom you cheered for, the comet was a good or evil omen.

However, it can be argued that cometary physics started in the late sixteenth century with Tycho Brahe's observations of "the great comet of 1577" (C/1577 V1). From his observations, of which a detail is shown in Figure 30.2, he found that the comet had a smaller daily parallax than the Moon. This means that it is a celestial object and not an atmospheric one, which was one of the leading ideas (although disputed already by Seneca the Younger at the beginning of the current era). Brahe also realized that the comet's tail is always pointing away from the Sun, as is clear from the drawing in Figure 30.2.

In the late seventeenth/early eighteenth century, after *Newton* [1687] published his theories in his *Principia* with which the observations of comets could be linked to their elliptical orbits, *Halley* [1705] used this to compute the orbits of comets and showed that the appearance of the bright comet in 1682 (now known as "Halley's comet") was periodic in nature. This explained the orbits of comets, but not the nature of these objects. *Bertier* [1760], a Jesuit priest, wrote a six volume book, *Physique des Comètes*, in which the brightness was explained in the following way [from *Bonamy*, 1760, 112, sic]:

> *Les Cometes, ajoute le P. Berthier, ne sont autre chose que ces tourbillons d'éther qui deviennent lumineux, parce que leurs couches inférieures, ayant plus de vîtesse que les supérieures, frottent celles-ci avec une violence extrême. Ces tourbillons ne durent qu'un certain tems, au bout duquel ils se décomposent & disparoissent.*

which basically says that "ether whirlwinds become luminous because they rub together due to their different speeds, and these whirlwinds only last a certain time and disappear."

Space Research Institute, Austrian Academy of Sciences, Graz, Austria

Electric Currents in Geospace and Beyond, Geophysical Monograph 235, First Edition.
Edited by Andreas Keiling, Octav Marghitu, and Michael Wheatland.

Figure 30.1 Bayeux Tapestry, Scene 32: men staring at Halley's Comet; Scene 33: Harold at Westminster. Bayeux Tapestry 32-33 comet Halley Harold.jpg via Wikimedia Commons.

Slowly, many new developments were made in the study of comets, up until the "modern cometary age" started in the 1950s. For further reading on the history of cometary physics, see for example *Festou et al.* [2004], *Heidarzadeh* [2008], *Krishna Swamy* [2010], or the ESA website (http://sci.esa.int/rosetta/54198-harbingers-of-doom-windy-exhalations-or-icy-wanderers/).

Comets are also know as "hairy stars" because of the nebulous cloud around the nucleus, which is formed by the sublimation of volatiles that are contained in the nucleus. When observed by a telescope, this cloud gives the comet a hairy appearance, setting it aside from regular stars. This cloud was named "coma," which was derived from the Greek word *kome* (κόμη), which means "hair," from which also the word "comet" originates.

30.2. WHIPPLE, OORT, BIERMANN, AND ALFVÉN

In the early 1950s, some important insights about comets were obtained. *Whipple* [1950] combined all spectral observations and published the now well-known "dirty snowball" model of comets. In this model, the nucleus of the comet is seen as a conglomeration of volatile ices (such as H_2O and CO_2) bound in a solid rocky body of meteoritic material. Actually, this model was developed to explain the non-Keplerian behavior of comets (specifically comet 2P/Encke).

It had been shown by earlier work that there were no comets with an original hyperbolic orbit [*Strömgren*, 1914, 1947], and that thus comets could not be of interstellar origin. *Oort* [1950] posited that there is a cloud of comets around our solar system between 50,000 and 150,000 AU distance from the Sun, from which the long-period comets arise. This region is now called the Oort cloud.

Interestingly, it was found that cometary tails, which consist of ions (apart from the dust tails, which consist of dust), often show acceleration of condensations, which exceed the gravitational pull of the Sun by a factor μ of 1 to 2 orders of magnitude. *Öpik* [1964] used observations of comet 1P/Halley from 1910, taken from the analysis by *Bobrovnikoff* [1930], and showed that the usage of a correct coordinate system is necessary to estimate the acceleration of the condensations, as often unreasonable values of $\mu > 1000$ were published. It was generally accepted, though, that this acceleration was produced by light pressure from the solar radiation [e.g., *Wurm*, 1943].

(a) (b)

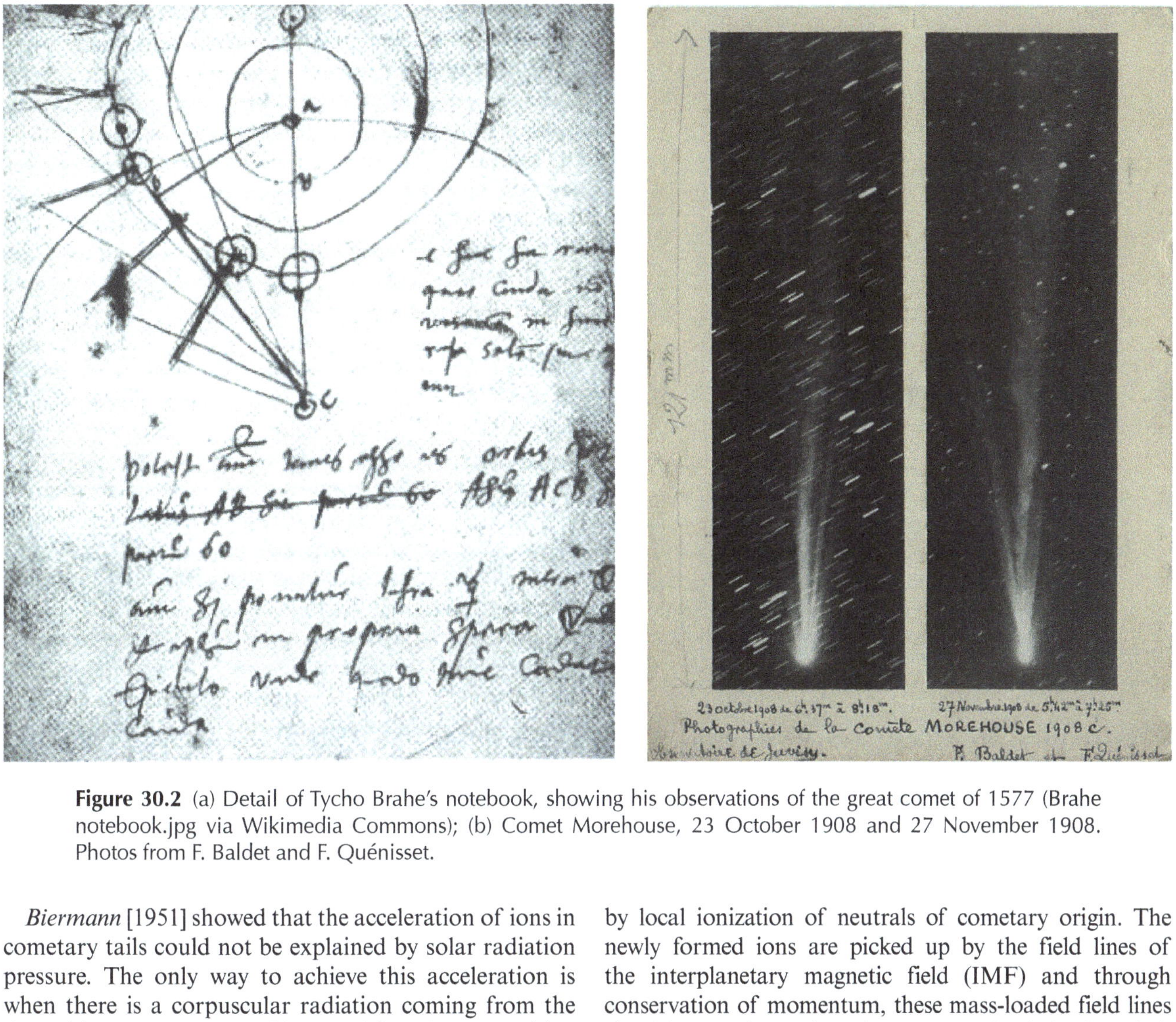

Figure 30.2 (a) Detail of Tycho Brahe's notebook, showing his observations of the great comet of 1577 (Brahe notebook.jpg via Wikimedia Commons); (b) Comet Morehouse, 23 October 1908 and 27 November 1908. Photos from F. Baldet and F. Quénisset.

Biermann [1951] showed that the acceleration of ions in cometary tails could not be explained by solar radiation pressure. The only way to achieve this acceleration is when there is a corpuscular radiation coming from the Sun with equal amounts of ions and free electrons with a density of 10^3 to $10^5\,cm^{-3}$ and a velocity of $10^3\,km/s$.

Taken together, this leads to a model of a cloud of dirty snowballs far out in the solar system, some of which get perturbed and enter the inner solar system on a parabolic orbit. Heating the comet leads to outgassing and consequently ionization. The emitted volatiles are ionized and blown away by corpuscular radiation from the Sun (i.e., the solar wind).

It took another 6 yr for a complete model of the interaction of the solar wind with an outgassing comet and the creation of cometary tails. Using the relatively new theory of magnetohydrodynamics (MHD) [*Alfvén*, 1942], it was shown by *Alfvén* [1957] how the solar wind with frozen-in magnetic field would drape itself around the active nucleus, (see Fig. 30.3AA).

In the model by *Alfvén* [1957], the solar wind magnetoplasma reaches the active nucleus and becomes mass-loaded by local ionization of neutrals of cometary origin. The newly formed ions are picked up by the field lines of the interplanetary magnetic field (IMF) and through conservation of momentum, these mass-loaded field lines should move slower than the non-mass-loaded solar wind. The magnetic field gets draped around the nucleus as the "far away" parts of the field lines continue with nominal solar wind speed, thereby stretching the field into a tail-like structure behind the nucleus as seen from the Sun, as in Figure 30.3AA.

30.3. THE COMET'S HAIR

The most noticeable feature of a comet is its hair, or the so-called cometary rays that seem to emerge from the bright center of the comet and are then bent backward toward the tail [e.g., *Wurm*, 1963]. A picture of comet Morehouse (C/1908 R1) in Figure 30.2 shows the ray-like structures in the coma (first discussed by *Eddington* [1910]), but drawings in *Rahe* [1968] Figure 1 show "fountain-like structures of the uprushing material in the inner coma region" as observed from telescopes.

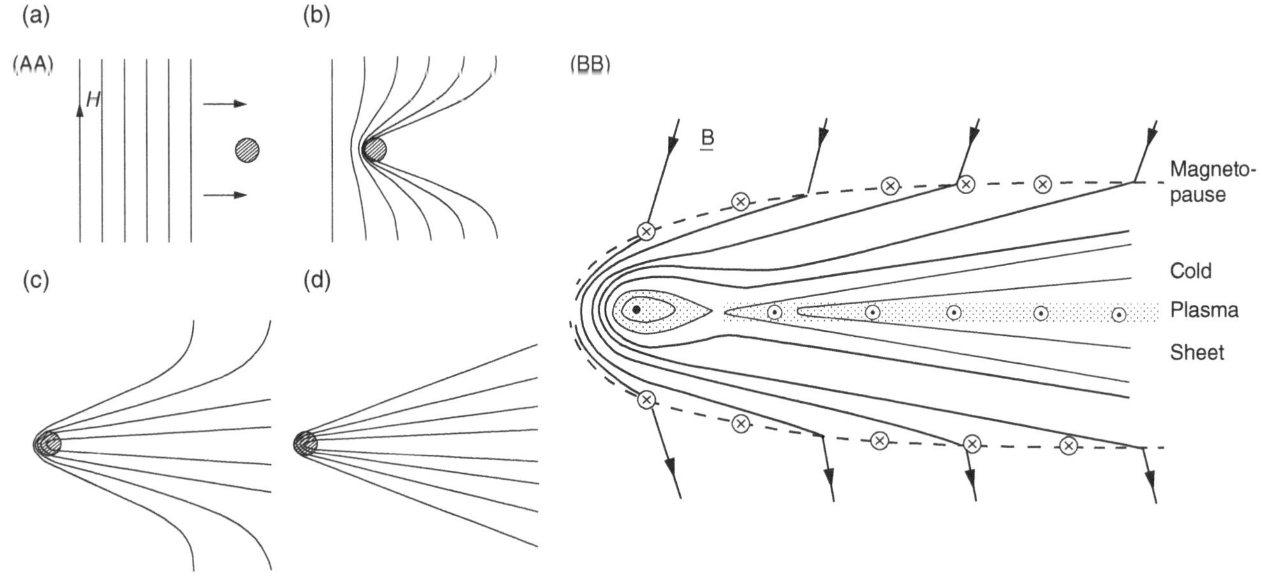

Figure 30.3 (AA) Magnetic field line draping around a cometary nucleus according to *Alfvén* [1957]. A beam with a frozen-in magnetic field H approaches the head of a comet. (a) The field is deformed; (b) and (c) Final state when the beam has passed. (d) Viewed perpendicular to the field of the beam. (BB)The magnetic field configuration around comet 21P/Giacobini-Zinner, as observed by the ICE spacecraft. The draped magnetic field lines are kinked where they are connected to the solar wind IMF, which leads to a current sheet on the "magnetopause." The oppositely directed magnetic fields in the top and bottom half of the tail are separated by a current sheet. From *Heubner* [1990].

Ray formation physics took off in the 1960s, after *Alfvén* [1957] explained the creation of the ion tail and thereby the bend-back of the cometary rays, but not the brightness variations that are seen. Observations, once more of comet Morehouse, showed that the rays are probably created by ionization of H_2O, CO, and CO_2 close to the nucleus, which occurs in a discontinuous fashion [*Rahe and Donn*, 1969]. Bursts were seen to last for up to 20 hr, and the connection between the nucleus and the tail seems to remain intact until the source of the plasma stops.

Wolff et al. [1985] consider the cometary rays to be channeled outflow of cometary ions along magnetic flux tubes. In this case, flux tubes are entering the cometary ionosphere through the Kelvin-Helmholtz instability and are able to penetrate the ionopause. At that point, there are two symmetric flux tubes with their foot points rooted near the nucleus, extended along the comet's tail. Variations in the location and the brightness of the rays are then explained through differences in the interplanetary magnetic field direction and outgassing rate of the comet. Later, *Ip* [1994] added the possibility that modification of the electron heat conduction could also be the cause for variations along the flux tubes.

30.4. CURTAINS AROUND THE NUCLEUS

The magnetic field line draping model explains well how the cometary ion tail is formed and why it is always (almost) pointing radially away from the Sun. However, the end stage in Alfvén's model, Figure 30.3AA.d only shows the situation deep inside the coma and tail. For a more global view, Figure 30.3AA.c is more realistic, as it shows that the field lines are inherently still connected to the IMF (unless reconnection takes place). A fuller model is given in Figure 30.3BB from *Heubner* [1990], which is deduced from measurements of the International Cometary Explorer (ICE) as it flew through the tail of comet 21P/Giacobini-Zinner [*Smith et al.*, 1986].

As can be seen, the mass-loaded field lines are held-up in front of the nucleus, whereas the parts of the field lines farther away travel with solar wind speed, thus stretching and draping the field lines. This leads to the creation of a "magnetopause" around the comet, where the slowly moving cometary plasma couples to the faster moving solar wind plasma, resulting in a kink in the field lines. This structure of the so-called induced magnetosphere [*Podgorny et al.*, 1980] is very similar to that near Venus [*Luhmann et al.*, 2004; *Zhang et al.*, 2008]. On the surface of the magnetopause, perpendicular currents are flowing allowing for the magnetic field lines to kink. The upper and lower part of the cometary tail in Figure 30.3BB have oppositely directed magnetic field and are by necessity separated by a cross tail current sheet. The cross tail current is carried by the cold picked-up plasma. As in the Earth's magnetotail, the cross tail current closes through the magnetopause currents [e.g., *Stern*, 1994].

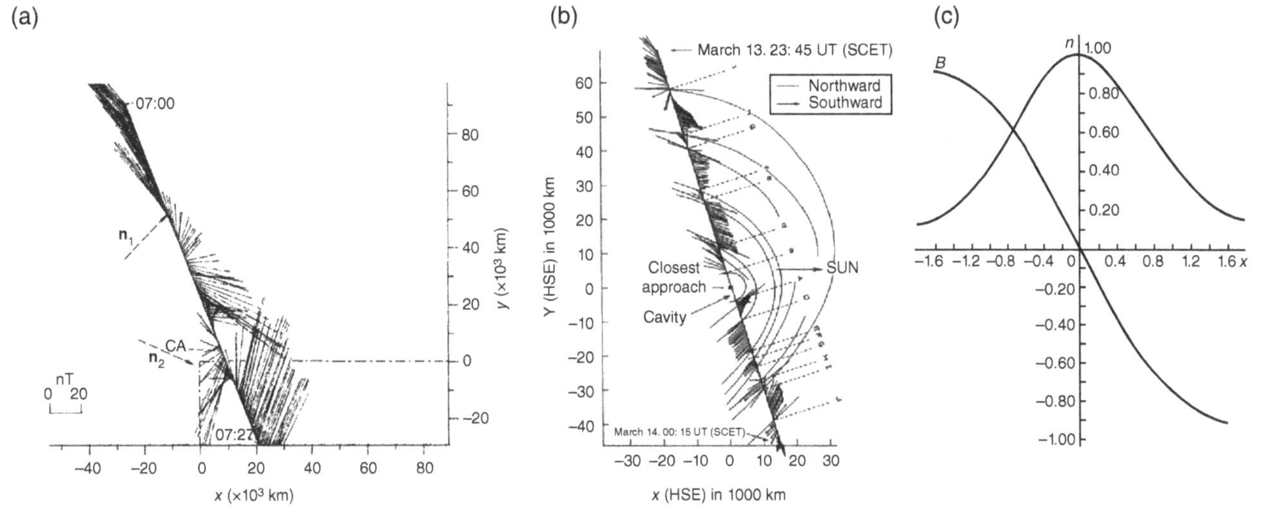

Figure 30.4 (a) Half-hour interval around closest approach of the VEGA 1 flyby of comet 1P/Halley. The magnetic field direction in the *XY*-plane is shown as a hedgehog plot along the orbit of the spacecraft. The rotations of the field just after closest approach are similar to rotations 2.75 hr earlier (not shown) and assumed to be from the same draped solar wind (from *Riedler et al.* [1986]); (b) Half-hour interval around closest approach of the Giotto flyby of comet 1P/Halley. The rotations of the field are clearly visible. The curved lines show how the different regions in ingress and egress are connected by nested draping (from *Raeder et al.* [1987]); (c) The magnetic field (B) and plasma density (n) for a Harris current sheet (from *Harris* [1962]).

Another fact that does not come into both images in Figure 30.3 is that the solar wind IMF does not have a steady direction. The well-known Parker spiral [*Parker*, 1958] of the IMF is only obtained after averaging over relatively long time periods [e.g. *Borovsky*, 2010; *Ness and Wilcox*, 1964] and, even in a perfect spiral, there will be sectors where $B_x > 0$ and $B_x < 0$, for example, pointing sunward or antisunward. This means that the magnetic field of different directions will be transported toward the comet, and will result in so-called nested draping. The result is that layers of magnetic field of different directions will be wrapping around the nucleus like an onion. Very nice examples are shown in Figure 30.4a from *Riedler et al.* [1986] and Figure 30.4b from *Raeder et al.* [1987], who studied the VEGA 1 flyby on 6 March and the Giotto flyby of comet 1P/Halley on 13 and 14 March 1986, respectively. Looking at the layering of the magnetic field directions for ingress and egress, *Raeder et al.* [1987] were able to identify and connect layers with different polarity.

This means that the induced magnetosphere around a comet consists of a complex geometry of magnetic fields, as if a large amount of different kinds of curtains has been draped around the nucleus. One of the consequences of these rotations of the magnetic field is that there will be current sheets generated in the coma.

30.5. HARRIS AND CO.

In a region where oppositely directed magnetic fields meet, a current sheet will be generated, by necessity of Ampère's law:

$$\nabla \times \mathbf{B} = \mu_0 \mathbf{J} + \frac{1}{c^2}\frac{\partial \mathbf{E}}{\partial t}. \tag{30.1}$$

This situation often occurs in space physics, for example, in planetary magnetotails or in nested magnetic field draping. *Harris* [1962] published an analytical model for the magnetic field $\mathbf{B}(z)$ and plasma pressure $p(z)$ profile given by:

$$\mathbf{B}(z) = B_e \tanh\left(\frac{z}{z_0}\right)\hat{\mathbf{e}}_x, \tag{30.2}$$

$$p(z) = p_0 \operatorname{sech}^2\left(\frac{z}{z_0}\right),\ p_0 = \frac{B_x^2}{2\mu_0}. \tag{30.3}$$

Also, the structure needs to be in pressure balance such that:

$$p(z) + \frac{B^2(z)}{2\mu_0} = \text{constant}. \tag{30.4}$$

In this model the plasma density $\rho(z)$ is arbitrary but has to comply with equations (30.2)–(30.4). One solution for the density is given by *Epstein* [1930], where it varies from ρ_0 in the center of the current sheet to ρ_e at very large z:

$$\rho(z) = \rho_0 \operatorname{sech}^2\left(\frac{z}{z_0}\right) + \rho_e \tanh^2\left(\frac{z}{z_0}\right). \tag{30.5}$$

Figure 30.4c shows the magnetic field and plasma density for a so-called Harris current sheet. Naturally, generalizations on this basic model have been performed for the presence of a guide field [e.g., *Yoon and Lui*, 2008] or for relativistic plasmas [e.g., *Balikhin and Gedalin*, 2008].

30.6. THE COMET HAIR'S MAKEUP

Naturally, it is not the magnetic fields that are visible in the pictures of comets, but it is the sunlight reflected from the dust coming from the nucleus and emission lines from the neutral gas and plasma.

When a comet enters the region inside the orbit of Jupiter, the solar irradiation is strong enough to start heating the nucleus and for the volatiles to start subliming. In order to model the outgassing of a comet at a rate of Q_n (molecules s^{-1}), a spherical expansion is assumed where the gas moves away from the nucleus at a velocity of V_e (m/s). The neutral gas escaping from the comet will get ionized by solar UV radiation and/or collisions with the solar wind at a rate of ν (s^{-1}). Using this, the radial dependence of the neutral gas density N_n can be formulated by the *Haser* [1957] model:

$$N_n(r) = \frac{Q_n}{4\pi V_e r^2} \exp\left\{\frac{-\nu r}{V_e}\right\}. \tag{30.6}$$

The number of ions that are implanted per second into the solar wind is then given by $Q_i = \nu N_n$ (see also, *Huddleston et al.* [1990]). It is this heavy ion plasma (H_2O, CO, CO_2), in combination with the solar wind plasma (H, He), which, in combination with the emitted neutrals, makes up the coma. Depending on the activity stage of the comet and the distance of the spacecraft from the nucleus, this means that different kinds of plasma compositions can be measured. In the early stage of the Rosetta mission [*Glassmeier et al.*, 2007a], with comet 67P/Churyumov-Gerasimenko (67P/CG) between 3.6 and 2.0 AU distance from the Sun, there was a clear mixture of cometary ions and solar wind ions measured by the Rosetta Plasma Consortium (RPC) [*Carr et al.*, 2007] Ion Composition Analyzer (ICA) [*Nilsson et al.*, 2007]. At a later stage, with comet 67P/CG only 1.4 AU distant from the Sun, the same instrument solely measured cometary ions [*Edberg et al.*, 2016a] at a distance between 300 and 1500 km from the nucleus.

30.7. PICKING UP CURRENTS ALONG THE WAY

The freshly created plasma, through ionization of the escaping neutrals from the cometary nucleus, is picked up by the solar wind magnetic field. The ions get accelerated by the motional electric field of the solar wind, given by $\mathbf{E} = -\mathbf{v}_{sw} \times \mathbf{B}$, and start to gyrate around the magnetic field at the specific cyclotron frequency $\omega_{ci,e} = q_{i,e}B/m_{i,e}$, which can lead to the generation of cyclotron and mirror-mode waves [e.g., *Gary*, 1992; *Volwerk et al.*, 2013, 2014, 2016a]. *Ip and Axford* [1980] and *Goertz* [1980] showed that in the pick-up process the guiding centers of the ions and the electrons are displaced by their respective gyroradii along the motional electric field (but see also *Cloutier* [1970]). This leads to a so-called pick-up current:

$$\mathbf{J}_{pu} = \frac{\dot{\rho}\mathbf{E}}{B^2} = \frac{m_i \dot{n}_i \mathbf{v}_{SW}}{B}. \tag{30.7}$$

where $\dot{\rho}$ is the mass-loading rate by the ionization and pick up of the newborn ions and $\dot{n}_i$ is the number density rate. The plasma is then decelerated by the $\mathbf{J} \times \mathbf{B}$ force.

Although this straightforward derivation of the pick-up current works well, *Vasyliunas* [2016] showed that there is a problem in the framework of MHD, in which currents $\mathbf{J}$ are determined by $\nabla \times \mathbf{B}$ and not the other way around as in the derivation above. *Vasyliunas* [2016] states that the pick-up current does not exist as a distinct kind of electric current. However, the pick-up process main effect is to slow down the flow of the magnetoplasma (see also section 30.2), which produces a disturbance of the magnetic field, which by Ampère's law creates a current $\mathbf{J}$.

A full derivation of the electrodynamics of the pick-up process can be read in *Vasyliunas* [2016], but the "Root of the inconsistency" (Vasyliunas's section 2.3) is that in the derivation above, where $\mathbf{J}$ determines $\mathbf{B}$, an additional current created by polarization drift of the plasma, given by:

$$\rho \frac{\partial \mathbf{v}}{\partial t} = \mathbf{J}_{pol} \times \mathbf{B} \Rightarrow \mathbf{J}_{pol} = \frac{c^2}{v_A^2} \frac{\partial \mathbf{E}}{\partial t}, \tag{30.8}$$

is neglected (v_A is the Alfvén velocity). This current is mentioned by *Goertz* [1980] (his equation (5)) but is not carried along further in that paper.

Doing the full derivation leads to an expression for the averaged current $\langle \mathbf{J}_\perp \rangle$ (over timescales longer than the ion gyro period but shorter than the long-term evolution of the fields) of the form:

$$\langle \mathbf{J}_\perp \rangle = \left(\dot{\rho}\frac{\mathbf{E}}{B^2}\right)\left[\frac{v_A^2}{c^2 + v_A^2}\right], \tag{30.9}$$

which is smaller by a factor $v_A^2/c^2 \ll 1$ than the current in equation (30.7). The temporal development of the current is determined by the time that the Lorentz force can balance the momentum loading by the pick-up process. As soon as a situation is reached in which $\partial v/\partial t \to 0$ (see equation [30.8]), the neglected term disappears and the pick-up current reduces to the "traditional" value given in equation (30.7). The timescale over which this happens is the Alfvén travel time of the system $t_A \approx L/v_A$ with a characteristic length much greater than the electron inertial length $L \gg \lambda_e$. In well-known pick-up regions, for example, around Io and Europa, $t_A \approx 10$ s, and around comet 67P/CG during the time intervals shown later (section 30.10.4), the Alfvén velocity is $v_A \approx 20-100$ km/s and the obstacle size varies from a few hundred up to more than 1500 km (no bow shock was seen during the far dayside excursion [*Mandt et al.*, 2016]), which results in similar values for t_A, on the order of tens of seconds.

The above discussion holds for the MHD framework, in which processes are looked at over timescales greater than the gyro period of the heaviest ion. However, in the case of a weakly outgassing comet interacting with unshocked solar wind, the gyroradii and periods can be very large. This means that when the spacecraft remains close to the comet in the source region of the pick-up ions, the newly formed and accelerated ions will constitute a cross-field current, which is discussed in section 30.10.1.

30.8. OF DAYS LONG PAST

In the 1970s, ground-based observations of comets had indicated that there are strong magnetic fields around comets, for example, for comet Kohoutek (C/1973 E1, 10973 XII, 1973f), a field strength $\geq 100\gamma$ (now called nT; *Ip and Mendis* [1975]) was derived. *Hyder et al.* [1974] noticed a helical structure moving down the tail of the comet, at the estimated Alfvén velocity. They assumed that this structure was created by a kink-mode instability generated by a current flowing along the field lines down the tail and estimated the necessary current to be $I \approx 2 \times 10^7$ A. However, *Ip and Mendis* [1975] corrected this estimate to $I \approx 2 \times 10^6$ A.

These same helical waves were also studied by *Ershkovich and Heller* [1977] for comets Kohoutek (C/1973 E1), Morehouse (C/1908 R1), and Arend-Roland (C/1956 R1). In this case, the cometary tail was modeled as a plasma cylinder separated by a tangential discontinuity from the solar wind, the magnetic field is along the axis of the cylinder, and the plasma is ideal and compressional. The dispersion relation for helical oscillations of this system, in the form $\exp\left[i\left(kz + m\phi - \omega t\right)\right]$ was determined. For typical plasma parameters, it was found that the phase velocity of the helical waves is close to the Alfvén velocity of the tail plasma and the authors concluded that the waves are generated by a Kelvin-Helmholtz instability.

The interest in these tail currents was because of the observed strong CO^+ features and their temporal behavior close to the nucleus of several comets (e.g., *Wurm* [1961], see also the brightness variations discussed in section 30.3). It was posited that such currents would have to close through the "head" of the comet, and thereby could be an extra source for ionization in the coma. They made great efforts to decrease the ionization time, to be well below the solar UV photo-ionization time ($t_{UV} \approx 10^6$ s) by maximizing the current reaching times of $t_I \approx 5 \times 10^4$ s.

Observations showed the back-bending of so-called streamers on the dayside of the comet (see, e.g., *Wurm and Mammano* [1967], and the discussion in section 30.3) when they interact with the solar wind (still referred to as the corpuscular radiation of the Sun). *Ip and Mendis* [1976] discussed how this folding of the tail rays generates a strong tail field through accumulation of magnetic flux. Through flux conservation they estimated the field strength B_t of the tail to be:

$$B_t \approx \frac{B_0 v_s t_f}{h_t}, \tag{30.10}$$

where $B_0 \approx 5$ nT and $v_s \approx 300$ km/s are the solar wind magnetic field and velocity and $h_t \approx 5 \times 10^4$ km is the half-width of the tail. This leads to $B_t \approx 1500$ nT. They realized that this folding leads to two regions with oppositely directed field, similar to the Earth's magnetotail, and thus that there needs to be a cross-tail current through Ampère's law (equation [30.1]), with $I_t \approx 10^8$ A for a tail length of $l_t \approx 10^6$ km. *Ip* [1979] assumed that (part) of this current can suddenly be channeled through the cometary atmosphere, in a similar way to what happens in the Earth's tail during a substorm [*Boström*, 1974]. This field-aligned current can then, again, be used to increase the ionization rate near the nucleus.

The magnetic field strength B_t determined by *Ip and Mendis* [1976] is unrealistically high at ≈ 1500 nT. Indeed, *Mendis and Alfvén* [1974] came to the conclusion that the field could be 100 – 1000 nT using an estimate $B = \sqrt{\mu_0 \rho v^2}$, equating the magnetic energy to the solar wind kinetic energy. Both papers basically assume that the creation of the cometary tail is similar to that of the Earth's tail. However, there are differences between the Earth's magnetotail and a cometary tail; for one, the cometary tail remains relatively constant in diameter, and does not flare like the Earth's.

Based on the presence of the helical waves in the cometary tail (see above) *Ershkovich* [1976] states that the tail magnetic field should be calculated through pressure balance over the tangential discontinuity separating the tail plasma (subscript i) from the solar wind plasma (subscript e).:

$$\frac{B_i^2}{2\mu_0} + p_i = \frac{B_e^2}{2\mu_0} + p_e, \tag{30.11}$$

and it is concluded that the magnetic field strength B_i along the tail, well away from the stagnation point, should be of the same order as the solar wind magnetic field B_e.

Podgorny et al. [1982] and *Dubinin et al.* [1982] returned to the results by *Öpik* [1964] and used the observed acceleration of the condensations (see also section 30.3) to get an estimate of the magnetic field strength in the tail. The former paper shows that the large variations $7 \leq B_{tail} \leq 75$ nT that were found using this acceleration are caused by the uncertainties in the plasma density. The latter shows that the magnetic field is strongest at the boundary of the tail and decreases toward the center, in agreement with an induced magnetosphere as proposed by Alfvén's model.

It took until 1986 before detailed in situ measurements became available from probably the most famous of comets: 1P/Halley.

30.9. THE HALLEY ARMADA

The first comet to be studied in situ with a fleet of six spacecraft was 1P/Halley. Figure 30.5a shows the geometry of the 1986 flybys by VEGA 1 (6 March, *Riedler et al.*, [1986]), SUISEI (8 March, *Mukai et al.*, [1986]), VEGA 2 (9 March, *Riedler et al.* [1986]), SAKIGAKE (11 March, *Saito et al.* [1986]), Giotto (14 March, *Neubauer et al.*, 1986]), and ICE (25 March, *Wenzel et al.* [1986], formerly known as ISEE -3). Not all spacecraft entered into the cometary magnetosheath and only VEGA 1, 2 and Giotto performed close flybys of the nucleus.

Before the arrival of this flotilla, controlled experiments were performed with the AMPTE spacecraft on 27 December 1984 and 18 July 1985. Through the release of 2 kg barium vapor, which ionized quickly through solar UV radiation, an artificial comet was created. The first phenomenon to be observed by the magnetometer was the creation of a diamagnetic cavity, which was predicted to exist at comets theoretically [*Ip and Axford*, 1982; *Pilipp*, 1971] and experimentally [*Haerendel*, 1973; *Luehr et al.*, 1986]. The electrons that are set free directly interact with the solar wind magnetic field, but the ions overshoot their equilibrium and an electric field is set up. This field brakes the ions and lets the electrons **E** × **B**-drift. These electrons crate an electric current, which shields the field-free region of the barium cloud.

30.9.1. Diamagnetic Cavity

Only the Giotto spacecraft [*Reinhard*, 1986], with its closest approach (CA) to the nucleus, observed the diamagnetic cavity around comet 1P/Halley. The observations by the Giotto magnetometer [*Neubauer et al.*, 1986] clearly showed a field-free region around CA. The duration of this signature was ~124.5 s, which corresponded to a physical width of ~8513 km. Figure 30.5b shows the transition of Giotto into the diamagnetic cavity, where over a very sharp boundary (C) the magnetic field strength decreases from ~20 nT to almost zero. From the behavior of the magnetic field components, it was deduced that there are not only currents perpendicular to the magnetic field (taking care of the shielding of the cavity) but there are also strong field-aligned currents in alternating directions.

The change in magnetic field is carried by currents, which will result in a **J** × **B** force on the plasma. *Neubauer* [1988] assumed that the magnetic field is draped and the strength varies with radial distance r from the nucleus and that the curvature radius is equal to r, and found for the radially directed force:

$$F_r = (\mathbf{J} \times \mathbf{B})_r = -\frac{B}{\mu_0 r}\frac{d}{dr}(rB), \tag{30.12}$$

which results in a force $F_r = 5.7\ (7.3) \times 10^{-16}$ N/m² for the inbound (outbound) crossing of the cavity. When the authors assumed stationarity of the boundary in the comet's frame, a thickness of the boundary was deduced of $d \approx 24.5$ km and was recognized as a tangential discontinuity. Assuming that the radial force F_r from equation (30.12) is representative for the boundary (ionopause), the full force equation has to be zero:

$$-\frac{B}{\mu_0 r}\frac{d}{dr}(rB) - \frac{dp}{dr} + m(0.5n\nu_{\text{in}} + q)v_{\text{n}} = 0, \tag{30.13}$$

where ν_{in} is the ion-neutral collision frequency and v_n is the neutral velocity and q photo-inonization rate. Various assumptions have been made in solving equation (30.13): *Ip and Axford* [1987] and *Cravens* [1986] used negligible pressure gradients and *Cravens* [1986] also neglected the curvature force. *Neubauer* [1988] assumed $dp/dr = 0$ and solved for the frictional force based on the **J** × **B**-force alone. With $v_n = 830$ m/s, they obtained $(0.5n\nu_{\text{in}} + q) = 2.2 \times 10^7$ m^{-3}s^{-1}, which is below the theoretical estimates for the sum of these two quantities, and a possible extra source of mass-loading through energetic electrons may have to be invoked (see, e.g., recent Rosetta discoveries by *Feldman et al.* [2015], for extra ionization through photo-dissociation electrons).

In order to model the diamagnetic cavity of comet 1P/Halley, *Israelevich et al.* [2003] realized that the interaction of the solar wind with the outgassing comet is not unlike the creation of Alfvén wings (see, e.g., *Drell et al.* [1965]). Using the data from the Giotto crossing of the diamagnetic cavity and a numerical model describing the cavity boundary, they find a distributed current density with a maximum $j \approx 0.9$ μA/m².

30.9.2. Nested Draping and Reconnection

As mentioned in section 30.4 and Figure 30.4a and b, the changing direction of the IMF (mainly 180°, i.e. field reversals) over time leads to a nested draping of magnetic

(a)

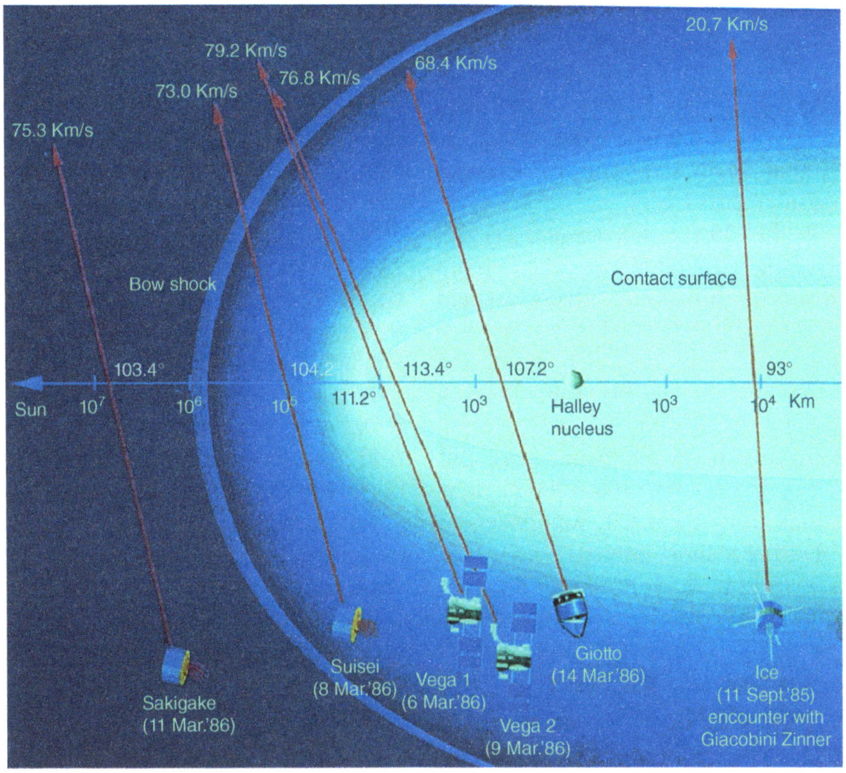

(b)

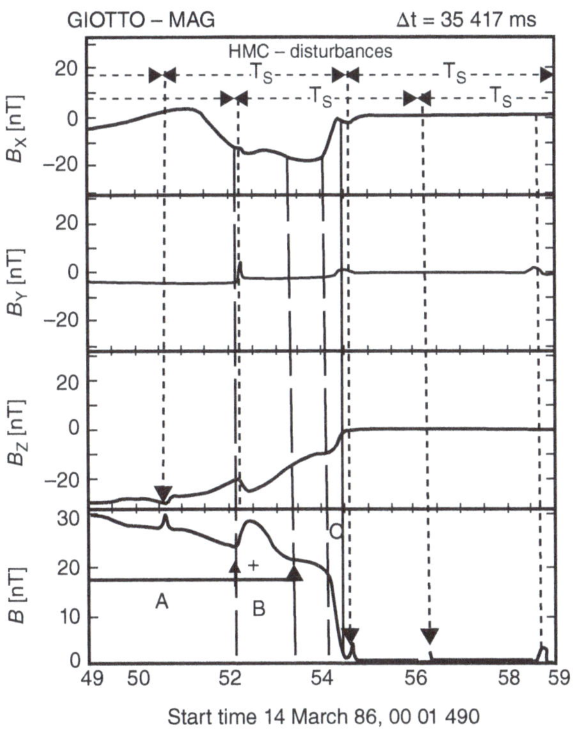

Figure 30.5 (a) The geometry of the six flybys of comet 1P/Halley. The distances along the Halley-Sun-line are on a logarithmic scale. (b) The Giotto magnetic field data near closest approach showing the fast transition of the spacecraft into the diamagnetic cavity. From *Neubauer* [1988].

field around the outgassing nucleus. For VEGA 1 observations, *Schwingenschuh et al.* [1987] observed that rotations of the field around CA were similar to rotations observed ~2.75 hr earlier (not shown in Fig. 30.4a), and assumed them to be created by the same IMF. A consequence of the field rotations is that current sheets are generated. These should be observable in high-enough resolution density measurements (see, e.g., *Volwerk et al.* [2016b], discussed in section 30.10). *Raeder et al.* [1987] stated that "field reversals, convected into the pile-up region are not destroyed by diffusional effects or field line merging but persist even close to the ionopause."

During the VEGA 1 flyby, there was much less nested draping (see, e.g., *Volwerk et al.*, [2014], Fig. 3). Well inside the coma of comet 1P/Halley, the VEGA 1 plasma instrument did no longer observe solar wind ions. But near CA to the nucleus, a burst of 100–1000 eV ions with a duration of 5 min, 0719–0724 UT, was observed [*Verigin et al.*, 1987]. During a short interval before CA, 0711–0724 UT, the magnetic field draping direction was opposite to that observed over the main part of the cometosheath. *Verigin et al.* [1987] showed a model (see also Fig. 30.6a), which showed the topology of the magnetic field along the orbit of VEGA 1, and they inferred a reconnection site located before CA, which accelerates the ions to the observed energies.

In a later paper, *Kirsch et al.* [1989] discussed observations from Giotto, where they state that the nested draping, that is, the fast 180° rotations of the field direction, are "considered as an indication of the field line merging process in the front side of Halley's cometosheath." A further study of the ions showed that during the magnetic field rotations, there is a peak in the high-energy ions, coinciding with a drop in the low-energy ions. The pitch angle of the high-energy ions is shown to be near zero degrees, that is, field aligned. However, *Kirsch et al.* [1989] estimated the energization of the ions through reconnection and found that the maximum energy would be $E_i \approx 22$ keV assuming acceleration of the ion over a length of 50,000 km, well below the observed energies of 97–145 keV (see Fig. 30.6b). These are indirect indications that magnetic reconnection could be taking place in the current sheets generated by nested draping. Probably, the reconnection rate is strongly reduced, due to high plasma densities ρ around the comet, and the rate being proportional to $1/\sqrt{\rho}$ [e.g., *Parker*, 1957; *Sweet*, 1956] for slow and $\propto 1/\ln S$ for fast reconnection [*Petschek*, 1964], where S is the Lundquist (or magnetic Reynolds) number.

30.9.3. Halley's Global Current Structure

Israelevich et al. [1994] used the magnetic field data from Giotto to infer the structure of the induced magnetosphere of comet 1P/Halley. The draping direction of the magnetic field should reflect the direction of the IMF. Ideally, the draping should be viewed in a IMF-related coordinate system X_{IMF} along the solar wind direction $\mathbf{v}_{sw}$, Y_{IMF} in the $\mathbf{v}_{sw} \times \mathbf{B}$ direction, and Z_{IMF} completing the triad. Unfortunately, for the Giotto flyby, the IMF direction is not known, only the $\mathbf{v}_{sw}$ determined by *Coates et al.* [1987] is assumed to be constant over the flyby. Based on symmetry assumptions in the induced magnetosphere with respect to the XY_{IMF}- and XZ_{IMF}-plane, *Israelevich et al.* [1994] were able to find the $X_{IMF}Y_{IMF}Z_{IMF}$ coordinate system. Transforming the Giotto data to this system, they obtained a global view of the induced magnetosphere.

The observations were basically in the $X_{IMF} = 0$-plane. *Israelevich and Ershkovich* [1994] expanded the magnetic field observations with the "mirror points" that are determined by the symmetry assumptions, and from there used a linear gradient model to fill up the whole magnetic field of the induced magnetosphere with a square grid of 10^3 (10^4) km for high(low) resolution. This model was then used to determine the currents in comet 1P/Halley's induced magnetosphere through $\mathbf{J} = \nabla \times \mathbf{B}/\mu_0$ and the accuracy of the result was determined by $\nabla \cdot \mathbf{B}$. The j_x currents in the XY_{IMF} plane are shown in Figure 30.7a. The authors state that "it is worth mentioning that the calculated current density refers to currents creating the magnetic barrier and, thereby responsible for magnetic field lines draping about the comet." The grid that is used to calculate the currents has too little spatial resolution to resolve the currents at the diamagnetic cavity. Overall, the magnetospheric currents are on the order of tens of nA/m^2.

30.9.4. A Tail Tale

The only spacecraft to cross the tail of a comet was ICE during its encounter with comet 21P/Giacobini-Zinner, as shown in Figure 30.7b. The magnetic field measurements showed a clear draping pattern of the magnetic field around the comet, creating the ion tail. This structure is reminiscent of the Earth's magnetotail. The crossing can be described as the spacecraft passing through two regions of strong and oppositely directed magnetic field (the "lobes"), which are separated by a high-β plasma (current) sheet [*Slavin et al.*, 1986a,b]. The current layer was found to be ~200 km thick. *McComas et al.* [1987] determined the current density inside the current sheet and used that to calculate the $\mathbf{J} \times \mathbf{B}$-force. They found that this force is negative throughout the current sheet, which means that it reaccelerates the plasma tailward. Naturally, this tail configuration, like at Earth, is prone to instabilities and can lead to reconnection in the current sheet. This leads to so-called (tail) disconnection events, which were studied as early as the late 1970s, for example, by *Niedner and Brandt* [1978]. These authors

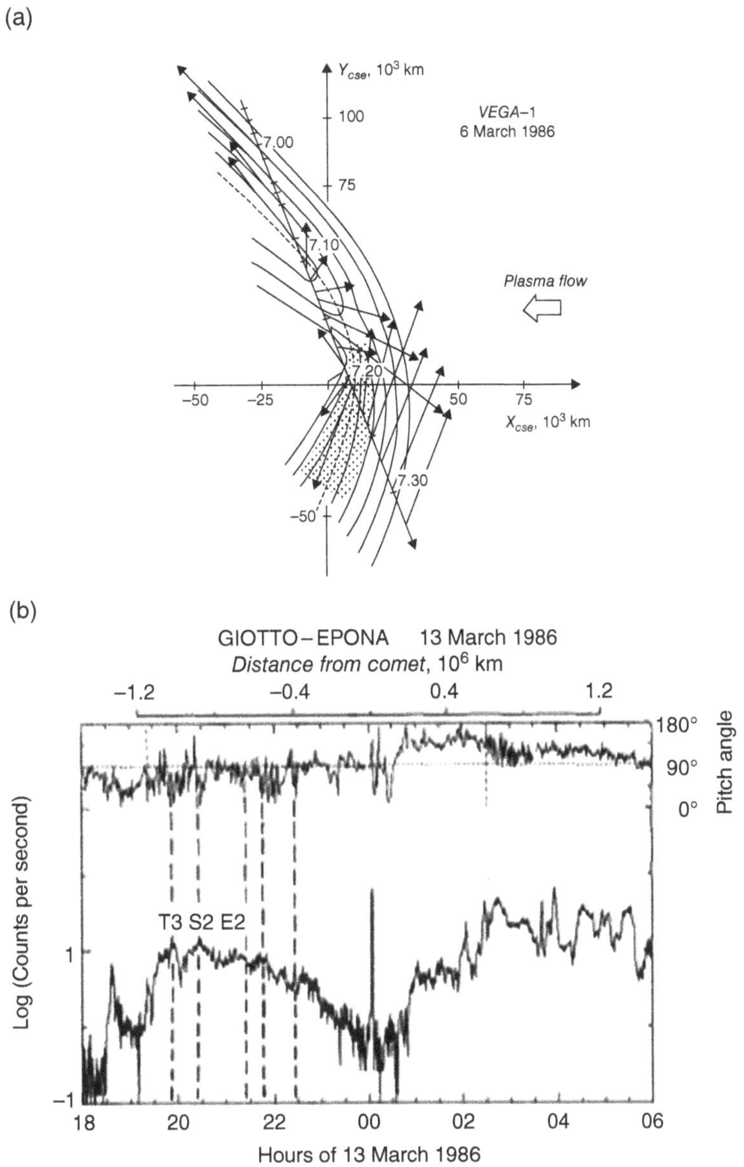

Figure 30.6 (a) VEGA 1 flyby of comet 1P/Halley. A model of the draped magnetic field is drawn around the nucleus based on the magnetic vectors along the orbit of the spacecraft. Magnetic reconnection is assumed to take place just before CA and to accelerate ions in the downward direction of the figure (from *Verigin et al.* [1987]); (b) Energetic ion counts (E_{H2O} = 97 – 145 keV) and ion pitch angle for the Giotto flyby of comet 1P/Halley. The peaks in the ion counts coincide with zero pitch angle, in regions where the magnetic field rotates over 180° (from *Kirsch et al.* [1989]).

discussed, in the frame of Alfvén's draping model, the reaction of the induced magnetosphere to the crossing of a sector boundary in the solar wind. The change of the magnetic field direction after the crossing leads to a boundary between the oppositely directed magnetic field in the draping/pile-up region in front of the comet. Magnetic reconnection takes place at this boundary, detaching the inner part of the tail from the pile-up region, which then moves away down the tail (see their Fig. 8). More recent observations of disconnection events is discussed in section 30.10.4.

The length of a cometary tail can be very extensive, easily up to 100 million km, which means that the ICE flyby discussed above at ~7800 km down the tail was relatively close by. An extreme case of cometary tail identification was done by *Jones et al.* [2000] using the Ulysses magnetometer data. On 1 May 1996, an unusual magnetic field structure was observed in combination

(a)

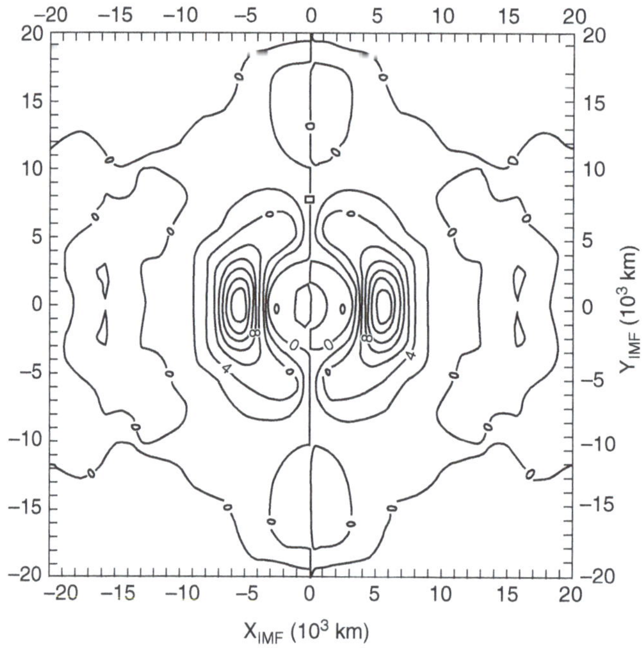

(b)

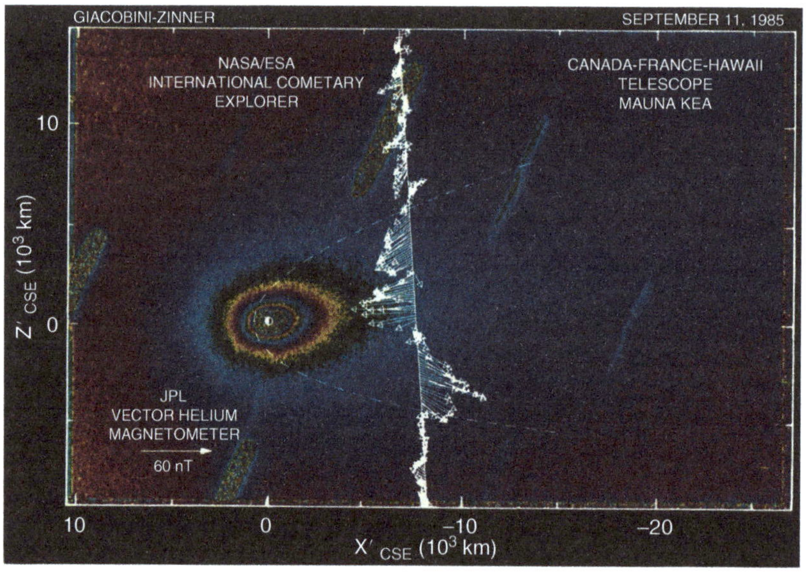

Figure 30.7 (a) Contour plot of the X-component of the magnetospheric currents in the XY_{IMF} plane for a high-resolution grid. The contour lines are labeled in nA/m² (from *Israelevich and Ershkovich* [1994]); (b) The ICE flyby through comet 21P/Giacobini-Zinner's ion tail. The magnetic field data along the orbit are superimposed on a ground-based telescope image taken in a band around the 7025 Å- H_2O^+-line (from *Slavin et al.* [1986a]).

with anomalously low solar wind density. At that time, Ulysses was at a position that was radially away from the Sun, where comet Hyakutake (C/1996 B2) was located about 8 days earlier. The magnetic field signature was consistent with that of a cometary tail with nested draping, showing symmetry around the center of the signature. The distance between Ulysses and the location of comet Hyakutake was a staggering 3.8 AU (i.e., 550 million km) and the width of the tail signature was estimated to be at least 7 million km.

30.10. ROSETTA, ARE YOU BETTER?

All interesting results discussed so far were obtained by quick flybys of comets with a duration on the order of hours, and thus only "snapshots" were obtained. This situation changed drastically with the arrival of Rosetta [*Glassmeier et al.*, 2007a] at comet 67P/Churyumov-Gerasimenko (67P/CG) on 6 August 2014. This spacecraft was to orbit around the comet and follow it through perihelion (13 August 2015) and end its (extended) mission with landing on the Ma'at region on 30 September 2016.

Rosetta was packed with a full plasma suite called the Rosetta Plasma Consortium (RPC) [*Carr et al.*, 2007]. With these instruments, enhanced with the neutral gas data from Rosina [*Balsiger et al.*, 2007], Rosetta was in a perfect state to investigate the development of the plasma environment and induced magnetosphere of an increasing-activity comet, up to perihelion and decreasing-activity as it moved away from the Sun again.

30.10.1. A Comet's Song

One of the first remarkable observations by the magnetometer RPC-MAG [*Glassmeier et al.*, 2007b] during the first months of the mission was the presence of low-frequency, between 25 and 75 mHz with the main peak at ≈ 40 mHz, large-amplitude, $\delta B/B \approx 1$, waves. These waves were dubbed "the singing comet" [*Richter et al.*, 2015]. It was found that the magnetic energy W_B in the waves varied with radial distance from comet 67P/CG, and weakly correlated with the radial dependence of the neutral gas density N_n as $W_B \propto N_n^{-\gamma}$ with $\gamma = 5.62 \pm 0.45$.

The waves are not generated by the ion-cyclotron instability created by the freshly picked-up water ions [e.g., *Gary*, 1992] as the cyclotron frequency is orders of magnitude smaller than that of the observed waves. Also, in the early stage of the mission, the local magnetic field strength was still very low; little draping, pileup, and solar wind slowdown had occurred, which makes the gyroradius of the newly created ions very large: $\rho_L \approx 2 \times 10^4$ km. This large ρ_L means that in the pick-up region the water ions can be considered nonmagnetized, that is, the motion of the ions is dominated by the acceleration of the motional electric field and not by the Lorentz force.

At a pick-up flux of at least $10^{10}\,\mathrm{m^{-2}s^{-1}}$, which was estimated by RPC-ICA [*Nilsson et al.*, 2015], the unmagnetized ions will constitute a cross-field current of $j \approx 4.8\,\mathrm{nA/m^2}$. *Richter et al.* [2015] estimated that, if this cross-field current becomes unstable, it can generate waves with a wave number given by Ampère's law (equation [30.1]) $k_\perp = \mu_0 \delta j / \delta B \approx 6.6 \times 10^{-6}\,\mathrm{m^{-1}}$. Using a beam-instability [*Chang et al.*, 1990; *Sauer et al.*, 1998], a resonant frequency $f \approx 38$ mHz was obtained.

A further theoretical investigation into the origin of these waves was performed by *Meier et al.* [2016], introducing a modified Weibel instability [*Weibel*, 1959], using the observed plasma parameters by RPC and assuming a homogeneous cold background plasma (i.e., the solar wind). The unmagnetized pick-up ions were considered to be a current across the field, and this leads to waves with maximum growth perpendicular to the magnetic field. In Rosetta's rest-frame, the waves have a frequency of $f \approx 40$ mHz, as observed by RPC-MAG. Also a maximum wavelength of ≈ 980 km and a phase velocity of ≈ 40 km/s was determined. However, an apparent smaller wavelength, due to Doppler shift, $\lambda \approx 200$ km was measured.

In a later paper, *Richter et al.* [2016] discussed 2-point measurements of these waves, using both the measurements of RPC-MAG and of the ROMAP magnetometer on Philae [*Auster et al.*, 2007] during the descent phase on 12 November 2014. The orbits after Philae separation and before landing are shown in Figure 30.8a. Both magnetometers detected the same waves with frequencies around ~20–50 mHz and an amplitude up to ~3 nT. With knowledge of the location of the spacecraft, it is possible to determine the projected wave number and the dispersion relation. *Richter et al.* [2016] found $\langle k_p \rangle = (2.3 \pm 0.2) \times 10^{-5}\,\mathrm{m^{-1}}$ and a phase velocity $v_p = 6.1 \pm 0.8$ km/s for a mean wavelength of $\lambda = 278$ km. This is in agreement with the wavelength determined from the modified Weibel instability.

The singing "disappeared" when comet 67P/CG moved closer to Sun, increasing the production rate. It is unclear, as yet, if it truly disappears or if it is just hidden in the much larger amplitude waves and turbulence, up to $\Delta B \approx 200$ nT. After perihelion passage, the activity of the comet slowly reduced again, and therewith the strong waves and turbulence and the signature of the singing comet reappeared in the RPC-MAG data [*Glassmeier*, 2016].

30.10.2. 67P/CG'S Unordinary Cavity

Observations of the diamagnetic cavity expected around an active comet are scarce from earlier flybys. With the arrival of Rosetta at comet 67P/CG, this changed with continuous flybys of the nucleus over the period of more than 2 yr. The first detection of the diamagnetic cavity at comet 67P/CG was on 26 July 2015 [*Goetz et al.*, 2016b]. A clear, field-free region with sharp boundaries was observed by RPC-MAG. Surprisingly, this happened at a larger distance, ≈ 170 km, from the nucleus as compared with that obtained from steady-state numerical models of the comet [e.g., *Koenders et al.*, 2015]. An explanation was found in the possibility of Kelvin-Helmholtz waves on the ionopause of comet 67P/CG.

Recently, however, *Hansen et al.* [2016] showed that the production rate of the comet assumed in the simulations

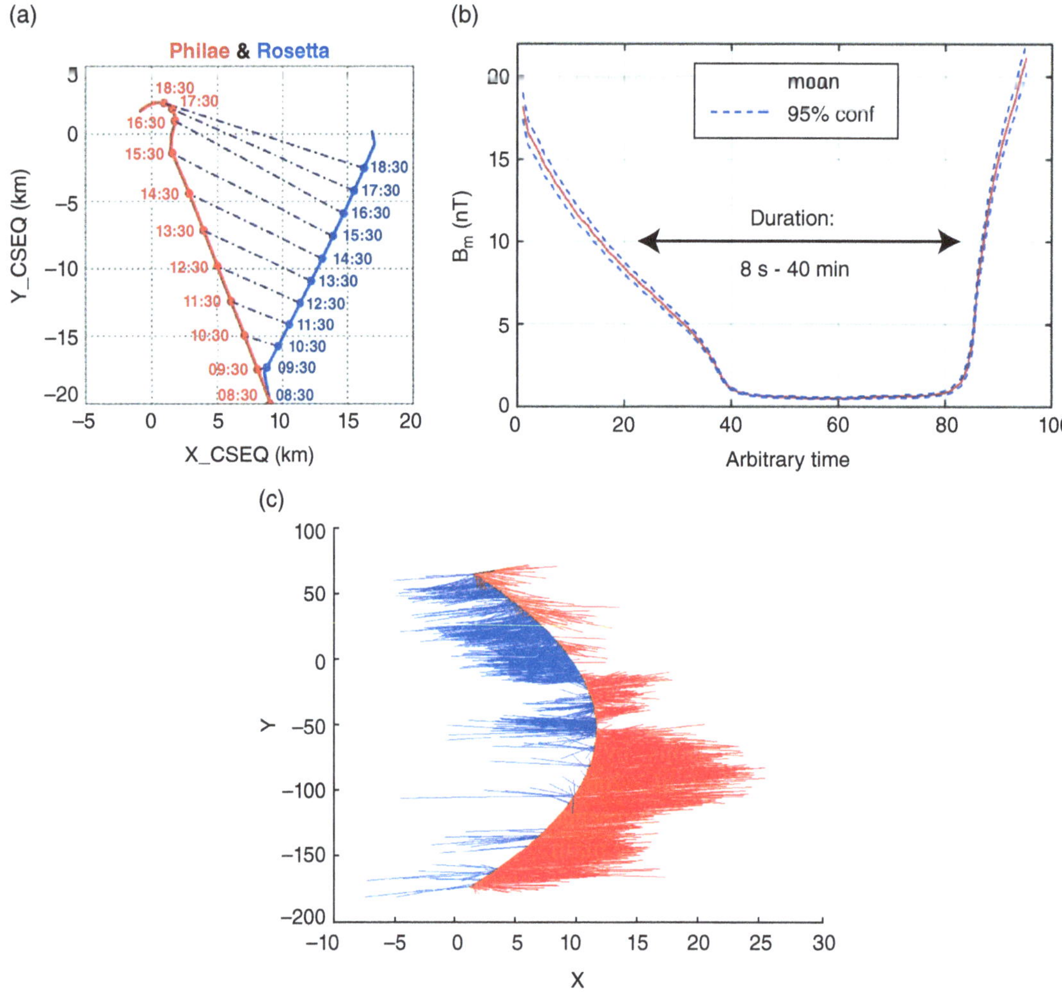

Figure 30.8 (a) The locations of Rosetta and Philae during the descent of the lander. The dash-dotted lines show the connection between the two spacecraft along which the wave characteristics are determined (from *Richter et al.* [2016]); (b) A superposed-epoch analysis of 665 diamagnetic cavity observations by Rosetta, showing the asymmetric and fast decreases and increases of the magnetic field strength (from *Goetz et al.* [2016a]); (c) A hedgehog plot of the magnetic field along the orbit in the *XY*-plane (axes are not scaled equally) of Rosetta during June 2015. The red quivers indicate sunward and the blue shows antisunward directed magnetic field. There is a clear draping signature of the field, with interspersed nested draping (see also, *Volwerk et al.* [2016b]).

by *Koenders et al.* [2015] ($Q = 5 \times 10^{27}$ s^{-1}) was much too low and should be around $Q \approx 3.5 \times 10^{28}$ s^{-1}. This will move out the boundary of the diamagnetic cavity, however, not enough to let Rosetta enter it along its orbit. Variations in the outgassing rate, pushing the boundary even further out, and instabilities on its surface are still needed.

A statistical study was performed on the RPC-MAG data from April 2015 to February 2016, in which a total of 665 cavity-events were found by Goetz *et al.* [2016a]. A superposed-epoch analysis [*Chree*, 1913] was performed on these structures, with a time interval variation from 8 s up to 40 min., and the result is shown in Figure 30.8b. *Goetz et al.* [2016a] described the cavities as follows: "Any features that may be seen in this analysis will be common features of most cavity crossings. First the magnetic field decreases from 20 to 5 nT, then the decrease steepens until the field is close to zero. In the cavity the field remains zero until a slight upward trend leads over into the transition region where the magnetic field again increases to 20 nT." There seems to be a correlation between the distance of the cavity boundary, r_c, from the nucleus and the outgassing rate $r_c \propto Q^{0.678(0.621)}$ before (after) July 2015, which is comparable to the theoretical estimate, $\propto Q^{0.75}$ by *Cravens* [1987] and *Ip and Axford* [1987].

Naturally, such a strong change in the magnetic field strength needs electric currents, which can be determined by equation (30.1), which gets approximated by:

$$|j| = \frac{|B|}{\mu_0 L}, \quad (30.14)$$

where L is the characteristic length over which the field changes. Based on observations at comet 1P/Halley the thickness of the cavity boundary is $L \approx 25$ km [*Neubauer*, 1988], a value that agrees well with the numerical simulations for comet 67P/CG by *Rubin et al.* [2012] and *Koenders et al.* [2015]. A superposed epoch analysis of 665 entries/exits of the diamagnetic cavity shows that the magnetic field strength just outside the boundary is $|B| > 20$ nT. From equation (30.14), *Goetz et al.* [2016a] then obtained currents on the order of 1 μA/m^2. It is interesting that this value is very similar to the currents determined for the artificial comet experiment by AMPTE [*Luehr et al.*, 1988] and for the cavity around comet 1P/Halley.

30.10.3. The Inner Curtains

Solar wind magnetic field draping around comet 67P/CG also takes place and *Koenders et al.* [2016] discussed one Rosetta flyby on 28 March 2015, where the structure of the magnetic field direction clearly shows the telltale bidirectional field structure on either side of the nucleus, which fits well with the numerical model they presented.

Volwerk et al. [2016b] studied May, June, and July 2015 in order to find nested-draping and associated current sheets. Figure 30.8c shows the June flyby, where the axes are not equally scaled to enhance the clarity of the field line draping. The main pattern is that for $Y_{CSEQ} \leq -50$ where the field vectors are mainly red, indicating sunward-directed field, whereas, on the other side, the field vectors are mainly blue, pointing away from the Sun. However, there is a mixing of the colors, which means that there is a certain amount of nested draping. An example of such a magnetic field rotation is shown in Figure 30.9a (see also *Volwerk et al.* [2016b]). In the gray box there is a strong rotation of the cone angle of the magnetic field, accompanied by a minimum in (low-pass filtered) total magnetic field, filBm, and a peak in the electron density N_e measured by the mutual impedance probe (MIP) [*Trotignon et al.*, 2006]. These are the signatures one expects for crossing a current sheet. Assuming that the magnetic field moves together with the plasma (i.e., is frozen in) at a velocity given by the ion and electron spectrometer (IES) [*Burch et al.*, 2006], $v \approx 10$ km/s, a thickness of the rotation region can be obtained and thereby the current density of the embedded current sheet, using equation (30.14). This results in $j_{cs} \approx 85$ nA/m^2 for the rotation in Figure 30.9a, but stronger current sheets up to ~250 nA/m^2 have also been found [*Volwerk et al.*, 2016b].

30.10.4. Solar Wind Impact

The solar wind is variable, both in the magnetic field direction responsible for nested draping, and with magnetic explosions on the surface of the Sun regularly sending coronal mass ejections (CMEs) into interplanetary space. These are fast, high-density, magnetic flux ropes [e.g., *Möstl et al.*, 2014], which can interact with the planets, creating for example strong disturbances in the Earth's magnetosphere [e.g., *Zhang et al.*, 2007]. They also interact with comets and lead to interesting phenomena.

Vourlidas et al. [2007] observed the interaction of a CME with comet 2P/Encke with the SECCHI HI-1 instrument on STEREO-A. The movie from the camera showed that as the CME passed over the comet and its tail there was a disconnection event (see http://www.nasa.gov/mission_pages/stereo/news/encke.html). *Jia et al.* [2007] modeled this interaction with an MHD code and showed that a sudden 180° rotation of the IMF started reconnection first at the front side of the comet and then in the current sheet of the comet's tail, leading to a disconnection event. The cutoff part of the tail was transported away with the solar wind, and slowly a new tail grew behind the comet. *Jones and Brandt* [2004] showed multiple ICME interactions with comet 153P/Ikeya-Zhang, where possible shocks were observed traveling along the cometary tail, and disrupting it, leading to disconnection events. However, even though disconnection events can be generated by IMF reversals, there is no one-to-one correlation, as has been shown by *Delva et al.* [1991], using VEGA-1/2 data during the comet 1P/Halley flybys.

Rosetta did not enter the tail region of comet 67P/CG (except for one tail-excursion, ~1000 km behind the nucleus in March 2016), however, interactions with CMEs or with corotational interaction regions (CIRs) were observed by RPC on the dayside.

CIRs are structures in the solar wind where a region of fast wind catches up with a region of slow wind. *Edberg et al.* [2016b] studied the interaction of CIRs with the plasma environment around comet 67P/CG and found that the CIR compresses the comet's magnetosphere and there is an increased plasma production because of enhanced particle impact ionization and possible charge exchange. Some plasma observations during three events, a sudden reduction of the piled-up magnetic field and an increase of energetic electrons, can be interpreted as the comet's ionospheric response to tail disconnection events driven by the CIR.

(a)

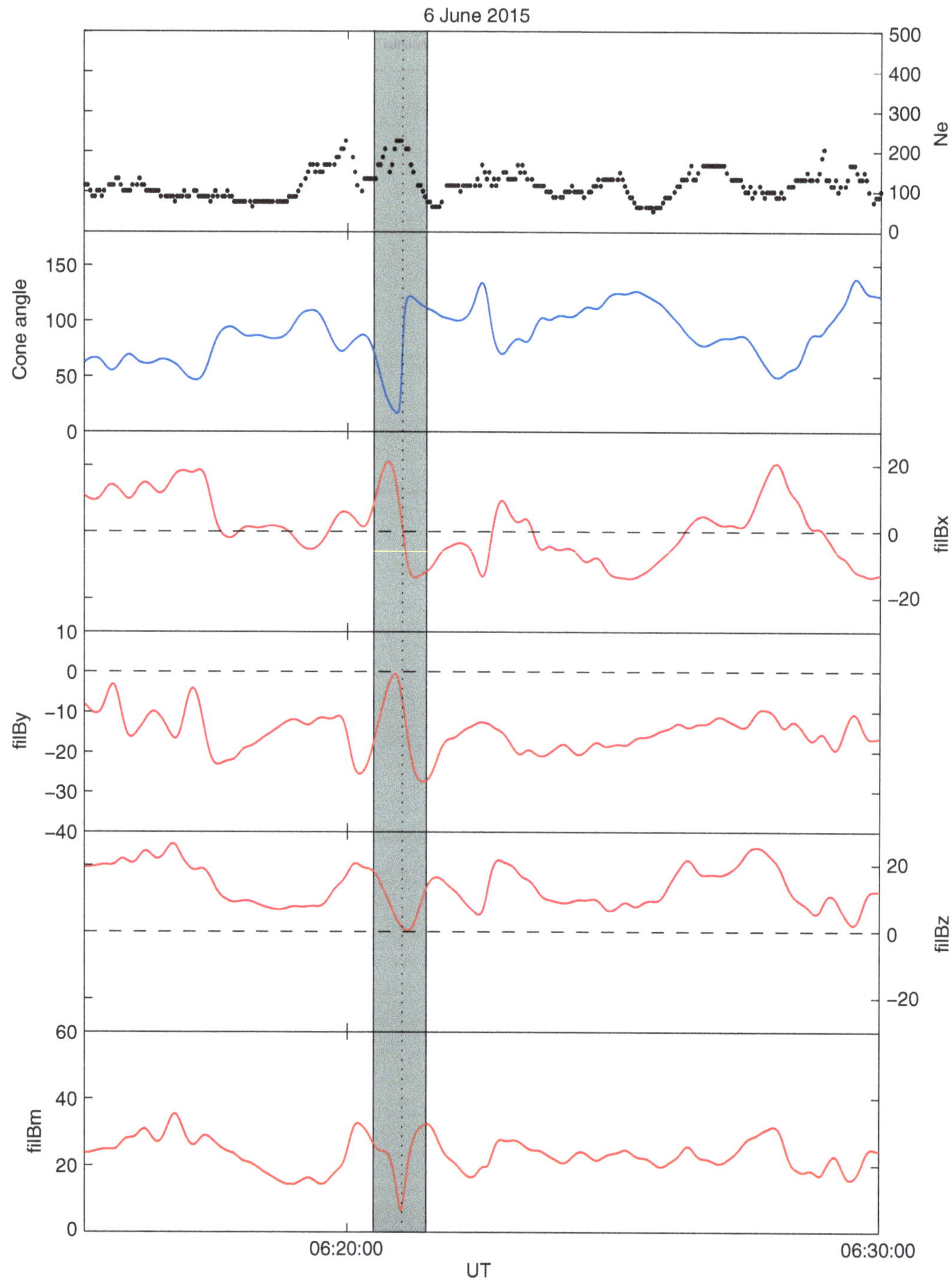

Figure 30.9 (a) The crossing of a current sheet on 6 June 2015. In the gray box the cone angle of the magnetic field changes strongly over more than 100°. This rotation of the field coincides with a minimum in total magnetic field and a maximum in electron density, indicating the crossing of a current sheet (see also *Volwerk et al.* [2016b]); (b) The signature of a magnetic flux rope in 67P/CG's magnetosheath. The data in minimum variance coordinates show the characteristic signature of a flux rope. From *Edberg et al.* [2016a].

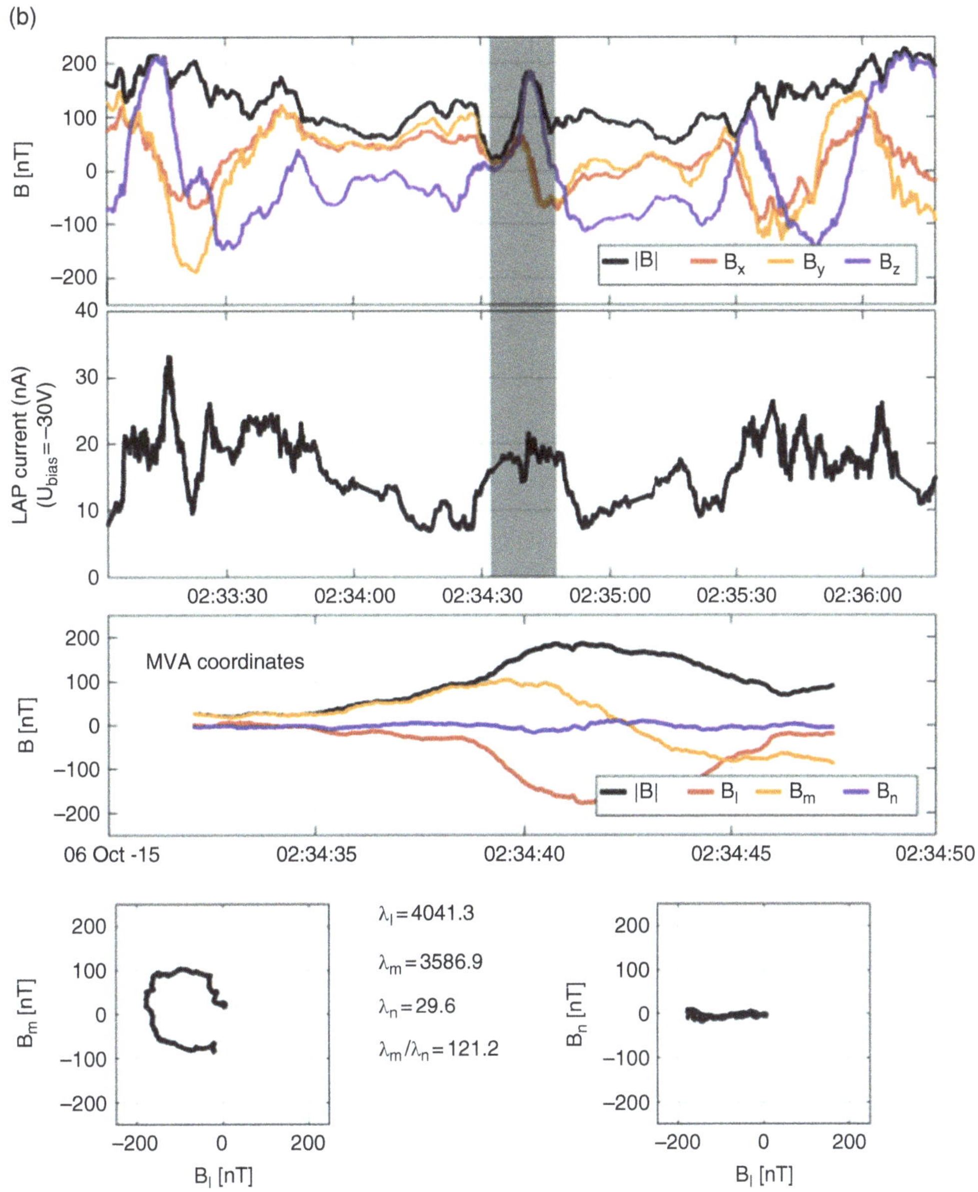

Figure 30.9 (Continued)

An interaction with a CME on 5–6 October 2015 was studied by *Edberg et al.* [2016a]. With Rosetta at a distance of ~800 km from the nucleus, the coma was compressed and solar wind ions were detected, which had been absent for months, even during the 1500 km dayside excursion of the spacecraft. Both magnetic field and plasma were compressed. Also, large spikes were observed in the MAG data, which were interpreted as flux ropes created by reconnection of the CME magnetic field and that draped around the nucleus. In Figure 30.9b, an example of such a structure is shown in minimum variance coordinates [*Sonnerup and Scheible*, 1998], which has the classical characteristics of a flux rope, which means that there is a strong current flowing along the center of the tube. Using the model by *Vršnak et al.* [2015], the axial current can be estimated by:

$$I_{\parallel} = \frac{0.519\pi}{\mu_0} dB_c \approx 10^5\,\mathrm{A}, \tag{30.15}$$

where $d \approx 100$ km is the diameter of the flux tube and $B_c \approx 200$ nT is the core field, given by *Edberg et al.* [2016a].

30.11. AND THUS IT ENDS

The complex interaction between a time variable solar wind and an outgassing cometary nucleus gives rise to a whole zoo of interesting plasma phenomena. Unfortunately, a size-limited review paper cannot discuss everything in full detail, but for the interested reader there is an extensive literature list in the references.

Considering the publications of the last few years, both with the old data and the new Rosetta observations, a lot of discoveries can still be made. Comparisons between strongly outgassing (1P/Halley) and weakly outgassing (67P/Churyumov-Gerasimenko) comets need to be made for various processes, as well as the temporal development during slowly increasing/decreasing activity.

ACKNOWLEDGMENTS

It was only possible to write this review because of the excellent work that has been performed by my predecessors in the early days of cometary science, some of whom are still working in the field. I would like to thank the ROSETTA team in general and the RPC team specifically for all the time and effort that has been put into this mission, which resulted in outstanding data. Special thanks to Tom Broiles, Magda Delva, Niklas Edberg, Anders Eriksson, Karl-Heinz Glassmeier, Charlotte Götz, Pierre Henri, Geraint Jones, Christoph Koenders, Christian Möstl, Hans Nilsson, Ingo Richter, and Bruce Tsurutani for discussions most clarifying and fruitful (sometimes in liquid form). I also would like to thank the two referees for their useful comments on additions to improve this chapter. I also appreciate Mike Wheatland for proofreading the manuscript.

REFERENCES

Alfvén, H. (1942), Existence of electromagnetic-hydrodynamic waves, *Nature*, *150*, 405–406.

Alfvén, H. (1957), On the theory of comet tails, *Tellus*, *9*, 92–96.

Auster, H.-U., I. Apathy, G. Berghofer, A. Remizov, R. Roll, K.-H. Fornaçon, K.-H. Glassmeier, et al. (2007), Romap: Rosetta magnetometer and plasma monitor, *Space Sci. Rev.*, *128*, 221–240; doi:10.1007/s11214-006-9033-x.

Balikhin, M., and M. Gedalin (2008), Generalization of the Harris current sheet model for non-relativistic, relativistic and pair plasmas, *J. Plasma Phys.*,*74*, 749–763; doi:10.1017/S002237780800723X.

Balsiger, H., K. Altwegg, P. Bochsler, P. Eberhardt, J. Fischer, S. Graf, A. Jäckel, et al. (2007), ROSINA: Rosetta orbiter spectrometer for ion and neutral analysis, *Space Sci. Rev.*, *128*, 697–712; doi:10.1007/s11214-006-8335-3.

Bertier, E. L. (1760), Physique des Comètes, L'Imprimerie Royale, Paris.

Biermann, L. (1951), Kometenschweife und solar Korpuskularstrahlung, *Zeitschr. Astrophys.*, *29*, 274–286.

Bobrovnikoff, N. T. (1930), Halley's comet in its apparition of 19091911, *Lick Obs. Public.*, *17*, 305.

Bonamy, P.-N. (1760), Suite de la clef ou Journal historique sur les matiere du tems, Ganeau, Paris.

Borovsky, J. (2010), On the variations of the solar wind magnetic field about the Parker spiral direction, *J. Geophys. Res.*, *115*, a09101; doi:10.1029/ 2009JA015040.

Boström, R. (1974), Spectral analysis, in *Magnetospheric Physics*, edited by B. M. McCormac, D. Reidel Publishing Co., Dordrech, 45–59.

Burch, J. L., R. Goldstein, T. E. Cravens, W. C. Gibson, R. N. Lundin, C. J. Pollock, J. D. Winningham, and D. T. Young (2006), RPC-IES: The ion and electron sensor of the Rosetta plasma consortium, *Space Sci. Rev.*, *128*, 697–712; doi:10.1007/s11214-006-9002-4.

Carr, C., E. Cupido, C. G. Y. Lee, A. Balogh, T. Beek, J. L. Burch, C. N. Dunford, et al. (2007), RPC: The Rosetta plasma consortium, *Space Sci. Rev.*, *128*, 697–712; doi:10.1007/s11214-006-9136-4.

Chang, C., H. Wong, and C. Wu (1990), Electromagnetic instabilities attributed to a cross-field ion drift, *Phys. Rev. Lett.*, *65*, 1104–1107; doi:10.1103/PhysRevLett.65.1104.

Chree, C. (1913), Some phenomena of sunspots and of terrestrial magnetism at Kew observatory, *Phil. Trans. R. Soc. Lond.*, *212*, 75–116.

Cloutier, P. A. (1970), Dynamics of the interaction of the solar wind with a planetary atmosphere, *Radio Sci.*, *5*, 387–389; doi:10.1029/ RS005i002p00387.

Coates, A. J., A. D. Johnstone, M. F. Thomsen, V. Formisano, E. Amata, B. Wilken, K. Jockers, J. D. Winningham, H. Borg, and D. A. Bryant (1987), Solar wind flow through the comet P/Halley bow shock, *Astron. Astrophys.*, *187*, 55–60.

Cravens, T. E. (1986), Ion distribution functions in the vicinity of comet Giacobini-Zinner, *Geophys. Res. Lett.*, *13*, 275–278.

Cravens, T. E. (1987), Theory and observations of cometary ionospheres, *Adv. Space Res.*, *7*, 147–158; doi:10.1016/0273-1177(87)90212-2.

Delva, M., K. Schwingenschuh, M. B. Niedner, and K. I. Gringauz (1991), Comet Halley remote plasma tail observations and in situ solar wind properties, Vega-1/2 IMF/plasma observations and ground-based optical observations from 1 December 1985 to 1 May 1986, *Planet. Space Sci.*, *39*, 697–708; doi:10.1016/0032-0633(91)90063-G.

Drell, S. D., H. M. Foley, and M. A. Ruderman (1965), Drag and propulsion of large satellites in the ionosphere: An Alfvén propulsion engine in space, *J. Geophys. Res.*, *70*, 3131–3145; doi:10.1029/JZ070i013p03131.

Dubinin, E. M., P. L. Izrailevich, and I. M. Podgorny (1982), Distribution of the magnetic field in Halley's comet tail obtained from the plasma cloud kinetics, *Sov. Astron.*, *26*, 608–611.

Edberg, N. J. T., A. I. Eriksson, E. Odelstad, E. Vigren, D. J. Andrews, R. Johansson, J. L. Burch, et al. (2016b), Solar wind interaction with comet 67P: Impacts of corotating interaction regions, *J. Geophys. Res.*, *121*, 949–965; doi:10.1002/2015JA022147.

Edberg, N. J. T., M. Alho, M. André, D. J. Andrews, E. Behar, J. L. Burch, C. M. Carr, et al. (2016a), CME impact on comet 67P/Churyumov-Gerasimenko, *Mon. Not. R.. Astron. Soc., 462*, S45–S56; doi:10.1093/mnras/stw2112.

Eddington, A. S. (1910), The envelopes of comet Morehouse (1098 c), *Mon. Not. R. Astron. Soc., 70*, 442–458; doi:10.1093/mnras/70.5.442.

Epstein, P. S. (1930), Reflection of waves in an inhomogeneous absorbing medium, *Proc. Nat. Acad. Sci., 16*, 627–637.

Ershkovich, A. I. (1976), On the origin of the magnetic field in type-I comet tails, *Mon. Not. R.. Astron. Soc., 177*, 649–652; doi:10.1093/mnras/177. 3.649.

Ershkovich, A. I., and A. B. Heller (1977), Helical waves in type-1 comet tails, *Astrophys. Space Sci., 48*, 365–377; doi:10.1007/BF00648123.

Feldman, P. D., M. F. A'Hearn, J.-L. Bertaux, L. M. Feaga, J. W. Parker, E. Schindhelm, A. J. Steffl, et al. (2015), Measurements of the near-nucleus coma of comet 67P/Churyumov-Gerasimenko with the Alice far-ultraviolet spectrograph on Rosetta, *Astron. Astrophys., 583*, A8; doi:10.1051/0004-6361/201525925.

Festou, M. C., H. U. Keller, and H. A. Weaver (2004), A brief conceptual history of cometary science, in *Comets II*, edited by M. C. Festou, H. U. Keller, and H. A. Weaver, 3–16, University or Arizona Press, Tucson.

Gary, S. P. (1992), The mirror and ion cyclotron anisotropy instabilities, *J. Geophys. Res., 97*, 8519–8529.

Glassmeier, K.-H. (2016), Interaction of the solar wind with comets: A ROSETTA perspective, *Phil. Trans. R. Soc. A.*, submitted.

Glassmeier, K.-H., H. Boehnhardt, D. Koschny, E. Kührt, and I. Richter (2007a), The Rosetta mission: Flying toward the origin of the solar system, *Space Sci. Rev., 128*, 1–21.

Glassmeier, K.-H., I. Richter, A. Diedrich, G. Musmann, U. Auster, U. Motschmann, A. Balogh, et al. (2007b), RPC-MAG the flux-gate magnetometer in the ROSETTA plasma consortium, *Space Sci. Rev., 128*, 649–670.

Goetz, C., C. Koenders, I. Richter, K. Altwegg, J. Burch, C. Carr, E. Cupido, et al. (2016b), First detection of a diamagnetic cavity at comet 67P/Churyumov-Gerasimenko, *Astron. Astrophys., 588*, A24; doi:10.1051/0004-6361/201527728.

Goetz, C., C. Koenders, K. C. Hansen, J. Burch, C. Carr, A. Eriksson, D. Frühauff, et al. (2016a), Structure and evolution of the diamagnetic cavity at comet 67P/Churyumov-Gerasimenko, *Mon. Not. R. Astron. Soc.*, in press.

Goertz, C. K. (1980), Io's interaction with the plasma torus, *J. Geophys. Res., 85*, 2949–2956.

Haerendel, G. (1973), Results from barium cloud release in the ionosphere and magnetosphere, *Space Res., 2*, 601–617.

Halley, E. (1705), *Astronomiae Cometicae Synopsis (A Synopsis of the Astronomy of Comets*, 1705), London, Oxford.

Hansen, K. C., K. Altwegg, J.-J. Berthelier, A. Bieler, N. Biver, D. Bockelée-Morvan, U. Calmonte, et al. (2016), Evolution of water production of 67P/Churyumov-Gerasimenko: An empirical model and a multi-instrument study, *Mon. Not. R. Astron. Soc., 462*, in press.

Harris, E. G. (1962), On a plasma sheet separating regions of oppositely directed magnetic field, *Nuovo Cimento, 23*, 115–121.

Haser, L. (1957), Distribution d'intensité dans la tête d'une comète, *Bull. Soc. R. Sci. Liege, 43*, 740–750.

Heidarzadeh, T. (2008), *A History of Physical Theories of Comets, from Aristotle to Whipple*, Springer, New York.

Heubner, W. F. (1990), *Physics and Chemistry of Comets*, Springer Verlag, Berlin.

Howarth, D. (1977), *1066 The Year of Conquest*, Penguin Books, New York.

Huddleston, D. E., A. D. Johnstone, and A. J. Coates (1990), Determination of comet Halley gas emission characteristics from mass loading of the solar wind, *J. Geophys. Res., 95*, 21–30.

Hyder, C. L., J. C. Brandt, and R. G. Roosen (1974), Tail structures far from the head of comet Kohoutek. *I, Icarus, 23*, 601–610; doi:10.1016/ 0019-1035(74)90028-1.

Ip, W.-H. (1979), Currents in the cometary atmosphere, *Planet. Space Sci., 27*, 121–125; doi:10.1016/0032-0633(79)90041-2.

Ip, W.-H. (1994), On a thermodynamical origin of the cometary ion rays, *Astrophys. J., 432*, L143–L145; doi:10.1086/187532.

Ip, W.-H., and D. A. Mendis (1975), The cometary magnetic field and its associated electric currents, *Icarus, 26*, 457–461; doi:10.1016/0019-1035(75) 90115-3.

Ip, W.-H., and D. A. Mendis (1976), The generation of magnetic fields and electric currents in cometary plasma tails, *Icarus, 29*, 147–151; doi:10.1016/ 0019-1035(76)90110-X.

Ip, W.-H., and W. I. Axford (1980), A weak interaction model for Io and the Jovian magnetosphere, *Nature, 283*, 180–183; doi:10.1038/283180a0.

Ip, W.-H., and W. I. Axford (1982), Theories of physical processes in the cometary coma and ion tails, in *Comets*, edited by L. L. Wilkening, 588–634, University of Arizona Press, Tucson.

Ip, W. H., and W. I. Axford (1987), The formation of a magnetic-field-free cavity at comet Halley, *Science, 325*, 418–419.

Israelevich, P. L., and A. I. Ershkovich (1994), Induced magnetosphere of comet Halley 2. magnetic field and electric currents, *J. Geophys. Res., 99*, 21,225–21,232.

Israelevich, P. L., A. I. Ershkovich, T. I. Gombosi, F. M. Neubauer, and O. Cohen (2003), Fine structure of the diamagnetic cavity boundary in comet Halley, *J. Geophys. Res., 108*, 1097; doi:10.1029/2002JA009622.

Israelevich, P. L., F. M. Neubauer, and A. I. Ershkovich (1994), The induced magnetosphere of comet Halley: interplanetary magnetic field during Giotto encounter, *J. Geophys. Res., 99*, 6575–6583.

Jia, Y.-D., M. R. Combi, K. C. Hansen, and T. I. Gombosi (2007), A global model of cometary tail disconnection events triggered by solar wind magnetic variations, *J. Geophys. Res., 112*, A05223; doi:10.1029/2006JA012175.

Jones, G. H., and J. C. Brandt (2004), The interaction of comet 153P/Ikeya-Zhang with interplanetary coronal mass ejections: Identification of fast ICME signatures, *Geophys. Res. Lett., 31*, L20805; doi:10.1029/ 2004GL021166.

Jones, G. H., A. Balogh, and T. S. Horbury (2000), Identification of comet Hyakutake's etremely long ion tail from magnetic measurements, *Nature, 404*, 574–576.

Kirsch, E., S. McKenna-Lawlor, P. Daly, A. Korth, F. M. Neubauer, D. O'Sullivan, A. Thompson, and

K. P. Wenzel (1989), Evidence for the field line reconnection process in the particle and magnetic field measurements obtained during the Giotto Halley encounter, *Ann. Geophys.*, *7*, 107–113.

Koenders, C., C. Götz, I. Richter, U. Motschmann, and K.-H. Glassmeier (2016), Magnetic field pile-up and draping at intermediately active comets: Results from comet 67P/Churyumov-Gerasimenko at 2.0 AU, *Mon. Not. R. Acad. Sci.*, *462*, S235–S241; doi:10.1093/mnras/stw2480.

Koenders, C., K.-H. Glassmeier, I. Richter, H. Ranocha, and U. Motschmann (2015), Dynamical features and spatial structures of the plasma interaction region of 67P/ChuryumovGerasimenko and the solar wind, *Planet. Space Sci.*, *105*, 101–116; doi:10.1016/j.pss.2014.11.014.

Krishna Swamy, K. S. (2010), *Physics of Comets*, 3d ed., World Scientific, Singapore.

Luehr, H., N. Kloecker, and M. H. Acuna (1988), The diamagnetic effect during AMPTE's tail releases: Initial results, *Adv. Space Res.*, *8*, 11–14.

Luehr, H., N. Kloecker, D. J. Southwood, M. W. D. Mier-Jedrzejowicz, W. A. C. Mier-Jedrzejowicz, R. P. Rijnbeek, M. Six, B. Haeusler, and M. Acuna (1986), In situ magnetic field observations of the AMPTE artificial comet, *Nature*, *320*, 708–711; doi:10.1038/320708a0.

Luhmann, J. G., S. A. Ledvina, and C. T. Russell (2004), Induced magnetospheres, *Adv. Space Res.*, *33*, 1905 1912; doi:10.1016/j.asr.2003.03. 031.

Mandt, K. E., A. Eriksson, N. J. T. Edberg, C. Koenders, T. Broiles, S. A. Fuselier, P. Henri, et al. (2016), RPC observation of the development and evolution of plasma interaction boundaries at 67P/Churyumov-Gerasimenko, *Mon. Not. Royal Astron. Soc.*, *462*, S9–S22; doi:10.1093/mnras/stw1736.

McComas, D. J., J. T. Gosling, S. J. Bame, J. A. Slavin, E. J. Smith, and J. L. Steinberg (1987), The Giacobini-Zinner magnetotail: Tail configuration and current sheet, *J. Geophys. Res.*, *92*, 1139–1152; doi:10.1029/ JA092iA02p01139.

Meier, P., K.-H. Glassmeier, and U. Motschmann (2016), Modified ion-Weibel instability as a possible source of wave activity at comet 67P/Churyumov-Gerasimenko, *Ann. Geophys.*, *34*, 691–707; doi:10.5194/angeo-34-691-2016.

Mendis, A., and H. Alfvén (1974), Magnetisation of comets, *Nature*, *248*, 36–37; doi:10.1038/248036a0.

Möstl, C., K. Amla, J. R. Hall, P. C. Liewer, E. M. De Jong, R. C. Coaninno, A. M. Veronig, et al. (2014), Connecting speeds, directions and arrival times of 22 coronal mass ejections from the Sun to 1 AU, *Astrophys. J.*, *787*, 119; doi:10.1088/0004-637X/787/2/119.

Mukai, T., W. Miyake, T. Terasawa, M. Kitayama, and K. Hirao (1986), Plasma observation by Suisei of solar-wind interaction with comet Halley, *Nature*, *321*, 299–303; doi:10.1038/321299a0.

Ness, N. F., and J. M. Wilcox (1964), Solar origin of the interplanetary magnetic field, *Phys. Rev. Lett.*, *13*, 461–464; doi:10.1103/PhysRevLett.13. 461.

Neubauer, F. M. (1988), The ionopause transition and boundary layers at Comet Halley from Giotto magnetic field observations, *J. Geophys. Res.*, *93*, 7272–7281; doi:10.1029/JA093iA07p07272.

Neubauer, F. M., K. H. Glassmeier, M. Pohl, J. Raeder, M. H. Acuna, L. F. Burlaga, N. F. Ness, et al. (1986), First results from the Giotto magnetometer experiment at comet Halley, *Nature*, *321*, 352–355.

Newton, S. I. (1687), *Philosophiae Naturalis Principia Mathematica*, Jussu Societalis Regiae ac typis Josephi Streatti, London.

Niedner, M. V., and J. C. Brandt (1978), Interplanetary gas: XXII, plasma tail disconnection events in comets, evidence for magnetic field line reconnection at interplanetary sector boundaries, *Astrophys. J.*, *223*, 655–670; doi:10. 1086/156299.

Nilsson, H., G. Stenberg Wieser, E. Behar, C. S. Wedlund, H. Gunell, M. Yamauchi, R. Lundin, et al. (2015), Birth of a comet magnetosphere: A spring of water ions, *Science*, *347*, aaa0571-1–aaa0571-4; doi:10.1126/science.aaa0571.

Nilsson, H., R. Lundin, K. Lundin, S. Barabash, H. Borg, O. Norberg, A. Fedorov, et al. (2007), Rpc-ica: The ion composition analyzer of the rosetta plasma consortium, *Space Sci. Rev.*, *128*, 671–695; doi:10.1007/s11214-006-9031-z.

Oort, J. H. (1950), The structure of the cloud of comets surrounding the solar system, and a hypothesis concerning its origin, *Bull. Astr. Inst. Netherl.*, *11*, 91–110.

Öpik, E. J. (1964), The motion of the condensation in the tail of Halley's comet June 5–8, 1910, *Zeitsschr. Astrophys.*, *58*, 192–201.

Parker, E. N. (1957), Sweet's mechanism for merging magnetic fields in conducting fluids, *J. Geophys. Res.*, *62*, 509–520; doi:10.1086/146579.

Parker, E. N. (1958), Dynamics of the interplanetary gas and magnetic fields, *Astrophys. J.*, *128*, 664–676.

Petschek, H. E. (1964), Magnetic field annihilation, *NASA Special Publication*, *50*, 425–439.

Pilipp, W. G. (1971), Expansion of an ion cloud in the Earth's magnetic field, *Planet. Space Sci.*, *19*, 1095–1119; doi:10.101 6/0032-0633(71)90107-3.

Podgorny, I. M., E. M. Dubinin, and P. L. Israelevich (1980), Laboratory simulation of the induced magnetospheres of comets and Venus, *Earth Moon Planet*, *23*, 323–338; doi:10.1007/BF00902047.

Podgorny, I. M., E. M. Dubinin, and P. L. Israelevich (1982), The estimates of the magnetic field in Halley's comet, *Moon Planets*, *27*, 135–138; doi:10.1007/BF00941563.

Raeder, J., F. M. Neubauer, N. F. Ness, and L. F. Burlaga (1987), Macroscopic perturbations of the IMF by P/Halley as seen by the Giotto magnetometer, *Astron. Astrophys.*, *187*, 61–64.

Rahe, J. (1968), The structure of tail rays in the coma region of comets, *Zeitschr. Astrophys.*, *68*, 208–213.

Rahe, J., and B. Donn (1969), Ionization and ray formation in comets, *Astron. J.*, *74*, 256–258.

Reinhard, R. (1986), The Giotto encounter with comet Halley, *Nature*, *321*, 313–318.

Richter, I., C. Koenders, H.-U. Auster, D. Frühauff, C. Götz, P. Heinisch, C. Perschke, et al. (2015), Observation of a new type of low frequency waves at comet 67P/Churyumov-Gerasimenko, *Ann. Geophys.*, *33*, 1031–1036.

Richter, I., H. Auster, G. Berghofer, C. Carr, E. Cupido, K. Fornaçon, C. Goetz, et al. (2016), Two-point observations of low-frequency waves at 67P/Churyumov-Gerasimenko during the descent of PHILAE: Comparison of RPCMAG and ROMAP, *Ann. Geophys.*, *34*, 609–622; doi:10.5194/angeo-34-609-2016.

Riedler, W., K. Schwingenschuh, Y. E. Yeroshenko, V. A. Styashkin, and C. T. Russell (1986), Magnetic field observations in comet Halley's coma, *Nature*, *321*, 288–289.

Rubin, M., K. C. Hansen, M. R. Combi, L. K. S. Daldorff, T. I. Gombosi, and V. M. Tenishev (2012), Kelvin-Helmholtz instabilities at the magnetic cavity boundary of comet 67P/Churyumov-Gerasimenko, *J. Geophys. Res.*, *117*, A06227.

Saito, T., K. Yumoto, K. Hirao, T. Nakgawa, and K. Saito (1986), Interaction between comet Halley and the interplanetary magnetic field observed by Sakigake, *Nature*, *321*, 303–307; doi:10.1038/321303a0.

Sauer, K., E. Dubinin, K. Baumgärtel, and V. Tarasov (1998), Low-frequency electromagnetic waves and instabilities within the Martian bi-ion plasma, *Earth Planet Space*, *50*, 269–278; doi:10.1186/BF03352113.

Schwingenschuh, K., W. Riedler, Y. Yeroshenko, J. L. Philips, C. T. Russell, J. G. Luhmann, and J. A. Fedder (1987), Magnetic field draping in the comet Halley coma: Comparison of Vega observations with computer simulations, *Geophys. Res. Lett.*, *14*, 640–643.

Slavin, J. A., B. A. Golberg, E. J. Smith, D. J. McComas, S. J. Bame, M. A. Strauss, and H. Spinrad (1986a), The structure of a cometary type I tail: Ground-based and ICE observations of P/Giacobini-Zinner, *Geophys. Res. Lett.*, *13*, 1085–1088; doi:10.1029/GL013i011p01085.

Slavin, J. A., E. J. Smith, B. T. Tsurutani, G. L. Sicsoe, D. E. Jones, and D. A. Mendis (1986b), Giacobini-Zinner magnetotail: ICE magnetic field observations, *Geophys. Res. Lett.*, *13*, 283–286; doi:10.1029/ GL013i003p00283.

Smith, E. J., B. T. Tsurutani, J. A. Slavin, D. E. Jones, G. L. Siscoe, and D. A. Mendis (1986), Ice encounter with Giacobini-Zinner: Magnetic field observations, *Science*, *232*, 382.

Sonnerup, B. U. Ö., and M. Scheible (1998), Minimum and maximum variance analysis, in *Analysis Methods for Multi-Spacecraft Data*, edited by G. Paschmann and P. Daly, 185–220, ESA, Noordwijk.

Stern, D. P. (1994), The art of mapping the magnetosphere, *J. Geophys. Res.*, *99*, 17,169–17,198; doi:10.1029/94JA01239.

Strömgren, E. (1914), Über der Ursprung der Kometen, Publ. Obs. Copenhagen, Copenhagen.

Strömgren, E. (1947), The short-period comets and the hypothesis of their capture by the major planets, Publ. Obs. Copenhagen, Copenhagen.

Sweet, P. A. (1956), The neutral point theory of solar flares, in *Proceedings of the International Astronomican Union Symposium on Electromagnetic Phenomena in Cosmical Physics*, Kluwer, Dordrecht, *123*.

Trotignon, J. G., J. L. Michau, D. Lagoutte, M. Chabassièreand G. Chalumeau, F. Colin, P. M. E. Décréau, et al. (2006), RPC-MIP: The mutual impedance probe of the Rosetta plasma consortium, *Space Sci. Rev.*, *128*, 713–728; doi:10.1007/s11214-006-9005-1.

Vasyliunas, V. M. (2016), Physical origin of pickup currents, *Ann. Geophys.*, *34*, 153–156; doi:10.5194/angeo-34-153-2016.

Verigin, M. I., W. I. Axford, K. I. Gringauz, and A. K. Richter (1987), Acceleration of cometary plasma in the vicinity of comet Halley associated with an interplanetary magnetic field polarity change, *Geophys. Res. Lett.*, *14*, 987–990.

Volwerk, M., C. Koenders, M. Delva, I. Richter, K. Schwingenschuh, M. S. Bentley, and K.-H. Glassmeier (2013), Ion cyclotron waves during the Rosetta approach phase: A magnetic estimate of cometary outgassing, *Ann. Geophys.*, *31*, 2201–2206.

Volwerk, M., G. H. Jones, T. Broiles, C. Carr, A. J. Coates, E. Cupido, M. Delva, et al. (2016b), Field line draping and current sheets in comet 67P/Churyumov-Gerasimenkos coma, *J. Geophys. Res.*, submitted.

Volwerk, M., I. Richter, B. Tsurutani, C. Götz, K. Altwegg, T. Broiles, J. Burch, et al. (2016a), Mass-loading, pile-up, and mirror-mode waves at comet 67P/Churyumov-Gerasimenko, *Ann. Geophys.*, *34*, 1–15.

Volwerk, M., K.-H. Glassmeier, M. Delva, D. Schmid, C. Koenders, I. Richter, and K. Szegö (2014), A comparison between VEGA 1, 2 and Giotto flybys of comet 1P/Halley: implications for Rosetta, *Ann. Geophys.*, *32*, 1441–1453; doi:10.5194/angeo-32-1441-2014.

Vourlidas, A., C. J. Davis, C. J. Eyles, S. R. Crothers, R. A. Harrison, R. A. Howard, J. D. Moses, and D. G. Socker (2007), First direct observation of the interaction between a comet and a coronal mass ejection leading to a complete plasma tail disconnection, *Astrophys. J.*, *668*, L79–L82; doi:10.1086/522587.

Vršnak, B., T. Rollett, A. M. Veronik, C. Möstl, U. V. Amerstorfer, M. Leitner, M. Temmer, et al. (2015), Heliospheric evolution of magnetic clouds, *Astrophys. J.*, submitted.

Weibel, E. S. (1959), Spontaneously growing transverse waves in a plasma due to an anisotropic velocity distribution, *Phys. Rev. Lett.*, *2*, 83–84; doi:10.1103/PhysRevLett.2.83.

Wenzel, K.-P., T. R. Sanderson, I. G. R. Cowley, S. W. H. Cowley, R. J. Hynds, S. J. Bame, R. D. Zwickl, E. J. Smith, and B. T. Tsurutani (1986), In-situ observations of cometary pick-up ions greater than 0.2 AU upstream of Comet Halley–ICE observations, *Geophys. Res. Lett.*, *13*, 861–864; doi:10.1029/GL013i008p00861.

Whipple, F. L. (1950), A comet model I. the acceleration of comet Encke, *Astrophys. J.*, *111*, 375–394.

Wolff, R. S., G. L. Siscoe, D. G. Sibeck, and M. M. Neugebauer (1985), Cometary rays: magnetically channeled outflow, *Geophys. Res. Lett.*, *12*, 749–752; doi:10.1029/GL012i011p00749.

Wurm, K. (1943), Die Natur der Kometen, *Mitt. Hamburger Sternw.*, *8*, Nr. 51.

Wurm, K. (1961), Structure and development of the gas tails of comets, *Astron. J.*, *66*, 362–367; doi:10.1086/108437.

Wurm, K. (1963), The physics of comets, in *The Moon, Meteorites And Comets*, edited by G. P. Kuiper and B. Middlehurts, 573–617, University of Chicago Press, Chicago.

Wurm, K., and A. Mammano (1967), Dissoziation und ionisation in kometen, *Icarus*, *6*, 281–291; doi:10.1016/0019-1035(67)90026-7.

Yoon, P. H., and A. T. Y. Lui (2008), Drift instabilities in current sheets with guide field, *Phys. Plasmas*, *15*, 072101; doi:10.1063/1.2938386.

Zhang, J., I. G. Richardson, D. F. Webb, N. Goplaswamy, E. Huttunen, J. C. Kasper, N. V. Nitta, et al. (2007), Solar and interplanetary sources of major geomagnetic storms (Dst ≤ -100 nT) during 19962005, *J. Geophys. Res.*, *112*, A10102; doi:10.1029/2007JA012321.

Zhang, T. L., M. Delva, W. Baumjohann, M. Volwerk, C. T. Russell, H. Y. Wei, C. Wang, et al. (2008), Induced magnetosphere and its outer boundary at Venus, *J. Geophys. Res.*, *113*, E00B20; doi:10.1029/2008JE003215.

INDEX

Electric Currents in Geospace and Beyond, Geophysical Monograph 235, First Edition.
Edited by Andreas Keiling, Octav Marghitu, and Michael Wheatland.